AGRICULTURAL MECHANICS

Fundamentals & Applications

Third Edition

In Memory of
Elmer L. Cooper

FFA

AGRICULTURAL MECHANICS

Fundamentals & Applications

Third Edition

ELMER L. COOPER, Ed. D.

Delmar Publishers

ITP An International Thomson Publishing Company

Albany • Bonn • Boston • Cincinnati • Detroit • London • Madrid
Melbourne • Mexico City • New York • Pacific Grove • Paris • San Francisco
Singapore • Tokyo • Toronto • Washington

NOTICE TO THE READER

Cover design courtesy Design Works

Delmar Staff
Publisher: Tim O'Leary
Acquisitions Editor: Cathy L. Esperti
Developmental Editor: Cathy L. Esperti
Senior Project Editor: Andrea Edwards Myers
Production Manager: Wendy A. Troeger
Marketing Manager: Maura Theriault

Printed in the United States of America

For more information, contact:

Delmar Publishers
3 Columbia Circle, Box 15015
Albany, New York 12212-5015

International Thomson Publishing Europe
Berkshire House 168-173
High Holborn
London, WC1V7AA
England

Thomas Nelson Australia
102 Dodds Street
South Melbourne, 3205
Victoria, Australia

Nelson Canada
1120 Birchmount Road
Scarborough, Ontario
Canada M1K 5G4

International Thomson Editores
Campos Eliseos 385, Piso 7
Col Polanco
11560 Mexico D F Mexico

International Thomson Publishing Gmbh
Konigswinterer Strasse 418
53227 Bonn
Germany

International Thomson Publishing Asia
221 Henderson Road #0510
Henderson Building
Singapore 0315

International Thomson Publishing - Japan
Kirakawacho Kyowa Building, 3F
2-2-1 Hirakawacho
Chiyoda-ku, 102 Tokyo
Japan

5 6 7 8 9 10 XXX 02 01 00 99 98 97

Library of Congress Cataloging-in-Publication Data

Cooper, Elmer L.
Agricultural mechanics: fundamentals & applications/Elmer L. Cooper. — 3rd ed.
p. cm.
Includes index.
ISBN 0-8273-6854-2 (textbook)
1. Agricultural mechanics. I. Title
S675.3.C66 1996
631.3—dc20 95-15152
CIP

CONTENTS

PREFACE

Agricultural Mechanics: Fundamentals & Applications, third edition, is written for students and individuals needing written materials on basic mechanical skills. It grew out of the need for an easy-to-read, easy-to-understand, and highly illustrated text on modern agricultural mechanics for high school and postsecondary programs.

This book addresses the specific needs of students enrolled in agriscience, production agriculture, ornamental horticulture, agribusiness, agricultural mechanics, and natural resources programs. The text starts with very basic and general information, such as career opportunities, and then provides competency-based instruction on basic mechanical skills and applications.

Agricultural Mechanics: Fundamentals & Applications, third edition, is highly illustrated to clarify the written word. It contains instructions on materials such as wood, sheet metal, structural steel, electrical wire, rust preventatives, wood preservatives, and paint and other finishes. It includes procedures utilizing hand tools, portable power tools, and stationary power equipment.

The selection and use of nails, screws, bolts, and glue are covered in detail. Processes include woodworking; metalworking; electric and gas welding; heating and cutting; small engine maintenance; electrical wiring; electronics; robotics; pneumatics; plumbing; concrete and masonry work; and construction.

In response to popular demand, seven additional units had been added to the book for the second edition. The titles of those new units are as follows:

1. Small Engine Adjustment and Repair (Unit 30)
2. Electronics in Agriculture (Unit 33)
3. Electric Motors, Drivers, and Controls (Unit 34)
4. Drainage and Irrigation Technology (Unit 36)
5. Hydraulic, Pneumatic, and Robotic Power (Unit 37)
6. Planning and Constructing Agricultural Structures (Unit 39)
7. Aquaculture, Greenhouse, and Hydroponics Structures (Unit 40).

The additional units address some of the more specialized and emerging areas of agricultural mechanics and, as such, make the text more comprehensive and useful for new and emerging programs in agricultural education.

This third edition has focused on additional technical information as needed throughout. The arc welding unit has been expanded to include MIG and TIG welding.

A competency-based approach is used throughout the text. This makes the text consistent in format, easy to use for individualized instruction, easy to teach from, simple for substitute teachers, and easy to assess student progress. Each unit is part of a section and includes (1) a statement of objective, (2) competencies to be developed, (3) a list of new terms, (4) a materials list, (5) highly illustrated text material, (6) student activities, and (7) a self-evaluation. All new terms are carefully defined in the text as well as in the glossary.

The appendices includes fifty-four project plans with bills of materials and construction procedures. The projects were carefully selected to match the skills covered in the text. Plans include some projects that have become classics in the field, some that cover targeted enterprises, and some that are new and innovative. The projects were selected because they are used in high school agriculture/agribusiness, agriscience, or related programs.

The appendices also includes thirty-six tables containing information for estimating, planning, selecting, purchasing, and building in agricultural mechanics. The project plans, tables, glossary, and index provide unique reference materials that, for many users, are alone worth the price of the text.

AGRICULTURE MECHANICS:

The third edition of Agricultural Mechanics Fundamentals & Applications has been carefully designed to enhance the study of mechanics and technology in agriscience programs. For best results, you may want to become familiar with the features incorporated into this text and accompanying learning tools.

ADDED FEATURES

- Addition of color.
- Safety issues and procedures are highlighted throughout.
- English-Spanish glossary with over 425 definitions. Provides an excellent study aid and reference.

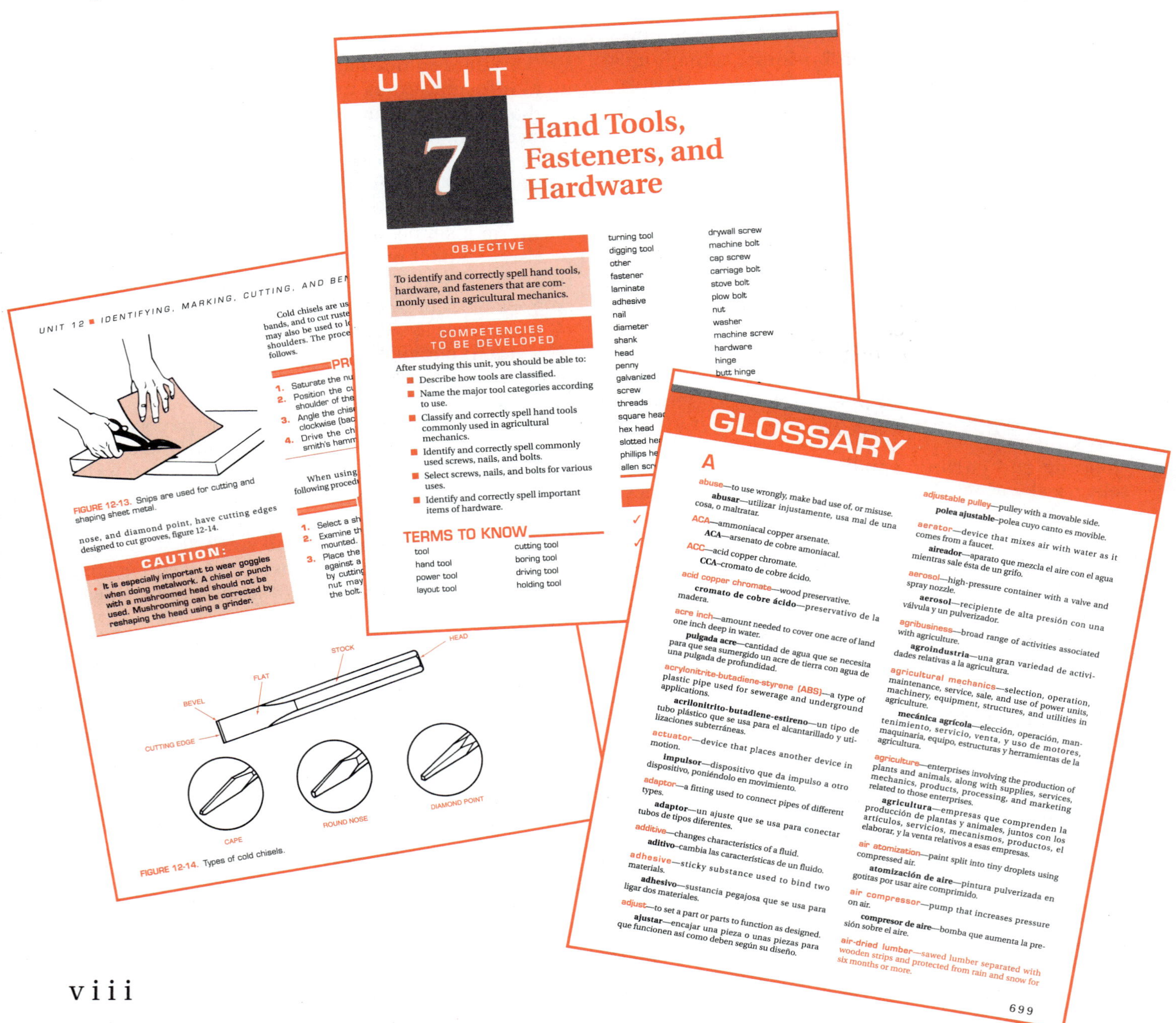

UNIT 12 ■ IDENTIFYING, MARKING, CUTTING, AND BEN

FIGURE 12-13. Snips are used for cutting and shaping sheet metal.

nose, and diamond point, have cutting edges designed to cut grooves, figure 12-14.

CAUTION:

- It is especially important to wear goggles when doing metalwork. A chisel or punch with a mushroomed head should not be used. Mushrooming can be corrected by reshaping the head using a grinder.

FIGURE 12-14. Types of cold chisels.

UNIT 7

Hand Tools, Fasteners, and Hardware

OBJECTIVE

To identify and correctly spell hand tools, hardware, and fasteners that are commonly used in agricultural mechanics.

COMPETENCIES TO BE DEVELOPED

After studying this unit, you should be able to:

- Describe how tools are classified.
- Name the major tool categories according to use.
- Classify and correctly spell hand tools commonly used in agricultural mechanics.
- Identify and correctly spell commonly used screws, nails, and bolts.
- Select screws, nails, and bolts for various uses.
- Identify and correctly spell important items of hardware.

TERMS TO KNOW

tool, hand tool, power tool, layout tool, cutting tool, boring tool, driving tool, holding tool, turning tool, digging tool, other, fastener, laminate, adhesive, nail, diameter, shank, head, penny, galvanized, screw, threads, drywall screw, machine bolt, cap screw, carriage bolt, stove bolt, plow bolt, nut, washer, machine screw, hardware, hinge, butt hinge

GLOSSARY

A

abuse—to use wrongly, make bad use of, or misuse.
abusar—utilizar injustamente, usa mal de una cosa, o maltratar.

ACA—ammoniacal copper arsenate.
ACA—arsenato de cobre amoniacal.

ACC—acid copper chromate.
CCA—cromato de cobre ácido.

acid copper chromate—wood preservative.
cromato de cobre ácido—preservativo de la madera.

acre inch—amount needed to cover one acre of land one inch deep in water.
pulgada acre—cantidad de agua que se necesita para que sea sumergido un acre de tierra con agua de una pulgada de profundidad.

acrylonitrite-butadiene-styrene (ABS)—a type of plastic pipe used for sewerage and underground applications.
acrilonitrito-butadieno-estireno—un tipo de tubo plástico que se usa para el alcantarillado y utilizaciones subterráneas.

actuator—device that places another device in motion.
impulsor—dispositivo que da impulso a otro dispositivo, poniéndolo en movimiento.

adaptor—a fitting used to connect pipes of different types.
adaptor—un ajuste que se usa para conectar tubos de tipos diferentes.

additive—changes characteristics of a fluid.
aditivo—cambia las características de un fluido.

adhesive—sticky substance used to bind two materials.
adhesivo—sustancia pegajosa que se usa para ligar dos materiales.

adjust—to set a part or parts to function as designed.
ajustar—encajar una pieza o unas piezas para que funcionen así como deben según su diseño.

adjustable pulley—pulley with a movable side.
polea ajustable—polea cuyo canto es movible.

aerator—device that mixes air with water as it comes from a faucet.
aireador—aparato que mezcla el aire con el agua mientras sale ésta de un grifo.

aerosol—high-pressure container with a valve and spray nozzle.
aerosol—recipiente de alta presión con una válvula y un pulverizador.

agribusiness—broad range of activities associated with agriculture.
agroindustria—una gran variedad de actividades relativas a la agricultura.

agricultural mechanics—selection, operation, maintenance, service, sale, and use of power units, machinery, equipment, structures, and utilities in agriculture.
mecánica agrícola—elección, operación, mantenimiento, servicio, venta, y uso de motores, maquinaria, equipo, estructuras y herramientas de la agricultura.

agriculture—enterprises involving the production of plants and animals, along with supplies, services, mechanics, products, processing, and marketing related to those enterprises.
agricultura—empresas que comprenden la producción de plantas y animales, juntos con los artículos, servicios, mecanismos, productos, el elaborar, y la venta relativos a esas empresas.

air atomization—paint split into tiny droplets using compressed air.
atomización de aire—pintura pulverizada en gotitas por usar aire comprimido.

air compressor—pump that increases pressure on air.
compresor de aire—bomba que aumenta la presión sobre el aire.

air-dried lumber—sawed lumber separated with wooden strips and protected from rain and snow for six months or more.

699

Fundamentals & Applications, 3E

ENHANCED CONTENT

The following updated and enhanced content addresses the evolving agriscience curriculum:

- Expanded coverage of MIG and TIG welding.
- Integrates agricultural mechanization with plant, animal, and environmental sciences to provide students with a broad view of the world of agriculture.
- Every unit combines theory with practice.
- Mathematical skill development is emphasized throughout.
- Each unit has been updated to include a wide variety of agricultural careers.

EXTENSIVE TEACHING/ LEARNING PACKAGE

The complete supplement package was developed to achieve two goals:

1. To assist students in learning the essential information needed to continue their exploration into the exciting field of agricultural mechanics
2. To assist instructors in planning and implementing their instructional programs for the most efficient use of time and other resources.

INSTRUCTOR'S GUIDE TO TEXT

The Instructor's Guide provides answers to the end-of-chapter questions, additional materials to assist the instructor in the preparation of lesson plans, and numerous suggestions for using the text. Also includes transparency masters.

WORKBOOK

Order #0-8273-6928-X
This comprehensive workbook tests students' knowledge and reinforces learning of text content.

WORKBOOK INSTRUCTOR'S GUIDE

Order #0-8273-6929-8
The *Instructor's Guide* provides answers to workbook exercises and additional guidance for the instructor.

TEACHER'S RESOURCE GUIDE

Order #0-8273-6895-X
This new supplement provides the instructor with valuable resources to simplify the planning and implementation of the instructional program. It includes transparency masters, motivational questions and activities, answers to questions in the text, and lesson plans to provide the instructor with a cohesive plan for presenting each topic.

COMPUTERIZED TESTBANK

Order #0-8273-7488-7
An all new computerized testbank (IBM and MAC compatible) with more than 500 new questions that will give the instructor an expanded capability to create tests.

AGRISCIENCE TESTMASTER CD-ROM

Order #0-8273-7496-0
This testbank contains over 3,000 questions correlated to the following Delmar titles:

Cooper/*AGRISCIENCE: Fundamentals & Applications, 2E*

Cooper/*AGRICULTURAL MECHANICS: Fundamentals & Applications, 3E*

Reiley/*INTRODUCTORY HORTICULTURE, 5E*

Herren/*THE SCIENCE OF AGRICULTURE: A Biological Approach*

Ricketts/*LEADERSHIP PERSONAL DEVELOPMENT AND CAREER SUCCESS*

Camp/*MANAGING OUR NATURAL RESOURCES, 3E*

Gillespie/*MODERN LIVESTOCK AND POULTRY PRODUCTION, 5E*

The CD-ROM also contains ESAgrade Electronic Gradebook and ESAtest III. These enable the instructor to create and administer on-line tests and keep track of grades.

DELMAR'S AGRISCIENCE LASERDISC PACKAGE

Order #0-8273-7300-7

Delmar's Agriscience Laserdisc Package offers instant accessibility to interactive video segments on four double-sided Level I laserdiscs directly corresponding to all the major content areas in agriculture. The discs can be obtained as a package or individually.

Disc 1: Animal Science
Order #0-8273-7301-5
Disc 2: Plant Science
Order #0-8273-7302-3
Disc 3: Business and Mechanical Technology
Order #0-8273-7303-1
Disc 4: Forestry and Natural Resources Management
Order #0-8273-7304-X

In addition, a correlation guide complete with barcodes correlates all barcodes contained on the laserdiscs with all of Delmar's top agriscience texts and lab manuals.

ABOUT THE AUTHOR

Dr. Elmer L. Cooper was a consultant and writer in Agricultural Mechanics and Agricultural Education. He taught agricultural mechanics in public schools for over 17 years, including grades 8 through 12, adults, and for one year, at the post-secondary level. Dr. Cooper served as State Supervisor of Agriculture/Agribusiness and Renewable Natural Resources in Maryland for four years. He had been a teacher educator at VA TECH and the University of Maryland for 12 years, and many outstanding teachers in the field have benefited from his teaching and guidance.

The content of this textbook was carefully selected from work with hundreds of agricultural mechanics teachers and visits to dozens of outstanding programs. Dr. Cooper's extensive experience qualifies him exceptionally well for selecting, organizing, and writing *Agricultural Mechanics: Fundamentals & Applications, third edition.*

ACKNOWLEDGMENTS

The author wishes to thank the many individuals who provided encouragement and assistance during the preparation of this text. He would particularly like to thank the many agricultural education teachers, students, and faculty at North Harford High School, Hereford High School, Linganore High School, Walkersville High School, the Institute of Applied Agriculture and the Department of Agricultural Engineering at the University of Maryland, and other educational institutions for providing the settings and/or being a subject in the photographs.

The expert contributions by agricultural instructors Steen Westerberg, Vernon Marshall, Charles Cramer, James Ferrant, Doug Herring, and Thomas Hawthorne are especially noteworthy. Similarly, contributions by Dr. Thomas Handwerker, Assistant Professor at the University of Maryland at Princess Anne, were invaluable. Appreciation is expressed to photographers Frederick Doepkens, Richard Kreh, and Ronald Seibel for their invaluable services.

The early approval and continuous support of the project by Dr. Clifford Nelson, then Chairman of the Department of Agricultural and Extension Education, University of Maryland, and H. Edward Reiley, consulting editor, is appreciated and gratefully acknowledged. In like manner, appreciation is expressed to his wife, Dollye Cooper, for her constant support, assistance, and love.

The continuous encouragement and periodic handling of typing, compilation, and mailing tasks by Cathie Galeano, and typing by Sandy Smith and the typists at TERP Service are gratefully acknowledged. The author would also like to express appreciation to the many schools, colleges, curriculum centers, businesses, and government offices across the nation for supplying photographs and illustrative materials for use in the text.

Reviewers who contributed their expertise to the third edition of *Agricultural Mechanics: Fundamentals & Applications* include:

Sam May
Bismark Vo-Tech
Bismark, ND

Ermanno Arizzi
Eureka Senior High School
Eureka, CA

Harrison A. Dixon
Gloucester High School
Gloucester, VA

Rick Gamblin
Miles, TX

Mike Rich
Holmes County High School
Bonifay, FL

Special thanks to reviewers of past editions:

Don Gates
Centralia Missouri High School
Centralia, MO

Danny Beck
LaMesa High School
LaMesa, TX

E. C. Conner
Parkview Senior High School
South Hill, VA

Daniel Lantis
Baker High School
Baker, MT

Raymond L. Holt
Texas Education Agency
Austin, TX

Bill Honig
California State Department of Education
Sacramento, CA

Exploring Careers In Agricultural Mechanics

UNIT 1

Mechanics in the World of Agriculture

OBJECTIVE

To determine how mechanical skills, concepts, and principles are used in agriculture and related occupations.

COMPETENCIES TO BE DEVELOPED

After studying this unit, you should be able to:

- Define agriculture and agricultural mechanics.
- Define occupation and describe an occupational cluster.
- Describe the role of mechanics and mechanical applications in society.
- Demonstrate knowledge of contributions made by mechanical application to the development of agriculture.
- Name eight inventors of important agricultural machines.

MATERIALS LIST

- ✓ Pencil
- ✓ Paper
- ✓ Encyclopedias

TERMS TO KNOW

agriculture
agriscience
agribusiness
renewable natural resources
occupational cluster
occupation
business
profit
employment
trade
mechanical
mechanic
technician
mechanics
agricultural mechanics
efficiency

Historically, the word agriculture meant to farm or to grow plants or animals. Today, agriculture may be defined as those activities concerned with plants and animals and the related supplies, services, mechanics, products, processing, and marketing related to plants, animals, and the environment. The term agriscience evolved during the current decade to clearly denote that agriculture is a science. Actually, modern agriculture covers so many activities that a simple definition is not possible. The United States Department of Education developed the phrase "agriculture/agribusiness and renewable natural resources" to refer to the broad range of activities associated with agriculture. **Agribusiness** refers to the network of commercial firms that have developed with or stem out of agriculture.

Renewable natural resources are the resources provided by nature that can replace or renew themselves. Examples of such resources are wildlife, trees,

FIGURE 1-1. The industry of agriculture provides interesting careers in production, management, science, education, finance, communication, government, conservation, and mechanics.

and fish as well as their natural surroundings. Some occupations in renewable natural resources are game trapper, forester, and waterman. A waterman uses boats and specialized equipment to harvest fish, oysters, and other seafood.

All jobs and types of work in the field of agriculture/agribusiness and renewable natural resources make up an occupational cluster. An occupational cluster is a group of related jobs. There are many occupations that are related to agriculture as illustrated in figure 1-1. Agricultural mechanics is only

FIGURE 1-1. *Continued*

one of the many careers that fall under the broad term of agriculture.

Agriculture depends on many nonfarm workers for goods and services. For every farm worker in the United States, there are said to be four workers in nonfarm agricultural jobs, figure 1-2. Nonfarm agriculturists provide machinery, equipment, fertilizer, feed, seed, money, research, and government services. Similarly, marketing, transporting, processing, and distributing farm products are done by nonfarm agriculturalists. This network of nonfarm agricultural workers helps the average U.S. farmer feed about 128 people per year, figure 1-3. On the most productive and efficient farms, however, one farmer may produce enough to feed more than 200 people.

FIGURE 1-3. The average American farmer produces enough food per year to feed about 128 people. (*Courtesy of U.S.D.A.*)

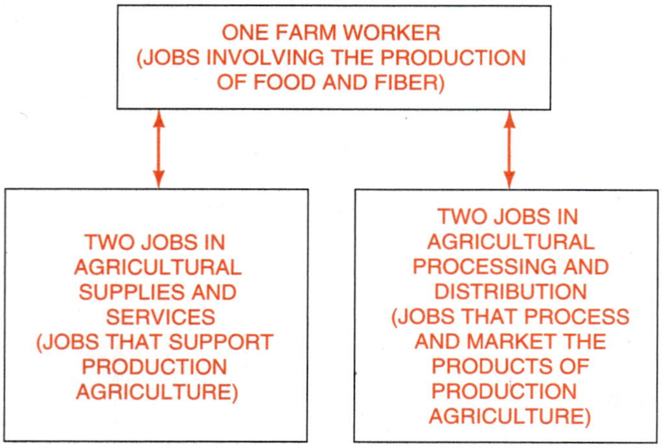

The term occupation means business, employment, or trade. It is the work a person does regularly to earn a living. Business generally refers to work done for profit. Profit refers to income made from the sale of goods or services after expenses have been taken out. An example of profit is the income made by a local machinery dealer or a local garden center minus expenses.

Employment means work done for pay. Employees are paid by the hour, day, week, month, or year. A small-engines mechanic employed at the local machinery dealer is probably paid by the hour, but the sales manager is probably paid an annual salary with bonuses based on sales volume.

Trade refers to specific kinds of work or businesses, especially those that require skilled mechanical work. Mechanical means having to do with a machine, mechanism, or machinery. A person who is specifically trained to perform mechanical tasks is a mechanic. A mechanic who uses high technology is generally called a technician. The mechanic must be skilled in the use of tools and machines. The mechanic must also be able to select appropriate materials, use a variety of processes, and analyze problems. The term mechanics is defined as the branch of physics dealing with motion, and the action of forces on bodies or fluids.

The term agricultural mechanics is the selection, operation, maintenance, servicing, selling, and use of power units, machinery, equipment, structures, and utilities used in agriculture. Hundreds of different jobs are available in agricultural mechanics. A total of 99 out of 305 occupational entries in *A Concise Handbook of Occupations* are mechanical in nature.

ONE FARM WORKER (JOBS INVOLVING THE PRODUCTION OF FOOD AND FIBER)

TWO JOBS IN AGRICULTURAL SUPPLIES AND SERVICES (JOBS THAT SUPPORT PRODUCTION AGRICULTURE)

TWO JOBS IN AGRICULTURAL PROCESSING AND DISTRIBUTION (JOBS THAT PROCESS AND MARKET THE PRODUCTS OF PRODUCTION AGRICULTURE)

FIGURE 1-2. The industry of agriculture now includes five people for each worker on a farm. There are about four nonfarm agricultural jobs for each farm job.

AGRICULTURE—A BASIC INDUSTRY

Agriculture is a very complex industry. The industry produces plant and animal products from which thousands of commodities are made. Since every person and many industries depend upon agriculture, it is said to be a basic industry. Some products of agriculture are food, oils, fiber, lumber, ornamental trees and shrubs, flowers, leather, fertilizers, feed, seed, and more. Basic agricultural products form the raw materials for many items of everyday living.

Fabrics for clothing, curtains, and floor coverings are made from oils such as corn oil, soybean oil, and cottonseed oil. Plastics of all kinds are also made from vegetable oils. Products from animals are used to make materials such as glue, leather, and paint.

Many medicines come from plants and animals. The manufacture of automobiles, furniture, airplanes, radios, stereos, and computers all depends on agriculture for certain raw materials. The construction of homes, boats, and factories all depends on agriculture for lumber, fiber, and other basic commodities. Most dwellings in America are surrounded by lawn, shrubs, or other plants for beautification. These are also agricultural commodities.

Agriculture is, indeed, a basic industry upon which all people depend. It is the backbone of the American society.

THE ROLE OF AGRICULTURAL MECHANICS

Mechanical applications are found throughout agriculture. Some of these applications are shown in figures 1-4 through 1-8. A few examples of jobs involving agricultural mechanics are:

- the engineer who designs tractors and other farm and ranch machines
- the forester who keeps chainsaws and other equipment going
- the builder of processing plants, farm buildings, and aquaculture facilities

FIGURE 1-5. Agricultural construction is an important segment of agricultural mechanics. (*Courtesy of U.S.D.A.*)

- the electrician who installs climate controls, silo unloaders, and milling equipment
- the soil conservationist who constructs terraces to control erosion
- the hardware store employee who must locate repair parts for agricultural tools and machines
- the air conditioning and refrigeration specialist in processing and storage facilities

FIGURE 1-4. Landscaping includes machinery and mechanical operations that require people in agricultural mechanics to keep machinery repaired and in top operating condition. (*Courtesy of U.S.D.A.*)

FIGURE 1-6. Workers in agricultural mechanics are responsible for the design, manufacture, testing, sales, and service of farm buildings, machinery, and equipment. (*Courtesy of S & R Photo Acquisitions*)

FIGURE 1-7. Many jobs in modern agricultural mechanics involve computerized operations. *(Courtesy of U.S.D.A.)*

- the designer and installer of field, turf, landscape, and golf course drainage and irrigation systems
- the lawn equipment service mechanic who repairs lawn tractors
- the welder who repairs farm machinery
- the mechanic who keeps the diesel trucks and machines going.

Processing plants for field crops, livestock, poultry, fruits, and vegetables all have machinery. Such machinery requires designers, engineers, operators, maintenance and repair personnel, and construction workers. Even people with jobs in finance, publications, and communications may need some knowledge of mechanics when their assignments deal with agriculture. All are likely to use computers and computer applications in their work.

FIGURE 1-8. Many jobs in agricultural mechanics are found in suburban or urban settings. Here, heat from the sun is trapped in the solar pond on the left and is then used to heat the production greenhouses on the right. *(Courtesy of U.S.D.A.)*

THE INFLUENCE OF MECHANIZATION

At the birth of the United States in 1776, over 90 percent of the American colonists were farmers, yet many of General Washington's troops at Valley Forge died for lack of food and clothing. Today, less than 10 percent of all Americans work in agriculture, yet there are generally food surpluses in America. The ratio of farm workers to nonfarm people in America approximately reversed itself in 200 years. In 1776, the farm-to-nonfarm ratio was approximately 9 to 1. Today, the ratio of agricultural workers to the remaining population is approximately 1 to 9. Mechanization has played a major role in this rise in production efficiency. **Efficiency** means ability to produce with a minimum waste of time, energy, and materials.

America provided the inventors for many of the world's most important agricultural machines. Cyrus McCormick invented the reaper in 1834 to cut small grain crops. Later, the combine was invented, which cut and threshed the grain in the field. Today, one modern combine operator can cut and thresh as much grain in one day as 100 persons could cut and bundle in one day in the 1830s.

Of great significance was Thomas Jefferson's invention of an iron plow to replace the wooden plows of the time. Later, in 1837, a man named John Deere developed a steel plow. This development permitted plowing of the rich prairie soils of the great American West. In 1793, Eli Whitney invented the cotton gin. The cotton gin removes the cotton seed from the cotton fiber. This paved the way for an expanded cotton and textile industry, In 1850, Edmund W. Quincy invented the mechanical corn picker. Joseph Glidden's development of barbed wire permitted establishment of ranches with definite boundaries. In 1878, Anna Baldwin invented the vacuum milking machine; and in 1904, Benjamin Holt invented the tractor.

Many of the early inventors worked alone or with one or two partners. They all could be considered workers in the area of agricultural mechanics. By the early 1900s, many people worked in factories or operated businesses. The companies that were formed to produce agricultural machinery or process agricultural products turned to invention also. For instance, mechanical cotton pickers were developed in the 1930s by several American companies.

Agriculture has become highly mechanized in the developed countries of the world. For the undeveloped countries, many engineers, teachers, and technicians have sought simple, tough, and reliable small machines to improve agriculture. In such countries, America's highly developed, complex, computerized, and expensive machinery will not do. Most countries do not have people trained for the variety of agricultural mechanics jobs that are needed to support America's agriculture.

Many features, such as rubber tires, have been standard equipment on American farms since the 1930s. Yet, a machine with rubber tires is useless if a tire is damaged and repair services are not available. This is the case in most undeveloped countries in Central and South America, Asia, and Africa. Much of the world cannot compete with American agriculture because the related agricultural products and services are not available to support the farm worker.

STUDENT ACTIVITIES

1. Define the Terms to Know in this unit.
2. Interview a Cooperative Extension Specialist for Agricultural Resources in your county or city. Ask the specialist to describe the different jobs people do in your locality that are regarded as agricultural or agriculturally related.
3. Look up "inventors" or "inventions" in an encyclopedia. Pick out the inventions that relate to agriculture and report your findings to the class.
4. Select three or five classmates to join you in a debate on the role of agriculture in society. One team should support the position that agriculture is the backbone of society. The opposing team should support the notion that it is not.
5. Consider an everyday product such as bread, milk, leather gloves, or a corsage for Mother's Day. Trace the production, processing, and marketing of the item from its source to finished product. List points along the way where agricultural mechanics are used.

SELF-EVALUATION

Multiple Choice. Select the best answer.

1. The production of plants and animals and the related supplies, services, mechanics, products, processing, and marketing defines
   a. horticulture
   b. renewable natural resources
   c. agricultural mechanics
   d. agriculture
2. Agribusiness is
   a. the same as agricultural mechanics
   b. limited to the sale of agricultural products
   c. business stemming from agriculture
   d. special work done by medical doctors
3. Examples of renewable natural resources are
   a. oil, gas, and coal
   b. fish, trees, and wildlife
   c. rubber, steel, and water
   d. air, soil, and minerals
4. The term *occupation* means
   a. business
   b. employment
   c. trade
   d. all of these
5. Agricultural mechanics stems mostly from
   a. physics
   b. biology
   c. medicine
   d. horticulture
6. Agricultural products come from
   a. soil and coal
   b. plants and animals
   c. iron ore and aluminum
   d. atomic fuel
7. Products of agriculture include
   a. leather seat covers
   b. paint
   c. flower arrangements
   d. all of these
8. Agricultural mechanics includes the occupation of
   a. garden tractor repairperson
   b. automobile mechanic

c. pile driver
d. systems analyst

9. Mechanization of agriculture has resulted in
a. decreased soil production
b. decreased farm expenses
c. increased production efficiency
d. increased numbers of farm workers

10. Cyrus McCormick invented the
a. steel plow
b. cotton gin
c. milking machine
d. reaper

11. In 1776, the ratio of farm workers to nonfarm people was approximately
a. 9 to 1
b. 1 to 1
c. 4 to 1
d. 1 to 9

12. Today, the ratio of agricultural workers to the remaining population is approximately
a. 9 to 1
b. 1 to 9
c. 4 to 1
d. 1 to 1

UNIT

2 Career Options in Agricultural Mechanics

OBJECTIVE

To determine how skills in agricultural mechanics may be used to earn a good living.

COMPETENCIES TO BE DEVELOPED

After studying this unit, you should be able to:

- List the major divisions in the agricultural cluster of occupations.
- Identify occupations in agriculture that require mechanical skills.
- Describe the relationship between mechanical applications and success in certain agricultural occupations.
- Conduct an in-depth study of one or more jobs in agricultural mechanics.
- Establish tentative personal goals for using agricultural mechanics skills.

MATERIALS LIST

✓ A computerized, microfiche, or other career information system
✓ *Occupational Outlook Handbook*

TERMS TO KNOW

FFA
off-the-farm agricultural jobs
occupational division
agribusiness and agricultural production
4-H
Boy Scouts of America

Many hats from which to choose! This is one way to state the career opportunities and options in agriculture. Figure 2-1 shows a young man with many hats stacked on his head and holding many tools. The picture suggests that young men and women who have agricultural skills have many good jobs available to them. Today, many jobs are found in urban areas; however, the serenity of rural areas also attracts many people to careers in agriculture, figure 2-2.

FFA members and agricultural education students have bright futures. Such students are preparing for an industry with many opportunities, figure 2-3. The advice is simple, "Plan ahead!"

AGRICULTURE IS NUMBER ONE

Agriculture is America's number one employer. Some estimate there are 23 million jobs in the broad field of

FIGURE 2-1. The career options available to agricultural education students provide "many hats" from which to choose.

agriculture. This is nearly 22 percent of America's total work force. A large proportion of these jobs are off-the-farm agricultural jobs. Off-the-farm agricultural jobs are those jobs requiring agricultural skills but not regarded as farming or ranching. The efficiency of this unique mix of farm and nonfarm agri-

FIGURE 2-2. The serenity of rural areas attracts many people to careers in agriculture. (*Courtesy of U.S.D.A.*)

A. Agricultural engineers and soil and water technicians seek better ways to use water wisely. (*Photo courtesy of Rick Parker*)

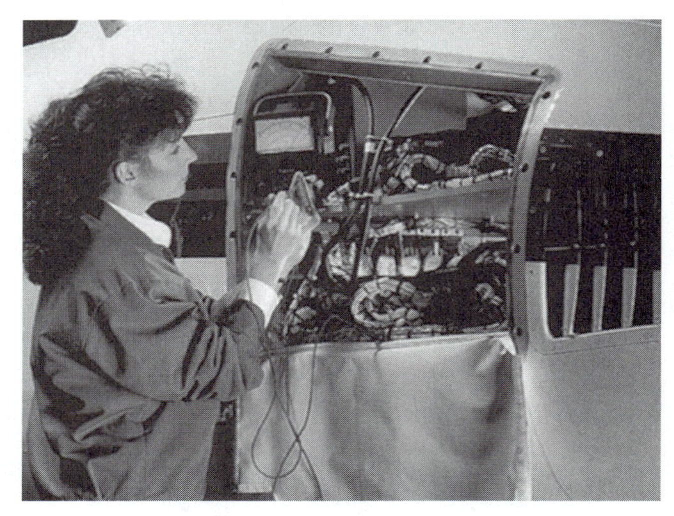

B. Repair of electronic equipment is a necessary part of modern agriculture. (*Photo courtesy of Cleveland Institute of Electronics, Inc.*)

C. Agriculture provides many jobs for those who prefer to work outdoors. (*Courtesy of U.S.D.A.*)

FIGURE 2-3. Many occupations are available in the field of agriculture.

cultural jobs permits the average American to use only 11 percent of his or her income on food. This is the lowest of any nation in the world.

The opportunities in agriculture are found at all levels of employment. The first level of employment is laborer; this level requires the least amount of preparation. A few laborer jobs require less than a high school diploma. Additional training is needed for the better paying jobs. Other job levels include semiskilled, skilled, managerial, and professional. Some professional jobs, such as veterinarian and scientist, require a doctor's degree. The doctor's degree requires seven or more years of college education. It is important to make a career choice as early as possible so the proper training can be obtained. Career goals should be set while in high school. High school courses, supervised agricultural experience, work experience abroad, and future schooling can then help prepare for the chosen career. Jobs are available in agricultural occupations at every work level.

Students enrolled in an agricultural education course have the edge on others who seek careers in agriculture. Agricultural education programs prepare the student specifically for the world of work, as well as for college. With so many areas inside the broad field of agriculture, choices must be narrowed. Choosing a division within agriculture generally permits more specializing and better job opportunities. The largest career areas in agriculture are in marketing and sales, scientists and engineers, figure 2-4. There are also many other areas to choose from. This suggests an excellent employment outlook for agricultural education graduates.

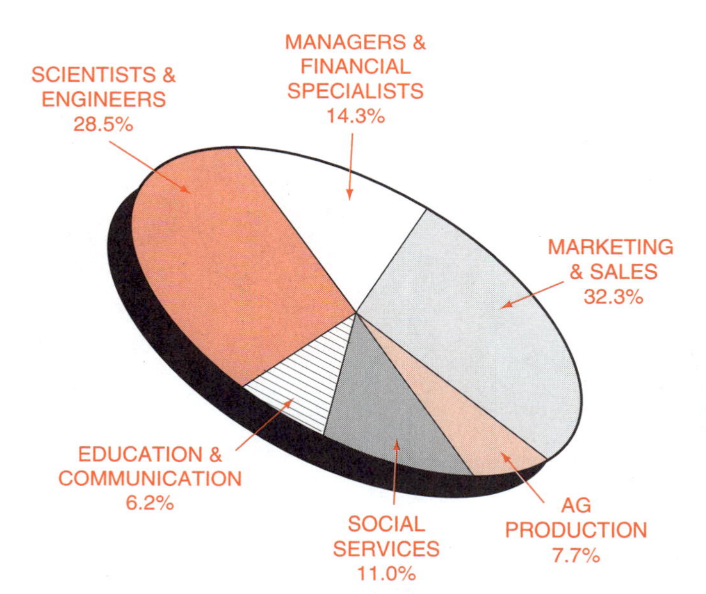

FIGURE 2-4. Employment opportunities for agriculture/agribusiness as reported by U.S.D.A. (*Courtesy of the U.S. Department of Education*)

AGRICULTURAL DIVISIONS

A number of United States government agencies have worked together to classify occupations. The National Center for Educational Statistics publishes *A Classification of Instructional Programs.* This book arranges all occupations into occupational clusters and divisions. An occupational division is a group of occupations or jobs within a cluster that requires similar skills. All jobs in agriculture are in one agricultural cluster grouped under the headings of 1) Agribusiness and Agricultural Production, 2) Agricultural Sciences, and 3) Renewable Natural Resources.

Divisions in Agribusiness and Agricultural Production

The agribusiness and agricultural production area contains eight divisions. They are:

- agricultural business and management
- agricultural mechanics
- agricultural production
- agricultural products and processing
- agricultural services and supplies
- horticulture
- international agriculture
- agribusiness and agricultural production, other.

Notice that agricultural mechanics is one of the divisions. Farm equipment operator is a part of agricultural mechanics. Many young people find satisfaction in a career operating heavy equipment, figure 2-5. There are many interesting jobs available in agri-

FIGURE 2-5. Many people make careers of operating the massive machinery used in agricultural occupations. (*Courtesy of Michael Dzaman*)

culture that use mechanical applications. Figure 2-6 shows how geothermal water is used to heat greenhouses. Figure 2-7 shows career areas in agriculture.

Divisions in Agricultural Sciences

The agricultural sciences have six divisions. They are:

- agricultural sciences, general
- animal sciences
- food sciences
- plant sciences
- soil sciences
- agricultural sciences, other.

Notice that each division under agricultural sciences has an *s* on science. This is because there are numerous sciences under each division.

FIGURE 2-6. Geothermal water at about 98°F is used to heat greenhouses. Pipes carry the hot water to the heater where the fan removes heat from the water and pushes warm air through the greenhouse. (*Courtesy of Rick Parker*)

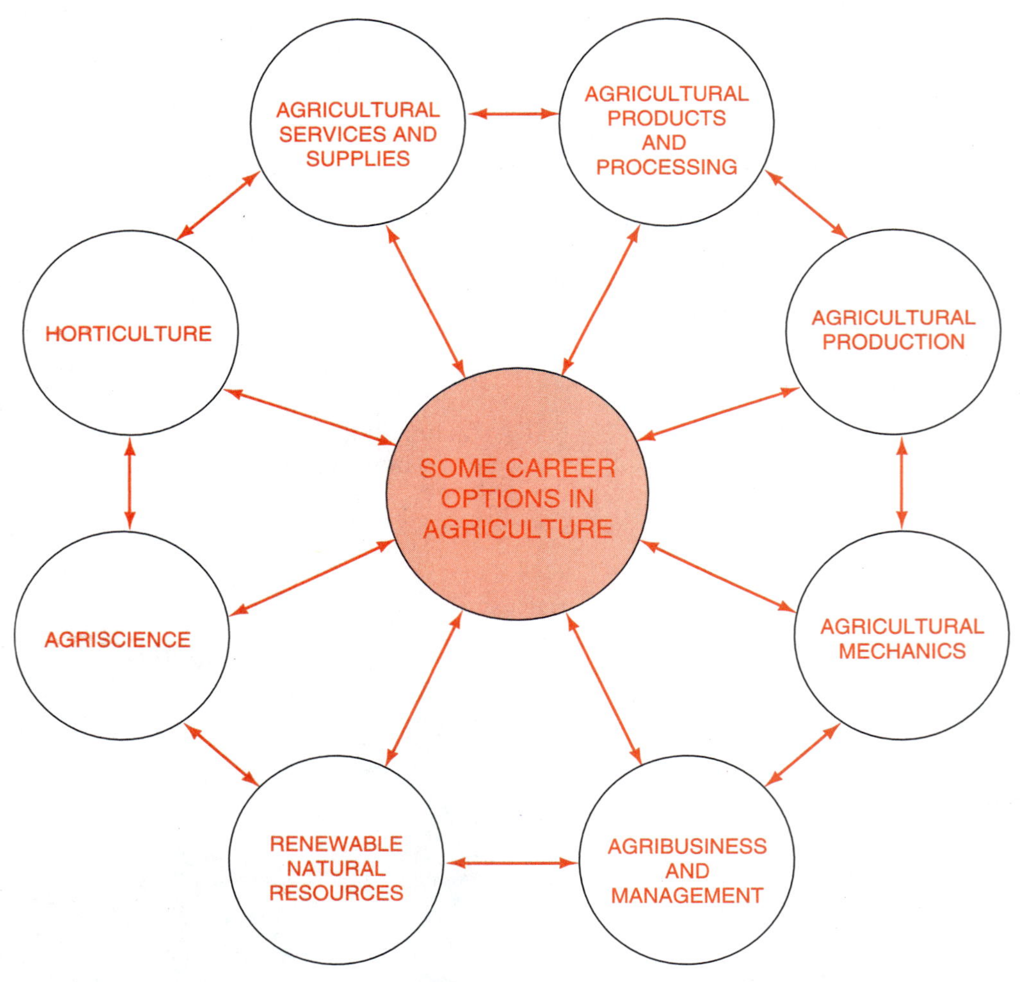

FIGURE 2-7. There are many career areas to choose from in agriculture. Each area relies on many others to be successful. For example, a greenhouse cannot be successful without people knowledgeable in mechanics and business.

American agriculture is based on scientific knowledge. Production, management, and mechanics all rely on information obtained through the scientific approach. Figure 2-8 shows laser technology being used in a land leveling operation.

FIGURE 2-8. Scientific application is an important part of agriculture. In this photo, laser technology is being used in a land leveling operation. (*Courtesy of* The National Future Farmer *magazine*)

Divisions in Renewable Natural Resources

The renewable natural resources area contains seven divisions. They are:

- renewable natural resources, general
- conservation and regulation
- fishing and fisheries
- forestry production and processing
- forestry and related sciences
- wildlife management
- renewable natural resources, other.

There are many jobs listed under the renewable natural resources division that require agricultural mechanics.

CAREER SELECTION

Agriculture teachers, guidance counselors, and librarians are sources of information regarding careers in the field of agriculture. Figure 2-9 provides a breakdown of the various divisions in agriculture.

There are specific job titles in agricultural mechanics. They are classified under the following categories:

- agricultural mechanics, general
- agricultural electrification, power, and controls
- agricultural mechanics, construction, and maintenance skills
- agricultural power machinery
- agricultural structures, equipment, and facilities
- soil and water mechanical practices
- agricultural mechanics, other.

Figure 2-10 shows some examples of jobs that are included in these classifications.

Farmers, ranchers, greenhouse operators, pesticide applicators, veterinarians, wildlife officers—all are better at their jobs if they have agricultural mechanics skills. Even those who simply use buildings, equipment, or materials find mechanical skills help them solve problems. As students plan courses in agricultural education, they should get as much experience in agricultural mechanics as possible. Such experiences should be stimulating and rewarding. They may occur in the classroom, shop, greenhouse, school farm, at home, in supervised occupational experience programs, or FFA.

In FFA, students may participate in agricultural mechanics or tractor operation contests. There are proficiency awards for farm and home safety and agricultural mechanics. Proficiency award winners receive up to $500 depending on the level of the award. Additionally, agricultural mechanics may be used as a speech topic in any organization.

Members of 4-H conduct many projects using agricultural mechanics skills. Some of these projects are wood science, electricity, tractor safety and maintenance, and automotive and small engines. The Boy Scouts of America has over 115 merit badge areas of which at least 25 involve agricultural mechanics skills. Some of these are camping, drafting, electricity, energy, engineering, farm mechanics, forestry, gardening, home repairs, landscape architecture, machinery, soil and water conservation, metalwork, plumbing, and woodwork.

MEETING THE CHALLEGE

Students having trouble choosing a specific career within agriculture should examine figure 2-11. This may help them set career goals. Most importantly, students should talk with the teacher, prospective employers, parents, grandparents, counselors, and workers to learn more about appropriate careers. Answering the following questions may be helpful in gaining insight into a career choice.

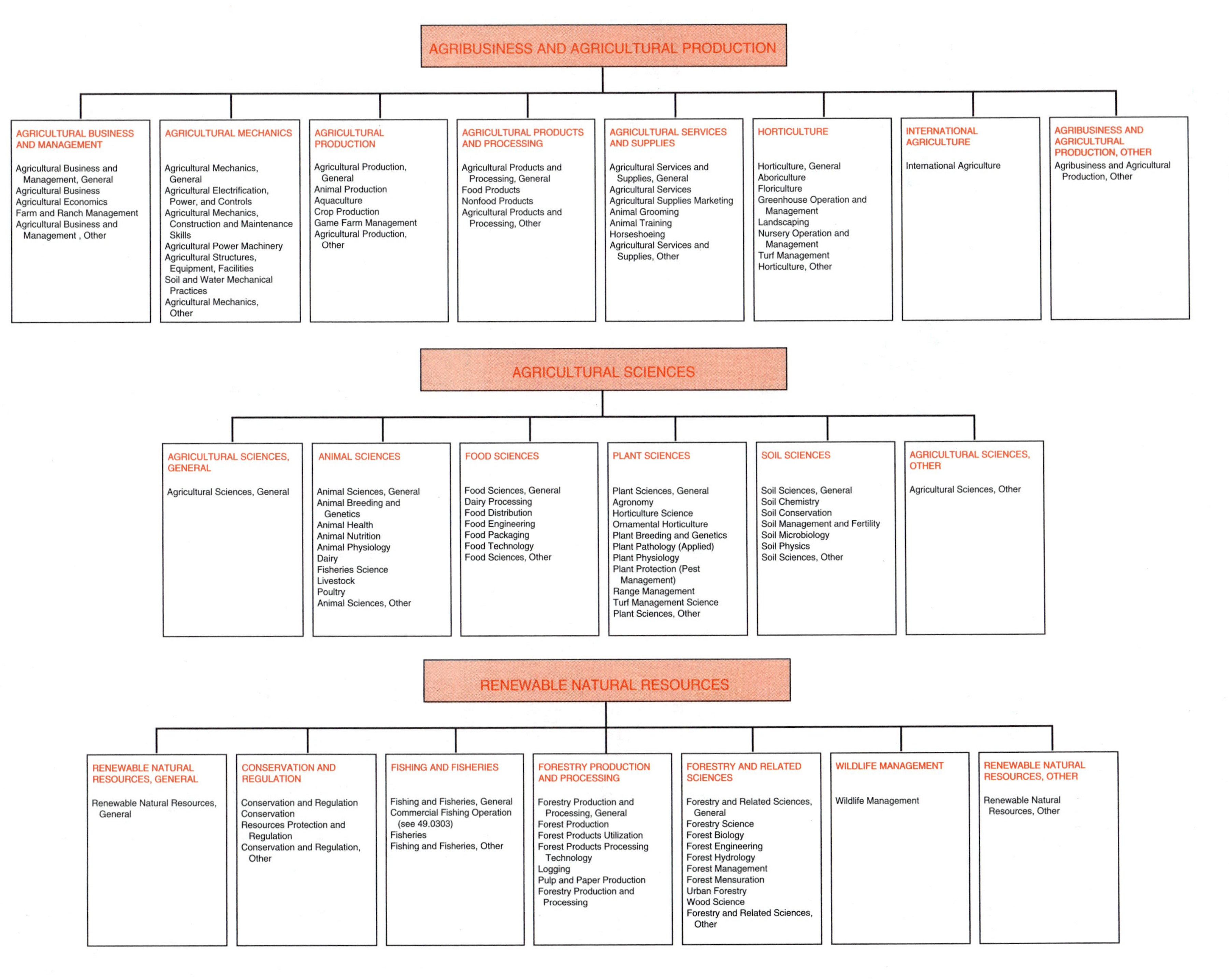

FIGURE 2-9. Agricultural programs as listed in *A Classification of Instructional Programs* (1981), National Center for Education Statistics, U.S. Department of Education, Washington, DC 20202. (*Courtesy of the U.S. Department of Education*)

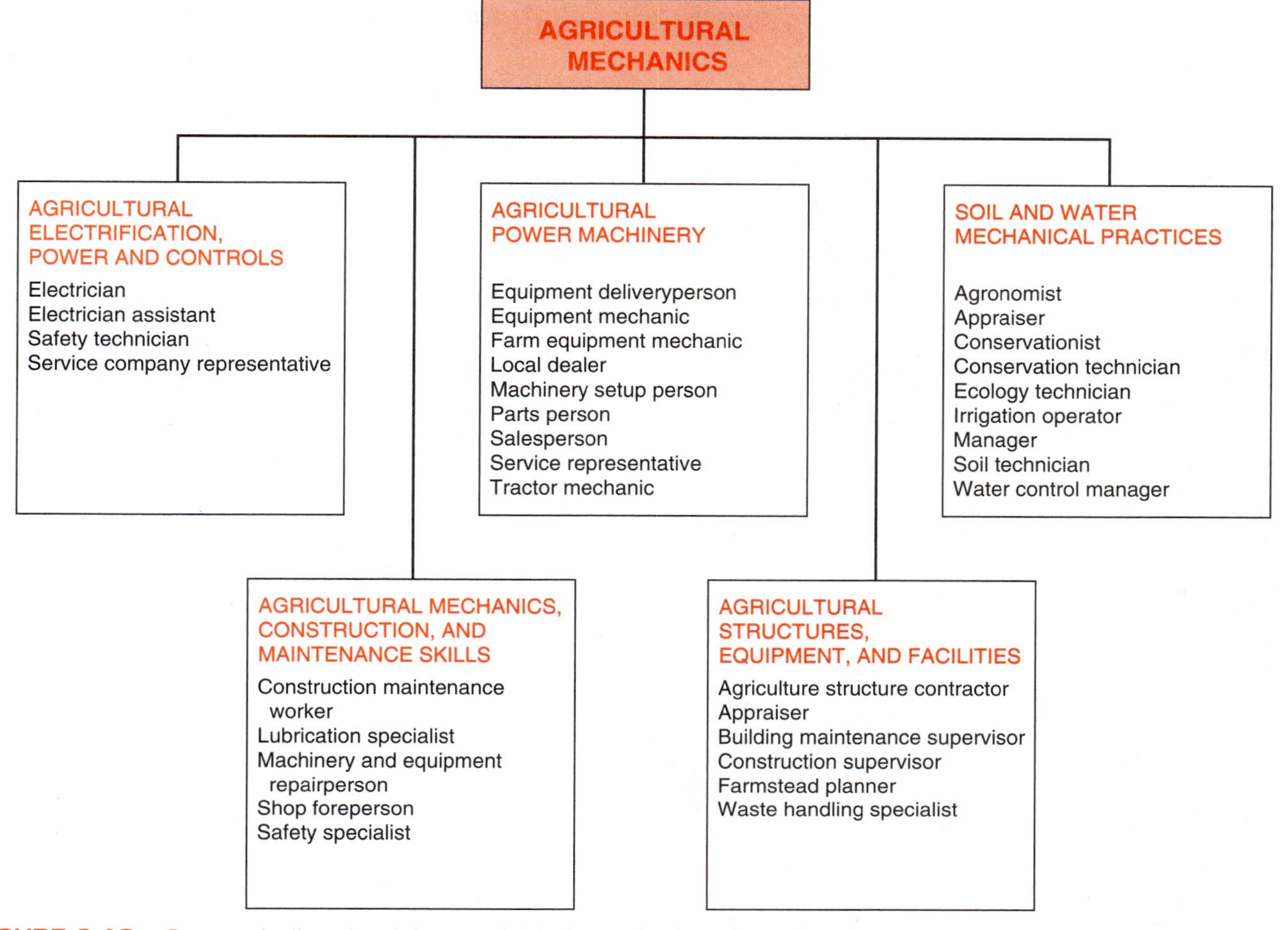

FIGURE 2-10. Some challenging jobs available in agricultural mechanics.

- Do you want to feel pride in a job well done?
- Do you want to help others by doing a job for them they cannot do themselves?
- Do you want to keep machinery operating efficiently from soil preparation through harvesting?
- Do you want to work on engines, build buildings, operate equipment, or install electrical equipment?
- Are you interested in welding, hydraulics, soil management, water management, or use of tools?

Agricultural mechanics is an area that can lead to great specialization or it may provide supplemental skills for most agricultural-related careers. It can lead to business and career opportunities locally or it can lead to international opportunities. Agricultural mechanics skills may be helpful in many career areas, or may be absolutely necessary, depending on the career choice. Students planning a career in this field need to get prepared.

To get prepared, students should:

- visit with people who have jobs in agriculture that use mechanical skills
- prepare a list of questions to ask workers about their jobs, figure 2-12
- talk to as many people as possible about agricultural mechanics
- plan to learn every skill possible in school, at home, and on the job
- learn what and why in the classroom
- learn how through shop and laboratory activities.

You may plan to go into business for yourself and be an employer. Or, you may wish to work as an employee. Either way, it will be helpful to develop good work and management skills. The experts say one needs the proper schooling, good grades, a serious work attitude, good communication skills, and leadership competencies to do well in today's competitive job market and business world.

FIGURE 2-11. Areas of career opportunities in American agriculture. (*Courtesy of* FFA New Horizons *magazine and the College of Agriculture, University of Illinois*)

The following is a list of questions which could be used when interviewing people about their occupation. This list is not complete and is intended to be used as a guide for developing questions.

1. Why did you pick this job?
2. How did you get started in your occupation?
3. How did you choose your place of training?
4. What educational, training, and other qualifications are there for the job?
5. If you should wish to change jobs, would the training contribute in any way?
6. Do you think this job would have a good future for me?
7. How could I get started in this career?
8. What is the salary range of this occupation?
9. What could a beginning person expect to make?
10. What are the fringe benefits?
11. Do you get paid vacations?
12. Do you have medical insurance?
13. Is there any chance of layoff? If so, how often?
14. What sort of planning does this business have for retirement?
15. What do you or don't you like about your job?
16. What are the advantages?
17. What are the disadvantages?
18. What are the hours and working conditions?
19. Do you ever have to work holidays? If so, which ones?
20. Do you ever work on weekends?
21. Is there a special uniform you must wear, or are you free to wear what you want? Does the company provide the uniform or does the employee?
22. What tools do you need?
23. Do you have to provide your own tools and equipment?
24. What are the physical requirements?
25. What do you do in this occupation?
26. How much traveling is involved?
27. What kinds of people do you work with?
28. Are there chances for advancement?
29. What are your responsibilities?
30. Do you belong to a union?
31. What's a typical day like for you in this job?
32. Is there any on-the-job training?
33. Has there ever been a time when you couldn't stand your job? If so, why and when?
34. Do you have to move if the company does?
35. What work experience did you have before you started to work in this occupation?
36. Who depends on your work? Upon whom do you depend?
37. Are there opportunities for advancement in this job? If so, what are the requirements for advancement?
38. How does your job affect your personal life?
39. What kinds of people do you meet?
40. Do you work mainly with people or things?
41. Do you work a lot with ideas?
42. Does your job offer opportunities to be creative?
43. Are people with your skills usually needed—even when business may be bad?
44. Is your work at all seasonal?
45. Could you briefly describe the personal qualities a person would need to do your job—strength, height, agility, ability to think rapidly, ability to make decisions, ability to deal with other people, etc.?
46. Would you recommend this kind of work for your children?
47. How do you spend your time after work?
48. If you could have any job in the world, what would you like to be?
49. Do you still go to school for special training?
50. When are people promoted? When are people fired?

FIGURE 2-12. Some questions for career exploration interviews.

STUDENT ACTIVITIES

1. Define the Terms to Know in this unit.
2. Check the *Dictionary of Occupational Titles* for agricultural mechanics job titles that interest you.
3. Ask your school guidance counselor to help you use the career information systems available in your school. Ask about computerized and/or microfiche systems. Use these or other systems to find information on three or more job titles of your choice in agricultural mechanics.
4. Use the above sources and the *Occupational Outlook Handbook* to prepare a report on the job in agricultural mechanics that interests you the most. Include the following parts in your report: a) job title, b) outlook for the future, c) working conditions, d) personal and educational requirements for success, e) special advantages of the job, and f) hazards and other disadvantages.
5. Consult the *Yellow Pages* of your local phone book. Record the names, addresses, and phone numbers of all agencies that hire people in the job of most interest to you.
6. Use the classified ads section of a large newspaper. Mark all jobs mentioned in which a knowledge of agricultural mechanics would be helpful.
7. Visit a business that hires people with agricultural mechanics skills. Ask the employer and some workers about the working conditions, travel requirements, employee benefits, skills needed, salary, and other items of interest to you.

SELF-EVALUATION

A. **Multiple Choice**. Select the best answer.

1. America's number one employer is
   a. agriculture
   b. chemicals
   c. oil
   d. steel
2. The estimated number of agriculture-related jobs in the United States is
   a. 5 million
   b. 11 million
   c. 23 million
   d. 44 million
3. The first level of employment in agriculture is
   a. skilled
   b. semiskilled
   c. professional
   d. laborer
4. The U.S. Department of Education classifies agricultural jobs by the number(s)
   a. 01
   b. 02
   c. 03
   d. all of these

B. **Matching**. Match the job classification in column I with the job title in column II.

Column I

1. agricultural electrification, power, and controls
2. agricultural mechanics, construction, and maintenance skills
3. agricultural power machinery
4. agricultural structures, equipment, and facilities
5. soil and water mechanical practices

Column II

a. farmstead planner
b. electrician assistant
c. machinery setup person
d. safety specialist
e. ecology technician

Using The Agricultural Mechanics Shop

UNIT 3

Shop Orientation and Procedures

OBJECTIVE

To recognize major work areas and use safe procedures when working in an agricultural mechanics shop.

COMPETENCIES TO BE DEVELOPED

After studying this unit, you should be able to:

- Identify major work areas in an agriculture mechanics shop.
- State school policies regarding shop procedures.
- State general safety precautions regarding the shop.
- Inform parents of shop policies and procedures.
- Sign a shop policy and procedures statement and a physical problems statement.
- Have parents sign a shop policy and procedures statement and a physical problems statement.
- Submit signed policy and procedures statement and physical problems statement to the teacher.

TERMS TO KNOW

hands-on experience
electricity
scale drawing
plan reading
farmstead layout
concrete
supervised occupational experience program
safe
efficient
policy
procedure

MATERIALS LIST

- ✓ Tape
- ✓ Drawing paper
- ✓ Pencil, eraser, and ruler
- ✓ Shop policies and procedures statement form
- ✓ Physical problems statement form

THE SHOP AS A PLACE TO LEARN

The agricultural mechanics or ag mechanics shop is a wonderful place to learn. Such shops may be found in schools, on farms, or in other agricultural businesses. Some people have shops in their basement or in special buildings, figure 3-1. Good lighting, adequate electrical power, and grounded electrical outlets are essential for safety and efficiency in all shop areas. A study bench and vise are basics for the shop. Some people have tools and equipment that are easy to move from place to place, figure 3-2.

Sometimes students and teachers refer to the agricultural mechanics shop in school as the ag mech shop or simply the ag shop. Students learn many things that help them now and in the future. Important skills for now include the use of basic hand tools and power equipment, figure 3-3. Most teenagers and young adults find many uses for tools, such as repair of bicycles, automobiles, stereos, appliances, garden equipment, and engines. Agricultural students use tools to help them in their agricultural projects and agricultural experience programs.

The modern agricultural mechanics shop has facilities for the serious student to learn skills that are useful for a lifetime. Many adults who are agricultural education graduates learned mechanical skills when in high school. They now use these skills at home, in their businesses, and for leisure.

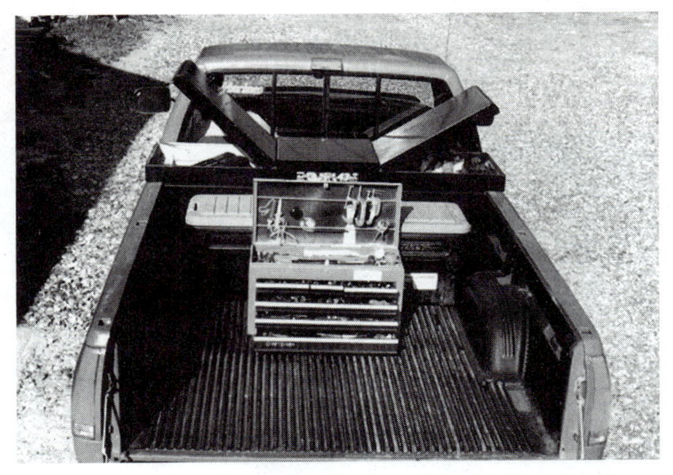

FIGURE 3-2. A large tool box can hold frequently used tools, fasteners, and lubricants. The portability of the tool box means the tools can be taken to the site for repairs to buildings and/or machinery. (*Courtesy of S & R Photo Acquisitions*)

Agricultural Shop Operations

Most agricultural mechanics shop operations fall under several broad categories:

- selection, care, and correct use of hand tools and power equipment

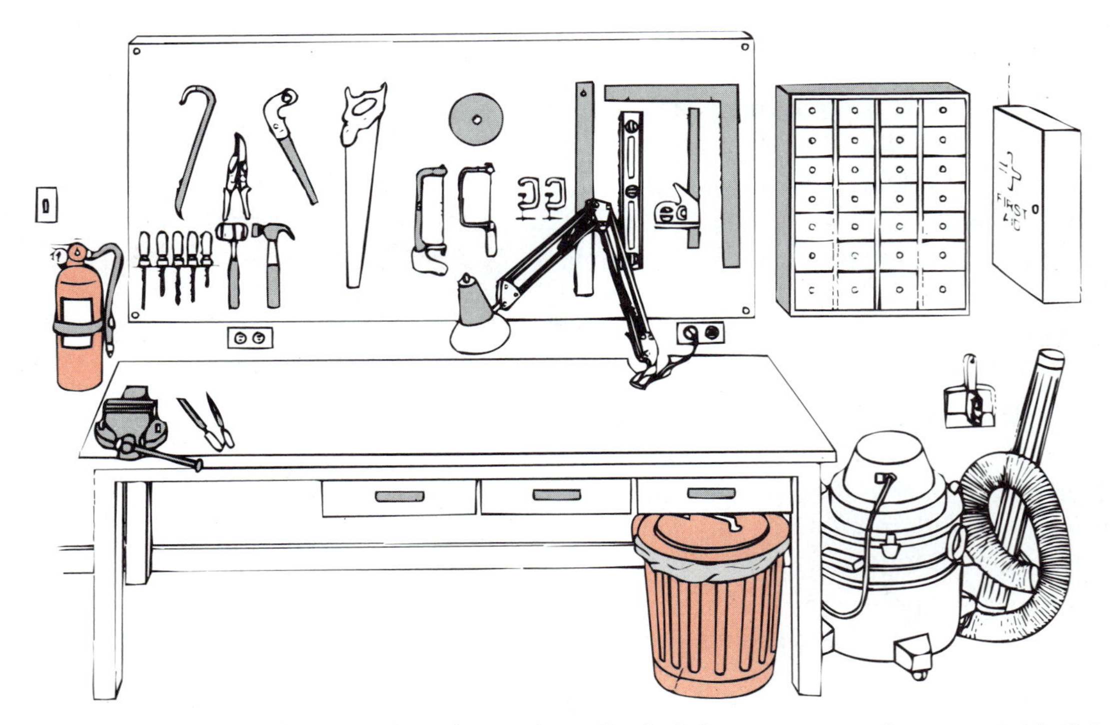

FIGURE 3-1. Many farm and nonfarm workers have shops in their basements or in a special building.

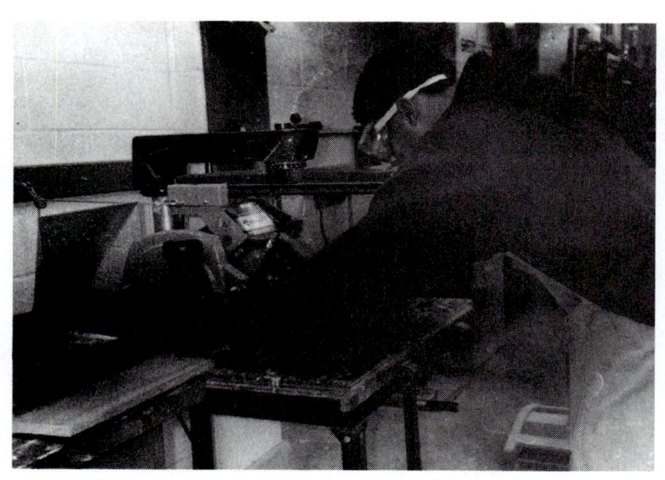

FIGURE 3-3. An agriculture instructor demonstrates the proper use of a radial arm saw.

FIGURE 3-4. The agricultural mechanics shop must have large, open areas for students to work on large projects.

- woodworking and carpentry
- use of sheet metal (metalworking)
- handling and fastening structural steel, including welding
- pipe selection and fastening (pipe fitting)
- rope work such as tying knots and splicing
- machinery maintenance and repair
- painting and finishing
- electrical wiring, motors, and applications
- hydraulic and pneumatic applications.

Many of these items deal with the use of materials for constructing equipment or buildings used in agriculture. Therefore, a large amount of space in a shop is used for material storage.

Agricultural Power Machinery

The agricultural mechanics shop should have large, open spaces for project construction and machinery repair, figure 3-4. Students typically build projects to develop skills. They also bring equipment from their homes, farms, or places of employment and repair it as a way of learning through hands-on experiences, figures 3-5 and 3-6. Hands-on experience simply means learning by doing an operation rather than just reading or talking about it. It is very important, however, that reading, discussion, and demon-

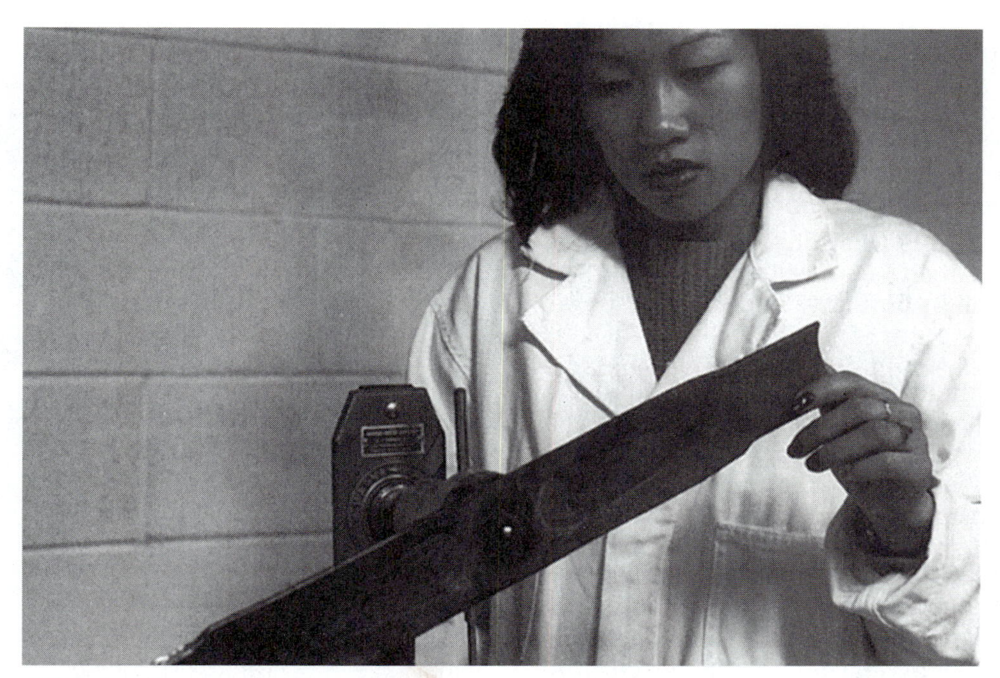

FIGURE 3-5. Rotary blade on a balancer. (*Courtesy of Emmons/*Turfgrass Science and Management, 2E, *Delmar Publishers, 1995)*

FIGURE 3-6. An agricultural student completes a tractor tune-up.

strations precede hands-on experiences for most activities.

Agricultural power and machinery experiences include the selection, management, adjustment, operations, maintenance, and repair of:

- gas and diesel engines
- trucks
- tractors
- field machines
- feed handling equipment
- crop storage and handling equipment
- special machines such as those used in horticulture
- other mechanical devices used in agriculture.

Agricultural Electrification

American agriculture is a very efficient industry. Because of this efficiency, the average American spends only 11 percent of income on food. American agriculture is more efficient than those of most other countries. This is due in part to the use of energy to operate machines.

Electricity is a form of energy that can produce light, heat, magnetism, and chemical changes. By using the energy of magnetism, engineers have developed electric motors and controls. Similarly, they create high-voltage current to provide spark for gasoline engines.

The agricultural student can benefit much from a knowledge of electricity and its use. This includes the use of electricity in the home, on the farm, and in all agricultural settings. Knowledge of electricity includes the selection, installation, and maintenance of wiring and electrical equipment, figure 3-7. Many of these skills may be learned in the agricultural mechanics shop.

Agricultural Buildings and Equipment

Many people in agriculture are around buildings and equipment in their work. This is true whether the person is an aquaculturalist, crop farmer, nursery operator, greenhouse operator, agricultural banker, or agriculture teacher. Many agricultural businesses are family businesses. Many of the construction and maintenance procedures on buildings and equipment are done by the owner, family members, or employees, figure 3-8. Therefore, agriculturists should have a thorough knowledge of building construction, and repair and maintenance of equipment.

Scale Drawing and Plan Reading. Most agricultural mechanics projects are too large to be drawn actual size on paper. Therefore, drawings must be scaled down (reduced) to make the dimensions proportionately smaller so the project can be planned on paper. Scale drawing is object representation on paper and uses a smaller dimension to represent a larger dimension, e.g., 1/4 inch on the paper = 1 foot on the object, or 1/4" = 1'. This means each foot of the actual object is represented by 1/4 inch on the drawing. The dimension ratio used must be indicated on the drawing. The actual dimensions of the object are stated on the drawing.

Scale drawings can also be made of a part or object that is too small to show details clearly. The actual dimensions of the part or object are increased on the drawing. A scale drawing that states 2" = 1" means the drawing is twice the size of the actual part or object.

FIGURE 3-7. Electric motor maintenance and repair are important skills for agricultural workers. *(Courtesy of U.S.D.A.)*

FIGURE 3-8. Equipment operators frequently are required to maintain and repair the equipment they operate. (*Courtesy of U.S.D.A.*)

Students build projects from plans developed by others, or they design their own projects personally. Either way, they need to know about scale drawing and plan reading.

Plan reading simply means using the scale drawing to build the project from the information given. The ability to do scale drawing and plan reading is basic to all construction, figure 3-9.

Farmstead Layout. Farmstead layout refers to the efficient arrangement of buildings on a farm. While arrangement has some relationship to construction, students typically study farmstead layout in the classroom rather than in a shop. Mistakes on a plan can be corrected quickly and easily. However, once construction is completed, mistakes are difficult and expensive to correct.

Functional Requirements of Buildings. Functional skills are needed to plan, install, or repair water systems, irrigation equipment, sprinkler systems, drainage pipe, and sewage systems, figure 3-10. This requires the agriculturalist to be a junior engineer or a jack-of-all-trades to keep operating costs down.

Concreting. Concrete is a mixture of portland cement, water, sand, and course aggregates (stones).

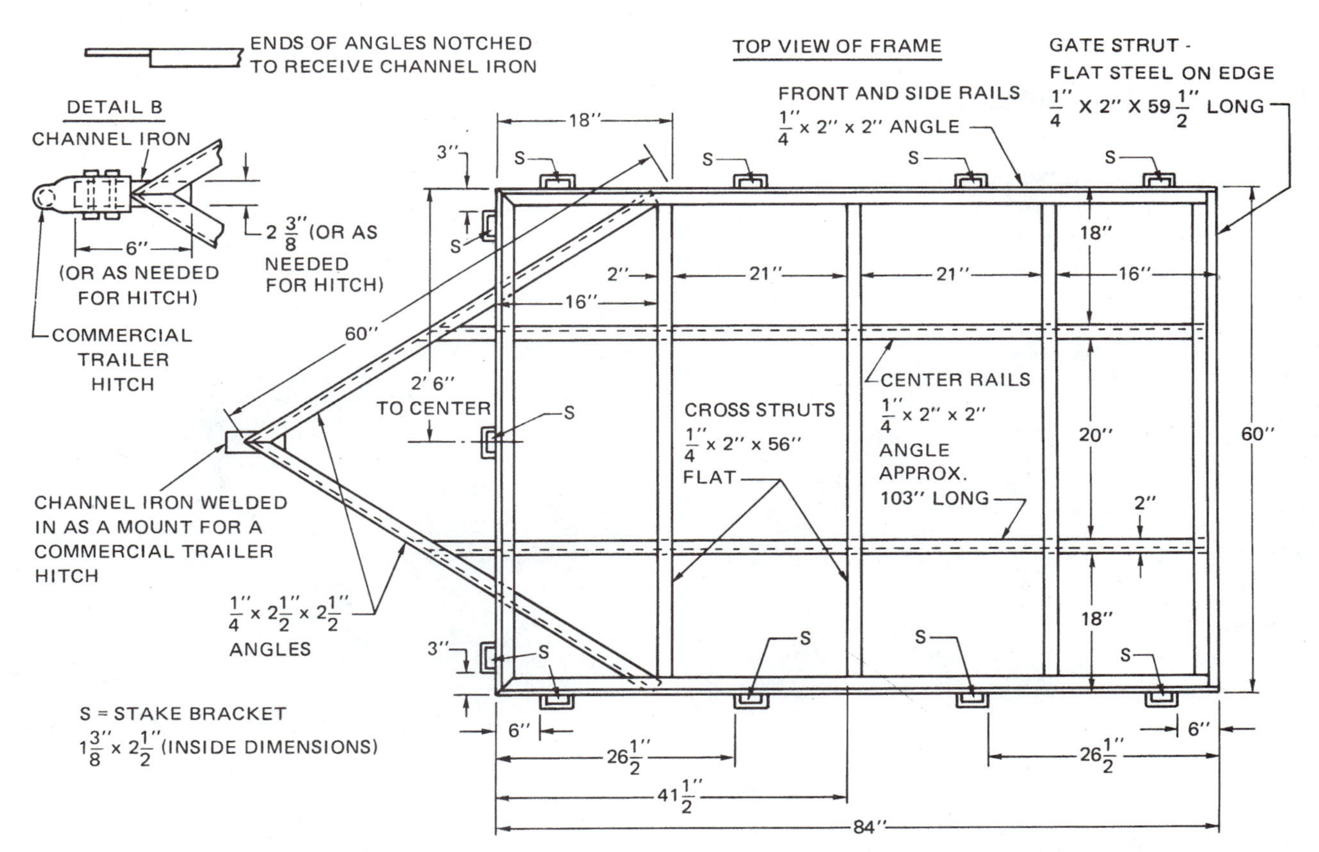

FIGURE 3-9. A plan or blueprint provides the worker with information on materials, dimensions, and details of construction.

FIGURE 3-10. This student is examining the impeller in a water pump. The next step will be to replace worn parts to restore the pumping capacity of the unit.

Concrete is used in agriculture for patios, landscape features, feeding floors, walks, roads, bridges, holding tanks, and other projects. Concrete can be poured to make objects or structures of nearly any shape. It is hard, durable, and resistant to abuse. Hence, much concrete is used in agriculture, figure 3-11.

Soil and Water Management

Mechanical skills needed for soil and water management are generally developed in the classroom or field. These include land leveling, land measurement, mapping, drainage, irrigation, terracing, contouring, and strip cropping, figure 3-12. However, some of the basic shop skills are useful in learning soil and water management practice.

MAJOR WORK AREAS IN THE AGRICULTURAL MECHANICS SHOP

Briefly, the major work areas in a modern agricultural mechanics shop are:

- tool storage
- materials storage
- woodworking
- finishing
- metalworking (including welding)
- electricity
- machinery repair
- spray painting.

In some schools, one or more of these areas are combined, such as tool storage and materials storage. In other schools, these areas are subdivided. For example, metalworking may be in three or four areas, such as cold metal, hot metal, sheet metal, and welding.

Work areas in the shop are generally not divided by definite barriers. Major walkways are often identified with strips of vinyl tape or painted lines. This helps students identify work areas and keep walkways clear of tools and materials, figure 3-13.

Large, open areas should be available inside the agricultural mechanics shop for students to build large projects and repair machinery that they use in their supervised agricultural experience programs, figure 3-14.

FIGURE 3-11. After pouring concrete, the worker must smooth the surface, provide construction joints, and finish the edges. *(Courtesy of S & R Photo Acquisitions)*

FIGURE 3-12. The bulldozer is used to construct terraces to control the flow of water on slopes. Proper installation of terraces and other soil and water controls help conserve both water and soil. *(Courtesy of U.S.D.A.)*

FIGURE 3-13. Shops with large, open areas and marked traffic lanes provide safe working conditions for agricultural students.

FIGURE 3-15. Large wood and metal projects are frequently built in areas outside the school shop. Typically, the area is secured by a woven wire fence and locked gates during nonschool hours.

The term Supervised Agricultural Experience Program (SAEP) refers to activities the student does outside the agricultural class or laboratory to develop agricultural skills.

Many school agricultural mechanics shops have fenced-in areas with concrete or blacktopped surfaces. Such areas are located just outside the shop and greatly increase the amount of space available for students to develop valuable shop skills, figure 3-15.

POLICIES FOR SAFE PROCEDURES IN THE SHOP

The school shop is similar to well-equipped shops used by contractors, farmers, or other agricultural businesses. Such home or farm shops must be equipped with basic hand tools, power equipment, and materials storage. Most procedures recommended for safe operation of a school shop also apply to home shops. Students are encouraged to think in terms of both situations as they study agricultural mechanics shop procedures and organization.

FIGURE 3-14. This farm machinery carrier is being constructed by students for use in their supervised occupational experience program.

Shop Size

The school shop must be big enough for twenty or more students to work at one time. Home and farm shops generally need work areas for only several people at one time. To be safe and efficient, the shop must be large enough to meet reasonable standards for space for each person in the shop. Safe means free from harm or danger. Efficient means able to produce with a minimum of time, energy, and expense. Even with the recommended amount of space per pupil, learning opportunities and personal safety are decreased without careful management.

Proper Instruction

Students must be properly instructed in the correct and safe way to use each tool. To avoid crowding, which creates serious hazards, students must be spread throughout the shop. The instructor generally has students working in different shop areas during the shop period.

Safety Policies and Procedures

Every school shop should have policies and procedures to improve instruction and help promote student safety. A policy is a plan of action or a way of management. A procedure is a method of doing things or a particular course of action. Figure 3-16 gives an example of a good policies and procedures statement for an agricultural mechanics shop. The instructor may add to or take away from this list according to established requirements in the school. It is suggested that the student discuss every item in

SHOP POLICIES AND PROCEDURES

Safe Conduct and Dress. I AGREE TO:

1. Purchase school insurance or bring a note signed by my parent(s) verifying that I have suitable personal accident and medical insurance.
2. Inform my teacher of any allergies or handicaps before using the shop.
3. Occupy my assigned place at the beginning of the period.
4. Wear safety glasses at all times when in the shop, use special goggles as needed for handling chemicals and caustic materials, and wear shields or helmet when conducting hazardous operations.
5. Wear prescribed protective clothing and not wear long neckties, loose sleeves, or other loose clothing.
6. Practice general cleanliness and orderliness at all times.
7. Not wear finger rings when handling molten metal or working on electrical systems.
8. Not loiter in the shop.
9. Not throw objects in the shop.
10. Avoid "horseplay" at all times in the shop.
11. Report all accidents including minor cuts, scratches, and splinters to the instructor promptly.
12. Remain in the shop or classroom at all times except when excused by the instructor.
13. Help with shop cleanup and storage duties until the job is finished.

Safe Machine Use. I AGREE TO:

1. Never operate power equipment unless specifically authorized.
2. Never operate machines unless the guards are in place.
3. Use all tools for the purpose intended and in the approved manner as taught by the instructor.
4. Stand to the side of grinding wheels, buffers, and blades while the machine is gaining speed, and out of the fire path when furnaces and other burners are being started.
5. Remove keys from chucks when adjustments are complete.
6. Fasten all work securely before drilling, milling, sanding, and other such operations.
7. Avoid talking to or otherwise distracting others using machines or doing hazardous activities.
8. Not leave machines unattended, nor repair, oil, clean, or adjust them while they are running.
9. Not force machines beyond the capacity for which they were designed.
10. Report to the teacher all electrical equipment that is not properly grounded or otherwise safe to use.
11. Report to the teacher all tools and machinery in need of repair, and any hazards that I observe.

Safe Materials Handling. I AGREE TO:

1. Obtain the teacher's permission and read the label before using any pesticides or other chemicals.
2. Handle, use and store pesticides and other chemicals properly.
3. Not use gasoline as a cleaner or solvent, and to handle all fuels in the prescribed ways.
4. Not permit water to be poured into acid or molten metal, nor permit grease or oil to come into contact with molten metal.
5. Handle and store all soiled materials, rags, paints, solvents, and flammable materials in the approved manner.
6. Lift or move heavy objects in the approved manner only.

I have read the above rules and discussed them in class with my instructor. I realize they are for my protection and I will do all I can to see that they are enforced. I will observe all precautions given by my instructor or others assigned to supervise my participation in school and school-related activities.

Signed ______________________ / __________
student's signature date

______________________ / __________
parent's signature date

FIGURE 3-16. Example of a shop policies and procedures statement.

the instructor's policy and procedure list in detail with the instructor and parents. Students and parents should sign the statement to indicate complete understanding. The signed statements may then be filed with other student records at the school.

Another important procedure to help protect students in the shop is to fill out and sign an allergies and physical problems statement, as shown in figure 3-17. Parents should sign these statements also to ensure proper medical treatment can be given in case of accident or other emergency.

ALLERGIES AND PHYSICAL PROBLEMS STATEMENT

The agriculture teachers and others responsible for providing treatment for me in case of an emergency should be aware of the following:

1. I am allergic to ______________________________
____
2. I should not be given the following medication
____
3. I have the following physical or personal problems
____
4. My family physician is ________________________
5. My physician's phone number is ________________
6. My parents' names are _________________________
7. My parents' phone numbers are _________________

Signed ______________________ / ________
student's signature date

______________________ / ________
parent's signature date

FIGURE 3-17. Example of an allergies and physical problems statement.

STUDENT ACTIVITIES

1. Define the Terms to Know in this unit.
2. Measure the length, width and other major dimensions of your agricultural mechanics shop. How many square feet are available per pupil in your class?
3. Make (or ask your teacher to provide) a drawing of your shop and any fenced-in area which may be used by students. Label the tool storage, lumber storage, metal storage, woodworking, metalworking, plumbing, electrical, and project areas.
4. List five projects which other agriculture students have constructed in the shop.
5. List five projects which you would like to make in the shop.
6. Refer to figure 3-16. Study each statement and list the number of each statement that is related to safety.
7. Read, discuss with your parents, sign, and obtain your parents' signatures on the policies and procedures statement for your shop. Return the signed document to your teacher.
8. Fill out, sign, and have your parents sign a personal allergies and physical problems statement for you. Return the signed statement to your teacher.

SELF-EVALUATION

A. Multiple Choice. Select the best answer.

1. The agricultural mechanics shop is also referred to as the
   a. ag mechanics shop
   b. ag mech shop
   c. ag shop
   d. all of these
2. The agricultural mechanics shop is a good place for serious students to learn skills that are useful
   a. now
   b. in the future
   c. both of these
   d. either of these
3. Agricultural mechanics includes
   a. woodworking and carpentry
   b. metalworking and welding
   c. pipe fitting and irrigation
   d. all of these
4. Hands-on experience means
   a. made by hand tools only
   b. a process of learning by doing
   c. a procedure requiring many people to help
   d. a wasteful method of education
5. Large, open spaces are needed in agricultural mechanics shops as compared with other shops for
   a. safe operation of stationary power equipment
   b. meeting fire code regulations
   c. storing materials for the school custodians
   d. student project work
6. Safety in the shop depends on
   a. students
   b. teacher(s)
   c. school shop designers
   d. all of these
7. Safety in the shop depends on
   a. students staying in assigned areas
   b. students wearing proper clothing
   c. use of safety glasses by all persons in the shop
   d. all of these
8. Which of the following enhances student safety in the shop?
   a. shop cleanliness and orderliness
   b. proper instruction
   c. machines that are kept in adjustment
   d. all of these

B. Matching. Match the definitions in column I with the correct term in column II.

Column I

1. energy source for agriculture
2. construction and repair
3. irrigation and drainage
4. woodworking, metalworking, welding, and pipe fitting
5. efficient arrangement of buildings
6. portland cement, water, sand, and stone
7. selection, management, adjustment, operation, maintenance, and repair
8. object representation on paper

Column II

a. agricultural shop operations
b. agricultural power and machinery
c. agricultural electrification
d. agricultural buildings and equipment
e. scale drawing
f. farmstead layout
g. concrete
h. soil and water management

UNIT

Personal Safety in Agricultural Mechanics

OBJECTIVE

To interpret safety colors and codes, protect the body against injury, and work safely in agricultural mechanics settings.

COMPETENCIES TO BE DEVELOPED

After studying this unit, you should be able to:

- State how to create a safe place to work.
- Recognize hazards in agricultural mechanics.
- List the types of parts and areas identified by various safety colors.
- Describe what each safety color means.
- Select appropriate protective clothing and devices for personal protection.

MATERIALS LIST

✓ Examples of shop protective clothing and devices

TERMS TO KNOW

safety
focal color
noise intensity
noise duration
decibel

Students sometime grow weary of teachers, parents, and others who constantly remind them to work and play safely. Yet, far more people are injured each year than need be to carry on the work of society.

Most people react to injury with anger at themselves or others who were responsible. This is because many injuries could have been avoided. People are willing to accept severe pain and hardship if they must, but they are resentful if they could have escaped the loss by being more careful.

Injury and disability are troublesome problems in society. Not only does the victim suffer, but others suffer as well. Friends of the injured are frightened and inconvenienced. Parents, guardians, spouses, and others frequently must take off work for visits to the doctor's office or hospital. They often forfeit income and must pay for extra expenses. Therefore, it is well worth the time to create safe places to work and to learn safety in everything. Safety means freedom from accidents. Some common causes of accidents are indicated in figure 4-1.

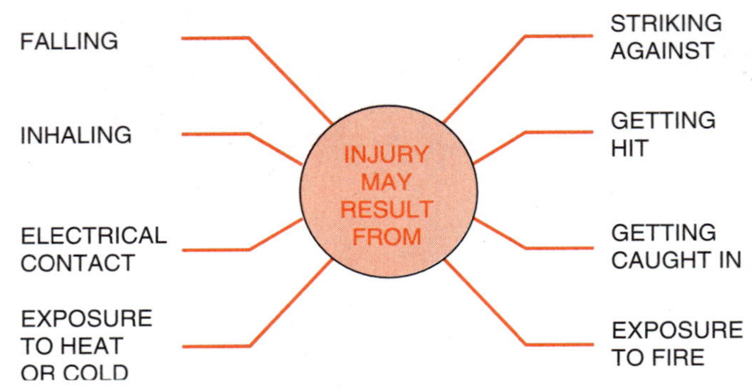

FIGURE 4-1. Some common causes of accidents.

Many accidents leave the victim partially or totally disabled. This carries a lifetime of regret. Many accidents are fatal. This results in needless loss of loved ones and bereavement to those left behind. Obviously, the thinking person will make every reasonable effort to work safely. This helps avoid the hardships caused by accidents. Accidents in farm populations involve machinery, drownings, firearms, falls, falling objects, burns, and others, figure 4-2.

THE SAFE PLACE TO WORK

Work in agricultural mechanics involves extensive contact with tools and machinery. Therefore, workers should be especially aware of the hazards that exist and take special precautions as needed. By taking the following precautions, a safer work place can be created.

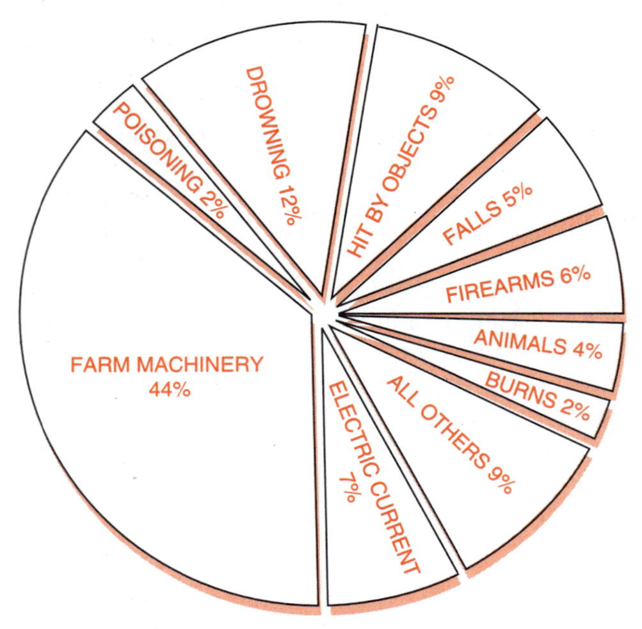

FIGURE 4-2. Many farm accidents are machinery related.

- Install all electrical devices according to the *National Electrical Code®*.
- Install all machinery according to the manufacturer's specifications.
- Keep all tools and equipment adjusted or fitted according to specifications.
- Use tools and equipment skillfully.
- Provide proper storage for tools, materials, fuels, chemicals, and waste materials.
- Keep work areas clean and free of tools, materials, grease, and dirt.
- Keep moving parts properly shielded.
- Manage all situations to avoid the likelihood of falling objects.
- Avoid areas where objects may fall.
- Avoid the flight path of objects that could be thrown by machines.
- Protect eyes, face, feet, and other parts of the body with protective clothing and devices.
- Move slowly enough to avoid creating hazards to self and others.
- Read and heed all precautions.

These precautions help make the work place safe. However, each person should have insurance to cover any personal injury or property damage that may occur.

SAFETY COLORS

National organizations have worked together to develop a safety color coding system for shops. The American Society of Agricultural Engineers and the Safety Committee of the American Vocational Association have published such a code. In developing the code, these agencies drew upon materials published by the American National Standards Institute (ANSI), the United States Department of Transportation (DOT), the National Safety Council (NSC), and the Occupational Safety and Health Act (OSHA).

Colors in the coding system are used to:

- alert people to danger or hazards
- help people locate certain objects
- make the shop a pleasant place to work
- promote cleanliness and order
- help people react quickly to emergencies.

Each color or combination of colors conveys a specific message, based on a standard code. Students need to memorize the message conveyed by each color; then the safety message will be understood in the shop. Shops must be properly painted for the color coding system to make its contribution to shop safety and efficiency.

The following descriptions show how each safety color is used to convey a safety message and help students work safely in the shop. The colors are listed in figure 4-3 and used as follows:

Red = Danger. Red is used to identify areas or items of danger or emergency such as safety switches and fire equipment.

Orange = Warning. Orange is used to designate machine hazards such as edges and openings. Orange is also used as background for electrical switches, levers, and controls.

Yellow = Caution. Yellow, like the amber traffic light, means be cautious. It is used to identify parts of machines, such as wheels, levers, and knobs that control or adjust the machine. Yellow and black stripes are used in combination to mark stairs, protruding objects, and other stationary hazards.

Blue = Information. Blue is used for signs if a warning or caution is intended. Such signs are made of white letters on blue background and carry messages such as "OUT OF ORDER" or "DO NOT OPERATE."

Green = Safety. Safety green is a special shade of green and indicates the presence of safety equipment, safety areas, first aid, and medical practice.

Black and Yellow Diagonal Stripes = Radioactivity. A black and yellow diagonal striped pattern is designated as the marking for radiation hazards.

NINE SAFETY COLORS

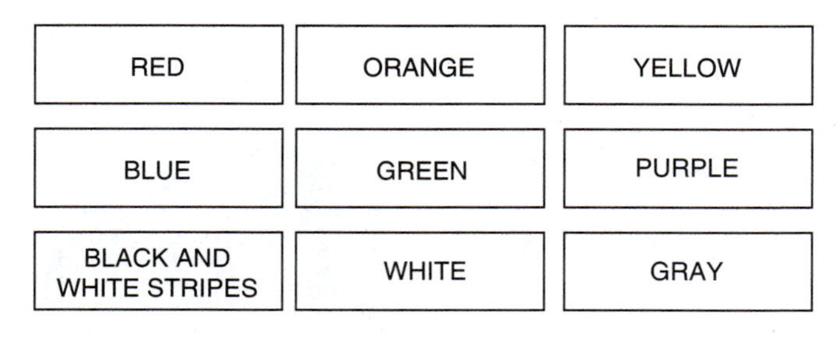

THREE FOCAL COLORS

FIGURE 4-3. The nationally accepted shop safety color-coding system uses nine safety colors and three focal colors to designate objects and areas.

White. White is used to mark off traffic areas. White arrows indicate the direction of traffic. White lines also mark work areas around objects in the shop. Yellow may be used in place of white to mark areas and lanes.

White and Black Stripes. White and black in alternate stripes or checkers are traffic markings. An example of such use is to mark traffic-stopping barricades.

Gray. Gray is used on floors of work areas in the shop. It is a restful color and provides good contrast for other safety colors. It is used to paint body areas of machines and may be used on the table tops if painting is desired.

FOCAL COLORS

The nationally accepted shop safety color coding system includes three focal colors. A **focal color** is used to draw attention to large items such as machines, cabinets, and floors. The focal colors provide contrast for the safety colors and create pleasant surroundings for people using the shop. The focal colors are ivory, vista green, and aluminum.

Ivory. Ivory is used to highlight or improve visibility of certain items. These items include tool storage chests, table edges, and freestanding vises and anvils.

Vista Green. Vista green is a special shade of green. It is used to paint bodies of machines, cabinets, and stationary tools such as vises. It is regarded as a pleasing color and contrasts with the safety colors.

Aluminum. Aluminum is used on waste containers such as those for scrap wood, scrap metal, and rags.

Some of these colors are used in combinations with other colors to mark pipes, hoses, and vents. Such markings identify the material in the lines as oxygen, natural gas, compressed air, air at atmospheric pressure, and water.

A properly color-coded shop is attractive and pleasant. The coloring system helps guide the user safely through many enjoyable experiences in agricultural mechanics.

PROTECTIVE CLOTHING AND DEVICES

Most work situations in agricultural mechanics require some type of body protection. The eyes, ears, hands, arms, feet, and legs are easily injured, figure

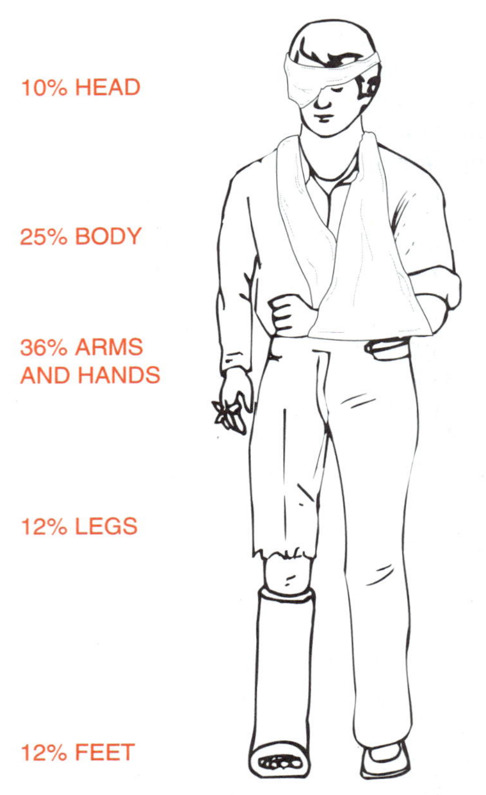

FIGURE 4-4. Studies show that any part of the body may be injured in the shop.

4-4. The best protection against injury is to prevent the accident from happening. The next best approach is to protect the body where injury may occur. Both lines of defense are needed to minimize injuries.

Safety Glasses and Face Shields

The face and eyes are regarded as the most critical parts of the body to be protected. This is because the eyes are so easily damaged and the face is easily disfigured. Flying objects striking the head can easily cause blindness or result in death. Acids, caustic chemicals, fertilizers, pesticides, solvents, molten metal, and hot water are all dangerous materials. Those who work around these materials should shield themselves against the possibility of accidental contact with them.

Safety Glasses and Goggles. Safety glasses and goggles offer minimum eye protection and are the first line of defense for the eyes, figure 4-5. Glasses and goggles should be the approved type with special impact-resistant lenses and side shields. They should fit the face and be kept clean for proper visibility. Single-piece goggles and cup goggles are used where special eye protection is needed against chemicals, flying objects, or damaging light. Special shaded lenses are needed when welding.

Many states have special laws requiring students to wear safety glasses or cover goggles upon entering and participating in shops and laboratories in the schools. Some states have laws requiring an approved face shield when an individual is:

- participating in, or exposed to, the immediate vicinity where hot molten metal or solder is being prepared, poured, or used in any form
- participating in, or exposed to, the immediate vicinity where milling, turning, sawing, shaping, grinding, sanding, cutting, or stamping of any solid materials are taking place
- participating in, or exposed to, the immediate vicinity where heat treating, tempering, or kiln firing of any material is taking place.

Approved cup goggles, helmets, or hand shields are required when in the areas of gas welding, electric arc welding, or weld flash exposure.

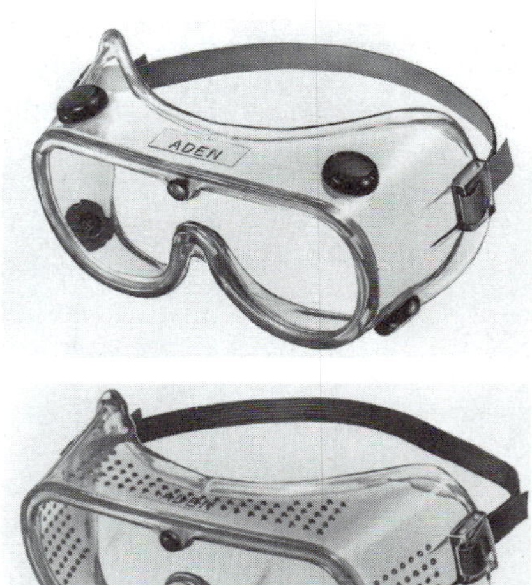

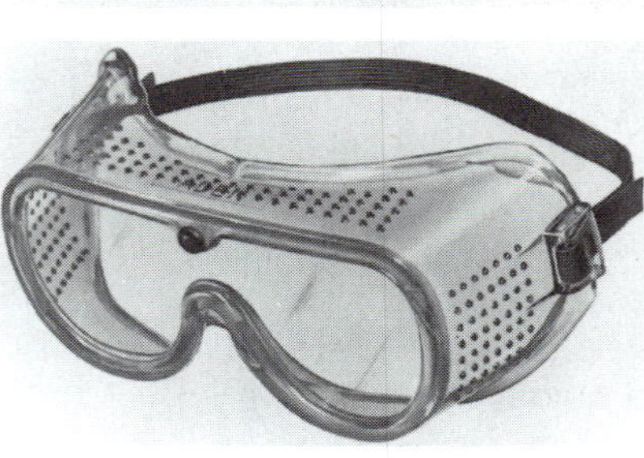

FIGURE 4-5. Safety glasses, goggles, or face shields should be worn when working around materials, processes, and machines. *(Courtesy of J. I. Scott Advertising/Marketing)*

Students must always wear safety glasses as minimum protection against eye injury. When students are in special hazard areas, special eye and face shields must be used, figure 4-6.

Hair Restraints. Serious accidents can occur if long hair becomes tangled in drill presses, saws, or other turning equipment. When hair is long or loose, it should be contained by wearing one of the following:

- a woolen hat
- a head band
- a hardhat
- a hairnet.

Protective Clothing

No part of the body is safe from injury in shop accidents. Suitable protective clothing that fits properly helps to prevent or reduce injuries. There should be no cuffs, strings, or ties for turning machinery to catch. Clothing should be fire resistant and provide protection from scrapes and abrasions. Protective clothing must be easily cleaned. Keeping clothing clean keeps clothing more fire resistant.

Coveralls. Coveralls are the most versatile and all-around item of clothing for agricultural mechanics. Coveralls protect the arms, body, and legs. They can be buttoned or zipped to the neck for maximum protection. Coveralls should fit well and be easily removable. Coveralls are safer if they do not have sleeve or pant cuffs. Cuffs are hazardous because flying sparks can be caught and trapped, resulting in the garment catching on fire. Bulky cuffs can catch in machinery and other objects.

FIGURE 4-6. Clear face shields provide protection to the entire face. They also provide extra protection to the eyes when worn over glasses.

Coveralls have the advantage of many pockets for pencil, pad, small tools, and objects the worker uses. The pockets should be covered with snap-down flaps to provide a smooth body cover that will not catch sparks and other objects. Coveralls should have an elastic waist, sleeves that button, and flaps that cover all buttons and zippers. When jackets, pants, or shirts are worn instead of coveralls, it is important to select clothing that fits snugly and has the same safety qualities as coveralls.

Aprons. Some shop teachers permit students to wear aprons in the shop or laboratory. Heavy cloth or leather aprons provide good protection for the front of the body and upper legs. Aprons are recommended only for limited shop work at benches or as additional covering over coveralls.

Vinyl or rubber aprons should be worn when liquids are used. Aprons are economical to buy and easy to store.

Shop Coats. Shop coats have the benefits of aprons plus additional body protection. The arms, body, and upper legs are protected for most work where the operator is standing. The shop coat is easy to put on. Therefore, it is used frequently by instructors and others who need to put on and remove protective clothing frequently throughout the day.

Footwear. Leather shoes with steel toes are recommended when working in the shop and when using machinery, figure 4-7. Many workers prefer shoes with 6-inch tops; others prefer the higher boot-type shoes. Leather is the preferred material for footwear because of its strength, durability, and comfort. The ability of leather to breathe explains why it is more healthful, cool, and comfortable than other materials. Farmers, ranchers, foresters, carpenters, plumbers, and all who handle heavy or hot objects need sturdy shoes. The popular soft, lightweight

FIGURE 4-7. Leather shoes with steel-reinforced toes provide excellent protection to the feet.

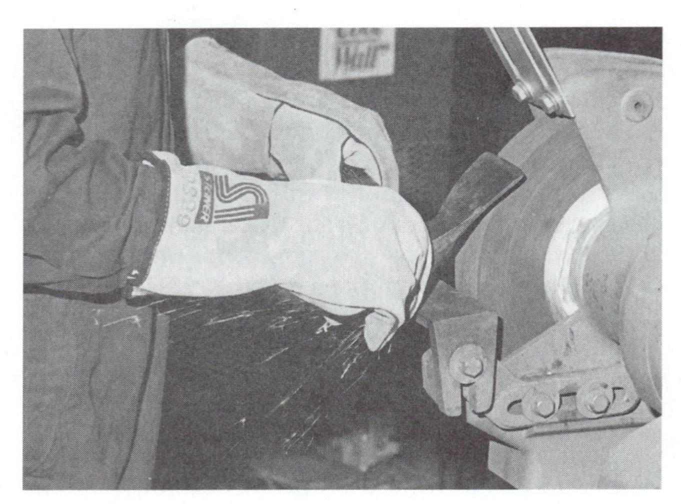

FIGURE 4-8. Leather gloves provide good protection for the hands.

FIGURE 4-9. Hard hats are worn where there is danger from falling objects or other sources of potential head injury. *(Photo courtesy of Rick Theriault)*

FIGURE 4-10. Face masks should be worn when dust is encountered in the work area.

vinyl or canvas shoes that many wear for casual activities are not safe for agricultural work.

Rubber boots are needed when working in water or using pesticides. Rubber is more durable than leather in extremely wet conditions and offers better protection against wet feet. Rubber will not absorb pesticides and can be washed free of such materials.

Gloves. Gloves are used to keep the hands warm as well as to protect them from excessive abra-

DURATION OF TIME PERMITTED AT VARIOUS SOUND LEVELS

Duration Per Day in Hours	Sound Level in dB
8	90
6	95
3	97
2	100
$1^1/_2$	102
1	105
$^1/_2$	110
$^1/_4$ or less	115
none	over 115

DECIBEL (DB) LEVELS OF COMMON SOUNDS AT TYPICAL DISTANCE FROM SOURCE

dB	Sound
0	Acute threshold of hearing
15	Average threshold of hearing
20	Whisper
30	Leaves rustling, very soft music
40	Average residence
60	Normal speech, background music
70	Noisy office, inside auto 60 mph
80	Heavy traffic, window air conditioner
85	Inside acoustically insulated protective tractor cab in field
90	OSHA limit—hearing damage on excess exposure to noise above 90 dB
100	Noisy tractor, power mower, all-terrain vehicle, snowmobile, motorcycle, in subway car, chain saw
120	Thunderclap, jackhammer, basketball crowd, amplified rock music
140	Threshold of pain—shot gun, near jet taking off, 50 hp siren (100')

FIGURE 4-11. The Occupational Safety and Health Act (OSHA) establishes maximum safe levels of noise, in decibels, for specified periods of time. Workers should not be exposed to a higher level of noise or noise for more than the recommended amount of time without wearing ear muffs or ear plugs. *(Courtesy of OSHA)*

sion, heat, liquids, or chemicals. Cloth gloves are suitable for warmth, but leather is needed where protection from heat, abrasion, or impact is needed, figure 4-8. Only rubber or vinyl gloves are suitable where liquids or chemicals are involved.

Hardhats. Hardhats are needed when working where objects are above head level or flying objects could be encountered, figure 4-9. Hardhats are made from special lightweight and impact-resistant materials. It is important that hardhats be approved by the Occupational Safety and Health Administration. Standards set by this agency help protect workers.

Masks and Respirators. Masks that cover the nose and mouth are needed to filter out particles of dust or spray paint, figure 4-10. Such materials irritate the nostrils and sinuses. Continuous inhaling of dust leads to lung diseases such as black lung and cancer. Effective dust masks are not expensive and should be worn when sanding, painting, welding, mixing soil, shoveling grain, or whenever dust is encountered.

Respirators that cover the nose and mouth and contain special filters are needed for certain jobs. When using pesticides, it is important to use the specific type respirator recommended by the pesticide manufacturer.

Ear Muffs and Ear Plugs. Ear protection is recommended when working in certain types and levels of noise. Equipment such as the radial arm saw, planer, router, chain saw, tractor, and lawnmower can produce noise that may damage the ears and cause a hearing impairment.

Ear muffs or plugs are recommended when the intensity, frequency, or duration of noise reaches certain levels. Noise intensity refers to the energy in the sound waves. Noise duration refers to the length of time a person is exposed to a sound.

Distance has a great effect on sound pressure or intensity. A person standing 5 feet from a machine can reduce the sound pressure to 25 percent by moving away another 5 feet. The decibel (dB) is the standard unit of sound. The Occupational Safety and Health Act (OSHA) has established a 90-dB noise level for an 8-hour period as the maximum safe limit. An 85-dB limit is safer. The sound level in decibels and duration of time are illustrated in figure 4-11. A sound level meter is used to determine noise levels.

Time is an important factor on the effect of noise on hearing. The ears can stand loud noises for a few minutes. That same noise may damage the ears if exposed for longer periods of time.

STUDENT ACTIVITIES

1. Define the Terms to Know in this unit.
2. Interview the school nurse or medical assistant to determine what kind of accidents have occurred in the school.
3. Search recent newspapers for articles on home, shop, farm, or work accidents and discuss them in class. Classify each accident or incident described in the newspaper articles as preventable or not preventable.
4. Make a bulletin board using newspaper clippings about home, shop, farm, or work accidents.
5. Make safety posters depicting the causes of accidental injury.
6. Contact your Cooperative Extension Service or other agencies and request current accident information regarding your community.
7. Study every machine and work area in your home shop or school shop, and list all hazards that should be corrected. If at home, correct them; if at school, assist your teacher in correcting them.
8. Examine your school agricultural mechanics shop for proper use of safety and focal colors. Ask your teacher what plans there are to correct any incorrectly painted areas.
9. Survey your agricultural mechanics shop for the following items: safety glasses for all students, goggles, face shields, welding helmets, protective clothing, respirators, hearing protectors, and fire extinguishers. Ask your teacher to acquire any of the items not on hand.

SELF-EVALUATION

A. Multiple Choice. Select the best answer.

1. Accidents among farm workers most often involve
   a. burns
   b. drowning
   c. falls
   d. machinery
2. For safety purposes, moving parts on machines should be
   a. labeled
   b. oiled
   c. painted
   d. shielded
3. Color coding is used in the shop to
   a. alert people to dangers and hazards
   b. make the shop a pleasant place to work
   c. help people react quickly to emergencies
   d. all of these
4. Which of the following is *not* regarded as a major type of accident that causes injury?
   a. assault and battery
   b. electrical contact
   c. falling
   d. inhaling
5. The national organization(s) that helped to develop safety color coding is / are the
   a. American Society of Agricultural Engineers
   b. American Vocational Association
   c. National Safety Council
   d. all of these
6. The safety color used to identify wheels, levers, or knobs that control or adjust machines is
   a. red
   b. yellow
   c. orange
   d. none of these
7. Fire equipment and safety switches are indicated by the color
   a. orange
   b. purple
   c. red
   d. bright green
8. The number of safety colors in the shop color-coding system is
   a. nine
   b. eight
   c. seven
   d. four
9. The number of focal colors in the shop color-coding system is
   a. one
   b. two
   c. three
   d. four
10. Suitable eye protection must be worn when working with
    a. chemicals
    b. grinding machinery
    c. welding equipment
    d. all of these
11. Protective clothing used in the shop must
    a. be fire resistant
    b. fit properly
    c. be clean
    d. all of these
12. The best item of protective clothing for agricultural workers is
    a. an apron
    b. a shop coat
    c. jeans
    d. coveralls
13. The length of time a person is exposed to sound is called
    a. noise intensity
    b. noise duration
    c. decibels
    d. sound pressure
14. Hearing damage may occur if excessively exposed to noise above
    a. 30 decibels
    b. 60 decibels
    c. 75 decibels
    d. 90 decibels

B. Matching. Match the meaning in column I to the correct color code in column II.

Column I

1. indicates warning
2. indicates danger
3. indicates information
4. indicates radioactivity
5. indicates caution
6. indicates safety
7. indicates traffic areas
8. indicates traffic markings

Column II

a. yellow
b. red
c. green
d. blue
e. black and yellow stripes
f. orange
g. black and white stripes
h. white

C. Completion. Fill in the blanks with the word or words that will make the following statements correct.

1. Many accidents leave the victim partially or totally ______________________.
2. All electrical devices should be installed according to the ________ ________ *Code*®.
3. All machinery should be installed according to the __________________ specifications.
4. The standard unit of sound is the ________.
5. Minimum eye protection is provided by wearing ______ ________ or __________.
6. The hazards of breathing dust and paint spray can be reduced by wearing a ________________.
7. The body parts most often injured in shop accidents are the ________ and ________.
8. When working with objects above head level, a ______________ should be worn.

UNIT

5

Reducing Hazards in Agricultural Mechanics

OBJECTIVE

To recognize and reduce hazards in agricultural mechanics settings, and react effectively in case of fire or other emergencies.

COMPETENCIES TO BE DEVELOPED

After studying this unit, you should be able to:

- Reduce hazards in agricultural mechanics.
- State the three conditions necessary for combustion.
- Match appropriate types of fire extinguishers to each class of fire.
- Use a fire extinguisher.
- Interpret labels on hazardous materials.
- Describe appropriate action in case of fire, accident, or other emergency.

MATERIALS LIST

✓ Different types of fire extinguishers
✓ Labels from agricultural chemicals

TERMS TO KNOW

fire triangle
fuel
combustion
heat
oxygen
extinguish
slow-moving vehicle
SMV
cardiopulmonary resuscitation

Fire, slow-moving vehicles, highway crossings, and chemicals create unique hazards in agricultural mechanics. Each may hold dangers to the worker and to others in the area. Fortunately, there are ways to reduce hazards and take action quickly if accidents occur.

REDUCING FIRE HAZARDS

The discovery of fire and how to create it was one of man's most important achievements. Fire is used to heat homes, cook food, generate electricity, melt ore to refine metals, heat metals to bend and form them, and cut metals. Yet, fire has not really been tamed. Fire breaks out of control at unexpected times causing injury and loss of property and lives. Burns are probably the most painful of all injuries.

It is known how to prevent uncontrolled fire, except in the case of volcano and lightning. Even lightning can be directed to a safe ground and thus reduce its destruction. The important thing is that most losses from fire can be prevented; however, it requires attention and knowledge of how fire works.

The Fire Triangle

To produce fire, three components must be present at the same time. These three components are fuel, heat, and oxygen. They are known as the fire triangle, figure 5-1.

Fuel is any combustible material that will burn. Combustible comes from the word combustion, which means to burn. Common fuels are gasoline, kerosene, diesel fuel, wood, paper, acetylene, and propane. Most materials will burn if they are made hot enough in the presence of oxygen.

Heat simply refers to a type of energy that causes the temperature to rise. If the temperature of a room is changed from 50 degrees to 70 degrees, it is done by using heat. Remember, most materials will burn if they are made hot enough in the presence of oxygen.

Oxygen is a gas in the atmosphere. It is not a fuel but must be present for fuels to burn. Oxygen is nearly always present except in airtight conditions. This fact is important to remember in fire safety and control.

Preventing Fires in Agricultural Mechanics

If any one of the three components of the fire triangle (fuel, heat, oxygen) is eliminated, fire will be prevented from starting; or it will be stopped if it has started. Therefore, to prevent, control, or stop fires:

- store fuels in approved containers
- store fuels away from other materials that burn easily
- store materials in areas that are cooler than their combustion temperature
- use fire only in safe surroundings
- put out fires by removing one or more elements in the fire triangle.

The prevention of fire goes hand-in-hand with safe use of equipment and efficient management of work areas. For instance, the proper use of a gas cutting torch decreases the likelihood of fire resulting from its use. Proper storage of materials decreases the chance of fire and keeps materials readily available when needed. Clean work areas prevent people from slipping or tripping, and damage to parts or projects. A clean work area also decreases the chance of a fire. Special paint booths provide a clean area for paint jobs and also decrease the likelihood of fire.

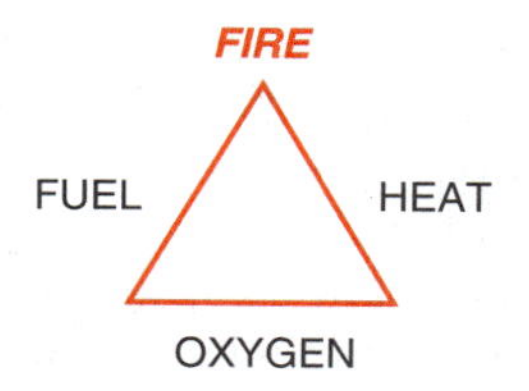

FIGURE 5-1. The fire triangle.

EXTINGUISHING FIRES

Fires are extinguished or put out by adding water to cool them, covering them to cut off the oxygen, or removing the fuel. This could mean wrapping a person whose clothing is on fire with a blanket. It could mean stopping fire in a field by raking grass and leaves out of the path of the fire or throwing soil on the fire to smother it. Fire at a gas torch or hose may be stopped by shutting off the gas at the cylinder. A burning container of paper may be extinguished by cooling it with water from a hose or bucket.

Classes of Fires

To effectively and safely put out a fire with a fire extinguisher, the class of fire must be known. This is determined by the material and the surroundings as follows:

- Class A—Ordinary Combustibles. Ordinary combustibles include wood, papers, and trash. Class A combustibles do not include any item in the presence of electricity or any type of liquid.
- Class B—Flammable Liquids. Flammable liquids include fuels, greases, paints, and other liquids, as long as they are not in the presence of electricity.
- Class C—Electrical Equipment. Class C fires involve the presence of electricity.
- Class D—Combustible Metals. Combustible metals are metals that burn. Burning metals are very difficult to extinguish. Only Class D extinguishers will work on burning metals.

Fire classification is based on how to safely and cheaply extinguish each type of material. Water is generally the cheapest material to use in fire control, but it may not be safe or effective. A firefighter can be electrocuted if the stream of water hits exposed electrical wires, plugs, appliances, or controls. Water is not suitable on fires involving petroleum products since the fuel floats to the top of the water and continues to burn.

FIGURE 5-3. Standardized symbols on fire extinguishers indicate the type of fire for which they are used.

Types of Fire Extinguishers

The proper fire extinguisher can put out a fire within seconds. However, such results occur only if the fire is extinguished when it first bursts into flames. The key is the proper extinguisher, used immediately, and in the proper way. This combination may make the difference between a mere frightful moment or a multimillion dollar fire loss with serious injuries and death.

Students should learn to recognize extinguishers by their type, figure 5-2. Common types of extinguishers are:

- water with pump or gas pressure (used for Class A fires)
- carbon dioxide gas (CO_2) (used for Class B and C fires)
- dry chemical (used for Class A, B, and C fires)
- blanket (used for smothering fires on humans or animals).

CARBON DIOXIDE FIRE EXTINGUISHER

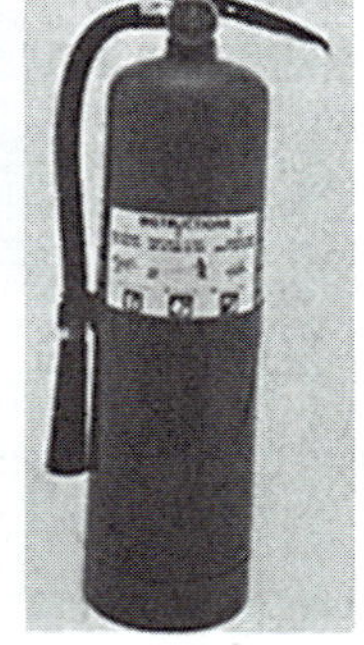

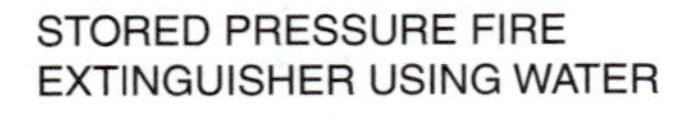

STORED PRESSURE FIRE EXTINGUISHER USING WATER

DRY CHEMICAL FIRE EXTINGUISHER

FIGURE 5-2. Types of fire extinguishers. *(Courtesy of Amerex Corporation)*

Extinguishers are marked according to the class or classes of fires on which they will safely work. Fire extinguisher labels contain standardized symbols to help the reader act quickly in an emergency, figure 5-3. The symbols are as follows:

- Green triangle—for Class A, ordinary combustibles
- Red square—for Class B, flammable liquids
- Blue circle— for Class C, electrical equipment
- Yellow star —for Class D, combustible metals.

Location of Fire Extinguishers

The location of fire extinguishers is very important. A few seconds lost in looking for the right type of extinguisher could allow a fire to rage out of control. Class A extinguishers should be placed in areas where Class A fires are likely to occur. Class B extinguishers should be placed in areas where Class B fires are likely to occur, and so on. Placing a water-type extinguisher in an area where an electrical fire is likely to occur is of little value. The extinguisher should not be used on an electrical fire.

Extinguishers should be placed in clean, dry locations near exits within easy reach. The extinguisher should be hung on the wall so the top of the extinguisher is not more than $3^1/_2$ to 5 feet above the floor, figure 5-4. The bottom of the extinguisher should be at least 4 inches above the floor. The extinguisher should be positioned so it can be removed quickly.

Everyone should be familiar with the locations and use of all types of extinguishers.

Using Fire Extinguishers

Generally, fire extinguishers are held upright and operated by a lever. However, some types are activated by inverting the tank, causing chemicals to mix inside the container. Sometimes the lever is blocked by a pin to prevent accidental discharge. Before operating an extinguisher, the instructions on the container should be read. For most extinguishers, a pin is pulled and a lever pressed.

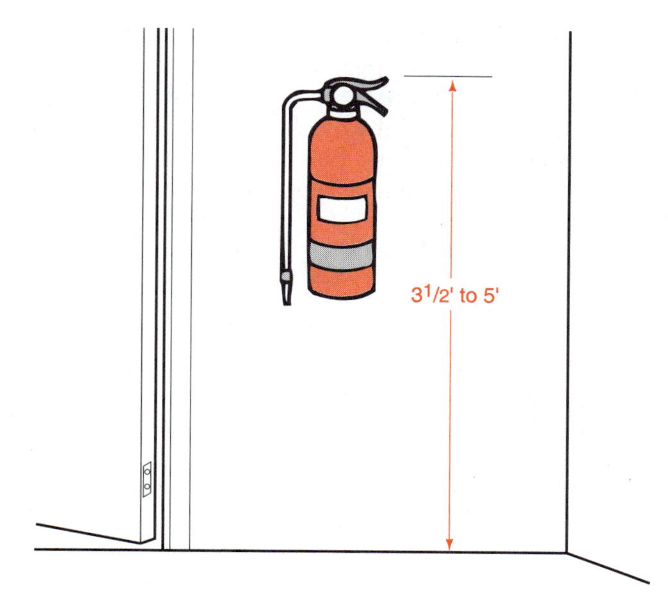

FIGURE 5-4. Fire extinguishers should be hung on the wall so the top of the extinguisher is not more than $3^1/2$ to 5 feet above the floor.

Before discharging an extinguisher, the firefighter should move to within 6 to 10 feet of the fire and direct the extinguisher nozzle toward the base of the fire, figure 5-5. The extinguisher will be empty in a matter of seconds, so any material that misses the fuel at the base of the fire will be wasted. A monthly inspection of all fire extinguishers should be made to ensure that the extinguishers are useable in case of an emergency, figure 5-6.

Always think before acting. Call for help immediately. Be sure any fire is completely out before leaving the area. The local fire department is the best source of help on fire safety and prevention.

MONTHLY FIRE EXTINGUISHER CHECK

- See that the proper class extinguisher is in the area of fire class risk.
- See that the extinguisher is in its place.
- See that there is no obvious mechanical damage or corrosive condition to prevent safe reliable operation.
- Examine or read visual indicators (safety seals, pressure indicators, gauges) to make certain the extinguisher has not been used or tampered with.
- Check nameplate for readability and lift or weigh extinguisher to provide reasonable assurance extinguisher is fully charged.
- Examine nozzle opening for obstruction. If equipped with shut-off type nozzle at the end of the hose, check the handle for free movement.

CHECK LIST

- ❑ Locate in a Proper Place
- ❑ Safety Seals
- ❑ Gauge or Indicator in Operable Range
- ❑ Proper Weight

FIGURE 5-6. Items to include in the monthly inspection of fire extinguishers.

SIGNS OF DANGER

There are many signs to warn of possible hazards. Stop, yield, caution, and crossing signs alert drivers to dangers on the highway. Danger, no trespassing, condemned, and keep out signs warn of dangers around old buildings.

FIGURE 5-5. Instruction for operating a fire extinguisher. *(Courtesy of Amerex Corporation)*

FIGURE 5-7. Safety warning labels on machinery are placed there by the manufacturer to alert operators of danger. It is important to follow the instructions to prevent personal injury or property damage. *(Courtesy of Anthony/*Farm & Ranch Safety Management*, Delmar Publishers, 1995)*

On the farm and in other agricultural mechanics settings, warning signs are found on machinery. Such signs may say not to remove a shield, not to overspeed, to keep bolts tight, or to position a part in a certain way, figure 5-7.

The Slow-Moving Vehicle Emblem

A very important sign for safety on the highway is the slow-moving vehicle (SMV) emblem, figure 5-8. This is a reflective-type emblem consisting of an orange triangle with a red strip on each of the three sides. It glows brightly when only a small amount of light hits it. Therefore, it is generally the first item to be seen on a vehicle. When operators of fast-moving vehicles know the meaning of the emblem, they have time to slow down before running up on slow vehicles. SMV emblems are required on all vehicles that travel a maximum of 25 miles per hour on public roads. It is important to have a SMV emblem on every piece of machinery on the highway. Drivers of automobiles and trucks always must be prepared to slow down when the SMV emblem comes into view.

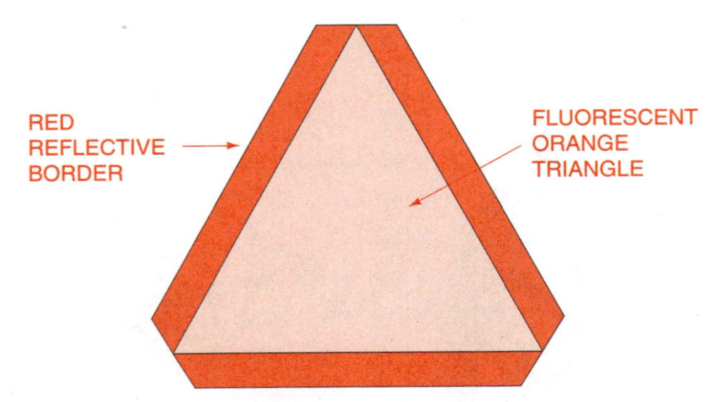

FIGURE 5-8. The slow-moving vehicle (SMV) emblem is commonly used on farm and industrial equipment, road trucks, and other vehicles that travel 25 miles per hour or less on public roads. *(Courtesy of the National Safety Council)*

Package Labels

All commercial products have a label to provide the user with certain information. Food product labels tell about nutrient content. Repair parts come with instructions for installation. Paint labels name the components, and clothing labels tell how they are to be cleaned.

Labels on hazardous products may be a matter of life or death in the event of accident. Common products such as kerosene and turpentine are poisonous if taken internally. Agricultural pesticides are products designed to be poisonous to pests; they may

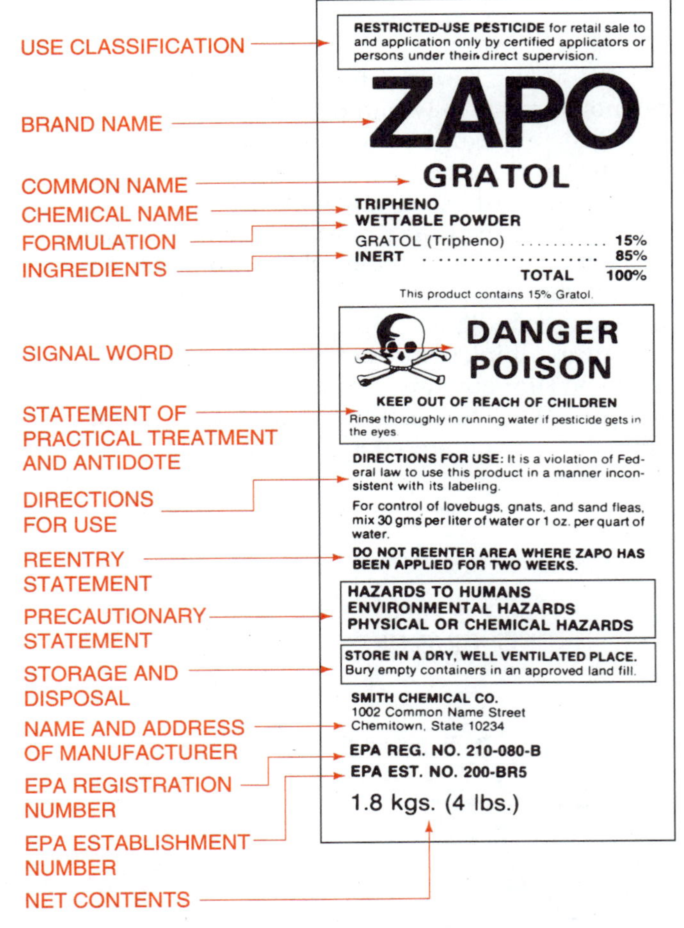

FIGURE 5-9. Agricultural pesticides and other hazardous chemicals have labels that warn the user of any hazards associated with the material. *(Courtesy of American Association for Vocational Instructional Materials)*

cause illness or death to humans if misused. If used according to the instructions on the label, pesticides are generally safe.

It is very important to keep materials in their original containers. Original containers are the correct type of container. They carry the label that describes the product, its hazards, and procedures to use in case of an emergency with the product.

Pesticide labels are legal documents. Label directions are required by law to be accurate and complete. Labels should be read again before handling, opening, mixing, using, or disposing of pesticides.

Pesticide labels have at least sixteen different items of information. Study the pesticide label shown in figure 5-9 to see how the sixteen items of information are displayed on the label. Pesticides should always be stored in a locked cabinet or area, figure 5-10.

EMERGENCIES OR ACCIDENTS

Quick action of the correct type can change the outcome of an emergency from a tragedy to simply a frightful moment. If fire should break out at school:

- notify the teacher
- keep everyone calm
- set off the fire alarm
- call the fire department
- clear the area
- use fire extinguishers if this seems logical under the circumstances.

FIGURE 5-10. Pesticides should always be stored in a locked cabinet or other locked area.

If an injury occurs, quick action is in order. Such action must be correct and based on good thinking; otherwise, additional injury may result. For many emergencies, the following procedures should be helpful.

PROCEDURE

1. Call or send for help.
2. Call for emergency police or ambulance service.

   NOTE: Some communities now have ambulance service and medical helicopters standing by that can reach an accident scene with paramedics and emergency equipment in minutes. These services often make the difference between life and death.
3. Do not move the victim unnecessarily.
4. Try to arouse the victim by talking.
5. Treat for shock. Keep the victim lying down. Elevate the victim's feet 8 to 10 inches if there are no signs of bone fractures or head or back injuries. Place a blanket under and over the victim to maintain body heat. If the victim complains of being thirsty, moisten a clean cloth and wet the victim's lips, tongue, and inside of the mouth.
6. If the victim is bleeding, stop the bleeding by wrapping or pressing clean cloth or gauze directly on the wound.
7. If the victim is not breathing, clear the air passage and find someone to administer rescue breathing or cardiopulmonary resuscitation. Cardiopulmonary resuscitation (CPR) is a first-aid technique to provide oxygen to the body and circulate blood when breathing and heart beat stops. CPR training is given to many teachers and students. The technique requires a special course. To save a victim's life, CPR must be started within 4 to 6 minutes of drowning, electrocution, suffocation, smoke inhalation, gas poisoning, or heart attack.
8. Do not move the victim if broken bones are suspected, unless problems with breathing, bleeding, or other life-threatening factors exist.

Qualified people should be called to give medical aid to accident victims. In schools, there are specific procedures for teachers and students to follow in case of injury or other emergency. Each student should learn these specific policies and procedures.

STUDENT ACTIVITIES

1. Define the Terms to Know in this unit.
2. List the types of fire extinguishers needed for classroom, shop, and laboratory. Check each fire extinguisher that is present for state of charge, type, proper hanger, and appropriate location. Ask your teacher to correct all problems regarding fire extinguishers.
3. Ask your teacher to arrange for a representative of the local fire department to visit your class and demonstrate appropriate fire safety practices.
4. Install slow-moving vehicle (SMV) emblems on pieces of school equipment that need them.
5. Install SMV emblems on pieces of equipment at home that need them.
6. As an FFA chapter project, conduct an SMV emblem sale and campaign to increase the use of the emblem in your community.
7. List on your class chalkboard all the signs of danger you and your classmates can name.

SELF-EVALUATION

A. Multiple Choice. Select the best answer.

1. Which is *not* part of the fire triangle?
   a. fuel
   b. combustion
   c. oxygen
   d. heat
2. A commonly used fuel is
   a. acetylene
   b. acetone
   c. oxygen
   d. magnesium
3. Fire can always be prevented or stopped by eliminating
   a. combustible gases in the area
   b. congestion in the shop
   c. improper storage of fuels
   d. any item in the fire triangle
4. Fire hazards associated with painting can be reduced by
   a. using a spray gun instead of a brush
   b. using newspaper to protect bench surfaces
   c. using a special paint booth
   d. painting with several people in the area
5. Effective fire control techniques include
   a. cooling a fire with water
   b. wrapping a blanket around a person whose clothes are on fire
   c. raking dead leaves and grass away from an advancing fire
   d. all of these
6. Fires are classified according to
   a. materials involved and techniques that safely extinguish them
   b. size and duration of the fire
   c. season of the year when the fire occurs
   d. the amount of material being burned
7. A green triangle on a fire extinguisher means the extinguisher can be used to put out burning
   a. metals
   b. liquids
   c. wood
   d. electrical wires
8. Most fire extinguishers will discharge when
   a. the pin is pulled and the lever is pressed
   b. the extinguisher is inverted
   c. either a or b, depending on the extinguisher
   d. none of these

9. SMV means
   a. small mechanical vehicle
   b. stop! moving vehicle
   c. slow-moving vehicle
   d. none of these
10. SMV emblems are required when
    a. vehicles are standing
    b. vehicles travel 25 miles per hour or slower
    c. vehicles travel 30 miles per hour or slower
    d. vehicles travel over 30 miles per hour
11. Pesticide labels are
    a. legal documents
    b. used only on insecticides
    c. used primarily on powdered chemicals
    d. generally written in two or more languages

B. **Matching.** Match the class of fire in column I with the burning material in column II.

Column I

1. Class A fire
2. Class B fire
3. Class C fire
4. Class D fire

Column II

a. paper and wood
b. flammable liquids
c. combustible metals
d. electrical equipment

C. **Matching.** Match the type of extinguisher in column I with the type of fire for which it is used in column II.

Column I

1. dry chemical extinguisher
2. water extinguisher
3. carbon dioxide gas extinguisher
4. foam extinguisher

Column II

a. Class B and C fires
b. Class A fires
c. Class A and B fires
d. Class A, B, and C fires

D. **Brief Answers.** Briefly answer the following questions.

1. Name four items of information provided on pesticide labels.
2. Name four signs that may be seen along highways that alert people of possible hazards.
3. Name three warning signs often found on machinery.
4. Describe the slow-moving vehicle emblem.
5. Name five things to do if a fire breaks out.
6. Name six emergency procedures used for accident victims.
7. Explain how to properly extinguish a fire using a fire extinguisher.

UNIT 6

Shop Cleanup and Organization

OBJECTIVE

To work cooperatively with classmates to clean the shop efficiently and store all tools and materials.

COMPETENCIES TO BE DEVELOPED

After studying this unit, you should be able to:

- Use shop cleaning equipment properly.
- Clean benches, machines, and floors.
- Store materials properly.
- Store tools properly.
- Do assigned tasks.
- Work cooperatively with others.

TERMS TO KNOW

silhouette
flammable materials cabinet
vertical rack
floor broom
dust mop
bench brush
scoop shovel
dust pan
cleanup wheel
cleanup skills checklist
cleanup assignment sheet

MATERIALS LIST

✓ Bench brush
✓ Floor broom
✓ Dust mop
✓ Vacuum cleaner
✓ Rag can
✓ Scrap wood box
✓ Scrap metal can
✓ Shop cleanup wheel chart
✓ Shop cleanup assignment sheet

A CLEAN AND ORDERLY SHOP

Each student should have a clear vision of what is meant by a clean and orderly shop. All students working under the direction of the teacher should help produce and maintain a clean and orderly shop. Some positive indicators of a properly cleaned shop are as follows:

- A signal is given to stop work and start cleanup at a specified time. A whistle is effective for a cleanup signal.
- Every student helps with cleanup.
- Benches are cleared and clean.
- Machines are clean.

- Paint brushes and spray equipment are properly cleaned and stored.
- Solvents, paints, and greases are properly stored.
- Tools are in their places.
- Lumber, metal, and other construction materials are stored.
- Projects and related materials are in approved places.
- Floor is clean and trash is in containers.
- Cabinets and storage areas are locked.
- Every job is checked for completeness.
- Every student is evaluated according to the quality of his or her cleanup contribution.
- Sinks and restrooms are clean and orderly.
- Students are waiting in an orderly manner for dismissal by the teacher.

REASONS FOR KEEPING THE SHOP CLEAN

A quick and efficient cleanup procedure is important to the safety of students. It adds greatly to the success of an agricultural mechanics shop program. There are good reasons for cleaning the shop after each class, every day. Some of these are related to personal safety, some to learning efficiency, and some to student comfort and convenience. A properly organized shop cleanup procedure is important for the following reasons:

- Each student's projects and possessions are stored properly. When projects are stored properly, they do not interfere with the work of other students and are not damaged by other students using the shop. Projects or project parts may be stored in drawers, lockers, storage cabinets, storage rooms, or fenced areas. In well-managed shops, project parts may be stored in containers by, on, or under large projects such as tractors, wagons, or machinery.
- All project parts are stored together. This enables students to see if any items need to be brought the next day to continue the project.
- Shop spaces are cleared so other classes can safely use the areas.
- Tools are returned to their proper places. Tools should be mounted on panels over colored outlines of each tool, figure 6-1. These outlines are called silhouettes; they make it easy to check for missing tools at the end of each class. Tools can also be easily checked to see if they are in need of repair.
- Each student learns to put tools and materials in their proper places and can expect to find them quickly when needed. This eliminates lost time looking for tools and materials.

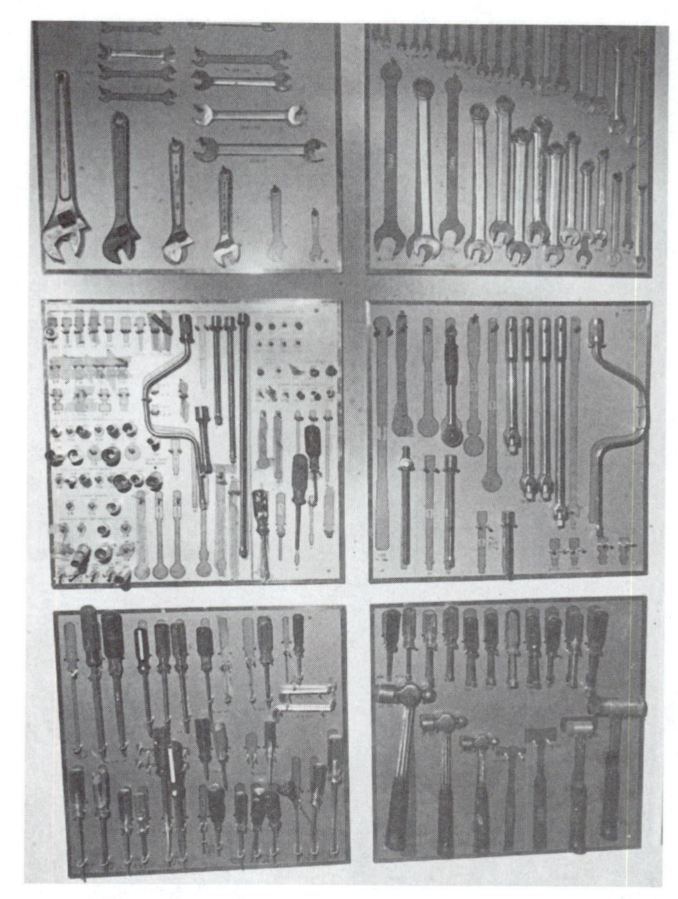

FIGURE 6-1. Tools should be mounted on panels over colored outlines of each tool.

- Paint materials and equipment are cleaned and stored to avoid wasted materials and ruined finishing materials.
- The hazards of fire and explosion are reduced by proper storage of materials.
- Students learn cooperation and teamwork.

EQUIPMENT AND CONTAINERS USEFUL FOR SHOP CLEANUP

It is important to have enough cleanup equipment and materials on hand so all students can participate in cleaning and storage activities. Each student must do his or her part to make the cleanup easier for everyone. It is also important to have storage containers for every type of materials used in the shop, figure 6-2. Many shops have excellent commercial flammable materials cabinets. Flammable liquids, such as grease, oil, and solvents, are stored in these cabinets, which are made of steel and close automat-

FIGURE 6-2. Storage containers are needed for different types of materials.

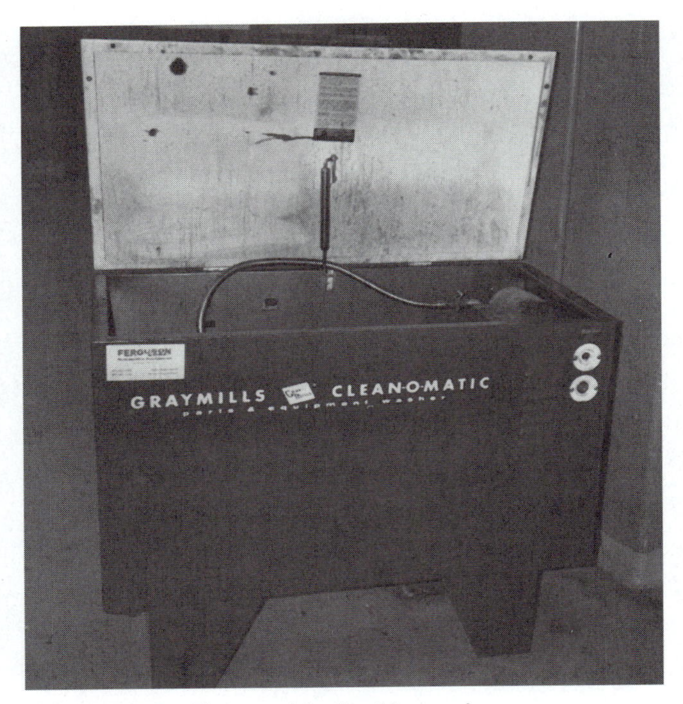

FIGURE 6-3. Flammable liquids such as grease, oil, and solvents must be stored in special cabinets that close automatically in the presence of fire.

ically in the presence of fire, figure 6-3. Such cabinets must meet all safety requirements.

Racks for lumber and metal provide safe and convenient storage for these materials, figure 6-4. Vertical racks permit the storage of both long and short items. On such racks, material can be reached with little moving of other items.

There are many items of equipment that are necessary to clean a shop quickly and efficiently and

FIGURE 6-4. Vertical racks for lumber, steel rods, bands, angle iron, and pipe make good use of wall space. These racks permit the storage of both long and short items. Material can be reached without moving other materials.

to store materials safely. These items include the following:

- floor brooms
- floor dust mops
- bench brushes
- shop vacuum cleaner(s)
- dust collection and chip removal system
- metal cans for rag storage
- large metal trash cans
- storage cabinets for combustible materials
- scoop shovels and dust pans to pick up dirt and trash
- varsol for cleaning up grease and oil spills
- sawdust to absorb liquids
- commercial material to sprinkle on the floor to control dust
- clean rags
- storage cabinets for tools and hardware
- storage racks for lumber and metal
- steel cans for metal used for practice welding
- suitable containers for scrap wood
- cabinets, lockers, fenced areas, etc., for project storage.

FIGURE 6-5. The shop vacuum cleaner makes cleanup easier and does not create dust.

The soft-bristled brush and shop vacuum cleaner are the standard stools for removing dirt, saw dust, and trash from benches and machines, figure 6-5. The floor broom and dust mop are important floor cleaning equipment. The dust pan and standard scoop shovel are commonly used to move the trash from the floor to the trash can.

TECHNIQUES FOR EFFECTIVE CLEANING

Effective cleaning skills cannot be taken for granted; they must be developed. Many students go through the motions of cleaning, but many do not do a good job. Shop cleanup tasks include:

- removing tools and materials from benches and floor before cleaning
- cleaning all paint brushes
- cleaning high areas such as racks, machines, and bench tops
- cleaning the shop starting at far sides and ends and working toward the trash collecting area(s)
- using brushes and brooms in short strokes and lifting intermittently (brushes and brooms should be tapped against the surface or floor frequently to shake out dirt particles)
- using a commercial dust-absorbing material, if available
- using a vacuum cleaner to clean machines whenever possible (the vacuum cleaner is very desirable because it does not create dust)
- putting oily rags in closed metal containers
- using sawdust or commercial materials to absorb liquids such as oil spills
- sweeping fine waste particles into floor drops of dust collection systems, if provided
- putting all trash in suitable metal containers
- placing trash containers in the proper place
- storing all cleanup equipment properly
- cleaning all sink areas and picking up paper towels
- helping others finish the cleanup.

SHOP CLEANUP SYSTEMS

Organization is the key to a clean shop. Without good organization, students who see the need for a clean and orderly shop soon lose their willingness to put the shop in good order. This is natural since the value of teamwork and fair play is learned at a very young age. Therefore, a system that involves every student on an equal basis is needed. Several systems have been developed to get the job done.

All-Pitch-In Method

For lack of a better term, the most simple system of shop cleanup is called the "all-pitch-in" method. With this system, the teacher announces cleaning time verbally, or by whistle, bell, or other device. Students start by putting their own materials away. They then do cleanup, arrangement, or storage tasks according to their knowledge, maturity, personal cleanup habits, and commitment to the program. This system generally fails because students lack the knowledge and skills necessary to do a good job at all cleanup tasks. Students and teacher soon discover that certain important cleanup tasks are left undone. This is due to lack of organization.

Cleanup-Wheel Method

Many shops use a cleanup wheel, figure 6-6. This system uses a chart shaped like a wheel. The teacher specifies all cleanup tasks in equal sections on the outer section of the wheel chart. Students are placed in groups or given a group number. Either names or group numbers are placed in the inner section of the wheel chart. Each group of students is assigned the task or tasks listed in line with their name or group number on the wheel chart. The tasks remain stationary at the edge of the chart. The wheel itself can be periodically rotated so each group has a chance to do all cleanup tasks.

For the cleanup-wheel method to work, some kind of cleanup skills checklist is needed. It is important that a student who is respected by his or her classmates and is a good judge of achievement be given this job. A check sheet with a format that works well is shown in figure 6-7. The checklist should be placed in a prominent location so students can see how they are being evaluated. Students with zeroes (0) and ones (1) should be encouraged to improve their cleanup skills.

When using a checklist, the foreperson's name is listed along with all others. During the shop cleanup time, the foreperson carries the checklist on a clipboard and evaluates all students except himself or herself. The teacher then evaluates how well the foreperson has done the evaluations.

This method helps students see how well they are developing their cleanup skills. It also helps develop management and supervisory skills. The checklist has only enough columns for a limited number of shop periods. New cleanup tasks should be assigned to students every two weeks. This ensures that everyone will do all jobs sometime during the year and learn how to do each one in a skillful manner.

Both the teacher and the students must realize that the shop cleanup skills are valuable. Future employment demands such skills be developed. Therefore, every effort should be made to teach and learn procedures and tasks used in effective cleanup operations.

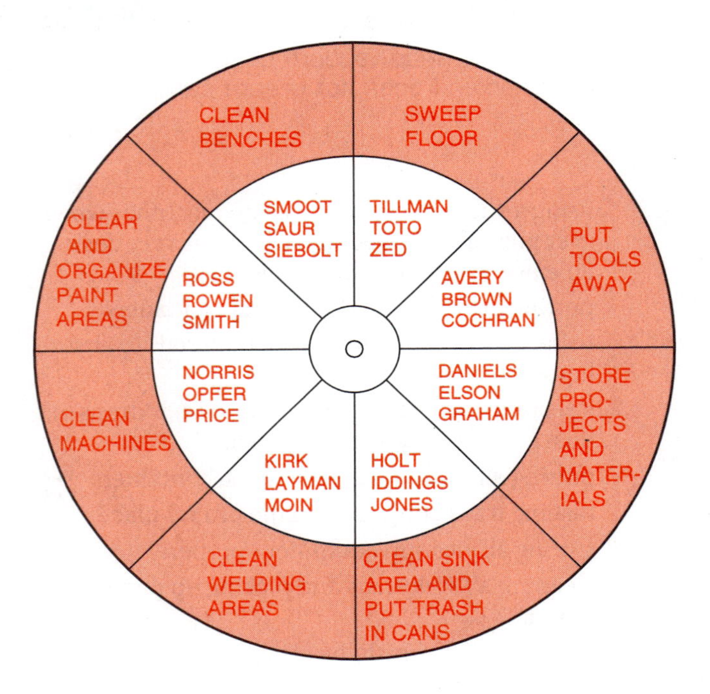

FIGURE 6-6. An example of a shop cleanup wheel chart. All sections in the wheel are of equal size. The hub and inner ring with students' names will rotate to change assignments. The outer section of the wheel does not rotate.

Assignment-Sheet Method

The assignment sheet method may be used in place of the cleanup wheel and the shop cleanup skills checklist, figure 6-8. One important advantage of the assignment sheet is that as many students as necessary can be assigned to any given task to get it done. The equal-sized groups necessary for the cleanup wheel are not needed for the assignment sheet. A disadvantage of assignment sheets is the time required to periodically reassign students.

With the shop cleanup assignment sheet, the teacher and class develop a list of cleanup tasks. Class members are then assigned to these tasks in numbers needed to get the job done. For instance, if it takes two people to sweep off benches, four to clean the welding areas, and five to sweep the floor,

SHOP CLEANUP SKILLS CHECKLIST

KEY:
3 = Done well
2 = Done satisfactorily
1 = Done poorly
0 = Job not done
ab = Student is absent

DATE 4/4/96

SHOP FOREPERSON Walker

DATE AND RATING

NAME	3/5	3/6	3/7	3/8	3/9	3/12	3/13	3/14	3/15	3/16	AVERAGE SCORE	COMMENTS
Avery	3	3	2	3	3	3	3	3	3	3	2.9	
Saur	3	2	1	1	1	ab	0	0	1	1	1.0	
Cochran	1	3	0	3	3	3	3	3	3	3	2.5	
Opfer	ab	3	3	2	2	2	2	2	3	2	2.1	
Walker (Foreperson)	JW	JW	JW	JW	JW	JW	JW	JW	JW	JW		

FIGURE 6-7. An example of a shop cleanup skills checklist. Notice in this example that the teacher initialed the box by the foreperson's name. This indicates the teacher observed the general conditions of the shop and approved the job done by the foreperson.

SHOP CLEANUP ASSIGNMENT SHEET

Key:
3 = Done well
2 = Done satisfactorily
1 = Done poorly
0 = Job not done
ab = Student is absent

DATE ______

SHOP FOREMAN ______

Task	Person(s) Responsible	Date and Rating								Average Score	Comments
	Date:										
Sweep Benches	Avery										
	Saur										
Clean Welding Areas	Opfer										
	Elson										
	Graham										
Sweep Floor	Holt										
	Iddings										
	Jones										
	Kirk										
Foreman –	Mozier										

FIGURE 6-8. An example of a shop cleanup assignment sheet. This is similar to the shop cleanup skills checklist except a task column is added.

the appropriate number of students can be assigned to each task. The teacher decides who will do the tasks although students can volunteer for various tasks. Assignments may be changed as often as desired. A new shop cleanup assignment sheet must be prepared each time a rotation is made.

A student foreperson and the check-off system are used to evaluate cleanup skills. The check-off, or evaluation, process is the same as described for the shop cleanup-wheel method.

Choosing a System

The shop cleanup system used is largely up to the teacher and students. Each method requires a cooperative attitude on the part of all. Each student must do the assigned task every day or the system will not work. This is because others must do the job for the negligent student. Jobs of absent students must also be covered by others assigned to the same task. To avoid this problem, several students may be designated as substitutes to do the tasks assigned to others who are absent on any given day.

The following must be provided if a shop cleanup system is to work well:

- assignment of every task to some person
- fair assignments based on desirability of the task and effort required to do it
- rotation of assignments so all students learn all tasks
- cooperation by all parties
- genuine honesty in evaluating performance
- a record of individual performance
- a clean and safe shop by the end of each period.

By working together, agricultural mechanics students can enjoy the benefits of a clean and safe shop. This provides satisfaction, good workmanship, cooperative effort, and personal gain.

STUDENT ACTIVITIES

1. Define the Terms to Know in this unit.
2. List the items in the school agricultural mechanics shop that are used for the storage of:
   - lumber and metal
   - fasteners such as nails and screws
   - flammable liquids
   - waste materials
   - tools.
3. Examine the tools and equipment in your school agricultural mechanics shop that are used to clean the shop. Learn to use each item to clean thoroughly.
4. Check all containers used in your agricultural mechanics shop to see if they are of the proper type and properly labeled. Report your findings to the teacher.
5. Help improve the cleanup equipment and facilities in your shop.
6. Volunteer for the shop cleaning job of your choice.

SELF-EVALUATION

A. Multiple Choice. Select the best answer.

1. Oily rags should be stored in a
   a. cardboard box
   b. plastic bag
   c. wooden box
   d. closed, metal can
2. A clean, organized shop reduces the chance of
   a. fire
   b. lost tools
   c. damage to projects
   d. all of these
3. Brushes and brooms work better if pushed
   a. in a continuous path
   b. and lifted intermittently
   c. back and forth
   d. in long strokes
4. Sawdust is useful in shop cleanup to
   a. absorb liquids on the floor
   b. reduce dust in the trash container
   c. condition bristles on floor brooms
   d. none of these
5. A recommended material for cleaning grease from the floor is
   a. water
   b. gasoline
   c. varsol
   d. sawdust
6. The foreperson's job in the cleanup process is
   a. supervision
   b. reward
   c. evaluation
   d. assigning jobs
7. The best item for cleaning nongreasy machines is a/an
   a. rag
   b. brush
   c. air gun
   d. vacuum cleaner
8. The shop cleaning method that gives the best control over the cleanup process is the
   a. all-pitch-in method
   b. cleanup-wheel method
   c. assignment-sheet method
   d. honor-system method
9. The main advantage of the shop cleanup assignment sheet over the shop cleanup wheel is the
   a. flexibility in assigning students to tasks
   b. ease in reassigning tasks
   c. use of a checklist for evaluation
   d. use of a foreperson for evaluations
10. Rotating shop cleanup duties
    a. enables everyone to learn the various cleaning tasks
    b. promotes fairness in assigning undesirable tasks
    c. involves every student on an equal basis
    d. all of these

B. Brief Answers. Briefly answer the following questions.

1. List fifteen signs of a properly cleaned shop.
2. List seven reasons for using a properly organized cleanup procedure.
3. List fifteen tasks included in shop cleanup.
4. Name two cleaning tools that are used to move trash from the floor to the trash cans.
5. What is the best cleaning tool to use when cleaning machines?

Section 3

Hand Woodworking And Metalworking

UNIT

Hand Tools, Fasteners, and Hardware

OBJECTIVE

To identify and correctly spell hand tools, hardware, and fasteners that are commonly used in agricultural mechanics.

COMPETENCIES TO BE DEVELOPED

After studying this unit, you should be able to:

- Describe how tools are classified.
- Name the major tool categories according to use.
- Classify and correctly spell hand tools commonly used in agricultural mechanics.
- Identify and correctly spell commonly used screws, nails, and bolts.
- Select screws, nails, and bolts for various uses.
- Identify and correctly spell important items of hardware.

TERMS TO KNOW

tool
hand tool
power tool
layout tool
cutting tool
boring tool
driving tool
holding tool
turning tool
digging tool
other
fastener
laminate
adhesive
nail
diameter
shank
head
penny
galvanized
screw
threads
square head
hex head
slotted head screw
phillips head screw
allen screw
drywall screw
machine bolt
cap screw
carriage bolt
stove bolt
plow bolt
nut
washer
machine screw
hardware
hinge
butt hinge
strap hinge
T hinge
screw hook and strap hinge
hasp
gusset
flush plate

MATERIALS LIST

- Hand tools commonly used in agricultural mechanics
- Common types of screws, nails, and bolts
- Items of hardware commonly used in agricultural mechanics

The correct use of hand tools is fundamental in agricultural mechanics. Large and efficient power tools are used to do most of the work in society. However, hand tools are used to do the small jobs and to do the work where large machines cannot function.

The use of hand tools is basic and essential for work in design, construction, maintenance, and repair. Hand tools are used by all who construct buildings, install landscape structures, wire computers, or repair tractors. In fact, all people in society use hand tools or pay others to do jobs that require their use. A **tool** is any instrument used in doing work. A **hand tool** is any tool operated by hand to do work, figure 7-1. This is contrasted with a **power tool**, which is operated by some source of power other than human power.

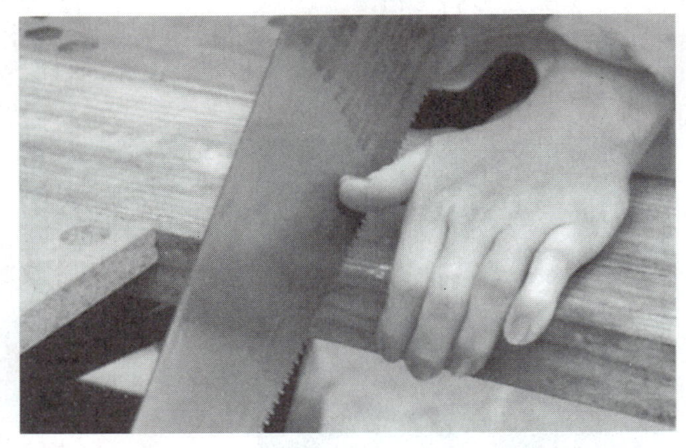

FIGURE 7-1. Hand tools are instruments powered by human hands to do work. *(Courtesy of Lewis/*Carpentry*, Delmar Publishers, 1995)*

TOOL CLASSIFICATIONS

There are various methods of classifying tools. Therefore, the student needs to be prepared to recognize tools by their various classifications. Tools may be classified according to who uses them, e.g., carpenters' tools, masons' tools, mechanics' tools, or machinists' tools. Tools from all such trades are used in agriculture. Therefore, classification of tools by use or function is more meaningful to workers in agricultural mechanics.

The major classifications of tools according to use or function are layout tools, cutting tools, boring tools, driving tools, holding tools, turning tools, digging tools, and other tools. Students should study tools and learn how to classify and correctly spell them. This is helpful when buying or using tools in the shop, home, or on the job.

Layout Tools (L)

A **layout tool** is a tool used to measure or mark wood, metal, and other materials, figure 7-2. Layout tools may be designated by using a capital *L*. Rules,

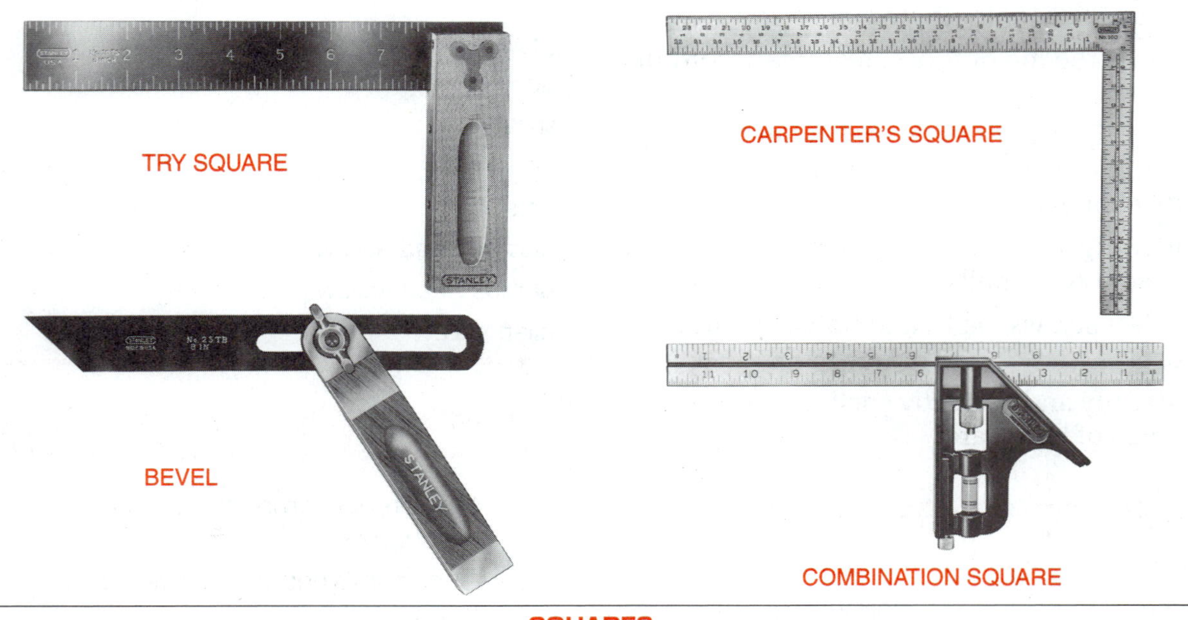

FIGURE 7-2. These common layout tools are used to measure and mark materials. *(Courtesy of Stanley Tools; inside caliper courtesy of Mac Tools Inc.)*

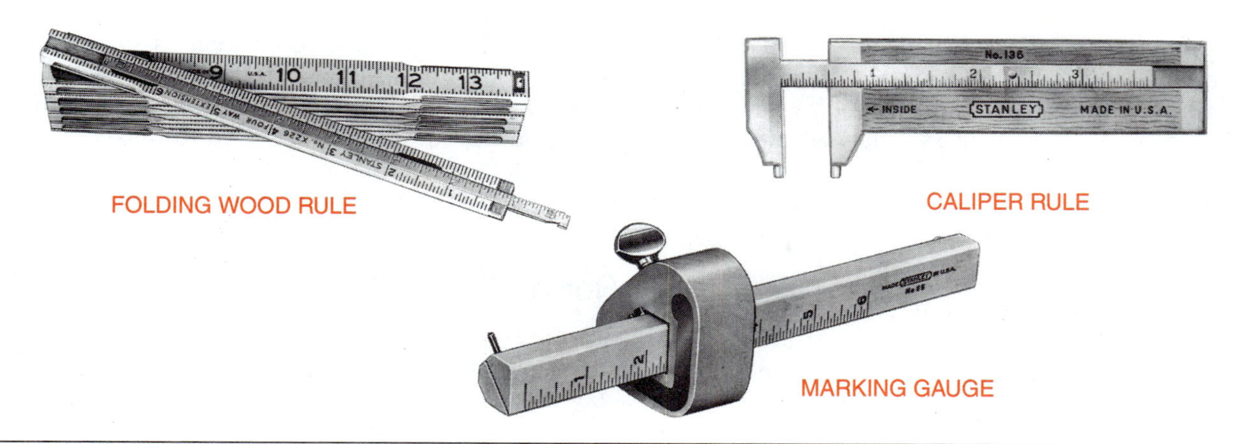

RULES

LEVELS

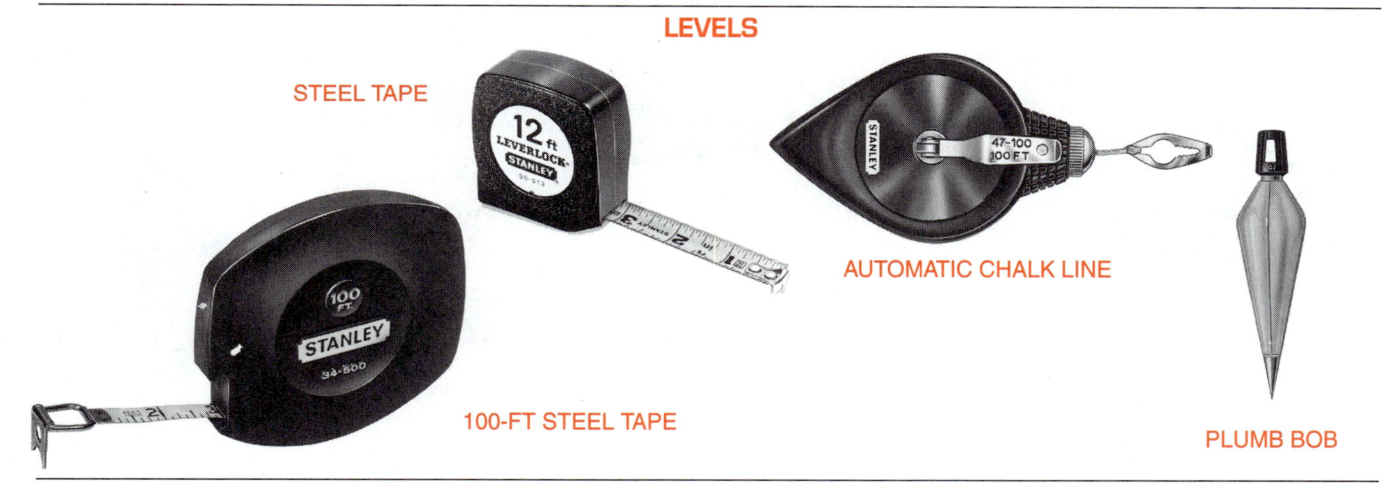

TAPES AND LINES

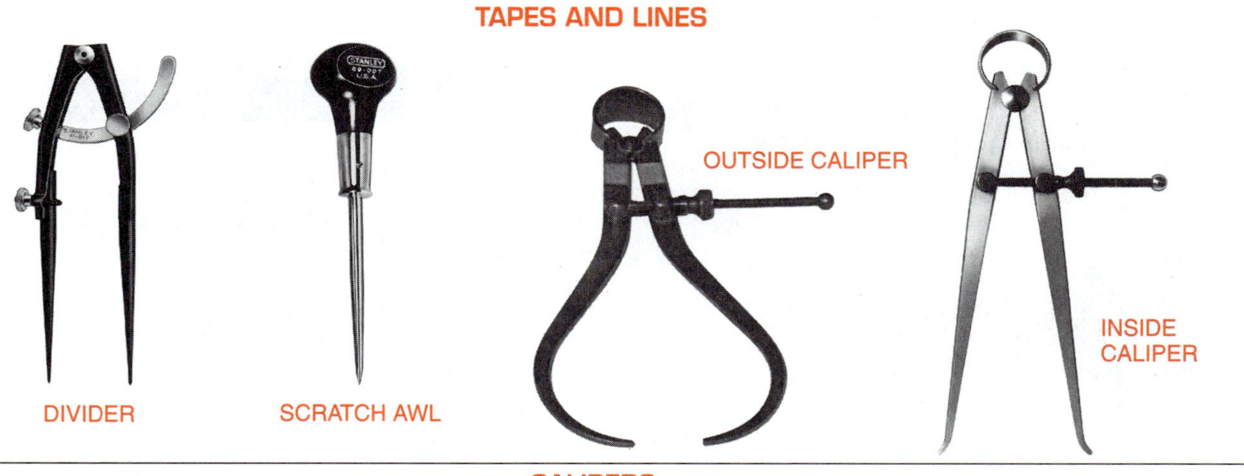

CALIPERS

FIGURE 7-2. *Continued*

squares, scratch awls, calipers, measuring tapes, and dividers are classified as layout tools. These tools are used to measure and mark materials before cutting or shaping is done. Measuring and marking are done when laying out work so other functions can follow according to plan.

Cutting Tools (C)

A cutting tool is a tool used to cut, chop, saw, or otherwise remove material. This permits the user to shape the material. A capital *C* may be used to designate cutting tools. Tools such as saws, chisels, hatchets, and planes are classified as cutting tools, figure 7-3. Since there are so many cutting tools in common use, those that are used to make holes are classified in a separate group and are known as boring tools. Boring tools are designated by the capital letter *B*.

Boring Tools (B)

A boring tool is a tool used to make holes or change the size or shape of holes. Boring tools include bits,

FIGURE 7-3. Cutting tools. *(Courtesy of Stanley Tools,* 1–27, 31–32; *Klein Tools Inc.,* 29, 33; *Mac Tools Inc.,* 34–40, 42)

15. BLOCK PLANE

17. JACK PLANE

16. SMOOTHING PLANE

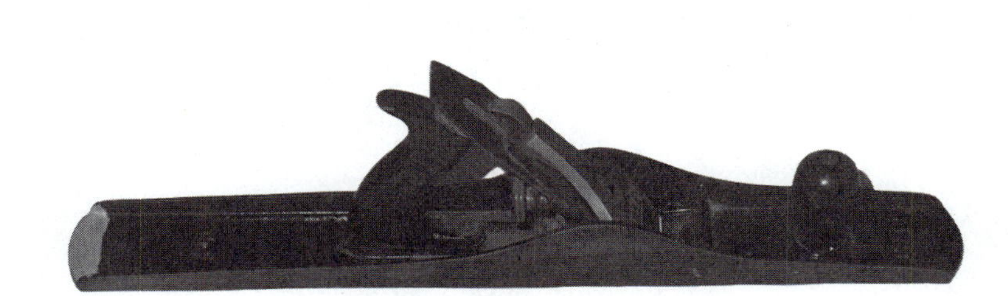

18. JOINTER PLANE

19. SURFORM

PLANES

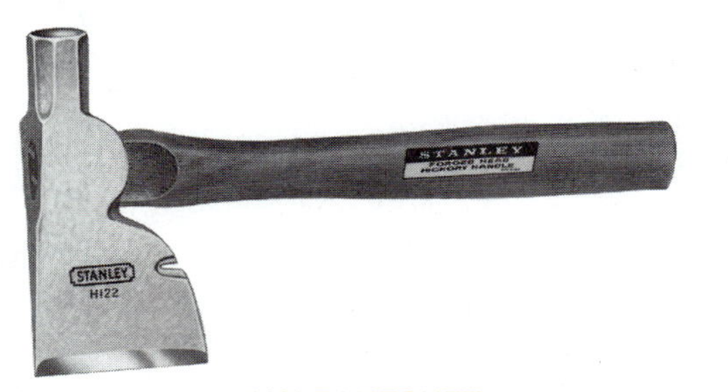

20. HALF HATCHET

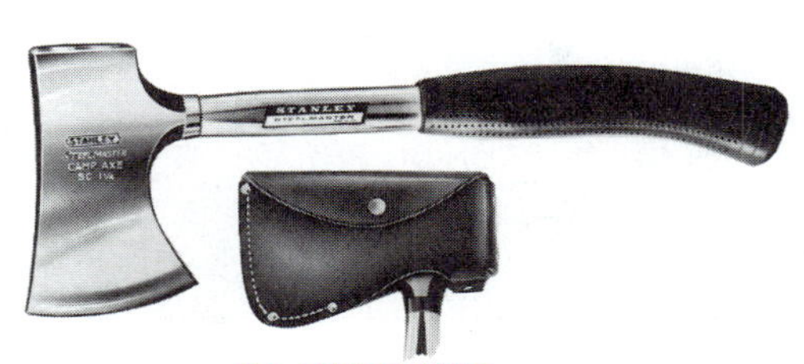

21. CAMP AXE

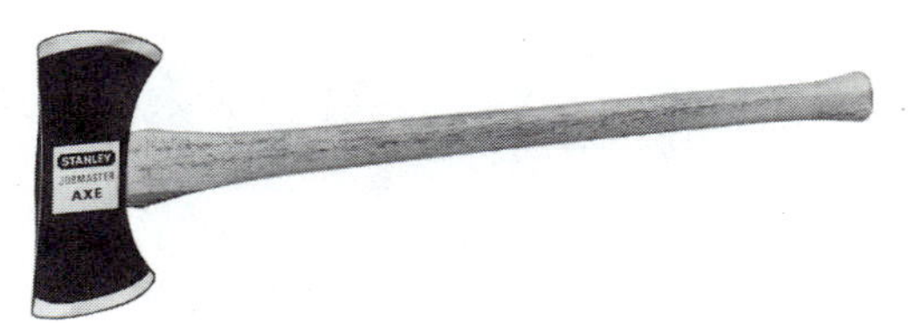

22. DOUBLE BIT AXE

23. SINGLE BIT AXE

24. SPLITTING WEDGE

AXES AND WEDGES

FIGURE 7-3. *Continued*

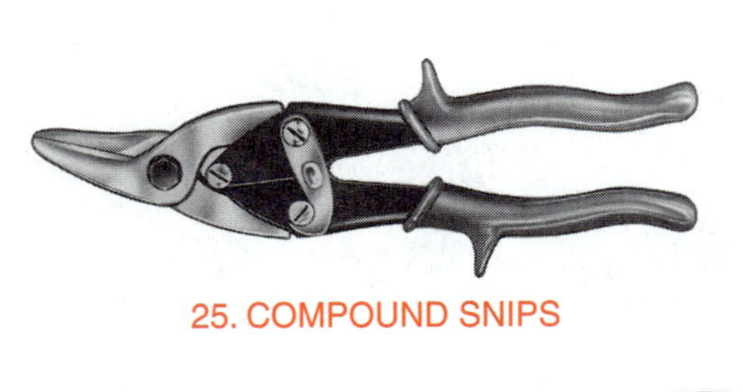

25. COMPOUND SNIPS

29. BOLT CUTTERS

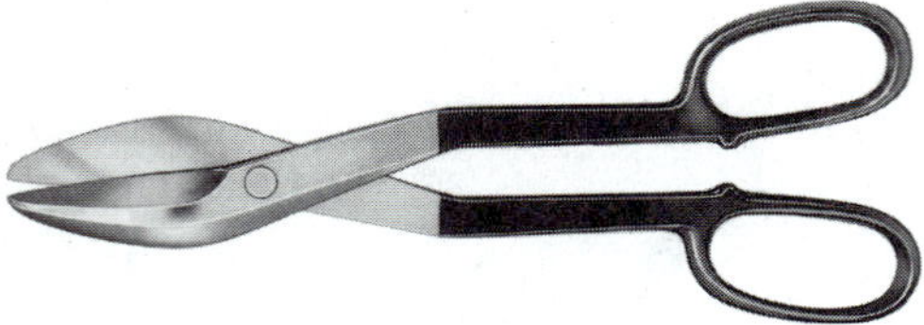

26. TIN SNIPS

30. PIPE CUTTERS

31. GLASS CUTTERS

27. DIAGONAL CUTTERS

32. UTILITY KNIFE

28. PRUNING SHEARS

33. NIPPERS

CUTTERS

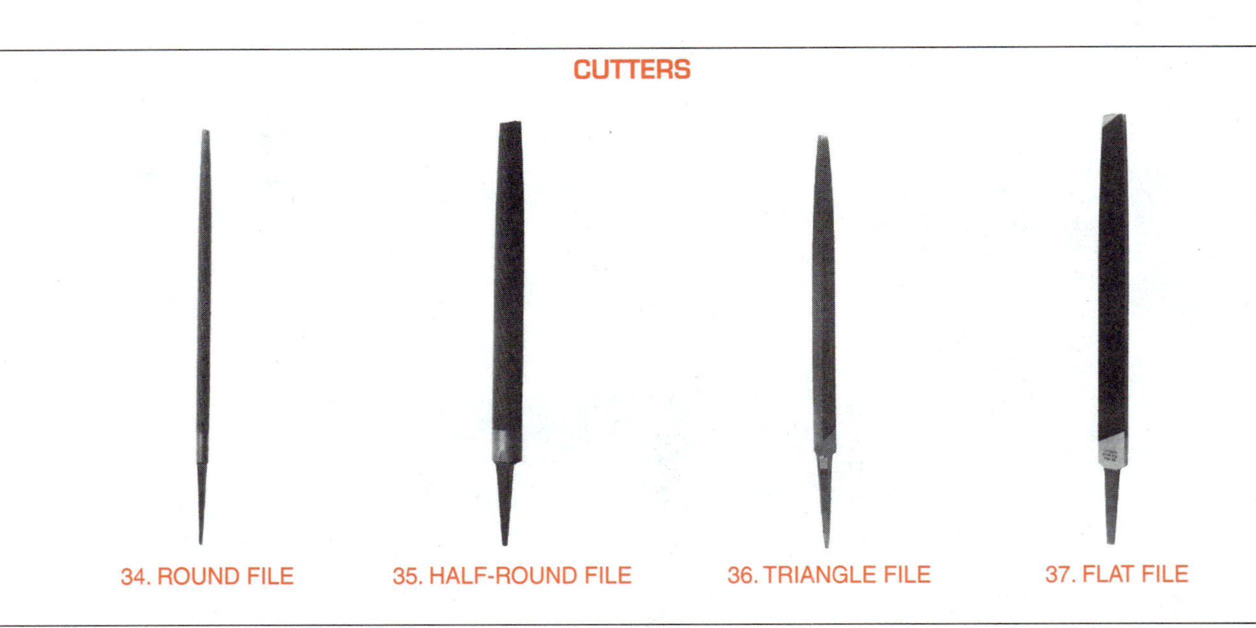

34. ROUND FILE

35. HALF-ROUND FILE

36. TRIANGLE FILE

37. FLAT FILE

FILES

FIGURE 7-3. *Continued*

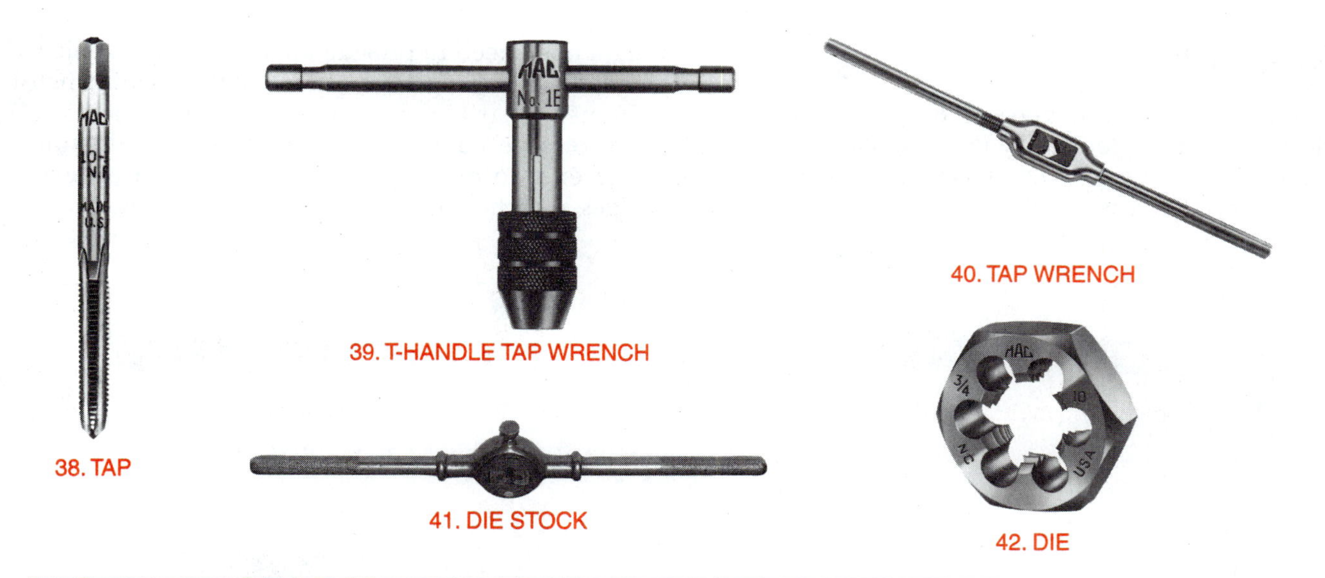

FIGURE 7-3. *Continued*

drills, reams, and the devices used to turn them, figure 7-4. In addition to many common hand tools, this group includes a large variety of specialized tool bits for power machines.

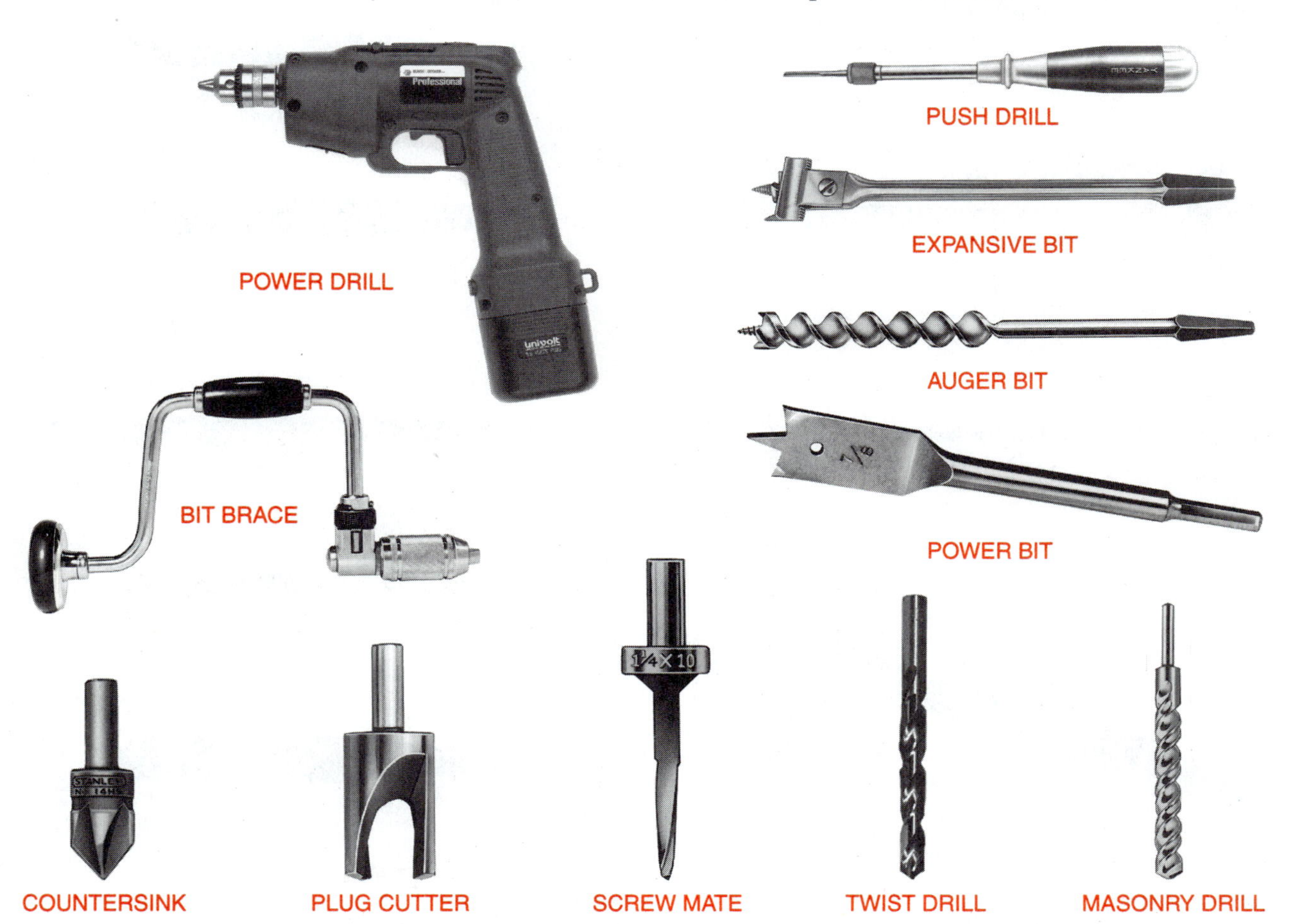

FIGURE 7-4. Boring tools. *(Courtesy of Stanley Tools)*

Driving Tools (Dr)

Another group of tools classified by their use is driving tools. These tools may be designated by the letters *Dr.* A driving tool is a tool used to move another tool or object, figure 7-5. Driving tools rely on their weight and speed to provide force to move an object when it is hit. Using a hammer to move a sliding belt tightener or striking a cold chisel to cut a rivet are used to force nails into wood, or stakes into soil. Some examples of driving tools are hammers, sledges, punches, and various auto body tools.

FIGURE 7-5. Driving tools. *(Courtesy of Stanley Tools)*

Holding Tools (H)

A holding tool is a tool used to grip wood, metal, plastic, and other materials, figure 7-6. They may be designated by the capital letter *H*. Holding tools are used to hold material while other tools are used to cut, shape, modify, or turn threaded items like screws. Holding tools are also used to grip objects such as bolts or pieces of wire. They may be used to bend or shape such objects as needed. Holding tools include vises, pliers, and clamps.

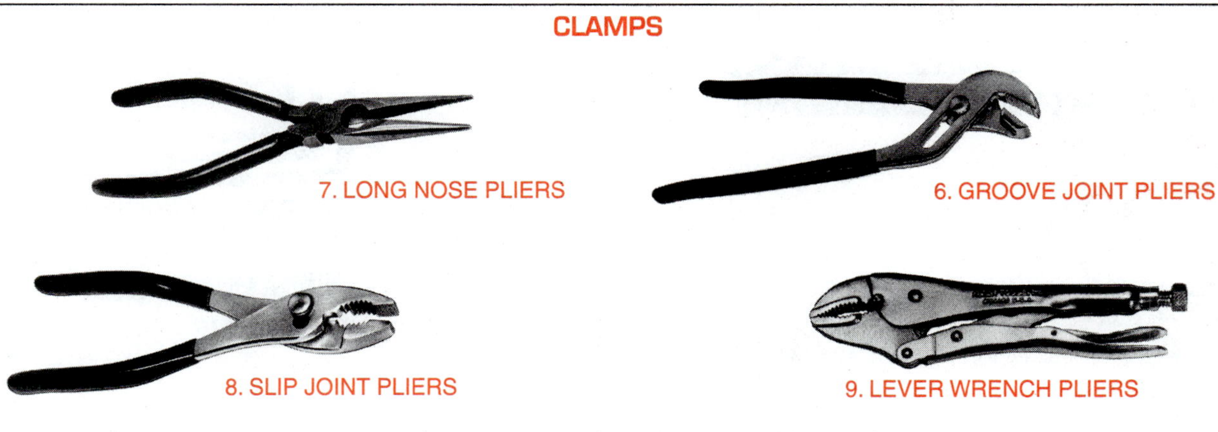

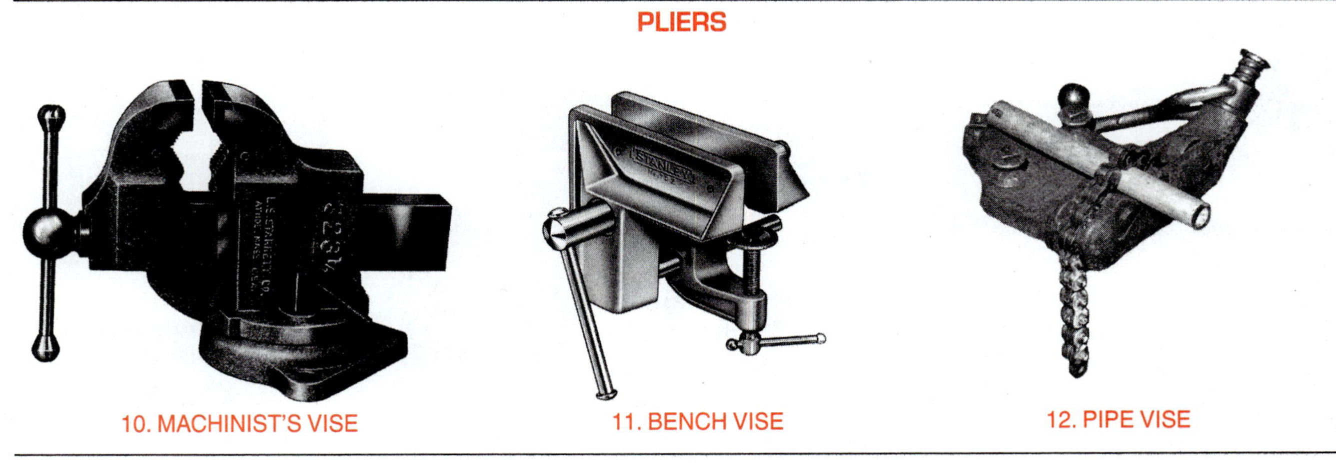

FIGURE 7-6. Holding tools. *(Courtesy of Stanley Tools,* 1–2, 4–8, 11; *Klein Tools Inc.,* 9; *The L. S. Starrett Company,* 10)

Turning Tools (T)

Engine and machinery mechanics make frequent use of turning tools. Turning tools are designated by the capital letter *T*. A turning tool is a tool used to turn nuts, bolts, or screws. Turning tools include wrenches, sockets, and screwdrivers, figure 7-7. The handles and extensions used with sockets are also classified as turning tools.

FIGURE 7-7. Turning tools. *(Courtesy of Stanley Tools; pipe wrench courtesy of Klein Tools Inc.)*

Digging Tools (D)

Many people working in agriculture use a group of tools known as digging tools (designated by the capital letter *D*). A digging tool is any device used to turn up, loosen, or remove earth, figure 7-8. Digging tools include shovels, mattocks, hoes, rakes, post hole diggers, and garden trowels.

Other Tools (O)

All tools do not fit into neat categories. Therefore, the term other is used to classify tools used in agricultural mechanics that do not fit into standard categories. These tools are designated by the capital letter *O*. Pry bars, scrapers, and masonry tools may be classified as other tools, figure 7-9.

FASTENERS

A fastener is any device used to hold two or more pieces of material together or in place. The carpenter creates a building out of individual pieces by using fasteners such as nails. By using fasteners such as bolts and nuts, the manufacturer assembles a machine from many parts. Fasteners are used to hang pictures and laminate panels. Laminate means to fasten two or more flat pieces together with an adhesive. An adhesive is a sticky substance such as glue. The most common fasteners used in agricultural mechanics are nails, screws, bolts, nuts, washers, and machine screws.

Nails

A nail is a fastener that is driven into the material it holds. There are many types of nails classified gener-

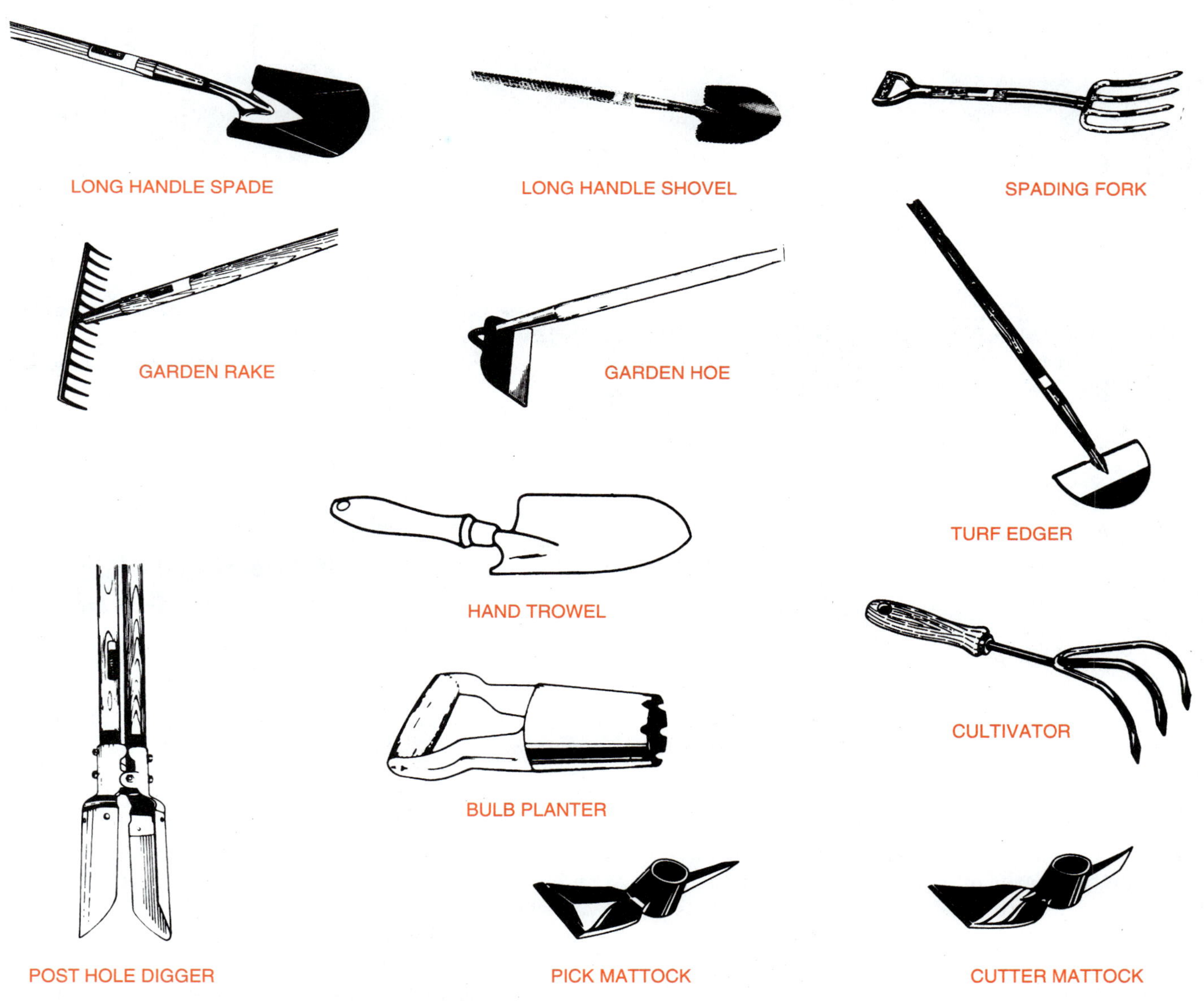

FIGURE 7-8. Digging tools. *(Courtesy of Ames Lawn and Garden Tools)*

FIGURE 7-9. Many tools do not fit into categories and are referred to as other tools. *(Courtesy of Stanley Tools,* 1–6; *Diamond Tool and Horseshoe Company,* 13; *Sears, Roebuck and Company,* 14–15; *Mac Tools Inc.,* 16)

ally by their use or form, figure 7-10. When buying nails, the purchaser must know the use for the nail, the desired length, and the desired thickness or diameter. Diameter refers to the distance across the center of a round circle or object. The shank is the long stem part of a nail or screw; the head is the enlarged part on top. Probably the most familiar nail is the common nail. It is a fairly thick shank and medium-sized head. Nails that are flat and tapered are called cut nails. These are cut from steel and used for wood flooring, or they may be hardened and used as fasteners in masonry materials.

Nail Lengths. The unit of measure used to designate the length of most nails is the penny, figure

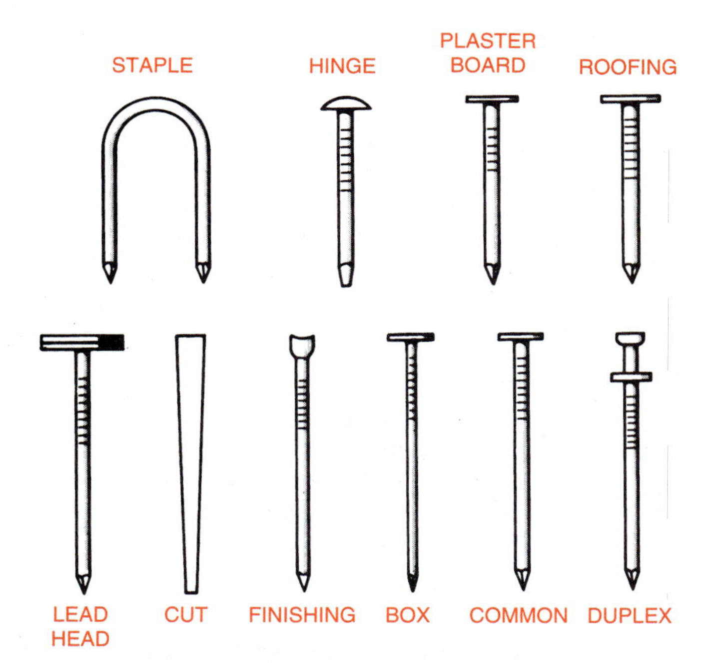

FIGURE 7-10. Types of nails.

7-11. The symbol for penny is the lowercase letter *d*. The term penny was originally used to indicate the number of English pennies needed to purchase 100 nails of a given size. The lengths of common nails, box nails, finishing nails, cut nails, and spikes are designated by penny. The size range for these nails are:

- common nail—2d to 60d
- box nail—2d to 40d
- finishing nail—2d to 20d
- cut nail—2d to 20d
- spike—16d to 12 inches.

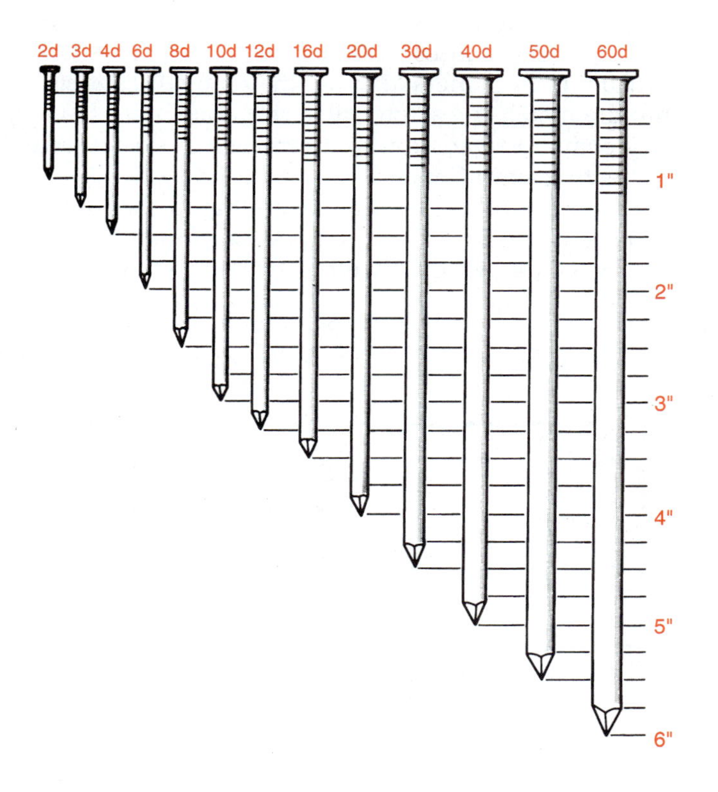

FIGURE 7-11. Length of nails by penny size.

Uses for Nails. Nails vary in thickness of shank and diameter of head according to their use. If the material being held is soft, a large-headed nail is needed. Otherwise, the material will pull over the head of the nail. If the material being held is heavy, a thick, strong shank is required. Following are some important types of nails and their uses.

- common nail—used for general construction; nailing sheeting, shiplap, and board fencing
- cut nail—used for nailing tongue-and-groove flooring; if hardened, used for nailing in masonry materials
- box nail—used for light household construction; nailing siding on buildings, nailing into the end grain of boards
- finishing nail—used for interior finishing of buildings; trim, cabinet, and furniture work, when countersinking is needed
- shingle nail—used for nailing wood and shingles
- roofing nail—used for nailing rolled roofing and composition shingles
- plaster board nail—used to attach plaster board to studs in buildings
- hinge nail—used to fasten hinges on doors and cabinets
- duplex nail—used for construction of forms for concrete work and for nailing insulators on wooden posts for electric fencing
- wire staple—used in wire fence construction
- lead head nail—used for nailing galvanized steel roofing and siding

Improved Nails. Changes in nail forms have come about with the development of new types of building materials. Some new materials require special fastening devices. For instance, soft insulating boards should be nailed with special nails having large, square heads.

Improved nails, sometimes referred to as thread nails, are basically the same as regular nails. They differ in that a portion of the nail shank is threaded with annular or helical threads, figure 7-12. The indentations in the threads along the nail shank provide grooves into which the wood fibers can expand. Thus, both friction and sheer resistance make it very difficult to remove the nail. Nails that must be driven

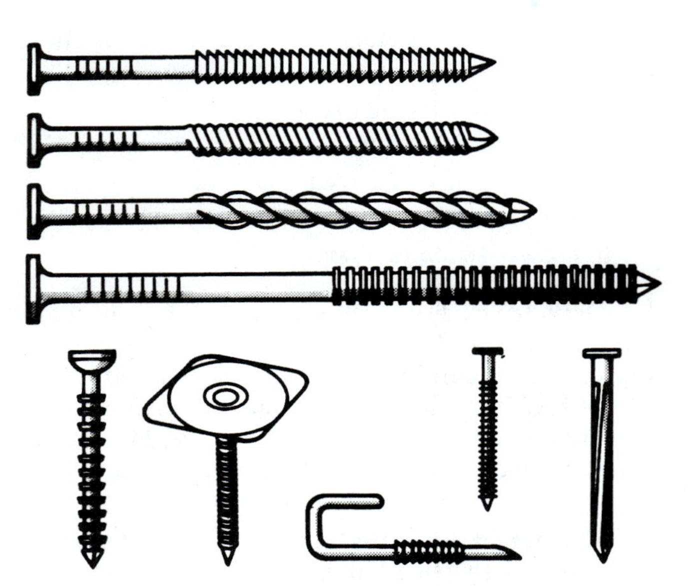

FIGURE 7-12. Improved nails have threaded shanks, and therefore, more holding power than regular nails.

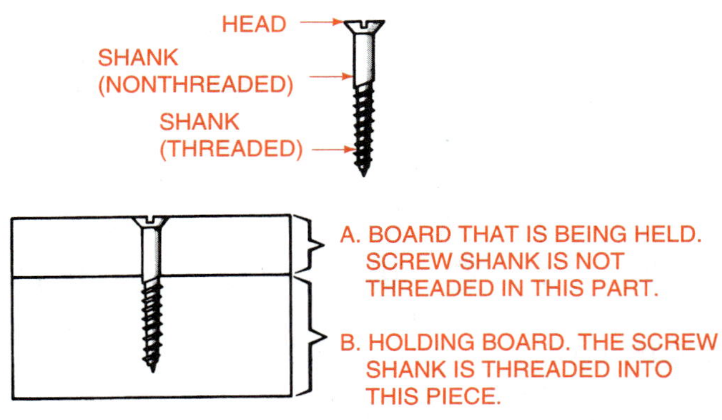

FIGURE 7-13. Screws hold and resist pulling by threading into the holding material as shown in B. The screw must not be threaded in the material being held. This permits the material in A to draw tightly against material B.

into very hard substances are heat treated to make them hard.

Nails are generally made of steel or aluminum. Steel nails will rust and must be galvanized if exposed to moisture. Galvanized means coated with zinc. Steel nails can be coated with a varnishlike cement to make them hard to remove from wood. Nails used with aluminum should be made of aluminum. This is necessary since steel nails will rust and galvanized nails will react with and destroy the aluminum that they touch.

Screws

A screw is a fastener with threads that bite into the material it fastens. The term threads refers to grooves of even shape and taper that wrap continuously around a shank or hole. Screws are made for use in wood, sheet metal, plastic, and any material solid enough to hold them. Screws generally cut threads into the material to which they hold, figure 7-13.

Classifications of Screws. Screws can be classified according to the material they hold. Wood screws have threads designed to bite into wood fibers, which draw the screw into the wood when turned. Sheet metal screws have threads that are wide enough to permit the thin metal to fit between the ridges of the threads. Cap screws are designed to thread into thick metal that has matching threads cut into the metal. Lag screws, also called lag bolts, have very coarse threads designed for use in structural timber or lead wall anchors.

Screws can be classified according to the metal they are made from or the finish used. Steel screws may be coated with a blued, galvanized, cadmium,

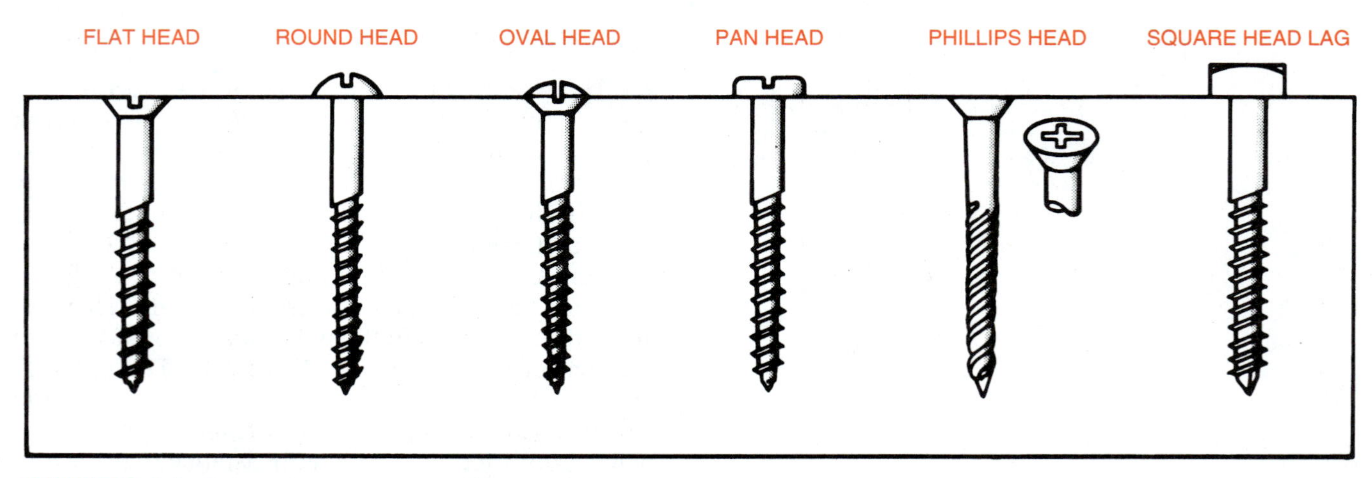

FIGURE 7-14. Screws can be classified by the shape of their heads.

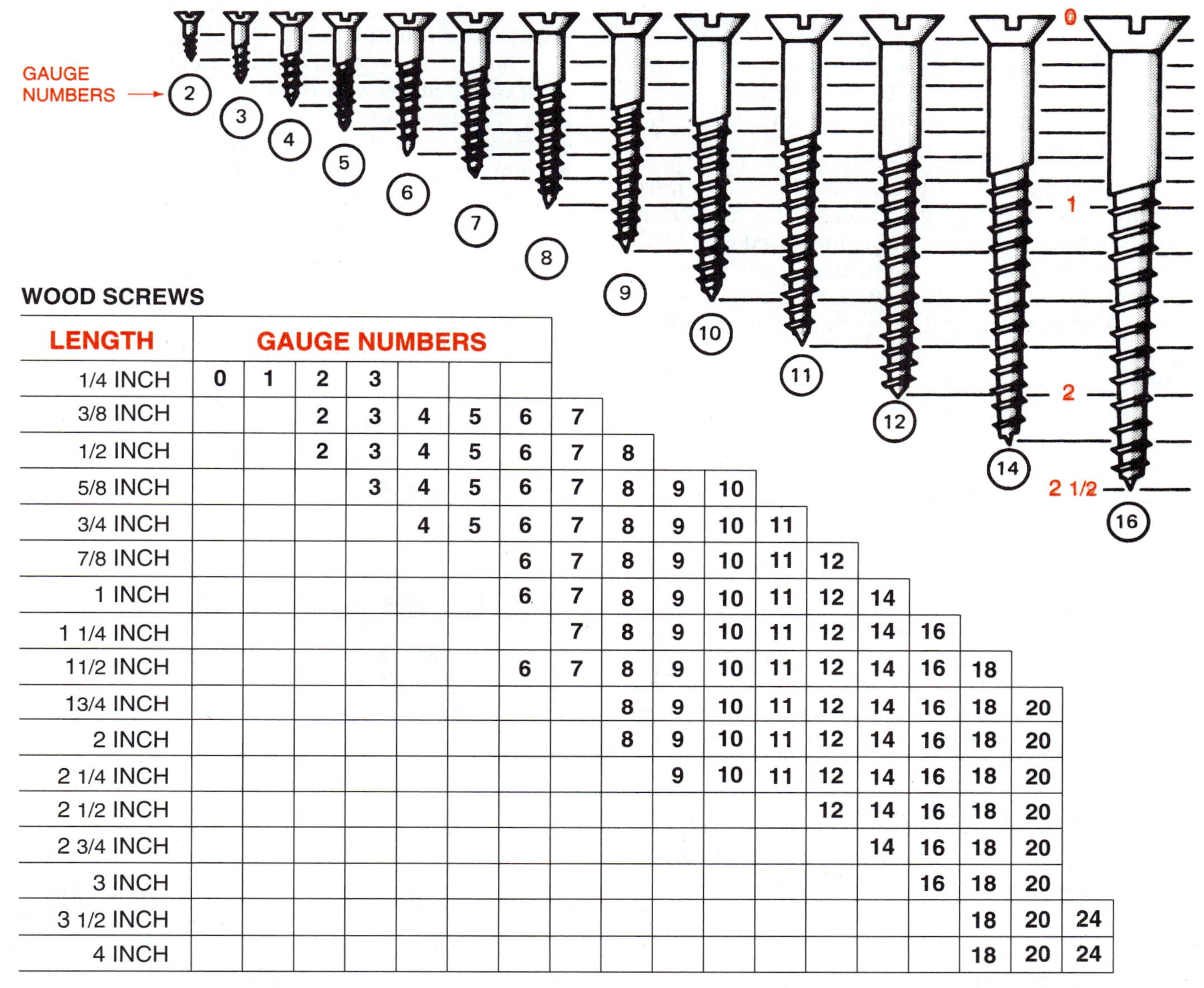

WOOD SCREWS

LENGTH	GAUGE NUMBERS																	
1/4 INCH	0	1	2	3														
3/8 INCH			2	3	4	5	6	7										
1/2 INCH			2	3	4	5	6	7	8									
5/8 INCH				3	4	5	6	7	8	9	10							
3/4 INCH					4	5	6	7	8	9	10	11						
7/8 INCH							6	7	8	9	10	11	12					
1 INCH							6	7	8	9	10	11	12	14				
1 1/4 INCH								7	8	9	10	11	12	14	16			
11/2 INCH							6	7	8	9	10	11	12	14	16	18		
13/4 INCH									8	9	10	11	12	14	16	18	20	
2 INCH									8	9	10	11	12	14	16	18	20	
2 1/4 INCH										9	10	11	12	14	16	18	20	
2 1/2 INCH													12	14	16	18	20	
2 3/4 INCH														14	16	18	20	
3 INCH															16	18	20	
3 1/2 INCH																18	20	24
4 INCH																18	20	24

When You Buy Screws Specify (1) Length, (2) Gauge Number, (3) Type of Head–Flat, Round, or Oval, (4) Material–Steel, Brass, Bronze, etc., (5) Finish–Bright, Steel Blued, Cadmium, Nickel, or Chromium Plated.

FIGURE 7-15A. Screws are classified by a gauge number according to the diameter of the screw. Length of screws is expressed in inches.

FIGURE 7-15B. Drywall screws can be purchased in a variety of lengths and types.

nickel, chromium, or brass finish. Solid brass screws are rust proof and are used where severe moisture problems destroy coated or plated screws.

Screws can be classified by the shape of their heads, figure 7-14. Flat head screws have tapered heads designed to fit into a hole countersunk in the material. The result is a screw head that is flush with the surface. Round head screws have heads that are half-round with a slot in the upper part. These heads or pan-head screws look like an upside-down frying-pan. They are used on sheet metal. A phillips head screw has a flat head with slots in the head that look like a plus sign (+). Lag screws have unslotted heads that are either square or hex shaped. (A square head has four equal sides; a hex head has six equal sides.) Lag screws are turned with a wrench instead of a screwdriver to obtain more leverage.

Screws can be classified by the type of tool needed to turn them. A slotted head screw requires a standard screwdriver. A phillips head screw requires a screwdriver shaped like a plus sign. A six-sided, hex, or allen wrench is used to turn allen screws.

Size of Screws. The size of the screw is specified by the diameter and length of shank. Diameter is expressed by gauge numbers that run from 2 through 24. Numbers 6, 8, and 10 are common sizes. Common lengths range from 1/4 inch through 4 inches, figure 7-15A.

Drywall Screws. The drywall screw has a flat phillips head, with a very thin shank. It is made from very tough steel. Screws can be purchased with black or galvanized finish and flat or pan heads. They can be driven with power tools without making drilled holes as needed with standard screws, figure 7-15B.

Bolts

A bolt is a fastener with a threaded nut, figure 7-16. Types of bolts that are often used in agricultural mechanics include machine bolts, carriage bolts, stove bolts, plow bolts, and special bolts. Bolts and nuts have national course, national fine (SAE), or metric threads.

Bolts are generally straight with round, square, or hex heads. However, some are bent, modified, or shaped for special uses. These include the eye bolt, hook or J bolt, U bolt, turn buckle, and plow bolt. A lag screw, when used with an expansion plug in a masonry wall, is known as an expansion bolt. A toggle is a collapsible winged nut that opens when inserted into a hollow wall. A toggle with bolt is called a toggle bolt.

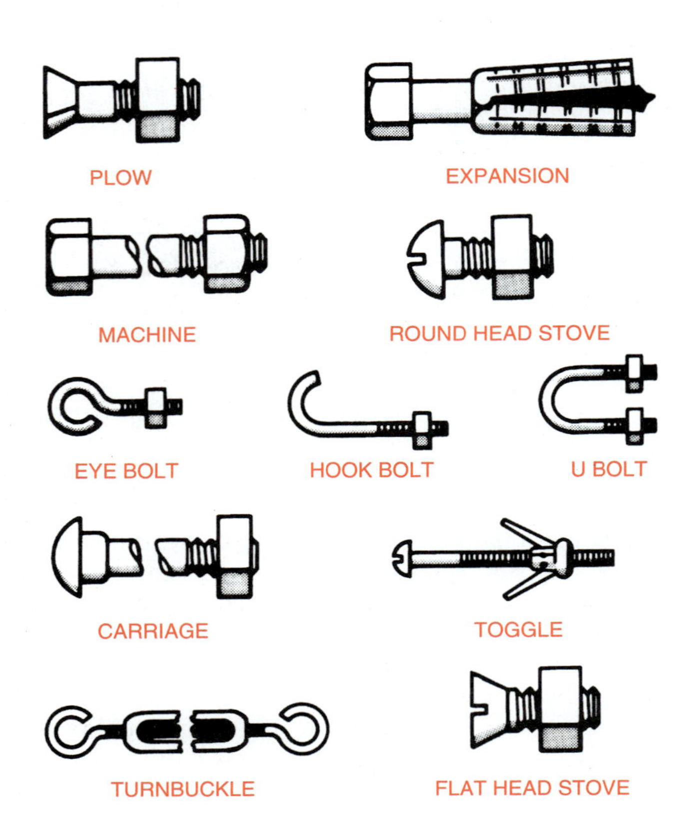

FIGURE 7-16. Common types of bolts.

The machine bolt is a fastener with a square head or hex head with threads on just the last one inch or more of the shank. Machine bolts are used extensively in engines and other machines. Special hardened machine bolts with extra strength and toughness have heads marked with short lines that point from the shoulders toward the center of the head. Three lines designate a very tough bolt and six lines designate a very, very tough bolt. The tougher the bolt, the more expensive it is to manufacture.

The cap screw looks like a machine bolt with the following exceptions. A cap screw

- usually is threaded over its entire length
- generally is 2 inches or shorter in length
- may have a special head such as one requiring an allen wrench
- threads into an object rather than a nut.

The carriage bolt is a fastener with a round head over square shoulders and is used with wood. The shoulders are drawn down into the wood to prevent the bolt from turning. The round, low profile of the head makes it almost flush with the surface of the wood. Carriage bolts are used for construction of wagon bodies, feeders, doors, and other wooden projects.

The stove bolt is a round headed bolt with a straight screwdriver slot. It is threaded the entire length. When purchased, stove bolts generally come with square nuts. They are generally available in sizes up to 3/8-inch diameter and up to 6 inches in length.

The plow bolt has a square tapered head. When placed in a hole designed for a plow bolt, the head is flush with the surface. The plow bolt is used to hold shares and other parts in place on tillage implements.

Nuts

A nut is a device with a threaded hole, figure 7-17. Nuts are the movable part of bolts that are used to fasten two items together. Nuts may be square with four sides or hexagonal with six sides. Hexagonal is also simply referred to as hex. Wing nuts have winged extensions for tightening with the fingers.

Special nuts are available with slots on one side to permit the use of cotter pins or keys to lock the nut in place. Another technique used to lock a nut in place is to tighten a second nut against the first. The second nut is then referred to as a lock nut. Some nuts are said to be self-locking. These have special design features that make them difficult to turn on or off a bolt.

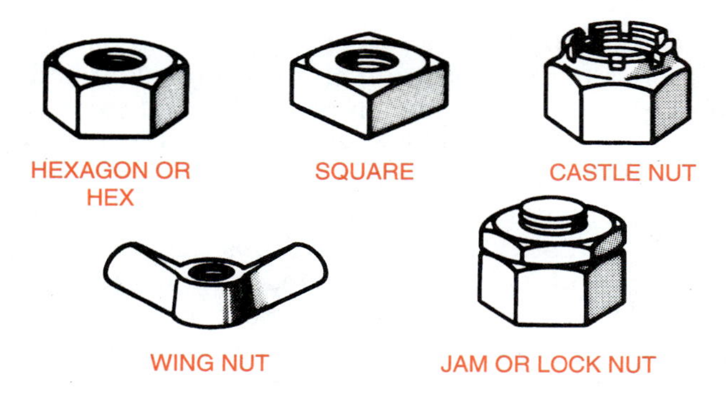

FIGURE 7-17. Types of nuts.

Nuts are available with several types of threads. The most popular threads used on bolts and nuts for agricultural applications are national coarse (NC), national fine (NF) or (SAE), and metric.

Washers

A washer is a flat device with a hole in the center and is used as part of a fastener, figure 7-18. Flat washers are used to prevent bolt heads or nuts from penetrating material. Special lock washers are used to prevent nuts or bolts from loosening due to vibration and use. Lock washers may be split and made with spring steel to exert back pressure when pressed. They may also have spring steel fingers that bite the surfaces that press against them. The spring pressure or biting capability of lock washers prevents nuts or bolts from loosening.

Washers are designated by the size of the bolt for which they are designed. For instance, a washer with a hole that is correct for fitting on a 3/8-inch bolt is called a 3/8-inch washer. Washers are sold by the piece or by the pound.

Machine Screws

A machine screw is a small bolt with a hex nut. Machine screws range in size from very tiny up to less than 1/4 inch in diameter, and up to several

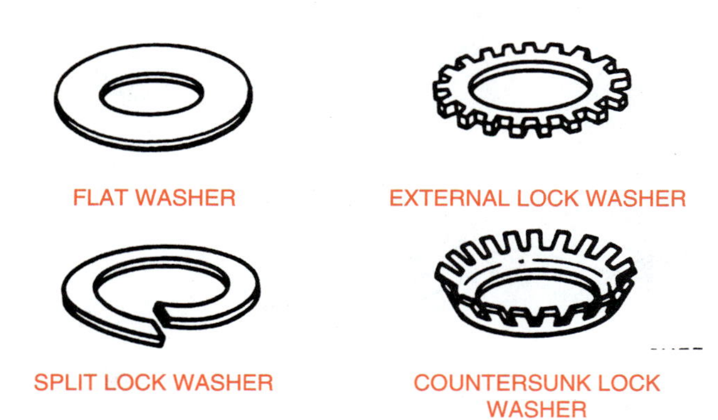

FIGURE 7-18. Types of washers.

DIAMETER NUMBER	THREADS PER INCH
4	40
6	32
6	40
8	32
8	36
10	24
10	32
12	24
12	28
1/4*	20
1/4	28

*Same as 1/4-inch stove bolt or cap screw with standard threads

FIGURE 7-19. Machine screw sizes and threads.

inches in length. The diameter of machine screws is expressed by a number, figure 7-19. The most common machine screws range from number 4 (very small) to number 12 (nearly 1/4 inch in diameter).

Threads on machine screws are designated by the number per inch. Finer threads have more threads per inch than coarse threads and are less likely to vibrate loose. On the other hand, coarse threads permit greater ease of starting and greater speed in applying a nut.

When buying machine screws, the diameter by number, the threads per inch, and the length must be specified. For example, a 4-40 × 1/2 machine screw is number 4 diameter, 40 threads per inch, and 1/2 inch long.

HARDWARE

The term hardware is used in agricultural mechanics for special fasteners. These include hinges, brackets, plates, and miscellaneous metal objects. Hardware is used to support doors, lock windows, secure sideboards on trucks, and anchor roofs. Whenever two objects need to be connected, hardware is generally used. Hardware, as the term suggests, is hard because it is usually made from steel or brass.

Hinges

A hinge is an object that pivots and permits a door or other object to swing back and forth or up and down. Small hinges used on cabinets and furniture are referred to as cabinet hinges. Hinges are classified according to their type and size. Some common types of hinges are butt hinges, strap hinges, T hinges, screw hook and strap hinges, and hasps, fig-

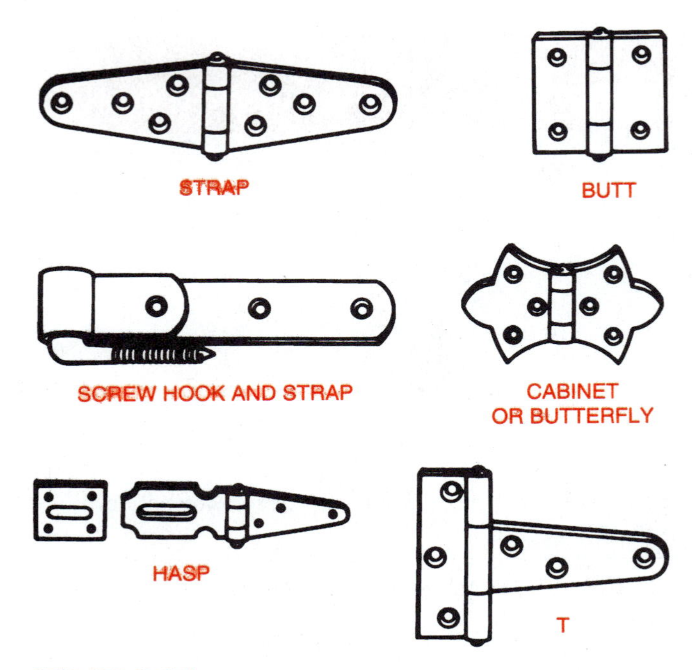

FIGURE 7-20. Types of hinges.

ure 7-20. All hinges consist of two parts plus a pin. The parts are fastened to the stationary object and the moving object. The pin is then inserted to permit movement between the parts, figure 7-21.

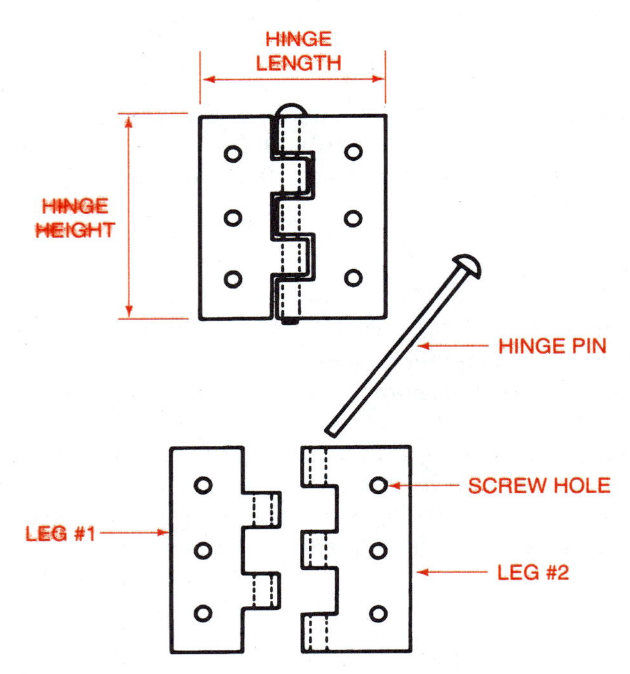

FIGURE 7-21. A hinge has three parts: two legs and a pin. Leg 1 is interlocked with leg 2 using the hinge pin. Hinges may be purchased with removable pins for use on doors intended for occasional removal. However, many hinge pins are not removable.

Butt Hinges. Butt hinges are used to mount doors in a butted position. This means the door and the strip it is mounted on are flush and form a smooth surface. Butt hinges are fairly short in length as compared to their height or the length of their pins. Butt hinges may be mounted on the surface or may be mounted in a concealed position between the door and its mounting strip. If the hinges are concealed or out of sight, they must be set in the material and have countersunk screw holes with flat head screws.

Some surface-mounted butt hinges are designed to be attractive. An example is the butterfly hinge. It is so named because it is shaped like a butterfly in flight.

Strap Hinges. Strap hinges are much longer than they are high. Their long, thin, straplike appearance gives them their name. The strap hinge is used where the hinge must reach across a long surface; it provides additional support for the door. This feature is useful where the door or its mounting strip is not

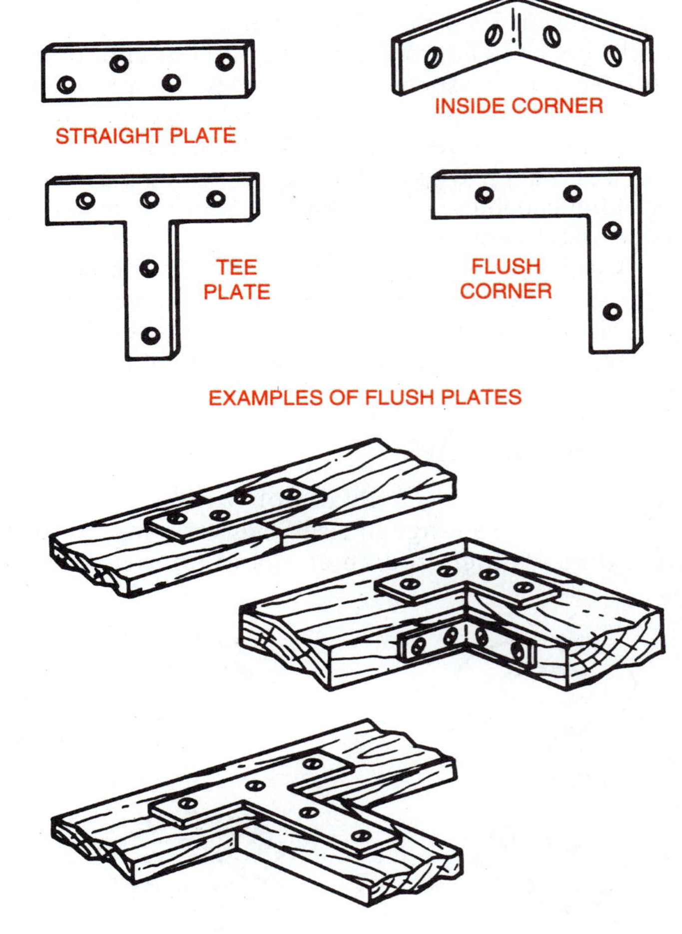

FIGURE 7-22. Flush plates are used to strengthen corners of frames and doors.

very strong. Strap hinges are also useful where excessive weight or pressure is used on a door.

T Hinges. The T hinge is made by combining the features of the butt hinge and the strap hinge. One leg of the hinge is short and high like a butt hinge. The other leg is narrow and long like a strap hinge. The T hinge is useful where the mounting strip for a door is thin but strong, and the door is flimsy and needs extra hinge support.

Screw Hook and Strap Hinges. The screw hook and strap hinge is also called a gate hinge. Two special screw hooks are screwed into a post or door jamb. The two straps are then mounted in the correct positions on the door. The door with hinge straps mounted is set or hung on the screw hooks. This completes the installation.

Hasps. The hasp is not really a hinge but is constructed like one and is sometimes classified with hinges. It is used to hold doors closed. One part of the hasp is attached to the door. The hinged part fits over a metal loop attached to the mating surface. A padlock or other device is placed in the loop to hold the door closed.

Brackets and Flush Plates

Hardware suppliers can provide many types of brackets, plates, door closers, door catches, and other hardware items for construction and maintenance. Special devices called gussets are available to reinforce joints in wood and metal. Wall brackets and special mounting channels with prefinished shelving are some hardware items that make construction fast and easy.

Some common devices used for strengthening corners of frames and doors are flush plates, figure 7-22. A flush plate is made of metal about 1/16 inch thick and 1/2 inch wide. They are drilled and have countersunk holes to receive flat head screws. Flush plates are available as straight, tee, inside, or flush corners.

STUDENT ACTIVITIES

1. Define the Terms to Know in this unit.
2. Study individual tools provided by the instructor and classify them according to their use.
3. Prepare a bulletin board with captions showing the major classes of tools: layout, cutting, boring, driving, holding, turning, digging, and other. Cut pictures of tools from old issues of magazines and catalogs and place them under their caption on the bulletin board.
4. Examine the tool storage area in your agriculture department. On a sheet of paper, list all tools that you see. Place a letter after each tool to indicate its classification as follows: L = Layout; C = Cutting; B = Boring; Dr = Driving; H = Holding; T = Turning; D = Digging; and O = Other.
5. Make up a board with samples of various types of nails attached; properly label the nails.
6. Make up a board with samples of screws and bolts attached; properly label them.
7. Visit a hardware store and examine various nails, screws, bolts, and other fasteners. Ask the store manager for manufacturer charts or tables that describe fasteners.
8. Ask your instructor if there are some activities you can do to improve the arrangement and storage of nails, screws, bolts, and other fasteners in the agriculture department.

SELF-EVALUATION

A. Multiple Choice. Select the best answer.

1. The use of hand tools is
   a. for those who cannot afford power tools
   b. for a limited number of highly specialized jobs
   c. primarily for engine and machinery mechanics
   d. the foundation of agricultural mechanics
2. Tools are generally classified according to
   a. use
   b. color
   c. construction
   d. origin
3. An example of a layout tool is/are
   a. a claw hammer
   b. outside calipers
   c. a hand saw
   d. a plug cutter
4. Saws are classified as
   a. kerf tools
   b. push tools
   c. flexing tools
   d. cutting tools
5. Taps and dies are classified as
   a. holding tools
   b. digging tools
   c. cutting tools
   d. turning tools
6. Wrenches are classified as
   a. turning tools
   b. digging tools
   c. other tools
   d. cutting tools
7. The lowercase letter *d* is used to designate sizes of
   a. lumber
   b. screws
   c. nails
   d. bolts
8. Which is *not* the name of a type of nail?
   a. lumber
   b. plaster board
   c. roofing
   d. duplex
9. The term *improved* means a nail
   a. is made of copper
   b. is easy to remove
   c. has a thick shank
   d. holds better
10. Screws are classified
    a. according to the material they hold
    b. by the metal from which they are made or the finish used
    c. by the shape of their heads
    d. by all of these
11. The number of a screw refers to its
    a. diameter
    b. length
    c. head type
    d. use
12. The difference between a bolt and a screw is
    a. a bolt has threads
    b. a bolt has a nut
    c. a screw has a slotted head
    d. a screw is suitable for use in wood
13. A bolt used in wood that has a round head over square shoulders is
    a. a stove bolt
    b. a machine bolt
    c. a carriage bolt
    d. none of these
14. How many sides are there on a hexagon nut?
    a. four
    b. six
    c. eight
    d. twelve
15. A 4-40 × 1/2 machine screw
    a. has 4 threads per inch
    b. has 40 threads per inch
    c. is 4 inches long
    d. comes four to the package
16. Which is *not* a type of hinge?
    a. butt
    b. strap
    c. T
    d. N
17. The hinge that contains a feature from each of the two other hinges is the
    a. butt hinge
    b. strap hinge
    c. T hinge
    d. N hinge
18. The best hinge to use if it is not to be seen is the
    a. butt hinge
    b. strap hinge
    c. T hinge
    d. N hinge

19. The best hinge for a very large and extra heavy door or gate is the
a. butt hinge
b. T hinge
c. N hinge
d. screw hook and strap hinge

20. Corners of frames and doors may be strengthened by using a
a. hasp
b. butt hinge
c. flush plate
d. hook and eye bolt

B. Brief Answers. Briefly answer the following questions.

1. What is a hand tool?
2. Name and correctly spell six layout tools.
3. Name and correctly spell six cutting tools.
4. Name and correctly spell six boring tools.
5. Name and correctly spell six driving tools.
6. Name and correctly spell six holding tools.
7. Name and correctly spell six turning tools.
8. Name and correctly spell six digging tools.

C. Identification. Name, correctly spell, and classify each of the following tools.

1.

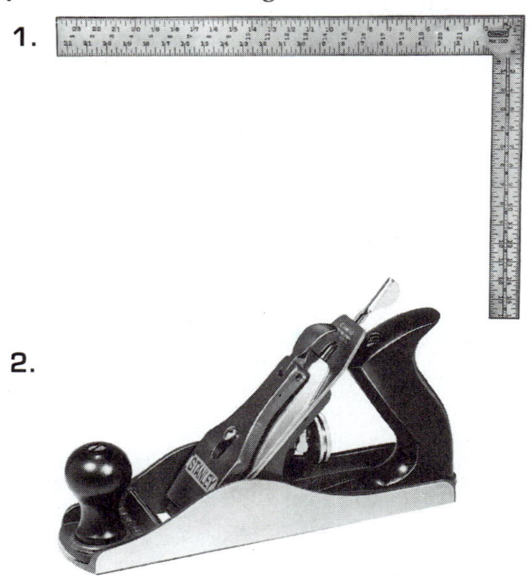

2.

3.

4.

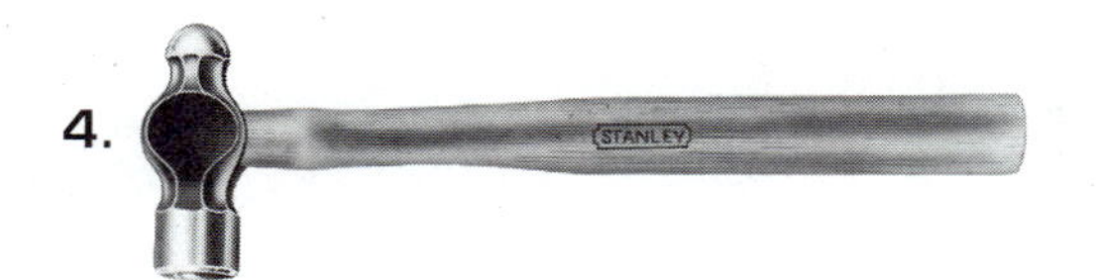

5.

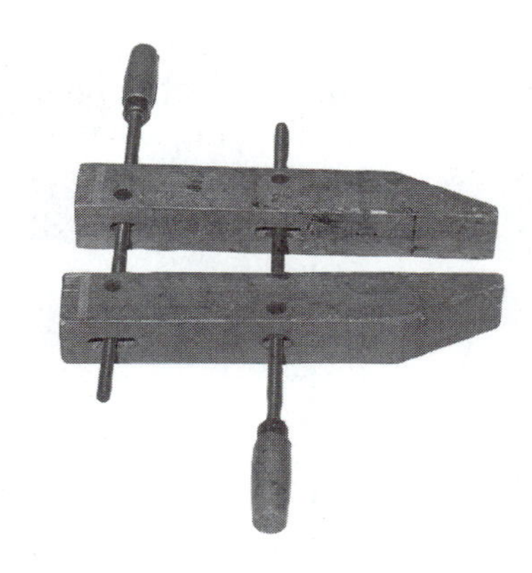

6.

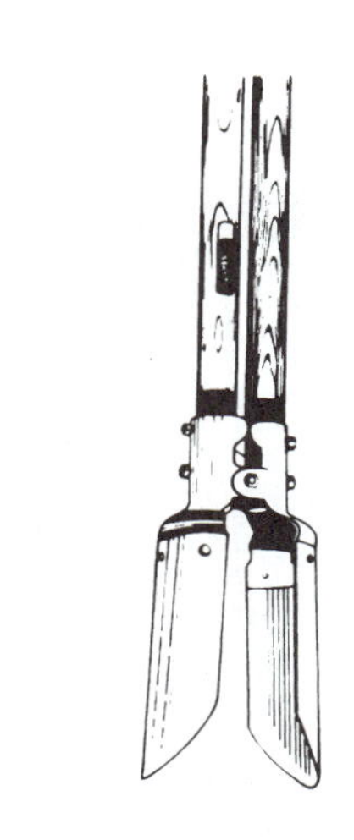

7.

8.

9.

10.

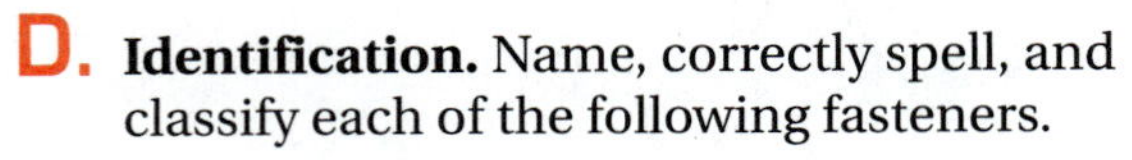

D. Identification. Name, correctly spell, and classify each of the following fasteners.

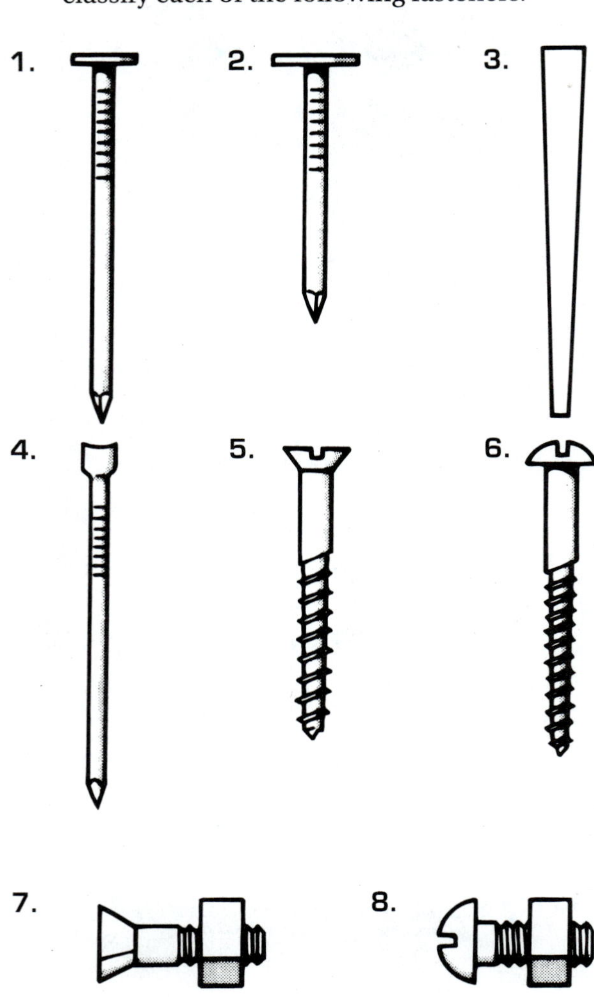

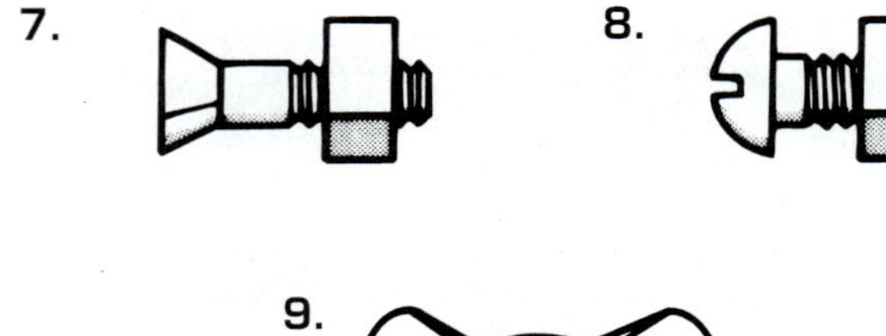

9.

E. Identification. Name, correctly spell, and classify each of the following items of hardware.

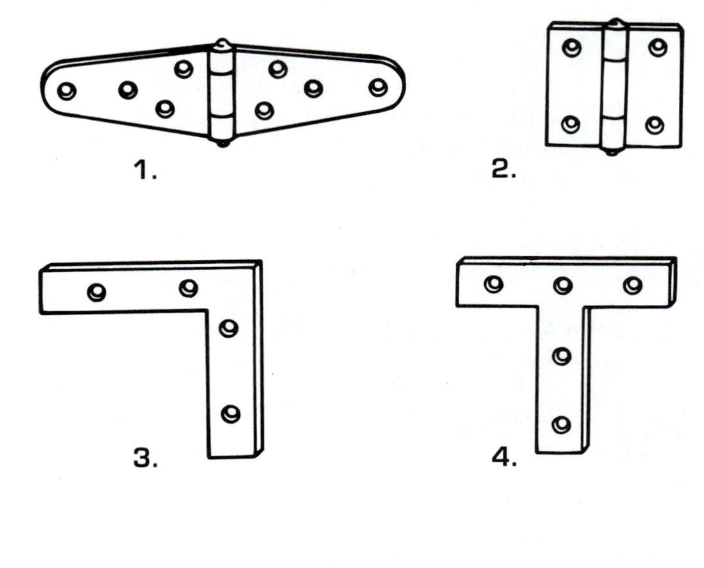

UNIT

Layout Tools and Procedures

OBJECTIVE

To select and use appropriate layout tools and procedures for wood and metalworking..

COMPETENCIES TO BE DEVELOPED

After studying this unit, you should be able to:

- Select appropriate tools for layout procedures in woodworking and metalworking.
- Use layout tools correctly and accurately.
- Make and use a pattern.
- Lay out wood and metal for cutting and shaping.

TERMS TO KNOW

layout
gradations
scale
aluminum
plastic
craftsman
English system
U.S. customary system
linear
English system
metric system
meter
centimeters
millimeters
kilometer
tape
folding rule
caliper
gauge
framing square
heel
try square
combination square
plumb
miter
sliding T bevel
bevel
right triangle
dividers
scribers
level
spirit level
line
chalk line
plumb line
plumb bob
pattern

MATERIALS LIST

- ✓ Examples of tools or tool parts made from steel, aluminum, wood, plastic, and cloth
- ✓ Examples of tools of poor quality that have been bent or broken with normal use
- ✓ A meter stick and a 12-inch ruler
- ✓ 12-foot tape (with English and metric numbers preferred)
- ✓ Steel or wooden scale
- ✓ Folding rule
- ✓ Inside calipers
- ✓ Outside calipers

- ✓ A variety of gauges (sheet metal, wire, drill, feeler, and spark plug)
- ✓ Framing square
- ✓ Combination square
- ✓ Try square
- ✓ Sliding T bevel
- ✓ Dividers and scribers
- ✓ Carpenter's level and line level
- ✓ Plumb line
- ✓ Chalk line
- ✓ Tag board and scissors
- ✓ Strig for 12-foot lines
- ✓ Hammer and three slender 12-inch stakes

Layout tools are used to guide the worker when cutting, sawing, drilling, shaping, or fastening. The term layout means to prepare a pattern for future operations. A truck driver uses a map to determine the correct route between two points. The map requires the use of many roads, numerous turns, and changes in direction to get to the destination. The situation is similar when workers design and construct projects, machinery, equipment, or buildings. They place marks on their work and then cut, shape, dig, or fasten according to the marks.

Many workers must use layout tools in performing their work. These workers include plumbers, electricians, designers, engineers, surveyors, farmers, landscapers, builders, and many others. The student will do well to learn how to use layout tools carefully. Layout tools have many uses.

When laying out work, it is important to choose the right tool. A second important factor is to use the tool correctly. For many jobs, there are several tools that will work, but usually one works better than others. For instance, a one-foot ruler or yardstick may be used to measure a house that is 60 feet long. However, a 100-foot tape is quicker and more accurate.

MATERIALS USED FOR LAYOUT TOOLS

Measuring devices are generally made of steel, aluminum, wood, plastic, or cloth. Steel tools are generally more durable than those made of other materials. When numbers and lines, called gradations, are stamped into steel tools, they stay readable for a long time. Numbers and gradations make up a scale on measuring tools. Another advantage of steel is it bends without breaking. Steel can stand the rough use found at construction sites. Steel is the first choice of material for calipers, dividers, and other slender tools that must measure very accurately and not bend easily.

Aluminum is the second choice of materials for use in many layout tools. Aluminum is a very tough metal that is light and durable. It is used extensively in levels and chalkers.

Wood is a cheap, relatively soft and light material. It is good for handles of layout tools. Wood is not used extensively in tools because it breaks easily and does not wear well. An exception is the wooden folding rule, which is compact, useful, and popular. However, it is very easily broken if not handled correctly.

Plastic is a term used for a group of materials made from chemicals and molded into objects. Some plastics are tough and light. They can be molded into any shape and are cheap to produce. A major disadvantage of plastics is their tendency to melt if touched by a hot object or flame. They may also be damaged by solvents found in shops. Some layout tools such as rulers are made of plastic, but the use of plastic is limited.

Cloth is a material used in some 50-foot tapes. It is cheap and lightweight. Cloth tapes are not very accurate as they stretch under stress and moisture. Any stretching of a measuring device reduces its accuracy.

Cost vs. Quality

Cheap tools are seldom a bargain. A cheap tool is likely to be inaccurate and unsatisfactory after little use. Good tools may seem expensive; however, they last a lifetime if used properly. High-quality tools are an excellent investment. Craftspeople use only good tools. A craftsperson is a skilled worker. Good tools permit the worker to become a good craftsperson if proper techniques are used. On the other hand, even skilled workers cannot do good work with poor tools.

SYSTEMS OF MEASUREMENT

The English system has always been the standard system of measurement in the United States. This system is now called the U.S. customary system, figure 8-1. It uses the inch, foot, yard, rod, and mile as units of linear measure. Linear means in a line. The English system uses units of various size, based on similarity to a certain familiar object or part of the human body, such as the foot or hand. Even today, horses are said to be so many "hands" high.

There are 12 inches in a foot, 3 feet in a yard, $16^{1}/_{2}$ feet in a rod, and 5,280 feet in a mile. The English system does not seem very logical.

LIQUID MEASURE	
2 cups	1 pint
2 pints	1 quart
4 quarts	1 gallon
3 1 1/2 gallons	1 barrel
2 barrels	1 hogshead
7 1/2 gallons of water	1 cu ft
U.S. Gallon	231 cu in

AVOIRDUPOIS WEIGHT	
16 ounces	1 pound
100 pounds	1 hundredweight
20 hundredweight	1 ton

DRY MEASURE	
2 pints	1 quart
8 quarts	1 peck
4 pecks	1 bushel

1 bushel contains 2150.42 cubic inches or approximately 1 1/4 cubic feet.

SQUARE MEASURE	
144 square inches	1 square foot
9 square feet	1 square yard
30 1/4 square yards	1 square rod
160 square rods	1 acre (43,560 sq ft)
640 acres	1 square mile
1 square mile	1 section

WEIGHT	
Gram	15.432 grains
Gram	.0353 ounce
Kilogram	2.2046 lb
Kilogram	.0011 ton (short)
Metric ton	1.1025 ton (short)
Grain	.064 gram
Ounce	28.35 grams
Pound	453.5 grams
Ton (Short)	907.18 kilograms
Ton (Short)	.907 metric tons
Ton (Short)	2,000 lb

LINEAR MEASURE	
12 inches	1 foot
3 feet	1 yard
5 1/2 yards (16 1/2 feet)	1 rod
320 rods	1 mile
1760 yards (5,280 feet)	1 mile

CUBIC MEASURE	
1728 cubic inches	1 cu ft
27 cubic feet	1 cu yd
128 cu ft (8' × 4' × 4')	1 cord
1' × 1' × 1"	1 bd ft

LIQUID DATA

1 gallon of water weighs 8.34 pounds
1 gallon of milk weighs 8.6 pounds
1 cubic foot of water is equal to 7.48 gallons
1 inch of rainfall amounts to 27,154 gallons per acre

FIGURE 8-1. The English system, now known as the U.S. Customary system, has long been the standard of weights and measures in the United States.

In the early times, people could function satisfactorily with approximate units of measure such as a hand to measure the height of a horse. However, as the technological age evolved, measurements had to be standardized. Can you imagine the results if the human foot was used as the method for measuring when constructing buildings?

Today, the foot and most other linear units of measure are very precise lengths. If layout tools are constructed properly, a foot is the same length on all tools. Variations may be observed on poor-quality tools and tools that measure long distances, such as tapes. The longer the instrument, the greater the problem of inaccuracy due to stretching, expansion, and contraction. The foot is represented by one mark (') and the inch by two marks ("). Hence, 2'6" means 2 feet and 6 inches.

The **metric system** of measurement has always been used for scientific work in the United States, figure 8-2. Only recently has the metric system been used for nonscientific work. Linear measurements in the metric system are based on the **meter**. The meter (m) equals 39.37 inches. The meter contains 100 **centimeters** (cm) and 1,000 **millimeters** (mm). One thousand meters equals a **kilometer** (km). Notice how the units of measure in the metric system relate to one another by multiples of ten. The metric system is mathematically logical and easy to use.

The millimeter, centimeter, meter, and kilometer are the most used metric units of linear measurement in agricultural mechanics. The inch, foot, yard, rod, and mile are the most useful units of linear measurement in the U.S. Customary system. The U.S. Customary system is still the most used system in the United States. However, the metric system is increasing in use and many layout tools use both systems, figure 8-3.

The Inch as a Unit of Measurement

The inch is the traditional unit of measurement for woodworking and metalworking in the United States. It must be divided into smaller units to be useful for most applications. Some fine rules or scales may have as many as 32 marks per inch. Each mark is 1/32 of an inch apart. One-sixteenth inch is more commonly used as the smallest unit on a rule. Lines of different lengths are used to show 1/2, 1/4, and 1/8 of an inch on many measuring devices, such as rules and squares, figure 8-4.

The Millimeter as a Unit of Measurement

The millimeter (mm) is slightly smaller than 1/16 of an inch. It is a very convenient unit for linear measurement without using fractions. It has the advan-

A COMPARISON OF THE INTERNATIONAL METRIC SYSTEM AND THE ENGLISH SYSTEM OF MEASUREMENT

1 Centimeter	= .3937 inch
1 Inch	= 2.54 centimeters
1 Foot	= 30.48 centimeters
1 Meter	= 39.37 inches
1 Meter	= 100 centimeters
1 Meter	= 1.094 yards
1 Meter	= 1000 millimeters
1 Millimeter	= .001 meter
1 Yard	= .9144 meter
1 Mile	= 1609.344 meters
1 Kilometer	= 1000 meters
1 Kilometer	= .62137 mile
1 Sq. Centimeter	= .155 Sq. inch
1 Sq. Decimeter	= 100 cu. centimeters
1 Cu. Centimeter	= .061 cu. inch
1 Cu. Decimeter	= 1000 cu. centimeters
1 Cu. Meter	= 100 liters
1 Fluid Ounce	= 29.54 milliliters
1 Liter	= 1000 cu. centimeters
1 Liter	= 1.057 quarts
1 Gallon	= 3.785 liters
1 Gram	= 15.43 grains
1 Ounce	= 28.35 grams
1 Kilogram	= 1000 grams
1 Kilogram	= 2.205 pounds
1 Pound	= 7000 grains
1 Pound	= .4536 kilogram
1 Kilogram	= 1000 milliters
1 Kilogram	= 1 liter

METRIC PREFIXES

Names of multiples of Metric units are formed by adding a prefix to "meter," "gram," or "liter." These prefixes are used also with units other than these three metric ones.

Prefix	Symbol	Power	Number	Name
tera	- T	10^{12}	1,000,000,000,000.	ONE TRILLION
giga	- G	10^{9}	1,000,000,000.	ONE BILLION
mega	- M	10^{6}	1,000,000.	ONE MILLION
kilo	- k	10^{3}	1,000.	ONE THOUSAND
hecto	- h	10^{2}	100.	ONE HUNDRED
deka	- dk	10^{1}	10.	TEN
—	-	10^{0}	1.	ONE

Prefix	Symbol	Power	Number	Name
—	-	10^{0}	1.	ONE
deci	- d	10^{-1}	0.1	ONE TENTH
centi	- c	10^{-2}	0.01	ONE HUNDREDTH
milli	- m	10^{-3}	0.001	ONE THOUSANDTH
*micro	- μ	10^{-6}	0.000001	ONE MILLIONTH
nano	- n	10^{-9}	0.000000001	ONE BILLIONTH
pico	- p	10^{-12}	0.000000000001	ONE TRILLIONTH

**The special case of one millionth of a meter is called a micron.*

CONVERSION TABLE

Multiply	To Obtain
Bushels by .	Cu. Feet
Bushels by	Pecks
Bushels by .04545	Cu. Yards
Centimeters by 0.3937	Inches
Centimeters by 0.01	Meters
Centimeters by 10	Millimeters
Cu. Centimeters by .00003531	Cu. Feet
Cu. Centimeters by .06102	Cu. Inches
Cu. Centimeters by .00001	Cu. Meters
Cu. Centimeters by .000001308	Cu. Yards
Cu. Centimeters by .0002642	Gallons
Cu. Centimeters by .001	Liters
Cu. Centimeters by .002113	Pints (Liq.)
Cu. Centimeters by .001057	Quarts (Liq.)
Cu. Feet by .0002832	Cu. Centimeters
Cu. Feet by 1728	Cu. Inches
Cu. Feet by 0.02832	Cu. Meters
Cu. Feet by 0.03704	Cu. Yards
Cu. Feet by 7.48052	Gallons
Cu. Feet by 28.32	Liters
Cu. Feet by 59.84	Pints (Liq.)
Cu. Feet by 29.92	Quarts (Liq.)
Cu. Feet by 1.25	Bushels
Cu. Inches by 16.39	Cu. Centimeters
Cu. Inches by .0005787	Cu. Feet
Cu. Inches by .00001639	Cu. Meters
Cu. Inches by .00002143	Cu. Yards
Cu. Inches by .004329	Gallons
Cu. Inches by .01639	Liters
Cu. Yards by 764600	Cu. Centimeters
Cu. Yards by 22	Bushels
Cu. Yards by 27	Cu. Feet
Cu. Yards by 46.656	Cu. Inches
Cu. Yards by 0.7646	Cu. Meters
Cu. Yards by 202.0	Gallons
Cu. Yards by 764.6	Liters
Cu. Yards by 1616	Pints (Liq.)
Cu. Yards by 807.9	Quarts (Liq.)
Fathoms by 6	Feet
Feet by 30.48	Centimeters
Feet by 12	Inches
Feet by 0.3048	Meters
Feet by 1/3	Yards
Gallons by 3785	Cu. Centimeters
Gallons by 0.1337	Cu. Feet
Gallons by 231	Cu. Inches
Gallons by .003785	Cu. Meters
Gallons by .004951	Cu. Yards
Gallons by 3.785	Liters
Gallons by 8	Pints (Liq.)
Gallons by 4	Quarts (Liq.)
Gallons Water by 8.3453	Lbs. of Water
Grams by 0.03527	Ounces
Grams by 0.03215	Ounces (Troy)
Grams by .002205	Pounds
Inches by 2.540	Centimeters
Kilometers by 100000	Centimeters
Kilometers by 3281	Feet
Kilometers by 1000	Meters
Kilometers by 0.6214	Miles
Meters by 100	Centimeters
Meters by 3.281	Feet
Meters by 39.37	Inches
Meters by .01	Kilometers
Meters by 1000	Millimeters
Meters by 1.094	Yards
Microns by .00001	Meters
Miles by 160900	Centimeters
Miles by 5280	Feet
Miles by 1.609	Kilometers
Miles by 1760	Yards
Miles per Hr. by 44.70	Centimeters per Sec
Miles per Hr. by 88	Feet per Min.
Miles per Hr. by 1.467	Feet per Sec.
Miles per Hr. by 1.609	Kilometers per Hr.
Miles per Hr. by 0.8684	Knots
Miles per Hr. by 26.82	Meters per Min.
Millimeters by 0.03937	Inches
Ounces by 2	Tablespoons (Liq.)
Ounces by 6	Teaspoons (Liq.)
Ounces by 3	Tablespoons (Dry)
Ounces by 9	Teaspoons (Dry)
Ounces by 28.349527	Grams
Ounces by 0.9115	Ounces (Troy)
Ounces (Fluid) by 1.805	Cu. Inches
Pounds by 16	Ounces
Pounds of Water by 0.01602	Cu. Feet
Pounds of Water by 27.68	Cu. Inches
Pounds of Water by 0.1198	Gallons

FIGURE 8-2. The metric system of measurement and factors for converting units in the English and metric systems.

CONVERSION TABLE (Cont'd.)

Multiply	To Obtain
Cu. Inches by 0.03463	Pints (Liq.)
Cu. Inches by 0.01732	Quarts (Liq.)
Cu. Meters by 10000	Cu. Centimeters
Cu. Meters by 35.31	Cu. Feet
Cu. Meters by 61.023	Cu. Inches
Cu. Meters by 1.308	Cu. Yards
Cu. Meters by 264.2	Gallons
Cu. Meters by 1000	Liters
Cu. Meters by 2113	Pints (Liq.)
Cu. Meters by 1057	Quarts (Liq.)
Kilometers by 1094	Yards
Liters by 1000	Cu. Centimeters
Liters by 0.03531	Cu. Feet
Liters by 61.02	Cu. Inches
Liters by .01	Cu. Meters
Liters by .001308	Cu. Yards
Liters by 0.2642	Gallons
Liters by 2.113	Pints (Liq.)
Liters by 1.057	Quarts (Liq.)
T (in) by W (ft) by L (ft) Board Feet	
Tablespoons (Liq.) by 0.5	Ounces
Tablespoons (Dry) by 0.3333	Ounces
Tablespoons by 3	Teaspoons
Teaspoons (Liq.) by 0.1666	Ounces
Teaspoons (Dry) by 0.1111	Ounces
Teaspoons by 0.3333	Tablespoons
Temp (C)+17.78 by 1.8	Temp (F)
Temp (F)–32 by $^{5}/_{9}$	Temp (C)
Tons (Long) by 2240	Pounds

FIGURE 8-2. *Continued*

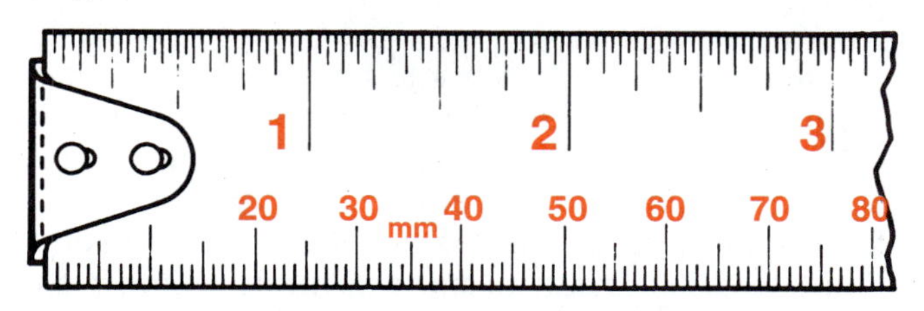

FIGURE 8-3. This tape has both U.S. Customary and metric units of measurement. The 1, 2, 3 values refer to inches. The 20, 30, 40 values refer to millimeters.

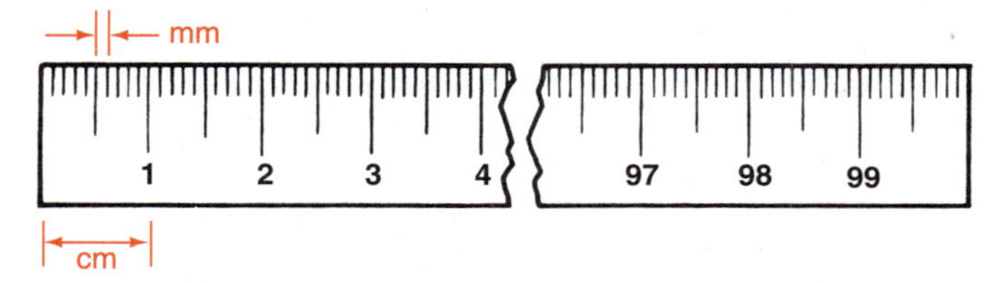

FIGURE 8-5. The marks that appear between each centimeter division are millimeters. A meter has 100 centimeters and 1,000 millimeters.

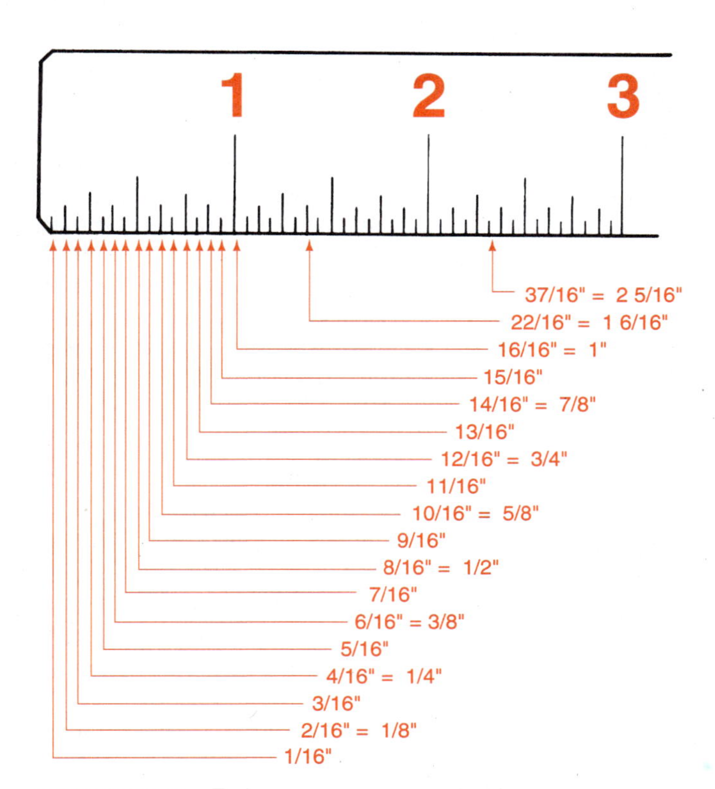

FIGURE 8-4. Rules, squares, and other measuring devices are divided into halves, quarters, eighths, and sixteenths of an inch. These divisions are frequently illustrated by lines of different lengths.

tage of being 1/1,000 of a meter and 1/10 of a centimeter. This means one meter plus 250 millimeters equal 1 1/4 meters. It is more convenient to write 1/4 meters using a decimal rather than a fraction. Using the decimal, 1 1/4 meters reads 1.250 meters. Similarly, 1 1/2 meters reads 1.500 meters, or simply 1.5 meters. On many metric rules, each centimeter division contains ten marks to represent millimeters, figure 8-5. Centimeters can be changed to millimeters by multiplying by 10. Simply adding a 0 to the centimeter's value obtains the millimeter value, e.g., 4 centimeters is 40 millimeters.

When working on small projects, measurements may be made entirely in millimeters to avoid the use of fractions or decimals. If a board is 1.5 meters in length, this value can be written as 1500 millimeters. To change meters to millimeters multiply by 1000.

MEASURING TOOLS

Measuring tools include tapes, rules, calipers, and other devices used to determine specific distances. They are used to measure length, width, height, depth, thickness, spacing, and clearances.

When measuring, the last number on the scale is read and then the fraction of inches or number of millimeters is added, depending on the scale. If very exact measurements are necessary, a thin tape is used or the measuring device is turned up on its edge to get an exact reading.

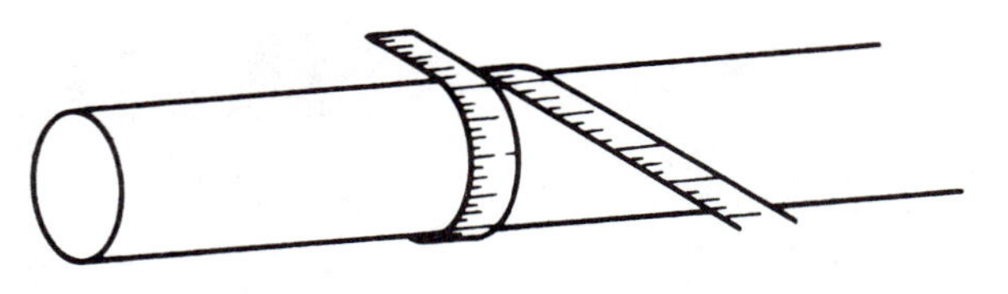

MEASURING THE CIRCUMFERENCE OF A ROUND OBJECT.

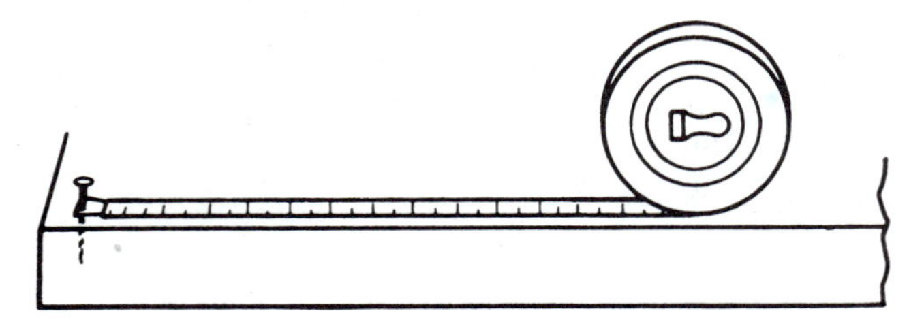

TAKING A MEASUREMENT STARTING AT A NAIL.

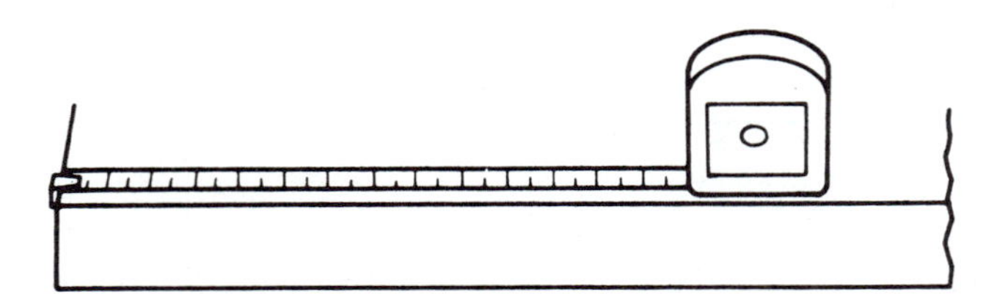

TAKING A MEASUREMENT STARTING FROM THE END OF AN OBJECT.

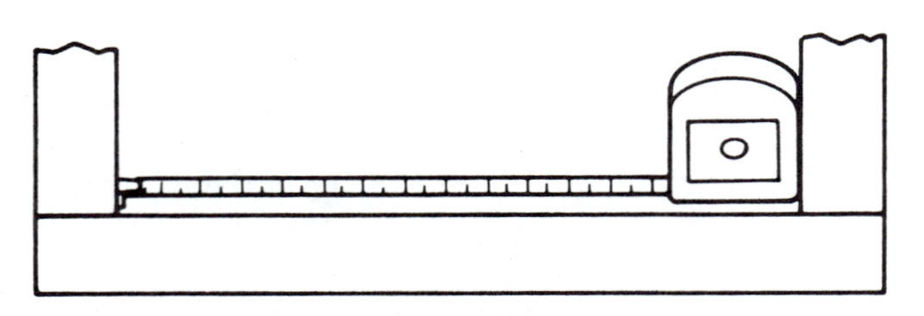

MEASURING THE DISTANCE BETWEEN TWO INSIDE SURFACES.

FIGURE 8-6. Tapes can be used to measure in various ways. The book provided on one end of most tapes provides for accurate measurement whether it is hooked onto an object or pushed against a surface.

Tapes

The term tape refers to flexible measuring devices that roll into a case. They range in length from several feet to hundreds of feet and are made of steel, cloth, or fiber materials. Some tapes are self-retracting; they have a spring inside to wind up the tape when released. Some have locks to hold the tape out while in use. Some have a button on the side to control the movement of the tape.

Most tapes have a hook on the end that slides a distance equal to the thickness of the hook. This permits the tape to provide accurate measurements whether hooked to an object or pushed against a surface. Figure 8-6 shows various ways of measuring using a tape measure. Students must handle tapes carefully as they are easily broken. Tapes break if pulled out too far, forced back into the case, or bent excessively.

Folding Rules

A folding rule is a rigid rule of 2 to 8 feet in length. It can be folded for handling and storage. At one time the folding rule was the standard measuring device used by carpenters. However, the compactness and length of modern tapes make them preferable to the folding rule for many jobs. Folding rules are made of wood, plastic, or metal.

Good quality wooden folding rules have spring-loaded buttons to keep them open; these rules are called spring-joint rules. A good wooden folding rule is made of tough wood and is easy to read. Some folding rules have a sliding metal insert in one end, figure 8-7. The insert permits accurate inside measurements and depth measurements.

Special care must be used in opening and closing folding rules. A careless twist can cause the rule to break. However, careful workers, who oil the joints periodically, may use the same folding rule day after day for many years without breakage.

Scales and Bench Rules

The term scale as used here means a rigid steel or wooden measuring device, figure 8-8. Scales are generally one to three feet in length and 3/4 inch to 1 inch in width. Wooden scales are sometimes called bench rules. They are about 1/4-inch thick. Metal scales are relatively thin and, therefore, better than wooden scales where very accurate work is required.

Scales and bench rules are handy in the shop. They are not used much on other jobs since they cannot be folded.

Calipers

A caliper is an instrument used to measure the diameter or thickness of an object. There are inside calipers and outside calipers, figure 8-9. Inside calipers are so named because they measure inside distances, such as the diameter of a cylinder. Outside calipers are used to measure the outside of round objects, such as pipes.

With the most economical caliper a rule or scale may be needed to measure its settings. Other calipers have their own dimension scales. Another type of caliper has a dial to indicate very small differences in the size of objects measured.

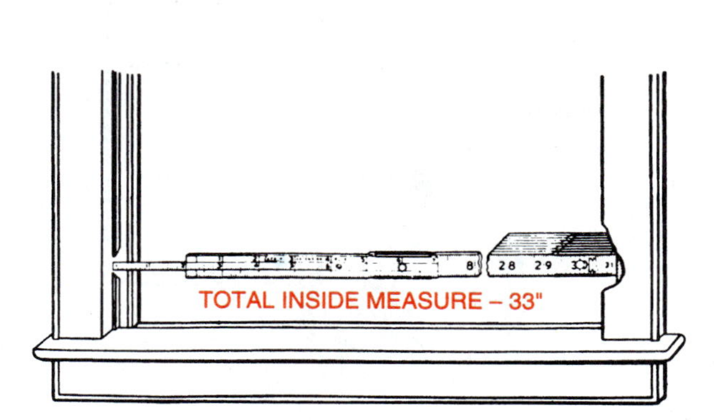

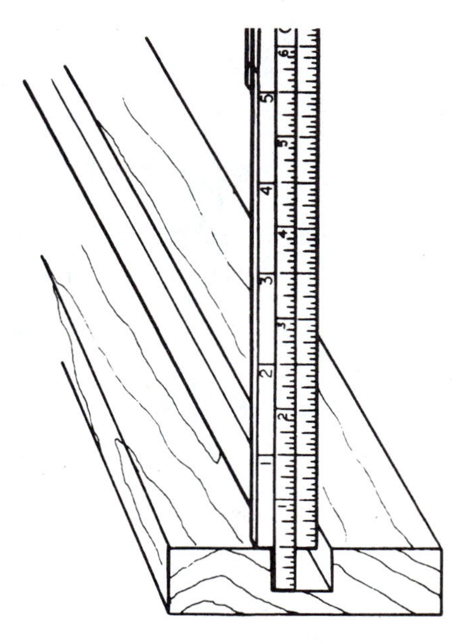

FIGURE 8-7. The insert in the end of a wooden folding ruler makes the ruler useful for accurate inside measurements. *(Courtesy of McDonnell & Kaumeheiwa, Use of Hand Woodworking Tools, Delmar Publishers, 1978)*

Gauges

A gauge is a device used to determine thickness, gap in space, diameter of materials, or pressure of flow. Examples are sheet metal gauges, wire gauges, drill gauges, spark plug gauges, feeler gauges, and pressure gauges..

SQUARES AND THEIR USES

Squares are used to guide the builder. A square is a device used to draw angles for cutting and check the cuts for accuracy. The squares most often used are the framing square, try square, combination square, and the sliding T bevel.

Framing Square. The framing square is a flat square with a body and tongue. It is usually made of steel. The body is 24 inches long and the tongue is 16 inches long. The heel is the place where the tongue and body meet to form a 90 degree angle.

The framing square is also called a carpenter's square and a steel square. It may contain several tables. One table found on many framing squares helps the user to calculate the board feet in a piece of lumber. Another table may help the user to lay out rafters. Special booklets and chapters in advanced texts provide more information on the use of these tables.

Framing squares have measurements that start at the inside of the heel and run out the body and tongue. Inside measurements of the tongue are 1 to 14 inches. Inside measurements of the body are 1 to 22 inches. Outside measurements of the tongue are 1 to 16 inches. Another set of measurements starts at the outside of the heel and runs out of the body and tongue. Outside measurements of the body are 1 to 24 inches. It is important to study the scales carefully before using a framing square. Figure 8-10 shows how the tongue and the body are used to measure different materials.

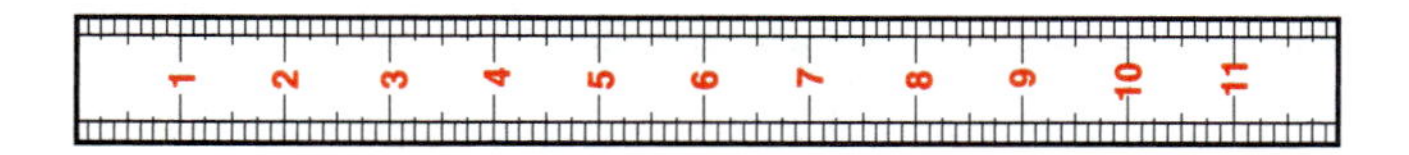

FIGURE 8-8. Wooden bench rules and steel scales are useful in the shop. They are not often used on job sites since they cannot be folded.

Try Square. The try square is so named because it is used to try or test the accuracy of cuts that have been made. It is also used to mark lines on boards in preparation for cutting. Try squares have wood, steel, or plastic handles and steel blades.

The try square is a good tool for marking lines on boards up to 12 inches wide. Small try squares are handy to mark lines across the edge or end of boards. They may be used to draw both 90 or 45 degree lines on boards, figure 8-11.

Combination Square. The combination square is so named because it combines many tools

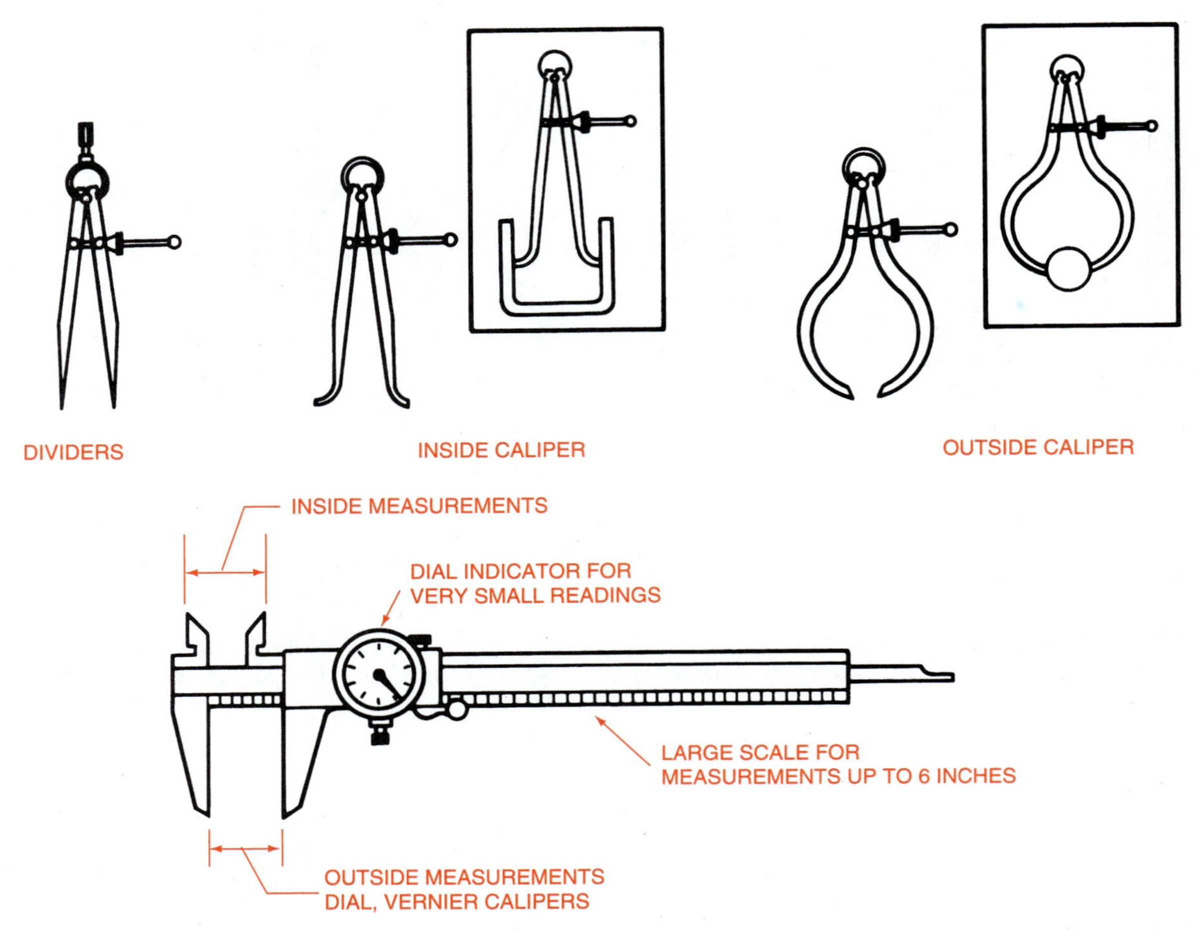

FIGURE 8-9. Outside calipers are used to measure the distance across the outside of objects. Inside calipers are used to measure the inside of objects.

into one, figure 8-12. It contains a bubble that permits the tool to be used as a level. Another bubble permits it to be used for plumbing objects. An object is plumb when it is vertical to the axis of the earth or in line with the pull of gravity. In other words, when something is straight up and down it is said to be plumb.

The combination square contains a removable

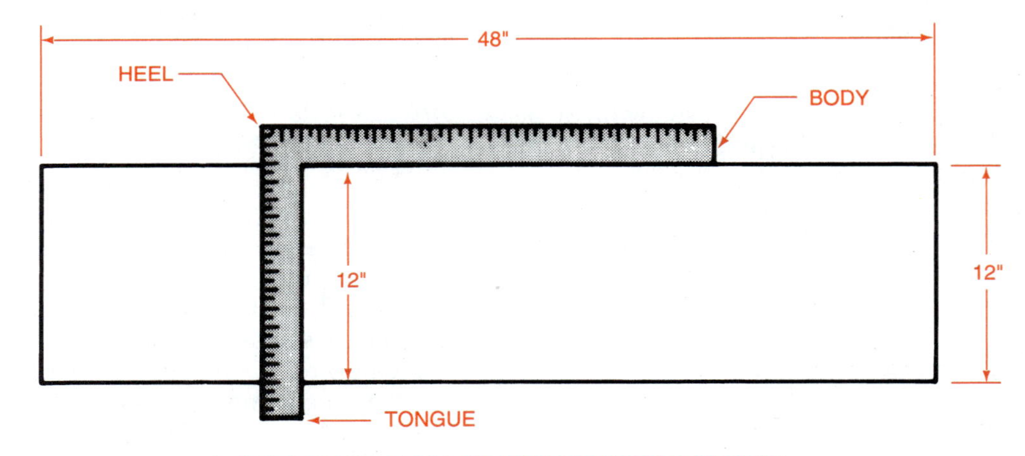

A. THE TONGUE OF THE FRAMING SQUARE IS USED TO SQUARE BOARDS UP TO 16 INCHES WIDE.

FIGURE 8-10. The tongue and the body of the framing square are used to measure materials of different size.

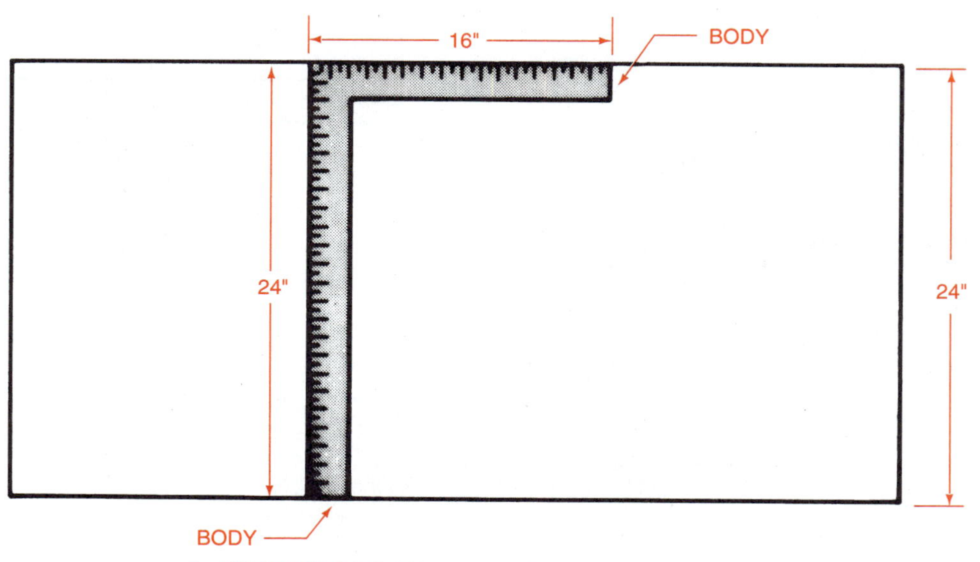

FIGURE 8-10. Continued

blade that can be used as a steel scale. When the blade is in place, it may be used as a depth gauge. The tool is very handy as an ordinary square and as a 45 degree miter square. A miter is an angle. Special heads make the tool useful for locating the center of round objects. The combination square is probably the most used square because of its many functions.

Sliding T Bevel. The sliding T bevel is a device used to lay out angles. It is also called a bevel square. A bevel is a sloping edge. If the corner is cut from a board, a bevel is created, figure 8-13. To use a sliding T bevel, first loosen the wing nut or thumb screw.

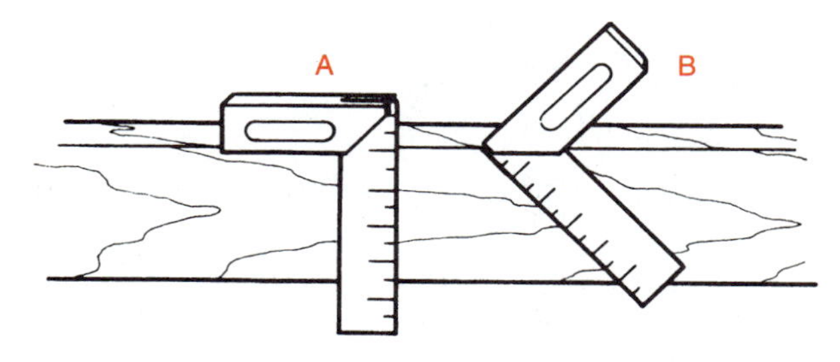

FIGURE 8-11. The try square can be used to draw 90 degree or 45 degree lines across the face, edge, or ends of boards. To mark a 90 degree angle, the try square is positioned as shown in A. To mark a 45 degree angle, the 45 degree should of the try square is used to check the exactness of the cut. *(Courtesy of McDonell & Kaumeheiwa, Use of Hand Working Tools, Delmar Publishers, 1978)*

Then adjust the blade to the desired angles and length, and retighten.

Using a Square to Mark a Board

A square is used when marking boards to be sawed exactly square or at 90 degrees to the edge. When using a square to mark a board the following procedure is used:

PROCEDURE

1. Measure and mark the length of board desired. The mark should be thin and at a 90 degree angle to the board's edge.
2. Place the handle of the square firmly against the edge of the board.
3. Move the square until the blade is against the mark.
4. Hold a sharp pencil or knife against the blade and draw a line through the mark and across the board, figure 8-14. The board is now ready to be sawed.

SQUARING LARGE AREAS

When working large areas, such as laying out a building, care must be taken when squaring corners. An error as small as one degree will cause serious construction problems. One way to establish 90 degree

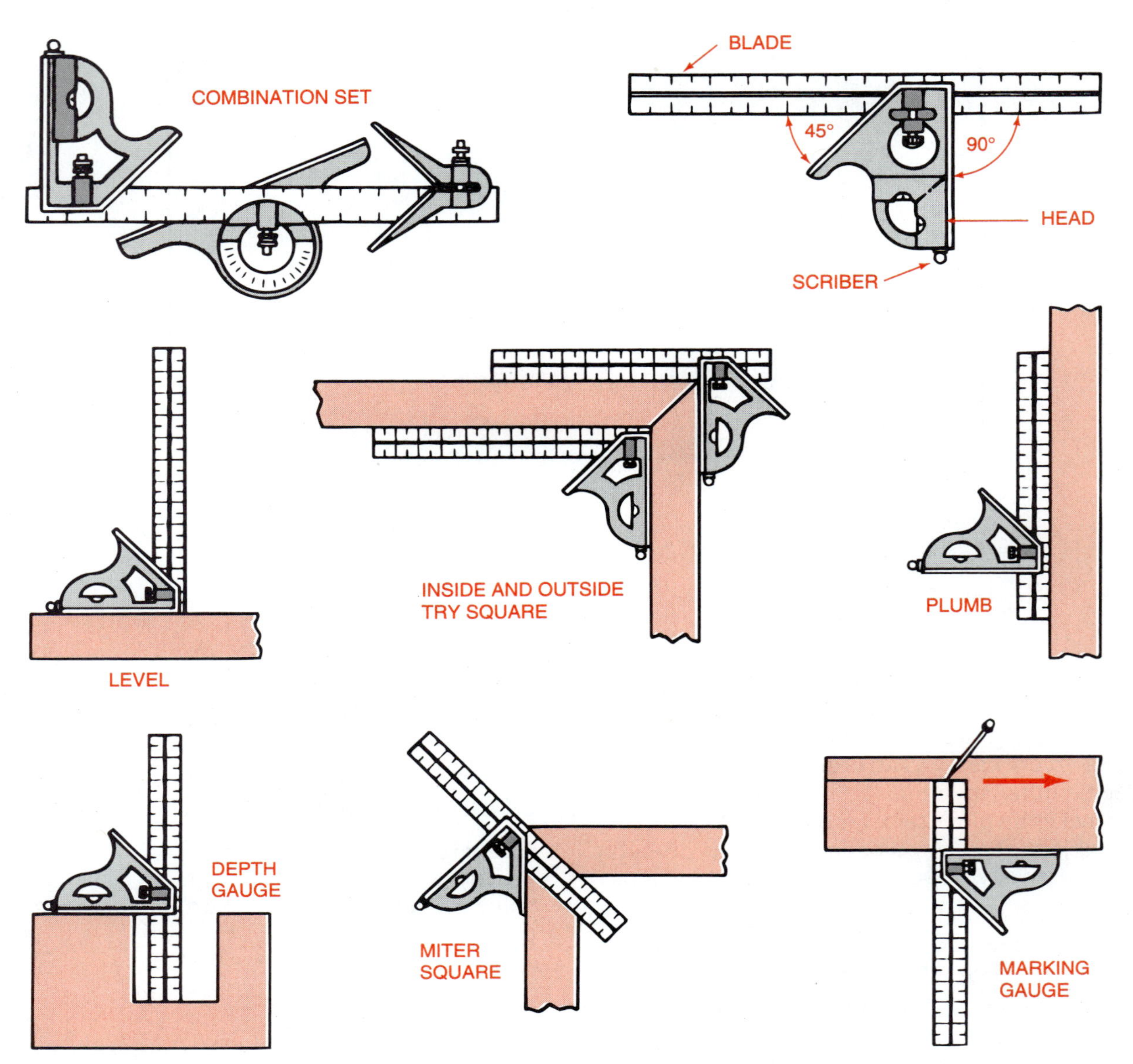

FIGURE 8-12. The combination square is many tools in one. It is a plumb, level, scriber, scale, depth gauge, marking gauge, miter square and regular scale.

corners is to create a right triangle with sides that are 3 feet, 4 feet, and 5 feet in length. Any right triangle with sides that are multiples of 3 feet, 4 feet, and 5 feet will work. A right triangle is a three-sided figure with one angle of 90 degrees.

To lay out a 90 degree corner for a large area, use 6, 8, and 10 feet, respectively, and the following steps.

PROCEDURE

1. Drive a thin stake at the point of the right angle or corner.
2. Lay the heel of a framing square against the stake.
3. Attach a piece of string and stretch it along the line of one leg of the square. Stake the other end of the string and call this line A.
4. Attach a second string at the point of the right angle and stretch it along the other leg of the square. Call this line B.
5. Measure along line A a distance of 6 feet. Tie a piece of string at this 6-foot mark.
6. Measure along line B and tie a piece of string at the 8-foot mark.

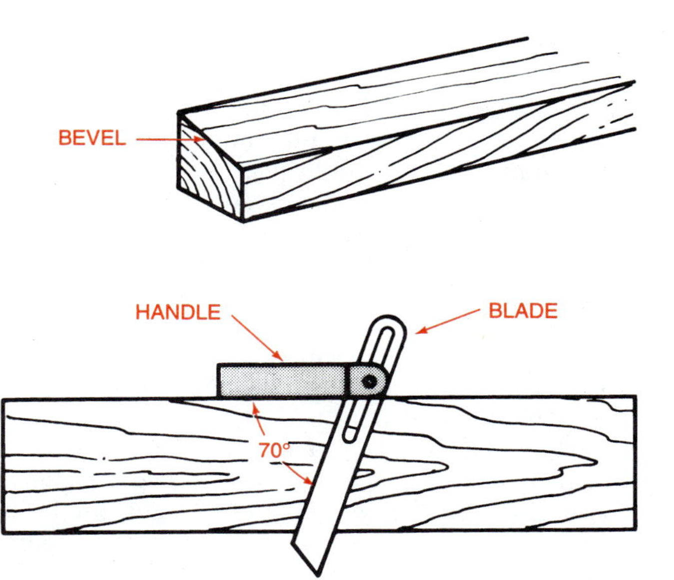

FIGURE 8-13. A bevel is a sloping edge made on boards by cutting down a corner. The sliding T bevel is used to lay out bevels and other angles.

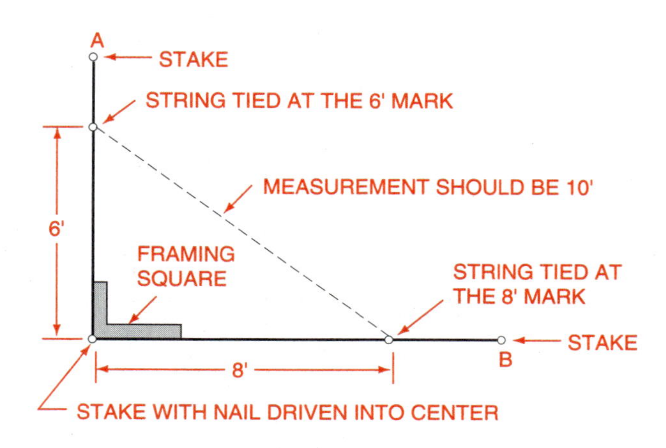

FIGURE 8-15. The 6, 8, 10 method is useful in establishing a square corner in a large area. The framing square provides a means to place the strings at approximate right angles to each other. Measurements of 6 feet, 8 feet, and 10 feet provide more accuracy in placing stakes A and B to create an angle that is exactly 90 degrees.

FIGURE 8-14. Holding the square against the mark, a line is drawn through the mark across the board.

7. Measure the distance from the string tied on line A and the string tied on line B. The two points should be exactly 10 feet apart. If they are not, move either line A or B in or out until the distance is exactly 10 feet (within 1/16 of an inch). The angle will then be square or 90 degrees, figure 8-15.

A useful procedure for determining if four corners of a rectangle are square is to measure the distance diagonally from corner to corner. If all four corners are square, the diagonals will be equal in length, figure 8-16. To be considered equal, two measurements must be exactly the same.

DIVIDERS AND SCRIBERS

Dividers and scribers are instruments used to make circles and other curved lines. They are also used to transfer equal measurements from a scale or piece of material to another item. Dividers and scribers are

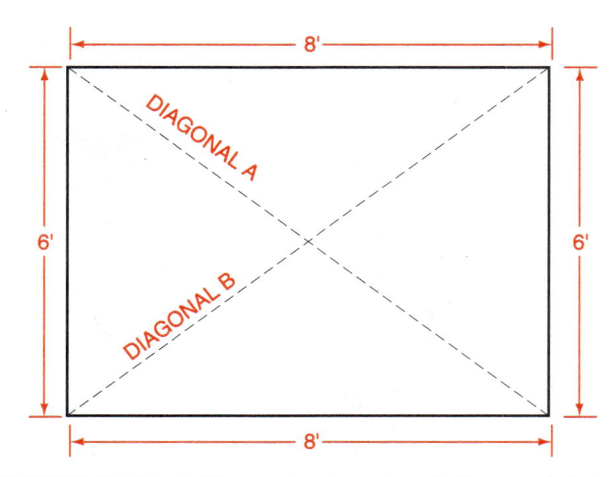

FIGURE 8-16. When the length of diagonal A is the same as the length of diagonal B, all corners are square if the figure is a rectangle. To be a rectangle, the two sides must be of equal length and the two ends must be of equal length.

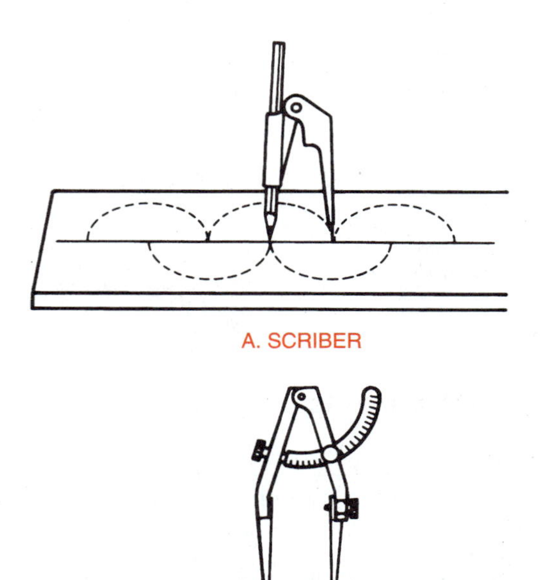

FIGURE 8-17. Both scribers and dividers are used to mark circles. Scribers have one steel leg and a pencil. Dividers have two steel legs, which are useful for marking dark metal.

very similar. One difference is that the scriber has one steel leg and a pencil for the other leg. The divider has two sharp steel legs, figure 8-17.

LEVELS

A level is a device used to determine if an object has the same height at two or more points. To determine this, a spirit level is generally used. A spirit level consists of alcohol in a sealed, curved tube with a small air space or bubble mounted in a bar of wood or aluminum. When both ends of the level are exactly the same height, the bubble flows to the high point of the tube. This point is indicated by the bubble fitting exactly between two lines marked on the glass. If the tube is mounted crossways in the bar, the level may be used to plumb an object.

Levels may be made to attach to a string or line stretched between two distant points. This is called a line level. When the string is pulled tight, the level indicates when both ends are the same height, figure 8-18. Line levels are useful for leveling the corners when building with block and setting stakes to gauge the depth of concrete.

LINES

Strong cotton or nylon string is used as a guide when laying block or flooring, cutting rafters, applying ceiling tiles and other jobs. A line is any thin material stretched tightly between two or more points.

A chalk line is a cord with chalk applied by rubbing or by drawing the line through powdered chalk. When stretched and snapped, the line leaves a thin trail of chalk to mark straight lines up to 50 feet or more.

A plumb line is a string with a round and pointed piece of metal called a plumb bob attached. The plumb line hangs perfectly vertical. It is useful when

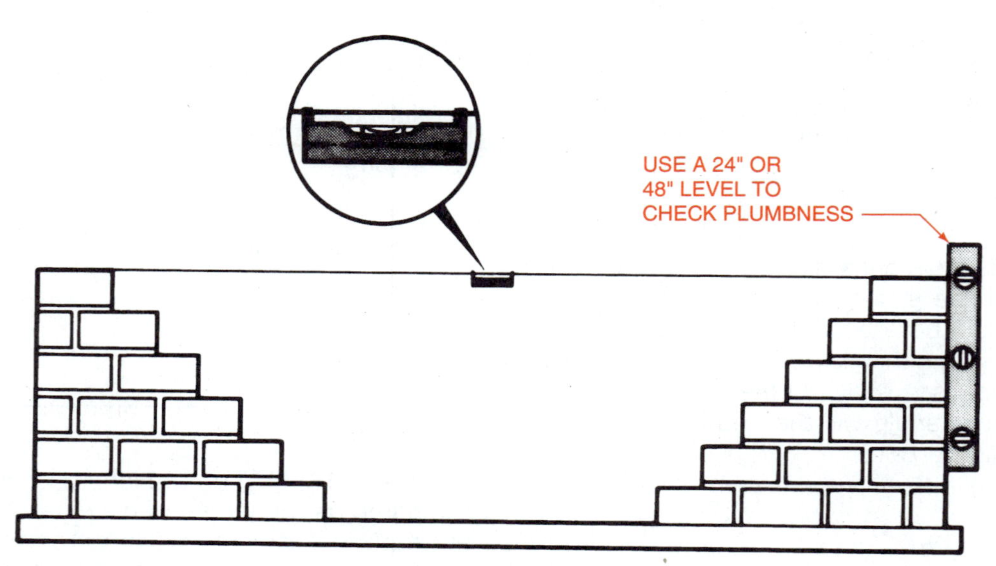

FIGURE 8-18. A level attached to a string is called a line level. It is used to check the levelness of two distant points. The corners of the walls shown are 16 feet apart. A carpenter's level may be used to check the wall for plumbness.

building greenhouses, sheds, barns, and other buildings to check for plumbness of walls and structures.

USING PATTERNS

Layout tools and instruments can be used to create almost any object. However, much time may be saved by copying the shapes of objects or designs rather than creating them. A pattern is useful for copying objects. A pattern is a model or guide from which to make an object.

To build a wall, a carpenter may cut one 2 × 4 inch piece of lumber exactly 7 feet 6 inches long with both ends square. This piece will then be used as a pattern to mark and saw several dozen pieces exactly like the pattern.

In the agricultural mechanics shop, students may wish to develop simple woodworking skills by making a decorative cutting board as a project. The teacher may have a pattern made of a thin piece of masonite cut the shape of a pig or other object. The masonite is laid on a piece of wood and the outline of the pattern is drawn. Within minutes, the project is ready to be cut.

Individual designs for a project or project part can be created. When creating designs, it is always wise to begin by making a pattern out of paper. Several techniques may be used to make a pattern.

- *Tracing a pattern.* If a picture of the desired design is available, the pattern can be made by tracing. This is done by placing thin tracing paper over the picture and drawing an outline of

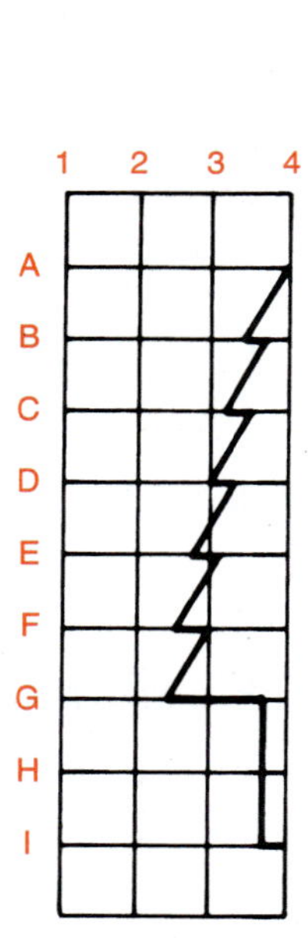

STEP 1: DRAW A GRID OVER THE ORIGINAL PICTURE. USE 1/4" SPACES BETWEEN LINES. NUMBER AND LETTER THE LINES IN THE GRID; THEN FOLD THE ORIGINAL IN HALF.

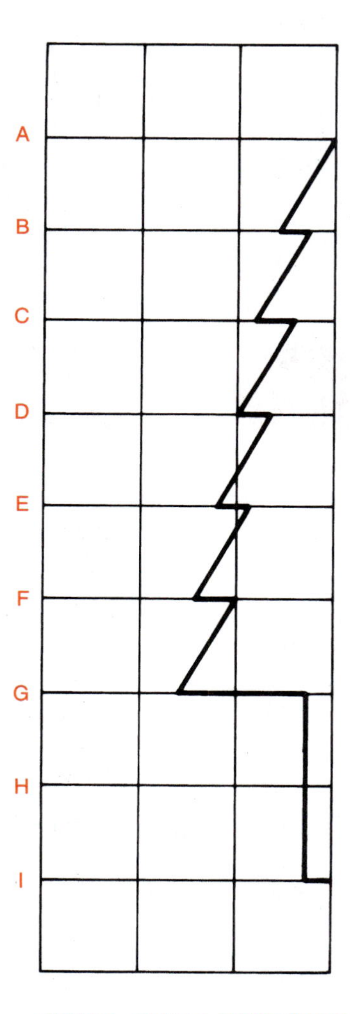

STEP 2: FOLD A PIECE OF TAG BOARD IN HALF. ADD A GRID USING 1/2" SPACINGS BETWEEN LINES. DRAW THE TREE ONTO THE ENLARGED GRID AND CUT OUT THE TREE AS OUTLINED.

STEP 3: THE PATTERN IS NOW READY FOR DRAWING THE OUTLINE ONTO WOOD OR METAL. THE GRID LINES ON HALF OF THE PATTERN MAY BE IGNORED.

FIGURE 8-19. A pattern of any size may be made from a drawing of an object such as a Christmas tree. In this example, a pattern is made that doubles the size of the original picture.

the object. The tracing paper is then placed over a piece of heavy paper, such as tag board. Both pieces are then cut out at one time.

- *Using carbon paper.* Carbon paper may be placed between a picture and a piece of heavy paper. The picture is then outlined with a ballpoint pen and the image is transferred to the heavy paper. The paper is then cut into a pattern.
- *Drawing a pattern.* A pattern can be made by using a pencil, ruler, compass, or other drawing instruments. A pattern can also be drawn free-hand without instruments. If the object is to be the same on both the left and the right, the paper should be folded in half to make the pattern symmetrical. One-half of the object is then drawn. By cutting the doubled paper, a pattern is obtained with both halves being exactly alike. This procedure is especially helpful if the pattern has curved lines.

Enlarging a Pattern

If a pattern is smaller than desired, an enlarged version can be made. First, a grid of vertical and horizontal lines is drawn over the original pattern or a copy of the pattern. The lines must all be spaced the same distance apart. One-quarter inch spacings work well for objects smaller than a standard sheet of paper.

A similar grid is then drawn on a piece of heavy drawing paper, but the space between lines in the grid is increased. This increase is in proportion to the desired increase in size of the pattern. For example, suppose the original is a picture in a book that is 6 inches high and 4 inches wide. If the pattern is to be twice as large, the desired height would be 12 inches and the width 8 inches. In this case, the distance between the lines of the grid in the pattern is doubled. Suppose the picture has identical left and right sides, such as a Christmas tree. After the grid on the paper is prepared, the paper is folded in half. Observe how the lines of the original picture cut across the grid. An outline of half of the tree is drawn onto the new grid using the corresponding lines and squares of the original to guide the drawing on the pattern. The pattern is then cut out and unfolded. It should be a perfect enlargement of the original, figure 8-19.

An alternate method is to use an overhead or opaque projector to project an enlarged pattern onto paper. The image is then marked with a pencil or pen.

STUDENT ACTIVITIES

1. Define the Terms to Know in this unit.
2. Obtain a steel tape, steel or wooden scale, combination square, try square, and framing square. Examine the tools and record the information needed to complete the following table.

Tool	Basic Units of Measure	Marks Per Unit	Length
a. Steel tape	Inch/Foot	16 per inch	12 inch
b. Steel or wooden scale	________	________	________
c. Combination square	________	________	________
d. Try square	________	________	________
e. Framing square	________	________	________

3. Using a ruler, draw ten straight lines 5 inches long on a piece of paper. Place short marks across each line at distances exactly as follows:
   a. Line 1: Distance between marks = 3"
   b. Line 2: = 4 1/2 inches
   c. Line 3: = 2 1/8 inches
   d. Line 4: = 2 3/4 inches
   e. Line 5: 3 7/8 inches
   f. Line 6: 15/16 inches
   g. Line 7: 7 centimeters
   h. Line 8: 84 millimeters
   i. Line 9: 7.3 centimeters
   j. Line 10: 7 centimeters and 3 millimeters.
4. Use a tape and measure the length and width of your classroom, your desk top, and the classroom door.

5. Examine a folding rule. Carefully unfold and refold the rule being careful to swing each section open and closed, using the spring joint as a hinge.
6. Examine inside calipers and outside calipers. Measure the inside and outside diameter of a round object, such as a can, pipe fitting, or piece of pipe.
7. Obtain a square of your choice and a board that is 6 inches to 10 inches wide. Draw a line across the board this is straight and square to the edge of the board.
8. Obtain a 12-inch, 24-inch, or 48-inch level.
   a) Place the level on the window sill in the classroom, shop or other room. Is the window level?
   b) With the level resting on the window sill, lift the right end of the level. Which way did the bubble move?
   c) Lift the left end of the level. Which way did the bubble move?
   d) Place the level in a vertical position against the classroom or shop door. Is the door plumb? Do all bubbles in the level read alike?
   e) With the level still in a vertical position against the door, hold the bottom of the level against the door but move the top away from the door. Which way did the bubble move?
   f) Hold the top of the level against the door and move the bottom away. Which way did the bubble move?
9. Use a combination square to:
   a) draw a 90 degree line on a board
   b) draw a 45 degree line on a board
   c) measure the width of a board
   d) measure the depth of a groove
   e) test the levelness of a bench.___________ ___________ ___________
10. Create a 90 degree corner in the lawn using the 6', 8', 10' method. You will need three stakes, a hammer, two lengths of string each 12 feet long, two string ties, a framing square and a 12-foot tape.
11. Use a scriber to create a 4-inch circle on a piece of paper. Use a divider to create a 3-inch circle on a scrap piece of wood or metal.
12. Prepare a pattern for some small object you would like to make out of wood or metal.

SELF-EVALUATION

A. Multiple Choice. Select the best answer.

1. On most rules and tapes used in the shop, the shortest line represents
   a. 1/8 inch
   b. 1/16 inch
   c. 1/64 inch
   d. 1 inch
2. Of the following, the smallest unit of metric measurement is
   a. meter
   b. millimeter
   c. kilometer
   d. centimeter
3. Tapes generally break due to
   a. frequent use
   b. use of spring retractors
   c. use outdoors
   d. forcing the tape back into the case
4. Wooden scales are sometimes called
   a. depth gauges
   b. bench rules
   c. marking gauges
   d. scribers
5. The best tool for measuring the outside diameter of a pipe in its middle section is the
   a. try square
   b. tape
   c. calipers
   d. dividers

B. Matching. Match the word or phrase in column I with the correct word or phrase in column II.

Column I
1. high-quality tools
2. English system
3. metric system

4. millimeter
5. centimeter
6. meter
7. kilometer
8. framing square
9. combination square
10. plumb line
11. line level
12. 90 degrees
13. body, tongue, and heel
14. sliding T bevel
15. 6', 8', 10'
16. scriber
17. glass tubes and alcohol
18. tracing paper
19. pattern enlargement
20. makes a circle

Column II

a. m
b. cm
c. mm
d. board foot table
e. km
f. good investment
g. yards, feet, inches
h. check high structures
i. meters, centimeters, millimeters
j. includes a sliding scale and level
k. requires tight string
l. used to lay out angles
m. right triangle
n. dividers
o. bubble
p. grid
q. framing square
r. right angle
s. contains a pencil
t. pattern making

C. Completion. Fill in the blanks with the word or words that will make the following statements correct.

1. When laying out work, it is important to choose the right tool and use the tool ____________.
2. The materials tools are made of include ______, ________, ________, ________, and ________.
3. Tools used to guide workers when cutting, sawing, drilling, shaping, or fastening are called ______________ tools.
4. The official system of measurement for non-scientific purposes in the United States is the ____________________ system.
5. The metric system of measurement is regarded as a logical system because its units of measure relate to each other in multiples of _______.
6. Use the following drawing and read the rule to the nearest one-eighth inch.

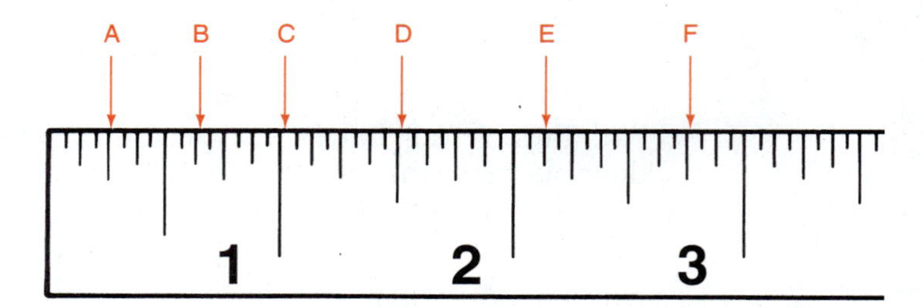

A ________
B ________
C ________
D ________
E ________
F ________

7. Use the following drawing and read the rule to the nearest $^1/_{16}$ inch.

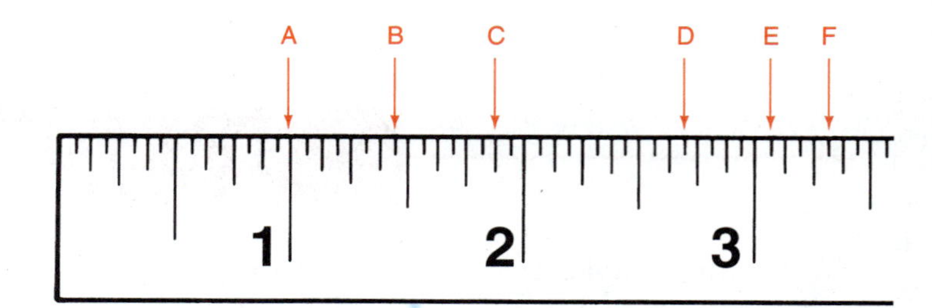

A ________
B ________
C ________
D ________
E ________
F ________

UNIT

Selecting, Cutting, and Shaping Wood

OBJECTIVE

To select wood and use tools and procedures in cutting and shaping wood.

COMPETENCIES TO BE DEVELOPED

After studying this unit, you should be able to:

- Name and correctly spell twelve species of lumber that may be used in agricultural mechanics shops.
- Select lumber for agricultural projects.
- Determine the most useful hand tools for cutting and shaping wood.
- Cut and shape wood.

TERMS TO KNOW

- wood
- annual rings
- species
- lumber
- warping
- kiln dried
- planer
- nominal
- handsaw
- cross cut
- rip
- grain
- crosscut saw
- rip saw
- kerf
- back saw
- miter box
- coping saw
- compass saw
- keyhole saw
- bore
- drill
- bit brace
- tang
- hand drill
- chuck
- push drill
- automatic drill
- clockwise
- counterclockwise
- plane
- plane iron
- chamfer
- file
- rasp
- file card
- dado
- rabbet
- perpendicular

MATERIALS LIST

- ✓ Twelve specified species of lumber
- ✓ Planed and rough lumber
- ✓ Dried and green lumber
- ✓ Samples of S2S, S4S, and sanded lumber
- ✓ Samples of standard cuts of lumber
- ✓ Combination square
- ✓ Handsaws of various kinds
- ✓ Boring tools
- ✓ Planes
- ✓ Files
- ✓ Miter box
- ✓ 3/4-inch wood chisel
- ✓ Sandpaper

The agricultural mechanics student should know some basic information about wood and its source, figure 9-1. Wood is the hard, compact fibrous material that comes from the stems and branches of trees. Nutrients flow up and down a tree trunk in bundles of tubelike tissues that are formed just inside the bark of trees each year. It is the yearly addition of new bundles that causes a tree to have annual rings. Annual rings are patterns caused by the hardening and disuse of the bundles; they form the grain in wood. Annual rings can be seen by looking at the end of a tree trunk or tree stump.

CHARACTERISTICS OF WOOD AND LUMBER

It is important to know the characteristics of wood from various species of trees, figure 9-2. The word species means plants or animals with the same permanent characteristics. Some woods are hard and some are relatively soft. Some woods resist rotting better than others. Some are very attractive while other woods are plain. Some woods are stronger and tougher than others.

When building with wood, it is important to select the correct type. Common species of trees used for fine furniture are oak, walnut, maple, cherry, and mahogany. Other woods commonly used for furniture are white pine, red cedar, and poplar. Wood used for building construction includes white pine, yellow pine, hemlock, fir, redwood, cypress, and oak. Each species has its advantages and shortcomings.

Lumber Grade and Wetness

Lumber is wood that has been cut into boards. Lumber is graded according to its appearance and soundness. The fewer and less obvious the scars, knots, or blemishes, the more the lumber costs. For instance, clear pine shelving costs more than knotty

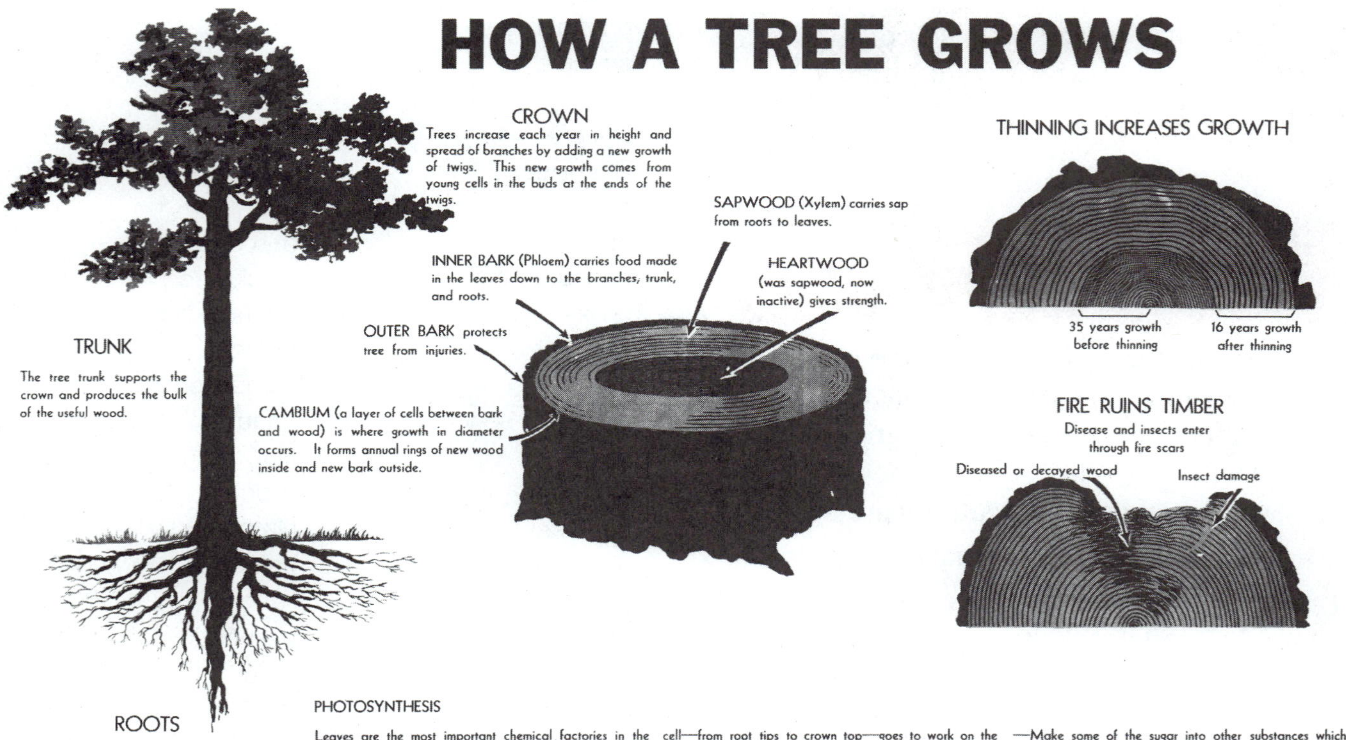

FIGURE 9-1. How a tree grows. *(Courtesy of U.S.D.A.)*

SPECIES	HARDNESS	KNOWN FOR	SOME MAJOR USES
Birch	Hard	Surface veneer for panels	Cabinets and doors
Cedar, red	Medium	Pleasant odor	Furniture, chests and closet linings
Cherry	Hard	Red grain	Fine furniture
Cypress	Medium	Rot resistance	Structural material in wet places
Fir and Hemlock	Soft	Light, straight, strong	Construction framing, siding, sheathing
Locust, black	Hard	Rot resistance	Fence posts
Maple	Hard	Light grain	Floors, bowling alleys, durable furniture
Mahogany	Medium	Reddish color	Fine furniture
Oak	Hard	Toughness, strength feeders, farm buildings	Floors, barrels, wagon bodies,
Pine, yellow	Medium	Wear resistance, tough	Floors, stairs, trim
Pine, white	Soft	Easy to work, straight	Shelving, siding, trim
Redwood	Soft	Excellent rot resistance	Yard posts, fences, patios, siding
Walnut, black	Hard	Brown grain	Fine furniture
Willow, black	Soft	Brown grain, easy to work, walnut look	Furniture

FIGURE 9-2. Characteristics of common woods.

pine shelving. Lumber with knots that are loose or missing costs less than lumber with tight knots. The lowest grade of lumber must simply hold together under ordinary handling.

When lumber is first sawed, it is said to be green or wet. This is due to the high moisture content. Wood with too much moisture will split, bend, cup, and twist if not dried evenly. This is called warping. Excessive moisture will evaporate from lumber if narrow strips of wood called stickers are placed between all boards in a stack. The process is called air drying and takes about a year. Moisture can be driven from lumber within several weeks if placed in special ovens called kilns. The lumber is then said to be kiln dried. Kiln-dried lumber must be used when constructing furniture or the wood pieces will warp or split. Warping can also be controlled by constructing an object quickly and then sealing the wood to prevent the escape of moisture.

Lumber Finish

Lumber is generally purchased with all sides and edges smooth. The planer is a machine that cuts lumber down to an exact size and leaves it smooth. Lumber may be purchased in the following four categories:

- **Rough**—not planed; delivered as it comes from the sawmill; width and thickness varies from piece to piece
- **S2S**—surfaced two sides; all pieces are the same thickness; edges are not planed; widths vary
- **S4S**—surfaced four sides; the sides and edges are planed to exact dimensions
- **Sanded**—width and thickness dimensions are exact on all pieces; all surfaces are sanded.

Many agricultural students buy rough, green lumber to make livestock feeders, wagon bodies, and other projects for the farm. The rough lumber is satisfactory for such uses and much cheaper than finished lumber.

Standard Lumber Sizes

Rough pieces of lumber, as they come from a sawmill, are not all the same size. What is called a 2 x 4 may be $2\,1/4" \times 4$ or $4\,1/4"$. When a 2×4 is planed, the actual size is only $1\,1/2" \times 3\,1/2"$. The name given to cuts of lumber reflects the nominal or approximate size. The real or actual sizes are $1/2$ inch less than the nominal sizes, figure 9-3.

CUTTING AND SHAPING WOOD WITH SAWS

Wood has become quite expensive, therefore, much care should be taken when measuring and cutting. A good rule is to measure twice and cut once. If a piece of wood must be recut because of an error, it generally means a new piece of wood must be used or it may mean all pieces must be recut and the project made smaller.

STANDARD LUMBER SIZES

1 x 2 (ACTUAL $\frac{3}{4}$ x $1\frac{1}{2}$)

1 x 3 (ACTUAL $\frac{3}{4}$ x $2\frac{1}{2}$)

1 x 4 (ACTUAL $\frac{3}{4}$ x $3\frac{1}{2}$)

1 x 5 (ACTUAL $\frac{3}{4}$ x $4\frac{1}{2}$)

1 x 6 (ACTUAL $\frac{3}{4}$ x $5\frac{1}{2}$)

1 x 8 (ACTUAL $\frac{3}{4}$ x $7\frac{1}{4}$)

1 x 10 (ACTUAL $\frac{3}{4}$ x $9\frac{1}{4}$)

1 x 12 (ACTUAL $\frac{3}{4}$ x $11\frac{1}{4}$)

2 x 2 (ACTUAL $1\frac{1}{2}$ x $1\frac{1}{2}$)

2 x 3 (ACTUAL $1\frac{1}{2}$ x $2\frac{1}{2}$)

2 x 4 (ACTUAL $1\frac{1}{2}$ x $3\frac{1}{2}$)

2 x 6 (ACTUAL $1\frac{1}{2}$ x $5\frac{1}{2}$)

2 x 8 (ACTUAL $1\frac{1}{2}$ x $7\frac{1}{4}$)

2 x 10 (ACTUAL $1\frac{1}{2}$ x $9\frac{1}{4}$)

2 x 12 (ACTUAL $1\frac{1}{2}$ x $11\frac{1}{4}$)

3 x 4 (ACTUAL $2\frac{1}{2}$ x $3\frac{1}{2}$)

4 x 4 (ACTUAL $3\frac{1}{2}$ x $3\frac{1}{2}$)

4 x 6 (ACTUAL $3\frac{1}{2}$ x $5\frac{1}{2}$)

6 x 6 (ACTUAL $5\frac{1}{2}$ x $5\frac{1}{2}$)

8 x 8 (ACTUAL $7\frac{1}{2}$ x $7\frac{1}{2}$)

FIGURE 9-3. The actual thickness and width of planed lumber is less than its nominal size.

Saws are used to cut boards to length and width. Saws are also used to cut curves, make holes, and cut panels to size. The more common saws are the handsaw, back saw, coping saw, and compass saw. The compass saw is very similar to the keyhole saw.

Handsaws

The handsaw is used to cut across boards or to rip boards and panels. The words cross cut mean to cut across the grain of a board. The word rip means to cut along the length of the board or with the grain. Grain in a board refers to lines caused by the annual rings in the tree.

The teeth of a handsaw determine how the saw should be used, figure 9-4. Teeth cut and filed to a point are designed to cut across the grain of boards. Such saws are called crosscut saws. Teeth filed to a knifelike edge are designed to cut with the grain. These saws are called rip saws.

Both crosscut and rip handsaws may be purchased in sizes 20 to 28 inches in length. The shorter saws are easier to use by smaller people. Shorter saws are frequently the choice for saws with small teeth designed for finer cuts.

Saws may be purchased with various tooth sizes. These range from six to fourteen teeth per inch on handsaws. The tooth size is designated by the number of tooth points per inch. For example, an eight-point saw has eight large teeth per inch of blade and is said to be a coarse saw. A twelve-point saw has twelve small teeth per inch and is considered a fine saw. When sawing a board, the saw removes wood and leaves an opening called a kerf.

When sawing wood, the material should rest on a solid work area or be held in a vise. Most carpenters use saw horses to support the material being sawed. The following procedure is recommended when using a crosscut saw or a rip saw.

PROCEDURE

1. Select a piece of 1" x 6" lumber, several feet long.
2. Use a square and draw a thin pencil mark across the board one inch from the end.
3. Place the stock (board) on a saw horse with the marks over the end to the right. This assumes the operator is right-handed. (Reverse if left-handed.) Hold the stock down on the saw horse with the left knee.
4. Grasp the saw handle in the right hand. The saw can be better controlled by placing the right forefinger along the outside of the handle.
5. Place the left hand on the board and use the thumb nail against the saw as a guide to start the cut.
6. Place the heel of the saw on the mark at the edge of the board and pull it toward your body to start the cut. Stand so your eyes are in line with the cut.
7. After the cut or kerf is started, push the saw forward, applying a light, downward pressure. Since the saw is designed to cut on the downward stroke only, apply no pressure on the return stroke. Operate the saw at a 45 degree angle (60 degree angle when ripping with a rip saw) to the surface of the board, figure 9-5.
8. Use long, slow strokes to complete the cut. The last few strokes should be very slow with no pressure on the saw.

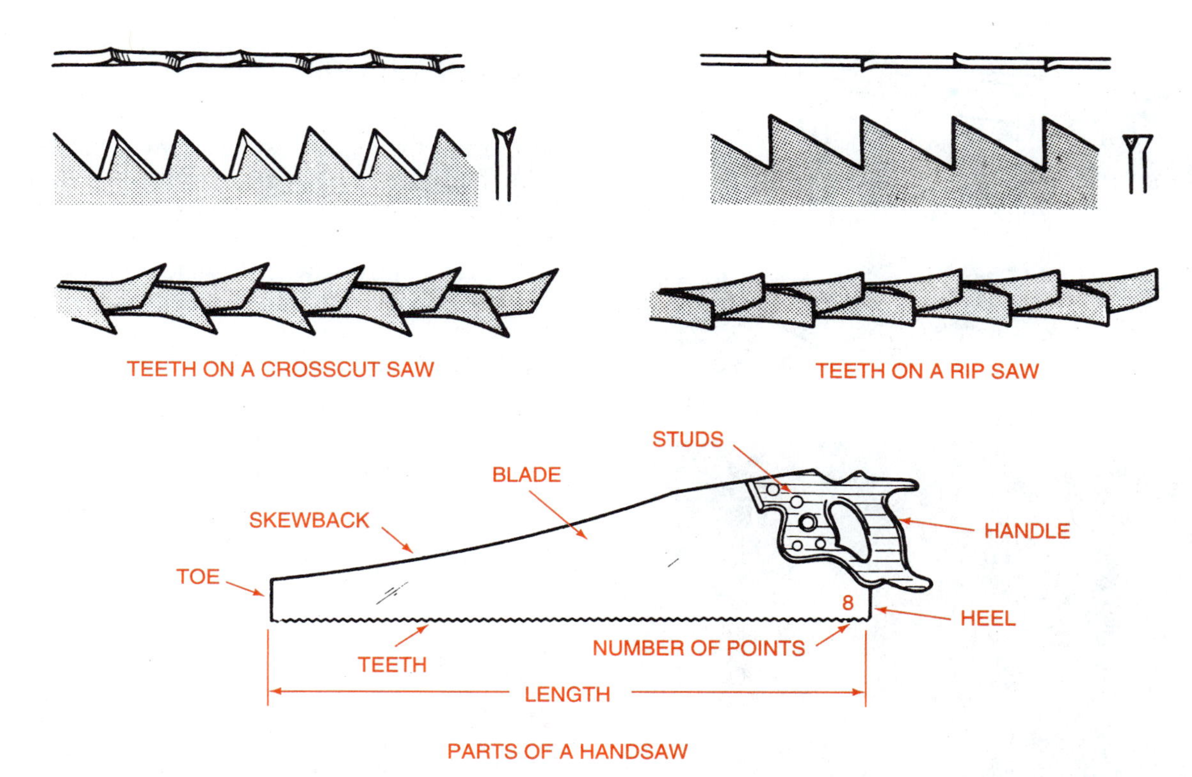

FIGURE 9-4. Handsaws are used for cross cutting or ripping, depending on the shape of the teeth. *(Courtesy of McDonnell & Kaumecheiwa,* Use of Hand Woodworking Tools, *Delmar Publishers, 1978)*

The Back Saw

A **back saw** is similar to a crosscut handsaw. It has very, very fine teeth and a stiff metal back from which it gets its name. These features make it an excellent tool for making very accurate cuts in smooth lumber. The back saw may be placed in a tool called a miter box to guide the saw, figure 9-6. A **miter box** is a device used to cut molding and other narrow boards at any desired angle. An adjustable

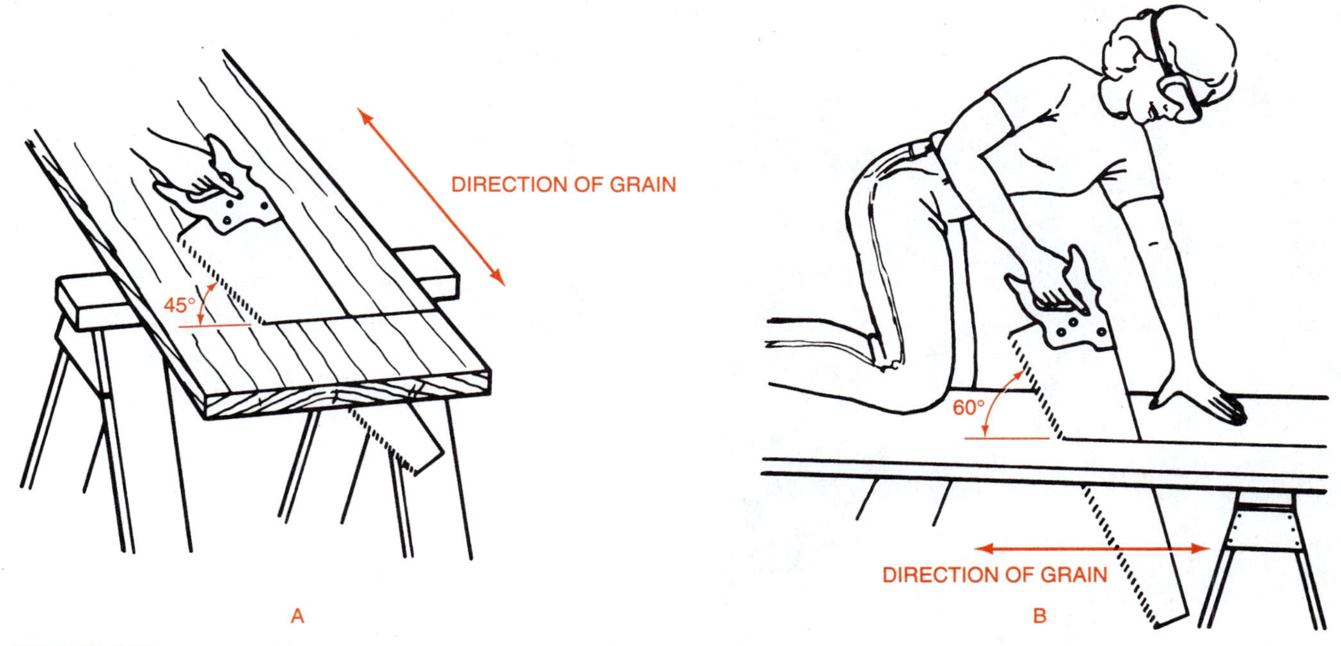

FIGURE 9-5. When using a crosscut saw, hold the saw at a 45 degree angle. When using a rip saw, hold the saw at a 60 degree angle. Start all cuts with short strokes, then use full, even strokes for good cuts. The saw kerf should be on the waste side of the line.

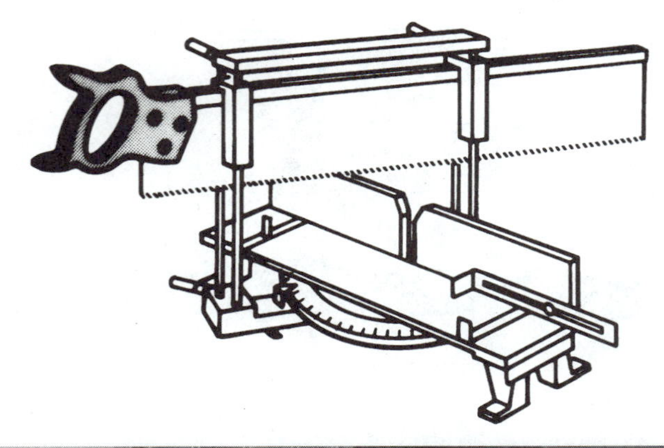

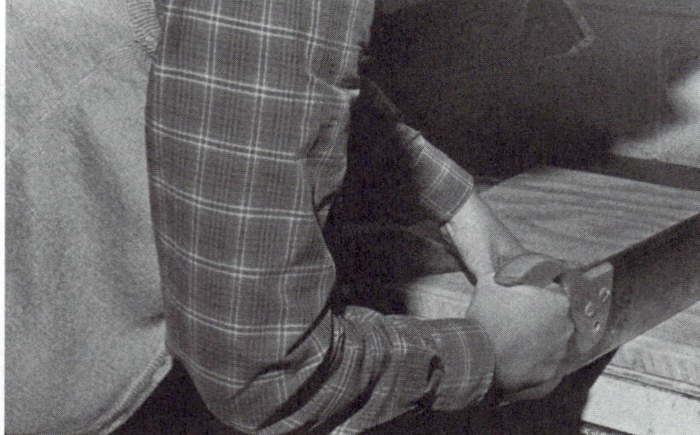

FIGURE 9-6. The back saw can be used alone or in a miter box. It is excellent for making very fine and accurate cuts.

miter box is operated by simply squeezing a lever and swinging the saw to the desired angle. The lever is released and the saw locks at that angle. The wood is inserted, held firmly, and then sawed.

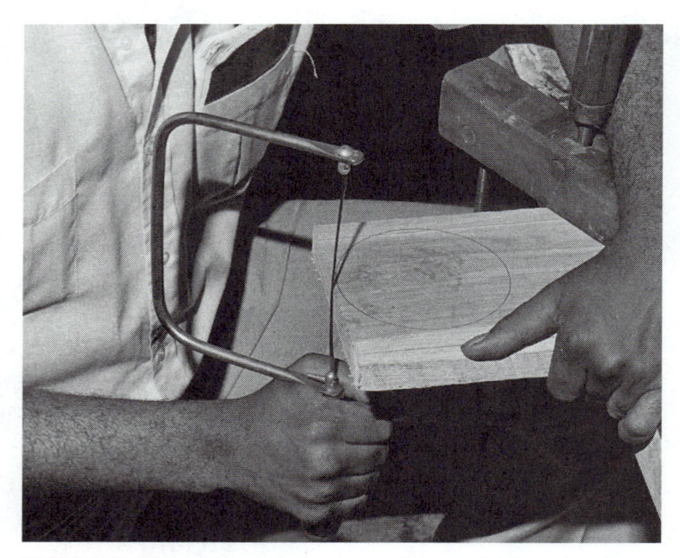

FIGURE 9-7. The coping saw is useful for cutting irregular shapes in wood. The blade can be removed from the frame, inserted into a hole, and reinstalled in the frame for the inside cutting.

The Coping Saw

The coping saw has a very thin and narrow blade supported by a spring steel frame. This design permits the blade to be removed from the frame, figure 9-7. This makes it possible to use the saw to cut large holes or other shapes that are totally surrounded by wood. The coping saw is useful for cutting any kind of irregular, curved cuts in wood or other soft materials.

Compass and Keyhole Saws

The compass saw and keyhole saw are designed for making cuts starting from a hole, figure 9-8. A hole about one inch in diameter is bored and the slim blade of the saw is inserted into the hole. The operator saws outward and follows the lines of the desired cut. The compass saw is similar to the keyhole saw, but its blade is wider at the base.

CUTTING AND SHAPING WOOD WITH BORING AND DRILLING TOOLS

The words bore and drill are often used to mean the same thing—to make a hole. However, the word boring generally refers to low speed, while drilling is a high-speed operation. The most popular hand tools for boring and drilling are the bit brace, hand drill, and push drill.

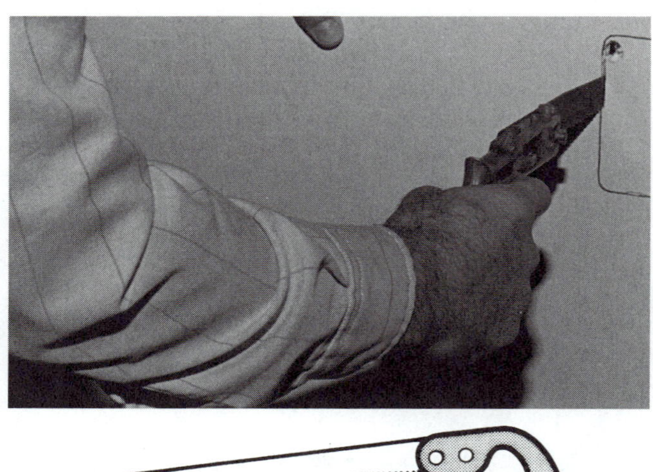

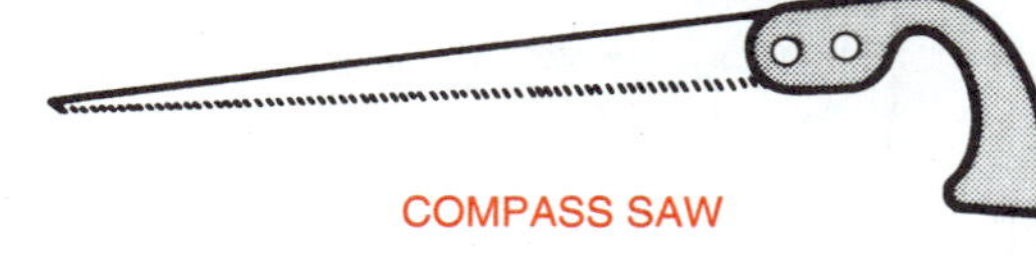

FIGURE 9-8. The compass and keyhole saws are used to make cuts starting from holes.

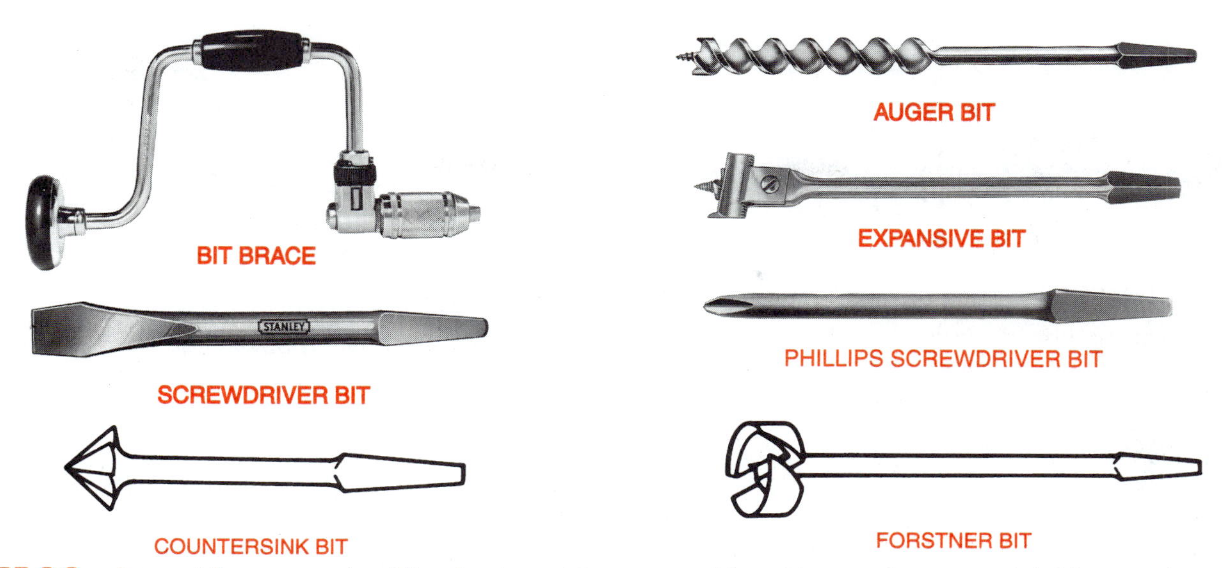

FIGURE 9-9. Auger bits, expansive bits, forstner bits, screwdriver bits, and countersink bits can be used with the bit brace. *(Photos courtesy of Stanley Tools)*

The Bit Brace

The bit brace has a large crank-type handle. This permits the worker to turn bits, which make holes up to 2 inches or more in diameter. Bit braces are also available with ratchet drives. A ratchet drive permits use of the tool in tight places, such as against a wall. The handle drives the bit forward and ratchets backward for another stroke.

The jaws of the bit brace receive auger bits, expansive bits, forstner bits, screwdriver bits, and countersink bits, figure 9-9. These bits all have a tang. A tang is a tapered shank with four corners. A bit with a tang will not slip in the jaws of a bit brace.

The best hand tool for making holes larger than 1/4 inch is the bit brace and auger bit, figure 9-10. Auger bits range in size from 1/4 inch to one or more inches. The size is designated by a number equal to the diameter of the bit, in sixteenths of an inch. For example, a number 6 auger bit is 6/16 inch or 3/8 inch in diameter.

The auger bit has a screw on its end to pull the bit into the wood, figure 9-11. Care must be taken not to damage this screw point. The bit also has four knifelike edges to cut into the wood. Moderate pressure is needed by the operator to keep the bit cutting. The cutting edges of the auger bit must be sharp. If the screw point is damaged or the cutting edges are not sharp, excessive pressure must be applied by the operator to make a hole. Both the screw point and the cutting edges will be ruined if the bit hits bolts, screws, or nails in the wood. They will also be damaged if they strike hard surfaces such as a vise, metal bench, or concrete floor.

FIGURE 9-10. The bit brace and auger bit is the most popular hand tool for boring holes larger than 1/4 inches in diameter.

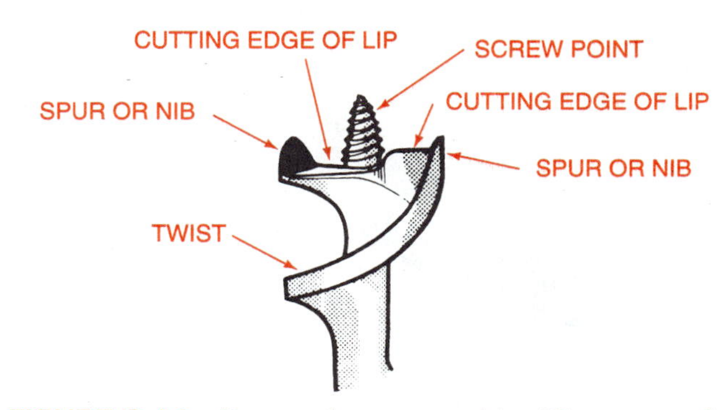

FIGURE 9-11. Parts of an auger bit. *(Courtesy of McDonnell & Kaumeheiwa,* Use of Hand Woodworking Tools, *Delmar Publishers, 1978)*

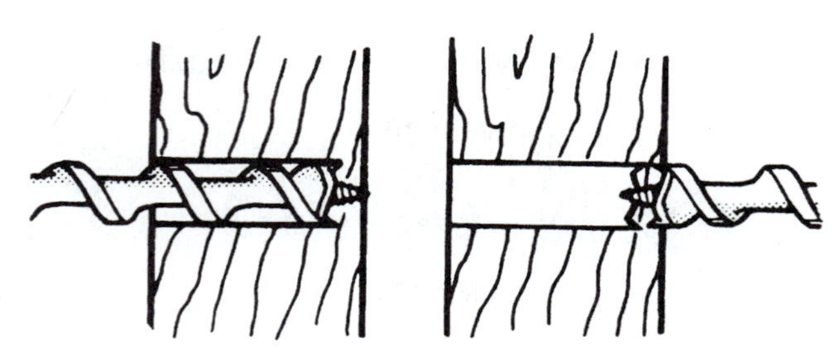

FIGURE 9-12. When the tip of the bit pierces the back side of the board, boring should be stopped. The hole should be completed by boring from the back side of the board. *(Courtesy of McDonnell & Kaumeheiwa,* Use of Hand Woodworking Tools, *Delmar Publishers, 1978*).

When boring a hole and the tip of the bit pierces the back side of the board, the operator should stop boring, figure 9-12. The hole should be completed by boring from the other side of the board. Completing a hole by boring from the back side of the board is a very important step when using an auger bit. This procedure prevents the bit from splitting the face of the board when it breaks through.

The Hand Drill

The hand drill is a device with gears that drive its bit much faster than the handle turns. This increased speed, however, means the turning power delivered to the bit is reduced. Therefore, the hand drill is useful only for making holes up to about 1/4 inch in diameter in soft materials.

The hand drill must be held carefully to prevent injury to the hands and to ensure the hole is straight. Care must be taken to keep hands and clothing away from the turning cogwheels of the tool. When cogwheels come together, they can catch cloth or skin and severely pinch them between the teeth of the cogwheels.

The bit of the hand drill is held in place by a chuck. A chuck is a device with jaws that open and close to receive and hold bits with smooth shanks,

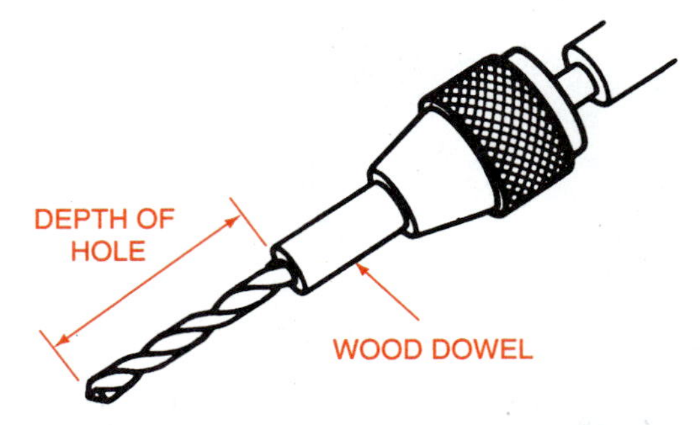

FIGURE 9-14. A wood dowel can be placed on the bit to act as a depth gauge to drill holes of uniform depth.

figure 9-13. Bits with smooth shanks include the twist drill, countersink bit, and combination wood drill and countersink. To drill holes of uniform depth, a wooden dowel can be placed on the bit to act as a depth gauge, figure 9-14. If a hole is to be drilled the whole way through a board, pressure would be eased on the drill when the bit starts to show on the back side of the board. Easing pressure prevents the wood from splitting.

To remove the bit from a hand drill, the chuck is held with one hand. The crank is then turned backwards with the other hand, figure 9-15.

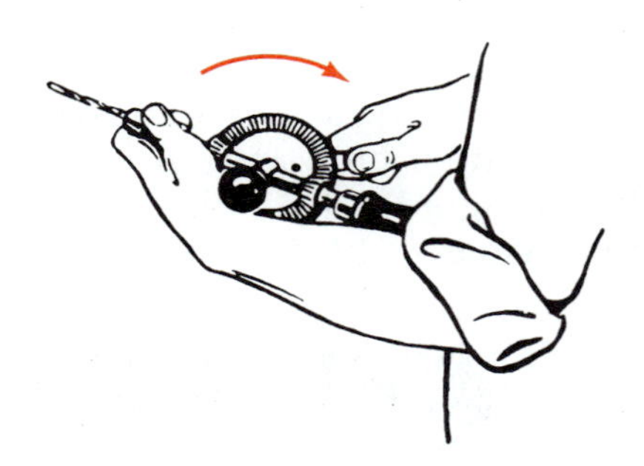

FIGURE 9-15. To remove a bit from the hand drill, hold the chuck with one hand and turn the crank backward with the other hand. *(Courtesy of McDonnell & Kaumeheiwa,* Use of Hand Woodworking Tools, *Delmar Publishers, 1978)*

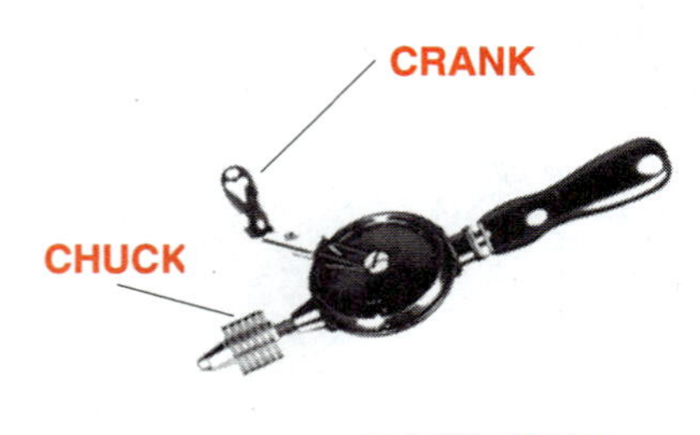

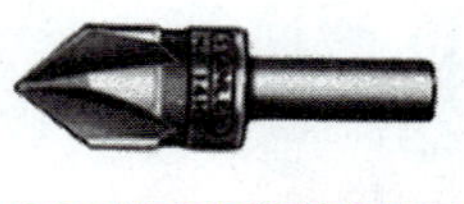

FIGURE 9-13. The chuck of a hand drill receives bits with smooth shanks. *(Courtesy of Stanley Tools)*

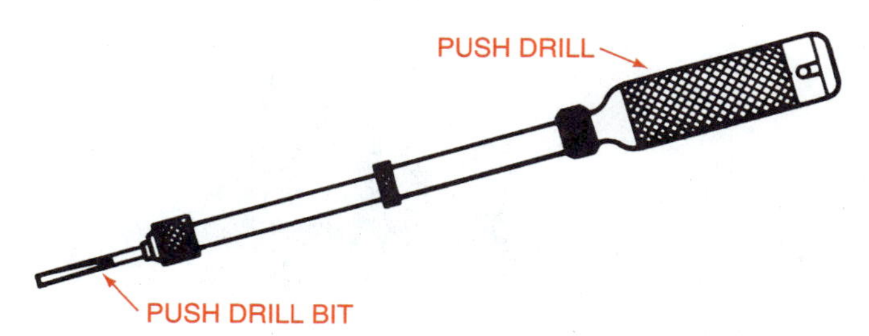

FIGURE 9-16. Push drills have spiral shanks that turn the bit when pressure is applied to the handle. The operator's up and down motion keeps the bit turning at a high rate of speed.

The Push Drill

The push drill is also called an automatic drill. It has a spiral-shaped shaft that turns clockwise when pushed against an object. Clockwise means turning in the direction the hands of a clock turn. A spring causes the shaft to turn counterclockwise or backwards when the pressure is released. Therefore, the pumping of the handle causes the bit of the drill to turn.

The push drill is very fast and handy for drilling holes up to 3/16 of an inch in diameter. It is generally used to make holes for wood screws. The bit for a push drill has a round shank with flat places on its end, figure 9-16. This enables the bit to lock quickly and securely in the chuck of the push drill. Bits are stored in the handle of most push drills, figure 9-17.

Downward pressure drives the bit of a push drill. Therefore, if the operator drills all the way through a piece of wood using normal pressure, the bit will split the wood when it breaks through. Such damage is avoided if the push drill is used only for making holes to install screws.

The procedure for using a push drill is as follows:

PROCEDURE

1. Mark the spot where the center of the hole will be.
2. Install the correct bit.
3. Place the point of the bit on the wood.
4. Hold the push drill at a 90 degree angle to the work.
5. Push the handle up and down until the hole is the desired depth.
6. Pull the drill bit out of the hole.

SHAPING EDGES

Some projects have all straight edges and square ends. In this case, a crosscut handsaw or back saw is used to cut the board to length. The task of cutting and shaping wood is greatly simplified if S4S lumber is used. When surfaced on four sides, it has smooth, straight edges from which to work. Planes are used to work the edges of boards.

A plane is a tool that shaves off small amounts of wood and leaves the surface smooth. The most common planes are the jointer, jack, smoothing, and block planes, figure 9-18. The jointer, jack, and smooth planes require the operator to use both hands, figure 9-19. The block plane is small enough to be used with one hand.

The cutting part of a plane is called the plane iron. Plane irons must be very sharp and properly adjusted, figure 9-20. The procedure for adjusting a plane iron follows.

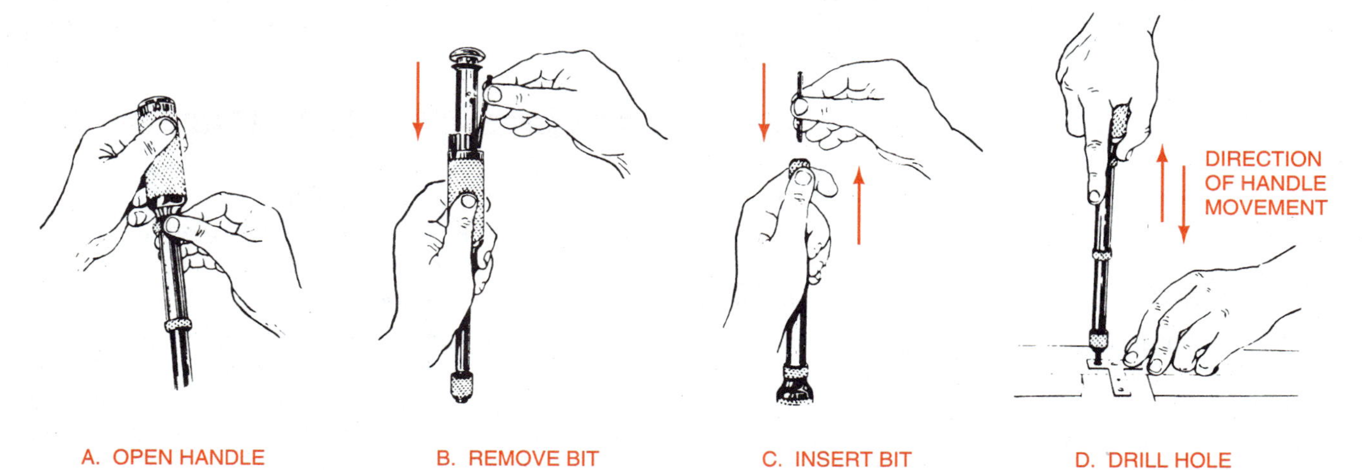

FIGURE 9-17. Bits are stored in the handles of most push drills. This feature, plus the easy push action of the tool, makes it a favorite for making holes for screws. *(Courtesy of McDonnell & Kaumeheiwa,* Use of Hand Woodworking Tools, *Delmar Publishers, 1978)*

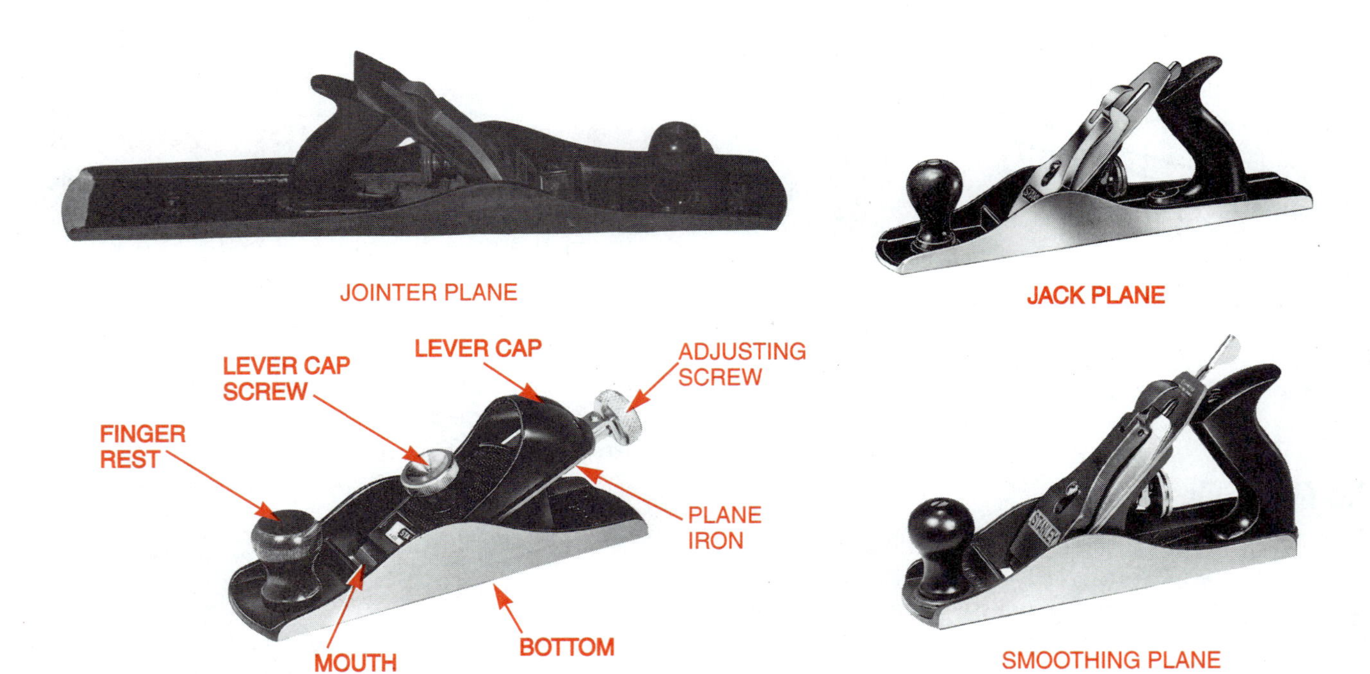

FIGURE 9-18. Common planes used for making straight, smooth edges. *(Courtesy of Stanley Tools)*

FIGURE 9-19. The jointer plane, jack plane, and smoothing plane require the operator to use both hands.

PROCEDURE

1. Hold the plane upside down and look down the bottom of the plane from front to rear.
2. Move the adjusting lever or adjusting screws to set the plane iron. The plane iron should project less than 1/16 of an inch through the bottom of the plane. Both sides should be projecting the same distance.
3. Make a trial cut by pushing the plane along the edge to be planed. Plane with the grain, i.e., push the plane in the direction that makes the smoothest cut.
4. Readjust the plane until it cuts smoothly and a thin curl of wood is produced by each stroke.

CUTTING A BOARD TO WIDTH

The procedure for cutting a board to width involves several tools and steps. The following procedure is used when cutting boards to width. The board used as an example in the procedure is to be cut to a width of 4 inches.

PROCEDURE

1. Set a combination square so its blade extends out from the handle 4 inches.
2. Place the handle of the square against a smooth edge of the board.
3. Place a sharp pencil at the end of the blade.
4. Move the pencil and the square along the entire length of the board making a line, figure 9-21. The pencil should be held straight up and down and the angle not changed at any point along the way.
5. Reset the square at 4 1/16 inches and draw a second line.

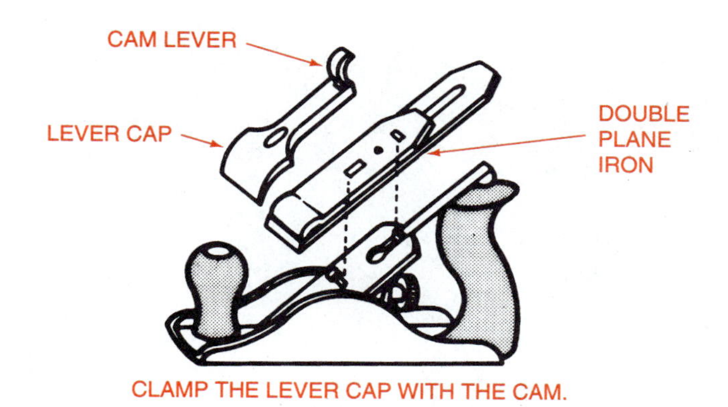

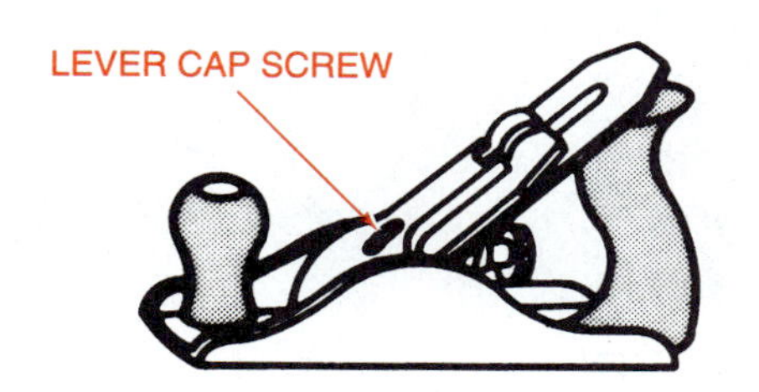

FIGURE 9-20. Adjusting the plane iron.

6. Use a hand rip saw to saw the board. Make the saw kerf on the outside or waste side of the second line. The kerf should just touch the line.
7. Use a plane to smooth the saw cut. Lift the plane on each back stroke to avoid dulling the blade.
8. Check the edge for squareness.
9. Plane off any remaining wood until the edge is straight, smooth, square, and exactly down to the first line draw.

FIGURE 9-21. When marking the width of a board, the pencil and combination square are moved along the board together.

MAKING FINAL CUTS IN CURVED EDGES

When cutting curved lines in wood and other soft materials, a coping saw is used. A cut is made about 1/16 inch outside the line. The saw must be kept moving exactly parallel to the line. After sawing out the project, a block plane, file, and sandpaper are used to work the wood down to the line. After using sandpaper no cutting tool should be used. This is because sand particles in the wood will dull cutting tools.

A block plane is useful for gently curving outside lines. It is also good for cutting chamfers, figure 9-22. A chamfer results when a corner between the edge and face of a board is cut. A chamfer is similar to a bevel except a bevel extends all the way across the edge of the board.

Using Surforms

A relatively new tool, called a Surform®, is available. Surform® is a trademark of Stanley Tool Works. The Surform® may be used instead of a plane or file, figure 9-23. It consists of several styles of handles and blades. The blades have sharp ridges that cut wood, plastic, or body fillers. Small curls of material are produced much like a plane.

Surforms® are handy since they do not need to be adjusted or sharpened. The blade is simply replaced. Surforms® come in a variety of shapes. Some are useful for cutting material to form straight and flat surfaces. Others are useful for rounding edges and smoothing curves. A variety of surfaces can be produced by simply changing the direction of

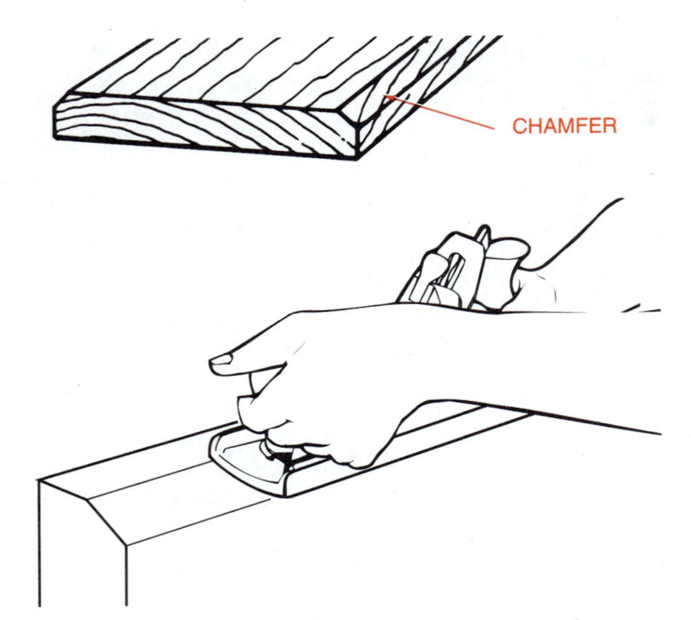

FIGURE 9-22. A plane is used to create a chamfer.

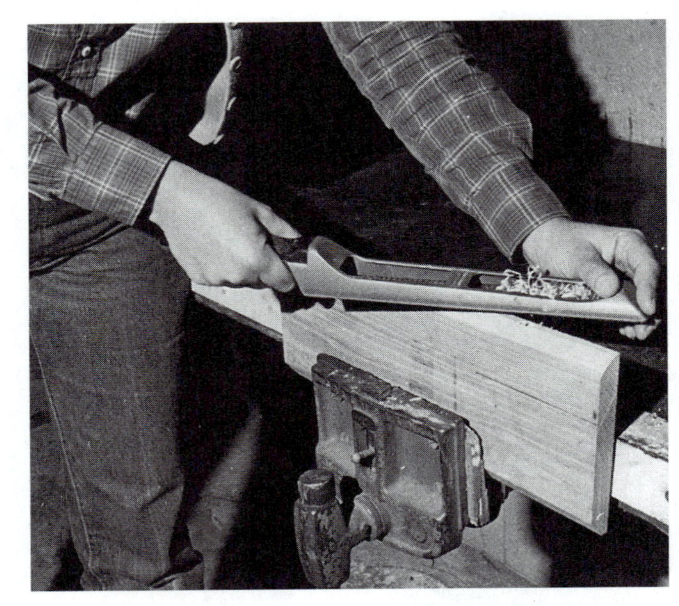

FIGURE 9-23. A Surform® can be used instead of a plane or a file to shape edges of boards.

stroke, figure 9-24. Surforms® will remove large amounts of material if pushed at a 45 degree angle to the material. Pushing the tool in a straight line in the direction of the work produces a fine, smooth cut.

Using Files

Various files are used to dress down and finish off curved cuts on boards. Files are handy for smoothing edges and shaping materials to odd shapes, figure 9-25. A file may be flat, round, half-round, square, or three-sided. Files have teeth that cut wood or metal. Files designed for cutting wood and soft metals have coarse teeth. Those designed for cutting steel have smaller teeth. Most files are hard enough to cut either wood or metal. The rasp is a type of file with very, very coarse teeth. It is designed to be used only on wood and other soft materials. Rasps are good for making rough cuts when a lot of material must be removed.

When filing wood, the file must not be pushed across the edge of the board as this would tear down the corner and leave it ragged. Best results are obtained when the file is pushed lengthwise along

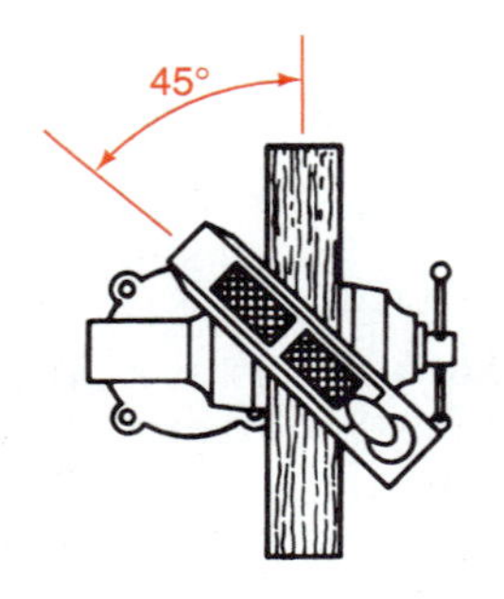

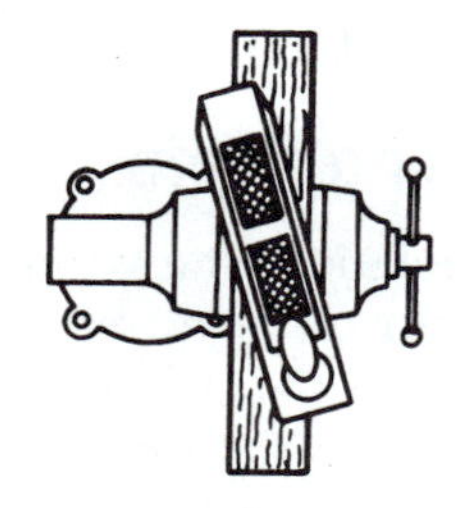

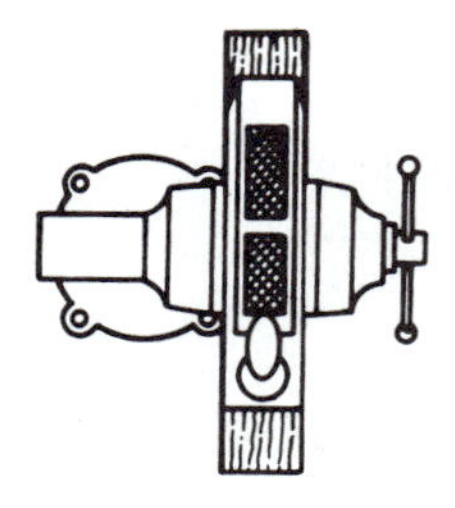

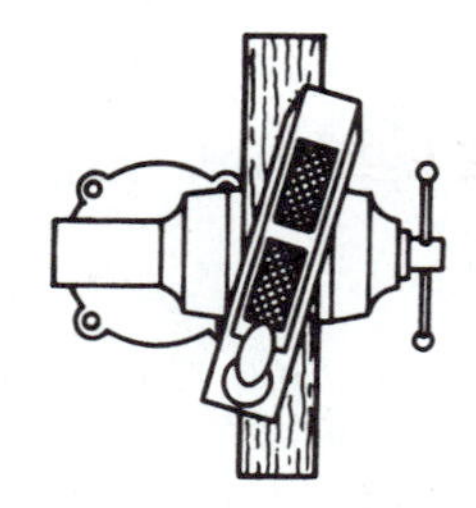

FIGURE 9-24. By changing the direction of the stroke, Surforms® can produce a variety of surfaces. The maximum amount of material is removed by pushing the tool at a 45 degree angle to the board. Less material is removed and a smoother surface obtained by reducing the angle. A smooth surface is obtained by pushing the tool in a straight line. An almost polished effect is obtained by pushing the tool at a slightly reversed angle.

FIGURE 9-25. Files are handy for smoothing edges and shaping materials to odd shapes.

FIGURE 9-27. A file wrapped with sandpaper can be used for finishing edges.

the board. File teeth cut only when the file is pushed forward. Therefore, it is advisable to apply pressure on the forward stroke only. The file should be lifted on the return stroke. The teeth of the file may become filled with the material being filed. Material may be removed and the file cleaned by tapping the handle up and down on a bench top. A special wire brush called a file card may also be used to clean files, figure 9-26.

A file has a sharp end called the tang. The tang is designed to go into a wooden or plastic handle. Files should not be used without handles. Serious injury to the hand can result from using a file without a handle. A file handle can be installed by driving the handle on the tang using a soft mallet. Another method is to place the tang in the handle and tap the handle up and down on the bench top until the tang is totally hidden in the handle.

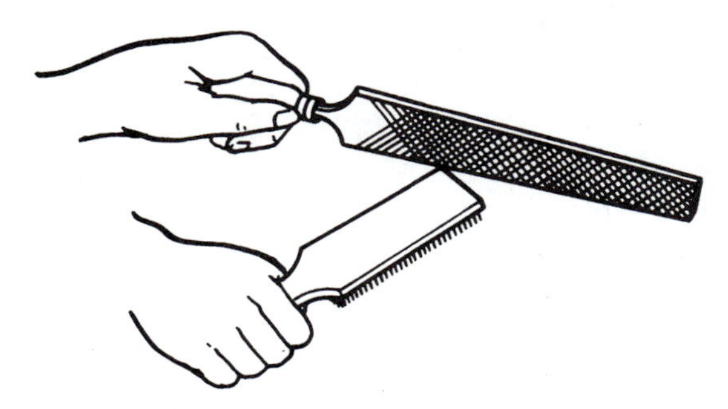

FIGURE 9-26. A wire brush called a file card is used to clean material from the teeth of a file.

Using Sandpaper

A small piece of sandpaper can be wrapped around a file for the finishing touches, figure 9-27. The file provides a thin, rigid core of the desired shape to support the sandpaper. The sandpaper is wrapped with the rough surface out.

To use the sandpaper and file, pinch the two lightly at both ends. Then, simply move the two as a unit. Sandpaper can be pushed across edges without splitting the corner of boards. Medium sandpaper should be used first and then very fine sandpaper can be used to finish.

CUTTING DADOS AND RABBETS

A dado is a square or rectangular groove in a board. The purpose of the dado is to receive the end or edge of another board to make a dado joint. A rabbet is a cut or groove made at the end or edge of a board to receive another board. The result is a rabbet joint when the two are fastened together. Dado and rabbet joints are used wherever strong 90 degree joints are needed, figure 9-28. To make a dado, the following procedure is used. (A rabbet is made using a similar procedure.)

PROCEDURE

1. Using a combination square, mark the face of the board with two lines 3/4 inch apart to indicate the position and width of the dado.
2. Use a square to mark down across the edge of the board with lines perpendicular to the lines on the face of the board. Perpendicular means at a 90 degree angle.

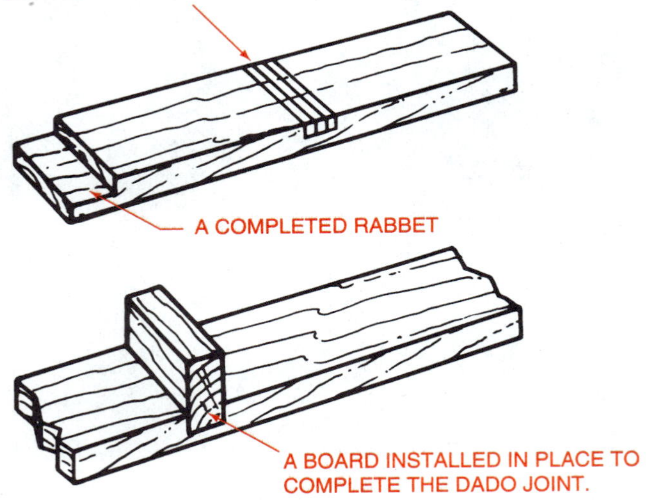

FIGURE 9-28. Dados and rabbets are made by making multiple saw cuts and finishing with a wood chisel and file. A groove made in the end and/or edge of a board to create a joint is called a rabbet. A groove made elsewhere in the board to create a joint is called a dado.

3. Set the blade on the combination square to extend a distance equal to one-half the thickness of the board. The distance will be 3/8 inch for a board that is 3/4-inch thick.
4. Lay the blade of the square against the edge of the board, and mark the bottom of the dado by drawing a line at the end of the blade.
5. Mark the depth and sides of the dado on the other edge of the board.
6. Carefully saw four kerfs into the dado, stopping with the kerf exactly at the bottom of the dado. This can be seen on both edges of the board. A backsaw or miter box saw should be used for this operation.
7. Use a wood chisel to remove the wood between the kerfs, figure 9-29. Be careful to chisel the wood to a uniform depth at the bottom of the kerf.
8. Smooth the bottom of the dado with a flat file. Be careful not to rock the file. Rocking will round the edges at the bottom of the dado and make a poor joint.

FIGURE 9-29. A wood chisel is used to remove the wood between the kerfs when making a dado.

STUDENT ACTIVITIES

1. Define the Terms to Know in this unit.
2. Examine the stump of a recently cut tree. Count the number of rings in the stump. How old was the tree? Examine the distance between the rings. Wide distances indicate a year of good growth, perhaps plenty of rain.
3. Examine samples of oak, walnut, maple, cherry, mahogany, white pine, red cedar, poplar, fir, redwood, and cypress. Record the color and degree of hardness of each.
4. Learn to identify and correctly spell the various species of wood.
5. Examine lumber that is warped and has split ends. Try to explain the reason for the warping.
6. Describe the steps in preparing lumber from log to sanded board.
7. Measure the width and thickness of five different types of finished lumber. Record the actual size and nominal size of each.
8. Obtain a board 18 inches long and 5 1/2 inches wide. Soft wood is recommended. Place five lines across the board 1/2 inch apart, near the end and square to the edge. Use a crosscut saw to saw off the

piece nearest the end. Check the cut with a try square. Is it exactly square when checked in both directions? Saw and check each of the other four cuts, trying to make perfect cuts. Report the results to your instructor.

9. Use a combination square and mark a line 1/2 inch from the edge of a board 1 to 2 feet long. Rip the board on the side of the line closest to an edge. Check the cut for straightness and squareness. What are the results?
10. Use a plane to improve the edge of the board sawed in activity 9. Now is the edge straight? Is it square?
11. Set a compass at one-half the width of the board used in activity 10. Place the steel point on the board so the pencil can be swung around and just touch both edges and an end of the board. Swing the compass in an arc and mark a rounded end on the board. Saw the end round with a coping saw and finish it with a file or Surform®.
12. Using the board from activity 11, bore a 1/2-inch hole with a brace and bit at the mark left by the steel compass point.
13. Using the board from activity 12, remove 1/2 inch from the nonrounded end by using a miter box. Cut a rabbet in the end that is 3/8 inch deep and 3/4 inch wide.
14. Measure and mark the board (used in activity 13) at 3 inches and 3 3/4 inches from one end. Cut a dado 3/8 inch deep and 3/4 inch wide.
15. Set a combination square so the blade extends 1/4 inch. Choose one of the four corners of the board you used in activity 14. Place the square and draw a line on the edge of the board 1/4 inch down from the corner you chose. The line should run the entire length of the board. Reposition the square and draw a line from one end to the other on the face of the board. Use a block plane or a smoothing plane to make a chamfer.
16. Discuss your workmanship in the preceding procedures with your instructor. Save the board for an activity in Unit 10.

SELF-EVALUATION

A. Multiple Choice. Select the best answer.

1. Grain in lumber is caused by the
   a. age of the board
   b. annual rings
   c. special drying techniques
   d. stain
2. Lumber is graded according to its
   a. appearance and soundness
   b. color and species
   c. strength and durability
   d. cost and length
3. A crosscut handsaw with very coarse teeth would have how many teeth per inch?
   a. 6
   b. 10
   c. 12
   d. 14
4. A crosscut handsaw with very fine teeth would have how many teeth per inch?
   a. 6
   b. 8
   c. 14
   d. 20
5. The wood removed by a saw blade leaves an opening called a
   a. bevel
   b. channel
   c. chamfer
   d. kerf
6. The back saw gets its name from
   a. its use as a back-up tool
   b. its fine teeth
   c. its stiff back
   d. its original use in making chair backs
7. A suitable tool for cutting curves is the
   a. compass saw
   b. coping saw
   c. keyhole saw
   d. all of these
8. A bit brace uses
   a. an auger bit
   b. an expansive bit
   c. a screwdriver bit
   d. all of these

9. The plane small enough to be used with one hand is the
   a. jointer plane
   b. jack plane
   c. smoothing plane
   d. block plane
10. The teeth on a file cut
    a. best when oiled
    b. only in soft materials
    c. only on the backward stroke
    d. only on the forward stroke
11. The groove cut across the end of a board to receive another board is called a
    a. dado
    b. rabbet
    c. miter
    d. ratchet
12. A file handle is necessary for
    a. operator protection
    b. file functioning
    c. compliance with the law
    d. none of these

B. Matching. Match the type of wood in column II to its major use in column I.

Column I
1. fence posts
2. reddish, fine furniture
3. barrels, wagon bodies, farm buildings
4. shelving, siding, trim
5. floors, stairs, trim
6. excellent rot resistance; patios
7. structural material in wet places
8. brown, fine furniture
9. floors and bowling alleys
10. chests and closet linings
11. construction framing, siding, sheathing
12. surface veneer for cabinets and doors

Column II
a. red cedar
b. cypress
c. fir and hemlock
d. black locust
e. mahogany
f. maple
g. oak
h. white pine
i. yellow pine
j. redwood
k. black walnut
l. birch

C. Completion. Fill in the blanks with the word or words that will correctly complete the statement.

1. A piece of lumber that is planed and measures $1^1/_2$" × $^1/_2$" is called a ______ by ______.
2. A rip saw is designed to cut __________ the grain and the crosscut saw is designed to cut __________ the grain.
3. When boring a hole with a twist drill, it is very important to ___________ the pressure when the bit is coming through the reverse side.
4. Two advantages of Surform® tools over planes are they do not need to be ________ or ________.

D. Brief Answers. Briefly answer the following questions.

1. Explain why "measure twice, cut once" is a good rule when cutting wood.
2. What does the designation S2S for lumber mean?
3. Explain the difference between boring and drilling.

UNIT

Fastening Wood

OBJECTIVE

To use nails, screws, bolts, and glue in assembling wood.

COMPETENCIES TO BE DEVELOPED

After studying this unit, you should be able to:

- Drive nails.
- Set screws.
- Use bolts.
- Use glue.
- Assemble wood parts.

MATERIALS LIST

- ✓ Samples of common wood joints (butt, lap, dado, rabbet, and dovetail)
- ✓ Common and finishing nails: 4d, 8d, 10d
- ✓ Scraps of 2 × 4s in hard and soft woods
- ✓ Claw hammers, nail set
- ✓ Stapling gun and staples
- ✓ Flat head wood screws (#8 × 1 1/4 inch)
- ✓ Hand drill and drill bit set; countersink
- ✓ Push drill and drill bits
- ✓ Bit brace and auger bit (3/8 inch recommended)
- ✓ Standard screwdriver to fit #8 screws
- ✓ Hinge set with screws
- ✓ Liquid glue
- ✓ Assortment of dowel pins
- ✓ Glue
- ✓ Clamps
- ✓ Clean, soft rags

TERMS TO KNOW

joint
butt joint
lap joint
member
assembly
dado joint
rabbet joint
miter joint
dovetail joint
setting
nail set
toe nail
end nailing
flat nailing
clinch
staple
countersink
shank hole
clearance hole
pilot hole
anchor hole
glue
dowel

Wood and wood products are attractive, strong, and long lasting. They are used for framing, siding, walls, doors, cabinets, counters, and furniture. Wood is used extensively on farms for construction of buildings, fences and machinery such as wagons and trailers. Wood is easy to fasten. Common materials for fastening wood include nails, screws, bolts, and glue.

When fastening wood, a joint that is strong enough to do the job should be chosen. A joint is the union of two materials. Joints may be secured with nails, screws, bolts, or glue. The most popular joints are butt, lap, dado, rabbet, miter, and dovetail, figure 10-1.

The butt joint is formed by placing two pieces end-to-end or edge-to-edge in line or at a 90 degree angle. Butt joints may be strengthened by applying thin wood or metal plates at corners or across flat surfaces where parts meet.

The lap joint is formed by fastening one member face-to-face on another member of an assembly. *Member* means piece; *assembly* means pieces fastened together. The members may be offset or may be cut so the two form a flat surface. In other words, the members are set into each other. Lap joints are stronger than butt joints.

The dado joint is a rectangular groove cut in a board. The end or edge of another board is then inserted in this groove. The rabbet joint is similar to a dado except that it occurs at the end or edge of a board.

The miter joint is formed by cutting the ends of two pieces of lumber at a 45 degree angle. The two pieces are then joined to form a 90 degree angle.

The dovetail joint is formed by interlocking parts of two pieces. This is the strongest joint. However, it is also the most difficult to make. Dovetail joints are used extensively in high-quality and expensive furniture.

FASTENING WITH NAILS

Nailing is the fastest way to fasten wood. However, it is the least rigid and has the least strength compared to the other methods of fastening wood. Carpenters use nails to build frames of houses, attach siding, and trim out house interiors. Farmers and others in agricultural mechanics find that proper nailing techniques are a very useful skill. A few nails, carefully placed and carefully driven, make quick and effective repairs. Where extensive nailing is done, electric and air-driven automatic nailing machines speed the process.

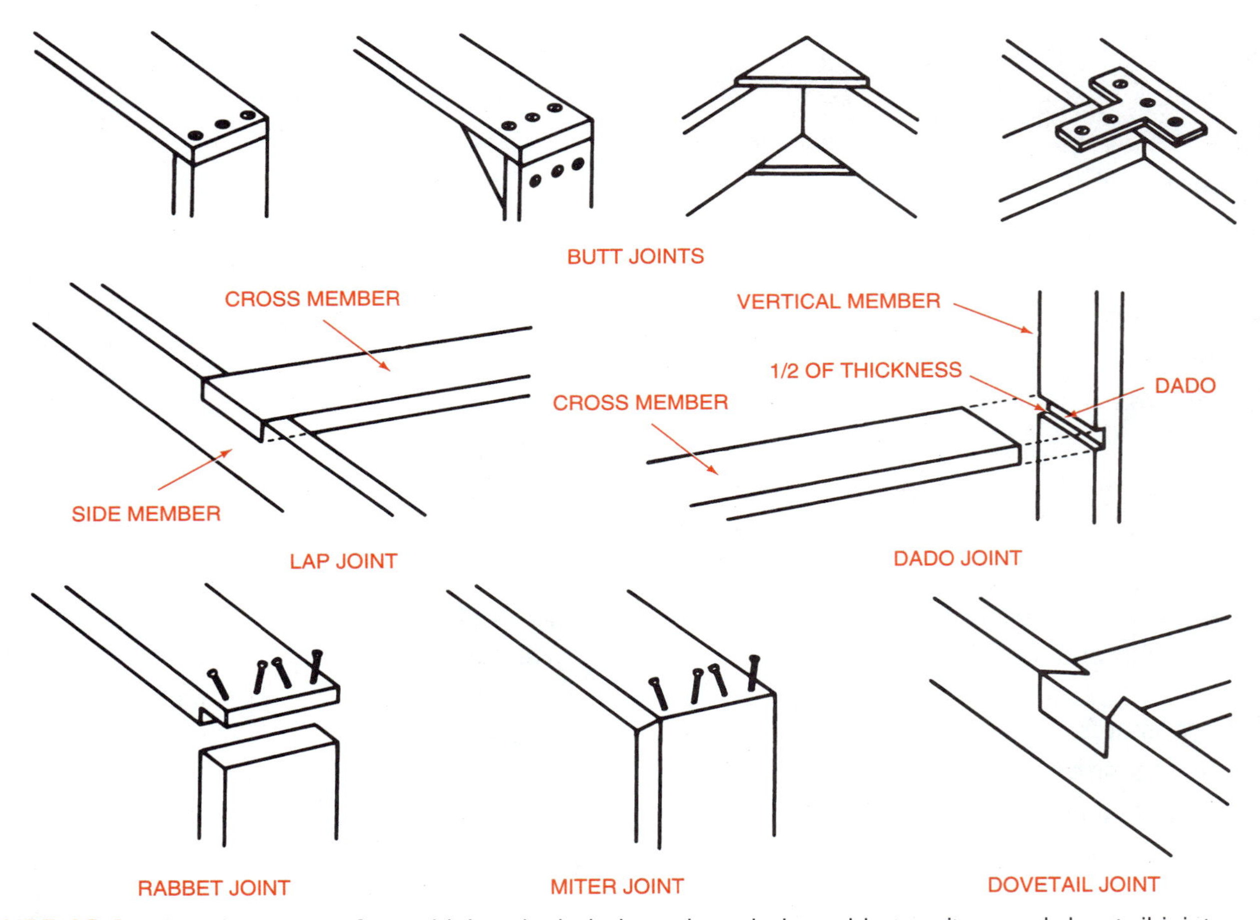

FIGURE 10-1. Popular types of wood joints include butt, lap, dado, rabbet, miter, and dovetail joints.

Driving Nails

Nails can be driven into flat surfaces easily if the material is soft. To drive a nail, the following procedure is used.

PROCEDURE

1. Hold the nail between the thumb and index finger and place the point of the nail on the material, figure 10-2A. The fingers should be placed high on the nail to permit them to be knocked free rather than smashed if the hammer accidentally hits them.
2. Hold a claw hammer in the other hand with the hand near the end of the handle, figure 10-2B.
3. Keep your eyes focused where you want the hammer to hit.
4. Tap the nail with the hammer until it stands up itself.
5. Use the wrist and arm to deliver firm blows using the weight of the hammer to drive the nail.
6. Drive the nail until the head is flush with the wood and the two pieces of wood are tightly fastened.

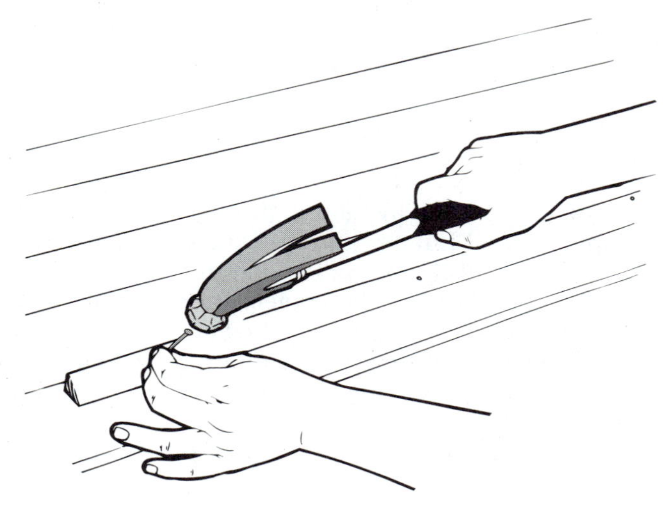

A. Start a nail by holding it with one hand and tapping it with the hammer held in the other hand.

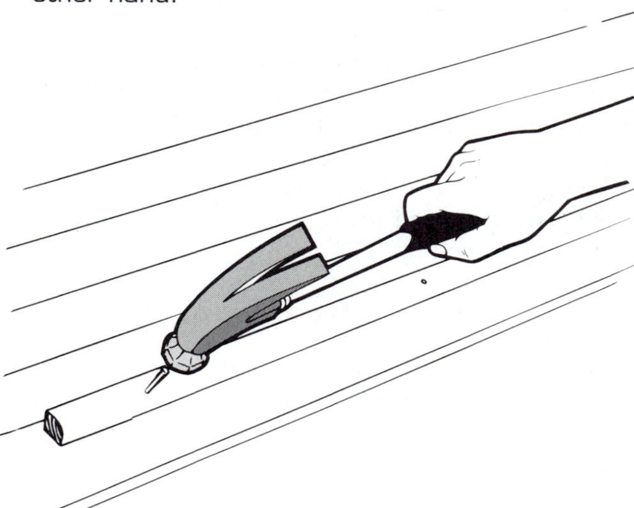

B. Drive the nail with long, even strokes. The hammer handle should be parallel with the work when the head strikes the nail.

FIGURE 10-2. Proper technique for driving a nail.

Match the hammer to the size of the nail. Common weights of claw hammer heads are 7, 13, 16, and 20 ounces. The 7-ounce hammer is excellent for brads, tacks, and small finishing nails. The 13-ounce hammer is good for light general nailing. For most farm applications, the 16- or 20-ounce hammer is preferred because it is heavy enough to drive large nails and spikes. Splitting of lumber is decreased when the end of the nail is filed to create a blunt end.

Pulling Nails. Claw hammers and ripping bars are used to pull nails and to rip boards from surfaces. When pulling a nail with a claw hammer, place a block under the hammer to prevent breaking the hammer handle, figure 10-3.

Setting Nails

When using finishing nails, the head of the nail is hidden in the wood. This is done by setting the nail. Setting a nail means driving the head below the surface, figure 10-4. To set a nail, the nail is driven

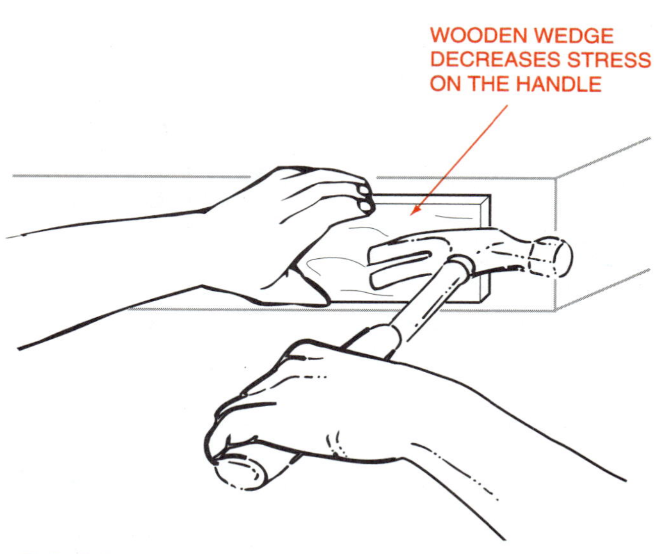

FIGURE 10-3. When a claw hammer is used to pull a nail, a block is placed under the hammer head to prevent breaking the hammer handle.

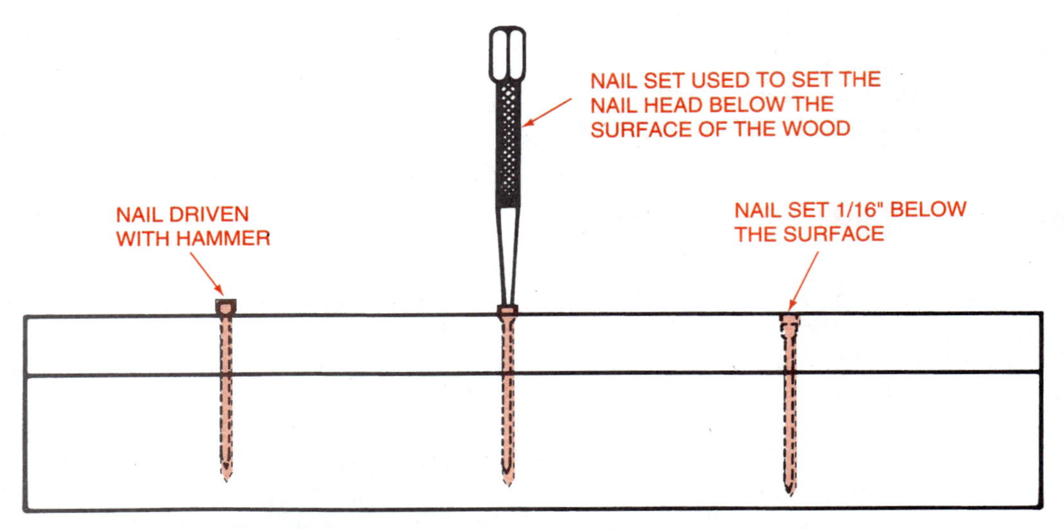

FIGURE 10-4. When appearance is important, finishing nails are used and the heads set below the surface. The holes are then filled with putty before the wood is finished.

with a hammer until the head touches the wood. A **nail set** is then used to drive the head below the surface about 1/16 inch. A nail set is a punchlike tool that has a cupped end rather than the flat or pointed end of a punch.

Toe Nailing

When two large pieces of wood must be fastened at right angles, they may be toe nailed. To **toe nail** means to drive a nail at an angle near the end of one piece and into the face of another piece, figure 10-5. Toe nailing is done extensively in framing using 2 × 4 or 2 × 6 lumber and 8d common nails.

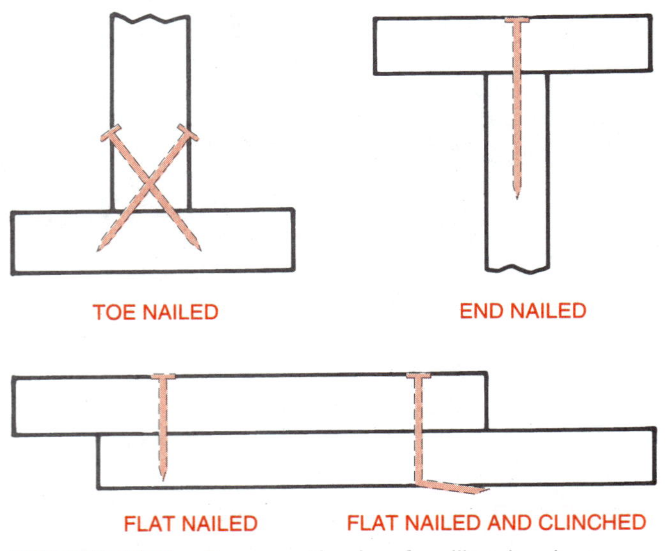

FIGURE 10-5. Four methods of nailing lumber.

End Nailing

End nailing is done by nailing through the thickness of one piece of lumber and into the end of another piece, figure 10-5. The nail is driven with the grain into the end of the receiving piece. Common nails or spikes are used for end nailing; box nails can be used for end nailing thin materials. End nailing has poor holding power. The technique is most useful when weight is permanently resting on the assembly. When end nailing small projects, it is advisable to use glue as well.

Flat Nailing

Flat nailing means to fasten two flat pieces to each other. One flat piece may also be nailed to a thicker piece. When two thin pieces are nailed together, the nails must be clinched. To **clinch** means bending the nail over and driving the flattened end down into the wood, figure 10-5. If splitting is likely, the nail should be clinched across the grain. Clinching results in a very strong nailed joint. To clinch nails, use the following procedure.

PROCEDURE

1. Choose a common nail that is about 1 inch longer than the thickness of the two boards.
2. Drive the nail through the boards until the head is flush with the surface of the wood. Be careful to support the wood so there is a space under the area where the nail comes through. If a space is not provided, the nail will damage the surface or the support, or the pieces will be nailed to the bench or saw horse.

3. Turn the two boards over and place the head of the driven nail on a hard surface. A steel vise works well. When nailing large assemblies, a heavy sledge hammer can be held against the nail head.
4. Hit the end of the nail sideways until it is bent flat to the surface.
5. Drive the nail until it is embedded in the wood.

Stapling

A staple is a piece of wire with both ends sharpened and bent to form two parallel legs. Staples are generally used to attach fencing or electrical wiring to wood. Staples are driven until they just touch the wire. If they are driven too tight, they will cause a short in electrical wire. It is also possible to weaken or break fence wire if staples are driven tightly against the wire.

Staples are also used to fasten ceiling tile and back panels on furniture. This type of staple is driven with a staple gun.

FASTENING WITH SCREWS

Screws hold better than nails. They are installed with a screwdriver, which is a turning tool with a straight tip, phillips tip, or special tip. The phillips screwdriver end is shaped like a plus sign (+). Screws may be driven quickly and easily with power screwdrivers or variable speed drills. Screws are used extensively in doors, windows, wall systems, and furniture. The flat head screw is the one used more extensively in woodworking. Its head is countersunk and flush with the surface after it is installed. The phillips head is preferred because it is easier to keep the screwdriver bit in the head of the screw.

Preparing Wood for Flat-Head Screws

A hole must be drilled in the wood before a screw is inserted. Holes made in wood to receive screws must be the correct diameter and depth; otherwise, the screw will not hold properly. Screws have a head, shank, and core with threads—all require exact holes, figure 10-6. The flat head must have a hole with tapered sides that fit. The hole for the head is called the countersink. The shank requires a hole that will permit it to drop through. The hole for the shank is called a shank hole or clearance hole. The core requires a hole small enough to permit the threads to screw into the wood. The hole for the core is called a pilot hole or anchor hole. When drilling holes for screws, it is recommended that the drill sizes be determined using a drill size chart, figure 10-7.

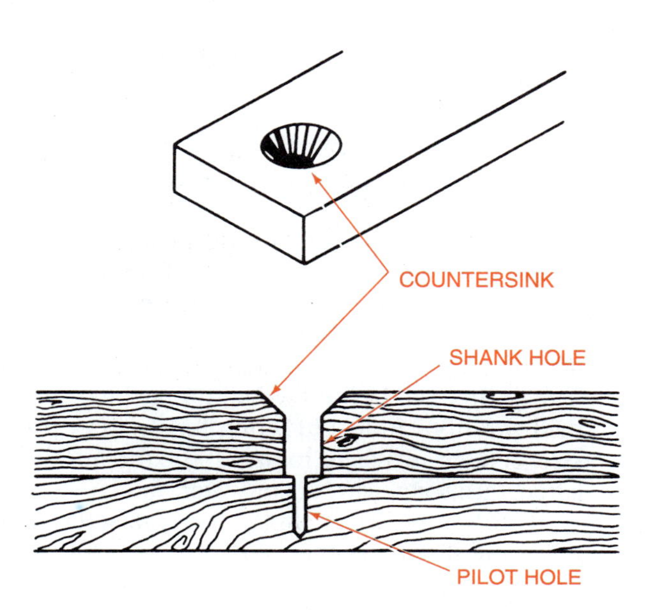

A. WOOD PIECES PROPERLY DRILLED TO RECEIVE A FLAT-HEAD WOOD SCREW.

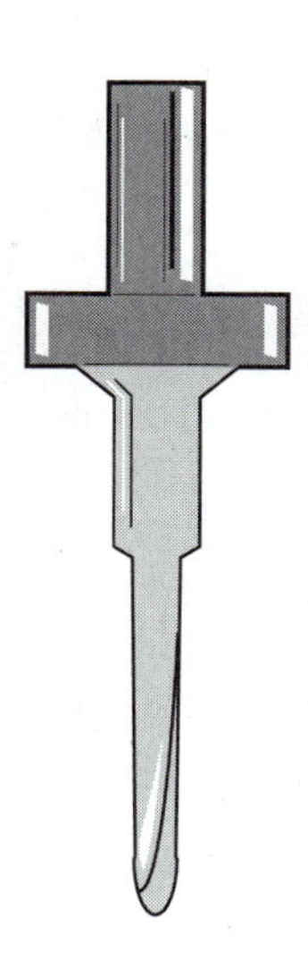

B. A COMBINATION DRILL THAT MAKES THE PILOT HOLE, SHANK HOLE, AND COUNTERSINK IN ONE OPERATION MAY BE PURCHASED FOR THE MORE POPULAR SIZES OF WOOD SCREWS.

FIGURE 10-6. Holes must be predrilled for screws.

Screw Gauge	0	1	2	3	4	5	6	8	10	12	14	16	18
Shank Hole	1/16	5/64	3/32	7/64	7/64	1/8	9/64	11/64	3/16	7/32	1/4	17/64	19/64
Pilot Hole	1/64	1/32	1/32	3/64	3/64	1/16	1/16	5/64	3/32	7/64	7/64	9/64	9/64

FIGURE 10-7. Chart of drill sizes for wood screws. *(Courtesy of M. Huth,* Introduction to Construction, *Delmar Publishers, 1980)*

Attaching Wood with Screws

When using screws to fasten two pieces of wood together, the following procedure is recommended.

PROCEDURE

1. Measure the thickness of the material to be attached.
2. If possible, use a screw three times as long as the thickness of the board being attached. If the screw will reach all the way through the second board, use a shorter screw.
3. Determine the appropriate screw spacing and mark the spots for all screws.
4. The diameter of the screws should look in balance with the spacing. Close screws should be smaller in diameter. The most frequently used screw sizes are numbers 6, 8, 9, and 10.
5. Use a chart to determine the drill size for the shank and pilot holes. An alternative method is to hold different drills under the screw shank until you have one which appears to be the same diameter. For the pilot hole, the drill must be the size of the screw core, not the diameter of the threads.
6. Insert the pilot hole drill into a hand drill chuck. Adjust the length of the exposed drill until it equals the length of the screw. The bit may be too long. If so, mark the drill with a sliver of masking tape or use a piece of wooden dowel to limit the exposed length of the drill.
7. Place two pieces of wood together and drill the pilot hole through both pieces of wood.
8. Install the shank hole drill in a hand drill. Use the shank hole drill to enlarge the hole in the first piece.

   NOTE: The practice of enlarging a hole with a drill is acceptable in wood only. If attempted in metal, it may damage the bit.

9. Use a countersink tool to create a countersink that exactly fits the screw head. The screw should drop through the piece to be attached. The head of the screw should be perfectly level with the surface of the board.

   NOTE: Steps 7, 8, and 9 can be done in one operation using a combination drill designed for this purpose.

10. Screw the two pieces together. Use a screwdriver that fits the screw slot. Turn the screw until it is snug. Do not over tighten.
11. Drill pilot holes for all additional screws.
12. Drill the shank holes for all additional screws. The depth of the shank hole is very critical. If it is drilled deeper than the thickness of the board, the screw will not hold properly.
13. Countersink all holes.
14. Set all screws.

    NOTE: A project looks better if the screws are evenly tightened and all of the screw slots are aligned.

Installing Round-Head Screws

The procedure for using round-head screws is similar to the procedure for using flat-head screws. Countersinking is simply eliminated. When tightening screws, screws should not be forced. If all holes are the proper depth and diameter, the screw should be easy to turn. In hard wood, it may be necessary to drill the pilot hole 1/64 inch larger than the chart size. Screws are easier to turn if the threads are lubricated with soap, wax, oil, or water.

If the screwdriver slips out of the screw slot, the following should be checked:

- be sure the screwdriver fits the screw, figure 10-8
- be sure the holes are the proper depth and diameter
- lubricate the screw threads
- check the condition of the end of the screwdriver—reshape if necessary.

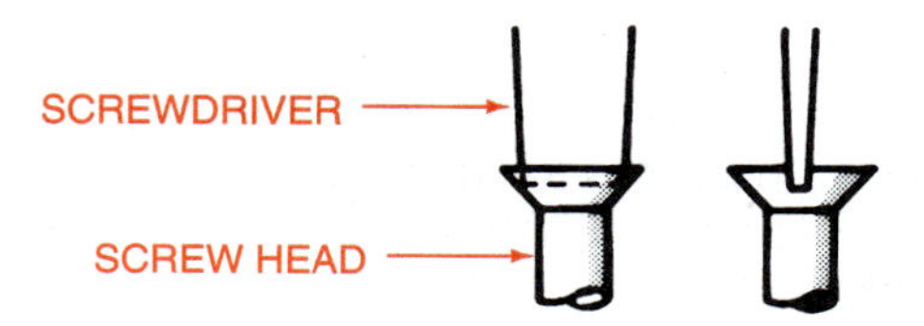

FIGURE 10-8. The screwdriver should be the same width as the screw head. It is very important that the end of the screwdriver be squared off so it fits the screw slot exactly.

A standard screwdriver should have a flat end with square edges. If the end is rounded, it should be reshaped with a file or another screwdriver should be used.

If a phillips screwdriver end is damaged or the end does not fit the screw properly, another screwdriver should be used. Damaged or worn phillips screwdrivers should be discarded or reshaped for other purposes.

Attaching Metal with Screws

Metal objects, such as hinges, require short screws. This is due to the thinness of the material being attached. When setting screws to attach hinges, the following procedure is recommended.

PROCEDURE

1. Hold the door in place.
2. Hold the hinge in place.
3. Draw a circle in one hole of the hinge.
4. Make a pilot hole with a push drill, hand drill, or pointed awl pushed by hand.

   NOTE: The shank hole is already drilled in the hinge.

5. Install one screw.
6. Repeat the process on the other end of the hinge.
7. Install the second hinge using two screws.
8. Check the door for proper closing and clearance.
9. Reset one or more screws if necessary.
10. Install all remaining screws.

FASTENING WITH BOLTS

Bolts can be the strongest method of fastening wood—with the possible exception of glue. The strength of bolts and the addition of large washers make them especially useful at high stress points. To install bolts, use the following procedure.

PROCEDURE

1. Select the type of head needed. The carriage bolt is designed specifically for wood. It leaves a fairly smooth head.
2. Drill a hole in both members the size of the bolt.
3. Place a washer on the bolt if a machine bolt is being used.
4. Insert the bolt.
5. Add a flat washer and nut.
6. Tighten the nut until the members are tight.

Lock washers are not needed when using bolts in wood. The wood creates back pressure on the nut. When tightening bolts, crushing the wood fibers or drawing the head below the surface of the wood should be avoided.

FASTENING WITH GLUE

Gluing is used extensively as a fastener of wood. Glue is a sticky liquid used to hold things together. It is used in the manufacture of plywood and particle boards. It is used extensively in manufacturing furniture. Even the rafters of wooden barns and large gymnasiums require glue to laminate the pieces into arched shapes. Glue is often used in combination with nails, screws, or bolts to provide extra strong joints.

There are many kinds of glue. Some examples are resorcinol, urea, polyvinyl, epoxy, contact cement, casein, and animal glues. The white liquid polyvinyl glue combines the advantages of several other glues. Polyvinyl fills cracks, dries colorless, sets quickly, and gives good results if protected from moisture.

Another important adhesive for agricultural uses is epoxy. Epoxy produces a strong, waterproof bond on wood, plastics, ceramics, metals, and other materials. Epoxy materials are also available for difficult masonry repairs. The disadvantage of epoxy is that it is supplied as two materials. These materials must be mixed in the correct proportions and used promptly once mixed.

Using White Polyvinyl Glue

A properly glued wood joint will be as strong as the wood itself. When gluing two pieces of wood together with polyvinyl glue, the following procedure is used.

PROCEDURE

1. Cut the two pieces of wood so all surfaces match perfectly.
2. Put the two dry pieces of wood together to

FIGURE 10-9. Bar clamps can be used to draw the pieces of wood together until the glue sets. The clamps should be only moderately tightened to avoid glue-starved joints.

check that the fit is good. Modify the pieces to make them fit, if needed.

3. Drill all screw or bolt holes.
4. Obtain all clamps that may be needed.
5. Apply a small bead of glue on the mating surfaces of both pieces of wood.
6. Spread the glue evenly with a flat object.
7. Put the two pieces together and nail, screw, bolt, or clamp the material, figure 10-9.
8. Check the joint and project to be certain every part is correctly aligned.
9. Retighten all clamps.

FIGURE 10-10. When the correct amount of glue is used, only small droplets of excess glue will be squeezed from the joint. These droplets are removed with a putty knife after the glue dries.

10. Remove all glue runs when slightly dry with a putty knife or wood chisel, figure 10-10.
11. Wipe all glue marks a second time using a clean wet rag. Finally, wipe the wood with a dry rag.

    NOTE: If all glue is not removed, the wood will not stain evenly. If too much water is used in the cleanup process, the grain of the wood will get rough and will need sanding before finishing.

12. Clamps may be removed after 1/2 hour. For better results, leave the clamps on for about 12 hours.

Glue will not adhere to paint, grease, or wax. Therefore, it is best to cut, file, sand, or otherwise clean the areas to be glued.

Glue guns that melt a glue stick and apply a thick substance that does not require clamping are very useful for arts and crafts projects.

Using Dowel Pins

A dowel is a round piece of wood. Dowels are generally sold in 36-inch lengths. They range from 1/16 inch to 1 inch or larger in diameter. Popular sizes are 1/8, 3/16, 1/4, 5/16, 3/8, 7/16, 1/2, 5/8, 3/4, 7/8, and 1 inch. An important use of dowels is to strengthen glue joints by extending from one part to the other, figure 10-11.

FIGURE 10-11. Dowel pins that extend into the two pieces of wood greatly strengthen glue joints. The dowel pins should be covered with a film of glue before installation.

Dowels are also sold in lengths of about 2 inches. These lengths have tapered ends that make them easy to use when gluing parts together. They have spiral grooves that permit glue to make an excellent bond between the dowel and the board.

STUDENT ACTIVITIES

1. Define the Terms to Know in this unit.
2. Obtain two pieces of scrap wood 2 × 4 or 2 × 6, each about 12 inches long, cut from pine, spruce, hemlock, redwood, poplar, or other soft wood. Place one piece on top of the other to double the thickness. Practice driving 4d, 8d, and 10d common nails and finishing nails. Try different sizes of hammers.
3. Drive several nails part way into a piece of wood. Place a block of wood (about 2/3 the height of the nail) beside a nail. Place the hammer claws under the head of the nail with the hammer head on the block. Pull the nail. Practice this skill being careful to use wooden blocks to avoid breaking the hammer handle. Why does the use of a block under the hammer reduce strain on the handle?
4. Repeat activity 3 using oak, locust, maple, or other hard wood. What differences do you observe? Nails are easier to drive into hard wood if a hole smaller than the nail is drilled through the first piece.
5. Use a nail set to drive the heads of several finishing nails below the surface.
6. Practice toe nailing a 2 × 4 to a 2 × 4.
7. Study some pieces of furniture and identify butt, lap, dado, rabbet, and dovetail joints.
8. Attach two small boards together using glue and two flat head screws. Be careful to provide the correct countersink, shank hole, and pilot hole.
9. Take out the board you used to cut a dado in Unit 9. Cut a 3/4-inch piece of wood several inches wide to fit into the dado groove. Glue and clamp the board to form a dado joint. Save this board for an activity in Unit 11.
10. Make a sanding block with a hold-down cleat like the one shown in figure 10-12.

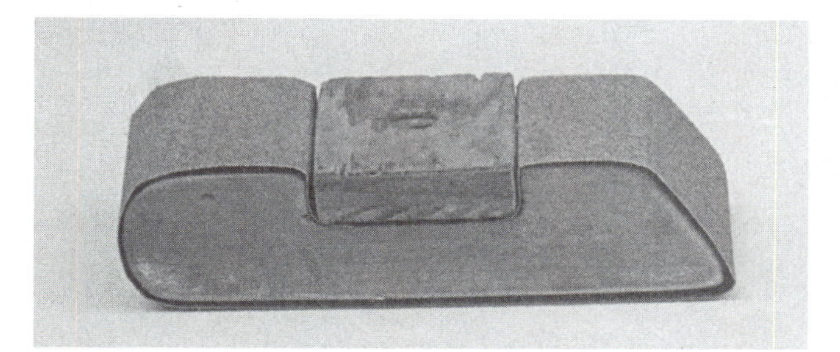

FIGURE 10-12. This sanding block works well because it fits comfortably in one hand. It uses a strip of sandpaper 2 inches wide and the length of a standard sheet of sandpaper. The cleat is held down with a wood screw to keep the paper in place. This sanding block is made from a piece of soft wood 3/4" × 2" × 4". The dado (groove) is 3/8 inch deep and 1 1/2 inches wide. The cleat is 3/8" × 1 3/8" × 2". The screw is No. 8 × 3/4 inch long.

SELF-EVALUATION

A. Multiple Choice. Select the best answer.

1. Two pieces of wood joined together is called
   a. an angle
   b. a joint
   c. a lap
   d. a splice
2. The tool used to push nail heads below the surface of wood is a nail
   a. driver
   b. press
   c. punch
   d. set
3. The strongest nailing method is
   a. clinching
   b. end nailing
   c. flat nailing
   d. toe nailing
4. An advantage of epoxy glue is
   a. ease of use
   b. no mixing
   c. waterproof bond
   d. none of these

B. Matching. Match the word or phrase in column I that best matches the word or phrase in column II.

Column I

1. groove in end of board
2. interlocking extensions
3. groove in middle of board
4. face-to-face
5. edge-to-edge
6. flat head of screw
7. smooth stem of screw
8. threaded core of screw
9. combination drill
10. shank hole for No. 8 screw
11. shank hole for No. 10 screw
12. pilot hole for No. 8 screw
13. pilot hole for No. 10 screw

Column II

a. butt joint
b. dado joint
c. dovetail joint
d. lap joint
e. rabbet joint
f. three holes at once
g. countersink
h. shank hole
i. pilot hole
j. 11/64 inch
k. 3/64 inch
l. 5/64 inch
m. 3/32 inch

UNIT

11 Finishing Wood

OBJECTIVE

To prepare wood for finishing and to apply attractive and durable finishes.

COMPETENCIES TO BE DEVELOPED

After studying this unit, you should be able to:

- Sand wood.
- Remove dents.
- Fill holes.
- Select finishing materials (except paint).
- Select solvents for thinning.
- Apply stain and clear finishes.
- Clean brushes.

TERMS TO KNOW

finish
putty
glazing compound
pliable
plastic wood
shellac
varnish
spar varnish
marine varnish
tack rag
polyurethane
satin
wood filler

MATERIALS LIST

- ✓ Sandpaper
- ✓ Steel wool
- ✓ Heating iron
- ✓ Clean, soft rags
- ✓ Putty, glazing compound, or plastic wood
- ✓ Clean paint brush (1 inch)
- ✓ Sealer (shellac suggested)
- ✓ Finishing material (varnish suggested)
- ✓ Stain
- ✓ Paste furniture wax
- ✓ Paint thinner
- ✓ Shellac thinner

Wood objects and buildings that were properly finished have remained durable and beautiful for centuries. There are many, many ways to finish wood to protect the surface.

Wood can be truly beautiful if finished so the grain shows. If painted, wood offers only its shape, strength, and paint color. The secret to an attractive wood project is to prepare the wood for finishing. This means removing dents, leveling high spots, removing glue stains, and sanding to a smooth surface.

This unit covers the finishing of wood with stain and clear finishes. Painting will be covered in another unit.

PREPARING WOOD FOR FINISHES

Soft wood is easily damaged until a finish is applied. A finish is a chemical layer that protects the surface of a material. Lumber should always be handled carefully. Walking on clean boards or laying them on dirty benches or floors should be avoided. Sanded boards should be protected with cloth or paper. Pencil marks should be made lightly and then erased. If pencil marks are sanded, they become embedded in the wood.

Removing Dents

If wood is accidentally dented, it may be repaired unless the wood fibers are cut. The procedure is as follows.

PROCEDURE

1. Heat a soldering copper to about 400°F or a clothes iron to its hottest setting.
2. Place a damp, soft, clean cloth over the dent.
3. Apply the hot iron to the cloth until steam rises from the cloth, figure 11-1.
4. Let it steam for about 5 seconds, but do not burn the wood.
5. Remove the iron.
6. Remove the cloth.
7. Examine the dent. The steam should have caused the compressed wood fibers to expand. The previous dent should be level.
8. If there is still a small dent, redampen the cloth and repeat steps 2 through 7.
9. Sand the area when the surface dries.

FIGURE 11-1. To remove a dent in wood, lay a wet cloth over the dent. Press the flat area of a hot iron on the cloth. The steam from the water in the cloth should cause the compressed fibers of the wood to expand, thus eliminating the dent.

Filling Holes

A word of caution is in order about filling holes. Some students expect to correct mistakes with filling materials. It is very difficult to restore the looks of a project after bad cuts are made or it has been inadequately sanded. In this regard, wood filling materials work well only in holes where dowels or nails are set below the surface, or where a puncture or blemish is small in diameter.

Holes made by installing finishing nails may be filled with caulking, putty, glazing compound, or plastic wood. Caulking, putty, and glazing compound are used to hold window panes and fill cracks and crevices before painting.

Putty and Glazing Compound. Putty and glazing compound are soft filling materials containing oils to keep them pliable. Pliable means it will give if push or pulled. To use either putty or glazing compound to fill holes, the procedure is as follows:

PROCEDURE

1. Seal the wood with a sealer or primer.
2. Take a small amount of putty or compound from the can. Work it in the palm of your hand until it is soft.
3. If you wish to color the material, work in some oil base tinting pigment to obtain the desired color.
4. Place a bit of the material on a screwdriver tip and press it into the hole, figure 11-2A.
5. Smooth and level the material by wiping the surface firmly with a flexible putty knife or your finger, figure 11-2B.
6. Permit the material to firm up overnight.
7. Apply the desired finishing materials to the wood.

Plastic Wood. Plastic wood is a filler product that is soft when it comes from a sealed can but dries very quickly. It has the advantage of becoming very hard in a matter of minutes. Plastic wood cannot be colored. However, it can be purchased in colors to match various woods such as pine, oak, walnut, and mahogany. To apply plastic wood, the following procedure is recommended.

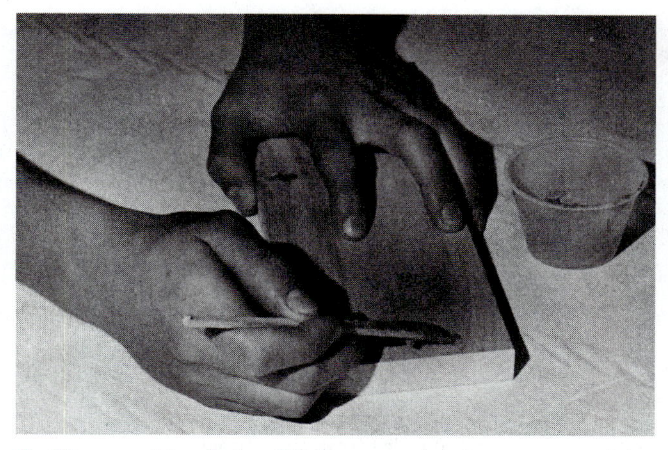

A. Place a bit of the filling material on a screwdriver tip or other implement and press it into the hole.

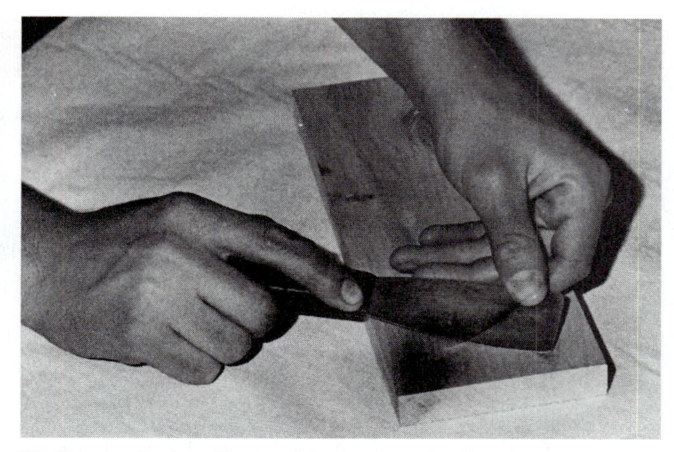

B. Smooth the filler with a putty knife.

FIGURE 11-2. Holes in wood may be filled with putty, glazing compound, or plastic wood.

PROCEDURE

1. Select a can or tube of plastic wood of the color to match the finished color of the wood.
2. Put a small amount of plastic wood on the tip of a screwdriver and close the can or tube.
3. Quickly press the plastic wood into the hole.
4. Smooth the surface by drawing the flat part of the screwdriver blade back over the plastic wood.
5. Add more plastic wood and smooth out if necessary.
6. Recheck to be sure the can or tube is tightly closed.
7. Scrape off any plastic wood on the surface around the hole.
8. After 3 to 5 minutes, sand the area thoroughly with fine sandpaper. Otherwise, every spot touched by plastic wood will leave an ugly stain on the wood.
9. Apply the desired finishing materials to the wood.

Putty Stick. A more recent product for filling holes is the putty stick. Putty sticks may be purchased in dozens of shades to match any color of wood finish. The procedure for using a putty stick follows.

PROCEDURE

1. Apply a stain (if desired) and a final finish on the wood.
2. Select a putty stick to match the finish of the wood.
3. Rub the stick back and forth over the hole until the hole is filled with putty.
4. Smooth the putty so it is level with the surface of the wood.
5. Rub the area with a cloth or paper towel to remove all excess putty. The job is then complete since all finishes are applied before using the putty stick, and putty is a good, protective finish.

Selecting the Finish

Wood products must be carefully prepared before applying stains and other finishing materials. Paint and opaque stains require less sanding than penetrating stains and clear finishes. The type of finish must be selected before the project is completed.

If expensive wood with an attractive grain is used, a penetrating stain with a clear finish is desirable. If the grain is not attractive, an opaque stain or paint can be used. This requires less time than preparing the wood for a finish which lets the grain show.

Sanding

All edges and ends of boards should be worked down with a file or other cutting tool to the desired shape. The wood is then sanded with medium sandpaper and finally with fine or very fine sandpaper. Figure 11-3 gives guidelines for selecting sanding materials.

For best results, always do the final sanding with the grain; otherwise, ugly sanding marks will be left in the wood, figure 11-4.

MATERIAL	RECOMMENDED USE	SPECIAL COMMENTS
Flint	Painted or pitchy surfaces; jobs where clogging is a problem	Low cost; paperback abrasive wears off quickly; use where paper clogs easily
Garnet	Hand sanding of wood	Durable; low cost; paper backed
Emery	Hand sanding of metal to remove rust or smooth rough areas	Expensive; cloth backed; disintegrates when wet
Aluminum Oxide	Machine sanding of wood or metal; finishing of bronze and steel	Fast cutting; long lasting
Silicone Carbide (wet and dry)	Wet: remove automotive-type finishes; feather edges of chipped paint. Dry: finish brass, copper, aluminum; sand plastics, glass and ceramics	Has a waterproof backing to permit rinsing of paint from the abrasive; very effective when used with water; durable; expensive
Steel Wool	Remove old finishes; smooth and polish fine wood finishes; polish brass	Coarse and medium grades work well with paint removers. Very fine is used to cut the gloss and provide a dead smooth surface after applying wood finishes

FIGURE 11-3. Guidelines for selecting sanding materials and steel wool.

Faces of boards should be sanded with a power sander or sandpaper wrapped around a block of wood. This helps keep the surface of the wood level. It also spreads the use across a large area of the sandpaper. A sanding block with a beveled edge permits sanding in corners. Again, start with medium sandpaper and end with very fine. After final sanding, the wood should be wrapped in cloth or paper to protect the surface from dirt and damage until the finish is applied.

APPLYING CLEAR FINISHES

Clear finishes protect the wood and let the beauty and color of the wood show through. An excellent finish is obtained by simply rubbing boiled linseed oil into the wood. The same is true for paste-type furniture wax. However, since four to ten coats are required, most people use finishes that are easier to apply.

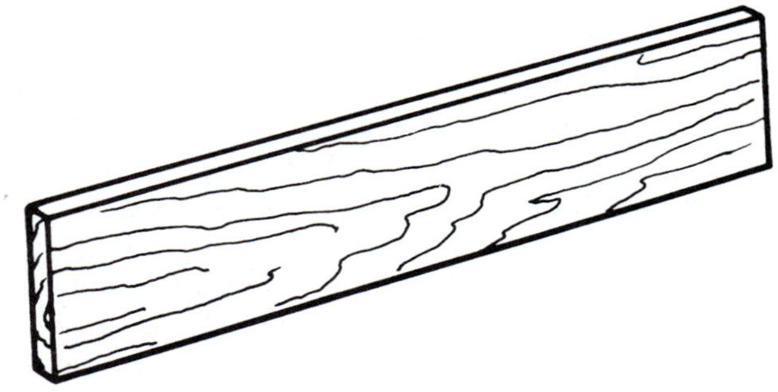

The procedures given for applying finishes are general instructions. The student is reminded to always read the label and follow the specific directions given with the product.

FIGURE 11-4. Final sanding should always be done with the grain.

Shellac and Varnish

Shellac is a product made from the secretion of the lac insect and is used to seal wood. After sealing with shellac, varnish is used to provide a protective finish. Varnish is a clear, tough, water-resistant finish made from various oils. Special varnishes such as spar varnish and marine varnish are designed for use on objects exposed to a great deal of moisture. The procedure for a durable shellac-varnish finish is as follows.

PROCEDURE

1. Sand the wood until all marks are absent, the wood is extremely smooth, and the grain shows in all areas.
2. Wipe the wood with a rag dampened with shellac thinner (denatured alcohol). A rag dampened with a solvent is called a tack rag.
3. Choose white shellac for a clear finish. Choose orange shellac for an amber finish.
4. Thin the shellac according to the directions on the container.
5. Apply the shellac quickly with a clean, pure bristle brush. Do not brush excessively and do not go over the surface a second time.

6. Hang up the project or support it with small nails until dry.
7. Clean the brush with alcohol and then wash it with warm water and soap. Wrap the bristle end in a paper towel to permit the bristles to dry and maintain the brush shape.
8. Allow 12 hours or more for the project and brush to dry.
9. Rub the dried shellac lightly with 000 grade steel wool or very fine sandpaper. The wood surface should become very smooth.
10. Choose a work area that is clean and especially free of dust in the air or on work areas.
11. Wipe the project with a rag dampened with turpentine, varsol, or paint thinner.
12. Drive four small nails in the bottom of the project to serve as temporary legs. An alternate method is to drive a nail with a head in the bottom of the project and then use the nail to suspend the project on a wire for drying.
13. Select a good quality, perfectly clean, pure bristle brush. Brushes ranging from 1 inch to 2 1/2 inches are recommended for shop projects and furniture.
14. Open the can of varnish and stir it slowly with a clean object. Avoid causing bubbles due to vigorous stirring.
15. Rewipe the project to remove dust.
16. Apply the varnish by starting in hard-to-reach areas and moving toward the easier-to-reach areas.
17. Frequently check previously varnished areas. Smooth out any areas where the varnish runs.
18. Do not skip any areas; cover the end grain and undersides as well as visible areas.
19. Set the project in a dust-free area to dry at least 12 hours (24 or more in damp weather).
20. When thoroughly dry, rub with 000 steel wool or sand lightly with very fine sandpaper.
21. Apply a second coat of varnish. Runs may be harder to control on the second coat, so extra care is needed when applying the varnish.
22. Rub again with 000 steel wool.
23. Restore the shine with a paste-type furniture wax. Wipe on a thin coat of wax and allow it to dry to a haze—about 10 minutes. Polish the surface with a clean, soft, dry rag. The wax provides a hard, waterproof, and durable finish.

Polyurethane

Polyurethane is a clear, durable, water-resistant finish. It requires no separate sealer. It seals the pores of the wood and provides a very durable finish. Polyurethane can be purchased in a high gloss form or *satin* form. The satin form looks like a hand-rubbed varnish finish without the need for steel woolling or sanding between coats. A satin finish has a low sheen and is regarded by many as more desirable than a very shiny or glossy finish.

Polyurethane dries quickly and is applied like varnish. Two or three coats are recommended. Turpentine, varsol, or paint thinners are satisfactory thinners and brush cleaners. Once dried, polyurethane is not easily removed even with chemical paint and varnish removers.

Lacquers

Lacquer is a clear finish that requires its own special thinner. It seals wood and provides a good protective finish. Lacquer and lacquer thinners are identified by their bananalike odor. Lacquer thinner softens paint, so it is important not to apply lacquer over other finishes.

Lacquer is popular because it is a clear finish which dries almost instantly if sprayed. A special slower drying lacquer is available for brushing. The procedure for using the brush-type lacquer follows.

PROCEDURE

1. Sand and clean the wood properly.
2. Use a clean, pure bristle brush.
3. Wipe the wood with a rag dampened with lacquer thinner.
4. Quickly apply a coat of lacquer. Move forward into new wood and do not brush over previously coated areas.
5. Replace the lid on the can promptly.
6. Clean the brush immediately in lacquer thinner followed with soap and water.
7. Allow the finish to dry thoroughly.
8. Apply a second coat.
9. After drying, rub paste wax into a pad of 000 steel wool. Rub the finish briskly with the wax and steel wool pad. Before the wax dries, polish the surface vigorously with terrycloth or a coarse towel.

APPLYING WOOD STAINS

Wood may be stained to many shades of black, brown, red, blonde, and white. Stains may be purchased in forms that penetrate the wood and enhance the grain. Another type of stain colors the surface of the wood and hides the grain. A third type is thick and fills the pores of open grained woods. This is called wood filler.

Penetrating oil stains are popular and easy to use. To apply penetrating oil stains, use the following procedure.

PROCEDURE

1. Sand and prepare the wood for finishing.
2. Stir the stain thoroughly.
3. Use a brush or soft, clean cloth to cover all parts of the wood with stain.
4. Allow about 5 minutes for the stain to penetrate the wood.
5. Wipe off the excess stain with clean, soft, absorbent rags.
6. Wipe and lightly polish the wood until no stain comes off when touched.
7. Set the project aside to dry for at least 24 hours.
8. Clean the brush in turpentine, varsol, or paint thinner followed by soap and warm water.
9. Clean stain off your hands with the brush cleaner.
10. Apply a clear finish as described previously.

SOLVENTS AND THINNERS

Many problems can be traced to a lack of knowledge about solvents and thinners. The five solvents that are used extensively in the shop and home are denatured alcohol, turpentine, paint thinners, varsol, and lacquer thinner, figure 11-5.

Denatured Alcohol

Denatured alcohol or wood alcohol is sold as shellac thinner. It will thin shellac and finishes that contain shellac. For the most part, alcohol is a single product solvent in the finishing room. However, it may be used elsewhere in the shop as a cleanup solvent when overhauling brake cylinders and other metal hydraulic parts.

Turpentine, Paint Thinners, and Varsol

These products are all suitable solvents or thinners for oil-based stains, paints, and varnishes. While some cost more than others, most product labels indicate equally good results from all three. However, varsol is especially desirable as a solvent in the shop. Varsol is safer than most solvents from the standpoint of fire hazard. Secondly, it is a good inexpensive grease solvent and can be rinsed away with water.

THINNER/ SOLVENT*	USED WITH	CLEANS BRUSHES USED IN
Denatured Alcohol	Shellac (both white and orange) Sealers that contain shellac	Shellac Sealers that contain shellac
Turpentine and Paint Thinners	Oil base stains and wood fillers Oil base (alkyd) paints Varnish Polyurethane	Oil base stains, fillers, and paints Varnish Polyurethane
Varsol	Oil base stains, wood fillers, and alkyd paints Solvent for grease and oil	Oil base stains, wood fillers, and alkyd paints
Lacquer Thinner**	Brush lacquer Spray lacquer Finishes containing lacquer	Lacquer and finishes containing lacquer Oil base stains, fillers, and alkyd paints (Also softens hard lacquer, oil base stains and fillers, and alkyd paints in brushes)
Water	Latex paints and other latex products	Latex paint and products (when water is warm and used with a detergent) Other thinners (when water is warm and used with soap)

***Caution:** Protect eyes with safety devices and use all solvents in well-ventilated area.
****Caution:** Lacquer thinner is especially hazardous because it evaporates quickly and dissolves the natural oils from skin. Avoid excessive contact with skin.

FIGURE 11-5. Thinners and solvents commonly used with finishing materials.

Lacquer Thinner

Lacquer thinner is a fairly universal solvent. Not only does it thin lacquer, but it softens hard paint as well. This makes it useful for cleaning paintbrushes used for lacquer as well as other materials.

Lacquer thinner evaporates quickly. This creates problems with fumes in poorly ventilated areas. Extra care is needed to avoid breathing lacquer thinner fumes. Similarly, special care must be taken to avoid sparks or fire when using lacquer or any other finishing material.

USE AND CARE OF BRUSHES

High-quality brushes make it possible to apply high-quality finishes. Without good brushes, high-quality finishes are impossible. Unfortunately, a good brush can be ruined the first time it is used. Following is the procedure for correctly using a brush.

PROCEDURE

1. Choose the right brush for the material you are using. Nylon and other man-made bristle brushes work best with latex paints. Natural bristle brushes are best for most other stain, varnish, and painting products.
2. Dip only about 1/2 inch of the brush into the material.
3. Brush back and forth to keep the bristles together. Do not bend or spread the bristles excessively.
4. Once a job is started, alternate dipping the brush into the material and applying it so the material does not dry in the brush.
5. When the job is finished, clean the brush promptly with the proper solvent and then wash it with soap and warm water until all evidence of the finishing material is gone.

After use, brushes must properly cleaned. The cleaning procedure should leave the brush perfectly clean. When dry, the bristles should be soft, compact, and ready for reuse. Like a good tool, a good brush, properly used and properly cleaned, can provide many years of excellent service. The procedure for cleaning a brush follows and is shown in figure 11-6.

PROCEDURE

1. Remove as much material from the brush as possible. Rags, newspapers, or paper towels may be used to wipe the brush.

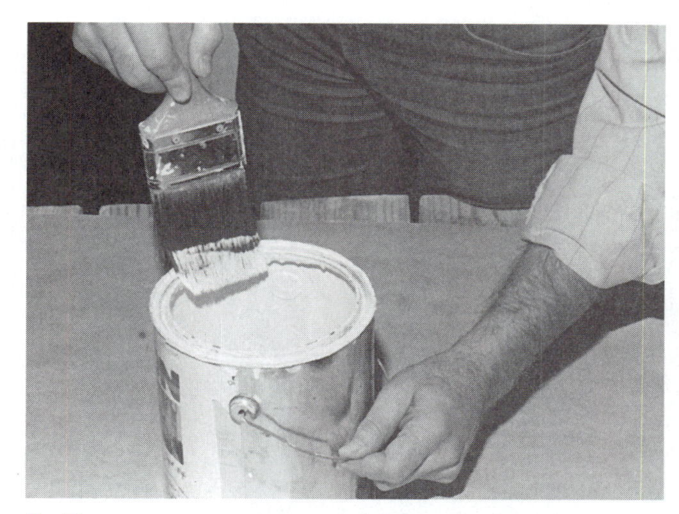

A. Remove excess material from the brush.

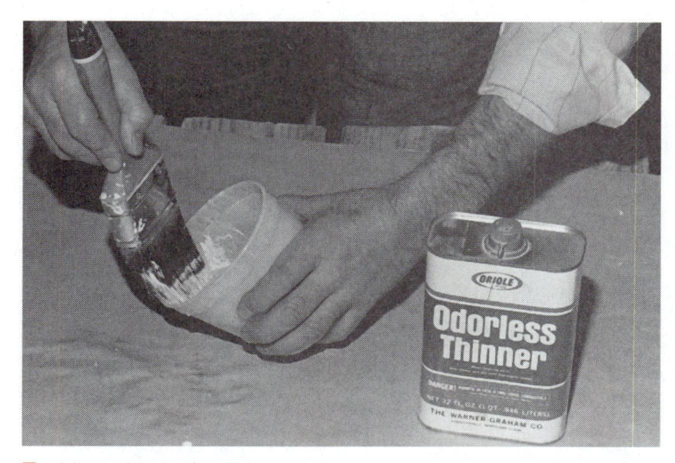

B. Use thinner to dissolve material in the brush.

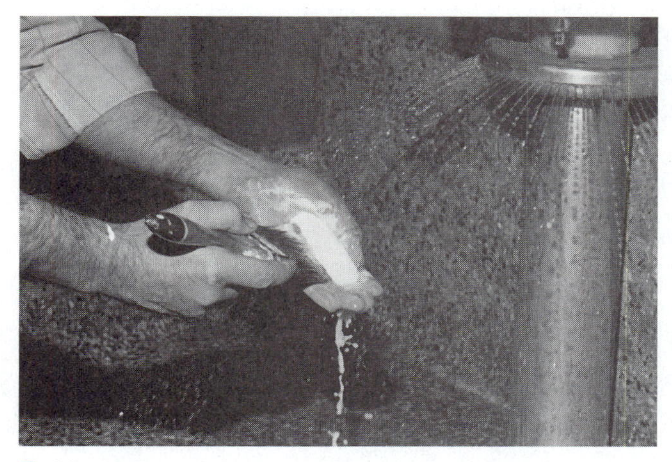

C. Use soap and warm water to remove the thinner and dissolved material from the brush.

FIGURE 11-6. Brushes must be cleaned properly after use.

D. A brush that is clean and ready to dry.

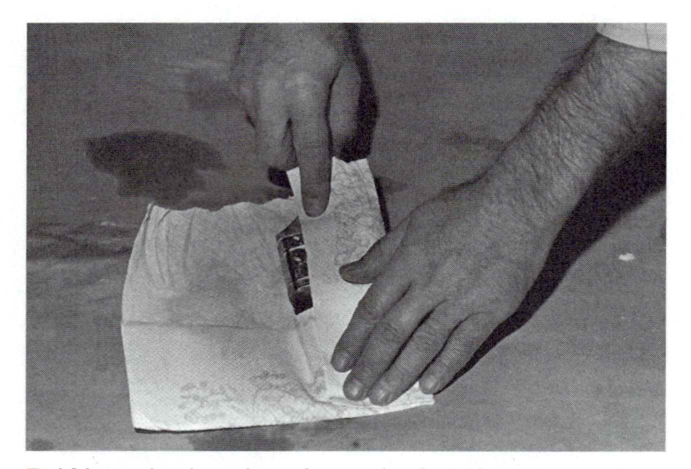

E. Wrap the brush to keep the bristles compact and in shape while drying.

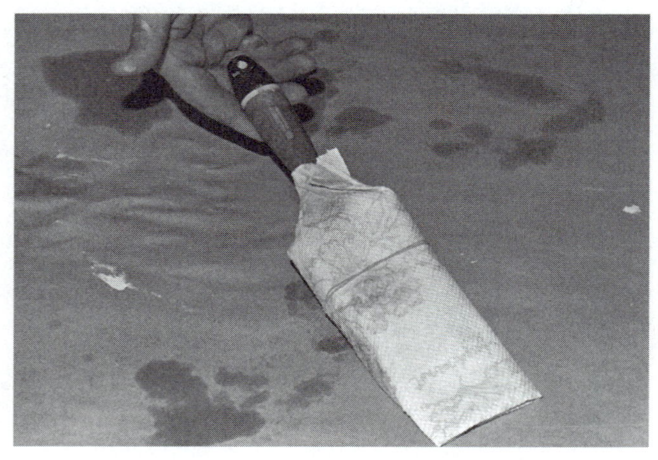

F. A clean brush, properly wrapped and ready to be left to dry.

FIGURE 11-6. *Continued*

2. Pour about $1/2$ inch of solvent for the product being used into a container just a little bigger than the width of the brush. Using a small container reduces the amount of solvent needed.
3. Insert the brush and push it up and down bending the bristles slightly, but not flattening them.
4. Pour the used solvent into an empty solvent can marked "used."
5. Pour some more clean solvent into the container and again flush the brush.
6. Pour the solvent into the "used" solvent can. Replace the caps on all solvent cans. After several days the used solvent will become clear and usable again for cleaning brushes.
7. Wipe all excess solvent from the brush with a rag or paper towel.
8. Place a bar of soap in the palm of your hand and run a small stream of warm water on the soap.
9. Rub the brush on the soap and in the water until all solvent and material are flushed away and the soap creates a good lather.
10. Rinse all soap from the brush. The bristles should feel like clean hair.
11. Lay a paper towel on a bench.
12. Position the brush on the left lower quarter of the towel.
13. Wrap the brush one or two turns, fold the paper down over the top of the brush, and continue the wrap until all of the towel is used. Twist the towel around the handle of the brush.
14. Lay the brush in a warm place to dry. It will take about 24 hours.
15. The properly cleaned brush will not show any evidence of previous use, as noted by color of material used or any bristle stiffness.

STUDENT ACTIVITIES

1. Define the Terms to Know in this unit.
2. Take out the board that you used for activity 9 in Unit 10 and perform the following procedures.
   - Drill two $^3/_{16}$ inch holes about $^1/_4$ inch deep.
   - Fill the holes with natural-colored plastic wood.
   - Use a file and sandpaper to smooth all edges of the board.
   - Sand all surfaces, remove all marks, and create clean, smooth surfaces on the board.
   - Apply stain to the back side (side without the dado).
   - Apply a sealer, such as shellac.
   - Apply one or two coats of finish, such as varnish.
   - Apply wax.
3. Examine the finished board. What procedures did you do well? What errors did you make? How can the errors be avoided in the future?

SELF-EVALUATION

A. **Multiple Choice**. Select the best answer.

1. A dent may be removed from wood with
   a. a special tool
   b. a grinder
   c. sandpaper
   d. steam from a wet rag
2. Holes in wood may be filled with
   a. glazing compound
   b. plastic wood
   c. putty
   d. all of these
3. Oil base stains, varnish, and alkyd paints should be thinned with
   a. alcohol
   b. lacquer thinner
   c. turpentine
   d. water
4. Wood may be colored by applying
   a. polyurethane
   b. stain
   c. varnish
   d. wax
5. The material used to fill holes after wood has a final finish is
   a. glazing compound
   b. plastic wood
   c. putty
   d. putty stick
6. Plastic wood is recommended for
   a. filling nail holes
   b. hiding bad cuts
   c. removing sanding marks
   d. all of these
7. Final sanding should be with
   a. coarse sandpaper
   b. the grain
   c. medium sandpaper
   d. a file
8. The correct sequence when applying finishes is
   a. wax, shellac, varnish, stain
   b. stain, shellac, varnish, wax
   c. stain, varnish, wax, shellac
   d. shellac, varnish, wax, stain
9. A clear, durable, water-resistant finish that is brushed on and requires no separate sealer is
   a. polyurethane
   b. shellac
   c. stain
   d. wax
10. The solvent that dissolves the greatest number of finishing products is
    a. alcohol
    b. lacquer thinner
    c. turpentine
    d. varsol

B. Matching. Match the finishing materials in column I with the thinner/solvent they are used with in column II.

Column I

1. oil base stain and varnish
2. latex paint
3. shellac
4. lacquer

Column II

a. denatured alcohol
b. lacquer thinner
c. turpentine and varsol
d. water

C. Brief Answers. Briefly answer the following questions.

1. What is a rag dampened with solvent called?
2. List the steps to follow when cleaning a varnish brush.

UNIT

12 Identifying, Marking, Cutting, and Bending Metal

OBJECTIVE

To identify, mark, cut, and bend cold metal.

COMPETENCIES TO BE DEVELOPED

After studying this unit, you should be able to:

- Identify metals.
- Mark metal.
- Cut and file metal.
- Bend square, round, and flat steel.
- Form sheet metal.

TERMS TO KNOW

ferrous
nonferrous
alloy
malleable
casting
ductile
malleable cast iron
anneal
tempering
galvanized steel
corrosion
solder
scratch awl
scriber
soapstone
dividers
center punch
hacksaw
single cut file
double cut file
drawfiling
stock
snips
shears
cold chisel
eye

MATERIALS LIST

- ✓ Samples of
  - cast iron
  - wrought iron
  - mild steel (various shapes)
  - tool steel
  - stainless steel
  - galvanized steel
  - aluminum
  - copper
  - lead
- ✓ Scratch awl
- ✓ Scriber
- ✓ Assorted files
- ✓ Soapstone
- ✓ Chalk
- ✓ Dividers
- ✓ Center punch
- ✓ Hack saw
- ✓ Metal snips
- ✓ Cold chisel
- ✓ 10-inch or 12-inch adjustable wrench
- ✓ Machinist's vise
- ✓ Anvil
- ✓ Magnet
- ✓ Combination square with scriber
- ✓ Stove bolt with nut ($^1/_4$" x 2")
- ✓ $^1/_8$" x 1" x 6" flat steel
- ✓ Blacksmith's or heavy ball pein hammer

Metals form much of the structures in the world around us. Skyscraper buildings became a reality only after it became possible to make structural steel. The transportation industry depends on metal for vehicles, rails, and reinforcement for concrete roads. Even the high technology of electronics and computers relies extensively on copper wire, silver connectors, and other metal components.

Most of agricultural machines and equipment are made of steel and other metals. A successful career in many areas of agricultural mechanics depends upon a knowledge of metals and their uses in agriculture.

IDENTIFYING METALS

All metals can be classified as either ferrous or nonferrous. Ferrous metals come from iron ore. Nonferrous metals do not contain iron. The distinction is necessary because ferrous metals are used differently than nonferrous metals.

Most metals are not used in their pure metallic state. They are usually combined with one or more other metals. The combination of two or more metal elements is called an alloy. Alloys have characteristics that make them different from the original metals that were used to form the alloy. The alloy is made to improve the strength or some other quality of the original metals. For example, carbon is added to iron to make steel, which is tougher and more flexible than iron. Similarly, nickel and chromium may be mixed with iron to make nonrusting stainless steel. Figure 12-1 gives the characteristics and major uses of some common metals and alloys.

METAL	ORIGIN	CHARACTERISTICS	MAJOR USES
Cast iron	Iron ore	Forms into any shape; brittle	Machinery parts; engine blocks
Wrought iron	Iron ore	Malleable; tough; rust-resistant	Decorative fences; railings
Mild steel	Iron ore	Malleable; ductile; tough	Structural steel
Tool steel	Iron ore	High carbon; heat treatable; expensive	Tools; tool bits
Stainless steel	Iron ore, nickle, and chromium	Very corrosion resistant; bright appearance; hard; tough	Food handling equipment; milktanks; restaurant equipment
Galvanized steel	Steel; zinc	Zinc coated steel	Water tanks; towers; fencing; roofing; siding
Aluminum	Ore	Light; tough; relatively soft; good electrical conductor; silver-white color	Roofing; siding; truck bodies; automobiles; electric wires and cables
Copper	Ore	Tough; malleable; corrosion resistant; excellent heat and electrical conductor; reddish brown color	Pipe; electrical wire and cables; rain spouts and gutters; electrical equipment; bronze; brass
Brass	Copper and zinc	Soft; malleable; corrosion resistant	Water valves; boat accessories; ornaments
Bronze	Copper, zinc, and tin	Soft; malleable; corrosion resistant	Ornaments
Lead	Ore	Soft, very heavy; bluish gray	Batteries; cable coverings; shot; solder
Tin	Ore	Very malleable; corrosion resistant; silver color	Plating; bronze; solder

FIGURE 12-1. Common metals and alloys.

Ferrous Metals

Ferrous metals are mostly iron. Iron is refined from iron ore. Iron ore is particles of iron mixed with other minerals found in abundance throughout the world. Because of the abundant supply, iron is relatively inexpensive and, therefore, used extensively. However, ferrous metals rust very easily and must be protected by coatings that resist corrosion. Ferrous metals are easy to identify because they are magnetic and give off sparks when ground on an emery wheel.

Iron is the most useful metal for making tools and machinery. Its usefulness is increased with the addition of carbon to make steel. Further, when combined with certain other metals, iron becomes a very corrosion-resistant material called stainless steel.

Cast Iron. Cast iron is a type of iron that is grainy and not easily shaped or bent. It is not malleable. Malleable means workable. Parts with odd shapes may be made by pouring molten iron into a mold called a casting. When the metal cools and the form is removed, the item is said to be made of cast iron. Pure cast iron will break before it will bend.

Cast iron can be treated so it is somewhat workable. This is done by special heat treatments. The result is iron with a ductile outer layer. Ductile means that the metal can be bent slightly without breaking. The combination of the cast iron core and ductile metal outer layer is called malleable cast iron. Malleable cast iron is used extensively in farm and factory machinery.

Wrought Iron. Wrought iron is almost pure iron. It is very malleable. It can be bent, shaped, welded, drilled, sawed, and filed. Wrought iron also resists rust. These qualities make it a popular material for fences and other ornamental uses.

Mild Steel. Mild steel is the workhorse of metals. It is made by adding small amounts of carbon with iron. Mild steel is tough, strong, ductile, and malleable. It is rolled into many shapes such as flat bands, angles, channels, tees, I-beams, rods, and pipe, figure 12-2. Mild steel is used extensively in the agricultural mechanics shop.

Tool Steel. Tool steel contains a specific amount of carbon which permits it to be hardened. Tool steel can be annealed. Anneal means to heat a metal to the proper temperature and then slowly cool it. The process of annealing softens and toughens steel.

Tool steel can be hardened by heating to the proper temperature and then rapidly cooling the steel. The degree of hardness is determined by controlling the temperature of the metal and the speed of cooling after heating. Only tool steel can be tempered. Tempering is the process of carefully controlled reheating and cooling of steel after it has been shaped. Tempering results in a specified degree of hardness, relieves stress, and prevents cracking in steel. A tempered product is hard, tough, and not easily bent.

Tool steel can generally be identified by the exploding, sparkling nature of the sparks given off when the metal is ground, figure 12-3.

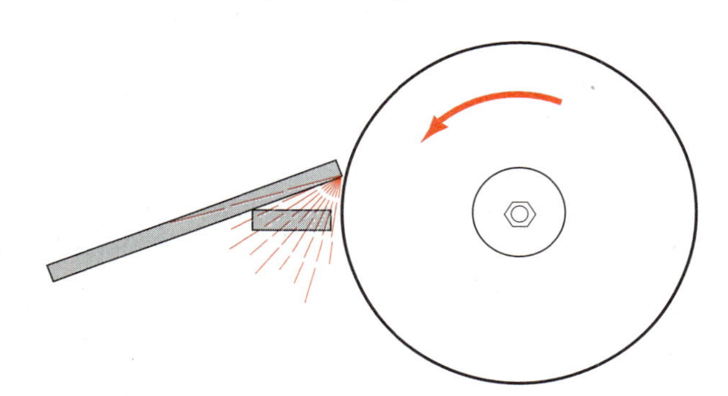

FIGURE 12-3. The carbon content of steel may be judged by the brightness and explosiveness of the sparks produced by a grinding wheel.

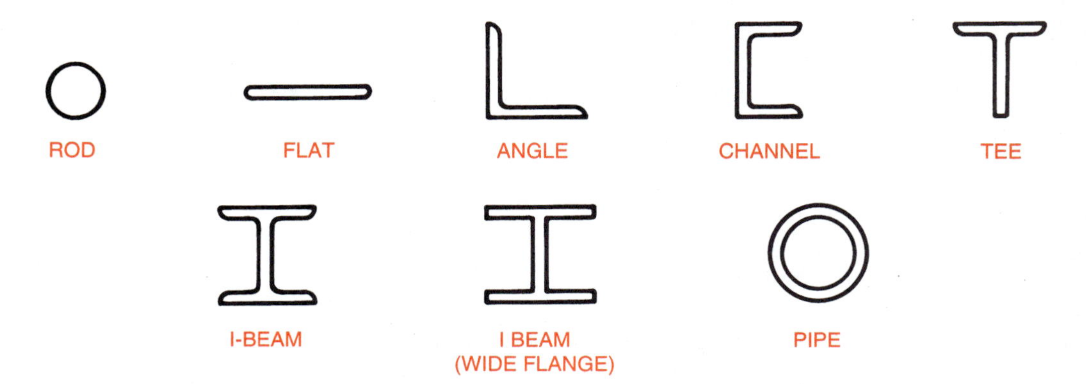

FIGURE 12-2. Mild steel is formed into many shapes for use in agriculture and industry.

Stainless Steel. Stainless steel is made by adding nickel and chromium to steel. It is very tough, will not rust, and resists corrosion of all types. Stainless steel is used extensively for milk tanks, kitchen equipment, and factory equipment where food is being processed and sanitation is necessary.

Plated Steel. Steel may be made rustproof by applying a coating of metal such as tin or zinc. At one time tin was used on rolled steel roofing material. This gave rise to the term *tin roof.* When a relatively thick coating of tin was used, the roof stayed bright for many years without rusting. When rust started to appear, the roof was painted. Today, most steel for roofs and siding is coated with zinc. Zinc-coated steel is called galvanized steel. The rust resistance of galvanized steel depends upon the thickness of the zinc coating. The thickness is rated in ounces of zinc per square unit of metal surface.

Galvanized steel sheets, pipe, buckets, and tanks must be handled with care in the shop. When heated or welded, the zinc gives off a poisonous gas. Therefore, such operations must be done with great care in a well-ventilated area. Also, galvanized buckets and containers can be ruined if the zinc finish is damaged by cracking or chipping, by dropping, or other abuse.

Nonferrous Metals

Nonferrous metals and their alloys do not contain iron. They are more expensive than ferrous metals on a per weight basis and their supply is rather limited. However, nonferrous metals are needed and used extensively in industry and agricultural mechanics. Nonferrous metals include aluminum, copper, lead, tin, and zinc. Some nonferrous metals are light in weight.

Aluminum. Aluminum is identified by its toughness, light weight, and silver-white color. The combination of toughness and light weight makes aluminum the first choice for truck bodies, aircraft, ladders, equipment, containers, roofing, and siding.

Aluminum is a very good electrical conductor. Therefore, aluminum wire is used for high-voltage lines and often used when copper wire becomes expensive. Aluminum resists corrosion. Corrosion is the reaction of metals to liquids and gases that cause them to deteriorate or break down. Aluminum is often painted to provide choices in color when used in doors, siding, gutters, awnings, and toys.

Pure aluminum is very soft and malleable. Other materials, such as silicon, manganese, and magnesium, are added to give aluminum more strength and other qualities. Special techniques must be used to weld aluminum.

Copper. Copper is an excellent electrical conductor and does not corrode much in air or water. Therefore, copper is used extensively in electrical wiring, electric motors, and plumbing systems.

Copper makes up the bulk of the metal in the alloys brass and bronze. Brass is made of 60 to 90 percent copper; the rest is zinc. Bronze is copper and zinc plus about 10 percent tin. These alloys have been used for centuries. They are used extensively today for their decorative and corrosion-resistant qualities.

Lead. Lead is a very soft and heavy metal. It is used in the manufacture of batteries, lead pipe, wire coverings, and in combination with tin to make solder. However, lead solder used in drinking water systems may contribute to lead poisoning and is not permitted by building codes. Solder is a soft alloy used to join many kinds of metal. Lead is also used in washers on roofing nails to provide a watertight seal between the nail and roofing.

Tin. Tin is used to coat steel for temporary rust protection. It is also a component of solder and bronze.

Tin has long been used to coat steel used in cans, but the coating generally provides protection for only a short period of time. Steel, coated with tin, does not rust until it is discarded outdoors. The tin coat then wears away quickly and the steel rusts.

MARKING METAL

Steel is the most commonly used metal in agricultural mechanics. Therefore, most metalworking skills will focus on steel. The student should realize, however, that most procedures used in working with steel apply to other metals as well.

Tools for marking metal and wood are discussed in the unit on layout tools and procedures. The main difference between marking wood and marking metal is the instrument used for marking. Metals are either very shiny or very dark in color. Therefore, a lead pencil mark does not show up well. The scratch awl, scriber, file, soapstone, dividers, and center punch are tools used to mark metal.

Scratch Awl. The scratch awl, or simply awl, is a tool with a wood or plastic handle and a pointed steel shank. It is held in the hand and used like a pencil. The awl is used to scratch the surface of metal to make a mark, figure 12-4.

Scriber. The name scriber is generally given to very small, metal, sharp-tipped markers. Many combination squares have a scriber in the handle that can be removed and used for marking metal.

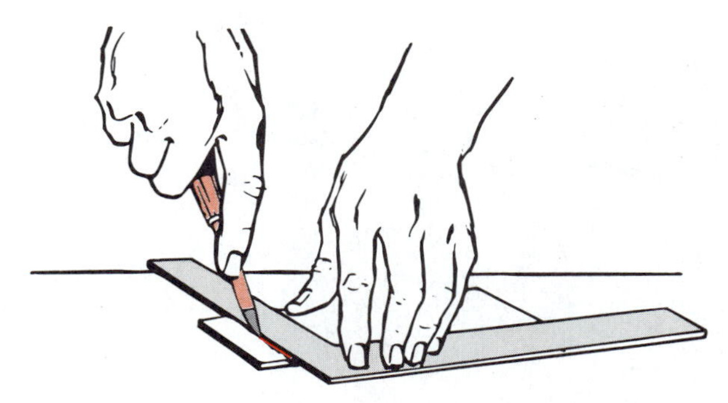

FIGURE 12-4. Marking metal with a scratch awl.

Files. Any file will work for marking metal. The sharper the edge or corner, the finer and more accurate the mark can be. To mark with a file, the sharp corner of the file is pushed over the flat side or the corner of the metal. The file works better on structural steel than it does on plate or sheet steel.

Soapstone and Chalk. Effective markers are made by cutting soapstone into thin pieces resembling pencils. Soapstone is a soft, gray rock that shows up well on most metals, figure 12-5. It can be wiped off, but not easily. Soapstone can be sawed to a knifelike edge, if desired. This sharp edge makes a narrow line on metal.

Common chalk can also be used to mark metal. However, chalk will not leave a narrow mark and it brushes off easily.

Dividers. Dividers have two steel legs with sharp points that are used to make arcs and circles on metal. The tool is also useful for transferring measurements from one item to another. A third important use is to divide any measurement into two equal parts, hence the name dividers.

FIGURE 12-5. Soapstone can be used to mark metal.

Center Punch. The center punch is a steel punch with a sharp point. The center punch is used to make a small indentation or hole in metal. This hole is used to help start a twist drill. The procedure for using the center punch is as follows.

PROCEDURE

1. Place the point of the center punch exactly where the center of the hole should be.
2. Hold the punch straight.
3. Give the punch a light blow with a steel hammer.
4. Check to see if the dent is exactly where the center of the hole should be.
5. Place the punch in the dent and give the punch one solid blow with the hammer.

CUTTING METAL

Metal can be cut or formed by sawing, shearing, filing, or grinding. Since metals are harder to work than wood, the choice of tools is somewhat more limited for working metal. The hacksaw, file, snips, and cold chisel are tools used to cut metal.

The Hacksaw

The tool used most often for cutting metal is the hacksaw, figure 12-6. The term hacksaw means a device that holds a blade designed for cutting metal.

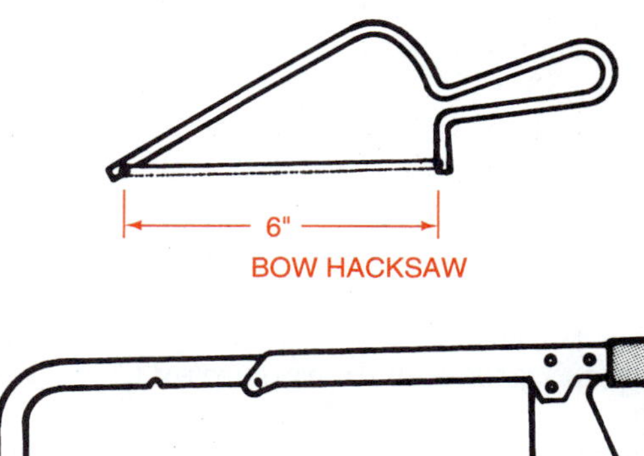

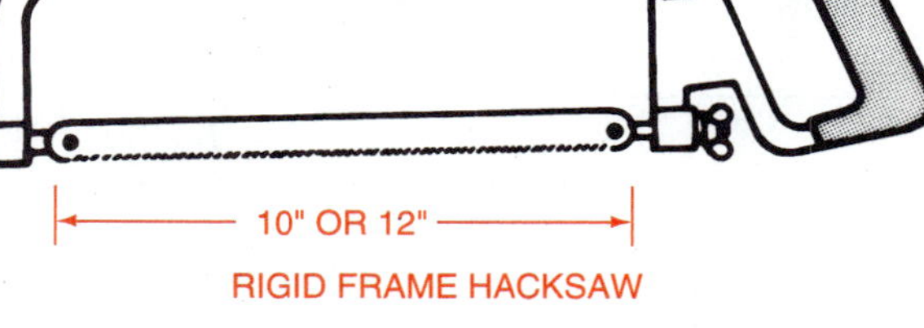

FIGURE 12-6. Commonly used hacksaws. *(Courtesy of Stanley Tools)*

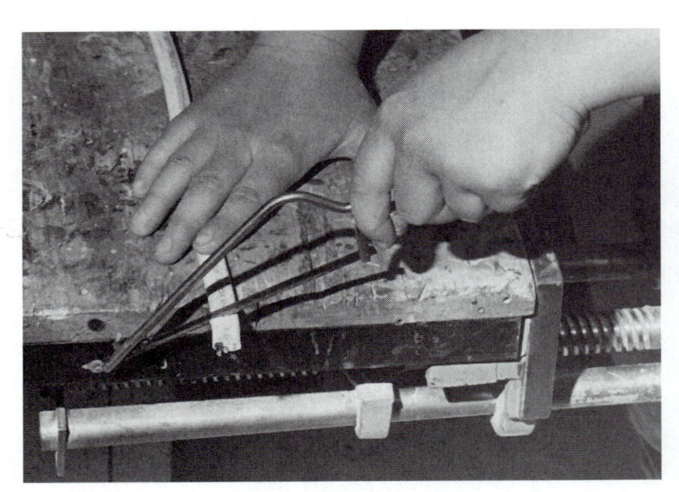

FIGURE 12-7. The bow-shaped hacksaw is often used by electricians to cut heavy wire.

FIGURE 12-9. Place the metal to be sawed 1/2 inch from the vise jaws. *(Courtesy The L. S. Starrett Company)*

Such tools may be as simple as a handle that grips a common hacksaw blade. Another type of hacksaw is shaped like a miniature bow saw and holds a 6-inch blade with very fine teeth. Electricians favor such saws for cutting heavy copper and aluminum wire, figure 12-7.

When using a hacksaw, a blade should be selected that is fine enough to allow at least three teeth at a time on the metal, figure 12-8. Standard hacksaw blades come in 10-inch or 12-inch lengths. They may be bought with 14, 18, 24, or 32 teeth per inch. The blade with 32 teeth per inch has very small teeth and should be used to cut very thin metal. The thicker the metal, the coarser the teeth should be so the blade will cut faster.

Hacksaws are designed to cut on the forward stroke. Therefore, blades must be installed with the teeth pointing away from the handle. When using hacksaws, a small notch should be filed to help start the cut. The blade is pulled backwards and then pushed forward with slight pressure. After starting the cut, long full strokes with pressure on the forward stroke should be used. No pressure is used on the backward stroke.

Metal must be held tightly when sawing to prevent it from flexing or twisting and to prevent damage to the hacksaw blade. The best way to secure metal for sawing is to use a metal vise. The metal should be placed in the vise so the cut will be only 1/2 inch from the vise jaws, figure 12-9. The metal should be positioned so that three or four teeth are cutting at one time.

A used hacksaw blade should not be replaced after the cut has been started. A new blade will have a wider set in its teeth. When a new blade is forced down into the kerf left by an old blade, the teeth are damaged. To avoid such damage, a cut with the new blade should be started on the opposite side of the metal.

Band saws, power hacksaws, and saber saws all may be used to saw metal. However, the correct blades must be used. The use of power saws with hacksaw blades is discussed in another unit.

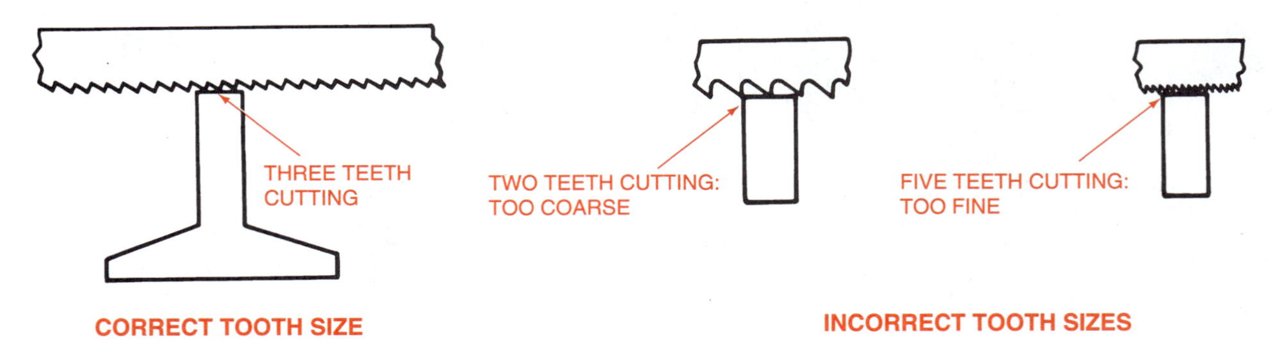

FIGURE 12-8. When sawing with a hacksaw, three teeth at a time should be on the metal. If less than three teeth are in use, the teeth will be stripped off the blade. If more than three teeth are being used, a lot of pressure will be needed to push the saw forward.

Files

Files are classified by length, shape, design, teeth, and coarseness of teeth, figure 12-10. Files used in the agricultural mechanics shop are generally 8 inches to 12 inches long. The shop usually has one or more of each of the various file shapes (triangular, half-round, round, and flat) so any kind of filing job can be done.

If a file has straight teeth going in one direction, it is said to be a single cut file. If a file has teeth going in two directions, it is a double cut file. A double cut file cuts faster than a single cut file. Single cut and double cut files can be purchased in bastard (coarse), second cut (medium), and smooth (fine) grades.

The rasp cut file has raised, sharp, individual teeth. Rasps are used for shaping wood, horses' hooves for shoeing, and soft metal. They are not suitable for steel.

The curved tooth file has teeth that follow a half-round pattern. The teeth are very coarse. The curved tooth file is designed to cut soft metals such as aluminum and copper.

Filing Metal. When filing metal, every effort should be made to secure the metal firmly, figure 12-11. Filing will cause metal to vibrate if it is not held tightly. A file cuts on the forward stroke only. Filing may be done across metal or along its length.

To file, the point of the file is grasped between the thumb and index finger of one hand. The handle is gripped with the other hand. The file is pushed forward while light pressure is applied. The file should be slightly lifted and brought back to the starting position.

Drawfiling. Drawfiling is done by placing the file at a 90 degree angle to the metal and pushing or

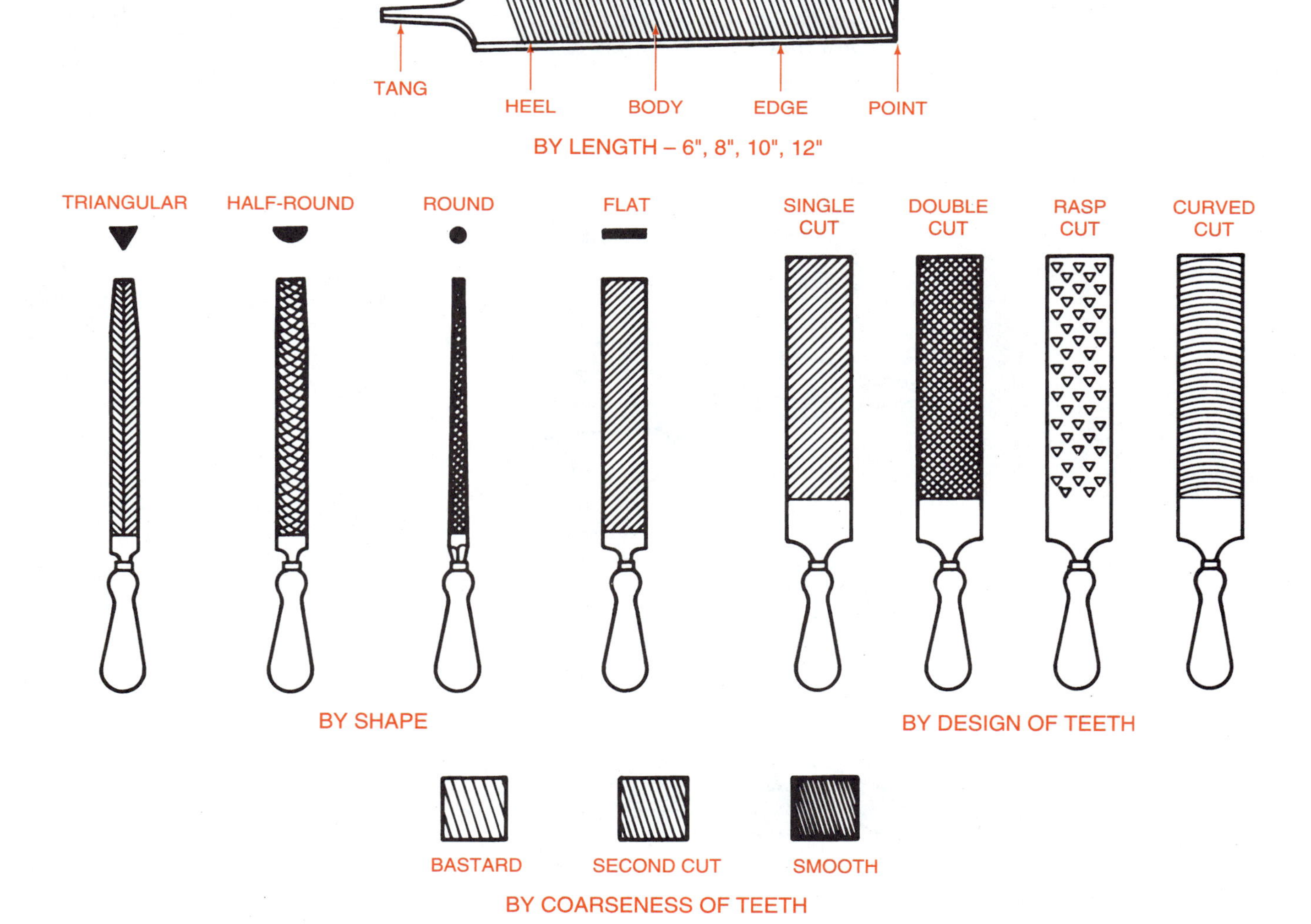

FIGURE 12-10. Files are classified according to the length and shape of the file, the design of the teeth, and the coarseness of the teeth.

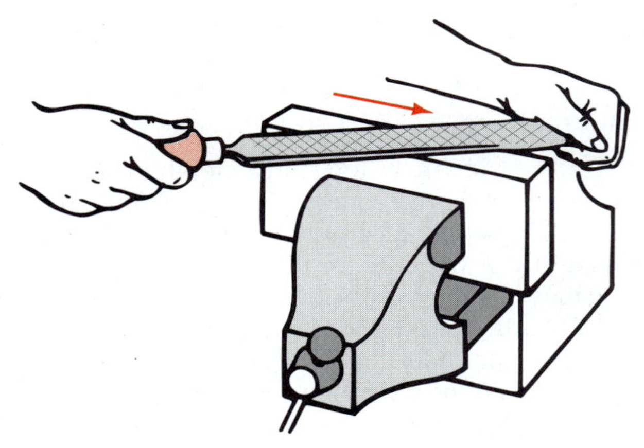

FIGURE 12-11. When filing metal, the metal should be held tightly in a vise.

pulling the file in this 90 degree position. Pressure is applied with a push or pull movement of the file over the length of the stock. Stock refers to a piece of material such as wood or metal.

Cleaning Files. Files are cleaned by tapping the handle sharply on the bench top. Another cleaning method is to use a file card or wire brush when particles are difficult to remove.

Snips and Shears

The terms snips and shears are terms used for the same tool. Snips or shears are large scissorlike tools for cutting sheet metal and fabrics. They may be purchased for cutting straight, left-hand, or right-hand curves. Regular snips are designed like scissors. Snips for heavy cutting have compound handles and are also called aviation snips, figure 12-12.

To cut sheet metal, a pair of snips that are sharp and free of nicks is selected. A piece of scrap should be cut first to be sure the snips are heavy enough. The snips are used like scissors. For snips to work well, the metal must curl or lift up and out of the snips as the cut progresses. Since curled metal is sharp and will cut, it is advisable to wear gloves when handling sheet metal. If cutting is difficult when using regular snips, let the lower handle rest on the bench top. Exert extra pressure by pushing down on the top handle, figure 12-13.

Cold Chisels

A cold chisel is a piece of tool steel shaped, tempered, and sharpened to cut mild steel when driven with a hammer. A regular cold chisel has a wide cutting edge. Special chisels, such as the cape, round

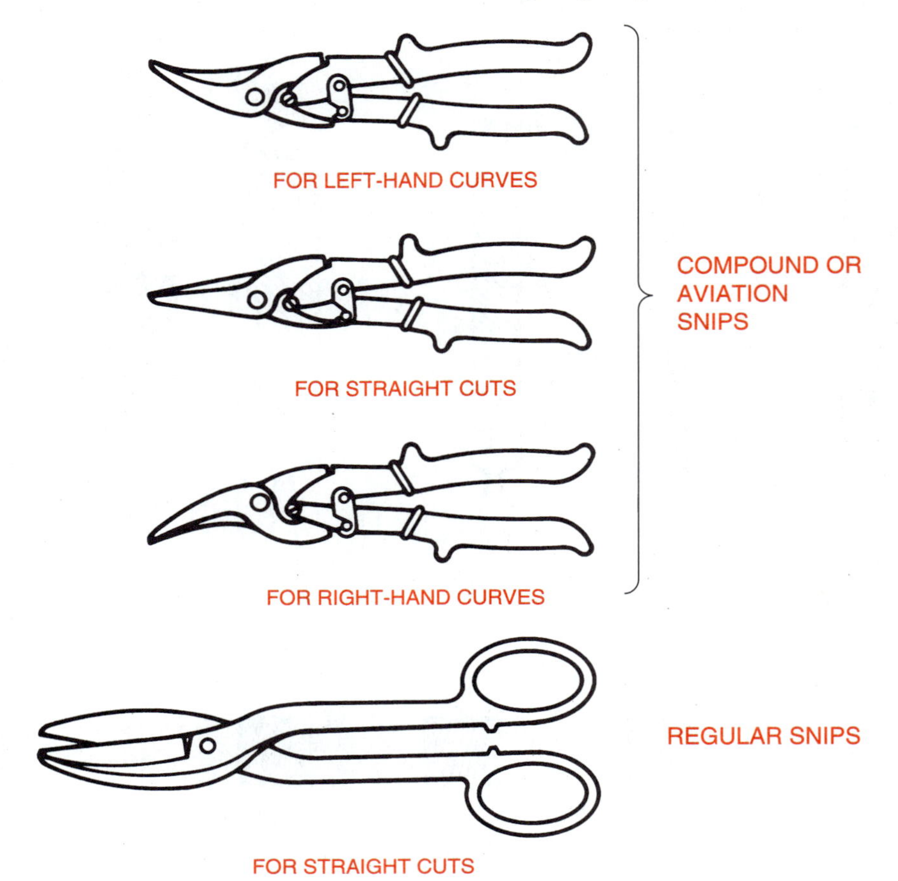

FIGURE 12-12. Several types of snips are available. Compound or aviation snips provide extra leverage for heavy cutting.

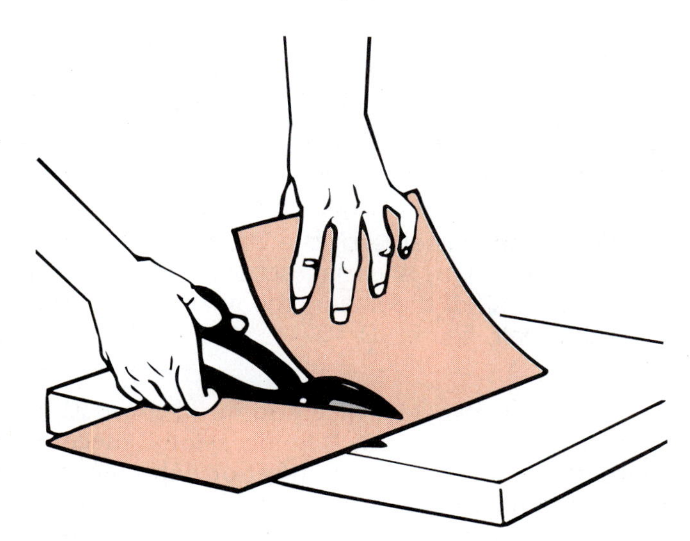

FIGURE 12-13. Snips are used for cutting and shaping sheet metal.

nose, and diamond point, have cutting edges designed to cut grooves, figure 12-14.

CAUTION:

- **It is especially important to wear goggles when doing metalwork. A chisel or punch with a mushroomed head should not be used. Mushrooming can be corrected by reshaping the head using a grinder.**

Cold chisels are used to cut mild steel rods and bands, and to cut rusted nuts and bolts. A cold chisel may also be used to loosen nuts that have rounded shoulders. The procedure for loosening such nuts follows.

PROCEDURE

1. Saturate the nut with a rust solvent.
2. Position the cutting edge of the chisel on a shoulder of the nut.
3. Angle the chisel so it will turn the nut counterclockwise (backwards) when struck.
4. Drive the chisel with a ball pein or blacksmith's hammer.

When using the cold chisel to cut nuts, the following procedure is recommended.

PROCEDURE

1. Select a sharp chisel.
2. Examine the way in which the bolt and nut are mounted.
3. Place the chisel on the nut so it will be driven against a solid structure. The nut can be split by cutting downward parallel to the bolt. The nut may also be cut into a right angle to the bolt.

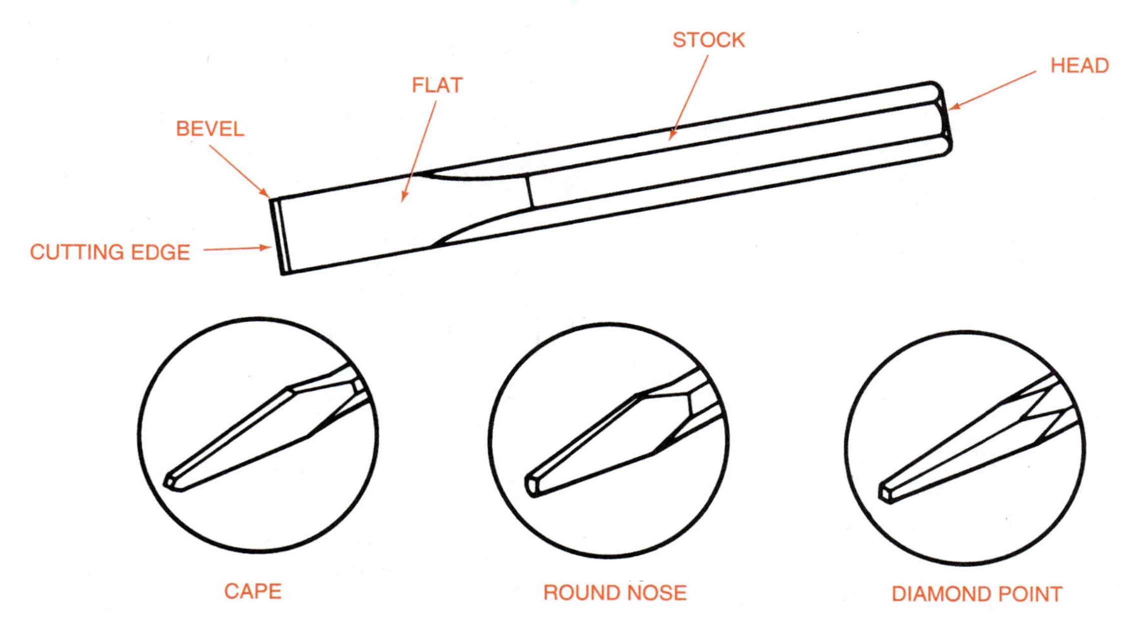

FIGURE 12-14. Types of cold chisels.

4. Cut through the nut as far as possible.
5. Change the position of the chisel and drive it in such a way as to turn the nut off the bolt. The nut should either turn off or split off the bolt.

Sometimes it is easier to cut a bolt than to cut a nut. This may be true when the bolt is 3/8 inch or smaller. For best results, the bolt should be tight in its hole. The procedure for cutting a bolt follows.

PROCEDURE

1. Place a sharp chisel at the base of the nut and at a right angle to the bolt.
2. Drive the chisel so it is forced under the nut. The lifting action of the chisel should cause the bolt to be exposed under the nut.
3. Continue driving until the bolt is cut.

The hacksaw is the best tool for cutting mild steel. However, there are times when the cold chisel may be preferred. When cutting rod or flat iron, the following procedure is recommended.

PROCEDURE

1. Mark the metal where it is to be cut.
2. Place the steel in a heavy vise. The cutting mark should be even with the jaws of the vise.
3. Lay a heavy cold chisel on top of the moveable jaw of the vise. The cutting edge of the chisel is to rest against the steel to be cut. If cutting flat iron, the cutting edge of the chisel should be placed on the corner of the iron.
4. Hold the chisel so it rests at about a 30 degree angle to the bench top. The flat of the chisel should be parallel to the bench top.
5. Drive the chisel until the metal is cut. The metal is cut by shearing action. The vise acts as one-half of the shear; the cold chisel is the other half, figure 12-15.

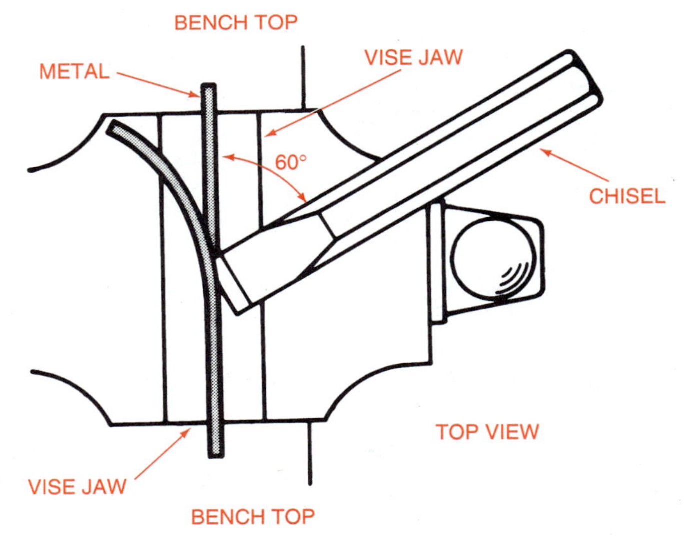

FIGURE 12-15. When cutting with a cold chisel, drive the chisel so it acts as the second half of a pair of shears.

Always use special face and hand protection when using driving tools. The occasional piece of metal that may fly from a chisel or hammer may become embedded in an eye or other parts of the face. Hands may be bruised or seriously injured by a chisel that slips or a hammer that misses its mark. Therefore, every precaution should be taken to protect the body.

BENDING METAL

Mild steel can be bent cold in sizes up to 1/2 inch square, 1/2 inch round and 3/16" × 1" flat. A vise anchored to a heavy bench makes the bending of cold metal possible. Metal can be twisted or bent for ornamental purposes such as for porch railings. Metal may also be bent at various angles to fit some object or to finish an assembly. Round metal may be bent to form an eye. An eye is a piece of metal bent into a small circle.

Bending Flat Metal

To bend a flat piece of metal up to one inch wide, or a rod up to 1/2 inch in diameter, the following procedure is recommended.

PROCEDURE

1. Mark the metal where the bend is to occur.
2. Place the metal in a vise with the mark at the top of the vise jaws. It is best to have the longest part of the metal extending above the vise.
3. Tighten the vise securely.
4. Hold a heavy ball pein or blacksmith's hammer in one hand.
5. Grasp the metal near its end with the other hand.
6. Push the metal with one hand while hitting it with the hammer near the vise. Coordinate the movements so the metal bends sharply at the vise and not above it, figure 12-16.

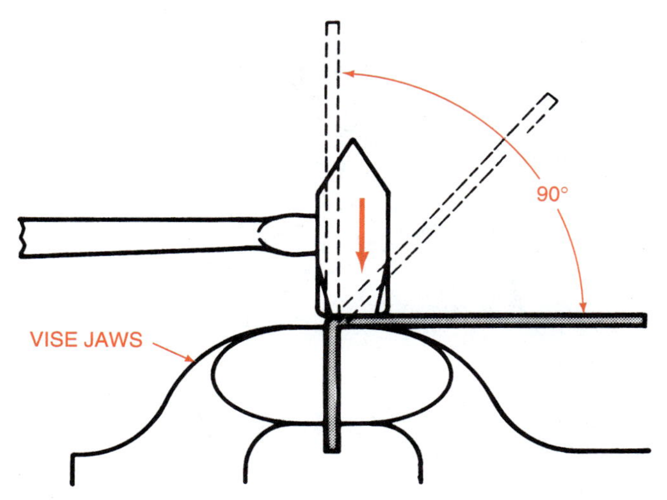

FIGURE 12-16. Metal should bend sharply at the vise.

CAUTION:

- **Wear a glove or hold the chisel with a cloth to protect the hand.**

Rounding Metal

Metal can be rounded by using a vise and a piece of round stock or pipe. The procedure for rounding metal is as follows.

PROCEDURE

1. Select a piece of round stock or pipe with a diameter equal to the desired bend.
2. Place the end of the material to be bent tightly between the round stock and a vise jaw, figure 12-17A.
3. Using a hand and a hammer, bend the metal around the round stock.
4. Open the vise and move the material around the round stock, figure 12-17B.
5. Tighten the vise.
6. Bend and hammer the material to fit the round stock.
7. Repeat the above steps until the part is bent to the desired shape.
8. Reposition the metal as needed and use the closing power of the vise to tighten and shape the bend.

Using an Anvil. An anvil is a heavy steel object that is used to help cut and shape metal, figure 12-18. Metal may be rounded by holding the long part of the metal in one hand and bending the short end of the metal over the horn of an anvil, figure 12-19.

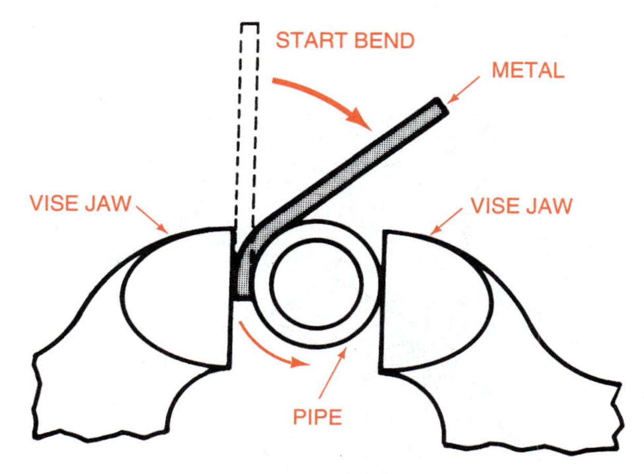

A. PLACE METAL TIGHTLY BETWEEN THE PIPE AND THE JAWS OF THE VISE.

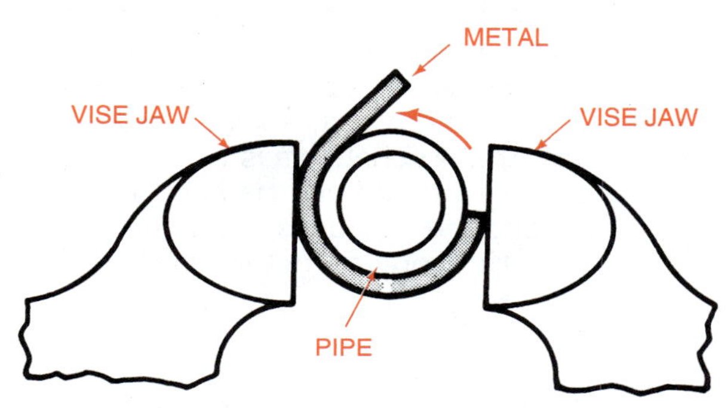

B. MOVE THE MATERIAL AROUND THE ROUND STOCK AND CONTINUE THE BENDING PROCESS.

FIGURE 12-17. Rounding metal using a pipe and a vise.

Twisting Metal

Another useful procedure in agricultural mechanics is the twist. To make a 90 degree twist in flat iron, the following procedure is used.

PROCEDURE

1. Mark the metal where the twist should start.
2. Measure up the metal a distance equal to 1 1/2 times the width of the metal and make a second mark.
3. Place the metal in a vise with the twist mark at the top of the vise jaws.
4. Take a 10-inch or 12-inch adjustable wrench and adjust it to fit the thickness of the metal.
5. Place the wrench on the metal so the jaws extend the width of the metal. Move the wrench up the metal until the bottom of the wrench is on the upper line.

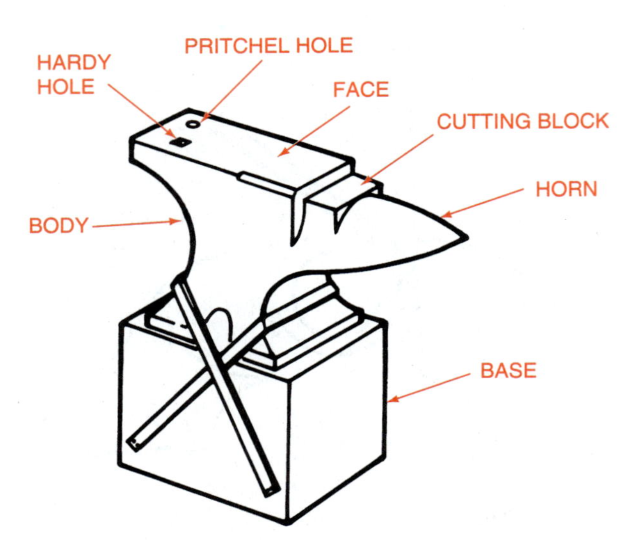

FIGURE 12-18. The parts of an anvil. The tapered square shank of an anvil chisel, or hardy, fits into the hardy hole. Metal is placed on the cutting edge of the hardy and is struck with a hammer to cut the metal. The pritchel hole is used to help punch holes in metal. Red hot metal is placed over the hole and the round, tapered part of the punch is driven through the metal and into the hole.

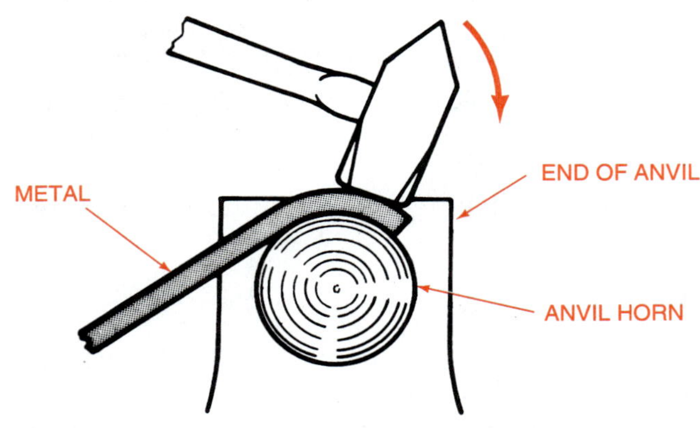

FIGURE 12-19. Using the horn of the anvil and a blacksmith hammer to round metal.

6. Hold the metal straight up in the air with one hand while turning the wrench with the other, figure 12-20.
7. Stop when the upper section of the metal is at 90 degrees to the lower section.

It may be necessary to move the wrench up or down the metal. Move it up if the metal starts to shear at the vise. Move it down if the bend is too long or gentle.

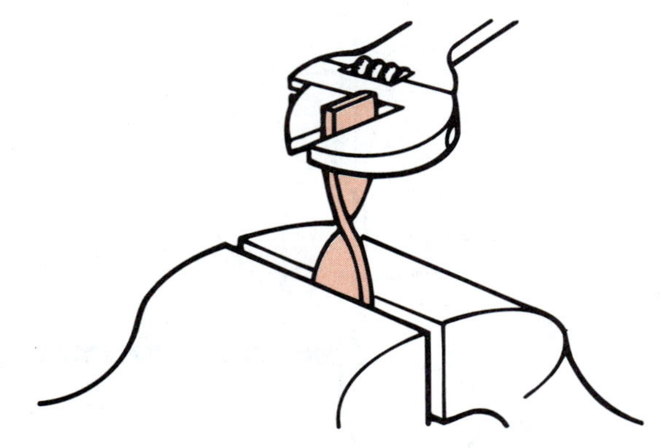

FIGURE 12-20. Twisting flat iron using an adjustable wrench.

Bending Sheet Metal

Sheet metal is relatively easy to bend. Commercial benders are available. However, most agricultural shops use hand tools for this task. To bend sheet metal without special tools, the following procedure is used.

PROCEDURE

1. Mark the metal where the bend should occur.
2. Find a bench with a sharp 90 degree angle at the edge. If the edge is not sharp, place a 90 degree piece of angle iron over the edge and clamp the angle to the bench top.
3. Place the sheet metal so the metal to be bent extends over the bench. Be sure the bend line is directly over the edge of the bench.
4. Push the metal down along the edge to start the bend.
5. Complete the bend by tapping the metal with a hammer along the length of the bend line, figure 12-21.

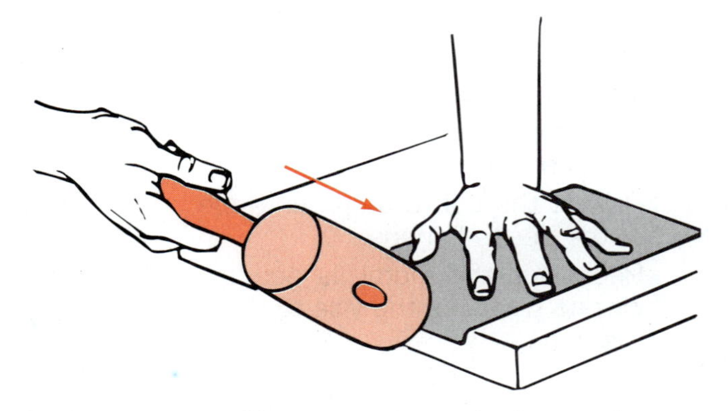

FIGURE 12-21. Sheet metal may be bent using a hammer or mallet and the edge of a bench top.

For small projects, place an angle iron or heavy piece of flat steel of the suitable size in a vise. Form the metal using the steel to support the metal as it is hammered and shaped, figure 12-22.

CAUTION:

- **Use goggles for all metalworking procedures.**

CAUTION:

- **Warn others to stay away from the area, since the end of the bolt and the nut can become a projectile if driven hard.**

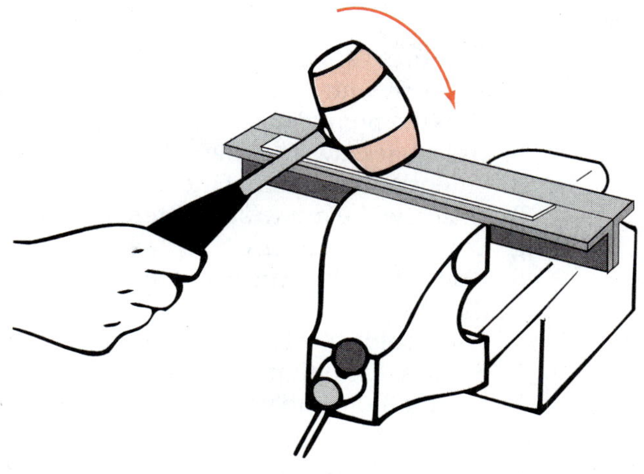

FIGURE 12-22. Two pieces of angle iron in a vise make a good devise for bending sheet metal.

STUDENT ACTIVITIES

1. Define the Terms to Know in this unit.
2. With the use of a magnet, classify the following items as ferrous or nonferrous. Determine what metal or metals are in the items.
   - Penny
   - Quarter
   - Door knob
   - Door hinge
   - Framing square
   - Automatic chalker
   - Piece of electric wire
   - Wrench
   - Steel water pipe
3. Break a piece of cast iron that has been discarded. Describe the appearance of the metal exposed by the break.
4. Examine a combination square. Does it have a scriber? If so, remove, examine, and replace it.
5. Examine five or six different files. Indicate the shape and type of teeth for each file.
6. Cut and bend flat steel.
   a. Obtain a piece of $^1/_8$" × 1" flat mild steel, 6 inches long.
   b. Measure 1 inch in from one end.
   c. Use a square and scratch awl to make a line across the metal.
   d. Saw off 1 inch using a hacksaw. Check the work for squareness.
   e. Again, measure 1 inch from the end.
   f. Square a line across the metal with soapstone. (If soapstone is not available, use chalk).
   g. Shear off 1 inch using a vise and cold chisel.
   h. Again, measure 1 inch in from one end. Square across using a scriber or scratch awl.
   i. Measure in $1^1/_2$ inches from the line, and square a line across.
   j. Place a 90 degree twist between the two lines using a vise and adjustable wrench.
   k. Secure 1 inch of the piece of metal in the vise and make a 90 degree bend.

7. Cut bolts.
   a. Obtain a 1/4-inch stove bolt, 2 inches or more in length, and a 1/4-inch nut.
   b. Screw the nut on the bolt about 1/2 inch.
   c. Grip the nut tightly in a machinist's vise with the bolt in a horizontal position.
   d. Use a hacksaw and saw off the part of the bolt that extends beyond the nut.
   e. Smooth up the end of the bolt with a fine file.
   f. Unscrew the nut to straighten the threads.
   g. Screw the nut back on the bolt about 1/2 inch.
   h. Place the bolt vertically in the vise with the nut resting against the top of the vise. Tighten the vise.
   i. Use a cold chisel to shear off the nut. Drive the chisel slowly.
   j. Replace the bolt in the vise leaving about 1/2 inch extended above the vise.
   k. Shear off the end with a cold chisel.
8. Shear and bend sheet metal.
   a. Lay out a piece of sheet metal 4 inches square.
   b. Cut out the 4-inch square sheet metal with snips.
   c. Use a combination square and scratch awl to lay out a line 1 inch from one side of the sheet metal. Extend the line from edge to edge.
   d. Set a dividers with the legs 1 inch apart. Place one leg of the dividers on the line 1 inch from the end. Make an arc across the corner of the metal.
   e. Repeat step d on the other end of the line.
   f. Round the two corners with snips.
   g. Using a hammer and the edge of the bench, make a 90 degree bend on the line laid out in step c.

SELF-EVALUATION

A. Multiple Choice. Select the best answer.

1. When using a hacksaw, the number of teeth cutting at one time should be
   a. one
   b. two
   c. three
   d. five
2. Another name for snips is
   a. aviation
   b. combination
   c. scissors
   d. shears
3. Standard hacksaw blades come in lengths of
   a. 6 and 10 inches
   b. 7 and 10 inches
   c. 9 and 11 inches
   d. 10 and 12 inches
4. Choice of teeth per inch for hacksaw blades are
   a. 14, 18, 24, and 32
   b. 14, 20, 26, and 32
   c. 14, 24, and 36
   d. none of these
5. A hacksaw cuts on
   a. the backward stroke only
   b. the forward stroke only
   c. both the forward and backward strokes
   d. the stroke recommended by the manufacturer
6. Single cut and double cut refer to
   a. teeth design on files
   b. teeth on a hacksaw blade
   c. speed designations for blades
   d. width of saw kerfs
7. Placing a file at a 90 degree angle to the metal and pushing or pulling is
   a. burnishing
   b. drawfiling
   c. push filing
   d. none of these
8. It is very important to wear face protection when using a chisel because
   a. a chip of metal may hit you
   b. the hammer head is likely to come off
   c. chisels frequently break in half
   d. sight is improved by good face shields

B. Matching. Match the shape of the steel in column I with the correct name in column II.

Column I	Column II
1.	a. channel
2.	b. tee
3.	c. pipe
4.	d. rod
5.	e. I-beam
6.	f. flat
7.	g. angle

C. Matching. Match the description in column I with the type of metal in column II.

Column I

1. malleable iron
2. flats, I-beams, pipe
3. zinc on steel
4. any shape; brittle
5. electrical wire; pipe
6. batteries
7. alloy of copper, zinc, and tin
8. plating for steel cans
9. alloy of copper and zinc
10. light, tough, roofing, and siding
11. food equipment, shiny
12. chisels, punches

Column II

a. cast iron
b. wrought iron
c. mild steel
d. tool steel
e. stainless steel
f. galvanized steel
g. aluminum
h. copper
i. brass
j. bronze
k. lead
l. tin

D. Completion. Fill in the blanks with the word or words that will make the following statements correct.

1. A soft, gray rock that is used to mark metal is ____________.
2. A tool with a handle and a pointed steel shank used to mark metal is the __________.
3. Ferrous metals are easily identified with a __________.
4. Thin metal may be bent into curves and circles by using a piece of __________ in a vise.
5. A 90 degree twist may be made in flat iron by using a vise and an __________ __________.

UNIT

Fastening Metal

OBJECTIVE

To fasten metals using procedures frequently used in agricultural mechanics.

COMPETENCIES TO BE DEVELOPED

After studying this unit, you should be able to:

- Drill holes in metal.
- Tap threads in holes.
- Cut threads on bolts and pipe.
- Fasten metal with bolts and screws.
- Fasten metal with rivets.
- Solder sheet metal.
- Sweat copper pipe.

TERMS TO KNOW

soldering
brazing
welding
high-speed drill
jobbers-length drill
tap
tap wrench
die
die stock
tap and die set
screw plate
taper tap
plug tap
bottoming tap
bolt threads
pipe fittings
teflon tape
pipe joint compound
rivet
pop rivet
solder
soldering
flux
tinning
50-50 solder
hollow-core solder
flux core solder
soldering copper
sal ammoniac
sweating

MATERIALS LIST

- ✓ Center punch
- ✓ Set of twist drills
- ✓ Drill press or power hand drill
- ✓ Tap and die set or selected parts
- ✓ Oil for cutting
- ✓ Rags
- ✓ Pipe cutter
- ✓ Tubing cutter
- ✓ Pipe wrenches, adjustable wrench, and screwdriver
- ✓ Pipe reamer
- ✓ Pipe dies and stock
- ✓ A length of steel pipe—1/2 inch, 3/4 inch, or 1 inch diameter and 1 foot or more in length
- ✓ A pipe fitting for the length of steel pipe
- ✓ A length of copper tubing or pipe—3/8 inch or 1/2 inch in diameter and 6 inches or more in length
- ✓ Copper fitting for the length of copper tubing or pipe

- Metal countersink
- Sheet metal screws
- Stove bolt, 5/16" × 1/2" long
- Steel rivets, round head and countersink head
- 50-50 solder and flux
- Metal fruit or vegetable can
- One foot of #12, #14, or #16 stranded, covered copper wire
- Hot rolled band iron—1/2" × 1" × 18" long
- Cold rolled round stock, 3/8 inch diameter × 3 inch long
- Soldering copper and/or soldering gun
- Teflon tape or pipe compound

Metals are fastened with various devices such as screws, bolts, and rivets. Fasteners are needed to assemble machines, equipment, tractors, tools, and buildings. Screws are used for things such as holding the metal panels on appliances and stampted sheet metal parts on machinery. Bolts are used to hold metal parts together on engines and transmissions. Rivets may be used to hold the rain gutters and downspouts together on garages or barns. Rivets are also used in certain large bridges and steel buildings.

Metal pieces are also fastened together by melting a second metal between two pieces. This procedure is called soldering or brazing, depending on the material used. The copper pipes in houses, schools, and other buildings are held together with solder.

A third important way to hold metals together is to simply melt them together. This is called welding. A filler rod may be used to add metal to strengthen the bond.

MAKING HOLES IN METAL

Holes are drilled in metal to permit screws to pass through a shank hole in the first piece. A pilot hole is needed in the second piece to accommodate the threaded part of the screw. In sheet metal, the pilot hole in the second piece is drilled so that the solid core of the screw will pass through. The threads of the screw hold by straddling the metal.

Holes are also drilled in metal to receive bolts and rivets. When threads are needed in a hole, the hole must be of a very specific size; otherwise, it cannot be threaded.

FIGURE 13-1. A drill press is commonly used to drill holes in metal. *(Courtesy of S & R Photo Acquisitions)*

Tools for Drilling Holes

Holes are usually drilled in metal with a portable electric drill or power drill press, figure 13-1. Some farm shops may still have a blacksmith's post drill. This is a hand-powered machine that is very effective for drilling metal.

Only high-speed twist drills are recommended for drilling metal. A high-speed drill is a drill that is hardened for use in metal. Such drills may be purchased in regular length, or longer length known as jobbers-length drills. It is especially important to use eye protection when drilling wood or metal.

Drilling Holes in Metal

To drill a hole in metal, the following procedure is used.

CAUTION:

- **It is especially important to wear eye protection for all metalworking procedures.**

PROCEDURE

1. Use a center punch. Make a small dent to guide the center of the drill.
2. Secure the metal by clamping it to the drill press table, holding it in a metal vise, or resting it against the post of the drill press, figure 13-2.
3. Use a power drill that turns at the correct speed for the size of the drill being used (fast for small drills and slow for large drills), figure 13-3.

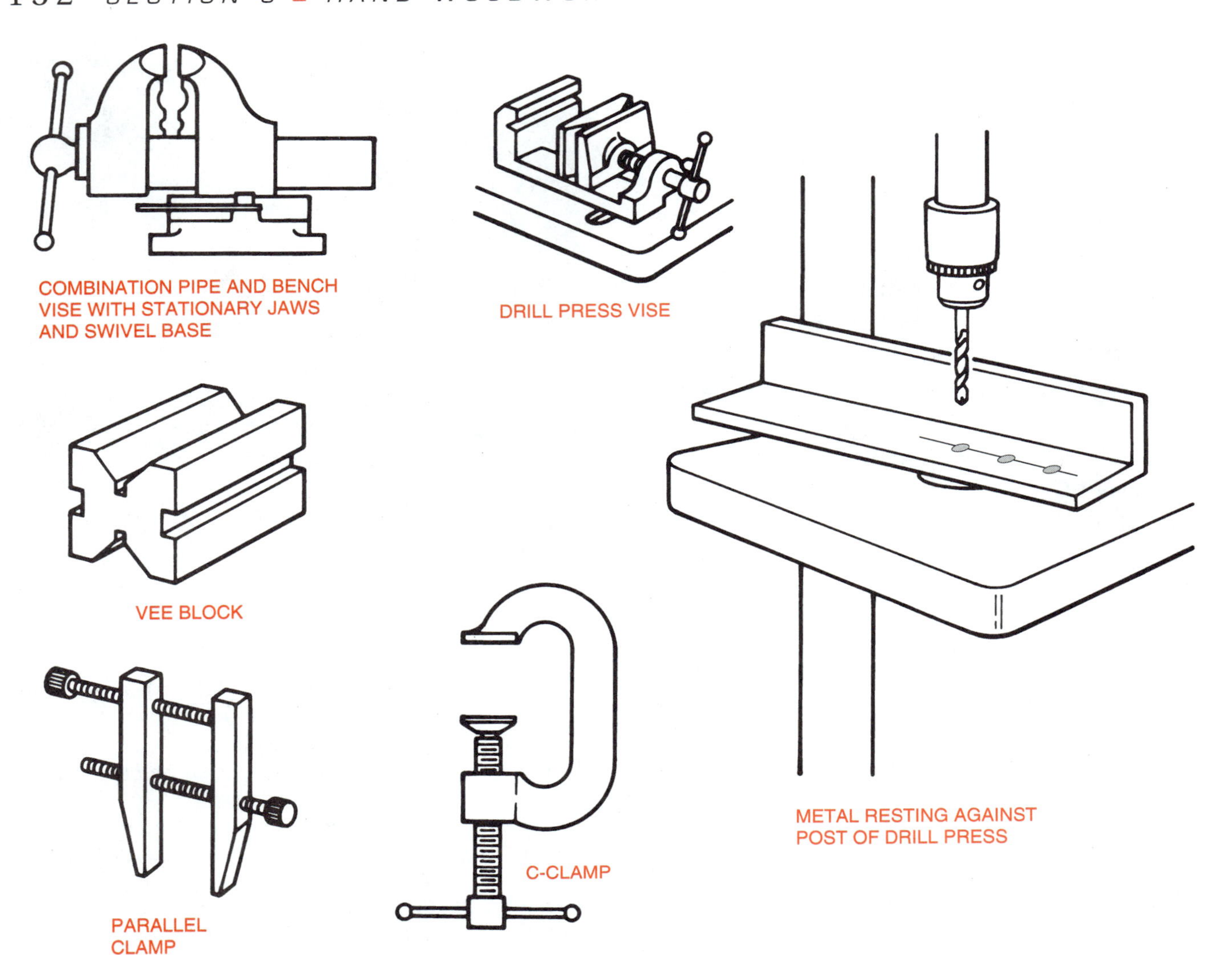

FIGURE 13-2. Some devices for holding metal while it is being drilled.

DRILL SPEEDS IN RPM FOR STEEL	
Drill Size	**Recommended Drill Speed**
1/8	2100–2450
3/16	1400–1600
1/4	1000–1200
5/16	850–950
3/8	700–800
7/16	600–700
1/2	450–600
9/16	425–550
5/8	400–500
11/16	375–450
3/4	350–400
13/16	325–375
7/8	300–350
15/16	275–325
1	250–300

FIGURE 13-3. Selecting the correct speed for the size of drill being used ensures long drill life and good results.

4. Select a high-speed twist drill of the desired size.

   NOTE: For holes larger than 3/8 inch in diameter, first drill a small hole (pilot hole) about the size of the dead center of the twist drill. Proceed to drill with the full-sized bit.

5. Place the drill bit in the drill chuck. Tighten the chuck by using the key in at least two holes in the chuck (three holes are preferred), figure 13-4.

CAUTION:

- **Be sure to remove the key from the chuck after tightening.**

6. Apply cutting oil to carry heat away from the drill point. Cutting oil is a special oil that does not have additives used in oils designed for engines.

FIGURE 13-4. Tighten the chuck using the chuck key in at least two holes. This will prevent the chuck from slipping and damaging the drill shank.

FIGURE 13-5. Threads are cut into a hole with a tap.

7. Apply firm, even pressure to the drill when starting and drilling.
8. Keep the drill perpendicular to the metal.
9. A uniform ribbon of metal is cut when the drill cuts steadily.
10. Use very little pressure as the drill breaks through the metal.
11. Be alert to the possibility of the bit seizing the metal as the drill breaks through. Hold portable drills very firmly at this point.
12. Avoid any side pressure on the drill bit. Small drill bits are brittle and break very easily.
13. After drilling the hole, remove any metal burrs with a file.

Punching Holes in Metal

Holes may be made in metal with punches. Soft, thin metal, like aluminum in storm windows, is punched rather than drilled. Punching is faster than drilling.

Steel may be punched if it is very thin. Steel up to 1/2 inch thick can be punched with hand tools if the metal is red hot. Blacksmith's tools are needed to heat and punch steel. Heavy stationary punches driven by hand, hydraulics, or electricity may be used to punch cold steel. These machines are used when many holes of the same size are needed.

THREADING METAL

Threads may be cut into a hole in metal with a tap, figure 13-5. A tap is a hardened, brittle, fluted tool that cuts threads into the hole. A tap is turned with a tap wrench. Threads are cut onto a rod or bolt with a die, figure 13-6. A die is turned with a die stock. A set of tools for making threads is called a tap and die set or a screw plate, figure 13-7. Common types of bolt threads are National Coarse (NC), National Fine (NF/SAE), and metric.

Tapping Threads

When tapping threads in a hole, the following procedure is used.

FIGURE 13-6. Threads are cut onto a rod with a die.

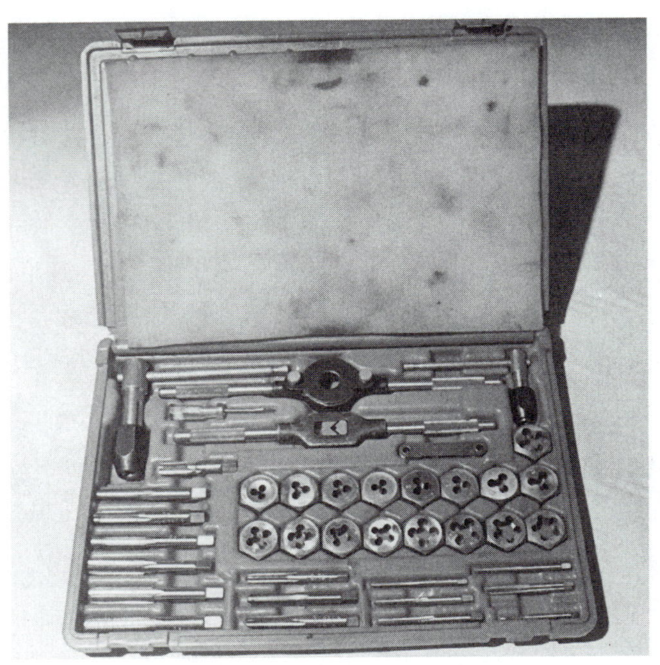

FIGURE 13-7. A set of taps and matching dies and the handles to turn them is called a tap and die set.

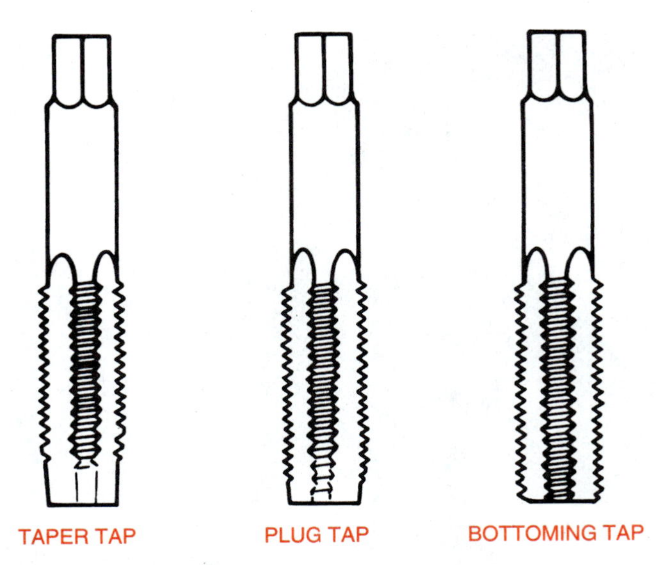

FIGURE 13-8. When tapping threads to the bottom of a blind hole, first use a taper tap, then a plug tap, and finally a bottoming tap.

PROCEDURE

1. Select the bolt that is to be threaded into the hole.
2. Determine the outside diameter of the bolt. Machine screw diameters are stated as Numbers 2, 4, 6, 8, 10, or 12. Other bolts are 1/4 inch, 5/16 inch, 3/8 inch, 7/16 inch, 1/2 inch, etc. Metric bolts have diameters ranging from about 3 to 20 millimeters.
3. Determine the type of threads on the bolt by using a screw pitch gauge. Most bolt threads are National Coarse (standard), National Fine (SAE), or standard metric. Metric bolts are also available with metric fine threads.
4. Select a taper tap of the correct size and thread type. The taper tap will start threads and complete them if the tap can extend through the material. If the hole does not go through the metal, a plug tap is used, followed by a bottoming tap. These taps will cut threads to the bottom of a blind hole, figure 13-8.
5. Center punch the location of the hole to be drilled.
6. Obtain a drill of the exact size specified for the tap selected, or use a tap and drill size chart, figure 13-9. When selecting a drill size, the exact number or fractional size is needed when the hole is to be threaded.
7. Carefully drill the hole. The hole must be drilled at a 90 degree angle to the surface of the metal. Do not permit the drill bit to wobble.
8. Place the tap in a tap wrench.
9. Place the end of the tap in the hole. Be careful to keep the tap in line with the hole during the entire tapping process, figure 13-10.
10. Apply enough cutting oil (if available) or motor oil to cover the threads of the tap and the hole in the metal.
11. Apply downward pressure on the tap and turn the tap forward (clockwise) about one-half turn. Do not exert any sideways pressure or cause the tap to wobble. The cap will break very easily.
12. Turn the tap backward (counterclockwise) one-quarter turn to break the chip of metal that forms. Add oil to the threads if dry.
13. Turn the tap forward one or two turns.
14. Turn the tap backward to break the chip. Add oil.
15. Repeat steps 13 and 14 until the tap is through the hole or to the bottom of a blind hole. Keep the tap and metal covered with oil.
16. Remove the tap by turning it backward.
17. Wipe oil and metal chips from the parts and work area using a cloth or paper towel.
18. Dispose of oily materials in the proper metal container.

NATIONAL COARSE THREADS (NC)			NATIONAL FINE THREADS (NF)		
Size of Bolt or Screw and Tap	Size of Drill to Use	Threads Per Inch	Size of Bolt or Screw and Tap	Size of Drill to Use	Threads Per Inch
#1	#53 or 1/16"	64	#1	1/16"	72
#2	#50	56	#2	#50	64
#3	#47 or 5/64"	48	#3	#45	56
#4	#43	40	#4	#42 or 3/32"	48
#5	#38	40	#5	#37 or 7/64"	44
#6	#36 or 7/64"	32	#6	#33	40
#8	#29	32	#8	#28 or 9/64"	36
#10	#25	24	#10	#21 or 5/32"	32
#12	#16	24	#12	#14	28
1/4"	#6 or 13/64"	20	1/4"	#3 or 7/32"	28
5/16"	1/4"	18	5/16"	17/64"	24
3/8"	5/16"	16	3/8"	21/64"	24
7/16"	23/64"	14	7/16"	25/64"	20
1/2"	27/64"	13	1/2"	29/64"	20
9/16"	31/64"	12	9/16"	33/64"	18
5/8"	17/32"	11	5/8"	37/64"	18
3/4"	21/32"	10	3/4"	11/16"	16
7/8"	49/64"	9	7/8"	13/16"	14
1"	7/8"	8	1"	15/16"	14

FIGURE 13-9. Taps and appropriate drill sizes.

CAUTION:

- **Do not brush metal chips away with bare hands as the chips are sharp and will cause injury.**

Threading a Rod or Bolt

The principles involved in threading a rod or bolt are similar to tapping threads. There is one major difference, however. Some dies are adjustable, but taps are not adjustable. With an adjustable die, the threads on a rod can be cut to a greater or lesser depth as desired. This adjustment permits a rod to be threaded to fit a nut perfectly. The procedure for cutting threads on a rod follows.

PROCEDURE

1. Clamp a rod of cold rolled steel in a vise (hot rolled steel will not thread satisfactorily).
2. Bevel the end of the rod with a file, figure 13-11.
3. Select a die of the correct size and type of thread.
4. Mount the die in a stock (handle).
5. Place the die over the rod with its tapered side toward the rod.
6. Oil the die and rod.

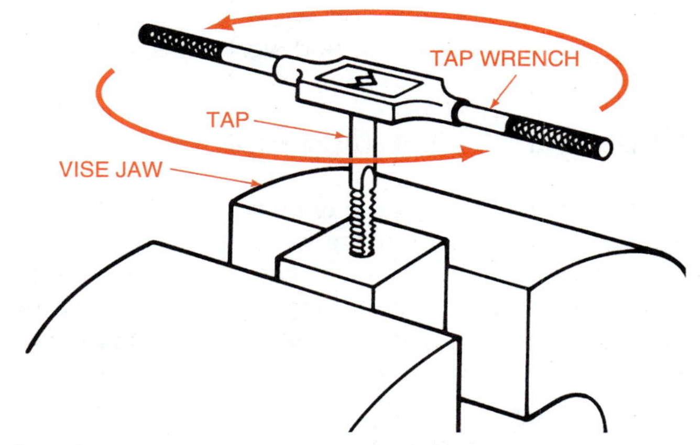

FIGURE 13-10. The tap must be kept in line with the hole and turned carefully. Keep the tap and the area being threaded covered with oil so tapping proceeds smoothly. Turn the tap forward about one-half turn, then turn it backward one-quarter turn to break the chip. Proceed until the tap turns freely, indicating the entire hole is threaded.

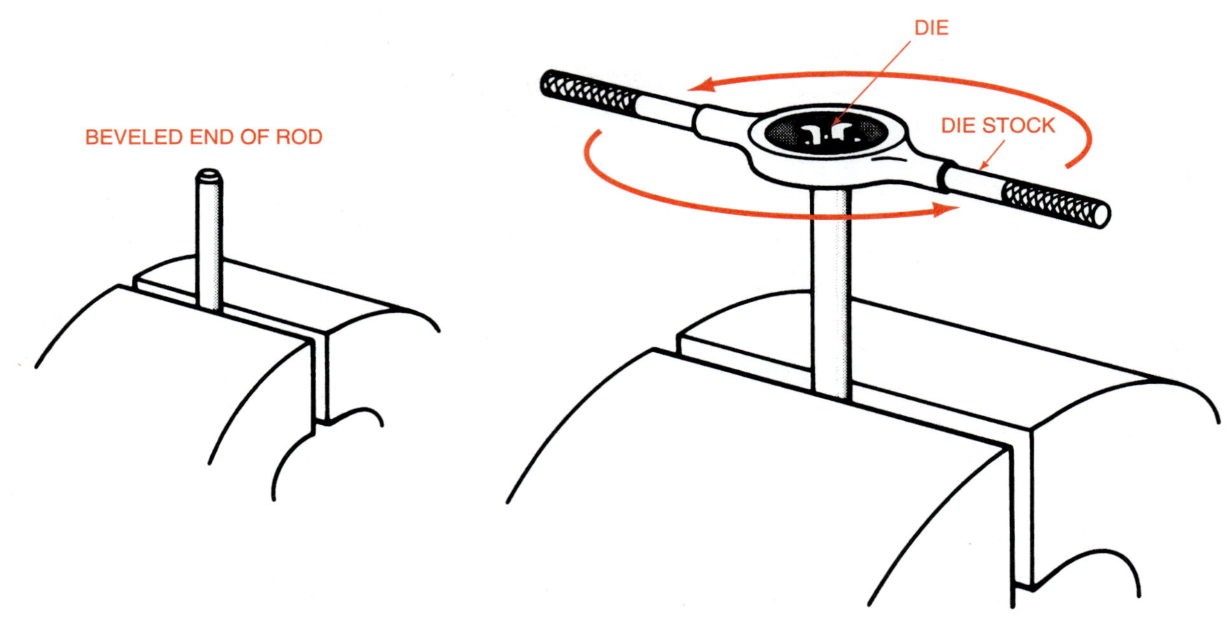

FIGURE 13-11. Bevel the end of the rod with a file. The die must be the correct size.

7. Apply pressure on the die and turn forward about one-half turn.
8. Back up one-quarter turn to break the chip.
9. Repeat steps 7 and 8 until a short length of rod has been threaded. Add oil as needed to keep the die lubricated and cutting smoothly. If the die is adjustable, check to see if the nut threads on properly. If not, adjust the die to obtain a good fit before cutting threads over the entire area.
10. Continue cutting threads until the desired threaded length is obtained.
11. Remove the die and wipe all chips and oil from the rod, tools, and work area.

Cutting and Threading Pipe

Cutting Pipe. Pipe is cut with pipe cutters that consist of one steel cutting wheel plus two rollers. Some pipe cutters have three cutting wheels and no rollers. A small version of the pipe cutter is used for cutting copper and steel tubing. When using pipe or tubing cutters, the following procedure is used.

PROCEDURE

1. Unscrew the cutter handle until the pipe fits between the rollers and wheels.
2. Slide the cutter onto the pipe until the cutter wheel(s) is on the mark.
3. Tighten the handle until the cutter wheel starts biting into the pipe.
4. Roll the pipe cutter around the pipe one or two times. The addition of oil will help the cutting process.
5. Repeat steps 3 and 4 until the pipe is cut. Do not tighten the handle more than one-half turn before rotating the cutter each time.
6. Insert a pipe reamer in the end of the pipe. Turn the reamer until the inside ridge is removed.

Threading Pipe. Pipe threads are not the same as bolt threads. Bolt threads are straight and can be cut the entire length of a rod. Pipe threads are tapered and can only be cut a short distance on the end of a piece of pipe, figure 13-12. The die gets tight

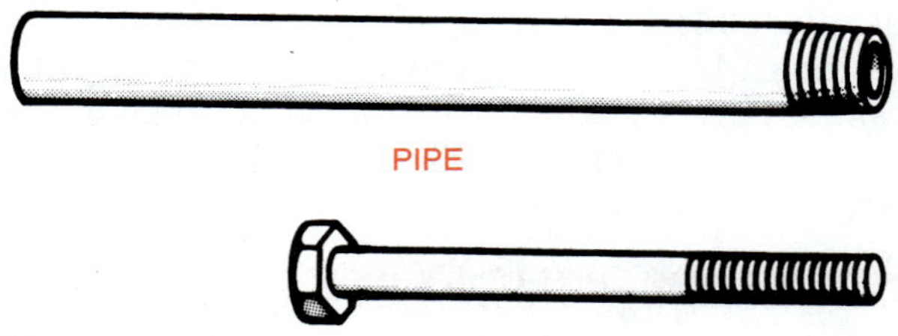

DIAMETER OF THREADED AREA IS LESS THAN THE DIAMETER OF THE PIPE. THE THREADS TAPER TO THE END OF THE PIPE.

PIPE

BOLT

DIAMETER OF THREADED AREA IS UNIFORM.

FIGURE 13-12. Pipe threads are tapered. Approximately one inch only can be threaded at the end of the pipe. In contrast, bolt threads are not tapered and can be cut along the entire length of the bolt.

and binds if too many threads are attempted. The tapered threads permit pipe connections to be gas and liquid tight.

Pipe fittings screwed onto pipe with appropriate joint compound create joints that hold air, steam, and liquids under pressure without leaking. Pipe fittings are hollow connectors designed to attach to pipe. Threaded fittings for steel pipe require some material to seal the areas between threads if they are to be used for gases or liquids. Materials for this purpose are teflon tape and pipe joint compound, also referred to as pipe dope. The procedure for threading pipe follows.

PROCEDURE

1. Select the correct die size for the pipe to be threaded.
2. Place the die in the stock.
3. Tighten the pipe in a pipe vise with 8 to 10 inches extending beyond the pipe vise. A regular vise may crush the pipe.
4. Slide the tapered side of the die onto the end of the pipe.
5. Apply cutting oil to the die and pipe as needed.
6. Turn the die until the end of the pipe starts to extend through the die.
7. Remove the die. Check to see if a pipe fitting will thread on and become tight after 4 to 6 turns. If not, cut more threads on the pipe.
8. Use a cloth to wipe all oil and chips from the pipe, tools, and work area. Remove all chips and oil from the inside of the pipe. Such materials interfere with the use of the pipe if they are not properly removed.

FASTENING METAL USING BOLTS AND SCREWS

Flat pieces of metal are fastened quickly and easily with bolts. One of two methods is generally used. A nut is used with the bolt, or the bolt is screwed into the second piece of metal after threads are tapped.

Using a Bolt and Nut

When using a bolt and nut, the following procedure is recommended.

PROCEDURE

1. Drill or punch a hole through both pieces of metal. The holes should be the same diameter as the bolt. Oversized holes may be used if there is a poor fit.
2. Select the desired type of bolt.
3. Countersink to accommodate the bolt head, if using a flat head stove bolt.
4. Insert the bolt.
5. Add a lock washer.
6. Thread on the nut.
7. Line up all parts.
8. Tighten the nut until the lock washer is flattened.

Using Bolts without Nuts

Cap screws, machine bolts, and machine screws may be used without nuts. The procedure follows.

PROCEDURE

1. Select the desired size and type of bolt.
2. Drill a hole in the first piece of metal the diameter of the bolt.
3. Drill the second piece of metal with the exact drill specified for the bolt size and thread type.
4. Tap threads into the second piece.
5. Place a lock washer on the bolt.
6. Push the bolt through the first piece.
7. Screw the bolt into the second piece.
8. Tighten the bolt until the lock washer is flattened.

Using Sheet Metal and Dry Wall Screws

Sheet metal screws are used to attach thin pieces quickly. The procedure follows.

PROCEDURE

1. Select the desired type and size of sheet metal screw.
2. Drill or punch a shank hole in the first piece of metal.
3. Drill or punch a pilot hole in the second piece.
4. Screw the two pieces together, but do not tighten the screw.
5. Align the parts.
6. Tighten the screw snugly. Be careful not to overtighten. If the screw is overtightened, the pilot hole may be stripped and the screw will not hold.
7. If the material is soft or the metal is thin, self-tapping sheet metal screws and dry wall screws may cut their own threads.

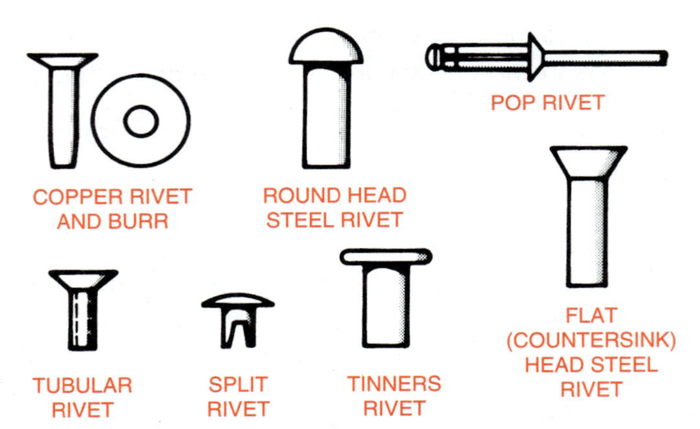

FIGURE 13-13. Rivets may be solid, tubular, or split and are available in different head styles. They are generally made of steel, aluminum, or copper.

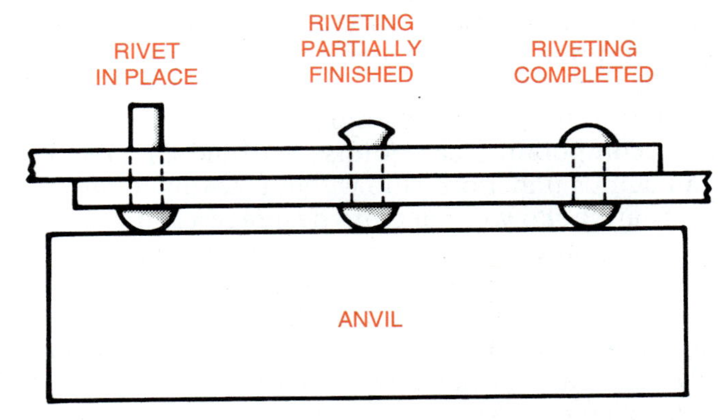

FIGURE 13-14. Riveting procedure using a ball pein hammer.

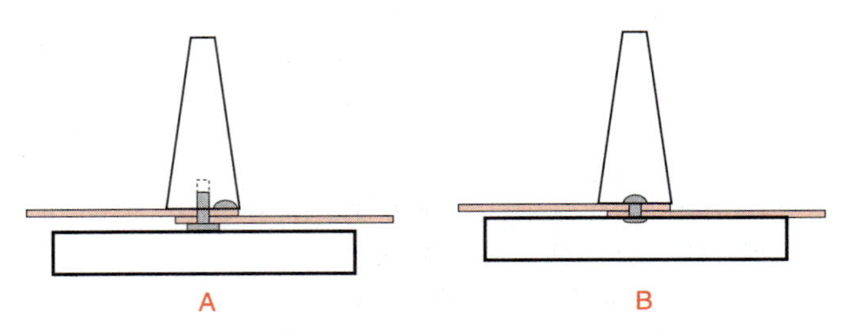

FIGURE 13-15.

A. When using a rivet set, place the rivet set over the rivet and tap the pieces of metal together.

B. Then place the cupped part over the rivet and drive the rivet set to form the mushroom-shaped head.

FASTENING WITH RIVETS

A rivet is a fastening device held in place by spreading one or both ends. A special tool called a rivet set makes it easy to install rivets. Many shops do not have a rivet set and the peining method is used.

Most rivets are purchased with flat, round, or countersunk heads; they may be solid, tubular, or split, figure 13-13. Tubular and split rivets have the advantage of being easy to rivet. This means they can be spread or shaped by hammering. They will not hold as well as solid rivets. Steel rivets, copper rivets, and pop rivets are commonly used.

The Steel Rivet

The round head steel rivet is used to fasten structural and sheet steel in bridges, buildings, and truck bodies. An important use of steel rivets on the farm is to attach mower knives. For this application the flat head rivet is used. To install a steel rivet, the following procedure is used.

PROCEDURE

1. Mark the spot and center punch the metal.
2. Drill a hole in both pieces of metal the exact size of the rivet.
3. Determine the length of the rivet. The length should be the thickness of the two pieces of metal plus the diameter of the rivet. For example, when fastening two pieces of metal each 1/8 inch thick with a rivet of 1/8 inch diameter, the rivet length should be 3/8 inch (1/8" + 1/8" + 1/8" = 3/8").
4. Cut the rivet to length with a hacksaw.
5. Insert the rivet through the two pieces of metal.
6. Place the head of the rivet on a solid steel object such as an anvil or metal vise.
7. Place a rivet set, or other object with a hole the size of the rivet, over the protruding rivet end.
8. Drive the metal down tight against the rivet head.
9. Use a ball pein hammer to create a rivet head on the extended end. This is done by striking or peining the rivet stem on the edges until they are rounded down and mushroomed against the metal, figure 13-14. A rivet set may also be used to form the riveted end, figure 13-15.

The Copper Rivet

Copper rivets with a tight fitting washer called a burr are used to fasten harness and other leather products. A burr is as large as the head so soft materials are held without pulling the rivet through the material.

The copper rivet with burr is installed much like a steel rivet. The rivet is inserted in the material; the washerlike burr is then added to the rivet. The burr and material are hammered together. Finally, the end is riveted to keep the burr in place.

The Pop Rivet

The pop rivet is so named because it pops when installed. It is especially useful in situations in which it is difficult to hold one end of the rivet and hammer the other end. An example of this is repairing sheet metal on an automobile.

When purchased, the pop rivet appears to be mostly stem. However, during installation, the special pop rivet tool pulls the stem until its ball-like head forces the end of the rivet to expand, figure 13-16. This enlargement holds the rivet in place. As the stem is pulled further by the tool, it breaks with a popping sound. The stem pulls out of the hollow rivet and is thrown away. The riveting process is then complete.

FASTENING METALS WITH SOLDER

Solder is a mixture of tin and lead or other metals. The joining of two materials with solder is known as soldering. Solder is frequently used to join thin metal, to make electrical connections, and to join copper tubing and pipe. The following general procedure is used when soldering.

PROCEDURE

1. Remove all dirt and corrosion from the metal with sandpaper or emery cloth. The metal should be shiny and bright.
2. Coat the tip of the soldering copper with solder. This process is called tinning. Tinning is done by cleaning the soldering copper with steel wool, sandpaper, or a file. The copper is heated and flux is applied. After the flux melts, solder is applied to the copper until it covers the tip. Flux is a material that removes tarnish or corrosion, prevents corrosion from developing, and acts as an agent to help the solder spread over the metal. Three fluxes that are commonly used are acid, rosin, and sal ammoniac.
3. Apply heat to the metal and apply flux.
4. Apply additional heat to the metal being soldered until that metal will melt solder.
5. Tin the metal by applying solder from a wire spool or a solid bar.
6. Apply heat to both pieces of metal until the solder melts and flows between them.
7. Wipe off excess solder with a cloth or sponge dampened with water while the solder is still molten.
8. Do not permit movement in the joint until the solder solidifies or hardens.

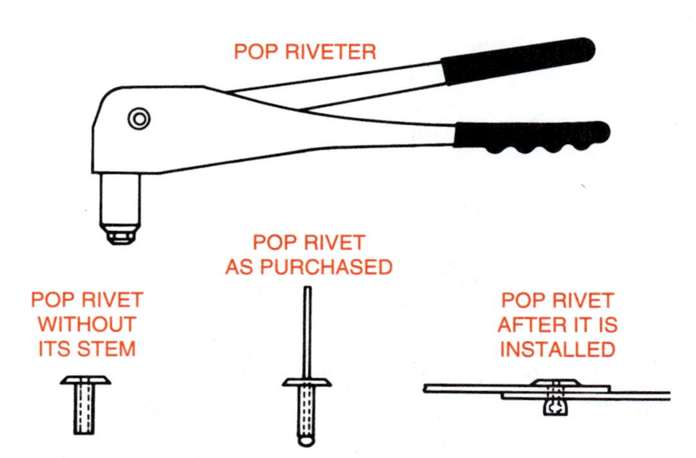

FIGURE 13-16. Pop rivets are installed from one side of the metal only. A built-in stem with a round ball is pulled into a hollow rivet. The ball causes the rivet to spread on the opposite side of the metal.

When soldering, several points should be kept in mind. It is important to select the appropriate flux and solder for the metal being soldered. Directions on the labels of commercial flux products as to the metals they will clean should be followed. Most soldering jobs in agricultural mechanics can be done with solder composed of 50 percent tin and 50 percent lead. This solder is called 50-50 solder. Other solders are 30-70, 40-60, 60-40, and 80-20. Notice that each combination of numbers equals 100. Different solders are recommended for various special jobs.

Solder may be purchased with flux inside. It is referred to as hollow core solder or flux core solder. Many people find the flux in the core of the solder makes soldering more convenient and easy.

Tinning a Soldering Copper

A soldering copper is a tool used for soldering and consisting of a wooden or plastic handle, steel shank, and copper tip, figure 13-17. It may also have an electric heating element. The procedure for tinning a soldering copper is as follows.

PROCEDURE

1. Remove burned residues with sandpaper, emery cloth, steel wool, or a file.
2. Heat the soldering copper until it will melt the flux or make smoke when rubbed in sal

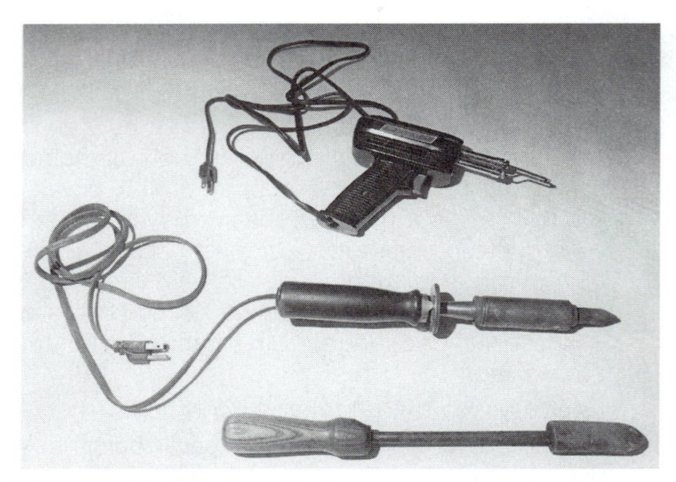

FIGURE 13-17. A soldering copper consists of a wooden or plastic handle, steel shank, and copper tip.

ammoniac. Sal ammoniac is a cube or block of special flux for cleaning soldering coppers, figure 13-18.

3. Rub the heated copper in sal ammoniac, or apply paste flux with a small wooden stick. The copper should get shiny.
4. Unroll about 6 inches of solder from the spool and touch the end to the copper. The solder should melt and flow over all flat surfaces of the copper if the copper is clean, fluxed, and the correct temperature.
5. If the solder does not melt, heat the copper to a higher temperature. Then repeat steps 3 and 4.
6. Brush the copper lightly with a cloth dampened with water. The soldering copper should be a shiny silver color. The silver color indicates the copper is tinned.

FIGURE 13-18. A block of sal ammoniac is used to clean the soldering copper.

Soldering Sheet Metal

To solder sheet metal, a large soldering copper is used. A soldering copper may be heated using a gas torch, a gas furnace, or an electric element. Electric units are sometimes called soldering irons; however, the tip of the unit is made of copper, not iron. When soldering sheet metal, the following procedure is used.

PROCEDURE

1. Clean the area to be soldered using an abrasive or a file.
2. Heat each piece of metal with the flat part of a tinned soldering copper.
3. Add flux to the metal as it heats up.
4. As the metal gets hotter, keep the flat part of the soldering copper flat on the metal near one edge.
5. Touch the solder wire to the metal, not the soldering copper.
6. Watch the point when the solder melts and begins to flow.
7. Move the soldering copper slowly across the metal, figure 13-19.
8. Follow the soldering copper with the solder as needed to tin the area to be soldered.
9. Repeat steps 2 through 8 on the second piece of metal.
10. Lay the two tinned surfaces face-to-face on a piece of scrap wood if support is needed.
11. Heat both sheets of metal with the soldering copper until the solder starts to flow.
12. Feed in additional solder to fill the joint as needed.

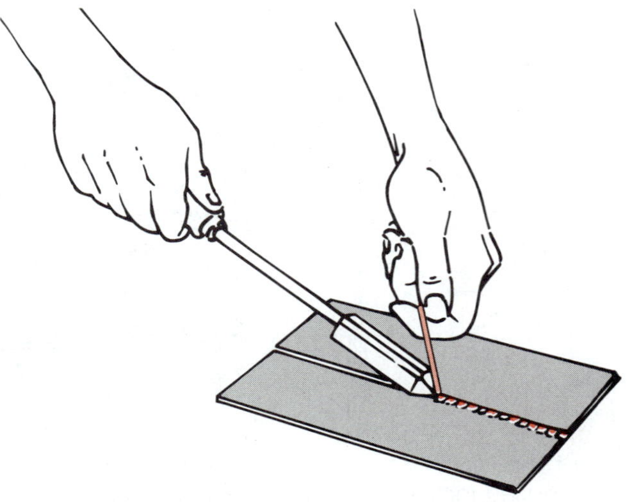

FIGURE 13-19. When soldering, the soldering copper is moved slowly across the metal.

13. Wipe off excessive molten solder with a damp cloth.
14. Hold the two pieces together with a piece of scrap wood until the solder cools. The solder is cool when it changes from a shiny to a dull finish.

Some of the steps in this procedure must be done at the same time, or they must be done in close sequence. Therefore, exact instructions from an experienced teacher may be needed to learn the technique.

Wood is recommended to support or hold the metal because wood does not carry heat away from the metal. Metal table tops, vises, and metal tools interfere with heat control. The control of heat is the heart of the soldering process.

Soldering Stranded Electric Wires

Copper wires for low-voltage wiring are soldered by first removing about 1 inch of insulation. The soldering procedure follows.

PROCEDURE

1. Twist the two bare ends of the wire together.
2. Insert one of the wires in a vise and gently tighten the vise jaws against the insulation.
3. Hold a hot, tinned soldering copper under the twisted wires, figure 13-20.
4. Hold rosin core solder on the top of the wire.

NOTE: Acid flux is not recommended for electrical connections.

CAUTION:

- **Do not use an acid flux as this interferes with the function of the splice.**

5. Wait for the heated wire to cause the rosin to flow and clean the wire.
6. Watch the solder melt and run into and around the strands of wire.
7. Remove the heat.
8. Apply several thicknesses of vinyl electrical tape to the joint.

An electric soldering pencil or gun works well for soldering copper wires. These special soldering tools provide the correct amount of heat for electrical applications.

Soldering Copper Pipes and Fittings

The process of soldering a piece of copper pipe into a fitting is called sweating, figure 13-21. The ability to sweat copper fittings is useful for repairing and installing water lines at home, on the job, or on the farm.

Common pipe fittings include elbows, tees, couplings, unions, and adaptors. Fittings are used to change the direction of a pipe line. They are also used to adapt one type of pipe to another, such as copper to plastic.

Fittings are used to connect pipe to faucets, valves, and other fixtures. To sweat a fitting onto a piece of copper pipe, the following procedure may be used.

PROCEDURE

1. Use very fine sandpaper or steel wool to polish about 3/4 inch at the end of the pipe.
2. Place the sandpaper or steel wool across the end of your index finger and polish the inside of the fitting.
3. Apply a paste flux to the polished surfaces of the pipe fitting.

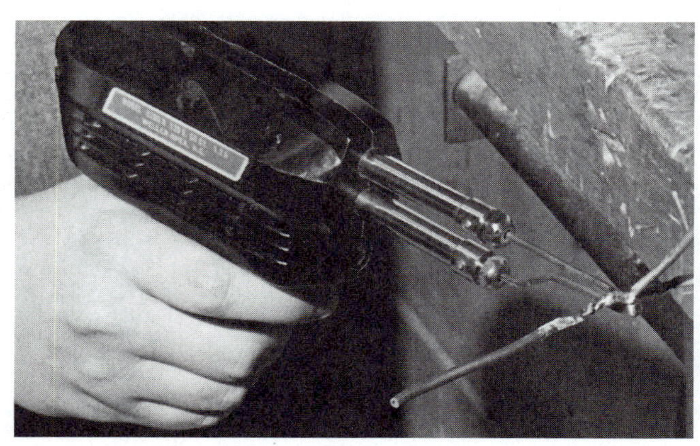

FIGURE 13-20. When soldering stranded electrical wire, a hot, tinned soldering copper is held under the twisted wires.

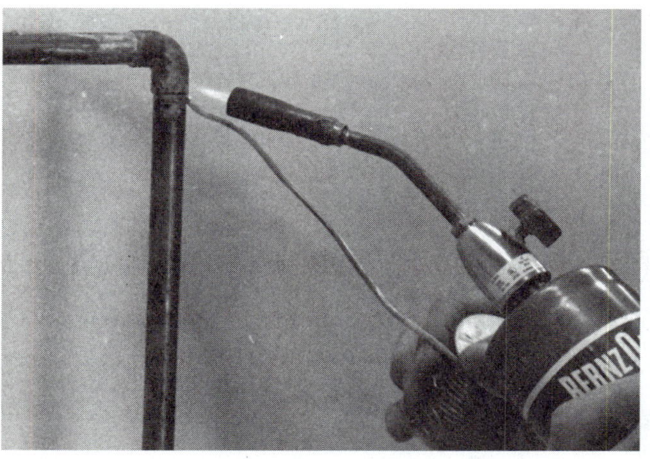

FIGURE 13-21. Soldering a piece of copper pipe into a fitting is called sweating.

4. Press the fitting onto the pipe.
5. Use a propane or other gas torch to heat the outside of the fitting near the opening where the pipe is inserted.
6. Heat the fitting 1/2 inch in from the opening. Use the tip of the inner flame of the torch to heat the underside of the fitting. Heat will rise to the upper areas.

NOTE: It is important to heat the fitting rather than the pipe to draw the solder into the joint.

7. Hold a length of lead-free solid core solder at the joint between the fitting and pipe. Be sure the solder is of the correct type for water lines.
8. Continue to heat the pipe until the solder starts flowing into the joint.
9. Move the solder around the joint to permit solder to fill all areas between the pipe and fitting.
10. Remove the heat and solder when solder is seen all the way around the joint and the joint is full of solder.
11. Before the solder hardens, wipe around the joint with a damp cloth.
12. Cool the joint.
13. Turn on the water to pressurize the system.
14. Check for leaks.

When using a propane torch several precautions are important. First, wear goggles and insulated gloves. Second, work only in areas that are free of paper, grease, fuel, and other materials that burn easily. Third, point the torch away from you when you turn it on and light it. Fourth, use approved fire safety procedures and materials.

Whenever any flame is used, be sure there is a fire extinguisher nearby. To reduce fire hazards, place sheet metal behind pipes where a torch is used. When finished with the torch, it is best to extinguish the flame. Certain propane torches have high flame triggers. When the propane unit is set down, the flame becomes low and creates less of a fire hazard.

STUDENT ACTIVITIES

1. Define the Terms to Know in this unit.
2. Make a bench stop using the following procedure.
   a. Cut a piece of band iron 1/4 inch thick × 1 inch wide × 5 inches long.
   b. Make the ends square; slightly round all corners to remove sharp edges.
   c. Center punch the metal 1/2 inch from each end and each side.
   d. Drill a 5/16 inch hole through the metal at the two punch marks.
   e. Tap National Coarse (NC) threads in each hole to receive a 3/8 inch threaded rod.
   f. Cut two pieces of round cold rolled steel 3/8 inch in diameter and 1 1/2 inches long. Remove all sharp corners with a file.
   g. Thread each rod down 1/4 inch.
   h. Screw the two threaded rods into the 1/4" × 1" × 5" piece. The rods should thread in until tight. The end of the rods should be flush with the bottom of the metal strip.

NOTE: The bench stop is useful to stop lumber when sanding, planing, or nailing on a bench. To use the stop, drill two holes in the bench top 3/8 inch or larger in diameter. The holes are drilled so the two rods drop into them. The bench stop is then ready for use.

3. Cut and thread a piece of steel pipe.
   a. Obtain a piece of steel pipe 1 foot or more in length. Pipe with an inside diameter of 1/2 inch, 3/4 inch, or 1 inch is recommended.
   b. Cut off any existing threads with a pipe cutter.
   c. Use a pipe reamer to remove any inside ridge.
   d. Cut threads on the pipe.
   e. Apply teflon tape or pipe joint compound.
   f. Screw on a pipe fitting such as a tee or elbow.
   g. Have the teacher check your work.
   h. Remove the fitting, wipe off any pipe joint compound, and store the pipe.

4. Bolt and rivet two pieces of band iron.
   a. Cut two pieces of band iron $1/4" \times 1" \times 6"$.
   b. Mark one piece A and the other B.
   c. Center punch piece A at $1\frac{1}{2}$ inches, 3 inches, and $4\frac{1}{2}$ inches from one end and along the center line of the metal.
   d. Lay A over B.
   e. Drill a hole through both pieces near one end with a 1/8-inch drill.
   f. Redrill A with a 5/16-inch drill.
   g. Redrill B with a 1/4-inch drill.
   h. Tap 5/16 inch NC threads into the hole in B.
   i. Screw the two pieces together with a $5/16" \times 1/2"$ stove bolt, machine bolt, or cap screw.
   j. Drill a 1/4-inch hole through both pieces at the center and end marks.
   k. Install a round head rivet in the center hole.
   l. Countersink the third hole and install a flat head steel rivet.
5. Solder sheet metal.
   a. Obtain a clean steel fruit can and its lid.
   b. Cut the outer corrugations off the lid so only the flat center remains.
   c. Turn the empty can upside down.
   d. Solder the round piece from the lid onto the flat round area on the bottom of the can.
   e. Punch several holes in the corrugations outside the patched areas with a 4d nail.
   f. Close the holes by soldering.
6. Solder electric wires.
   a. Obtain two pieces of number 12, 14, or 16 stranded and insulated copper wire about 4 inches long.
   b. Remove 1 inch of insulation from one end of each wire.
   c. Cross the two bare wires and twist them around each other to form a smooth splice.
   d. Solder the joint.
   e. Tape the joint.
7. Sweat copper pipe.
   a. Obtain a short length of 3/8- or 1/2-inch copper pipe.
   b. Sweat a copper fitting onto the pipe.
   c. Have the instructor inspect the job.
   d. Remelt the solder and remove the fitting with pliers.
   e. Wipe off the hot solder with a damp cloth.

SELF-EVALUATION

A. **Multiple Choice**. Select the best answer.

1. Metal may be fastened by
   a. bolts
   b. rivets
   c. screws
   d. all of these
2. The process of joining metal by melting a different metal between two pieces is known as
   a. gluing
   b. soldering
   c. washing
   d. welding
3. Holes are usually made in heavy metal by using
   a. an auger bit
   b. a forge
   c. a punch
   d. a high-speed twist drill
4. When drilling, the lightest pressure should be placed on the drill when
   a. breaking through
   b. midway drilling
   c. starting the hole
   d. none of these

5. Threads are cut onto a rod with a
   a. die
   b. ream
   c. stock
   d. tap
6. When tapping threads, always start with a
   a. bottoming tap
   b. plug tap
   c. taper tap
   d. die
7. The first tool to use in drilling a hole in metal is the
   a. center punch
   b. drill
   c. file
   d. ream
8. When cutting threads, oil is used to
   a. clean the tool
   b. harden the threads
   c. lubricate the tool
   d. soften the metal
9. Pipe threads differ from bolt threads in that
   a. pipe threads are tapered, bolt threads are not
   b. oil is needed to cut pipe threads but not bolt threads
   c. pipe is threaded with a die, a bolt is not
   d. all of these
10. If you cannot get to both sides of sheet metal, the rivet to use is the
   a. copper rivet and burr
   b. pop rivet
   c. split rivet
   d. solid steel rivet

B. Matching. Match the word or phrase in column I with the correct word or phrase in column II.

Column I

1. drill
2. pop
3. solder
4. countersink rivet
5. tin and lead
6. N.C.
7. N.F. or S.A.E.
8. sal ammoniac
9. teflon tape
10. sweating

Column II

a. rivet
b. solder
c. jobbers-length
d. National Fine
e. seal pipe threads
f. flux
g. 50-50
h. mower knives
i. National Coarse
j. soldering copper pipe

C. Completion. Fill in the blanks with the word or words that will make the following statements correct.

1. The screw used to fasten thin metal is called a ________ ________ screw.
2. The metal used in the ends or tips of all soldering irons, guns or pencils is ________________.
3. The process of covering the tip of the soldering copper with solder is called ____________.
4. When soldering electrical wire or connections, ________ core solder must be used.
5. After cleaning with an abrasive, metal must be treated with a ____________ before soldering.

Power Tools In The Agricultural Mechanics Shop

UNIT

Portable Power Tools

OBJECTIVE

To select and safely use major portable power tools in agricultural mechanics.

COMPETENCIES TO BE DEVELOPED

After studying this unit, you should be able to:

- State recommended procedures for using portable power tools.
- Write a description of the uses of portable power tools.
- Name and properly spell the names of common portable power tools used in agricultural mechanics.
- Identify and spell the names of the major parts of portable power tools.
- Safely operate a portable power drill, belt sander, disc sander/grinder, finishing sander, sabre saw, reciprocating saw, and circular saw.

TERMS TO KNOW

saw horse
trestle
ground-fault interrupter (GFI)
double insulated
air tool
compressed air
duty cycle
continuous duty
variable speed
reversible
hammer
cordless
pilot hole
finishing sander
orbital sander
belt sander
disc sander
grinders
power handsaw
sabre saw
bayonet saw
blind cut
reciprocating saw
tiger saw
portable circular saw
bushing
router

MATERIALS LIST

✓ Examples of portable electric and air-driven tools

✓ Unlabeled diagrams of portable power tools:

Drill
Belt sander
Disc sander
Grinder
Finishing sander
Sabre saw
Reciprocating saw
Circular saw

NOTE: Before proceeding with this unit, review the material in Unit 4 on personal safety.

Portable power tools save labor and are relatively inexpensive to buy. Most people prefer to invest in portable power tools rather than stationary ones for agricultural use. Since work on the farm and in other agricultural settings requires that tools be taken to the job, portable tools are especially useful and efficient.

The use of battery-operated power tools permit fast and efficient use of tools without the need for electricity at the work site. Recharge batteries immediately after use to keep the batteries healthy and always ready for use.

When using portable power tools, it is important that the work be well secured. There are several ways to do this. If possible, place the work in a vise or some other kind of holding device. If large pieces are being used, saw horses, also called trestles, are recommended. A saw horse or trestle is a wood or metal beam or bar with legs. It is used for the temporary support of materials. Large flexible panels require special handling to prevent the binding of saws and equipment. If two or three 2 × 4s are placed on saw horses, flexible panels can then be placed across them for support. Large panels can be cut safely when supported in this manner.

SAFETY PRECAUTIONS

Several safety problems are inherent with portable electric tools. When working with power tools, sharp blades, bits, and abrasives should be used to reduce the pressure needed to make the tool function. Reduced pressure decreases the likelihood of the tool binding or the hand or tool slipping.

Wet areas are dangerous areas for using electrical power tools. Work areas should always be set up in dry locations. Wooden floors are the best, but concrete is safe if it is dry. Shoes with rubber soles or rubber boots reduce electrical hazards when power tools are being used. A ground-fault interrupter (GFI) should be used in circuits when the tool operator is working with wet hands, feet, or body. A GFI breaks the electrical circuit when the operator is threatened by electrical shock.

Another important safety consideration in using portable power tools is proper balance and footing. It is important to stay balanced at all times. Excessive reaching which may cause loss of balance or control should be avoided.

Eye and face protection are very important when using power tools. Under no circumstances should power tools be used without wearing safety glasses or goggles. Face shields are also recommended.

SAFETY RULES

1. Plug into an outlet protected by a ground-fault interrupter when possible and at all times when wet conditions exist.
2. If the tool housing is metal, be sure to use a three-wire grounded power source.
3. Be certain the power cord, switch, and all electrical parts are in good condition.
4. Be certain that blades, bits, and other cutting units are clean and sharp.
5. Support work carefully to avoid any tendency for it to bend, buckle, or bind when power tools are used.
6. Keep power tools clean and free of dirt.
7. Keep vent holes free of dirt to permit ventilation and cooling of the motor.
8. Wear complete body protection.
9. Exercise care to avoid accidential electrocution caused by working in wet areas or by cutting power cords.
10. Do not force power tools to cut, drill, or otherwise work faster than designed to work.
11. Hold every power tool firmly and in control at all times.
12. In school, always obtain the instructor's permission before using a power tool.
13. If there is any question about the condition of a power tool, check with the instructor before proceeding.
14. Announce to others around you when you are ready to start a power tool.
15. Lay the power tool down safely so it will not damage cutting parts or injure others.
16. Report any faulty condition of the power tool and its cutting parts to the instructor as soon as it is noticed.

FIGURE 14-1. Basic safety rules for using electrical power tools.

In addition to eye protection, coveralls and other protective clothing are recommended when using power tools. It is important to avoid all loose clothing. Leather shoes are recommended as a minimum. Steel-toed shoes are needed when heavy materials are being handled.

Figure 14-1 lists some basic safety rules to follow when using electrical power tools.

Power Cords

The electrical cord of a power tool must be in good condition. The cord should be checked for broken insulation, broken plugs, bare wires, or other evidence of cord damage. Power cords often break, short out, or become electrical hazards due to breaks in the insulation where the cord leaves the power

FIGURE 14-2. A three-prong, grounded plug.

tool. It is advisable not to turn the first wind of the power cord too sharply if cords are wrapped around tools for storage.

All motors must have some method for protecting the operator against electrical shock in case of an internal electrical problem. Motors with metal housings should have a special ground wire and a plug with a ground prong, figure 14-2. The cords of such tools contain three wires: two wires carry current for the motor and the third wire is a safety ground wire to ground the housing of the tool. The ground wire connects to the longer prong on a 120-volt plug. This ground prong should never be broken off or damaged. If a tool in the agriculture shop is not grounded properly, the instructor should be notified immediately.

Motors with plastic housings generally are double insulated and do not need the special ground wire and ground prong. Double-insulated tools use two-wire, nongrounded cords. The electrical parts are insulated or separated from the user by special insulation inside the motor and by the plastic motor housing. Some electric motors have internal parts insulated in two different internal locations.

Extension Cords. In most cases an extension cord is needed with a portable power tool. Before a power tool is used, it should be determined if the tool is meant to be used with a three-wire extension cord. It is important that a three-wire grounded extension cord be used with tools that have metal bodies.

Extension cords should always be checked before use. Cords with frayed insulation or damaged ends should not be used. Extension cords longer than needed to reach the job should be used. If short cords are used, it is possible for tools to be pulled from benches if the cord is moved by passersby. Workers must always be aware of the location of the power cord. Otherwise, electrical shock may result due to damage to the cord by sawing, drilling, sanding, binding, or crushing.

Air-Driven Tools

Some agricultural mechanics shops have air tools. An air tool is a tool that is powered by compressed air. Compressed air is air pumped under high pressure and carried by special hoses. Compressed air provides pressure for spray guns; it also drives portable tools such as drills.

Compressed air can be dangerous if not handled properly. It is particularly hazardous when air hoses are being coupled and uncoupled, figure 14-3. High pressure air lines must not be uncoupled when people are nearby. Streams of air must never be directed toward a person's face or body. Dust and dirt may be driven into the eyes, or the compressed air may damage ear drums and other organs of the body. Serious accidents of many types have been reported from the misuse of compressed air. Compressed air is not recommended for cleanup purposes unless the pressure is less than 30 pounds per square inch.

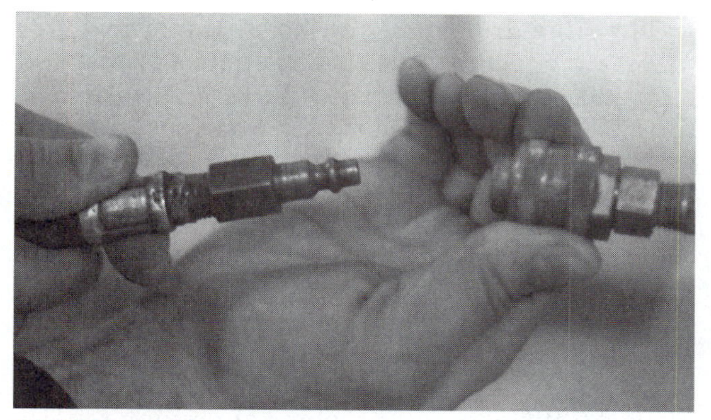

FIGURE 14-3. Couple and uncouple air hoses with care. The short burst of air during coupling or uncoupling can cause injury to the operator or bystanders.

PORTABLE DRILLS

A portable drill is a small tool that can be easily moved to the work. The major parts of a portable drill are the:

- power cord
- handle
- motor housing
- gear chuck
- vents
- trigger switch
- trigger switch lock
- reversing switch
- chuck wrench or key.

These parts are shown in figure 14-4.

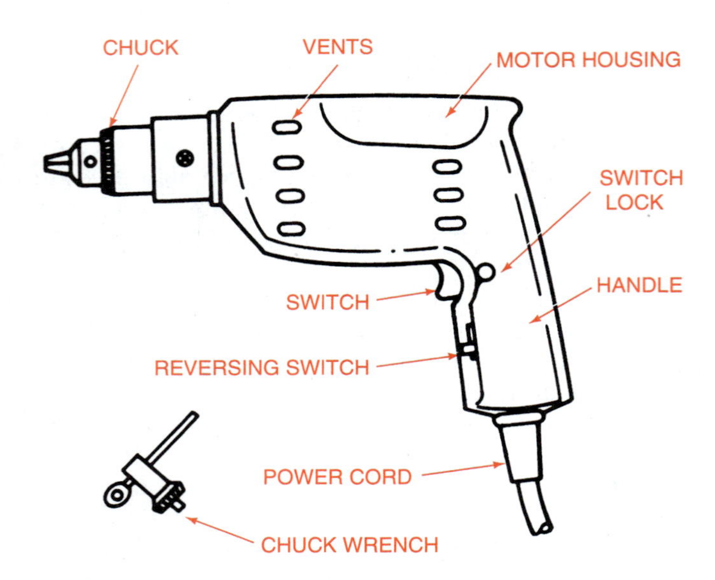

FIGURE 14-4. Major parts of a portable power drill.

Portable drills may be classified by the chuck size. A chuck is the device used to hold a drill or tool bit in the machine. Common portable electric drill sizes used in agricultural mechanics are $^1/_4$ inch, $^3/_8$ inch, and $^1/_2$ inch. The $^1/_4$-inch and $^3/_8$-inch drills usually turn quite fast. The $^1/_2$-inch drills run more slowly. This type of drill has reduction gears that generate more torque than is provided by the drill motor itself.

Another way of classifying drills is by their power rating. Power drills typically draw from two to five amperes of electricity. A 115-volt motor using five amperes of electricity develops approximately $^1/_2$ horsepower. Such a drill is considered a powerful portable drill.

Portable drills may also be rated by duty cycle. Duty cycle refers to the amount of time a motor can run versus the time it needs to cool off. Continuous duty means a tool can be used all the time for a 6- or 8-hour day. Most drills, however, are not continuous duty cycle drills. Therefore, it is possible to overwork and overheat a power drill. When a drill gets too warm to hold comfortably, it is time to stop work and let it cool off. If overused, it is possible to burn out the motor. This same principle applies to all portable electric power tools.

Types of Drills

Drills may be single speed or variable speed. Variable speed means the speed of the motor can be controlled by the operator. Variable speed drills are useful for special purposes such as running slowly enough to drive screwdriver bits. These bits are used to install and remove screws. A variable-speed drill will turn fast for small drill bits and more slowly as needed for larger ones.

Some drills are reversible. Reversible means they will run backward as well as forward. A variable speed, reversible drill is useful to back screws out as well as drive them in.

Another feature of some drills is their ability to hammer. The capacity to hammer means a drill will turn a bit and also provide a rapid striking action on the bit to speed up drilling in masonry materials. This feature is especially useful for workers who must make holes in brick, concrete, or stone walls, or who must install plugs and anchors in such walls.

Some drills are cordless. A cordless drill means it contains a rechargeable battery pack to drive the unit when it is not plugged into an electrical outlet. Cordless drills operate for only a few hours. The tool must then be plugged into an electrical outlet or rechargeable unit to recharge the battery. The cordless drill is handy because it can be used in the field or any other location even though no electrical wiring exists.

Uses for Drills

Portable power drills have many uses, figure 14-5. The most basic is the drilling of holes. Power drills can turn many different types of bits. Some power drills can turn screws in and out. Masonry drill bits permit the drilling of holes in brick, block, or stone walls. Hole saws are driven by portable electric power drills. Some people use sanding discs and polishing heads on portable electric drills. When using a portable power drill, the following procedure is recommended.

FIGURE 14-5. The portable power drill may be used to drive drills, wood bits, hole saws, sanding discs, polishing bonnets, screwdriver bits, and other accessories. *(Courtesy of Lewis/Carpentry, 1995, Delmar Publishers)*

PROCEDURE

1. Use only straight-shank bits.
2. When tightening a drill chuck, place the key into one hole and tighten the chuck securely. Place the key into a second hole and, again, tighten securely.

CAUTION:

- **Always remove the chuck key from the chuck after tightening a drill bit. Otherwise, the chuck key will be thrown when the drill is started.**

3. Always center punch metal to help start a bit.
4. Hold materials in a vise or other secure device.
5. Use slow-turning drills for large bits.
6. Use even pressure on the drill.
7. Ease off the pressure when the drill is breaking through the material.
8. Hold the drill so as to avoid binding the drill bit.
9. Keep operator positioned so that balance is always maintained.
10. Always remove the drill bit from the chuck when finished.
11. Store the portable power drill in its own case or in a special storage rack.

When drilling large holes, a pilot hole is used. A pilot hole is a small hole drilled in material to guide the center point of larger drills. By drilling a pilot hole, the bit stays exactly where planned and cuts with less power and pressure.

If a drill bit is not cutting, check the reversing switch to see if it is turned on. If the drill is turning counterclockwise, the bit will not cut. If the drill is turning clockwise and is not cutting, the drill bit is dull and must be sharpened.

PORTABLE SANDERS

Three types of portable sanders are used in agricultural mechanics. They are the portable belt sander, portable disc sander, and portable finishing sander. Belt sanders and disc sanders are used for coarse sanding; finishing sanders are used for the last operation before applying finishes.

Power sanders do the same work as hand sanding. However, power sanders remove wood or other materials faster and easier than hand sanding. It is the speed that makes power tools desirable for sanding or grinding. However, the best work often can be done by hand sanding. Hand sanding permits the operator to control the sanding process better.

Effective Sanding

When sanding, it is important to sand with the grain of the wood for the fine work. However, crossgrain sanding may be useful

- if boards are uneven.
- if extremely rough boards are encountered.
- if very difficult finishes must be removed.

Generally, coarse sandpaper is used first, then medium sandpaper, and finally, fine or very fine sandpaper is used to complete the job.

Sometimes, workers choose to use a belt or disc sander for the rapid removal of material. A finishing sander or hand sanding is then used to complete the job. The finishing sander moves forwards and backwards. Therefore, the sanding can always be with the grain of the wood.

A slight variation of the forward and backward movement is provided by an orbital sander. The orbital sander moves in a circular pattern. This results in a faster cut, but still leaves a fine finish. After using a finishing sander, it is desirable to hand finish using fine or very fine sandpaper. This process finishes wood to the point of maximum smoothness and beauty.

Safe Use of Portable Belt Sanders

The belt sander is a tool with a moving sanding belt, figure 14-6. Major parts include the:

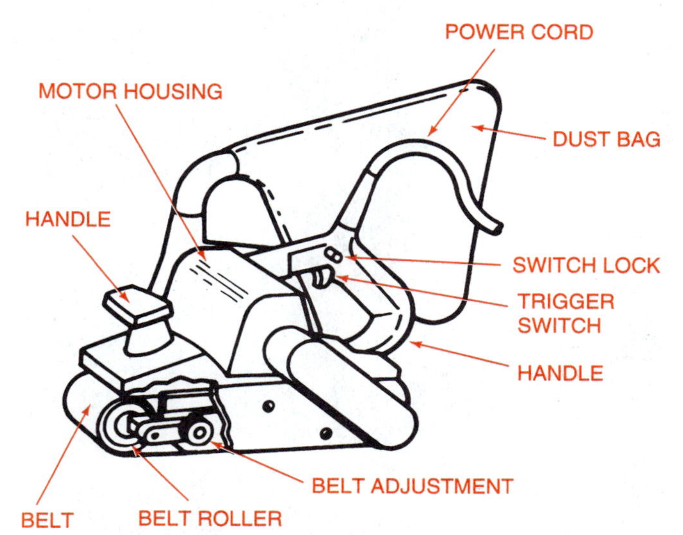

FIGURE 14-6. Major parts of a portable belt sander.

- motor housing
- handles
- belt
- belt rollers
- belt adjustment
- trigger switch
- switch lock
- power cord
- dust bag.

Each part should be learned by the student.

The belt sander is often used in agricultural mechanics shops. It is a relatively safe tool to use, but care must be taken as with all power tools. When using the belt sander, the work should be secured well and both hands used on the machine to manage it very carefully, figure 14-7.

When using belt sanders, the following general procedures are recommended for safe use and effective operation.

FIGURE 14-7. When using a belt sander, both hands are used to manage the machine so sanding is controlled.

PROCEDURE

1. Wear suitable face protection and protective clothing.
2. Check the power cord and extension cords for safety.
3. Install a sanding belt of suitable coarseness.
4. Lay the sander on its side when not in use. This prevents the sander from running off the bench if accidently turned on.
5. Be sure the dust bag is empty or nearly so before starting to sand.
6. Always start the machine while holding it slightly above the material.
7. Keep the power cord out of the way of the belt.
8. After turning on the sander, touch the work with the front part of the belt first, then slowly settle the rest of the belt down onto the work.
9. Operate the machine with two hands at all times.
10. Sand with the grain. Move the machine from one end of the board to the other in a straight path; then move it slightly sideways and draw the machine back over new area. Gradually work across the board by slightly overlapping the forward and backward passes.
11. Keep the machine in motion. If permitted to sand in one spot, it will cut a depression in the wood.
12. The final movement is to lift the machine off the work while it is still running.
13. Examine the work carefully. If necessary, resand in order to create a perfectly level surface that is smooth.
14. Install a fine sanding belt and resand. This resanding leaves the work in its smoothest possible form using the belt sander.
15. Use a finishing sander or hand sand to obtain the degree of fineness desired.

Safe Use of Portable Disc Sanders and Grinders

Some portable tools may be used only as sanders or only as grinders. Others are designed to be used as either, figure 14-8. By simply changing the sanding disc to a grinding wheel, many disc sanders become grinders. The manufacturer's instructions should be followed regarding any single tool. It is important to use the proper guards and wear protection with either the sanding disc, grinding wheel, or wire brush.

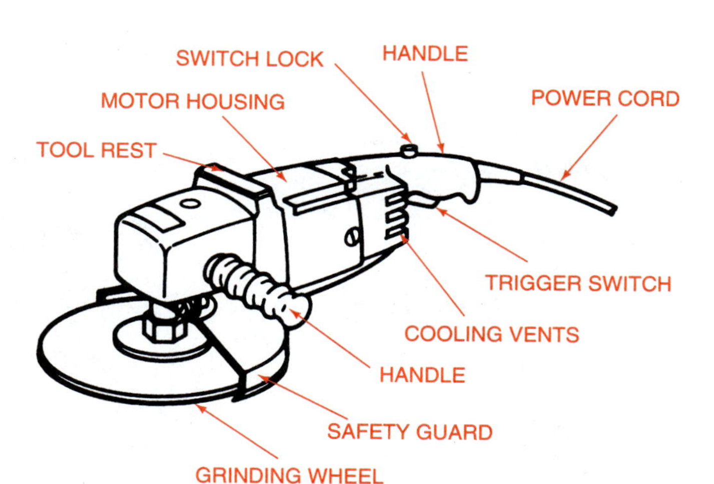

FIGURE 14-8. Major parts of a portable sander/grinder.

Some important parts of portable sanders and/or grinders are the:

- motor housing
- handles
- power cord
- switch
- switch lock
- cooling vents
- wheel
- spindle
- safety guard
- tool rest.

A disc sander is a tool with sanding materials, called grit, on a revolving plate. Disc sanders can be used for sanding wood or metal. However, only certain types of sanding discs are suitable for metal. Discs made from aluminum oxide may be used for sanding wood or metal. Discs made from flint paper are usable only for wood sanding.

Grinders have rigid grinding wheels instead of flexible discs. They cut metal only. Grinders may be used to shape metal, grind down welds, and remove metal as needed. Portable grinders may also be used to turn wire brushes used for cleaning metal. In agricultural mechanics, wire brushes are often used to remove rust and scaling paint.

Some special safety precautions are in order when using a grinding wheel. Some of these are as follows.

- Wear a face shield.
- Always check the grinding wheel for cracks or damage before use. Do not use a wheel that shows any sign of damage.
- Be sure to use wheels that are designed for the machine.
- Tighten the wheel securely and carefully.
- Never use a grinding wheel that is less than one-half of its original diameter.
- When preparing to grind small pieces, secure them in a vise, if possible.
- Do not grind metal in areas of combustible gases or materials.
- Hold the machine with both hands at all times.
- Do not discharge sparks against persons, clothing, or other combustible materials.

The procedure for using a sander or grinder is as follows.

PROCEDURE

1. Select the correct sanding disc, grinding wheel, or wire brush for the job.

CAUTION:

- **Be sure the wheel or disc is rated to turn at speeds higher than that of the machine.**

2. Install the appropriate guard for the job being done.
3. Wear appropriate face protection and protective clothing.
4. Be sure the work is properly secured.
5. Keep the power cord out of the way of the machine.
6. Grip the machine firmly with both hands and turn on the switch.
7. Settle the turning sanding disc, grinding wheel, or wire brush, onto the work slowly.
8. Touch the work gently with the wheel. This is to avoid the wheel catching the work and throwing metal particles toward the operator, bystanders, or flammable materials.
9. Do not apply pressure to the machine. The weight of the machine is generally sufficient to sand, grind, or brush properly. Keep the wheel clean and sharp. Keep a fresh abrasive disc on the machine to grind or sand quickly and efficiently.
10. After turning the switch off, do not lay the machine down until it has completely stopped.
11. Most machines have a rest or a flat spot to rest upon. Do not lay the machine down on its disc or wheel.
12. Remove the grinding wheel or sanding disc and store the machine properly after use.

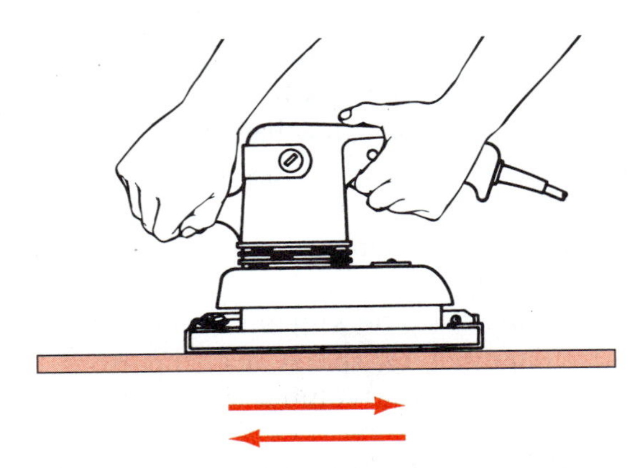

FIGURE 14-9. The pad on a straight-line finishing sander moves backward and forward in a straight line.

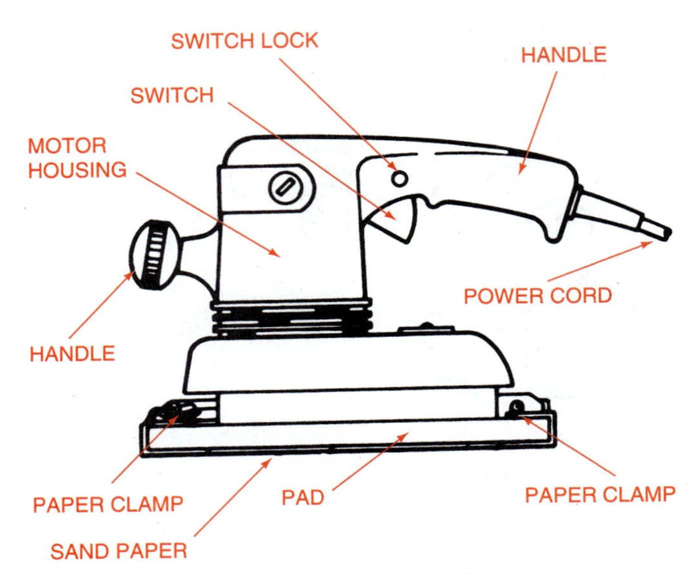

FIGURE 14-10. Major parts of a finishing sander.

Safe Use of Finishing Sanders

The finishing sander is a tool with a small sanding pad driven in a forward-backward or circular pattern. A straight-line sander moves backward and forward in a straight line, figure 14-9. An orbital sander is a finishing sander that moves in a circular pattern. The finishing sander is generally the last power tool used on a project. It cuts slowly but gives the work a very smooth finish.

Important parts of the finishing sander are the:

- motor housing
- handles
- switch
- switch lock
- power cord
- paper clamps
- pad
- sandpaper.

These parts are shown in figure 14-10.

The recommended procedure for using a finishing sander is as follows.

PROCEDURE

1. Use appropriate face and body protection.
2. Check the sander to determine if it is a straight-line or orbital type. Some sanders can be set to operate in either mode. Remember, straight sanding is the smoothest sanding.
3. When starting with rough work, use coarse sandpaper for the first sanding. Switch to medium sandpaper, and then to fine sandpaper for the final sanding.
4. To prepare sandpaper, cut the paper to the appropriate size so it fits the machine. Precut sandpaper can also be purchased.
5. To install sandpaper, loosen the clamps on the pad, insert the paper, and close the clamps.
6. Proceed to sand applying only slight pressure on the sander. The sander should be in constant movement over the work.
7. Remove the dust from the work frequently to keep the material being removed from clogging the sandpaper.
8. Store the machine properly when finished.

PORTABLE SAWS

Portable saws are useful for carpentry projects and repair activities on the farm and in other agricultural settings. Their use will speed jobs and permit the operator to move tools to the job quickly and easily.

Portable saws include the sabre saw, reciprocating saw, and circular saw. The action of sabre and reciprocating saws is up and down or back and forth. They are compact, portable, and useful for cutting curves in plywood, paneling, dry wall, and other sheet materials. The circular saw is generally known simply as a power hand saw. The power hand saw has a circular blade and is used extensively for cutoff work.

Safe Use of Sabre Saws

The sabre saw is also referred to as a bayonet saw. It is used primarily to cut curves or holes in wood, metal, cardboard, and similar materials, figure 14-11. Sabre saws are compact. The saw blade motion is up and down rather than circular and continuous. This feature makes sabre saws less hazardous than

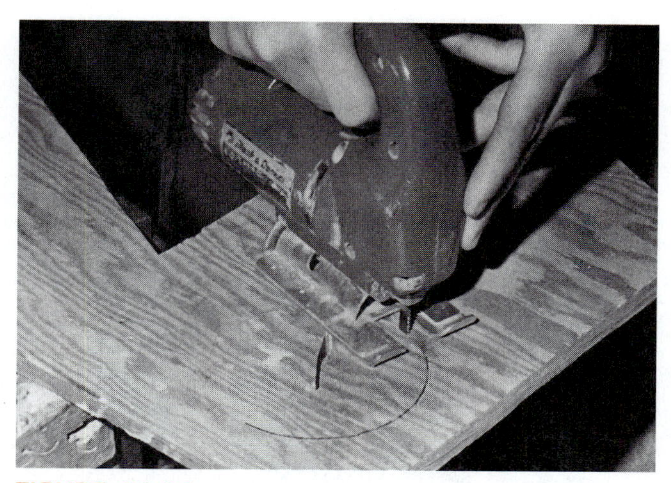

FIGURE 14-11. The sabre saw is a convenient tool for making curved and irregular cuts.

circular saws; they are dangerous nonetheless if not handled properly. The major parts of the sabre saw are the:

- motor housing
- base
- blade
- toe
- blade retention screw
- handles
- switch
- power cord.

These parts are shown in figure 14-12. A safe procedure for using the portable sabre saw follows.

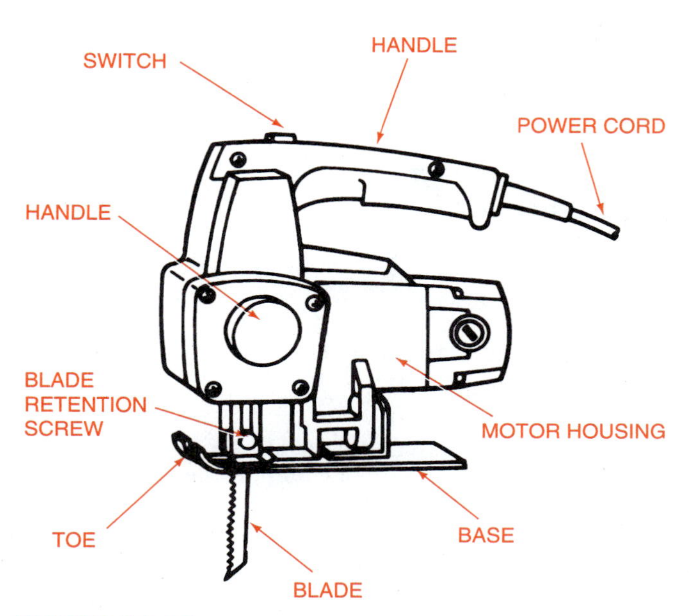

FIGURE 14-12. Major parts of a sabre saw.

PROCEDURE

1. Wear proper eye protection and protective clothing.
2. Select the correct blade for the job. Narrow blades permit shorter turns than wider blades but break more easily. Blades are available for cutting wood, metal, cardboard, and other material. Follow the manufacturer's instruction for the selection and use of blades.
3. Insert the blade and tighten the blade retention screw firmly. Do not apply so much pressure that the threads on the screw are stripped.
4. Adjust the base for the work being done. Some saws have bases that tilt to make angle cuts. Almost all saws have a depth control.
5. Carefully secure all material being cut.
6. Start the cut at the edge of the board. When making inside cuts, bore a hole inside the circle, insert the blade, and proceed.
7. A blind cut is a cut made by piercing a hole with the saw blade. To start blind cuts, place the toe of the base against the material with the blade free to start without striking the material. Start the saw, and carefully roll the saw backward to permit the blade to saw a hole gradually into the material.
8. Use firm pressure and uniform forward movement when operating the saw.
9. Make turns slowly and carefully and give the saw an opportunity to cut as it is directed into the curve.
10. Near the end of the cut take extra care to support the work. Also, reduce the rate of speed and pressure on the saw.
11. When finished, remove the blade and lightly retighten the blade retention screw to prevent accidental loss of the screw.
12. Store the tool properly.

Safe Use of Reciprocating Saws

The reciprocating saw, also referred to as tiger saw, is held and operated much like a portable drill. The action is at the end rather than underneath the tool. The main parts of the reciprocating saw are the:

- motor housing
- blade retention screw
- blade
- shoe
- vents

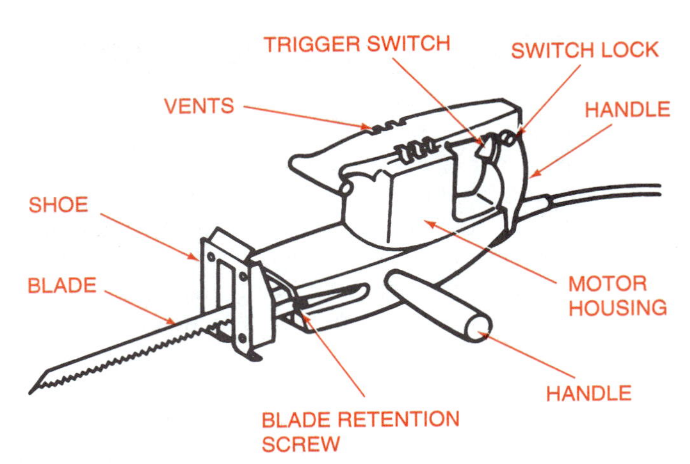

FIGURE 14-13. Major parts of a reciprocating saw.

- handles
- power cord
- trigger switch
- switch lock.

These parts are shown in figure 14-13. Safe operating procedures for using the reciprocal saw include the following.

PROCEDURE

1. Wear eye protection and protective clothing.
2. Select the correct blade for the job. Blades are available for cutting wood, plastic, metal, and other materials. Follow the manufacturer's recommendations for selecting blades for the reciprocal saw.
3. Select the speed for the job. Some saws have high speeds for woodworking and low speeds for metalworking.
4. To make a blind or plunge cut, rest the saw on its shoe and gradually tilt the blade forward into the work.
5. Handle the saw as needed to make cuts in a fashion similar to the sabre saw.
6. Be careful not to bind, pinch, or crowd the blade when sawing.
7. Operate the saw with the shoe against the work at all times.
8. Remove the blade for storage.
9. Keep extra blades on hand for a variety of jobs.
10. Be especially careful not to wear loose fitting clothing. Observe other general safety precautions when using the reciprocating saw.
11. Always use the manufacturer's recommendations whenever there are questions of use and safety.

FIGURE 14-14. The circular saw is used extensively for ripping and cutoff work. Special blades are available for cutting concrete, brick, and stone.

Safe Use of Circular Saws

The portable circular saw, also called a power hand saw, is a lightweight, motor-driven, round-bladed saw. It is perhaps the most popular saw used by people doing woodworking in agricultural mechanics. This saw is useful for many building and repair jobs on the farm and in other agricultural settings, figure 14-14. The major parts of the portable circular saw are the:

- motor housing
- handles
- power cord
- trigger switch
- switch lock
- guard
- guard lift lever
- retractable guard
- blade
- spindle
- base
- angle adjustment lock
- angle scale.

These parts are shown in figure 14-15.

The portable circular saw can be a very useful tool. However, its high speed and tendency to kick back make it a very dangerous tool. The user is cautioned to pay attention to detail and hold the tool very carefully when using it. A safe procedure for using the power hand saw follows.

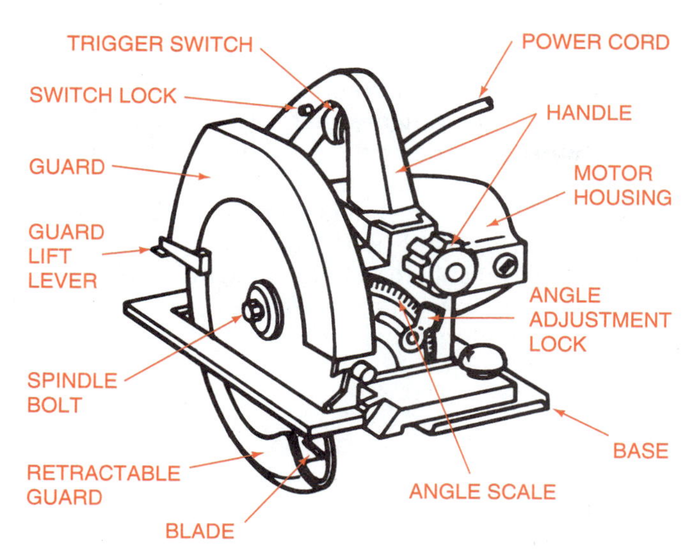

FIGURE 14-15. Major parts of a portable circular saw.

PROCEDURE

1. Always use a face shield and protective clothing.
2. Choose an appropriate blade for the work being done. Special blades include crosscut, rip, combination, hollow ground, and safety blades.
3. Install the blade with care. Use the securing washers in the appropriate sequence. Use the correct wrench to tighten the spindle bolt. If a bushing is used for the blade to make it fit the spindle, be sure the bushing is in place. Install the blade so the bottom saw teeth point toward the front of the saw.
4. Support the work with trestles or other solid materials to avoid pinching or binding of the blade when sawing.

CAUTION:

- **Do not attempt to saw a board between two trestles. The board should be placed so the waste piece can drop off without binding the saw.**

5. Adjust the blade for depth so that only ¼-inch, or the length of a saw tooth, extends below the material.
6. Hold the saw securely with both hands at all times.
7. Start the saw while it is positioned near the work. Be careful that the blade is not touching another object or clothing.
8. Move the saw steadily into the work. If the saw stalls, back it up to clear its teeth from the material and correct the cause of the stalling.
9. When sawing, watch the line ahead of the saw and move the saw so its reference mark stays on the line.
10. Near the end of the cut, reduce the pressure and release the switch as the cut is being finished.
11. When ripping, use the ripping guide if provided on the saw and if it is long enough for the cut being made.
12. For bevel cuts, adjust the angle of the base to the angle of the cut desired. Handle the saw the same as for other crosscuts or rip cuts.
13. When making a pocket cut, place the front of the saw base on the board, turn on the saw, and gradually lower the saw into the work.
14. When cuts are finished, be sure the saw blade stops completely before placing the saw at rest.
15. When the saw is not in use, unplug it to avoid accidental starting. Later model saws have a special switch or locking device which must be pushed before the switch will engage.
16. If the blade is gummy, remove the gum with alcohol or other solvent. If the blade is becoming dull, sharpen or replace it. If the retractable guard does not move freely and cover the blade at all times except when in use, correct the problem.
17. Properly store the tool.

POWER ROUTERS

Power routers may be needed for specialized jobs in agricultural mechanics. A router is a power tool used to cut grooves and ornamental shapes on faces and edges of wood and other soft materials. It may be useful in agricultural mechanics for making rabbet, dado, and dovetail joints, installing butt hinges, creating decorative edges on wood, making signs, and other decorative work. Routers are generally $^1/_2$ to 1 horsepower and turn bits at very high speeds, figure 14-16. The major parts of a router are the:

- base
- see-through chip deflector
- depth-of-cut lock
- depth indicator

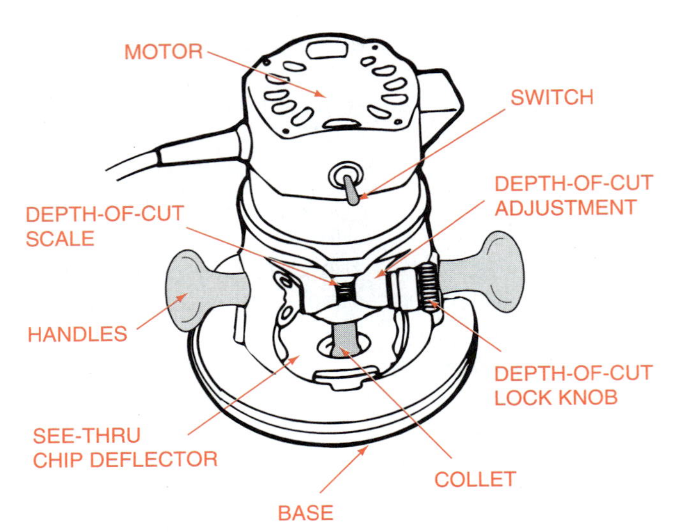

FIGURE 14-16. Major parts of a power router.

- handles
- motor
- switch
- depth-of-cut adjuster
- collet.

Safe Use of Routers

The router has a smooth base that is moved across the material by the operator as the protruding bit turns and creates a groove or cut. The groove or cut is determined by the type, shape, and size of the bit. Bits are made of high speed steel and may be carbide-tipped for extra resistance to dulling. Bits that are dull, rusted, or covered with gum from the wood are dangerous and should not be used. Clean, sharp bits will provide clean even cuts if the router is held firmly to the wood, figure 14-17. To create a straight cut or a cut along desired curved lines, some kind of guide or jig is generally needed. Some guides may come with the router. Others may be purchased from suppliers or devised by the craftsman as needed. Guides are available to help do straight line, circular, or contour routing, figure 14-18.

Some recommended procedures for the safe use of routers follow.

PROCEDURE

1. Thoroughly read the operator's manual and follow its instructions.
2. Clothe yourself with safety goggles and protective clothing.
3. Secure the work firmly before starting.
4. Select a sharp bit of the type needed for the job.
5. With the router unplugged, install the bit according to the manufacturer's instructions.
6. Install appropriate guides or jigs.
7. Set the depth carefully. Do not set deeper than the capacity of the bit.
8. Set up a piece of scrap wood for a trial cut.
9. Plug the router into a safe extension cord.
10. Set the router on its base with the bit clear of the wood, near the starting point, and turn it on.

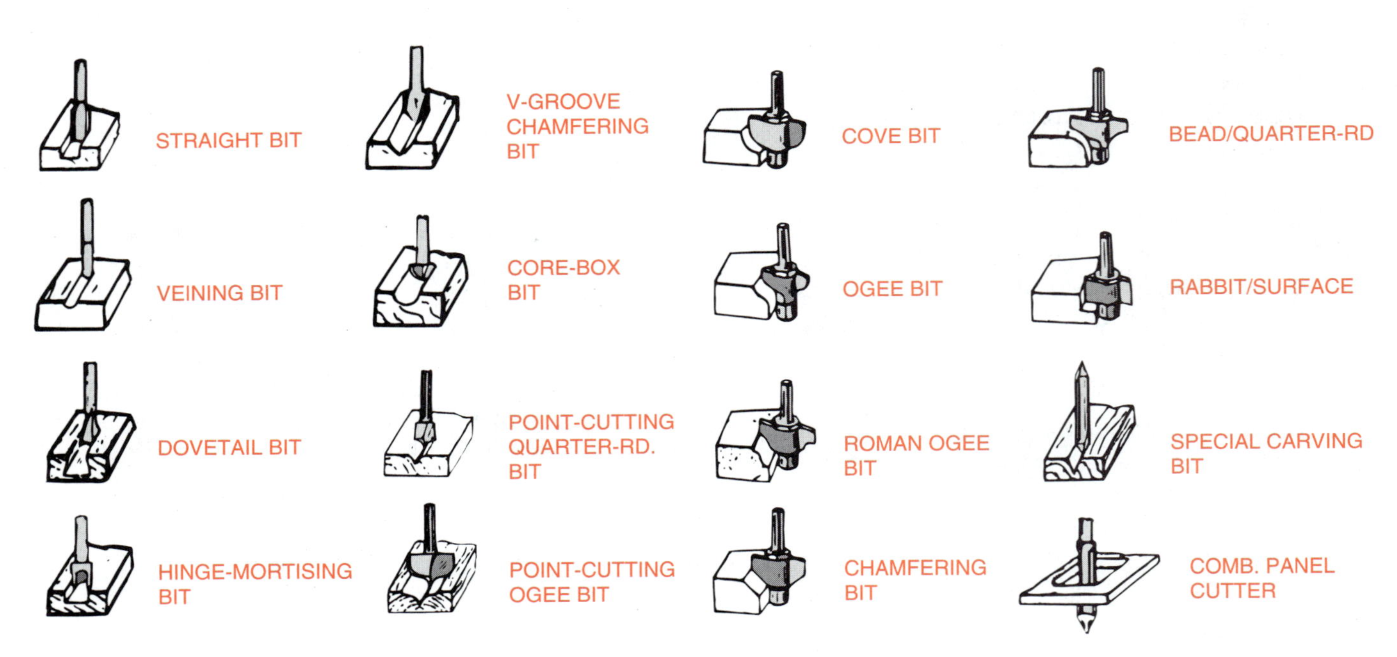

FIGURE 14-17. Common types of router bits.

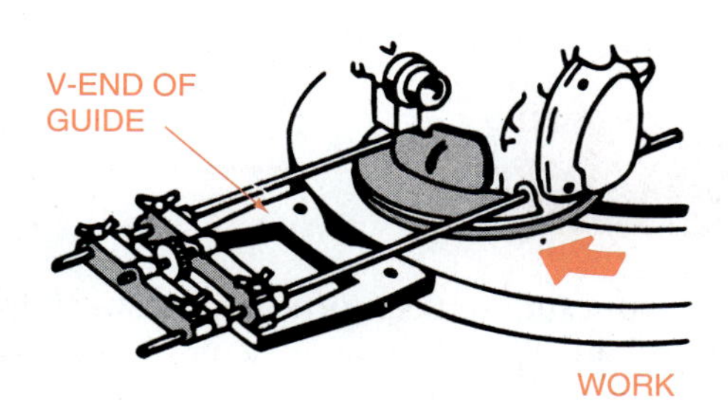

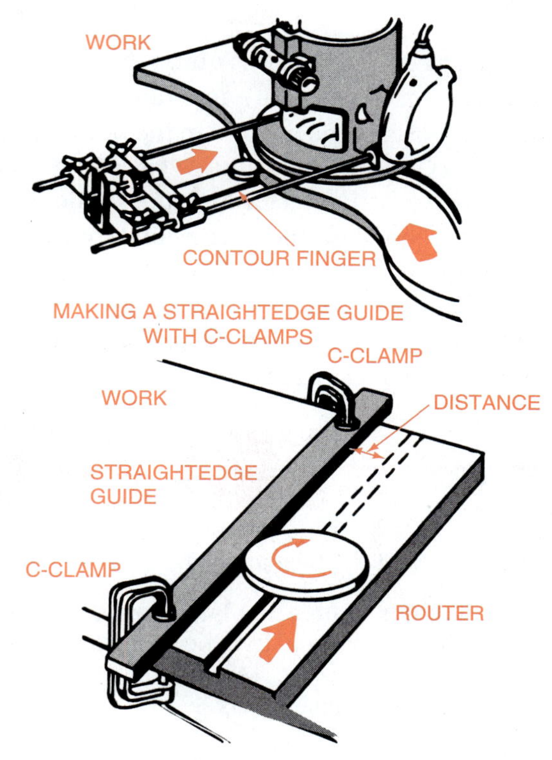

FIGURE 14-18. Edge guides are useful for making both straight and curved router cuts.

11. Move the router slowly into the wood and adjust the pace so as to produce tiny, cleanly severed chips. Move slowly enough into the wood that the motor maintains high speed, yet fast enough to keep cutting without heating the bit or glazing the wood.
12. Follow the guide or jig carefully.
13. Avoid putting side thrust on the bit. Small bits are easily broken and others easily ruined by forcing the bit rather than simply moving it into the wood and letting it cut its way as it progresses.
14. Finish the cut carefully and turn off the router.
15. Place the router in an appropriate stand so it cannot roll and the bit cannot be touched. The bit turns at 20,000+ RPMs and cannot be totally guarded.
16. Examine the cut and adjust the machine or technique as needed.
17. Repeat the process while performing the work on the project.
18. Unplug, clean, and store the router and accessories properly.

Every effort must be made to maintain a safe work environment when using power tools. A moment of carelessness or neglect can permit or cause an accident which can lead to a lifetime of regrets. Power tools are built to save labor and perform tasks quickly and accurately, and they are a real asset to the responsible user.

STUDENT ACTIVITIES

1. Define the Terms to Know in this unit.
2. Identify and correctly spell the names and parts of all power tools described in this unit.
3. Examine each portable power tool in the school agricultural mechanics shop. Determine its size classification, general condition, and the accessories available for use.
4. Study the operator's manual for each power tool in the shop.
5. Use each power tool described in this unit, under the close supervision of the instructor.
6. Examine all portable power tools you have at home. Work with your parent(s) to arrange for any repairs needed on any of the tools.
7. Obtain a three-wire extension cord for the power tools at home if needed.
8. Arrange to have all bits, blades, and other accessories at home sharpened as needed.

SELF-EVALUATION

A. **Multiple Choice**. Select the best answer.

1. A power tool that is double insulated has parts internally insulated in two locations or has a
   a. plastic motor housing
   b. three-wire cord
   c. special carrying case
   d. continuous duty cycle
2. Turning the first wrap of a power cord tightly around a power tool can cause the
   a. housing to break
   b. tool to use more electricity
   c. warranty to be voided
   d. insulation in the cord to break
3. When using power tools, electrical shock hazards may be reduced by wearing
   a. gloves
   b. rubber-soled shoes
   c. safety glasses
   d. coveralls
4. Compressed air is especially dangerous when
   a. coupling and uncoupling hoses
   b. room temperatures are high
   c. portable tools are used
   d. air lines are long
5. Most portable power drills have chuck sizes of
   a. $^1/_8$, $^1/_4$, or $^1/_2$ inch
   b. $^1/_8$, $^3/_{16}$, or $^1/_2$ inch
   c. $^1/_4$, $^3/_8$, or $^1/_2$ inch
   d. $^1/_4$, $^3/_8$, or $^9/_{16}$ inch
6. A variable speed tool is one that will run
   a. a long time without damage
   b. backwards as well as forward
   c. in hot locations
   d. at different speeds
7. A feature that makes some drills especially good for drilling in masonry materials is its
   a. reversible action
   b. variable speed
   c. capacity to hammer
   d. continuous duty cycle
8. When drilling metal with a power drill
   a. use a center punch
   b. make a pilot hole
   c. use a cordless drill if appropriate
   d. all of these
9. Routers are handy for
   a. crosscutting boards
   b. making holes
   c. setting screws
   d. creating grooves and decorative edges
10. A router bit rotates at
   a. 100 RPM
   b. 1,000 RPM
   c. 20,000 RPM
   d. 200,000 RPM

B. **Matching.** Select the word or phrase in column I that best matches the word or phrase in column II.

Column I

1. circular pattern
2. finishing sandpaper
3. flint paper
4. aluminum oxide
5. sabre saw
6. reciprocating saw
7. power hand saw
8. metal housing
9. bushing
10. cordless
11. vent holes
12. crosscut

Column II
- a. good for sanding only wood
- b. good for sanding metal
- c. three-wire, grounded cord
- d. type of saw blade
- e. tiger saw
- f. fine or very fine
- g. ventilation
- h. battery pack
- i. bayonet saw
- j. circular saw
- k. blade insert
- l. orbital sander

C. Completion.

Completion. Fill in the blanks with the word or words that will make the following statements correct.

1. When using portable power tools, it is important to have the work well ________.
2. Another word for saw horse is __________.
3. Under no circumstance should one use power hand tools without protecting the eyes with safety _______ or ___________.
4. A three-wire power cord and plug with a ground prong is used on power tools with __________ bodies.
5. The type of saw with the blade extending from its end is called a ____________ saw.
6. Sanding belts should run _________ the grain of the wood.
7. A good accessory to use on portable grinders to remove rust from metal is the ________.
8. Grinding wheels that are ________ should be discarded.
9. When sawing large panels, ________ are used on trestles to provide extra support to the material.
10. The portable power tool that has a retractable guard is the _____ _____.

D. Identification.

Identification. Identify the parts of the following power tools.

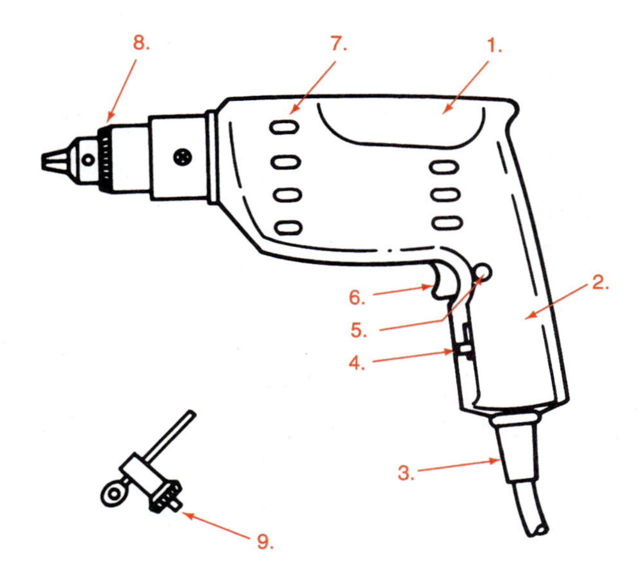

1. Portable drill.

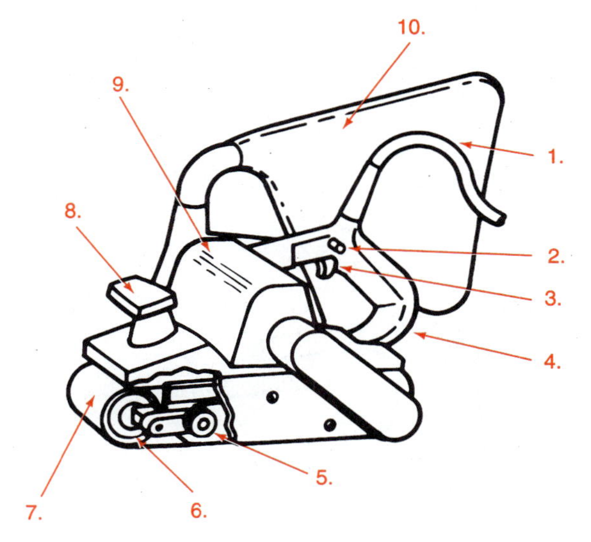

2. Belt sander.

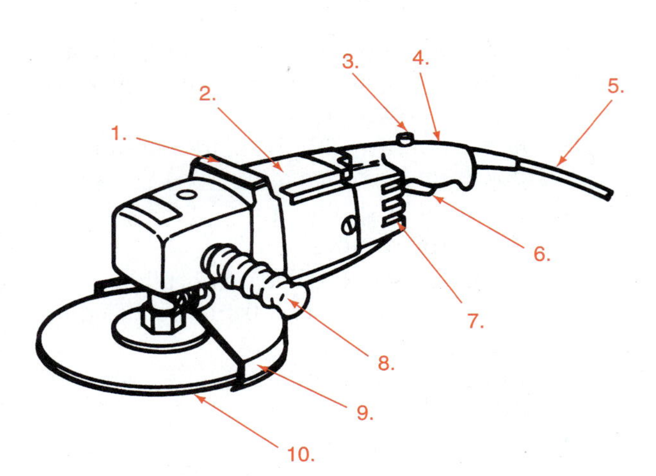

3. Portable grinder.

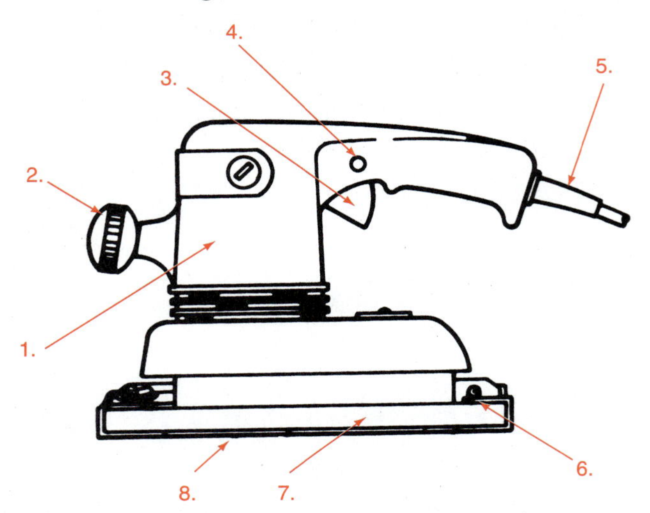

4. Finishing sander.

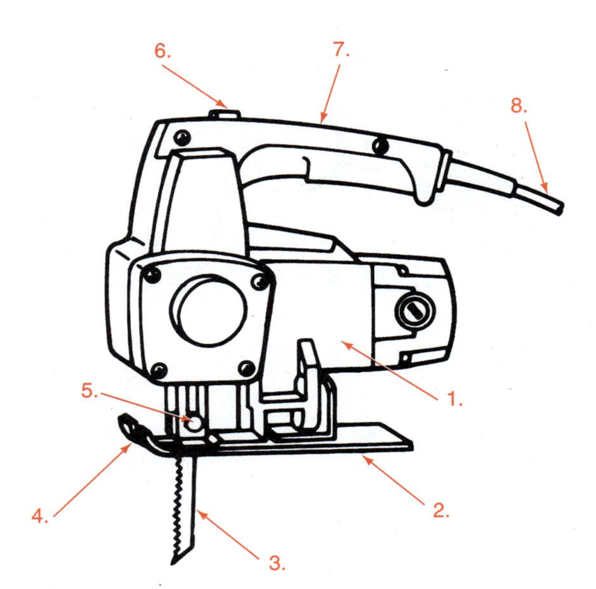

5. Sabre saw.

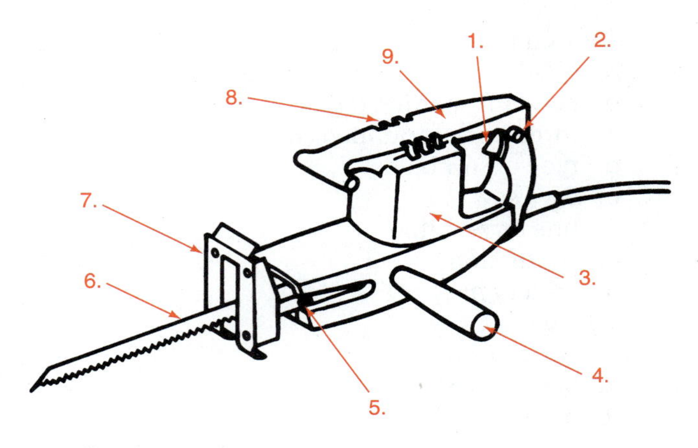

6. Reciprocating saw.

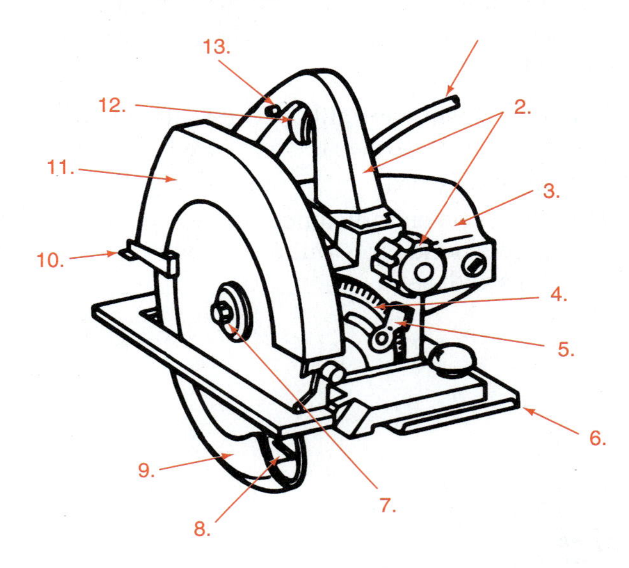

7. Circular saw.

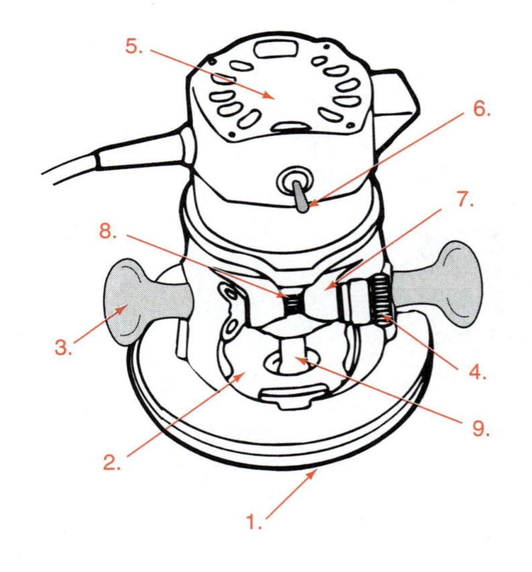

8. Power router.

UNIT

15 Woodworking with Power Machines

OBJECTIVE

To use stationary power woodworking machines in a safe manner.

COMPETENCIES TO BE DEVELOPED

After studying this unit, you should be able to:

- State basic procedures for using stationary power woodworking machines.
- Identify and properly spell major parts of specified machines.
- Operate a band saw.
- Operate a jig saw.
- Operate a table saw.
- Operate a radial arm saw.
- Operate a jointer.
- Operate a planer.
- Operate a sander.

TERMS TO KNOW

- power machine
- stationary
- band saw
- blade guide
- tilting table
- miter gage
- rip fence
- jig saw
- scroll saw
- reciprocal
- pulley
- table saw
- bench saw
- tilting arbor
- parallel
- push stick
- dado heads
- molding heads
- radial arm saw
- pivot
- cutoff saw
- jointer
- dress
- rough lumber

MATERIALS LIST

- ✓ Band saw
- ✓ Jig saw
- ✓ Radial arm saw
- ✓ Table saw
- ✓ Jointer
- ✓ Planer
- ✓ Bench brush and/or shop vacuum
- ✓ Roller stand or helper
- ✓ 1 piece full dimension (rough) lumber, 2" × 4" × 14"

NOTE: Before proceeding with this unit, review the material in Unit 4 on personal safety.

The use of electricity has improved the quality of life for all who use it wisely. Electricity provides a way of moving energy from place to place. Electrical energy is generated from water power, wind power, or fuels and is then sent over long distances by power lines to be received in homes, shops, and industries where it is put to use. In the agricultural mechanics shop, electrical energy is converted by relatively small motors to power in other forms used by shop machines. It is the power that makes machines hazardous to the careless operator. However, the careful operator uses the energy to get work done safely, quickly, and with ease.

SAFETY WITH POWER MACHINES

A power machine is a tool driven by an electric motor, hydraulics, air, gas engine, or some force other than, or in addition to, human power. Some tools such as metal shears may use levers and cams to increase human power. Such machines develop so much force that they, too, should be regarded as power machines.

Large power tools are stationary and should be placed in permanent locations in the shop. The word stationary means having a fixed position.

To reduce the likelihood of injury when working with stationary power machines and to ensure efficient use of the machines, the following are recommended.

- Plan the location of each machine carefully.
- Firmly anchor each machine to the floor.
- Have a licensed electrician provide electrical hookups if needed.
- Use a stripe or narrow line to mark the safety zone around each machine.
- Follow the manufacturer's recommendations for the installation, use, adjustment, and repair of each machine.
- Keep guards and shields in place on each machine at all times.
- Keep blades, knives, and bits sharp.

Safety Precautions

Safety rules are important and should be followed when using all shop machines. Some important precautions are:

1. Wear goggles and/or face shield.
2. Wear protective clothing that is not loose or baggy.
3. Walk—do not run around machines.
4. Only the operator is to be in the safety zone around a machine.
5. Do not use a machine without the instructor's permission.
6. Do not use a machine unless it is in good working order.
7. Perform only the procedures on a machine for which you have had instruction.
8. Do all operations slowly and cautiously.
9. Use a push stick to help push or guide small pieces.
10. Get help with large pieces of stock.
11. If the machine is worked too hard, an overload protector should stop the motor. If this happens, notify the instructor and the instructor will correct the problem.
12. Turn off the machine before leaving it.
13. Unplug the machine or switch off the circuit breaker when changing blades or doing repairs.
14. Clean woodworking machines with a brush or vacuum cleaner. Never clean the machines directly with the hand.

NOTE: Before continuing with this unit, learn these safety precautions.

BAND SAW

The band saw is a power tool with saw teeth on a continuous blade or band, figure 15-1. The band saw will cut straight or curved lines and different kinds of materials. It is fast cutting and versatile when the proper blade and speed are used. Careful operation is absolutely necessary.

The blade of a band saw is flexible and is stretched over flat wheels. The face of each wheel is covered with flat rubber called a tire. The rubber protects the teeth of the saw blade. Generally, the wheel diameter determines the size of the saw. A band saw with 16-inch wheels is classified as a 16-inch band saw. This means the saw can be used to cut to the center of stock up to nearly 32 inches wide.

A blade guide, with carefully adjusted rollers, supports the blade. The blade guide and other parts of the guide and the table assembly of a band saw are shown in figure 15-2. The rollers provide blade support as the operator pushes material into the teeth of the blade. In addition, material may be turned gradually to make curved cuts. The guides keep the blade from bending, breaking, or moving off

FIGURE 15-1. The band saw gets its name from the fact that cutting is done by a thin, continuous band of steel that has teeth along its entire length. (*Courtesy of Delta International Machinery Corporation*)

the wheels during normal use. Blades are available in various types and widths. The more narrow the blade, the shorter the curve which may be cut. Blades used in school shops are generally 1/4 inch to 1/2 inch in width.

Most band saws have a tilting table. The table can be set at various angles so that cuts of 45 to 90 degrees may be made.

Most band saw tables also are equipped with a miter gage. A miter gage is an adjustable, sliding device to guide stock into a saw at the desired angle. Some band saws have a rip fence. A rip fence is a guide that helps keep work in a straight line with the saw blade. It can be adjusted to a distance as close to, or as far from, the blade as the table size permits.

Band saws may have a speed control. This permits the saw to be used for cutting metal as well as wood. A slow-running saw with an appropriate blade is needed to cut metal.

Safe Operation of Band Saws

Band saws are adaptable to many cutting jobs. Therefore, specific instructions are needed for various operations. However, the following procedure is recommended for use of all band saws.

PROCEDURE

1. Obtain the instructor's permission to use the band saw.
2. Put on a face shield and protective clothing.

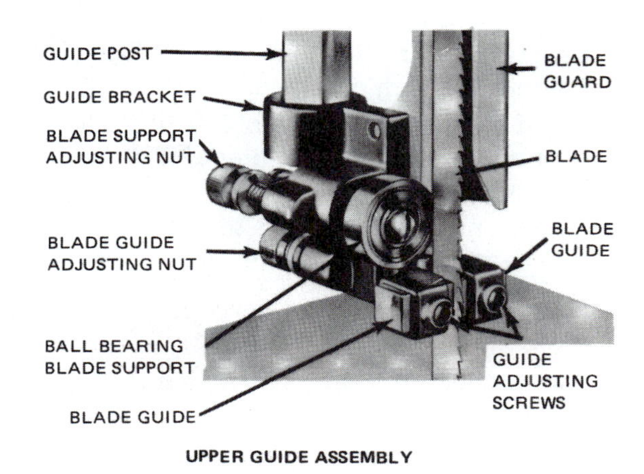

FIGURE 15-2. Guide and table assemblies for a typical band saw. (*Courtesy of Delta International Machinery Corporation*)

3. Check to see if the saw is equipped with a suitable blade for the job.
4. Be sure the blade is tight and the guide is properly adjusted.
5. Move the upper guide assembly down until it is within 1/8 inch of the top of the stock.
6. Place the miter gage or the rip fence in position and adjust as needed.

CAUTION:

- **Never use both the miter gage and the rip fence at one time as the stock will bind and create a serious hazard.**

7. Clear the table and blade of materials. Keep hands out of the work area. Switch the machine on and off again.
8. If the blade seems to be tracking correctly, stand back from the blade and turn the machine on and let it run.
9. Slowly and carefully push the stock to be cut into the teeth of the blade.

CAUTION:

- **Keep fingers and hands away from the front of the blade.**

10. When straight cuts are made, move the stock in a straight line.
11. When curved cuts are made, rotate the stock slowly as the saw cuts.
12. Try to arrange all cuts so you can cut all the way into and out of the stock.

CAUTION:

- **If you must back out of a cut, turn the machine off to do so.**

13. Make sure the blade has stopped before leaving the machine.

JIG SAW

The jig saw is also known as a scroll saw and is designed for sawing curves, figure 15-3. It is a poor tool for straight cutting. The saw cuts by means of the reciprocal action of the blade. Reciprocal means back and forth. The reciprocal action makes it a relatively safe tool if proper procedures are followed. However, the operator may receive serious injuries if the machine is not properly adjusted and if the work is not moved carefully into the blade. The major parts of a jig saw are shown in figure 15-4. Jig saws may be mounted on a bench or on a stand.

Blades for the jig saw are generally $6\,^1/_2$ inches long. They resemble a coping saw blade except they do not have pins in the ends. Blades range in thickness from 0.05 inch to 0.25 inch. When buying blades, the specifications are stated by length, thickness, and width. For example, a $6\,^1/_2" \times 0.020" \times 0.110"$ 20T blade means that it is $6\,^1/_2$ inches long, 0.02 inch thick, 0.11 inch wide and has 20 teeth per inch. The number of teeth per inch indicates the size of the teeth and the speed at which the blade will cut. The narrower the blade, the shorter the turn the blade can make.

FIGURE 15-3. The jig saw will cut very short curves. (*Courtesy of Delta International Machinery Corporation*)

The tension assembly of a jig saw consists of a sliding sleeve, a spring, and a pistonlike rod with a chuck. This upper chuck holds the upper end of the blade; a lower chuck holds the other end. The lower chuck is driven by a crank mechanism in the base. The crank pulls the blade down and a spring pulls it up. Blades must be installed with their teeth pointing down towards the table.

The jig saw's hold-down foot puts spring tension on the material being cut. This tension prevents the

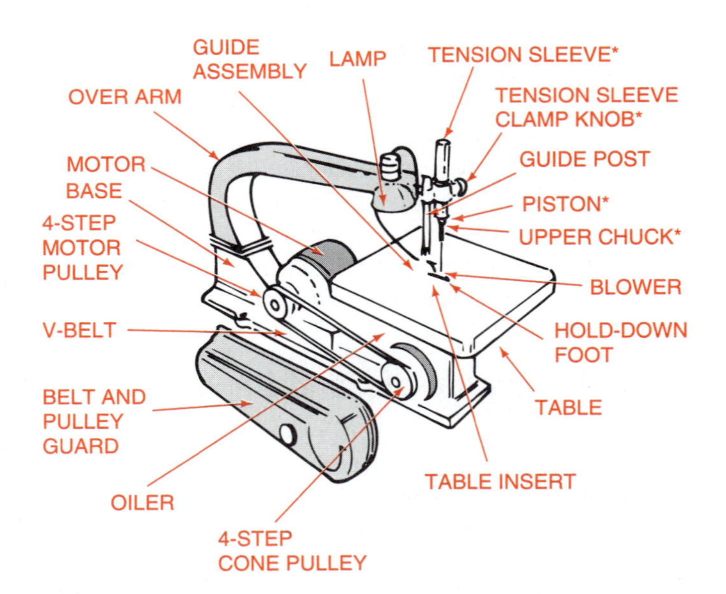

FIGURE 15-4. Major parts of the jig saw. (*Courtesy of Delta International Machinery Corporation*)

material from being lifted from the table as the blade rises. Since the teeth point downward, the blade cuts on the downward stroke.

Many jig saws have special pulleys. A pulley is a round device attached to a shaft and designed to fit and drive or be driven by a belt. Many jig saws have pulleys with two or more steps. A motor with a large pulley drives a machine with a small pulley faster than the speed of the motor. Conversely, if the motor has a pulley smaller than the machine, the machine runs slower than the motor.

Safe Operation of Jig Saws

Jig saws may be used to cut wood, plastic, or metal. However, the correct blade and machine speed must be used. The following procedure is recommended when installing a blade and using a jig saw.

PROCEDURE

1. Obtain the instructor's permission to use the machine.
2. Put on a face shield and protective clothing.
3. Unplug the motor.
4. Select an appropriate blade and determine the appropriate speed to run the saw.
5. Place the belt on the correct pulley steps to achieve the desired speed.
6. Attach the blade in the lower chuck.
7. Rotate the motor pulley until the blade is at its highest point.
8. Attach the blade in the upper chuck.
9. Loosen the tension sleeve clamp knob.
10. Lift the tension sleeve until moderate spring tension is placed on the blade. Retighten the tension sleeve clamp knob.
11. Rotate the pulley by hand to be sure the blade will go all the way up without buckling. Readjust, if needed.
12. Adjust the hold-down foot so it puts slight pressure on the stock.
13. Plug in and turn on the machine.
14. Push the work slowly into the front of the blade. Keep fingers to the side of the blade.
15. Make the saw kerf on the waste side of the line.
16. Rotate the stock as the saw cuts so it follows the curves.
17. Avoid backing out of cuts. Saw out across waste wood to the edge if necessary.
18. To make a cut inside a circle, first bore a hole in the stock inside the area to be cut. Remove the saw blade and reinstall it through the hole in the stock. Saw inside the circle.
19. Turn off the machine when finished.

TABLE SAW

The table saw is known also as a bench saw. A table saw or bench saw is a type of circular saw with either a tilting arbor or a tilting table for bevel cuts. A tilting arbor is a motor, belt, pulley, shaft, and blade assembly that moves as a unit, figure 15-5. The operator can adjust the arbor for depth or angle of cut.

Table saws are classified by the diameter of the blade. Most saws are 8 inches, 10 inches, or 12 inches. Saws with 8-inch blades make beveled cuts in wood up to 2 inches thick. The 8-inch saw is recommended for sawing small pieces. Its blade, guides, and guards work well for fine cutting and are easier to adjust than is the case with larger saws. Agricultural mechanics shops are often equipped with the larger, more powerful 10-inch or 12-inch saws. These are more suitable for cutting large boards and rough-sawed lumber.

Most saws have a hand wheel on the left side which is used to adjust the tilt of the blade from 45 to 90 degrees. In addition, there are slots in the table top for a miter gage. The miter gage is also adjustable from 45 to 90 degrees. This permits crosscutting of boards at the desired angle.

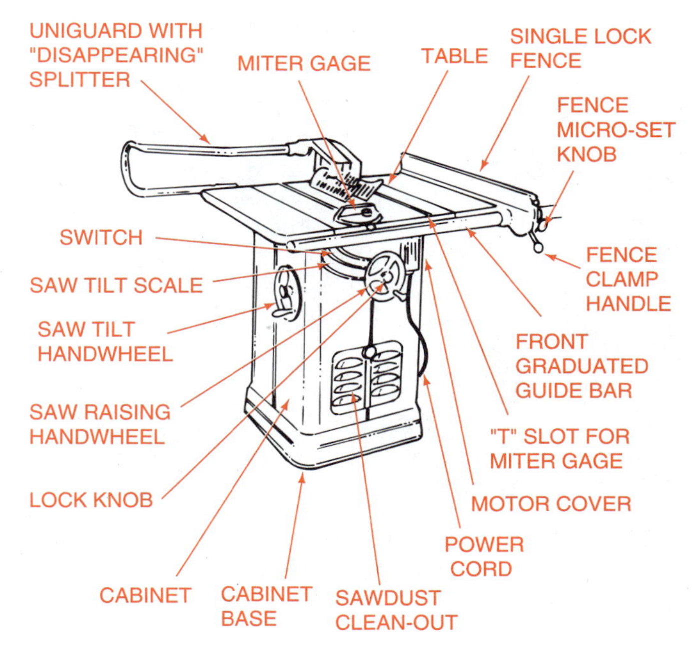

FIGURE 15-5. A tilting arbor table saw. *(Courtesy of Delta International Machinery Corporation)*

A hand wheel on the front of the saw controls depth of cut. The wheel is near the operator and is easy to reach. The saw blade should be adjusted so it is never higher above the board than a distance equal to the depth of the teeth.

Table saws are equipped with a rip fence. The rip fence slides on the front and back edges of the table. It has a scale to indicate the width of cut that will be made at any given setting. When the fence clamp is tightened, the fence guides lumber through the saw for cuts with parallel edges. **Parallel** means two edges or lines are the same distance apart at all points along the length of the object. Saws should be equipped with antikickback devices for ripping, figure 15-6.

All bench saws are provided with a blade guard, figure 15-7. The guard shields the hands of the operator from the dangerous cutting blade. It also provides some protection from splinters, sawdust, sparks, or metal bits that can be thrown by the blade.

The blade guard should always be kept in place. The guard will be lifted by approaching stock which glides under the guard. This provides maximum safety for the operator.

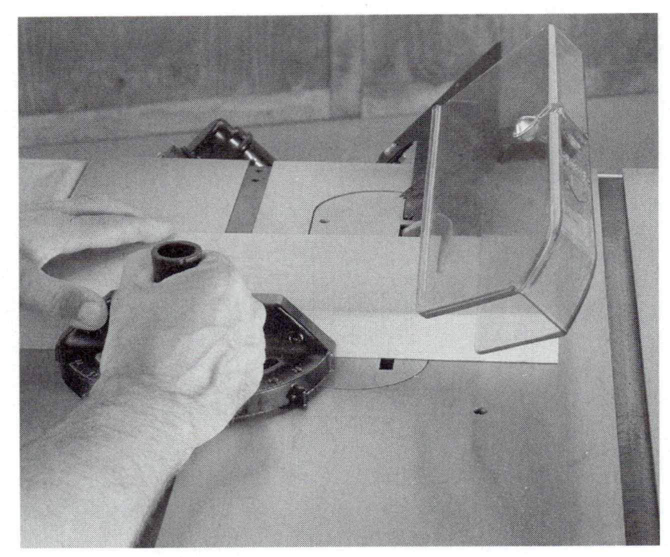

FIGURE 15-7. Blade guards of various types are available for table saws. Here a special see-through design permits the operator to see the blade, yet protected from direct contact with it. *(Courtesy of Delta International Machinery Corporation)*

Sawing Large or Small Pieces

Two people are needed to safely manage pieces of material over three feet long or two feet wide. If a helper is not available, adjustable roller stands to help support the material should be used.

Pieces of wood being ripped to 3 inches or less should be pushed with a push stick. A **push stick** is a wooden device with a notch in the end to push or guide stock on the table of a power tool. Every effort should be made to keep hands several inches from the sides of a saw blade. The operator should never place the fingers or hands in front of the blade.

FIGURE 15-6. Setting the position of the rip fence on a table saw. For an exact measurement, the distance is checked between the ripping fence and a tooth that is pointing toward the ripping fence. Notice the antikickback device on the splitter located behind the saw blade.

Selecting the Correct Saw Blade

Saw blades are classified generally by their diameter. Hence, a blade that has a diameter of 8 inches is called an 8-inch blade.

Blades are also classified by the type of teeth. A blade may be a rip blade, crosscut blade, or combination blade. It may have large teeth for fast, aggressive cutting or small teeth for smooth cutting. All blades must be sharpened according to their design. Blades with carbide tips on their teeth stay sharp longer because of their extra hardness, figure 15-8.

Blades are also designed and classified according to their unique function. For instance, **dado heads** are special blades that can be adjusted to cut kerfs from 1/8 inch to 3/4 inch wide in a pass. Similarly, **molding heads** are blades that hold knives to shape wood into moldings of various types.

Care is needed when installing a blade on a saw. The hole in the blade must be the exact size of the shaft. A suitable bushing can be used if the hole is too large. Further, the slot in the plate in the saw table must be suitable for the blade being used.

Safe Operation of Table Saws

Table saws are used for many sawing jobs. Instruction is needed for each special use. However, the following procedure is advisable when using table saws.

FIGURE 15-8. Many types of blades are available for table saws. The carbide tip blade has become very popular because it stays sharp longer than regular steel teeth. (*Courtesy of Sears, Roebuck, and Co.*)

PROCEDURE

1. Obtain the instructor's permission to use the saw.
2. Put on a face shield and protective clothing.
3. Unplug the motor.
4. Install the correct blade for the job with the teeth pointing toward the direction of rotation.
5. Adjust the blade to the correct angle. Use the degree scale on the saw or use a sliding T-bevel to establish the desired angle.
6. Adjust the height of the blade until only the teeth extend above the board to be sawed.
7. If ripping, move the miter gage off the saw table. Always use the miter gage or the rip fence, but never both at the same time. Set the rip fence for the desired width of cut. This is checked by measuring from the fence to the tip of a tooth closest to the fence.
8. If crosscutting, move the rip fence to the right edge of the table or remove it. Adjust the miter gage to the desired angle. The correctness of the angle is determined by using the degree scale on the gage or by using a sliding T-bevel.
9. Check to see that the guard assembly is in place and the saw is ready.
10. Arrange for a helper or set up stands if a long or large piece of lumber is to be sawed.
11. Plug in the motor.
12. Stand to the side of the blade's path and turn on the motor.
13. Push the work toward the blade with a slow and even movement.
14. Hold the work firmly so as to support both pieces on both sides of the blade.
15. Push the final work through with a push stick.
16. Turn off the machine.

RADIAL ARM SAW

The **radial arm saw** is a power circular saw which rolls along a horizontal arm. The saw can be raised or lowered. The arm will also pivot up to 45 degrees to the left and right.

The many movements of the radial arm saw make it capable of numerous cutting operations. These same movements make it a very dangerous tool. Whole books are written on using the radial arm saw. Only the simpler and more popular uses are described in this unit.

The radial arm saw has many moving parts and major assemblies, figure 15-9. To **pivot** means to turn or swing on. The arm pivots on the column, the yoke pivots under the arm, and the motor and blade assembly pivot at the bottom of the yoke. Each pivot point has a scale to indicate the position of the assembly. Each point also has a lock to hold the assembly once it is set. The major parts of the radial arm saw are shown in figure 15-10.

Special Safety Precautions

When using a radial arm saw it is important to:

- Ask the instructor before attempting any operation.
- Always wear a face shield.
- Unplug or lock the switch in the off position when making adjustments.
- Always pull the saw toward the operator when sawing—never push the saw back into the wood.
- Always have all guards in place.
- Always have one hand on the saw handle when the saw is turning or about to be turned on (it is easy for the saw to accidentally roll forward).
- Never turn the saw on when the blade is touching wood.
- Never stop a turning blade by pushing a piece of wood into it.

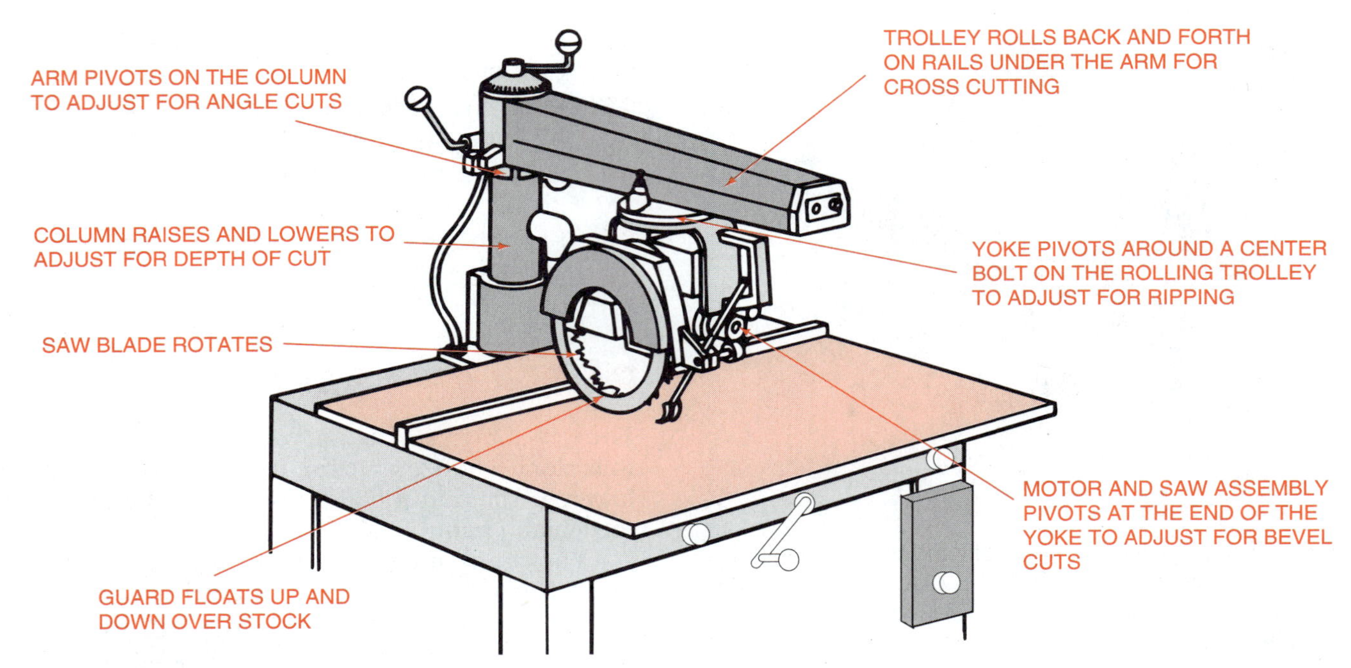

FIGURE 15-9. Moving assemblies on a radial arm saw.

Safe Operation of Radial Arm Saws

The most popular use of the radial arm saw is for cutoff work, including squaring boards, cutting them to length, cutting them at angles and bevels, cutting dados, and cutting rabbets.

Checking the Table for Levelness. The radial arm saw cannot make accurate cuts unless the table is level and parallel to the arm. The following procedure may be used to check the table.

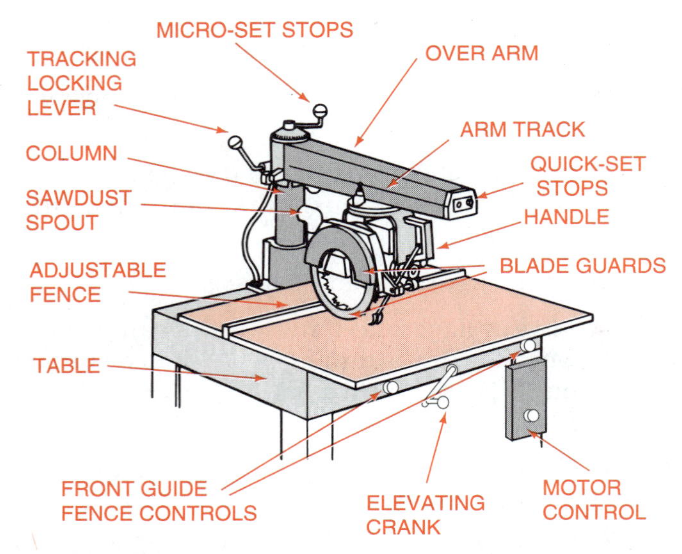

FIGURE 15-10. Major parts of a radial arm saw.

PROCEDURE

1. Obtain the instructor's permission to check the table.
2. Put on a face shield and protective clothing.
3. Unplug the machine or lock the switch in the "off" position.
4. Loosen the column clamp handle.
5. Turn the elevating crank until the points of the teeth on the blade lightly touch the table.
6. Tighten the column clamp handle.
7. Loosen the trolley lock.
8. With the saw stilll turned off, grasp the saw handle and slowly pull the assembly across the table. The teeth should just touch the table at all points. If they do not, report the condition to the instructor. Do not use the saw until the problem is corrected.

Crosscutting Boards. The following procedure is recommended for crosscutting.

PROCEDURE

1. Obtain the instructor's permission to use the saw.
2. Put on a face shield and protective clothing.
3. Mark the board where the saw will start the cut.

FIGURE 15-11. When using the radial arm saw for cutoff work, it is important to hold the work securely and pull the saw slowly into the wood. (*Courtesy of Delta International Machinery Corporation, formerly Rockwell Power Tool Division*)

4. Elevate the saw until the teeth will just reach through the board and touch the table top.
5. Position the board on the table tightly against the fence so the saw will cut on the waste side of the mark.
6. Grasp the saw handle with one hand.
7. Turn on the saw with the free hand.
8. Hold the board with the free hand, being careful to position the hand so the hand and arm are not in line with the blade.
9. Stand out of line of the path of the blade.
10. Very slowly and firmly pull the saw forward until the cut is complete, figure 15-11.
11. Push the saw back to the column.
12. Maintain the hold on the handle of the saw while you turn the saw off and lock the switch using the other hand.
13. Do not touch the lumber until the blade stops.

Crosscutting Bevels. The saw can be swung to the left or right to crosscut at angles. A beveled cut may also be made at any angle within the swing radius of the arm. When cutting a bevel, the following additional steps are required.

PROCEDURE

1. Elevate the saw about 2 inches above the table.
2. Loosen the bevel latch.
3. Swing the saw assembly to the desired angle as shown on the bevel scale.
4. Tighten the bevel latch.
5. Lower the saw to the table.
6. Proceed to crosscut.

Making Dados and Rabbets. Dados and rabbets are wide grooves cut only part way through the board. To cut dados and rabbets, the following procedure is used.

PROCEDURE

1. Mark the board with a sharp pencil to show exactly where the grooves start and stop.
2. Mark the end of the board to indicate how deep the cut should be.
3. Install a dado blade if available.
4. Place the board on the saw table and pull the saw out so the blade is against the end of the board.
5. Elevate the blade until the points of the teeth touch the mark which indicates the depth of cut on the board.
6. Push the saw back and position the board so the cut will remove the desired material.
7. Turn on the saw.
8. Make the pass.
9. Push the saw back.
10. If a dado blade is not being used, move the board $1/8$ inch and make another pass.
11. Repeat until the groove is finished.
12. Turn off the saw.

Many other types of cuts can be made with the radial arm saw. It is recommended that other books or bulletins on the subject be read if additional operations are desired.

CUTOFF SAWS

A cutoff saw is a motor-driven circular blade that may be adjusted for angle cuts and is fed down into the material being cut, figure 15-12A. A cutoff saw for wood may be called a motorized miter box or compound miter box by some manufacturers. These are equipped with wood-cutting blades and are used for making quick, accurate cuts in moldings, floor

A. CUTOFF SAW FOR WOOD

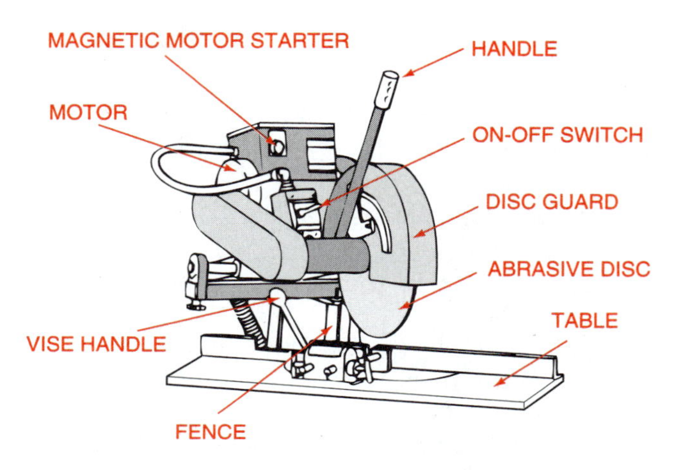

B. CUTOFF SAW FOR METAL

FIGURE 15-12. When equipped with appropriate blades, cutoff saws are fast and efficient for cutting wood, metal, or masonry metals. (*Courtesy of Delta International Machinery Corporation*)

boards, and framing. Similarly, by using the proper reinforced abrasive blades, cutoff saws are useful for cutting metal angles, bars, and rods, figure 15-12B. By using shatter-proof abrasive or diamond blades, cutoff saws may be used to cut masonry materials.

Safe Use of Cutoff Saws

Cutoff saws have a spring-loaded mechanism to hold the motor and saw up and clear of the stock in its parked position. The blade is shielded by a guard, but care must be taken to stay clear of the blade, even when not in use. Material to be sawed must be held firmly or clamped in place to avoid any movement while the cut is being made. The saw may be set to cut any angle from 45 to 90 degrees.

It is important to follow good safety practices when using cutoff saws.

PROCEDURE

1. Put on protective clothing and face shield.
2. Unplug the machine and install the proper blade for the material being cut.
3. Set the saw for the desired angle of cut.
4. Stand to the side of the blade to avoid any thrown particles.
5. Grasp the handle with one hand and turn on the saw with the other.
6. Slowly and carefully lower the saw into and through the stock.
7. Permit the saw to slowly rise to its parked position.
8. Turn off the saw and wait for the saw to stop turning.
9. Release the handle and remove the stock.

Cutoff saws are stationary power tools that are light enough to move to the job. They are relatively low in cost and perform cutoff operations quickly and efficiently. Therefore, they are widely used throughout the construction industry.

JOINTER

The jointer is a machine with rotating knives used to straighten and smooth edges of boards and to cut bevels, figure 15-13. The jointer is potentially a very dangerous tool. The knives can inflict severe cuts, and lumber may be thrown if not handled properly.

The size of a jointer is determined by the length of the knives. Most school shops have either 6-inch or 8-inch jointers. However, jointers are available up to $15^3/_4$ inches or more.

The main parts of the jointer are shown in figure 15-14 and include the base, front infeed table, front table adjusting hand wheel, rear outfeed table, rear table adjusting hand wheel, rabbeting ledge, depth scale, knife assembly or cutterhead, guard, fence, fence clamp, and tilt scale.

All knives are installed so their cutting edges extend to the same height and leave the board smooth and even as the cutter head rotates. The rear outfeed table is adjusted so it is level with the cutting edges of the knives; the table is locked at this level. Any adjustments of the rear outfeed table should be made by the instructor. The fence may be moved across the table to any point. Its position determines the maximum possible width of cut. In addition, the fence can be set at any angle from 45 to 90 degrees.

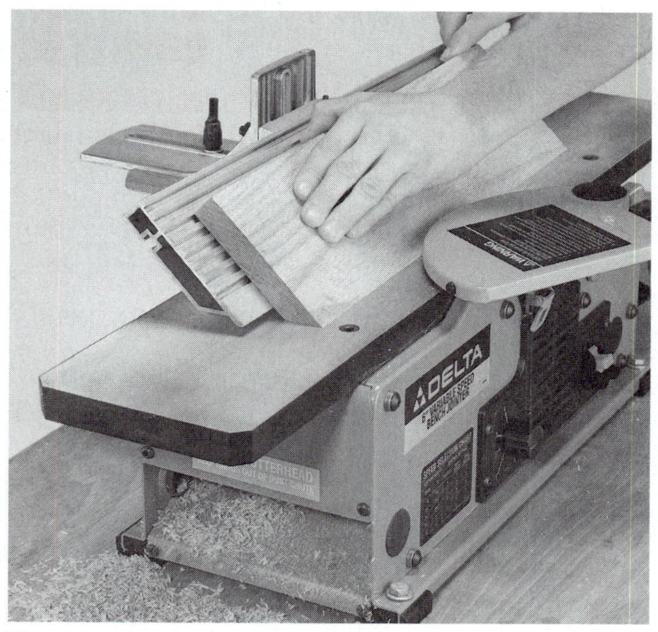

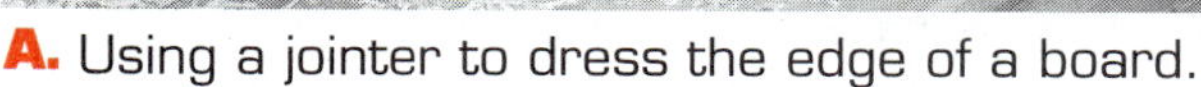

A. Using a jointer to dress the edge of a board.

B. Using a jointer to bevel the edge of a board.

FIGURE 15-13. A jointer is used to dress and bevel the edges of boards. *(Courtesy of Delta International Machinery Corporation)*

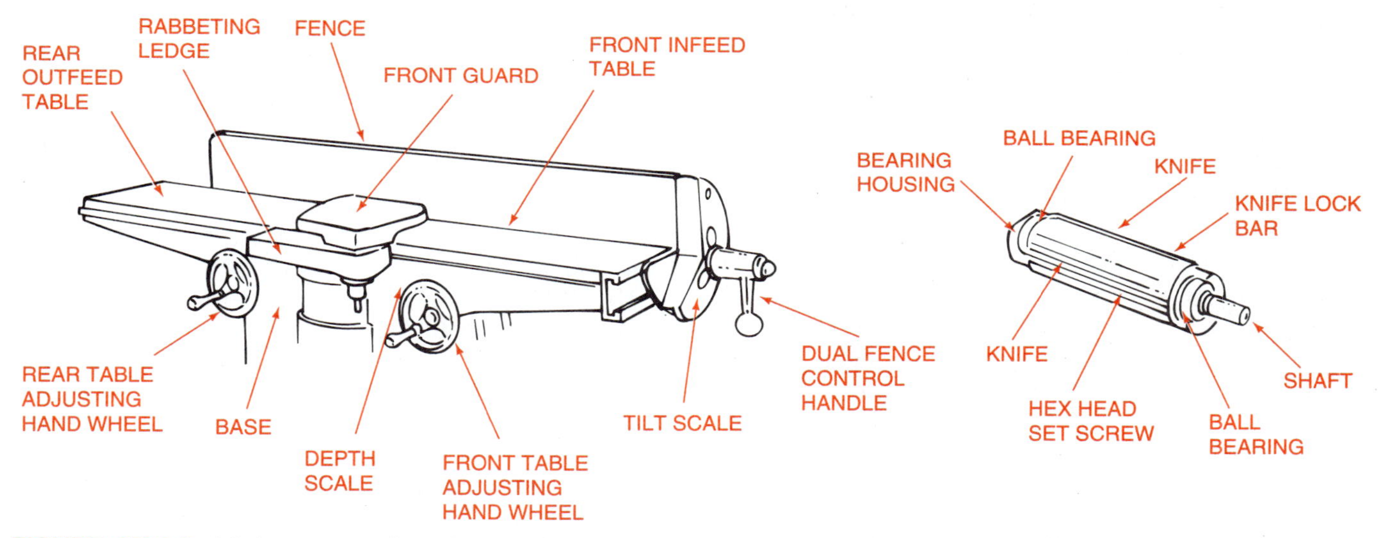

FIGURE 15-14. Major parts of a jointer. *(Courtesy of Delta International Machinery Corporation, formerly Rockwell Power Tool Division)*

This position determines the angle of the edge which is created.

The height of the front infeed table determines the depth of cut. If the front infeed table is exactly level with the knives, the depth of cut will be 0 inches. The table is lowered to increase the depth of cut. The depth of cut is indicated by a scale on the side of the jointer.

Safe Operation of Jointers

The jointer is designed to remove a small amount of wood ($^1/_{16}$ inch or less) at a time. Generally, the operator should joint one edge smooth, then use a table saw to cut the board to $^1/_8$ inch wider than needed. The board is then jointed down to the final width. This leaves the board with both edges smooth.

Strict attention must be given to the correct procedure to be used with the jointer.

PROCEDURE

1. Obtain the instructor's permission to use the machine.
2. Put on a face shield and protective clothing.
3. Check to see that the guard is covering the knives.
4. Adjust the fence to the proper angle (usually 90 degrees).
5. Adjust the front infeed table for a 1/16-inch cut. Only 1/16 inch per cut should be removed. Never cut deeper than 1/8 inch in one pass.
6. Be sure the board is 12 inches or more in length. Arrange for a helper if the board is longer than 4 feet.
7. Stand aside and turn on the machine.
8. Place the board on its edge against the fence on the front infeed table.
9. Grip the board by placing the hands over the top edge.

CAUTION:

- **Keep fingers high on the board and not over the cutterhead.**

10. Advance the board forward against the fence with moderate downward pressure.
11. When the leading hand is nearly over the knives, reposition it on the board over the rear outfeed table.

CAUTION:

- **Keep hands away from the cutter knives.**

12. When most of the board has passed over the knives, transfer the second hand so it is also over the rear table.
13. Continue moving the board to finish the pass with decreasing pressure near the end. Use a push stick to complete the pass for boards that are less than 4 inches wide. Do not use the jointer on boards less than 2 inches wide.
14. Reverse the board from end to end each time another pass is made. This decreases the tendency to create a wedge-shaped board after many passes.
15. Make the last pass with the grain to get a smooth cut.

FIGURE 15-15. The planer dresses surfaces and leaves the board at the desired thickness. *(Courtesy of Delta International Machinery Corporation, formerly Rockwell Power Tool Division)*

PLANERS

The planer (also known as a thickness planer or surface planer) is a machine with turning knives that dress the sides of boards to a uniform thickness. To dress means to remove material and leave a clean surface.

The planer is an excellent tool to convert low-cost, locally grown lumber into smooth, uniform materials. Planers are used to dress rough lumber to a desired thickness and smooth surface. Lumber is rough when it comes from a sawmill. Rough lumber may be thicker on one end than on the other. By making successive passes through the planer, a board can be dressed down to the desired thickness, figure 15-15.

The planer is also an excellent tool to level and smooth wide pieces made by gluing boards together. Agriculture students can make attractive projects such as gates, wagon bodies, and bench tops from rough lumber at a reasonable cost if a planer is available.

The planer is a massive machine with many parts. It includes an adjustable bed with smooth rollers to support lumber, corrugated rollers to draw lumber into the machine, and a rotating cutter head with knives that dress the lumber. Once a board is started in the machine, the machine is self-feeding. The parts of the planer are shown in figure 15-16. However, most parts of the planer are shielded and normally hidden from view.

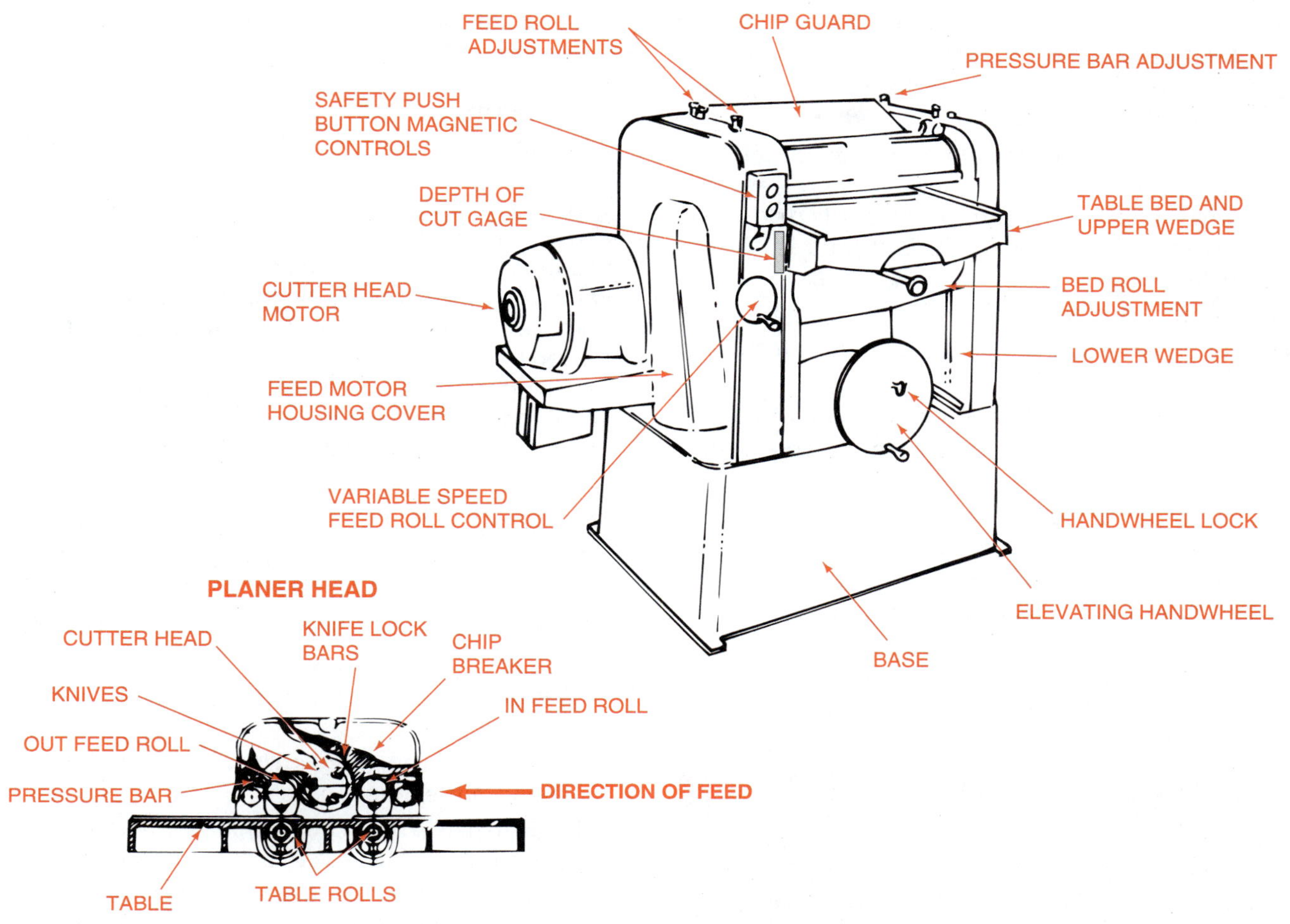

FIGURE 15-16. Major parts of the planer. (*Courtesy of Delta International Machinery Corporation, formerly Rockwell Power Tool Division*)

Safe Operation of Planers

Several critical adjustments on the planer must be made by the instructor or power tool specialist. These include installing knives, adjusting the feed rollers, and adjusting the smooth rollers.

CAUTION:

- **Students are not to make these adjustments.**

If used carefully, the planer will safely dress large volumes of lumber. If used carelessly, the knives can be ruined by the first board planed after newly sharpened knives are installed. Careless use can also cause the operator to be seriously injured. The planer has a large motor capable of throwing massive pieces of lumber. Great care must be exercised when using the planer.

When using a planer, it is important that no metal or loose knots be in the lumber. The cutter heads may throw these items, or the planer may be damaged by them.

The planer generates large volumes of dust and wood chips. Therefore, a vacuum dust and chip removal system should be used to reduce the dust created by planing.

The following procedure is recommended for dressing rough lumber with a planer. This same procedure is used for edge-planing materials that are 2 inches thick or more. Lumber should be reasonably straight for edge planing.

PROCEDURE

1. Obtain the instructor's permission to use the planer.
2. Put on a face shield and protective clothing.
3. Check the machine and be sure the chip guard and other shields are in place.
4. Sort the lumber according to its general thickness.

CAUTION:

- **Never plane lumber that has grit or nails, staples, or other metal in it.**

5. Identify the thickness of the thickest end of the thickest piece of lumber.
6. Measure the thickest part of the thickest board.
7. Turn the thickness adjustment wheel until the thickness scale reads 1/8 inch less than the thickest piece to be planed. For instance, if the board is 2 inches thick, the planer should be set at 1 7/8 inches.
8. Arrange for a helper.
9. Clear bystanders out of the path in front and back of the planer.
10. Set the feed control at its slowest rate.
11. Stand to the side and turn on the planer motor.
12. Have the helper place the thinner end of the board on the front end of the bed.
13. Lift the other end of the board until it is slightly higher than the end on the planer bed.
14. While staying beside the board and keeping the end high, push the board slowly forward into the planer.

CAUTION:

- **When the thick part of the lumber is in the planer, the feed rollers will pull the lumber forward. Otherwise, the operator must push the lumber.**

15. When the lumber is halfway through the planer, the helper should move to the back of the machine to pull the lumber through and complete the first pass.
16. Run additional boards through the planer.
17. Raise the bed 1/8 inch with the thickness adjustment wheel.
18. Repeat steps 12 through 17 until both sides of all pieces are smooth and within 1/8 inch of the desired thickness.
19. Make one final pass on each side of the boards, going with the grain, and removing 1/16 inch per final pass.
20. Turn off the machine and wait for the machine to stop.

NOTE: The last pass on each side of the board should be made so it is planed with the grain and removes only 1/16 inch per pass. This technique results in a smooth finish. The depth of each cut may vary from as little as 1/32 inch to 1/8 inch. The adjustments should be planned so the final pass leaves the board at the desired thickness.

SANDERS

Some agricultural mechanics shops may have stationary sanders, figure 15-17. For procedures in sanding, refer to Portable Sanders in Unit 14. When using portable sanders, the machine is moved to the material. To use stationary sanders, the material is moved to the machine.

Sanders may be equipped with belts or discs suitable for use on wood or metal. The procedure for using a stationary sander is as follows.

PROCEDURE

1. Obtain the instructor's permission to use the sander.
2. Wear safety glasses or other eye protection.
3. Wear a dust mask.
4. Use a belt or disc that is not clogged with particles.
5. Turn on the dust collector (if so equipped) and the sander motor.
6. Grip the material firmly with fingers well away from the surface to be sanded.
7. On belt sanders, place the work on the belt lightly and move it against the stop.
8. On belts, move the material back and forth in a sideways motion. This helps prevent the belt from over heating and becoming clogged with particles.
9. On discs, use the half of the disc moving downward toward the table. Again, keep the stock moving on the table.
10. Use only moderate pressure to sand.

FIGURE 15-17. Using a stationary sander.

11. Reduce pressure before lifting or pulling the work from the belt or disc.
12. When doing freehand sanding of faces, ends, edges, and corners, use light pressure and keep the work moving at all times.
13. Turn off the machine when finished.

STUDENT ACTIVITIES

1. Define the Terms to Know in this unit.
2. Identify and correctly spell the names of major parts of each machine in this unit.
3. Examine each stationary power woodworking machine in your shop and make a list of items that need maintenance or repair. Submit the list to your instructor for review and appropriate action.
4. Properly clean a band saw, table saw, radial arm saw, jointer, planer, and other woodworking machines in the shop.
5. Do the following Power Working Machine Exercise.

POWER WOODWORKING MACHINE EXERCISE

(Courtesy of Douglas Hering, Frederick High School, Frederick, MD)

Purpose: The purpose of the Power Woodworking Machine Exercise is to obtain experience quickly. It requires the use of many power tools yet the material used is generally scrap material in an agricultural mechanics shop.

Note to the Teacher: Students should be instructed on the use of each machine before operations are performed.

Power Machine to Use	Procedures
Radial Arm Saw or Portable Circular Saw	1. Obtain a rough 2 × 4 that is at least 2 inches thick and 4 inches wide. Cut it to a length of 14 inches, figure 15-18.
Planer	2. Plane the 2 × 4 to a thickness of $1^{1}/_{2}$ inches.
Jointer	3. Joint one edge of the 2 × 4 one time or until smooth. (Jointer should be set at a depth of $^{1}/_{16}$ inch.)
Table Saw	4. Place the jointed edge of the 2 × 4 against the fence. Rip the 2 × 4 to a width of $3\,^{5}/_{8}$ inches.
Jointer	5. Joint the rough side of the 2 × 4 to a width of $3\,^{1}/_{2}$ inches.
Jointer	6. Joint the top edges of the 2 × 4 at 45 degree angles, $^{1}/_{8}$ inch deep. This is done by setting the jointer for a $^{1}/_{16}$ inch cut and making 2 passes.
Radial Arm Saw	7. Cut the 2 × 4 to a length of 12 inches.
	8. On the bottom side of the 2× 4, mark and label the angles shown in figure 15-19 using a try square and pencil.
	9. Cut the pieces labeled "A" from the 2 × 4.
Band Saw	10. Cut th epiece labeled "B" from the 2 × 4.
	11. On the bottom side of the 2×x 4, starting at the pointed end, make pencil marks at 3, 6, and 9 inches down the center. Locate the exact center so that cross marks can be drawn, figure 15-20.
Portable Hand Drill	12. Drill a $^{1}/_{2}$-inch hole through the 2× 4 at the cross mark closest to the point.
Drill Press or Portable Power Drill	13. Drill two 1-inch holes through the other two cross marks with a spade bit.
	14. Connect the edges of the two 1-inch holes with a straight edge and pencil. Label the area inside as "C," figure 15-21.
Sabre Saw or Jig Saw	15. Cut "C" out by sawing on the two pencil lines.
Router (optional)	16. Shape the inside edge of the top side of the 2 × 4, moving the router clockwise around the hole.
Disc Sander or Belt Sander	17. Sand the point of the 2 × 4 project lightly.
	18. Put your name on the bottom side of the 2×x 4 project in pencil.
	19. Turn your project in to your instructor to be graded, figure 15-22.

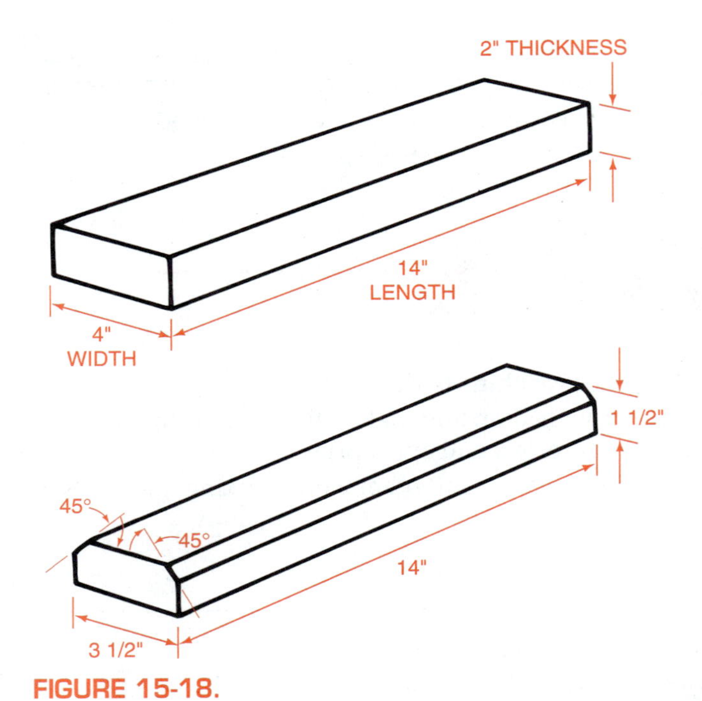

FIGURE 15-18.

FIGURE 15-19.

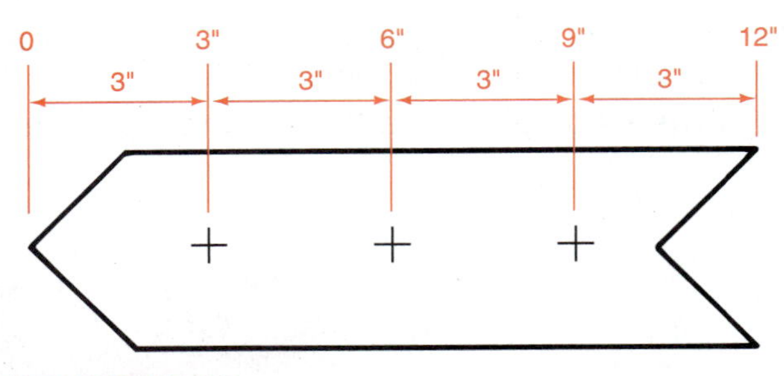

FIGURE 15-20.

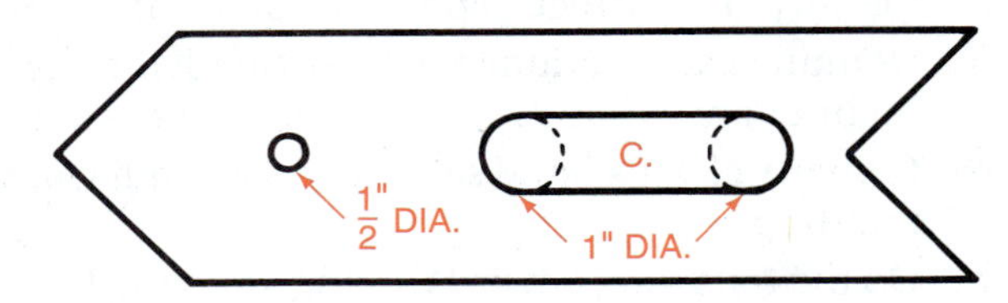

FIGURE 15-21.

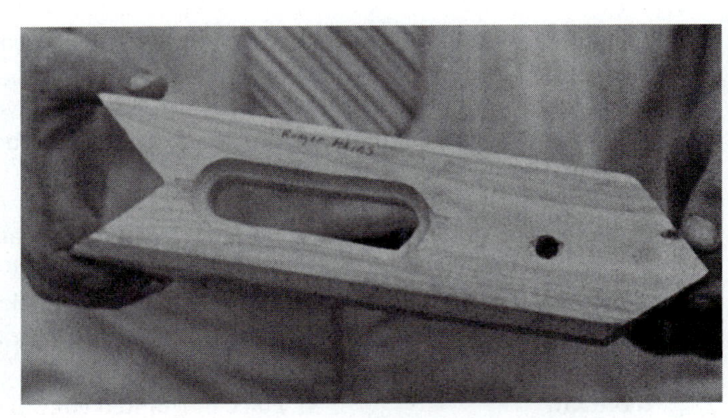

FIGURE 15-22. The completed project for the exercise should look like the piece shown in this illustration.

SELF-EVALUATION

A. Multiple Choice. Select the best answer.

1. The number of people permitted in the safety zone around a machine is
   a. one
   b. two
   c. three
   d. any number
2. Woodworking machines should be cleaned with
   a. a brush
   b. the hand
   c. a rag
   d. an air gun
3. If a machine is worked too hard, the electric motor should stop because of
   a. burnout
   b. general fatigue
   c. overload protection
   d. voltage drop
4. When using a band saw, the operator should avoid
   a. backing out of cuts
   b. crosscutting
   c. cutting metal
   d. sawing curved lines
5. The band saw blade is held in position when cutting by
   a. wheels
   b. tires
   c. levers
   d. guides
6. Jig saws are best for cutting
   a. very short curves
   b. straight lines
   c. rabbets
   d. dados
7. The blade on a table saw should extend how far above the work?
   a. 1 inch
   b. 2 inches
   c. 3 inches
   d. none of these
8. Small pieces of wood should be moved on a saw table by a
   a. bare hand
   b. gloved hand
   c. push stick
   d. hammer handle
9. Table saw blades are classified by the
   a. type of teeth
   b. size in diameter
   c. unique function
   d. all of these
10. The most popular use of the radial arm saw is
    a. ripping
    b. cutoff work
    c. curve cutting
    d. dado cutting
11. Minimum protection when using power machines starts with
    a. steel-toed shoes
    b. leather apron
    c. finger guards
    d. face shield
12. When cutting with the radial arm saw, the operator should
    a. move the wood into the saw
    b. pull the saw into the wood
    c. push the saw into the
    d. any of the above are safe
13. An adjustment that should be made on a jointer by the instructor is the
    a. rear outfeed table
    b. front infeed table
    c. fence
    d. miter gage
14. The maximum safe depth per cut by a jointer is
    a. 1/2 inch
    b. 1/4 inch
    c. 1/8 inch
    d. 1/16 inch
15. The power machine that generates large volumes of wood chips is the
    a. band saw
    b. bench saw
    c. jointer
    d. planer
16. For a good job when planing lumber, the operator should
    a. make each final pass while planing with the grain
    b. make each final pass with a shallow cut
    c. end up with the desired thickness
    d. all of these

B. Matching. Match the word or phrase in column I with the correct word or phrase in column II.

Column I

1. miter gage
2. rip fence
3. radial arm saw
4. many moving assemblies
5. tilting arbor
6. rotating cutterhead
7. dado head
8. planer
9. sander
10. jig saw

Column II

a. table saw
b. parallel to the blade
c. guides angle cuts
d. saw kerf $^{1}/_{8}$ inch to $^{3}/_{4}$ inch wide
e. jointer
f. dress sides of lumber to a uniform thickness
g. also called a cutoff saw
h. radial arm saw
i. special need for dust mask
j. reciprocal action

C. Completion. Fill in the blanks with the word or words that will make the following statements correct.

1. Electrical wiring for shop machines should be installed by a __________ __________.
2. The safety zone around a machine should be marked with a __________.
3. Bits, knives, and blades should be kept __________.
4. Do only the procedures on a power machine for which you have had __________.
5. The table saw is also known as a _____ _____.

UNIT

16 Metalworking with Power Machines

OBJECTIVE

To safely use stationary power machines for metalworking in agricultural mechanics.

COMPETENCIES TO BE DEVELOPED

After studying this unit, you should be able to:

- State basic procedures for using stationary machines for metalworking.
- Identify and properly spell major parts of specified machines.
- Operate a drill press.
- Operate a grinder.
- Operate power metal-cutting saws.
- Operate a power shear.
- Operate a metal bender.

TERMS TO KNOW

- drill press
- grinding wheel
- grit
- face
- dress
- coolant
- horizontal
- horizontal band saw
- shear

MATERIALS LIST

- ✓ Drill press
- ✓ Grinder and wire wheels
- ✓ Power hacksaw and/or metal-cutting band saw
- ✓ Metal shear
- ✓ Metal bender
- ✓ Drill bits, 1/8 inch and 1/2 inch
- ✓ Length of 1/8" x 1" flat steel (longer than 6 inches)

NOTE: Before proceeding with this unit, review Unit 4, Personal Safety in Agricultural Mechanics, and the topic Safety with Power Machines in Unit 15.

Power is used in a number of ways for metalworking in agricultural mechanics. It is used to cut, make holes, shape, and sharpen metal. These functions will be discussed under the following headings of drill press, grinder, metal-cutting power saws, power shears, and metal benders.

DRILL PRESS

The drill press is a stationary tool used to make holes in metal and other materials. Its design and structure permit it to drive large drills and apply heavy pressure on bits. When material can be placed on a drill press, this tool is preferred over the portable power drill for drilling large holes. The drill press is also preferred for precision drilling or when many holes are needed.

Drill presses are available as floor models or bench models. The parts are similar for both. The drill press consists of a base, column, table, and head. The head is an assembly consisting of the motor, switch, drive belt, speed control, shaft, quill, chuck, and feed. Parts of the drill press are shown in figure 16-1.

Safe Operation of the Drill Press

The motor of the drill press drives a shaft, which, in turn, drives the chuck. The speed of the chuck is controlled by the use of step pulleys, variable speed pulleys, or a variable speed motor. When the operator moves the feed handle, a gear moves the quill up and down. This movement raises and lowers the chuck to permit drilling, figure 16-2. The movement of the quill may be limited by the depth stop, or the quill may be locked for special functions.

The table slides up or down on the column. This permits thick or thin material to fit on the machine. The table may be swung to one side so large objects can rest on the base in a drilling position. Some drill presses have tables that tilt. Tilting tables permit materials to be supported for angle drilling.

The following procedure is recommended for the safe use of the drill press.

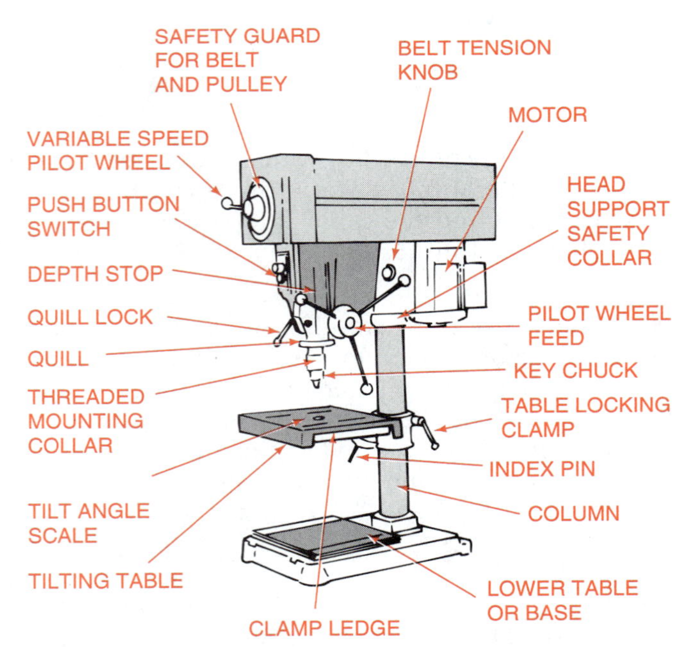

FIGURE 16-1. Major parts of the drill press. *(Courtesy of Delta International Machinery Corporation, formerly Rockwell Power Tool Division)*

FIGURE 16-2. When the operator moves the feed handle, a gear moves the quill up and down. This action raises and lowers the chuck allowing drilling to take place. *(Courtesy of Delta International Machinery Corporation)*

PROCEDURE

1. Obtain the instructor's permission to use the machine.
2. Wear a face shield and protective clothing. Leather gloves are recommended when drilling sharp or irregular-shaped metal.
3. When drilling wood, use a sharp pencil to mark where holes are to be drilled. A cross (+) or caret (^) is useful to mark an exact location. Use a scratch awl to mark metal.
4. When drilling metal, use a center punch to aid in starting the drill.
5. Use only straight shank drills in gear chucks, and taper shank drills in taper chucks.
6. Select the correct drill by observing the size stamped on the shank, or use a drill gage to determine drill size. Use a pilot hole for holes larger than $^3/_8$ inch.
7. Tighten gear chucks securely by inserting the chuck wrench into at least two of the three holes, twisting the wrench as tightly as possible in each.

CAUTION:

- **Be sure to remove the chuck wrench after tightening.**

8. Check table alignment to prevent damage to the table by the drill bit. To do this, move the feed handle until the drill passes down through the hole in the center of the table and then return it. Also, a flat piece of wood may be used to protect the table.
9. Clamp flat material to the table; clamp round material in a V-block.

CAUTION:

- **The long end of the material should be on the operator's left and against the drill press column to prevent any rotation.**

10. Hold metal or clamps with a gloved left hand.
11. Turn on the motor.
12. Grasp the feed handle and lower the drill to start the hole.
13. Apply steady pressure while drilling.
14. Add cutting oil to cool the drill when drilling steel.
15. Ease off and use very little pressure as the drill breaks through the metal.
16. Raise the drill.
17. Turn off the machine.
18. Use a file to remove burrs from the metal.
19. Use a bench brush to remove metal chips from the project and work area.
20. Use a cloth to clean oil from the project and work area.

NOTE: Refer to Unit 10, Fastening Wood, and Unit 13, Fastening Metal, for applications in which drilling is needed.

GRINDER

A grinder removes metal by abrasive action. Grinders are available in many types and sizes. They are used to sharpen tools, shape metal, prepare metal for welding, and remove undesirable metal.

Most shop grinders consist of a pedestal, a double-shafted motor, switch, wheels, guards, tool rests, and safety shields, figure 16-3. Some grinders have a light to improve vision and a water pot for cooling metal. Small grinders have wheels that are 6 inches or 7 inches in diameter and 1 inch wide. Large grinders generally have wheels that are 12 inches in diameter and 2 inches wide.

Grinders may be equipped with grinding wheels or wire brush wheels. A grinding wheel is a wheel made of abrasive cutting particles formed into a wheel by a bonding agent. The cutting particles are called grit.

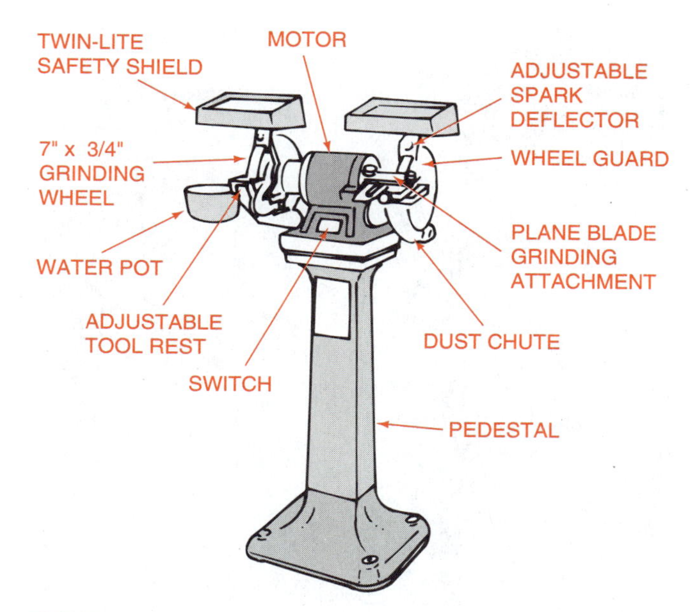

FIGURE 16-3. Major parts of the grinder. (*Courtesy of Delta International Machinery Corporation, formerly Rockwell Power Tool Division*)

When selecting a grinding wheel it is important to use the correct wheel for the job. The wheel must have the correct speed rating for the motor. Failure to select the proper wheel can result in a serious accident. The wheel may fly into pieces, hurling rocklike fragments into the face and body of the operator and bystanders.

When replacing wheels, obtain specifications from the operator's manual or from the wheel on the machine. Bushings that fit the shaft must be used when the hole in the wheel is larger than the shaft.

A coarse or very coarse texture is recommended for wheels used to shape metal for sharpening, welding, or construction. A medium texture wheel is good for sharpening axes, mower blades, and other tools. Fine textured wheels are not generally recommended for agricultural mechanics except for power oil stone or wet stone grinders. Fine wheels cut slowly and overheat the metal easily.

Grinding wheels are manufactured to be used on one surface only. The surface intended for use is called the face.

Wire wheels are useful for removing rust and dirt from machinery parts, figure 16-4. Such cleaning is useful before welding or painting. Wheels with large, stiff wire are called coarse wire wheels. Coarse wheels cut aggressively and can be dangerous. Care must be exercised to use a tool rest to support metal being brushed. Otherwise, the wheel may drag the metal down between the wheel and the guard. Such accidents may result in injury to the operator, a damaged object, or a broken machine. Some instructors

FIGURE 16-4. Grinder equipped with a wire wheel. Wire wheels on grinders are useful for removing rust and welding slag.

prefer to remove the tool rest when objects to be brushed are large. Fine wire wheels are good for moderate cleaning jobs. They are softer than coarse wheels and safer for the operator.

Safe Operation of Grinders

In addition to the points previously mentioned, the following procedure is recommended for the safe operation of a grinder:

PROCEDURE

1. Obtain the instructor's permission to use the grinder.
2. Wear a face shield.
3. Wear close-fitting leather gloves and a leather apron. Small work pieces should be held with lever lock pliers.
4. Check the wheels. Use the machine only if the wheels are clean and free of nicks, chips, and cracks.

CAUTION:

- **If any of these conditions are not met, have the instructor inspect the wheels and correct the problem(s).**

5. Adjust the tool rest so its top surface is level with the center of the motor shaft and within 1/16 inch of the wheel. For bevel grinding, set the tool rest at the appropriate angle.

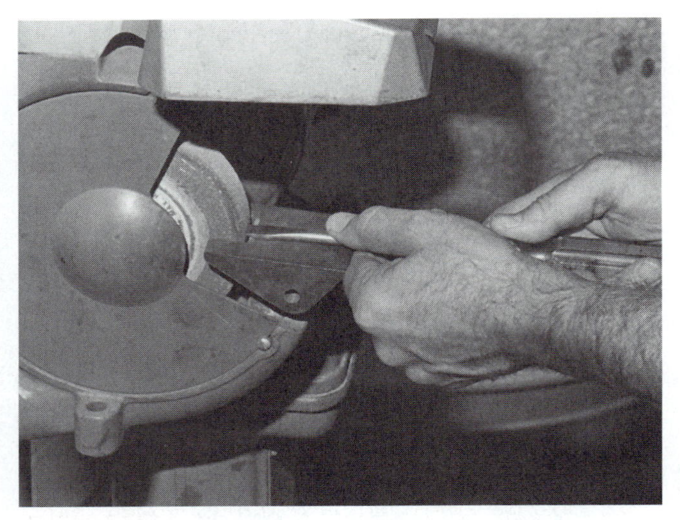

FIGURE 16-5. The tool rest helps position and guide small objects being ground. The tool rest should be adjusted so it is 1/16 inch or less from the wheel and level with the motor shaft.

6. Hold the metal to be ground firmly on the tool rest, figure 16-5.
7. Move the metal back and forth or in a curved motion as needed. The movement will help keep the stone clean and avoid overheating of the metal.

CAUTION:

- **Do not use excessive pressure on the metal. The speed of grinding is determined by the grit and speed of the wheel, not by pressure.**
- **Never grind with the side of the wheel.**
- **Grinding creates heat, therefore, avoid handling the metal with bare hands.**

8. Turn off the machine when finished and wait for it to stop running.

Dressing the Grinding Wheel. After proper instruction, the student may dress a grinding wheel. To dress a wheel means to remove material so the wheel is perfectly round with the face square to the sides and sharp abrasive particles exposed. Dressing a wheel removes clogged abrasives and slight bulges that cause the wheel to get out of balance, figure 16-6. The procedure for dressing a wheel follows.

PROCEDURE

1. Obtain the instructor's permission.
2. Wear a face shield. Leather gloves and leather apron are also recommended.

FIGURE 16-6. When dressing a grinding wheel, material is removed to leave the wheel perfectly round. The face is left square with the sides of the wheel.

3. Wear a dust-type filter respirator.
4. Obtain a wheel-type grinding wheel dresser.
5. Turn on the grinder.
6. Place the dresser on the tool rest of the grinder.
7. Slowly rock the dresser wheels forward until they touch the grinding wheel.
8. Apply firm and even pressure on the dresser and move it back and forth across the wheel for about 30 seconds.

CAUTION:

- **Be prepared for abrasive particles to fly off the wheel. The wheel is cleaned, balanced, and squared by removing particles.**

9. Remove the dresser and turn off the grinder.
10. Readjust the tool rest so it is within 1/16 inch of the wheel. Then rotate the wheel by hand to check for roundness, figure 16-7A.
11. Use a combination square or the tool rest to determine if the face of the wheel is square to the side, figure 16-7B.
12. Repeat steps 5 through 11 if additional dressing is needed.

Applications for grinders will be found in units on metalworking, welding, tool fitting, and others. The grinder is a basic tool and is found in most shops.

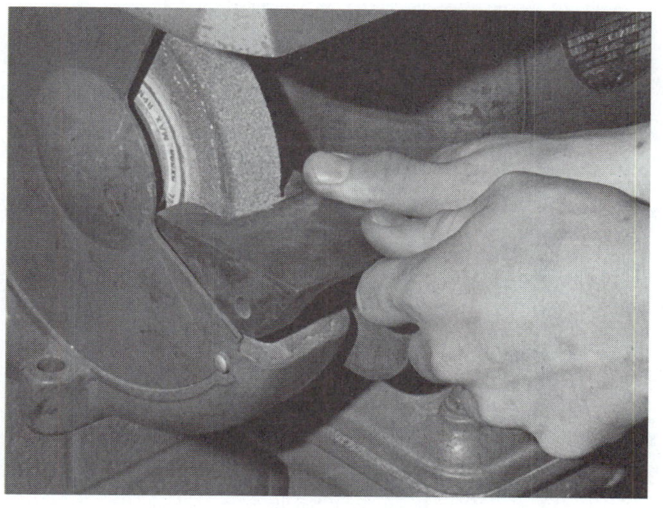

A. Checking a wheel for roundness using the tool rest.

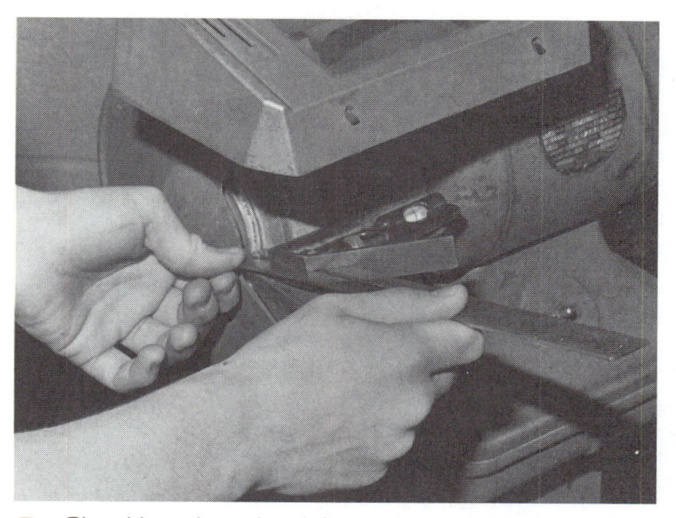

B. Checking the wheel for squareness of the face using a combination square. The tool rest may also be used if it has not been damaged.

FIGURE 16-7. Checking the grinding wheel.

METAL-CUTTING POWER SAWS

Metal-cutting power saws are reciprocating hacksaws, band saws, or thin grinding-type wheels that cut metal. The hacksaw and band saw are used most often in agricultural mechanics.

Hacksaws

The power hacksaw has a reciprocating movement that operates like a hand hacksaw, figure 16-8. However, it is motor driven and cuts much faster. Fast-cutting saws have a system to pump a coolant onto the cutting area. A coolant is a liquid used to cool parts or assemblies.

FIGURE 16-8. The power hacksaw has a reciprocating movement that works like a hand hacksaw. The blade is hard and, therefore, brittle and easily broken. The metal being cut must be clamped carefully to eliminate blade breakage due to binding.

The power hacksaw consists of a motor-driven frame mounted on a stand. The frame holds a rigid blade which is 3/4 to 1 inch wide and 12 inches to 18 inches long. Blades are available with fine to coarse teeth. Very hard blades are needed to cut hard steel. Such blades are brittle and break easily if the frame is dropped or if the saw binds. Extra care is needed to prevent blade breakage.

The power hacksaw stand is narrow. Therefore, metal must be supported to avoid tipping the machine. Sturdy models can support the weight of heavy steel. However, long stock must be supported by stands or a helper.

The power hacksaw is very useful in the agricultural mechanics shop. However, an expensive blade can be broken in an instant if the machine is not used properly. The following procedure is recommended for using a power hacksaw.

PROCEDURE

1. Obtain the instructor's permission to use the saw.
2. Wear a face shield, leather gloves, and a leather apron.
3. Check the machine to be certain it has the proper blade and that the blade is tight.
4. Place the frame in the raised position.
5. Adjust the vise on the machine to hold the metal at the desired angle.

CAUTION:

- **If the vise does not hold the metal so it is rigid, the blade will break.**

6. Position the metal in the vise on the machine and tighten securely.
7. Turn on the machine.
8. Lower the frame slowly and carefully until the blade is on the stock and starting to cut.
9. Turn on the coolant, if it does not turn on automatically.
10. Stay near and watch the machine while it is cutting.

CAUTION:

- **Do not put pressure on the blade or otherwise interfere with the machine while it is running.**

11. Switch off the machine when the cut is finished if it does not turn off automatically.
12. Remove all scrap metal and clean up all metal dust and coolant.

Horizontal Band Saw

The word horizontal means flat or level. The horizontal band saw has a blade that saws parallel to the ground. Generally, however, it is constructed like an upright band saw. It has a band-type blade that travels on wheels and moves through rollers and guides, figure 16-9. Since blade movement is forward at all times, it cuts continuously. As a result, it cuts faster than a power hacksaw.

The procedure for cutting with a horizontal band saw is similar to the procedure described for a hack-

FIGURE 16-9. The horizontal band saw is especially useful for cutting large pieces of structural metal, such as round stock, flats, angles, I beams, and channels. (*Courtesy of Wells Manufacturing*)

saw. However, the manufacturer's instructions should be followed. Machines vary in minor details of design and function.

POWER SHEARS

Sheet metal is easy to cut with hand shears. Therefore, power sheet metal shears may not be seen in many agricultural mechanics shops. To shear means to cut at an angle. Generally, shearing is done with two movable blades.

The cutting of flat, angle, and other structural steel is difficult by hand, and slow by power saw. The cutting of these materials is fast and clean when a shear is used, figure 16-10. Some shears have features that permit them to bend, cut, and punch metal. To operate a hand power shear, the following procedure is recommended.

PROCEDURE

1. Obtain the instructor's permission to use the machine.
2. Put on a face shield, leather gloves, and a leather apron.
3. Raise the handle of the shear.
4. Insert the metal to be cut or punched.
5. Carefully align the cut mark with the cutting edge of the stationary blade.
6. Support the metal so it is level.
7. Lower the handle until the cut is complete.

CAUTION:

- **Keep hands away from the cutting shears or punch.**

8. Store the handle so it cannot accidentally fall and operate the shear.

CAUTION:

- **The weight of the handle alone is sufficient to close the shears with enough power to cut off fingers.**

The procedure described for use of a hand power shear also applies to hydraulic shears. In the case of hydraulic shears and punches, power is delivered by cylinders driven by hydraulic fluid.

METAL BENDERS

Bending metal by hand is discussed in Unit 12. However, some shops have hand-operated benders. These are useful for bending small round and bar stock. By using cams, pins, and levers, such tools permit fast and accurate bends, figure 16-11.

FIGURE 16-10. Metal shear for cutting flat and round stock. Special knives for some shears make them capable of cutting angle iron.

FIGURE 16-11. The hand metal bender is useful for bending round, flat, and angle stock. (*Courtesy of Strippit/Di-Acro Houdaille*)

Sheet metal benders are available for making rain gutters, heating and air-conditioning ducts, and other commercial uses, figure 16-12. However, sheet metal bending in agricultural mechanics is generally limited to repair work. Such occasional work may be done with hand tools.

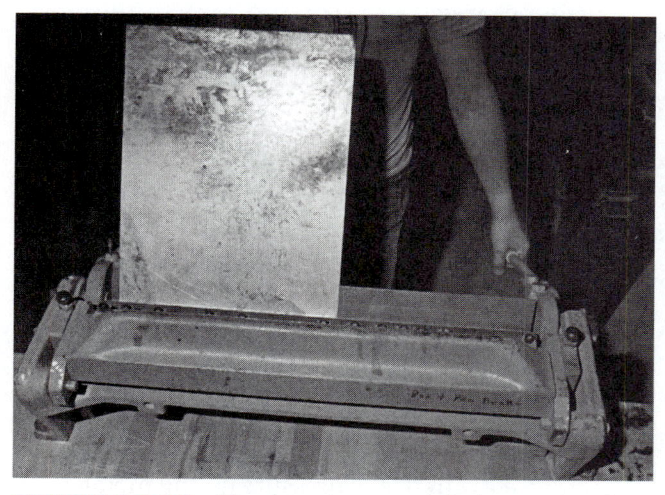

FIGURE 16-12. The hand-operated sheet metal bender is useful for making angle bends and seams.

STUDENT ACTIVITIES

1. Define the Terms to Know in this unit.
2. Examine the grinders in the shop to determine if the wheels need dressing. Ask the instructor to demonstrate how to dress a wheel.
3. Check the wheels on any grinders you have at home. Dress them if needed.
4. Do the following sawing, brushing, grinding, drilling exercise.
   a. Use a metal-cutting power saw to cut a 6-inch piece of 1/8 × 1" flat steel.
   b. Clean the 1/8 × 1" × 6" piece with a power wire brush.
   c. Round the corners and edges of each end slightly with a power grinder.
   d. Measure in one inch from one end and center punch on the center line.
   e. Drill the center punched hole with a 1/8-inch drill bit.
   f. Redrill the hole with a 1/2-inch drill bit.
   g. Remove all burrs or rough spots with a power wire wheel.
   h. Measure 1 1/2 inches from the end without a hole.
   i. Make a 90 degree bend at the line with a metal bender.
   j. Submit your project to the instructor.

SELF-EVALUATION

A. Multiple Choice. Select the best answer.

1. After installing a drill in a gear chuck, the next important thing is to
   a. start the motor
   b. remove the chuck wrench
   c. place the table off center
   d. check the belt for tightness
2. Round stock is best held for drilling by a
   a. C-clamp
   b. helper
   c. vise
   d. V-block
3. A grinding wheel may fly apart when running if the wheel does not have
   a. a coarse texture
   b. a coolant device
   c. a clean surface
   d. adequate speed rating for the motor
4. Grinding wheels are cleaned and restored to roundness with a
   a. grit cutter
   b. screwdriver
   c. wheel dresser
   d. any one of these
5. A dangerous act is to grind
   a. with heavy pressure on the metal
   b. using the side of the wheel
   c. without wearing a face shield
   d. all of these
6. Blades for power hacksaws are generally
   a. less than 12 inches long
   b. 12 inches to 18 inches long
   c. 1/2 inch to 3/8 inch wide
   d. none of these
7. A critical step to prevent breaking of power hacksaw blades is
   a. buy brittle blades
   b. clamp the work securely
   c. cool the blade frequently
   d. provide some slack in the blade
8. Gloves that are worn while doing metalwork should be made of
   a. asbestos
   b. cotton
   c. leather
   d. all are recommended
9. Power metal shears can cut
   a. angle stock
   b. flat stock
   c. round stock
   d. all of these

B. Matching. Match the word or phrase in column I with the correct word or phrase in column II.

Column I

1. chuck
2. quill
3. center punch
4. taper shank
5. shear
6. grinder wheel
7. used to shape metal
8. used to sharpen small tools
9. hydraulic

Column II

a. type of drill bit
b. bonded grit
c. coarse grinding wheel
d. medium grinding wheel
e. power for some shears
f. tighten in two holes
g. helps start drill bit
h. to cut at an angle
i. controlled by the feed handle

C. Completion. Fill in the blanks with the correct word or words that will make the following statements correct.

1. The _______ _______ of the drill press permits stock to be supported for angle drilling.
2. Grinding wheels may be checked for squareness by using a _______ or a _______ _______.
3. The liquid used to cool metal-cutting saw blades is called a _____.
4. Removing material so the grinding wheel is perfectly round with the face square to the sides and the abrasive particles exposed is called _______________ the wheel.

Section 5

Project Planning

UNIT

17 Sketching and Drawing Projects

OBJECTIVE

To use simple drawing techniques to create plans for personal projects.

COMPETENCIES TO BE DEVELOPED

After studying this unit, you should be able to:

- Identify common drawing equipment.
- Match basic drawing symbols with their definitions.
- Distinguish between pictorial and three-view drawings.
- Use common drawing techniques to represent ideas.
- Read and interpret a drawing.
- Make a three-view drawing of a given object.

TERMS TO KNOW

represent
sketch
drawing
dimension
pictorial drawing
three-view drawing
protractor
border line
object line
hidden line
dimension line
extension line
break line
center line
leader line
border
title block
full scale
scale
show box

MATERIALS LIST

- ✓ Unlined paper, 8 1/2" × 11"
- ✓ Lined paper, 8 1/2" × 11"
- ✓ Ruler, 12-inch
- ✓ Pencil
- ✓ Soft eraser
- ✓ Sample drawing blocks
- ✓ Optional:
  - a. Drawing board
  - b. T-square
  - c. Right triangle
  - d. Tape

Many objects are represented by other things. To represent means to stand for or to be a sign or symbol of. For instance, a map uses lines to represent roads in a geographic area, such as the community or state, figure 17-1. Books and bulletins use words to describe and create images in the mind that stand for ideas or objects. A photograph represents a person or object.

REPRESENTING BY SKETCHING AND DRAWING

A sketch is a rough drawing of an idea, object, or procedure. A drawing is a picture or likeness made with a pencil, pen, chalk, crayon, or other instrument.

Sketching and drawing are used in agricultural mechanics to put ideas on paper. For example, in a

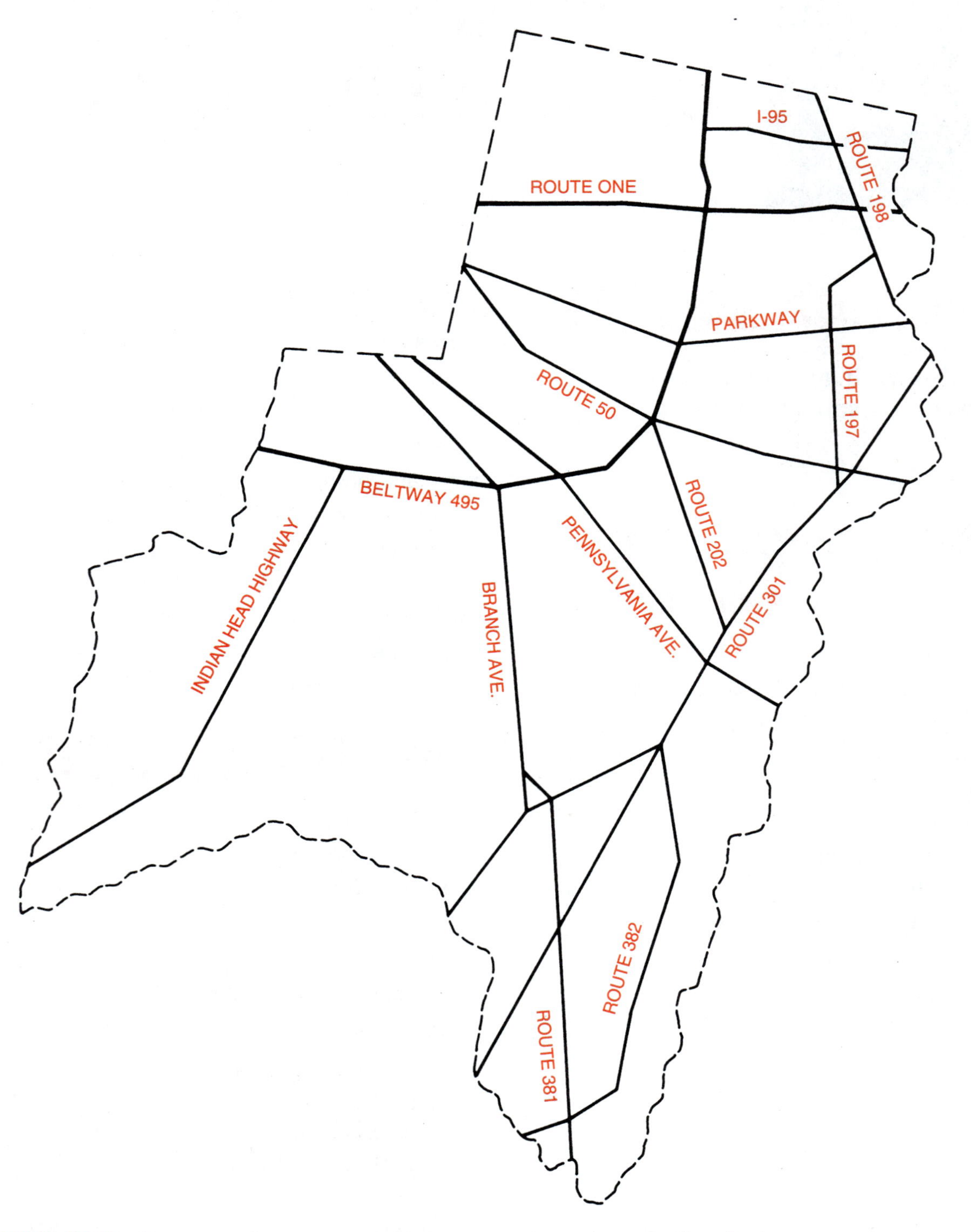

FIGURE 17-1. On maps, lines are used to represent roads and other features.

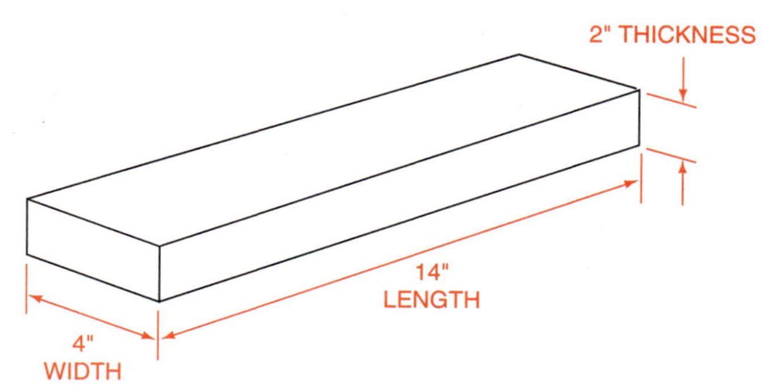

FIGURE 17-2. On drawings, lines show edges and corners of objects. Dimensions indicate length, width, and thickness.

building project an accurate sketch records the details of construction and indicates how the finished product should look. A sketch is needed to determine the amount of lumber, nails, bolts, and other materials to be used. Any construction project will be more efficient if there is an accurate sketch or carefully drawn plan to follow.

The ability to make sketches and simple drawings is a valuable skill. It is a way to record ideas for present and future use. In this way, a project can be planned in every detail before investing time and materials in construction. As the plan develops, some ideas may prove unworkable. Since the discovery is made before lumber or metal are cut, other workable ideas can be substituted.

A sketch or plan can be simple; it does not have to be done with specialized drawing equipment. However, a sketch must be complete. It must include lines to represent edges and corners of sections or parts of an object. Dimensions must be included to indicate the size of each part of the object. A dimension is a measurement of length, width, or thickness, figure 17-2.

Perspective

Figure 17-2 is a pictorial drawing. A pictorial drawing shows all three dimensions at once. That is, it shows the item turned so the front, side, and top are in view. Pictorial drawings are very useful, but it takes a good deal of training and experience to draw them accurately.

Almost anyone can make clear drawings by showing one view at a time. This method is called three-view drawing. It will be discussed in detail later in this unit.

Drawing Instruments

Sketches and simple drawings can be prepared with instruments found in most homes. These instruments include a sharp lead pencil, eraser, 12-inch ruler, compass, and protractor, figure 17-3. A protractor is an instrument for drawing or measuring angles.

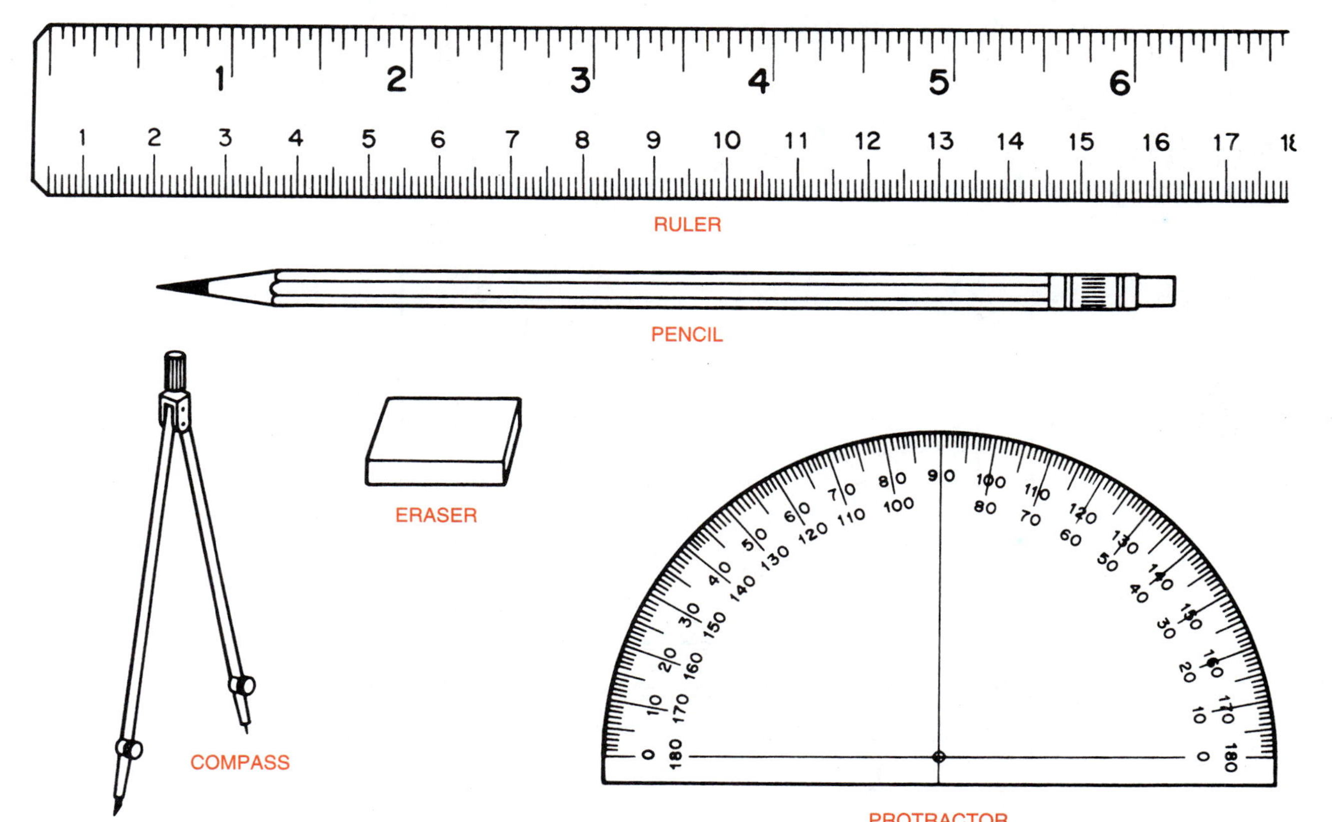

FIGURE 17-3. Instruments for making sketches and simple drawings.

If much drawing is to be done, several additional items will help save time. These items include a small drawing board with a T-square, a 30° × 60° × 90° plastic triangle, a plastic scale, and some masking tape, figure 17-4.

To use these tools, place the T-square across the bottom of the drawing board. Then place an 8 1/2" × 11" piece of paper on the board above the T-square. Adjust the paper so it rests on the T-square with its base parallel to the base of the board. Using masking tape, attach the paper to the board as shown. The T-square can now be used to make horizontal lines. Place the base of the triangle on the T-square. Using the triangle, lines can be drawn at 30°, 60°, or 90° angles to the T-square. Use of the scale will be discussed later.

Symbols Used in Drawing

Hundreds of figures have been adopted by architects and engineers for use as symbols in drawing. Only a few of these are needed for preparing sketches and simple drawings, however.

Several types of lines have specific meanings, figure 17-5. They are as follows:

- border line—a heavy, solid line drawn parallel to the edges of the drawing paper
- object line—a solid line showing visible edges and form of an object
- hidden line—a series of dashes that indicates the presence of unseen edges
- dimension line—a solid line with arrowheads at the ends to indicate the length, width, or height of an object or part
- extension line—a solid line showing the exact area specified by a dimension
- break line—a solid, zigzag line used to show the illustration stops but the object does not
- center line—a long-short-long line used to indicate the center of a round object
- leader line—a solid line with an arrow used with an explanatory note to point to a specific feature of an object.

These lines and symbols help the person who is drawing to communicate with the person who will use the plan. Learning to recognize these lines and symbols is the first step in learning to read plans. Similarly, learning to use them is the first step in learning to draw plans.

ELEMENTS OF A PLAN

Paper of any size may be used for drawing. However, it is advisable to start with 8 1/2" × 11" paper. This is standard notebook size, is readily available, and easy to handle.

Border. When making a drawing, it is useful to draw a heavy line all around and close to the outer edges of the paper. This is called a border. The lines used to make a border are called border lines. It is suggested that borders be made 1/2 inch in from the edge of the paper.

Title Block. A few items of information about the total drawing are necessary. These include:

- the name of the person who prepared the drawing

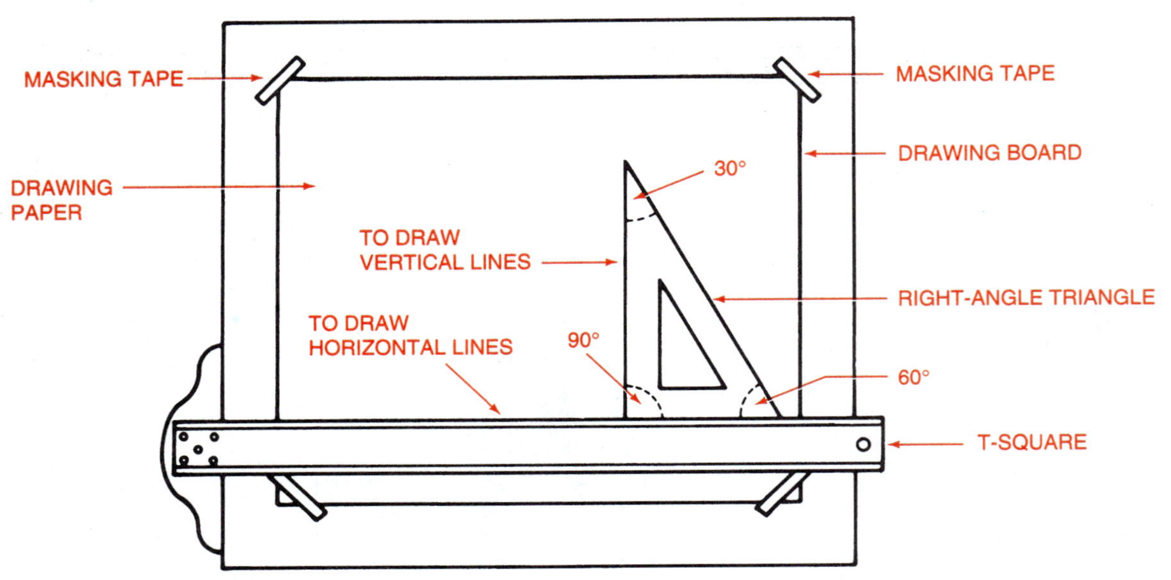

FIGURE 17-4. A drawing board with paper taped in place. The T-square is helpful for drawing horizontal lines. A right angle triangle is useful for drawing vertical lines.

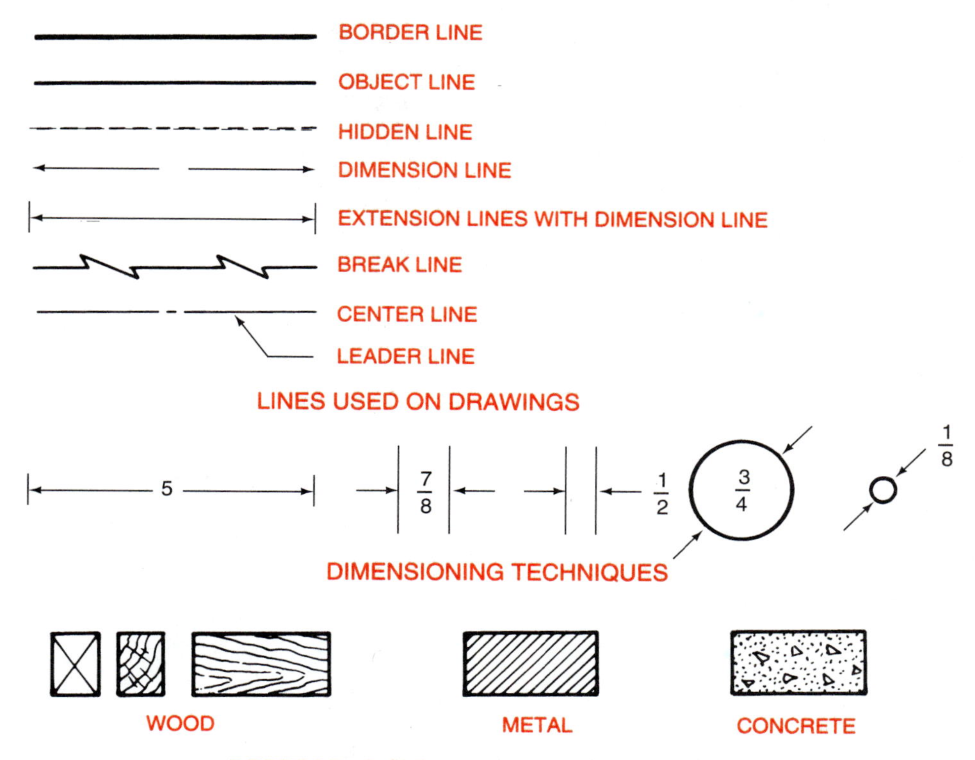

FIGURE 17-5. Some symbols used on drawings.

- the date when the drawing was completed
- the name of the drawing
- the scale of the drawing.

The section of a drawing reserved for information about the drawing in general is called the title block. For simple drawings, a $^1/_2$-inch line drawn above the border line at the bottom of the paper works well. The information can be printed within these lines. The addition of very light guidelines drawn $^1/_8$ inch from the top and bottom lines of the title block makes lettering easier.

Refer to figure 17-6. Note that a border line has been drawn around the edges of the paper. The title block indicates that the drawing is a sample block, drawn to full scale by Bill Brown on January 1, 1986. Full scale means that the drawing is the same size as the object it represents. In other words, 1 inch on the paper represents 1 inch on the block.

Views. The sample block is represented in this drawing by two views. The upper left-hand drawing is a top view of the block. In other words, the viewer is looking down on the block from above. The dimensions indicate that the block is 2 inches long and 2 inches wide. The inner circle indicates there is a hole in the center of the block. The dimension indicates it is 1 inch in diameter. The outer circle shows that a larger hole goes part way through the block. The size of the outer hole is not shown. However, remember that the object is drawn to full scale. This means the drawing is the same size as the block. Measuring the outer circle on the drawing shows it to be $1^1/_4$ inches in diameter. Therefore, the hole in the block is the same size or $1^1/_4$ inches.

The lower drawing in figure 17-6 shows a front view of the block. This view shows that the height of the block is $1^1/_2$ inches ($^3/_4$ inch + $^3/_4$ inch = $1^1/_2$ inches). It also shows that the larger hole is $^3/_4$ inch deep and the smaller hole goes through the remaining $^3/_4$ inch of the block. A note is provided just above the title block on the right to specify the material from which the block is to be made. Enough information is provided in figure 17-6 to make the block correctly. Therefore, it is a good drawing.

Most objects require three views in a drawing to provide complete information. For example, consider the bookend shown in figure 17-7. The top view is

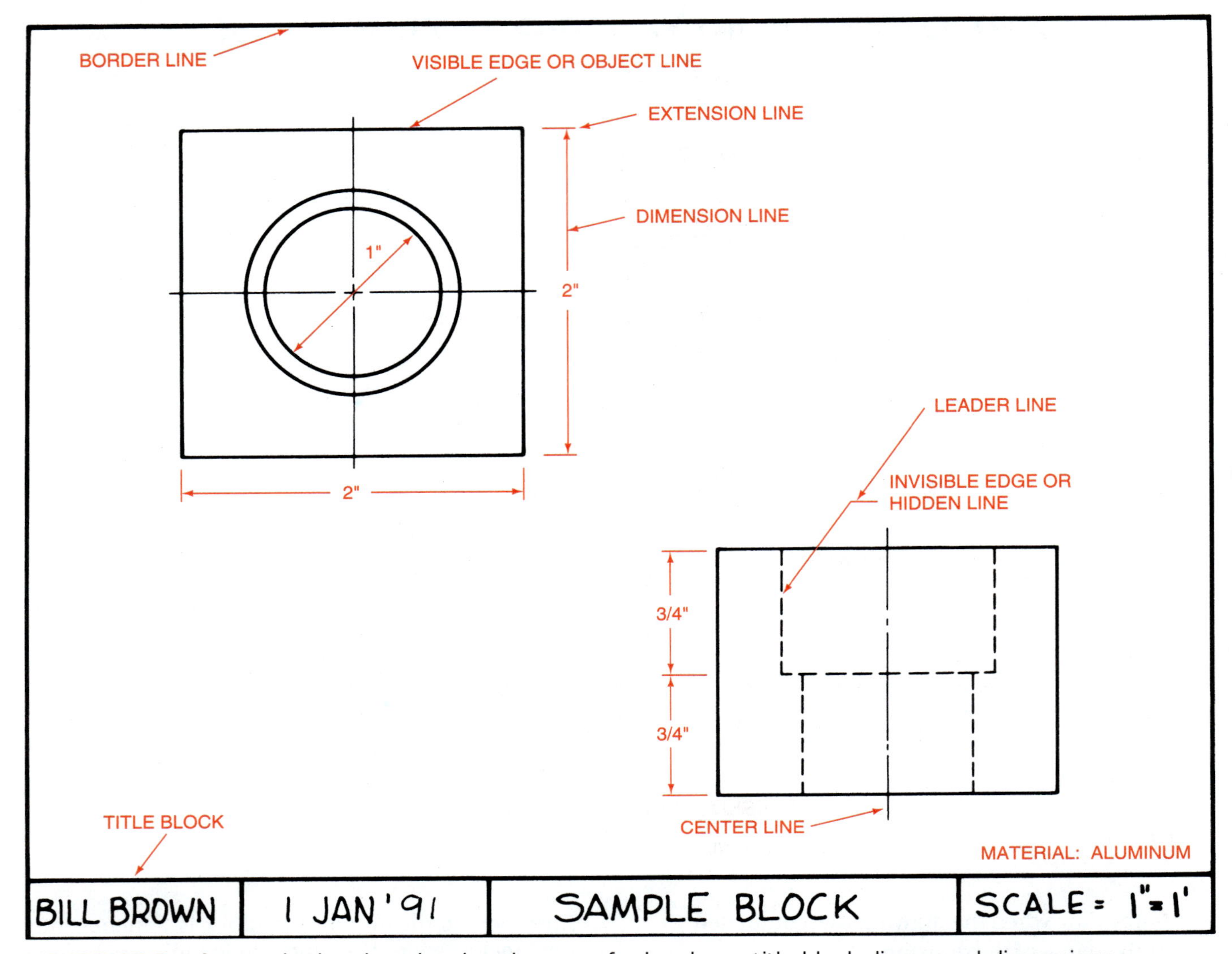

FIGURE 17-6. A sample drawing showing the use of a border, a title block, lines, and dimensions

at the upper left. The front view is at the lower left. However, the object is not recognized as a common bookend until the end view is seen. In this case, the end view gives the best indication of the shape of the object. End views are generally placed to the right of the front view and in line with it.

The views are placed in the drawing with the front view in the lower left, the top view above it, and the end view to the right of the front view. This arrangement allows dimensions placed on the drawing to represent more than one view, figure 17-8. By using this technique, dimensions are not repeated and the drawing is less cluttered, figure 17-9.

SCALE DRAWING

The word scale as used here means the size of a plan or drawing as compared to that of the object it represents. In figure 17-6, the drawing was made as large as the object. Since the size of the object was rather small, the drawing fit on the paper. However, most objects are much larger than a sheet of 8 1/2" × 11" paper. Therefore, the scale to use in preparing the drawing must be determined.

Determining the Scale

A piece of 8 1/2" × 11" paper, turned sideways, with borders and title block drawn, has 10 inches of horizontal drawing space and 7 inches of vertical drawing space. The three views must be planned to fit into this space without crowding.

A good place to start in deciding upon the scale is to assume that 1 inch equals 1 foot (1" = 1'). It is recommended that 1 inch of space for dimensions be allowed on every side of each view. Therefore, 1 inch on the left of the front view, 1 inch between views, and 1 inch on the right of the end view is reserved. The original 10 inches of horizontal space left after

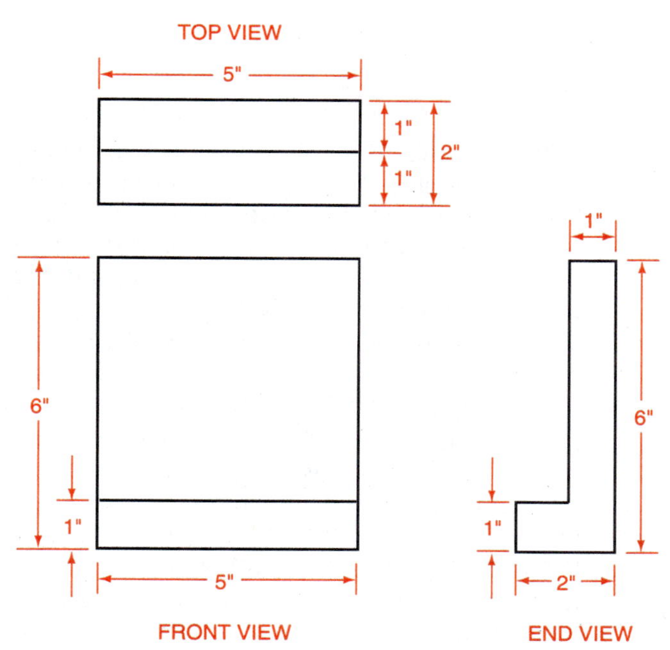

FIGURE 17-7. A three-view drawing with repeated dimensions. The names of the views do not appear on a final drawing.

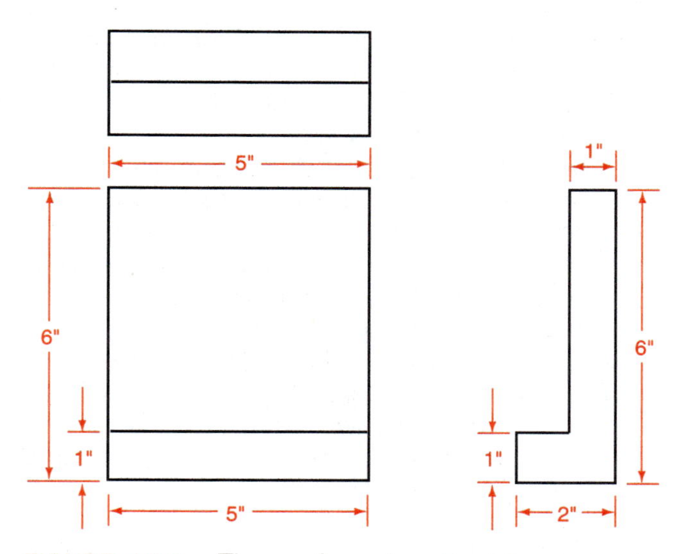

FIGURE 17-8. The preferred method for presenting three-view drawings in which each dimension is shown only once.

the borders are drawn in, reduced by 3 inches for dimensions, leaves 7 inches of horizontal space for the two views.

A three-view drawing has a front view and a top view, one over the other. Three spaces of 1 inch each are allowed for dimensions. This reduces the original 7 inches of vertical space by 3 inches, leaving 4 inches of vertical drawing space.

This procedure indicates that a scale of 1" = 1' is adequate to draw objects with front end horizontal

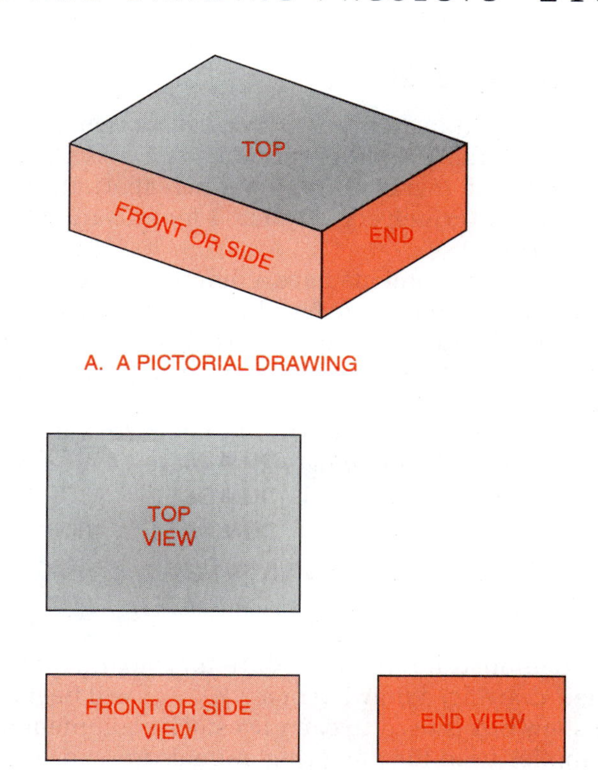

FIGURE 17-9. Two ways to represent the same object in drawings.

dimensions that add up to 7 feet or less and front and top vertical dimensions that total no more than 4 feet. Larger objects require a larger scale, such as 1" = 2' or 1" = 3', and so on. For example, consider a three-view drawing of a wagon body with hay racks. The dimensions are:

- Length = 16'
- Width = 8'
- Height = 8'

Remember that there are 7 inches of horizontal drawing space and 4 inches of vertical drawing space. With a scale of 1" = 4', 1 inch of drawing represents 4 feet of wagon (divide each dimension by 4):

- L = 16' actual dimension, 4" scale value
- W = 8' actual dimension, 2" scale value
- H = 8' actual dimension, 2" scale value

For the horizontal views, the dimensions on the drawing would be:

4" (front view) + 2" (end view) = 6"

Since there are 7 inches of horizontal drawing space, the scale is usable for the horizontal views. The extra inch will simply be extra space.

For the vertical views, the dimensions would be:

2" (front view) + 2" (top view) = 4"

Since there are 4 inches of vertical drawing space, the scale will work.

Using a Scale

A scale is an instrument with all increments shortened according to proportion, figure 17-10. On a $^{1}/_{2}$ scale, 1-inch marks are only $^{1}/_{2}$-inch apart. It can be used to let $^{1}/_{2}$ inch equal 1 inch, 1 foot, 1 yard, 1 mile, and so on.

Some instruments have a different scale on each edge. A triangular scale has three sides but six scales, with each side showing two scales. Some instruments show different scales at each end for a total of four scales per side, or 12 scales per instrument.

An expensive scale should never be used as a straightedge for drawing. However, low-cost plastic scales may be used for this purpose.

MAKING A THREE-VIEW DRAWING

To demonstrate the procedure for making a three-view drawing, a drawing of a show box will be made. A show box is used to carry livestock equipment and supplies used for preparing animals for showing. It can also be used to store and transport tools and materials for other purposes. The plans could be used to construct a trunk, chest, or toy box.

The show box will be 4 feet long, 2 feet wide, and $1^{1}/_{2}$ feet high. The scale will be one inch equals 1 foot or 1" = 1'. This can also be written one inch equals 12 inches or 1 = 12 or simply $^{1}/_{12}$.

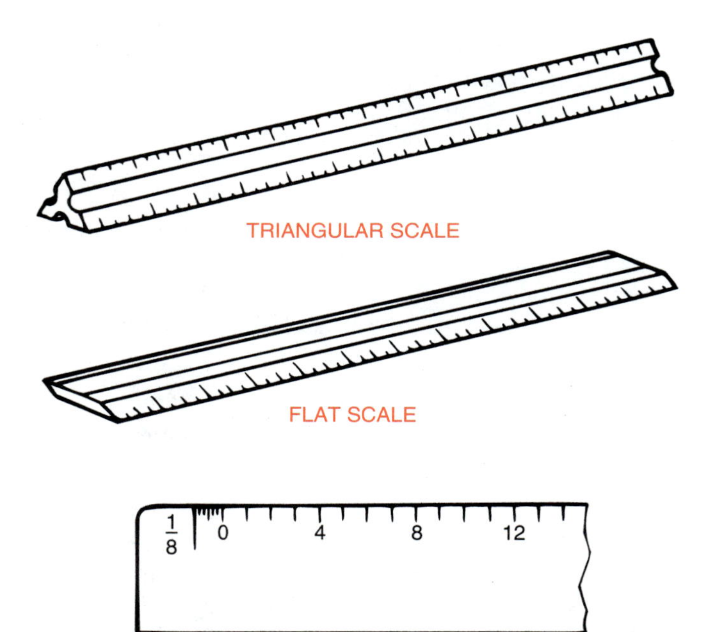

FIGURE 17-10. Triangular and flat scales. Notice that $^{1}/_{8}$ inch is used to represent 1 inch in the bottom example.

Blocking in the Views

PROCEDURE

1. Select a plain 8 $^{1}/_{2}$" x 11" piece of heavy paper. Regular drawing paper is preferred, but mimeograph paper will do. Turn the paper lengthwise in front of you, figure 17-11. Draw a $^{1}/_{2}$-inch border around the paper. Add a $^{1}/_{2}$-inch title block.
2. In the title block, letter your name and the date. For the name of the project, print "Show Box"; for scale, Scale = 1" = 1'.
3. Make a light mark 1 inch in from the left margin and draw a very light line through it from top to bottom and parallel to the left border. This becomes the object line for the left side of the top and front views.
4. Make a light mark 1 inch up from the title block and draw a very light line through it from border to border and parallel to the title block. This becomes the bottom object line.
5. The length of the box is 4 feet. The scale is 1" = 12", or 1" = 1'. Measure across 4 inches from the left object line and draw a very light vertical line from top to bottom. This becomes the object line on the right side of the top and front views.
6. Measure 1 inch to the right and draw a very light vertical line from top to bottom. This becomes the right side of the end view.
7. Measure another 2 inches to the right and draw another very light vertical line. This becomes the right side of the end view.
8. The show box is 1 $^{1}/_{2}$ feet high. So, start at the bottom object line and measure up 1 $^{1}/_{2}$ inches. Draw a very light line to form the top object lines of the front and end views.

   NOTE: Do not extend the lines into the spaces around the views.

9. Measure up 1 inch from the top object line and draw a light line to form the bottom object line of the top view.

   NOTE: Do not extend the line into the spaces around the views.

10. The show box is 2 feet wide. So, measure up another 2 inches and draw a light line to form the upper object line of the top view.
11. Now all three views are blocked in and drawn to scale, figure 17-12. Erase excess

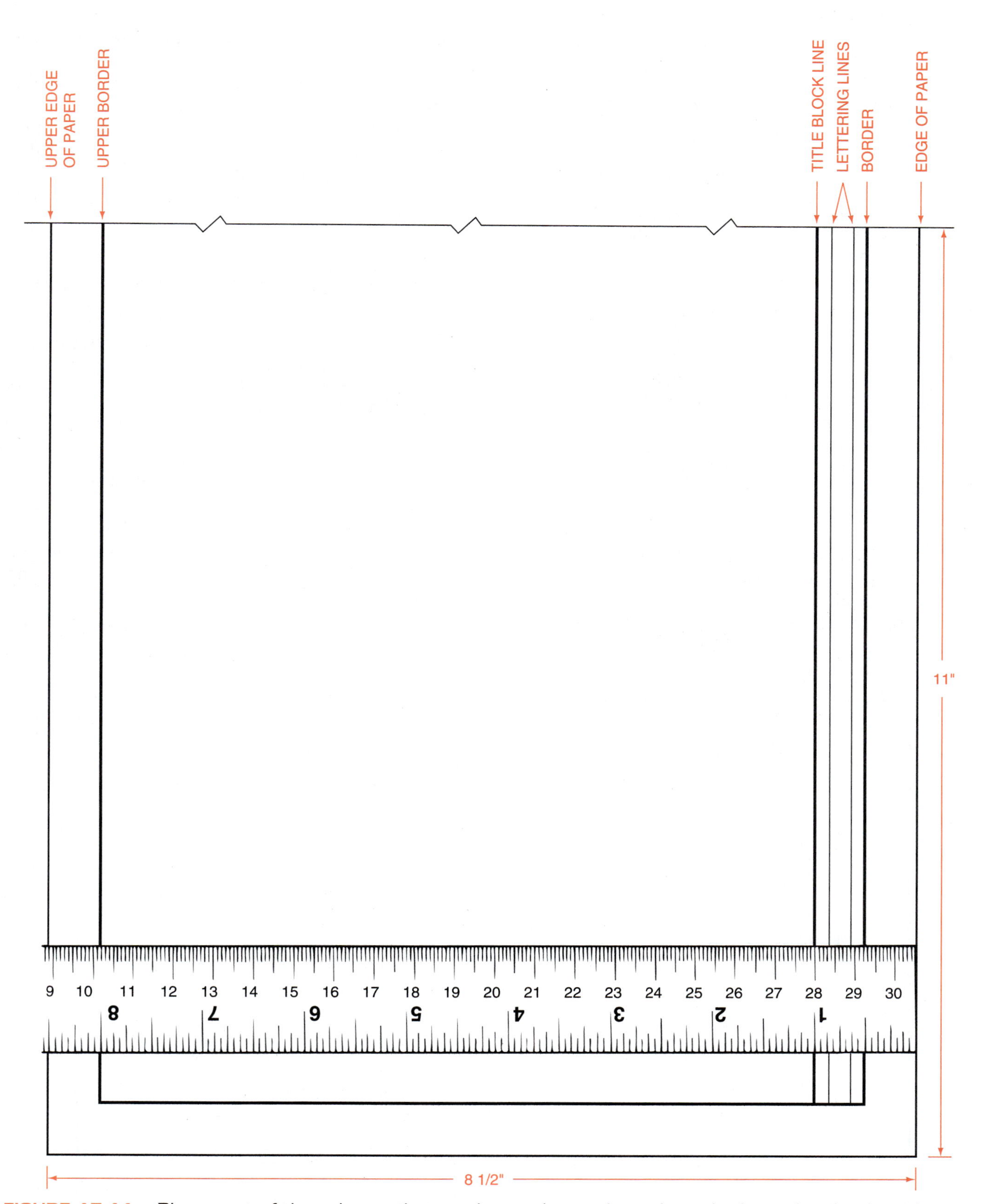

FIGURE 17-11. Placement of the ruler so that marks can be made to draw the lower border, lettering guidelines, title block line, and upper border line at the same time.

lines that extend beyond the views, figure 17-13.

NOTE: Remember that you have planned for 1 inch more of space around views. Therefore, when blocking in views, final lines generally stop before they touch the borders.

All three views are now blocked in. This means the outside object lines are drawn. Next, the details of each view must be added.

Completing the Front View

The most prominent view is generally the front view. Therefore, the front view should be drawn first. It will probably tell the viewer more about the project than either the top or the end view.

Assume that the show box will be made of $^3/_4$-inch plywood. The sides will be rabbetted and the ends set in. The bottom will be hidden by the sides and ends. The lid will fit on top of the sides and ends.

Before drawing, it must be decided how far apart two lines should be on paper to represent the thickness of $^3/_4$-inch plywood at $^1/_{12}$ scale. For convenience, round off the 3/4-inch thickness to 1 inch. Therefore, $^1/_{12}$ of an inch on the paper represents lumber that is $^3/_4$ inch thick. Rulers are laid out in eighths and sixteenths, not twelfths. Therefore, either $^1/_{16}$ or $^1/_8$ of an inch must be chosen to represent an inch or a special $^1/_{12}$ drawing scale must be obtained. For sketching purposes, $^1/_8$ inch will be used to represent the $^3/_4$-inch plywood to better show the detail of the joints.

To draw in the details of the front view, the steps are as follows.

PROCEDURE

1. Measure in $^1/_8$ inch from the left side of the front view and draw a broken vertical line from top to bottom. This line is a hidden line because the edge of the end panel cannot be seen, figure 17-14.
2. Measure in $^1/_8$ inch from the right side of the front view and draw a broken vertical line from top to bottom. This line represents the edge of the right-hand end panel.
3. Measure up from the bottom $^1/_8$ inch and draw a broken horizontal line between the two vertical broken lines. This line represents the hidden top edge of the bottom panel.
4. Measure down $^1/_8$ inch from the top of the front view and draw a horizontal solid line

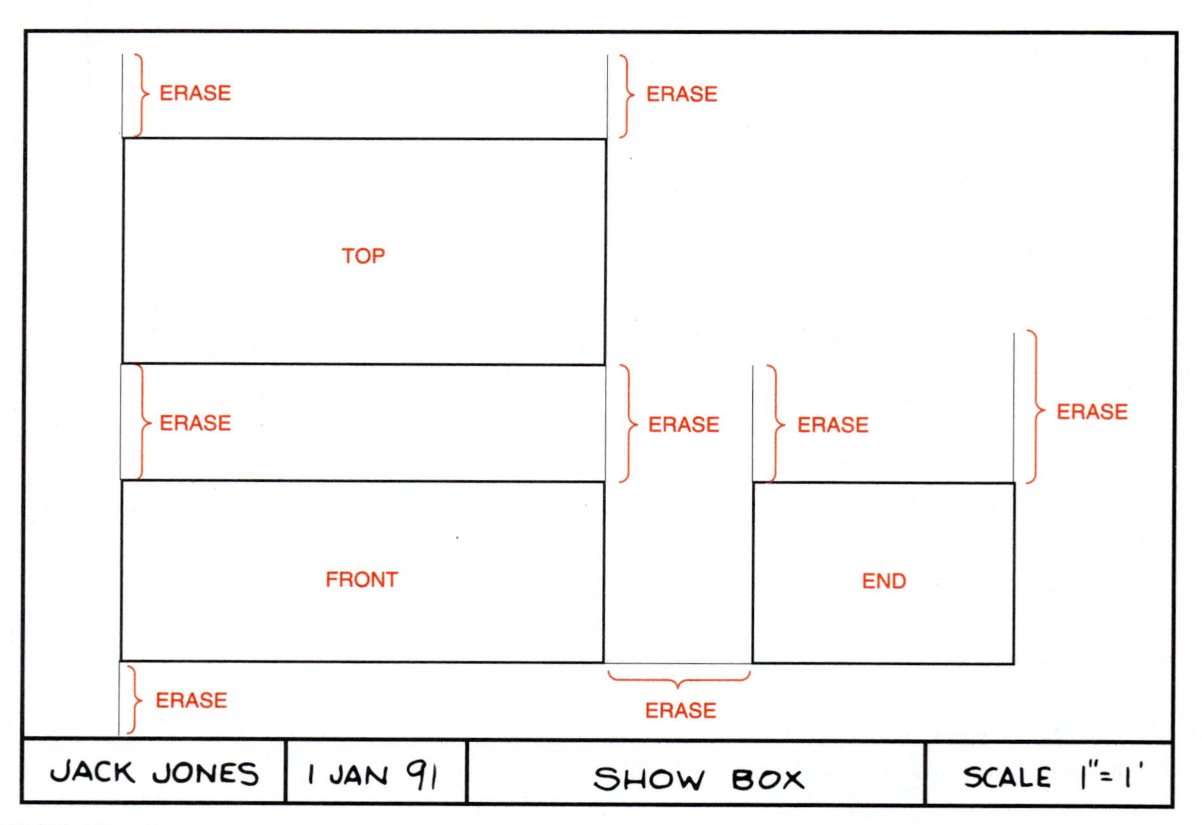

FIGURE 17-12. Blocked-in, three-view drawing of the show box before excess lines are erased.

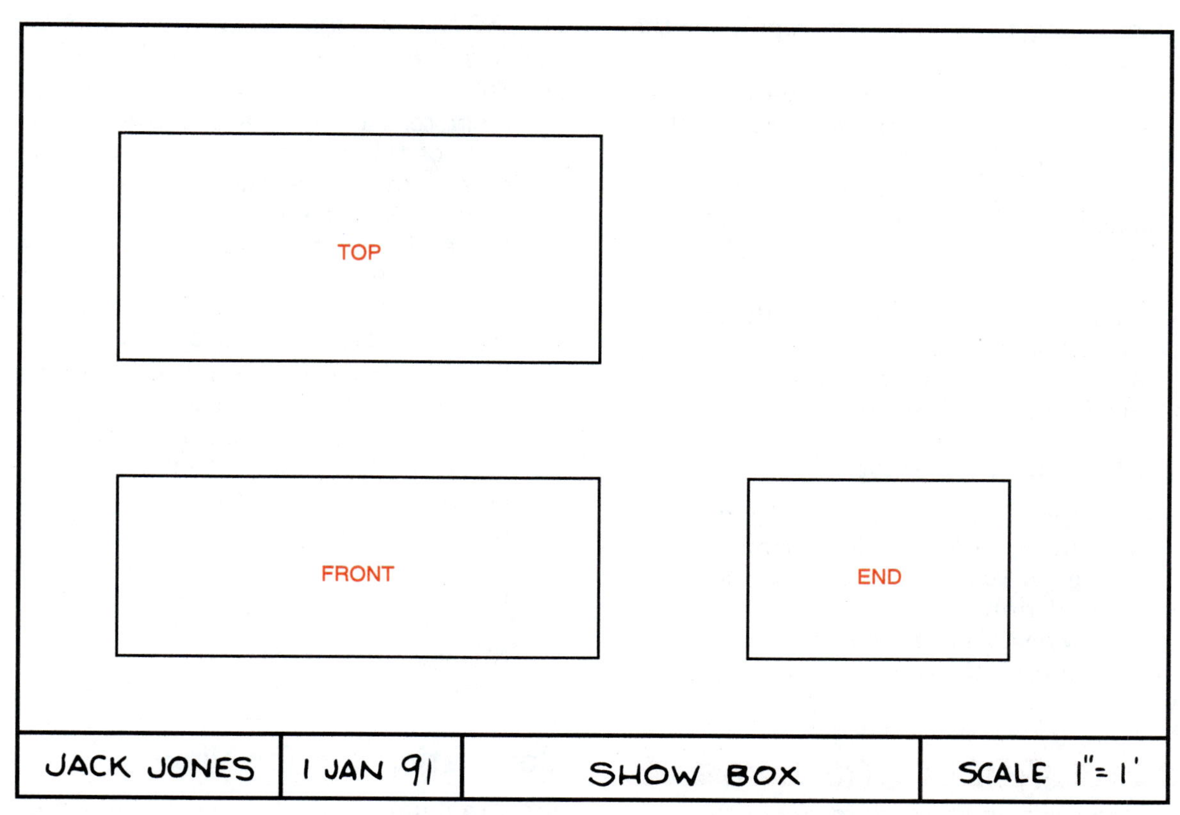

FIGURE 17-13. Blocked-in, three-view drawing of the show box after excess lines are erased.

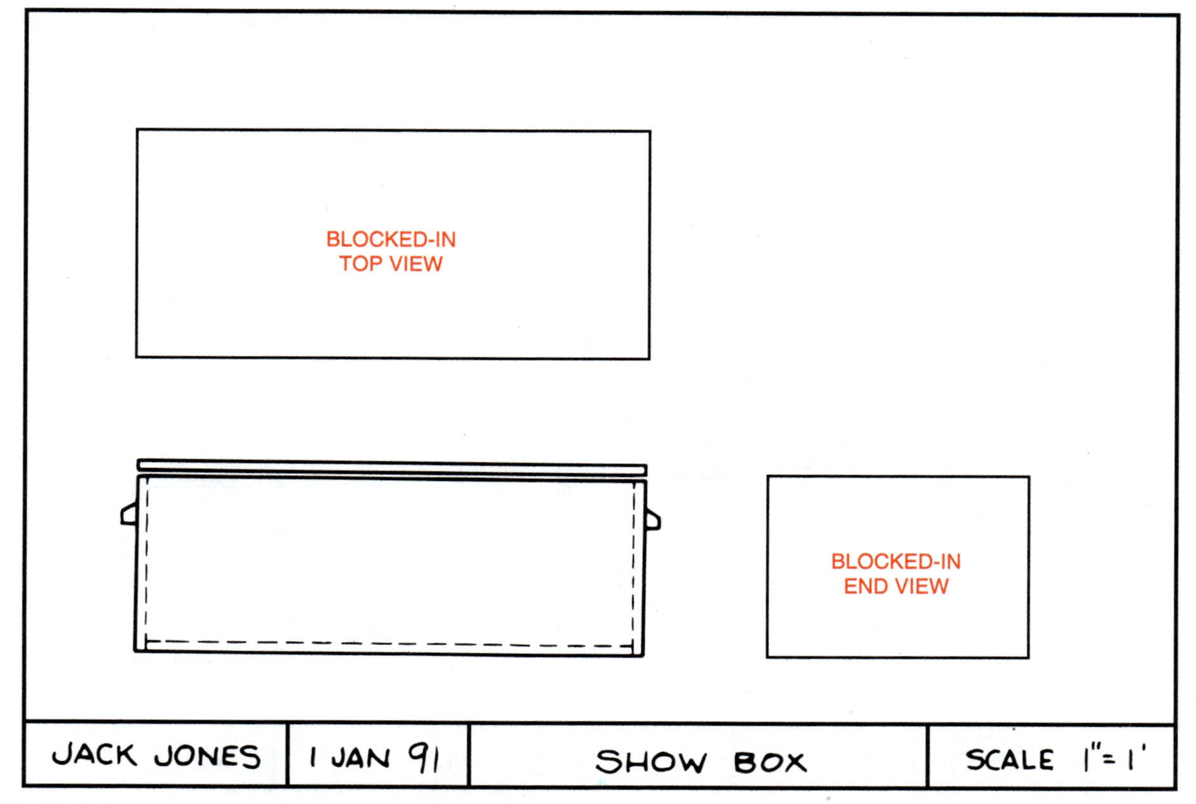

FIGURE 17-14. Completed front view of the show box.

from solid line to solid line to represent the thickness of the lid.

5. Erase any parts of the broken vertical lines that show inside the two lines representing the edge of the closed lid.
6. Show where handles will be attached, 12 inches up from the bottom of the chest.

This completes the front view unless the exposed heads of nails or screws used to attach the front to the ends and bottom are to be shown. A hasp or other locking device could also be pictured.

Completing the End View

With experience in drawing, some of the parts of the top and end views can be drawn at the same time the front view is drawn. Many of the lines can be drawn by placing the ruler on the measurement and drawing the lines in two views without moving the ruler.

Refer to figure 17-15. The steps in drawing the end view follow.

PROCEDURE

1. Place the ruler on the bottom line of the lid in the front view. The ruler should extend across the end view also. Draw the bottom line of the lid in the end view. (Use a solid line.)
2. The end view of the sides needs special treatment to show the rabbet joints. That is, 3/8 inch of the plywood end grain is seen; the other 3/8 inch is hidden. To show this, measure in 1/16 inch from the left-hand side of the end view and draw a solid vertical line. Measure another 1/16 inch and draw a vertical broken line. Repeat the procedure on the right-hand side of the end view.
3. Place the ruler on the hidden top edge of the bottom panel in the front view. draw in the hidden top edge of the bottom panel in the end view. (Use a broken line.)
4. Draw in the handle. Notice that the handle is on the same level in both the front and end views. The handle consists of a 1" x 2" x 6" hollowed strip of wood. Since the scale is 1" = 12", the handle is 1/2" long on the drawing. Measure to the center of the end and then outward 1/4 inch in each direction.

Completing the Top View

Refer to figure 17-16. The steps in drawing the top view follow.

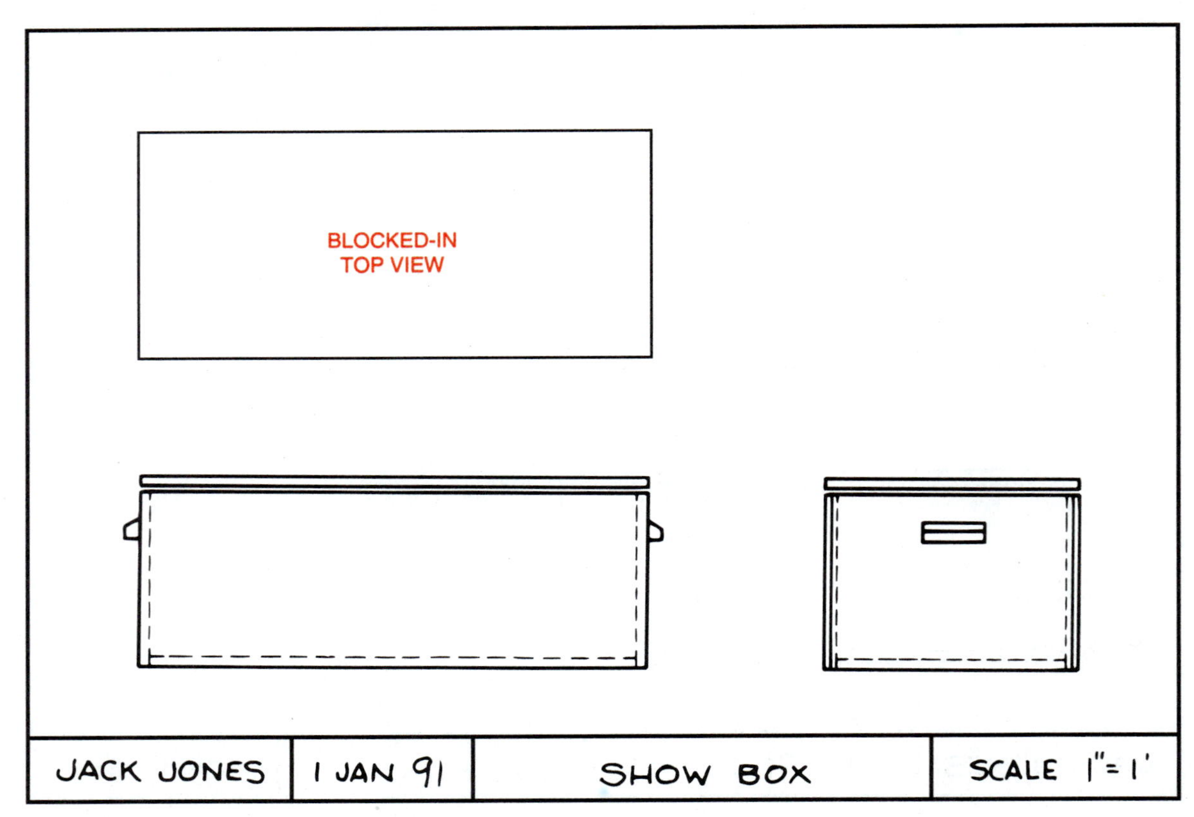

FIGURE 17-15. Completed front and end views of the show box.

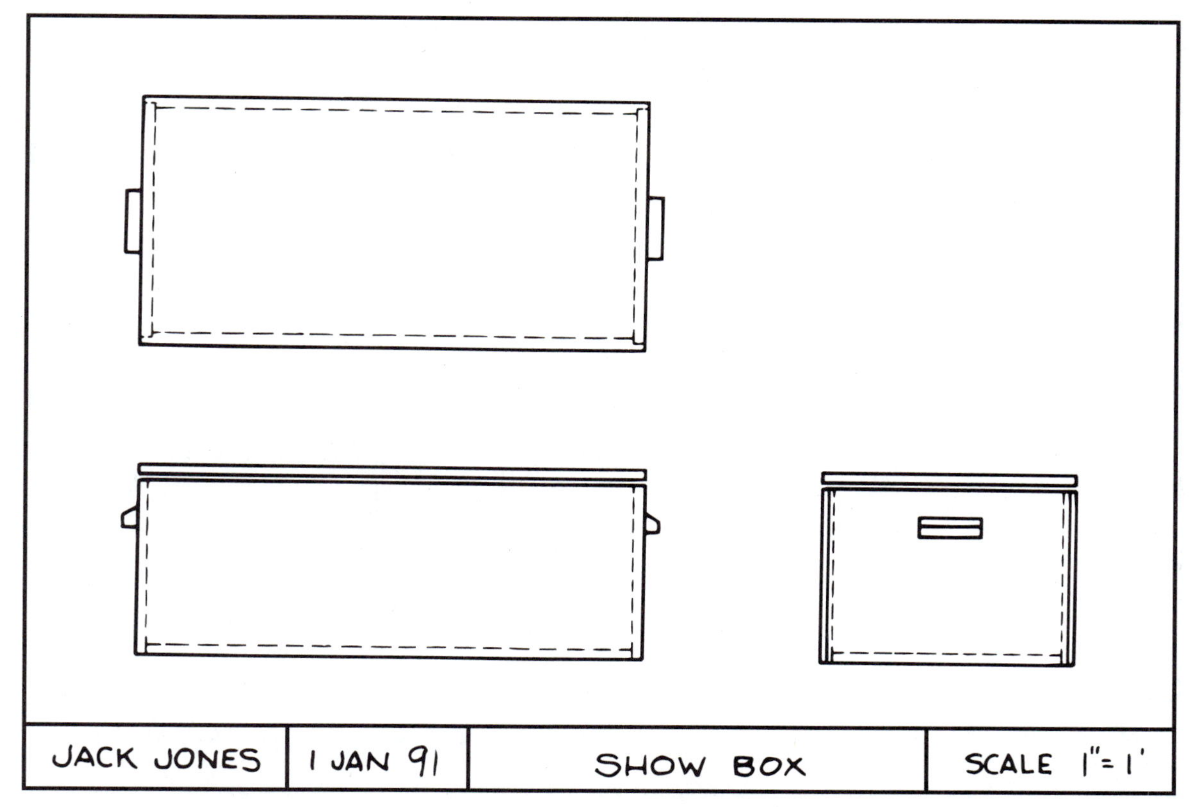

FIGURE 17-16. All three views of the show box completed without dimensions.

PROCEDURE

1. Place the ruler on the hidden inner edge of the left end panel of the front view and extend it over the top view (Use a broken line.)
2. Repeat step one on the right side of the view.
3. Measure 1/8 inch up from the bottom and 1/8 inch down from the top lines and draw the inner edges of the sides with broken lines.
4. Show the rabbet joints correctly in each corner. That is, show how the side panels are sawed and the end panels fit into the rabbet grooves.
5. Add the handles on each end by locating the center and measuring out 1/4 inch in each direction. Assume a thickness of 1-inch wood or 1/8 inch on the paper.

Dimensioning the Drawing

The final step to complete the drawing is to add dimensions, figure 17-17. The dimensions are the exact measurements, more exact than the drawing itself. The drawing focuses on details of the assembly. The dimensions tell the exact length, width, and height of the object. The dimensions also indicate the thickness of materials and other construction details as needed.

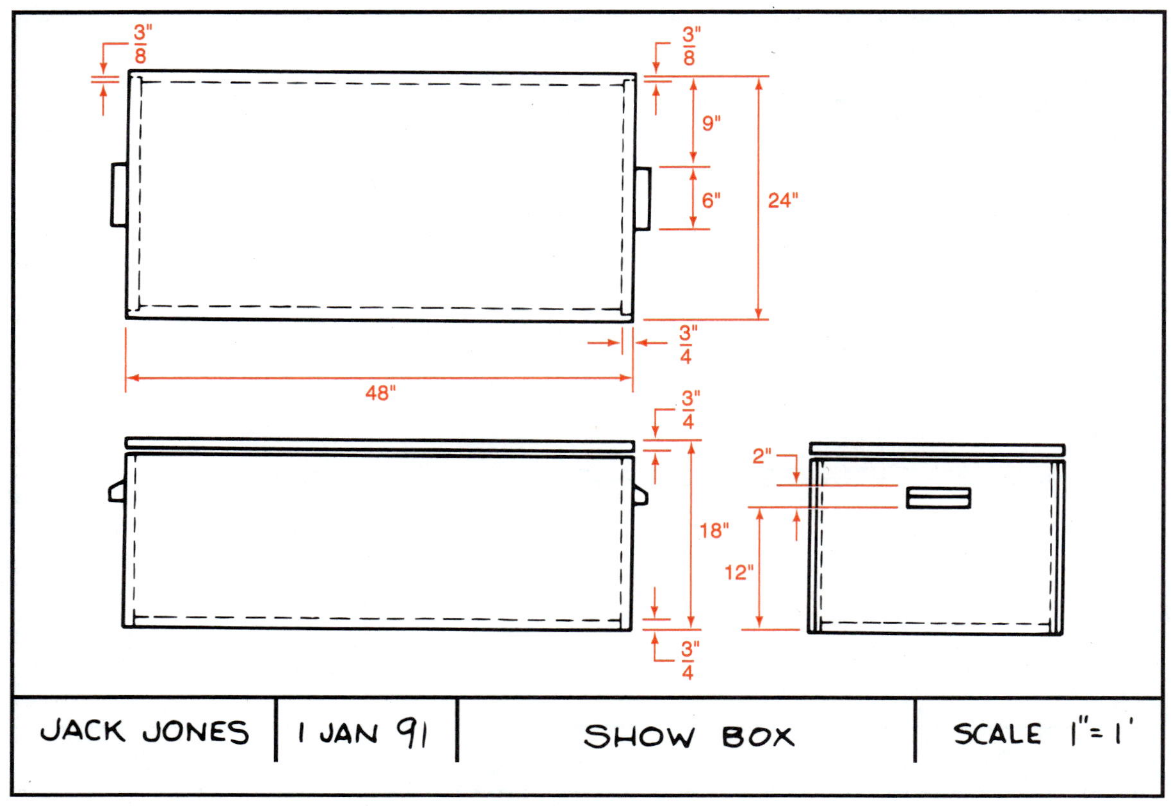

FIGURE 17-17. Completed drawing of the show box with dimensions.

STUDENT ACTIVITIES

1. Define the Terms to Know in this unit.
2. Study agricultural mechanics project plans to visualize how individual parts look and how they are fastened together.
3. Practice printing the alphabet in capitals and lower-case letters. Follow the form provided in figure 17-5. Use lined paper.
4. Using a ruler, make twenty arrows following the form used in figure 17-5 for dimension lines.
5. Practice making neat figures for dimensions by numbering from 0 through 9. Follow the form provided in figure 17-5.
6. Practice measuring accurately by drawing lines on a sheet of paper the length of each figure stated below. Make some lines solid and some broken. Submit the paper to your instructor for evaluation.
   a. 1 inch
   b. 1 1/2 inches
   c. 1 3/4 inches
   d. 3 1/8 inches
   e. 3 5/8 inches
   f. 1 1/16 inches
   g. 1 3/16 inches
   h. 1 1/4 inches
   i. 1 9/16 inches
   j. 1 3/8 inches

7. Ask the instructor to provide blocks or other objects for practice drawing. Draw one or more of such objects.
8. Convert the following project dimensions to the measurements they would be on paper, first using a 1/12 scale (1" = 12"); then using a $^{1}/_{2}$ scale (1" = 2").
   a. 12 inches = _____ _____
   b. 18 inches = _____ _____
   c. 36 inches = _____ _____
   d. 6 inches = _____ _____
   e. 3 inches = _____ _____
9. If a drawing instrument with $^{1}/_{12}$ and $^{1}/_{2}$ scales is available, draw lines on a sheet of paper representing the length of each item in activity 8.

SELF-EVALUATION

A. **Multiple Choice.** Select the best answer.

1. A simple plan must
   a. be done with specialized equipment
   b. be done by a draftsperson or engineer
   c. have dimensions
   d. all of these
2. The triangle recommended as an aid for simple drawing is
   a. 30°, 60°, 90°
   b. plastic
   c. three sided
   d. all of these
3. Which scale is best for making a three-view drawing on a sheet of 8 $^{1}/_{2}$" × 11" paper if the project is 4 feet long, 2 feet wide, and 1 $^{1}/_{2}$ feet high?
   a. 1" = 12"
   b. 1" = 6"
   c. 1" = 3"
   d. 1" = 1"
4. Which scale is best for making a drawing of the front view only on an 8 $^{1}/_{2}$" × 11" sheet of paper if the project is 4 feet L, 2 feet W, and 1 $^{1}/_{2}$ feet H?
   a. 1" = 12"
   b. 1" = 6"
   c. 1" = 3"
   d. 1" = 1"
5. The process of first drawing outside object lines for all three views is called
   a. blocking in
   b. pictorial drawing
   c. scale drawing
   d. sketching

B. **Matching.** Match the terms in column I with those in column II.

Column I
1. dimension
2. pictorial
3. object line
4. border line
5. hidden line
6. extension line
7. leader line
8. center line
9. title block
10. scale

Column II
a. $^{1}/_{2}$" from edge of paper
b. use with dimensions
c. round objects
d. three views in one
e. targets special information
f. measurement
g. name, date, project, scale
h. proportion
i. broken line
j. visible edges

C. Completion.

Completion. Fill in the blanks with the word or words that make the following statements correct.

1. The view placed in the upper left corner of a plan is the __________.
2. The view placed at the lower left corner of a plan is the __________.
3. The view placed in the lower right corner of the plan is the __________.

D. Brief Answers.

Brief Answers. Briefly answer the following questions.

1. What is a drawing?
2. What is a sketch?
3. What is a protractor?
4. Using a scale of 1" = 12", convert the following measurements.
   a. 48 inches = _____
   b. 6 inches = _____
   c. 9 inches = _____
   d. 1 inch = _____

UNIT

Figuring a Bill of Materials

OBJECTIVE

To state the use and format of a bill of materials and make all calculations needed to develop a bill of materials.

COMPETENCIES TO BE DEVELOPED

After studying this unit, you should be able to:

- Define terms associated with a bill of materials.
- State the components of a bill of materials.
- Record dimensions of structural metals and lumber.
- Calculate board feet.
- Calculate costs included in a bill of materials.
- Prepare a written bill of materials.

MATERIALS LIST

✓ Paper
✓ Pencil

TERMS TO KNOW

bill of materials
item
rounded up
tongue and groove
galvanize
cadmium
board foot

COMPONENTS OF A BILL OF MATERIALS

A bill of materials is a list and description of all the materials to be used in constructing a project. For each component, the bill of materials includes the following:

- item or part name
- number of pieces
- type of material
- size of pieces
- description of parts
- total feet
- unit cost
- cost

A bill of materials is used to determine the material that must be assembled for a project and its estimated cost. To determine the amount of lumber needed, the student must be able to compute board feet. The student must also be able to combine the costs of numerous pieces and types of materials to arrive at a total estimate of the cost of the project.

Several formats are used for bills of materials. These vary with the project or building being planned. One format for listing items is shown in figure 18-1.

The term item means a separate object. In a bill of materials, the term may refer to a part of a structure such as the side, end, bottom, or top. In a trailer body, an item may be "brackets" for attaching cross members to beams, "leg screws" for attaching pockets for side boards, "boards" for the bottom, "steel" for the sides, or "paint" for the whole project.

An item in a bill of materials may be paint, rafters, doors, nails, hinges, or 2 × 4s. The important thing to remember in making a bill of materials is that items of different types, materials, sizes, or costs must *not* be mixed and listed as one item.

The number of pieces is helpful since it can be used to buy materials efficiently. It is also needed in the calculation of total feet and total cost. For example, consider a plan to build a small storage building. The door will be made from boards that are 1" × 6". These boards can be bought in 6-, 8-, 10-, 12-, 14-, and 16-foot lengths.

The plan indicates that the door is 6 feet high and 3 feet wide, figure 18-2. This knowledge enables the worker to determine the number of pieces as follows: 36 inches (width of door) divided by 5 1/2 inches (width of a finished 1" × 6") = 6.5 boards. This calculation indicates that 7 boards, each 6 feet long, are needed to make the 36-inch door. The number 6.5 is rounded up to 7 boards. When a number is rounded up, it means the next higher whole number is used.

The plan also shows cross boards at the top and bottom of the door to hold the vertical boards together. Each cross piece is 36 inches (or 3 feet) long or equal to the distance across the 36-inch door. Finally, the plan calls for a 1" × 6" diagonal brace. Assume this brace needs to be 5 feet 2 inches long.

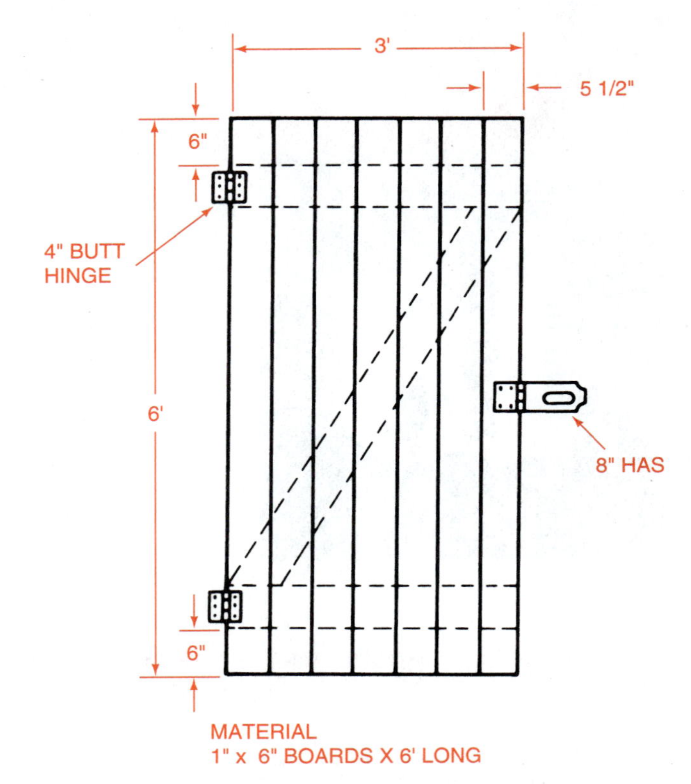

FIGURE 18-2. Plan of a storage shed door.

The bill of materials may be simplified by combining smaller items. The total number of 1" × 6" pieces required to make the door can now be stated as follows:

7 pieces—6 feet long
2 pieces—3 feet long
1 pieces—5 feet 2 inches long

Since the wooden pieces are all 1" × 6" boards, the items listed can be combined. To combine the three items into one item in the bill of materials, the two 3-foot pieces are cut from one 6-foot piece. The 5-foot-2-inch piece can also be cut from a 6-foot piece. Therefore, to make the door, nine pieces of 1" × 6" × 6' lumber are needed.

To make the door weathertight and strong, tongue and groove (T&G) lumber should be used. Tongue and groove boards have a tonguelike edge sticking out on one side and a groove cut into the other, which permits them to lock together.

ITEM	NO. OF PIECES	TYPE OF MATERIALS	SIZE	DESCRIPTION	TOTAL FEET	UNIT COST	COST

FIGURE 18-1. Format for a bill of materials.

ITEM	NO. OF PIECES	TYPE OF MATERIALS	SIZE	DESCRIPTION	TOTAL FEET	UNIT COST	TOTAL COST
Door	9	Boards	1" × 6" × 6'	S4S T&G Pine	54	$0.30	$16.20
Hinge	1 pr	Steel	4"	Butt	N/A	$4/pr	4.00
Hasp	1	Steel	8"	Galvanized	N/A	$3.50	3.50
Nails	1 lb	Steel	6d	Common	N/A	$0.75/lb	0.75
Screws	24	Steel	No. 10 × 3/4	Cd finish	N/A	$0.06 ea	1.44
Paint	1	Oil Base	Quart	Red	N/A	$6.50	6.50
							$32.39

FIGURE 18-3. Lumber, hardware, and paint requirements in a bill of materials.

Assume the cost of the boards will be $.30 per linear foot. The lumber requirement would be included in the bill of materials as shown for "door" in figure 18-3. The plan also calls for a pair of 4-inch butt hinges and an 8-inch hasp. These items of hardware, along with nails and screws, are also included on the bill of materials shown in figure 18-3.

Some additional explanation is in order regarding the sample bill of materials. The lumber was priced by the linear foot since tongue and grooved boards are used. This means that each running foot, or each foot of length in the board, costs $.30. The amount of cost of a linear foot is influenced by the width and thickness. The total number of linear feet is 54; thus, 54 × $.30 = $16.20.

Rough lumber from the sawmill is sold by the board foot. Calculating board feet requires computation of the volume of the board. Calculating board feet will be discussed later. The *S4S* under "Description" is a symbol designating "surfaced 4 sides" or four surfaces of the board have been planed. The symbol S4S means the boards are planed or surfaced on four sides, while the symbol *S2S* would mean the boards were surfaced on two sides.

Since the hasp in the example is galvanized, the finish on the screws is specified as cadmium (Cd). To **galvanize** is to coat a metal with zinc. **Cadmium** is a similar metal used for rust-resistant plating of fasteners and other steel products. "No. 10 × 3/4" size screws" means the screws are number 10 diameter and 3/4" long. "N/A" means "not applicable" and is used when requested information does not apply to the situation.

If the door is a single project, the paint should be added to the bill of materials. However, the door could be part of a whole building under construction. If so, paint would appear as a single item for the entire building unless the door was painted a different color.

UNITS OF MEASURE

A bill of materials provides much information in a small space. Therefore, standard abbreviations are helpful, figure 18-4.

It is easier to plan and build projects with a knowledge of the sizes and weights of lumber and steel products. These are given in figures 18-5 through 18-8.

STANDARD ABBREVIATIONS

" or in = inch
' or ft = foot
yd = yard
mi = mile
ea = each
@ = at
N/A = not applicable
pt = pint
qt = quart
gal = gallon
LF = linear foot
BF = board foot
S1S = surface 1 side
S2S = surface 2 sides
S3S = surface 3 sides
S4S = surface 4 sides
no. or # = number
in^2 or sq in = square inch
ft^2 or sq ft = square foot
yd^2 or sq yd = square yard
square = 10' × 10' or 100 square feet
NC = national coarse threads
NF = national fine threads
NPT = national pipe threads
d = penny (nails)
lb = pound
oz = ounce
Cwt = hundredweight (100 pounds)

FIGURE 18-4. Useful abbreviations for making a bill of materials.

LUMBER

(All available in 6', 8', 10', 12', 1', and 16' lengths)

1" × 6"	2" × 6"
1" × 8"	2" × 8"
1" × 10"	2" ×10"
1" × 12"	2" × 12"
2" × 4"	4" × 6"

PLYWOOD

(Interior and Exterior Grades)

1/4" × 4' × 8'	5/8" × 4' × 8'
3/8" × 4' × 8'	3/4" × 4' × 8'
1/2" × 4' × 8'	1" × 4' × 8'

STRUCTURAL STEEL

(Standard length = 20')

Flat Band Iron—1/8", 3/16", 1/4", 5/16", 3/8", and 1/2" thick —1/2", 3/4", 1", 1 1/2", 2", 3", and 4" wide

Angle Iron—1/8", 3/16", 1/4", 5/16", and 3/8" thick —1/2" × 1/2", 3/4" × 3/4", 1" × 1", 1 1/2" × 1 1/2", 2" × 2", and 3" × 3" (width of legs)

Round (Hot Rolled, Cold Rolled, or Tool Steel) —1/4", 5/16", 3/8", 1/2", 5/8", 3/4", 1", 1 1/2", and 2" (diameter)

BLACK OR GALVANIZED STEEL PIPE

(Standard Length = 21')

—1/4", 3/8", 1/2", 3/4", 1", 1 1/4", 1 1/2", 2", 2 1/2", and 3" (inside diameter)

FIGURE 18-5. Common sizes and units of lumber and structural metals.

CALCULATING BOARD FEET

Lumber in the form of trees, logs, or rough lumber is measured and sold by board feet. Sometimes dressed lumber is sold by the board foot as well.

A board foot is an amount of wood equal to a board 1 inch thick, 1 foot wide, and 1 foot long; or 1 board foot is 1" × 12" × 12". A board foot is also 144 cubic inches (cu in or in^3). That is, 1" × 12" × 12" = 144 cu in, figure 18-9. Cubic measures are determined by multiplying thickness by width by length. To convert cubic inches back to board feet, divide by 144. Hence, 144 cubic inches divided by 144 = 1 board foot.

Two Methods for Calculating Board Feet

Two methods or formulas for calculating board feet are especially useful. Each method is useful for certain sizes of lumber.

Small Pieces of Lumber. Remember that 1 board foot equals 144 cubic inches. With this in mind, the board feet in small or odd-sized pieces of lumber can be calculated. The formula is:

$$BF = \frac{\text{Thickness (in) x Width (in) x Length (in)}}{144}$$

$$\text{Therefore, } BF = \frac{T" \text{ x } W" \text{ x } L"}{144}$$

What is the number of board feet in a piece of wood 1 inch thick, 2 inches wide, and 24 inches long? The answer is calculated as follows:

$$BF = \frac{T" \text{ x } W" \text{ x } L"}{144}$$

$$BF = \frac{1 \text{ x } 2 \text{ x } 24}{144}$$

$$BF = \frac{48}{144}$$

$$BF = .33$$

Consider a wooden nail box with two sides that are each 1" × 4" × 14", two ends that are each 1" × 4" × 9", and a bottom that is 1" × 9" × 12 1/2". How many board feet are needed for the project? The answer is obtained by the steps shown in figure 18-10.

Large Pieces of Lumber. There is a better formula for large pieces of lumber, such as those used in wagon bodies, feed bunks, or buildings. The formula is:

$$BF = \frac{T" \text{ x } W" \text{ x } L'}{12}$$

In this formula, the length is expressed in feet instead of inches and the set is divided by 12 instead of 144. The formula is derived by dividing both the top and bottom by 12. Hence

$$BF = \frac{1" \text{ x } 6" \text{ x } 24"}{144} = \frac{1" \text{ x } 6" \text{ x } 2'}{12} = 1$$

ANGLES—BAR SIZES

Size in Inches	Weight Per Ft (Lbs)
1 × 1 × 1/8	.80
3/16	1.16
1/4	1.49
1 1/8 × 1 1/8 × 1/8	.90
1 1/4 × 1 1/4 × 1/8	1.01
3/16	1.48
1/4	1.92
1 3/8 × 7/8 × 1/8	.91
3/16	1.32
1 1/2 × 1 1/4 × 3/16	1.64
1 1/2 × 1 1/2 × 1/8	1.23
3/16	1.80
1/4	2.34
1 3/4 × 1 1/4 × 1/8	1.23
1 3/4 × 1 3/4 × 1/8	1.44
3/16	2.12
1/4	2.77
2 × 1 1/4 × 3/16	1.96
1/4	2.55

Size in Inches	Weight Per Ft (Lbs)
2 × 1 1/2 × 1/8	1.44
3/16	2.12
1/4	2.77
2 × 2 × 1/8	1.65
3/16	2.44
1/4	3.19
5/16	3.92
3/8	4.70
2 1/4 × 1 1/2 × 3/16	2.28
2 1/2 × 1 1/2 × 3/16	2.44
1/4	3.19
5/16	3.92
2 1/2 × 2 × 3/16	2.75
1/4	3.62
5/16	4.50
3/8	5.30
2 1/2 × 2 1/2 × 3/16	3.07
1/4	4.10
5/16	5.00
3/8	5.90
1/2	7.70

ANGLES—STRUCTURAL

Size in Inches	Weight Per Ft (Lbs)
3 × 2 × 3/16	3.07
1/4	4.10
5/16	5.00
3/8	5.90
1/2	7.70
3 × 2 1/2 × 1/4	4.50
5/16	5.60
3/8	6.60
1/2	8.50
3 1/2 × 3 × 1/4	5.40
5/16	6.60
3/8	7.90
1/2	10.20
3 1/2 × 3 1/2 × 1/4	5.80
5/16	7.20
3/8	8.50
7/16	9.80
1/2	11.10
4 × 3 × 1/4	5.80
5/16	7.20
3/8	8.50
7/16	9.80
1/2	11.10
5/8	13.60
4 × 3 1/2 × 1/4	6.2
5/16	7.7
3/8	9.1
7/16	10.6
1/2	11.9
4 × 4 × 1/4	6.6
5/16	8.2
3/8	9.8
7/16	11.3
1/2	12.8
5/8	15.7
3/4	18.5
5 × 3 × 1/4	6.6
5/16	8.2
3/8	9.8
7/16	11.3
1/2	12.8
5 × 3 1/2 × 3/16	8.7
3/8	10.4
7/16	12.0
1/2	13.6
5/8	16.8
3/4	19.8

Size in Inches	Weight Per Ft (Lbs)
3 × 3 × 3/16	3.71
1/4	4.90
5/16	6.10
3/8	7.20
7/16	8.30
1/2	9.40
3 1/2 × 2 1/2 × 1/4	4.90
5/16	6.10
3/8	7.20
1/2	9.40
5 × 5 × 3/8	12.3
1/2	16.2
5/8	20.0
3/4	23.6
6 × 3 1/2 × 5/16	9.8
3/8	11.7
1/2	15.3
6 × 4 × 5/16	10.3
3/8	2.3
7/16	14.3
1/2	16.2
5/8	20.0
3/4	23.6
7/8	27.2
6 × 6 × 3/8	14.9
7/16	17.2
1/2	19.6
5/8	24.2
3/4	28.7
7/8	33.1
1	37.4
7 × 4 × 3/8	13.6
1/2	17.9
5/8	22.1
8 × 4 × 1/2	19.6
8 × 6 × 1/2	23.0
5/8	28.5
3/4	33.8
1	44.2
8 × 8 × 1/2	26.4
5/8	32.7
3/4	38.9
7/8	45.0
1	51.0

FIGURE 18-6. Weight per foot of angle steel. To determine the cost per foot, multiply the weight in pounds per foot by the price per pound.

CONCRETE REINFORCING BARS

Bar Sizes Old (Inches)	Bar Sizes New (Numbers)	Weight Per Ft (Lbs)	Diameter (Inches)
1/4	2	.167	.250
3/8	3	.376	.375
1/2	4	.668	.500
5/8	5	1.043	.625
3/4	6	1.502	.750
7/8	7	2.044	.875
1	8	2.670	1.000
1	9	3.400	1.128
1 1/8	10	4.303	1.270
1 1/4	11	5.313	1.410

ROUNDS

Size in Inches	Weight Per Ft (Lbs)	Size in Inches	Weight Per Ft (Lbs)
c1/4	0.167	2 1/4	13.52
5/16	0.261	2 3/8	15.06
3/8	0.376	2 1/2	16.69
7/16	0.511	2 5/8	18.40
1/2	0.668	2 3/4	20.20
9/16	0.845	3	24.03
5/8	1.043	3 1/4	28.21
3/4	1.502	3 1/2	32.71
7/8	2.044	3 3/4	37.55
1	2.670	4	42.73
1 1/8	3.379	4 1/4	48.23
1 1/4	4.173	4 1/2	54.08
1 3/8	5.05	4 3/4	60.25
1 1/2	6.01	5	66.76
1 5/8	7.051	5 1/4	73.60
1 3/4	8.18	5 1/2	80.78
1 7/8	9.39	5 3/4	88.29
2	10.68	6	96.13
2 1/8	12.06		

CHANNELS—STRUCTURAL

Depth of Channel Inches	Weight Per Ft (Lbs)	Thickness of Web	Width Flange (Inches)
3	4.1	0.170	1.410
	5.0	0.258	1.498
	6.0	0.356	1.596
4	5.4	0.180	1.580
	7.25	0.320	1.720
5	6.7	0.190	1.750
	9.0	0.325	1.885
6	8.2	0.200	1.920
	10.5	0.314	2.034
	13.0	0.437	2.157
7	9.8	0.210	2.090
	12.25	0.314	2.194
	14.75	0.419	2.299
8	11.5	0.220	2.260
	13.75	0.303	2.343
	18.75	0.487	2.527
9	13.4	0.230	2.430
	15.0	0.285	2.485
	20.0	0.448	2.648
10	15.3	0.240	2.600
	20.0	0.379	2.739
	25.0	0.526	2.886
	30.0	0.673	3.033
12	20.7	0.280	2.940
	25.0	0.387	3.047
	30.0	0.510	3.170
13	31.8	0.375	4.000
	50.0	0.787	4.412
15	33.9	0.400	3.400
	40.0	0.520	3.520
	50.0	0.716	3.716
18	45.8	0.500	4.000

FIGURE 18-7. Weight per foot of round and channel steel. Multiply the weight in pounds per foot by the price per pound to get the cost per foot.

GAUGE	THICKNESS (INCHES)	WEIGHT/SQ. FT (POUNDS)
No. 18	0.0478	2.00
No. 16	0.0598	2.50
No. 14	0.0747	3.125
No. 12	0.1046	4.375
No. 11	0.1196	5.000
No. 10	0.1345	5.625
No. 8	0.1644	6.875
No. 7	3/16	7.500

FIGURE 18-8. Weight per square foot of sheet metal. Multiply the weight in pounds per square foot by the price per pound to determine the cost per square foot.

This formula can be used to determine the number of board feet in the door shown in figure 18-11. The math is easier in the second formula because expressing the length in feet makes the figures smaller. Remember that lumber must be bought in lengths of even numbers of feet. Calculations must be based on the length of the lumber to be purchased, not just on the length to be used in the project. To ignore the waste lumber means the final cost of the project will be higher than calculated.

PRICING MATERIALS

Lumber is priced by the thousand board feet. The Roman numeral for one thousand is a capital *M*. For

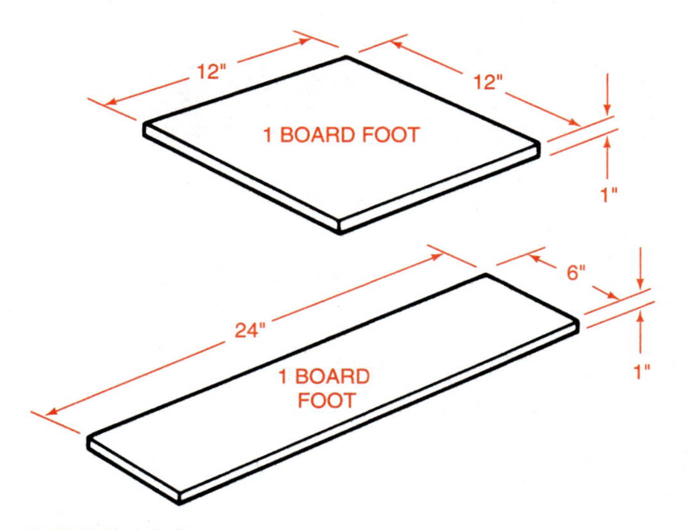

FIGURE 18-9. A board foot is any volume of wood that equals 144 cubic inches.

example, lumber dealers may describe the price of lumber as \$300/M. This means \$300 for 1,000 board feet. In this example, to determine the price of 1 board foot, simply divide \$300 by 1,000:

$$\frac{\$300}{1{,}000} = \$.30$$

Wood products sold in sheets are sold by the square foot or panel. Examples of such products are plywood, pressed wood, chip board, wall paneling, and insulation board. These panels generally are sold in 4' × 8' sheets of various thicknesses. A $^3/_4$' × 4' × 8' piece of exterior-grade plywood may sell for \$.50 per square foot. However, an entire 4' × 8' sheet will probably have to be purchased for \$16 (4' × 8' = 32 sq ft × \$.50/sq ft = \$16).

ITEM	NO. OF PIECES	SIZE	BF PER PIECE		TOTAL BF
1. Sides	2	1" × 4" × 14"	0.39		0.78
2. Ends	2	1" × 4" × 9"	0.25		0.50
3. Bottom	1	1" × 9" × 12$^1/_2$"	0.78		0.78
				Grand Total	2.06

Step 1. Sides: $\frac{1" \times 4" \times 14"}{144} = \frac{56}{144} = 0.39 \times 2 \text{ pcs.} = 0.78$

Step 2. Ends: $\frac{1" \times 4" \times 9"}{144} = \frac{36}{144} = 0.25 \times 2 \text{ pcs.} = 0.50$

Step 3. Bottom: $\frac{1' \times 9" \times 12^1/_2"}{144} = \frac{112.5}{144} = 0.78$

FIGURE 18-10. Calculation of board feet for small pieces.

ITEM	NO. OF PIECES	SIZE	BF PER PIECE	TOTAL BF
1. Vertical	7	1" × 6" × 6'	3	21
2. Cross	2	1" × 6" × 3'	1.5	3
3. Diagonal	1	1" × 6" × 6'	3	3
			Grand Total	27

Step 1. Vertical: $\frac{1" \times 6" \times 6'}{12} = \frac{6 \times 6}{12} = \frac{36}{12}$ = 3 × 7 pcs. = 21 BF

Step 2. Cross: $\frac{1" \times 6" \times 3'}{12} = \frac{6 \times 3}{12} = \frac{18}{12}$ = 1.5 × 2 pcs. = 3 BF

Step 3. Diagonal: $\frac{1' \times 6" \times 6'}{12} = \frac{6 \times 6}{12} = \frac{36}{12}$ = 3 × 2 pcs. = 3 BF

Grand Total = 27 BF

FIGURE 18-11. Calculation of board feet for large pieces.

STUDENT ACTIVITIES

1. Define the Terms to Know in this unit.
2. List the eight components recommended for a bill of materials.
3. Calculate the number of board feet in each item below.
   a. 1 piece 1" × 6" × 12'
   b. 1 piece 1"× 12" × 12'
   c. 1 piece 2" × 6" × 8'
   d. 5 pieces 2" × 4" × 12'
   e. 1 piece 1" × 3" × 18'
   f. 3 pieces 1" × 12" × 14"
4. Calculate the cost of materials in each item below.
   a. 4 pieces 2" × 6" × 12' at a cost of $.50/BF
   b. 20 pieces 2" × 4" × 16" at a cost of $.40/BF
   c. 12 pieces 1" × 10"× 12' at a cost of $.55 per LF
   d. 1 piece 1/4" × 1" × 1" angle steel 20 feet long at a cost of $.30 per pound. (Use Figure 18-6.)
   e. 1 piece 3/4-inch black pipe 21 feet long at $.45 per foot

SELF-EVALUATION

A. Multiple Choice. Select the best answer.

1. A board with all four surfaces planed is specified
   a. S2S
   b. S4S
   c. S2E
   d. S2+2
2. Wood products for building may be sold by the
   a. board foot
   b. linear foot
   c. square foot
   d. all of these
3. Boards that lock together are
   a. T&G
   b. BF
   c. LF
   d. S4S
4. Rough lumber is generally sold by the
   a. T&G
   b. BF
   c. LF
   d. piece
5. A board foot is based on the
   a. number of cubic inches
   b. length and width only
   c. length × width × thickness divided by 50
   d. none of these
6. Sheet steel is sold by the pound. To determine the cost per square foot
   a. divide the cost per sheet by 12
   b. multiply length by width of sheet
   c. multiply price per pound by weight per square foot
   d. none of these

B. Matching. Match the terms in column I with those in column II.

Column I
1. steel, angle
2. dressed lumber
3. logs
4. pipe
5. $500/M
6. plywood

Column II
a. priced by the square foot
b. priced by BF and LF
c. priced by the pound
d. priced by board feet
e. priced by linear foot
f. price per 1,000 board feet

C. Completion. Fill in the blanks with the word or words that make the following statements correct.

1. A bill of materials is a ___________.
2. The components of a bill of materials for agricultural mechanics projects are
   a. ___________
   b. ___________
   c. ___________
   d. ___________
   e. ___________
   f. ___________
   g. ___________
   h. ___________
3. In a board 1" × 4" × 12'
   a. the 1" refers to the ________
   b. the 4" refers to the ________
   c. the 12" refers to the _______
4. Five items that might be in a bill of materials for a trailer are
   a. ___________
   b. ___________
   c. ___________
   d. ___________
   e. ___________
5. The two formulas used to calculate board feet are
   a. ___________
   b. ___________

D. Brief Answers. Briefly answer the following questions.

1. Which of the two formulas used to calculate board feet is best for small boards?
2. Calculate board feet.
   a. 1 piece 1" × 10" × 16' = __________ BF
   b. 1 piece 2" × 8" × 6' = __________ BF
   c. 1 piece 2" × 6" × 8' = __________ BF
   d. 2 pieces 2" × 4" × 8' = __________ BF

3. Calculate board feet and costs of a project with the following parts if lumber costs \$.40 per board foot.

	Board Feet	*Cost*
2 sides, each 1" × 6" × 18" =	_____	_____
2 ends, each 1" × 6" × 10" =	_____	_____
1 bottom, 1" × 10" × 16 1/2" =	_____	_____
8 screws, \$.07 each =	_____	_____
1 pint of paint, \$4.50 =	_____	_____
Total =		_____

4. Calculate the cost of each item if the price of steel is \$.50 per pound.

a. 1 piece 1/8" × 1" × 20' angle steel
(pounds per foot = .80)
Cost = _____

b. 4 pieces 1/4" × 2" × 2" × 1'
(pounds per foot = 3.19)
Cost = _____

c. 2 pieces 1/2" × 10' concrete reinforcing bar
(pounds per foot = .67)
Cost = _____

UNIT

19 Selecting, Planning, and Building a Project

OBJECTIVE

To select and plan projects that develop the woodworking and metalworking skills needed in agricultural jobs.

COMPETENCIES TO BE DEVELOPED

After studying this unit, you should be able to:

- Determine the skills you now have in agricultural mechanics.
- Select projects that require the use of woodworking tools.
- Select projects that require the use of metalworking tools.
- Select projects that are appropriate for developing basic skills.
- Modify plans of projects to meet personal needs.

MATERIALS LIST

- ✓ Lead pencil
- ✓ Eraser
- ✓ Ruler
- ✓ Graph paper, 1/4-inch

TERMS TO KNOW

project
scope
hands-on
graph paper

This unit will show how to plan a project for construction in the agricultural mechanics shop. A project is a special activity planned and conducted to aid learning. At school, project work is supervised by an instructor. Shop construction projects are fun. They develop eye-hand coordination and skills for jobs. In addition to personal development, there is a usable product when the project is completed.

PURPOSES OF PROJECTS

The project method of learning has been used in American schools for about a century. It stimulates students to think and learn. It helps students develop good and safe work habits. Any project a student builds for himself or herself stimulates interest and a desire to do a good job.

Students may design their own projects or they may modify plans from another source. Either way, the design phase requires thinking—to seek solutions and to solve problems. The project may stimulate the student to buy materials and get to know local suppliers. It can provide experience in shopping for the best prices and in looking for other ways to cut costs. These are good business skills.

School projects help to develop construction skills. These include measuring, cutting, drilling, assembling, and finishing. A good project will require the use of a wide variety of shop skills.

A good project will also help students judge their own achievement. When students are finished, they can decide what they did well. They can also see where they made mistakes and decide how to avoid them the next time.

Finally, the project will leave a visible reminder of students' efforts, figure 19-1. If they did well, the project will show it. If they were careless, lacked skill, or did not complete the project, this too will show. Students like to take home well-made projects. These are a source of pride for students and their families.

SCOPE OF PROJECTS

The scope of a project refers to its size and complexity. Students in agriculture should select and plan projects that help them learn. Learning is the most important reason for doing a shop project. Therefore, woodworking projects are appropriate when woodworking is being studied. Similarly, metalworking projects are appropriate when metalworking is being studied. The project provides an opportunity for hands-on experience. Hands-on means actually doing something, rather than reading or hearing others tell about it. It means practicing an activity to develop a skill.

Shop projects should be relatively simple during the first course in agriculture. At this time the students are learning basic skills. Many tools and equipment will be used for the first time. Work may progress slowly. In addition, there may be competition for tools in the class. However, in completing these projects, the students will develop skills to help them move on to more complex projects in the future.

Projects such as nail boxes, tool boxes, feed scoops, bird houses, stools, funnels, bench stops, foot scrapers, and gate staples have been favorites of the first-year students and instructors for years, figures 19-2 and 19-3. They have been popular because they require skills appropriate to the experience of beginning students.

For second-year students, more complex projects are desired. Such projects typically are larger, require more skills, and take longer to complete. Examples are jack stands, car ramps, sawhorses, shop stands, storage units, yard gates, feed carts, pickup truck racks, and picnic tables, figures 19-4 through 19-9.

Advanced students should choose more challenging projects. Advanced classes frequently are smaller. This may permit larger projects to be built. Such projects should be designed to provide additional skill development in carpentry, metal shaping and bending, welding, pipe cutting and bending,

FIGURE 19-1. A properly constructed shop project can be rewarding. This agriculture student won a $100 welding award for designing and building this utility trailer.

FIGURE 19-2. Tool boxes make excellent projects for beginning agriculture students. They provide practice in appropriate skills in woodworking and encourage the student to acquire and organize personal tools.

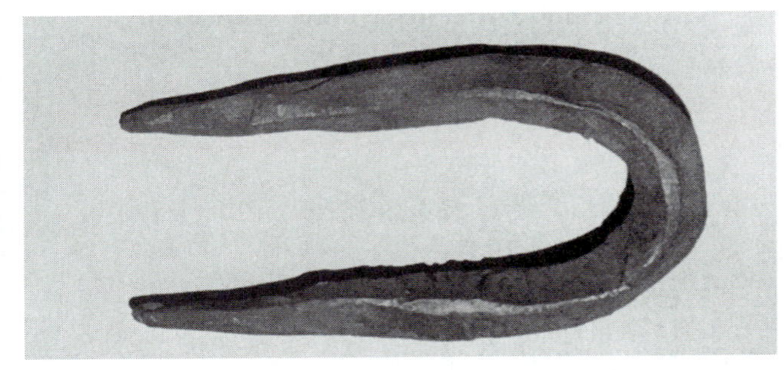

FIGURE 19-3. The gate staple is a good project for learning hot metalworking skills.

A. Automobile jack stand.

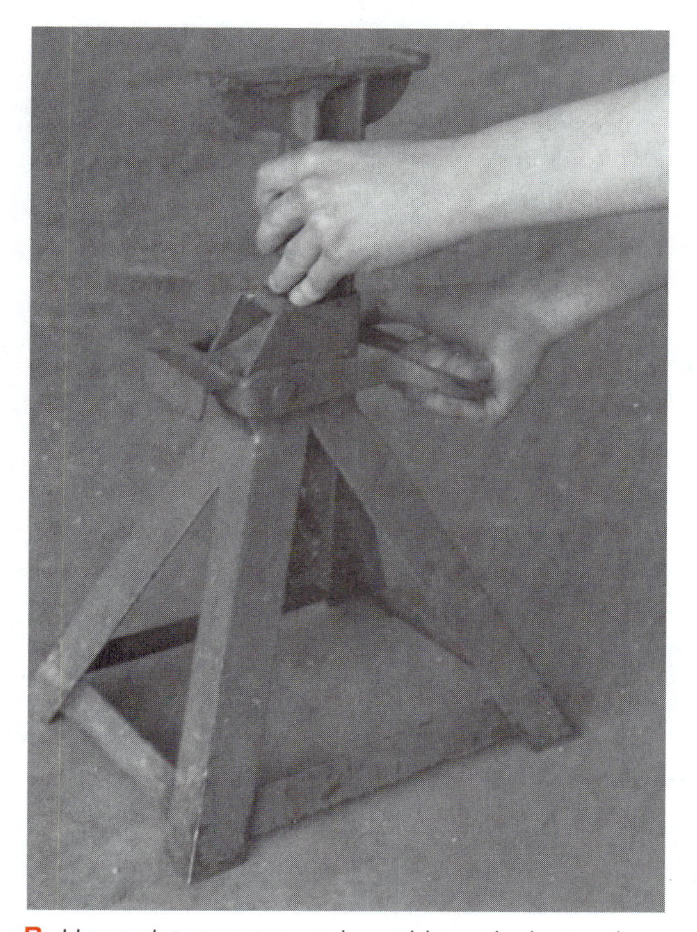

B. Heavy-duty tractor and machinery jack stand.

FIGURE 19-4. Jack stands for automobiles and machinery develop skill in metalworking and welding.

FIGURE 19-5. A car ramp project combines experiences in metalworking and woodworking. When making projects such as jack stands and ramps, special care must be taken to use designs that provide for extra strength. Careful design and quality workmanship are needed to assure the safety of users.

FIGURE 19-6. The lightweight (spider) trestle was made in an agricultural mechanics shop and has been in service for over thirty years. It features long, slender legs and lightweight construction.

nailing, bolting, and painting, figure 19-10. Good projects for developing such skills include loading chutes, farm and utility trailers, wagon bodies, truck bodies, horse trailers, cattle guards, cattle feeders, post hole diggers, and portable buildings, hydroponics units, and aquaculture structures.

If farm tractor and machinery maintenance are being taught, farm machinery and tractors that need repair should be brought to the school shop from home or work. Appropriate repair and maintenance operations include engine tuneup, horsepower checks, minor engine repair, clutch and brake adjustments, servicing wheel bearings, replacing electrical wiring, repairing frames, replacing broken parts, and painting.

Whether for beginning or advanced students, projects must be matched with level of experience.

FIGURE 19-7. Storage units may be built for the home agricultural library or other areas of the home or farm.

FIGURE 19-8. A rack made from pipe is useful for lumber storage in shops at home or school.

FIGURE 19-9. Picnic tables are popular home improvement projects. Good plans are available for picnic tables made of lumber, pipe, or steel.

FIGURE 19-10. Wagons and trailers can be relatively simple if only the low sideboards are constructed. However, if the entire running gear and body are constructed, it is a challenging project for the advanced agricultural mechanics student.

Before proceeding too far with a project, students should discuss their ideas with their instructor.

SELECTING A PROJECT

Projects should be selected carefully. Projects that are unsuitable may not be completed or may be poorly constructed. Either way, they do not achieve their intended purpose. The following steps are recommended for selecting a shop project.

PROCEDURE

1. Take a personal inventory of agricultural mechanics skills (competencies), figure 19-11.
2. Discuss the results of the inventory with your instructor.
3. Ask your instructor to suggest projects that will help you develop appropriate skills.
4. Search books, project bulletins, and departmental files for project ideas and plans.
5. Examine sample projects and look for details of design.
6. Ask parents, guardians, friends, and employers for ideas and suggestions.
7. Make a tentative decision and obtain approval from your instructor.

AGRICULTURAL MECHANICS COMPETENCY INVENTORY

INSTRUCTION: Check the column that best describes your personal level of experience in each area of competency.

	Level of Experience			
	1 Have Done	2 Am Familiar With	3 Will Learn This Year	4 No Skill; Will Not Learn This Year
1. Practice Safety and Shop Organization				
a Work safely with hand tools	❑	❑	❑	❑
b. Work safely with power tools	❑	❑	❑	❑
c. Follow safety rules in shop	❑	❑	❑	❑
d. Maintain safe work areas	❑	❑	❑	❑
e. Eliminate fire hazards	❑	❑	❑	❑
f. Select and use fire extinguishers	❑	❑	❑	❑
g. Interpret product labels	❑	❑	❑	❑
h. Act correctly in an emergency	❑	❑	❑	❑
i. Store materials correctly	❑	❑	❑	❑
j. Clean the shop effectively	❑	❑	❑	❑
2. Identify and Fit Agricultural Tools				
a Identify tools and equipment	❑	❑	❑	❑
b. Sharpen hand tools	❑	❑	❑	❑
c. Sharpen twist drills	❑	❑	❑	❑
d. Fit handles for hammers and axes	❑	❑	❑	❑
e. Fit handles for hoes and shovels	❑	❑	❑	❑
f. Dress grinding wheels	❑	❑	❑	❑
3. Maintain and Service a Home Shop				
a Clean and maintain an orderly shop	❑	❑	❑	❑
b. Observe safety regulations	❑	❑	❑	❑
c. Place and use fire fighting equipment	❑	❑	❑	❑
d. Inventory tools and equipment	❑	❑	❑	❑
e. Plan shop layout	❑	❑	❑	❑
f. Select shop site	❑	❑	❑	❑
4. Apply Farm Carpentry Skills				
a. Select lumber for a job	❑	❑	❑	❑
b. Operate power saws	❑	❑	❑	❑
c. Operate a jointer	❑	❑	❑	❑
d. Operate a planer	❑	❑	❑	❑
e. Saw to dimension	❑	❑	❑	❑
f. Bore holes	❑	❑	❑	❑
g. Glue wood	❑	❑	❑	❑
h. Drive and remove nails	❑	❑	❑	❑
i. Set screws and install bolts	❑	❑	❑	❑

FIGURE 19-11. Agricultural mechanics competency inventory.

	Level of Experience			
	1 Have Done	2 Am Familiar With	3 Will Learn This Year	4 No Skill; Will Not Learn This Year
5. Properly Use Paint and Paint Equipment				
a. Select paints or preservatives	❑	❑	❑	❑
b. Compute area for painting	❑	❑	❑	❑
c. Prepare wood surfaces	❑	❑	❑	❑
d. Prepare metal surfaces	❑	❑	❑	❑
e. Apply paint with a brush or roller	❑	❑	❑	❑
f. Clean and store paint brushes	❑	❑	❑	❑
g. Mask areas prior to painting	❑	❑	❑	❑
h. Apply paint with a spray can	❑	❑	❑	❑
i. Clean and store spray gun	❑	❑	❑	❑
6. Operate an Arc Welder				
a. Practice safety in arc welding	❑	❑	❑	❑
b. Prepare metal for welding	❑	❑	❑	❑
c. Determine welder settings	❑	❑	❑	❑
d. Select electrodes	❑	❑	❑	❑
e. Operate AC and DC welders	❑	❑	❑	❑
f. Strike an arc	❑	❑	❑	❑
g. Run flat beads	❑	❑	❑	❑
h. Make weld joints	❑	❑	❑	❑
i. Weld horizontally	❑	❑	❑	❑
j. Weld vertically	❑	❑	❑	❑
k. Weld overhead	❑	❑	❑	❑
l. Cut with arc	❑	❑	❑	❑
m. Weld cast iron	❑	❑	❑	❑
n. Hard surface steel	❑	❑	❑	❑
7. Operate an Oxyacetylene Welder				
a. Check for leaks	❑	❑	❑	❑
b. Turn equipment on and off	❑	❑	❑	❑
c. Adjust equipment	❑	❑	❑	❑
d. Change cylinders	❑	❑	❑	❑
e. Cut with cutting torch	❑	❑	❑	❑
f. Choose proper tips	❑	❑	❑	❑
g. Braze thin metal	❑	❑	❑	❑
h. Run beads	❑	❑	❑	❑
i. Make butt welds	❑	❑	❑	❑
8. Perform Skills in Hot and Cold Metal Work				
a. Identify metals	❑	❑	❑	❑
b. Cut with hand hacksaw	❑	❑	❑	❑
c. Cut with cold chisel	❑	❑	❑	❑
d. Bend metal	❑	❑	❑	❑
e. Cut with tinsnips	❑	❑	❑	❑
f. Drill holes	❑	❑	❑	❑

FIGURE 19-11. Agricultural mechanics competency inventory. (Cont.)

	Level of Experience			
	1 Have Done	2 Am Familiar With	3 Will Learn This Year	4 No Skill; Will Not Learn This Year
g. Cut threads	❑	❑	❑	❑
h. Use files	❑	❑	❑	❑
i. Solder metal	❑	❑	❑	❑
j. Use power hacksaw	❑	❑	❑	❑
k. Heat metal with torch	❑	❑	❑	❑
l. Operate gas forge	❑	❑	❑	❑
9. Perform Skills in Concrete and Masonry Work				
a. Build and prepare forms	❑	❑	❑	❑
b. Treat forms	❑	❑	❑	❑
c. Reinforce concrete	❑	❑	❑	❑
d. Test aggregates for impurities	❑	❑	❑	❑
e. Mix concrete	❑	❑	❑	❑
f. Pour concrete	❑	❑	❑	❑
g. Embed bolts	❑	❑	❑	❑
h. Finish concrete	❑	❑	❑	❑
i. Protect concrete while curing	❑	❑	❑	❑
j. Trowel concrete	❑	❑	❑	❑
k. Remove forms	❑	❑	❑	❑
l. Lay concrete blocks	❑	❑	❑	❑
m. Drill holes in concrete	❑	❑	❑	❑
10. Operate and Maintain Farm Engines				
a. Read and follow operator's manual	❑	❑	❑	❑
b. Start and operate engines	❑	❑	❑	❑
c. Change oil and oil filters	❑	❑	❑	❑
d. Service air and fuel filters	❑	❑	❑	❑
e. Maintain battery water	❑	❑	❑	❑
f. Maintain and operate small gas engines	❑	❑	❑	❑
g. Identify engine components and systems	❑	❑	❑	❑
h. Remove and connect battery cables	❑	❑	❑	❑
i. Charge batteries	❑	❑	❑	❑
j. Read battery hydrometer	❑	❑	❑	❑
k. Test batteries	❑	❑	❑	❑
l. Troubleshoot fuel problems	❑	❑	❑	❑
m. Troubleshoot ignition problems	❑	❑	❑	❑
11. Perform Minor Tuneup and Repair of Farm Engines				
a. Clean, gap, and replace spark plugs	❑	❑	❑	❑
b. Adjust carburetor mixture and speed crews	❑	❑	❑	❑
c. Install gap breaker points	❑	❑	❑	❑
d. Set dwell with meter	❑	❑	❑	❑
e. Install condensers	❑	❑	❑	❑
f. Time engines using timing light	❑	❑	❑	❑
g. Disassemble and reassemble distributors	❑	❑	❑	❑

FIGURE 19-11. Agricultural mechanics competency inventory. (Cont.)

	Level of Experience			
	1 Have Done	2 Am Familiar With	3 Will Learn This Year	4 No Skill; Will Not Learn This Year
h. Measure compression	❑	❑	❑	❑
i. Adjust valves	❑	❑	❑	❑
j. Set float levels in carburetors	❑	❑	❑	❑
k. Rebuild carburetors	❑	❑	❑	❑
l. Operate engine analyzers	❑	❑	❑	❑
m. Test and replace coils	❑	❑	❑	❑
n. Clean and install brushes	❑	❑	❑	❑
o. Clean commutator bars	❑	❑	❑	❑
12. Perform Electrification Skills				
a. Use safety measures in electrical wiring	❑	❑	❑	❑
b. Select correct fuse sizes	❑	❑	❑	❑
c. Replace fuses	❑	❑	❑	❑
d. Make splices	❑	❑	❑	❑
e. Repair electrical cords	❑	❑	❑	❑
f. Wire on-off switches	❑	❑	❑	❑
g. Select wire sizes	❑	❑	❑	❑
h. Know electrical terminology such as; volts, amps, watts, ohms	❑	❑	❑	❑
i. Attach wires to terminals	❑	❑	❑	❑
j. Solder splices	❑	❑	❑	❑
k. Install wire nut connectors	❑	❑	❑	❑
l. Install light fixtures	❑	❑	❑	❑
m. Install electric motors	❑	❑	❑	❑
n. Wire buildings and structures	❑	❑	❑	❑
o. Wire three-way switches	❑	❑	❑	❑
p. Wire four-way switches	❑	❑	❑	❑
13. Learn Plumbing Skills				
a. Repair leaky faucets	❑	❑	❑	❑
b. Assemble pipe and pipe fittings	❑	❑	❑	❑
c. Thread pipe	❑	❑	❑	❑
d. Measure and cut pipe	❑	❑	❑	❑
e. Cut plastic tubing/pipe	❑	❑	❑	❑
f. Install plastic tubing/pipe	❑	❑	❑	❑
g. Flare copper tubing	❑	❑	❑	❑
h. Cut copper tubing	❑	❑	❑	❑
i. Ream pipe	❑	❑	❑	❑
j. Install fixtures on plastic pipe	❑	❑	❑	❑
k. Sweat joints on copper pipe	❑	❑	❑	❑
l. Adjust air control mechanism on pressure system	❑	❑	❑	❑
m. Install pressure pumps	❑	❑	❑	❑
n. Repair pumps	❑	❑	❑	❑

FIGURE 19-11. Agricultural mechanics competency inventory. (Cont.)

	Level of Experience			
	1 Have Done	2 Am Familiar With	3 Will Learn This Year	4 No Skill; Will Not Learn This Year
o. Install water heaters	❑	❑	❑	❑
p. Cut soil pipe	❑	❑	❑	❑
q. Caulk joints on soil pipe	❑	❑	❑	❑
r. Lay soil pipe	❑	❑	❑	❑

FIGURE 19-11. Agricultural mechanics competency inventory. (Cont.)

PREPARING A PLAN

Good plans for projects are often available in textbooks, special plan books, magazines, and agricultural mechanics shop files. Many project plans are presented in the appendix of this text. A good plan with detailed drawings should be used. Such plans are generally well thought out and tested through actual construction.

If an exact plan is not available, a plan that is close to the need can be modified. If nothing is available, then an original design can be made.

Detailed information on how to make a simple project plan and a bill of materials is presented in previous units. However, project ideas can be tried out quickly on graph paper. Graph paper is paper laid out in squares of equal size. Paper with 1/4-inch squares is recommended for preparing views of projects, figure 19-12A. With graph paper, the blocks can be counted as a means of measuring. A ruler can then be used as a straightedge to draw lines. On plain paper, however, every line requires two or more measurements to establish length and position, figure 19-12B.

To design on paper, a tentative scale is first tried. If it doesn't work, another scale is tried. Graph paper with 1/4-inch squares has 34 blocks one way and 44 blocks the other. To draw a project view on a sheet of this graph paper, follow these steps.

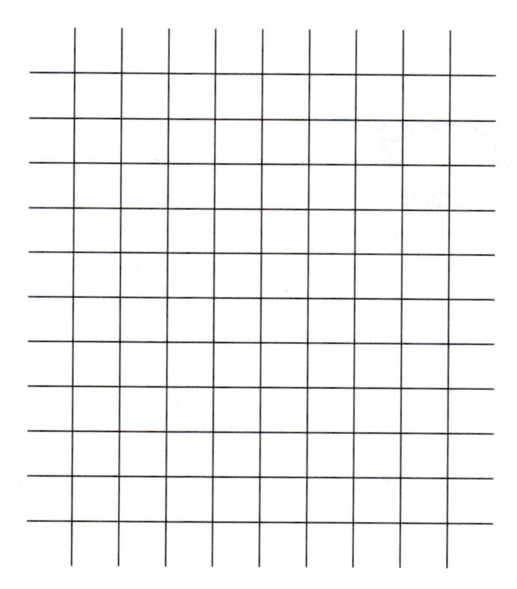

A. GRAPH PAPER WITH 1/4-INCH SQUARES IS EXCELLENT FOR PROJECT PLANNING. ON GRAPH PAPER, DISTANCES ARE DETERMINED BY COUNTING SQUARES AND USING A RULER TO CONNECT POINTS.

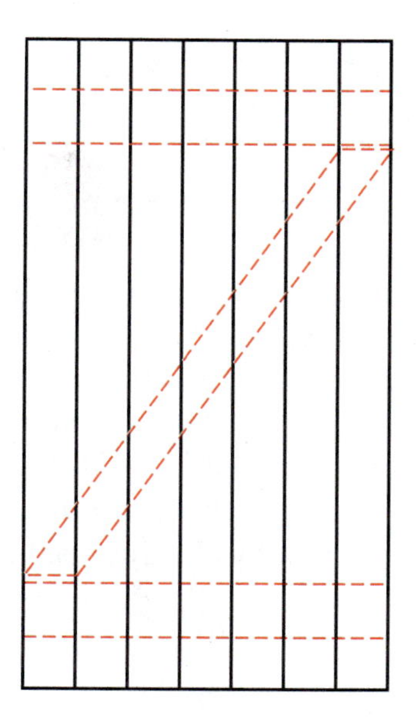

B. A PROJECT DRAWN ON PLAIN PAPER REQUIRES MORE TIME AND EFFORT THAN ONE DRAWN ON GRAPH PAPER. ON PLAIN PAPER, EVERY LINE REQUIRES TWO OR MORE MEASUREMENTS TO ESTABLISH LENGTH AND POSITION.

FIGURE 19-12. Graph paper makes plan drawing easier.

PROCEDURE

1. Determine the longest dimension of the project. This may be the length or the height.
2. Turn the graph paper horizontally for long projects, vertically for high ones.
3. Divide 44 (maximum number of squares in the long direction) by the longest dimension of the project. The result tells how many feet or inches one square must represent.
4. Choose a scale that permits a one-view drawing on the paper with some room for dimensions and notes.
5. Count squares and draw lines to complete one view. Include dimensions.
6. Draw additional views on additional sheets of paper.
7. Modify the plan until it is what you want.
8. Review the plan with your instructor.
9. Draw better plans after obtaining the instructor's approval.

After the plan is drawn, a bill of materials can be prepared and the materials purchased. It is best not to change the design after construction has started. However, if errors in design become obvious during construction, design modifications must be made.

DESIGN

Design work is not easy. It generally requires a great deal of experience to be done well. When designing projects, the following principles and facts should be kept in mind:

- Base designs on engineered plans—that is, start with tested plans.
- Use materials of extra strength and size if safety is a factor. For example, car ramps and jack stands should be extra sturdy. Failure of these objects could cause injury or death.
- Structural steel generally holds up better than wood.
- Welding is stronger than riveting.
- Gluing is generally the most secure method of fastening wood.
- Bolts are more secure than screws; screws are more secure than nails.
- Crimps and folds increase the strength of sheet metal panels.
- Tongue and groove, dowels, and splines all increase the strength of joints in wood.
- Both wood and metal must be protected by preservative, paint, or other protective film if they are to be exposed to weather.

Examples of project plans are presented in Appendix A.

STUDENT ACTIVITIES

1. Define the Terms to Know in this unit.
2. Take a personal agricultural mechanics skills inventory.
3. Write down the skills that you need to develop through project construction.
4. Discuss project ideas with your instructor.
5. Search books, bulletins, files, and other sources for project ideas.
6. Select a project that fits your shop experience.
7. Make or modify plans as needed.

SELF-EVALUATION

A. Multiple Choice. Select the best answer.

1. The project method of learning is used because it
   a. develops body coordination
   b. develops skills for jobs
   c. provides a project when finished
   d. all of these
2. The project method has been used in the public schools for about
   a. ten years
   b. two decades
   c. thirty years
   d. one century
3. The project method of learning usually
   a. causes students to think
   b. causes students to work carefully
   c. creates personal interest
   d. all of these
4. A suitable project must
   a. provide practice in skills taught in class
   b. earn student money
   c. be completed in four weeks
   d. be paid for totally by student funds
5. Projects that are not carefully chosen will
   a. result in student failure
   b. never be completed
   c. not achieve the educational purposes
   b. be too costly
6. Good project ideas may come from
   a. books
   b. teachers
   c. parents or guardians
   d. all of these
7. Good sample projects have the benefit of
   a. showing details of design
   b. showing mistakes to avoid
   c. eliminating the need for a bill of materials
   d. all of these
8. In general, preferred projects for first-year students are
   a. those already designed with plans published
   b. those designed by the student
   c. those modified by the student
   d. all of these
9. Graph paper for drawing permits
   a. easy drawing of angles
   b. quick measurement
   c. easy three-view drawing
   d. economy of cost

B. Matching. Match the terms in column I with the terms in column II.

Column I
1. hands-on
2. nail box
3. feed cart
4. farm trailer
5. graph paper
6. nailing
7. gluing
8. steel
9. paint
10. weld

Column II
a. second- or third-year project
b. third- or fourth-year project
c. first-year project
d. doing
e. strongest wood joint
f. weakest wood joint
g. strongest metal joint
h. stronger than wood
i. 1/4-inch squares
j. protects from weather

C. Completion. Fill in the blanks with the word or words that make the following statements correct.

1. Building shop projects helps students develop buying skills because __________.
2. Shop skills developed by building suitable projects include ______, ______, ______, ______ and _____.
3. Students in agriculture should select and plan projects that __________.

Tool Fitting

UNIT

20 Repairing and Reconditioning Tools

OBJECTIVE

To restore worn, damaged, or abused tools to good working condition.

COMPETENCIES TO BE DEVELOPED

After studying this unit, you should be able to:

- Remove rust from metal tools.
- Apply rust-inhibiting materials to metal surfaces.
- Repair split wood handles.
- Replace broken wooden handles.
- Reshape screwdriver tips.
- Reshape the heads of driving and driven tools.

TERMS TO KNOW

abuse
rust
mushroomed
tool fitting
recondition
saddle soap
neat's-foot oil
wedge
eye
ferrule
tang
crown
torque

MATERIALS LIST

- ✓ Tools that need repair or reconditioning
    a. Tools with rust showing
    b. Driving tool with a split wooden handle
    c. Driving tool with a broken wooden handle
    d. Fork, rake, hoe, or shovel with a broken handle
    e. Tools with mushroomed heads
    f. Screwdriver that needs fitting
    g. Leather tools or protective clothing that need reconditioning
- ✓ Replacement wooden handles for previous items
- ✓ Saddle soap or neat's-foot oil
- ✓ Stiff bristle scrub brush
- ✓ Soft cloths
- ✓ Paste finishing wax
- ✓ Varsol or kerosene
- ✓ Wire brush or wire wheel
- ✓ Waterproof silicon carbide paper, 220 and 400 grit
- ✓ Lightweight lubricating oil, No. 10 or lighter
- ✓ Mediumweight lubricating oil, No. 20 or 30
- ✓ Wood glue
- ✓ Vise, hand tools, and electric drill
- ✓ Assorted steel rivets
- ✓ Wood and metal handle wedges

Modern farms and agribusinesses make large investments in tools. The most obvious use of tools is to build, repair, or adjust objects and machines. However, tools need repair too.

This unit provides instruction in repairing tools. The student is advised to review Units 14 and 16 on the use of power drills and grinders. These tools are useful in performing the tasks described in this unit.

A high-quality tool will last for years if used and maintained properly. Normal use does cause tools to lose their shape, however. Also, tools are frequently abused. To abuse is to use wrongly, make bad use of, or misuse. Such practices necessitate tool repair.

Metal tools rust if they get wet and are not dried off quickly. Rust is a reddish brown or orange coating that results when steel reacts with air and moisture. It creates pits in the surface of steel.

Wood handles may split or break, and metal handles bend out of shape, leaving tools useless until repairs are made. Tools such as punches, chisels, wedges, and sledges get mushroomed faces. Mushroomed means a spread or pushed-over condition caused by being struck many times and occurs over a long period of use. After much use, screwdrivers get rounded tips and cannot grip screws properly. Screw heads are then damaged by attempts to use such screwdrivers.

The term tool fitting means to clean, reshape, repair, or resharpen a tool. To recondition means to do what is needed to put the tool into good condition. This unit will cover cleaning and protecting tool surfaces as well as reshaping and otherwise repairing tools. The next unit will cover tool sharpening. Many students in agricultural mechanics start or add to their tool holdings by restoring tools that others have discarded.

PROTECTING TOOL SURFACES

Most tool parts are made of wood, plastic, leather, or steel. Plastic is tough and is not affected by moisture. It is too soft to restore by grinding or buffing. Therefore, plastic parts must be handled carefully at all times. It is important to avoid cuts, nicks, or other damage to plastic.

Restoring Leather Parts

Leather is used in a few tools. It may be rolled very tightly to form a mallet head. Layers of leather may be stacked and glued on wooden chisel handles. This cushions the rest of the tool when it is struck with a mallet. Leather is also used in protective clothes such as gloves and aprons.

When leather becomes stiff or dry, it can be restored by rubbing with saddle soap. Saddle soap, when mixed with water, will clean, soften, and preserve leather. The procedure follows.

PROCEDURE

1. Brush off all dirt with a stiff bristle brush.
2. Moisten a sponge or soft cloth with water.
3. Rub the cloth onto the surface of the saddle soap until suds develop into a lather.
4. Rub the sudsy mixture into all leather surfaces until the leather is clean and soft.
5. Remove suds with a damp cloth.
6. Let leather dry.
7. Rub with a soft, dry cloth or brush with a soft bristle brush.

Another product used to soften, restore, and protect leather is neat's-foot oil. Neat's-foot oil is a light yellow oil obtained by boiling the feet and shinbones of cattle. Simply apply neat's-foot oil to clean leather surfaces with a lamb's wool pad or soft cloth.

Other products may be used to recondition leather. They contain various mixtures of silicone, lanolin, or other oils, figure 20-1.

Restoring Wooden Surfaces

Wooden handles frequently become dried out, porous, and rough. Repeated use of such handles results in sore hands and sometimes splinters. Handles that dry out become loose and dangerous.

Wooden handles and other parts are easy to restore if treated as soon as drying is noticed. The procedure follows.

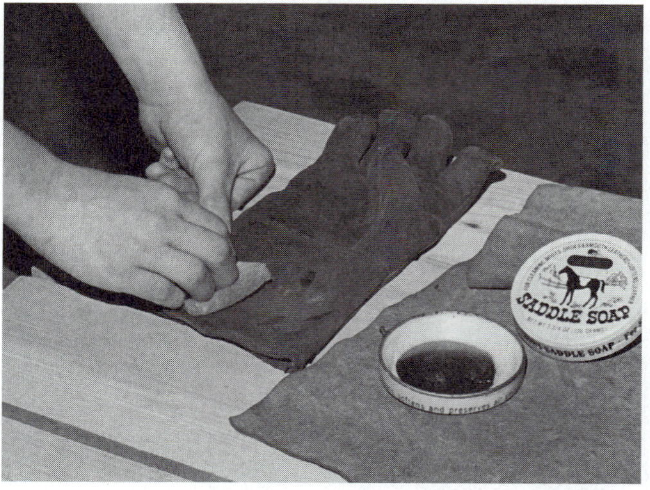

FIGURE 20-1. Leather may be restored with saddle soap, neat's-foot oil, or other commercial products containing silicone or lanolin.

PROCEDURE

1. Sand wooden parts with very fine sandpaper to remove all rough areas.
2. Rub all wood with a soft cloth dipped in boiled linseed oil, figure 20-2.

FIGURE 20-2. Linseed oil or paste finishing wax will restore and protect wooden parts.

3. After the oil dries, rub the wood briskly with a soft, dry cloth.
4. Repeat steps 2 and 3 several times if the wood is very porous.

Another product that may be used to restore wooden tool handles is paste finishing wax. The procedure is similar to that for linseed oil.

PROCEDURE

1. Before applying the wax, sand the wood thoroughly.
2. Rub on a thin coat of wax and let it dry.
3. Once the wax is dry, polish it with a soft cloth.
4. Repeat the process if needed to obtain a smooth surface that will prevent moisture from entering the wood.

Restoring Metal Surfaces

Metal surfaces may be quite difficult to restore if they are badly neglected. However, a few standard procedures can stop rust and restore a neglected tool to service.

Removing Grease and Oil. Grease and oil on tool handles are dangerous as well as messy. Tools are likely to slip out of the hand or be difficult to control when greasy.

To remove grease or oil, wipe the tool with a clean cloth dipped in kerosene, varsol, or other commercial solvent for grease and oil. After cleaning with a solvent, dry the tool with a clean cloth or paper towel.

CAUTION:

- **Do not use gasoline to remove grease and oil from tools. Gasoline is very volatile. Its explosive fumes will travel to sparks, flames, or other sources of fire.**

Removing Dirt. Tools such as shovels, hoes, and rakes may become caked with mud or concrete. When dry, these materials can be removed by tapping the tool with a metal object. The vibration generally causes the material to drop off. Any remaining material should be removed with a scraper, wire brush, or wire wheel.

Removing Rust. Light rust can be removed with varsol, kerosene, or other solvents. However, once pitting starts, a wire brush, wire wheel, steel wool, emery cloth, or aluminum oxide paper must be used, figure 20-3. These materials will dislodge deep rust but may scratch polished surfaces. A smooth final surface may be obtained by using 400 grit silicon carbide paper. Use water to help keep the paper clean while sanding.

Immediately after sanding, dry the tool thoroughly and coat all smooth metal surfaces with a light oil such as No. 10. Rough surfaces should be coated with mediumweight oil such as No. 20 or 30. If the tools are not treated in this manner, the water will cause rusting to start again within a few hours.

Protecting Steel from Rust. As noted before, rust occurs when steel is exposed to moisture and air. Therefore, rust is prevented by keeping air or moisture away from steel. Special rust-resisting primers and paints are the products most frequently used to protect steel from rust. However, many tool surfaces must remain smooth and unpainted in order to cut well. In addition, smooth, unpainted surfaces permit material to slide over metal more easily. Table tops on power machines are examples. Unpainted steel edges and surfaces are protected by rubbing with a good grade of lubricating oil. Paste wax also works satisfactorily on unpainted steel. However, wax is useful only if the steel is free of rust when it is applied.

REPAIRING DAMAGED TOOLS

Tools may be repaired by welding or gluing of broken parts. Wooden handles on files, wood chisels,

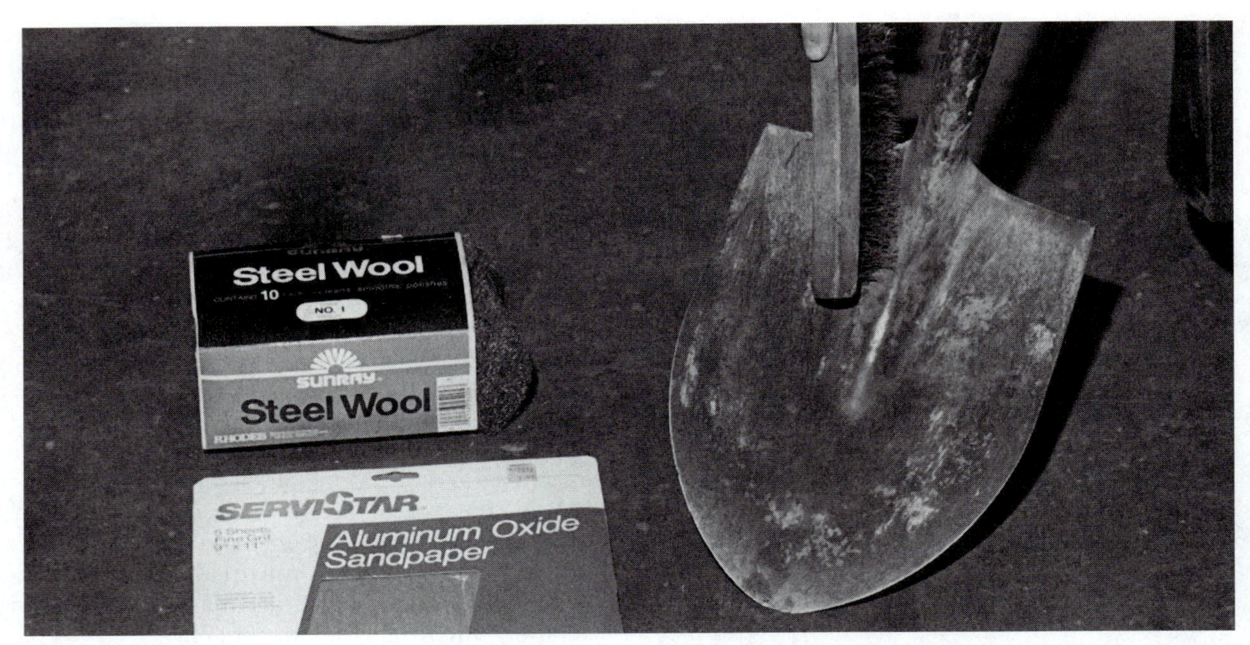

FIGURE 20-3. When pitting is evident on steel, rust must be removed by using a wire brush, wire wheel, steel wool, emery cloth, silicon carbide paper, or other abrasives.

hammers, axes, shovels, and the like can be replaced when broken. Sometimes it is less expensive to replace the tool than it is to replace the handle. However, for most tools, replacing handles is the most economical step.

Repairing Wooden Handles

Split Handles. Wooden handles may split when exposed to excessive strain. If the split is clean, it can be repaired with wood glue, figure 20-4. The procedure follows.

PROCEDURE

1. Position the tool in the vise so the split can be forced open.
2. Force open the split and spread glue on both surfaces.
3. Close the split and clamp the handle in several places.

FIGURE 20-4. This split handle is strong, but may cause splinters if the split area is not taped. Split wooden handles may be repaired by gluing.

4. Wipe off all excess glue.
5. After the glue dries, thoroughly sand the handle.
6. Treat the handle with boiled linseed oil and polish with a clean cloth when dry.

Loose Handles. Loose handles on hammers, hatchets, sledges, and axes are very dangerous. Handles have a wooden wedge and one or two metal wedges driven into the end. A wedge is a piece of wood or metal that is thick on one end and tapers down to a thin edge on the other. Check frequently to ensure that wedges are securely in place. If they are not, drive them into place. Even when handles are properly attached, they can still loosen due to drying. To reduce this problem, soak the end of the handle in the head of the tool with linseed oil about once a year, figure 20-5. An alternate method for treating tool heads with linseed oil follows.

PROCEDURE

1. Clamp the head of the tool in a vise with the end of the handle pointing upward. Brush enough boiled linseed oil over the end of the handle and wedges to saturate all parts. Keep them saturated for 12 to 24 hours if possible.
2. Permit the oil to dry while the head of the tool is still clamped in the vise.
3. Remove the tool from the vise and wipe a coat of boiled linseed oil on the rest of the handle.

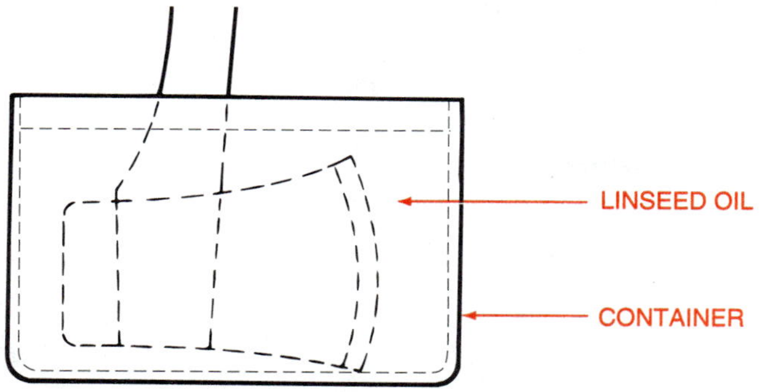

FIGURE 20-5. Treatment with linseed oil for 12 to 24 hours prevents drying and loosening of wooden handles.

4. Polish the handle with a dry cloth after the oil dries.

Replacing Wooden Handles

As mentioned before, most wooden handles are secured in the heads of tools by the use of wedges. The hole in the head of the tool is called the eye. This hole is smaller on the side where the handle enters than on the opposite side. Therefore, once the handle is inserted, it can be wedged out to fill the larger portion of the hole. As long as the wedges stay in place, the head remains tight on the handle.

Removing Broken Handles. A wooden handle generally breaks off flush with the head of the tool. When this occurs, the wooden core and metal wedges are still lodged securely in the tool head. To remove the broken piece of handle, the following procedure is recommended.

PROCEDURE

1. Place the head securely in a vise.
2. Use a 1/4-inch or 3/8-inch metal cutting drill bit and drill numerous holes into the wooden core.
3. Drive the remaining honeycomb of wood out of the head using a large punch or rod.

CAUTION:

- **Do not attempt to remove wooden pieces from metal tools by burning them out. The heat will cause the metal to lose its hardness.**

Attaching Handles to Hammers and Axes. When purchasing a replacement handle it is important to buy the correct size. The end of the handle that is shaped to go into the head should be slightly longer than the head is deep and should have the same shape as the hole in the head. It should also be slightly larger in cross section than the smaller part of the eye, or hole into which it is to be inserted.

Handles must be shaped to fit a head exactly. Start by placing the handle in a vise. The end to be worked on should be exposed so that it can be sawed, filed, and shaped. The procedure for attaching handles to hammers and axes follows and is also illustrated in figures 20-6 and 20-7.

PROCEDURE

1. Place the head against the end of the handle to see if it will start on.

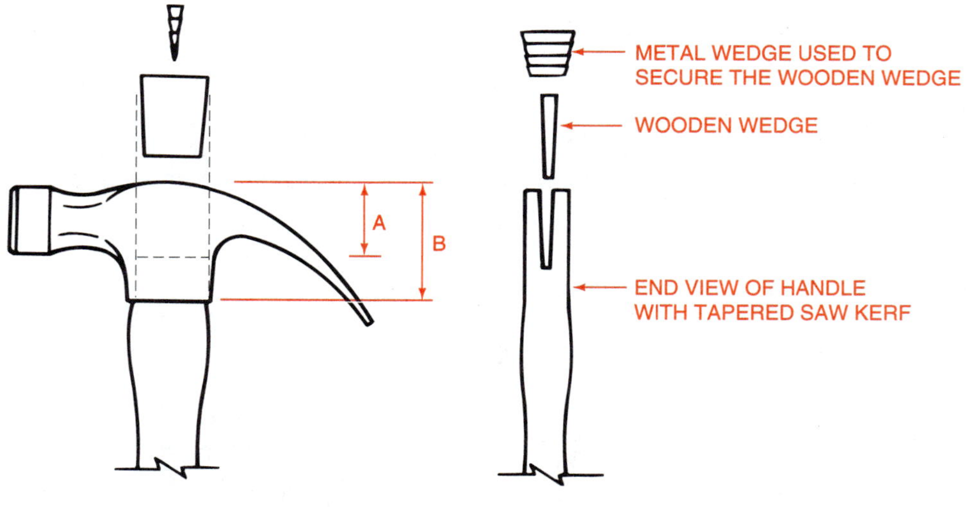

FIGURE 20-6. A handle is attached to a tool by wedging.

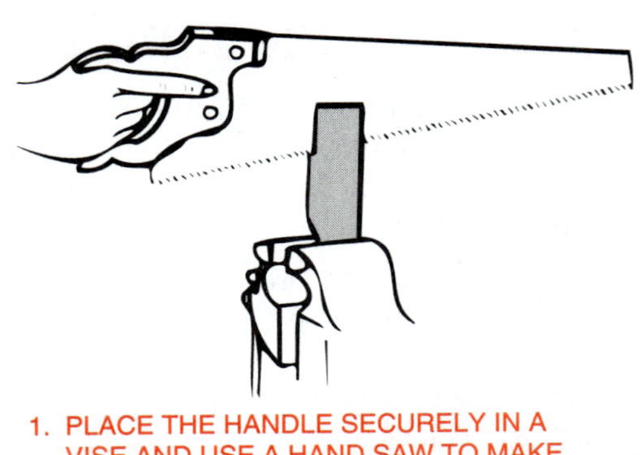

1. PLACE THE HANDLE SECURELY IN A VISE AND USE A HAND SAW TO MAKE A KERF ACROSS THE LONGEST CENTER LINE OF THE HANDLE.

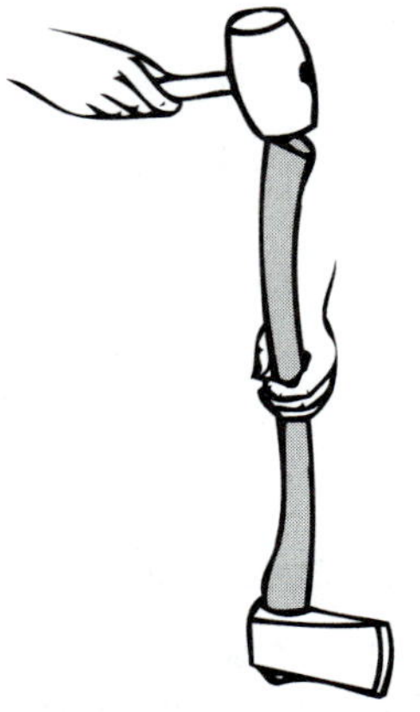

2. DRIVE THE HANDLE INTO THE HEAD WITH A WOODEN MALLET.

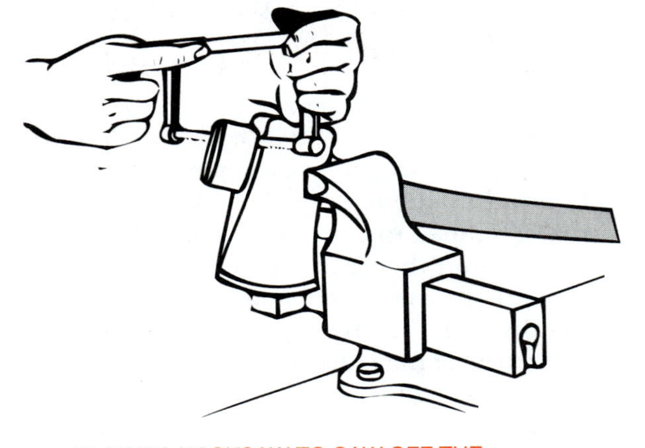

3. USE A HACKSAW TO SAW OFF THE EXCESS HANDLE FLUSH WITH THE HEAD.

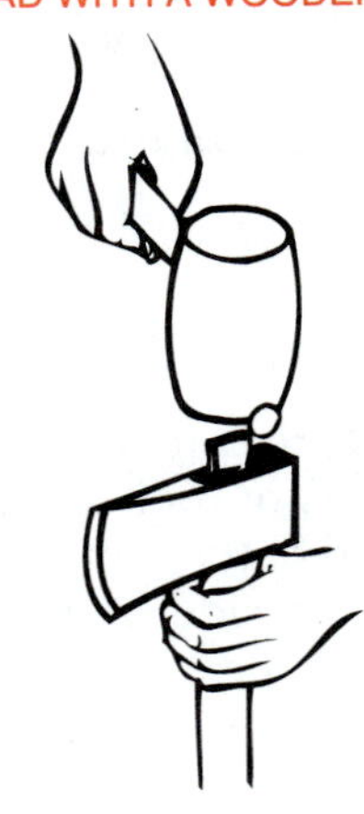

4. DRIVE IN A WOODEN WEDGE SO THE HANDLE SPREADS AND FILLS THE HEAD.

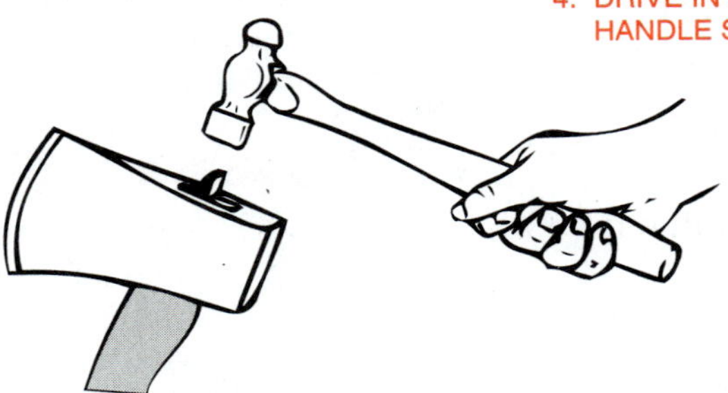

5. DRIVE ONE OR TWO METAL WEDGES IN AT RIGHT ANGLES TO THE WOODEN WEDGE TO HOLD THE WOODEN WEDGE IN PLACE.

FIGURE 20-7. Procedure for attaching a wooden handle to an axe or hammer after the handle is shaped to fit the head.

2. Observe where wood must be removed to shape the handle to enter the head.
3. Use a wood rasp or coarse file to shape the handle.

CAUTION:

- **Do not remove too much wood from the end. This will help get the head on, but it will also damage the wedged fit.**

4. Try the head frequently as wood is removed to avoid a loose fit.
5. Work the handle down until the head slides on snugly. It should come to rest about $1/2$ inch from the enlarged part of the handle.
6. Mark the handle on both sides of the head.
7. Remove the head.
8. Reposition the handle vertically in the vise.
9. Use a hand saw to make a kerf across the longest center line of the handle. The kerf

CAUTION:

- **Do not drive the head. It will bind and be very hard to remove.**

should extend two-thirds of the distance between the two marks made in step 6.

10. Reposition the handle in the vise and squeeze the end until the saw kerf is completely closed.
11. Run the saw down through the kerf again. When released, the kerf will be wider at the end than further down in the handle.
12. Now make a wooden wedge as wide as the oval hole in the head and thick enough to spread the handle when driven in.
13. Slide the handle into the head and drive it in securely with a wooden mallet.
14. Grip the handle with the vise just below the head.

CAUTION:

- **A steel hammer will ruin the end of the handle.**

15. Use a hacksaw to saw off the excess handle flush with the head.
16. Drive the wooden wedge so the handle spreads and fills the head.
17. Use a hacksaw to saw off the excess wooden wedge.

CAUTION:

- **Drive the wedge evenly or it will split.**

18. Drive one or two metal wedges in at right angles to the wooden wedge. These will hold the wooden wedge in place.
19. Place the tool, head down, into a metal or plastic container about the size of the head.
20. Add several inches of boiled linseed oil and brush oil around the handle and head.
21. The handle and head should soak for several days to seal the wood.
22. Remove the tool, rub oil on all parts of the handle, dry, and polish.

Attaching Handles to Rakes, Hoes, and Forks. Rakes, hoes, and forks are driven into their handles. The handles are fitted with metal collars to prevent splitting of the wood. This metal collar is called a ferrule. The tool has a metal finger called a tang, which is driven into the ferrule, figure 20-8. In some handles, a hole is provided in the ferrule for a nail or rivet. The nail or rivet is driven through the hole into the handle and tang to help hold the parts together. Other handles rely on friction only to keep the parts together.

Attaching Handles to Shovels. Shovel handles are exposed to tremendous forces when digging. Therefore, a metal tube extends up the handle for extra support, figure 20-9. Some shovel handles have a single bend and others have a double bend. Shovel handles are made to fit specific shovels. Therefore, it is best to take the shovel to the store when selecting a replacement handle. To attach a shovel handle, the following procedure is suggested.

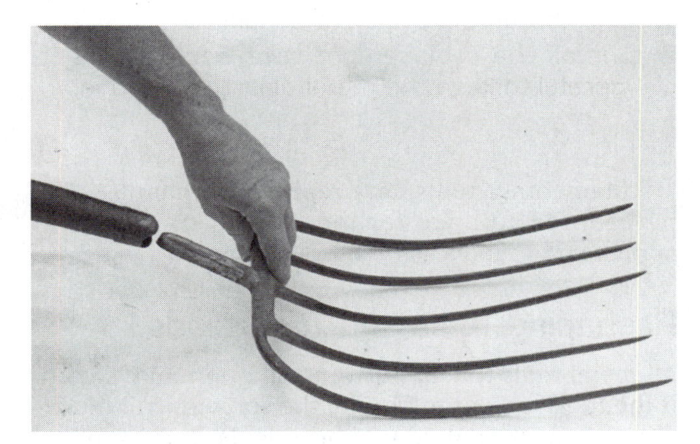

FIGURE 20-8. Forks, hoes, and rakes have a tang that is held in the handle with a nail or rivet or by friction between the wood and metal.

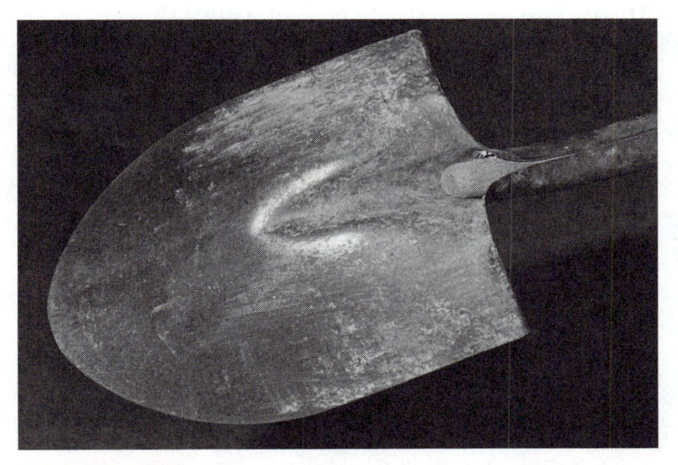

FIGURE 20-9. Shovels have a split metal tube that is tightened around the handle for support. A rivet is installed through the metal and handle to hold the handle in place.

PROCEDURE

1. Grind off the head of the rivet that holds the handle.
2. Remove the rivet with a drift punch.
3. Spread the metal tube and drive out the remains of the old handle.
4. Use a mallet to drive the new handle securely into place.

CAUTION:

- **If the handle drives very hard, consider removing some wood to improve the fit.**

5. Use a vise to close the metal around the handle and secure the assembly while drilling.
6. Insert a drill through the hole in the metal and drill through the wooden handle.
7. Install the replacement rivet securely, being careful to keep the metal tight to the wood.

Many other tools have replaceable handles. The procedures outlined for the tools in this unit will apply to other tools as well.

Reshaping Tools and Tool Heads

All metal tools that are driven will eventually flatten. If the tools are not reground, the surfaces will mushroom. A mushroomed tool head is dangerous. When struck, fragments of steel are likely to break off and fly into the face and body of the operator or bystanders.

Driven tools, such as punches, chisels, and wedges, are designed with a slight bevel around the crown. The crown is the end of a tool that receives the blow of a hammer. The bevel and the crown together make up the head of the tool, figure 20-10.

The procedure to reshape the head of a driven tool follows.

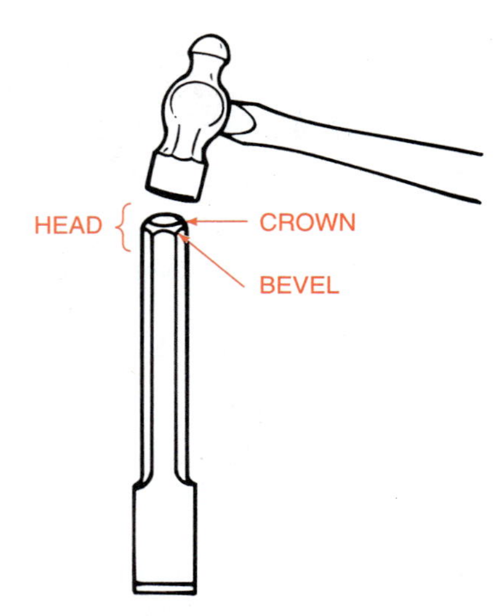

FIGURE 20-10. Driven tools, such as punches, chisels, and wedges, have a beveled edge around the crown to resist flattening when struck.

PROCEDURE

1. Examine the head of a similar tool that is in good condition.
2. With the good tool as a guide, use a medium grinding wheel to grind a taper from each flat surface to the crown.

CAUTION:

- **Use a face shield and leather gloves when grinding.**

3. Finish the taper by twirling the tool slightly so that all corners are slightly rounded. Round tools will have an even taper in a circular pattern around the crown.

The heads of axes, sledges, and hammers can also be reshaped using the procedures just described. However, these tools do not have much of a taper around the crown. Instead, they have rounded edges. It is important to keep the edges slightly rounded to avoid mushrooming of the head.

Reshaping Screwdrivers. The tips of standard screwdrivers can be reshaped. Phillips and other crossheaded screwdrivers must be discarded when their tips become worn. Standard screwdriver tips must be reground often to maintain a shape that will grip screws securely. Properly shaped standard screwdriver tips have a flat end and parallel sides. The width and thickness of the tip are determined by the size of the slot in the screw to be used.

When reshaping a standard screwdriver tip, the following procedure is recommended.

PROCEDURE

1. Obtain a screw of the size you wish to drive with the screwdriver.
2. Check to see that the screwdriver tip is as wide as the screw head. If it is not, obtain a wider screwdriver.
3. Check the face of the grinding wheel to be sure it is flat.
4. Check to see that the grinder's tool rest is properly adjusted.

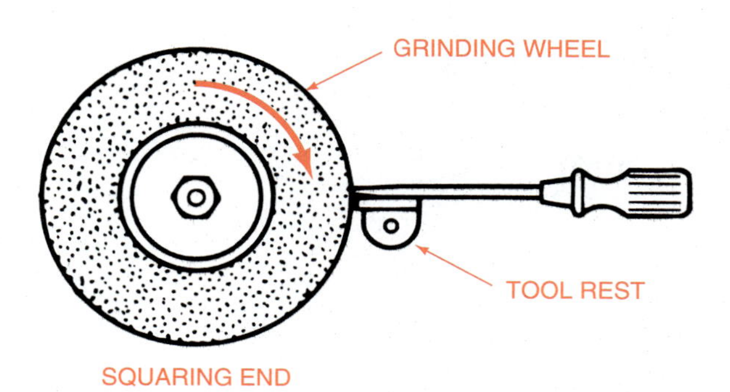

A. TO SQUARE THE END OF A SCREWDRIVER, SLOWLY PUSH THE TIP OF THE SCREWDRIVER ACROSS THE TOOL REST AND INTO THE GRINDING WHEEL.

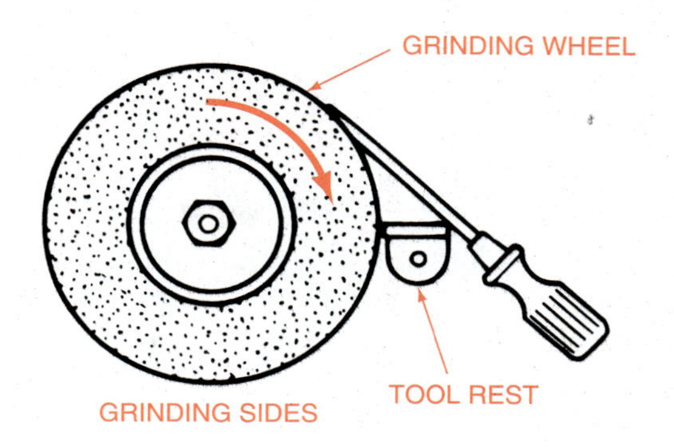

B. TO FLATTEN THE SIDES OF A SCREWDRIVER, HOLD THE SCREWDRIVER DIAGONALLY ACROSS THE TOOL REST WITH THE SIDE FLAT AGAINST THE WHEEL WHILE GRINDING.

FIGURE 20-11. Shaping the tip of a standard screwdriver.

5. Slowly push the flat end of the screwdriver tip across the tool rest and into the grinding wheel, figure 20-11A.
6. Grind the tip until it is flat, square, and even.

CAUTION:

- **Avoid overheating the tip of the screwdriver.**

7. Place the tip into the screw slot. It should reach to the bottom of the slot and fit the sides snugly.
8. If the tip is too thick, grind the flat sides of the tip slightly to narrow the screwdriver blade, figure 20-11B.

CAUTION:

- **Be careful to maintain the flat end. The flat surfaces of screw and screwdriver must fit perfectly. Any taper in the tip will cause it to twist out of the screw slot when torque is applied. Torque is a twisting force.**

9. Dress the edges of the tip so they are equal to the width of the screw head.

STUDENT ACTIVITIES

1. Define the Terms to Know in this unit.
2. Bring hand tools from home that need cleaning, rust protection, or repairing.
3. Remove dirt and rust from tools.
4. Repair wooden handles on tools.
5. Replace handles on driving and digging tools.
6. Reshape the heads of driven tools such as chisels and punches.
7. Reshape screwdriver tips.

SELF-EVALUATION

A. **Multiple Choice**. Select the best answer.

1. A common problem with wooden handles is
   a. rotting
   b. splitting
   c. rusting
   d. fatigue
2. Leather parts should be reconditioned with
   a. neat's-foot oil
   b. saddle soap
   c. lanolin products
   d. any of these
3. Hammer handles become loose if
   a. the handle dries out
   b. linseed oil is not applied occasionally
   c. the handle is not properly installed
   d. all of these
4. Wooden handles should be treated with
   a. linseed oil
   b. shellac
   c. paint
   d. all of these
5. A solvent recommended for removing grease and light rust is
   a. gasoline
   b. varsol
   c. turpentine
   d. water
6. After water touches an unprotected steel surface, rusting starts within
   a. hours
   b. days
   c. weeks
   d. months
7. Rust may be prevented from forming on metal surfaces by applying
   a. oil
   b. water
   c. saddle soap
   d. any of these
8. Split wooden handles are best repaired with
   a. nails
   b. screws
   c. glue
   d. tape
9. Driving tool handles are held in place by
   a. bolts
   b. nails
   c. screws
   d. wedges
10. When standard screwdrivers have rounded tips they
   a. work better
   b. need replacing
   c. slip out of screw slots
   d. should be heated and reshaped

B. **Matching.** Match the terms in column I with those in column II.

Column I

1. mushroomed
2. linseed oil
3. tool fitting
4. neat's-foot oil
5. mallet
6. ferrule
7. wedged handle

Column II

a. damaged head
b. hammer
c. reconditioning
d. treat handles
e. rakes and hoes
f. drive handles
g. leather

C. **Completion.** Fill in the blanks with the word or words that will make the following statements correct.

1. When a tool head has mushroomed, it has ________.
2. The crown is the part of the tool that _______.
3. When reconditioning leather, a sponge may be used to apply ________.
4. Shovel handles are held in place with a _____.
5. A mushroomed condition is corrected by ________.
6. To obtain a smooth final surface, metal tools should be sanded with _____.___
7. A good tool for removing rust is a _____ _____.

UNIT

Sharpening Tools

OBJECTIVE

To keep cutting tools sharp.

COMPETENCIES TO BE DEVELOPED

After studying this unit, you should be able to:

- Examine tools and determine the design of the cutting edges.
- Select appropriate procedures for sharpening tools.
- Sharpen knives.
- Sharpen wood chisels and plane irons.
- Sharpen cold chisels and center punches.
- Sharpen axes and hatchets.
- Sharpen twist drills.
- Sharpen rotary mower blades.
- Sharpen digging tools.

MATERIALS LIST

- ✓ Grinder with medium and fine grit wheels
- ✓ Eye or face protection and protective clothing
- ✓ Heavy-duty portable sander with fine aluminum oxide discs
- ✓ Large, flat file with medium or fine teeth
- ✓ Machinist's vise
- ✓ Assorted hand tools
- ✓ Oil bench stone (one fine side and one coarse side)
- ✓ Ax stone
- ✓ Tool fitting gauge
- ✓ Tools requiring repair and sharpening
    - a. Pocket or pruning knife
    - b. Plane iron and/or wood chisel
    - c. Cold chisel
    - d. Center punch
    - e. Ax or hatchet
    - f. Twist drill
    - g. Rotary mower blade
    - h. Digging tools

TERMS TO KNOW

bench stone	true
hand stone	whet
inclined plane	serrated
convex	arbor
concave	ax stone
hollow ground	field use
tool steel	high-speed drill
anneal	clockwise
temper	balance
draw the temper	

A properly sharpened tool of good quality can make cutting, drilling, and digging jobs relatively easy. Such tools result in work of excellent quality when used by skilled craftsmen. Sharp tools are safer than dull ones because they require less pressure. A craftsman should not attempt a job with a dull tool, because the resulting quality of work will be disappointing.

TOOLS USED FOR SHARPENING

Tools in the agricultural mechanics shop are generally sharpened with files, bench stones, hand stones, sanders, and grinders. The use of files and grinders is covered elsewhere in this text. A **bench stone** is a sharpening stone designed to rest on a bench. A **hand stone** is one designed to be held in the hand when in use. Sharpening stones include the common 6" × 2" × 1" bench stone, hand-held slip stones, and power-driven wheels. Bench stones are generally mounted in a wooden box which holds the stone during use and storage, figure 21-1. The box also helps keep the stone clean and oil-filled. Many stones require either water or oil to keep them clean and to ensure that the abrasive materials cut during use.

Files needed for sharpening tools include flat, round, triangular, and special styles. Examples of special files are auger bit files and chain saw files. Coarse files are useful for sharpening large objects such as blades for rotary mowers. Fine-toothed files are used for the majority of sharpening jobs.

CUTTING EDGES

Always observe the design and shape of the edges of cutting tools when they are new. The tool is designed to do a specific job. The nature of that job determines the shape of the cutting edge. For example, shaving should remove all evidence of hair. Hair is relatively soft. Therefore, a razor blade has a very thin, sharp edge. This edge does a good job in shaving but is too fragile for most other cutting jobs.

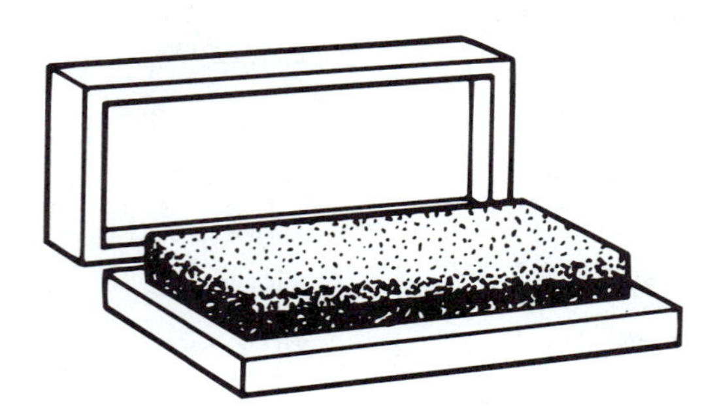

FIGURE 21-1. An oil-filled bench stone mounted in a wooden box.

In contrast, steel is tough and hard. Therefore, tools used to cut steel must be tough and hard. Their edges must be thick and strong.

Wood is relatively soft, so wood chisels and plane irons can be ground to fine, sharp edges. However, this means that they nick, break, or dull quickly when they are used on metal objects or come in contact with dirt or metal surfaces.

Tool edges are usually designed with either single or double inclined planes, figure 21-2. An **inclined plane** is a surface that is at an angle to another surface. It is one of the six simple machines. Wood chisels and plane irons have single inclined planes.

Other tools are designed with edges formed from two inclined planes. These tools must be sharpened from both sides. Some examples are knives, axes, hatchets, and cold chisels.

The term **convex** means curved out. An edge that is convex is strong because it has extra steel to back it up. However, the bulging nature of a convex edge means it requires more force to make it cut. Axes and similar tools with convex edges get their force from their speed and weight. Wedges get their force from a driving tool.

The term **concave** means hollow or curved in. Wood chisels and plane irons are ground to a concave edge. Their edges are very sharp and require little force to make them cut. Certain saws may be ground so the teeth are wider at the points than they are at their base. Such blades are said to be **hollow**

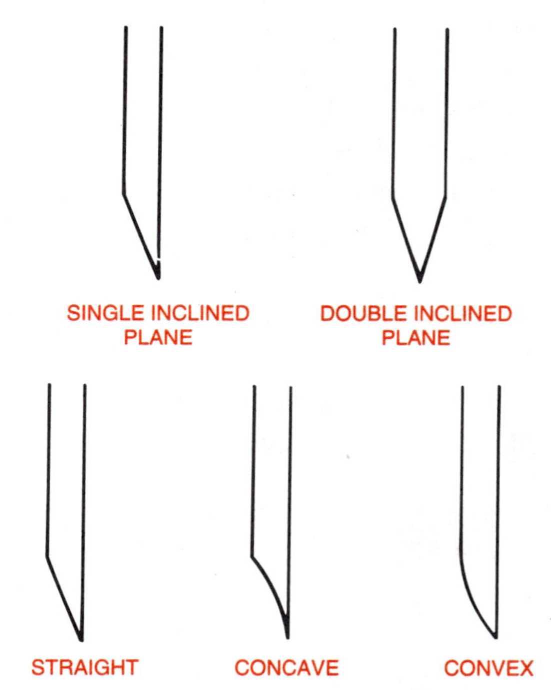

FIGURE 21-2. Tool edges usually have single or double inclined planes. These planes may also be straight, concave, or convex.

ground. The teeth are wider at their points than the thickness of the blade behind the teeth. Therefore, the teeth have no set and the blade makes a very smooth cut.

Preserving Temper

Most tools are made from tool steel. Tool steel is steel with a specific carbon content that allows the tool to be annealed and tempered. To anneal is to heat and then cool the steel slowly so as to make it soft and malleable. To temper is to heat the steel and then cool it more quickly so as to control the degree of hardness. After tools are properly tempered, excessive heating will draw the temper, or modify it, to render the steel soft and useless until it is retempered.

TESTING FOR SHARPNESS

Knives, plane irons, and chisels should be sharpened to a very fine edge. To check for sharpness, place the tool blade in a flat position on a narrow piece of soft wood. Raise the back slightly and push the cutting edge across the wood. The edge should catch the wood and shave off a fine sliver. If this does not happen, the tool is probably not very sharp.

CAUTION:

- **Any test of sharpness using the hands or body should be avoided.**

SHARPENING TOOLS

When sharpening tools, first remove all metal burrs or rough spots. Burrs should be ground or filed away without changing the shape of the edge. If deep nicks are present, straighten the entire cutting edge. The edge should be ground to the proper shape and angle.

CAUTION:

- **Grind slowly to avoid drawing the temper. A guide for sharpening tools is given in figure 21-3.**

When using grinders, be sure to wear a face shield and protective clothing. Check the grinding wheel(s) for soundness. True and clean the wheel, if needed, and then proceed to sharpen tools.

Tools that require a very sharp edge must be whetted. To whet means to sharpen by rubbing on a stone.

TOOL	SIDES TO SHARPEN	SHAPE OF THE EDGE	RECOMMENDED ANGLE
Ax/hatchet	Both	Convex	20°
Center punch	Point	Straight	60° to 75°
Cold chisel	Both	Straight	60° to 75°
Knife	Both	Straight	As original
Hoe/shovel	One	Straight	As original
Plane iron	One	Concave	29°
Rotary mower blade	One	Concave	As original or 45°
Scissors or snips*	One	Concave	80°
Twist drill	Both	Straight	118° to 120°
Wood chisel	One	Concave	25° to 29°

*Scissors and snips have two moving blades. Each blade is sharpened on one side only at an 80° angle.

FIGURE 21-3. A guide for sharpening tools.

Knives

Knives with smooth edges may be sharpened in the shop. Those with serrated edges must be sharpened by the manufacturer or a business with special sharpening equipment. Serrated means notched, like the edge of a saw. Most schools have policies prohibiting students from carrying knives in school. Therefore, it is especially important that students obtain permission from their instructor before fitting and sharpening knives in the shop.

To sharpen most knives, the following procedure is suggested.

PROCEDURE

1. Use a fine grit oil or water stone grinder.
2. Remove the nicks and grind away some of the thickness of the blade along the edge. Grind slowly to avoid overheating.

CAUTION:

- **Hold the knife with the cutting edge away from the operator, figure 21-4. Move the knife back and forth to create an even grinding mark behind the edge.**

3. Use an oil stone to whet the blade to a sharp edge. Whetting is done by holding the knife flat to the stone with the back of the blade slightly raised. Draw the knife, edge first, across the oiled stone; flip the knife over and push it in the opposite direction, figure 21-5.

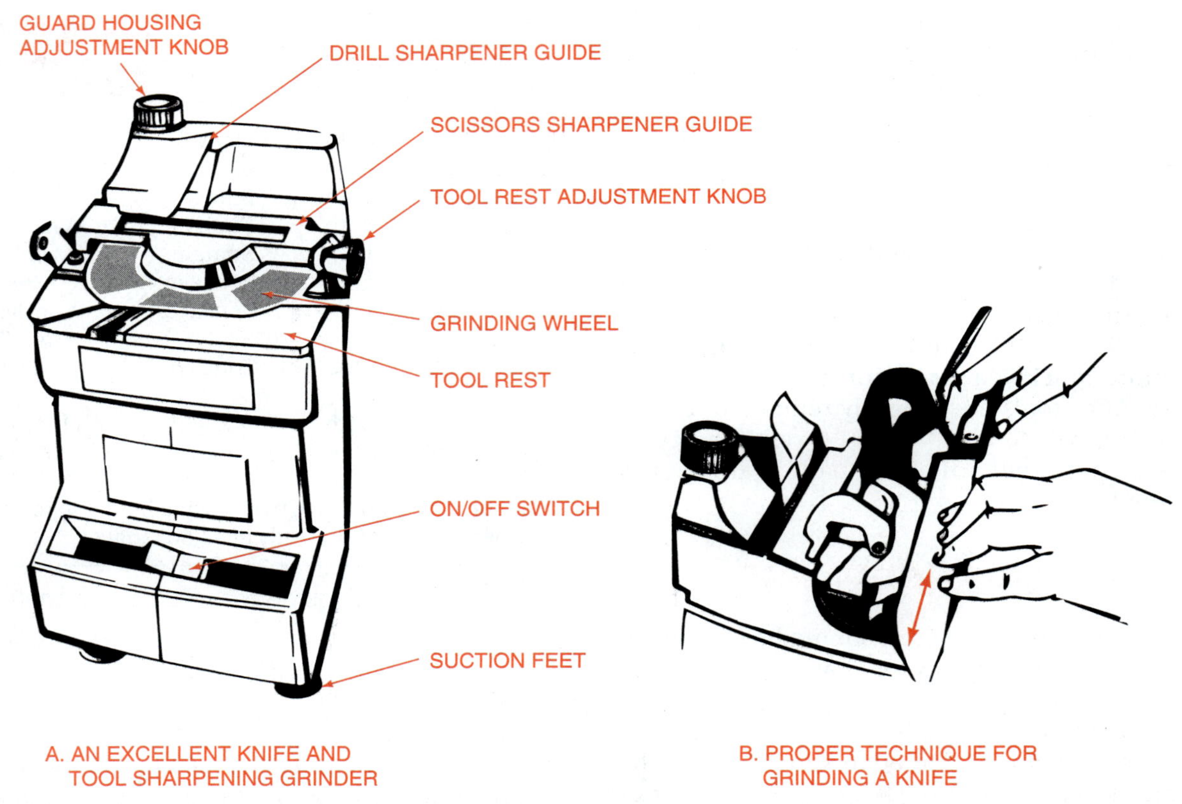

FIGURE 21-4. **A.** When grinding knives, use an oil or water stone grinder. **B.** Hold the knife with the cutting edge away from the operator and move it back and forth across the wheel to create an even taper to the blade.

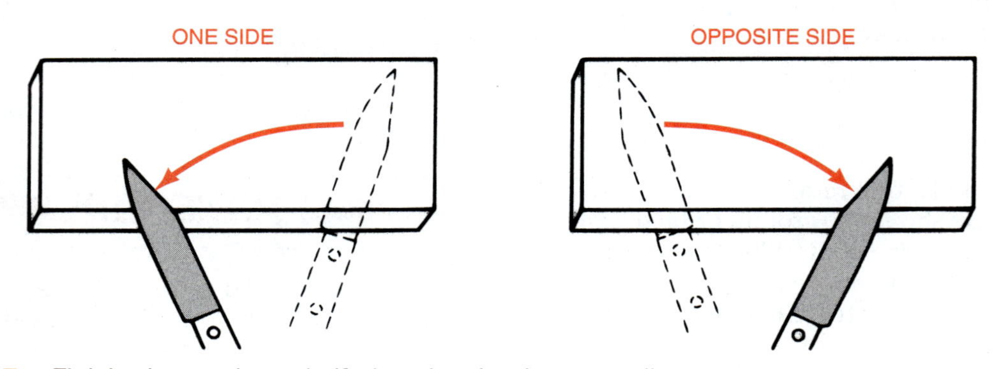

FIGURE 21-5. Finish sharpening a knife by whetting it on an oil stone.

Continue to alternate sides until the edge is even and sharp.

4. Some stones have one coarse and one fine side. Always make the final strokes on the fine side.
5. When finished, wipe the oil slurry from the stone. Cover the stone with an even coat of light oil and store it in its wooden box or original carton.

Plane Irons and Wood Chisels

The edges of plane irons and wood chisels are identical in type and shape. Both are sharpened on one side only to a 29 degree concave angle. (A tool sharpening gauge, figure 21-6, is used to check the angles.)

To sharpen a plane iron or wood chisel, grind the end until it is square and free of nicks. A special oil grinder is best, but dry-wheel grinders may also be used, with special care. Use the following procedure.

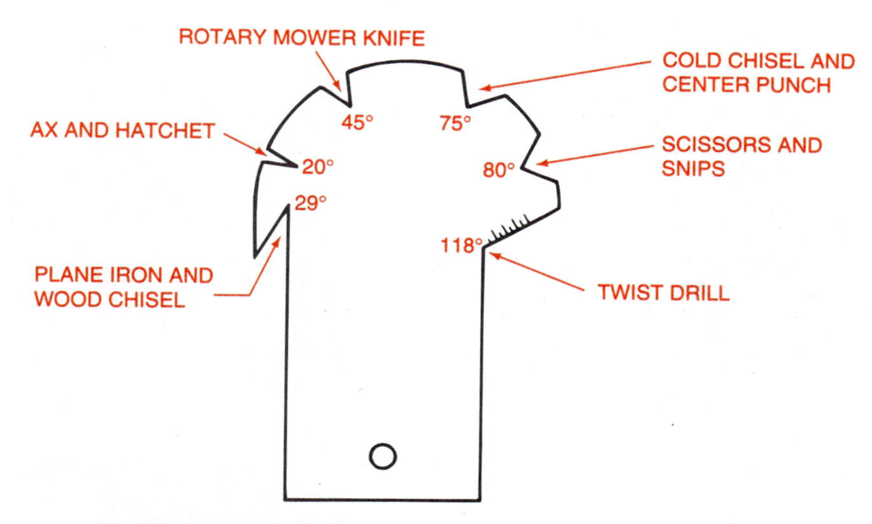

FIGURE 21-6. A tool sharpening gauge greatly improves the accuracy of tool sharpening.

PROCEDURE

1. Loosen the tool rest on a grinder with a medium or fine wheel. Angle the tool rest so that the plane iron lies on the curve of the wheel, figure 21-7. Tighten the tool rest.
2. To grind the plane iron, move it back and forth on the tool rest. After grinding, the cutting edge should be square to the side of the plane iron.
3. Check the angle of the edge with a 29 degree gauge, figure 21-8. It can also be compared to another plane iron that is properly sharpened. Resharpen if necessary.
4. The next step is to put a fine edge on the plane iron by whetting. To whet a plane iron, place the ground section flat against the surface of an oiled stone, then raise the heel off the ground area until it just lifts from the stone, figure 21-9A. Push the iron across the stone to sharpen the edge.
5. Lay the blade on its back and push it across the stone to remove the wire edge, figure 21-9B. Alternate these two steps until the blade has a fine, sharp edge.

FIGURE 21-7. Sharpening a plane iron using an oil grinder.

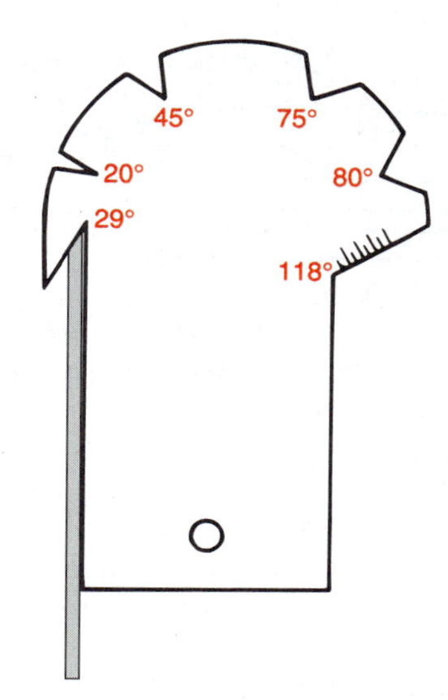

FIGURE 21-8. Checking the blade angle of a plane iron using a tool sharpening gauge.

Cold Chisels and Center Punches

The end of a cold chisel is relatively thick, because it must withstand heavy blows from a hammer to cut steel. However, the thickness tapers at a second angle to a sharp edge. This secondary taper forms a 60 to 75 degree angle.

A. LIFT THE HEEL SLIGHTLY AND PUSH THE PLANE ACROSS THE STONE.

B. LAY THE PLANE IRON FLAT ON THE STONE AND PUSH IT FORWARD TO REMOVE THE WIRE EDGE.

FIGURE 21-9. Whetting a plane iron using an oil bench stone.

When sharpening a cold chisel, select a medium grit grinding wheel. If the chisel edge has nicks, hold the chisel horizontally on the tool rest and ease the edge against the wheel. Move the chisel back and forth across the wheel and grind the end until all nicks are removed.

To resharpen, use the following procedure.

PROCEDURE

1. Hold the chisel on the tool rest. The chisel should be pointed upward and leaning toward the wheel. Form the cutting edge by making one pass across the wheel.
2. Then rotate the chisel half a turn and make one pass on the other side.
3. Alternate passes until a sharp edge is formed, figure 21-10.
4. Check the edge for squareness.
5. Use a tool fitting gauge to check the edge for the correct angle.
6. Regrind the chisel until a perfect edge is formed.

FIGURE 21-10. Sharpening a cold chisel using a grinder. Here the operator is grinding the edge to a 60 degree angle.

Center punches are also sharpened at angles ranging from 60 degrees to 75 degrees. A center punch is sharpened by holding it at an angle to the wheel and twirling it as it is ground. The twirling action should create a round, even point.

Axes and Hatchets

Axes are large and hard to grind on standard grinding machines equipped with the proper guards. For this reason some shops use grinding wheels mounted on a belt-driven arbor. Another method is to clamp the ax in a vise and use a portable disc sander, figure 21-11. The edge of an ax is ground to a 20 degree angle.

The ax is slowly ground away by many sharpenings. This causes the angle to increase and the area behind the edge to get thicker. This problem is reduced by removing metal from as far back as an inch from the edge.

While cutting wood the ax may be touched up or resharpened slightly with an ax stone. An ax stone is a small round stone held in the hand for field use. Field use means the area where an operation takes place. In this case field means the forest or wood pile.

Hatchets may be sharpened with a disc sander or grinder. The common shop grinder with a medium wheel may also be used to sharpen a hatchet. First remove any small nicks by pushing the blade horizontally across the tool rest and into the wheel. Make one pass across the wheel. Examine the edge to see if all small nicks are removed. Make another pass if needed.

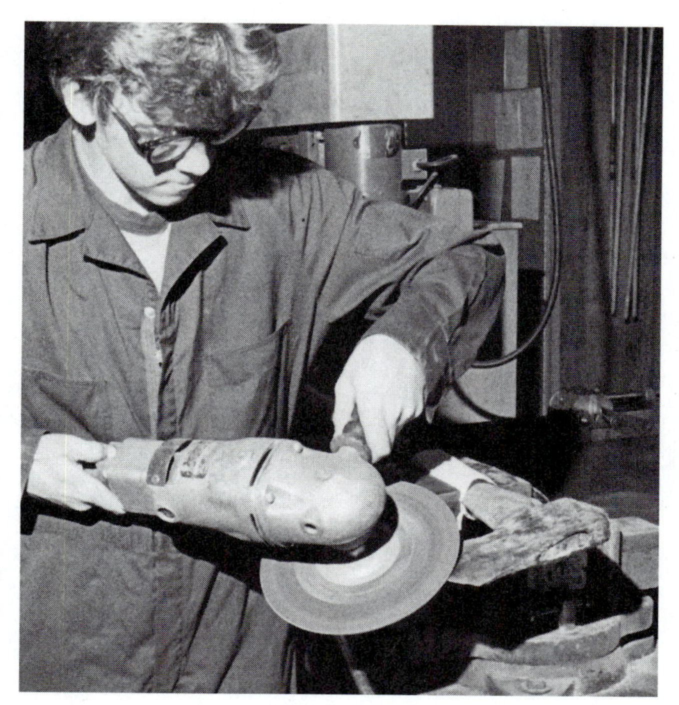

FIGURE 21-11. A disc sander can be used to sharpen an ax.

In some cases, it may be better to leave large nicks in the blade than to grind away so much metal that the hatchet or ax becomes excessively thick. Generally, large nicks occur at the end of the blade. If so, the end of the blade may be rounded in order to grind the nick into a sharp edge. Needless to say, large nicks ruin an ax or hatchet. They should be avoided by never cutting wood that may contain metal. Another possible source of nicks is chopping through wood and into stones in the ground.

To sharpen a hatchet, use the following procedure.

PROCEDURE

1. Draw a line on each side of the blade parallel to and 1/2 to 5/8 of an inch from the edge to serve as a guide.
2. Hold the head, edge up, against the wheel. Use a gloved hand to move the head quickly up and down 1/2 an inch or more as the tool is moved slowly across the wheel, figure 21-12. The up-and-down motion creates a wide grinding band and maintains the correct angle.
3. Reverse the position of the tool and make a similar pass on the other side.
4. Repeat the process until a good edge is obtained.
5. Remove grinding marks with a hand stone or fine flat file.

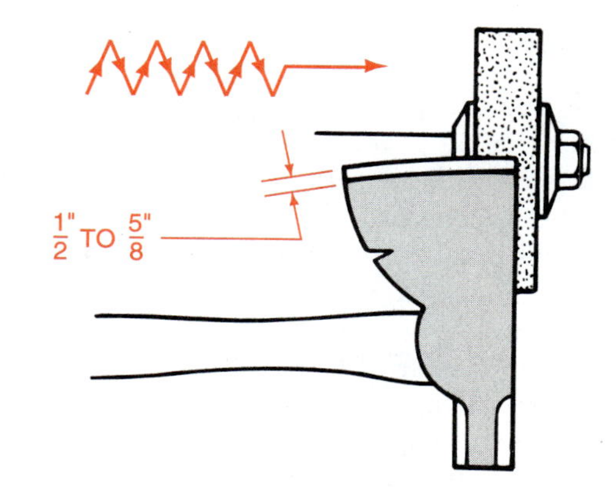

FIGURE 21-12. When using the grinder to sharpen a hatchet, move the hatchet across the wheel with quick up-and-down movements (shown by the arrows) to produce a long taper on the edge.

A large, sharp, medium-cutting, flat file provides an alternate and equally effective way to sharpen a hatchet or ax, figure 21-13. The tool to be sharpened is first clamped securely in a machinist's vise, the end of the file is then placed on the flat of the tool, and metal is removed by filing.

CAUTION:

- **Use extreme care to prevent fingers or hand from slipping into the cutting edge. Direct the file away from the cutting edge for safety reasons.**

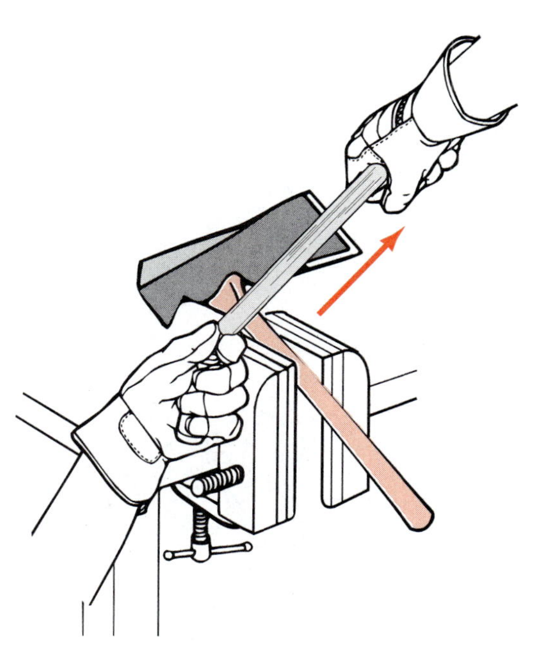

FIGURE 21-13. When using a file to sharpen a hatchet, wear heavy leather gloves and direct the file away from the cutting edge.

Twist Drills

Twist drills stay sharp for a long time when drilling wood or plastic. However, when drilling metal they quickly become dull. Although aluminum is soft, it dulls drill bits rather quickly. Drilling steel has a moderate dulling effect on high-speed drills. A **high-speed drill** is a twist drill made and tempered specially to drill steel.

To sharpen a twist drill requires familiarity with its design. The cutting tip of a drill consists of a dead center and two cutting lips, figure 21-14.

The dead center must be exactly centered after the drill is sharpened. The dead center and two cutting lips must touch metal before the rest of the drill end. Otherwise, the drill will ride on its rounded end rather than cut into the metal.

The two cutting lips start at the two ends of the dead center. They are ground at 59 degrees to the center line of the drill shank and together form a 118 degree angle when new. When resharpening a twist drill, a 118 degree or 120 degree gauge is used to test the accuracy of the angle.

The metal recedes or falls back from the cutting tip to the heel. The resulting angle is 12 degrees on a new drill and should be kept as close to that as possible when resharpening. This provides clearance for the cutting lips and also gives adequate support to keep the lips from breaking.

Twist drills should be sharpened with a special jig to obtain the correct angles. However, with practice, they can be sharpened freehand by using the following procedure.

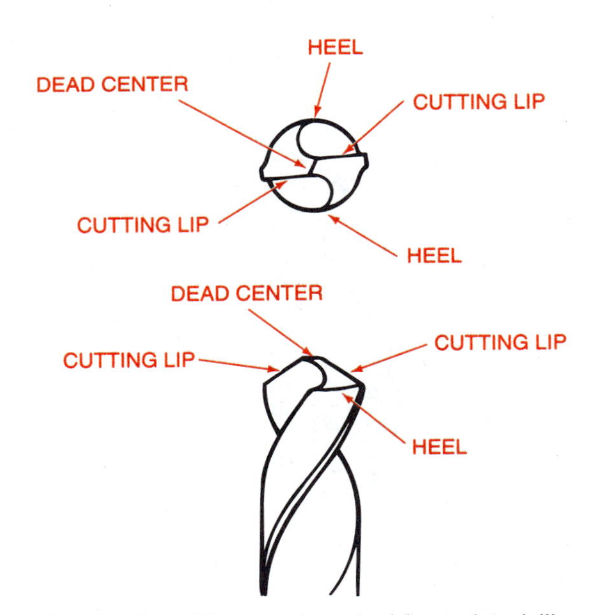

FIGURE 21-14. The cutting tip of a twist drill.

FIGURE 21-15. Sharpening a twist drill on a grinder. (*Courtesy of Delta International Machinery Corporation, formerly Rockwell Power Tool Division*)

PROCEDURE

1. Redress a fine grinding wheel if needed to make it true.
2. Set the tool rest horizontal to the center of the wheel.
3. Study the shape of a new drill that is about the size of the one to be sharpened.
4. Hold the drill between your thumb and index finger with the tip of the drill exposed about 1 inch, figure 21-15.
5. Place the back of your index finger on the tool rest with your thumbnail up and the drill at an angle to the wheel. The lip on the left should be visible and parallel to the stone.
6. Touch the lip against the grinding wheel and lower the opposite end of the drill as you give it a slight **clockwise** twist.
7. Rotate the drill half a turn so the other lip is visible.
8. Repeat step 6 to sharpen the second lip.

Use a tool sharpening gauge to check the angle of the lips, position of the dead center, and clearance of the heels, figure 21-16. If the angle of the lips is less than 118 degrees, the bit will cut too fast. If the angle is flatter (more than 118 degree to 120 degree), it will cut too slowly. If correct, test the tool by drilling a piece of mild steel. If the drill does not cut well, it must be resharpened.

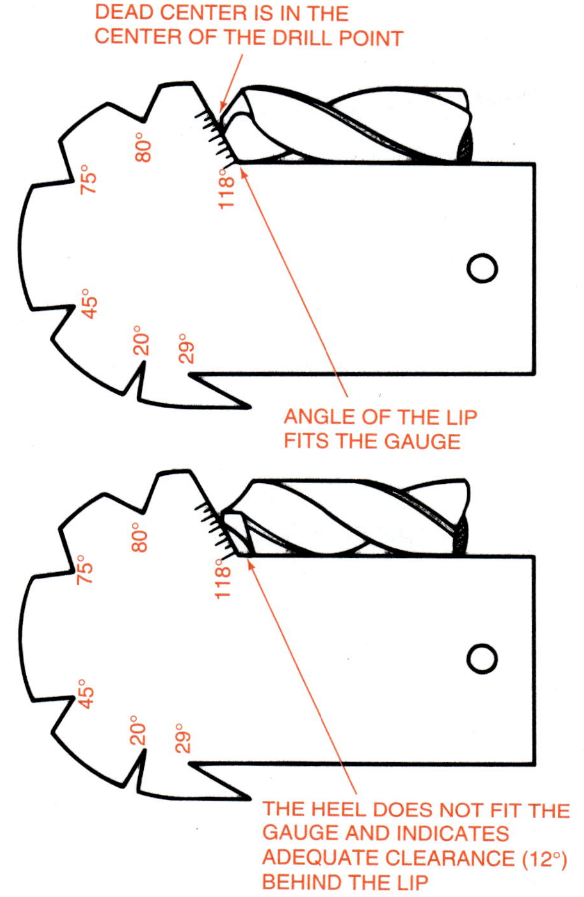

FIGURE 21-16. A tool sharpening gauge is used to check the angle of the lips, position of the dead center, and clearance of the heels of a resharpened twist drill.

Rotary Mower Blades

Blades for both lawn and field rotary mowers need frequent sharpening. Sharp blades require less power, do a better looking job, and damage the plants less.

When sharpening blades, try to maintain the original angle. If this is not known, grind the blades to a 45 degree angle. One side of the blade is kept perfectly flat; the other side is ground to a cutting edge.

When grinding mower blades, remove the same amount of metal from both ends. This keeps the blade in balance. Balance means the weight is equally distributed on both sides of the center. Therefore, if one end is badly nicked, grind both ends equally until the nick is removed. Failure to do so will cause the blade to be out of balance and the mower to vibrate. This causes damage to the shaft, bearing, and mower body.

To sharpen a rotary mower blade, first remove the nicks by grinding or filing the cutting edge, then clamp the blade in a machinist's vise. Restore the flat side by using a large, medium-cutting, flat file. Restore the cutting edge using a flat file or aluminum oxide disc on a heavy-duty portable sander.

Mower blades may be sharpened on a stationary grinder. Since the blades are hard to position properly on a grinder, this is a less desirable method.

CAUTION:

- **Whenever sanders or grinders are used, care must be taken to avoid overheating, which can cause the tool to lose its hardness.**

Digging Tools

The edges of digging tools, such as hoes, shovels, spades, and spading forks, get ragged and dull with use. The same is true for scoop shovels. To reshape these tools, first observe the side on which they were ground before. If metal is curled back, place the tool on an anvil and tap the bent metal back into place with a steel hammer. Restore the cutting lip with a file or grinder, figure 21-17. All shovels are sharpened on the inside. The outside or back remains straight.

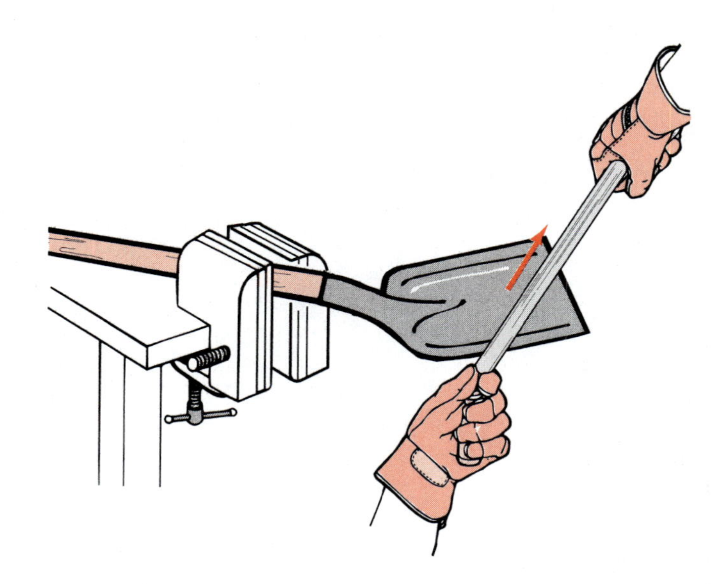

FIGURE 21-17. Restoring the cutting edge of a tool.

STUDENT ACTIVITIES

1. Define the Terms to Know in this unit.
2. Copy the data in figure 21-3 into your notes.
3. Use a grinder to sharpen a
   a. knife
   b. plane iron or wood chisel
   c. cold chisel
   d. center punch
   e. hatchet
   f. twist drill
4. Use a bench stone to whet a knife and plane iron.
5. Use a file to sharpen a hatchet, rotary mower blade, and shovel.
6. Use a portable power sander to sharpen an ax.
7. Do the tool sharpening gauge exercise. (A drawing of the tool sharpening gauge is shown in figure 21-18.)

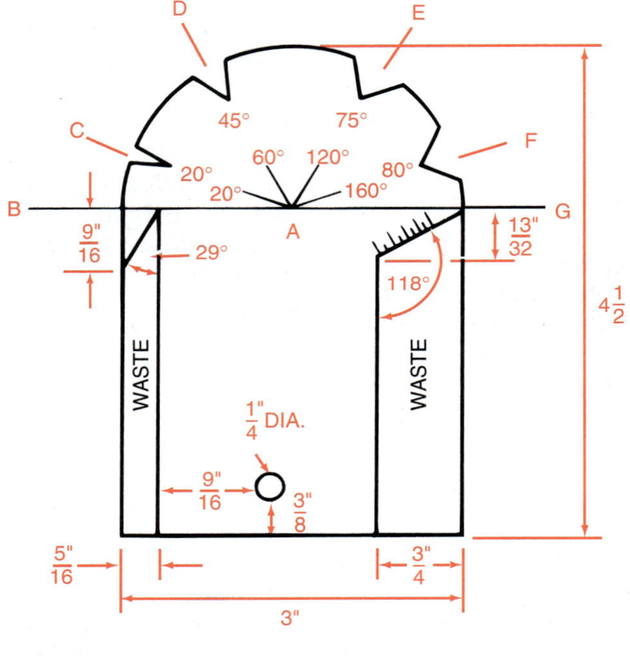

20°	AX, HATCHET
45°	ROTARY MOWER BLADE
75°	COLD CHISEL, CENTER PUNCH
80°	SCISSORS, SNIPS
29°	PLANE IRON, WOOD CHISEL
118°	TWIST DRILL

FIGURE 21-18. Plan for a tool sharpening gauge.

SELF-EVALUATION

A. Multiple Choice. Select the best answer.

1. The reason for learning tool sharpening skills is that
   a. better work is possible with sharp tools
   b. sharp tools are easier to use
   c. sharp tools are safer
   d. all of these
2. A 6" × 2" × 1" oil stone is called
   a. an ax stone
   b. a bench stone
   c. a common stone
   d. a slip stone
3. The best tool to sharpen an ax is a
   a. grinder
   b. belt sander
   c. portable disc sander
   d. all of these
4. When sharpening tools with a grinder, the tool rest should be
   a. removed
   b. placed at a 90 degree angle
   c. placed to support the tool
   d. pushed down out of the way
5. A properly ground drill must have the
   a. proper angles
   b. proper clearance
   c. proper centering
   d. all of these

B. Matching. Match the terms in column I with those in column II.

Column I

1. concave
2. convex
3. serrated
4. anneal
5. temper
6. whet
7. dead center

Column II

a. make soft
b. curved out
c. curved in
d. sharpen
e. drill part
f. notched
g. make hard

C. Brief Answers. Briefly answer the following questions.

1. List the number of sides to be sharpened and the recommended angle for grinding each of the following.
   a. ax
   b. cold chisel
   c. hatchet
   d. knife
   e. mower blade
   f. plane iron
   g. scissors
   h. shovel
   i. wood chisel

Gas Heating, Cutting, Brazing, and Welding

UNIT

22 Using Gas Welding Equipment

OBJECTIVE

To use heating, cutting, and gas welding equipment safely.

COMPETENCIES TO BE DEVELOPED

After studying this unit, you should be able to:

- Identify major parts of propane and oxyacetylene welding equipment.
- Change oxygen and acetylene cylinders.
- Turn on and adjust oxyacetylene controls.
- Light and adjust oxyacetylene torches.
- Shut off and bleed oxyacetylene equipment.
- Check for leaks in gas equipment.

TERMS TO KNOW

weld
fusion
gas
compress
flammable
apparatus
manifold
rig
oxyacetylene
torch
cylinder
valve
regulator
gauge
hose
crack the cylinder
seat
purge the lines
carbonizing flame
neutral flame
oxidizing flame
tip cleaner
bleeding the lines

MATERIALS LIST

- ✓ Oxyacetylene welding outfit
- ✓ Variety of welding tips
- ✓ Welding gloves and apron
- ✓ No. 5 shaded goggles
- ✓ Tip cleaners
- ✓ Spark lighter

Many gases will burn. This quality makes them both dangerous and beneficial. The burning qualities of gases are used in agricultural mechanics to heat, cut, and weld metals. To weld means to join by fusion. Fusion means melting together.

A gas is any fluid substance that can expand without limit. This expandable nature also means that a gas can be compressed. To compress means to apply pressure to reduce in volume. For example, an air compressor uses the force of a pump to compress great volumes of air into a small tank.

A compressed gas is dangerous simply because it is under pressure and is always trying to get free. It is like a spring in a compressed state. If a compressed gas is flammable (meaning that it burns easily) there is the additional danger that it may explode or burn out of control. Fortunately, suitable equipment and techniques are available for using compressed, combustible gases safely.

The most popular gases for heating, cutting, and welding metals are propane and acetylene. Before these gases will burn, they must be mixed with oxygen from the air or pure oxygen from tanks. The student is referred to Unit 5, Reducing Hazards in Agricultural Mechanics, for information on how to prevent and control fires.

FIGURE 22-1. Cylinders must be fastened securely in place at all times.

GENERAL SAFETY WITH COMPRESSED GASES

Some general precautions in handling compressed gases are:

- Wear safety goggles or a face shield at all times.
- Obtain the instructor's permission before using compressed gases.
- Store fuel gas cylinders separately from oxygen cylinders.
- Keep gas cylinders upright and chained securely at all times, figure 22-1. They should be stored outdoors or in well-ventilated, fire-safe areas.
- Do not bump or put pressure on pipes, connections, valves, gauges, or other equipment connected to compressed gas cylinders.
- When connections are opened or cylinders changed, check thoroughly for leaks before using the equipment.
- Never use equipment exposed to oil or grease. Spontaneous or instant fires may result.
- Follow specific procedures for turning systems on and off.
- Work only in areas that are free of materials that burn.
- Never use gas-burning equipment without approved fire extinguishers in the area.
- Always wear leather gloves and apron when using gas-burning equipment.
- Screw the caps on all cylinders that do not have regulators or other apparatus attached. Apparatus means objects necessary to carry out a function.
- All equipment or cylinders that may discharge gas should be pointed away from the operator, other people, and clothing. Fire will follow a gas stream.
- Never leave clothing where it can become saturated by oxygen or fuel gases.
- If gas equipment catches fire, immediately turn off the gas at the tanks. If this is not practical, or if this action does not extinguish the gas fire, evacuate the area and call for help.
- Learn to recognize the odors of combustible fuels.
- Protect gas cylinder storage areas with locked chain link fences or concrete enclosures.

GAS-BURNING EQUIPMENT FOR AGRICULTURAL MECHANICS

Shops generally use propane and/or acetylene and oxygen for heating, cutting, and welding. Propane and acetylene are fuels. Oxygen is not a fuel, but it must be present for other fuels to burn. Some shops have manifolds to which many cylinders and welding outfits may be attached. A manifold is a pipe with two or more outlets. Most agricultural mechanics shops have one or more portable oxyacetylene rigs. A rig is a self-contained piece of apparatus assembled to conduct an operation. Oxyacetylene is a shortened version of the words oxygen and acetylene. It refers to equipment and processes where the two gases are used together.

Shops may also have propane and oxygen torches and propane furnaces for heating and soldering. Propane and oxygen torches are desirable as cutting units since propane gas is generally less expensive than acetylene.

Oxyacetylene Equipment

Major Parts. A portable oxyacetylene rig includes a cart, cylinders, valves, regulators, gauges, hoses, and torch assemblies, figure 22-2. A torch is an assembly that mixes gases and discharges them to support a controllable flame, figure 22-3. A gas cylinder is a long round tank with extremely thick walls built to hold gases under great pressure. Valves and regulators are devices that control or regulate the flow of the gas; a gauge measures and indicates the pressure in the hose, tank, or manifold, figure 22-4. Hoses are flexible lines that carry the gases. They are rubber reinforced with nylon or other material to withstand high pressures and heavy use.

The gas flow and control components of a portable oxyacetylene rig are:

Acetylene Side

- acetylene cylinder
- cylinder valve
- red cylinder pressure gauge
- regulator
- hose pressure gauge
- red hose
- acetylene valve on torch

Oxygen Side

- oxygen cylinder
- cylinder valve
- green cylinder pressure gauge
- regulator
- hose pressure gauge
- green hose
- oxygen valve on torch
- torch mixing chamber
- torch tip

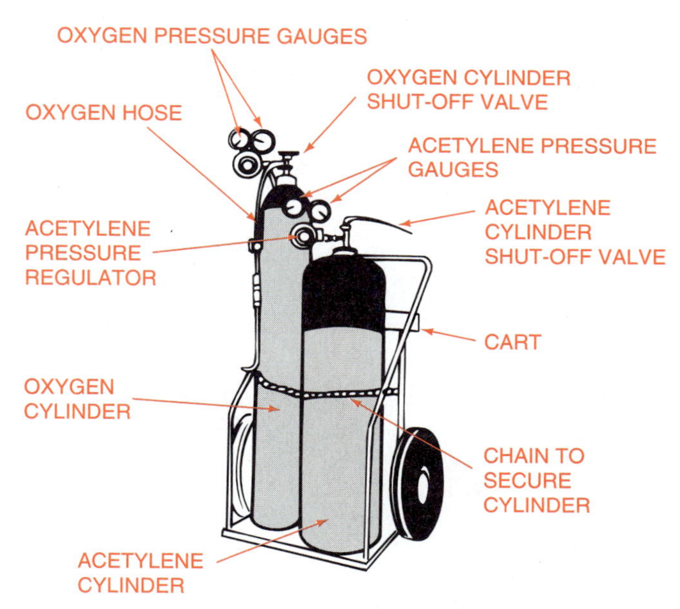

FIGURE 22-2. Portable oxyacetylene rig.

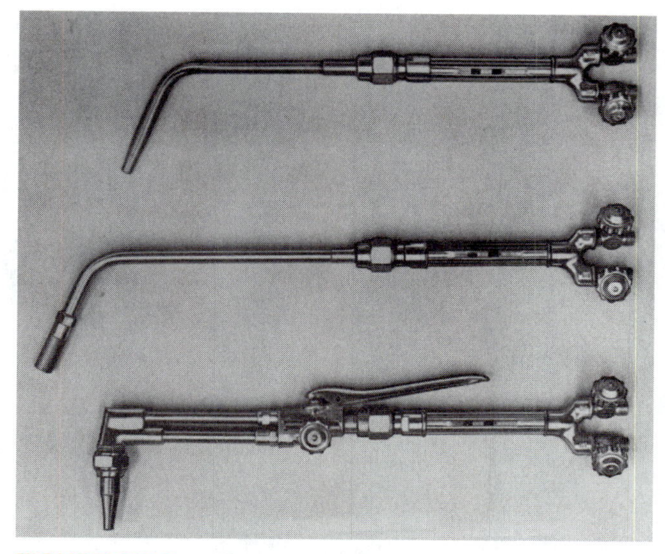

FIGURE 22-3. Torches designed for brazing and welding (top), heating (center), and cutting (bottom). (*Courtesy of Victor Equipment Company*)

Setting Up. When setting up oxyacetylene equipment, one must be aware of certain design features. The valve on most acetylene cylinders is protected with a high collar. The valve is turned on and off with a handle. On cylinders with removable handles, care must be taken to leave the handle in place when turned on in case the gas must be turned off in an emergency. Acetylene equipment is color coded red. Another important distinction is that all acetylene couplings have left-hand threads and notched nuts.

Oxygen cylinders generally have valves that are not protected during use. However, they have heavy caps that screw in place over the valves to protect them when not in use. Oxygen equipment is color coded green. All oxygen couplings have right-handed threads.

WORKING PRESSURE GAUGE 0-200 psi
CYLINDER PRESSURE GAUGE 0-4000 psi
OXYGEN CYLINDER INLET FITTING
OXYGEN HOSE OUTLET FITTING
OXYGEN REGULATOR ADJUSTING SCREW
A. OXYGEN REGULATOR

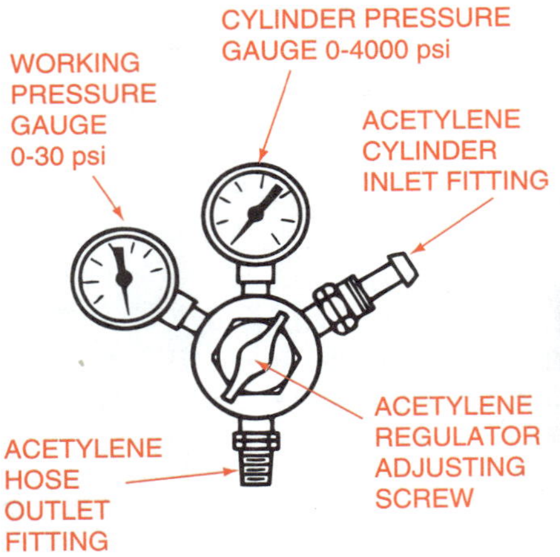

B. ACETYLENE REGULATOR

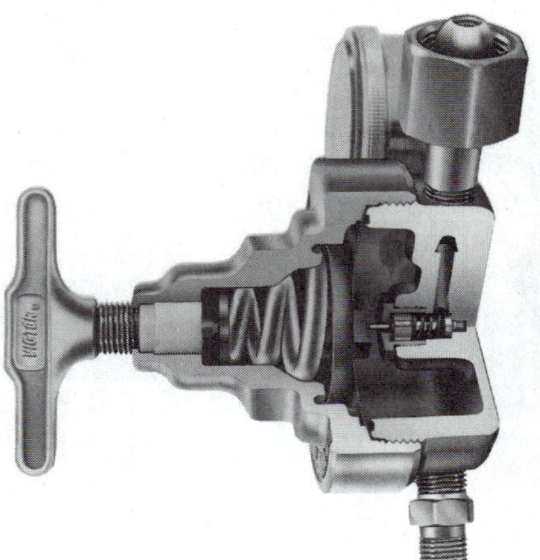
C. CROSS-SECTIONAL VIEW OF A GAS PRESSURE REGULATOR

FIGURE 22-4. Oxygen and acetylene regulators and gauges. (*Courtesy of Victor Equipment Company*)

Before attaching regulators to cylinders, crack the cylinder by turning the gas on and off quickly to blow any dust from the opening.

CAUTION:

- **When attaching regulators, hoses, gauges, or torch connections, the connectors must be threaded in the correct direction. The red color coding and opposite threads for acetylene are means of preventing the accidental mixing of acetylene and oxygen, except in the torch itself.**

All connections are gastight when screwed on properly. Some parts such as regulators have metal-to-metal fittings. These draw together snugly and must be gastight. Some brass fittings used for connecting hoses are also metal-to-metal types. Gauge fittings have tapered threads. These are wrapped with teflon tape to ensure a gastight fit when screwed into regulators or pipe fittings.

CAUTION:

- **Oxygen and acetylene valves on torches seal with light finger pressure. Overtightening will damage the seats of the valves. A seat is the point where the moveable part of a valve seals off the gas.**

Turning On the Acetylene and Oxygen. After all components are assembled, the following sequence is used to turn on the gases.

CAUTION:

- **The area must be properly ventilated. Wear protective goggles and clothing.**

PROCEDURE

1. Close the acetylene valve on the torch.
2. Close the oxygen valve on the torch.
3. Turn the acetylene regulator handle counterclockwise until no spring tension is felt.
4. Turn the oxygen regulator handle counterclockwise until no spring tension is felt.
5. Open the oxygen cylinder valve slowly until the pressure gauge responds. Open the valve all the way.

CAUTION:

- **Do not stand in front of the gauges when gas is being turned on.**

6. Open the acetylene cylinder valve slowly half a turn.

CAUTION:

- **This valve is never opened more than half a turn so that it can be turned off quickly in an emergency.**

7. Open the oxygen torch valve an eighth of a turn. Turn the oxygen regulator handle clockwise until the pressure gauge reads 10 psi

(pounds per square inch) pressure (the final desired pressure will vary with the specific torch equipment being used). Close the oxygen valve on the torch.

8. Open the acetylene torch valve an eighth of a turn. Turn the acetylene regulator handle clockwise until the pressure gauge reads 5 psi. (The final desired pressure will vary with the equipment being used.) Close the acetylene valve on the torch.

Steps 7 and 8 purge the lines and set the regulators at safe starting pressures. To purge the lines is to remove undesirable gases. The unit is now pressurized and ready for use.

Testing for Leaks. A leak test should be performed when equipment is first set up, when cylinders are changed, or if the odor of acetylene is present when the unit is not in use. To test for leaks, put a small amount of water in a small jar or can. Add a sliver of nondetergent hand soap. Use a 1-inch paint brush to produce a soapy lather; use the brush to apply the soap solution gently around each fitting and point where gas may escape. If a leak exists, it will cause bubbles in the solution.

If a leak is found, tighten the fitting. If this does not stop the leak, turn off both gases. Disassemble the joint and correct the problem. Turn on the gases again and recheck the entire system with the soap solution.

Lighting and Adjusting Torches. Once the system is pressurized, attention can be given to selecting torch parts. This is followed by lighting and adjusting the torch according to the work to be done.

Two types of torches are available: welding torches, figure 22-5, and cutting torches, figure 22-6. Both include a body or handle with hose connections and valves to control the oxygen and the fuel. Welding tips with mixing chambers are screwed onto the handle. Tips of different sizes are available. A cutting assembly with another set of valves may also be attached to the handle.

To light and adjust a welding or cutting torch, use the following procedure.

PROCEDURE

1. Put on leather gloves and goggles with a No. 5 shaded lens, figures 22-7.
2. Open the acetylene valve an eighth of a turn.
3. Use a spark lighter to ignite the torch, figure 22-8.

CAUTION:

- **Do not point the torch directly into the lighter.**

4. Open the acetylene valve slowly until the flame is 1/4 inch off the tip of the torch. Increase or decrease the regulator pressure until the flame just touches the tip. At this point, the acetylene valve on the torch is open several turns. This will be the correct pressure for the tip being used. The flame is called a carbonizing flame. A carbonizing flame is one with an excess of acetylene, figure 22-9. It is cooler than other types of flames.
5. Turn the oxygen valve on slowly and watch the inner flame shorten. Continue to add oxygen

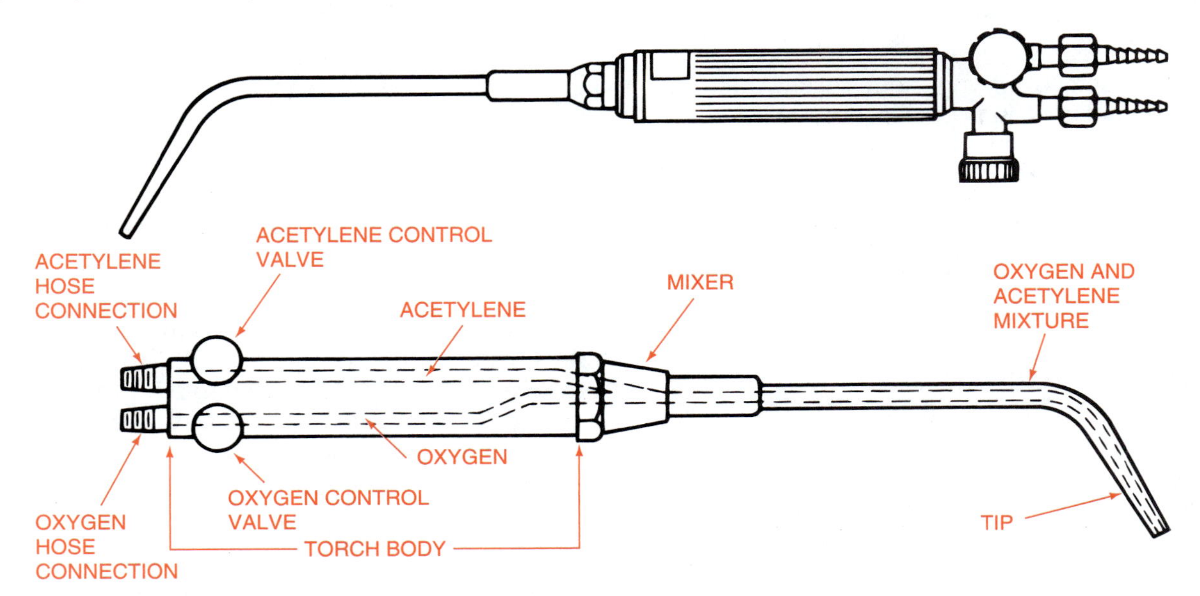

FIGURE 22-5. Exterior and interior views of a welding torch.

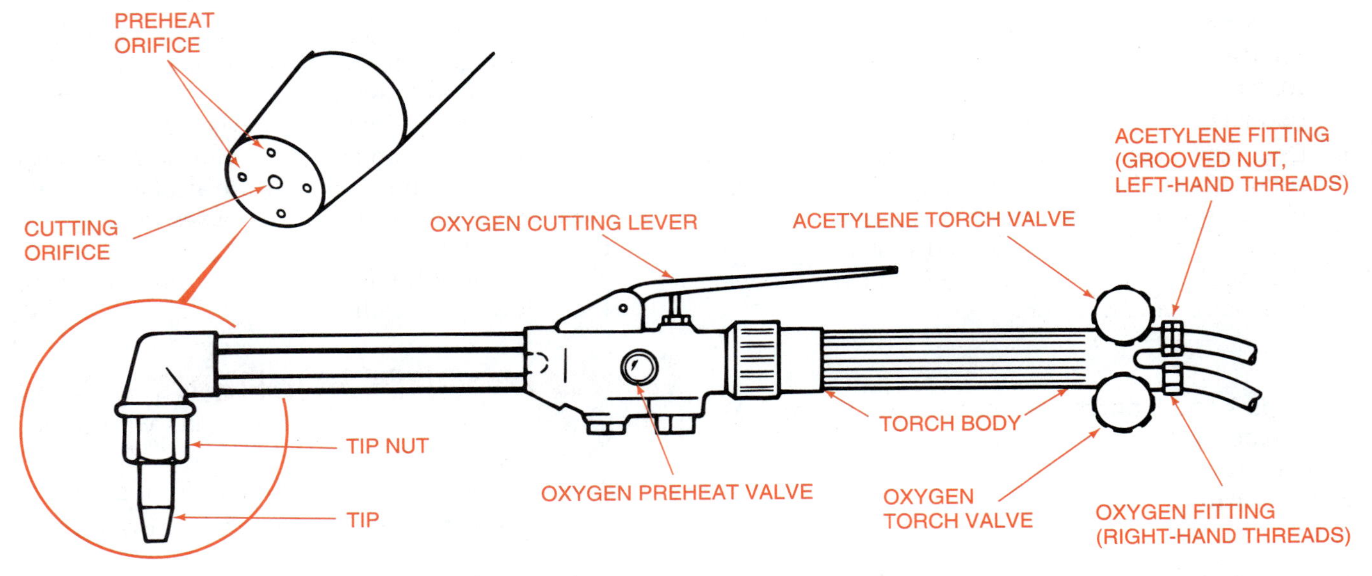

FIGURE 22-6. An oxyacetylene cutting torch.

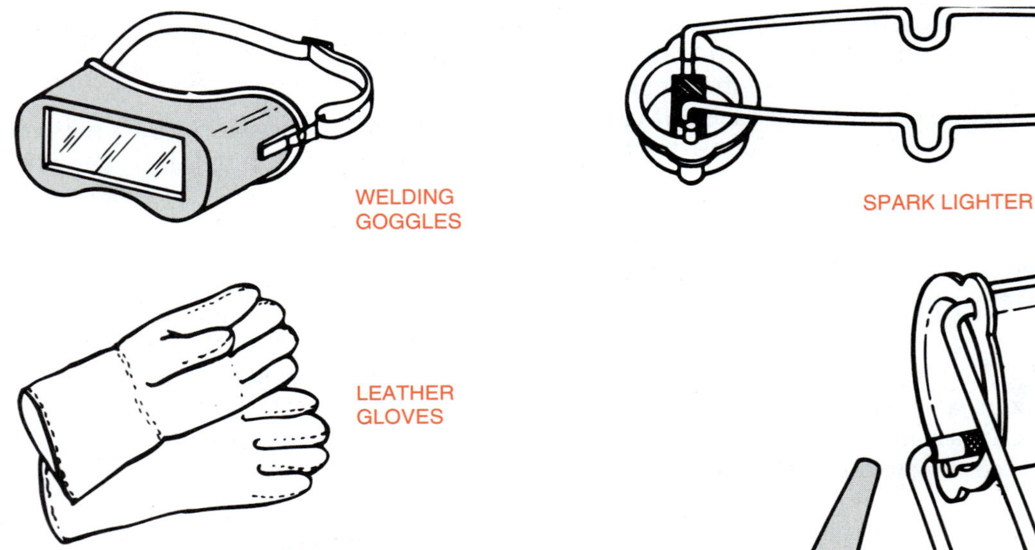

FIGURE 22-7. Before lighting a gas torch, the operator must put on the proper protective clothing.

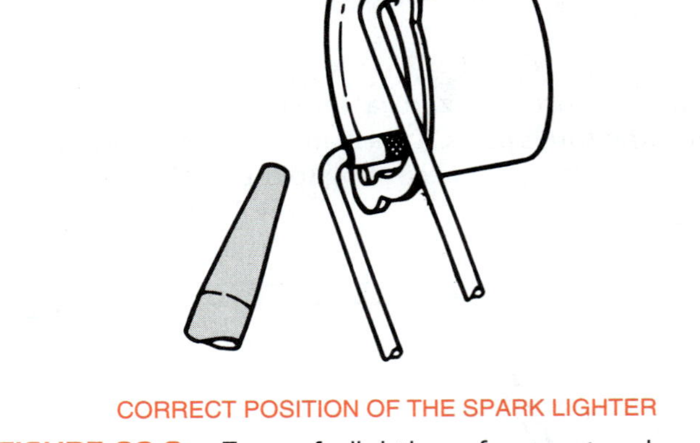

FIGURE 22-8. For safe lighting of a gas torch, use a spark lighter.

until the long inner flame just fits the cone. This is a neutral flame, or one with a correct balance of acetylene and oxygen. A neutral flame is correct for heating, cutting, and welding.

6. If additional oxygen is added, the cone becomes sharp and the flame noisy. This is an oxidizing flame, or one with an excess of oxygen. It is the hottest type of flame, but not recommended except for special applications. Additional instructions for lighting and adjusting cutting torches are provided in Unit 23.

If the torch cannot be adjusted to produce a neutral flame, clean the tip, figure 22-10. Tip cleaners are rods with rough edges designed to remove soot, dirt, or metal residue from the hole in the tip. Use a tip cleaner that is equal to or smaller in size than the hole in the tip. If molten metal is fused to the tip, remove it with fine emery cloth.

Shutting Off Torches. Torches become a fire hazard if the proper procedure is not followed when

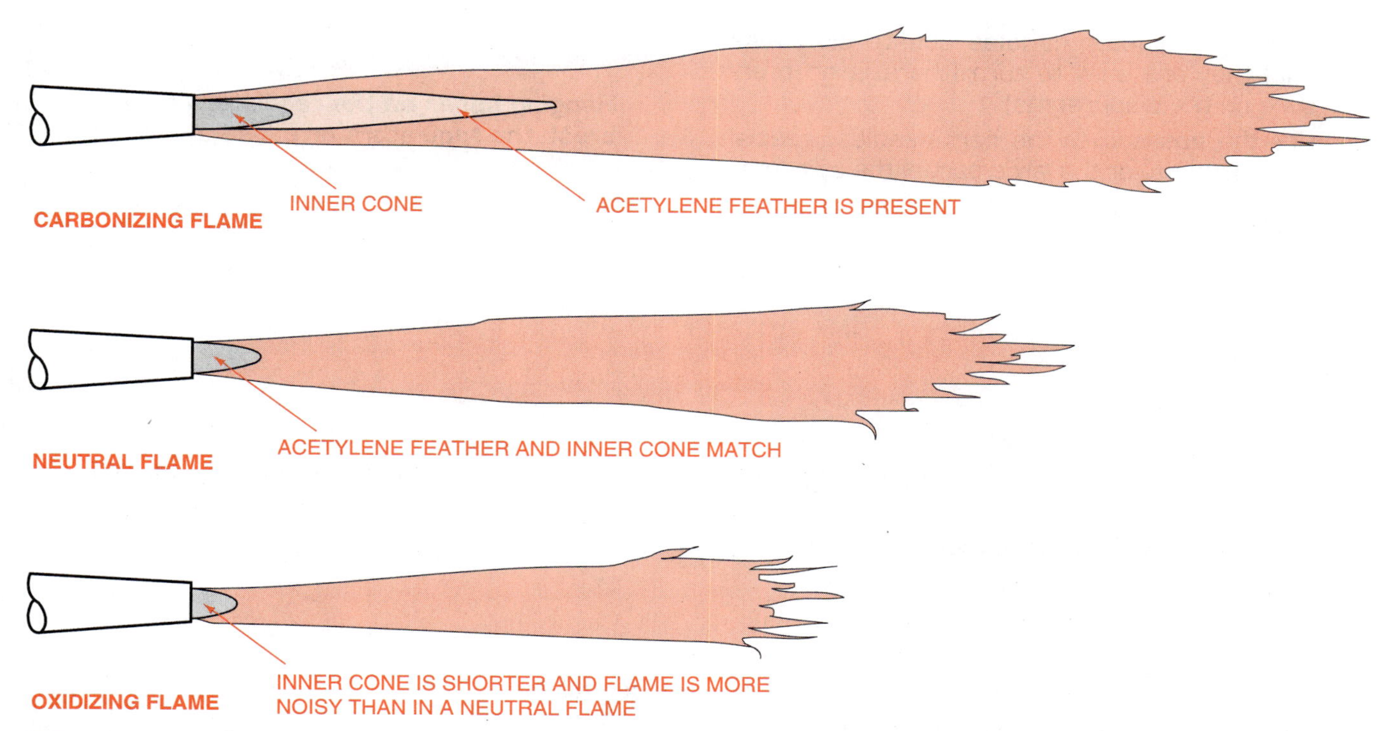

FIGURE 22-9. Carbonizing, neutral, and oxidizing flames. The neutral flame is correct for heating and cutting.

shutting them off. Improper procedures may cause excessive popping of the torch, soot from unburned gases, or carbon deposits in the tip. To shut off a torch correctly, first close the acetylene valve on the torch, then close the oxygen valve.

CAUTION:

- **Use gentle pressure on the torch valves as they have soft internal parts that may be damaged.**

Bleeding Lines. When the torch is not in use, it is important to close every point where gas may escape. Gas should also be removed from all lines and equipment. This practice is called bleeding the lines. To bleed the lines use the following procedure.

PROCEDURE

1. Turn off the acetylene at the cylinder.
2. Turn off the oxygen at the cylinder.
3. Open the acetylene valve at the torch until both regulators return to zero. Close the acetylene valve at the torch.
4. Open the oxygen valve at the torch until both regulators return to zero. Close the oxygen valve at the torch.

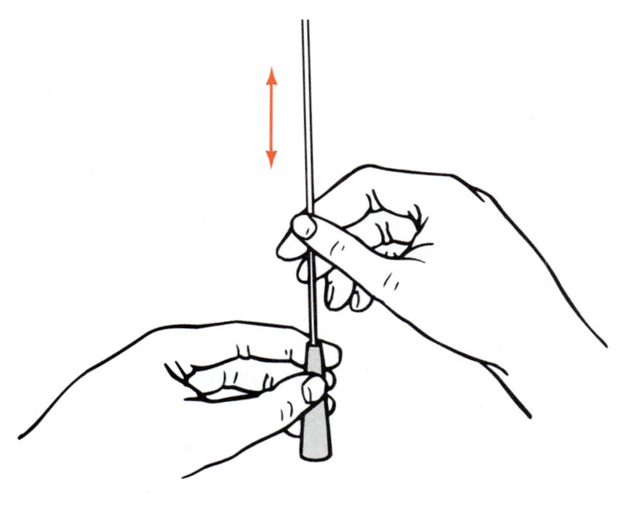

A. CLEANING A TIP WITH A STANDARD TIP CLEANER

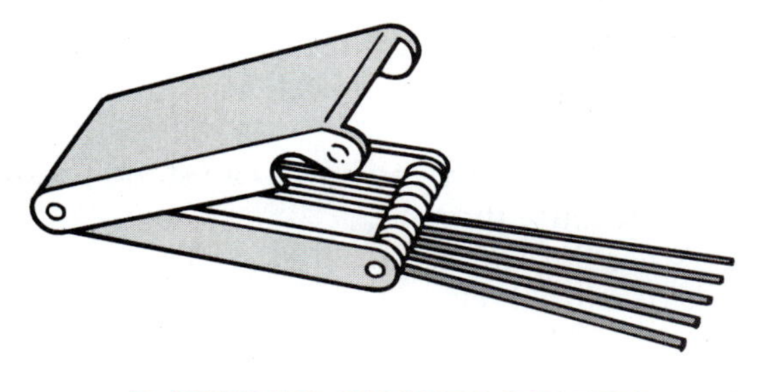

B. STANDARD SET OF TIP CLEANERS

FIGURE 22-10. Use a tip cleaner that is equal to or smaller in size than the hole in the tip.

5. Turn all regulator handles counterclockwise until they are easy to turn (there is no pressure on the diaphragms).
6. Coil the hoses over the cart handles, special hose hangers, or the large part of the cylinders.
7. If portable, store the rig in a suitable place.

CAUTION:

- **Hanging hoses on the regulators may break the regulators or cause a leak.**

STUDENT ACTIVITIES

1. Define the Terms to Know for this unit.
2. Study the diagrams of all equipment pictured in this unit. Memorize all the names of all equipment components.
3. Attach regulators to oxygen and acetylene cylinders.
4. Turn on the oxygen and acetylene.
5. Test the apparatus for gas leaks.
6. Clean the tip of the torch.
7. Light the torch and adjust the gas flow to achieve a neutral flame.
8. Shut off the torch and bleed the lines.
9. Store the apparatus properly.

SELF-EVALUATION

A. Multiple Choice. Select the best answer.

1. Gas can be compressed with a
   a. cylinder
   b. lever
   c. pump
   d. valve
2. Acetylene may be dangerous because it is
   a. compressed
   b. flammable
   c. explosive
   d. all of these
3. Which gas is not a fuel used for torches?
   a. acetylene
   b. oxygen
   c. propane
   d. none of these
4. Oxygen and acetylene hoses can stand pressure because they are
   a. color coded
   b. extra thick
   c. made of steel
   d. reinforced
5. Oxygen hoses and related equipment are color coded
   a. green
   b. ivory
   c. orange
   d. red
6. Acetylene hoses and related equipment are color coded
   a. green
   b. ivory
   c. orange
   d. red
7. The acetylene pressure to light a torch should be
   a. 5 psi
   b. 15 psi
   c. 25 psi
   d. 50 psi
8. Gas leaks are checked with
   a. compressed air
   b. flame
   c. soapy water
   d. teflon

B. Matching. Match the terms in column I with the terms in column II.

Column I

1. weld
2. gas
3. compress
4. oxyacetylene outfit
5. tank
6. neutral flame
7. goggles
8. teflon

Column II

a. pressure
b. cylinder
c. balanced
d. tape
e. expands without limit
f. join by fusion
g. rig
h. No. 5

C. Completion. Fill in the blanks with the word or words that make the following statements correct.

1. Gas cylinders must be securely ________.
2. Never use gas-burning equipment without approved _______ _______ in the area.
3. When no apparatus is attached, a _____ should be screwed onto oxygen cylinders.
4. The presence of acetylene and propane is detected by ________.
5. The term oxyacetylene comes from _____ and _____.
6. The valve on acetylene cylinders is protected by a _______.
7. Connectors for acetylene hoses have _____ threads.
8. Connectors for oxygen hoses have _____ threads.
9. Removing gas from oxyacetylene equipment is known as _______.
10. Hanging hoses on regulators may result in _______.

UNIT 23 Cutting with Oxyfuels

OBJECTIVE

To use oxygen and selected fuels to cut steel with a flame torch.

COMPETENCIES TO BE DEVELOPED

After studying this unit, you should be able to:

- Write the names and characteristics of common fuels used for cutting.
- State and apply recommended safety practices for using oxyfuels.
- Select appropriate pressures for using oxygen and common fuel gases for cutting.
- Cut steel with oxyfuels.
- Pierce steel with oxyfuels.

TERMS TO KNOW

oxyfuel	base metal
oxyfuel cutting	fusion welding
kerf	backfire
slag	flashback
brazing	slag box
brass	pierce

MATERIALS LIST

- ✓ Oxyacetylene cutting rig and welding area
- ✓ Welding safety equipment and protective clothing
- ✓ Two pieces of steel, 1/4" to 1/2" x 2" x 12"
- ✓ Plate steel, 1/4" to 1/2" x 6" x 12"
- ✓ Black pipe up to 3-inch diameter, 1 to 2 feet long
- ✓ Soapstone

Combustible gases combined with oxygen have been used since the 1800s for welding metals. Over the years, scientists have experimented to find the gases that work best. Unfortunately, no one gas is the safest and best for all jobs. A study of gas heating, cutting, and welding procedures is an important part of modern agricultural mechanics.

OXYFUEL PROCESSES

The term oxyfuel refers to the combination of pure oxygen and a combustible fuel gas to produce a flame. Oxyfuels are used for welding, brazing, cutting, and heating metals.

General Principles

Oxygen and Fuel Gases. These gases are stored under pressure in tanks or cylinders. They are released as individual gases through carefully designed valves, regulators, and hoses. The gases are mixed as they flow through torch assemblies. They burn as the mixture is discharged through carefully engineered tips.

The gas flame from an oxyfuel burns with intense heat. The temperature may range from 5000°F to 6000°F. These temperatures are hot enough to melt most metals and permit cutting and fusion welding.

Cutting, Brazing, and Heating. Cutting, brazing, and heating differ from welding. Oxyfuel cutting is a process in which steel is heated to the point that it burns and is removed to leave a thin slit called a kerf.

To burn, hot steel must combine with oxygen. Slag is a product formed during the process. Slag is a good insulator. It hinders the cutting process because it forms a layer between the burning steel and the torch flame.

Cutting torches are designed to send a forceful stream of oxygen into preheated cherry red steel. This oxygen supports the combustion. The force of the oxygen stream drives the slag out of the area and permits the heat from the torch to keep the steel burning. As a cutting torch moves forward, the oxygen stream pushes slag out to form the kerf. Torches can be adjusted to cut kerfs that are nearly as straight and clean as the cut produced by a saw.

Heating. In heating, the temperature may be raised enough to soften metal for bending or shaping. No melting takes place. The job may call for heat in a very narrow band to aid in making a sharp bend in a specific place. The heat may also be spread equally over a large area to reduce stress from welding. Different tips for oxyfuel torches are available for these purposes.

Brazing. Brazing is the process of bonding with metals and alloys that melt at or above 840°F. Brass is a mixture of copper and zinc. It is an example of an alloy used for brazing. The process of brazing is similar to soldering. When brazing, the base metal (the main piece of metal) is heated until the brass melts, flows, and bonds to it. The base metal is not melted, and the metals do not mix during brazing.

Fusion Welding. Steel may be joined by fusion welding. Fusion welding is joining metal by melting it together. An oxyacetylene welding torch may be used to heat two pieces of steel until the metal from each runs together to form a joint. When properly done, such joints are as strong as the base metal itself.

CHARACTERISTICS OF OXYFUELS

Oxygen

Oxygen is not a fuel and it will not burn. However, it combines with other substances and causes them to burn. According to the fire triangle, fuel plus heat plus oxygen equals fire. Oxygen is necessary if other gases are to be used in torches, figure 23-1.

Oxygen must be 99.5 percent pure to support the combustion of iron. When iron is hot enough, it

FUEL 1 CUBIC FOOT	TOTAL O_2 REQUIRED	O_2 SUPPLIED THROUGH TORCH*	% TOTAL THROUGH TORCH	CU FT O_2 PER LB OF FUEL
Acetylene—5589°F/1470 Btu	2.5	1.3	50.0	18.9
MAPP R gas—5301°F/2406 Btu	4.0	2.5	62.5	22.1
Natural gas—4600°F/1000 Btu	2.0	1.9	95.0	44.9
Propane—4579°F/2498 Btu	5.0	4.3	85.0	37.2
Propylene—5193°F/2371 Btu	4.5	3.5	77.0	31.0

*Balance of total oxygen demand is entrained in the fuel-gas flame from the atmosphere.
Source: Airco Welding Products.

FIGURE 23-1. Cubic feet of oxygen needed per cubic foot of fuel burned.

burns in oxygen the way wood or paper burns in air. This principle makes cutting with a torch possible. To burn in the presence of oxygen, iron must be heated to between 1600°F and 1800°F. This is about 800° below the melting temperature of iron.

Acetylene

The fuel most suitable for welding is acetylene. This gas produces a cleaner weld than most other fuel gases. It also produces a more controllable flame. However, acetylene gas is unstable and, therefore, it is hazardous. Acetylene must be handled very carefully.

Acetylene is more expensive than other oxyfuels. Therefore, it is used in welding only when its burning characteristics are needed. In such cases, the safety hazards must be tolerated. For high-volume cutting and heating, less expensive fuels are generally preferred, figures 23-2 and 23-3. However, since acetylene performs most functions well, it is the choice of fuel for many agricultural mechanics shops.

CAUTION:

- **Acetylene must not be used at pressures greater than 15 psi. The lower the pressure, the more stable the gas is and the safer it is to use. Acetylene containers must not be subjected to electrical shock, rough handling, or excessive heat. Do not use acetylene in copper lines or in lines with any grease or oil residues.**

Propane and Natural Gas

Propane gas is available in cylinders and tanks. Natural gas is piped into buildings in urban and suburban areas through public utility lines. Both gases are used extensively for general heating and may be used for torch heating and cutting, but not welding. For torch heating and cutting, however, both propane and natural gas consume large volumes of

FUEL	NEUTRAL FLAME TEMP °F	PRIMARY FLAME BTU/FT³	SECONDARY FLAME BTU/FT³	TOTAL HEAT BTU/FT³
Acetylene	5589	507	963	1470
MAPP® gas	5301	517	1889	2406
Natural gas	4600	11	989	1000
Propane	4579	255	2243	2498
Propylene	5193	438	1962	2371

Source: Airco Welding Products.
Key: MAPP® gas—A multipurpose industrial fuel gas consisting of a mixture of methylacetylene and propadiene.
BTU/ft³—British thermal units per cubic foot of fuel. A BTU is a small unit of heat.

FIGURE 23-2. Heating values of major industrial fuel gases when burned with pure oxygen.

APPLICATION	ACETYLENE	MAPP® GAS	PROPYLENE
Cutting			
Under 3/8 in thick	100	95	90
5/8 in to 5 in thick	95	100	95
Over 5 in thick	80	100	95
Cutting dirty or scaled surfaces	100	95	80
Cutting low-alloy specialty steels	100	90	80
Piercing	100	100	85
Welding	100	70	0
Braze welding	100	90	70
Brazing	100	100	90

Source: Airco Welding Products.

FIGURE 23-3. Average performance ratings of selected oxyfuel flames.

cylinder oxygen. This fact may offset other price advantages offered by these gases.

MAPP® Gas

MAPP® gas is a formulated mixture of methylacetylene and propadiene gases. (MAPP® is a trade name of Airco, Inc.) The gas is reported to have many of the advantages of acetylene but to be more stable and, therefore, safer. Its high temperature flame is suitable for brazing, cutting, heating, and metallizing.

CUTTING STEEL WITH OXYFUELS

Review the previous unit on using gas-burning equipment.

CAUTION:

- **Use every precaution in handling oxygen and fuel gases, setting up apparatus, and performing cutting and heating tasks.**

The Flame

A description of a neutral flame and the procedure for obtaining it are given in Unit 22. However, it is more difficult to determine when a flame is neutral on a cutting torch than on a welding torch. A cutting torch must be adjusted to obtain a neutral flame both with and without the oxygen jet, figure 23-4.

To obtain a neutral flame with a cutting torch, the following procedure is recommended.

PROCEDURE

1. Wear appropriate protective clothing and goggles with a No. 5 shaded lens.
2. Check the area and remove all fire hazards.
3. Set up the oxyacetylene equipment in a safe manner and check for leaks.
4. Set the pressure gauges at the recommended pressure for the tip being used. These will vary with the fuel and tip, figure 23-5.
5. Light the torch using acetylene only.
6. Increase the acetylene pressure until the

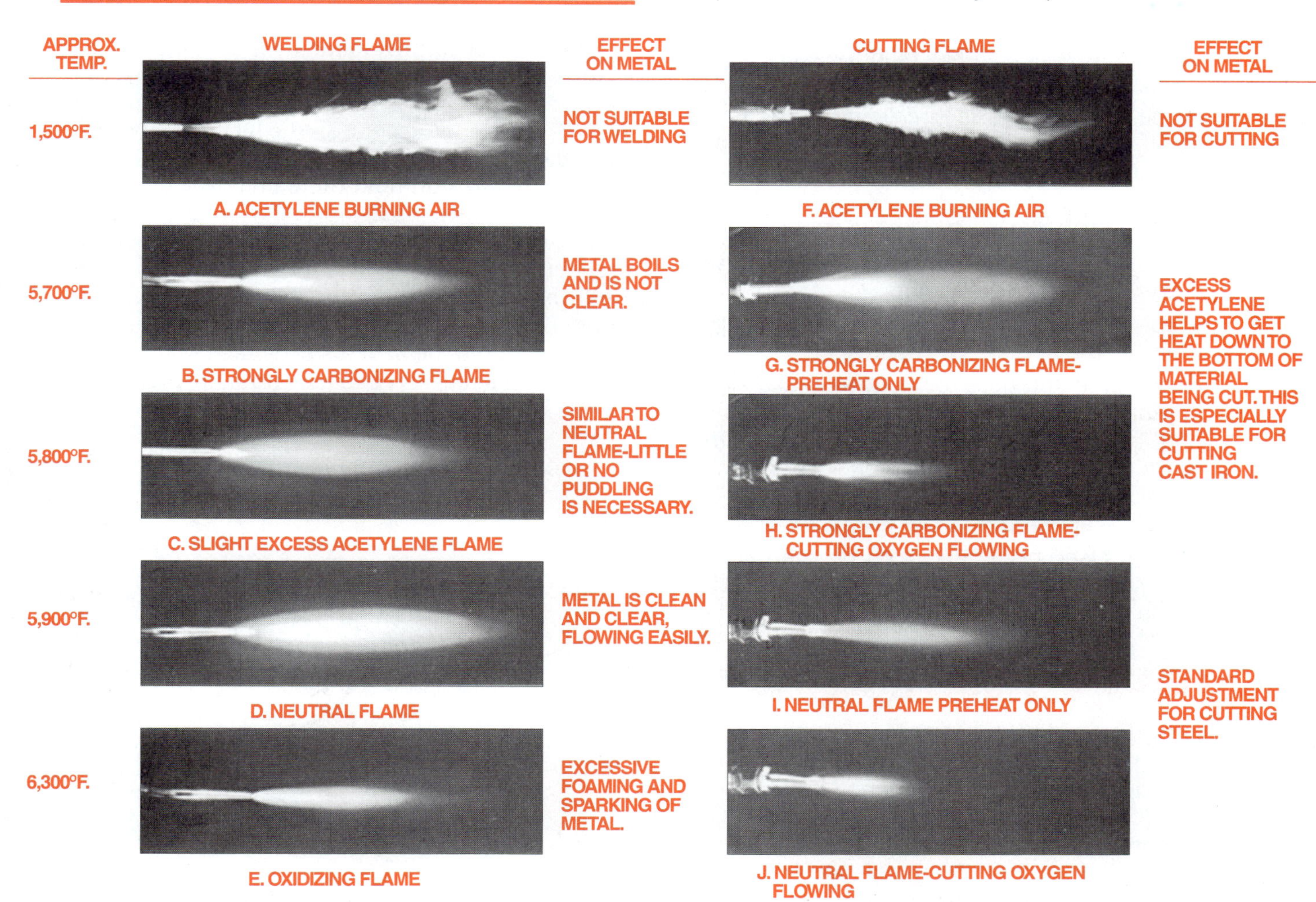

FIGURE 23-4. Oxyacetylene flame adjustments. (*Courtesy of Airco Welding Products, adapted*)

CUTTING GUIDE			
Metal Thickness (in.)	Airco Cutting Tip Size	Pressure PSI	
		Oxygen	Acetylene
1/8	00	30	1 1/2
1/4	0	30	3
3/8	1	30	3
1/2	1	40	3
3/4	2	40	3
1	2	50	3
1 1/2	3	45	3
2	4	50	3
3	5	45	4
4	5	60	4
5	6	50	5
6	6	55	5

FIGURE 23-5. Oxygen and acetylene pressures vary according to the size of the cutting tip. (*Courtesy of Airco Welding Products*)

flame just starts to leave the tip when the acetylene torch valve is opened several turns.

7. Close the acetylene torch valve until the flame touches the tip.
8. With the oxygen preheat valve closed, open the oxygen (O_2) torch valve several turns.
9. Slowly open the O_2 preheat valve until the acetylene feathers just match the inner cones of the preheat flames.
10. Press the oxygen cutting lever and observe whether the flames stay neutral.
11. If the flames do not stay neutral, make slight adjustments in the oxygen preheat valve and/or acetylene torch valve until the flame remains neutral both with and without the oxygen lever pressed.
12. If a neutral flame cannot be obtained, clean the tip. Use fine emery cloth on the outside and the appropriate tip cleaners for the inside of the holes.

CAUTION:

- **Touching the tip against the work, overheating, incorrect torch adjustment, a loose tip, a dirty tip, or damaged valves may all cause backfire. Backfire is a loud snap or popping noise which generally blows out the flame. The cause must be corrected before relighting.**

CAUTION:

- **Backfire sometimes causes a flashback. A flashback is burning inside the torch that causes a squealing or hissing noise. When this occurs, quickly turn off the torch oxygen valve and then the torch acetylene valve. If fire is suspected in the hoses, rush to close the acetylene valve and then the oxygen valve at the tanks. After a flashback, only an experienced operator can determine if the torch is safe to relight.**

Cutting the Steel

After obtaining a neutral torch flame, position the work for cutting. This is done by placing the metal to be cut over a slag box. A slag box is a metal container of water or sand placed to catch hot slag and metal from the cutting process. The work should be weighted or clamped so that it will not slip from the work area.

For general cutting, mark the line of the cut with soapstone. Place one gloved hand on the metal near the torch head. This permits control of the tip clearance. The other hand controls the handle and oxygen lever, figure 23-6.

To start the cut, hold the flame over the corner and edge of the metal, figure 23-7. Hold the torch at a slight angle away from the edge. The cones of the preheat flames should not quite touch the metal. Hold the torch steady until the edge of the metal turns cherry red. Press the oxygen lever and move the torch across the metal at a steady rate. Maintain the flame cones about 1/8 of an inch from the metal.

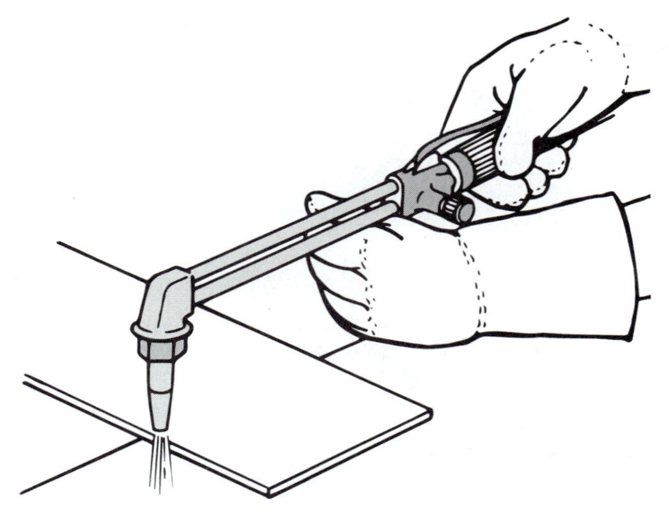

FIGURE 23-6. One gloved hand controls the tip clearance and the other controls the oxygen lever.

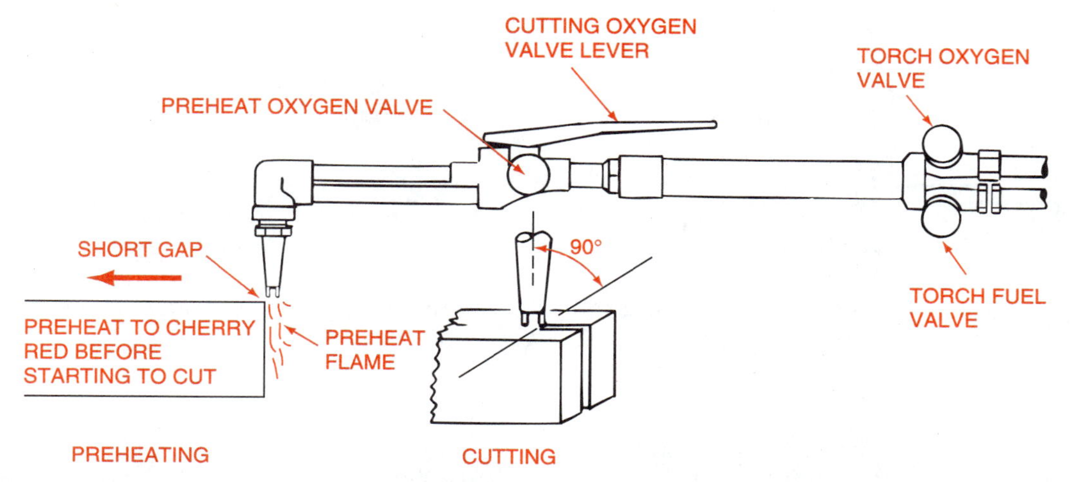

FIGURE 23-7. Torch positions when preheating and cutting. (*Courtesy of Airco Welding Products*)

Different techniques may be used to control speed and clearance. When making short cuts, it is helpful to slide the torch over the gloved hand positioned near the torch head. In this case, one hand controls the clearance and the other controls the oxygen lever and movement across the metal.

When making cuts longer than about 2 inches, both hands grip the torch and slide over the metal. Special devices such as wheeled trolleys may be used to maintain the correct clearance, figure 23-8. A rod with a sliding center point can be attached to a torch to aid in cutting perfect circles, figure 23-9. A piece of steel angle or similar device can be used to guide the torch when making straight cuts, figure 23-10. A smoother cut will be obtained by placing the guide metal on the waste piece since most of the slag will collect on the piece with the guide metal.

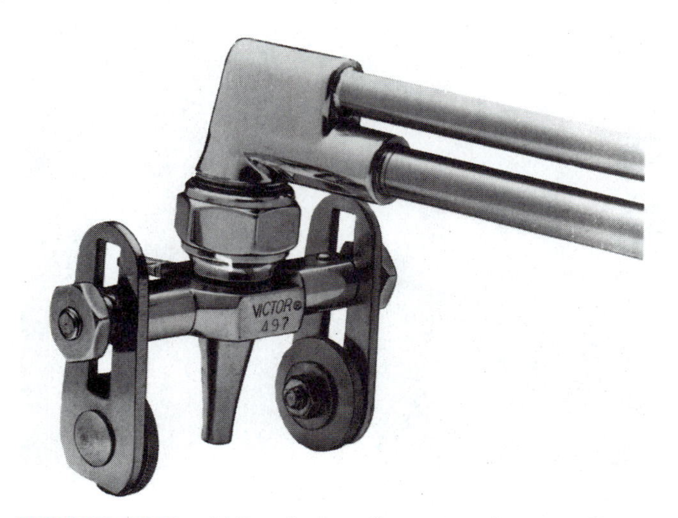

FIGURE 23-8. Wheeled trolleys may be attached to cutting tips to maintain the proper distance between the tip and the metal for long cuts. (*Courtesy of Victor Equipment Company*)

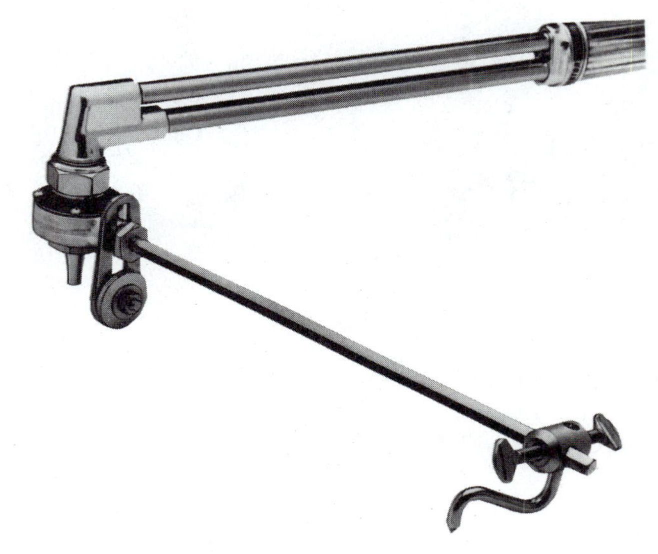

FIGURE 23-9. A rod with a sliding center point attached to a torch aids in cutting perfect circles. (*Courtesy of Victor Equipment Company*)

Improving the Cut

An examination of the metal along a cut reveals how the cut was made. Marks left by the flame provide clues to the preheat procedure, speed, and pressure.

Preheat. If the preheat flame is too hot or the torch travels too slowly, the surface melts before the metal is heated through. This leaves a melted or rounded appearance along the top, figure 23-11. The tip may be raised slightly to reduce the preheat. Increasing the speed slightly may also correct the problem.

Clearance. Clearance is the distance from the torch tip to the metal. Generally the clearance is correct when the tips of the primary flames are almost level with the surface metal.

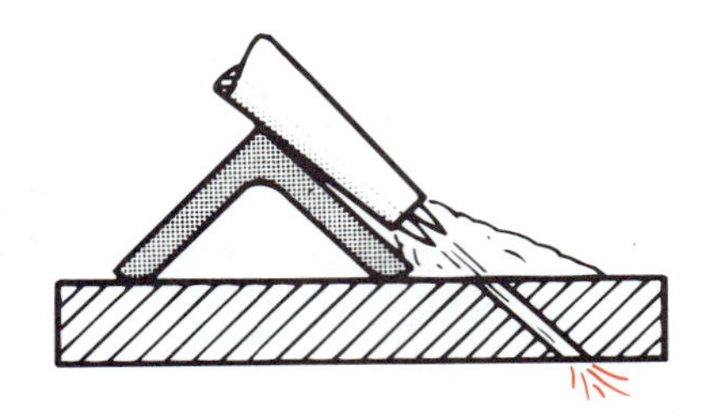

FIGURE 23-10. Steel angle used to guide the torch when cutting straight lines. (*Courtesy of Larry Jeffus,* Welding Principles and Applications, Third Edition, *Delmar Publishers, 1995*)

FIGURE 23-11. Excessive preheating and/or traveling too slowly across the metal result in a melted top edge. (*Courtesy of Larry Jeffus,* Welding Principles and Applications, Third Edition, *Delmar Publishers, 1995*)

FIGURE 23-12. When the torch travels too fast across the metal, rough edges and uncut metal result. (*Courtesy of Larry Jeffus,* Welding Principles and Applications, Third Edition, *Delmar Publishers, 1995*)

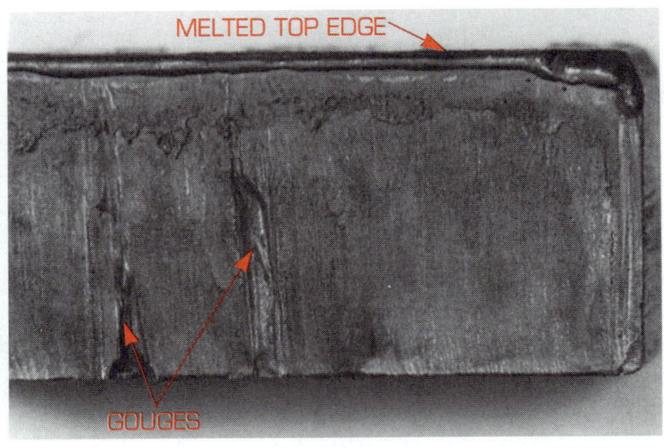

FIGURE 23-13. Moving the torch too slowly across the metal results in a melted top edge and gouges along the cut. (*Courtesy of Larry Jeffus,* Welding Principles and Applications, Third Edition, *Delmar Publishers, 1995*)

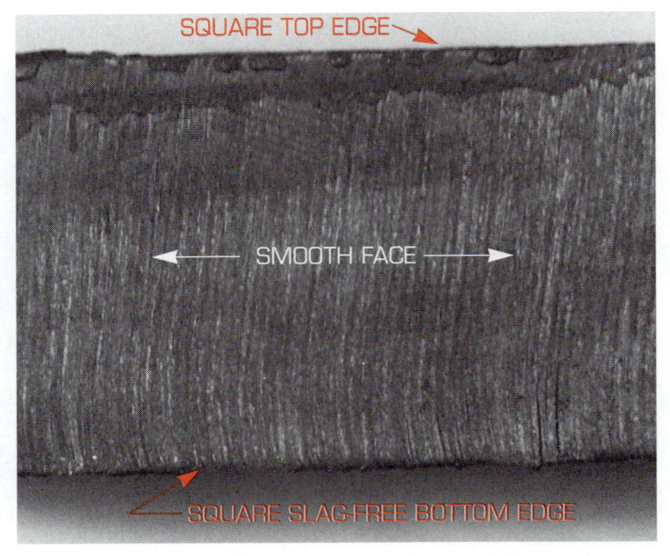

FIGURE 23-14. A correct cut is straight and square with a smooth face. Drag lines bend backward slightly at the bottom. (*Courtesy of Larry Jeffus,* Welding Principles and Applications, Third Edition, *Delmar Publishers, 1995*)

Speed. Moving the torch too fast across the metal results in an incomplete cut and rough edges. Incomplete cuts generally occur at the bottom and end of the cut, figure 23-12. On the other hand, moving too slowly results in a melted top edge and leaves gouges where the cutting stream has wandered, figure 23-13.

Pressure. If the oxygen pressure is too high, the result may be a dish shape in the kerf near the top. On the other hand, if the pressure is too low, the cut may not be complete at the bottom.

The Correct Cut. A correct cut is straight and square with a smooth face. Drag lines bend backward slightly at the bottom. If preheat, clearance, speed, and pressure are all correct, then cutting will be fast, clean, and accurate, figure 23-14. Such a combination requires a steady hand and plenty of practice. While a good cut is being made, there is a smooth, even sound and a steady stream of sparks from the bottom of the kerf, figure 23-15.

Piercing Steel

A flame cutting torch can also be used to pierce steel. To pierce means to make a hole. The operator must take special care to avoid being burned by molten slag and metal during the procedure.

CAUTION:

- **Use full face, head, shoulder, and body protection when piercing steel.**

FIGURE 23-15. During a good cut, there is a steady stream of sparks from the bottom of the kerf. (*Courtesy of Larry Jeffus,* Welding Principles and Applications, Third Edition, *Delmar Publishers, 1995*)

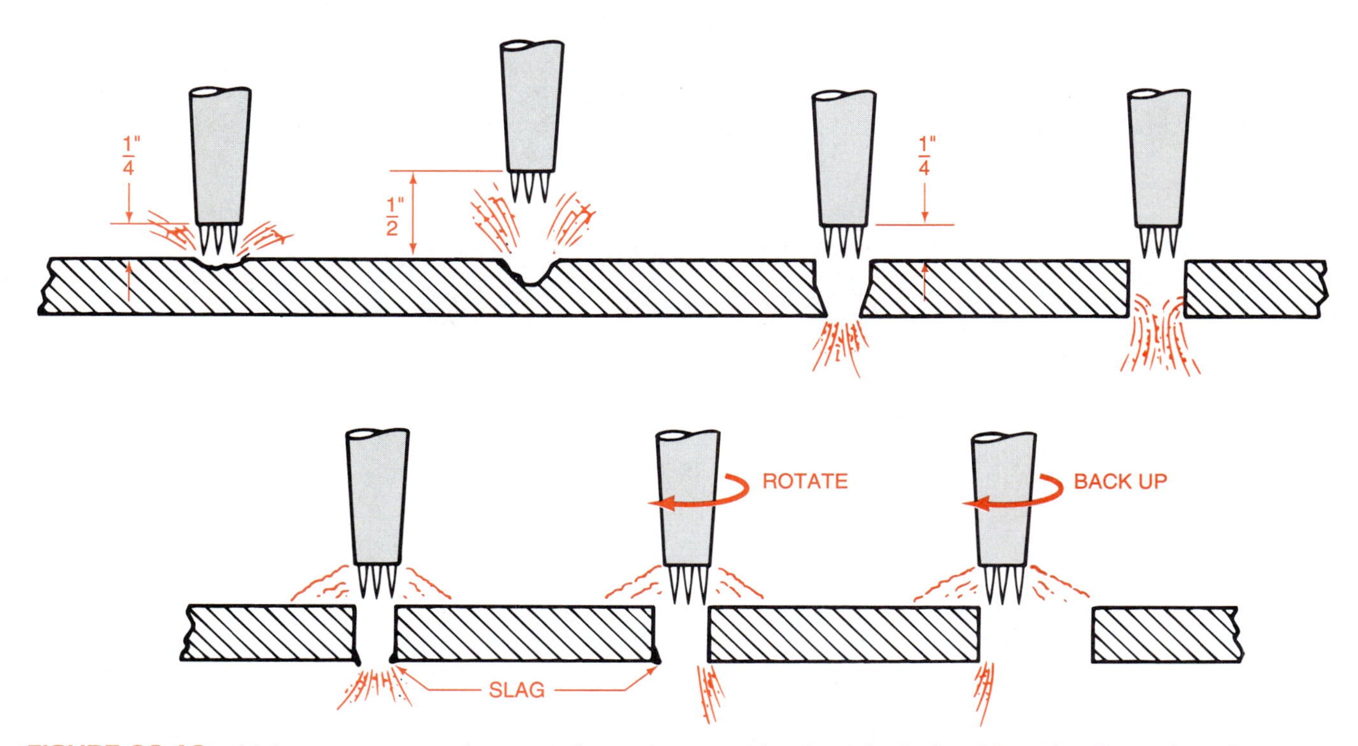

FIGURE 23-16. Using an oxyacetylene torch to pierce and enlarge a hole. After the flame breaks through the metal, the torch is rotated to enlarge the hole or backed up slightly before proceeding with the cut. (*Adapted from Larry Jeffus,* Welding Principles and Applications, Third Edition, *Delmar Publishers, 1995*)

To pierce steel, the cutting torch is held above the mark at the normal preheat distance. When the spot becomes cherry red, raise the torch 1/2 inch or more to reduce the hazard of molten metal; slowly press the oxygen lever.

Move the tip sideways and into a circular motion until the hole breaks through, figure 23-16. Enlarge the hole by cutting around the edges. If an inside cut is desired, proceed with the cut from inside the hole.

Cutting Pipe

Cutting pipe is much like piercing and cutting thin plate. To cut pipe up to 3 inches in diameter, first pierce a hole in the top of the pipe, then cut a kerf to the left side, followed by one to the right side, figure 23-17. Rotate the pipe and repeat the process to cut the underside.

To cut large pipe, the torch is held at a right angle to the pipe. It is then moved around the pipe to make the cut. An alternate method is to rotate the pipe in steps or continuously, figure 23-18.

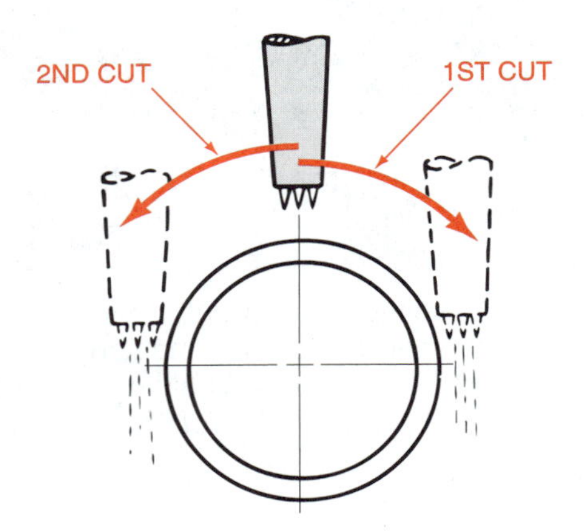

FIGURE 23-17. To cut pipe up to 3 inches in diameter, cut across the top, then rotate the pipe and cut the remaining section. (*Courtesy of Larry Jeffus,* Welding Principles and Applications, Third Edition, *Delmar Publishers, 1995*)

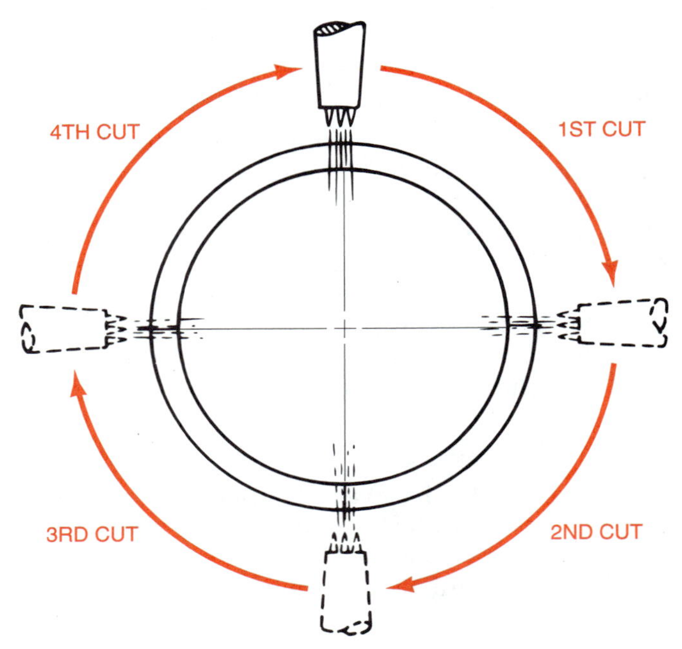

FIGURE 23-18. To cut large pipe, move the torch around the pipe, or hold the torch in position and rotate the pipe. (*Courtesy of of Larry Jeffus,* Welding Principles and Applications, Third Edition, *Delmar Publishers, 1995*)

STUDENT ACTIVITIES

1. Define the Terms to Know for this unit.
2. Light a cutting torch and adjust it to a neutral flame.
3. Obtain a piece of steel 1/4" × 2" × 12" or longer. Square six lines across the metal at 1/2-inch intervals. Use soapstone to mark the lines. Practice cutting and inspect each cut. Discuss your progress with your instructor.
4. Obtain a piece of scrap plate steel. Draw a straight line about 1/2 inch in from an edge. Practice making a long straight cut.
5. Obtain a piece of scrap steel 1/4" × 2" × 12". Practice piercing holes in the strip.
6. Practice cutting off rings about 1/2 inch wide from the end of black pipe.

CAUTION:

- **Do not use galvanized pipe. The melted zinc coating gives off poisonous fumes.**

SELF-EVALUATION

A. **Multiple Choice**. Select the best answer.

1. Combustible gases were first used for welding
   a. in the American space exploration program
   b. in the 1800s
   c. in the early 1900s
   d. during World War II
2. The temperatures from gas flames using oxyfuels are
   a. 150° to 200°F
   b. 400° to 500°F
   c. 1000° to 2000°F
   d. 5000° to 6000°F
3. The result of burning iron in the presence of pure oxygen is
   a. brass
   b. propane
   c. slag
   d. weld
4. When cutting steel, the oxygen stream
   a. aids in keeping the steel hot
   b. drives out slag
   c. supports combustion
   d. all of these
5. The fuel with the best qualities for welding and cutting is
   a. acetylene
   b. MAPP®
   c. natural gas
   d. propane
6. A mixture of gases with excellent qualities for cutting is
   a. acetylene
   b. MAPP®
   c. natural gas
   d. propane
7. A cutting torch must be adjusted so that it is neutral when
   a. cutting
   b. preheating
   c. the oxygen lever is down
   d. all of these
8. Correct oxygen and fuel pressure will vary with
   a. the tip
   b. the job
   c. the fuel
   d. all of these
9. When a correct torch cut is in progress, there will be a
   a. smooth, even sound
   b. spray of sparks
   c. slightly dished kerf
   d. all of these
10. When piercing, the clearance is increased after preheating to
    a. increase the force of the oxygen stream
    b. introduce more air into the process
    c. provide time for heat to move through the metal
    d. reduce the hazard from molten metal

B. **Matching.** Match the terms in column I with those in column II.

Column I
1. fusion
2. brazing
3. propane
4. oxygen
5. fire triangle
6. acetylene
7. neutral flame
8. backfire
9. flashback
10. hot galvanize

Column II
a. poisonous fumes
b. results in fire
c. maximum safe pressure is 15 psi
d. cone and feather are one
e. pop or snap at tip
f. bonding
g. oxyfuel not suitable for welding
h. supports combustion
i. fire inside the torch
j. melting metals together

C. **Completion.** Fill in the blanks with the word or words that make the following statements correct.

1. When cutting, slag should be caught in a _____.

2. When cutting, the edge of the metal should be preheated until it is ______.
3. When cutting, the flame cones should be about ______ from the metal.
4. When using proper equipment with a clean tip, most cutting problems can be traced to incorrect ______, ______, ______, or ______.
5. Cutting pipe is much like cutting ______.

UNIT

24 Brazing and Welding with Oxyacetylene

OBJECTIVE

To braze and weld safely with oxyacetylene equipment.

COMPETENCIES TO BE DEVELOPED

After studying this unit, you should be able to:

- Explain the nature and uses of braze welding.
- Prepare metal for welding.
- Select fluxes for welding.
- Identify joints commonly used in welding.
- Braze and braze weld butt, lap, and fillet joints.
- Fuse weld mild steel with and without filler rod.

TERMS TO KNOW

soldering
brazing
capillary action
filler rod
braze welding
fusion welding
oxide
oxidation
impurity
flux
play the flame
tinning
bead
puddle
butt weld
tacking
fillet weld
root
solidify
burn-through
puddle
pushing the puddle
back-stepping

MATERIALS LIST

- ✓ Oxyacetylene rig, welding table, and protective clothing
- ✓ Welding torch with No. 1-4 tips
- ✓ Spool of 50-50 solder
- ✓ Rosin flux
- ✓ Brazing flux
- ✓ Brazing rod, about 1/16 inch in diameter
- ✓ Welding rod, about 1/16 inch in diameter
- ✓ Steel plate (not galvanized), 16 gauge × 6" × 16"
- ✓ Steel plate (not galvanized), 1/8" × 6" × 10"
- ✓ Copper pipe or tubing, 3/8-inch or 1/2-inch, or sheet copper
- ✓ Power wire wheel or grinder
- ✓ Fine emery cloth

BASICS OF SOLDERING, BRAZING, AND FUSION WELDING

The term *welding* may be defined as uniting metal parts by heating or compression. Early American blacksmiths welded wagon rims and other parts by heating and hammering. Parts were heated to a certain color, and a joint was formed by pounding the parts together using a hammer and anvil.

Soldering, brazing, and braze welding have much in common. All three processes are done without melting the base metal. All three use a metal or alloy as the bonding agent.

Soldering

The term soldering means bonding with metals and alloys that melt at temperatures *below* 804°F. The most common soldering material for non-plumbing use is 50-50 solder, consisting of 50 percent tin and 50 percent lead. Solder used for water lines cannot contain lead because of its toxic qualities. Other alloys are available for special applications, figure 24-1. Tin-lead soldering is generally done with electric irons, soldering guns, and propane torches. Common applications of soldering are to join electrical wires, join copper gutters and spouts, sweat copper pipes, and fasten thin tin-plated steel. The basic procedure for soldering is covered in Unit 13, Fastening Metal.

Brazing

The term brazing means bonding with metals and alloys that melt at or above 840°F when capillary action occurs, figure 24-2. The term also refers to joining parts that are fitted extremely well. Only a very thin layer (.025 inch or less) of alloy is needed to fill the void between parts, figure 24-3. This small spacing allows the alloy to be drawn into the joint by capillary action. Capillary action is the rising of the surface of a liquid that is in contact with a solid.

ALLOY	USE ON
Tin-Lead	Copper and copper alloys Mild steel Galvanized metal
Tin-Antimony	Copper and copper alloys Mild steel
Cadmium-Silver	High strength for copper and copper alloys Mild steel Stainless steel
Cadmium-Zinc	Aluminum and aluminum alloys

FIGURE 24-1. Alloys used for soldering. (*Courtesy of Larry Jeffus,* Welding Principles and Applications, Third Edition, *Delmar Publishers, 1995)*

BASE METAL	BRAZING FILLER METAL
Aluminum	BAISi, aluminum silicon
Carbon Steel	BCuZn, brass (copper-zinc) BCu, copper alloy BAg, silver alloy
Alloy Steel	BAg, silver alloy BNi, nickel alloy
Stainless Steel	BAg, silver alloy BAu, gold base alloy BNi, nickel alloy
Cast Iron	BCuZn, brass (copper-zinc)
Galvanized Iron	BcuZn, brass (copper-zinc)
Nickel	BAu, gold base alloy BAg, silver alloy BNi, nickel alloy
Nickel-copper Alloy	BNi, nickel alloy BAg, silver alloy BCuZn, brass (copper-zinc)
Copper	BCuZn, brass (copper-zinc) BAg, silver alloy BCuP, copper-phosphorus
Silicon Bronze	BCuZn, brass (copper-zinc) BAg, silver alloy BCuP, copper-phosphorus
Tungsten	BCuP, copper-phosphorus

KEY: B = Brazing, Al = Aluminum, Si = Silicon, Cu = Copper, Zn = Zinc, Ag = Silver, Ni = Nickel, Au = Gold, P = Phosphorus

FIGURE 24-2. Base metals and the common brazing filler metals used for each. (*Courtesy of Larry Jeffus,* Welding Principles and Applications, Third Edition, *Delmar Publishers, 1995)*

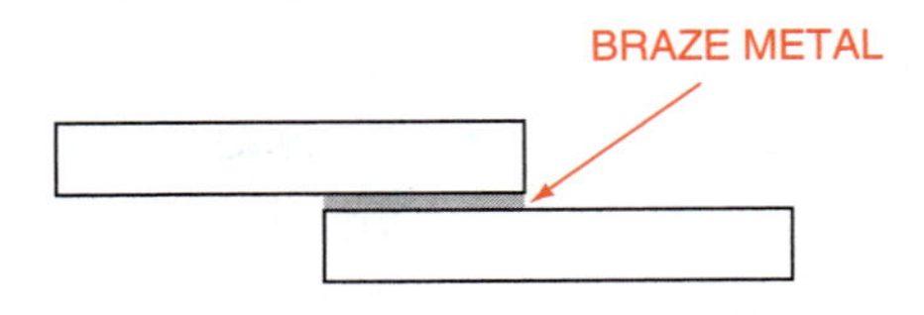

FIGURE 24-3. A very thin layer of braze metal joins the two pieces of base metal.

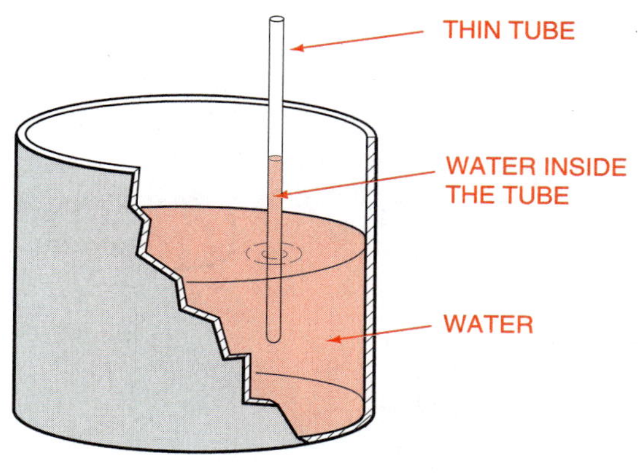

FIGURE 24-4. Capillary action pulls water up a thin tube.

Capillary action may be observed by placing a thin tube in a container of water, figure 24-4.

The temperature of properly prepared metal must first be raised to the melting point of the brazing alloy to be used. The brazing alloy is then added by means of a long, thin metal rod called a filler rod. The most popular filler rods for brazing and braze welding are copper-zinc alloys. The alloy is drawn between the parts and spreads throughout the spaces that are narrow enough for capillary action to work. This principle makes brazing an attractive method for repairing broken castings.

Braze Welding

Braze welding refers to bonding with alloys that melt at or above 840°F when capillary action does *not* occur. Here the alloy is bonded to each part, and the void between or around the part is filled with the melted alloy, figure 24-5. Since many applications require brazing as well as braze welding, the term brazing is frequently used for both operations.

Advantages of Soldering and Brazing

The advantages of soldering and brazing are that they (1) are low temperature processes, (2) permit easy disassembly, (3) allow different metals to be joined, (4) can be done at high speed, (5) do little damage to parts, (6) permit easy realignment of parts, and (7) permit parts of various thicknesses to be joined.

Fusion Welding

The term fusion welding means joining parts by melting them together. The base metals are melted and mixed together to form a continuous piece of the same kind of metal. Fusion welding can be done with or without a filler rod. A sawed cross section of a skillfully made fusion weld will show such complete fusion that it looks like a single piece of metal.

BRAZE METAL

FIGURE 24-5. Braze welding bonds pieces together and fills the voids between pieces with the braze metal.

Fusion welding is done extensively on steel. The strength of steel is such that only fusion welding is acceptable for most work. Oxyfuels are used only for welding on very thin steel. Most other fusion welding of steel is done using an electric arc welder as the source of heat.

Preparing Metals for Welding

Metals must be clean before soldering, brazing, or welding can take place. This means that oxides and other impurities must be removed from the metal. An oxide is the product resulting from oxidation of metal. Oxidation means combining with oxygen. Rust on iron and steel is an oxide. The dull coatings that form on brass and copper are also oxides.

Metal may be cleaned for soldering or welding by using a wire brush, wire wheel, grinder, sander, steel wool, file, or other mechanical process.

After cleaning metal by mechanical means, chemicals are used to remove any impurities that remain. An impurity is any product other than the base metal.

The term flux is given to any chemical used to clean metal. Fluxes remove tarnish and corrosion caused by oxidation. They also prevent corrosion from developing. Additionally, they promote wetting or movement of molten alloys on metal. Finally, fluxes aid in capillary action which moves molten alloys into very small cracks. Most fluxes work best when hot. If the metal gets too hot, however, the flux will burn and must be cleaned off before proceeding. Fluxes are available as solids, pastes, powders, liquids, sheets, rings, and washers, figure 24-6. They may also be in the hollow core or on the surface of filler rods.

Using Flux. In order to control the heat needed to manage the fluxing action, the following procedure is recommended.

PROCEDURE

1. Obtain a piece of copper pipe, tubing, or sheet. The copper should be clean but not shiny.
2. Set up, light, and adjust a small torch.
3. Put a pea-sized spot of rosin flux on the copper.

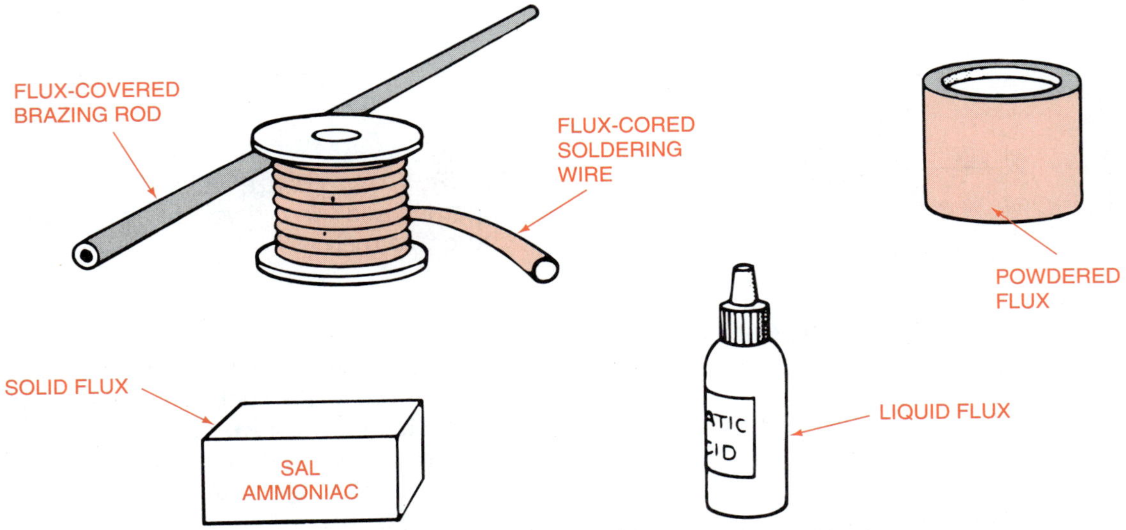

FIGURE 24-6. Fluxes come in many forms. (*Adapted from Larry Jeffus,* Welding Principles and Applications, Third Edition, *Delmar Publishers, 1995)*

4. Heat the flux and surrounding area by playing the flame on the rosin. To play the flame means to alternately move a flame into and out of an area to achieve temperature control.
5. Observe the point when the copper suddenly becomes shiny under the flux. Remove the heat. The shiny surface indicates that the flux has reacted with the oxide and removed it, figure 24-7.
6. Heat an area near the shiny spot. The flux should flow to the newly heated area and clean it, too.
7. Observe the effect of too much heat by continuing to heat the clean areas. Notice that too much heat will burn the flux and the area will turn dark or black.
8. Remove the burnt flux with steel wool or fine sandpaper.

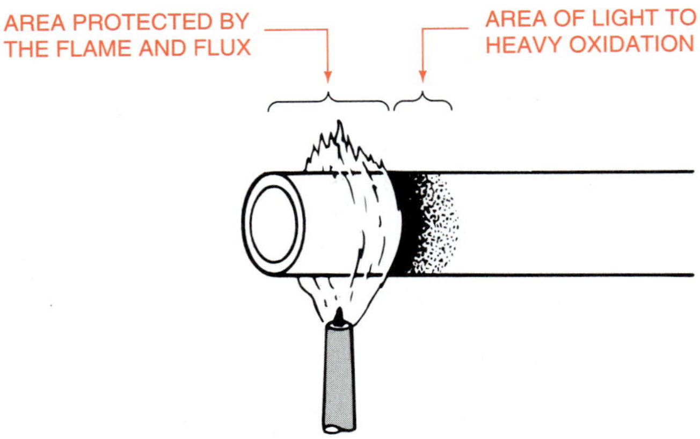

FIGURE 24-7. Fluxing copper pipe.

9. Reapply the flux and reheat. Note that the fluxing action can be repeated only after mechanical removal of the burnt flux. It is important to recognize fluxing action. It is necessary to learn how to control the fluxing process. Otherwise, soldering or brazing fillers will not bond to the metal.

Tinning. Tinning is the key to success in soldering and brazing. The term tinning means to bond filler material to a base metal. The term probably originated from the tinlike appearance of solder when it is applied to a base metal. Before two pieces of metal can be joined by soldering, brazing, or braze welding, both pieces must be tinned. Most unsuccessful attempts at soldering fail at this point. Learn to tin before attempting to solder, braze, or braze weld.

CAUTION:

- **Never braze or weld galvanized steel without special training and special precautions. Galvanized coatings give off toxic fumes when melted or heated above soldering temperatures.**

The following procedure is suggested for tinning.

PROCEDURE

A. Tinning with Solder

1. Obtain a piece of copper pipe or sheet.
2. Clean the metal with fine emery cloth.

3. Obtain a spool of tin-lead solder, paste flux, and a damp cloth.
4. Light and adjust an oxyfuel torch with a small tip and neutral flame.

   NOTE: A propane torch without oxygen may also be used.

5. Heat the copper until the fluxing action starts.
6. Touch the end of the solder to the hot metal. Feed the solder down to the metal as it flows across the fluxed area.
7. Use the torch to heat an enlarged area and attract flux and solder across the metal.
8. Play the torch over the tinned area to get a thin, smooth, and shiny surface.
9. To improve the appearance and to remove excess solder, wipe the solder-covered area quickly with a damp cloth. Note that the damp cloth removes excess solder and leaves a thin coating of solder bonded to the copper.

B. Tinning with Braze Filler

1. Obtain a piece of nonrusted steel plate approximately 1/8" × 2" × 6".
2. Clean the plate carefully with fine emery cloth.
3. Obtain a brazing filler rod and can of brazing flux.
4. Light and adjust an oxyfuel torch using a small tip and neutral flame.
5. Place the plate on firebricks.
6. Heat the center of the plate until it is dull red.

CAUTION:

- **Do not permit the steel to melt.**

7. Heat the end of a braze filler rod by placing it in the flame as the base metal is heating.

   NOTE: The operator controls the torch with one hand. The other hand holds the filler rod like a pencil, about 12 inches from the end.

8. With the tip of the filler rod heated, dip it into the can of flux, figure 24-8.
9. Place the flux-covered end of the rod into the flame and touch the hot base metal.
10. When the flux reaches the correct temperature, fluxing action occurs (the chemical flows across the hot metal and removes impurities).

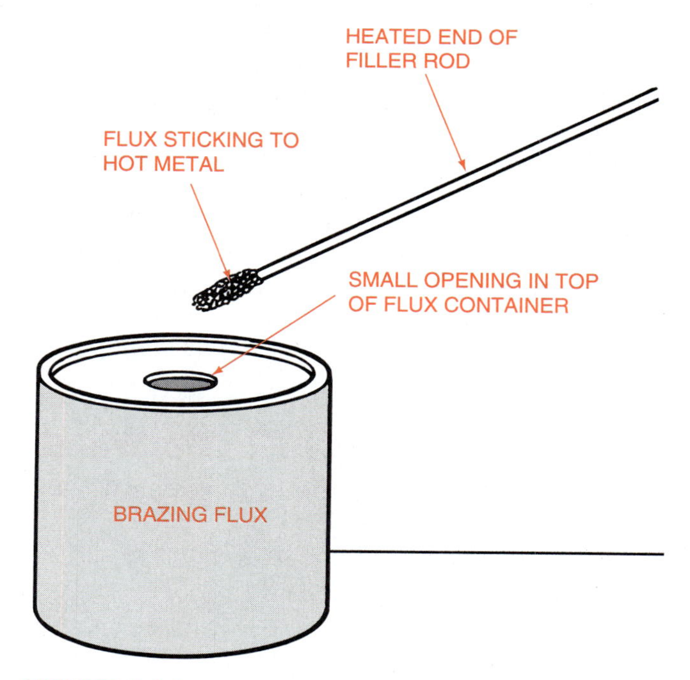

FIGURE 24-8. The heated end of a filler rod can be used to transfer flux from the container to the base metal.

11. Once fluxing occurs, watch for the end of the filler rod to melt. Then play the flame on the filler metal, flux, and plate, causing the filler to flow as you create the correct temperature on the base metal.
12. Play the heat over the entire area until the filler material flows into a thin layer of bonded metal.

CAUTION:

- **Do not overheat the filler material and base metal or the metals will turn dark from oxidation and will require mechanical cleaning again.**

If difficulty is encountered with the process, clean the metal completely and start again. The critical task is to keep the filler rod heated very nearly to its melting point. Then, when it is needed for fluxing or filling, the operator can easily control these processes. Quick melting of the filler rod is achieved by touching the base metal with the rod while playing the torch on the rod.

Controlling Heat. Controlling heat is the key to successful soldering and brazing. Metals are excellent heat conductors. Therefore, heat applied in one place soon moves to surrounding areas. The larger the piece of metal, the faster the heat gets away from the heated area. This must be kept in mind when joining two pieces of metal. If the torch directs an equal amount of heat onto two pieces of different size, the smaller piece will get hot faster. When this occurs, the flux will burn on the small piece before the large piece is up to fluxing temperature.

The operator must manipulate the torch to heat pieces of different size equally. The following procedure is helpful.

PROCEDURE

1. Obtain one piece of steel 1/8" × 2" × 2" and one 16 gauge x 2" × 2".
2. Place the two pieces end to end on a firebrick.
3. Light and adjust the torch.
4. Play heat on the joint and in a circular area covering both pieces until both are dull red.

The flame must be directed to the thick piece more than to the thin one. Practice with pieces of various sizes until you can make both pieces turn dull red at the same time. This skill is essential for soldering, brazing, and welding.

BRAZING AND BRAZE WELDING PROCEDURES

Brazing and braze welding are generally done in agricultural mechanics using an oxyfuel torch. Acetylene is perhaps the most widely used fuel. However, acetylene has the disadvantage of producing a very hot inner flame and a relatively cold outer flame. For the brazing and braze welding, a hot outer flame is useful for heating large areas uniformly. Such fuel gases as MAPP®, propane, butane, and natural gas have the advantage of providing more uniform heating.

Heating of large areas is necessary when entire assemblies, such as pipe fittings or large castings, are being bonded. A good way to control heat when brazing large assemblies is to first heat the basic assembly until it is near the melting point of the braze filler. It is fairly easy then to add the additional heat required to melt the braze metal and fill the voids in specific locations.

Torch Tips. Tips used for oxyfuel welding have a single hole. They are identified by a number which designates the size of the hole. Tip numbers range from 000 (the smallest) to 10 or larger. A small tip is used for thin metal. The tip size must be larger as the thickness and mass of the metal increases. The torch manual should provide proper guidelines for tip selection.

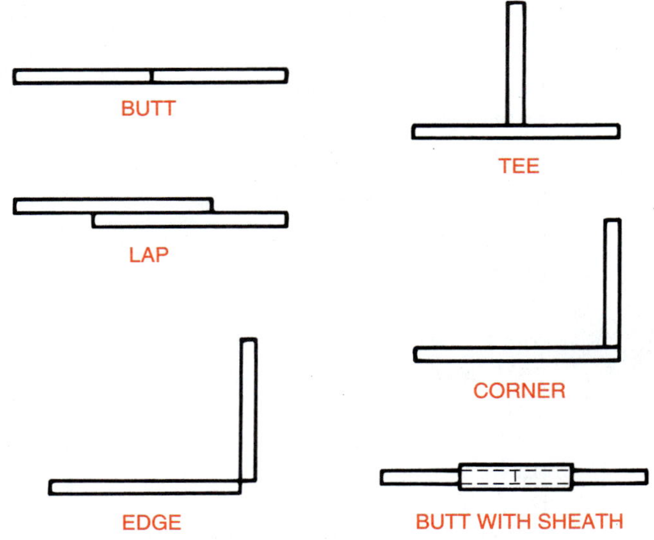

FIGURE 24-9. Some common types of joints used to fasten metal.

Torch Adjustment. A good general rule for torch adjustment is that most soldering, brazing, and welding is done with a neutral flame. Only specialized applications required slightly carburizing or slightly oxidizing flames.

Brazing

Since brazing works by capillary action, it requires that parts fit together very closely. Some common types of joints used to fasten metal are shown in figure 24-9. The following procedure is suggested for brazing flat material in a lap joint.

PROCEDURE

1. Obtain three perfectly flat and clean pieces of nongalvanized steel plate 1/8" × 2" × 2".
2. Clean a 1-inch strip along one edge of each of two pieces.
3. Place these two pieces on firebricks so that clean, 1/2-inch strips overlap one another. Use the third piece to support the top pieces, figure 24-10. If possible, clamp the assembly or hold the top piece in place with a brick.
4. Obtain brazing rod and flux.
5. Light and adjust the torch.
6. Apply heat to the clean joint until both pieces are dull red.
7. Add flux to the joint by way of the heated filler rod.

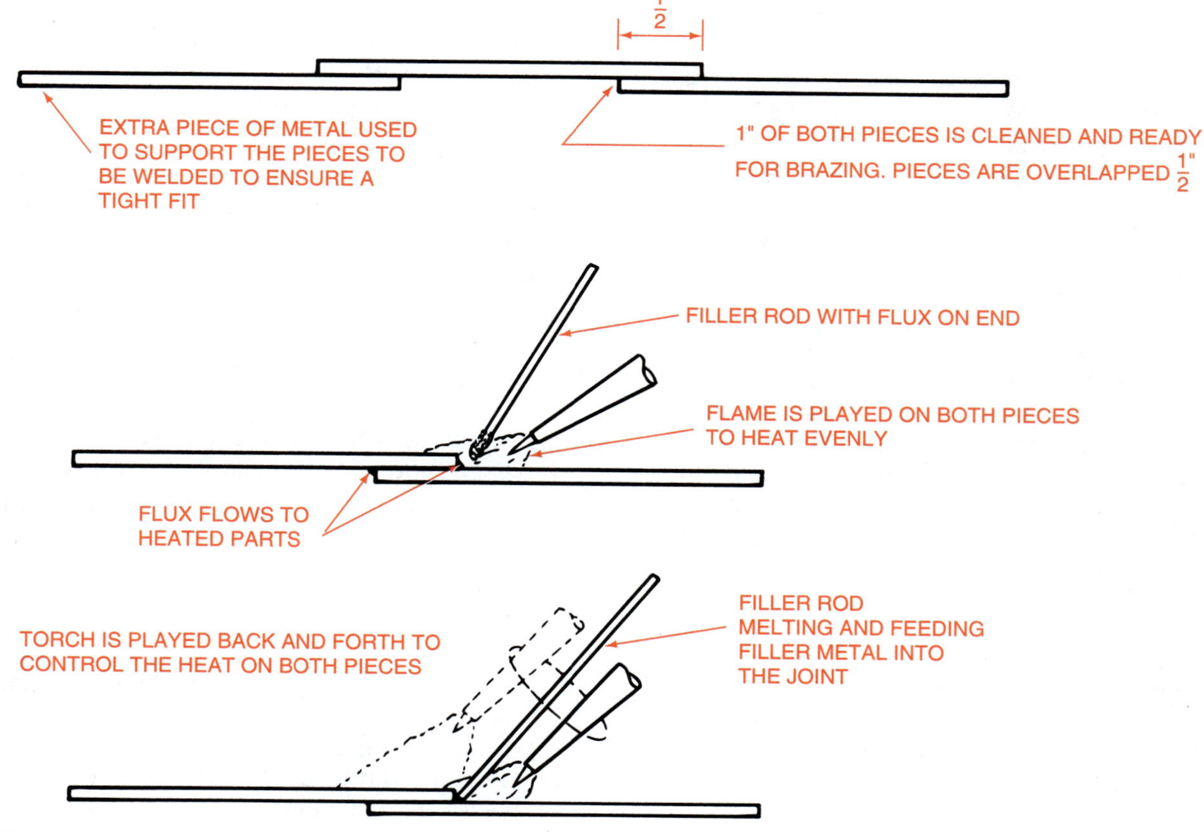

FIGURE 24-10. Brazing a lap joint.

8. Continue to raise the temperature of the assembly and move the filler rod in and out of the flame.
9. When both pieces are dull red, touch the metal with the rod and feed filler into the joint.

NOTE: Bend a loop into the cold end of the brazing rod before using.

10. Play heat more on the top piece and back from the edge to draw the filler into the joint.
11. Let the assembly cool.
12. Turn the assembly over and examine it for completeness of brazing. Filler metal should be present on the back edge of the joint.

Cast iron assemblies of irregular shapes may be brazed. Appropriate fluxes must be used to correct the presence of free graphite in some iron. Additionally, a large torch is needed to heat large assemblies. Brazing may be preferred over fusion welding of cast iron since there is less likelihood of heat distortion and breaking of the casting.

Braze Welding

When braze welding, many of the principles of brazing apply. The major difference is that a perfect fit is not needed. Still, good fits generally result in stronger joints.

Braze welding is useful on metal that is 16 gauge or thinner. Thicker metal can be braze welded, but electric arc welding is faster and easier and generally the preferred process for thicker metals.

Running a Bead. Most welding requires the ability to run a bead. A bead is a continuous and uniform line of filler metal. To run a bead with a braze filler rod, the following procedure is suggested.

PROCEDURE

1. Obtain a clean piece of 1/8" × 2" × 6" steel.
2. Clean both sides with fine emery cloth.
3. Light and adjust the torch.
4. Position the metal on a flat firebrick. The bead will be run across the metal 1/2 inch in from one end.
5. Heat the end of a filler rod and a spot on one edge of the plate about 1/2 inch from the end.

6. Dip the hot rod into flux and touch the fluxed rod to the heated spot.
7. When the rod starts to melt, move it in and out of the flame to add filler metal as needed.
8. The bead is formed by depositing filler material along a fluxed path across the metal.
9. Run additional beads across the metal at 1-inch intervals until you can run complete, straight, and even beads.

NOTE: When running a bead, it helps to move the torch in a circular pattern.The inner cone is kept about 1/8 inch from the molten puddle. A puddle is a small pool of liquid metal. Heat is controlled by the clearance of the cone, the angle of the flame, and the movement of the rod into and out of the puddle. The filler rod must be dipped into the flux frequently. This provides flux to cover the new bead as it forms.

Braze Welding Butt Joints. A butt weld is essentially a bead laid between two pieces of metal set edge to edge. This may be end to end or side to side. When braze welding a butt joint, it is important that the filler metal is down between the pieces and bonds the edges at all points. To achieve this, the two pieces are placed with a slight gap between them. To braze weld a butt joint, the following procedure is suggested.

PROCEDURE

1. Cut two pieces of nongalvanized steel 16 gauge × 2" × 2".
2. Clean the edges and faces with fine emery cloth.
3. Place them edge to edge on a flat firebrick with a gap between them of about 1/32 inch.
4. Weight the two pieces with firebricks to help hold them in position.
5. Heat a spot at one end of the joint. Flux the area and deposit a spot of filler material.
6. Repeat step 5 at the other end of the joint. These small welds are called tacks. The process of making a small weld to hold two pieces together temporarily is called tacking.
7. Lay a bead over the entire length of the joint, figure 24-11.

Braze Welding Fillets. A fillet weld is a weld placed in a joint created by a 90 degree angle. The joint may be in the form of an L or a T.

When making a fillet weld, the torch must be played on the flat and vertical pieces with care. If both pieces are equal in thickness and mass, the heat is applied equally. However, if one piece is thicker, that piece must receive more heat, figures 24-12 and 24-13.

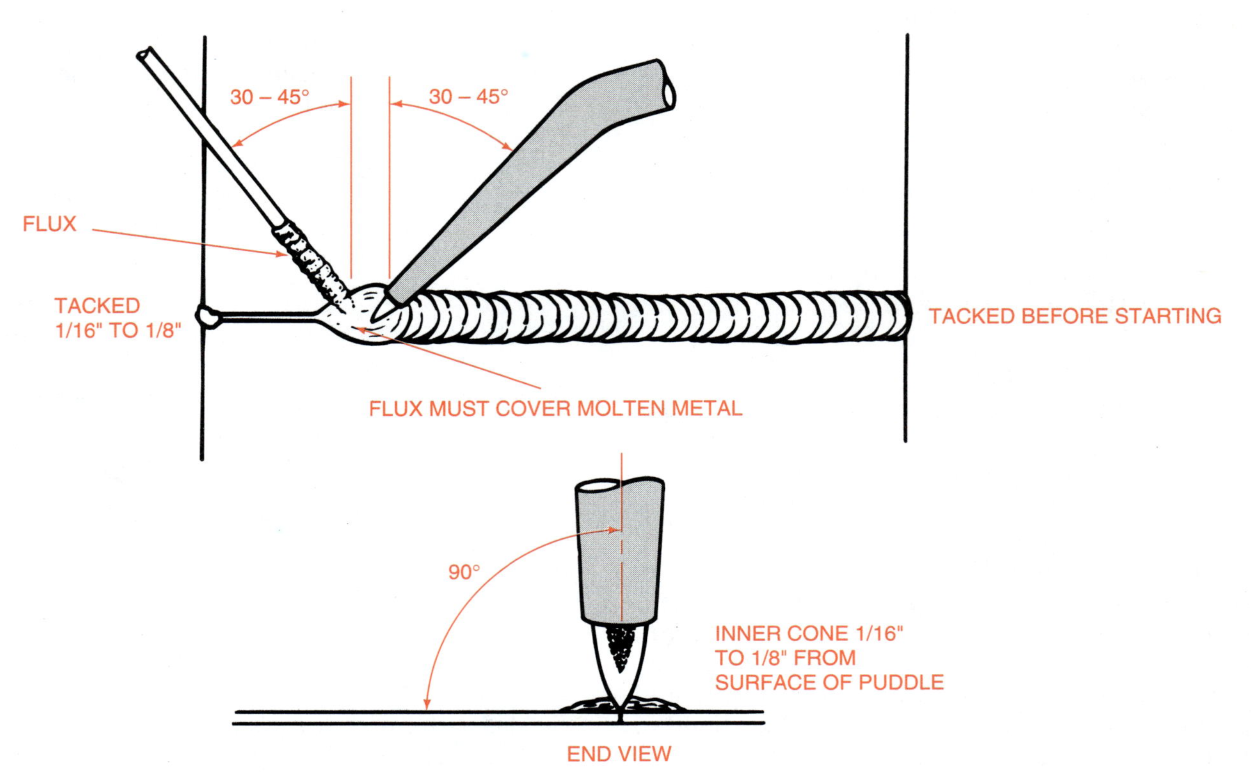

FIGURE 24-11. Braze welding a butt joint.

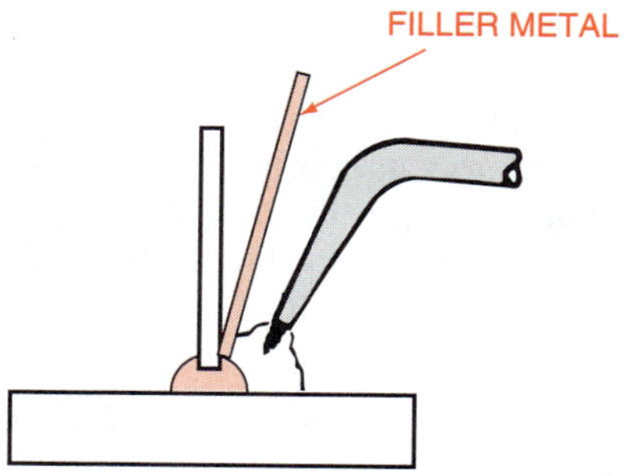

FIGURE 24-12. Torch and rod positions required to balance heat when the base is thicker than the vertical piece.

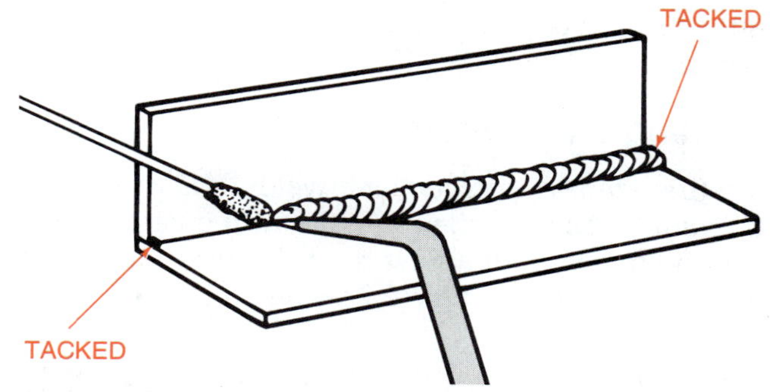

FIGURE 24-14. Making a fillet weld on pieces of equal thickness.

Special care must be taken to ensure that sufficient heat reaches the root of the weld. The root is the deepest point in a weld. In this case, it is the place where the two pieces melt to form an angle. A braze welded fillet should have the filler material under the vertical piece and a bead along the 90 degree corner, figure 24-14.

FUSION WELDING WITH OXYACETYLENE

Fusion welding of steel with oxyacetylene is generally limited to thin metal or small jobs where portability of equipment is a factor. Electric arc welders are now so versatile that they account for most fusion welding done in agricultural mechanics. Still, it is useful to be able to do small jobs and weld thin metal with oxyacetylene.

Factors Affecting the Weld

The Tip. Selection of the torch tip will depend upon the width and penetration of the bead desired. The larger the tip, the more heat the flame produces. The heat from a given tip is controlled by the distance the cone is held from the puddle and the angle of the tip to the work. The greatest melting ability is obtained when the angle is 90°. The ideal angle is 45°. The ideal cone distance is $^1/_8$ to $^1/_4$ inch from the puddle.

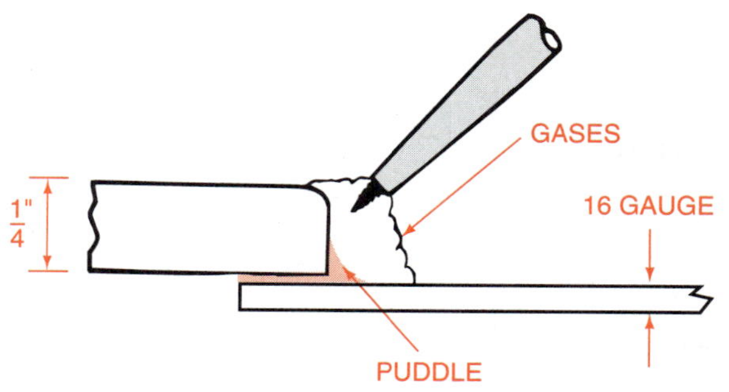

FIGURE 24-13. Torch position for directing heat to a vertical surface with a greater mass than the base.

The Rod. The filler rod for fusion welding should be made of the same material as that being welded. Remember, fusion welding is a melting and mixing of base metals. The filler rod adds volume so the joint will be filled and slightly reinforced, or strengthened. The larger the rod, the greater is its tendency to cool the molten puddle, increase the size of the bead, and reduce penetration.

CAUTION:

- **Always bend one end of the filler rod to form a loop. The loop identifies the unheated end and decreases chances of those in the area being hit by the sharp end or burned by the hot end.**

Controlling the Process

The gases in the air will contaminate, or pollute, an unprotected welding puddle. Fortunately, the presence of the torch flame prevents air from getting to the molten puddle. However, this makes it necessary to keep the torch on or nearly on the puddle at all times. To complete a weld properly, the torch is lifted slowly so the puddle can cool and solidify without contamination. To solidify is to turn from a liquid to a solid.

During the welding process, some components of the base metal burn and produce sparks. Changes in welding temperature can be detected by observing the amount of sparks given off. Hence, the alert welder can detect when a burn-through is about to happen and reduce the heat to prevent it. Burn-through is the process whereby the cool side of metal becomes molten and a hole opens in the surface due to excessive heat from the opposite side. A burn-through is generally preceded by an increase in spark activity.

Pushing the Puddle

The first step in oxyacetylene welding is to control the puddle. This is called puddling. To create a bead by puddling without a filler rod is called pushing the puddle and is done by the following procedure.

PROCEDURE

1. Obtain a clean piece of 16 gauge × 2" × 6" steel.
2. Light and adjust a torch with a number 2 or 3 tip, using 3 psi acetylene pressure and a neutral flame.
3. Place the strip of steel on a firebrick, and weight it with another firebrick.
4. Start at one end of the metal with the torch tip turned toward the plate and a gap of 1/8 inch from cone to metal.
5. When the metal starts to melt, move the torch in a small circular pattern down the plate to create a bead.
6. Lift the tip slowly at the end of the bead. Make additional practice beads until the metal is covered.

The puddle should move ahead as the torch moves down the plate. A bead will form as the flame passes. If the size of the pool changes, speed up or slow down, figure 24-15. The size of the pool should be kept constant.

The ideal torch angle is 45°. As the angle of the torch increases up to 90°, the bead gets wider and wider. If the angle is too flat, such as 30°, the bead will be too narrow, figure 24-16.

Making A Corner Weld without Filler Rod

A corner weld is one made on edges laid together to form a corner, figure 24-17. To practice making a corner weld, use the following procedure.

PROCEDURE

1. Obtain two pieces of clean 16 gauge × 1 1/2" x 6" steel.
2. Place the pieces on their edges on a firebrick to form a tentlike shape with a 60 degree angle.
3. Tack weld the joint at both ends and two places in between.
4. Make a weld bead along the entire corner.
5. Evaluate your speed and torch angle by comparing your results with figure 24-16.
6. Open the joint, examine the penetration, and discuss the results with your instructor.
7. Grind off the rough edges and save the metal for the next exercise.

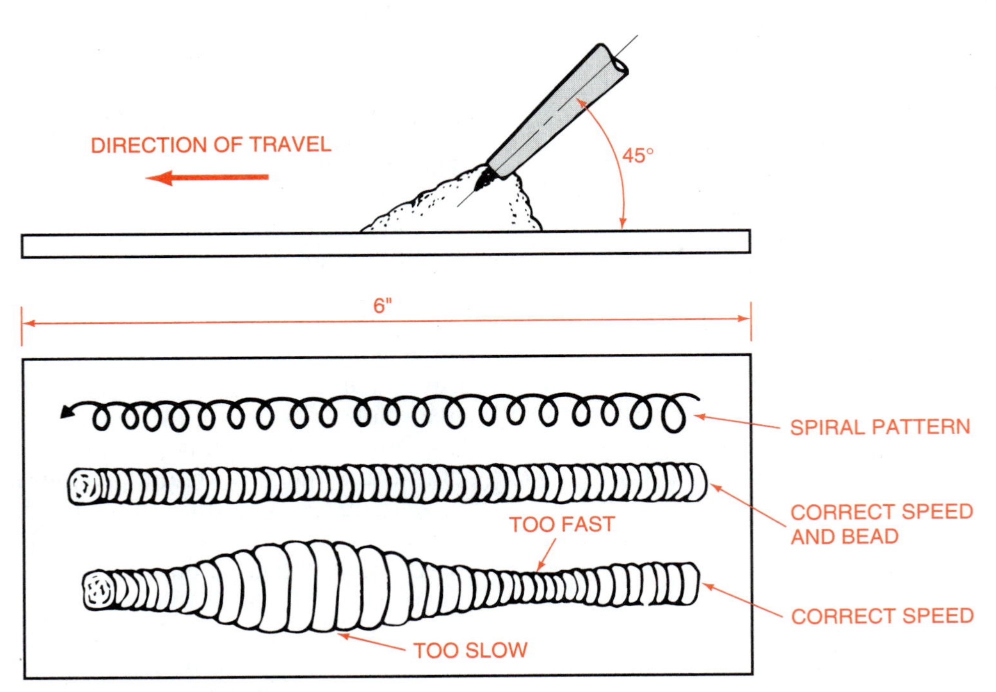

FIGURE 24-15. The torch speed affects the size of the bead.

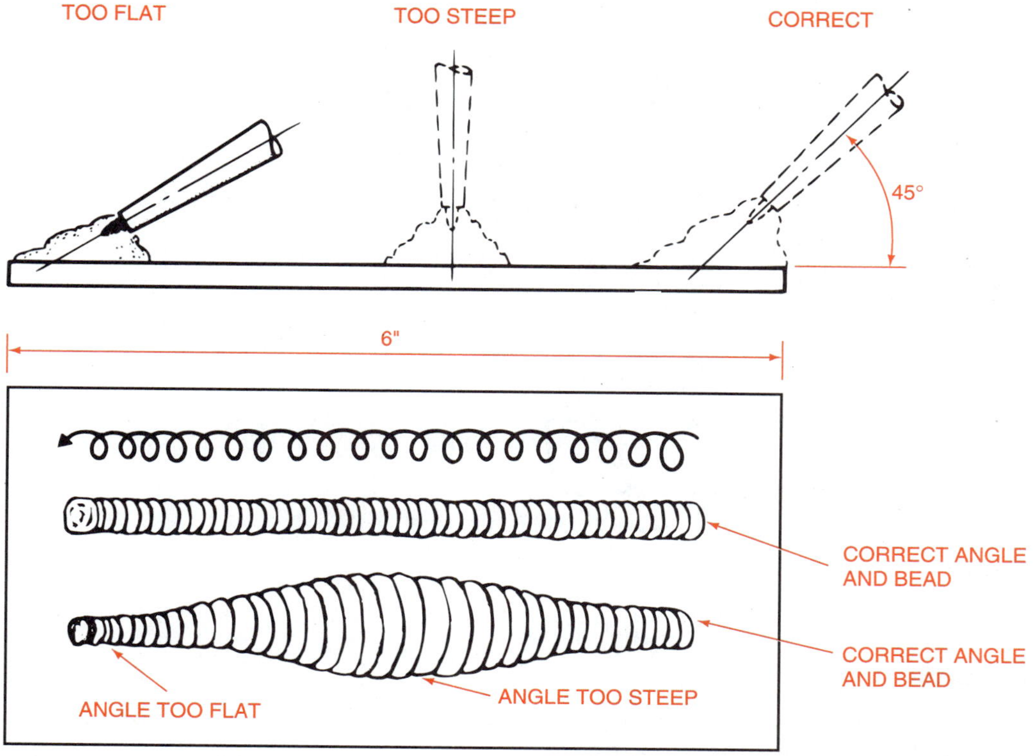

TOO FLAT
TOO STEEP
CORRECT
45°
6"
CORRECT ANGLE AND BEAD
CORRECT ANGLE AND BEAD
ANGLE TOO FLAT
ANGLE TOO STEEP

FIGURE 24-16. The torch angle also affects the shape of the bead.

Laying a Bead with Filler Rod

The movements for laying a bead with a filler rod in fusion welding are similar to those used in brazing. Practice laying a bead with a filler rod in a butt weld as follows.

INNER CONE 1/16 TO 1/8 INCH FROM SURFACE OF PUDDLE
45°
TACKED
TACKED
60°

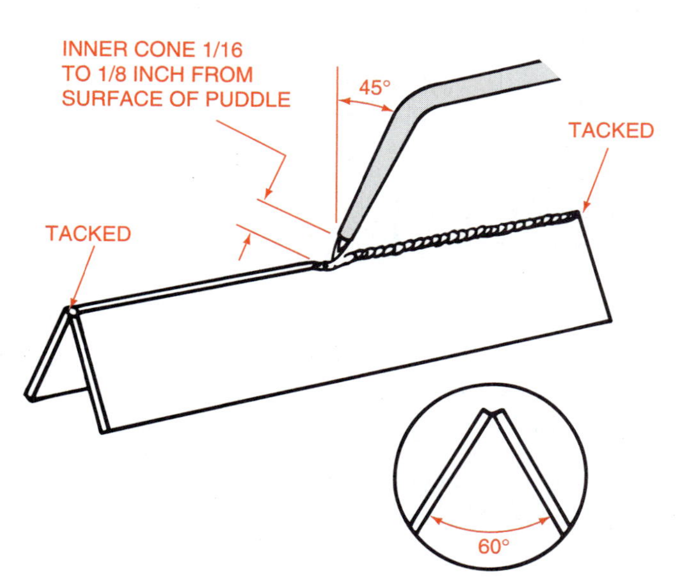

FIGURE 24-17. Making a corner weld without a filler rod.

PROCEDURE

1. Use the two pieces of 16 gauge × 1 1/2" ×6" metal left from the previous exercise.
2. Place the two pieces edge to edge on a firebrick to form a 6-inch-long butt joint.
3. Adjust the pieces until there is a 1/16-inch gap between them. Weight them with firebricks.
4. Tack weld at both ends of the joint and at two places in between.
5. Point the torch toward one end of the joint at an angle of 45°.
6. Establish a molten puddle with a 1/16- to 1/8-inch cone clearance.
7. Add filler metal to the front edge of the puddle.
8. Rotate the flame and move the rod in and out to create an even bead, figure 24-18.
9. Complete the bead by slowly raising the torch while filling the puddle as you withdraw the torch.
10. Discuss the resulting weld with your instructor.

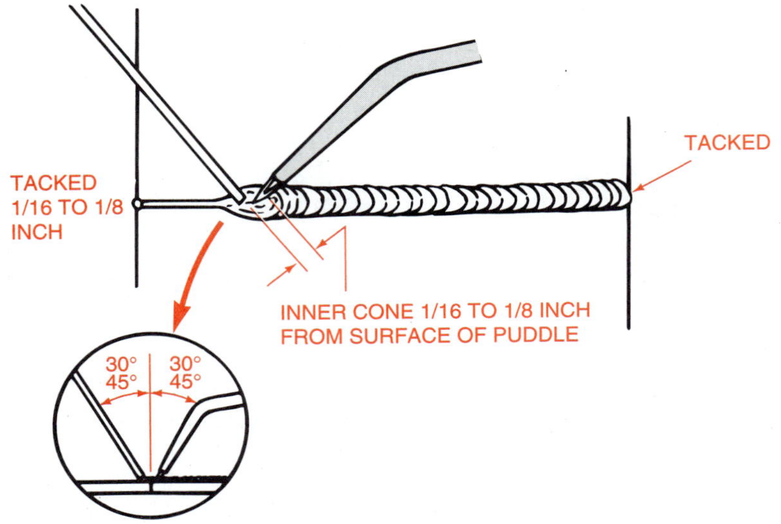

FIGURE 24-18. Laying a bead to form a butt weld.

The Finished Weld

The completed weld should be thoroughly fused to the base metal throughout the joint, figure 24-19. The weld metal should penetrate to the root of the joint with a small amount extending below the surface to assure a full section weld. At the face of the weld, there is usually a buildup known as "reinforcement." This reinforcement should be slight, blending smoothly with the base metal surfaces. It is especially important to avoid undercutting or overlapping at the juncture of the weld and base metal.

Bending or distortion of the metal can be reduced by back-stepping. Back-stepping is making short welds in the backward direction as the process

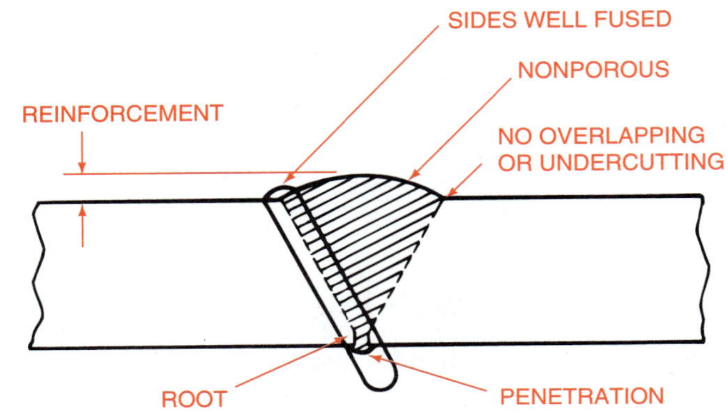

FIGURE 24-19. Characteristics of a proper fusion weld.

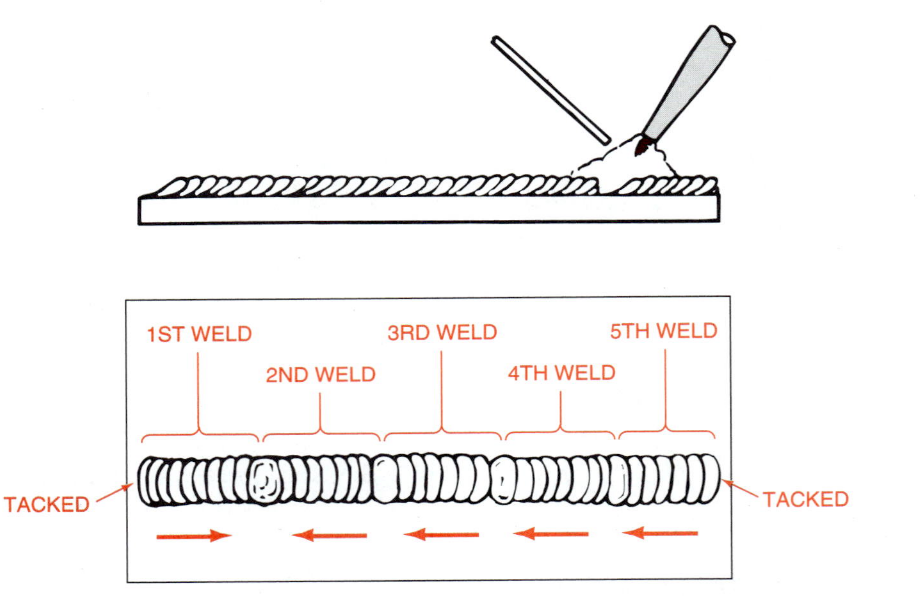

FIGURE 24-20. Distortion can be controlled by back-stepping. The first weld is made from left to right, but all subsequent welds are made from right to left.

progresses in the forward direction, figure 24-20. Repeat the previous exercise using the back-step method.

Making a Fillet Weld

A fillet weld is made by laying a bead in a 90 degree angle formed by two pieces of metal. To make a fillet weld, the following procedure is recommended.

PROCEDURE

1. Obtain two pieces of clean 16 gauge × 1½" × 6" steel.
2. Place one piece flat on firebricks. Set the other piece on its edge on the first piece to form a 90 degree angle.
3. Prop the assembly with firebricks.
4. Tack weld both ends and the middle.
5. Lay a bead along the entire joint, figure 24-21.

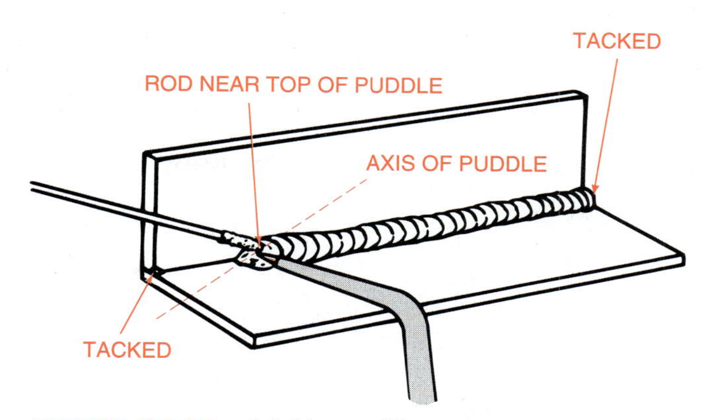

FIGURE 24-21. Making a fillet weld.

Other joints can be welded using the procedures outlined in this unit. The student is encouraged to learn well all basic procedures presented.

STUDENT ACTIVITIES

1. Define the Terms to Know for this unit.
2. Using a torch and flux, heat a piece of copper and practice fluxing copper without overheating the flux.
3. Using a torch, flux, 50-50 solder, and copper, tin the copper surface.
4. Using an oxyfuel torch, brazing filler rod, flux, and 1/8" × 2" × 6" steel plate, tin the plate without burning the flux.
5. Prepare an oxyfuel torch, a piece of 1/8" × 2" × 2" steel and a piece of 16 gauge × 2" × 2" steel. Place the pieces side by side and heat both at the same time to an even red color.
6. Prepare a torch, flux, brazing filler rod, and 1/8" × 2" × 6" steel plate. Practice laying braze weld beads across the plate at 1/2-inch intervals.
7. Prepare a torch, flux, brazing filler rod, and three pieces of 1/8" × 2" × 2" steel plate. Use one plate for support and braze a 1/2-inch lap joint between the other two.
8. Prepare a torch, flux, brazing filler rod, and two pieces of 16 gauge × 2" × 2" steel. Butt weld the two pieces together.
9. Prepare a torch, flux, brazing filler rod, and two pieces of 16 gauge × 1" × 2" steel. Lay one piece flat on a firebrick. Bend the other piece slightly and place it on its side on top of the first piece. Braze weld the two pieces to form a fillet weld.
10. Prepare a torch, welding flux, and one piece of 16 gauge × 1 1/2" × 6" steel. "Push the puddle" to form three beads across the steel without filler rod.
11. Prepare a torch, welding flux, and two pieces of 16 gauge × 1 1/2" × 6" steel. Prop the two pieces edge to edge to form a 60 degree angle. Make a corner weld over the ridge of the assembly without a filler rod.
12. Prepare a torch, flux, 1/16-inch welding filler rod, and two pieces of 16 gauge × 1 1/2" × 6" metal. Use the filler rod to butt weld the two pieces together.
13. Prepare a torch, flux, 1/1- inch welding filler rod, and two pieces of 16 gauge × 1 1/2" × 6" steel. Join the two pieces to form a T using a fillet weld.
14. Prepare a torch, flux, 1/16-inch welding filler rod, and two pieces of 16 gauge × 1 1/2" × 6" steel. Weld a lap joint.

SELF-EVALUATION

A. Multiple Choice. Select the best answer.

1. Welding is
   a. uniting
   b. heating
   c. fusion
   d. all of these
2. Brazing is much like
   a. soldering
   b. painting
   c. fusion welding
   d. arc welding
3. The most popular soldering alloy is
   a. aluminum silicon
   b. copper-zinc
   c. silver
   d. tin-lead
4. The most popular brazing alloy is
   a. aluminum
   b. copper-zinc
   c. silver
   d. tin-lead
5. Rust is a form of
   a. dirt
   b. galvanize
   c. oxidation
   d. weathered paint
6. Metal may be cleaned before brazing with a
   a. brush
   b. flux
   c. emery cloth
   d. all of these
7. Two pieces lying flat and end to end may be joined by a
   a. butt weld
   b. corner weld
   c. fillet weld
   d. lap weld
8. Bonding solder or braze material to a piece of metal is called
   a. brazing
   b. fusion
   c. soldering
   d. tinning
9. The best distance between cone and puddle when fusion welding is
   a. $^1/_2$ inch
   b. $^1/_4$ inch
   c. $^1/_8$ inch
   d. $^1/_{32}$ inch
10. The best torch angle for flat welding is
    a. 30°
    b. 45°
    c. 70°
    d. 90°

B. Matching. Match the terms in column I with those in column II.

Column I
1. soldering
2. brazing process
3. fusion welding
4. bead
5. tacking
6. flux
7. small tip
8. large tip
9. emery cloth
10. galvanize

Column II
a. joins electrical wires
b. fastens temporarily
c. used before fluxing
d. capillary action
e. size 000
f. size 5
g. mixes metals
h. line of filler metal
i. toxic when heated
j. removes oxides

C. Brief Answers. Briefly answer the following questions.

1. What are three advantages of brazing over fusion welding?
2. What is the difference between brazing and braze welding?
3. Why is back-stepping used?

Arc Welding

UNIT 25 Selecting and Using Arc Welding Equipment

OBJECTIVE

To select electric arc welders, equipment, and materials needed for welding in agricultural mechanics.

COMPETENCIES TO BE DEVELOPED

After studying this unit, you should be able to:

- Describe the shielded metal arc welding process.
- Distinguish types of electric welding machines.
- Select suitable supplies and equipment for shielded metal arc welding.
- Recognize color and numerical code markings on electrodes.
- Select electrodes for use in agricultural mechanics.

MATERIALS LIST

- ✓ A variety of welding machines
- ✓ A safe and fully equipped welding area
- ✓ Electrodes for identification

TERMS TO KNOW

arc welder
arc
weldor
electrodes
shielded metal arc welding (SMAW)
arc welding
stick welding
slag
duty cycle
ampere (A)
amp
conductor
volt (V)
voltage
watt (W)
transformer
alternating current (AC)
60-cycle current
generator
direct current (DC)
polarity
straight polarity (SP)
reverse polarity
electrode holder
ground clamp
chipping hammer
National Electrical Manufacturers Association (NEMA)
end marking
spot marking
group marking
American Welding Society (AWS)
tensile strength
carbon arc torch

Heat for arc welding is obtained by using electricity. In this process electric current flows from a transformer connected to lines from a power plant. Another source of electricity for welding is electric motors or gas engines driving special generators. Regardless of the source of the electrical energy, a machine that produces current for welding is known as an arc welder. An arc is the discharge of electricity through an air space. A person who welds is known as a weldor.

ARC WELDING PROCESS AND PRINCIPLES

Shielded Metal Arc Welding

This unit focuses on a welding process that uses flux-coated metal welding rods called electrodes. The process is called shielded metal arc welding (SMAW). Some weldors call the process arc welding or stick welding. The term *shielded* refers to the gaseous cloud formed around the weld by the burning flux. The gases help the metal electrode burn evenly as it mixes with the base metal, figure 25-1. The flux also removes impurities from the base metal. The flux and impurities float to the top of the weld to form a layer called slag.

Advantages of Arc Welding. Arc welding is used extensively in agricultural mechanics. Some of the reasons for its popularity are as follows:

- Electricity is relatively inexpensive as a source of heat for welding.
- Electric welders suitable for farm welding are relatively inexpensive.
- Welders are available that work on ordinary 230-volt household or farmstead wiring.
- Engine-driven portable welders are available.
- Arc welding is fast and reliable.
- Agricultural students and workers can become good arc weldors quickly.
- An arc welder can be used for heating, brazing, and hardsurfacing as well as welding.

Temperature. The arc from a welder has a temperature of about 9000°F. The exact temperature varies with the length of the arc, size of the electrode, and amperage setting. Typical amperage settings for welders used in agricultural mechanics range from about 20 amps to 225 amps. Upper limits for industrial and commercial welders are higher, figure 25-2.

Duty Cycle. Welding machines get hot from use. The design of the machine determines how long it can operate. The duty cycle is the percentage of time that a welder can operate without overheating. Stated another way, it is the number of minutes out of 60 that a welder can operate at full capacity. A 20 percent duty cycle welder should weld only 12 minutes out of every hour, or 20 percent of 60 minutes. Welders are available with duty cycles ranging from 20 percent to 100 percent.

Duty cycle ratings are based on a midrange setting. The duty cycle is shorter if the welder is used at higher settings. Similarly, it is longer when the welder operates at lower settings.

Electricity for Welding

Three terms are basic to understanding how electricity is used for welding. They are amperes, volts, and watts. An ampere (A) or amp is a measure of the rate of flow of current in a conductor. A conductor is any material that permits current to move through it. Amperes of electricity flowing through a wire can be compared to gallons per minute of water flowing through a pipe.

A volt (V) or voltage is a measure of electrical pressure. Volts in a wire can be compared to pounds per square inch of water pressure in a pipe.

A watt (W) is a measure of energy available or work that can be done. For example, a 100-watt bulb gives out a specific amount of light. To figure the number of watts consumed, multiply volts by

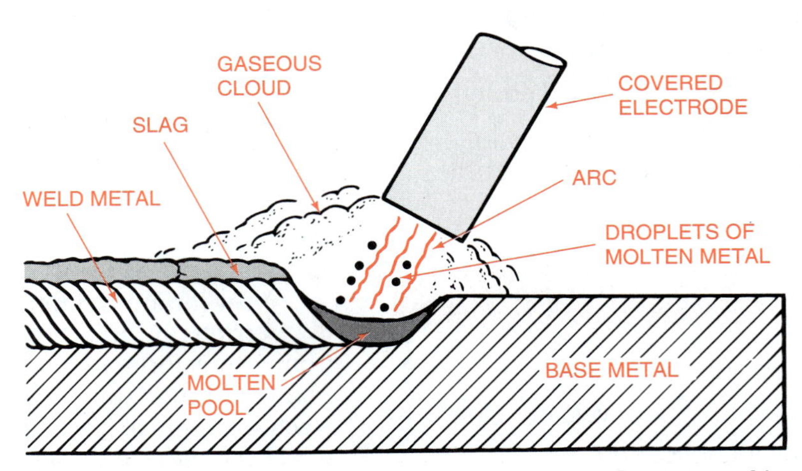

FIGURE 25-1. Components of the shielded metal arc welding process. (*Courtesy of Larry Jeffus,* Welding Principles and Applications, Third Edition, *Delmar Publishers, 1995)*

MACHINE MODEL	TYPE (60 Hertz)	NEMA RATING CURRENT (amperes)	ARC VOLTS	DUTY CYCLE	OUTPUT CURRENT RANGE (amperes)	TYPE	REQUIRED POWER SYSTEM	BULLETIN NUMBER
AC-225-S	K-1170	225	25	20%	40-225	AC	1 phase	E320
AC-250	K-1051	250	30	30%	35-300	AC	1 phase	E330
AC/DC-250	K-1053	250	30	30%	40-250#	AC & DC	1 phase	E330
TM-300	K-1103	300	32	60%	30-450	AC	1 phase	E340
TM-400	K-1105	400	36	60%	40-600	AC	1 phase	E340
TM-500	K-1108	500	40	60%	50-750	AC	1 phase	E340
TM-300/300	K-1104	300	32	60%	45-375#	AC & DC	1 phase	E340
TM-400/400	K-1107	400	36	60%	60-500#	AC & DC	1 phase	E340
TM-500/500	K-1110	500	40	60%	75-625#	AC & DC	1 phase	E340
TM-650/650	K-1126	650	44	60%	75-750#	AC & DC	1 phase	E340
R3R-300	K-1284	300	32	60%	45-375	DC	3 phase	E351
R3R-400	K-1285	400	36	60%	60-500	DC	3 phase	E351
R3R-500	K-1286	500	40	60%	75-625	DC	3 phase	E351

DC ranges, AC ranges same as straight AC models.
Source: Lincoln Electric Company.

FIGURE 25-2. The range of welding machines available allows the consumer to select one with the performance characteristics needed for a variety of jobs.

amperes: W = V × A or W = VA. This is called the West Virginia (W.VA.) formula to make it easier to remember. If any two values are known, the third can be calculated, figure 25-3.

Welding machines put out a high amperage and relatively low voltage. It is the low voltage that permits the welder to be a relatively safe machine while putting out so much energy. It requires about 60 volts to push current through the human body. At low to moderate settings, some welders put out less than 60 volts. However, reasonable caution to prevent electrical shock must be exercised at all times.

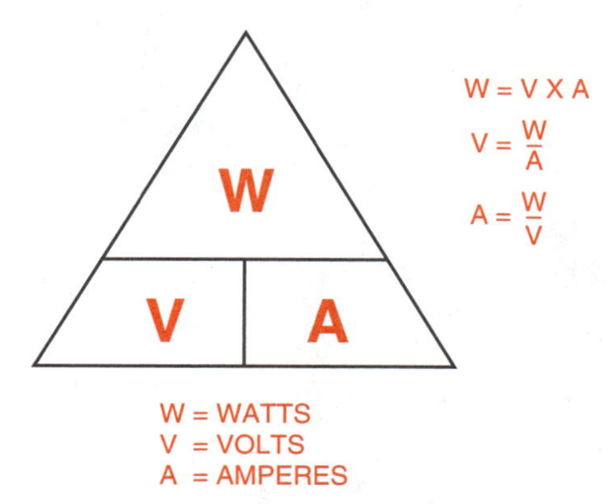

FIGURE 25-3. The third electrical value can be calculated when two out of three values are known. (*Courtesy of Larry Jeffus,* Welding Principles and Applications, Third Edition, *Delmar Publishers, 1995*)

Alternating and Direct Current. Some welders are transformers that receive current directly from utility power lines. Transformers convert high voltage and low amperage to low voltage and high amperage. These are called alternating current welders. Alternating current (AC) is current that reverses its direction of flow frequently. In the United States, power plants produce electricity that reverses its direction of flow 60 times per second. This electricity is referred to as 60-cycle current.

Other welders are really generators, driven by an electric motor or a gas or diesel engine. A generator is a machine that produces direct current. Direct current (DC) flows in one direction only in accordance with how the welder is set. DC welders can be set for straight or reverse polarity. Polarity refers to the direction of flow of electricity in the welding circuit. Certain electrodes and certain jobs work better with current flowing a certain way. Straight (negative) polarity (SP) is the term given to DC current flowing in one direction, while reverse (positive) polarity (RP) is direct current flowing in the opposite direction. Most DC welders have a switch to operate at either straight or reverse polarity as needed for any given electrode or job. Otherwise, polarity on DC welders can be switched by reversing the welding cables at the machine.

WELDING EQUIPMENT

AC Welders

AC welders rated at 180 and 225 amperes are popular for farm use, figure 25-4. Such welders can be purchased for just over $100 and are adequate for most farm welding jobs. They utilize standard 220 volt current. They do not draw excessive current, so that electricity bills are usually not greatly affected by them. Such machines are also available as AC/DC welders. Large farming operations may benefit from larger and more versatile welders for specialized welding.

DC Welders

Some DC welders are driven by electric motors. Such welders must be located near a suitable electrical outlet. Other welders are driven by gasoline or diesel engines. These units may be transported to the job by truck or trailer. Some machines are both DC and AC, figure 25-5. These machines can be used as welders and as portable generators to operate lights and equipment, figure 25-6. The control panel on such machines includes controls to operate the engine as well as the welder, figure 25-7.

Welding Cables

The diameter of electrical wire is stated in terms of gauge. The lower the gauge number, the larger the size of the wire. Wire sizes range from 18 for lighting circuits in automobiles to 0 or lower for large service cables. The gauge used for electrical wiring in homes is generally 14 for lights and 12 for heavier loads.

A welder puts out high amperage, so it needs large electrical cables. Cables carry current from the welder to the electrode, across the arc, and back to the welder. The cable that comes with a welder as

FIGURE 25-5. Diesel engine-drive AC/DC welder and generator. (*Courtesy of Miller Electric Manufacturing Company of Appleton, WI*)

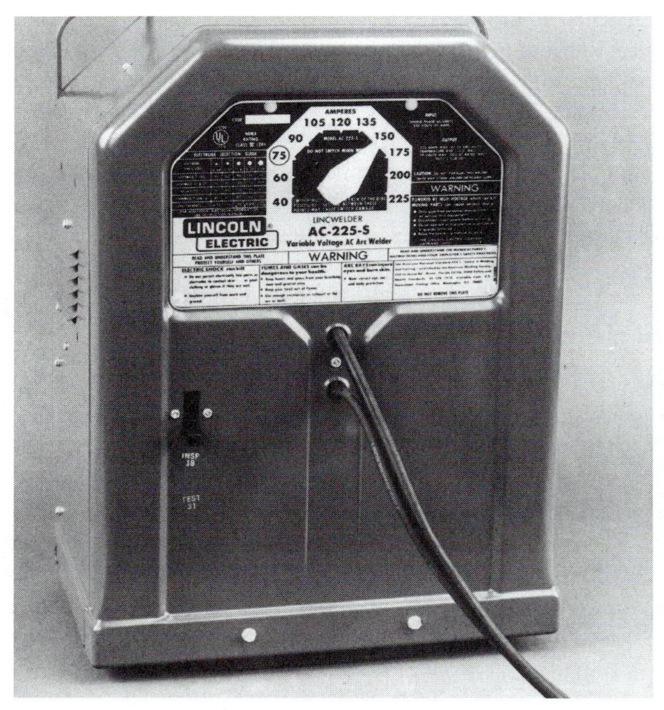

FIGURE 25-4. AC welders rated at 180 and 225 amperes are popular for farm use. (*Courtesy of Lincoln Electric Company*)

FIGURE 25-6. Some welders are portable and can act as generators to produce electricity for lights and power tools. (*Courtesy of Miller Electric Manufacturing Company of Appleton, WI*)

FIGURE 25-7. Control panel of an engine-driven welder/generator. *(Courtesy of Miller Electric Manufacturing Company of Appleton, WI)*

original equipment is the correct diameter for its length. If longer cables are used, they may also need to be of larger diameter, figure 25-8. If the cables are too small they will not deliver as much current as the welder is designed to provide. Under these conditions, the force of the arc is not as intense as the machine is capable of delivering.

CAUTION:

- **Careful attention should be given to cable care. Always protect cables from forceful impact and contact with sharp objects that may damage their coverings. These coverings insulate the conductors and prevent dangerous shocks and electrical shorts.**

Electrode Holders and Ground Clamps

A welder has two cables: one ends in an electrode holder and the other ends in a ground clamp. An electrode holder is a spring-loaded device with insulated handles used to grip a welding electrode, figure 25-9. Welding cables have a copper or aluminum core. This core is clamped or soldered to the metal part of the electrode holder. An insulated sleeve

LENGTH OF CABLE		COPPER WELDING LEAD SIZES								
AMPERES		100	150	200	250	300	350	400	450	500
ft	m									
50	15	2	2	2	2	1	1/0	1/0	2/0	2/0
75	23	2	2	1	1/0	2/0	2/0	3/0	3/0	4/0
100	30	2	1	1/0	2/0	3/0	4/0	4/0		
125	38	2	1/0	2/0	3/0	4/0				
150	46	1	2/0	3/0	4/0					
175	53	1/0	3/0	4/0						
200	61	1/0	3/0	4/0						
250	76	2/0	4/0							
300	91	3/0								
350	107	3/0								
400	122	4/0								

LENGTH OF CABLE		ALUMINUM WELDING LEAD SIZES								
AMPERES		100	150	200	250	300	350	400	450	500
ft	m									
50	15	2	2	1/0	2/0	2/0	3/0	4/0		
75	23	2	1/0	2/0	3/0	4/0				
100	30	1/0	2/0	4/0						
125	38	2/0	3/0							
150	46	2/0	3/0							
175	53	3/0								
200	61	4/0								
225	69	4/0								

FIGURE 25-8. Copper and aluminum welding lead sizes (gauges) for various current loads and cable lengths. *(Adapted from Larry Jeffus,* Welding Principles and Applications, Third Edition, *Delmar Publishers, 1995)*

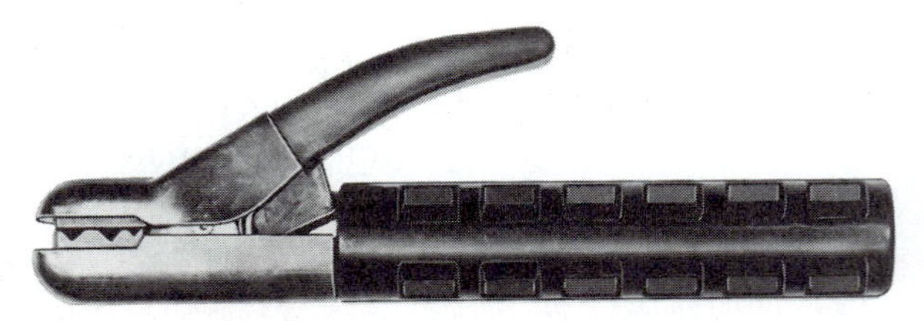

FIGURE 25-9. Electrode holder. (*Courtesy of Lincoln Electric Company*)

slides over the connection to protect the operator from electric shock. All other parts of the holder are insulated except the jaws.

A ground clamp is a spring-loaded clamp attached to an electrical cable, figure 25-10. It is not insulated and is not a shock hazard. It carries current between the welding table or project and to the welder.

Welding Table and Booth

A booth and metal table for welding are convenient and efficient, figure 25-11. The proper table permits the operator to stand or sit comfortably.

The booth must have curtains to protect other workers from blinding light flashes, figure 25-12. The booth should also have panels for the storage of welding tools and equipment. Exhaust equipment must be provided to remove the fumes resulting from welding. The entire area must be fire-resistant and free of flammable materials.

CAUTION:

- **It is essential to have a fire blanket and appropriate fire extinguishers in the welding area.**

Welding is often done on the floor of a shop or outdoors. In such situations, portable curtains must be set up to protect others in the area from welding flashes, figure 25-13. It is important that all combustible materials be removed before welding in such areas.

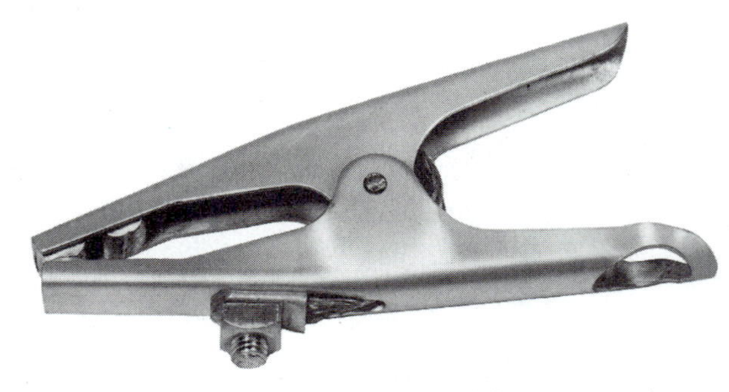

FIGURE 25-10. Ground clamp. (*Courtesy of Lincoln Electric Company*)

FIGURE 25-11. A welding booth and welding table are convenient and efficient.

Equipment for Cleaning Welds

Welding areas should be equipped with a large grinder with a wire wheel. The grinder is used extensively in preparing metal for welding, and the wire wheel is useful for cleaning beads after welding.

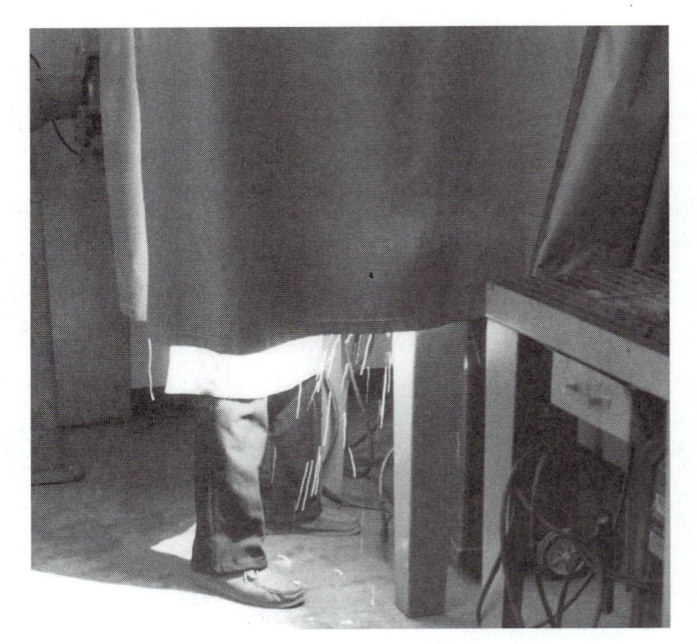

FIGURE 25-12. Curtains must cover the openings of a welding booth to protect workers in the area from light flashes.

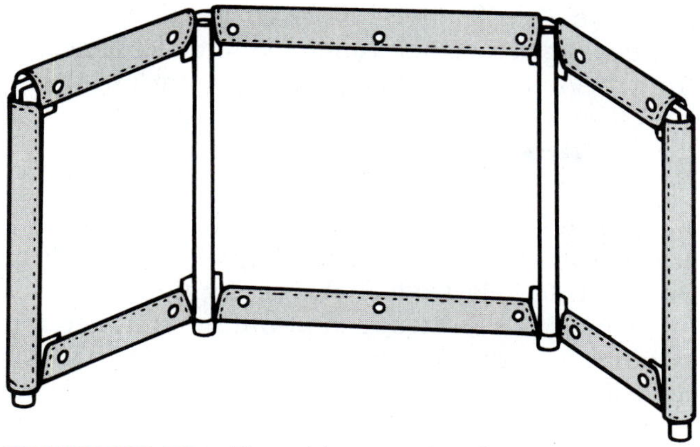

FIGURE 25-13. Portable curtains for use around welding done on the shop floor or outdoors.

Tongs or pliers of various types are useful for handling hot metal. Hand wire brushes and chipping hammers are recommended for removing slag. A chipping hammer is a steel hammer with a sharp edge and/or point.

ELECTRODES

Arc welding technology has developed to a high level. A wide variety of electrodes is available to enable a weldor to do many different jobs, including welding in all positions.

Welding positions include flat, horizontal, vertical up, vertical down, and overhead. (These are described in Unit 26.) Welding in agricultural mechanics may call for any of these positions. Some electrodes are suitable for all-position welding. Others are only suitable for one or two positions. The parts of an electrode are shown in figure 25-14.

NEMA Color Coding

The National Electrical Manufacturers Association (NEMA) developed a system to permit manufacturers to mark their electrodes by color codes. When used, the markings are placed on three areas of an electrode: (1) the exposed end of the metal rod, (2) the exposed surface of the metal rod, and (3) the flux near the exposed rod. The color markings are given specific names. Color on the end is called an end marking; color on the bare surface is called a spot marking; and color on the flux is called a group

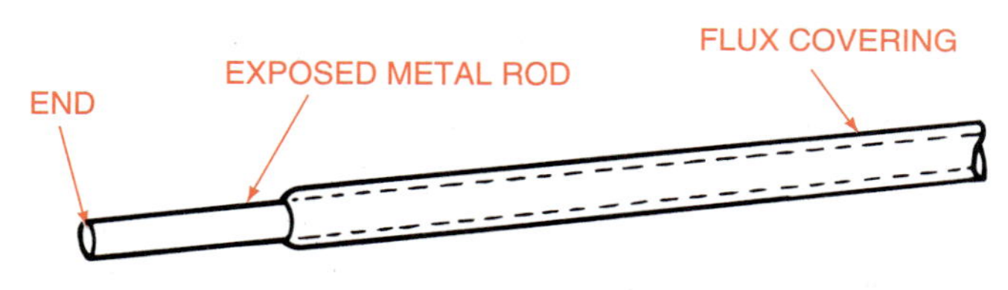

FIGURE 25-14. Parts of an arc welding electrode

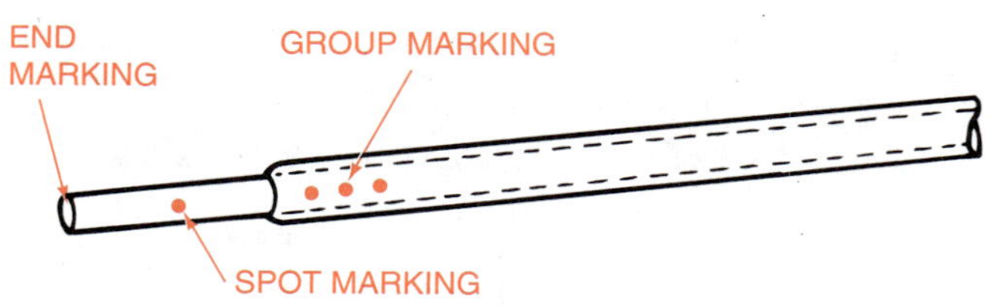

FIGURE 25-15. Locations on an electrode used in the color coding system of the National Electrical Manufacturers Association (NEMA).

marking, figure 25-15. Thus electrodes are identified by the colors of their end, spot, and group markings. Most manufacturers simply stamp the AWS classification number on each electrode, rather than use the color coding system.

AWS Numerical Coding

The American Welding Society (AWS) is an organization of individuals and agencies that support education in welding processes. The society has developed a numerical system for coding electrodes. The code condenses a great deal of information into a four- or five-digit number for mild steel electrodes. The number is preceded by the letter "E" to indicate that the number describes an electrode.

The first two digits (or the first three if it is a five-digit number) refer to the tensile strength of one square inch of weld. Tensile strength refers to the amount of tension or pull the weld can withstand. The number represents thousands of pounds of tensile strength per square inch. For example, E60 stands for 60,000 pounds per square inch (psi) of tensile strength.

The two right-hand digits in the AWS number refer to the type of welding the electrode is capable of doing. The third digit (or fourth if it is a five-digit number) refers to the welding positions for which the electrode is suited. The fourth digit (or fifth if a five-digit number) refers to the depth of penetration and/or welding current, figures 25-16 and 25-17. For

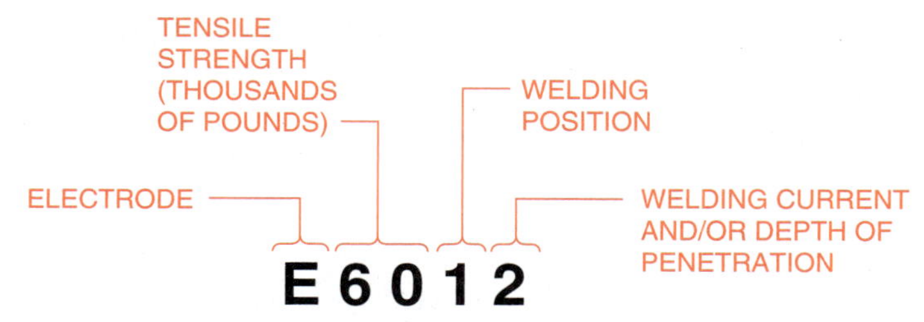

FIGURE 25-16. Components of a typical AWS classification number for electrodes. A five-digit AWS number is used if three digits are required to express the tensile strength. (*Adapted from Larry Jeffus,* Welding Principles and Applications, Third Edition, *Delmar Publishers, 1995*)

THE SECOND DIGIT FROM THE RIGHT INDICATES WELDING POSITION	
E _ _ 1 _	Usable in all directions
E _ _ 2 _	Usable in flat and horizontal positions only
E _ _ 4 _	Usable for vertical down only

THE RIGHT-HAND DIGIT INDICATES TYPE OF WELDING	
E _ _ _ 0	DC reverse polarity only
E _ _ _ 1	AC and DC reverse polarity
E _ _ _ 2	AC and DC straight polarity
E _ _ _ 3	AC and DC
E _ _ _ 4	AC and DC
E _ _ _ 5	DC reverse polarity
E _ _ _ 6	AC and DC reverse polarity
E _ _ _ 8	AC and DC reverse polarity

FIGURE 25-17. The last two digits in the AWS electrode numbering system specify the position of welding and the type of welding current.

complete identification it is more useful to read the last two digits together as defined in tables available from the American Welding Society, figure 25-18.

Electrode Selection

All-purpose, all-position, AC/DC mild steel electrodes are the choice for most shops. Both the E6013 and the E6011 electrode fit this description.

The E6013 electrode is popular for beginning weldors. It is relatively inexpensive and easy to use. It deposits metal quickly and leaves an attractive bead. The slag is easy to remove. However, the E6013 electrode has shallow penetration. Therefore, it is recommended primarily for thin metal, 1/8 inch or less.

The E6011 electrode is perhaps the most commonly used in agricultural mechanics. It is not easy for the beginner to make an attractive weld using the E6011. However, the deep penetration it provides generally results in a stronger weld than can be obtained with the E6013. Commercial welding shops

ELECTRODE CLASSIFICATION

AWS Classification	Type of Covering	Capable of Producing Satisfactory Welds in Position Shown[a]	Type of Current[b]
E60 Series Electrodes			
E6010	High cellulose sodium	F, V, OH, H	DCEP
E6011	High cellulose potassium	F, V, OH, H	AC or DCEP
E6012	High titania sodium	F, V, OH, H	AC or DCEN
E6013	High titania potassium	F, V, OH, H	AC or DC, either polarity
E6020 }	High iron oxide	H-fillets	AC or DCEN
E6022[c] }	High iron oxide	F	AC or DC, either polarity
E6027	High iron oxide, iron powder	H-fillets, F	AC or DCEN
E70 Series Electrodes			
E7014	Iron powder, titania	F, V, OH, H	AC or DC, either polarity
E7015	Low hydrogen sodium	F, V, OH, H	DCEP
E7016	Low hydrogen potassium	F, V, OH, H	AC or DCEP
E7018	Low hydrogen potassium, iron powder	F, V, OH, H	AC or DCEP
E7024	Iron powder, titania	H-fillets, F	AC or DC, either polarity
E7027	High iron oxide, iron powder	H-fillets, F	AC or DCEN
E7028	Low hydrogen potassium, iron powder	H-fillets, F	AC or DCEP
E7048	Low hydrogen potassium, iron powder	F, OH, H, V-down	AC or DCEP

a. The abbreviations, F, V, V-down, OH, H, and H-fillets indicate the welding positions as follows:
F = Flat
H = Horizontal
H-fillets = Horizontal fillets
V-down = Vertical down
V = Vertical }
OH = Overhead } {For electrodes 3/16 in. (4.8 mm) and under, except 5/32 in. (4.0 mm) and under for classifications E7014, E7015, E7016, and E7018.

b. The term DCEP refers to direct current, electrode positive (DC reverse polarity). The term DCEN refers to direct current, electrode negative (DC straight polarity).

c. Electrodes of the E6022 classification are for single-pass welds.

Source: Reproduced from AWS A5.1-81, American Welding Society, with permission.

FIGURE 25-18. Specifications for covered carbon steel arc welding electrodes.

can choose from many types and suppliers of electrodes, figure 25-19.

The experienced weldor needs at least a pound or more of each additional electrode in the shop.

See AWS A5.1-81. Specifications for Carbon Steel Covered Arc Welding Electrodes

MANUFACTURERS	AWS CLASSIFICATION						
	E6010	E6011	E6012	E6013	E6020	E6022	E6027
Aga de Mexico, S.A.	AGA C10 C12	AGA C11	AGA R12	AGA R10 R11	—	—	—
Airco Welding Products	AIRCO 6010	AIRCO 6010 6011C 6011LOC	AIRCO 6012 6012C	AIRCO 6013 6013C 6013D	AIRCO 6020	—	EASY ARC 6027
Air Products and Chemicals, Inc.	AP6010W	AP6011W	AP6012W	AP6013W	—	—	—
A-L Welding Products, Inc.	AL6010	AL6011	AL6012	AL6013	—	—	—
Alloy Rods, Allegheny International, Inc.	AP100 SW610	SW14	SW612	SW15	—	—	—
Applied Power, Inc.	—	No. 130 Red-Rod	—	No. 140 Production Rod	—	—	—
Arcweld Products Limited	Easyarc 10 Easyarc	Arcweld 230	Arcweld 387 Satinarc 11	Arcweld 90	—	—	—
Bohler Bros. of America, Inc.	Fox CEL	—	—	Fox OHV Fox ETI	Fox UMZ	—	—
Brite-Weld	Brite-Weld E6010	Brite-Weld E6011	Brite-Weld E6012	Brite-Weld E6013	—	—	Brite-Weld E6027
Canadian Liquid Air Ltd.	LA6010	LA6011P	LA6012P LA6013P	LA6013	—	—	—
Canadian Rockweld Ltd.	R60	R61	R62 R63A	R63	R620	—	R627
C-E Power Systems, Combustion Engineering, Inc.	—	—	—	—	—	—	—
Century Mfg. Co.	— 324	331	— 313	331	—	—	—
Champion Hobart, S.A. de C.V.	6010	6011	6012 Ducto P60	6013 Versa-T	—	—	—
CONARCO, Alambres y Soldaduras, SA	CONARCO 10 CONARCO 10P	CONARCO 11	CONARCO 12 CONARCO 12D	CONARCO 13A CONARCO 13	—	—	—
Cronatron Welding Systems, Inc.	— 6011	Cronatron	— 6013	Cronatron	—	—	—
Electromanufacturas, S.A.	West Arco XL-610M ZIP 10T	West Arco ACP611	West Arco FP-612 ZIP 12 SW-613 SW-10	West Arco SW-613M SUPER	—	—	West Arco ZIP 27

FIGURE 25-19. Carbon steel covered arc welding electrodes. (*Reproduced from AWS A5.0-83, American Welding Society, with permission*)

	AWS CLASSIFICATION						
MANUFACTURERS	E6010	E6011	E6012	E6013	E6020	E6022	E6027
ESAB	OK 22.45	OK 22.65	—	OK 46.00 OK 43.32 OK 50.10 OK 50.40	—	—	—
Eureka Welding Alloys	Eureka 6010	Eureka 6011	Eureka 6012	Eureka 6013	—	—	Eureka 6027
Eutectic Corporation	—	—	—	DynaTrode 666	—	—	—
Hobart Brothers Company	Hobart 10 60AP	Hobart 335A	Hobart 12 12A	Hobart 413 447A	—	Hobart 1139	Hobart 27H
International Welding Products, Inc.	INWELD 6010	INWELD 6011	INWELD 6012	INWELD 6013	INWELD 6020	—	INWELD 6027
Kobe Steel, Ltd.	KOBE 6010	KOBE 6011 RB-26D	KOBE TB-62 B-33	ZERODE 44 RB-26	—	— 27, B-27	ZERODE 27 AUTOCON
Latamer Company, Inc.	Latco E6010	Latco E6011	Latco E6012	Latco E6013	Latco E6020	—	Latco E6027
The Lincoln Electric Company	Fleetweld 5P	Fleetweld 35 35LS 180	Fleetweld 7	Fleetweld 37 57	—	—	Jetweld 2
Liquid Carbonic, Inc.	RD 704D RD 610P	RD 504D RD 611P	RD 604	RD 613	—	—	—
Murex Welding Products	SPEEDEX 610 611LV	Type A 611C GENEX M	Type N 13	Type U U13	F.H.P.	—	SPEEDEX 27
Oerlikon, Inc.	OERLIKON E6010	—	OERLIKON E6012	OERLIKON E6013	OERLIKON E6020	—	—
Sodel	—	SODEL 11	—	SODEL 31	—	—	—
Techalloy Maryland, Inc. Reid-Avery Div.	RACO 6010	RACO 6011	RACO 6012	RACO 6013	RACO 6020	—	RACO 6027
Teledyne Canada HARFAC	6010	6011	6012	6013	—	—	—
Teledyne McKay	6010	6011	6012	6013	—	—	—
Thyssen Draht AG	Thyssen Cel 70	—	Thyssen Blau Thyssen A5 Thyssen Grun	Thyssen Grun T SH Blau SH Gelb B SH Gelb T SH Gelb S SH Gelb R SH Grun TB SH Lila R Union 6013	SH Gelb	SH Tiefbrand	—

FIGURE 25-19. Carbon steel covered arc welding electrodes. (Cont.)

MANUFACTURERS	AWS CLASSIFICATION						
	E6010	E6011	E6012	E6013	E6020	E6022	E6027
WASAWELD	WASAWELD E6010	WASAWELD 6011	WASAWELD E6012	WASAWELD E6013	WASAWELD E6020	—	WASAWELD E6027
Weld Mold Company	—	—	—	—	—	—	—
Weldwire Co., Inc.	WELDWIRE 6010	WELDWIRE 6011	WELDWIRE 6012	WELDWIRE 6013	WELDWIRE 6020	WELDWIRE 6022	WELDWIRE 6027
Westinghouse Electric Corporation	ZIP 10 XL 610A	ACP ZIP-11R	ZIP 12	SW	DH 620	—	ZIP 27 ZIP 27M

MANUFACTURERS	AWS CLASSIFICATION						
	E7014	E7015	E7016	E7016-1	E7018	E7018-1	E7024
Aga de Mexico, S.A.	—	—	—	—	AGA B10	—	AGA RH10
Airco Welding Products	EASY ARC 7014	—	7016 7016M	—	EASY ARC 7018C 7018MR CODE ARC 7018MR	—	EASY ARC 7024 7024D
Air Products and Chemicals, Inc.	AP7014W	—	AP7016W	—	AP7018W	—	AP7024W
A-L Welding Products, Inc.	AL7014	—	—	—	AL7018 Nuclearc 7018	—	AL7024
Alloy Rods, Allegheny International, Inc.	SW151P	—	70LA-2	Atom Arc 7016-1	Atom Arc 7018 SW-47	Atom Arc 7018-1	7024
Applied Power, Inc. Hy-Pro-Rod	No. 146	—	—	— Marq-Rod	No. 7018	—	—
Arcweld Products Limited	Easyarc 14	—	Arcweld 312	—	Easyarc 328 Easyarc 7018MR	—	Easyarc 12
Bohler Brothers of America, Inc.	—	—	Fox EV47	—	Fox EV50	—	Fox HL 180 Ti
Brite-Weld	Brite-Weld 7014	—	—	—	Brite-Weld E7018	—	Brite-Weld E7024
Canadian Liquid Air, Ltd.	LA7014	—	—	—	Super Arc 18 LA7018, AA7018, LA7018B	— —	LA7024 LA24HD
Canadian Rockweld, Ltd.	R74	—	Tensilarc 76	—	Hyloarc 78	—	R724

FIGURE 25-19. Carbon steel covered arc welding electrodes. (Cont.)

MANUFACTURERS	AWS CLASSIFICATION E7014	E7015	E7016	E7016-1	E7018	E7018-1	E7024
C-E Power Systems, Combustion Engineering, Inc.	—	—	—	—	CE7018	—	—
Century Mfg. Co.	331 363	—	—	—	331 327	—	—
Champion Hobart, S.A. de C.V.	—	—	724	—	7018	718	MULTI-T
CONARCO, Alambres y Soldaduras, S.A.	CONARCO 14	CONARCO 15	CONARCO 16	—	CONARCO 18	CONARCO 18-1	CONARCO 24
Cronatron Welding Systems, Inc.	—	—	—	—	Cronatron 7018	—	—
Electromanufactures, S.A.	West Arco ZIP 14	—	West Arco WIZ 16	—	West Arco WIZ 18	—	West Arco ZIP 24
ESAB	OK 46.16	—	OK 53.00 OK 53.05	OK 53.68	OK 48.00 OK 48.04 OK 48.15	OK 48.68 OK 55.0	OK Femax 33.65 OK Femax 33.80
Eureka Welding Alloys	Eureka 7014	—	—	—	Eureka 7018	—	Eureka 7024
Eutectic Corporation	—	—	—	—	EutecTrode 7018	—	—
Hobart Brothers Company	Hobart 14A	—	—	—	Hobart 718 718LMP	—	Hobart 24 24H
International Welding Products, Inc.	INWELD 7014	INWELD 7015	INWELD 7016	—	INWELD 7018	—	INWELD 7024
Kobe Steel, Ltd.	KOBE RB-14	—	KOBE LB-52 LB-52U LB-26Vu ZERODE-52	—	KOBE LB-52-18 LTB-52A	—	KOBE RB-24 ZERODE 50F FB-24
Latamer Company, Inc.	Latco E7014	Latco E7015	Latco E7016	—	Latco E7018	—	Latco E7024
The Lincoln Electric Company	Fleetweld 47	—	—	—	Jetweld LH70 LH73 LH78 LH75	Jetweld LH75	Jetweld 1 3
Liquid Carbonic, Inc.	RD 714	—	—	—	RD 718	—	RD 724
Murex Welding Products	SPEEDEX U	—	HTS HTS-18 HTS-180	—	SPEEDEX HTS-MR HTS-M-MR-718	—	SPEEDEX 24 24D

FIGURE 25-19. Carbon steel covered arc welding electrodes. (Cont.)

MANUFACTURERS	AWS CLASSIFICATION						
	E7014	E7015	E7016	E7016-1	E7018	E7018-1	E7024
Oerlikon, Inc.	—	—	—	OERLIKON Spezials Extra Tenacito WZ	OERLIKON E7018	OERLIKON E7018-1	OERLIKON Ferromatic Ferrocito R
Sodel	—	—	—	—	—	SODEL 328	SODEL 314
Techalloy Maryland, Inc. Reid-Avery Div.	RACO 7014	RACO 7015	RACO 7016	—	RACO 7018	RACO 7018-1	RACO 7024
Teledyne Canada HARFAC	7014	—	7016	—	—	7018-1	7024
Teledyne McKay	7014	—	7016	—	7018 XLM	7018-1 XLM	7024
Thyssen Draht AG	SH Multifer 130	Thyssen K50 SH Grun K45 (6015)	SH Kb F Thyssen K50R SH Grun K50W Thyssen Kb Spezial SH Grun K70W Thyssen K90S	—	SH Grun K70	Thyssen 120K	Thyssen Rot R160 Thyssen Rot R160S SH Multifer 180 Thyssen Rot AR160
Weld Mold Company	—	—	—	—	WELD MOLD 7018	—	—
Weldwire Co., Inc.	WELDWIRE 7014	WELDWIRE 7015	WELDWIRE 7016	WELDWIRE 7016-1	WELDWIRE 7018	WELDWIRE 7018-1	WELDMIRE 7024
Westinghouse Electric Corporation	ZIP 14	—	LOH 2	—	WIZ 18	—	ZIP 24

MANUFACTURERS	AWS CLASSIFICATION			
	E7024-1	E7027	E7028	E7048
Aga de Mexico, S.A.	—	—	—	—
Airco Welding Products	—	—	EASY ARC 7028	—
Air Products and Chemicals, Inc.	—	—	—	—
A-L Welding Products, Inc.	—	—	—	—
Alloy Rods, Allegheny International, Inc.	7024-1	—	—	—
Applied Power, Inc.	—	—	—	—
Arcweld Products Limited	—	—	Super 28	—
Bohler Bros. of America, Inc.	—	—	Fox HL 180 Kb	—

FIGURE 25-19. Carbon steel covered arc welding electrodes. (Cont.)

MANUFACTURERS	AWS CLASSIFICATION E7024-1	E7027	E7028	E7048
Brite-Weld	—	—	—	—
Canadian Liquid Air, Ltd.	LA7024	—	LA7028 LA7028B	LA7048B
Canadian Rockweld, Ltd.	—	R727	Hyloarc 728	—
C-E Power Systems, Combustion Engineering, Inc.	—	—	—	—
Century Mfg. Co.	—	—	—	—
Champion Hobart, S.A. de C.V.	—	—	—	—
CONARCO, Alambras y Soldaduras, S.A.	—	—	CONARCO 28	CONARCO 48
Cronatron Welding Systems, Inc.	—	—	—	—
Electromanufacturas, S.A.	—	—	—	—
ESAB	—	—	OK 38.48 OK 38.65 OK 38.85 OK 38.95	OK 53.35
Eureka Welding Alloys	—	—	—	—
Eutectic Corporation	—	—	—	—
Hobart Brothers Company	—	—	Hobart 728	—
International Welding Products, Inc.	—	—	INWELD 7028	—
Kobe Steel, Ltd.	—	—	—	KOBE LB-26V LB-52V ZERODE 6V
Latamer Company, Inc.	—	—	Latco E7028	—
The Lincoln Electric Company	Jetweld 1	—	Jetweld LH3800	—
Liquid Carbonic, Inc.	—	—	RD 728	—
Murex Welding Products	—	—	SPEEDEX 28	—
Oerlikon, Inc.	—	—	OERLIKON E7028	—
Sodel	—	—	—	—

FIGURE 25-19. Carbon steel covered arc welding electrodes. (Cont.)

MANUFACTURERS	AWS CLASSIFICATION E7024-1	E7027	E7028	E7048
Techalloy Maryland, Inc. Reid-Avery Div.	—	—	RACO 7028	RACO 7048
Teledyne Canada HARFAC	—	—	—	—
Teledyne McKay	—	—	—	—
Thyssen Draht AG	—	SH Multifer 200	SH Multifer 150 K11 Thyssen Rot BR160	—
WASAWELD	—	—	—	—
Weld Mold Company	—	—	—	—
Weldwire Co., Inc.	WELDWIRE 7024-1	WELDWIRE 7027	WELDWIRE 7028	WELDWIRE 7048
Westinghouse Electric Corporation	—	—	—	—

Source: Reproduced from AWS A5.0-83, American Welding Society, with permission.

FIGURE 25-19. Carbon steel covered arc welding electrodes. (Cont.)

These include special electrodes for welding cast iron and stainless steel. Such electrodes are handy for specialized repair jobs.

Other useful electrodes include the hardsurfacing types. One type deposits material that resists soil abrasion. Layers deposited on parts such as plowshares, landsides, cultivator shovels, and bulldozer blades extend the life of the part. Another type of hardsurfacing electrode leaves a deposit on metal chains, drags, or bars to resist metal-to-metal wear.

Some shops have arc welders but do not have gas cutting and welding torches. Such shops may benefit from special arc welding electrodes designed for cutting and/or brazing. Additionally, the carbon arc torch may be a useful addition to the arc welder for heating. The carbon arc torch is a device that holds two carbon sticks and produces a flame from the energy of an electric welder.

STUDENT ACTIVITIES

1. Define the Terms to Know for this unit.
2. Examine the welders in your school agricultural mechanics shop. Classify them according to AC or DC type and determine if the DC welders are motor driven or engine driven.
3. Examine the electrodes used in your agricultural mechanics shop. Is there an AWS code stamped on the flux? Are they color coded? If so, determine what the code(s) mean.
4. Identify the tools and equipment in the welding area of your shop.

SELF-EVALUATION

A. Multiple Choice. Select the best answer.

1. Voltage is a measure of
   a. rate of current
   b. electrical pressure
   c. available energy
   d. all of these
2. A welder that gets its energy directly from a utility power plant is
   a. an alternator
   b. a generator
   c. a rectifier
   d. a transformer
3. When welding, the flux on an electrode
   a. forms a gas
   b. creates slag
   c. shields the weld
   d. all of these
4. Shielded metal arc welding is also called
   a. brazing
   b. electrode welding
   c. oxyacetylene welding
   d. stick welding
5. Suitable arc welders for farm use are
   a. fairly easy to use
   b. relatively inexpensive
   c. reliable
   d. all of these
6. The output of a welder is relatively
   a. low voltage and high amperage
   b. high voltage and low amperage
   c. high voltage and high amperage
   d. low voltage and low amperage
7. Types of polarity on DC welders include
   a. reversed
   b. straight
   c. both of these
   d. none of these
8. The proportion of time that a welder can operate without overheating is known as its
   a. AC/DC
   b. AWS classification
   c. duty cycle
   d. voltage drop
9. A chipping hammer is used to
   a. prepare edges for welding
   b. remove scale from steel
   c. remove slag
   d. temper beads
10. An organization of people and agencies interested in promoting welding is the
    a. AWS
    b. EPA
    c. SAE
    d. WPA

B. Completion. Fill in the blanks with the word or words that will make the following statements correct.

1. The color markings on an electrode will be _____, _____, or _____ markings.
2. The E in electrode code stands for _____.
3. In the American Welding Society electrode code, the first two (or three) numbers refer to _____; the next number refers to _____; and the final number refers to _____.
4. Three types of welding electrodes other than those used for welding mild steel are those used for _____, _____, and _____.

UNIT 26 Arc Welding Mild Steel and MIG/TIG Welding

OBJECTIVE

To use arc welding equipment and procedures in cutting and welding.

COMPETENCIES TO BE DEVELOPED

After studying this unit, you should be able to:

- Use safety equipment and protective clothing for arc welding.
- Strike an arc and run beads.
- Make butt and fillet welds.
- Make flat, horizontal, and vertical welds.
- Weld pipe.
- Pierce holes and cut with electrodes.
- Identify major parts of MIG and TIG welding outfits.
- Explain and do MIG and TIG welding.

TERMS TO KNOW

pad
crater
stringer bead
weaving
pass
root pass
horizontal welds
vertical down
vertical up
spot welding
spot welder
metal inert gas (MIG)
tungsten inert gas (TIG)

MATERIALS LIST

- ✓ Welding booth
- ✓ Electric welder, 180 amperes or larger
- ✓ Protective clothing: welding helmet, clear goggles, leather gloves, welding jacket, leather apron, and leather leggings
- ✓ Chipping hammer
- ✓ Pliers
- ✓ Wire brush
- ✓ Bucket with water
- ✓ Fire extinguisher and fire blanket
- ✓ Welding pads, steel, 1/4" × 2" × 6", 1/4" × 4" × 6", 1/8" × 1" × 6"
- ✓ Grinder with a coarse abrasive wheel and a wire wheel
- ✓ Soapstone markers
- ✓ Electrodes, 1/8-inch E6011 and E6013

The term *arc welding* as used in this unit refers to shielded metal arc welding (SMAW). In this type of welding, an electric welder and flux-covered electrodes are used. Before starting to weld, the operator must be thoroughly familiar with shop safety and fire prevention.

SAFETY PROCEDURES

Fire Protection

The temperature of the electric arc (about 9000°F) creates a very real danger of burns as well as fires. Nevertheless, with reasonable care and use of the proper equipment, the welding area is no more dangerous than many other areas of the shop.

The welding area should be equipped with metal benches. These serve both for fire protection and to electrically ground the work. Welding booths should be constructed of fireproof or fire-resistant materials, such as metal sheets or concrete block. Other materials such as pressed wood panels, plywood, and special fire-resistant canvas are also used.

Fire extinguishers suitable for Class A, B, and C fires, safety equipment, and a first-aid kit should be within easy reach at all times, figure 26-1. A wool fire blanket is another important piece of fire control equipment, figure 26-2. If human hair or clothing catches on fire, the fire blanket is used to wrap the victim and smother the fire.

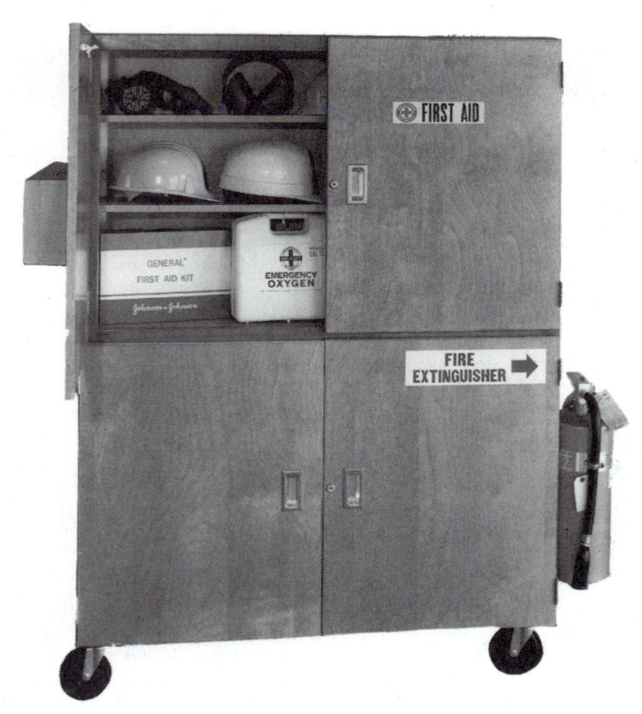

FIGURE 26-1. Fire extinguishers suitable for Class A, B, and C fires, safety equipment, and a first-aid kit should be within easy reach of every weldor.

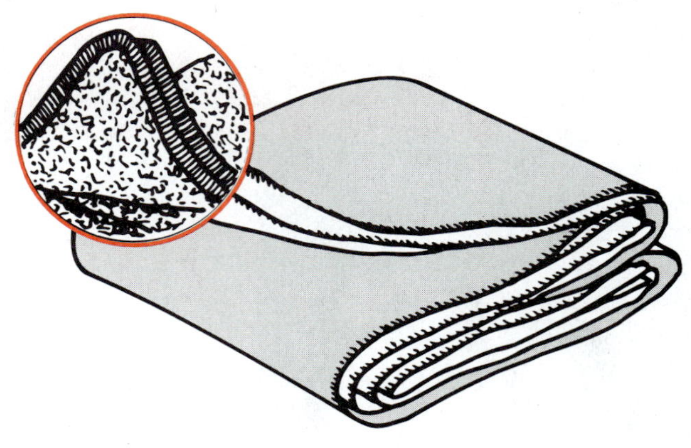
FIGURE 26-2. A wool fire blanket is the first line of defense if clothing catches on fire. The victim is wrapped in the blanket to smother the fire quickly.

Buckets of water are frequently used to receive sparks and cool metal in the welding areas. Cooling of metal used for welding practice reduces the chances of personal burns. The water may also be useful in extinguishing accidental fires when electric shock is not a hazard. However, any water spilled on the floor of the work area must be removed because it will create an electric shock hazard.

All grease, oil, sawdust, paper, rags, and other flammable materials must be removed from areas where welding is done. Good housekeeping is a major factor in reducing fire hazards.

Personal Protection

CAUTION:

- **The human eye must not be exposed to direct light from a welding arc. The welding arc contains light rays that burn, even from a distance. Fortunately, it is easy to protect the eyes from these rays. Any material that blocks the light also stops the damaging rays. The weldor's face and eyes are protected by using a face shield with a dark viewing glass. The glass must be classified as a No. 10 shade or higher, figure 26-3.**

The No. 10 shade lens must always be worn when welding, but is too dark to see through except when the welding arc is burning. Therefore, the operator must either wear safety glasses under the helmet or use a helmet with a flip-up lens for processes such as chipping, figure 26-4. When the flip-up lens helmet is not used, clear safety glasses *must* be worn

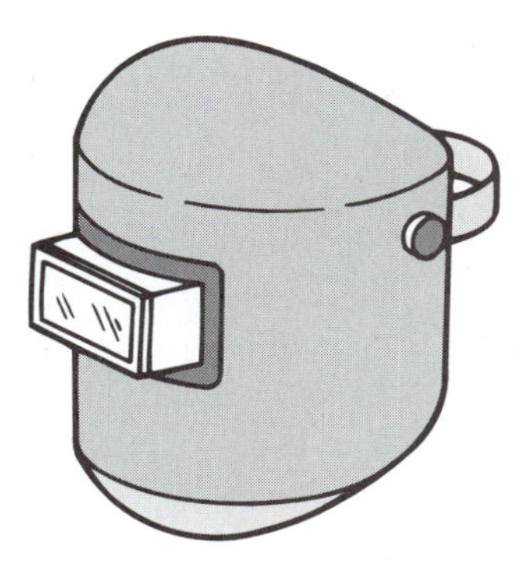

FIGURE 26-3. A welding helmet with a No. 10 shaded lens protects the eyes from damaging welding light rays.

under the helmet. When not welding, the operator may exchange the helmet for another type of clear face and eye protective covering, figure 26-5. Prescription eye glasses are not approved safety glasses unless ordered as such.

CAUTION:

- **Attempting to chip hot slag from welds without eye protection is likely to result in eye injury.**

Fire-resistant coveralls and high leather shoes are recommended as standard clothing when welding. In addition, leather gloves are needed to protect

FIGURE 26-4. Some welding helmets have a flip-up shaded lens. When this lens is up, the weldor can see through the clear lens, which backs up the shaded lens for no welding operations. *(Courtesy of J. I. Scott Company)*

FIGURE 26-5. When using a welding helmet that does not have a flip-up shaded lens, the weldor should wear standard safety glasses or goggles under the helmet.

hands and wrists. All skin areas must be covered or "sunburning" will occur from the rays of the arc. For many welding operations, protective clothing such as a leather apron, sleeves, jacket, or pants may be needed to protect the body from sparks and hot metal, figure 26-6. Further, suitable safety glasses or goggles are essential when chipping, cleaning, and examining welds.

SETTING UP

When preparing to weld, the appropriate electrode must be selected. For the beginning weldor this is likely to be an E6011 or E6013 electrode. As described in the preceding unit, the best single choice is probably the E6011. The E6013 is easier to use but is recommended primarily for thin metal.

The next step is to check the welding area to eliminate fire hazards. Gather all the necessary materials, including a piece of $^1/_4$" × 4" × 6" plate steel for practice, wire brush, chipping hammer, pliers to hold hot metal, and bucket of water to cool the metal. The practice metal is called a **pad**. It may be thicker and larger than specified, if desired. To improve handling, an electrode may be bent into a U shape and the ends welded to the pad to provide an insulated handle.

The welding machine should have suitable welding cables. One cable will end in an electrode holder and the other in a ground clamp. The metal pad to be used for practice is clamped to the welding table using the ground clamp, a vise-grip, or a spring

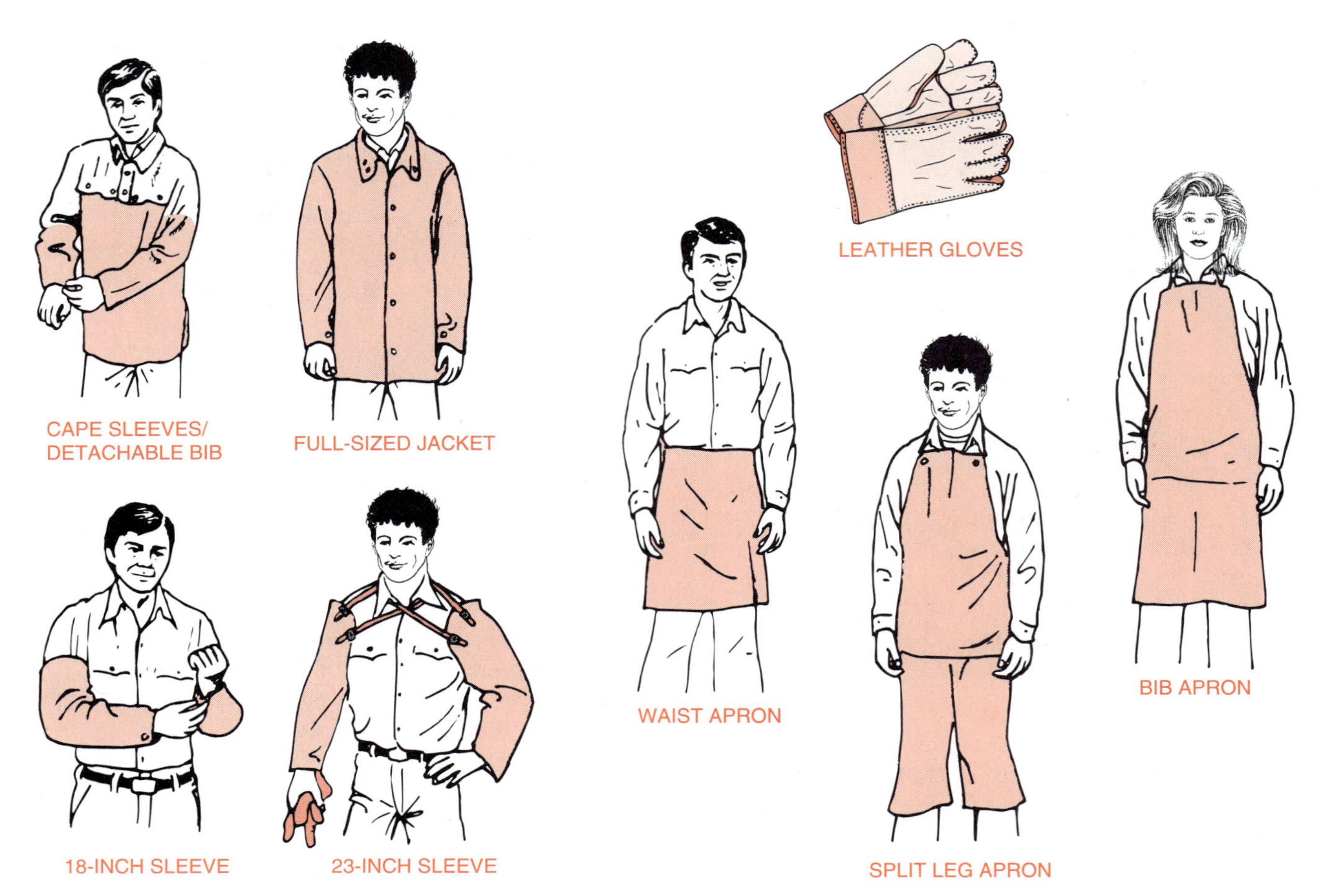

FIGURE 26-6. Certain welding jobs require the operator to wear leather gloves and other protective clothing such as an apron, sleeves, jacket, and/or pants.

clamp. When an extra clamp is used, the ground clamp is attached to either the practice metal or the table.

The operator must wear a welding helmet or use a face shield with a No. 10 lens. It is recommended that tight-fitting coveralls, leather shoes, and leather gloves be worn. Pants must not have cuffs as these will catch burning steel in the form of sparks.

Once dressed, the operator is ready to set the welder. Most welders have a rotary switch or lever that moves a pointer to the desired ampere setting. On other welders, there is a selection of sockets into which the cables can be plugged. The desired amperage setting is obtained by plugging the cables into the correct pair of sockets. For practice welding, electrodes 1/8 inch or 5/32 inch in diameter are to be used. Appropriate welder settings range between 70 and 220 amperes, depending on the welder and the electrode selected, figure 26-7. A recommended starting setting is 125 amperes for 1/8-inch E6011 or E6013 electrodes. The amperage must then be adjusted according to the individual machine.

ELECTRODE SIZE	CLASSIFICATION					
	E6010	E6011	E6012	E6013	E7016	E7018
3/32 in	40-80	50-70	40-90	40-85	75-105	70-110
1/8 in	70-130	85-125	75-130	70-125	100-150	90-165
5/32 in	110-165	130-160	120-200	130-160	140-190	125-220

FIGURE 26-7. Welding amperage ranges for various electrode classifications.

STRIKING THE ARC

Before striking an arc, the metal should be placed on the welding table. It should be located so the operator can reach across the metal and weld in a comfortable position. The metal should be clean and free of grease, oil, or rust. The following procedure is suggested for striking an arc.

PROCEDURE

1. Place a 1/8-inch E6011 electrode in a 90° position in the electrode holder.
2. Close all curtains in the welding booth.
3. Warn others that you are about to start welding by saying "Cover up!" This means they should place shielding in front of their eyes if they can see your work.
4. Turn on the welder.
5. Position the tip of the electrode about 1/4 inch above the practice metal and lean it slightly in the direction you plan to move. Right-handed weldors move from left to right.
6. Drop the face shield or cover lens over your eyes.
7. Strike the arc by quickly lowering, touching, and lifting the electrode, figure 26-8. The action is similar to striking a match. The lift should only be about 1/8 inch.
   a. If you do not lift the electrode in time, it will stick to the metal. If it does, use a quick whipping action to free it or release the electrode holder.

CAUTION:

- **Do not turn the welder off with the electrode frozen in the circuit as the welder would be damaged. First, release the electrode holder from the electrode.**

   b. If the flux becomes cracked or broken on the end of the rod, lay the electrode aside for a more experienced welder to use.

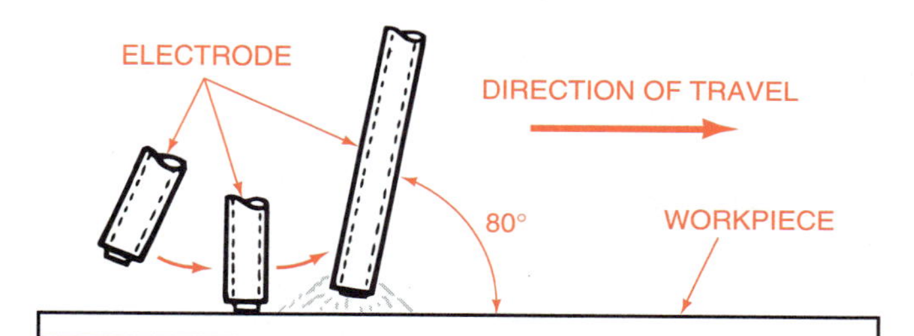

FIGURE 26-8. Striking the arc.

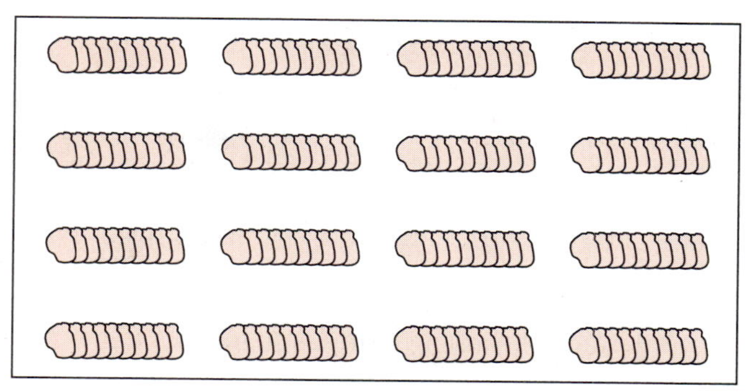

FIGURE 26-9. Practice striking and maintaining an arc until you can make a short bead of 1/2 to 1 inch long. The practice beads should be made in parallel lines across the metal. *(Adapted from Larry Jeffus,* Welding Principles and Applications, *Third Edition, Delmar Publishers, 1995)*

   Obtain another electrode for your next attempt to strike an arc.
   c. If the arc goes out, try again; keep the electrode closer to the metal after lifting it.
8. Feed the electrode very slowly as the weld metal burns away, keeping it about 1/8 inch from the metal. As you feed the electrode, move very slowly and evenly across the plate (generally from left to right for right-handed weldors).
9. Lift the electrode slowly to break the arc after traveling about 1 inch.

   NOTE: As you practice this technique, strike the arc and hold a slightly longer arc for a second or two before lowering the rod and moving across the metal. This practice preheats the metal and results in a better bead.

10. Continue practicing steps 5 through 9 until you can start and stop the arc at will. When you stop the bead, lift the electrode slowly to permit the crater to fill. The crater is a low spot in metal where the force of a flame has pushed out molten metal.
11. Make 16 strikes with 1-inch beads. Compare your results with figure 26-9. Discuss your welds with your instructor.

The method described is called the scratch method of striking an arc. Some people find it easier simply to touch the metal and raise the electrode. This is known as the tapping method. Both methods are acceptable. Striking an arc and keeping it going takes a lot of practice. However, it is well worth the time and patience required, since it is the starting point for arc welding.

RUNNING BEADS

A bead is produced by handling the electrode in a way that results in a proper mix of filler and base metal. A bead is called a **stringer bead** if it is made without weaving. **Weaving** means moving the electrode back and forth sideways to create a bead that is wider than the electrode would normally make. After learning to strike and maintain an arc, the student should practice making stringer beads.

Making stringer beads helps the student learn to:

- start a bead at the desired location
- hold the electrode at about an 80 degree angle
- feed the electrode at an even rate
- move across the metal in a straight line
- move across the metal at an even rate of speed
- stop or terminate a weld at the desired spot and in an acceptable manner.

Most of these skills must be developed at the same time. This means that attention to technique is important. It is helpful if someone who can make suggestions for correcting the student's technique watches the student practice.

Angle of the Electrode

Experienced weldors vary the angle of the electrode according to the electrode being used and the job. However, the beginner should lean the electrode slightly in the direction of travel. A 75 degree to 80 degree angle, 10 degrees to 15 degrees from the vertical (straight up) position, is suggested.

Arc Length

The arc length indicates the skill with which the electrode is fed as it burns away. The sound of the arc is a good guide. An arc that is the correct length sounds like bacon frying. If the arc is too long, a booing sound is heard. If the arc is too short, the electrode sticks to the metal.

Movement Across the Metal

The ability to weld in the intended direction is not easy. Looking through a dark lens, it is difficult to see a soapstone mark, a groove in the metal, another welded bead, or the molten puddle. However, the arc lights the surrounding area. With experience, it becomes possible to observe the welding process with ease.

Speed

To achieve a uniform rate of travel across the metal requires practice in arm movement. It requires the weldor to be in a comfortable position that permits such movement. The best way to control the movement is to watch the welded metal solidifying behind the puddle. When the electrode is moving at an even rate of speed, the weld material forms evenly spaced semicircles behind the puddle.

Amperage Setting

The correct amperage can be obtained by observing the welding process and the weld that results. If the amperage is too low, it will be difficult to strike the arc and keep it running. Low amperage results in a narrow, stringy bead. If the arc seems to struggle or start and stop, it is advisable to try a higher amperage. Increase the setting by 10 to 15 amperes and try again.

Learn to observe the shape of the puddle as it is forming. Long, narrow semicircles indicate the heat is too low. Semicircles that are wider than they are long indicate that the correct heat is being used. Gourd-shaped marks indicate that the heat is too high, figure 26-10.

Other symptoms or indicators of excessive heat are:

- the electrode covering turns brown
- the bead does not have clear markings
- the puddle burns through the plate
- it is difficult to manage the puddle
- the arc is very noisy.

When these conditions exist, decrease the current setting by 10 to 15 amperes and try again.

Making a Practice Bead

To practice running beads. it is best to make short beads about $^{3}/_{4}$ inch apart on the practice pad. This

AMOUNT OF HEAT DIRECTED AT WELD | WELD PUDDLE

TOO LOW

CORRECT

TOO HOT

FIGURE 26-10. The amount of heat directed on the welding puddle affects the shape of the puddle. *(Courtesy of Larry Jeffus,* Welding Principles and Applications, Third Edition, *Delmar Publishers, 1995)*

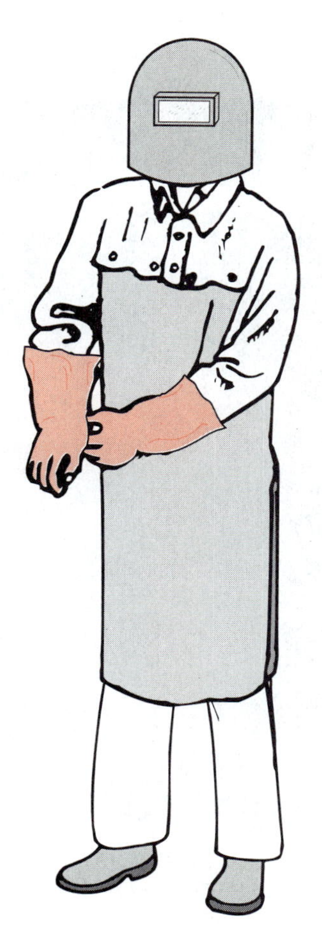

FIGURE 26-11. This student is wearing the proper protective clothing and is ready to weld in any position.

is far enough apart to evaluate the individual beads and also permit most efficient use of the metal.

It is advisable to use both hands to hold the electrode holder, or use the other hand to steady the wrist of the hand holding the electrode holder. It is important to wear proper protective clothing so hot slag cannot reach the skin and a comfortable position can be maintained, figure 26-11. To practice running beads, the following procedure is suggested.

PROCEDURE

1. Clamp a 1/4" × 4" × 6" welding pad in a suitable position on the table.
2. Set the welder at 110 amperes and turn on the welder.
3. Insert a 1/8-inch E6011 electrode in the holder at 90 degrees.
4. Cover up and remind others to do so too.
5. Strike the arc and quickly move the electrode to the starting point near the edge of the pad.
6. Watch the puddle, feed the electrode, move from left to right across the pad, observe

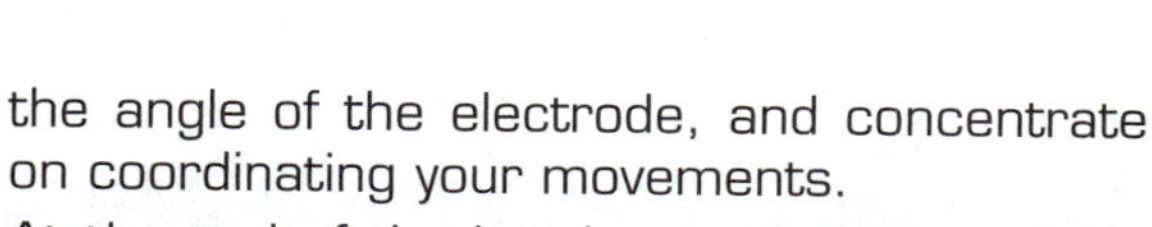

the angle of the electrode, and concentrate on coordinating your movements.

7. At the end of the bead near the edge of the plate, lift the electrode gradually until the crater fills and the arc goes out.
8. Put on clear goggles or a clear face shield (if you are not wearing safety glasses or goggles under your helmet), and remove the slag by striking the edges of the bead with a chipping hammer.
9. Compare your results with a welding chart or sample beads provided by the teacher, figure 26-12.
10. Repeat the process until you can strike the arc and run a good bead over and over again. Readjust the amperage setting if necessary. As the pad accumulates heat from welding, the amperage may need to be lowered to prevent overheating of the electrode and metal.

Weaving

After learning to run stringer beads, it is useful to practice running wider beads by weaving. Many patterns of weaving are used by experienced weldors to help control the heat in the puddle. These patterns are generally practiced as welding in different positions is learned. The beginning weldor should use a circular pattern, which provides good control of the bead width, figure 26-13.

WELDING JOINTS

The object of welding is generally to fasten metal pieces together. There are many types of joints, but most are variations of the butt joint and fillet joint. In a butt joint, the pieces are placed end to end or edge to edge. In a fillet joint, the two parts come together to form a 90 degree angle.

Butt Welds

Butt welds require the proper preparation. There must be a gap between the pieces to be welded of about the thickness of the electrode core. If the metal is thicker than the electrode core, it must be ground down. Grinding may be done on one or both sides of the metal. Grinding on both sides will produce stronger welds. The thickness of the unground metal should be no more than the diameter of the electrode core. For example, when welding with a 1/8-inch electrode, the top and/or bottom of the metal is ground at a 30 degree angle to leave a 1/8-inch thickness, figures 26-14 and 26-15.

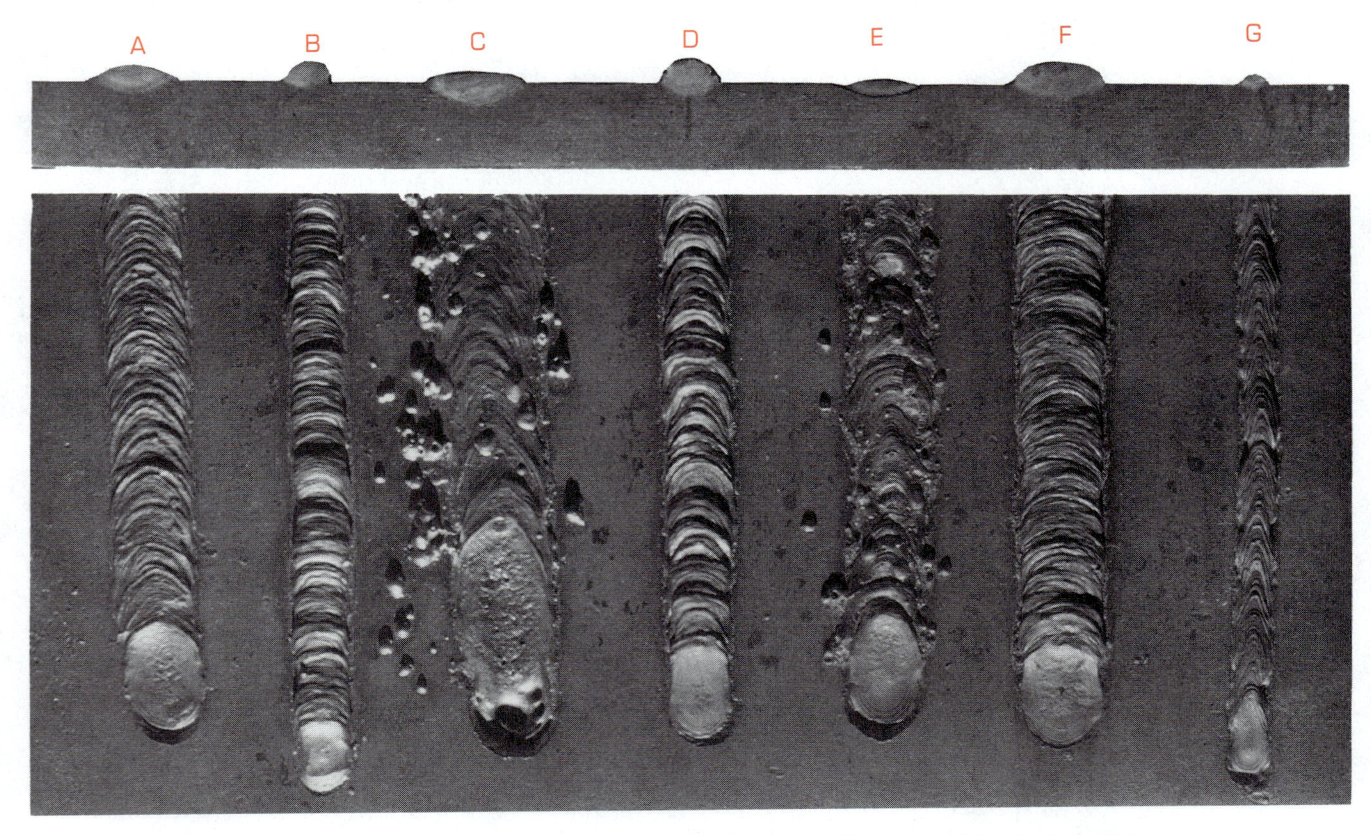

A. PROPER CURRENT, TRAVEL SPEED, AND ARC LENGTH
B CURRENT TOO LOW
C. CURRENT TOO HIGH
D. ARC LENGTH TOO SHORT
E. ARC LENGTH TOO LONG
F. TRAVEL SPEED TOO SLOW
G. TRAVEL SPEED TOO FAST

FIGURE 26-12. A careful study of welding beads will reveal the errors made while welding. In this way, the weldor can correct techniques and develop proficiency. *(Courtesy of Lincoln Electric Company)*

THIS WEAVE PATTERN RESULTS IN A NARROW BEAD WITH DEEP PENETRATION.

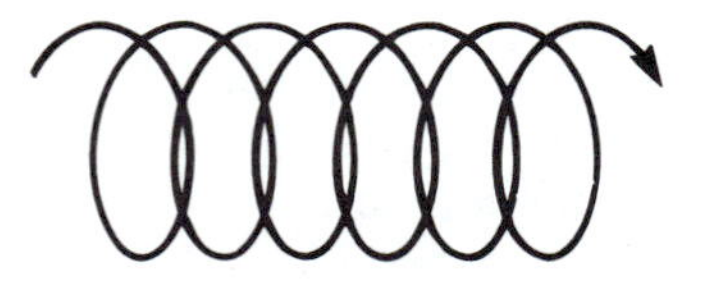

THIS WEAVE PATTERN RESULTS IN A WIDE BEAD WITH SHALLOW PENETRATION.

FIGURE 26-13. Weaving patterns are used to create wider beads and help control the heat when welding. *(Courtesy of Larry Jeffus,* Welding Principles and Applications, Third Edition, *Delmar Publishers, 1995)*

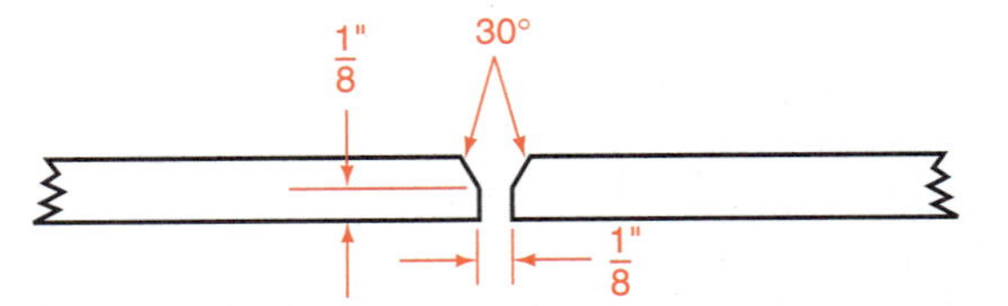

FIGURE 26-14. Before welding a butt joint, the metal pieces are ground and properly positioned. When welding with a 1/8-inch electrode, the top and/or bottom of the metal is ground at a 30° angle to leave a 1/8-inch thickness for the root weld.

When welding, one bead or layer of filler metal is called a pass, figure 26-16. The first pass made in a joint is called the root pass. The root pass is the most important pass in a weld.

When making a butt weld, the following procedure is suggested.

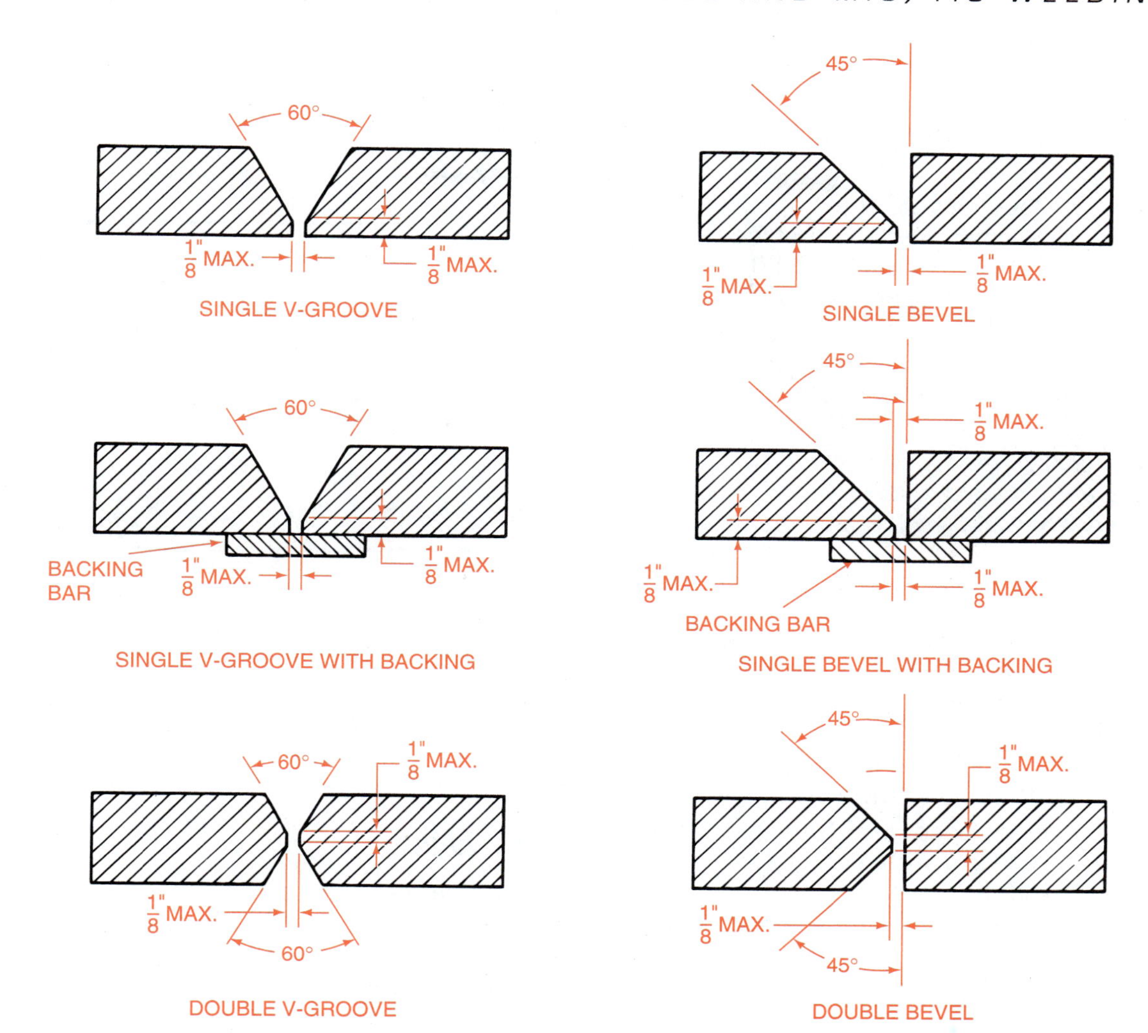

FIGURE 26-15. Typical butt joint preparations. *(Courtesy of Larry Jeffus,* Welding Principles and Applications, Third Edition, *Delmar Publishers, 1995)*

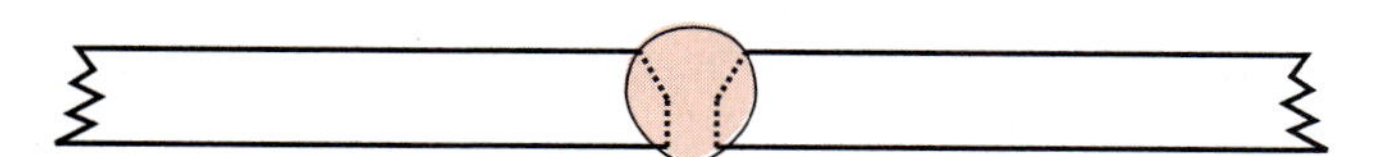

FIGURE 26-16. Butt weld using one heavy pass.

PROCEDURE

1. Obtain two pieces of 1/4" x 2" x 6" flat steel.
2. Grind one edge of each piece so the matching surfaces are 1/8 inch thick.
3. Position the two pieces 1/8 inch apart.
4. Set the welder at 110 amperes and obtain 1/8-inch E6011 electrodes.
5. Wear protective clothing.
6. Tack weld each end of the joint with about 1/4 inch of weld.
7. Straighten the joint so the two pieces are flat.
8. Strike the arc and keep a long arc as you move to the edge of the metal. Shorten the arc to the correct length and proceed to weld. Rotate the electrode in a circular pattern to create an even bead about 1/4 inch wide.
9. Chip the weld to remove all slag. Use a chipping hammer blade or point and a stiff wire brush to remove the slag.
10. Examine the weld for evenness and penetration. The answer should be "yes" to the following questions.
    - **a.** Do the sides blend in evenly with the base metal?
    - **b.** Are all the semicircles of the bead evenly spaced?
    - **c.** Does the weld go all the way through to the bottom?
    - **d.** Does the weld start at the starting edge?
    - **e.** Does the weld fill the groove at the starting edge?

f. Does the weld go all the way to the finishing edge?

g. Does the weld fill the groove at the finishing edge?

11. If the bead does not fill the gap, run one additional wide bead over the first bead, figure 26-17. If thicker metal is used, two or more beads may be needed to complete the weld, figure 26-18.

CAUTION:

- ***All* slag must be removed between beads to ensure a solid weld with no voids. Voids are caused when slag is trapped inside the weld.**

Prepare another butt weld, correcting any problems observed in the first weld. To make an internal check, the procedure follows.

PROCEDURE

1. Saw 1 inch off the end of the welded piece.
2. Examine the cross section of the weld. Voids that are open or filled with slag indicate weak parts of the weld, figure 26-19.
3. Saw off another inch-long section of the weld.
4. Clamp one side of the 1-inch section in a heavy vise with the welded section exposed.
5. Bend the welded section by driving the exposed piece with a heavy hammer. The welded section should be as tough and strong as the rest of the metal. If the welded section breaks, determine what steps are needed to improve the weld.

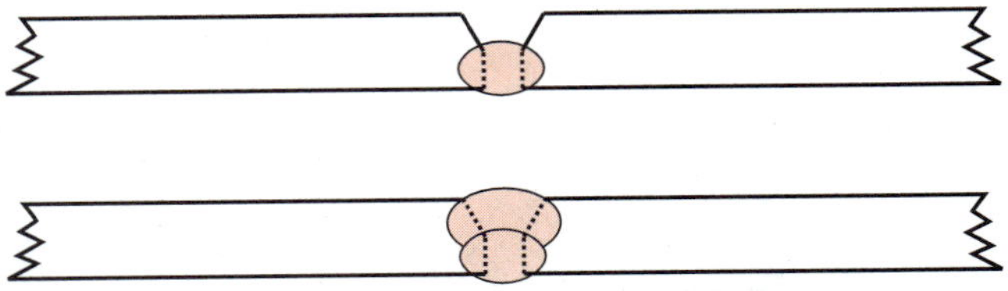

FIGURE 26-17. Butt weld using two passes.

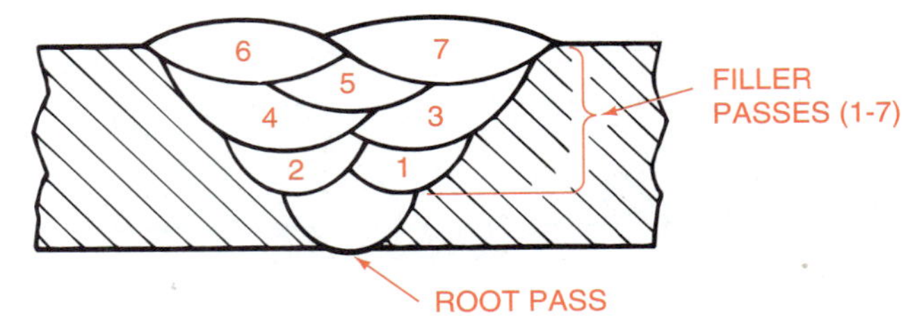

FIGURE 26-18. Butt weld using many passes. *(Courtesy of Larry Jeffus,* Welding Principles and Applications, Third Edition, *Delmar Publishers, 1995)*

Fillet Welds

Most of the principles and procedures that apply to butt welds also apply to fillet welds. To practice making fillet welds, prepare the piece that will be vertical so the weld metal will fuse both pieces completely. It is not necessary to grind 1/8-inch thick metal. However, thicker pieces need to have one or both edges beveled to permit welds to penetrate the entire thickness. A 1/8-inch electrode should penetrate about 1/8 inch deep. Therefore, metal that is thicker than 1/8 inch needs to be ground on one or both sides.

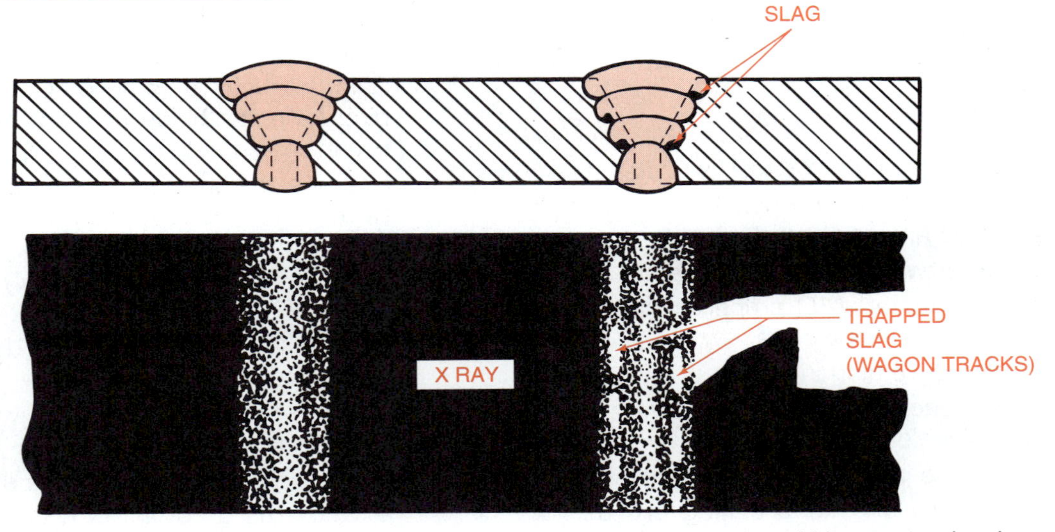

FIGURE 26-19. Slag trapped between passes will show on a cross-section of X-ray examination. *(Courtesy of Larry Jeffus,* Welding Principles and Applications, Third Edition, *Delmar Publishers, 1995)*

FIGURE 26-20. Making the first pass of a fillet weld

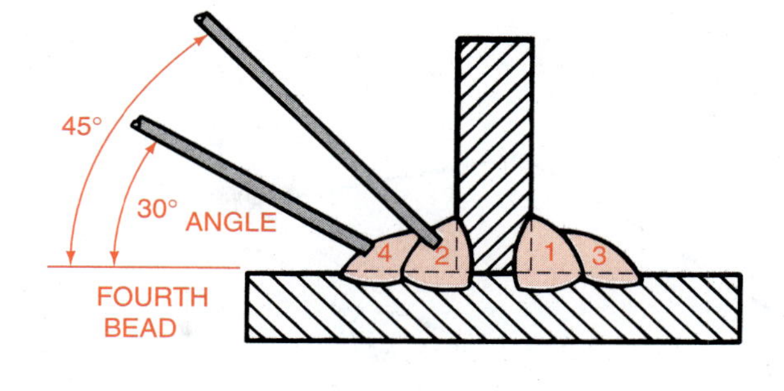

The recommended procedure for making a fillet weld follows.

PROCEDURE

1. Obtain one piece of $\frac{1}{8}$" × 1" × 6" flat steel.
2. Lay the piece on its edge on the piece of metal used in the butt welding exercise described previously; or use a new piece of $\frac{1}{8}$" × 2" × 6" or $\frac{1}{4}$" × 2" × 6" metal for the base piece.
3. Clamp the assembly to the table or steady the parts with firebricks.
4. Select a $\frac{1}{8}$-inch E6011 electrode and set the welder at 110 amperes.
5. Tack weld both ends of the fillet joint.
6. Square the vertical piece so it is at a 90 degree angle to the bottom piece.
7. Hold the electrode at a 45 degree angle to the table top and lean it in the direction of travel.
8. Run a bead from end to end along the joint, figure 26-20. Watch the puddle to be sure it penetrates both pieces equally. To create more penetration in one piece or the other, direct the end of the electrode more to that piece. Adjust the angle of the electrode as needed to obtain equal penetration. Adjust the amperes if needed. Check for undercut or overlap in the sides of bead, figure 26-21.
9. Run a bead on the other side of the vertical piece. Alternating passes from one side to the other reduces the problem of distortion.
10. Chip and examine the head.
11. Make additional passes following the steps shown in figure 26-22.

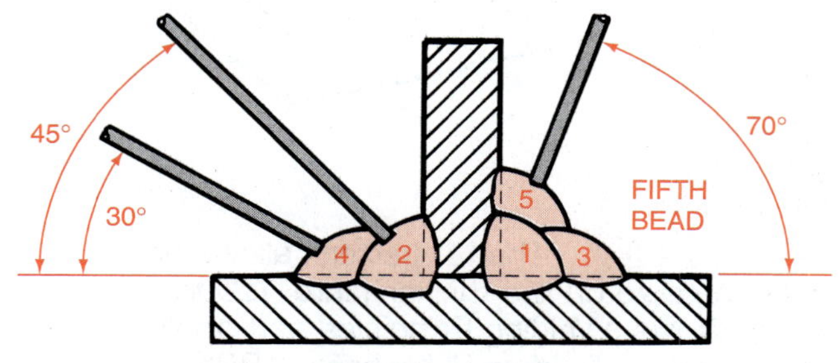

FIGURE 26-22. Alternating passes to control distortion.

It is important to remember that heat rises in metal. The weldor must make allowances for this fact in certain types of welds. For example, when making a fillet weld to form a lap joint, the "J" weave is useful to help control heat. When welding a thin piece of metal on top of a thicker piece, the J pattern concentrates more heat on the thicker bottom piece, figure

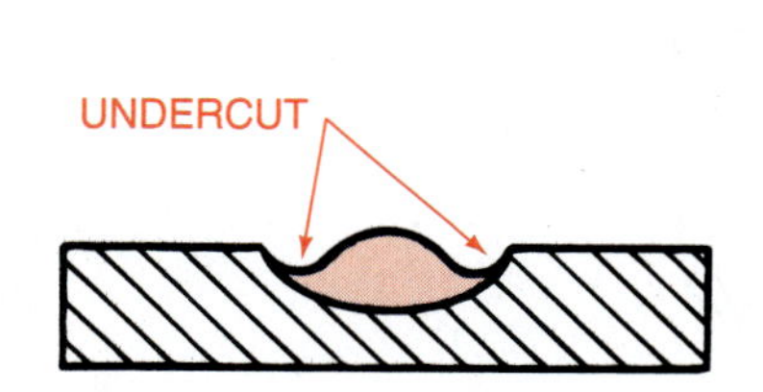

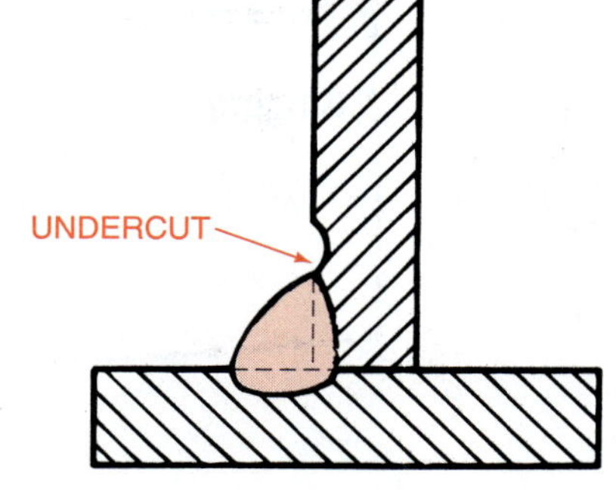

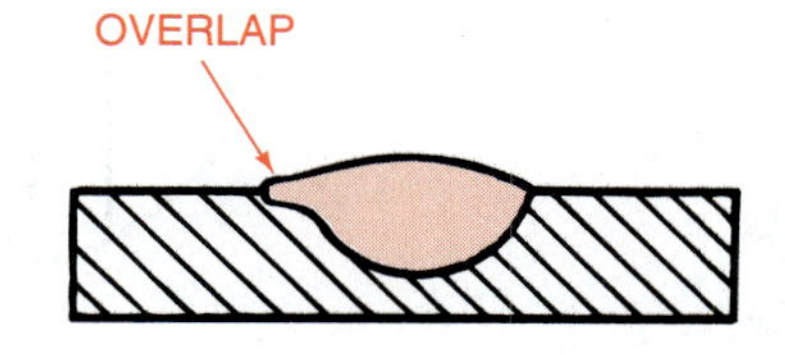

FIGURE 26-21. Undercut and overlap should be avoided in welding beads.

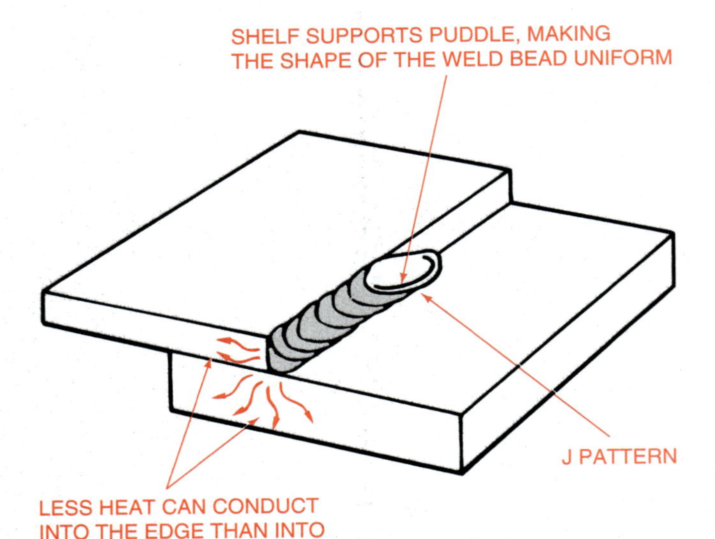

FIGURE 26-23. The "J" pattern allows heat to be concentrated on the thicker plate. *(Courtesy of Larry Jeffus,* Welding Principles and Applications, Third Edition, *Delmar Publishers, 1995)*

26-23. This permits proper weld penetration without overheating the vertical piece.

WELDING IN VARIOUS POSITIONS

Many welding jobs in agricultural mechanics require welding in a variety of positions. Fortunately, modern welding electrodes are available that make welding in all positions possible. The weldor should have reasonable skill in flat or down hand welding before attempting to weld in other positions. Even the experienced weldor will try to position materials so they can be welded in a down hand manner.

Horizontal Welds

Welds made by moving horizontally across a vertical piece of metal are called horizontal welds. When making horizontal welds, the metal is placed so that it is comfortable to run a bead across the surface. The metal should be at eye level so welding from a sitting position is recommended. To make a horizontal bead, the following procedure is suggested.

PROCEDURE

1. Obtain a 1/8" or 1/4" × 4" × 6" welding pad.
2. Weld two small pieces on the edge of the pad so the pad can be set on edge and clamped securely to the table.
3. Put on helmet, gloves, arm and shoulder protection, and a leather apron.
4. Sit in front of the table.
5. Select a 1/8-inch E6011 electrode and set the welder at 110 amperes.
6. Hold the tip of the electrode near the left side of the pad. Lean it in the direction of travel at an angle of about 70 degrees to the plate. Tilt the electrode so it points slightly upward at an 80 degree angle to the plate, figure 26-24.
7. Run a bead across the plate at a rate of speed that crates a proper bead without sagging, undercutting, or overlapping.
8. Adjust the amperes if needed and run additional beads to develop the skill.

Vertical Welds

Welds made by moving downward across a vertical piece of metal are called vertical down welds. Welds made by moving upward across the metal are called vertical up welds. Vertical down welds are easy to make. However, they are shallow in penetration and so should be made only on materials that are 1/8 inch or less in thickness.

To make a vertical down weld, set up a practice pad as described in the section on horizontal welds. A sitting position for welding is recommended. The

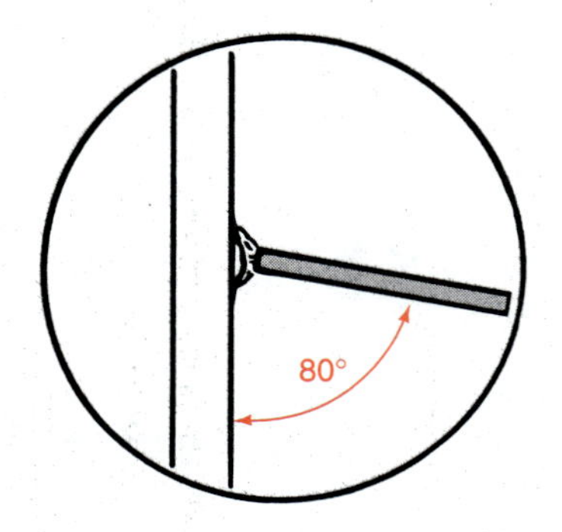

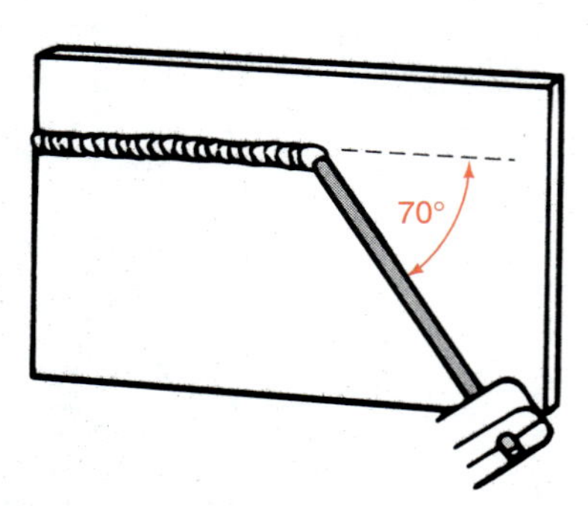

FIGURE 26-24. Electrode angles for horizontal welding.

electrode is held with the tip pointing upward at an 80 degree angle to the plate. Strike the arc at the top of the plate and let the force of the arc push the electrode downward across the plate at an even rate. Weaving is not recommended for this procedure. Adjust the amperes as needed to obtain an even bead.

Vertical up welding is rather difficult to manage, because heat builds up in the metal as the weld progresses up the metal. Special movements of the electrode are used to control the heat and permit individual parts of the puddle to cool. Unless this is done carefully, the weld metal will sag or drop out of the puddle. The shape of the puddle indicates the temperature of the surrounding base metal, figure 26-25.

A shorter arc, lower amperage, steeper electrode angle, and more rapid movement of the electrode may all be helpful in controlling the puddle. To make a vertical up weld on a flat area, the following procedure is recommended.

PROCEDURE

1. Set up a practice pad as outlined under horizontal welding.
2. Start at the bottom of the pad and work upward.
3. Use a wide weaving pattern, moving the tip of the electrode from side to side. Some weldors create the bead by making a series of very short horizontal beads as they progress upward. Other weldors use a figure eight pattern. Others use a wide, crescent-shaped, back-and-forth motion.
4. If the bead gets too hot and is not controllable, stop. Chip and clean the weld. Then start again. Make corrections by reducing the amperage and using more sideways motion to permit parts of the puddle to cool more quickly. Make additional passes, if needed, after the root pass is cool and all slag chipped, figure 26-26.

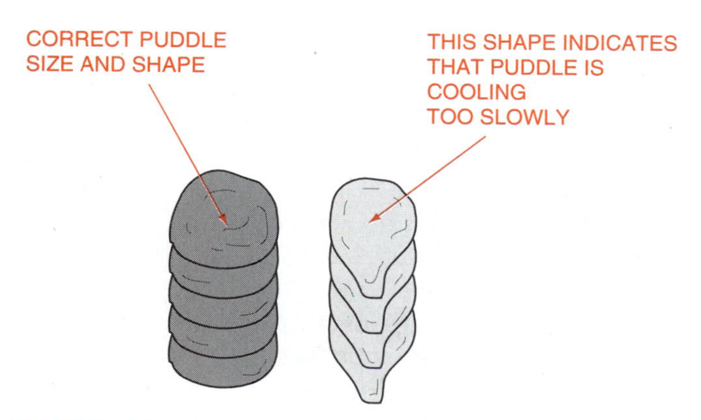

FIGURE 26-25. The shape of the puddle indicates the temperature of the surrounding base metal. *(Adapted from Larry Jeffus,* Welding Principles and Applications, Third Edition, *Delmar Publishers, 1995)*

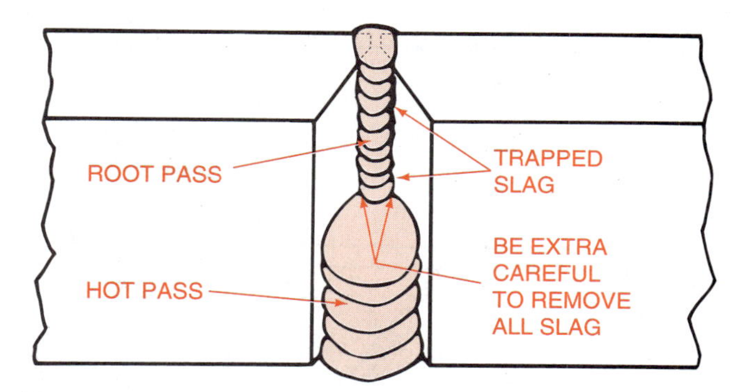

FIGURE 26-26. Making a vertical up weld on a flat area. *(Courtesy of Jeffus,* Welding Principles and Applications, Third Edition, *Delmar Publishers, 1995)*

To make a vertical up fillet weld, a T pattern is helpful. In this pattern, the electrode is directed first at the root of the weld. The movement is to one side, then to the other side, and finally back to the root to start the sequence again, figure 26-27.

Overhead Welding

Overhead welding is not a difficult procedure. However, it can be dangerous without protective clothing. The metal is positioned above the weldor and the weldor assumes a sitting position, if possible.

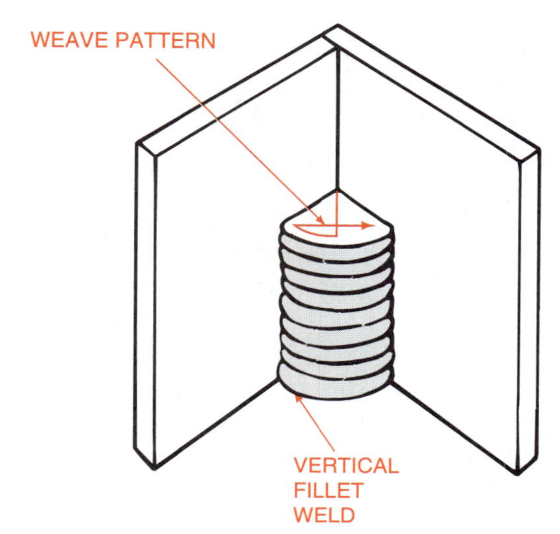

FIGURE 26-27. Making a vertical up fillet weld. *(Courtesy of Larry Jeffus,* Welding Principles and Applications, Third Edition, *Delmar Publishers, 1995)*

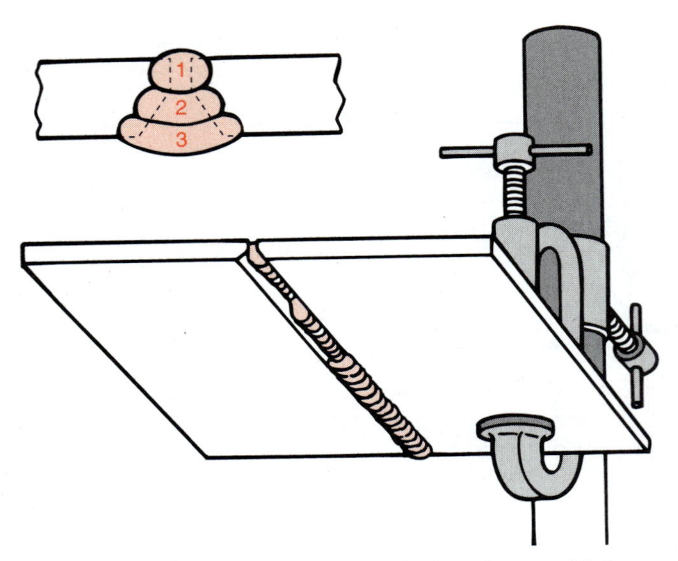

FIGURE 26-28. A welding pad showing multiple passes in the overhead position.

The electrode is held nearly straight up and down. Therefore, some weldors prefer to place the electrode so it extends out the end of the holder. The holder is then gripped like a pencil.

When welding overhead, use an E6011 electrode, a normal amperage setting, electrode angle of 10 to 15 degrees from vertical and normal speed. Welds may be single pass or multiple pass according to the need, figure 26-28.

WELDING PIPE

Welding of pipe is done extensively in agricultural mechanics. Pipe may be used to make gates, trailers, wagon sides, athletic equipment, and other projects. Used pipe is a cheap source of construction material in many communities.

Pipe is made of thin metal. Therefore, pipe welding requires the use of electrodes that produce shallow penetration. The E6013 electrode is a good choice. It is a fast-filling electrode that makes it suitable when fits are not perfect.

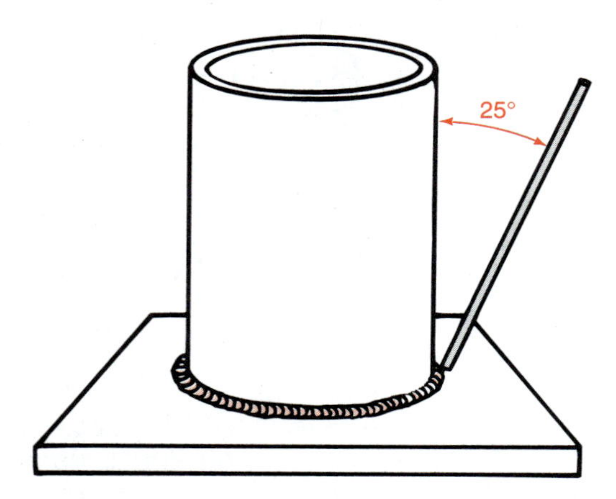

FIGURE 26-29. Making a fillet weld on pipe.

Pipe is generally welded without grinding. As the electrode proceeds around a pipe, the rate of travel must be faster at the electrode holder than at the electrode tip. It is important to maintain the same relationship of electrode-to-weld-surface as used for straight areas, figure 26-29. Welding of pipe ends is easier if the ends are first flattened. This permits the weldor to make straight passes.

CAUTION:

- **Do not weld galvanized pipe due to the hazard of poisonous fumes except under the supervision of an expert.**

PIERCING, CUTTING, AND SPOT WELDING

Piercing and cutting with arc welders is not as fast and clean as with an oxyacetylene torch. However, it is useful procedure for those who have arc welders but not an oxyacetylene torch.

Piercing

To pierce a hole with an arc welder, the following procedure is recommended.

PROCEDURE

1. Clamp a piece of 1/4" × 2" × 6" steel so it projects 4 inches from the table top.
2. Place a bucket half filled with water or a sandbox under the metal.
3. Put on protective clothing, including leather shoes and a leather apron and/or leggings.
4. Determine the correct amperage for welding and increase this value by 50 percent. Set the amperage at the increased value.
5. Select a 1/8-inch E6011 electrode or a special cutting electrode.
6. Strike the arc above the metal and hold a long arc (about 1/4 to 3/8 inch) until the metal is molten.
7. Rotate the tip of the rod to create an enlarged molten puddle.
8. Quickly push the electrode into the molten puddle to create a hole. If the rod does not go through, quickly withdraw the electrode or whip it loose if it sticks. Start again but make the molten puddle deeper before thrusting the electrode through the metal.

9. Keeping the arc burning, rotate the tip of the electrode around the hole to enlarge it as needed.
10. If a smooth hole is required, quickly place the hole in the metal over the hole in an anvil. Then drive a blacksmith's punch into the hole while the metal is still very hot.
11. Smooth the underside of the piece with a grinder as needed.

NOTE: If additional amperage gives better results, use a higher setting. However, excessive amperage burns and overheats the rod and creates a heavier load on the machine.

Cutting

To cut with an arc welder, the following procedure is recommended.

PROCEDURE

1. Use the same piece of metal as in the piercing exercise.
2. Use an E6011 electrode or a special cutting electrode.
3. Set the amperage 50 percent higher than for welding.
4. Strike the arc and direct a long arc at the edge of the metal until it becomes molten.
5. Use a quick up-and-down chipping movement to gouge out the molten metal and create a kerf across the metal.
6. Smooth the cut with a grinder, if needed.

Spot Welding

Electricity can be used to heat two pieces of metal in a small area between two electrodes. When the heat is sufficient to cause the metal to melt, welding occurs. The process is called spot welding and the device used to spot weld is called a spot welder, figure 26-30. A spot welder is used when sheet metal must be fastened in many places near the edges.

To operate a spot welder, two pieces of sheet metal are placed between the electrodes. The length of the electrode holders determines how far from the edge of the metal a spot weld can be made. The operator provides pressure to bring the electrodes into contact with the metal, and the machine provides the heat. The machine shuts off automatically when the weld is complete. The actual welding takes only a split second once it starts.

FIGURE 26-30. A spot welder.

MIG WELDING

Metal inert gas (MIG) welding became popular when manufacturers began using thin-gauge, high-strength, low-alloy (HSLA) steels. Manufacturers insisted that the only correct way to weld HSLA and other thin-gauge steel was with MIG (or similar gas metal arc welding [GMAW] system). Once the MIG welder was in place, it was easy to see that it provided clean, fast welds for all applications.

MIG welding is ideal for exhaust system work, repairing mechanical supports, installing trailer hitches and truck bumpers, and any other welds that would be done with either an arc or gas welder. In addition, it is possible to weld aluminum castings like cracked transmission cases, cylinder heads, and intake manifolds.

Safety in MIG Welding

MIG welding involves the use of electrical equipment as well as gas welding equipment. Therefore, a study of both gas and electrical arc welding should be completed before doing MIG and TIG welding. Some important safety practices to observe when MIG welding are:

1. Carefully follow the manufacturer's recommendations provided with the welder.
2. Check all welding cables to be sure they are in good repair and properly connected. Be sure the equipment is properly grounded.
3. Wear leather gloves, body jackets, and chaps for protection against burns.
4. Use an approved helmet with a minimum #11 shaded lens for non-ferrous and #12 for ferrous metals.

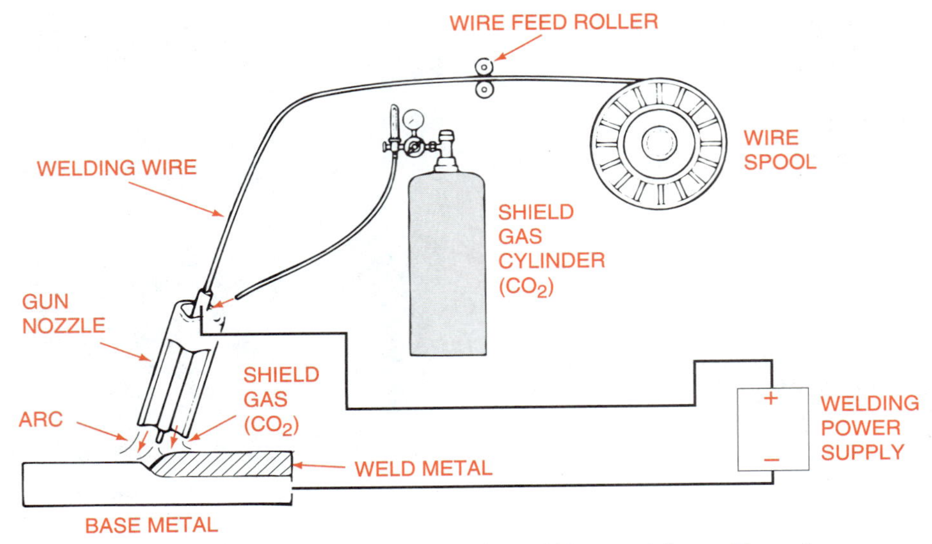

FIGURE 26-31. The principle of MIG welding. *(Courtesy of Toyota Motor Corp.)*

5. When the electrical switch is on, never touch electrical connections or the welding wire.
6. Never weld in wet locations or with wet hands, feet, or clothing.
7. Be sure there are no matches or other flammable materials in your pockets.They could ignite.
8. Handle hot metal with pliers or tongs.
9. Weld only in well-ventilated places.
10. Have only competent professionals do repair work on welding equipment.
11. When finished welding, be certain the welding equipment is turned off and safely stored.

Principles and Characteristics

The MIG welding method uses a welding wire that is fed automatically at a constant speed as an electrode. A short arc is generated between the base metal and the wire, and the resulting heat from the arc melts the welding wire and joins the base metals together. Since the wire is fed automatically at a constant rate, this method is also called semiautomatic arc welding. During the welding process (figure 26-31), either inert or active gas shields the weld from the atmosphere and prevents oxidation of the base metal. The type of inert or active gas used depends on the base material to be welded. For most steel welds, carbon dioxide (CO_2) is used, figure 26-32. With aluminum, either pure argon gas or a mixture of argon and helium is used, depending on the alloy and the thickness of the material. It is even possible to weld stainless steel by using argon gas with a little oxygen (between 4 and 5 percent) added.

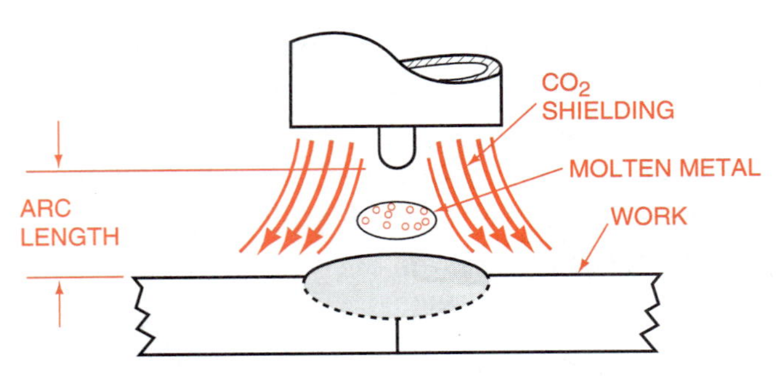

FIGURE 26-32. Carbon dioxide (CO_2) protects the molten metal from contamination by the atmosphere.

MIG welding is sometimes called carbon dioxide arc welding. Actually, MIG welding uses a fully inert gas such as argon or helium as a shield gas. Since carbon dioxide gas is not a completely inert gas, it is more accurately called MAG welding (metal active gas). The term *MIG* is used to describe all gas metal arc welding processes. In fact, many welders on the market can use carbon dioxide (a semiactive gas) or argon (inert gas) by simply changing the gas cylinder and the regulator.

MIG welding uses the short circuit arc method, which is a unique method of depositing molten drops of metal onto the base metal. Welding of thin sheet metal can cause welding strain, blow holes, and warped panels. To prevent these problems, it is necessary to limit the amount of heat near the weld. The short circuit arc method uses very thin welding rods, a low current, and low voltage. By using this technique the amount of heat introduced into the panels is kept to a minimum and penetration of the base metal is quite shallow.

As shown in figure 26-33, the end of the wire is melted by the heat of the arc and forms into a drop, which then comes in contact with the base metal and

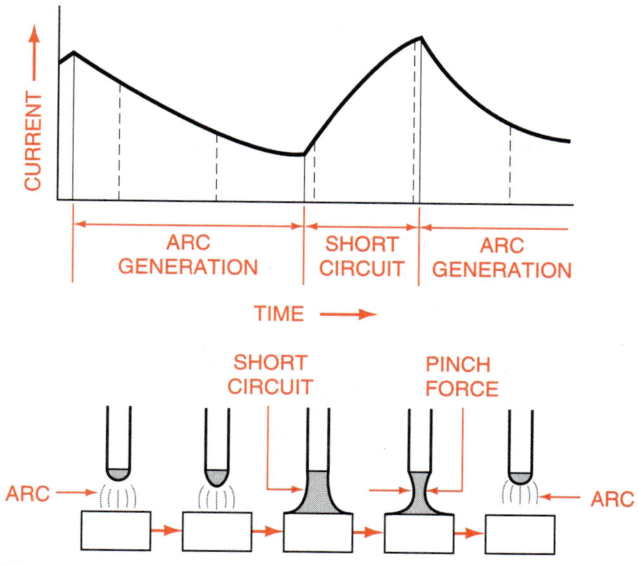

FIGURE 26-33. How the short arc method operates.

creates a short circuit. When this happens, a large current flows through the metal and the shorted portion is torn away by the pinch force or burnback, which reestablishes the arc. That is, the bare wire electrode is fed continuously into the weld puddle at a controlled, constant rate, where it short-circuits, and the arc goes out. While the arc is out, the puddle flattens and cools, but the wire continues to feed, shorting to the workpiece again. This heating and cooling happens on an average of 100 times a second. The metal is transferred to the workpiece with each of these short circuits. Generally, if current is flowing through a cylindrically shaped fluid (in this case molten metal) or current is flowing through an arc, the current is pulled toward the weld. This works as a constricting force in the direction of the center of the cylinder. This action is known as the pinch effect, and the size of the force is called the pinch force, as shown in figure 26-34.

In summary, the MIG process works like this:

- At the weld site the wire undergoes a split-second sequence of short-circuiting, burn-back, and arcing, figure 26-35.

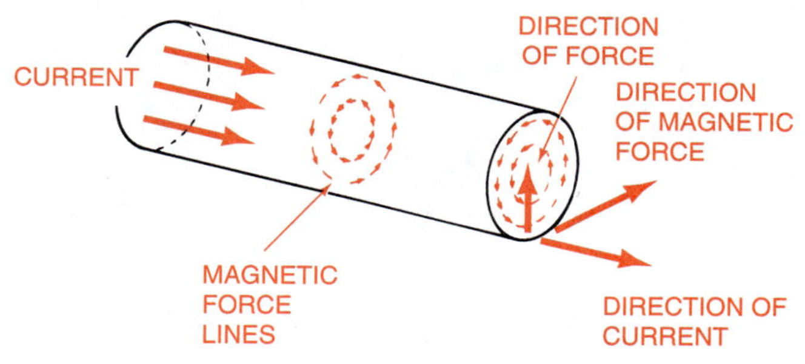

FIGURE 26-34. Typical pinch force and how it is formed.

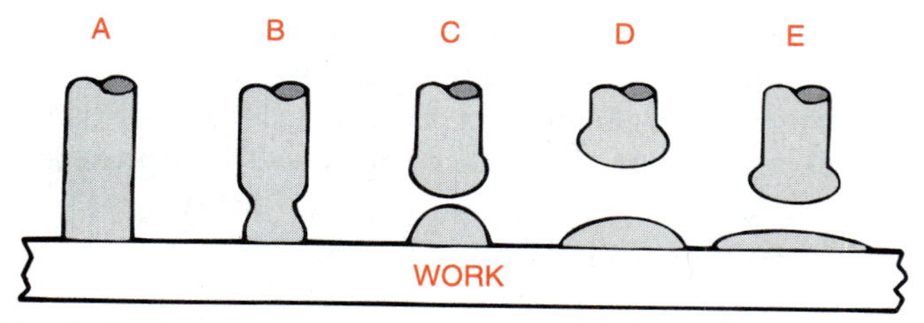

FIGURE 26-35. Here is the typical action of the welding wire as it burns back from the work during the MIG welding process.

- Each sequence produces a short arc transfer of a tiny drop of electrode metal from the tip of the wire to the weld puddle.
- A gas curtain or shield surrounds the wire electrode. This gas shield prevents contamination from the atmosphere and helps to stabilize the arc.
- The continuously fed electrode wire contacts the work and sets up a short circuit. Resistance heats the wire and the weld site.
- As the heating continues, the wire begins to melt and thin out or neck down.
- Increasing resistance in the neck accelerates the heating in this area.
- The molten neck burns through, depositing a puddle on the workpiece and starting the arc.
- The arc tends to flatten the puddle and burn back the electrode.
- With the arc gap at its widest, it cools, allowing the wire feed to move the electrode closer to the work.
- The short end starts to heat up again, enough to further flatten the puddle but not enough to keep the electrode from recontacting the workpiece. This extinguishes the arc, reestablishes the short circuit, and restarts the process.
- This complete cycle occurs automatically at a frequency ranging from 50 to 200 cycles a second.

MIG Welding Equipment

Most MIG welding equipment is semiautomatic. This means the machine's operation is automatic, but the gun is hand-controlled. Before starting to weld, the operator sets:

- Voltage for the arc
- Wire speed
- Shielding gas flow rate and presses the power button.

Then the operator has complete freedom to concentrate entirely on the weld site, the molten puddle, and whatever welding technique is used.

Regardless of the type of MIG equipment used, it will comprise the following basic components (figure 26-36):

- Supply of shielding gas with a flow regulator to protect the molten weld pool from contamination
- Wire/feed control to feed the wire at the required speed
- Spool of electrode wire of a specified type and diameter
- MIG type of welder machine connected to an electrical power supply
- Work cable and clamp assembly
- Welding gun and cable assembly that the welder holds to direct the wire to the weld area.

MIG spot welding is termed consumable spot welding because the welding wire is consumed in the weld puddle. Consumable spot welds can be made in a variety of methods and in all positions using various nozzles supplied with this option.

When spot welding different thicknesses of materials, the lighter gauge material should always be spotted to the heavy material.

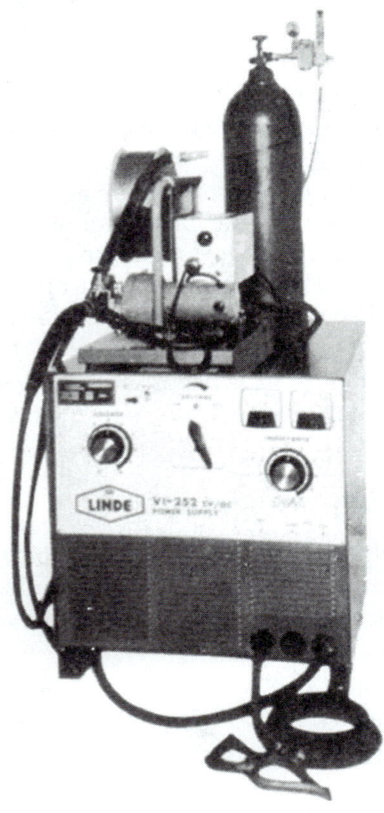

FIGURE 26-36. Basic components of MIG setup.

Spot welding usually requires greater heat to the weld than continuous or pulse welding. It is best to use sample materials when setting the controls for spot welding. To check a spot weld, pull the two pieces apart. A good weld will tear a small hole out of the bottom piece. If the weld pulls apart easily, increase the weld time or heat. After each spot is complete, the trigger must be released and then pulled for the next spot. MIG spot has the advantage of an easily grindable crown. The procedure does not leave any depression requiring a fill.

The pulse control allows continuous seam welding on the material with less chance of burn-through or distortion. This is accomplished by starting and stopping the wire for preset times without releasing the trigger. The weld "on" and "off" time can be set for the operator's preference and metal thickness.

The burn-back control on most MIG gives an adjustable burn-back of the electrode to prevent it from sticking in the puddle at the end of the weld.

In MIG welding, the polarity of the power source is important in determining the penetration to the workpiece. DC power sources used for MIG welding typically use DC reverse polarity. This means the wire (electrode) is positive and the workpiece is negative. Weld penetration is greatest using this connection.

Weld penetration is also greatest using CO_2 gas. However, CO_2 gives a harsher, more unstable arc, which leads with increased spatter. So when welding on thin materials, it is preferable to use argon/CO_2 (figure 26-37). If the material is super thin, weld in straight polarity. In straight polarity, the wire (electrode) would be negative and the workpiece would be positive. This would put more heat in the wire, providing less penetration. The disadvantage of using straight polarity would be a high ropelike bead, requiring more grinding.

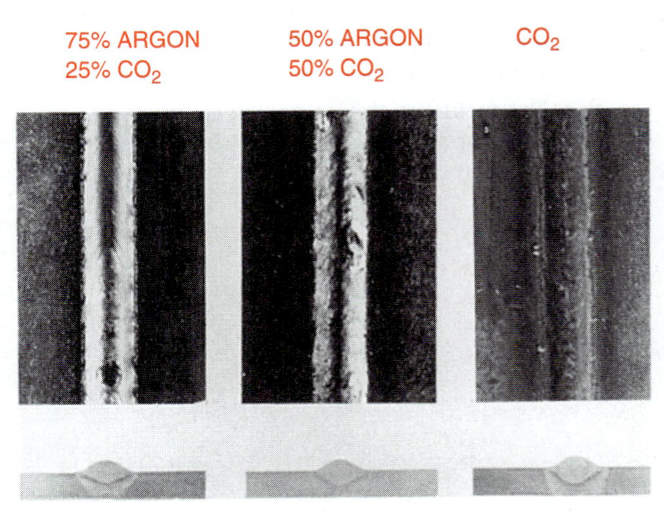

FIGURE 26-37. Typical weld penetration.

Voltage adjustment and wire feed speed must be set according to the diameter of the wire being used. It should be noted that when setting these parameters, manufacturers' recommendations should be followed to reach approximate settings. When rough parameters are selected, change only one variable at a time until the machine is fine tuned for an optimum welding condition. MIG welders can be tuned in using both visual and audio signals.

MIG Weld Defects

Defects in MIG welds and their causes are summarized in Table 26-1. Proper welding techniques ensure good welding results. If welding defects should occur, think of ways to change the method of operation to correct the defect.

When making any MIG repairs, the materials and panels must be similar enough to allow mixing when they are welded together. The melting and flowing of metals can be accomplished by many methods, depending on the materials being joined. The combinations of cleanliness of the welded area, the mixing of proper metals, and the right heat application will result in a good MIG weld.

TIG WELDING

Tungsten inert gas (TIG) welding, another form of GMAW, has somewhat limited applications. MIG welders lay down weld beads at the average of 25 inches per minute. TIG welding is much slower, with weld speeds ranging between 5 to 10 inches per minute. However, this slower speed gives much more control, and the end result is the best looking weld obtainable. A TIG unit can be used to repair cracks in aluminum cylinder beads, reconstruct combustion chambers, and other automotive components that need to be welded.

Like MIG welding, TIG welders use an inert gas such as argon or helium to surround the weld area and prevent oxygen and nitrogen in the atmosphere from contaminating the weld, figure 26-38. Instead of having a wire feed welding electrode like MIG units, TIG machines use a tungsten electrode with a very high melting point (about 6900°F) to strike an arc between the welding gun and the work.

Since the tungsten electrode has such a high melting point, it is not consumed during the welding process, so a filler rod must be used when welding thicker materials. Because the torch is held in one hand and the filler rod is held in the other (figure 26-39), TIG welding is similar in some ways to oxyacetylene welding.

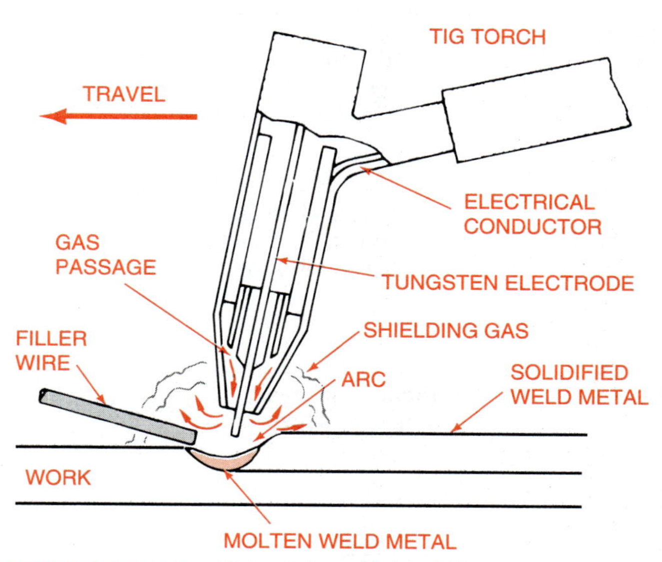

FIGURE 26-38. Principles of the TIG process. If filler metal is required, it is fed into the pool from a separate filler rod.

Safety in TIG Welding

When doing TIG welding, all safety practices listed under "Safety in MIG Welding" should be applied. Some major safety procedures to apply when TIG welding are:

1. Utilize all safety procedures that apply for MIG welding.
2. Wear hearing protection with pulsed power and high-current settings to protect against sound waves from the arc during high-current pulses.
3. Never touch the tungsten electrode with the filler rod or body parts, since the tungsten electrode carries electrical current.
4. Be careful to keep the TIG high- frequency unit setting within the limits prescribed by the manufacturer.

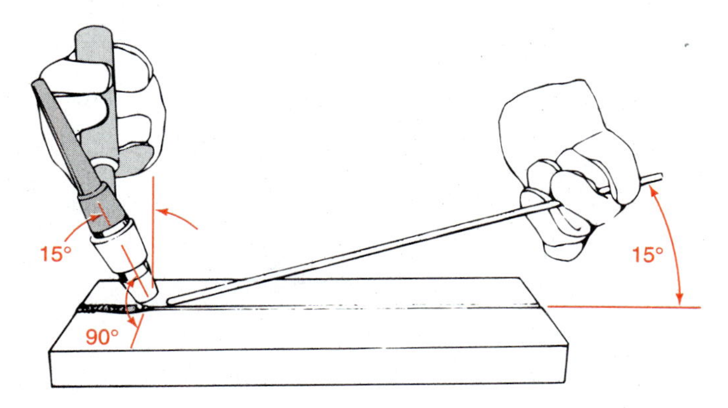

FIGURE 26-39. Proper position of the torch and filler rod for manual TIG welding.

WELDING PRECAUTIONS			
Defect	**Defect Condition**	**Remarks**	**Main Causes**
Pores/Pits	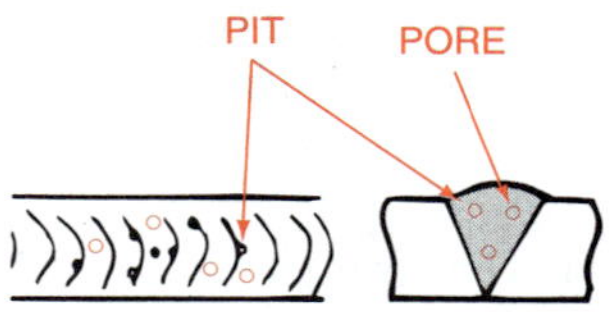 PIT PORE	There is a hole made when gas is trapped in the weld metal.	1. There is rust or dirt on the base metal. 2. There is rust or moisture adhering to the wire. 3. Improper shielding action (the nozzle is blocked or wind or the gas flow volume is low). 4. Weld is cooling off too fast. 5. Arc length is too long. 6. Wrong wire is elected. 7. Gas is sealed improperly. 8. Weld joint surface is not clean.
Undercut	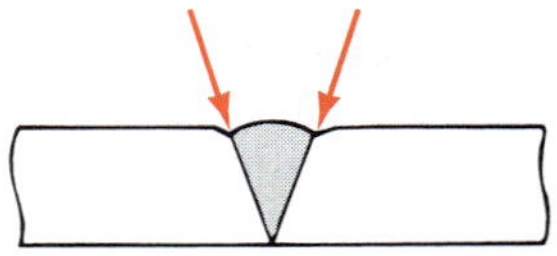	Undercut is a condition where the overmelted base metal has made grooves or an indentation. The base metal's section is made smaller and, therefore, the weld zone's strength is severely lowered.	1. Arc length is too long. 2. Gun angle is improper. 3. Welding speed is too fast. 4. Current is too large. 5. Torch feed is too fast. 6. Torch angle is tilted.
Improper Fusion	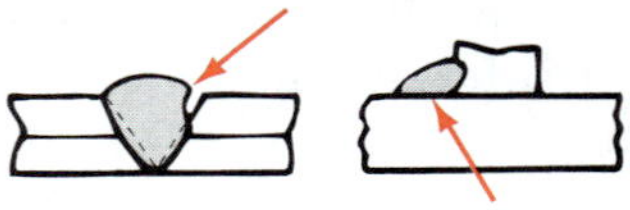	This is an unfused condition between weld metal and base metal or between deposited metals.	1. Check torch feed operation. 2. Is voltage lowered? 3. Weld area is not clean.
Overlap	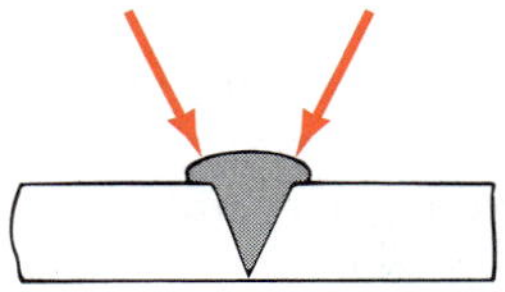	Overlap is apt to occur in fillet weld rather than in butt weld. Overlap causes stress concentration and results in premature corrosion.	1. Welding speed is too slow. 2. Arc length is too short. 3. Torch feed is too slow. 4. Current is too low.
Insufficient Penetration	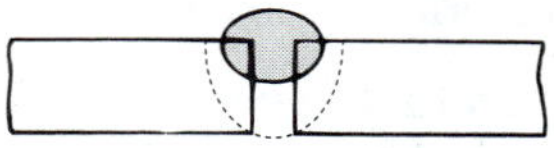	This is a condition in which there is insufficient deposition made under the panel.	1. Welding current is too low. 2. Arch length is too long. 3. The end of the wire is not aligned with the butted portion of the panels. 4. Groove face is too small.
Excess Weld Spatter	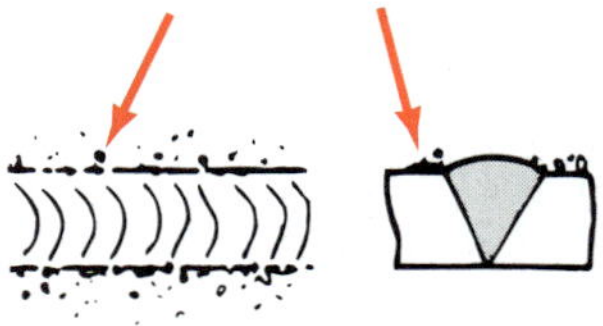	Excess weld spatter occurs as speckles and bumps along either side of the weld bead.	1. Arc length is too long. 2. Rust is on the base metal. 3. Gun angle is too severe.
Spatter (short throat)	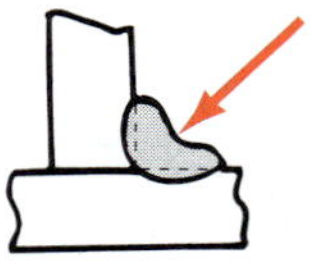	Spatter is prone to occur in fillet welds.	1. Current is too great. 2. Wrong wire is selected.

TABLE 26-1

WELDING PRECAUTIONS			
Defect	**Defect Condition**	**Remarks**	**Main Causes**
Vertical Crack		Cracks usually occur on top surface only.	1. There are stains on welded surface (paint, oil, rust).
The Bead Is Not Uniform		This is a condition in which the weld bead is misshapen and uneven rather than streamlined and even.	1. The contact tip hole is worn or deformed and the wire is oscillating as it comes out of the tip. 2. The gun is not steady during welding.
Burn Through		Burn through is the condition of holes in the weld bead.	1. The welding current is too high. 2. The gap between the metal is too wide. 3. The speed of the gun is too slow. 4. The gun-to-base metal distance is too short.

TABLE 26-1 Continued.

TIG Equipment

Three major components make up a TIG welding machine: the power supply, the welding gun, and the gas cylinder with flowmeter, as shown in figure 26-40. Most TIG units have a sophisticated power supply system (figure 26-41) that can supply current to the electrode as alternating current (AC), direct current straight polarity (DCSP), and direct current reverse polarity (DCRP). The type of current used depends on the type of material to be welded as well as the desired shape and penetration of the weld bead.

Welding with AC is primarily used for nonferrous metals like aluminum. The two DC currents are used primarily for welding ferrous metals like steel and cast iron. With DCSP, electron flow at the arc is from the electrode to the workpiece. The electrons strike the weld at a high rate of speed, which builds heat at the weld, resulting in a narrow weld with a high degree of penetration.

With DCRP, the electron flow is reversed, and heat builds up in the electrode instead of the weld.

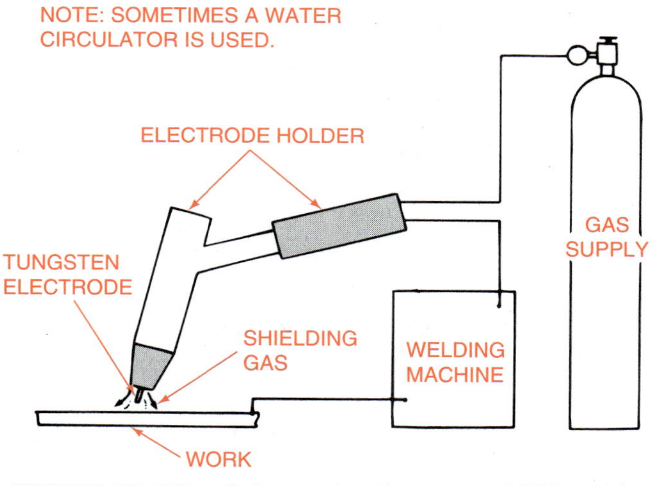

FIGURE 26-40. Schematic of a manual TIG welding outfit.

FIGURE 26-41. Typical TIG welding machine.

To handle the extra heat, a larger diameter electrode is needed for DCRP welding. The resulting weld bead tends to be wide, with minimum penetration.

The welding gun holds the tungsten electrode with a collet that screws into the body of the gun, figure 26-42. A ceramic cup blows the shielding gas around the weld area. The cups are available in a variety of sizes and flow rates to provide optimum shielding for different types of welds. Some guns also have a screened gas nozzle to eliminate troublesome turbulence, which can interfere with the shielding gases' ability to surround the weld.

Some TIG welders have a trigger on the gun to control current and gas flow. Other models have a foot control. When using filler rod, TIG welding becomes a two-handed procedure, and a foot-operated trigger is a handy thing to have. Electrodes are available in diameters of 1/16, 3/32, and 1/8 inch. The correct electrode diameter is based on the thickness of the material and the amount of current used. When DCRP or DCSP welding is used, the electrode must be ground to a point. For AC welding, the tip is rounded off.

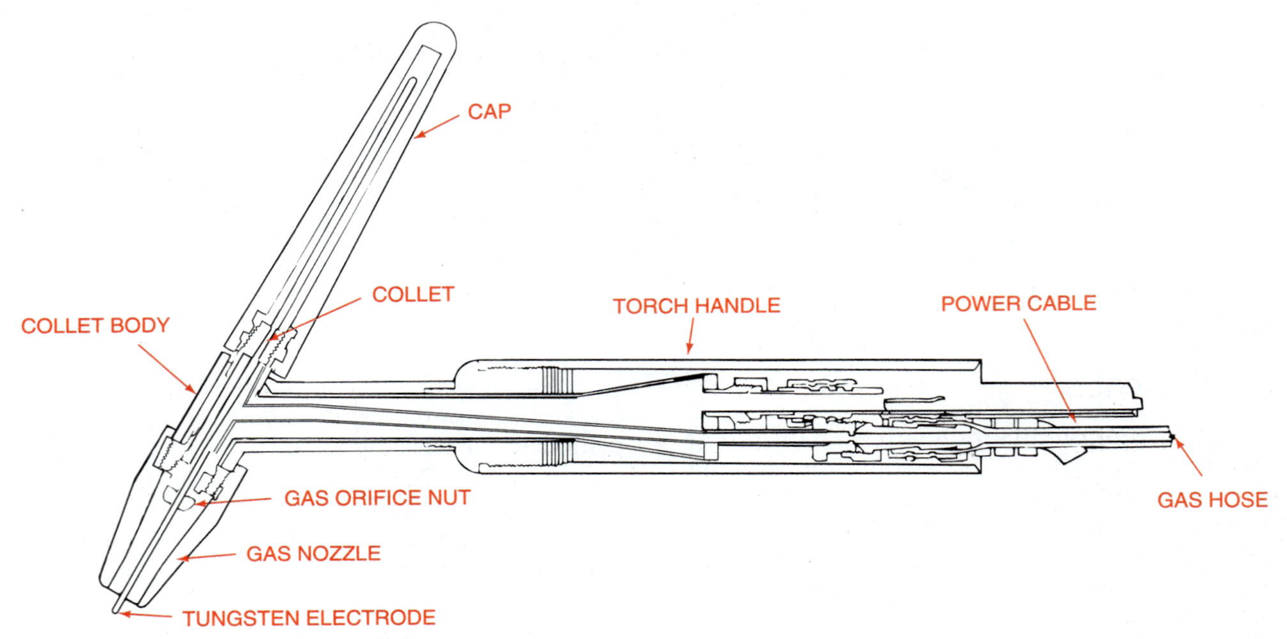

FIGURE 26-42. Cross section and parts of a TIG welding torch.

STUDENT ACTIVITIES

1. Define the Terms to Know for this unit.
2. In the welding area of the shop, examine all the protective clothing that is available.
3. Examine all fire protective equipment in the welding area. Report any irregularities to your instructor.
4. Ask your instructor to provide samples of weld showing good beads and poor beads due to improper welding.
5. Using a 1/4" × 4" × 6" welding pad, make 16 beads 1 inch long (four rows of four beads each).
6. Using a 1/4" × 4" × 6" welding pad, fill the pad with practice beads 3/4 inch apart.
7. Using two pieces of 1/4" × 2" × 6" steel, make a butt weld 6 inches long with an even bead.
8. Make an internal check of the bead from activity 6 by examining a cross section of a 1-inch piece sawed from the assembly. Bend open the 1-inch weld and analyze the bead.
9. Weld a 1/8" × 1" × 6" piece of steel to a 1/4" × 2" × 6" piece with a fillet weld on each side.
10. Run a straight 6-inch horizontal weld across a 1/8" or 1/4" × 4" × 6" welding pad.

11. Make a 2-inch vertical down butt weld.
12. Make a 2-inch vertical up butt weld.
13. Make a 2-inch vertical up fillet weld.
14. Pierce a 1-inch hole in a $^{1}/_{4}$" × 2" × 6" pad.
15. Cut three 1-inch strips from the pad used in activity 13, using an electric welder and electrode.
16. Observe MIG and TIG welding demonstrations.
17. Do MIG and TIG welding.

SELF-EVALUATION

A. Multiple Choice. Select the best answer.

1. The temperature of an electric welding arc is about
   a. 400°F
   b. 840°F
   c. 1800°F
   d. 9000°F
2. Welding tables should be made of
   a. concrete
   b. masonite
   c. metal
   d. wood
3. Fire extinguishers for welding areas should be suitable for
   a. Class A fires
   b. Class B fires
   c. Class C fires
   d. all of these
4. Burning clothes on a human should be extinguished with
   a. a fire blanket
   b. a fire extinguisher
   c. sand
   d. any of these
5. Water in a welding area is useful for
   a. receiving sparks from piercing
   b. extinguishing fires
   c. cooling metal
   d. all of these
6. Injury to eyes can result from
   a. chipping without goggles
   b. viewing welding without shielding
   c. welding with less than a No. 10 lens
   d. all of these
7. If only one kind of electrode for all arc welding is to be purchased, the best choice is an
   a. E6010
   b. E6011
   c. E6013
   d. E7018
8. For most welding in agricultural mechanics the best electrode size is
   a. $^{1}/_{16}$ inch
   b. $^{1}/_{8}$ inch
   c. $^{3}/_{16}$ inch
   d. $^{1}/_{4}$ inch
9. Correct arc length is approximately
   a. $^{1}/_{8}$ inch
   b. $^{1}/_{4}$ inch
   c. $^{3}/_{8}$ inch
   d. $^{1}/_{2}$ inch
10. When welding, the operator sees by
    a. daylight
    b. fluorescent light
    c. light from the arc
    d. all of these
11. The appearance and strength of a bead are influenced by
    a. amps
    b. angle
    c. speed
    d. all of these
12. The recommended position of the weldor for horizontal and vertical welding is
    a. standing
    b. sitting
    c. the most comfortable one
    d. laying flat
13. The recommended weave pattern for the beginning weldor doing down hand welding is
    a. circular
    b. figure eight
    c. J
    d. T
14. A second pass should never be done if
    a. the first was a poor weld
    b. the joint was ground
    c. the slag has been removed
    d. the slag has not been removed

15. In metal, the most rapid movement of heat is
    a. down
    b. equal in all directions
    c. horizontal
    d. up
16. MIG welding is especially useful for welding
    a. aluminum castings
    b. thin steel
    c. exhaust system parts
    d. all of these
17. The welding process that uses an automatic wire feed is
    a. MIG
    b. oxyacetylene
    c. stick
    d. TIG

B. **Matching.** Match the items in column I with those in column II.

Column I

1. lens for welding
2. lens for chipping
3. leather
4. E6013
5. E6011
6. root pass
7. vertical up
8. vertical down
9. J weave
10. electrode for pipe

Column II

a. protective clothing for weldors
b. electrode for thin metal
c. electrode for most welding
d. first pass
e. procedure results in a weld with shallow penetration
f. clear
g. relatively difficult
h. useful technique when welding metal of different thickness
i. E6013
j. No. 10

C. **Completion.** Fill in the blanks with the word or words that make the following statements correct.

1. The _____ _____ weld is the most difficult weld to make.
2. To make a hole with an electrode is called _____.
3. The best amperage setting for cutting is _____ percent higher than for welding.
4. The correct motion for cutting is _____.
5. MIG stands for _____ _____ _____.
6. TIG stands for _____ _____ _____.

Painting

UNIT

27 Preparing Wood and Metal for Painting

OBJECTIVE

To prepare wood and metal for painting.

COMPETENCIES TO BE DEVELOPED

After studying this unit, you should be able to:

- Prepare unpainted wood for painting.
- Prepare previously painted wood surfaces for repainting.
- Steam clean machinery.
- Prepare unpainted metal for painting.
- Remove rust and scale from metal surfaces.
- Feather chipped paint on metal surfaces.
- Mask tractors and other machinery for spray painting.
- Estimate materials for paint jobs.
- Paint buildings and machinery.

TERMS TO KNOW

waterproof
paint
pigment
vehicle
paint film
wood preservative
creosote
pentachlorophenol
zinc napthenate
cuprinol
new wood
new work
seal
sealer
rot
warp
air-dried lumber
old work
sandblasting
primer
caulk
steam cleaner
feather
mask

MATERIALS LIST

- ✓ Paint can label
- ✓ Varsol
- ✓ Rags
- ✓ Putty
- ✓ Scraper and wire brush
- ✓ Assortment of emery cloth and sandpaper
- ✓ Silicon carbide paper, 200 and 400 grit
- ✓ Window sash in need of repainting
- ✓ Wooden project to be painted
- ✓ Wooden project to be repainted
- ✓ Piece of machinery that needs repainting
- ✓ Portable power wire wheel
- ✓ Steam cleaner

Wood, steel, and concrete are the primary materials used to construct agricultural buildings and equipment. Wood may rot and steel will rust if not protected from moisture. Concrete is not damaged by moisture in the air but may need special treatment to make it waterproof. Waterproof means that water cannot enter the material.

Most wood exposed to the weather must be painted or treated with a preservative. Paint is a substance consisting of pigment suspended in a liquid known as a vehicle. A pigment is a solid coloring substance. The liquid is called a vehicle because it carries the pigment.

Paint must be applied in thin layers so some of the vehicle can evaporate. As the vehicle evaporates, the paint is said to be drying. After a paint dries, the material that is left is called a paint film. Wood and other materials must be properly prepared or the paint film will peel off rather than wear off. In such cases, the paint film first dries out excessively. Hairline cracks then start to appear. The film then breaks and starts to curl away from the wood. Finally, the paint film separates from the wood and leaves the surface exposed, figure 27-1.

PRESERVATIVES

Wood may be covered with a liquid that kills wood-rotting fungus and wood-eating insects such as termites. Such liquids are called wood preservatives. Creosote, pentachlorophenol, and zinc naphthenate are popular wood preservatives. Creosote is black and generally shows through if covered with paint. It also has a strong odor. Therefore, it is generally used for posts and poles in outside applications. Pentachlorophenol is sold in a nearly clear vehicle, does not bleed through, and makes an excellent undercoat for painting materials, figure 27-2. Zinc naphthenate, popularly known as cuprinol, is another nearly clear preservative used extensively for exterior surfaces such as shingles.

PREPARING WOOD FOR PAINTING

Wood that has never been painted or sealed is called new wood. Wood used in original construction is frequently referred to as new work. To seal wood means to apply a coating that fills or blocks the pores so no material can pass through the surface. The coating is called a sealer.

A sealer prevents moisture from entering the wood and causing it to rot or warp. To rot means to decay or break down into other substances. To warp means to bend or twist out of shape. A sealer also keeps moisture and natural liquids inside the wood. This prevents the wood from drying out and cracking.

Preparing New Wood

New wood is easy to prepare for painting. See Unit 11, Finishing Wood, for procedures to prepare wood for clear or stained finishes. The procedures are the same to prepare wood for painting.

New wood must be dry when it is painted, that is, there should be no indication of moisture on the surface and it should have been air dried. Air-dried lumber means that the boards are separated by wooden strips when they are stacked and are then protected from rain and snow for six months or more. Air drying permits the excessive natural moisture to leave the wood. If the wood contains excess moisture, that moisture will tend to lift the paint as it

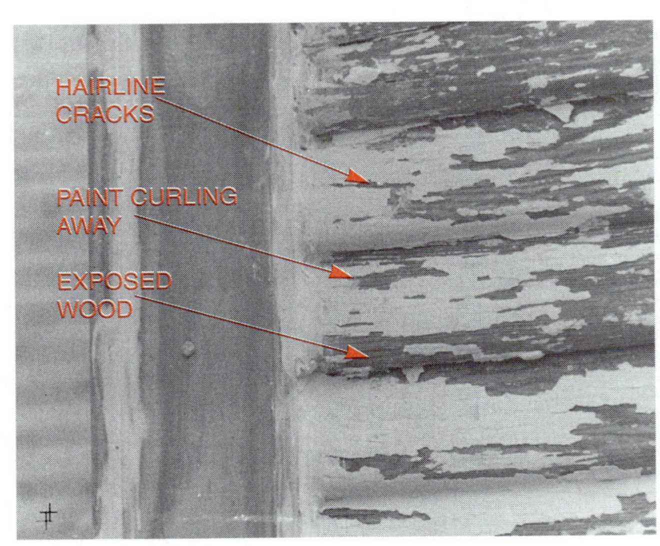

FIGURE 27-1. An advanced stage of paint peeling from siding on a building.

FIGURE 27-2. Creosote and pentachlorophenol are two commonly used wood preservatives for home and farm applications.

tries to escape to the drier surrounding air.

The wood must be free of grease, wax, dirt, and other substances or, again, the paint will not adhere to the wood. Grease and oil are removed by wiping the wood with a cloth soaked in varsol. Wax is removed by washing the wood with a solution of household ammonia. Dirt from shoes and other sources is removed by using a stiff bristle brush.

Planing of rough lumber greatly reduces the amount of paint needed to cover the surfaces. No sanding is needed except to remove objectionable roughness or to remove grease, wax, and other residues.

Bark must be removed before painting or planting posts and poles. Bark holds moisture and also serves as a home for disease organisms and insects that destroy wood. Posts may be dried by standing them on end around a tree trunk or other vertical support so air can flow around each one. Bark can then be removed with a draw knife or axe. After the bark is removed, the post must be left until all surfaces are thoroughly dry before painting is attempted.

Preparing Old Work

Wood that was previously painted is called old work. All loose paint must be removed from old work. This is done with a scraper, wire brush, and sandpaper, figure 27-3. The job is made easier with the use of abrasive flaps, wire wheels, and discs attached to a portable power drill, figure 27-4. If a smooth paint job is needed, the paint must be removed by a chemical paint remover or a torch and a scraper, or by complete sanding.

FIGURE 27-3. A wire brush is useful for removing loose paint.

FIGURE 27-4. To speed up the preparation of old work, a power drill can be used with a wheel and abrasive flaps, wire wheel, or sanding discs.

Paint manufacturers recommend the washing of previously painted surfaces before repainting. This can be done with a cloth and household detergent in water. The detergent removes grease, dirt, and weathered paint. A uniform, clean surface is left to which the new paint will adhere.

High pressure water cleaners are frequently used to clean floors, walls, siding, and concrete surfaces. Care must be used to avoid injury to by-standers or stripping of paint if repainting is not intended. Sandblasting where fine sand particles are thrown by compressed air is especially useful for cleaning metal and masonry.

A final step in preparing old work is to fill all holes, caulk all cracks, and seal any problem spots. The holes and cracks must be sealed before filling or liquids in the filling material will be absorbed by the dry wood. This means that the filler will shrink and loosen, causing the paint to crack and peel around the filler.

A primer is a special paint used to seal bare wood. Primer should be applied to the wood with a brush so the material can be worked into holes, cracks, and crevices. A thin coat is recommended because the job of the primer is to penetrate the wood as it is applied. When the primer dries, the pores are sealed and only a thin film is left on the surface. Knots or other problem areas that seem to absorb too much of the primer should be sealed with a primer containing shellac.

After priming, all cracks must be filled and the joints between materials must be caulked. Caulk is a material that stretches, compresses, and rebounds as

FIGURE 27-5. Caulking between two different materials, such as brick and wood, permits them to expand and contract without breaking the paint film.

materials expand and contract. This characteristic permits it to bridge the air gap between two materials such as brick and wood, figure 27-5.

Holes must be filled and window glass secured with putty or glazing compound on outdoor work. Plasterlike fillers may be used for indoor surfaces. Putty and glazing compound are soft materials containing oils that help keep them pliable over a long period of time, figure 27-6. Pliable means it will give if pushed or pulled. Putty is workable and may be pushed into holes and formed into special shapes. An example is the triangular bead of putty used to seal glass in a window, figure 27-7. Priming followed by applying new putty or glazing compound are important steps in repainting window sashes.

Scraping, brushing, washing, sanding, priming, caulking, and filling should prepare any old work for painting.

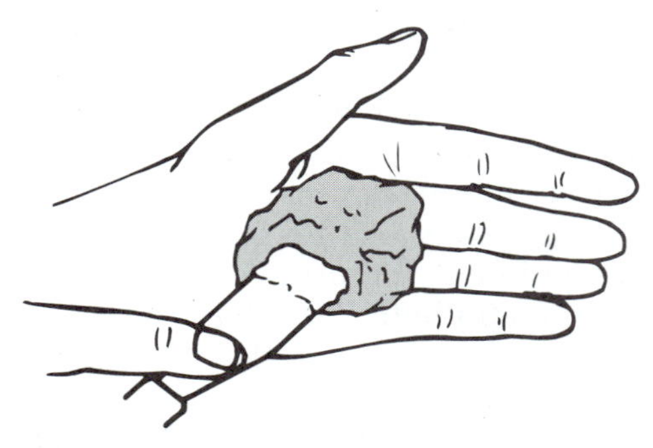

FIGURE 27-6. Putty and glazing compound are soft materials that contain oils to keep them pliable over long periods of time. Here the material is being worked in the hand to make it warm and pliable for application.

FIGURE 27-7. Window glass is sealed to wood or metal frames by applying a bead of putty and smoothing it to a triangular shape. Here paint is being applied to keep the putty and wood from drying out.

FIGURE 27-8. A steam cleaner removes grease and dirt from machinery, making it easier to repair or prepare the machinery for painting.

STEAM CLEANING MACHINERY

Tractors, engines, and machinery may be cleaned quickly and thoroughly with a steam cleaner. A steam cleaner is a portable machine that uses water, a pump, and a burner to produce steam, figure 27-8. Special chemicals in solution are added to the steam to dissolve grease and remove paint, dirt, manure, and other materials from metal, wood, and concrete.

A steam cleaner is powerful enough to strip paint from the metal if the appropriate chemical solution and pressure are used. This combination of heat and chemicals can make the steam cleaner hazardous in the hands of careless operators.

CAUTION:

- **Steam cleaners can cause serious burns to the operator or bystanders if not used carefully. The controls of the steam cleaner will vary with the manufacturer. Read the operator's manual before using a steam cleaner.**

A general procedure for operating a steam cleaner follows.

PROCEDURE

1. Cover all electrical parts on the machinery to be cleaned with plastic. Food bags, garbage bags, or vinyl film can be used. Wrap the parts with string or rubber bands to hold the plastic in place.
2. Place the machine at least 15 feet from any building.
3. Plug the motor into a three-wire grounded electrical extension cord attached to a proper outlet.

CAUTION:

- **Do not allow the plug to get wet. Place the plug up on the steam cleaner or other dry surface to keep it out of water that may accumulate in the area.**

4. Attach a garden hose to the water inlet of the cleaner and to a cold water faucet. Make all connections watertight.
5. Fill the fuel tank of the cleaner with kerosene or other recommended fuel.
6. Mix the cleaning chemical(s) in the solution tank according to the manufacturer's instructions.

CAUTION:

- **Cleaning materials are powerful chemicals that may burn eyes and skin. Use the proper goggles and protective clothing.**

7. Position the steam cleaner so you can observe all gauges and controls while operating the machine.

CAUTION:

- **Have a helper to watch the gauges and manage the controls of the machine as you steam clean. The helper should wear the same protective clothing as the operator.**

8. Unwrap and lay out the steam hose.
9. Hold the steam handle with the nozzle pointed toward concrete or free air.

CAUTION:

- **Never point the nozzle toward a person. Do not allow water or steam to contact electrical cords, wires, motors, outlets, or lights.**

10. Open the water faucet.
11. Turn on the switch to start the steamer motor.
12. Turn on the fuel valve.
13. Adjust the fuel valve to create the steam pressure recommended by the manufacturer.
14. When steam comes from the nozzle, turn on and adjust the solution valve to obtain the desired cleaning power.
15. Direct the steam to the high parts of the machine and move downward. The nozzle is held 8 to 12 inches from surfaces needing gentle cleaning and 2 to 4 inches from surfaces needing harsh cleaning.

CAUTION:

- **The nozzle should be positioned so steam and dirt are not forced into bearings or electrical components by the steam and water.**

16. Reclean the entire machine lightly to wash off all loose dirt.

Shutting down and storing the steam cleaner properly is very important. If stored incorrectly, the machine may freeze up, which will cause extensive damage. Some general recommendations for a safe shutdown are:

PROCEDURE

1. Close the solution valve.
2. Turn off the fuel.
3. Hold the nozzle until cold water comes out.
4. Turn off the motor.
5. Unplug the extension cord at the building, then unplug the steam cleaner.

CAUTION:

- **Since the surrounding area is wet, electric shock is a real hazard.**

6. Disconnect the garden hose from the machine and hose down the work area.
7. If the machine will be exposed to freezing temperatures, straighten the steam hose. Use compressed air to push water from the inlet side through the coils and out through the steam nozzle on the hose.
8. Store the steam cleaner with all cords and hoses protected.
9. Remove all plastic coverings from the clean machine.
10. If the machine has an engine, start the engine and let it warm up to help the drying process.
11. Dry the machine quickly and thoroughly.

Steam cleaning may produce moisture in the engine systems that will prevent starting. When this occurs, the trouble can usually be corrected by drying the spark plug wires and the inside of the distributor cap. A common hair dryer is effective for drying out electrical systems.

CAUTION:

- **Since steam cleaning may dissolve grease in bearings, these should be checked and repacked if needed.**

PREPARING MACHINES FOR PAINTING

Machines used in shops, buildings, and fields all need to be painted to prevent rusting. The paint film on a machine is damaged by mechanical contact, normal wear, and weathering. Such damage results in chipping, cracking, peeling, flaking, and fading of paint.

New Metal

Painting must follow new construction or repairs, such as welding, that leave metal exposed. To prepare new metal, the following steps are suggested.

PROCEDURE

1. Chip all welds and wire brush them thoroughly.
2. Wire brush or sand (using emery cloth) to remove all dirt or rust from the metal.
3. Steam clean or wipe down with varsol to remove grease.
4. Clean the metal with a commercial solvent designed to be a preparatory solvent.
5. Prime immediately.

Previously Painted Metal

Previously painted metal surfaces include castings, parts, and assemblies, and smooth surfaces, such as guards, hoods, and fenders, where an automobile-type finish is used.

All surfaces must first be cleaned of dirt, mud, manure, and grease. Steam cleaning is the fastest and easiest method. If steaming is not possible, then high-pressure water cleaning machines may be used. Or, materials can be removed by hand with a solvent, hose, water, scrapers, and brushes.

Heavy grease deposits are removed by scraping, followed by treatment with varsol or another commercial grease solvent. Generally, high-pressure water must then be used to remove the grease and solvent from the metal. Any grease or oil film left will prevent paint from holding to the metal.

Scaling paint and rust can be removed and rough or pitted areas can be smoothed by wire brushes, wire wheels, disc sanders, or emery paper. A combination of hand and power methods is commonly used, figure 27-9.

Smooth surfaces such as hoods and fenders need to be stripped of all paint if they are badly chipped. Otherwise, every chip will show in the new paint. Metal can be stripped using a paint remover or by sanding. If only a few chips exist, wet-type sandpaper can be used to **feather** the chipped areas. To feather means to sand so the chipped edge is tapered and no roughness can be felt between the painted and unpainted areas, figure 27-10. If a chip in a painted surface is feathered properly, the place where the chip starts and stops cannot be felt.

FIGURE 27-9. When cleaning metal prior to painting, a power sander can remove rust and smooth rough or pitted areas.

FIGURE 27-10. Chipped areas on a painted surface are sanded smooth or feathered so that no roughness can be felt between the painted and sanded areas.

Feathering is done using the following procedure.

PROCEDURE

1. Tear a piece of 200-grit silicon carbide waterproof paper into quarters.
2. Pour about one quart of water into a bucket or pan.
3. Dip the paper into the water.
4. Grip the paper between the thumb and index finger so the balls of the fingers serve as a backing pad for the paper.
5. Sand across the chipped area in all directions until a feathered edge is obtained. Dip the paper into water frequently to wash the pigment from the paper.
6. Finish with 400- or 600-grit paper to remove sanding marks and leave a smooth surface.
7. Wash the part thoroughly with water to remove all pigment.

CAUTION:

- **Paint pigment remains suspended in water, so the part or the area must be thoroughly washed and rinsed with clean water to avoid leaving a paint residue after the water dries.**

8. Dry the part thoroughly or rusting will start immediately.

MASKING FOR SPRAY PAINTING

A good-looking spray job requires careful masking of all glass, chrome, and other areas that should not be painted. To mask means to cover so paint will not touch. Masking may be done with tape or paper, figure 27-11.

Small items such as a light lens, distributor cap, and chrome trim are generally covered with masking tape. High-quality masking tape is required for good results. A sharp pocket knife is used to trim off the

FIGURE 27-11. Applying masking tape over newspaper to provide a tight masking on the grip surfaces of a steering wheel.

tape that extends beyond the part. For larger parts, paper is cut so that it is about 1/4 inch short of covering the area. Then 1/2-inch or 3/4-inch masking tape is used to finish covering the exposed area and tape the paper in place. Newspaper is frequently used for masking because of its availability and low cost. Very small areas such as grease fittings may be coated with a heavy film of general-purpose grease. When the paint job is dry, the grease and paint are wiped from the fitting.

ESTIMATING PAINT JOBS

Flat Surfaces

For most types of paint, one gallon will cover 400 to 500 square feet or about 100 square feet per quart. On flat surfaces such as doors, it is easy to calculate the square footage. For example, if two shed doors cover an opening 12 by 20 feet, the square footage is determined by multiplying length (20 feet) by height (12 feet). The amount is 240 square feet. A quart of paint will cover 100 square feet, so 240 is divided by 100 to get 2.4, which is rounded to 3. Thus, three quarts of paint will do the job. However, paint generally is cheaper by the gallon. For example, a quart may cost $7 while a gallon costs $21. In that case, three quarts of paint in quart cans will cost as much as four quarts in a one-gallon can. A good rule to use is: When the estimate indicates that two quarts will not be enough for the job, then buy a gallon.

If the job called for two coats of paint, then the decision to buy a full gallon would be an easy one. If the gallon did not cover the entire area with a second coat, an additional quart of paint could be purchased.

When estimating paint for a whole building, the square footage for all four sides should be added up. The estimate for a shed 15 feet wide by 30 feet long with an average height of 11 feet follows:

Front = 11' x 30' = 330 square feet
Back = 11' x 30' = 330 square feet
End = 11' x 15' = 165 square feet
End = 11' x 15' = 165 square feet
Total = 990 square feet

A gallon of paint is expected to cover 400 to 500 square feet. Dividing 990 square feet by 400 equals 2.5. If the building were very smooth two gallons might cover it. However, an extra quart or two may be needed to finish the job. If it is a long distance to the paint store, it would be wise to buy three gallons. This would eliminate the chance of running out of paint and avoid having to interrupt the painting once the job is started.

For buildings with shed-type roofs, the average height of the four sides is satisfactory for estimating purposes. Buildings with gable ends require an additional amount of paint to cover the areas in the triangles of the roof. The area *(A)* of a triangle is determined by multiplying the length of the base *(b)* by the height *(h)* of the triangle and dividing by 2. Thus $A = \frac{1}{2} bh$.

Spray Painting Machinery

Machinery has many irregular surfaces. When spraying, a lot of paint is wasted. For instance, the cone of spray coming from the gun may be 12 inches wide and the part being sprayed may be only 3 inches wide. Therefore, only one fourth of the paint is being used. The rest is escaping into the air and settling on other surfaces.

Experience shows that a gallon of machinery enamel will paint a small farm tractor. Generally, a quart of paint is sufficient to paint the trim or apply a second color to a tractor. One quart of enamel thinner may be sufficient to thin five quarts of enamel paint, but it may be wise to purchase a gallon and have some available for cleanup purposes. A cheaper solvent such as varsol should be used first to clean the spray gun and equipment. Therefore, the estimate for a tractor paint job would be as follows:

Main color	1 gallon
Trim color	1 quart
Enamel thinner	1 gallon
Cleanup solvent	1 gallon

Estimates for other jobs can be made by comparing them with the example given. Some useful guidelines for purchasing paint are as follows:

- One quart of spray enamel will be sufficient for one piece of lawn or garden equipment.
- A gallon of enamel is needed for a small farm tractor.
- If the paint is custom mixed, be sure to buy enough paint for the job. A second batch may not match the original batch perfectly.
- If two quarts are not likely to do the job, buy a gallon.
- Estimate on the high side (overestimate).
- To run out of paint during a spray paint job results in loss of time and money. Cleaning the spray equipment twice is time-consuming and requires extra solvent. A second trip for paint also costs time and money.
- When buying paint, purchase an extra amount with the understanding that unopened cans may be returned.

STUDENT ACTIVITIES

1. Define the Terms to Know for this unit.
2. Examine some buildings and machinery for signs of chipped paint and paint that has scaled or blistered.
3. Examine the label on a can of paint. Write down the ingredients listed as pigments. Write down the ingredients listed as the vehicle. Record the percentage of the total volume made up by each ingredient.
4. Examine a window with loose and missing putty. Note the cross-sectional view of the putty. Clean, prime, and apply new putty to the window. Repaint the window when the putty hardens.
5. Prepare a wooden project for painting.
6. Prepare a wooden project or building for repainting.
7. Steam clean a piece of machinery.
8. Wire brush and clean a piece of machinery for spray painting.
9. Sand a piece of metal with chipped paint. Feather the chipped area so that its presence cannot be felt with bare fingers.
10. Mask a machine for spray painting.
11. Estimate the amount of paint needed to apply one coat to the sides and ends of a building with the following dimensions: L = 40 feet, W = 24 feet, H = 12 feet. Assume that there are no windows.

SELF-EVALUATION

A. Multiple Choice. Select the best answer.

1. The solid substance that gives color to paint is
   a. film
   b. pigment
   c. thinner
   d. vehicle
2. The vehicle in paint is a
   a. film
   b. liquid
   c. pigment
   d. solid
3. The material that helps steam to dissolve grease is
   a. electricity
   b. fuel
   c. pressure
   d. a chemical solution
4. The first step in starting a steam cleaner is to connect the
   a. electricity
   b. fuel
   c. solution
   d. water
5. The first step in shutting down a steam cleaner is to
   a. stop the motor
   b. shut off the fuel
   c. close the solution valve
   d. turn off the water
6. A hazard generally *not* associated with the steam cleaner is
   a. chemical damage
   b. electrical shock
   c. falling objects
   d. steam burns
7. Steam cleaning will not
   a. cause hard starting
   b. dry out bearings
   c. increase paint preparation time
   d. strip paint
8. Scaling paint is removed by
   a. brushes
   b. scrapers
   c. wire wheels
   d. all of these
9. A one-gallon can of paint generally costs about the same as
   a. two quart cans
   b. three quart cans
   c. five quart cans
   d. none of these

10. One gallon of paint is generally expected to cover
    a. 400 to 500 square feet
    b. 300 to 400 square feet
    c. 200 to 300 square feet
    d. 100 to 200 square feet

B. Matching. Match the terms in column I with those in column II.

Column I

1. new work
2. old work
3. film
4. seal
5. creosote
6. pliable
7. chipped paint
8. waterproof paper
9. tape and paper
10. square feet

Column II

a. dry paint
b. preservative
c. purpose of a primer
d. correct by feathering
e. previously painted
f. silicon carbide
g. length × height
h. masking
i. putty
j. never painted

C. Brief Answers. Briefly answer the following questions.

1. What is a reasonable estimate of the amount of paint thinner and cleanup solvent needed to paint a small farm tractor?
2. How many quarts of paint would be needed to give a 10-foot by 10-foot door one coat?

UNIT

Selecting and Applying Painting Materials

OBJECTIVE

To select and apply paint in agricultural settings.

COMPETENCIES TO BE DEVELOPED

After studying this unit, you should be able to:

- Name major types of paint and paint components.
- Select paint for wood and metal.
- Apply paint with brushes.
- Spray paint with aerosol containers.
- Select and use spray painting equipment.
- Clean painting equipment.

TERMS TO KNOW

formulate
chalking
trim and shutter paint
enamel
gloss
semigloss
flat finish
water-base
oil-base
latex
alkyd
epoxy
interior
exterior
hiding power
lead
titanium dioxide
iron
zinc
aluminum
calcium
magnesium
silicon
asphalt
undercoater
compatible
drop cloth
fingering
loading the brush
paint roller
nap
roller cover
aerosol
propellant
spray gun
air atomization
regulator
extractor
siphon system
pressure feed system
CFM
viscosimeter
viscosity
sag
run
tacky

MATERIALS LIST

- ✓ Labels from oil-base and latex paints
- ✓ Paint brushes
- ✓ Paint thinner
- ✓ Latex paint
- ✓ Oil-base machinery enamel
- ✓ Spray aerosols with enamel

- ✓ Spray gun and equipment
- ✓ Paint room or outdoor area for painting
- ✓ Goggles, respirators, fire extinguishers, and other safety materials
- ✓ Cloths, waste cans, paint storage cabinet, drop cloths, and other painting equipment
- ✓ Projects to paint

Painting may be regarded as both a luxury and a necessity. Changing the colors of the surroundings just for fun is a luxury, adding enjoyment to life. However, most objects made of steel must be painted to prevent them from rusting. Similarly, wood that is exposed to high moisture levels must be painted or otherwise protected or it may rot. In these cases, painting is a necessity. Thus, there are two major reasons for painting: to improve appearance and to preserve.

The student is referred to Unit 11, Finishing Wood, for a discussion of thinners and solvents, procedures for painting with brushes, and brush cleaning and storage. Unit 27, Preparing Wood and Metal for Painting, should also be reviewed before continuing with this unit.

SELECTING PAINT

The chemistry of paint is complex. Therefore, the recommendations of the manufacturer must be relied upon heavily when choosing paints. However, some basic materials used in paints must be recognized to choose the right paint for the job and to buy paints that are a good value for the money.

Formulation

Paint is formulated for specific jobs. Formulate means to put together according to a formula. For example, paint for fences may be formulated so a tiny bit of the film washes off each time it rains. This keeps the paint bright. The process is called chalking. In addition, fence paints must not contain any toxic chemicals, such as lead, that might poison livestock when they lick painted surfaces. Fence paint should not be used to paint windows on a brick house, because the dull pigment that washes off will run down over the brick and cause streaks. A special type of paint called trim and shutter paint is formulated to stay bright without chalking.

Types of Finishes

Paint with a gloss or semigloss finish is referred to as enamel. Gloss means shiny. Paint with a slight shine is called semigloss. Most enamel is formulated to be used inside or outside. It is especially tough and useful on toys, boats, vehicles, tools, and machinery. It is also recommended for doors, trim, and baseboards in houses. Enamels are resistant to wear. Thus, they are useful on the floors of porches, patios, and other heavy-wear areas.

Some paints dry with a flat finish. Flat finish means dull or without shine. A flat finish paint is preferred for walls that are not generally touched or exposed to friction or moisture.

Vehicle

The vehicle is the liquid portion of the paint. The vehicle in most paints is either water or oil. Paints containing water are called water-base paints. Those containing oil are called oil-base paints.

The word latex on a paint label generally indicates a water-base paint. Most latex paints can be thinned with water, and brushes used with latex can be cleaned with water and detergent. The term alkyd indicates an oil-base paint. Such paint must be thinned with an oil, various petroleum products, or turpentine.

In recent years, epoxy paints have been added to the traditional line of paint materials. Epoxy is a synthetic material with special adhesive and wear-resistant qualities. Epoxy paints may be more difficult to apply, but their excellent durability makes them desirable for use in barns, milking parlors, public rest rooms, commercial kitchens, and other hard-use areas.

Interior Versus Exterior

The word interior on a label means the paint will not hold up if exposed to weather. The term exterior designates a paint that will withstand moisture and outside weather conditions. Both water-base and oil-base paints may be formulated for interior or exterior use.

Pigments

Pigments give color and hiding power to paint. Hiding power is the ability of a material to create color and mask out the presence of colors over which it is spread. Many materials are used for paint pigments. In the past, lead was used extensively. However, lead is a metal that has a cumulative toxic effect and stays in the body once it is ingested.

A high-quality pigment used in many paints is titanium dioxide. In addition, compounds of iron, zinc, and aluminum are associated with high-quality paints. Therefore, as a rule of thumb, the higher the proportion of these pigments in a paint, the better the quality.

CAUTION:

- **Lead paint has been found to act as a slow poison to livestock that lick it. The same is true of children who put objects painted with lead paint in their mouths. Laws have been passed to stop future use of lead paint, but old paint containing lead remains a hazard.**

Low-quality pigments provide little or no hiding power. Some low-quality pigments are calcium, magnesium, and silicon compounds. Asphalt in paints provides excellent black hiding power, but it wears away faster than the high-quality pigments.

Primers and Undercoaters

Primers and undercoaters are used to prepare surfaces for high-quality top coats. They stick to and seal surfaces. They must be compatible with the material being painted as well as with the top coats that follow. Compatible means they go together with no undesirable reactions. Primers and undercoaters will not stand weather, wear, or exposure. They must be covered with a proper top coat. It is generally wise to use the manufacturer's recommended primer or undercoat with any paint or finishing material.

Rust Resistance

Some paints consist of a vehicle and pigments that enable them to adhere well to rusted surfaces. Such paints work well if all loose rust is removed before the paint is applied. The primer and top coat must be compatible for best results.

USING BRUSHES

Brushes are useful for small paint jobs. They are especially adapted to use on irregular areas such as window sashes and building trim. The cost of equipment is low for both brush and roller painting. Therefore, both types of applicators are used extensively for home and farm projects. However, many commercial painters rely on the speed and professional results obtainable with modern spray painting equipment.

A new paint job can do almost as much to lift the spirits as a new purchase. However, a poorly done job may be ugly and messy, and may interfere with the proper operation of the object that is painted. For example, improperly painted windows may not open due to sticking. The recommended procedure for applying paint with brushes follows.

PROCEDURE

1. Prepare the object or area as outlined in the previous unit.
2. Mask all parts that are not to be painted. Cover areas under surfaces to be painted with drop cloths. A drop cloth is any material used to protect floors, furniture, shrubbery, and other objects from paint spatter or droppings, figure 28-1.
3. Use masking tape to make a straight line where paint should stop, figure 28-2. When the tape is removed, a straight paint line will be left, giving the job a professional appearance.
4. Mix paint thoroughly. Unopened oil-base paints should be shaken by machine. Once opened, they should be divided between two containers and stirred thoroughly, figure 28-3. Pouring the paint back and forth between containers is generally the final step in mixing.

FIGURE 28-1. Most paint jobs require the use of drop cloths to protect shrubbery, walks, blacktop, floors, or furniture.

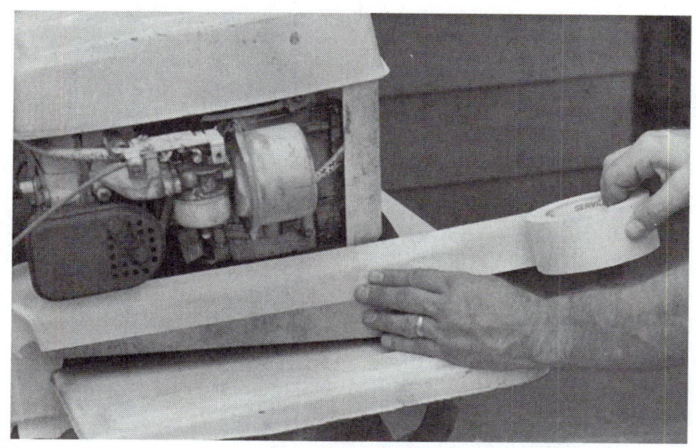

FIGURE 28-2. Use masking tape to make a straight line where paint should stop.

FIGURE 28-3. Oil-base paints should be divided between two containers, stirred thoroughly, and then poured back and forth between containers.

NOTE: Latex paints should not be shaken. They are mixed by stirring only.

5. Use a piece of screen or cheesecloth to strain paints containing lumpy particles, figure 28-4.
6. Use pure bristle brushes for oil paints and nylon or other artificial brushes for latex paints. Select a brush that has extra long and thick bristles and one that matches the job in size and shape—round, beveled, tapered, or flat.
7. Paint with the flat side of a flat brush. Painting with the edge causes fingering or dividing of the bristles, figure 28-5.

FIGURE 28-4. Use a piece of screen or cheesecloth to strain paints containing lumpy particles.

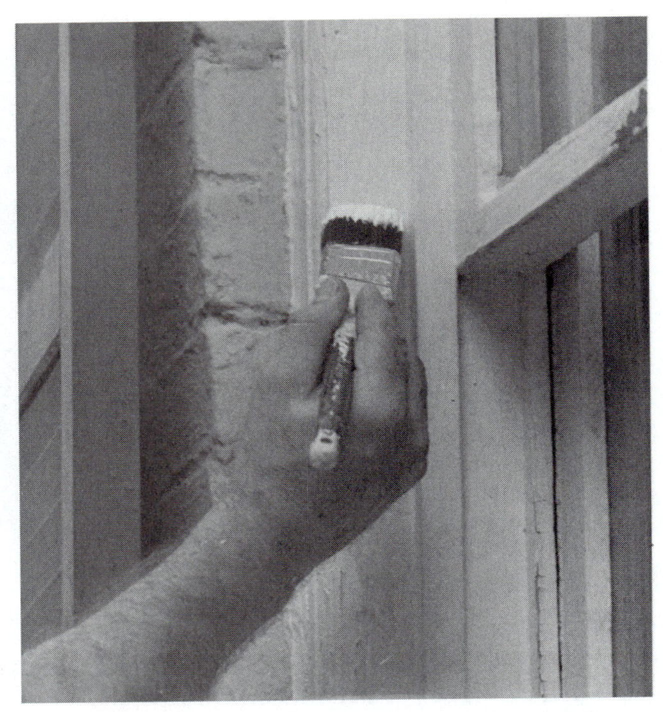

FIGURE 28-5. Use the flat side of a brush when painting. Painting with the edge will cause the bristles to divide into fingerlike clusters.

8. Dip about one third the length of the bristles into the paint. Touch the bristles to the side of the paint container to remove the excess paint and prevent dripping. This is called loading the brush.
9. Touch the loaded brush to the work in several places in a small area to deposit paint. Then smooth the paint by brushing back and forth to fill in the area. End all strokes in the paint rather than on unpainted surfaces, figure 28-6.

CAUTION:

- **Latex paints dry quickly and should not be brushed more than necessary to spread.**

10. Paint adjacent sections before the paint dries. This reduces the marks that result from uneven blending of paint layers.
11. Stir the paint frequently.
12. If it is necessary to stop painting briefly, wrap the bristles in a plastic food bag to prevent the paint from drying in the brush.
13. At the end of the work day, clean latex paint brushes with warm water and detergent. Wrap them in a paper towel and place them

FIGURE 28-6. End all brush strokes in the freshly painted area.

in a warm place to dry. This is necessary to maintain brush shape and compactness. Oil-base paint brushes may be suspended in linseed oil or thinner until the next day.

14. When restarting the paint job, wipe excess thinner from the bristles with rags or paper towels. Proceed with painting.
15. When the painting is finished, clean oil paint from brushes using varsol or thinner. Wash the brush thoroughly with soap and water to remove all paint and thinner residue. Clean bristles will squeak like clean hair. Wrap the brush in a paper towel to form the bristles in their proper shape. The bristles will then dry to provide a brush that is like new.

USING ROLLERS

Paint can be poured into a wide, shallow pan and applied with rollers more quickly and with less work than if brushes are used. A **paint roller** is a cylinder that turns on a handle. However, rollers tend to throw off tiny droplets of paint which spatter the surfaces below. Therefore, drop cloths are especially important when applying paint with a roller. Some suggestions for using rollers to apply paint follows:

- Use a 7-inch roller for narrow or uneven surfaces such as fence boards and siding. A 9-inch roller can be used for large, level surfaces such as plywood, wall sections, and ceilings.
- Use a roller with a short nap for smooth surfaces and one with a longer nap for rougher surfaces. **Nap** refers to the soft, woolly, thread-like surface of the roller cover. The **roller cover** is the hollow, fabric-covered cylinder that slides onto the roller handle.

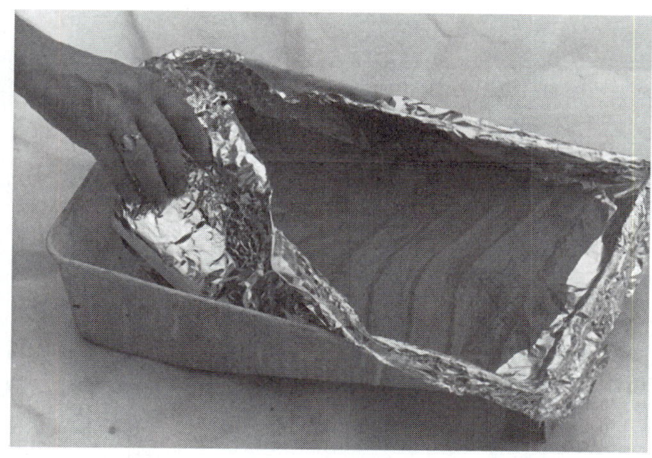

FIGURE 28-7. Cleanup time is saved by lining the paint tray with aluminum foil. After painting, the foil is removed and discarded.

- Line the paint pan with aluminum foil. The foil is discarded after painting, leaving the pan clean for the next job, figure 28-7.
- Use small rollers or sponges on bevels or grooves in siding or on rounded surfaces such as rails and posts, figure 28-8.
- Paint corners and edges first, then roll the larger areas. When painting large areas, apply paint in a diagonal or cross pattern, figure 28-9. Then, roll it again in the direction of movement across the work. Finish the surface by rolling toward the painted area—not the unpainted area.
- Use a roller with a very long nap on chain-link and woven-wire fences, figure 28-10. Paint both sides of the fence and roll in several directions for complete coverage of all joints.
- Stir the paint frequently, especially when the pan is refilled.
- When work is stopped, temporarily or even for

FIGURE 28-8. Use small rollers or sponges on bevels or grooves in siding or on rounded surfaces such as rails and posts.

FIGURE 28-9. Paint corners and edges first. Then roll the large areas by first applying paint in a cross pattern.

several days, store the roller in an airtight plastic or foil bag or wrapper, figure 28-11.

- When the job is finished, discard low-cost roller covers. High-quality covers are generally made with a washable core and may be cleaned. However, the process requires large amounts of solvent. Therefore, cleaning may be practical only when using latex paints, which are removed with warm water and detergent. If all paint is not removed, the roller cover will not be usable again.

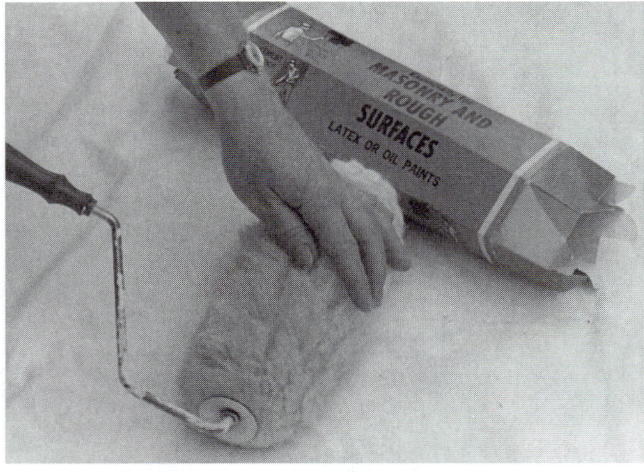

FIGURE 28-10. Use a roller with a very long nap for painting wire fences.

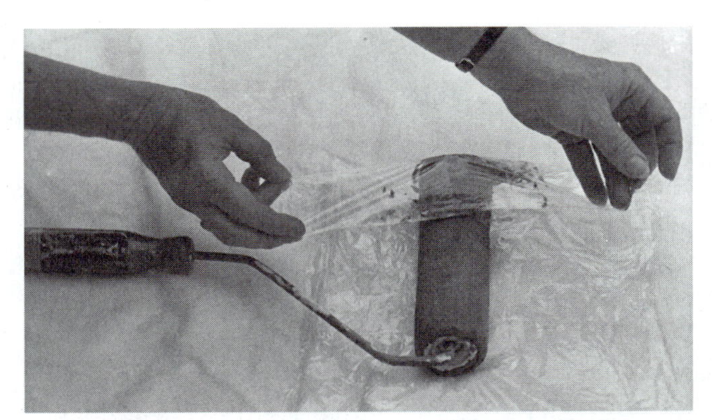

FIGURE 28-11. A used roller may be stored for several days without cleaning by wrapping it in an airtight material such as plastic.

USING AEROSOLS

Small paint jobs can be done quickly at a low cost with aerosol spray cans. An aerosol can is a high-pressure container with a valve and spray nozzle, figure 28-12. The container is partially filled with paint, which is put under pressure with compressed gas. The gas is called a propellant, since it pushes or propels the paint toward the object being painted.

When painting with aerosols, the following procedure is suggested.

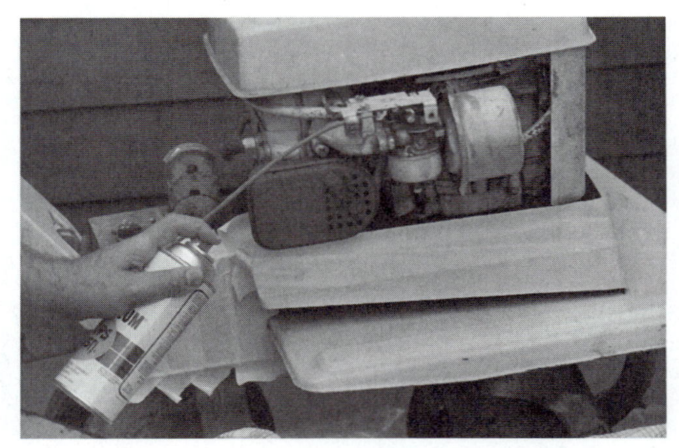

FIGURE 28-12. Aerosol spray paint is especially useful on small jobs and on areas that are hard to paint with brushes.

PROCEDURE

1. Degrease and sand the object to be painted.
2. Wash all dust and residue from the object.
3. Choose a clean, well-ventilated area in which to paint.
4. Cover or mask all adjacent areas.
5. Wipe areas to be painted with a tack rag and solvent.
6. Shake the can for several minutes.

7. Practice spraying on a piece of cardboard before starting on the project.
8. Hold the can 10 to 12 inches from the object. Start the spray in the air and then move it across the surface to be painted. Spray with rapid, uniform strokes.
9. Spray on a mist coat and let it become tacky or sticky. Gradually build a full cover by successive overlapped strokes to create an even paint film.

CAUTION:

- **Do not apply too much paint at one time on vertical surfaces or it will develop unsightly runs.**

10. When finished, invert the can and press the nozzle until there is no more paint in the spray. This leaves the nozzle free of paint so that it can be used another time.
11. If additional coats are needed for good coverage, follow the directions on the can regarding sanding and drying time.
12. Allow to dry thoroughly before heavy use. Paint may take 30 days to achieve maximum hardness.

A great selection of paints, varnishes, and other coatings is available in aerosol cans. Therefore, their use is encouraged for touch-up work on machinery and full paint jobs on small tools and equipment. Paint in aerosols is generally more expensive than an equal amount of paint in a regular can. However, the time saved and the attractive appearance obtained make the aerosol spray can the preferred method for many small jobs. Paint should be purchased in aerosols that use environmentally safe propellants.

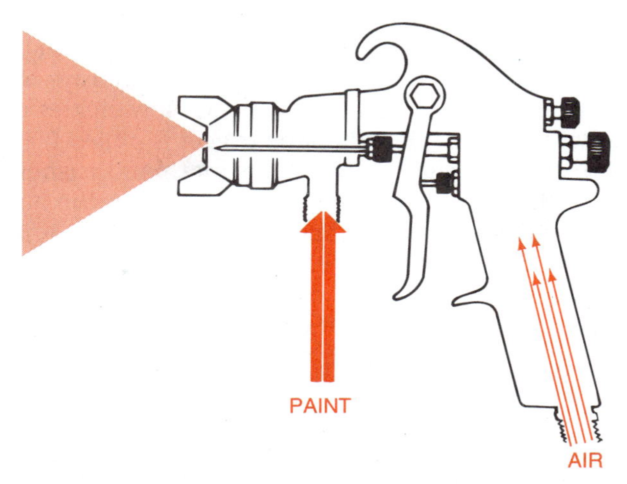

FIGURE 28-13. An air spray gun uses compressed air to split paint into tiny droplets. *(Courtesy of Binks Manufacturing Co.)*

SPRAY PAINTING EQUIPMENT

Spray Guns

A spray gun is a device that releases paint in the form of a fine spray. Spray guns may be the air atomization type or they may be airless. Air atomization means that the paint is split into tiny droplets using compressed air, figure 28-13.

A typical air spray gun consists of a gun body, fluid control knob, pattern control knob, air valve, paint needle valve, packing nuts, fluid nozzle, air nozzle, and trigger, figure 28-14. Air pressure is kept

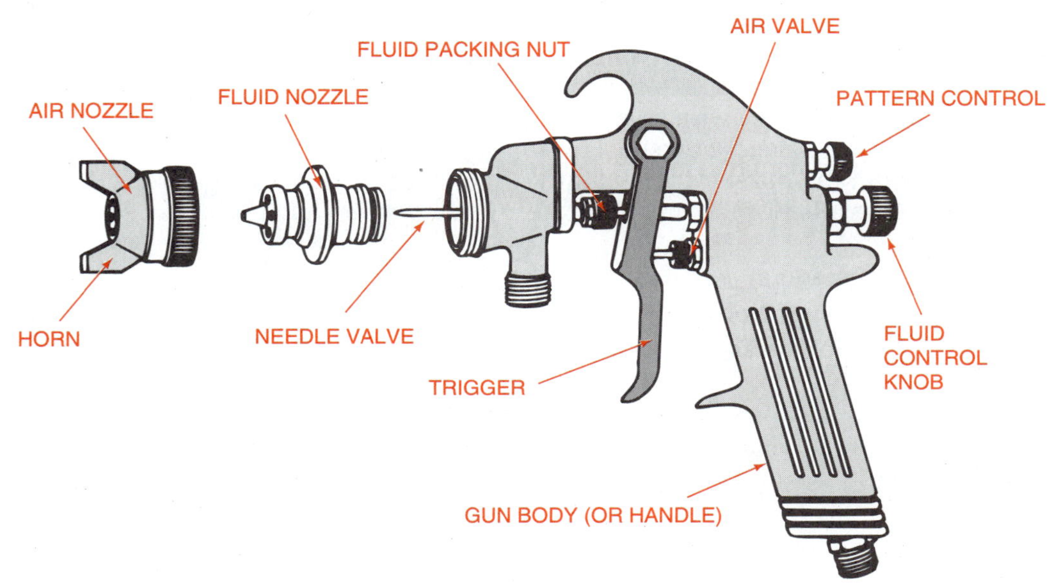

FIGURE 28-14. Parts of an air spray gun. *(Courtesy of Binks Manufacturing Co.)*

constant by a regulator. When the trigger is pulled a short distance, air flows through the passageways of the gun and out the air nozzle. The air flow creates a vacuum in the nozzle. When the trigger is pulled further, the fluid needle valve opens and paint rushes out the fluid nozzle. Holes are positioned in the air nozzle to discharge air at angles that force the paint into a fan or cone pattern, depending on the adjustment.

Airless spray guns are similar in design to air spray guns. However, no air is used in the gun. Instead, the paint is pumped through hoses under high pressure. The pressurized paint then rushes through the nozzle and is atomized as it is released into a spray pattern.

CAUTION:

- **Never point an airless gun at any part of your body or at anyone else. If the trigger is pulled, the paint is discharged at sufficient pressure to penetrate the skin. If this occurs, get medical attention immediately.**

Fluid Systems

Air spray guns may be connected to paint supplies in one of three ways: (1) a siphon feed cup, (2) a pressure feed cup, or (3) a pressure feed tank. All three systems must include a regulator and extractor in the air line. The regulator keeps the air pressure at a set level. The extractor removes moisture and oil from the compressed air. Moisture and oil in the air interfere with the spray pattern and ruin the paint job.

The siphon system, figure 28-15, has a small vent hole in the cap of the cup. This hole must be kept open at all times to permit atmospheric pressure on the paint. Paint may cover the hole if the cup is tipped too far. As a result, there will be a vacuum in the cup and the gun will not work. The hole may be reopened by using a small piece of wire or a stiff broom straw. The siphon-type system works by creating a low pressure area in the nozzle where paint can move in by force of atmospheric pressure.

The pressure feed cup and pressure feed tank systems operate with the paint under air pressure at all times. The pressure pushes the paint through the system. These systems do not use a vent hole in the cap. The pressure feed cup holds only one quart of paint; the pressure feed tank may hold five gallons or more. The pressure tank system is good for large jobs such as barns. The paint tank can remain on the ground while the operator is painting the sides or roof. The operator needs to carry only the gun and hoses.

Air Supply

Good spray painting systems permit fast and efficient painting. However, the initial cost is relatively high. For shop and farm use, a fairly large compressor is needed to operate a good spray gun. It is a good idea to buy the spray gun, compressor, air hose, regulator, extractor, and related equipment from one source, figure 28-16. The supplier should provide

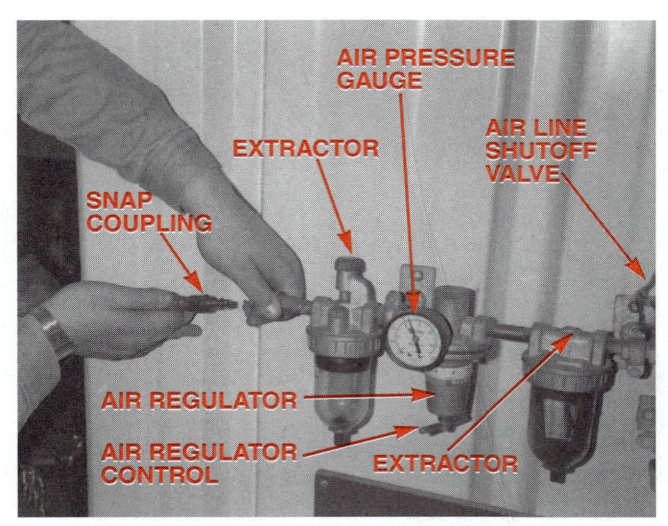

FIGURE 28-16. To prepare for spray painting, an air hose is attached between the gun and other parts of the spray system. This system includes a snap coupling, oil and water extractors, a pressure regulator with a gauge, a shutoff valve, an air line, tank, and compressor.

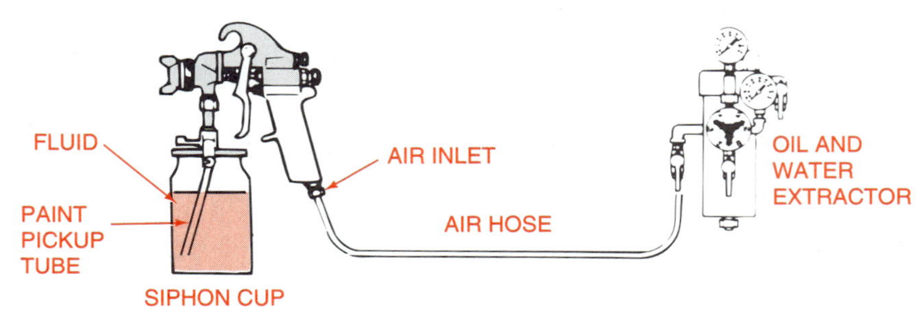

FIGURE 28-15. A siphon cup air spray system. *(Courtesy of Binks Manufacturing Co.)*

assurance that the components work well together to provide an efficient system.

It is especially important that the compressor provide as many cubic feet per minute (CFM) of air as the gun requires. The air must also be delivered to the gun at the required pressure. To achieve this requires adequate sizing of the motor, compressor, electric wiring, air pipe, and hoses.

SPRAY PAINTING PROCEDURES

Preparing the Surface

Work should be prepared for painting as outlined in the previous unit. Expensive equipment and a skillful painter cannot make paint attractive and permanent if the surface has not been prepared properly. All surfaces must be clean of blistered paint, rust, dirt, oil, and wax. Glossy surfaces must be sanded to create a dull finish to which paint can stick.

Choosing a Location

Gun spraying is not recommended indoors unless a commercial spray booth is available. Paint spray is a fire and explosion hazard. In addition, the spray will settle on surfaces and ruin ventilation equipment. Spraying may be done outdoors if the temperature is 50°F or above and there is no wind.

Following Safety Rules

Spray painting can be hazardous. Some safety rules for spray painting are:

- Mix paints and solvents in well-ventilated areas.
- Spray paint outdoors or in special booths with recommended electrical fixtures and ventilation.
- Never smoke or have any source of fire in the vicinity.

CAUTION:

- **All automobiles and other objects with good paint jobs should be at least 100 feet from the painting area. Otherwise, particles of paint will drift, settle on them, and ruin their appearance.**

- Wear a respirator or paint mask when painting.
- Use ladders that are in good condition. Do not use ladders with rungs or sides cracked or damaged.
- To avoid electrocution, be especially careful to keep aluminum ladders away from all electrical wires.

Preparing to Paint

Nothing is more frustrating than to have the project ready, spray equipment set up, paint in the gun, and then find that the gun will not spray paint. Such problems occur from the improper preparation of paint and equipment. Paint is prepared for spraying as follows.

PROCEDURE

1. Thoroughly mix and stir the paint until no streaks can be seen. If more than one container of paint is to be used, mix them together to assure a uniform color.
2. Thin the paint at room temperature according to the manufacturer's instructions. Since all thinners and paints are not compatible, mix a small amount of paint and thinner to see if they form a smooth paint with uniform color.
3. To be safe, use a viscosimeter to confirm that the paint is thin enough. A viscosimeter is an instrument used to measure the rate of flow of a liquid. A liquid's tendency to flow is called its viscosity.
4. Using a commercial paint strainer, strain the paint into a clean container. Strainers are available from paint and automotive parts stores.
5. Pour the thinned paint into a clean sprayer cup or tank.

Adjusting the Spray Gun

Fluid and air pressures should be set as low as possible, yet do the job. Pressure requirements vary with the design of the gun, the viscosity of the paint, and the temperature of the paint. All guns should be adjusted according to the manufacturer's instructions. However, some general recommendations for adjusting spray guns follow.

PROCEDURE

1. Set up all equipment.
2. Put on safety glasses and a paint respirator.
3. Set up a sheet of cardboard to use as a practice surface.
4. Turn on the air and set the regulator at the level specified for the gun.
5. Open the pattern control and fluid control knobs.
6. Prepare to make a pass by pointing the gun off to the left side of the panel. Pull the trigger slowly until air is heard coming from the

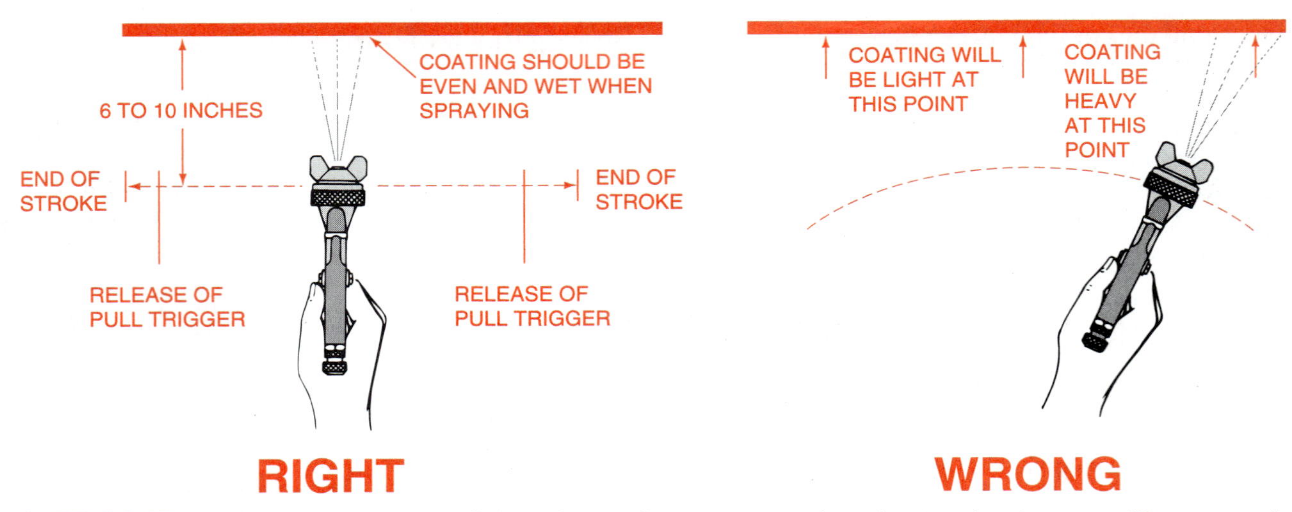

FIGURE 28-17. Moving the gun parallel to the surface assures that the coating is even. *(Courtesy of Binks Manufacturing Co.)*

nozzle, if using an air-type gun. Pull the trigger further until a faint paint fan is seen.

7. Position the gun with the nozzle 6 to 10 inches from the work and perpendicular to it. Slowly move the gun across the surface, maintaining the same distance from it, to create an even spray band, figure 28-17.
8. Readjust the pattern control to obtain a fan-shaped pattern.
9. Adjust the fluid control knob to permit the desired amount of paint to flow when the trigger is pulled all the way.
10. Adjust the distance from the object and the rate of travel until a good, even pattern is obtained. Each stroke should overlap the previous one by 50 percent.

Operating the Spray Gun

Spray guns must be held the correct distance from the work, figure 28-18. The gun stroke is made by moving the gun parallel to the work.

The closer the gun is held to the work, the faster the gun must be moved to prevent sags and runs. A sag occurs when a large area of paint shifts downward because it has been applied too thickly. A run is a narrow stream of paint flowing downward due to excessive buildup. On the other hand, holding the gun too far from the work causes dry spray and excessive spray dust. These result in a flat, dull, and rough finish.

When painting tractors and machinery, many irregular surfaces must be sprayed. Under these conditions the spray pattern should be narrow. This permits paint to reach parts that are further than the ideal distance from the nozzle. Large smooth areas, such as hoods and fenders, require a broad fan and should be removed from the machine being painted. The shape of the fan is controlled by adjusting the pattern control knob.

A. NOZZLE TOO FAR FROM THE SURFACE. DUSTING RESULTS IN A ROUGH PAINT FILM WITH A DULL FINISH.

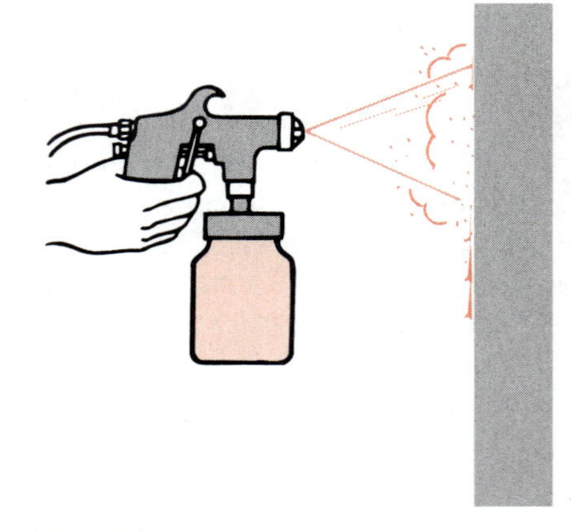

B. NOZZLE IS TOO CLOSE. A THICK AND UNEVEN FILM RESULTS IN SAGS AND RUNS.

FIGURE 28-18. When the spray nozzle is held at an incorrect distance from the object being painted, an undesirable effect is obtained.

Modern machinery enamels are designed to be sprayed in two or more layers at one time. To do this, spray on a thin layer and let it become tacky. Tacky means sticky. By the time a machine such as a tractor or baler has been given one coat, the paint applied at the beginning of the job will be tacky. Go back to the starting point and apply a heavy top coat. The paint layer should shine but not run or sag. When a run or sag occurs, the entire part or panel must be wiped clean with solvent and then repainted. Another method of correcting sags or runs is to allow the part to dry thoroughly, then sand the area and repaint.

Correcting Faulty Spray Patterns

Faulty spray patterns can be caused by incorrect pressure, clogged air passages, loose packing nuts, dry packings, thick paint, and unstrained paint.

After each use, a drop of light machine oil should be placed on each needle in a spray gun where it enters the packing nut. This procedure helps eliminate some problems with faulty patterns. Other recommendations for correcting faulty patterns are given in figure 28-19.

Cleaning Spray Equipment

Cleaning spray guns and equipment requires a great deal of solvent. Varsol is recommended for oil paints because it is low in cost. Another advantage of varsol is that the fire hazard is not as great as it is with other solvents.

To clean the gun, cup, and other equipment, the following procedure is suggested.

PROCEDURE

1. Pour all unused paint back into the container and reseal it.
2. Wipe most of the paint from the interior of the paint can and gun parts with paper towels or clean rags.
3. Pour about 1 inch of varsol into the cup.
4. Attach the cup to the gun and shake the gun so the varsol can dissolve the paint residue.
5. Spray a piece of cardboard until the cup is empty. Cardboard absorbs the thinner.
6. Wipe all visible surfaces, both inside and outside of the gun and cup, with clean varsol on a cloth.
7. Repeat steps 3 to 6 one or two times using paint thinner until all traces of paint are absent from the thinner.
8. Remove the air nozzle and paint nozzle.
9. Use a cloth with thinner to wipe away any remaining evidence of paint.

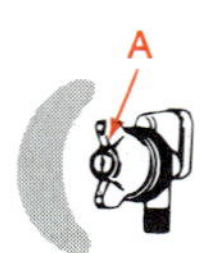

CAUSE
Dried material in side port "A" restricts passage of air through port on one side. Results in full pressure of air from clean side of port in a fan pattern in direction of clogged side.

REMEDY
Dissolve material in side port with thinner. Do not use metal devices to probe into air nozzle openings.

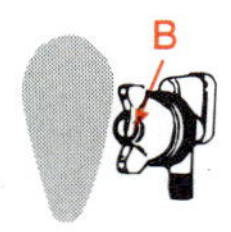

CAUSE
Dried material around the outside of the fluid tip at position "B" restricts the passage of atomizing air at one point through the center ring opening of the air nozzle. This faulty pattern can also be caused by loose air nozzle, or a bent fluid nozzle or needle tip.

REMEDY
If dried material is causing the trouble, remove air nozzle and wipe off fluid tip, using rag wet with thinner. Tighten air nozzle. Replace fluid nozzle or needle if bent.

CAUSE
A split spray pattern (heavy on each end of a fan pattern and weak in the middle) is usually caused by: (1) atomizing air pressure too high (2) attempting to get too wide a spray with thin material, (3) not enough material available.

REMEDY
(1) Reduce air pressure. (2) Open material control "D" to full position by turning to left. At the same time turn spray width adjustment "C" to right. This reduces width of spray but will correct split spray pattern.

SPITTING

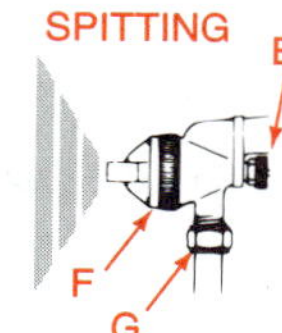

CAUSE
- Air entering the fluid supply.
- Dried packing or missing packing around the material needle valve that permits air to get into fluid passageway.
- Dirt between the fluid nozzle seat and body or a loosely installed fluid nozzle.
- A loose or defective swivel nut, siphon cup, or material hose.

REMEDY
- Be sure all fittings and connections are tight.
- Back up knurled nut "E," place two drops of machine oil on packing, replace nut, and finger tighten. In aggravated cases, replace packing.
- Remove air and fluid nozzles "F" and clean back of fluid nozzle and nozzle seat in the gun body, using a rag wet with thinner. Replace and tighten fluid nozzle using wrench supplied with the gun. Replace air nozzle.
- Tighten or replace swivel nut "G."

CAUSE
A fan spray pattern that is heavy in the middle, or a pattern that has an unatomized "salt-and-pepper" effect indicates that the atomizing air pressure is not sufficiently high, or there is too much material being fed to the gun.

REMEDY
Increase pressure from air supply. Correct air pressures as discussed elsewhere in this manual.

FIGURE 28-19. Faulty spray patterns and how to correct them. *(Courtesy of Binks Manufacturing Co.)*

10. Dry all parts and reassemble the gun.

CAUTION:

- **The spray gun will not work well the next time unless every trace of paint is removed. A good rule to follow is that you should not be able to tell the color of the paint used previously when examining the gun.**

11. Clean all other equipment using a cloth and solvent.
12. Discard all paper, cloths, used strainers, and containers in an approved metal trash container.
13. Store all equipment properly.
14. Turn off the air pressure and/or compressor.

Spray painting can be very satisfying. However, care must be taken to prepare the work and paint carefully. Absolute cleanliness is required to avoid spray gun problems. Since paint and thinners are very flammable, there is always a danger of fire. Every precaution must be exercised to keep fire and electrical sparks away from the spray painting area.

STUDENT ACTIVITIES

1. Define the Terms to Know in this unit.
2. Examine the paint on a wooden building, a wooden fence, and a piece of used farm machinery or garden equipment. What evidence is there to indicate a need for repainting?
3. Examine the labels on a can of oil-base paint and a can of latex paint. List the ingredients and the percentage of each. What is recommended as a thinner for each?
4. Prepare a wooden project for painting. Brush paint the project.
5. Prepare a metal project for painting. Spray paint the project using an aerosol spray container.
6. Prepare a piece of machinery for spray painting. Use a spray gun to paint the machine.
7. Use a pan and roller to paint a fence, room, or building.

SELF-EVALUATION

A. Multiple Choice. Select the best answer.

1. Painting is not a necessity when it
   a. is done for personal preference
   b. preserves wood
   c. prevents rot
   d. prevents rust
2. An example of a low quality pigment is
   a. aluminum
   b. calcium
   c. titanium
   d. zinc
3. Lumpy particles in paint should be
   a. ignored
   b. mashed up
   c. reduced by thinning
   d. removed by straining
4. Perfectly straight lines are obtained when painting by use of
   a. a steady hand
   b. masking tape
   c. a ruler and pencil
   d. a yardstick
5. Oil paints should not be applied with
   a. sponges
   b. nylon bristle brushes
   c. pure bristle brushes
   d. rollers
6. Latex paints should not be
   a. brushed excessively
   b. applied with rollers
   c. used outdoors
   d. used on wood

7. An aerosol for painting is
   a. not pressurized
   b. relatively expensive
   c. difficult to use
   d. too expensive for small jobs
8. All spray guns are
   a. airless
   b. pressurized
   c. siphon-type
   d. none of these
9. Painting may be hazardous because of
   a. explosive materials
   b. flammable materials
   c. compressed air
   d. all of these
10. Painting hazards are greatly increased by
   a. excessive ventilation
   b. smoking
   c. spraying outdoors
   d. using a respirator

B. **Matching.** Match the terms in column I with the correct definition in column II.

Column I

1. formulate
2. chalking
3. gloss
4. flat
5. latex
6. titanium dioxide
7. primer
8. lead
9. extractor
10. regulator

Column II

a. dull
b. thin with water
c. first coat
d. poisonous
e. removes water
f. controls pressure
g. put together
h. safe, high-quality pigment
i. shiny
j. wash off

C. **Completion.** Fill in the blanks with the word or words that make the following statements correct.

1. Gun spray painting should not be done indoors unless a _____ is available.
2. Aluminum ladders are a special hazard around _____.
3. Dry paint in a spray gun will _____.
4. Spray guns should be lubricated with thin machine oil on each _____ where it enters the _____.
5. The tendency of a liquid to flow is known as its _____.

Small Gas Engines

UNIT

29 Small Engine Maintenance

OBJECTIVE

To maintain and perform minor repairs on small engines.

COMPETENCIES TO BE DEVELOPED

After studying this unit, you should be able to:

- Practice appropriate safety precautions.
- Identify the major parts and systems of small engines.
- Describe the general operation of two- and four-cycle engines.
- Conduct recommended maintenance procedures on small engines.
- Solve minor small engine problems.
- Prepare small engines for storage.

TERMS TO KNOW

internal combustion engine
piston
crankshaft
reciprocating
horizontal shaft engine
vertical shaft engine
stroke
cycle
four-stroke cycle
four-cycle engine
two-stroke-cycle
two-cycle engine
poppet valve
port
reed valve
intake stroke
compression stroke
compression ratio
top dead center (TDC)
bottom dead center (BDC)
fire
ignition
power stroke
exhaust
exhaust stroke
revolution
momentum
maintenance
troubleshooting
repair
adjust
contaminant
saturate
dual
precleaner
shroud
score
torqued
speed indicator
RPM
leaner
richer
labor

MATERIALS LIST

- ✓ Protective goggles and coveralls
- ✓ General shop hand tools and wrenches
- ✓ Four-cycle engine in need of maintenance
- ✓ Two-cycle engine in need of maintenance
- ✓ Clean gasoline in an approved fuel can
- ✓ Oil of recommended type and grade for the engine
- ✓ Spark plug gauge
- ✓ Rags and approved waste disposal cans
- ✓ Varsol, kerosene, or approved parts cleaner

Gasoline and diesel engines have become the backbone of the world in motion. They power automobiles, trucks, tractors, diesel electric trains, and ships at sea.

Small gasoline engines are used on boats, recreational vehicles, snowmobiles, and bikes. Small engines run chain saws, portable grass cutters, leaf blowers, and lawnmowers, common in most neighborhoods. Small engines power garden tractors, tillers, bale throwers, and elevators on farms. The ability to keep engines working is a valuable skill for most workers in agriculture.

CAUTION:

- **Do not use gasoline as a parts cleaner due to the fire hazard.**

SAFETY FIRST

There is some danger inherent in using all machines. Gasoline is an added hazard in gasoline engines. Suggestions for avoiding fire and injury around small engines follow:

- Wear safety glasses and leather shoes.
- Keep hands and feet a safe distance from all moving parts.
- Do not smoke around oil, solvents, or gasoline. Do not smoke near the openings of gasoline tanks and containers.
- Stop the engine before refueling.
- Do not spill gasoline on hot engine parts.
- Whenever possible, handle gasoline outdoors.
- Disconnect the spark plug cable when making any adjustments on machines to prevent the possibility of accidental engine starting.
- Keep all shields in place.
- Do not operate engines above the specified speed.
- Do not overload engines or force equipment beyond the designed capacity.

HOW AN ENGINE RUNS

An engine requires fuel, oxygen, and heat to run. It harnesses the energy of fuels such as gasoline, diesel fuel, propane gas, or kerosene. The fuel mixes with air, which provides the oxygen. The mixture is ignited by a spark, which provides the heat.

Internal combustion engines have been around for about a hundred years. An internal combustion engine is a device that burns fuel inside a cylinder to create a force that drives a piston, or cylindrical device within the cylinder. The moving piston pushes a connecting rod, which turns a crankshaft, figure 29-1. A crankshaft is a device that converts circular motion to linear motion, or vise versa. In this way, the reciprocating, or back-and-forth, motion of the piston is changed to circular motion.

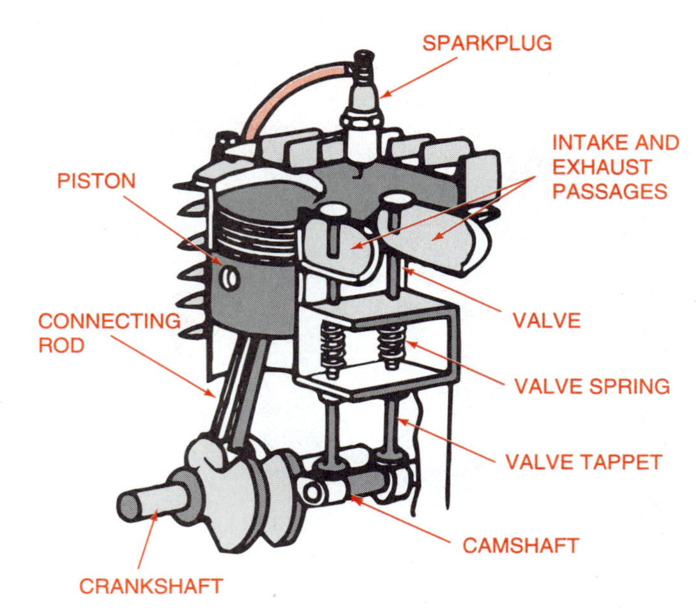

FIGURE 29-1. In an internal combustion engine, burning fuel pushes a piston, which is attached to a connecting rod. The connecting rod turns a crankshaft.

All of an engine's parts and systems exist to convert the energy of burning fuel to circular motion. Once circular motion is obtained, a system of gears, sprockets, chains, belts, and pulleys is used to drive machines, figure 29-2.

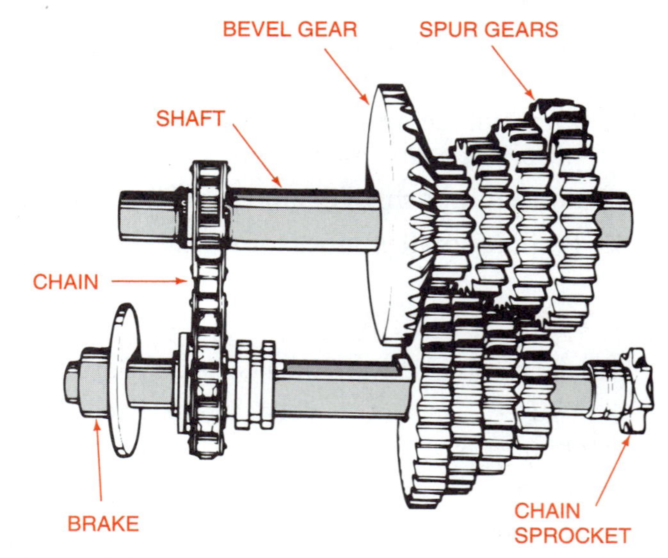

FIGURE 29-2. Once circular motion is obtained, a system of gears, sprockets, chains, belts, and pulleys is used to drive machines. *(Courtesy of Tecumseh Products Company, Engine and Gear Service Division)*

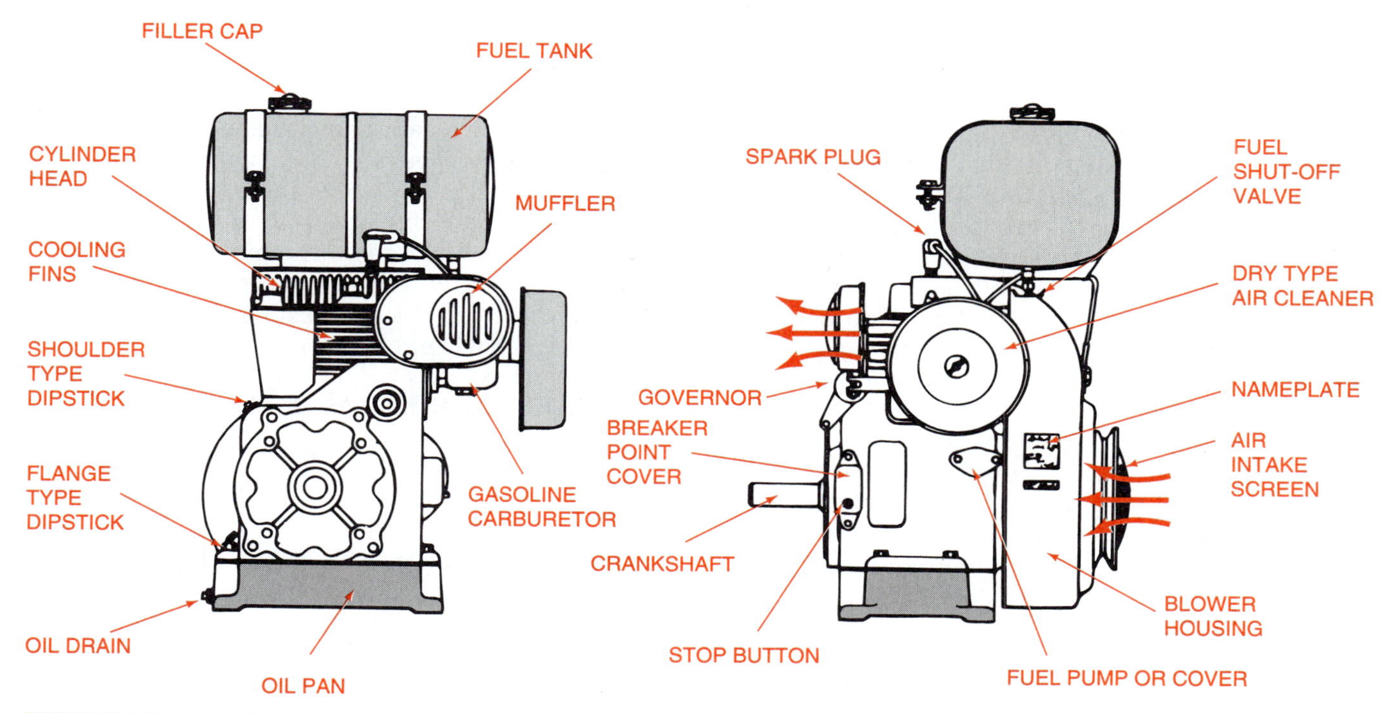

FIGURE 29-3. A four-cycle horizontal shaft engine. (*Courtesy of G. Stephenson,* Small Gasoline Engines, 4th ed., *Delmar Publishers, 1984*)

Engine Classification

Engines can be classified by the normal position of the crankshaft. If the crankshaft is placed sideways in an engine, the engine is classified as a horizontal shaft engine, figure 29-3. If the crankshaft is in an up-and-down position, the engine is known as a vertical shaft engine, figure 29-4. Engines designed to operate in all positions, such as those used on chain saws and weed trimmers, may be regarded as multiposition shaft engines.

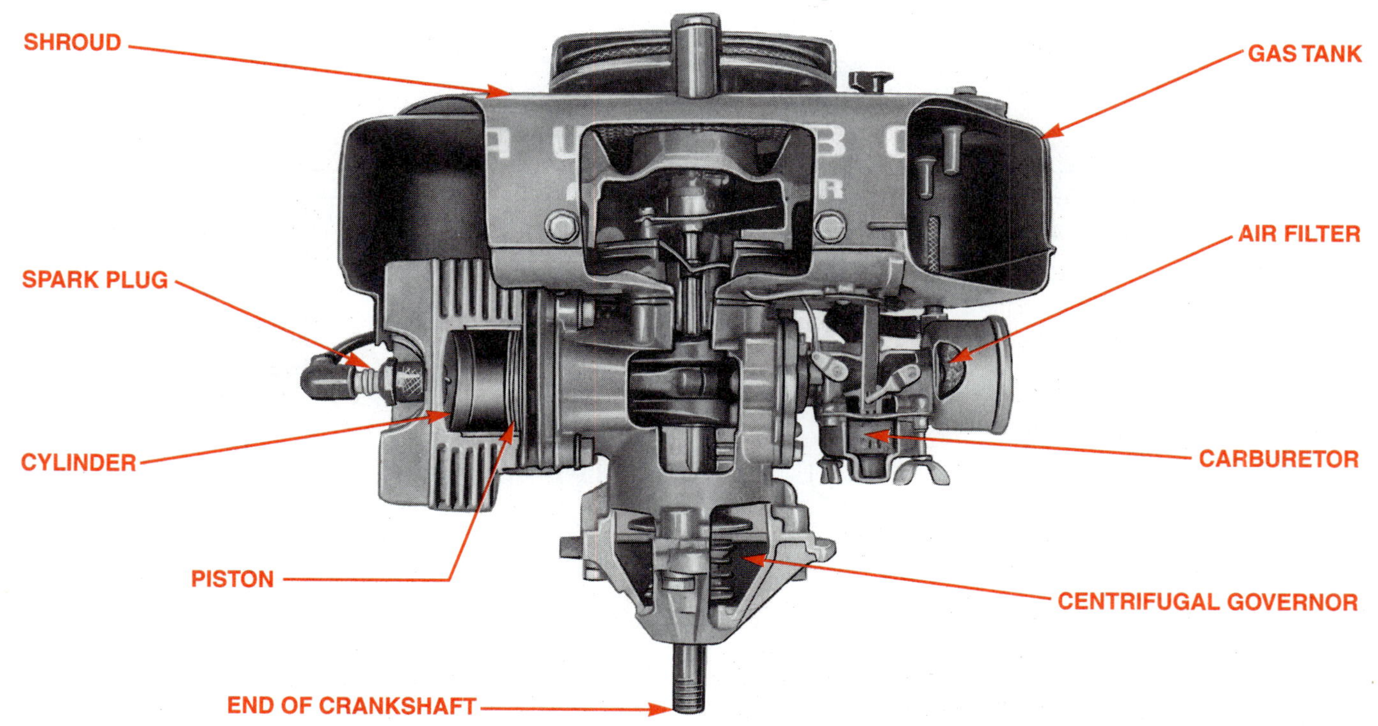

FIGURE 29-4. A two-cycle vertical shaft engine. (*Courtesy of G. Stephenson,* Small Gasoline Engines, 4th ed., *Delmar Publishers, 1984*)

Engines are also classified according to the number of strokes in a cycle. A stroke is the movement of the piston from top to bottom or from bottom to top. The top of the stroke refers to the position of the piston when it is most distant from the crankshaft. A cycle is all the events that take place as an engine: (1) takes in air and fuel, (2) compresses the air-fuel mixture, (3) burns the mixture, and (4) expels the burned gases.

In a four-stroke-cycle engine, one event is accomplished during each stroke. This means that four strokes are required to complete a cycle. The name is shortened to four-cycle engine. In a two-stroke-cycle engine, all four functions are completed in just two strokes. It is called a two-cycle engine for short. Both four-cycle and two-cycle engines are available with either horizontal or vertical crankshaft.

Engine Parts

The major parts of engines are shown in the cross-section views in figures 29-3 through 29-6. On the four-cycle engine, figure 29-3, the major parts are the oil pump, connecting rod, piston, cylinder, fins, cylinder head, spark plug, exhaust valve, intake valve, intake port, blower fan, carburetor, gas (fuel) tank, muffle, valve lifters, flywheel, camshaft drive gear, crankcase (also called the assembly), camshaft, and crankshaft.

The two-cycle engine has most of the same parts as the four-cycle engine. The main difference is in the valve systems. A four-cycle engine has two poppet valves per cylinder. A poppet valve controls the flow of air and gases by moving up and down. It is pushed open by a cam and closes by spring action. (A cam is a raised area on a shaft.) On the other hand, a two-cycle engine has one or more intake ports and exhaust ports per cylinder. A port is a special hole in the cylinder wall closed by the piston moving over the opening. A two-cycle engine also has a reed valve. A reed valve is a flat, flexible plate that lets air and fuel in but will not let the mixture out.

Operation of a Four-Cycle Engine

The four strokes of a four-cycle engine are shown in figure 29-5.

Intake Stroke. The intake stroke starts with the piston in its top position (most distant from the crankshaft). As the piston moves toward the crankshaft, it creates a vacuum in the cylinder. During this movement, the intake valve is open and a proper mixture of air and fuel from the carburetor rushes in (hence the term intake stroke). It is the function of the carburetor to supply a mixture of air and fuel in the correct proportion. The intake valve closes at the end of the stroke.

Compression Stroke. At the bottom of the intake stroke and with both valves closed, the crankshaft moves across the bottom of its swing. The movement of the engine parts pushes the piston upward and squeezes or compresses the air-fuel mix-

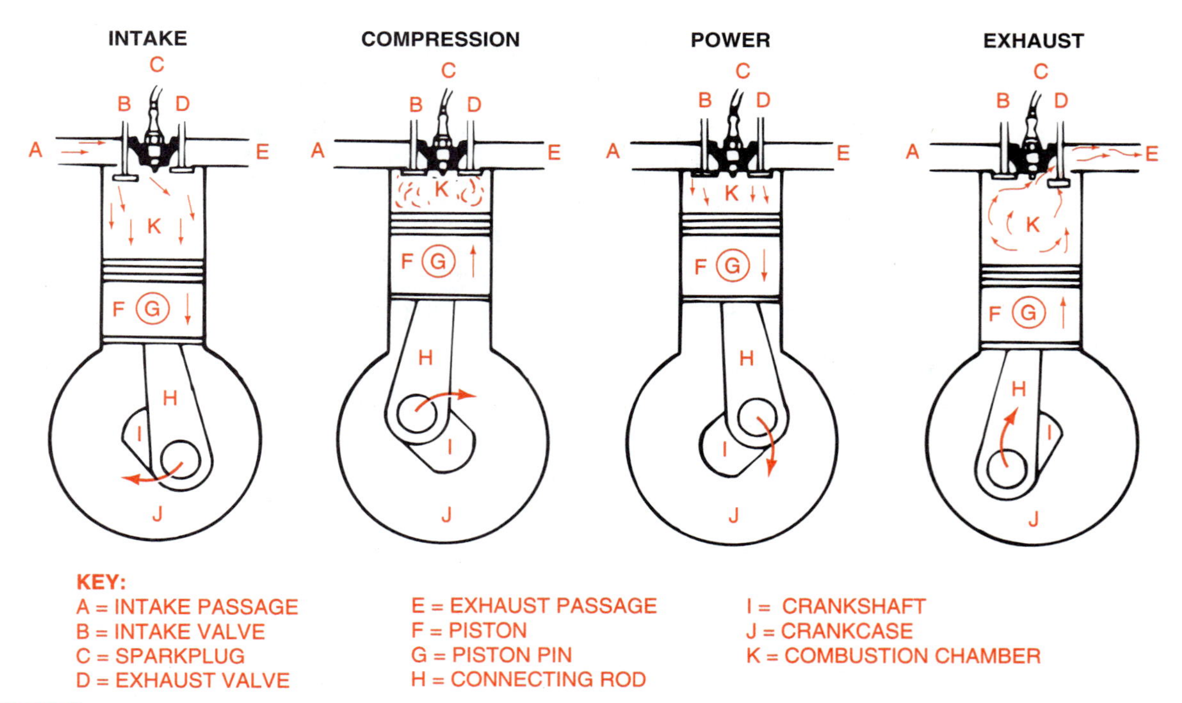

FIGURE 29-5. Operation of a four-stroke-cycle engine. (*Courtesy of The Pennsylvania State University, Department of Agricultural Education*)

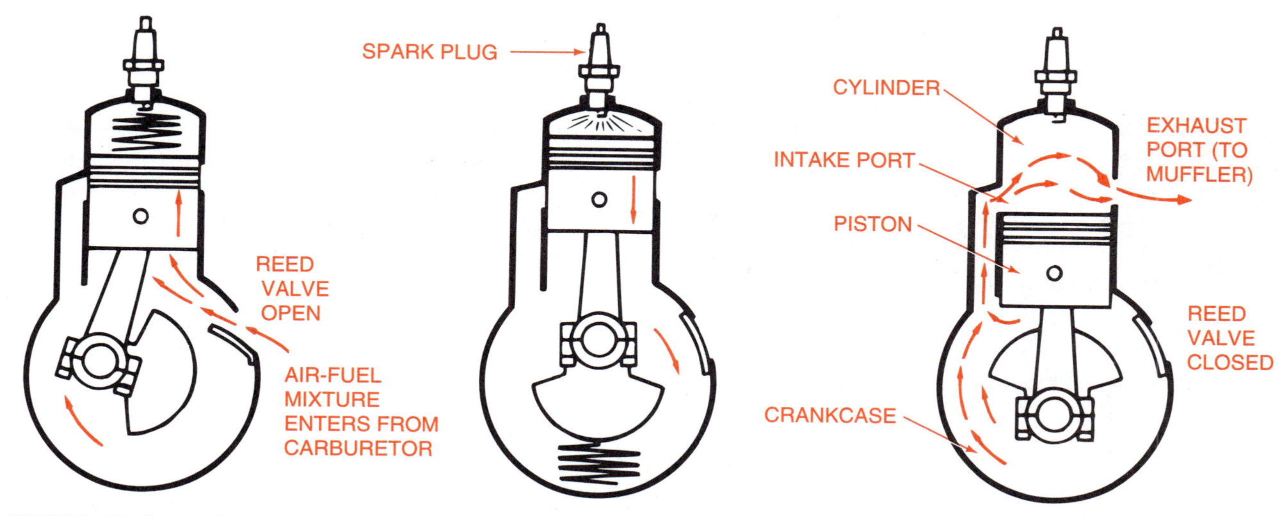

FIGURE 29-6. Operation of a two-stroke-cycle engine. (*Adapted from* Small Gasoline Engines Student Handbook, *The Pennsylvania State University, Department of Agricultural Education*)

ture. This action gives rise to the term compression stroke. At the beginning of this stroke, the compression chamber and the cylinder are full of air and fuel. At the top of the compression stroke, the air and fuel mixture has been compressed into just the area in the head called the combustion chamber. The relationship between the volume of the cylinder plus combustion chamber at the beginning and end of this stroke is known as the compression ratio. Compression ratios for small gas engines are around 6 to 1. The ratio is written as 6:1.

Power Stroke. When the piston is at its highest point (greatest distance from the crankshaft), it is said to be at top dead center (TDC). When it is at its lowest point (closest to the crankshaft), it is said to be at bottom dead center (BDC). At or near TDC on the compression stroke, the spark plug fires. To fire means to make a spark jump across an air gap. This spark ignites the compressed air-fuel mixture. The process is called ignition. The burning fuel expands rapidly but evenly and drives the piston down. This stroke is thus known as the power stroke. It is the only stroke that receives any force from the fuel.

Exhaust Stroke. At or near the bottom of the power stroke, the exhaust valve opens. As the piston moves upward it pushes out the burned or exhaust gases. Hence the term exhaust stroke. After the exhaust stroke, the exhaust valve closes and the piston is at the top and ready for another intake stroke to start the cycle again.

Notice that the crankshaft made two complete turns or revolutions in the cycle. It is the momentum or turning force of the flywheel and other moving parts that carries the engine through the three nonpower strokes. In engines with more than one cylinder, one cylinder will be on the power stroke while the other cylinders are on their nonpower strokes.

Another important point about the four-cycle engine is the use of the crankcase as an oil reservoir. Oil is splashed on engine parts by a special dipper attached to the crankshaft. Further, oil is splashed by the crankshaft and connecting rod parts that pass through the oil as the engine runs. In addition, most four-cycle engines have an oil pump to pump oil through small holes and lines to the valves and bearings. This lubrication is needed to minimize friction and keep the engine parts operating smoothly.

Operation of a Two-Cycle Engine

As stated before, the two-cycle engine completes the intake, compression, power, and exhaust stages in two strokes. The crankcase of the engine does not contain oil. Instead, it is airtight and contains a reed valve to admit the air-fuel mixture from the carburetor when the piston moves away from the crankshaft. The mixture is compressed when the piston returns toward the crankshaft. The crankcase then holds the mixture under pressure until the cylinder is ready to receive it. Oil is mixed with the gasoline to lubricate the moving parts of a two-cycle engine. Hence, there is no need for oil in the crankcase.

CAUTION:

- **It is essential that the gasoline and the proper type of oil be mixed in exact proportions as recommended by the engine manufacturer. Since a two-cycle engine does not contain oil in the crankcase, it depends totally on the oil in the gasoline to lubricate all engine parts.**

The operation of a two-cycle engine is shown in figure 29-6. To begin a cycle, the piston moves upward and creates a vacuum in the crankcase of the engine. The air-fuel mixture from the carburetor rushes into the crankcase to fill the vacuum. When the piston stops its upward movement, the reed valve closes by its own spring action. The piston then moves down and puts pressure on the mixture in the crankcase. Now the engine is ready to start a normal cycle.

Intake and Exhaust. As the piston nears the bottom of its stroke, it uncovers the intake and exhaust port(s), figure 29-6(C). Since the air-fuel mixture in the crankcase is under pressure, it rushes through a passage to the intake port and enters the cylinder. This incoming gaseous mixture pushes air or exhaust out of the cylinder. Therefore, intake and exhaust functions occur with very little movement of the piston.

Compression Stroke. The cylinder now is filled with an air-fuel mixture. The piston moves upwards, closes the intake and exhaust ports, and compresses the air-fuel mixture trapped in the cylinder. At the same time, a new supply of air and fuel rushes into the crankcase, figure 29-6(A).

Power Stroke. At or near top dead center (TDC) the spark plug fires to ignite the mixture. The burning and expanding gases drive the piston downward through the power stroke. This same downward movement puts pressure on the new air-fuel mixture in the crankcase, figure 29-6(B). Thus the engine completes its cycle of intake, compression, power, and exhaust with only two strokes of the piston.

SMALL ENGINE MAINTENANCE

An understanding of what happens in a cylinder is useful for engine maintenance and troubleshooting. Maintenance is doing the tasks that keep a machine in good condition. Troubleshooting is determining what is wrong with a machine so that the problems can be corrected. Repairing means replacing a faulty part or making it work correctly. Adjusting means setting a part or parts so the machine functions as designed.

Maintenance procedures for small engines are concerned with keeping the engine and everything that enters the engine clean. All oil, fuel, and air must be clean when they enter the engine or the engine will not run properly for very long.

Dirt or water in the fuel will make the engine miss or stall. Dirt or other contaminants in the oil will cause excessive wear of all parts. A contaminant is any material that does not belong in a substance. Dust in the air that enters the cylinder will scratch the cylinder walls. The engine will then lose compression and power.

Air Cleaners and Filters

An air cleaner is attached to the carburetor. It is designed to remove dirt and dust from the air before the air mixes with fuel. An engine may be ruined in just a few hours of operation if the air cleaner is not working properly. This is especially true under dusty conditions. Most small engine air filters should be cleaned after every 25 hours of operation or more frequently under dusty conditions.

Oil Foam Type. To service an oil foam air cleaner, the following general procedure is suggested. This procedure is also shown in figure 29-7.

PROCEDURE

1. Remove the wing nut and cover.
2. Carefully remove the spongelike foam element from the base.

CAUTION:

- **Do not rip the foam.**

3. Wash the foam in kerosene, varsol, or water with detergent.
4. Squeeze out the excess liquid.
5. Dry the foam by squeezing it with a dry cloth.
6. Saturate the foam with the motor oil recommended for the engine. (To saturate means to add a substance until the excess starts to run out.)
7. Squeeze the foam to remove the excess oil.
8. Carefully reassemble the air cleaner.

Dry Element Type. Dry element cleaners are efficient and easy to maintain. To clean a dry element, simply remove the metal cover and element, figure 29-8. Tap the bottom of the element on a flat surface to dislodge large particles of dirt and chaff and then reinstall the element.

WING NUT
COVER
SEALING LIP
BASE

REMOVE THE WING NUT AND COVER; REMOVE FOAM ELEMENT FROM THE BASE

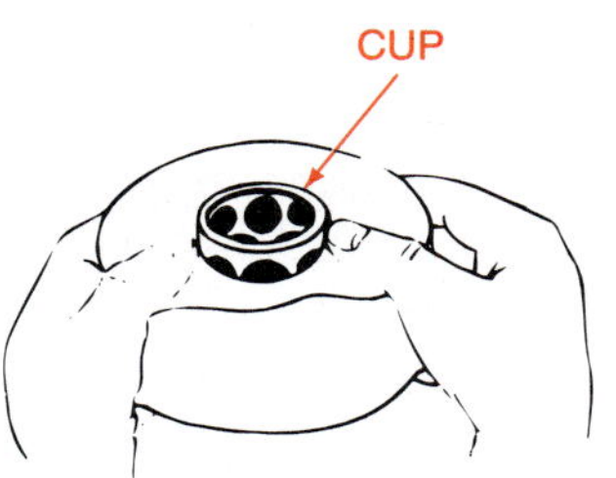

PUSH DOWN FOAM ELEMENT AND PULL OUT AIR CLEANER CUP

WASH THE FOAM IN KEROSENE, VARSOL, OR WATER WITH DETERGENT

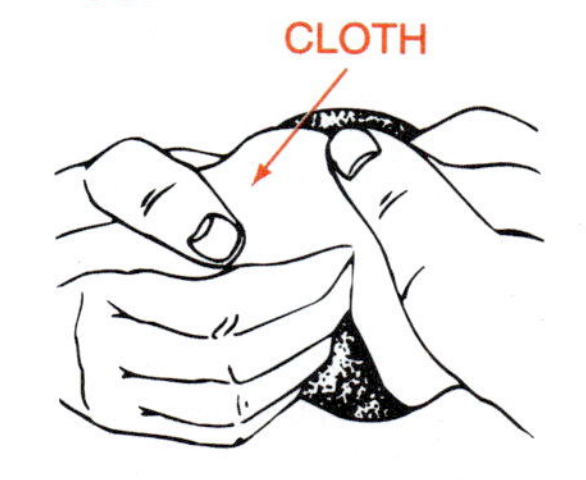

SQUEEZE OUT EXCESS LIQUID AND THEN DRY THE FOAM BY SQUEEZING IT WITH A DRY CLOTH

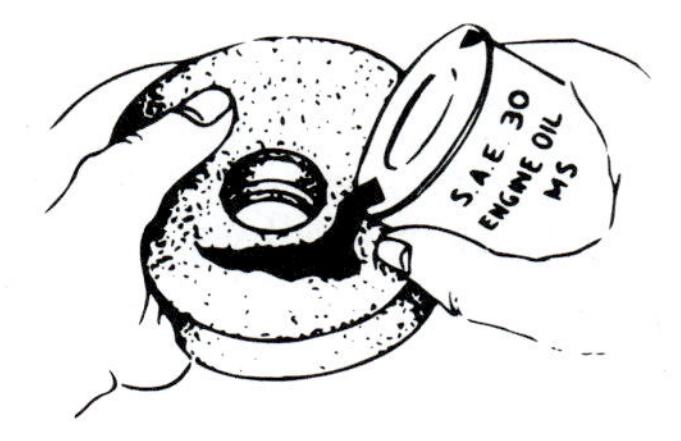

SATURATE THE FOAM WITH THE RECOMMENDED MOTOR OIL

SQUEEZE OUT THE EXCESS OIL

CAREFULLY REASSEMBLE THE AIR CLEANER

FIGURE 29-7. Servicing an oil foam air cleaner. (*Adapted from* Engine Maintenance Guide, *Briggs and Stratton Corporation*)

CAUTION:

- **Use of compressed air to clean the element may result in damage to the element.**

Replace the element at the time interval recommended by the engine manufacturer. Replacement is recommended whenever the element becomes damaged or oily, or appears to be restricting air flow. Any element that has even a small hole or slight tear must be replaced.

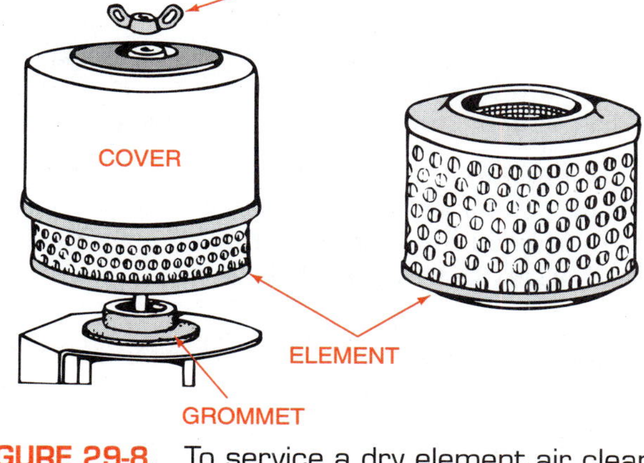

FIGURE 29-8. To service a dry element air cleaner, remove the metal cover and element. Tap the bottom of the element on a flat surface to dislodge the dirt. (*Courtesy of Briggs and Stratton Corporation*)

CAUTION:

- **When assembling the cleaner element be careful to center the element over its seat or grommet. Any crack or opening will permit dirt to bypass the filter.**

Dual Element Type. A dual element air cleaner has both a dry element and an oil foam cover, figure 29-9. Dual means two. The same procedures described for the oil foam and dry element cleaners are used here. The oil foam cover keeps most dirt

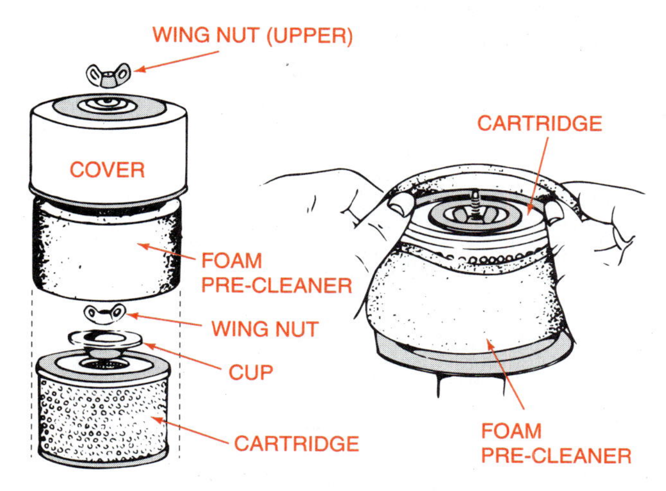

FIGURE 29-9. A dual element air cleaner has both a dry element and an oil foam cover. (*Courtesy of Briggs and Stratton Corporation*)

from entering the cleaner and so serves as a precleaner. The foam should be serviced every 25 hours. The dry element needs servicing less frequently. Some manufacturers recommend changing the dry element after about 100 hours of service.

Oil Bath Type. Oil bath cleaners are found on some older engines. They consist of a metal container with a metal mesh core. The following procedure is used to service this type of cleaner.

PROCEDURE

1. Remove the nut and top cover.
2. Lift out the wire mesh core.
3. Lift off the metal container.
4. Pour the oil and dirt out of the container. Wash the container in kerosene or varsol. Wipe dry.
5. Add motor oil to the mark on the container.
6. Flush the wire mesh core in kerosene or varsol.
7. Shake excess solvent off the core.
8. Reassemble the unit.

Crankcase Oil

It is important to keep the crankcase of a four-cycle engine filled to the correct level with the recommended type of oil. The correct oil level for most engines is indicated by a mark on a dip stick or by a filler plug located so that oil will fill to the level of the plug, figure 29-10.

Oil Functions and Classifications. Oil has four major functions: (1) to lubricate, (2) to cool, (3) to seal, and (4) to clean by carrying contaminants away from the engine. Oils for internal combustion engines are classified according to their performance under various operating conditions. Performance ratings are then designated by code letters in the classification. Most manufacturers of small gas engines call for oil with a service classification of SF, SE, SD, or SC. The service classification is clearly marked on every container of motor oil.

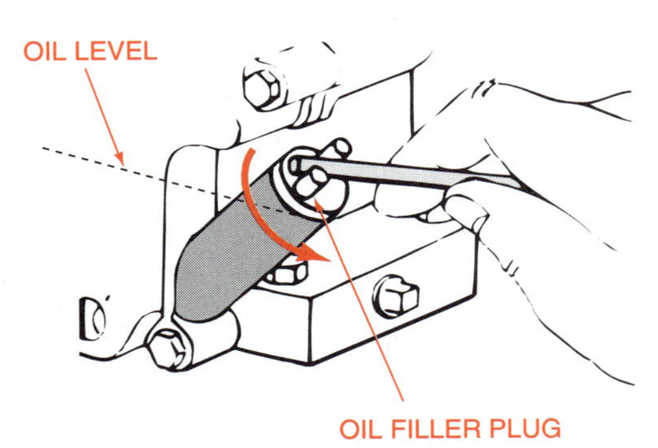

FIGURE 29-10. On some engines, oil fills to the level of the filler plug making oil level easy to check. (*Courtesy of Briggs and Stratton Corporation*)

CAUTION:

- **Using any oil other than the type recommended by the engine manufacturer may void the engine warranty and prove costly in other ways in the long run.**

Oil Viscosity. Engines need lighter weight oil as temperatures drop. Oil thickens as it gets cold so that it cannot get between tight-fitting moving parts. Therefore, the oil must be changed as the seasons change in many areas. It is important to use oil with the correct viscosity grade as rated by the Society of Automotive Engineers (SAE), figure 29-11.

CAUTION:

- **Do not overfill the crankcase or the gear reduction box.**

Oil Changes. After each 25 hours of engine use, the crankcase oil should be drained while the engine is warm, figure 29-12. Hot oil flows easily and carries the contaminants with it. The crankcase should then be refilled to the proper level using the recommended oil.

Additional attention may be needed if the engine has a reduction gear box. In this case, there is generally one hole for filling, one hole for checking the oil level, and one hole for draining the oil, figure 29-13. Each hole has a plug. Again, the manufacturer's recommendations for the oil type and change interval

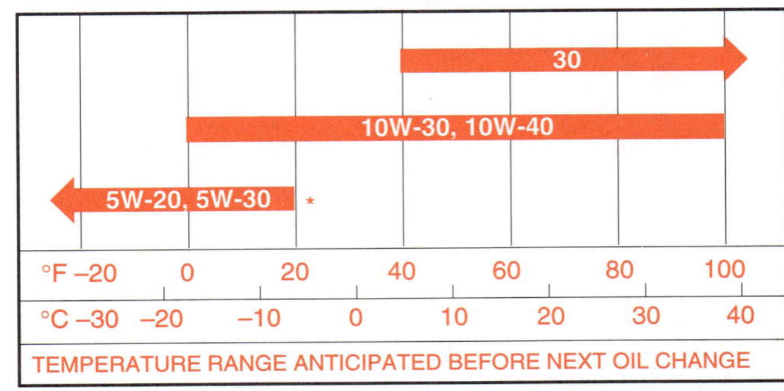

* If not available, a synthetic oil may be used having – 5W-20, 5W-30, or 5W-40 viscosity.

FIGURE 29-11. Recommended viscosity of oil for use in the crankcase of four-cycle engines under various temperature conditions. (*Courtesy of Briggs and Stratton Corporation*)

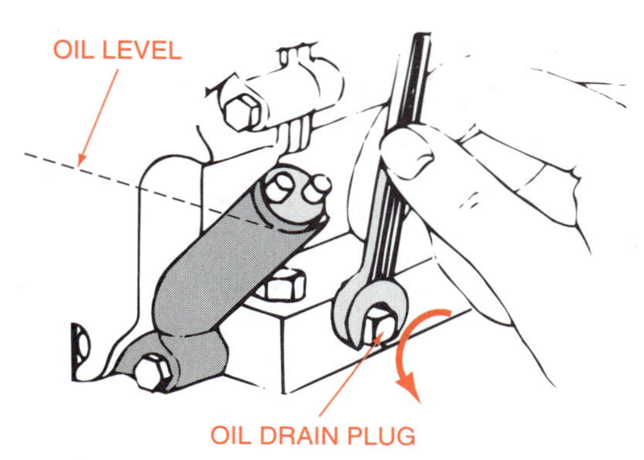

FIGURE 29-12. To change engine oil, remove the oil drain plug and drain the oil in the crankcase while the engine is warm. Replace and tighten the drain plug. Refill crankcase with new oil through the oil fill plug. (*Courtesy of Briggs and Stratton Corporation*)

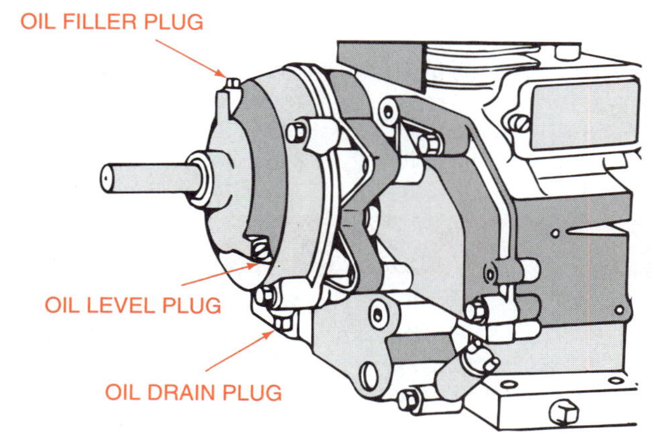

FIGURE 29-13. An engine equipped with a reduction gear box generally has one hole for filling, one hole for checking the oil level, and one hole for draining the oil. (*Courtesy of Briggs and Stratton Corporation*)

should be followed. The oil level can be checked between oil changes by removing the oil level plug. If oil does not drip out when the plug is removed, oil is added until it does.

Cooling System

The cooling system on air-cooled engines consists of:

- fins on the flywheel, which serve as a fan to move air
- fins on engine parts to transfer heat away from the parts
- shrouds to route air over these fins. A shroud is a cover. It is important to keep grass, chaff, and dirt from blocking any of the areas under an engine shroud or between fins, figure 29-14. Such materials may be removed by brushing or using compressed air.

FIGURE 29-14. Keep grass, chaff, and dirt out of the fins and airways under the shroud. (*Courtesy of Briggs and Stratton Corporation*)

Spark Plug

After 100 hours of operation, the spark plug should be serviced or replaced. To service the plug, scrape off all deposits using a pocket knife, then soak the plug in a commercial solvent. Wipe the plug dry.

File the electrodes with a point file to restore their flat surfaces. Regap the plug to .030 inch, or as recommended by the manufacturer. Blow out all dirt particles with compressed air. Replace the plug and tighten until the gasket is partially flattened.

CAUTION:

- **Do not sandblast plugs used in small engines. Sand left by blasting scores, or scratches, the cylinder walls and ruins the engine. Use of sandblasted plugs will void the warranty on some engines. In view of the low cost of a new plug and the time and problems involved in cleaning a plug, it is generally better to replace the plug than to clean it.**

Combustion Chamber

Carbon deposits build up on the tops of pistons, valves, and other surfaces in the combustion chamber. For this reason, it may be necessary to remove the cylinder head every 150 hours and scrape the carbon from all surfaces. An engine manual should be consulted for this work. Before scraping, make sure the piston is all the way up and the valves are closed. Use compressed air to blow away all carbon chips.

When reinstalling the head, clean all gasket surfaces with a smooth scraper. It is advisable to always replace the head gasket, figure 29-15. Head bolts should be tightened in the correct sequence, figure 29-16. All bolts on an engine should be torqued (tightened), according to the manufacturer's specifications. In the absence of specific information, refer to a general torque specifications table such as the one shown in figure 29-17.

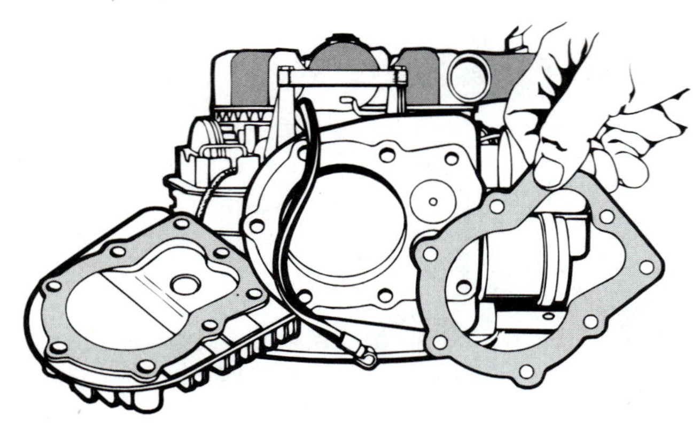

FIGURE 29-15. The head gasket must be replaced if it is damaged. (*Courtesy of Tecumseh Products Company, Engine and Gear Service Division*)

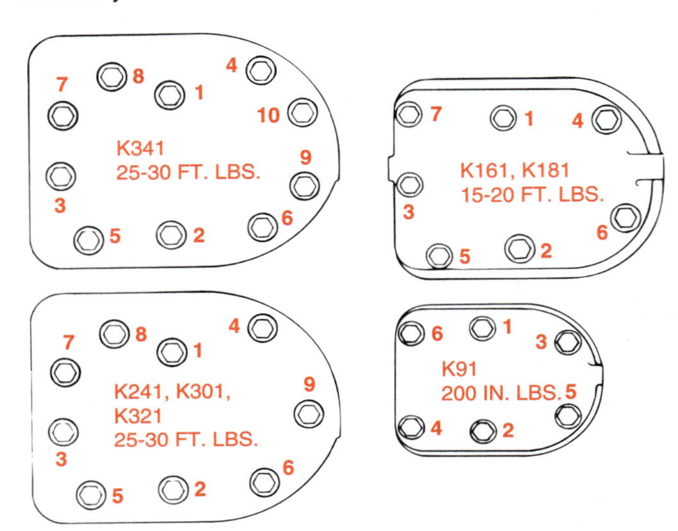

FIGURE 29-16. Recommended sequence for tightening head bolts. (*Courtesy of Kohler Co., Kohler Engines Division*)

All manufacturers provide specific maintenance schedules for their engines. While there are many similarities, it is wise to follow specific manufacturer recommendations, if available, figure 29-18.

TROUBLESHOOTING

Small engines are generally reliable, and adjustments are not normally required. If an engine does not start or if it stops during use, the simplest things should be checked first.

CAUTION:

- **Do not adjust the carburetor or other parts until the problem is diagnosed correctly. Suggested steps for troubleshooting, or diagnosing gas engine problems follow.**

CAST IRON OR STEEL			
Size	Grade 2	Grade 5*	Grade 8
8-32	20 in. lb.	25 in. lb.	
10-24	32 in. lb.	40 in. lb.	
10-32	32 in. lb.	40 in. lb.	
1/4-20	70 in. lb.	115 in. lb.	165 in. lb.
1/4-28	85 in. lb.	140 in. lb.	200 in. lb.
5/16-18	150 in. lb.	250 in. lb.	350 in. lb.
5/16-24	165 in. lb.	270 in. lb.	30 ft. lb.
3/8-16	260 in. lb.	35 ft. lb.	50 ft. lb.
3/8-24	300 in. lb.	40 ft. lb.	60 ft. lb.
7/16-14	35 ft. lb.	55 ft. lb.	80 ft. lb.
7/16-20	45 ft. lb.	75 ft. lb.	105 ft. lb.
1/2-13	50 ft. lb.	80 ft. lb.	115 ft. lb.
1/2-20	70 ft. lb.	105 ft. lb.	165 ft. lb.
9/16-12	75 ft. lb.	125 ft. lb.	175 ft. lb.
9/16-18	100 ft. lb.	165 ft. lb.	230 ft. lb.
5/8-11	110 ft. lb.	180 ft. lb.	260 ft. lb.
5/8-18	140 ft. lb.	230 ft. lb.	330 ft. lb.
3/4-10	150 ft. lb.	245 ft. lb.	350 ft. lb.
3/4-16	200 ft. lb.	325 ft. lb.	470 ft. lb.
ALUMINUM			
8-32	20 in. lb.	20 in. lb.	20 in. lb.
10-24	32 in. lb.	32 in. lb.	32 in. lb.
1/4-20	70 in. lb.	70 in. lb.	70 in. lb.
5/16-18	150 in. lb.	150 in. lb.	150 in. lb.

*Also Self Tapping Screws

FIGURE 29-17. Standard torque settings to be used when specific values are not specified. (*Courtesy of Kohler Co., Kohler Engines Division*)

Cold Engine Will Not Start

If the engine is cold and will not start, the problem may be that:

- no gas is in the tank—add gas.
- the fuel valve is shut off—turn on the fuel.
- the engine is flooded—leave it for about 10 minutes and then try it again. Flooding refers to an excessive amount of gasoline in the cylinder. Such a condition may be caused by applying the choke too long. A more difficult cause to correct is the float needle valve sticking in the carburetor. If this is the case, the carburetor must be disassembled and corrections made.
- one or more control knobs are slipping on their shafts and the controls are not moving—tighten or replace the knobs or move the control by hand.

FREQUENCY	TYPE OF MAINTENANCE	SERVICE LOG	DATE SERVICED
DAILY	Check Oil Level Replenish Fuel Supply Clean Air Intake Screen	**DAILY Before Starting Engine**	
FIRST 5 Hours	On new or rebuilt engine, change oil after first 5 hours	1 2 3 4 5 Change Oil	------------
EVERY 25 Hours	Service Precleaner Change Oil	6 … 25	------------
EVERY 50 Hours	Clean Cooling Fins and External Surfaces Inspect Air Cleaner Element Check Oil Level in Reduction Gear Unit Check Belt Tension—Motor Generator Set Clean Fuel Filter	26 … 50	------------
EVERY 100 Hours	Check Spark Plug Change Oil—Wet Type Clutch Check Battery Electrolyte Level	51 … 75 76 … 100	------------ ------------

ANNUALLY (or Every 500 Hours)
These services require the attention of a trained mechanic and the use of special tools and equipment. Your Kohler engine service dealer has the facilities, training and genuine replacement parts necessary to properly perform these services.

500 Hours	❏ Ignition Timing ❏ Breaker Points ❏ Valve/Tappet Clearance ❏ Starter Motor ❏ Alternator Connections ❏ Voltage Regulator ❏ Crankcase Breather ❏ Crankcase Vacuum ❏ Compression	500 Hours* 1000 Hours**	❏ Cylinder Heads Serviced—Constant Load & Speed *250 hours when leaded gasoline is used. ❏ Cylinder Heads Serviced—Constant Load & Speed **500 hours when leaded gasoline is used.

ENGINE: Model ______________ **Spec.** ______________ **Serial No.** ______________

FIGURE 29-18. Manufacturer's maintenance schedule for a single-cylinder engine. One block is crossed out for each hour of use. (*Courtesy of Kohler Co., Kohler Engines Division*)

Hot Engine Will Not Start

If the engine has been in use and will not restart, the problem may be that:

- no gas in the tank—refill.
- the implement is clogged—clear the implement.
- the engine is overheated—leave it for 10 minutes and try again.

Starting Problems Persist

The next step in troubleshooting is to determine if the cylinder is getting fuel and spark. The following procedure is recommended.

PROCEDURE

1. Remove and examine the spark plug. If the odor of gasoline is present on the plug or in the cylinder, the fuel system is probably working correctly. If the odor of gasoline is not present, trace the flow of fuel from the tank to the carburetor and look for fuel line blockages or restrictions.
2. To test for spark, attach the spark plug wire to the plug and lay the plug on its side on the engine head. Hold it down with a piece of rubber to avoid shock.
3. Crank the engine briskly and observe the spark at the gap. If there is no spark or if the spark is yellow, try a new plug. A good spark is blue in color.
4. If the engine still will not start or other problems are encountered, consult a troubleshooting guide, figure 29-19.

COMMON TROUBLES AND REMEDIES

The following charts list the most common troubles experienced with gasoline engines. Possible causes of trouble are given along with probable remedy. Paragraph references direct the reader to the correction portion of the handbook.

4-Cycle Engine Troubleshooting Chart

Cause	Remedy and Reference
ENGINE FAILS TO START OR STARTS WITH DIFFICULTY	
No fuel in tank.	Fill tank with clean, fresh fuel.
Shut-off valve closed.	Open valve.
Obstructed fuel line.	Clean fuel screen and line. If necessary, remove and clean carburetor.
Tank cap vent obstructed.	Open vent in fuel tank cap.
Water in fuel.	Drain tank. Clean carburetor and fuel lines. Dry spark plug points. Fill tank with clean, fresh fuel.
Engine overchoked.	Close fuel shut-off and pull starter until engine starts. Reopen fuel shut off for normal fuel flow.
Improper carburetor adjustment.	Adjust carburetor.
Loose or defective magneto wiring.	Check magneto wiring for shorts or grounds; repair if necessary.
Faulty magneto.	Check timing, point gap, and if necessary, overhaul magneto.
Spark plug fouled.	Clean and regap spark plug.
Spark plug porcelain cracked.	Replace spark plug.
Poor compression.	Overhaul engine.
No spark at plug.	Disconnect ignition cut-off wire at the engine. Crank engine. If spark at spark plug, ignition switch, safety switch or interlock switch is inoperative. If no spark, check magneto. Check wires for poor connections, cuts or breaks.
Electric starter does not crank engine.	See 12 volt starter troubleshooting chart.
ENGINE KNOCKS	
Carbon in combustion chamber.	Remove cylinder head and clean carbon from head and piston.
Loose or worn connecting rod.	Replacing connecting rod.
Loose flywheel.	Check flywheel key and keyway; replace parts if necessary. Tighten flywheel nut to proper torque.

FIGURE 29-19. Troubleshooting guide. (*Courtesy of Tecumseh Products Company, Engine and Gear Service Division*)

4-Cycle Engine Troubleshooting Chart (Cont.)

Cause	Remedy and Reference
Worn cylinder.	Replace cylinder.
Improper magneto timing.	Time magneto.
ENGINE MISSES UNDER LOAD	
Spark plug fouled.	Clean and regap spark plug.
Spark plug porcelain cracked.	Replace spark plug.
Improper spark plug gap.	Regap spark plug.
Pitted magneto breaker points.	Replace pitted breaker points.
Magneto breaker arm sluggish.	Clean and lubricate breaker point arm.
Faulty condenser (except on Tecumseh Magneto).	Check condenser on a tester; replace if defective.
Improper carburetor adjustment.	Adjust carburetor.
Improper valve clearance.	Adjust valve clearance to .010 cold.
Weak valve spring.	Replace valve spring.
ENGINE LACKS POWER	
Choke partially closed.	Open choke.
Improper carburetor adjustment.	Adjust carburetor.
Magneto improperly timed.	Time magneto.
Worn rings.	Replace rings.
Lack of lubrication.	Fill crankcase to the proper level.
Air cleaner fouled.	Clean air cleaner.
Valves leaking.	Grind valves and set to .010 cold.
ENGINE OVERHEATS	
Engine improperly timed.	Time engine.
Carburetor improperly adjusted.	Adjust carburetor.
Air flow obstructed.	Remove any obstructions from air passages in shrouds.
Cooling fins clogged.	Clean cooling fins.
Excessive load on engine.	Check operation of associated equipment. Reduce excessive load.
Carbon in combustion chamber.	Remove cylinder head and clean carbon from head and piston.
Lack of lubrication.	Fill crackcase to proper level.
ENGINE SURGES OR RUNS UNEVENLY	
Fuel tank cap vent hole clogged.	Open vent hole.
Governor parts sticking or binding.	Clean, and if necessary repair governor parts.
Carburetor throttle linkage or throttle shaft and/or butterfly binding or sticking.	Clean, lubricate, or adjust linkage and deburr throttle shaft or butterfly.
Intermittent spark at spark plug.	Disconnect ignition cut-off wire at the engine. If spark, check ignition switch, safety switch and interlock switch. If no spark, check magneto. Check wires for poor connections, cuts or breaks.
Improper carburetor adjustment.	Adjust carburetor.
Dirty carburetor.	Clean carburetor.

FIGURE 29-19. *Continued*

4-Cycle Engine Troubleshooting Chart (Cont.)

Cause	Remedy and Reference
ENGINE VIBRATES EXCESSIVELY.	
Engine not securely mounted.	Tighten loose mounting bolts.
Bent crankshaft.	Replace crankshaft.
Associated equipment out of balance.	Check associated equipment.
ENGINE USES EXCESSIVE AMOUNT OF OIL	
Engine speed too fast.	Using tachometer adjust engine RPM to spec.
Oil level too high.	To check level turn dipstick cap tightly into receptacle for accurate level reading.
Oil filler cap loose or gasket damaged causing spillage out of breather.	Replace ring gasket under cap and tighten cap securely.
Breather mechanism damaged or dirty causing leakage.	Replace breather assembly.
Drain hole in breather box clogged causing oil to spill out of breather.	Clean hole with wire to allow oil to return to crankcase.
Gaskets damaged or gasket surfaces nicked causing oil to leak out.	Clean and smooth gasket surfaces. Always use new gaskets.
Valve guides worn excessively thus passing oil into combustion chamber.	Ream valve guide oversize and install $^1/_{32}$" oversize valve.
Cylinder wall worn or glazed, allowing oil to bypass rings into combustion chamber.	Bore hole, or deglaze cylinder as necessary.
Piston rings and grooves worn excessively.	Reinstall new rings and check land clearance and correct as necessary.
Piston fit undersized.	Measure and replace as necessary.
Piston oil control ring return holes clogged.	Remove oil control ring and clean return holes.
Oil passages obstructed.	Clean out all oil passages.
OIL SEAL LEAKS	
Crankcase breather.	Clean or replace breather.
Old seal hardened and worn.	Replace seal.
Crankshaft seal contact surface is worn undersize causing seal to leak.	Check crankshaft size and replace if worn excessively.
Crankshaft bearing under seal is worn excessively, causing crankshaft to wobble in oil seal.	Check crankshaft bearings for wear and replace is necessary.
Seal outside seat in cylinder or side cover is damaged, allowing oil to seep around outer edge of seal.	Visually check seal receptacle for nicks and damage. Replace P.T.O. cylinder cover, or small cylinder cover on the magneto end if necessary.
New seal installed without correct seal driver and not seating squarely in cavity.	Replace with new seal, using proper tools and methods.
New seal damaged upon installation.	Use proper seal protector tools and methods for installing another new seal.
Bent crankshaft causing seal to leak.	Check crankshaft for straightness and replace if necessary.
Oil seal driven too far into cavity.	Remove seal and replace with new seal, using the correct driver tool and procedures.

FIGURE 29-19. *Continued*

4-Cycle Engine Troubleshooting Chart (Cont.)

Cause	Remedy and Reference
BREATHER PASSING OIL	
Engine speed too fast.	Use tachometer to adjust correct RPM.
Loose oil fill cap or gasket damaged or missing.	Install new ring gasket under cap and tighten securely.
Oil level too high.	Check oil level - Turn dipstick cap tightly into receptacle for accurate level reading. DO NOT fill above full mark.
Breather mechanism damaged.	Replace reed plate assy.
Breather mechanism dirty.	Clean thoroughly in solvent. Use new gaskets when reinstalling unit.
Drain hole in breather box clogged.	Clean hole with wire to allow oil to return to crankcase.
Piston ring end gaps aligned.	Rotate end gaps so as to be staggered 90 o apart.
Breather mechanism installed upside down.	Small oil drain holes must be down to drain oil from mechanism.
Breather mechanism loose or gaskets leaking.	Install new gaskets and tighten securely.
Damaged or worn oil seals on end of crankshaft.	Replace seals.
Rings not properly seated.	Check for worn, or out of round cylinder. Replace rings. Break in new rings with engine working under a varying load. Rings must be seated under high compression, or in other words, under varied load conditions.
Breather assembly not assembled correctly.	See section on Breather Assembly.
Cylinder cover gasket leaking.	Replace cover gasket.

Troubleshooting Carburetion

POINTS TO CHECK FOR CARBURETOR MALFUNCTION

Trouble	Corrections
Carburetor out of adjustment	3-11-12-13-15-19
Engine will not start	1-2-3-4-5-6-8-11-12-14-15-22
Engine will not accelerate	2-3-11-12
Engine hunts (at idle or high speed)	3-4-8-9-10-11-12-14-20-21-24
Engine will not idle	4-8-9-11-12-13-14-19-20-21-22
Engine lacks power at high speed	2-3-6-8-11-12-19-20-22-23
Carburetor floods	4-7-17-20-21-22-23
Carburetor leaks	6-7-10-18
Engine overspeeds	8-9-11-14-15-18-19
Idle speed is excessive	8-9-13-14-15-18-19-22-23
Choke does not open fully	8-9-15
Engine starves for fuel at high speed (leans out)	1-3-4-6-11-15-17-20-23
Carburetor runs rich with main adjustment needle shut off	7-11-17-18-20-22-23
Performance unsatisfactory after being serviced.	1-2-3-4-5-6-7-8-9-10-11-15-16-17-18-19-20-22-23

1. Open fuel shut-off valve at fuel tank—Fill tank with fuel.
2. Check ignition, spark plug and compression.
3. Clean air cleaner—Service as required.

FIGURE 29-19. *Continued*

Troubleshooting Carburetion (Cont.)

POINTS TO CHECK FOR CARBURETOR MALFUNCTION

4. Dirt or restriction in fuel system—Clean tank and fuel strainers, check for kinks or sharp bends.
5. Check for stale fuel or water in fuel—Fill with fresh fuel.
6. Examine fuel line and pick-up for sealing at fittings.
7. Check and clean atmospheric vent holes.
8. Examine throttle and choke shafts for binding or excessive play—Remove all dirt or paint, replace shaft.
9. Examine throttle and choke return springs for operation.
10. Examine idle and main mixture adjustment screws and "0" rings for cracks and damage.
11. Adjust main mixture adjustment screw; some models require finger-tight adjustment. Check to see that it is the correct screw.
12. Adjust idle mixture adjustment screw. Check to see that it is the correct screw.
13. Adjust idle speed screw.
14. Check for bent choke and throttle plates.
15. Adjust control cable or linkage, to assure full choke and carburetor control.
16. Clean carburetor after removing all non-metallic parts that are serviceable. Trace all passages.
17. Check inlet needle and seat for condition and proper installation.
18. Check sealing of welch plugs, cups, plugs and gaskets.
19. Adjust governor linkage.

SPECIFIC CARBURETOR CHECKS FOR FLOAT

20. Adjust float setting, if float type carburetor.
21. Check float shaft for wear and float for leaks or dents.

SPECIFIC CARBURETOR CHECKS FOR DIAPHRAGM

22. Check diaphragm for cracks or distortion and check nylon check ball for function.
23. Check sequence of gasket and diaphragm for the particular carburetor being repaired.

Troubleshooting 12 Volt Starters

Problem	Probable Cause	Fix
Does not function.	Weak or dead battery.	Check charge and/or replace battery.
	Corroded battery terminals and/or electrical connections.	Clean terminals and/or connections.
	Brushes sticking.	Free brushes. Replace worn brushes and those which have come in contact with grease and oil.
	Dirty or oily commutator.	Clean and dress commutator.
	Armature binding or bent.	Free armature and adjust end play, replace armature, or replace starter.
	Open or shortened armature.	Replace armature.
	Shorted, open or grounded field coil.	Repair or replace housing.
	Loose or faulty electrical connections.	Correct.
	Load on engine.	Disengage all drive apparatus and relieve all belt and chain tension.
	Electric starter cranks, but no spark at spark plug.	Disconnect ignition cut-off wire at the engine. Crank engine. If spark at spark plug, ignition switch, safety switch is inoperative. If no spark, check magneto. Check wires for poor connections, cuts or breaks.

FIGURE 29-19. *Continued*

Troubleshooting 12 Volt Starters (Cont.)

Problem	Probable Cause	Fix
Does not function.	Electric starter does not crank engine.	Remove wire from starter. Use a jumper battery and cables and attach directly to starter. If starter cranks engine the starter is okay; check solenoid, starter switches, safety switches and interlock switch. Check wires for poor connections, cuts or breaks.
Low RPM.	Unit controls engaged.	Insure all unit controls are in neutral or disengaged.
	Worn bearings in cap assemblies.	Clean bearings or replace cap assemblies.
	Bent armature.	Replace armature.
	Binding armature.	Free up armature. Adjust armature end play.
	Brushes not seated properly.	Correct.
	Weak or annealed brush springs.	Replace springs.
	Incorrect engine oil.	Ensure the correct weight of oil is being used.
	Dirty armature commutator.	Clean commutator.
	Shortened or open armature.	Replace armature.
	Loose or faulty electrical connections in motor.	Correct.
Motor stalls under load.	Shorted or open armature.	Replace armature.
	Shortened field coil.	Correct, or replace housing assembly.
Intermittent operation.	Brushes binding in holders.	Free up brushes. Replace worn brushes and those which have come in contact with grease and oil.
	Dirty or oily commutator.	Clean and dress commutator.
	Loose or faulty electrical connections.	Correct.
	Open armature.	Replace armature and interlock switch.
	Break in electrical circuit.	Disconnect ignition cut-off wire at the engine. Crank engine, if spark, check ignition switch, safety switch and interlock switch. Check wires for poor connections, cuts or breaks.
Sluggish disengagement of the drive assembly pinion gear.	Dirt and oil on assembly and armature shaft.	Clean drive assembly and armature shaft and relubricate shaft splines.
	Bent armature.	Replace armature.

Troubleshooting 120 Volt Starter System

NOTE: The power supply for the rectifier assembly must be 120 volts A.C. with a grounding type outlet. The A.C. circuit cable should be no smaller than #14 and should be fused with a 30 ampere fuse or a 25 ampere Fustate or Fusetron.

Trouble	Probable Cause	Remedy
Low RPM.	Incorrect oil viscosity.	Install correct amount of oil.
	Binding armature.	Free up armature. Adjust armature end play.
	Brushes not seated properly.	Correct.
	Weak or annealed brush springs.	Replace springs.
	Dirty armature commutator.	Clean commutator.
	Shortened or open armature.	Replace armature.

FIGURE 29-19. *Continued*

Troubleshooting 120 Volt Starter System (Cont.)

Problem	Probable Cause	Fix
	Low line voltage in A.C. circuit.	
Motor stalls under load.	Shorted or open armature.	Replace armature.
	Shorted field coil.	Correct.
Intermittent operation.	Brushes binding in holders.	Free up brushes. Replace worn brushes (see each starter) and those that have come into contact with grease and oil.
	Dirty or oily commutator.	Clean and dress commutator. Replace any brush that has come into contact with grease or oil.
	Faulty electrical connections in motor or rectifier assembly.	Correct.
	Open armature.	Replace armature.
Will not operate.	Armature binding.	Free up armature. Adjust armature end play.
	Brushes sticking in brush holders.	Free up brushes. Replace worn brushes (see each starter) and those which have come into contact with grease or oil.
	Dirty or oily commutator.	Clean and dress commutator. Replace any brush which has come into contact with grease or oil.
	Faulty electrical connections.	Correct.
	Open field coil.	Correct.
	"Blown" fuse in A.C. circuit.	Replace fuse. Check starting motor to determine if overload was caused by faulty motor. Check rectifiers for serviceability.
	Open or shorted armature.	Replace armature.
	2 rectifiers open in one heat sink.	Replace rectifier assembly.
Sluggish disengagement of the drive assembly pinion gear.	Dirt and oil on assembly and armature shaft.	Clean drive assembly and armature shaft and relubricate shaft splines.

FIGURE 29-19. *Continued*

Sheared Flywheel Keys

Many other problems can be diagnosed and corrected by the operator using common tools and a troubleshooting guide. If an engine has stalled abruptly, such as when a rotary mower blade strikes a root, the flywheel key is probably sheared slightly or completely. In this case, the operator's manual should be used for guidance in replacing the key. If service procedures are not covered in the operator's manual, it is advisable to consult a small engine technician or dealer.

Carburetor Adjustment

Small engine carburetors are adjusted correctly at the factory. They seldom get out of adjustment on their own. Unfortunately, carburetor settings are frequently changed by operators when engines do not start or when they run incorrectly due to other problems. When the real cause of the problem is corrected, the engine cannot perform well because the carburetor is out of adjustment. Typical carburetor adjustment points are shown in figure 29-20.

Idle Speed Adjustment. Most carburetors have an idle speed adjusting screw. The engine idle speed is increased when the screw is turned clockwise and decreased when the screw is turned counterclockwise. The engine idle speed should be set with a speed indicator. A speed indicator is a device used to measure the speed at which the crankshaft or other parts turn in revolutions per minute (RPM or r/min). It is important to keep the idle speed within the range specified by the manufacturer.

Air-Fuel Mixture. The air-fuel mixture must be carefully controlled if small engines are to run correctly. Most engines have one slotted head needle valve to adjust the air-fuel mixture at idle and one for intermediate and high speed operation. These valves are found on the carburetor and are called the idle mixture adjustment and the main mixture adjustment.

On most engines, these adjustment screws are turned clockwise to reduce the proportion of fuel and make the mixture leaner. A leaner mixture has less fuel and more air. Turning the screw counterclockwise increases the proportion of fuel to air, creating a richer mixture.

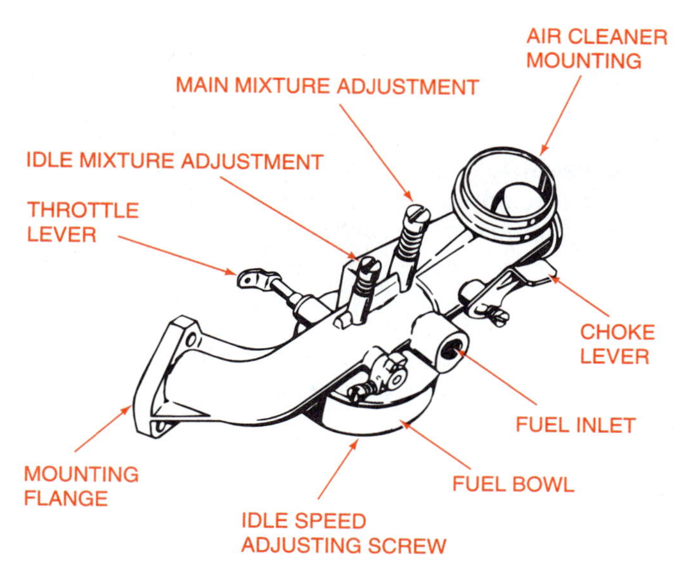

FIGURE 29-20. Typical carburetor adjustment points. (*Courtesy of Engine Service Association, Inc.*)

Adjusting the Idle Mixture. The operator's manual should be consulted before any attempt is made to adjust a small engine . However, in the absence of a manual, the following steps are generally followed to make an idle mixture adjustment.

PROCEDURE

1. Start the engine and run it until it is up to the recommended operating temperature.
2. Set the idle speed adjustment so the engine idles at the recommended speed.
3. Slowly turn the idle mixture adjustment screw clockwise until the engine slows down and starts to labor. To labor means to struggle or work hard to keep running.
4. Slowly turn the screw back counterclockwise until the engine runs smoothly and then decreases in speed.
5. Slowly turn the screw in the opposite direction to the point where the engine runs the most smoothly.
6. Reset the idle speed adjustment so the engine idles at the specified speed.

Adjusting the Main Mixture. When making a main mixture air-fuel adjustment, the following steps are commonly used.

PROCEDURE

1. Warm the engine up to operating temperature.
2. Set the idle speed to the manufacturer's specification.
3. Be certain the idle mixture is correctly adjusted.
4. Set the throttle on the engine at its high-speed operating position.
5. Slowly turn the main mixture adjustment screw clockwise until the engine labors.
6. Slowly turn the screw counterclockwise until the engine runs fast and then begins to slow down.
7. Slowly turn the screw in the opposite direction until the engine runs smoothly.
8. Put the engine under load to check its performance.
9. If the engine does not accelerate or pull well, turn the screw counterclockwise slightly to make the mixture richer.

CAUTION:

- **When adjusting a carburetor needle valve, the screw should be moved only 1/32 of a turn before observing the engine's performance. It takes a few seconds for the new valve setting to influence the performance of the engine.**

STORING SMALL ENGINES

Whenever a gas engine is to be stored, it is important to leave it in good condition. A little care when storing can save time when the engine is to be restarted.

When storing for a month or less, use the following procedure.

PROCEDURE

1. Close the fuel shut-off valve and drain the carburetor if it has a drain plug or drain plunger.
2. Fill the fuel tank with gasoline of the correct grade. If it is a two-cycle engine, fill the tank with gasoline and oil mixed in the proper proportion.
3. Store the engine in a dry place.
4. If the engine must be stored outside, cover it with plastic or a tarpaulin.
5. If the engine is supplied with a battery, be sure that the battery is fully charged.

When storing an engine for longer than a month, the following procedure is recommended.

PROCEDURE

1. Drain the fuel tank.
2. With the fuel shut-off valve open, run the engine until all fuel is burned.
3. For a four-cycle engine, drain the oil and refill with the correct oil for the weather when the engine will be put back into service.
4. Remove the spark plug. (For a two-cycle engine, turn the crankshaft until the piston covers the ports.) Pour 2 to 3 tablespoons of clean engine oil into the cylinder to prevent rust.
5. Reinstall the spark plug.
6. Turn the crankshaft slowly three or four times to distribute the oil.
7. Drain, clean, and reinstall the fuel filter bowl, if so equipped.
8. Remove the shrouds and guards so that the surface of the engine can be cleaned completely.

CAUTION:

- **When removing the shroud, be careful to avoid bending any carburetor or governor linkage. Also observe how the linkage is assembled so that you will be able to reassemble it correctly.**

9. Clean the exterior.
10. Remove all debris from between the fins.
11. Check to see that all linkages move freely.
12. Loosen all belts.
13. Clean and lubricate all chains.
14. Lubricate all fittings.
15. Reinstall all shrouds and guards.
16. Store the engine off the ground indoors. If it is stored outside, cover it to protect it from moisture.

STUDENT ACTIVITIES

1. Define the Terms to Know in this unit.
2. Name the parts of a four-cycle engine.
3. Name the parts of a two-cycle engine.
4. Obtain a used small engine and its operator's manual. Perform the following service procedures:
   a. Use an air gun or brush and small screwdriver to clean the fins and air passageways.
   b. Service the air cleaner
   c. Check the fuel tank for dirt, chaff, grass, or leaves. If dirty, drain the tank and remove all foreign matter. Refill with fresh gasoline after all service procedures are completed.

CAUTION:

- **Wear safety glasses and work in a well-ventilated area. Do not spill any gasoline. Wipe up oil spills immediately.**

   d. Change the oil in the crankcase.
   e. Service or replace the spark plug.

CAUTION:

- **The plug must be gapped using a wire gauge to the manufacturer's specifications.**

   f. If the engine has several hundred hours of use, ask your instructor if you should remove the head to clean the combustion chamber.

5. Study a small engine troubleshooting guide. Place a mark beside the procedures you can perform to correct engine problems.
6. Prepare a small engine for storage.

Stroke	Piston Movement	Intake Valve Position	Exhaust Valve Position	General Action/ Activity
Intake	Down	Open	Closed	Air-fuel mixture is drawn into the cylinder.
Compression				
Power				
Exhaust				

SELF-EVALUATION

A. Multiple Choice. Select the best answer.

1. An engine takes in air to provide
   a. compression
   b. fuel
   c. lubrication
   d. oxygen
2. Internal combustion engines have been around for about
   a. 20 years
   b. 25 years
   c. 50 years
   d. 100 years
3. Reciprocal means
   a. circular
   b. expanding
   c. explosive
   d. back and forth
4. On the power stroke of a four-cycle engine the
   a. fuel is compressed
   b. exhaust valve is open
   c. intake valve is open
   d. spark plug fires
5. A dry element air cleaner is serviced by
   a. washing in varsol
   b. tapping on a level surface
   c. blowing with compressed air
   d. dipping in detergent
6. If specifications are not available, the gap of a four-cycle small engine spark plug should be set at
   a. .030 inch
   b. .020 inch
   c. .010 inch
   d. .005 inch
7. Sandblasting of spark plugs is not recommended because
   a. the practice may void the warranty
   b. the practice may damage the cylinder wall
   c. the savings over the cost of a new plug is not worth the risk
   d. all of these
8. The troubleshooting procedure starts with
   a. checking the simple things
   b. draining and replacing the fuel
   c. removing the shroud
   d. servicing the spark plug

9. Before storing an engine, 2 or 3 tablespoons of oil are placed in the cylinder to
   a. prevent rust
   b. dissolve carbon
   c. prevent freezing
   d. mix with gasoline
10. A good spark at the spark plug gap is
   a. blue
   b. orange
   c. red
   d. yellow

B. Matching. Match the terms in column I with those in column II.

Column I

1. reed
2. stroke
3. hole
4. 6:1
5. TDC
6. BDC
7. two-cycle engine
8. SF, SE, SD, SC
9. exhaust
10. viscosity

Column II

a. valve in two-cycle engines
b. top dead center
c. mix oil with gasoline
d. oils for four-cycle engines
e. tendency to flow
f. one piston movement
g. burned gases
h. port
i. bottom dead center
j. compression ratio

C. Completion. Fill in the blanks with the word or words that will make the following statements correct.

1. The spongelike foam in an oil foam air cleaner may be washed in _____, _____, or _____.
2. Crankcase oil should be changed after every _____ hours of operation in small, four-cycle engines.
3. Most small engines for home and farm use are cooled by _____.

UNIT

Small Engine Adjustment and Repair

OBJECTIVE

To adjust and repair small engines.

COMPETENCIES TO BE DEVELOPED

After studying this unit, you should be able to:

- Identify tools for engine repair.
- Disassemble and reassemble a small engine.
- Clean carburetors.
- Replace and adjust ignition points.
- Replace, lap in, and adjust valves.
- Replace piston rings.
- Replace rod bearings.
- Service rope starters.

TERMS TO KNOW

overhaul
governor
air vane
carburetor
needle
jet
seat
battery
magneto
condenser
primary circuit
secondary circuit
high-tension wire
electromagnetic induction
ignition spark
discharge
electrolyte
distilled water
armature
head gasket
valve spring compressor
valve keeper
valve pin
head
stem
margin
face
lapped in
lapping compound
valve grinding
valve guide
valve stem clearance
cylinder
head
head gasket
piston
rings
inside micrometer
telescoping gauge
cylinder hone
ring expander
blow-by
piston ring compressor
wrist pin
micrometer
plastigage
torque
rope starter
wind-up starter
kick starter

MATERIALS LIST

- ✓ Protective goggles and coveralls
- ✓ General shop tools: socket and box-end wrenches, small engine wheel holder, wheel puller, valve grinder and compound, ring squeezer, feeler, and wire gages
- ✓ Clean gasoline in an approved can
- ✓ Oil of recommended type and grade for engine used
- ✓ Rags and approved waste disposal can
- ✓ Varsol, kerosene, or approved parts cleaner
- ✓ Small engine (3 to 5 horsepower preferred)

Small gasoline engines are the choice of power for lawn, garden, nursery, farm, recreation, and many modes of travel where low horsepower and portability is required, figure 30-1. A knowledge of maintenance of small engines is useful for all who use them. Further, skill in repairing small engines can save owners and operators time and money. There are many career opportunities for those with small engine repair skills.

Before proceeding, become thoroughly familiar with the material in Unit 29, Small Engine Maintenance. Reread the unit and pay special attention to Safety First and all other safety sections.

CAUTION:

- **Gasoline is very quick to vaporize, spread, ignite, and burn. Flash fires and explosions are constant threats to those repairing engines if extreme caution is not practiced at all times.**

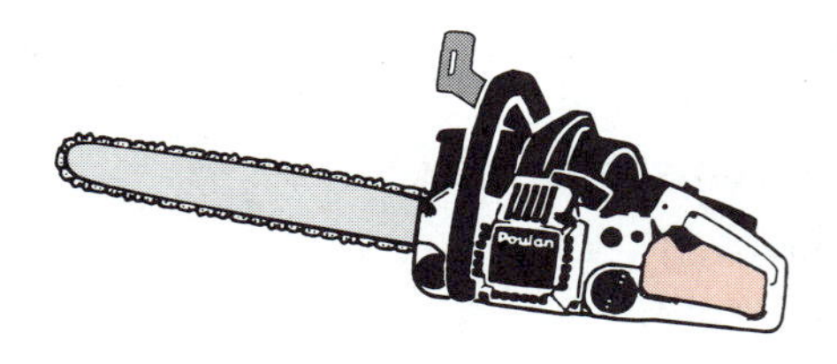

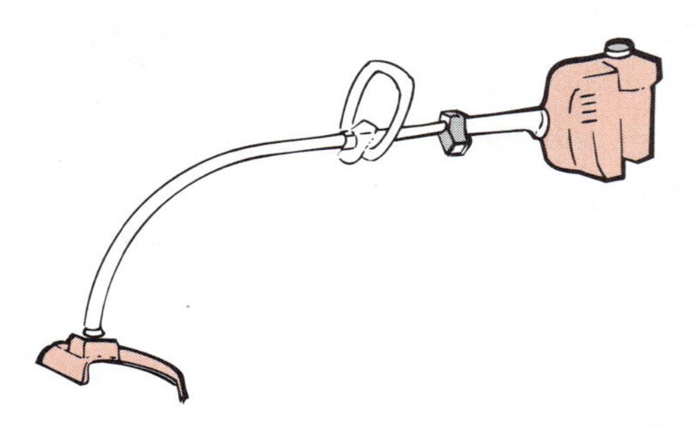

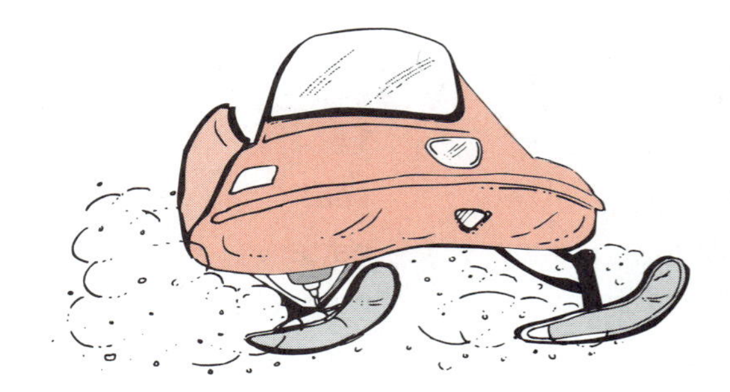

FIGURE 30-1. Small gasoline engines are used wherever a portable power source is required.

Some practices that should help reduce the fire hazards are:

- Work only in well-ventilated areas.
- Keep an approved fire extinguisher in the work area.
- Do not smoke, or use flame or electrical equipment in work area.
- Wear protective goggles at all times.
- Wear heavy, clean overalls.
- Wear leather shoes.
- Drain all fuel from tank and carburetor into an approved container out-of-doors.
- Store gasoline and other fuels in approved containers.
- Wipe up all oil and fuel spills immediately.
- Place rags that have fuel on them out-of-doors until dry.
- Store used rags in an approved metal container.
- Keep paper and other flammable material out of the work area.
- Read and observe all safety precautions found in the engine operator's and repair manuals.

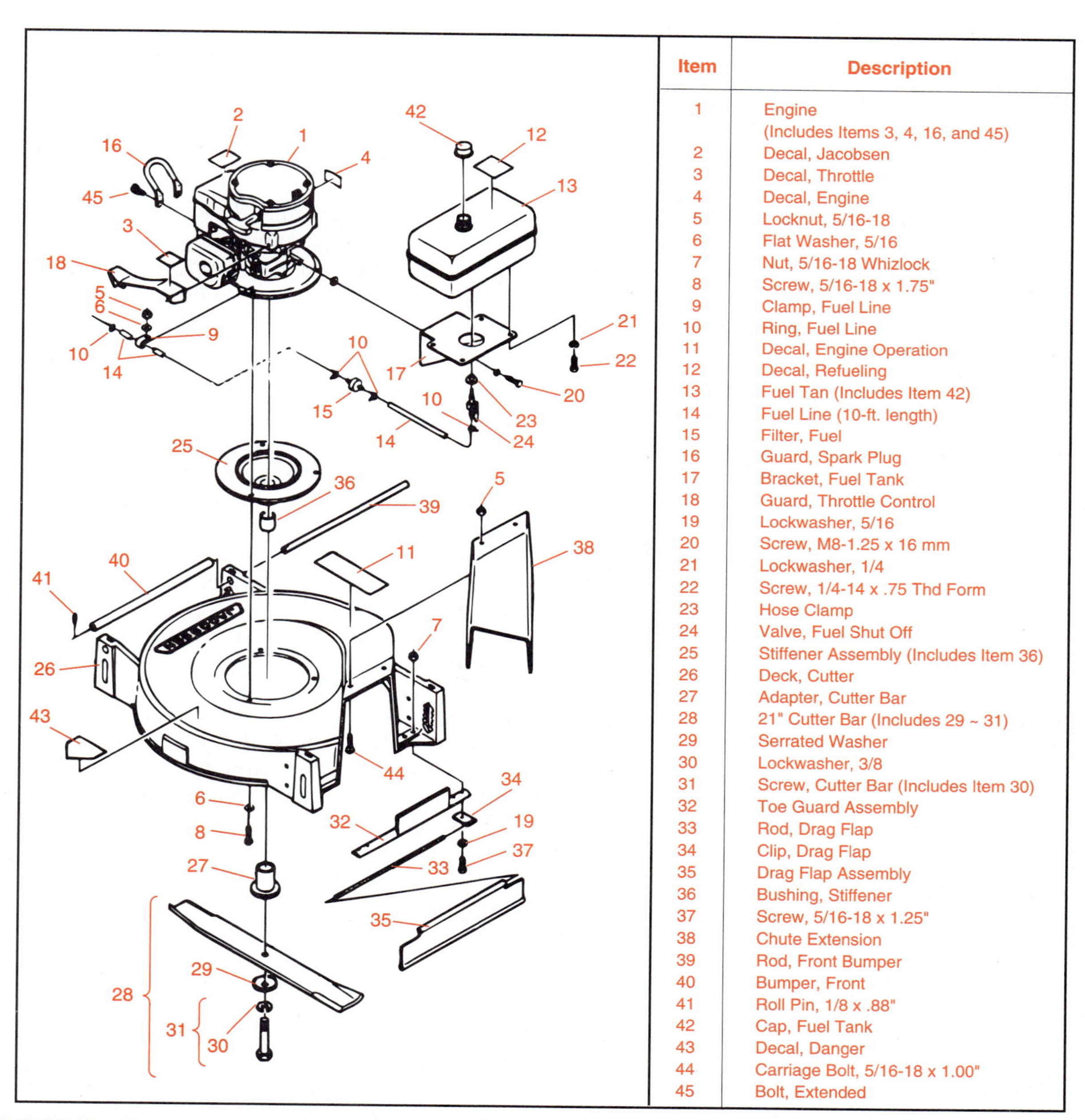

Item	Description
1	Engine (Includes Items 3, 4, 16, and 45)
2	Decal, Jacobsen
3	Decal, Throttle
4	Decal, Engine
5	Locknut, 5/16-18
6	Flat Washer, 5/16
7	Nut, 5/16-18 Whizlock
8	Screw, 5/16-18 x 1.75"
9	Clamp, Fuel Line
10	Ring, Fuel Line
11	Decal, Engine Operation
12	Decal, Refueling
13	Fuel Tan (Includes Item 42)
14	Fuel Line (10-ft. length)
15	Filter, Fuel
16	Guard, Spark Plug
17	Bracket, Fuel Tank
18	Guard, Throttle Control
19	Lockwasher, 5/16
20	Screw, M8-1.25 x 16 mm
21	Lockwasher, 1/4
22	Screw, 1/4-14 x .75 Thd Form
23	Hose Clamp
24	Valve, Fuel Shut Off
25	Stiffener Assembly (Includes Item 36)
26	Deck, Cutter
27	Adapter, Cutter Bar
28	21" Cutter Bar (Includes 29 ~ 31)
29	Serrated Washer
30	Lockwasher, 3/8
31	Screw, Cutter Bar (Includes Item 30)
32	Toe Guard Assembly
33	Rod, Drag Flap
34	Clip, Drag Flap
35	Drag Flap Assembly
36	Bushing, Stiffener
37	Screw, 5/16-18 x 1.25"
38	Chute Extension
39	Rod, Front Bumper
40	Bumper, Front
41	Roll Pin, 1/8 x .88"
42	Cap, Fuel Tank
43	Decal, Danger
44	Carriage Bolt, 5/16-18 x 1.00"
45	Bolt, Extended

FIGURE 30-2. Good reference materials are essential for successful engine repair. (*Courtesy of Jacobsen Textron*)

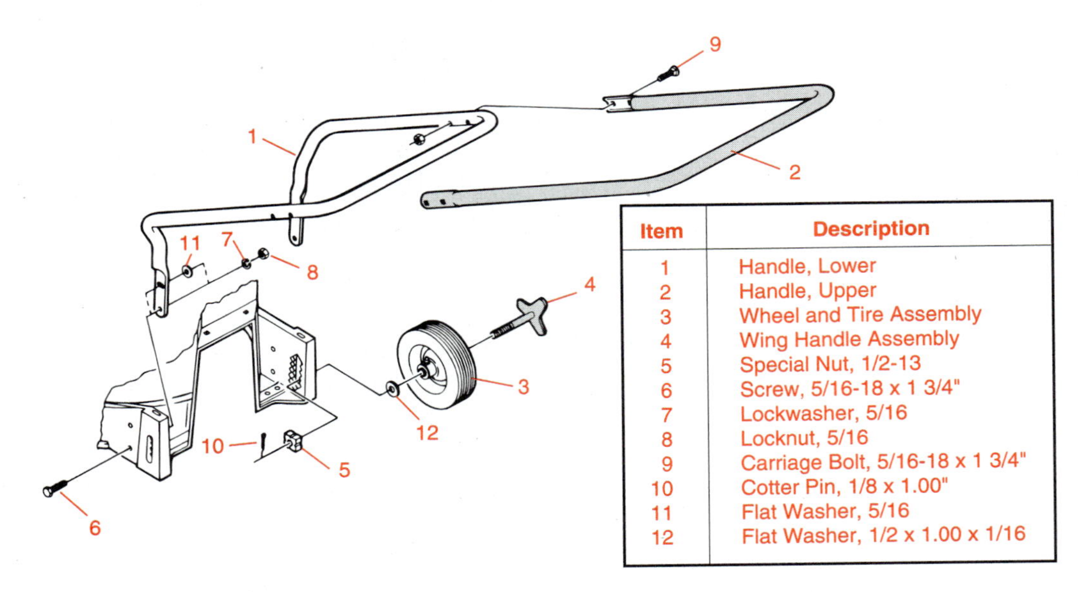

Item	Description
1	Handle, Lower
2	Handle, Upper
3	Wheel and Tire Assembly
4	Wing Handle Assembly
5	Special Nut, 1/2-13
6	Screw, 5/16-18 x 1 3/4"
7	Lockwasher, 5/16
8	Locknut, 5/16
9	Carriage Bolt, 5/16-18 x 1 3/4"
10	Cotter Pin, 1/8 x 1.00"
11	Flat Washer, 5/16
12	Flat Washer, 1/2 x 1.00 x 1/16

FIGURE 30-2 . *Continued*

TOOLS FOR ENGINE REPAIR

There is an endless list of tools used for mechanical work on engines and drive systems. However, a limited set of basic tools are adequate for small engine work.

Recommended Basic Tools

One set each of socket, open-end, and box wrenches ranging from $^1/_4$" to 1" will generally be sufficient for small engine work. Sockets in this size range may have $^1/_4$", $^3/_8$", or $^1/_2$" drive handles. Similar wrench sets in metric sizes will permit work on a wider range of engine makes.

Other general tools for small engine repair are a set of hex wrenches, 6" and 10" adjustable wrenches, slip joint pliers, long nose pliers, an 8- to 12-ounce ball pein hammer, chisels, and punches. Several sizes of standard and phillips head screwdrivers will round out the basic tools. These and other tools are pictured in Unit 7 Handtools, Fasteners, and Hardware. An individual will encounter the need for additional tools as the range of repair work increases.

Specialized Small Engine Tools

Small engines have parts such as valves, pistons, rings, and bearings that require specialized tools for repair. These tools will be introduced as the various repairs are discussed.

DISASSEMBLY

The careful observer and mechanically inclined person can repair many appliances, small machines, and small engines. Teenagers can learn valuable life skills by repairing simple things such as bicycles and toys. From there they can progress to lawn and garden equipment or to motorcycles, chain saws, boats, trucks, tractors, and automobiles.

References

It is important to obtain a suitable reference manual for an engine before disassembly. Step-by-step procedures and diagrams are provided in owner's manuals and shop or service manuals, figure 30-2. These are written by manufacturers. Additionally, text and reference books are available in popular book stores. Frequently, agricultural mechanics shops and reference files have suitable repair instructions for popular engines.

Work Area

The work area should be clean and free of flammable materials. All fuel and oil must be drained from the engine. Clean rags for wiping and small cans or trays with numerous dividers for small parts must be on hand. Assemble appropriate tools before starting the job.

Diagnosing the Job

Good engines may be ruined by disassembly by the unskilled person. Therefore, careful thought must be given to the engine problem before "tearing it down." Generally, proper cleaning and maintenance as described in Unit 29 will keep an engine in good running order. Periodic tune-ups and adjustments are also necessary to keep the engine running at top efficiency. Complete overhaul is seldom needed. Overhaul means complete disassembly with cleaning

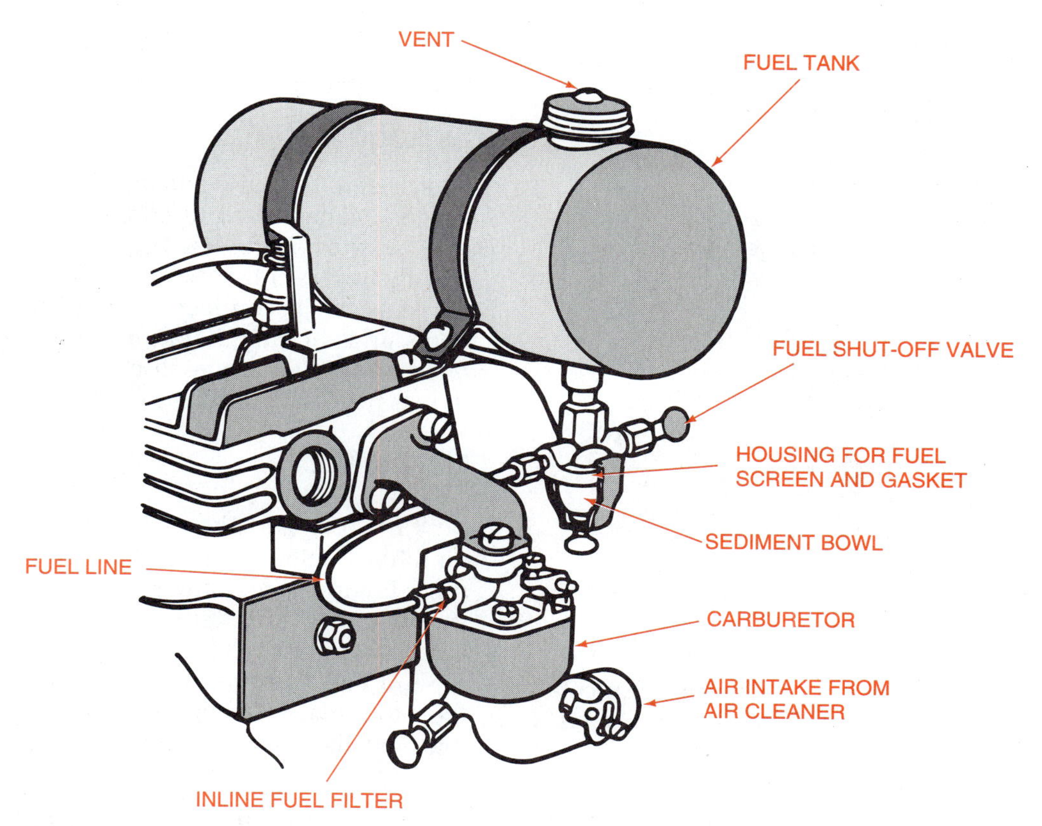

FIGURE 30-3. Fuel system on a four-cycle engine.

and reconditioning or replacement of most moving parts.

An engine component should only be removed if it needs service or if it prevents access to another component that needs service. Follow the manufacturer's instructions for disassembly. All components should be laid out in orderly fashion so they can be further disassembled if necessary, or reassembled without error or confusion.

FUEL SYSTEM

The fuel system consists of the fuel tank, fuel lines, shut-off valve, sediment bowl (optional), and carburetor, figure 30-3.

Fuel Tank and Lines

The fuel tank may be removed by disconnecting the fuel line and loosening the bracket bolts or screws. Check the inside for dirt or trash. If dirty, flush the tank, sediment bowl, and screen. Replace any inline fuel filters.

CAUTION:

- **Use only varsol or kerosene to clean fuel system parts. Wipe dry or air-dry all parts.**

Engine Speed Control

Carburetors generally have cables or wire links to the governor and throttle assembly and choke valve. A governor is a speed control device. To control engine speed, a lever, handlebar grip, or threaded shaft moves the cable or link and stretches a spring in the system between the cable and governor. The governor is linked to the throttle shaft of the carburetor, figure 30-4.

Governor

Spring tension in the linkage is countered by action of the governor. The engine speed is controlled by a balance of forces exerted by the cable in one direction and the governor in the other. As the cable pulls harder on the spring, the governor will permit the engine to run faster before leveling off to a constant speed. The governor controls engine performance by way of linkage to the throttle valve.

One type of governor consists of weights on a shaft turned by a gear driven by the engine. Centrifugal force causes the weights on a shaft to move outward and move a lever attached to the throttle shaft as the engine speed increases. This spreading force counteracts the action of the throttle cable on the spring.

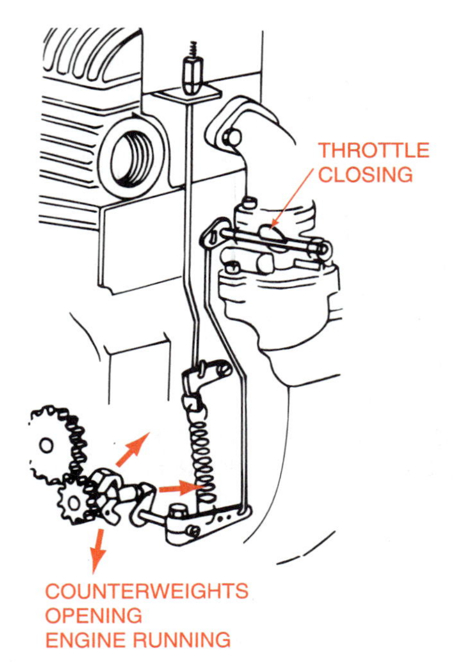

FIGURE 30-4. Governor action closing the throttle to stabilize engine speed. (*Courtesy of Briggs & Stratton Corporation*)

The air vane is another type of governor used on small engines. It is activated by the flow of air created by flywheel fins under the engine shroud. As a rule, the only service or repair needed on governors is to keep the parts clean.

Linkage

Fuel system linkage parts may be damaged if forced. Care is needed when disconnecting or reconnecting fuel system parts. During disassembly, observe which hole each link goes in to assure proper reassembly. Disconnect links by loosening components and rotating them until the link separates on one end. It is best not to disconnect any more linkage than is necessary to do the servicing.

Carburetor

Functions. The carburetor provides fuel and air to the engine in appropriate proportions and volume. Engine speed, power, and acceleration is controlled by the carburetor. To function effectively the carburetor must receive a continuous supply of clean air and fuel. Carburetors have several fuel flow systems to provide for engine idle and intermediate and high speeds, figures 30-5A and B.

Clean Air and Fuel. Clean air is assured if air cleaners and filters are serviced regularly as described in Unit 29. Clean fuel starts with fresh gasoline stored in proper storage cans or tanks.

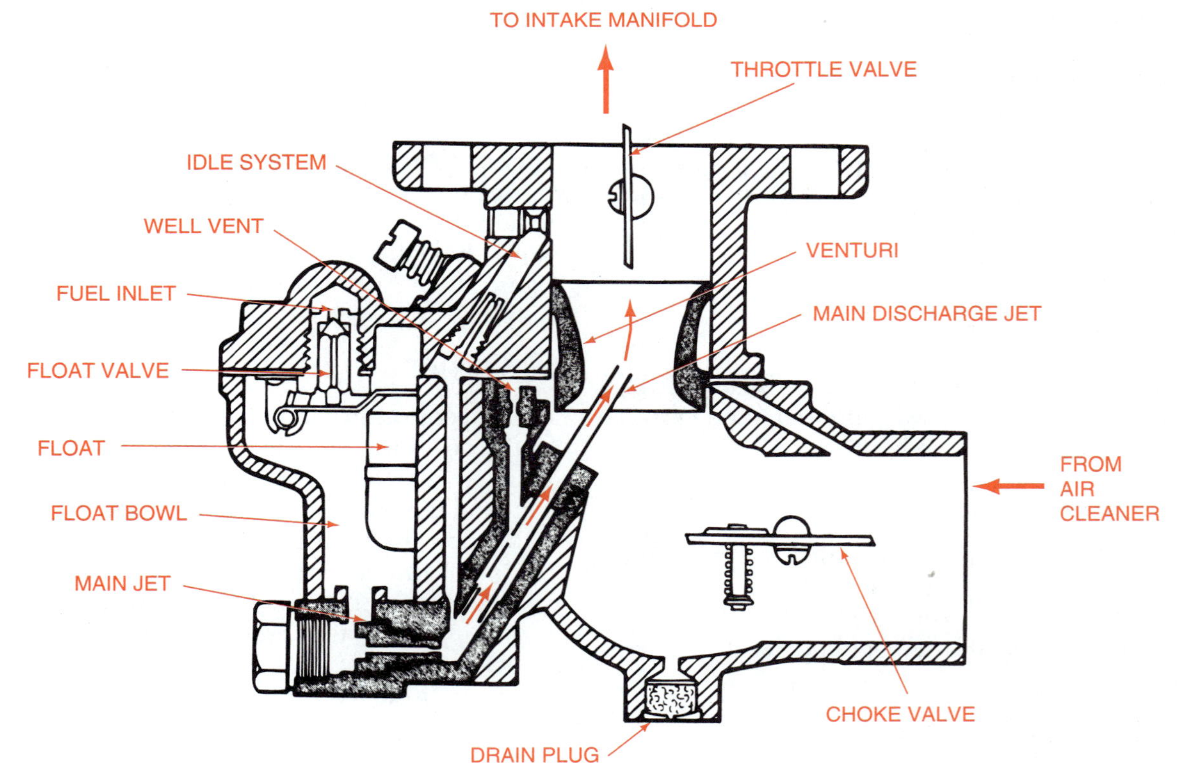

FIGURE 30-5A. Updraft carburetor requiring fuel by gravity flow or fuel pump (*Courtesy of G. Stephenson,* Small Gasoline Engines, *Delmar Publishers, 1984*)

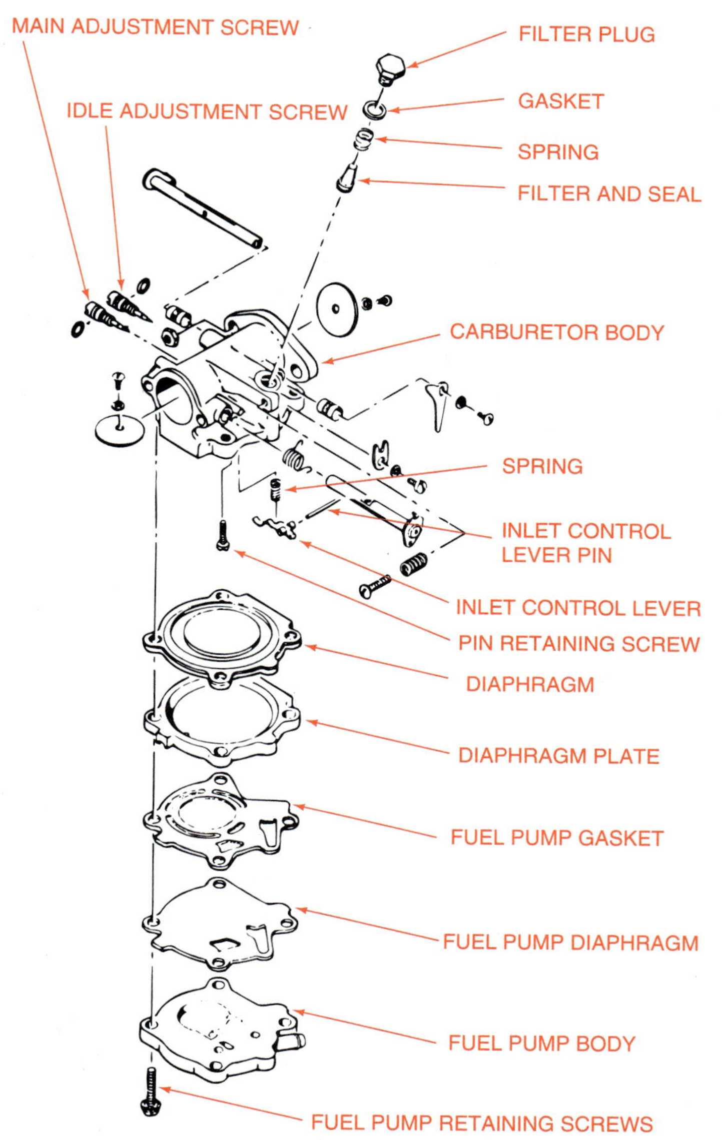

FIGURE 30-5B. Diaphragm carburetor with built-in fuel pump. (*Courtesy of G. Stephenson,* Small Gasoline Engines, *Delmar Publishers, 1984*)

Engine fuel tanks should be refilled after use to reduce condensation from temperature changes. Very small amounts of dirt or moisture can cause carburetor malfunction. Clean sediment bowls and inline fuel filters help assure a continuous supply of clean fuel to the carburetor.

Carburetor Overhaul. Carburetor overhaul is needed occasionally. When overhaul is necessary, an overhaul kit consisting mostly of gaskets and needle valves with matching jets, or seats, should be used. A needle is a long tapered shaft and a jet, or seat, is a hole shaped to receive the needle and control the flow of fuel. Follow instructions provided in the kit. A commercial carburetor solvent may be necessary if fuel deposits have accumulated in hidden passages.

Carburetor disassembly and reassembly must be done with careful attention to details. Use appropriate size screwdrivers and wrenches to avoid damage to screw slots, needle valves, and seats. Clean all parts in varsol, kerosene, or special carburetor solvent. A poorly done carburetor overhaul can cause the engine to run worse than before the overhaul.

Reassemble with new gaskets, seats, and needle valves. Tighten screws with moderate torque. Check the float level with the float gauge provided in the kit. Careful—do not turn needle valves tightly against their seats, or the needle will be damaged and make proper carburetor adjustment impossible. Use information provided with the overhaul kit, manufacturer's materials, or Unit 29, Small Engine Maintenance, for carburetor adjustments.

CAUTION:

- **Use extreme care to avoid fire with solvents.**

IGNITION

Gasoline engines require a high-voltage spark to ignite the mixture of air and fuel in the cylinder. The spark is achieved by using an electrical surge of 30,000 volts or more across an air gap between the two electrodes of a spark plug. To achieve such a spark, a battery or magneto, ignition switch (battery system), ignition coil, breaker points, condenser, and spark plug are needed, figure 30-6. A battery produces electricity by chemical action, while a magneto produces electricity by magnetism.

Primary Circuit

Low-voltage current produced by a battery or magneto enters the primary windings in the coil. The current flows to ground through a set of closed breaker points. Objectional surges of current are absorbed by a condenser. The condenser then releases stored current to boost power in the circuit when needed. These components constitute the low-voltage circuit, or primary circuit.

Secondary Circuit

The secondary circuit, or high-voltage circuit, consists of a large coil of very small diameter wire wound around an iron core, high-tension or high-voltage wire, and a spark plug. The circuit starts at the grounded terminal of the coil, flows through the high-tension coil wire and high-tension spark plug wire to the high-tension lead and spark plug. Current

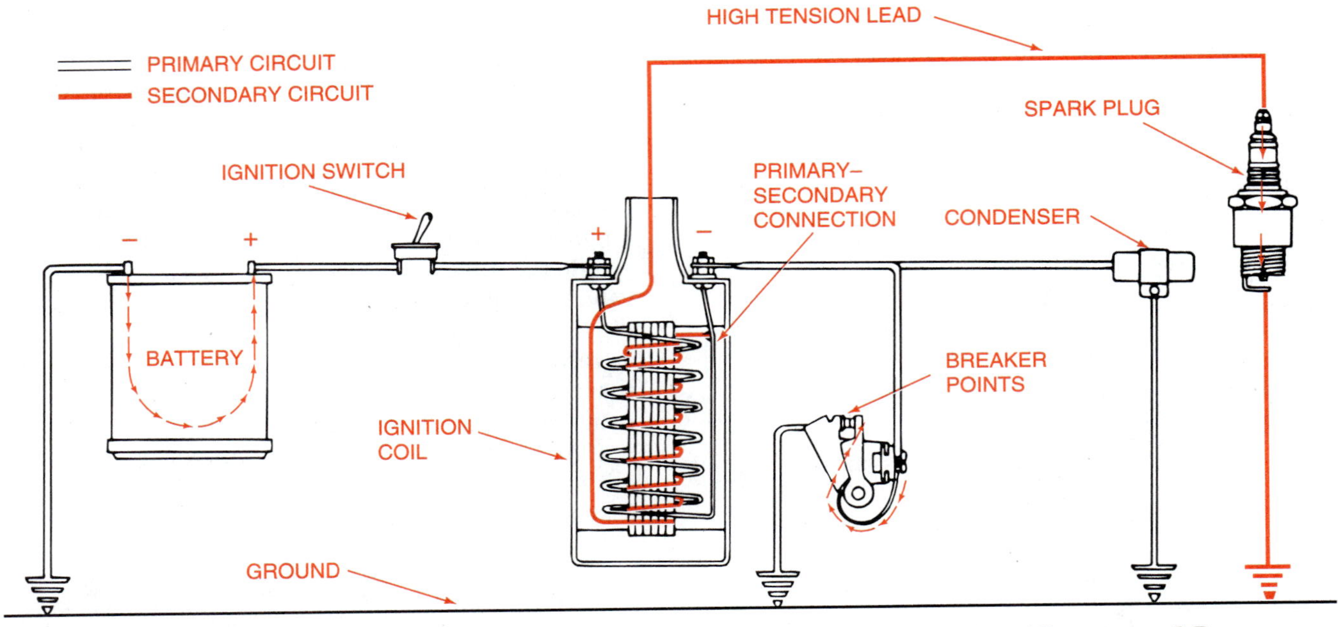

FIGURE 30-6. Battery ignition systems showing primary and secondary circuits. (*Courtesy of G. Stephenson,* Small Gasoline Engines, *Delmar Publishers, 1984*)

then flows through the spark plug and creates a high-voltage spark as it jumps the air gap of the spark plug and returns to its source through the metal engine parts and/or battery ground cable.

Electromagnetic Induction

The conversion of low-voltage to high-voltage current with a coil is called electromagnetic induction. The coil must consist of an insulated primary and a secondary winding around an iron core.

Low-voltage current is sent through the primary coil of wire. This creates a magnetic field around the primary and secondary windings. When the flow of current stops in the primary circuit, the magnetic field collapses and creates a high-voltage surge in the secondary circuit. (Note: Additional information on electromagnetism is provided in Unit 31, Electrical Principles and Wiring Materials.)

Creating Ignition Spark

An ignition spark is a hot electrical arc jumping across an air gap. In an ignition circuit, the ignition breaker points interrupt the flow in the primary circuit at the correct time to induce the high-voltage flow to the spark plug. Correct timing of the ignition system and valves with the crankshaft create the spark when the piston is at or near the top dead center on the compression stroke. For spark to occur, the spark plug electrodes and insulation must be clean and the gap set to specifications using an wire-type gauge, figure 30-7.

Battery Service

Batteries must be kept clean, dry, and free of any conductive material. Such material could permit current to flow from one terminal to the other and discharge or run down the battery.

Some batteries have electrolyte or acid reservoirs that are sealed and seldom need service. If the caps can be removed from the cells, however, the addition of distilled water, free of pollutants, may be needed periodically.

Magneto Systems

A magneto consists of an ignition coil, an iron core called an armature, and a permanent magnet. On some engines, the armature and coil assembly is

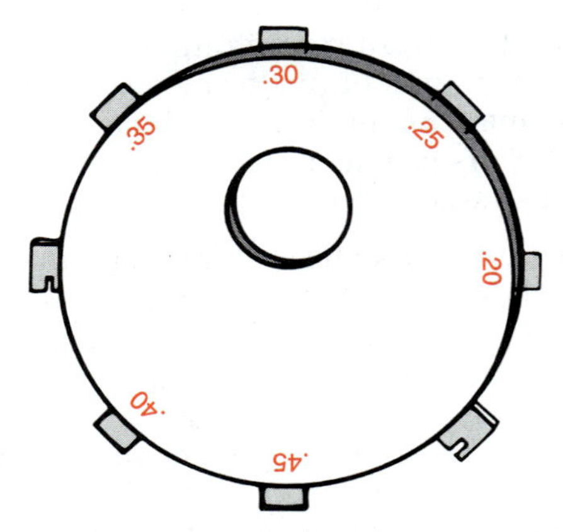

FIGURE 30-7. Wire-type spark plug gauge.

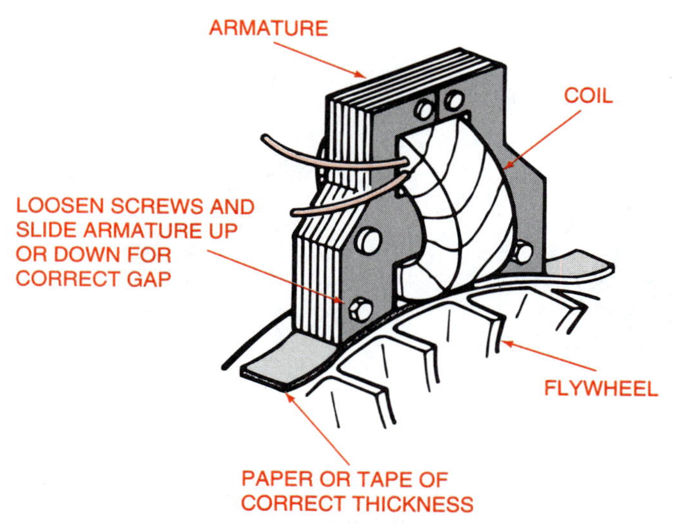

FIGURE 30-8. Magneto with paper gage to set correct clearance from the flywheel.

positioned so a magnet in the flywheel passes the armature at the instant a spark is needed in the cylinder. The passing of the magnet creates magnetism which, in turn, induces a high-voltage spark when the ignition points open. The air gap between the armature and flywheel is adjusted by loosening two screws and sliding the magneto. The air gap is approximately the thickness of a postcard, figure 30-8. Some magnetos are self-contained with their own rotating magnets.

Ignition Breaker Points

Flywheel Removal. The ignition breaker points may be in a special compartment that is easily accessible. However, most are hidden by the flywheel.

The flywheel is generally held in correct position with the crankshaft by a soft metal key in the keyway. This assures correct engine timing. The key will shear off if the engine is subjected to an abrupt stop such as a mower blade hitting an object. The shearing of a key decreases the chances of crankshaft breakage.

Appropriate pullers are needed to remove flywheels without breakage. To pull the flywheel, remove the crankshaft nut (left-hand threads), install the puller, tighten the puller to exert pulling force, and strike the center bolt of the puller with a steel hammer. The impact should cause the wheel to pop off of the tapered shaft, figure 30-9.

Examining Ignition Points. Points must be smooth, even, and perfectly aligned. If they are not, replace them. Buildup of metal on either surface of the points means the condenser should be replaced also, figure 30-10.

Replacing Ignition Points. To replace ignition points, first disconnect the primary wire. Remove the screw that holds the point assembly in place. Remove and install the new point assembly. Snug up but do not tighten the mounting screw until the point gap is set.

Adjusting the Point Gap. Turn the crankshaft until the moveable arm of points is pushed as far as the cam or plunger will push it. Place a clean feeler gauge of the specified thickness in the opening. The gap is generally around 0.020 inch. Move the stationary part of the assembly until the gauge exerts slight resistance when pulled. Tighten the hold-down screw. Recheck the gap by inserting the gauge. If the gauge does not have slight drag or if the moveable arm moves when the gauge is withdrawn, reset the points, figures 30-11 and 30-12.

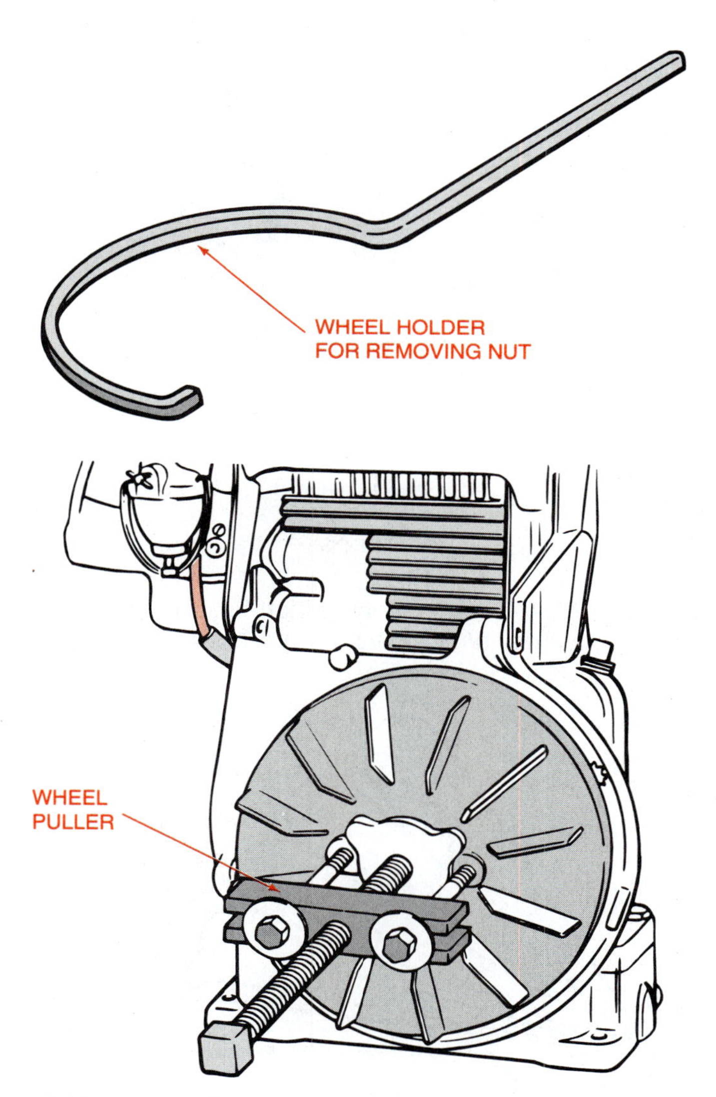

FIGURE 30-9. Pulling a flywheel. (*Courtesy of Kohler Company*)

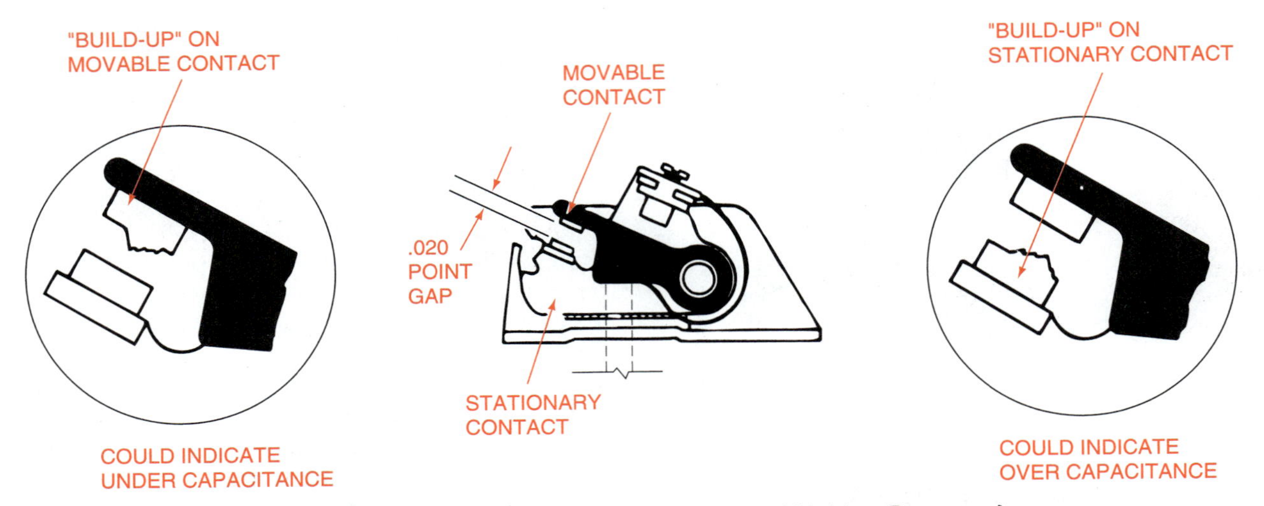

FIGURE 30-10. Buildup of metal on ignition points. (*Courtesy of Kohler Company*)

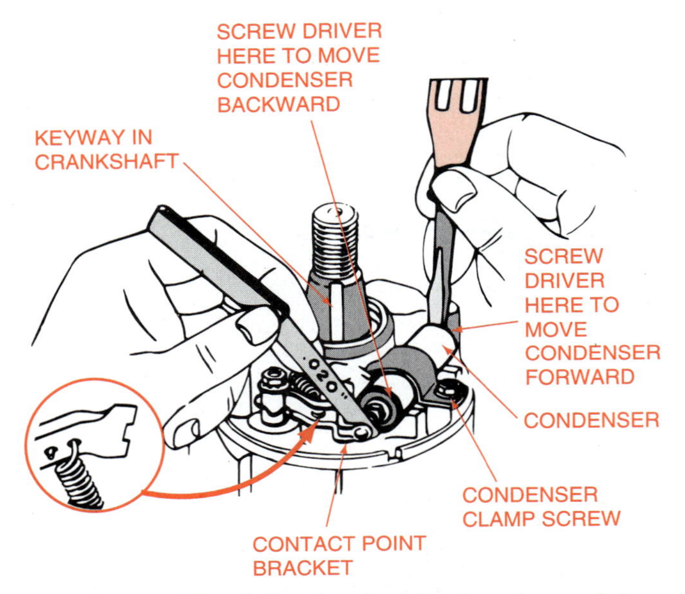

FIGURE 30-11. Adjusting ignition breaker point gap. (*Courtesy of Briggs & Stratton Corporation*)

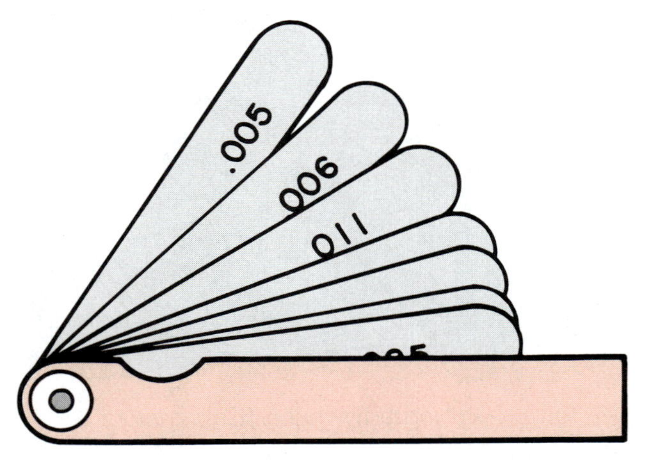

FIGURE 30-12. Feeler gauge.

HEAD AND HEAD GASKET

Frequently, compression leakage is caused by a leaking head gasket. The head gasket provides the seal between the head and the cylinder block. Carbon should be scraped from the combustion chamber and top of the piston whenever the head is removed. A new head gasket should be used when reinstalling the head.

THE BIG DECISION

Should the valves, piston rings, and bearings be reconditioned? Should they be replaced? Should they be left alone? For most engines they should be left alone.

Most engine problems are caused by dirt, trash, dirty fuel, incorrect adjustments, or worn parts in the ignition system. When all of these problems are corrected and the engine exhaust is blue in color, the piston rings probably need replacing. When rings are replaced, the piston, cylinder walls, and bearings should be evaluated and serviced if renewal is worth the time and cost. A new or factory rebuilt short block or engine may be cheaper than the time and material needed for overhaul.

VALVES

Cylinder valves control the entry of air and fuel into and the exit of exhaust gases from the cylinder. Valves are either intake or exhaust type. Valves, stems, and valve heads must fit the valve guides and valve seats according to specifications. Valves are removed from the engine using a valve spring compressor to remove the valve keepers or valve pin, figure 30-13.

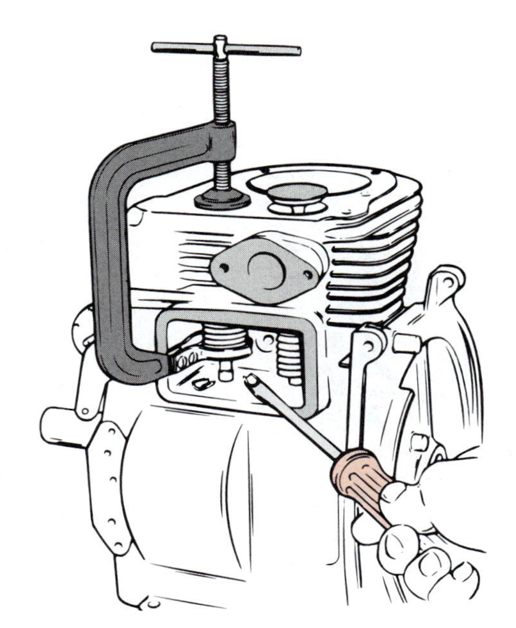

FIGURE 30-13. Removing valve keepers. *(Courtesy of Kohler Company)*

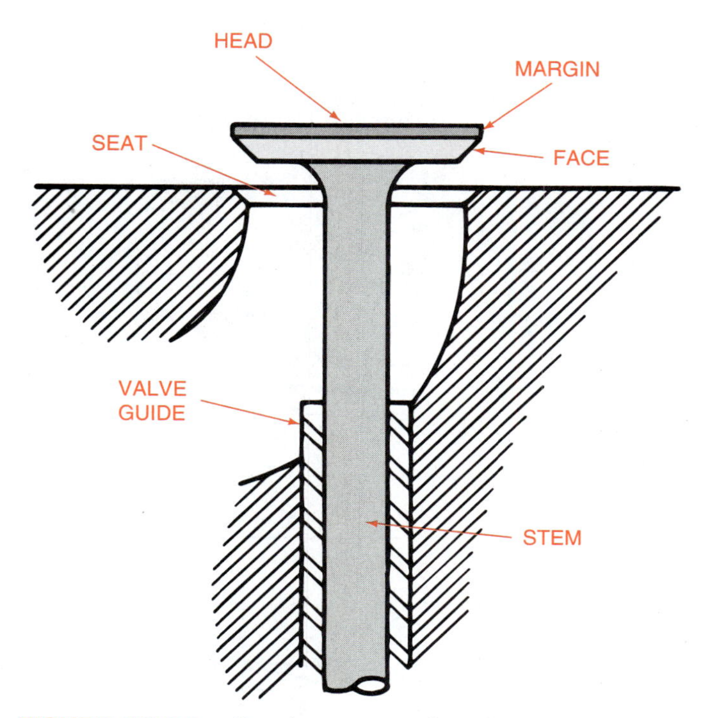

FIGURE 30-14. Good compressing depends on proper seating of valves.

Valve Heads and Seats

The flat part of a valve is called the head and the long, round section is the stem. The outer edge of the valve head is the margin, and the tapered section, the face, figure 30-14.

The valve face may be ground or redressed until the margin is half its original thickness. The face and seat must each be ground at matching 30-degree or 45-degree angles, according to specifications. Reground or new valves must be lapped in to fit the seat for a perfect seal. This is done by putting a gritty valve lapping compound on the valve seat and rotating the valve back and forth with a valve grinding tool, figure 30-15.

Valve Guides

Valve stems are held in alignment by a valve guide. Valve guides should be replaced and reamed to correct specifications when new valves are installed, figure 30-16.

Valve Clearance Adjustment

Valves must have a valve stem clearance of about 0.010 inches or as specified. The gap is achieved by grinding the end of the valve stem or by adjusting valve tappets. Valve clearance is measured using a feeler gauge with the valve closed and the cam out of the way, figure 30-17.

Cylinder, Piston, and Rings

In a gasoline engine, energy in fuel is converted into power and motion through combustion. The cylin-

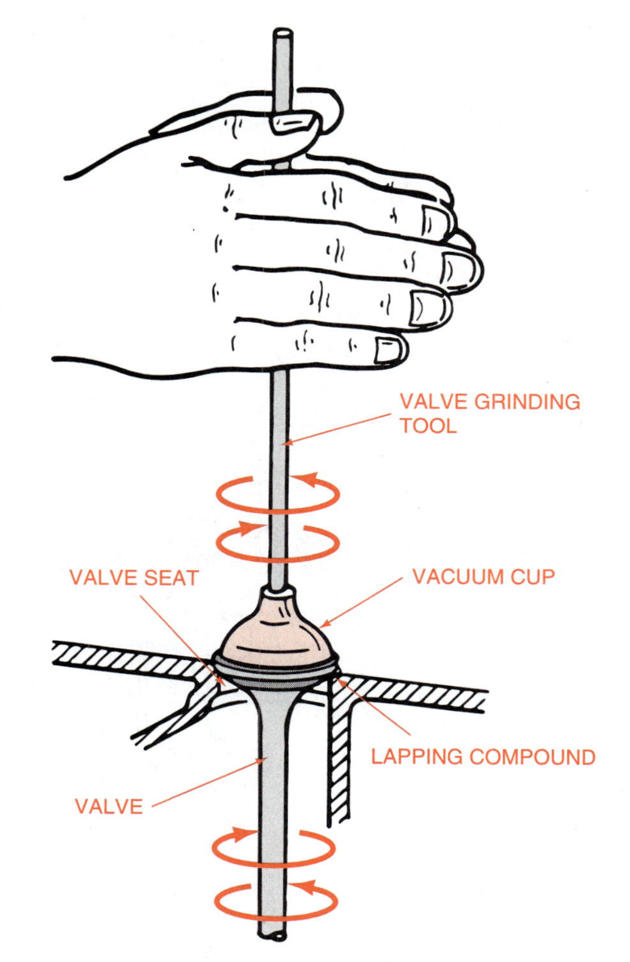

FIGURE 30-15. Lapping in a valve.

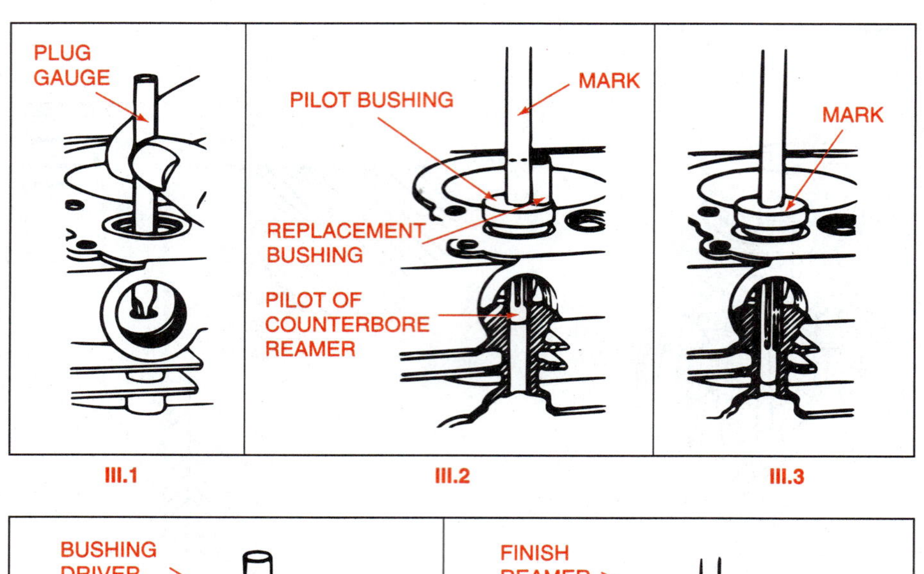

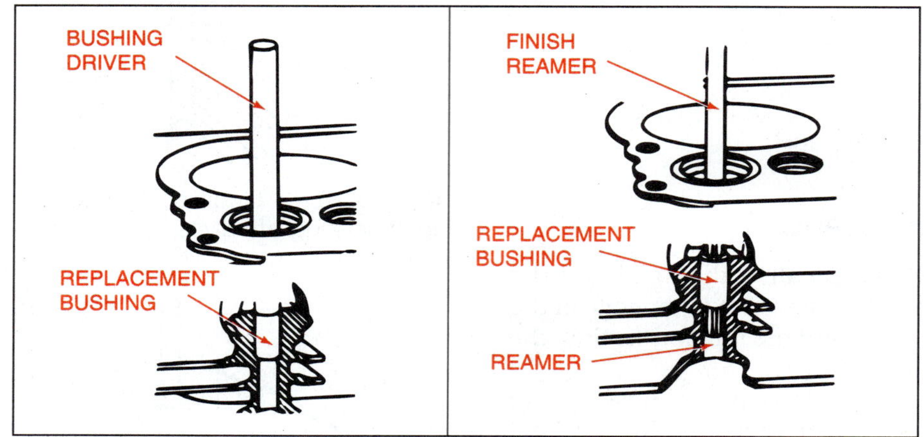

FIGURE 30-16. Reaming a valve guide. (*Courtesy of Briggs & Stratton Corporation*)

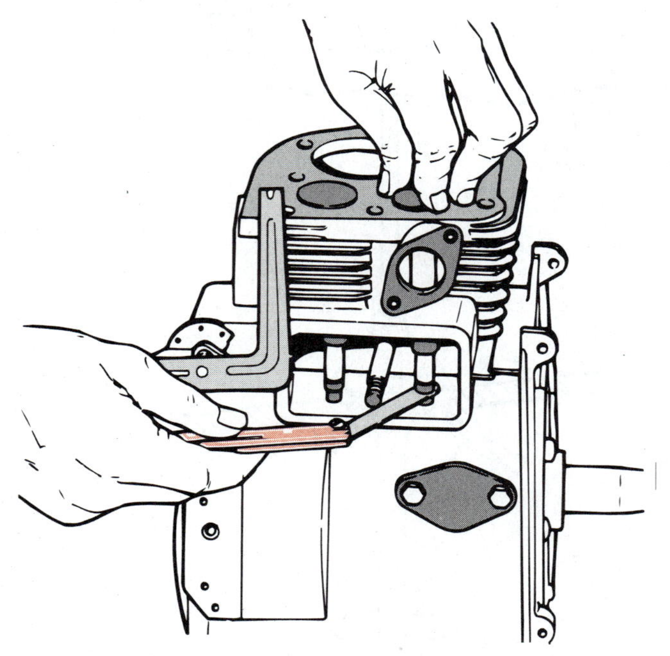

FIGURE 30-17. Checking valve stem clearance with a feeler gauge. (*Courtesy of Kohler Company*)

der, cylinder head, piston, piston rings, and valves are the major parts harnessing the energy of combustion in the engine.

The **cylinder** contains the piston of the engine. It is covered by a head attached with bolts. The **head** contains the spark plug and combustion chamber. A **head gasket** provides a seal between the cylinder and head.

The **piston** is the moving core of the cylinder and receives the force of combusting fuel. Piston **rings** complete the seal between the piston and cylinder wall.

The piston is removed by loosening the connecting rod bearing caps at the crankshaft journal. The piston is then pushed up and out of the cylinder by pushing the connecting rod.

Cylinder. After removal of the piston, rings, and connecting rod assembly, the cylinder must be analyzed for wear. An **inside micrometer** or **telescoping gauge** is used to determine the nature and extent of cylinder wear. Cylinders tend to wear to an elliptical out-of-round shape, figure 30-18.

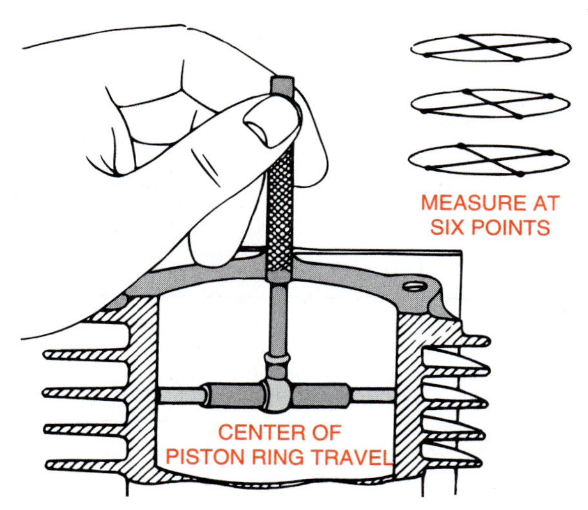

FIGURE 30-18. Measuring cylinder wear. (*Courtesy of Briggs & Stratton Corporation*)

Cylinders in cast iron blocks can be restored with a cylinder hone. A cylinder hone has spring-loaded carborundum stones, which are turned with a power drill. Cast iron blocks may also be rebored to accommodate slightly oversized pistons. Aluminum blocks must be replaced if wear exceeds specification limits.

Piston and Rings. Piston rings are removed and installed with a ring expander, figure 30-19. The ring grooves and piston are then cleaned with a soft wire brush to remove all carbon deposits. New piston rings are then installed and ring-gap and ring-groove clearances checked, figure 30-20.

Pistons are designed with various ring configurations and designs. Pistons typically have two compression rings designed to prevent compression blow-by or leakage. The oil ring or bottom ring is designed to prevent crankcase oil from getting past the piston. Oil getting past the piston rings is burned in the cylinder, figure 30-21. A piston ring compressor is used to force the rings into their grooves so the piston can be pushed down into the cylinder, figure 30-22.

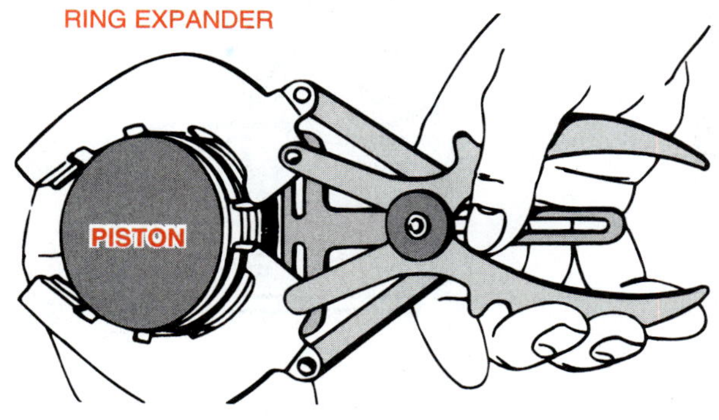

FIGURE 30-19. Using a ringer expander to remove rings. (*Courtesy of Briggs & Stratton Corporation*)

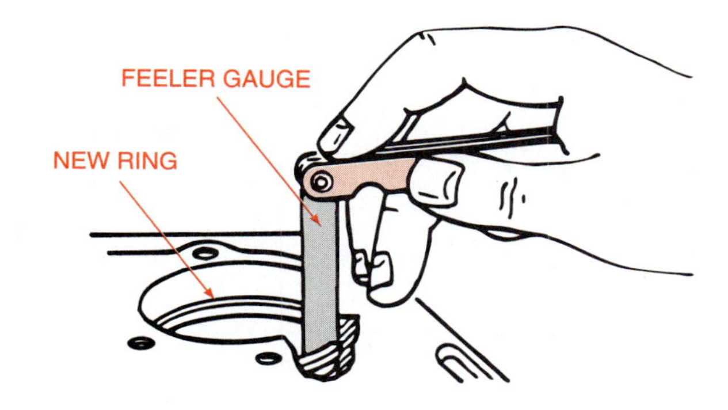

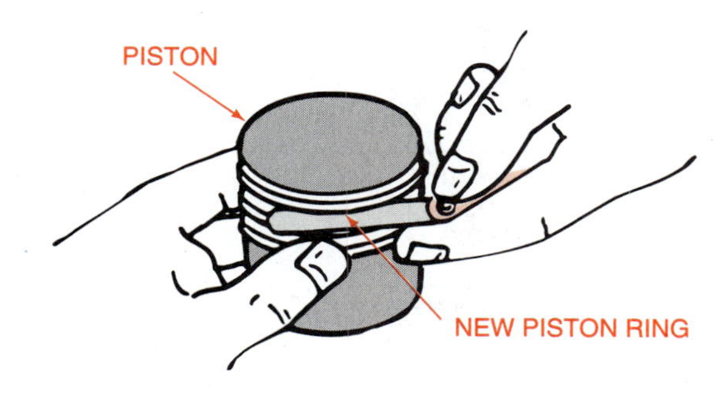

FIGURE 30-20. Checking ring-gap and ring-groove clearances.

BEARINGS

Bearings may need replacing where the crankshaft enters the engine crankcase, where the connecting rod attaches to the crankshaft, and the wrist pin where the connecting rod connects to the piston, figure 30-23.

Main bearings and oil seals may be replaced by pulling or driving them out of their seats. Special tools are available to reduce the likelihood of damaging these items during removal and replacement, figure 30-24.

The wrist pin may be replaced along with a matching bearing on the connecting rod. Another bearing is found on the connecting rod where it attaches to the crankshaft. This bearing must be fitted carefully to ensure correct clearance. The micrometer is used to measure outside surfaces of shafts to determine the nature and extent of wear on the shaft, figure 30-25.

Plastigage or thin feeler gauges may be used to measure connecting rod bearing clearance. Plastigage is a carefully designed material that flattens out uniformly when pressed. The material is laid

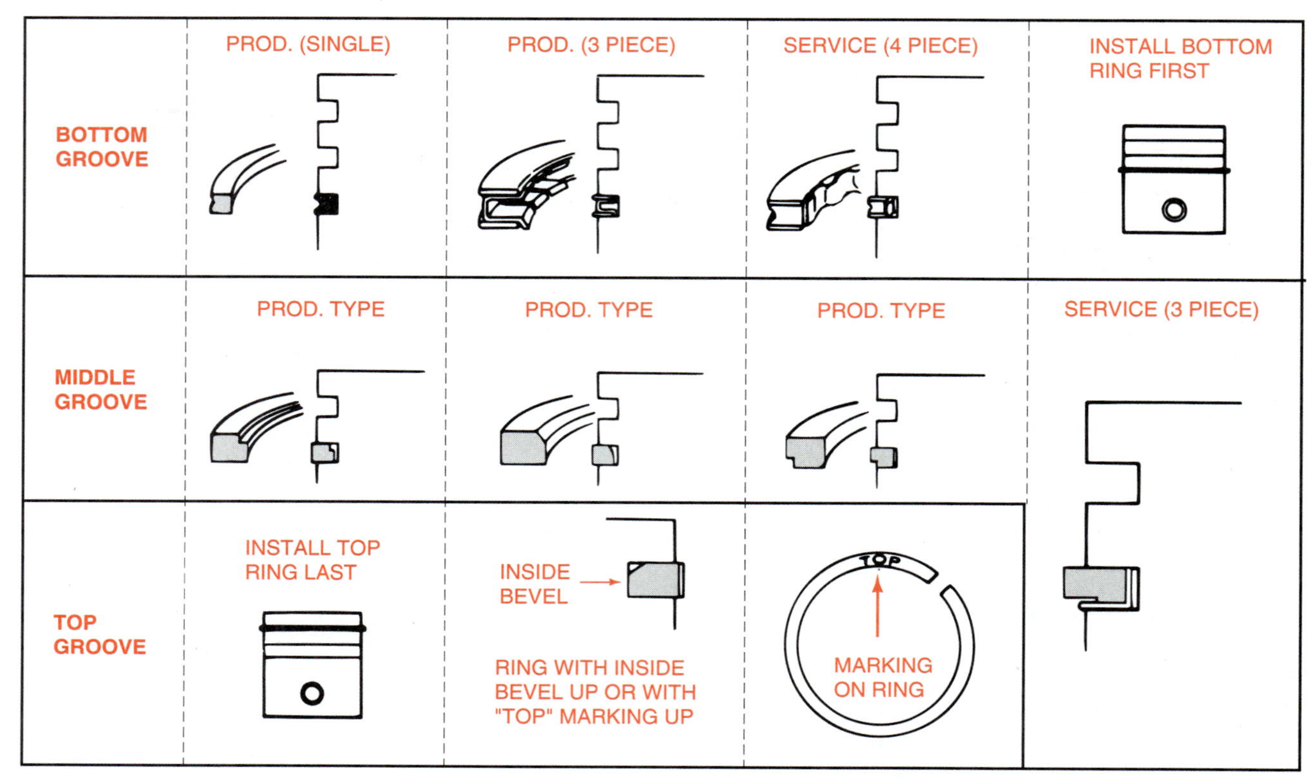

FIGURE 30-21. Types of rings. (*Courtesy of Kohler Company*)

across the bearing surface, and the bearing cap is attached and torqued. The bearing cap is removed again and the width of the plastigage measured with a special gauge. The gauge converts the measurement to thousandths of an inch of bearing clearance. Bearing clearance may be increased by mounting shims of specified thickness in the assembly, figure 30-26.

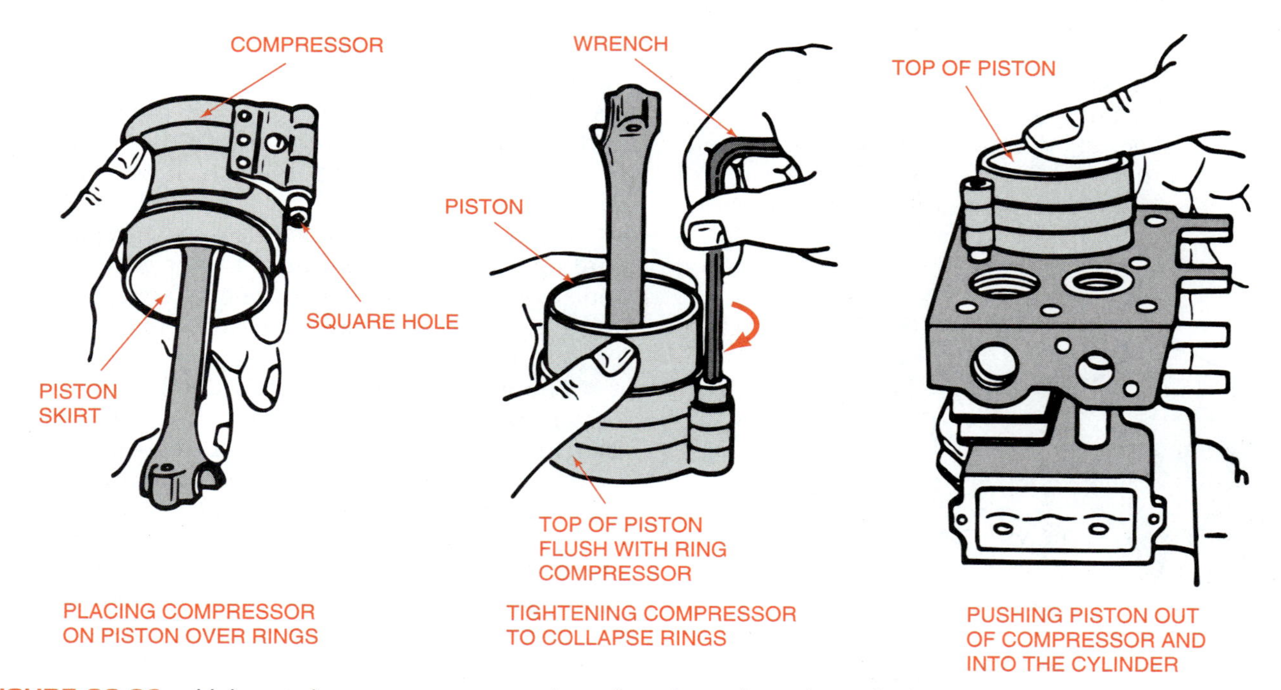

FIGURE 30-22. Using a ring compressor to install a piston into the cylinder. (*Courtesy of Briggs & Stratton Corporation*)

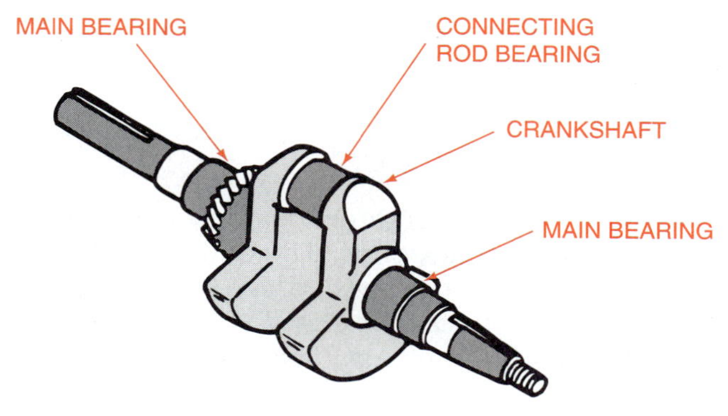

FIGURE 30-23. Crankshaft.

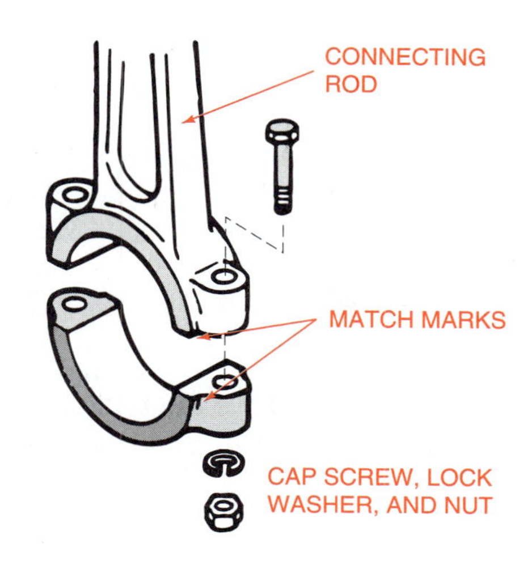

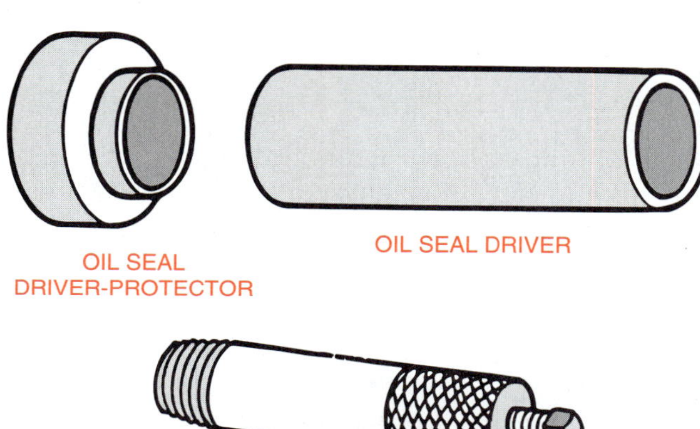

FIGURE 30-24. Tools for removing and installing main bearings and seals.

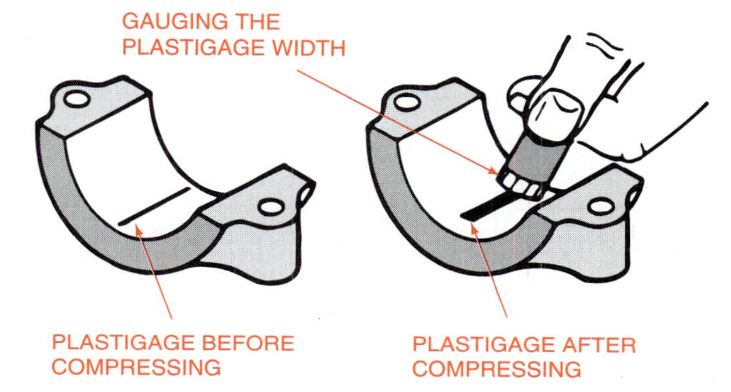

FIGURE 30-26. Connecting rod bearing fitting by using plastigage.

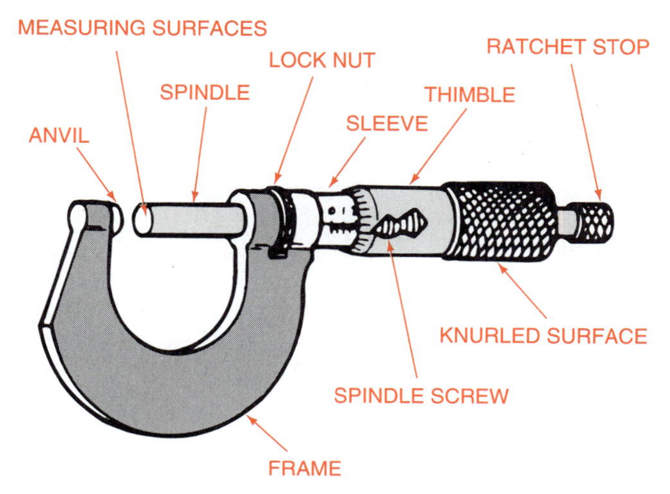

FIGURE 30-25. A micrometer is used to measure round surfaces. The drawing shows an outside 1-inch micrometer.

REASSEMBLY

Reassembly is done in reverse order of disassembly. All parts should be clean and all matching surfaces must be clear of grease, dirt, or carbon deposits. All gaskets should be replaced.

Torque. The extent of torque or twist on nuts and bolts is critical. Use a torque wrench to tighten bolts and screws to specification, figure 30-27.

FIGURE 30-27. Torquing a bolt to specifications.

Cylinder Head. Special care is needed when tightening a cylinder head. The head bolts must be torqued gradually and in the correct sequence. Failure to do so is likely to result in a warped head and leaking head gasket. (See example under Trouble Shooting in Unit 29, Small Engine Maintenance.)

ROPE AND WIND-UP STARTERS

Small engines are frequently started by hand or foot. The rope starter utilizes the pull of a rope wrapped around a pulley for turning power to start the engine. The wind-up starter uses a lever for a crank to twist a spring against a locked crankshaft. When the lock is disengaged, the coiled spring cranks the engine. The kick starter uses a foot-operated lever and cog mechanism to crank the engine.

Starter ropes need replacing and starter assemblies need to be cleaned and lubricated periodically. Follow instructions in the operator's manual for servicing starters.

CAUTION:

- **When servicing starters, be especially careful when disassembling and handling. The coiled spring, dogs, pawls, spurs, and clips can cut hands or be hurled into the face by spring release, figure 30-28.**

SUMMARY

Servicing of small engines can be enjoyable and productive. It can keep engines running long and efficiently. Further, engine servicing skills can lead to lifelong savings and excellent career opportunities in mechanics, sales, and service.

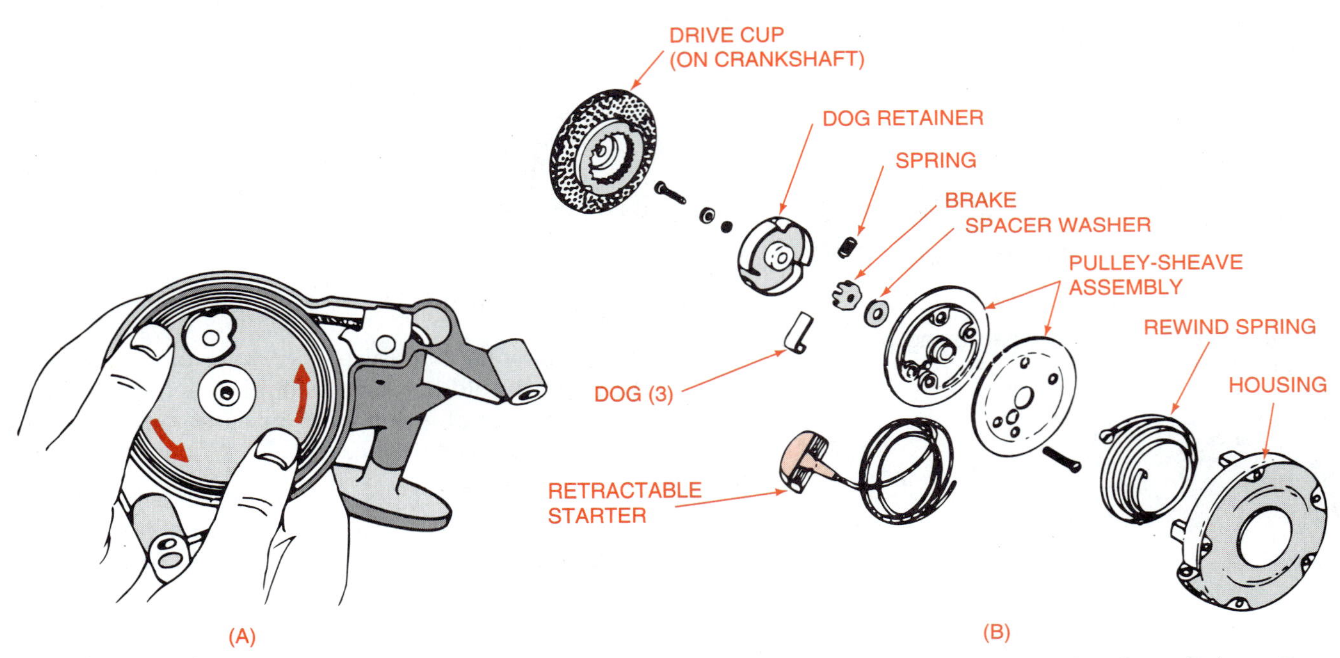

FIGURE 30-28. Rope starter assemblies. ([A] *Adapted from and used with permission from Briggs & Stratton Corporation; [B] Courtesy of Kohler Company*)

STUDENT ACTIVITIES

1. Define the Terms to Know in this unit.
2. Obtain a small engine repair manual.
3. Along with several other classmates, learn the exact names and identification of all parts of a typical small engine.
4. Sketch and label the major parts of an engine.
5. Study the disassembly and service procedures for some or all of the components discussed in this unit.
6. Write the correct and detailed procedure for overhauling a carburetor and replacing valves, rings, and bearings.
7. Research the procedures used to overhaul outboard boat engines.
8. Research the procedures used to overhaul chain saws or other small two-cycle engines.
9. Determine the procedures for repairing, troubleshooting, maintaining, or repairing the electrical system on a small engine with electrical starter and lights.
10. Prepare a teaching aid by cutting away one side of the major parts of a discarded small engine. Paint the exposed surfaces to highlight the various sections or internal parts.
11. Mount components from discarded engines on a wooden panel. Paint and label parts of each component.
12. Overhaul a small engine.

SELF-EVALUATION

A. Multiple Choice. Select the best answer.

1. Small gasoline engines are used extensively for
   a. lawn equipment
   b. nursery equipment
   c. recreation
   d. all of the above
2. Small engine work should be done
   a. where fuel is stored
   b. in a well-ventilated area
   c. with plenty of rags and paper nearby
   d. where a torch or welder is handy
3. Rags used to wipe up fuel should be
   a. placed out-of-doors to dry
   b. discarded promptly in a trash can
   c. piled up until several can be discarded at one time
   d. used until completely saturated
4. Clean fuel drained from an engine should be
   a. discarded down a sink drain
   b. discarded in a floor drain
   c. discarded with other trash
   d. stored in an approved container
5. Small engine overhaul is
   a. simple
   b. very expensive
   c. seldom needed
   d. a good substitute for regular maintenance
6. The following should *never* be used as a solvent
   a. gasoline
   b. kerosene
   c. varsol
   d. water
7. Parts or sections of a valve are
   a. stem, margin, and face
   b. stem, margin, and seat
   c. stem, head, and seat
   d. stem, face, and pushrod
8. Most valves are ground at
   a. 30 or 45 degrees
   b. 45 or 65 degrees
   c. 30 or 90 degrees
   d. 20 or 30 degrees

9. Blow-by occurs in the
   a. carburetor
   b. cylinder
   c. governor housing
   d. breaker point assembly
10. Connecting rod bearing clearance at the crankshaft may be gauged by a
   a. fine ruler
   b. micrometer
   c. piece of plastigage
   d. wire gauge

B. **Matching.** Match the terms in column I with those in column II.

Column I

1. governor
2. air vane
3. carburetor
4. needle valve
5. spark
6. battery
7. primary circuit
8. secondary circuit
9. electrolyte
10. breaker points

Column II

a. 30,000 volts or more
b. 6 or 12 volts
c. includes ignition points
d. speed control
e. includes spark plug
f. controls fuel flow
g. acid
h. provides air and fuel
i. approximately 0.020 inches
j. moved by air flow

C. **Completion.** Fill in the blanks with the word or words that will make the following statements correct.

1. Small gas engines are especially useful where low horsepower and _____ is required.
2. Gasoline is quick to vaporize, spread, ignite and _____.
3. _____ force causes the weights to move outward on a governor.
4. To reduce the problem of condensation, fuel tanks should _____.
5. A magneto must have a coil, iron core, and _____.
6. Small engines on mowers frequently have a _____ to protect the crankshaft against breakage due to impact caused by hitting objects.
7. A piston must have two types of rings. These are _____ and _____.
8. The connecting rod is attached to the piston by the _____.
9. Improper torquing of a cylinder head may cause _____ and _____.
10. A hand starter that utilizes an uncoiling spring to crank the engine is called a _____ starter.

Electricity And Electronics

UNIT

31 Electrical Principles and Wiring Materials

OBJECTIVE

To use principles of electricity and safety for planning simple wiring systems.

COMPETENCIES TO BE DEVELOPED

After studying this unit, you should be able to:

- Describe some basic principles of electricity and magnetism.
- Use safety practices with electricity.
- Steam clean machinery.
- Describe the relationship among volts, amperes, and watts.
- Select materials for electrical wiring.
- Design simple wiring systems.

MATERIALS LIST

- ✓ Horseshoe magnet
- ✓ Samples of wire, nonmetallic sheathed cable, armored cable, and conduit
- ✓ Electric meter dial

TERMS TO KNOW

electricity
filament
fluorescent lamp
resistance
conductor
insulator
amperes
volts
watts
ohm
Ohm's law
volt-ohm-milliampere meter (VOM meter)
milliampere
magnetism
permanent magnet
poles
north pole
attracts
south pole
repels
magnetic flux
magnetic field
reverses the polarity
electromagnetic
commutator
armature
field
generator
alternator
turbine
circuit
open circuit
short circuit
grounding
rod
wire
shock
ground fault interrupter (GFI)
entrance head
transformer
service drop
service entrance panel
meter
kilo
watthour
kilowatthour
branch circuits

fuse
circuit breaker
nonmetallic sheathed cable (Romex)
armored cable (BX)
conduit
electrical metallic tube (EMT)
strands
voltage drop
positive (hot) wires
neutral wires

Electricity is the major power source for stationary equipment in houses, farm and ranch buildings, and agribusinesses. It is the energy source commonly used for driving machinery, and for lighting, heating, and cooling. Some knowledge of electricity is essential for the safe use of electrical equipment. Understanding how to wire simple circuits and make minor electrical repairs is also useful. It is important to maintain electrical circuits and equipment properly to ensure their long life and safe operation.

PRINCIPLES OF ELECTRICITY

Heat and Light

Electricity is a form of energy that can produce light, heat, magnetism, and chemical changes. Light can be produced by heating a special metal element or filament in a vacuum tube called a bulb. The flow of electricity into the bulb must be carefully controlled so that the filament glows without burning out. In addition, electricity flowing through certain gases causes them to glow. A fluorescent lamp glows as a result of electricity flowing through a gas.

Heat is produced when electricity flows through metals with some difficulty. Any tendency of a material to prevent electrical flow is called resistance. If electricity flows easily, the metal is said to be a good conductor. Silver is an excellent conductor, copper is a very good conductor, and aluminum is a good conductor. Because silver is so expensive, copper is generally used in wiring systems; aluminum may be used when the price of copper is high. Aluminum is also used extensively in outside lines where heat buildup is not a problem. Applications include the high-voltage power lines that cross the countryside, as well as the overhead wires that stretch from building to building in residential areas or on the farm.

Heat in wires, switches, outlets, motors, and lights is not desirable. It wastes electrical energy, causes materials to deteriorate, heats up the surrounding areas, and may cause fires. Therefore, the proper design of wiring systems is important. In properly designed electrical systems, heat is kept within acceptable limits. In heating systems, when heat is generated by design, the heat is produced by using elements with just the right amount of resistance. Heat from electricity is clean and easy to control.

A material that provides great resistance to the flow of electricity is called an insulator. Examples of good insulators include rubber, glass, vinyl, and air. Anyone working with electricity must be aware that insulation is relative. For example, a very thin layer of rubber or vinyl will prevent electrical flow if the voltage is low, such as 12 volts for automobile lights. However, to prevent electricity at 30,000 volts in a spark plug wire from jumping to ground requires a much thicker layer of rubber.

Amperes, Volts, and Watts

Amperes, volts, and watts must be understood to design electrical circuits or install electrical materials safely. Amperes are a measure of the rate of flow of electricity in a conductor. Volts are a measure of electrical pressure. Watts are a measure of the amount of energy or work that can be done by amperes and volts.

The following relationships exist among amperes, volts, and watts:

$$\text{Watts} = \text{Volts} \times \text{Amperes}$$

$$\text{Volts} = \frac{\text{Watts}}{\text{Amperes}}$$

$$\text{Amperes} = \frac{\text{Watts}}{\text{Volts}}$$

For example, if a 200-watt light bulb operates at 120 volts, it will draw 1.67 amperes of electricity.

$$A = \frac{W}{V} \text{ or } 200 \div 120 = 1.67$$

A 5-ampere motor running on 120 volts will consume 600 watts of electricity.

$$W = V \times A \text{ or } 120 \times 5 = 600$$

A refrigerator motor rated at 3 amperes that consumes 360 watts of electricity should be plugged into a 120-volt circuit.

$$V = \frac{W}{A} \text{ or } 360 \div 3 = 120$$

These relationships are known as the West Virginia formula: W = VA. By remembering the basic formula, any one of the three quantities can be determined if the other two are known. Notice that the formulas used in the above illustrations are all from the basic W = VA formula.

Ohm's Law

A physicist named George Simon Ohm made a number of important discoveries about electric current. As a result, the unit used to measure a material's resistance to the flow of electrical current is known as the ohm. Ohm discovered that the flow of electricity through a conductor is directly proportional to the electrical or electromotive force that produces it. The relationship he discovered between electric current, electromotive force, and resistance is called Ohm's Law.

Electromotive force is measured in volts and is represented by a capital E in Ohm's Law. The rate of flow of electricity through a conductor is measured in amperes, represented by a capital I in Ohm's Law. The tendency of a material to prevent electrical flow, or resistance, is represented by a capital R.

E = Volts
I = Amperes
R = Resistance

Ohm's Law states that electromotive force is equal to amperes times resistance, or E = IR. From this basic formula, amperes (I) and resistance (R) can also be computed:

$$I = \frac{E}{R}$$

$$R = \frac{E}{I}$$

In any situation, one of the values can be computed if the other two are known. This knowledge is useful when designing, testing, or using electrical equipment.

Volts, amperes, and resistance can be measured using various types of meters. Voltage is measured with a voltmeter. Amperes are measured with an ammeter. Resistance in ohms is measured by an ohmmeter. A popular combination meter used in electronics is a volt-ohm-milliampere meter (VOM meter). A milliampere is a thousandth of an ampere.

Magnetism and Electricity

Electricity flowing through a conductor results in magnetism. Magnetism is a force that attracts or repels iron or steel, figure 31-1. Magnetism can be created by exposing iron or steel to magnetic forces. If iron or steel holds its magnetism, it is said to be a permanent magnet. Permanent magnets may be used to create or generate electricity.

Magnets have two ends, called poles. These are designated as the north pole and the south pole. The north pole of one magnet attracts the south pole of another magnet but repels, or pushes away, the north pole of another magnet. In a similar manner, the south pole of one magnet will repel the south pole of another but attract the north pole. In summary, like poles repel and unlike poles attract, or opposites attract, figure 31-2

Lines of magnetic force, often called magnetic flux, occur in patterns in the air between the two poles. The pattern is referred to as the magnetic field, figure 31-3.

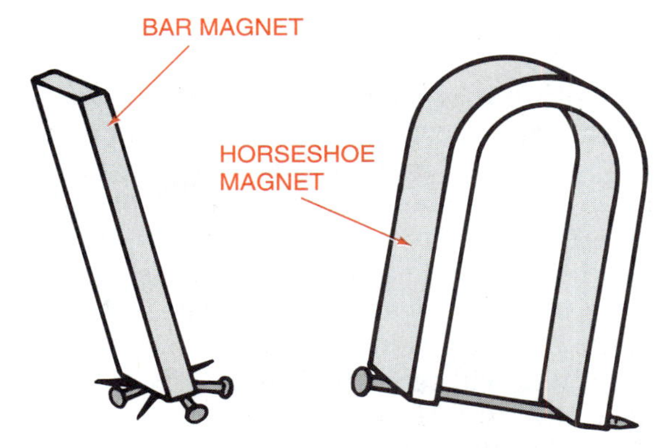

FIGURE 31-1. Magnetism is a force that attracts or repels iron or steel.

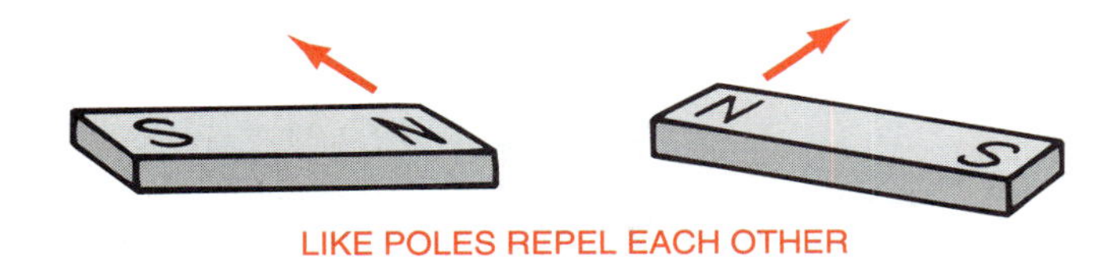

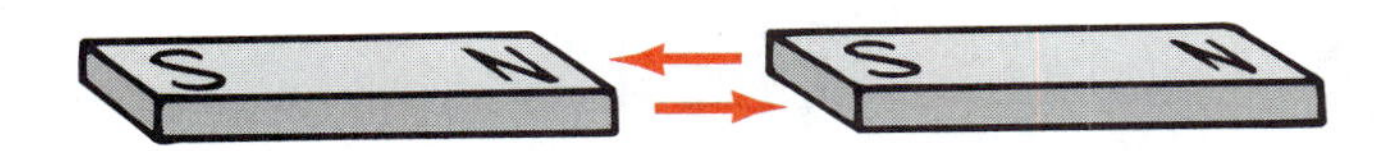

UNLIKE POLES ATTRACT EACH OTHER

FIGURE 31-2. Like poles repel and unlike poles attract.

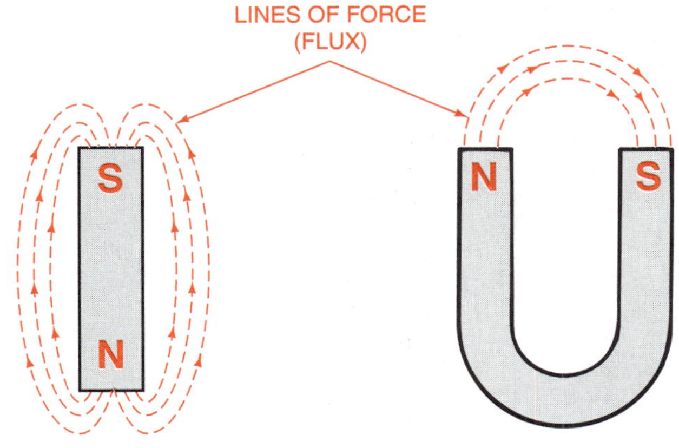

FIGURE 31-3. Lines of magnetic force occur in a pattern called the magnetic field.

Electric Motors

The principle of magnetism is the basis upon which electric motors operate. A strong magnet can be made by wrapping many coils of insulated wire around an iron core and passing electrical current through the wire, figure 31-4. When the current stops flowing, the magnetic field ceases to exist, figure 31-5. If the wires are switched to opposite battery terminals, the poles of the magnet will reverse, but the magnetic field will remain, figure 31-6. However, reversing the direction of current flow reverses the polarity of the magnet. The unit is called an electromagnet.

A motor may be made by bending the electromagnet into the shape of a horseshoe and suspending a permanent magnet on a bearing between the poles. When current flows, the unlike poles of the electromagnet and the permanent magnet will attract each other, causing the permanent magnet to make a half turn, figure 31-7. If the wires are reversed at the battery, the direction of the current and polarity of the magnet will be reversed, and the permanent magnet will make another half turn, figure 31-8. In a motor, the current is reversed by a part called a commutator. The commutator causes the motor to run continuously in the same direction. The rotating (turning) magnet is called the armature, and the magnetic forces around the armature are called the field.

Devices that produce electricity by magnetism are called generators and alternators. A generator produces direct current, which means the electricity

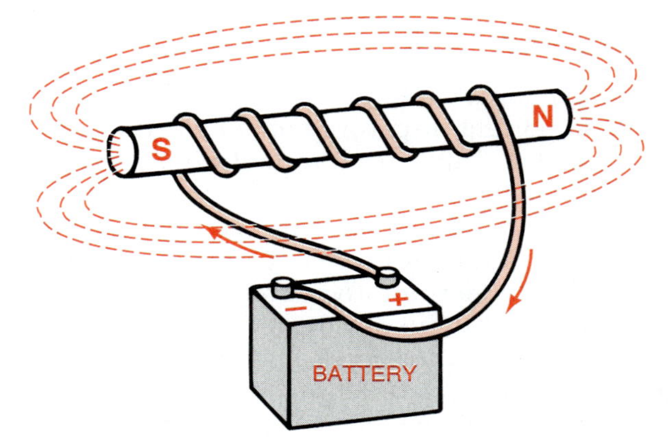

FIGURE 31-6. If the wires are switched to opposite battery terminals, the poles of the magnet reverse, but the magnetic field remains.

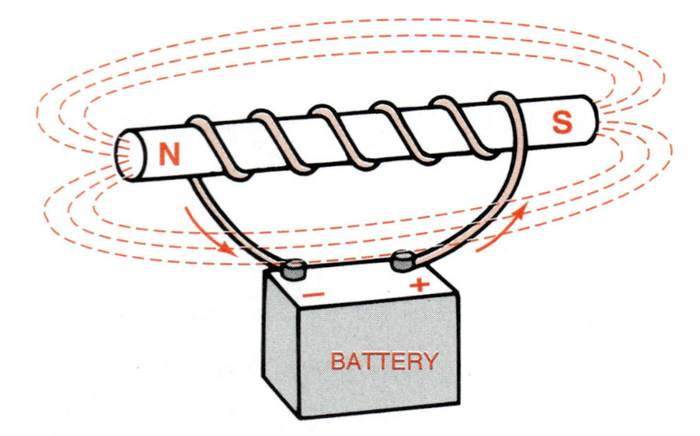

FIGURE 31-4. A strong magnet can be made by wrapping many coils of insulated wire around an iron core and passing electrical current through the wire.

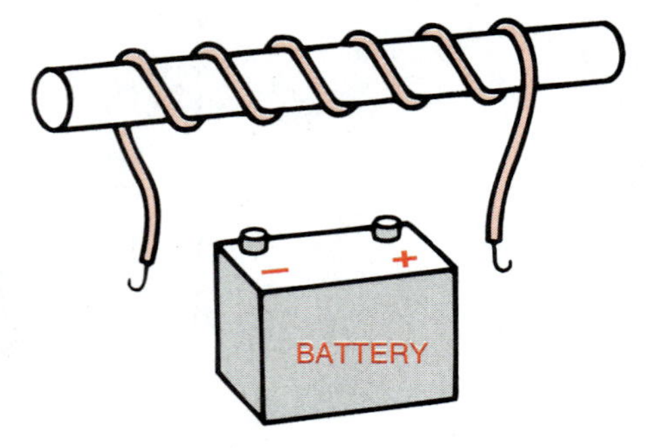

FIGURE 31-5. When the current stops flowing, the magnetic field ceases to exist.

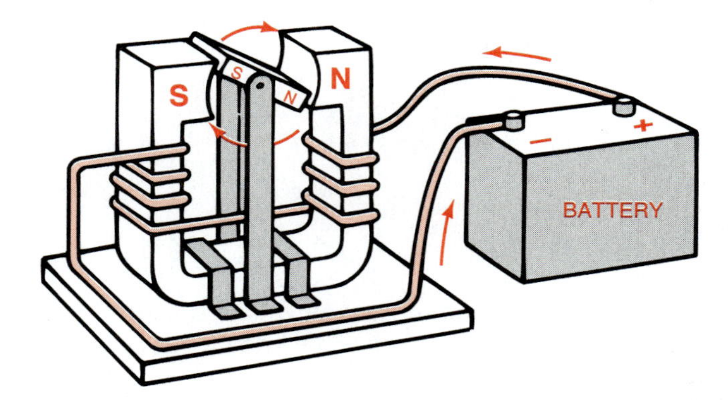

FIGURE 31-7. Unlike poles attract each other to move the permanent magnet a half turn.

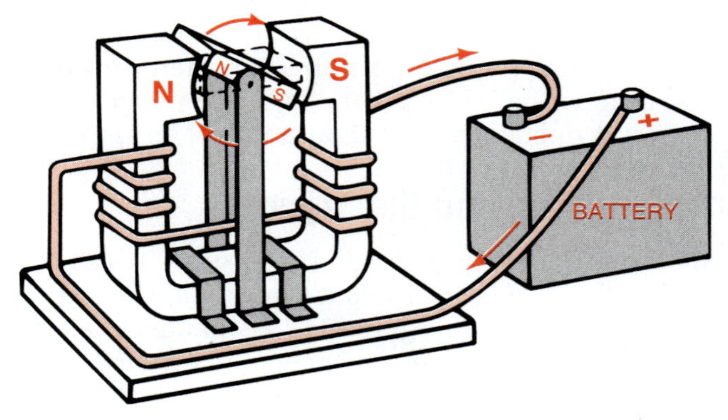

FIGURE 31-8. If the wires are reversed at the battery, the polarity of the electromagnet is reversed and the permanent magnet makes another half turn.

flows in one direction. An alternator produces alternating current, which reverses in direction, or alternates, each time the armature turns. Generators and alternators must be powered by human or animal power, water wheel, gas or diesel engine, turbine, or an electric motor getting electricity from a separate source. A turbine is a high-speed rotary engine driven by water, steam, or other gases.

Electric Motor Maintenance. Electric motors are designed for long life with little maintenance. Some have permanently lubricated bearings. With these motors, the only maintenance is an occasional cleaning with compressed air, figure 31-9. Under extreme conditions the motor may accumulate greasy dirt on inside parts. In those cases, disassembly and cleaning with an electric motor cleaner may be necessary, figure 31-10.

FIGURE 31-9. Electric motors need occasional cleaning with compressed air to remove dust and dirt from interior surfaces.

FIGURE 31-10. If motors are extremely dirty, they may need to be disassembled for cleaning.

Many motors have two oil cups or tubes with a wick to hold oil for the two shaft bearings. The operator's manual supplied by the manufacturer typically calls for about 10 drops of oil per year in these tubes. More oil than specified will cause dirty oil accumulations on interior and exterior parts. On the other hand, failure to use oil as specified by the manufacturer shortens the life of the motor due to excessive bearing wear. Therefore, it is important to follow the instructions provided, and to adjust oil volumes and intervals according to operating conditions.

Circuits

A source of electricity plus two wires connected to a light, heater, or motor are known collectively as a circuit. A circuit may be likened to a circle. If the circle is broken so that current cannot flow through it, it is said to be an open circuit.

All circuits must include an object with resistance that falls within a certain range. Otherwise, the electricity will flow through the circuit and back to its source too rapidly and blow fuses, burn wires, and drain batteries. Such a condition is called a short circuit.

Electricity can travel back to its source through the earth as well as through a wire. A circuit always includes a wire to carry the current back to its source. However, safety requires that an additional wire or conductor be provided to all metal parts in the system in case the current gets out of its circuit. Making this additional connection between a piece of equipment and the earth is called grounding. It is accomplished by driving a solid copper rod or a 1-inch diameter or larger galvanized steel pipe into soil that is always moist. The rod or pipe, called a ground rod, and the electrical equipment are connected by a conductor called a ground wire. Grounding equipment is a standard safety practice. In this way electricity that accidentally gets out of its circuit is channeled to the earth through a ground wire rather than through the body of a human or animal.

ELECTRICAL SAFETY

There are two deadly hazards associated with electric current: shock and fire. Shock refers to the body's reaction to an electric current. Shock can interfere with a normal heartbeat and result in the injury or death of the victim. Fire may result when electrical conductors overheat or when a spark is produced by an electrical current jumping an air space.

All people who work around electric current should observe safety practices at all times to avoid shock or fire. Some precautions advised by authorities are:

- Never disconnect or damage any safety device that is provided by the manufacturer or specified by electrical codes.

- Do not touch electrical appliances, boxes, or wiring with wet hands or wet feet.
- Do not remove the long ground prong from three-prong 120-volt plugs.
- Use ground fault interrupters in kitchen, bathroom, laundry, and outdoor circuits, or wherever moisture may increase shock hazard. A ground fault interrupter (GFI) is a device that cuts off the electricity if even very tiny amounts of current leave the normal circuit. This device reduces the likelihood of human shock injury. Ground fault interrupters must be installed in accordance with the instructions provided by the manufacturer.
- Immediately discontinue the use of any extension cord that feels warm or smells like burning rubber.
- Do not place extension cords under carpeting.
- Install all electrical wiring according to the specifications of the National Electrical Code.
- Use only double insulated portable tools or tools with three-wire grounded cords.
- If a fuse is blown or circuit breaker is tripped, determine and correct the problem before inserting a new fuse or resetting the breaker.
- Fuses and circuit breakers are designed to prevent the circuit from being overloaded. Do not create a fire hazard by installing higher capacity fuses or breakers than the system is designed to handle.
- Do not leave heat-producing appliances, such as irons, hair dryers, and soldering irons, unattended.
- Place all heaters and lamps away from combustible materials.
- Keep the metal cases or cabinets of electrical appliances grounded at all times.
- Do not remove the back of a television set. There is danger of shock by 20,000 or 30,000 volts even with the unit unplugged.
- Keep electrical motors lubricated and free of grease and dirt.
- Keep appliances dry to reduce shock hazard and prevent rust.
- Do not use any switches, outlets, fixtures, or extension cords that are cracked or damaged in any way.
- Follow manufacturer's instructions for installation and use of all electrical equipment.

ELECTRICAL WIRING

An electrical system must meet several conditions to be satisfactory. It must:

- be safe
- be convenient
- be expandable
- look neat
- provide sufficient current

Service Entrance

Electrical power comes to the home or farm by overhead or underground wires. The power company provides a transformer, service drop, and appropriate wiring to an entrance head. An entrance head is a waterproof device used to attach exterior wires to interior wires of a building. The transformer converts high voltage from the power lines to 240 volts for home and farm installations. The service drop is an assembly of electrical wires, connectors, and fasteners used to transmit electricity from the transformer to the entrance head and on to the service entrance panel. The service entrance panel is a box with fuses or circuit breakers where electricity enters a building. Wires feed from the service drop through the insulated entrance head, through cable or pipe to a meter box. The electric meter plugs into the meter box. The power continues on to the service entrance panel for distribution to branch circuits, figure 31-11.

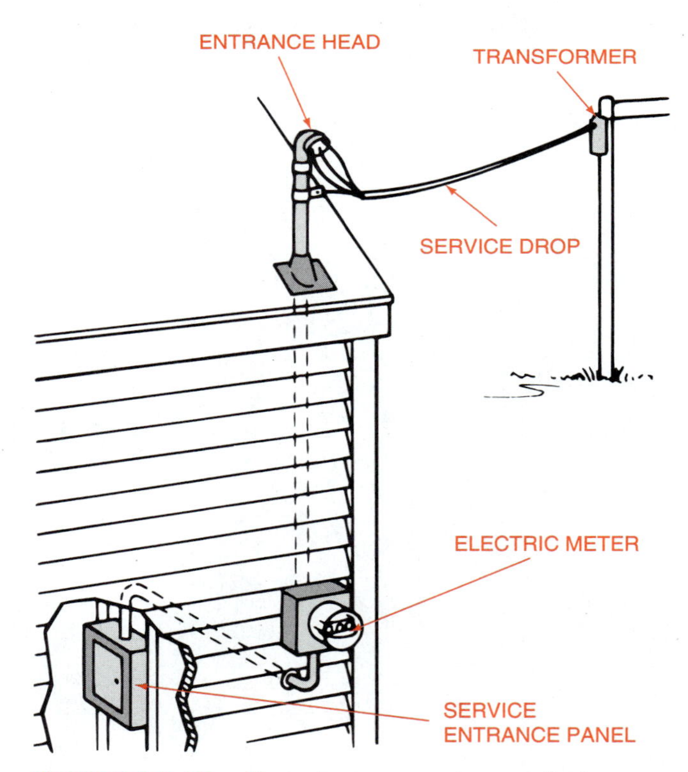

FIGURE 31-11. Electrical power is supplied to local homes and farms through a transformer, service drop, entrance head, meter, and service-entrance panel.

Meter

All electricity that passes through the system is measured by the meter, in kilowatthours, figure 31-12. Kilo means 1000. A watthour is the use of 1 watt for 1 hour. (A 100-watt bulb that burns for 1 hour will consume 100 watthours of electricity.) A kilowatthour is the use of 1000 watts for 1 hour.

The meter will convert any amount of electricity into kilowatthours. A power company representative reads the meter at regular intervals, and the customer is billed according to the number of kilowatthours used. When reading the meter, the last number passed by each pointer is read. These four numbers constitute the meter reading. Note that the first and third dials turn counterclockwise, and the second and fourth dials turn clockwise. In figure 31-12, the dials indicate that 0019 kilowatthours of electricity have passed through the meter. The third dial appears to read 2, but since the fourth dial has not completed its revolution, the third dial cannot quite have reached 2.

FIGURE 31-12. An electric meter.

Branch Circuits

Most circuits begin from the service entrance panel. They are called branch circuits since they branch out into a variety of places and for a variety of purposes, figure 31-13. A branch circuit generally includes only one motor or a series of outlets or a series of lighting fixtures, figure 31-14. It is important to provide the correct size of wire and fuse or circuit breaker for the load on each circuit.

A fuse is a plug or cartridge containing a strip of metal that melts when more than a specified amount of current passes through it. A circuit breaker is a switch that trips and breaks the circuit when more than a specified amount of current passes through it. A circuit breaker can be reset after it trips. A fuse must be replaced if it blows.

CAUTION:

- **Fuses and circuit breakers are used to protect circuits from damage and fire. Never destroy this protection by installing fuses or breakers with a larger ampere rating than is recommended for the size of wire being protected.**

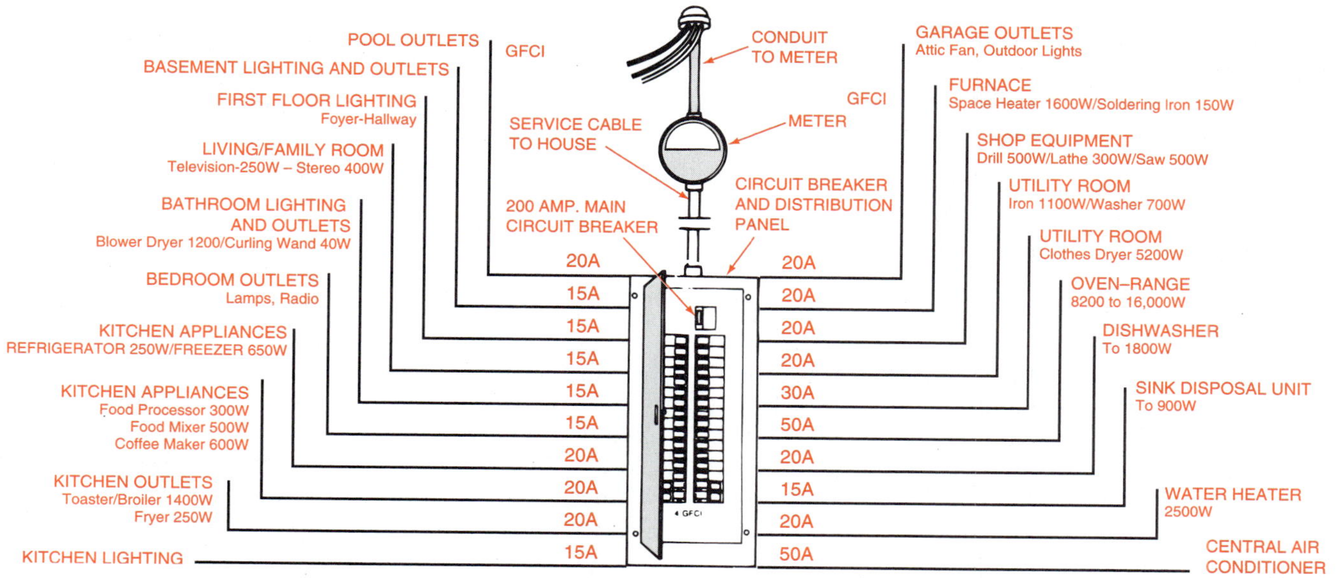

FIGURE 31-13. Branch circuits fan out from the service-entrance panel. (*Courtesy of American Association for Vocational Instructional Materials*)

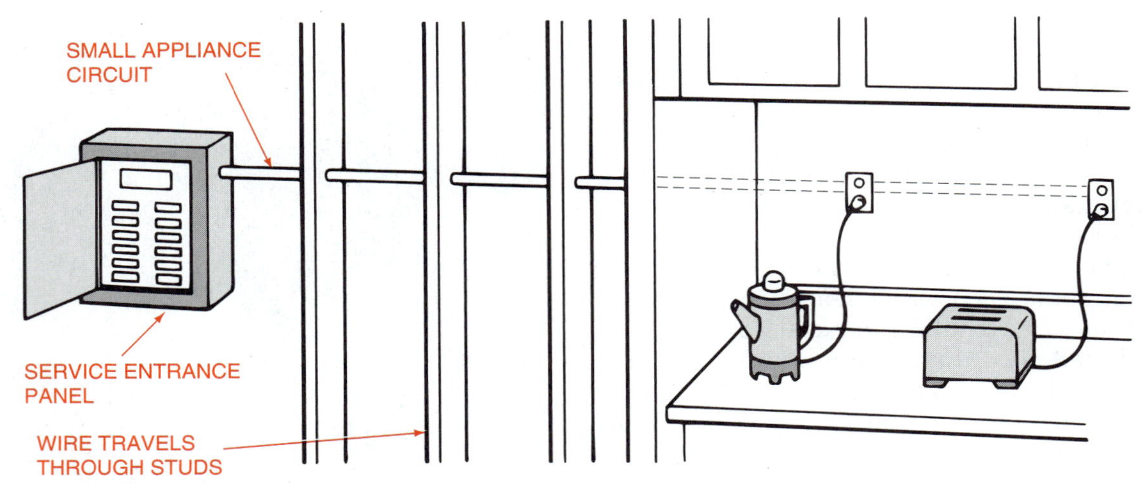

FIGURE 31-14. A branch circuit generally includes a series of outlets, a series of lights, or one electric motor.

When planning electrical wiring systems or reading blueprints, it is useful to know standard electrical symbols, figure 31-15. The symbols are used on drawings to indicate the location of outlets, receptacles, switches, and special appliances.

Types of Cable

Three systems of wiring are in general use for most home, farm, and commercial installations: (1) nonmetallic sheathed cable, (2) armored cable, and (3) conduit.

Nonmetallic sheathed cable, commonly called Romex, consists of copper or aluminum wires covered with paper, rubber, or vinyl for insulation and protection, figure 31-16. Certain types of nonmetallic sheathed cable are waterproof and suitable for burial in soil.

Armored cable, commonly called BX, consists of a flexible metal sheath with individual wires inside, figure 31-17. The wires are insulated with rubber or vinyl and wrapped in paper for protection. Armored cable provides good protection against mechanical damage. It is fairly easy to install in narrow and irregular building sections. However, its tendency to rust and develop short circuits makes it unsuitable for use in damp areas.

Conduit is tubing which contains individual insulated wires, figure 31-18. Such tubes are available in 1/2-inch, 3/4-inch, 1-inch or larger diameters. Conduit may be rigid or bendable. It is available in metal or plastic. The bendable type of metal conduit is known as electrical metallic tube (EMT). Special benders permit shaping EMT in bends of up to 90 degrees. Couplings and connectors are used to assemble conduit and attach it to boxes. Waterproof plastic fittings provide waterproof seals where needed for metal conduit. Conduit systems are used to provide the most protection for wiring and are required for most commercial jobs.

Wire Type and Size

Individual wires within cable or conduit may be aluminum or copper. Smaller sizes include No. 14 wire for 15-ampere circuits, No. 12 for 20-ampere circuits, and No. 10 for 30-ampere 120 volt circuits, figure 31-19. These ampere ratings apply to copper wire. For aluminum, it is necessary to use one wire size larger than is specified for copper wire. Wire size is designated by gauge or AWG (American Wire Gauge) number. The lower the gauge number, the larger the wire size.

Small wires called strands are placed together to form bundles for wire sizes No. 8 and larger in order to improve flexibility and conductivity. Electricity is carried on the outer surfaces of wire. Thus, stranded wire has more total surface area than solid wire of equal diameter and carries more current. Either solid wire or stranded wire may be placed in cables or conduit.

Voltage Drop

A common problem encountered in homes and on farms is voltage drop. Voltage drop refers to a loss of voltage as it travels along a wire.

Voltage drop causes lights to dim, heaters to put out less heat, and motors to put out less power and to overheat. The larger the wire, the less problem there is with voltage drop for a given amount of current. The longer the wire, however, the greater the problem.

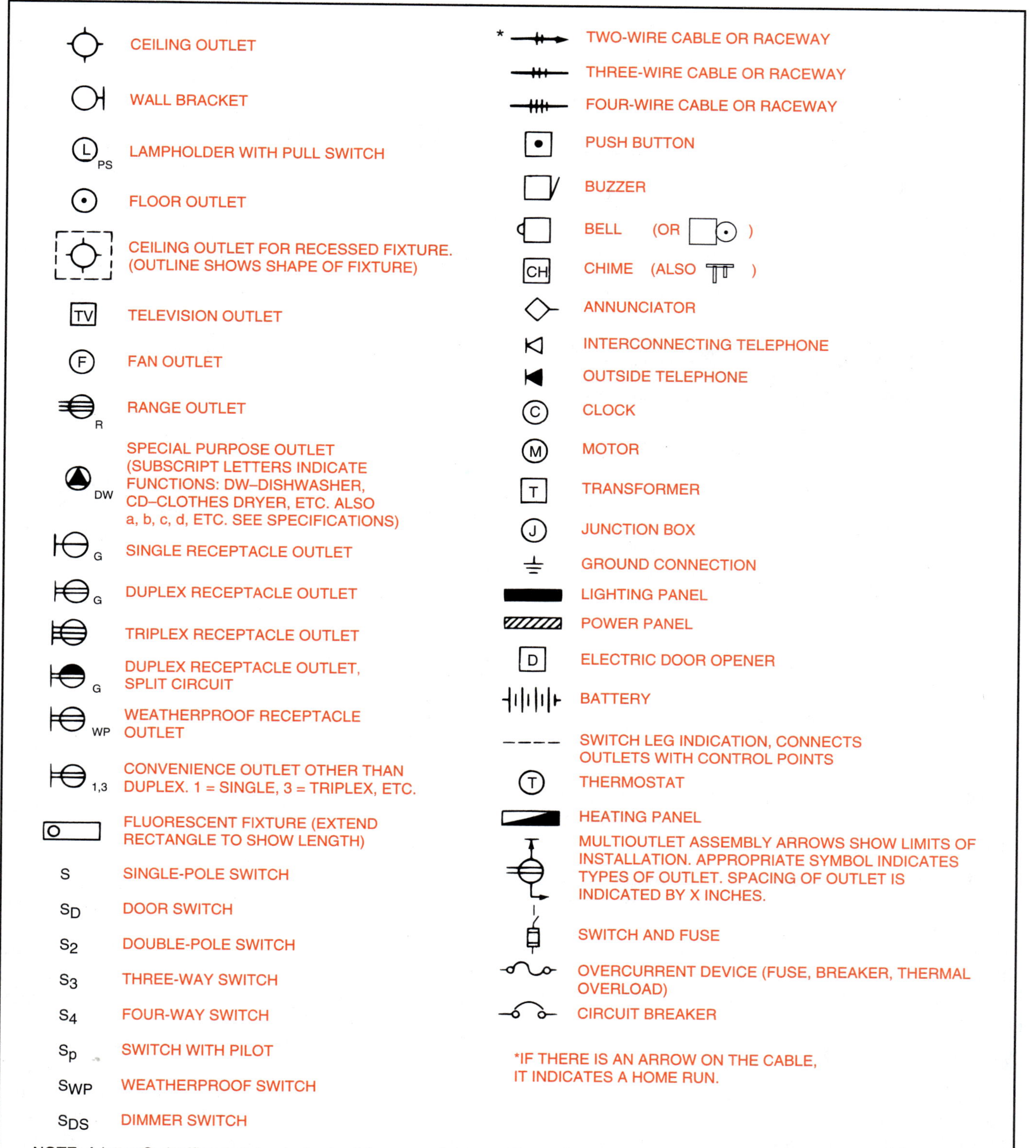

FIGURE 31-15. Standard symbols identify locations of outlets, switches, and equipment. (*Courtesy of R. Mullin,* Electrical Wiring—Residential, 12th ed., *Delmar Publishers, 1996*)

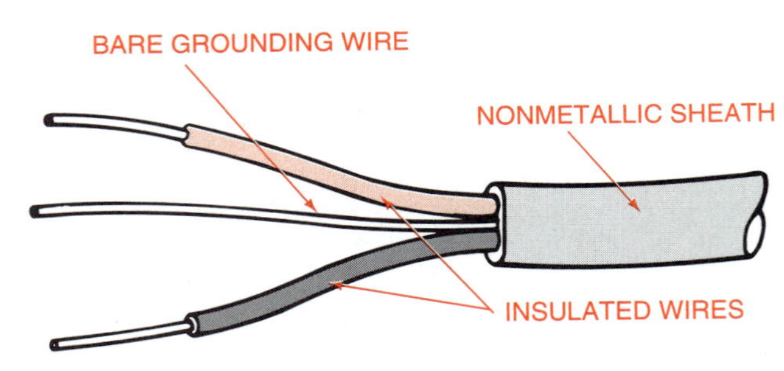

FIGURE 31-16. Nonmetallic sheathed cable.

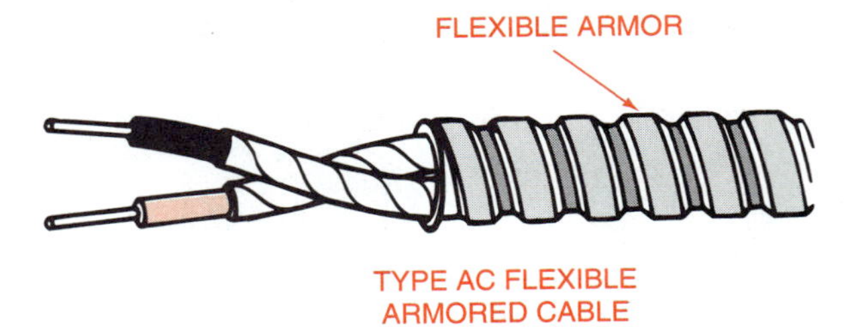

FIGURE 31-17. Armored cable.

FIGURE 31-18. Conduit. (*Courtesy of American Association for Vocational Instructional Materials*)

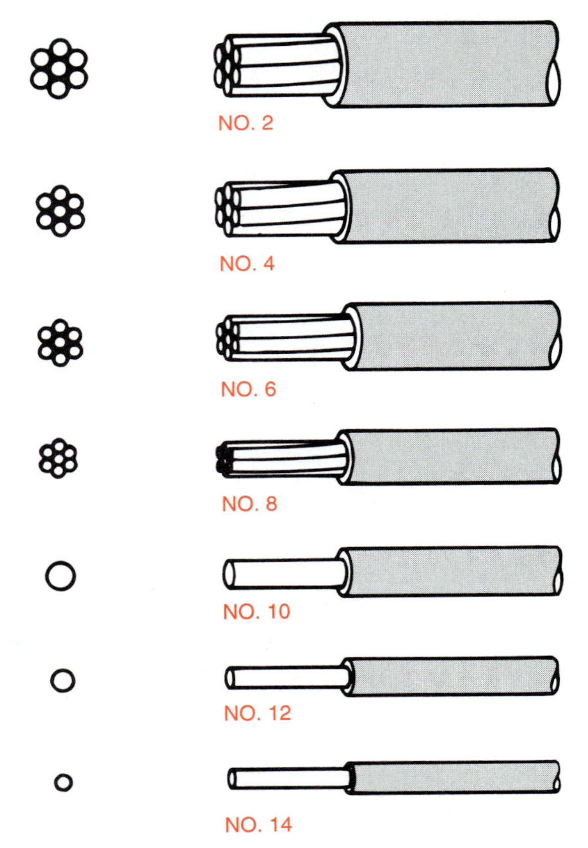

FIGURE 31-19. Common sizes of wire used in cables and conduit.

Allowance for long wires must be made by using either lighter loads or larger wires. For example, to obtain 20 amperes of current with standard voltage at the end of a circuit, the wire size must be increased according to the distance, figure 31-20. This is especially true when installing special circuits for motors.

Wire Identification

The type of outer covering, individual wire covering, cable construction, and number of wires all help determine where a cable can be used. The wire type is stamped on the outer surface of the wire or cable or cable covering. Some common types of wire and cable are:

- Type T—for use in dry locations
- Type TW—for use in dry or wet locations
- Type THHN—for use in dry locations with high temperature
- Type THW and THWN—for use in wet locations with high temperature
- Type XHHW—for high moisture and heat resistance
- Type UF—for direct burial in soil but not concrete

The *National Electrical Code*® and local building codes must be consulted for specific guidance in wire selection.

Individual wires are color coded to help identify their function in the circuit. Black wires, red wires, and blue wires are positive (hot) wires, which carry current to appliances. White wires are neutral wires, which carry current from the appliance back to the source. Green wires and bare wires are used to ground all metal boxes and appliances in the circuit.

Nonmetallic sheathed cable is generally stamped at regular intervals with a mark that identifies its type, wire size, and number of conductors. Typical markings and their meanings are as follows:

- 12-2—two strands of No. 12 wire, one black and one white
- 12-2 w/g—two strands of No. 12 wire plus a ground wire (w/g means "with ground"), one black, one white, and one green or bare

COPPER UP TO 200 AMPERES, 115-120 VOLTS, SINGLE PHASE, BASED ON 2% VOLTAGE DROP

Minimum Allowable Size of Conductor				Length of Run in Feet — Compare size shown below with size shown to left of double line. Use the larger size.																					
In Cable, Conduit, Earth			Overhead In Air*																						
Load in Amps	Types R, T, TW	Types RH, RHW, THW	Bare & Covered Conductors	30	40	50	60	75	100	125	150	175	200	225	250	275	300	350	400	450	500	550	600	650	700
5	12	12	10	12	12	12	12	12	12	12	10	10	10	10	8	8	8	8	6	6	6	6	4	4	4
7	12	12	10	12	12	12	12	12	12	10	10	8	8	8	8	5	6	6	6		4	4	4	4	3
10	12	12	10	12	12	12	12	10	10	8	8	8	6	6	6	5	4	4	4	4	4		4	4	3
15	12	12	10	12	12	10	10	10	8	6	6	6	4	4	4	4	4	3	2	2	1	1	1	0	0
20	12	12	10	12	10	10	8	8	6	6	4	4	4	4	3	3	2	2	1	1	0	0	00	00	00
25	10	10	10	10	10	8	8	6	6	4	4	4	3	3	2	2	1	1	0	0	00	00	000	000	000
30	10	10	10	10	8	8	8	6	4	4	4	3	2	2	1	1	1	0	00	00	000	000	000	4/0	4/0
35	8	8	10	10	8	8	6	6	4	4	3	2	2	1	1	0	0	00	00	000	000	4/0	4/0	4/0	250
40	8	8	10	8	8	6	6	4	4	3	2	2	1	1	0	0	00	00	000	000	4/0	4/0	250	250	300
45	6	8	10	8	8	6	6	4	4	3	2	1	1	0	0	00	00	000	000	4/0	4/0	250	250	300	300
50	6	6	10	8	6	6	4	4	3	2	1	1	0	0	00	00	000	000	4/0	4/0	250	250	300	300	350
60	4	6	8	8	6	4	4	4	2	1	1	0	00	00	000	000	000	4/0	250	250	300	300	350	400	400
70	4	4	8	6	6	4	4	3	2	1	0	00	00	000	000	4/0	4/0	250	300	300	350	400	400	500	500
80	2	4	6	6	4	4	3	2	1	0	00	00	000	000	4/0	4/0	250	300	300	350	400	400	500	500	600
90	2	3	6	6	4	4	3	2	1	0	00	000	000	4/0	100	250	250	300	350	400	500	500	500	600	600
100	1	3	6	4	4	3	2	1	0	00	000	000	4/0	4/0	250	250	300	350	400	500	500	500	600	600	700
115	0	2	4	4	4	3	2	1	0	00	000	4/0	4/0	250	300	300	350	400	500	500	600	600	700	700	750
130	00	1	4	4	3	2	1	0	00	000	4/0	4/0	250	300	300	350	400	500	500	600	600	700	750	800	900
150	000	0	2	4	2	1	1	0	000	4/0	4/0	250	300	350	350	400	500	500	600	700	700	800	900	900	1M
175	4/0	00	2	3	2	1	0	00	000	4/0	250	300	350	400	400	500	500	600	700	750	800	900	1M		
200	250	000	1	2	1	0	00	000	4/0	250	300	350	400	500	500	500	600	700	750	900	1M				

ALUMINUM UP TO 200 AMPERES, 115-120 VOLTS, SINGLE PHASE, BASED ON 2% VOLTAGE DROP

Minimum Allowable Size of Conductor				Length of Run in Feet — Compare size shown below with size shown to left of double line. Use the larger size																					
In Cable, Conduit, Earth			Overhead In Air*																						
Load in Amps	Types R, T, TW	Types RH, RHW, THW	Bare & Covered Conductors Single Triplex	30	40	50	60	75	100	125	150	175	200	225	250	275	300	350	400	450	500	550	600	650	700
5	12	12	10	12	12	12	12	12	10	10	8	8	8	8	6	6	6	6	4	4	4	4	3	3	3
7	12	12	10	12	12	12	12	10	10	8	8	6	6	6	6	4	4	4	4	3	3	2	2	2	1
10	12	12	10	12	12	10	10	8	8	6	6	6	4	4	4	4	3	3	2	2	1	1	0	0	0
15	12	12	10	12	10	8	8	8	6	4	4	4	3	3	2	2	2	1	0	0	00	00	00	000	000
20	10	10	10	10	8	8	6	6	4	4	3	3	2	2	1	1	0	0	00	00	000	000	4/0	4/0	4/0
25	10	10	10	8	8	6	6	4	4	3	2	2	1	1	0	0	00	00	000	000	4/0	4/0	250	250	300
30	8	8	10	8	6	6	6	4	3	2	2	1	0	0	00	00	00	000	4/0	4/0	250	250	300	300	350
35	6	8	10	8	6	6	4	4	3	2	1	0	0	00	00	000	000	4/0	4/0	250	300	300	350	350	400
40	6	8	10	6	6	4	4	3	2	1	0	0	00	00	000	000	4/0	4/0	250	300	300	350	350	400	500
45	4	6	10	6	6	4	4	3	2	1	0	00	00	000	000	4/0	4/0	250	300	300	350	400	400	500	500
50	4	6	8	6	4	4	3	2	1	0	00	00	000	000	4/0	4/0	250	300	300	350	400	400	500	500	600
60	2	4	6	6	4	3	3	2	0	00	00	000	4/0	4/0	250	250	300	350	350	400	500	500	600	600	700
70	2	2	6	4	4	3	2	1	0	00	000	4/0	4/0	250	300	300	350	400	500	500	600	600	700	700	750
80	1	2	6	4	3	2	1	0	00	000	4/0	4/0	250	300	300	350	350	500	500	600	600	700	750	800	900
90	0	2	4	4	3	2	1	0	00	000	4/0	250	300	300	350	400	400	500	600	600	700	750	800	900	1M
100	0	1	4	3	2	1	0	00	000	4/0	250	300	300	350	400	400	500	600	600	700	750	900	900	1M	
115	00	0	2	3	1	1	0	00	000	4/0	300	300	350	400	500	500	600	600	700	800	900	1M			
130	000	00	2	2	1	0	00	000	4/0	250	300	350	400	500	500	600	600	700	800	900	1M				
150	4/0	000	1	2	0	00	00	000	250	300	350	400	500	500	600	600	700	800	900	1M					
175	300	4/0	0	1	0	00	000	4/0	300	350	400	500	600	600	700	750	800	900							
200	350	250	00	0	00	000	4/0	250	300	400	500	600	600	700	750	900	900								

FIGURE 31-20. Size of copper and aluminum wire needed to carry given loads with 2 percent or less voltage drop. (*Courtesy of National Food and Energy Council, Inc., Columbia, MO*)

COPPER UP TO 400 AMPERES, 230-240 VOLTS, SINGLE PHASE, BASED ON 2% VOLTAGE DROP

Minimum Allowable Size of Conductor				Length of Run in Feet																					
In Cable, Conduit, Earth			Overhead In Air*	Compare size shown below with size shown to left of double line. Use the larger size.																					
Load in Amps	Types R, T, TW	Types RH, RHW, THW	Bare & Covered Conductors	50	60	75	100	125	150	175	200	225	250	275	300	350	400	450	500	550	600	650	700	750	800
5	12	12	10	12	12	12	12	12	12	12	12	12	12	12	10	10	10	10	8	8	8	8	8	6	6
7	12	12	10	12	12	12	12	12	12	12	12	10	10	10	10	8	8	8	8	6	6	6	6	6	6
10	12	12	10	12	12	12	12	12	10	10	10	10	8	8	8	8	6	6	6	6	4	4	4	4	4
15	12	12	10	12	12	12	10	10	10	8	8	8	6	6	6	6	4	4	4	4	4	3	3	3	2
20	12	12	10	12	12	10	10	8	8	8	6	6	6	6	4	4	4	4	3	3	2	2	2	1	1
25	10	10	10	12	10	10	8	8	6	6	6	6	4	4	4	4	3	3	2	2	1	1	1	0	0
30	10	10	10	10	10	10	8	6	6	6	4	4	4	4	4	3	2	2	1	1	1	0	0	0	00
35	8	8	10	10	10	8	8	6	6	4	4	4	4	3	3	2	2	1	1	0	0	0	00	00	00
40	8	8	10	10	8	8	6	6	4	4	4	4	3	3	2	2	1	1	0	0	00	00	00	000	000
45	6	8	10	10	8	8	6	6	4	4	4	3	3	2	2	1	1	0	0	00	00	00	000	000	000
50	6	6	10	8	8	6	6	4	4	4	3	3	2	2	1	1	0	0	00	00	000	000	000	4/0	4/0
60	4	6	8	8	8	6	4	4	4	3	2	2	1	1	1	0	00	00	000	000	000	4/0	4/0	4/0	250
70	4	4	8	8	6	6	4	4	3	2	2	1	1	0	0	00	00	000	000	4/0	4/0	4/0	250	250	300
80	2	4	6	6	6	4	4	3	2	2	1	1	0	0	00	00	000	000	4/0	4/0	250	250	300	300	300
90	2	3	6	6	6	4	4	3	2	1	1	0	0	00	00	000	000	4/0	4/0	250	250	300	300	350	350
100	1	3	6	6	4	4	3	2	1	1	0	0	00	00	000	000	4/0	4/0	250	250	300	300	350	350	400
115	0	2	4	6	4	4	3	2	1	0	0	00	00	000	000	4/0	4/0	250	300	300	350	350	400	400	500
130	00	1	4	4	4	3	2	1	0	0	00	00	000	000	4/0	4/0	250	300	300	350	400	400	500	500	500
150	000	0	2	4	4	3	1	0	0	00	000	000	4/0	4/0	4/0	250	300	350	350	400	500	500	500	600	600
175	4/0	00	2	4	3	2	1	0	00	000	000	4/0	4/0	250	250	300	350	400	400	500	500	600	600	600	700
200	250	000	1	3	2	1	0	00	000	000	4/0	4/0	250	250	300	350	400	500	500	500	600	600	700	700	750
225	300	4/0	0	3	2	1	0	00	000	4/0	4/0	250	300	300	350	400	500	500	600	600	700	700	750	800	900
250	350	250	00	2	1	0	00	000	4/0	4/0	250	300	300	350	350	400	500	600	600	700	700	750	800	900	1M
275	400	300	00	2	1	0	00	000	4/0	250	250	300	350	350	400	500	500	600	700	700	800	900	900	1M	
300	500	350	000	1	1	0	000	4/0	4/0	250	300	350	350	400	500	500	600	700	700	800	900	900	1M		
325	600	400	4/0	1	0	00	000	4/0	250	300	300	350	400	500	500	600	600	700	750	900	900	1M			
350	600	500	4/0	1	0	00	000	4/0	250	300	350	400	400	500	500	600	700	750	800	900	1M				
375	700	500	250	0	0	00	4/0	250	300	300	350	400	500	500	600	600	700	800	900	1M					
400	750	600	250	0	00	000	4/0	250	300	350	400	500	500	500	600	700	750	900	1M						

Conductors in overhead spans must be at least No. 10 for spans up to 50 feet and No. 8 for longer spans. See NEC, Sec. 225-6(a).
See Note 3 of NEC "Notes to Tables 310-16 through 310-19"

FIGURE 31-20. *Continued*

- 12-3—three strands of No. 12 wire, probably one black, one red or blue , and one white
- 12-3 w/g—same as 12-3 cable with the addition of a green or bare wire for grounding the circuit

The same system is used for all wire sizes. Letters such as T, TW, or THHN on the cable indicate its type. It is very important to install wire and cable with the correct type designations to accommodate heat, moisture, fungus, and other insulation-destructive factors.

ALUMINUM UP TO 400 AMPERES, 230-240 VOLTS, SINGLE PHASE, BASED ON 2% VOLTAGE DROP

Minimum Allowable Size of Conductor					Length of Run in Feet — Compare size shown below with size shown to left of double line. Use the larger size																					
	In Cable, Conduit, Earth		Overhead In Air*																							
			Bare & Covered Conductors																							
Load in Amps	Types R, T, TW	Types RH, RHW, THW	Single	Triplex	50	60	75	100	125	150	175	200	225	250	275	300	350	400	450	500	550	600	650	700	750	800
5	12	12	10		12	12	12	12	12	12	12	10	10	10	10	8	8	8	8	8	6	6	6	6	4	4
7	12	12	10		12	12	12	12	12	10	10	10	8	8	8	8	6	6	6	6	4	4	4	4	4	4
10	12	12	10		12	12	12	10	10	8	8	8	8	6	6	6	6	4	4	4	4	3	3	3	2	2
15	12	12	10		12	12	10	8	8	8	6	6	6	4	4	4	4	3	3	2	2	2	1	1	1	0
20	10	10	10		10	10	8	8	6	6	6	4	4	4	4	3	3	2	2	1	1	0	0	0	00	00
25	10	10	10		10	8	8	6	6	4	4	4	4	3	3	2	2	1	1	0	0	00	00	00	000	000
30	8	8	10		8	8	8	6	4	4	4	3	3	2	2	2	1	0	0	00	00	00	000	000	000	4/0
35	6	8	10		8	8	6	6	4	4	3	3	2	2	1	1	0	0	00	00	000	000	000	4/0	4/0	4/0
40	6	8	10		8	6	6	4	4	3	3	2	2	1	1	0	0	00	00	000	000	4/0	4/0	4/0	250	250
45	4	6	10		8	6	6	4	4	3	2	2	1	1	0	0	00	00	000	000	4/0	4/0	250	250	250	300
50	4	6	8		6	6	4	4	3	2	2	1	1	0	0	00	00	000	000	4/0	4/0	250	250	300	300	300
60	2	4	6	6	6	6	4	3	2	2	1	0	0	00	00	00	000	4/0	4/0	250	250	300	300	350	350	350
70	2	2(a)	6	4	6	4	4	3	2	1	0	0	00	00	000	000	4/0	4/0	250	300	300	350	350	400	400	500
80	1	2(a)	6	4	4	4	3	2	1	0	0	00	00	000	000	4/0	4/0	250	300	300	350	350	400	500	500	500
90	0	2(a)	4	2	4	4	3	2	1	0	00	00	000	000	4/0	4/0	250	300	300	350	400	400	500	500	500	600
100	0	1(a)	4	2	4	3	2	1	0	00	00	000	000	4/0	4/0	250	300	300	350	400	400	500	500	600	600	600
115	00	0(a)	2	1	4	3	2	1	0	00	000	000	4/0	4/0	250	300	300	350	400	500	500	600	600	600	700	700
130	000	00(a)	2	0	3	2	1	0	00	000	000	4/0	250	250	300	300	350	400	500	500	600	600	700	700	750	800
150	4/0	000(a)	1	00	2	2	1	00	000	000	4/0	250	250	300	300	350	400	500	500	600	600	700	750	800	900	900
175	300	4/0(a)	0	000	2	1	0	00	000	4/0	250	300	300	350	400	400	500	600	600	700	750	800	900	900	1M	
200	350	250	00	4/0	1	0	00	000	4/0	250	300	300	350	400	400	500	600	600	700	750	900	900	1M			
225	400	300	000			1	0	00	000	4/0	250	300	350	400	500	500	500	600	700	750	900	1M	1M			
250	500	350	000			0	00	000	4/0	250	300	350	400	500	500	500	600	700	750	900	1M					
275	600	500	4/0			0	00	000	4/0	250	300	400	400	500	500	600	600	750	900	1M						
300	700	500	250			00	00	000	250	300	350	400	500	500	600	600	700	800	900	1M						
325	800	600	300			00	000	4/0	250	300	400	500	500	600	600	700	750	900	1M							
350	900	700	300			00	000	4/0	300	350	400	500	600	600	700	750	800	900								
375	1M	700	350			000	000	4/0	300	350	500	500	600	700	700	800	900	1M								
400		900	350			000	4/0	250	300	400	500	600	600	700	750	900										

Conductors in overhead spans must be at least No. 10 for spans up to 50 feet and No. 8 for longer spans. See NEC, Sec. 225-6(a).
See Note 3 of NEC "Notes to Tables 310-16 through 310-19".

FIGURE 31-20. *Continued*

STUDENT ACTIVITIES

1. Define the Terms to Know in this unit.
2. Collect samples of cable and label them with the name, type of conductor, and type of insulation.
3. Use the formula W = VA to calculate the unknown value in the following problems.
   a. A 100-watt bulb in a 120-volt circuit draws _____ amperes.
   b. A 120-volt hair dryer is rated at 1200 watts. It will draw _____ amperes.
   c. A toaster draws 13 amperes at 120 volts. How many watts does it use?
   d. A circuit has three 100-watt lamp bulbs and a 1600-watt toaster. How many watts are used when all four appliances are in operation?
   e. Would a 15-ampere fuse carry the load described in item d? Why?
   f. A grain elevator motor has a 1-horsepower motor. A general rule of thumb is that 10 amperes at 120 volts equals 1 horsepower. How many amperes does this motor draw?
   g. A microwave oven draws 1400 watts at 12 volts on a 20-ampere circuit. It is 40 feet between the outlet and the service entrance panel. What size of copper wire is needed for the circuit? What size of aluminum wire?

4. Cover a horseshoe magnet with a piece of paper. Sprinkle iron filings over the paper. Observe the magnetic lines of force exhibited by the filings.
5. Read the electric meter at your home. Read it again two days later. How many kilowatthours of electricity did your family use in two days?
6. Clean an electric motor. Oil it according to instructions in the operator's manual.

SELF-EVALUATION

A. Multiple Choice. Select the best answer.

1. The major power source for stationary equipment in houses, farm buildings, and agribusiness is
   a. diesel fuel
   b. electricity
   c. gasoline
   d. steam
2. Electricity produces
   a. chemical changes
   b. heat and light
   c. magnetism
   d. all of these
3. A device that produces direct current by means of magnetism is
   a. a generator
   b. an alternator
   c. a turbine
   d. a pole
4. A device used to protect circuits that can be reset is a
   a. three-way switch
   b. ground fault interruptor
   c. fuse
   d. circuit breaker
5. A device in circuits to protect against human shock is a
   a. three-way switch
   b. ground fault interruptor
   c. fuse
   d. circuit breaker
6. Electricity is distributed to branch circuits by
   a. an electric meter
   b. an entrance head
   c. a service drop
   d. a service entrance panel
7. Tubes used to carry wires are called
   a. armored cable
   b. conduit
   c. nonmetallic sheathed cable
   d. pipe
8. A suitable wire for high temperature, high moisture locations is
   a. Type T
   b. Type THHN
   c. Type THW
   d. WVA
9. A cable consisting of No. 12 wire, one black, one red, one white, and a ground wire will be stamped
   a. 12-2
   b. 12-3
   c. 12-3 w/g
   d. 12-3 BRW
10. In order to run copper wire in a building 100 feet to a 10-ampere, 120-volt motor, and hold the voltage drop to 2 percent, the minimum size of wire must be
    a. No. 12
    b. No. 10
    c. No. 8
    d. No. 6
11. The first job in maintaining electric motors is to
    a. install them in series
    b. keep them clean
    c. change plugs frequently
    d. avoid frequent use
12. Motors that need lubricating
    a. require only small amounts of oil at one time
    b. must be oiled at least every three months
    c. have grease gun fittings
    d. generally need to be oiled in many places

B. Matching. Match the terms in column I with those in column II.

Column I

1. conductor
2. fluorescent
3. filament
4. insulator
5. amperes
6. volts
7. watts
8. magnet
9. armature
10. circuit

Column II

a. great resistance
b. electrical pressure
c. electrical energy or work
d. north and south poles
e. circle
f. rotating part of motor
g. rate of electrical flow
h. glowing element
i. glowing gas
j. carries electricity

C. Completion. Fill in the blanks with the word or words that will make the following statements correct.

1. Two deadly hazards of electricity are _____ and _____.
2. Electric meters measure the amount of electricity used and express it in _____.
3. Electrical circuits in a building are called _____ circuits.
4. Three types of cable used for wiring are _____, _____, and _____.
5. Strands of wire are gathered together in bundles to increase the _____ of the wire and therefore its conductivity.

UNIT

32 Installing Branch Circuits

OBJECTIVE

To run wires and safely install boxes, switches, outlets, and fixtures.

COMPETENCIES TO BE DEVELOPED

After studying this unit, you should be able to:

- Select electrical boxes, outlets, and switches.
- Install and replace switches, outlets, and fixtures.
- Install, extend, and modify branch circuits.
- Test electric circuits.

MATERIALS LIST

- ✓ Hand tools suitable for electrical wiring
- ✓ Cable ripper
- ✓ Wire stripper
- ✓ Test light
- ✓ 14-2 cable with ground
- ✓ 14-3 cable with ground
- ✓ Two switch boxes
- ✓ One octagon box
- ✓ One single-pole switch
- ✓ Two three-way switches
- ✓ One porcelain lamp holder
- ✓ One duplex receptacle
- ✓ Two switch covers
- ✓ One outlet cover
- ✓ Wire nuts
- ✓ Ground (gee) clip
- ✓ Electrical tape

TERMS TO KNOW

new work
fixture
wire nuts
receptacle
switch
duplex (double) receptacle
splitting the receptacle
single-pole switch
knockout
ground (gee) clip
continuity tester
continuity
three-way switch
switching (traveler) wires

A knowledge of electrical principles and wiring materials is important. The rules for the proper and safe installation of electrical devices and circuits given in the current National Electrical Code must be strictly observed at all times.

Cable and conduit are run through the floors, ceilings, and partitions of new buildings during construction. Such installations are called new work. Electrical codes may permit wiring to be placed in view in old work installations. This is due to the difficulty of concealing wires after a building is completed. In surface installations, however, the wiring must be protected from mechanical damage.

Utility buildings and other farm structures generally do not have finished interior walls. Such structures are relatively easy to wire at any time. However, the *National Electrical Code®* contains special requirements for such wiring.

WIRING BOXES

All wiring systems require the use of fuses or circuit breakers, and protected wires and boxes. The boxes are metal or plastic, figure 32-1. Boxes have several important functions:

- They hold the cable or conduit so stress cannot be placed on the wire connections.
- They are nailed, screwed, or clamped to the building to support switches, outlets, or fixtures. A fixture is a base or housing for light bulbs, fan motors, and other electrical devices.
- They contain all electrical connections made outside of fixtures.

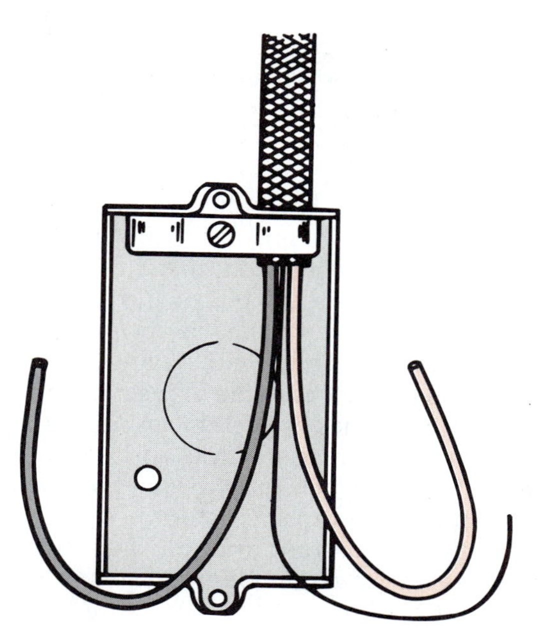

FIGURE 32-1. All wiring systems require the use of boxes.

Electrical boxes are available in rectangular or octagonal shapes and in various depths. Some switch and outlet boxes have removable sides so the boxes can be installed in gangs (series). Such boxes will hold multiple switches or outlets.

Large steel boxes are used for service entrances. They contain the main fuses or circuit breakers. They also contain fuses or circuit breakers for one or more branch circuits.

Certain basic rules apply to the wiring of all boxes. The *National Electrical Code®* and local building codes must be checked for specific regulations. Some basic requirements follow:

- The box must be fastened securely to the building.
- The cable or conduit must be clamped securely to the box.
- Cables running from box to box must run through the interior of the building's walls, floors, and ceilings, be secured by staples or clamps near each box, and secured as needed to prevent the cable from being accidentally caught and pulled.
- The box must be grounded if it is metal. The ground wire from the cable is attached to the box by a screw or by a grounding clip, figure 32-2.

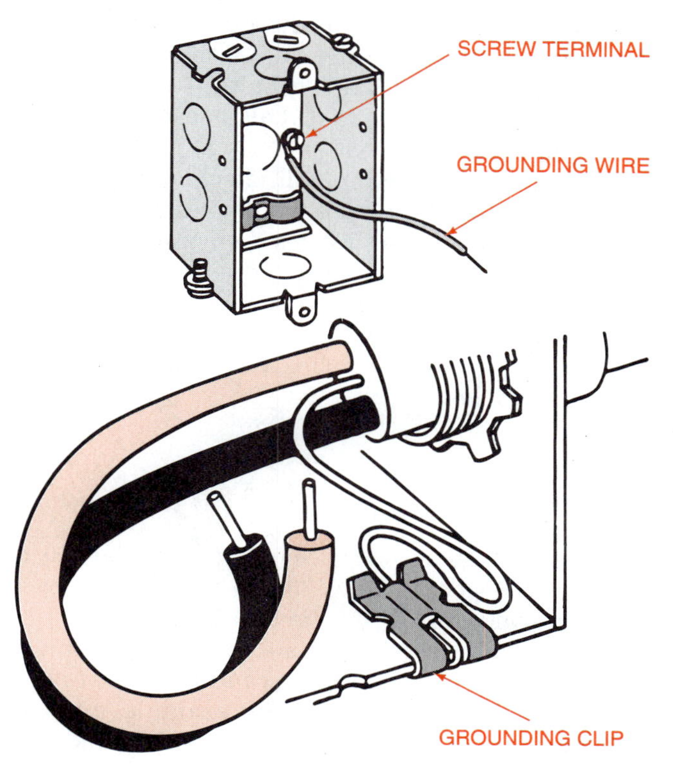

FIGURE 32-2. Ground wires are attached to metal boxes by means of a screw or grounding clip.

- Wires in boxes must be connected to each other by insulated solderless connectors called wire nuts. Bare areas are not permitted on wires except ground wires.
- Ground wires must be held together by a special metal clamp or a solderless connector.
- Wires must be attached to terminals of switches and receptacles by tightening no more than one wire under one screw or spring-clamp. A receptacle is a device for receiving electric plugs. A switch is a device used to stop the flow of electricity. If the receptacle or switch is equipped with special clamps, one wire may be inserted into each clamp provided.
- Positive or hot wires (black, red, or blue) must always be attached to yellow screws. Neutral wires (white) must always be attached to white screws. Ground wires (bare or green) are attached to green or uncolored screws.
- When a white wire must be used as a positive (hot) wire, the insulation showing in the box should be painted black or marked with black tape.

The most common type of receptacle is the duplex (double) receptacle wired so that both outlets are on the same circuit, figure 32-3. The current comes in on a black wire, flows from one screw through a metal strap to the other screw, and, when so wired, continues on to the next electrical box. The two receptacles in the duplex may be wired to two different circuits if the metal strap between the two screws is removed. This is called splitting the receptacle.

A standard switch box will safely hold four wires only. To determine the number of wires in a box, only the positive and neutral wires are counted, not the ground wires. In other words, one incoming cable with a black wire, a white wire, and a ground wire is counted as two wires. With an identical cable taking the current on to the next box, the total of four wires has been reached.

WIRING A SWITCH AND LIGHT

As stated, a switch is a device used to stop the flow of electricity. A single-pole switch is a switch designed to be the only switch in a circuit. The cable carrying current from the service entrance panel may come to an outlet box where a light is mounted or to one where a switch is mounted. For example, a power cable coming to a box where a light is mounted, and controlled by a single-pole switch, is shown in figure 32-4. To wire this circuit, cable with a white wire, black wire, and ground wire is used.

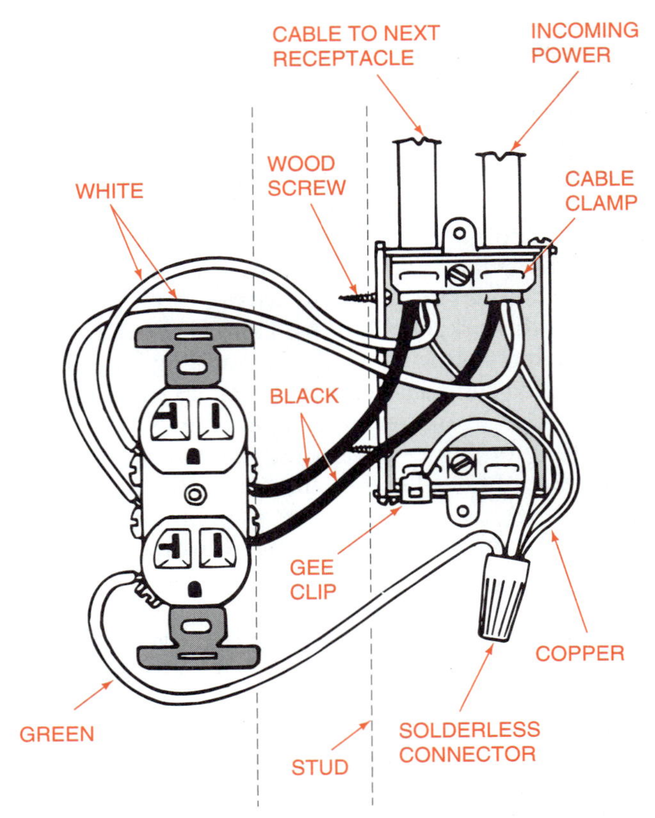

FIGURE 32-3. Typical duplex receptacle. (*Courtesy of The Pennsylvania State University, Department of Agricultural Education*)

The cable is prepared for insertion into the box by slitting 6 to 8 inches of the outside cable covering with a cable ripper, figure 32-5.

CAUTION:

- **Care must be taken to prevent damage to the insulation on the individual wires.**

The wires are then separated from the jacket and the excess jacket material is cut off.

Electrical boxes are provided with a knockout, or partially punched impression, that can be punched out and removed. The cable is pushed through this hole. Some boxes have cable clamps provided with the box, while others require the addition of a cable clamp, figure 32-4. The cable is inserted until $^{1}/_{16}$ inch of the cable jacket extends beyond the clamp. The clamp is then tightened and cable is run between the switch box and the light box. If there is more than one light in the circuit, a cable may be installed to carry current from one light box to the next. All cables must be clamped securely in the boxes and stapled or clamped to the structure within 12 inches of each box and at intervals of $4^{1}/_{2}$ feet or less.

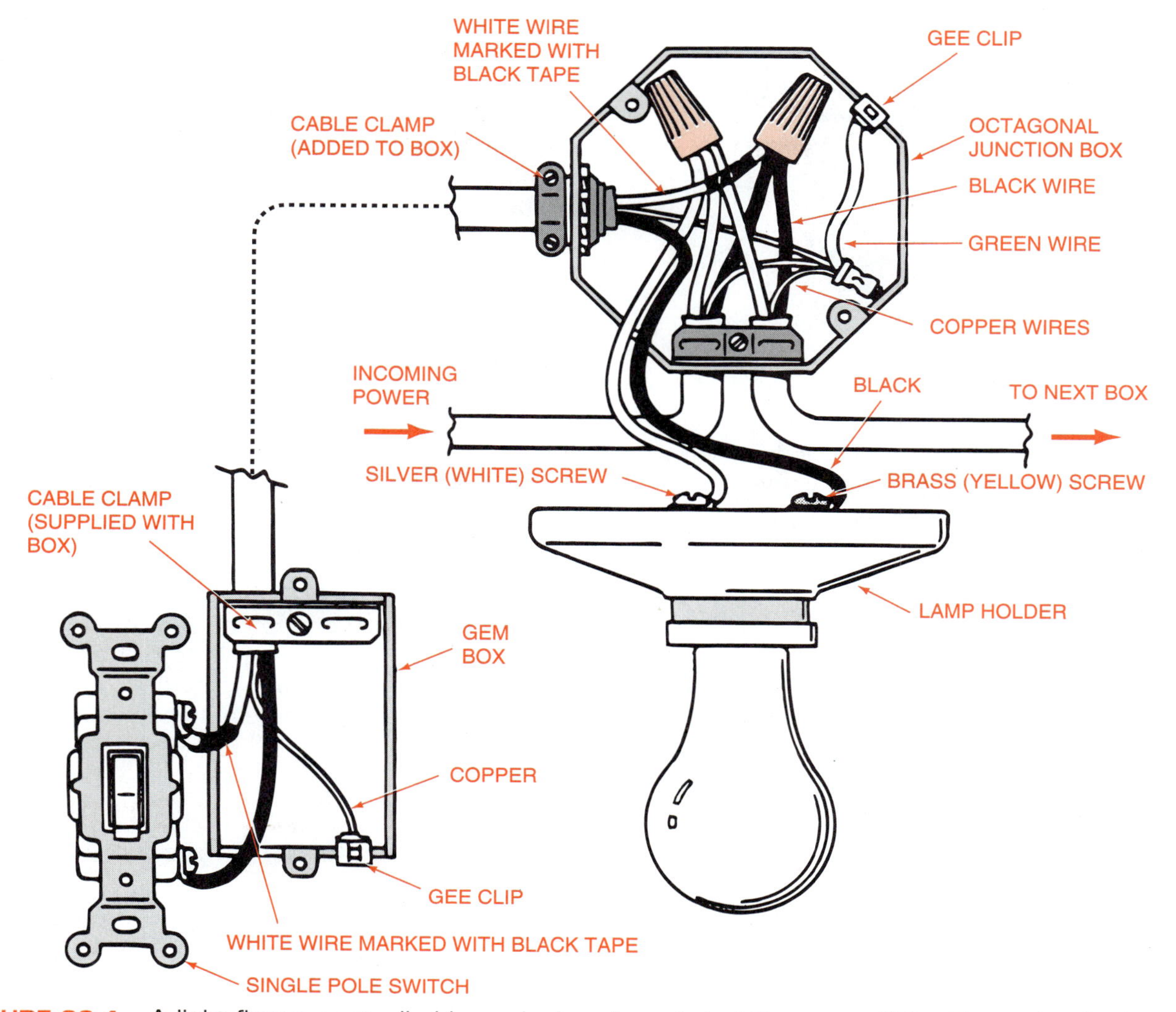

FIGURE 32-4. A light fixture controlled by a single-pole switch. (*Courtesy of The Pennsylvania State University, Department of Agricultural Education*)

To install the switch, a wire stripper is used to remove about 3/4 inch of insulation from the ends of individual wires, figure 32-6.

CAUTION:

- **Care must be taken not to nick the wire during this procedure.**

A round loop is then made in the end of each wire. Each loop is wrapped around a yellow screw in the direction the screw turns when being tightened. The screws are then tightened securely.

Switches are placed in hot wires only—never in neutral wires. Thus a piece of black tape is placed on the white wire to mark it as a hot wire. The ground wire is attached to the metal box with a ground (gee) clip. The switch is then screwed in place and a switch cover installed.

Wiring the lamp holder box is more difficult. However, the work progresses in a logical sequence according to the following steps.

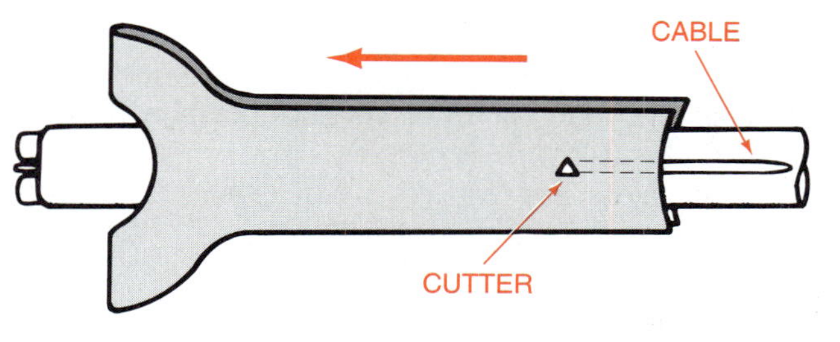

FIGURE 32-5. Seven inches of outer covering of nonmetallic sheathed cable is slit by a cable splitter. The individual wires are then pulled out of the jacket, and the loose jacket is cut off and discarded.

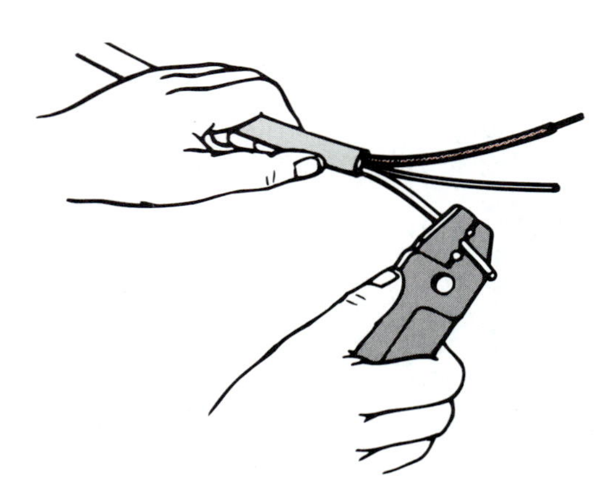

USING A WIRE STRIPPER

FIGURE 32-6. When preparing the ends of wires for screws, care must be taken to remove 3/4 inch of insulation without nicking the wire. Remove only 5/8 inch for wire nuts.

PROCEDURE

1. Strip the ends of all wires leaving 5/8 inch of wire exposed. The wires are now ready for connection with wire nuts.
2. Strip 5/8 inch from both ends of an 8-inch length of green wire and ground it to the box with a ground clip or screw.
3. Hold the four ground (green) wires so they form a bundle. Twist a wire nut onto the bundle and tighten it as much as possible with the hands. For large bundles, use a wire nut handle or pliers to tighten the wire nut.
4. Mark the white wire coming from the switch with black tape. This is now regarded as a black wire.
5. Attach the black wire from the switch to the yellow terminal of the fixture or lamp holder.
6. Cut an 8-inch piece of white wire and strip 5/8 inch of insulation from both ends. Attach one end to the white screw to the lamp holder.
7. Use a wire nut to connect the loose ends of the three white wires.
8. Use another wire nut to connect the ends of the three remaining black wires. Remember, the white wire with the black tape is treated as a black (hot) wire.
9. Check each connection for tightness by holding the wire nut and pulling hard on each of the wires. If one or more wires is loose, the entire bundle must be retightened.

The flow of electricity in this circuit proceeds from the service entrance box to the light box on the black wire of the incoming cable. It flows to the switch by way of the white wire marked with black tape. If the switch is in the "on" position, the current flows through the switch and through the black wire to the lampholder. Current flows through the bulb and back to the service entrance box by way of white (neutral) wires.

TESTING A CIRCUIT

After all lights and switches are wired, a circuit should be tested before turning on the circuit breaker or inserting the fuse. Use a continuity tester to be sure the circuit is not open. A continuity tester is a device used to determine if electricity can flow between two points. Continuity means connectedness. A circuit is open if there is a break or poor connection anywhere in it. The circuit should also be checked for shorts, or places where current (1) can get from the black wires to the ground wires or boxes or (2) can bypass the bulb and flow directly to the neutral wires. The final step is to test all boxes to ensure that they are properly grounded.

The following procedure may be helpful in testing a circuit.

PROCEDURE

1. Place switches in the "on" position.
2. Remove all bulbs.
3. Test the circuit by connecting to the wires at the fuse or circuit breaker box before the cable is wired into the fuse block or circuit breaker. Using an ohmmeter, touch one lead to the black wire and the other to the white wire of the cable. There should be no reading. This indicates no continuity. That is, there are no points where electricity can flow from the black wires to the white wires.
4. Install a good light bulb in the last fixture of the circuit. Repeat the test in step 3. There should be continuity through the filament of the bulb as indicated by a meter response. This indicates that current can flow through all black wires, through the bulb, and back through the white wires.
5. Test to see if each metal box in the circuit is grounded. Continuity is checked with an ohmmeter by touching one test lead to the box and the other to the ground wire. There should be a reading.

When all cover plates and fixtures have been installed, the cable can be wired into the fuse block or circuit breaker.

CAUTION:

- Do not add electricity to a circuit until a qualified electrician has checked it and all required inspections are complete.

WIRING LIGHTS WITH THREE-WAY SWITCHES

A three-way switch is a switch that, when used in a pair, permits a light or receptacle to be controlled from two different locations, figure 32-7. The wiring of a circuit with three-way switches will vary according to the location of the incoming power cable, the light, and the two switches. However, certain basic rules apply regardless of the location of components.

A three-way switch has three terminals. The current feeds into a common terminal identified by its dark screw. The current flows out through either one of the remaining two terminals identified by light-colored screws. The light-colored screw that carries the current is determined by the position of the switch at any point in time. The two wires attached to the light-colored screws are called the switching (traveler) wires. A wire running from the common terminal or dark screw of the second switch is attached to the yellow screw on the light fixture.

Basic rules for wiring three-way switches follow.

PROCEDURE

1. Connect the white wire of the incoming power cable to the silver terminal of the light fixture.
2. Connect the black wire of the incoming power

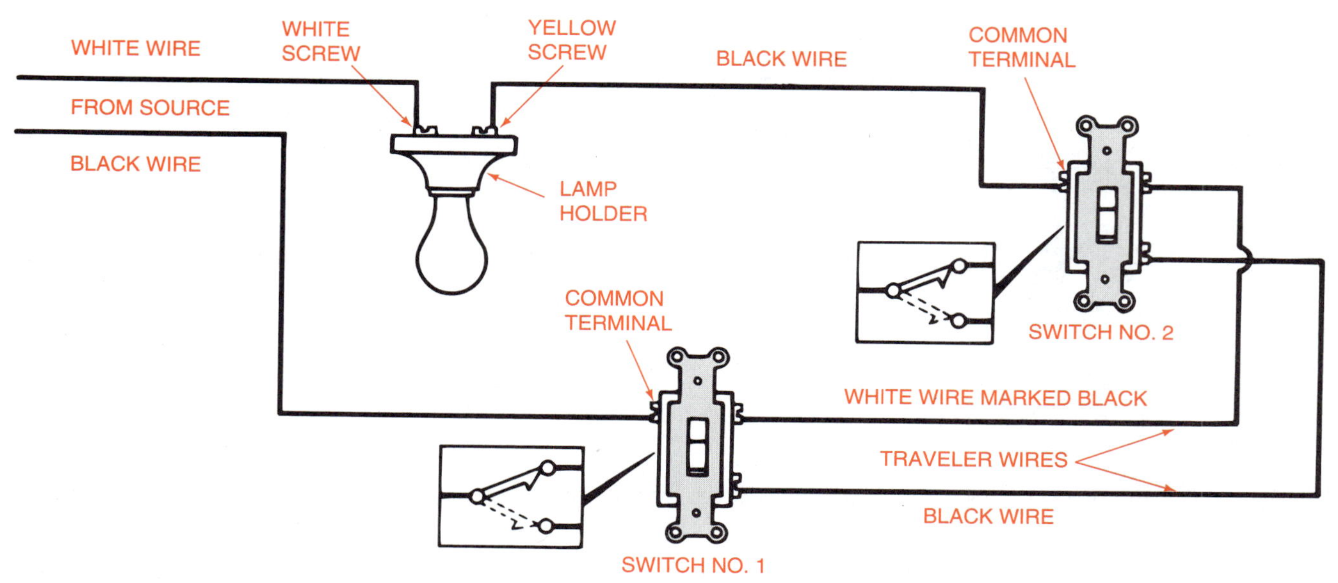

FIGURE 32-7. A three-way switch circuit.

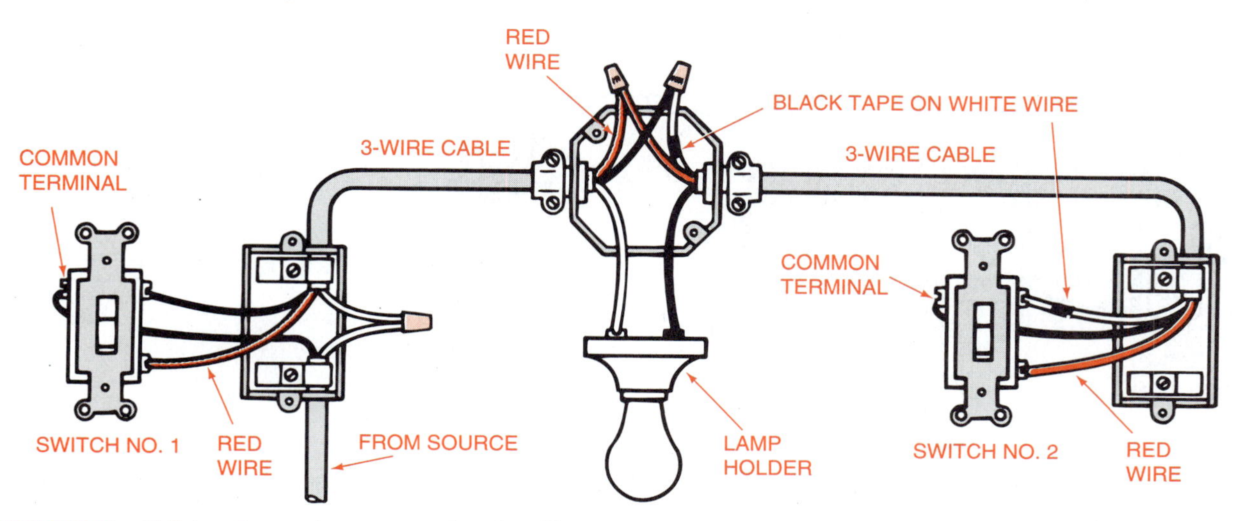

FIGURE 32-8. Wiring for a three-way circuit with current coming to one switch and operating a light located between two three-way switches.

cable to the dark screw (common terminal) of the switch.

3. Connect a black wire from the dark screw of the other switch to the yellow terminal of the light fixture.
4. Run a pair of traveler wires between the light-colored terminals of the two switches.

Three-wire cables are useful when wiring circuits controlled by three-way switches, figures 32-8 to 32-10. Such cables typically have a black, red, white, and ground wire. The white wire may be used as a hot wire if it is connected to a switch or to a black wire and is clearly marked with black tape.

Black and red wires are generally used as the traveler wires in a three-way switch circuit. If metal boxes are used, the ground wire grounds the boxes.

Wiring Lights Using Four-Way Switches

A four-way switch is a switch connected in the pair of traveler wires between two three-way switches, figure 32-11. Any number of four-way switches can be added to a three-way switch circuit to accommodate the total number of switch locations needed. The four-way switch permits the electricity to be changed from one traveler to the other depending on the position of the switch lever, figure 32-12.

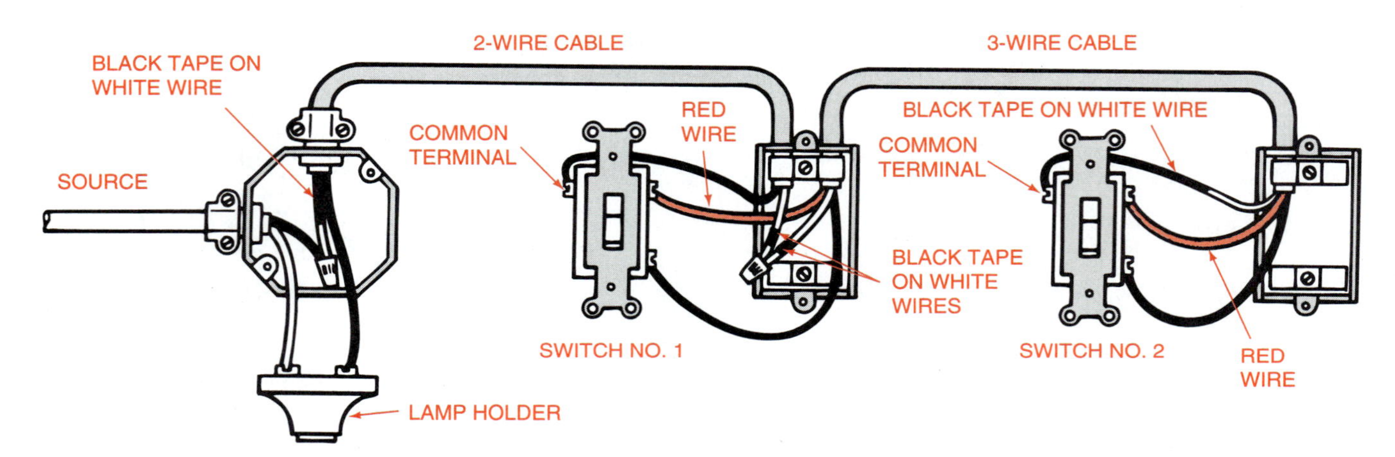

FIGURE 32-9. Wiring for three-way switches with current coming in at the light.

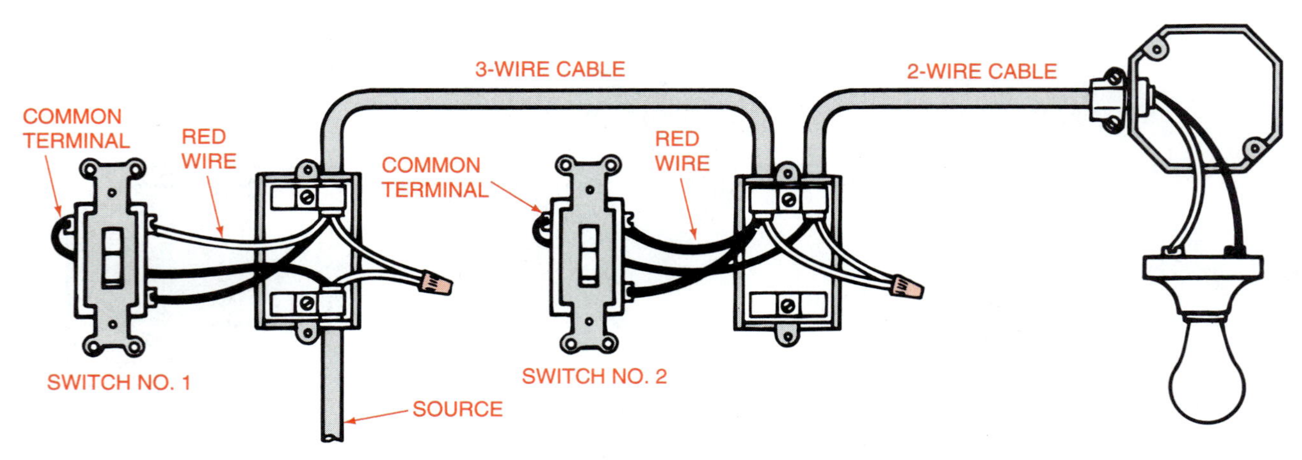

FIGURE 32-10. Wiring for three-way switches with the light at the end of the run and current coming into one switch.

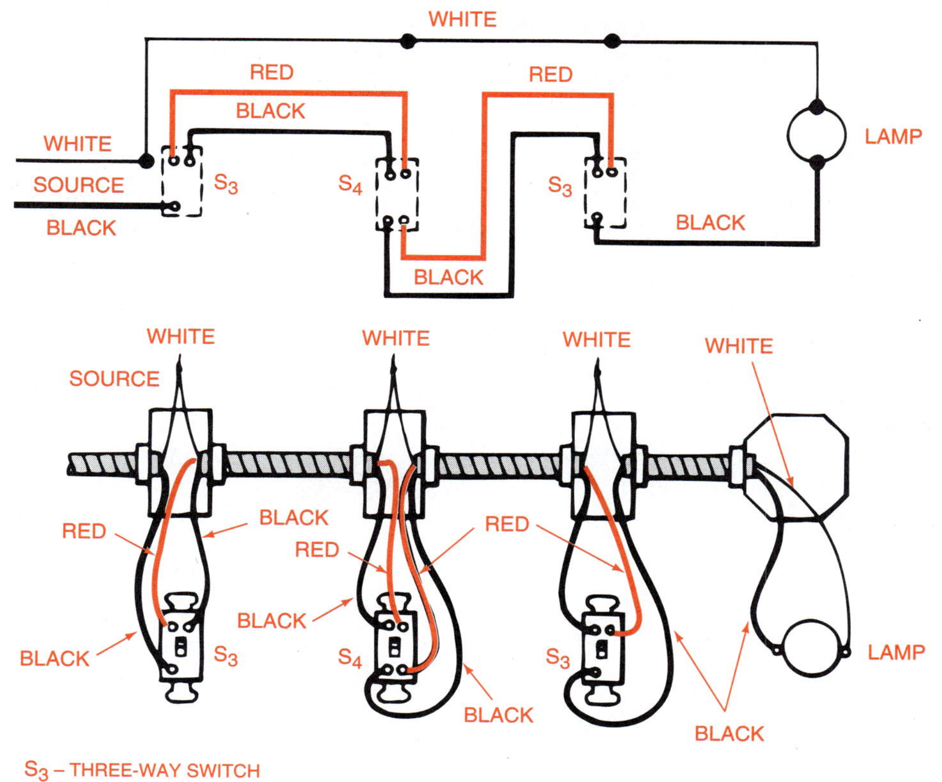

FIGURE 32-11. A circuit using a four-way switch. Four-way switches (S_4) can only be installed in the traveler wires between two three-way switches (S_3). (*Courtesy of R. Mullin,* Electrical Wiring—Residential, 12th ed., *Delmar Publishers, 1996*)

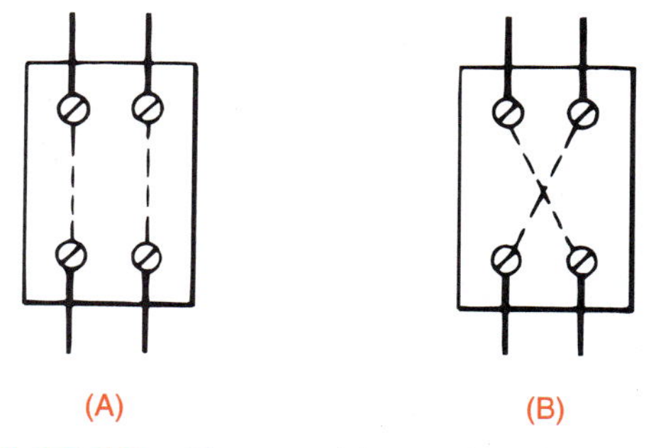

FIGURE 32-12. Two positions of a four-way switch. (*Courtesy of R. Mullin,* Electrical Wiring—Residential, 12th ed., *Delmar Publishers, 1996*)

STUDENT ACTIVITIES

1. Define the Terms to Know in this unit.
2. Wire a circuit with an outlet.
3. Wire a circuit with several outlets.
4. Wire a circuit with a light and single-pole switch.
5. Wire a circuit with a light and two three-way switches.

SELF-EVALUATION

A. Multiple Choice. Select the best answer.

1. All electrical connections in a circuit are made
   a. in boxes or fixtures
   b. by screws
   c. with solder
   d. with tape
2. All metal electrical boxes must
   a. be grounded
   b. be securely fastened
   c. secure the cable or conduit
   d. all of these
3. Neutral wires are attached to screws colored
   a. white
   b. green
   c. silver
   d. yellow
4. The device that receives electrical plugs is a
   a. box
   b. cap
   c. circuit breaker
   d. receptacle
5. White wires used as positive wires must be
   a. connected to black wires
   b. connected to fixtures
   c. stripped of all insulation
   d. taped or painted black
6. A properly wired circuit will be
   a. grounded
   b. open
   c. shorted
   d. all of these
7. In three-way switch circuits, electricity passes from one switch to the other by
   a. traveler wires
   b. neutral wires
   c. common terminals
   d. none of these
8. Three-way switch circuits usually include
   a. two-wire cables
   b. three-wire cables
   c. grounded boxes
   d. all of these

B. Matching. Match the terms in column I with those in column II.

Column I

1. old work
2. duplex
3. solderless connector
4. switch
5. positive or hot
6. red wire
7. neutral wire
8. circuit protector
9. electrical contact
10. switching wire

Column II

a. used in place of black wire
b. always white
c. traveler
d. fuse
e. extensions to existing systems
f. receptacle
g. continuity
h. stops current
i. black wire
j. wire nut

UNIT

33 Electronics in Agriculture

OBJECTIVE

To utilize electronic principles and applications in agricultural settings.

COMPETENCIES TO BE DEVELOPED

After studying this unit, you should be able to:

- Cite major historical landmarks in electronics development.
- State basic principles of electronics.
- Name typical applications of electronics in agriculture.
- Discuss strategies for maintaining electronics equipment.

MATERIALS LIST

- ✓ Publications with photos of electronics applications
- ✓ Poster materials
- ✓ Bulletin board materials
- ✓ Reference materials on electronics

TERMS TO KNOW

lodestone
conductance
resistivity
insulators
ohm
siemen
reciprocal
resistor
potentiometer (pot)
rheostat
series circuit
parallel circuit
series-parallel circuit
electrical meter
analog meter
digital meter
ammeter
volt meter
multimeter
inductance
henry (H)
inductor
capacitance
capacitor
plates
dielectric
cycle
hertz (Hz)
sine wave
oscilloscope
semiconductor
silicon
diodes
transistor
thyristor
integrated circuit
optoelectric device
photoconductive cell
photovoltaic cell (solar cell)
photodiode
phototransistor
light-emitting diode (LED)
power supply
rectifier
voltage regulator
filter
voltage multiplier
over-voltage protection circuits

fuses
circuit breakers
amplifier
video amplifier
oscillator
digital computer
digital
digits
binary system
binary
chips
bit
floppy disk (diskette)
magnetic tape
hard disk
mainframe
microcomputer
minicomputers
supercomputers
throughput
power takeoff (PTO)
infrared light sensor
biosensors
claw
viscosimeters
microprocessors
pallet

Electronics is certainly the most pervasive and influential technology of the twentieth century. Set your digital watch; use your computer; click on your television; listen to your stereo; start your car, truck, or tractor; ride a motorcycle or run a boat—you have used electronics, figure 33-1.

FIGURE 33-1. Electronic devices touch everyday occurrences in our lives. (*Courtesy of Radio Shack*)

Radio, television, audio and video recorders, computers, satellites, factory automation, space flight, stealth aircraft, and traffic control devices all rely on electronics, figure 33-2. Homes, schools, supermarkets, department stores, gas stations, banks, and offices are more efficient and productive because of electronic equipment.

DEVELOPMENT OF ELECTRONICS

Computers and telephones in homes, scanners and monitors in hospitals, audiovisual equipment in classrooms, electron microscopes in laboratories, electronic feeding and milking systems on farms, and communication satellites in space are electronic realities of today. However, it took thousands of years of searching and discovery to reach our present level of electronics technology.

The Ancients

As early as 600 B.C. Thales of Miletus, Greece, discovered that when a piece of amber was rubbed with a cloth, the amber would attract bits of feathers and pith from plants. Similarly, Thales believed there was a connection between electricity and magnetism. He was one of the seven wise men of Greece.

Some people of early civilizations tried electric shocks from electric eels in efforts to cure stiffening diseases of the body. This suggests a realization that somehow electronic impulses are used in the body to send messages from the brain to muscles. We know this today as the nervous system in the body.

Archaeologists have discovered strange pots in ancient Arabic ruins, which seem to indicate someone was electroplating jewelry thousands of years ago!

The first compass is believed to have been used on a cart used on the high Asian plains. The cart utilized a piece of lodestone in a wood spindle to keep an arm pointing north. It also had a counter that ticked off units of distance traveled. Lodestone is a magnetized piece of black iron oxide.

Pioneers from 1600 to 1900

From 1600 to 1900, many discoveries were made that make today's electronics possible. Some of these discoveries and pioneers include:

1. Numerous discoveries about static electricity and magnetism—William Gilbert

FIGURE 33-2. Electronics provide the communications core of our society. (*Courtesy of Lanier Worldwide, Inc.*)

2. The discovery that lodestones attract iron—William Gilbert
3. The discovery that only certain substances conduct electricity—Steven Gray
4. The observation that substances with like charges repel and opposite charges attract—Charles du Fay
5. The laws of attraction and repulsion—Charles Augustin de Coulomb
6. The development of the first battery—Allesandro Volta, figure 33-3
7. The demonstration that lightening is electricity—Benjamin Franklin
8. The recognition of different charges as negative and positive—Benjamin Franklin
9. The use of a battery to power the first electric light—Humphry Davy
10. The first machine to make electricity from mechanical energy—Michael Faraday
11. The first electric motor—Joseph Henry
12. The first telegraph—Samuel B. Morse
13. The telephone—Alexander Graham Bell
14. The incandescent light bulb—Thomas Edison

After the 1800s, research and development became more characterized as group activities and company projects. Whom to credit with new discoveries became more difficult as science entered the twentieth century.

Twentieth-Century Discoveries

The twentieth century was ushered in with Guglielmo Marconi's first transatlantic wireless message in 1901. From then on, discoveries and development mushroomed. A few highlights are:

1. The deForrest Audion (vacuum radio tube)—led to radios in homes
2. Superheterodyne receiver—gave the Allies a great advantage in World War I
3. Discovery of radio waves from outer space—beginning of radio astronomy
4. Television—developed in the 1930s; became a household item in the 1950s
5. Radar—a method of detecting distant objects by analysis of high-frequency radio waves received from their surfaces
6. Sonar—a system using transmitted and reflected acoustic waves to detect and locate submerged objects
7. Transistor—a three-terminal semiconductor device used for amplification, switching, and detection
8. Printed circuit—electrical conducting material placed on a flat board by printing techniques
9. Direct dialing—dial from a home telephone to anywhere in the United States and to other countries
10. Molecular electronics—placing of material only a millionth of an inch thick in electrical circuits
11. Radio/T.V. satellites—receive and forward electronic signals between points thousands of miles apart on Earth
12. Videocassette recording—records visual and audio signals on magnetic tape for playback on television
13. Laser—light energy focused in a very small and powerful beam
14. Computers—performs high-speed mathematical or logical calculations or processes and prints information
15. Electron microscope—views areas as small as one-billionth of an inch
16. Radiotherapy—treatment of diseases with radiation

FIGURE 33-3. The first battery was built by Volta, who demonstrated that certain metal and chemical combinations cause the flow of electricity.

PRINCIPLES OF ELECTRONICS

Magnetism and Amperes-Volts-Watts

Before proceeding with the remainder of this unit, the reader should review the material in Unit 31. There, some basic principles of electricity and magnetism are presented. Further, safety practices with electricity and volts, amperes, and watts are discussed. Elements of electrical wiring are covered in both Units 31 and 32. Additional information on electricity and electronics is also given in the units on arc welding and small engine technology.

Resistance

Resistance was defined earlier in the text as the tendency of a material to prevent electricity flow. Conductance is the opposite. It is a measure of how well a material allows electricity to pass through it.

The resistivity of a material is defined as the resistance of a wire made from a material that is 1 foot long and 1 mil (0.001 inch) in diameter at 20°C. Low-resistance materials are called insulators. The symbol for resistance is R.

Resistance is measured in ohms. An ohm is the amount of resistance that allows 1 ampere of current to flow when 1 volt is applied. The symbol for ohm is the Greek letter omega (Ω).

The ability of a material to carry electrons or electricity is conductance. The symbol for conductance is G. The unit for measuring conductance is the siemen. The symbol for siemen is S.

Resistance is the reciprocal, or opposite, of conductance. The relationship is illustrated by the following formulas:

$$R = 1/G \text{ and } G = 1/R$$

Example #1
Problem: What is the conductance of a 250-ohm resistor?
Formula: G = 1/R
Solution: G = 1/250 or G = 0.004 siemen

Example #2
Problem: What is the resistance of a conductor with a conductance of 600 siemens?
Formula: R = 1/G
Solution: R = 1/600 or R = 0.0017 ohm

Resistors

A resistor is a device that offers specific resistance to a current. Resistors are the most commonly used components in a circuit. They come in various sizes and shapes to meet various conditions, figure 33-4. They may be designed with fixed resistance values or may permit resistance to be varied.

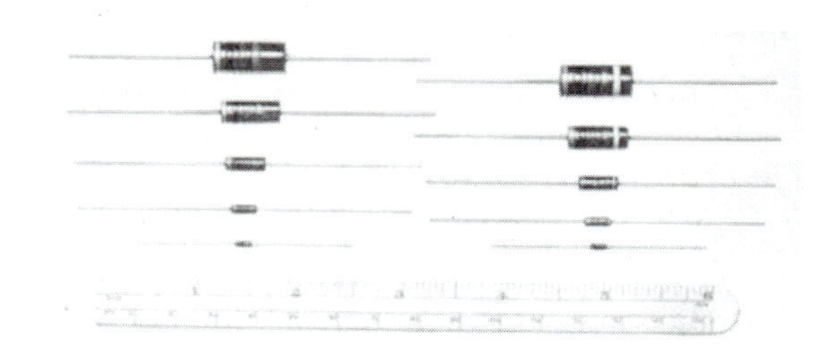

FIGURE 33-4. Resistors are built to meet various conditions. (*Courtesy of P. Crozier,* Introduction to Electronics, 2nd ed., *Delmar Publishers, 1987. Courtesy of the Allen-Bradley Company*)

Fixed Resistors. Fixed resistors are manufactured to deliver their rated resistance within certain tolerances above or below that rating. Lower tolerance resistors tend to cost more. Resistors are generally available with tolerance ratings of plus or minus 20, 10, 5, 2, and 1 percent.

Variable Resistors. Variable resistors have three terminals, figure 33-5. Two terminals form the ends of a resistor strip, and the third terminal attaches to a conductor on the end of a rotating shaft. When the shaft is rotated, the conductor moves to various points along the resistor to provide the desired level of resistance between the center terminal and each of two side terminals. A variable resistor used to control voltage is called a potentiometer (pot). A variable resistor used to control current is called a rheostat, figure 33-6.

Resistor Identification. The small size of some fixed resistors prevents printing the resistance value and tolerance on its case. Therefore, they are color coded. The industry uses a standard system established by the Electronics Industries Association (EIA).

The color code is represented by a system of colored bands. The first band is the band closest to an

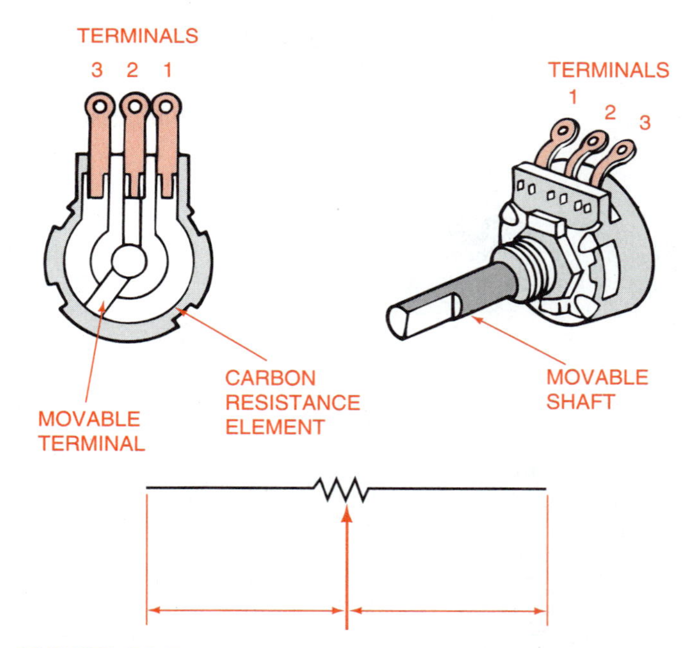

FIGURE 33-5. Variable resistors permit the selection of resistance required for the job. (*Courtesy of P. Crozier,* Introduction to Electronics, 2nd ed., *Delmar Publishers, 1987*)

FIGURE 33-6. The rheostat is a variable resistor used to control current. (*Courtesy of Vishay Intertechnology, Inc.*)

end of the resistor. The first band represents the first number of the resistance value, the second band represents the second number, and the third band represents the number of zeros to add to the first two digits. The fourth band represents the tolerance of the resistor, figure 33-7. There are certain exceptions to these color interpretations that the technician should also learn.

Use the top table in figure 33-7 to see why a resistor with bands that are brown, green, red, and silver would have a resistance value of 1500 ohms. Its tolerance level is 10 percent, figure 33-8. There is also a number system for use on larger resistors.

Types of Resistive Circuits. Resistive circuits may be of three types. The series circuit provides a single path for current flow. A parallel circuit provides two or more paths, and a series-parallel circuit combines the two, figure 33-9.

Checking Resistors. A resistor may be checked with an ohm meter. However, to check a resistor, it must be disconnected at both ends from the circuit.

Electrical Meters

An electrical meter is a device that measures the activities of invisible electrons and provides a visual interpretation of these activities. Meters may be analog or digital. The analog meter has a graduated scale with a pointer, figure 33-10. The digital meter provides a direct numerical readout, figure 33-11.

Some suggestions for effective use of meters are:

1. Contact meters observing correct polarity—red terminal positive and black terminal negative.
2. Before using an analog meter, adjust the pointer to zero on the scale.

Resistor Color Codes

Two-Significant-Figure Color Code

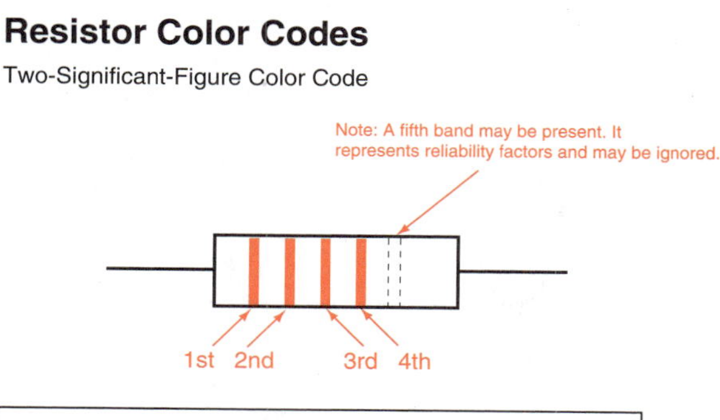

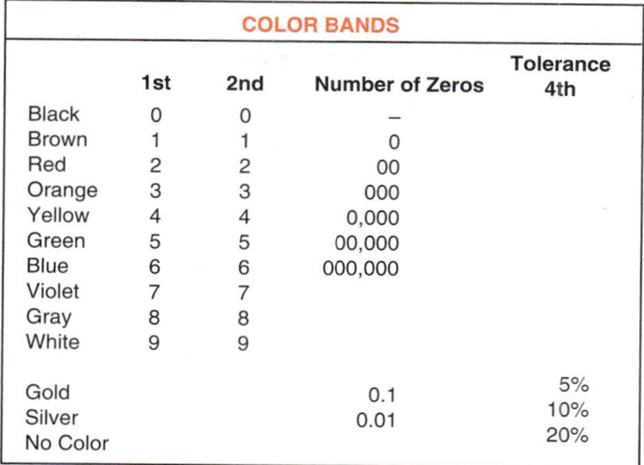

COLOR BANDS				
	1st	2nd	Number of Zeros	Tolerance 4th
Black	0	0	–	
Brown	1	1	0	
Red	2	2	00	
Orange	3	3	000	
Yellow	4	4	0,000	
Green	5	5	00,000	
Blue	6	6	000,000	
Violet	7	7		
Gray	8	8		
White	9	9		
Gold			0.1	5%
Silver			0.01	10%
No Color				20%

Three-Significant-Figure Color Code*

COLOR BANDS				
	1st	2nd	3rd	Multiplier 4
Black	0	0	0	–
Brown	1	1	1	0
Red	2	2	2	00
Orange	3	3	3	000
Yellow	4	4	4	0,000
Green	5	5	5	00,000
Blue	6	6	6	000,000
Violet	7	7	7	
Gray	8	8	8	
White	9	9	9	
Gold				0.1
Silver				0.01

All one-percent tolerance resistors.

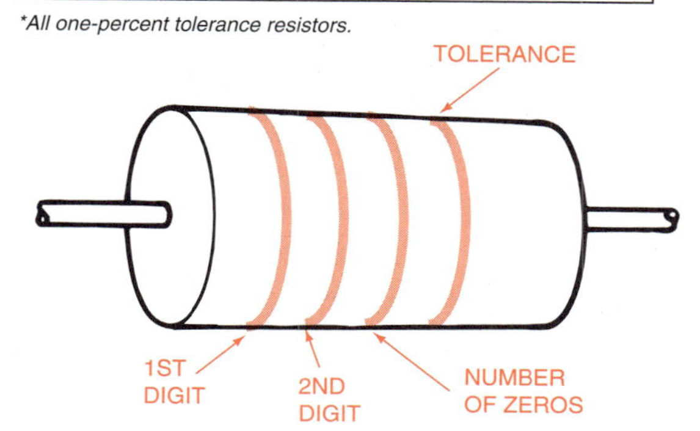

FIGURE 33-7. The color bands on a resistor indicate the resistance and tolerance. (*Courtesy of P. Crozier, Introduction to Electronics, 2nd ed., Delmar Publishers, 1987*)

3. To measure current (amperes) with an ammeter, open the circuit and place the meter into the circuit in series.

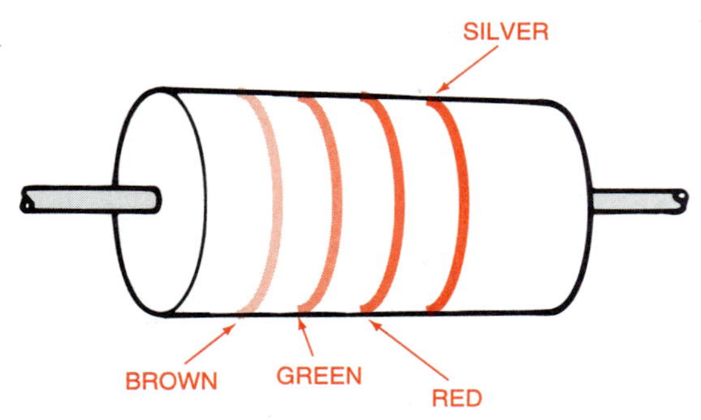

FIGURE 33-8. The color bands indicate this resistor has 1500 ohms of resistance and a tolerance of 10 percent. (*Courtesy of P. Crozier,* Introduction to Electronics, 2nd ed., *Delmar Publishers, 1987*)

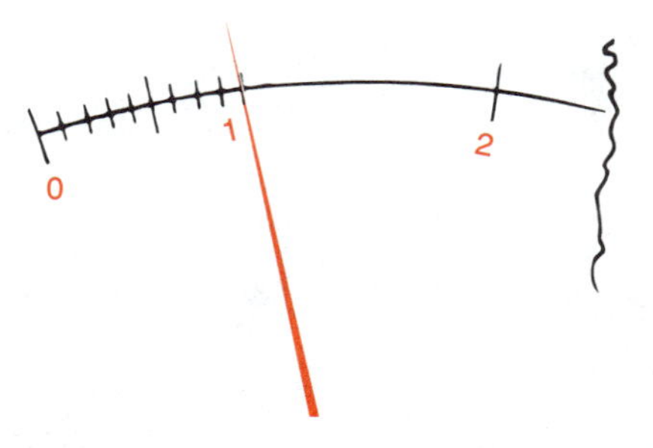

FIGURE 33-10. Example of an analog meter. (*Courtesy of P. Crozier,* Introduction to Electronics, 2nd ed., *Delmar Publishers, 1987*)

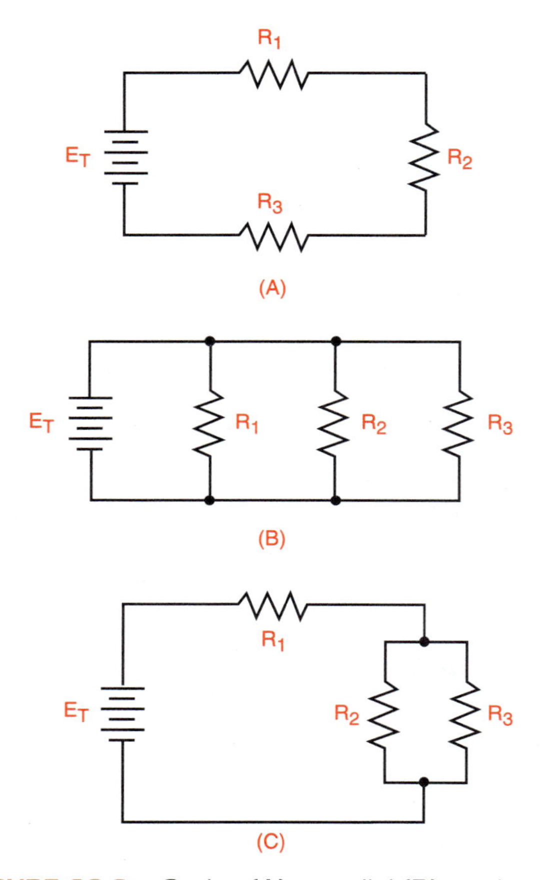

FIGURE 33-9. Series (A), parallel (B), and series-parallel (C) circuits. (*Courtesy of P. Crozier,* Introduction to Electronics, 2nd ed., *Delmar Publishers, 1987*)

4. To measure voltage with a volt meter, place the meter leads in the circuit in parallel.
5. To measure resistance, take the resistor, bulb, motor, or conductor out of the circuit. Zero the ohm meter, then connect the ohm meter prods or clips.

FIGURE 33-11. Example of a digital meter. (*Courtesy of P. Crozier,* Introduction to Electronics, 2nd ed., *Delmar Publishers, 1987. Courtesy of Beckman Industrial Corp.*)

CAUTION:

- **An error in connecting an ammeter or volt meter will damage the instrument!**

6. When using a multimeter, which is a combination volt-ohm-milliammeter, be sure the instrument is set for the appropriate function and scale before attaching the leads, figure 33-12.
7. Study the scale on the meter to be certain you understand what each mark and space represents, figure 33-13.
8. Amperes in an alternating current circuit may be measured without opening the circuit utilizing a clamp-on meter which measures magnetism and reads out in amperes, figure 33-14.

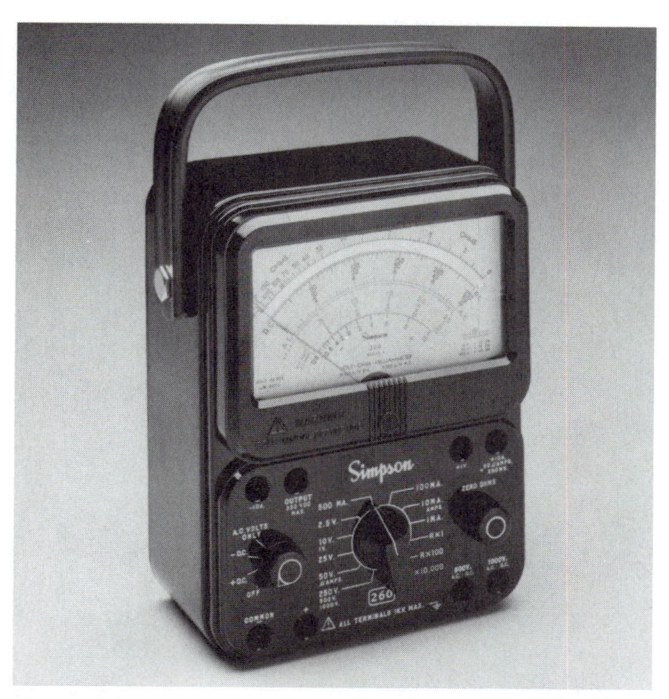

FIGURE 33-12. The volt-ohm-milliammeter (VOM) is a multimeter. (*Courtesy of Simpson Electric Company*)

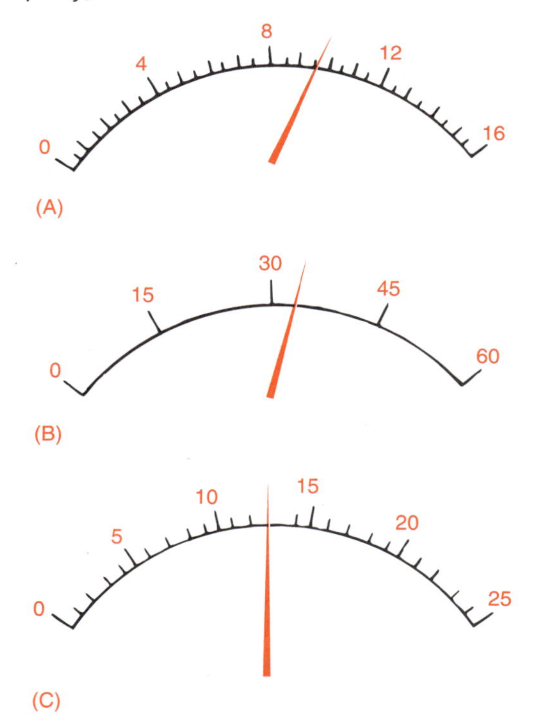

FIGURE 33-13. Scales must be read carefully to obtain accurate data from meters. (*Courtesy of P. Crozier,* Introduction to Electronics, 2nd ed., *Delmar Publishers, 1987*)

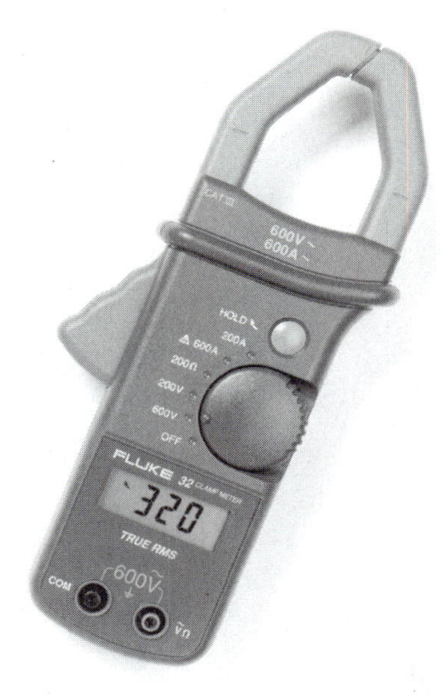

FIGURE 33-14. A clamp-on ammeter senses magnetism and converts the value to amps. (*Courtesy of The Fluke Corporation*)

Inductance

The ability to draw energy from a source and store it in a magnetic field is called inductance. For example, if current increases in a coil, the magnetic field expands. If the current decreases, the magnetic field collapses. However, the collapsing magnetic field induces a voltage back into the coil which maintains the current flow.

The Henry (H). The unit of inductance is called the henry (H). A henry is the amount of inductance required to induce an electromotive force (EMF) of 1 volt when the current in a conductor changes at the rate of 1 ampere per second. The symbol for inductance is L.

Inductors. An inductor is a device that provides inductance. Both fixed and variable inductors are available, figure 33-15.

Capacitance

The ability of a device to store electrical energy in an electrostatic field is capacitance. A capacitor is a device that possesses a specific amount of capacitance.

A capacitor is made of two conductors separated by an insulator. The conductors are called plates and the insulator is called dielectric, figure 33-16. Capacitors are generally made by rolling the plates and dielectric into a cylinder, figure 33-17.

CAUTION:

- **A capacitor can hold a charge indefinitely if not discharged. Therefore, to avoid electrical shock, never touch the leads of a capacitor until they have been touched together to discharge the unit.**

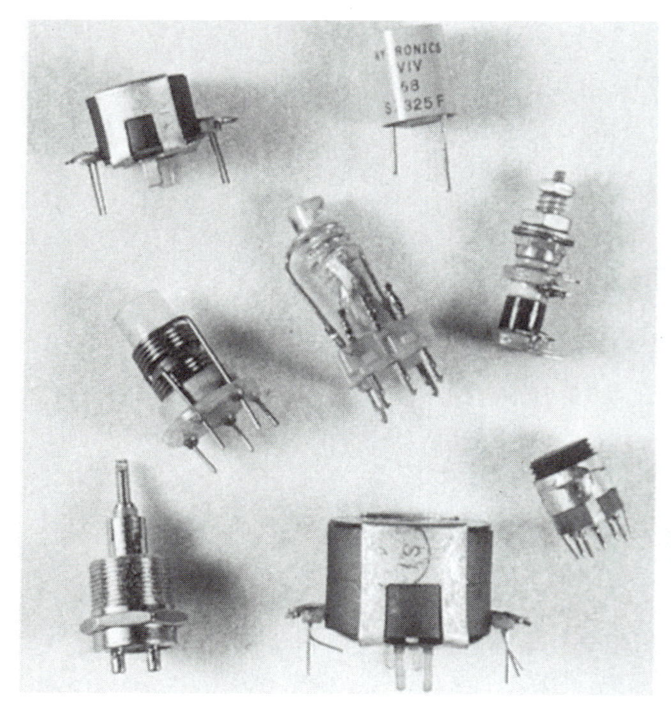

FIGURE 33-15. Some types of inductors. (*Courtesy of P. Crozier,* Introduction to Electronics, 2nd ed., *Delmar Publishers, 1987*)

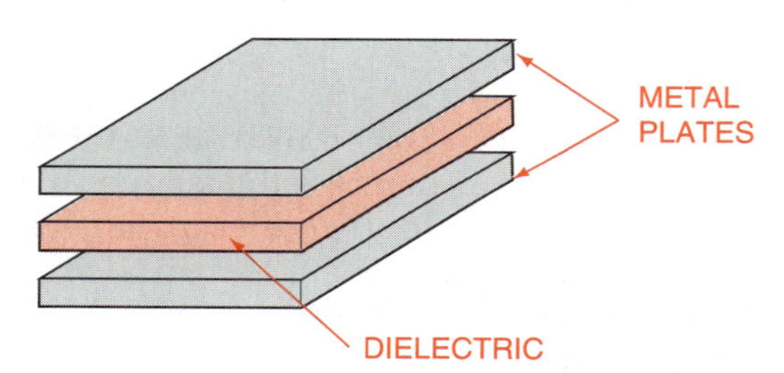

FIGURE 33-16. A capacitor consists of plates (conductors) separated by a dielectric (insulator). (*Courtesy of P. Crozier,* Introduction to Electronics, 2nd ed., *Delmar Publishers, 1987*)

Direct and Alternating Current

Current from batteries flows in one direction and is called direct current (DC). Direct current can also be produced with DC generators. On the other hand, alternating current (AC) reverses in direction of flow at rapid intervals caused by rotations of the armature of the generator or alternator creating the current.

Cycle. The flow of current through one positive thrust followed by flow through one negative thrust is called a cycle. One cycle is created by one turn of the armature in the generator, figure 33-18. One cycle per second is defined as a hertz (Hz). The wave created by the flow of current through one cycle is called a sine wave. AC current provided by power suppliers in the United States is generated at 60 cycles per second. Therefore, appliances operated on such current must be rated 60 Hz.

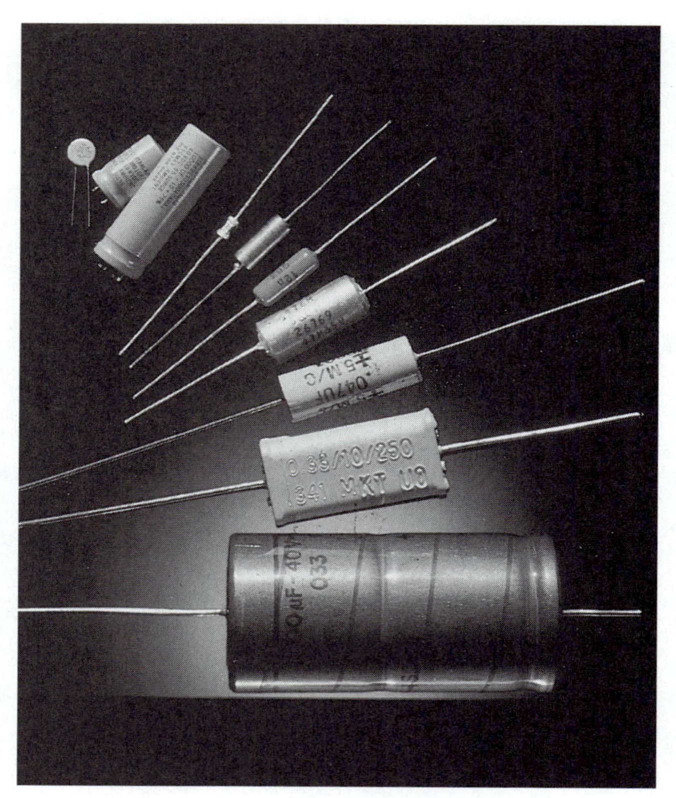

FIGURE 33-17. Various types of capacitors. (*Courtesy of Philips Components*)

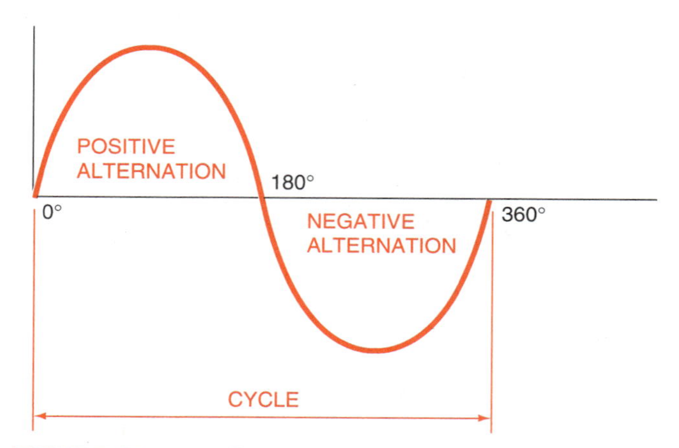

FIGURE 33-18. One turn of the generator armature produces one cycle. (*Courtesy of P. Crozier,* Introduction to Electronics, 2nd ed., *Delmar Publishers, 1987*)

Oscilloscopes. An oscilloscope is the most valuable piece of test equipment for working on electronic equipment and circuits. It provides a visual display of the following information: frequency and

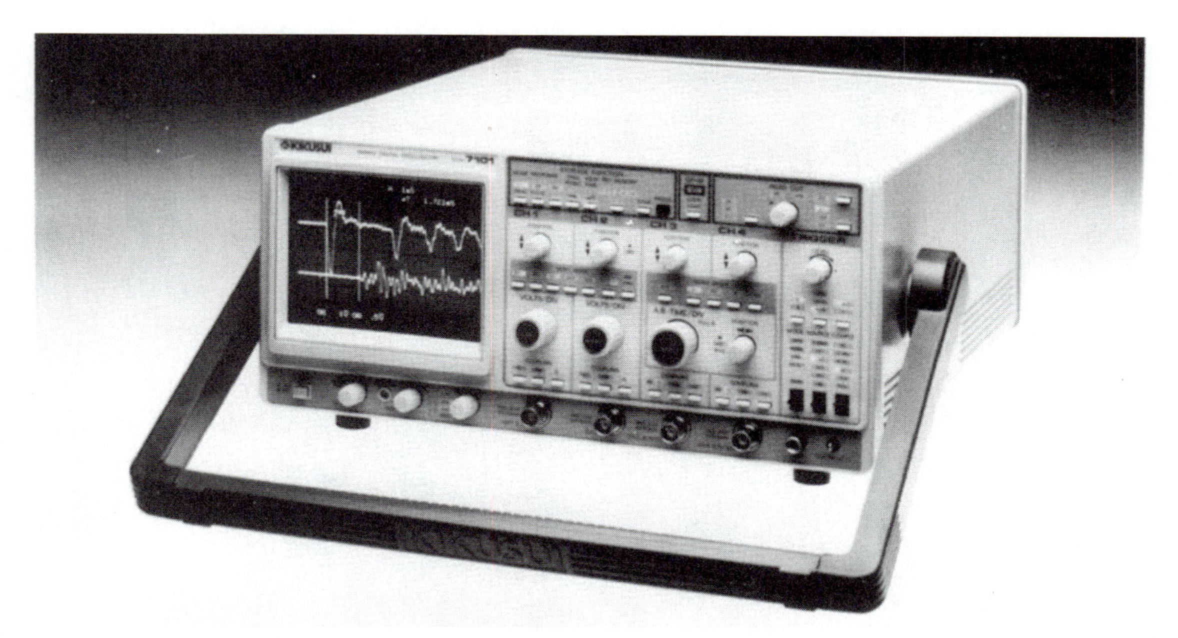

FIGURE 33-19. The oscilloscope shows visual displays of electronic signals. (*Courtesy of P. Crozier,* Introduction to Electronics, 2nd ed., *Delmar Publishers, 1987. Courtesy of Kikusui International Corp.*)

duration of signal, phase relationships, signal wave form, and amplitude (strength) of the signal, figure 33-19.

Semiconductors

A semiconductor is a material with characteristics that fall between those of insulators and conductors. Three pure semiconductor elements are carbon (C), germanium (Ge), and silicon (Si). Silicon and germanium are the most commonly used semiconductors. Silicon is the most used and is abundant in sand, quartz, agate, and flint.

Common uses of semiconductors include diodes (used to rectify), transistors (used to amplify), thyristors (used for switches), and integrated circuits (used to switch or amplify).

Semiconductors must undergo highly technical processes to modify their structure to achieve their performance in modern technology. The primary use of semiconductor devices is to control voltage or current for a desired result. Their advantages are, small size and weight, low power consumption, great reliability, instant operation, and economic mass production.

Diodes. Diodes are the simplest type of semiconductor. They allow current to flow in one direction only. A popular use of diodes is on automobile alternators to change AC current to behave like DC current since it permits current to flow in one direction only, figure 33-20.

Transistors. A transistor is a three-element, two-junction device used to control electron flow. By varying the amount of voltage applied to the three elements, the amount of current can be controlled for amplification, oscillation, and switching, figure 33-21.

Thyristors. A thyristor is a semiconductor device used for electronically controlled switches. They are used to apply power to a load or remove power from a load. Thyristors can also be used to regulate power or adjust power to a load. Examples are dimmer controls for lights and motor speed controls. Silicon-controlled rectifiers (SCRs) are the best known of the thyristors, figure 33-22.

Integrated Circuits. An integrated circuit is a miniature electronic circuit. The integrated circuit includes diodes, transistors, resistors, and capacitors. Their main advantage is smallness. An integrated circuit may be contained in a chip of semiconductor material $^1/_8" \times {}^1/_8"$ in size.

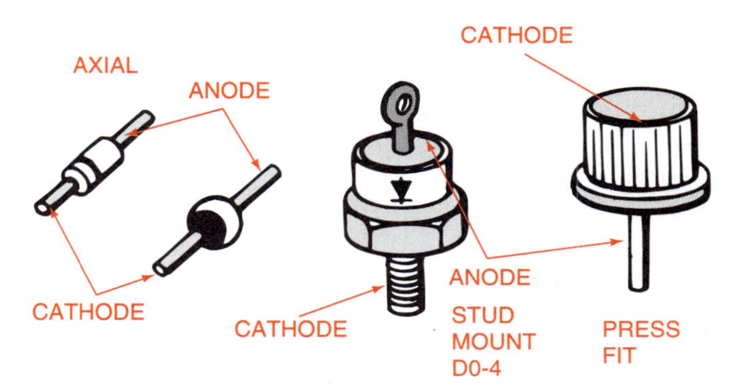

FIGURE 33-20. Common diode packages. (*Courtesy of P. Crozier,* Introduction to Electronics, 2nd ed., *Delmar Publishers, 1987*)

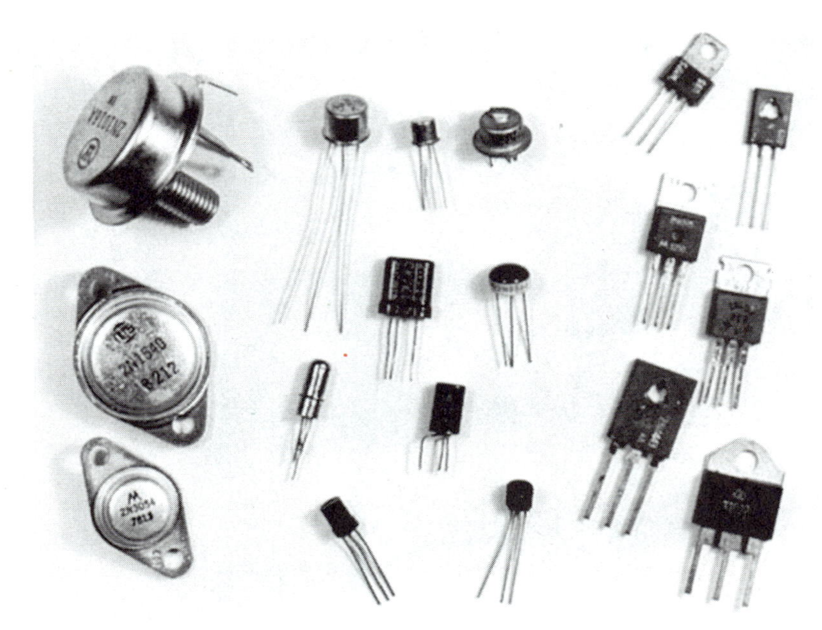

FIGURE 33-21. Typical transmitter packages. (*Courtesy of P. Crozier,* Introduction to Electronics, 2nd ed., *Delmar Publishers, 1987*)

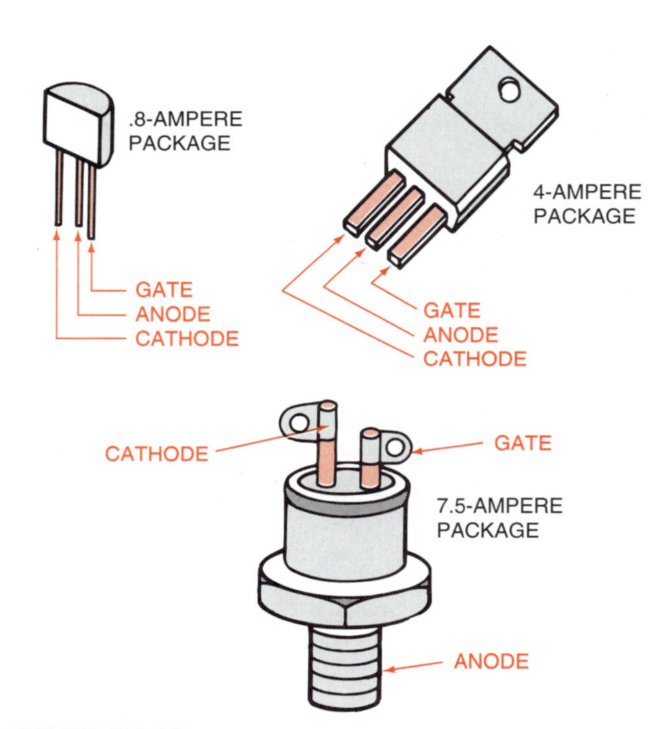

FIGURE 33-22. Silicon-controlled rectifiers, the best known of the thyristors. (*Courtesy of P. Crozier,* Introduction to Electronics, 2nd ed., *Delmar Publishers, 1987*)

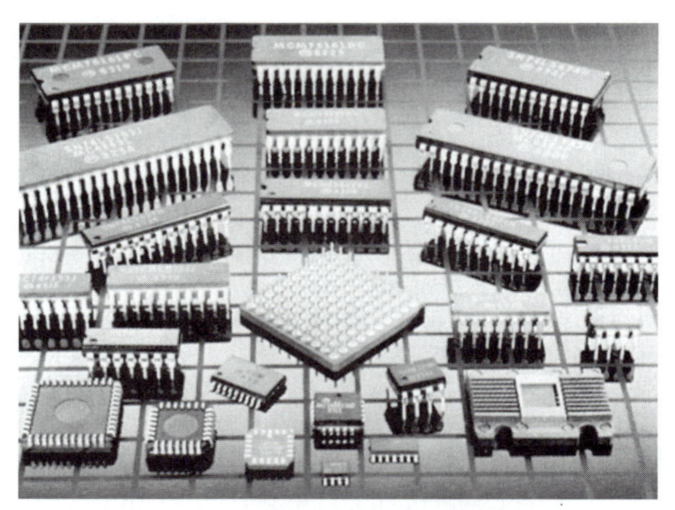

FIGURE 33-23. Examples of integrated circuits packages. (*Courtesy of P. Crozier,* Introduction to Electronics, 2nd ed., *Delmar Publishers, 1987. Courtesy of Motorola Semiconductor Products, Inc.*)

Computer systems, once the size of rooms, are now desktop accessories because of integrated circuits. Integrated circuits cannot be repaired, so faulty or damaged circuits are replaced in their entirety, figure 33-23.

Optoelectric Devices. An optoelectric device is designed to interact with light energy in the visible, infrared, and ultraviolet ranges. They include light-detecting, light-converting, or light-emitting devices. The manufacture of the semiconductor diode will determine the light wavelength of a particular device.

The photoconductive cell is a light-sensitive device. It is made from cadmium sulfide (CdS) or cadmium selenide (CdSe), figure 33-24. It is useful for low-light and relatively high-voltage applications. Photoconductive cells are popular automatic switches for all-night lights.

FIGURE 33-24. A photoconductive cell. (*Courtesy of P. Crozier,* Introduction to Electronics, 2nd ed., *Delmar Publishers, 1987*)

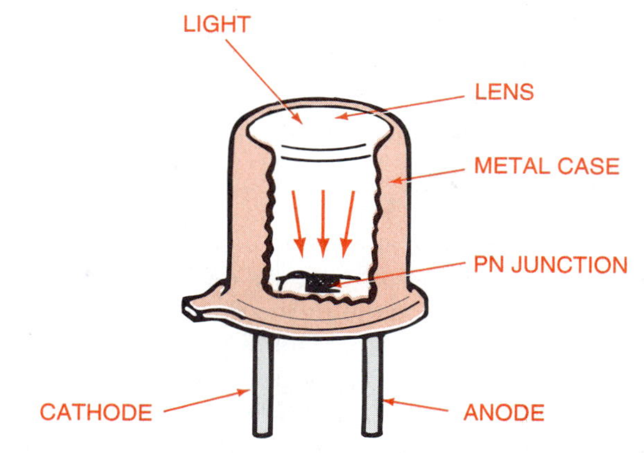

FIGURE 33-26. A photodiode package. (*Courtesy of P. Crozier,* Introduction to Electronics, 2nd ed., *Delmar Publishers, 1987*)

The photovoltaic cell (solar cell) converts light energy into electrical energy, figure 33-25. Solar cells are popular for powering pocket calculators, light meters for photography, sound track decoders for motion picture projectors, and battery chargers on satellites. The photodiode is similar to the photovoltaic cell except that it is used to control current flow, not generate it, figure 33-26.

The phototransistor is packaged like a photodiode except it has three leads. The third lead permits adjustment of the turn-on point. Phototransistors can produce higher output than photodiodes. They are used for phototachometers, photographic exposure controls, flame detectors, object counters, and mechanical positioners.

The light-emitting diode (LED) produces light when subject to a current flow, figure 33-27. These devices provide digital readouts for many types of electronic equipment.

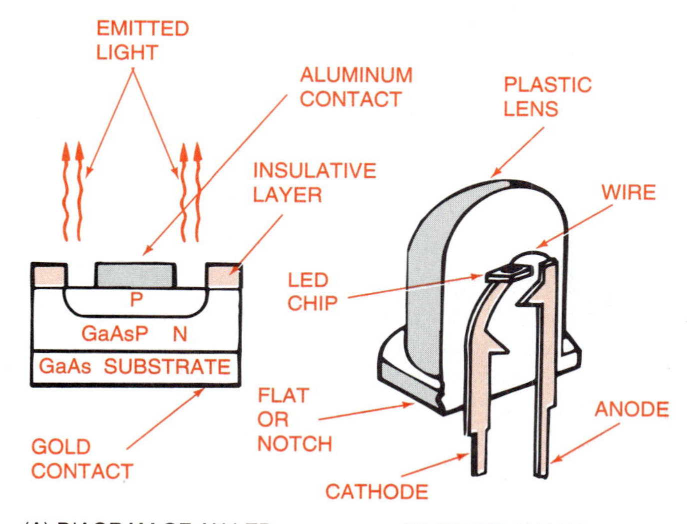

FIGURE 33-27. Construction of a light-emitting diode (LED). (*Courtesy of P. Crozier,* Introduction to Electronics, 2nd ed., *Delmar Publishers, 1987*)

Power Supplies. A power supply is a device that provides electricity of the proper type and amount to circuits. The main functions are to increase or decrease the incoming AC voltage by means of a transformer, convert AC to DC current by using a rectifier, and controlling voltage output with a voltage regulator. Further, a filter converts pulsating DC voltage to smooth voltage. DC voltage may be stepped up, if desired, with a voltage multiplier. Additionally, circuit protection devices, such as over-voltage protection circuits, fuses, and circuit breakers, may be used.

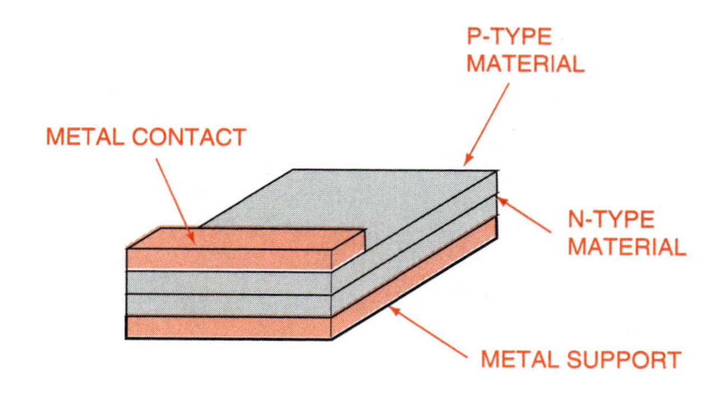

FIGURE 33-25. Construction of a solar cell. (*Courtesy of P. Crozier,* Introduction to Electronics, 2nd ed., *Delmar Publishers, 1987*)

Amplifiers. An amplifier is an electronic circuit used to increase the amplitude of an electronic signal. A circuit may be designed to function as a voltage amplifier or a current amplifier. A video amplifier is used to amplify pictorial information.

Oscillators. An oscillator is a nonrotating device for producing alternating current. Oscillators are used extensively in radios, televisions, communications systems, computers, industrial controls, and timekeeping devices. Most electronics equipment cannot function without oscillators.

Microcomputers

Computers have become the backbone of the communications revolution. As early as the 1930s, progress was being made in the development of the modern digital computer. The digital computer is a device that automatically processes data using digital techniques. Digital means using digits. Digits are the numbers 0 through 9. Computers record, store, and manipulate data using the binary number system. Known simply as the binary system, it contains only two digits, 0 to 1. Binary means two parts or components.

Computer Chips. Data are stored, processed, displayed, or printed in or from tiny electronic circuits in computers called chips. One chip holds hundreds of thousands of bits of information. The term bit comes from the *b* in *b*inary and the *it* in dig*it*. One bit is equal to one digit, one number, or a yes/no answer to a question. One computer can have many chips.

Computer Disks and Tapes. Computers can manage huge volumes of information by storing data on a flexible disk called a floppy disk (diskette). They can also use magnetic tape, like that used in tape recorders, to store data. Another popular device for storing data is a rigid disk or hard disk found with most microcomputers.

Types of Computers. The early computers occupied entire rooms. They operated on relatively large vacuum tubes that burned out quickly. Therefore, the early computers were large, heavy, generated a lot of heat, and failed frequently. All the computer capability of an institution was placed in a large unit called a mainframe computer.

With the development of transistors and computer chips, the microcomputer has revolutionized business, government, education, and the home. The microcomputer is a self-contained small computer suitable for desktop, portable, or on-board use, figure 33-28. It may work independently or in a system with other computers. Microcomputers are found in kitchen appliances, automobiles, trucks, tractors, boats, aircraft, offices, factories, and elsewhere.

While microcomputers are very popular, industry, universities, government, and other large systems still use minicomputers, mainframe computers, and super computers. Minicomputers are intermediate in capacity, while supercomputers have enormous capacity.

FIGURE 33-28. The microcomputer is a common accessory today. (*Courtesy of Radio Shack*)

ELECTRONICS IN AGRICULTURE

Electronics provide the mechanisms for high-technology applications in every conceivable area of agriculture/agribusiness and renewable natural resources.

In the Field

In land leveling, drainage, irrigation, and construction, laser equipment is essential for determining accurate elevations and sightings. In field planting and harvesting equipment, electronic sensors, lights, sound alarms, and digital readouts inform operators of proper equipment operation, figure 33-29.

Planting equipment utilizes electronics to monitor seed drop, seed spacing, and planting rate per

FIGURE 33-29. Laser equipment used to gauge elevation. (*Courtesy of Caterpillar, Inc.*)

acre. Similarly, fertilizer application rate and placement, acreage covered, ground speed, and rate per acre can be monitored, figure 33-30.

Grain combines have devices to monitor and/or control speed of the machine over the ground, reel speed, cylinder speed, engine speed, and throughput or rate of flow of grain and straw through the combine. Warning devices tell of header overload, cylinder clearance and load, walker area overload, excessively dirty or unthreshed grain, and grain tank fullness. Management information obtained by electronic devices include bushels per acre of yield, bushels per minute being harvested, and percent of moisture in the grain, figure 33-31.

Tractor performance is also monitored by electronic devices. Digital readouts may provide engine RPM, percent of wheel slippage, forward ground speed, axle load, drawbar load, power takeoff (PTO) speed, and PTO load. Infrared light sensors detect heat or motion.

Livestock and Dairy Operations

Feedlots and milking areas are becoming more high technology every year. Biosensors, or devices that measure processes of the body, are used on cattle and other animals to indicate body temperature, breeding status, feed requirements, disease problems, and supplying additional feed if appropriate, figure 33-32.

Heat sensors in the claw or cup of milking machines can signal the possibility of mastitis and other udder infections. Similarly, viscosimeters or flow meters can indicate thickening of milk or presence of blood clots which indicate mastitis problems. Various temperature-sensing devices can monitor milk cooling operations to assure proper cooldown for maximum milk quality.

Robots for putting on milkers are making milking a totally automated process. Cameras, visual imagery, computers, and various sensors all guide the activities of robots.

Humidistats gauge moisture buildup from livestock, thermostats sense temperature changes, and gas analyzers determine when ammonia is being produced from improper conditions in poultry and livestock manure. Such equipment typically controls heating, cooling, and ventilation in buildings.

Grain and Feed Processing

Special microprocessors of computers monitor and control the drying, grinding, and handling of grains

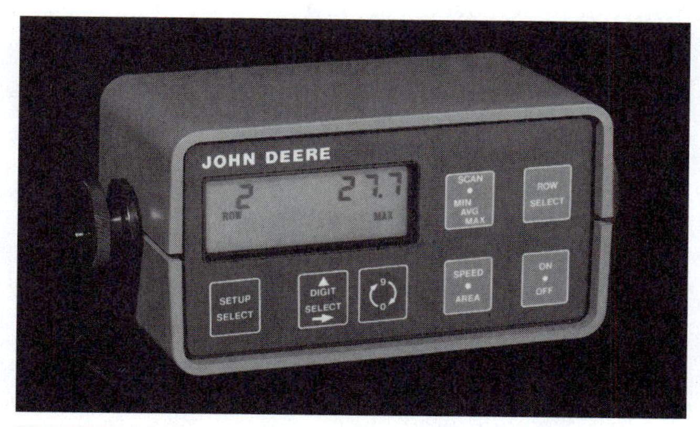

FIGURE 33-30. Monitoring planting procedures. (*Courtesy of Deere & Company, Moline, IL*)

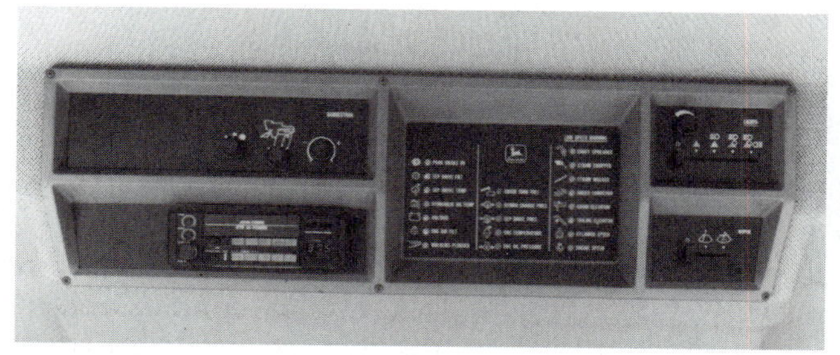

FIGURE 33-31. Monitoring harvesting procedures. (*Courtesy of Deere & Company, Moline, IL*)

FIGURE 33-32. Biosensors on livestock. (*Courtesy of E. Cooper,* Agriscience: Fundamentals and Applications, 2nd ed., *Delmar Publishers, 1986, and the National FFA*)

for human and livestock consumption. Computers determine proportions for rations and control the machinery that mix and transport the ingredients and final products.

Laboratories

Soil analysis, fertilizer ingredient proportioning, milk quality control, chemical analysis, pesticide monitoring, and other laboratory functions are now automated. Entire feed processing plants are typically fully automated today.

Aquaculture

Aquaculture uses electronics for depth finding, fish finding, and navigational aids. Automated water analysis equipment for fish culture, pollution monitoring, and water media management will be increasingly important in the future, figure 33-33.

FIGURE 33-33. Water analysis equipment for agriculture and water management. (*Courtesy of E. Cooper,* Agriscience: Fundamentals and Applications, 2nd ed., *Delmar Publishers, 1986*)

Food Distribution

Food distribution centers for modern supermarkets are highly computerized and automated. Such centers have huge buildings laid out in cells for pallet storage. A pallet is a portable platform for storing or moving cargo.

Food and household items needed by a given store are ordered by way of a computer. The computer directs the automatic forklift unit to the appropriate item in the warehouse, the item is picked up, and taken to and loaded on the appropriate truck at the dock. The process is repeated until all items are loaded. The computer automatically adjusts the inventory in the warehouse and the store to reflect the change.

Smart House

Homes are now being equipped to accommodate the physically handicapped. Verbal commands and other automated mechanisms can turn water on and off, open doors, lock windows and doors, and adjust bed positions. Such equipment can be designed to do almost any task that people with handicaps find difficult to do for themselves. The use of electronic devices can alert homeowners or the police of intruders and have become valuable home security devices.

STRATEGIES FOR MAINTAINING ELECTRONICS EQUIPMENT

Electronics components contain very tiny, complex, and complicated circuitry. Many components are difficult or impossible to repair once they are manufactured. Therefore, service of electronic devices should only be done by specially trained personnel. Such individuals have analytical and testing equipment to isolate the problem and indicate the faulty component. The faulty component or module can generally be replaced with ease.

General maintenance procedures for all electronic equipment should include:

1. Keep the item free of dust, dirt, metal objects, and liquid spills.
2. Wipe off routinely with a dry, clean cloth.
3. Lubricate equipment as specified by the manufacturer.
4. Maintain a cool, dry environment for computers, copiers, sound systems, and most electronic devices.
5. Read the operator's manual and follow the instructions carefully.
6. Call the supplier, factory representative, or qualified repair technician when needed—do not tamper with electronic devices.

SUMMARY

Electronics have become a major part of our everyday lives. Electronics permit us to communicate, automate many processes, and become more productive. They increase our health and safety, and improve our quality of life. The career outlook in electronics, and its applications in agriculture, is excellent.

STUDENT ACTIVITIES

1. Define the Terms to Know in this unit.
2. Ask your teacher to divide the class into groups of five. Each group should select a recorder and a chairperson. Have each group spend five minutes listing every activity or process in which electronics equipment is used. Have the recorders report the lists to the class.
3. Develop a collage depicting electronic devices.
4. Prepare and present a class report on one major discovery that contributed to today's electronics technology.
5. Develop an Electronics Hall of Fame poster. Include persons, discoveries, and dates for major landmark activities in the discovery and development of electricity and electronics.
6. Research the manufacturing and service procedures for a major type of electronic component, such as semiconductors, power supplies, or computer chips.
7. Design a security system for your home utilizing modern electronic devices.
8. Construct a chart or bulletin board illustrating uses of electronics in agriculture/agribusiness and renewable natural resources.

SELF-EVALUATION

A. Multiple Choice. Select the best answer.

1. Static electricity was described as early as
   a. 10,000 B.C.
   b. 5,000 B.C.
   c. 600 B.C.
   d. 1600 A.D.
2. Archaeologists have found evidence of electroplating dating to
   a. 1607 A.D.
   b. 1240 A.D.
   c. 4,000 B.C.
   d. thousands of years ago
3. The conductance (G) of a 100 ohm resistor is
   a. 0.01 siemen
   b. 0.10 siemen
   c. 10 siemens
   d. 100 siemens
4. The resistance (R) of a conductor with a conductance of 300 siemens is
   a. 0.0033 ohm
   b. 0.0013 ohm
   c. 0.0010 ohm
   d. 0.0001 ohm
5. Unless discharged after power is turned off, a serious electric shock can occur from a
   a. transistor
   b. rheostat
   c. potentiometer
   d. capacitor
6. Which substance is not used in semiconductor?
   a. carbon
   b. germanium
   c. silicon
   d. titanium
7. Devices that interact with light are called
   a. chips
   b. optoelectric
   c. silicon saddles
   d. transistors
8. LED stands for
   a. low exterior density
   b. limited emission device
   c. lithium emulsion drawndown
   d. light-emitting diode
9. One may convert AC current to DC with
   a. a transformer
   b. an oscilloscope
   c. a rectifier
   d. an optoelectric device
10. Data is processed in a computer by
    a. bitmaps
    b. chips
    c. disks
    d. transformers

B. Matching. Match the terms in column I with those in column II.

Column I

1. lodestone
2. Thales
3. Charles du Fay
4. Alessandro Volta
5. potentiometer
6. resistance
7. a henry (H)
8. a hertz (Hz)
9. binary
10. digits

Column II

a. discovered static electricity
b. opposite of conductance
c. variable resistor
d. unit of inductance
e. one cycle per second
f. observed that like charges repel
g. magnetized iron oxide
h. 0 to 9
i. developed the first battery
j. 0 or 1

C. Completion. Fill in the blanks with the word or words that make the following statements correct.

1. What does each of the following color bands on a resistor tell you?
   a. first band _____
   b. second band _____
   c. third band _____
   d. fourth band _____
2. Computers may be classified by capacity and size as _____, _____, and _____.
3. _____ are used on animals to indicate body temperatures, breeding status, and disease problems.
4. Two uses of electronics in agriculture are _____ and _____.

UNIT

34 Electric Motors, Drives, and Controls

OBJECTIVE

To install, maintain, and utilize motors and controls.

COMPETENCIES TO BE DEVELOPED

After studying this unit, you should be able to:

- List advantages of electric motor power.
- Discuss the types of electric motors.
- List factors to consider when selecting motors.
- Discuss mounts and drives for motors.
- Select and use motor controls.
- Maintain motors and controls.

MATERIALS LIST

- ✓ Pencil
- ✓ Pad
- ✓ Calculator

TERMS TO KNOW

ball bearing
needle bearing
bushing
frame
clutch
drive train
type
induction
centrifugal
horsepower
duty rating
temperature rise
pulley
step pulley
adjustable pulley
variable speed pulley
sprocket
transmission
gear
sensor
fustat
actuator
float switch
sump
manual
override
pilot light
timer
heat sensor
flow switch
pressure switch
gas analyzer
air pollution detector

Electric motors have become the backbone of modern technology. They provide the power for most portable power tools, household appliances, shop power equipment, stationary farm equipment,

FIGURE 34-1. Electric motors have become the backbone of modern technology.

processing equipment, and conveying equipment, figure 34-1. They drive massive machinery for mining and manufacturing, trains, and water and oil pumps. Electric motors are also essential components of portable machines such as autos, trucks, tractors, earth movers, boats, airplanes, mowers, tillers, combines, and others. They start gasoline and diesel engines, provide power for accessories, and pump fluid for hydraulic systems, figure 34-2.

Important information on electricity, electronics, and motors is presented in the previous units in this section of the text. Therefore, it is suggested that these be reviewed before proceeding.

FIGURE 34-2. Electric motors provide power for starting engines and running accessories. *(Courtesy of S & R Acquisitions)*

SELECTING ELECTRIC MOTORS

Electric motors should be of the proper type, size, and capacity for the job. Economical and efficient operation depends upon selection of the correct motor for the job.

Why Use Electric Motors

Electric motor use in domestic and agricultural settings is increasing. Electric motors are frequently chosen over other types of power because they are:

1. Adaptable—motors can be used in almost any location including underwater.
2. Automatic—motors are easily controlled by automatic devices.
3. Compact—a small unit develops a relatively large amount of power.
4. Dependable—electric motors selected specifically for the job seldom give trouble.
5. Economical—a small motor operating at less than 25¢ per hour can do the work of several humans.
6. Efficient—electric motors have efficiency ratings ranging up to 95 percent.
7. Long endurance—electric motors frequently run trouble-free for 20 to 30 years.
8. Low-maintenance—if protected from dust and dirt, motors require little or no maintenance.
9. Quiet—motors are generally quieter than the machinery they drive.
10. Safe—when properly installed, maintained, and used, motors are very safe to operate.
11. Simple to operate—very little training is needed to operate most motors.

Factors in Selecting the Correct Electric Motor

For most users, the motor comes with the equipment, building, or facility. Therefore, every motor has been selected by a manufacturer, engineer, or electrician. However, a knowledge of motor selection factors is useful when motors are replaced or equipment is being designed or made on the job.

Power Supply. What electrical power is available? Is it 120 volts or 230 volts? Single-phase or three-phase? Motors larger than 1/2 horsepower run more efficiently and cause less voltage drop problems if operating with 230 volts. Additionally, the use of three-phase current further reduces such problems.

New Wiring. Would the benefit of updated or modified wiring override the cost of new wiring from 220-volt or three-phase motors? Many farms, ranches, and businesses do not have three-phase current. However, 220-volt single-phase current is usually in the building, an adjacent building, or on a central meter pole on the property.

Power Requirements. How many horsepower are needed to run the machine? Is the machine easy to start? If not, the motor must have high torque starting capabilities. What speed is needed? Motors typically run at 1150, 1750, or 3450 rpm. The motor speed should match or be adaptable to that required by the machine. When replacing motors, the characteristics of the replacement motor should be as near as possible to those of the original motor. If the machine is being designed or rigged up, use a motor as large or larger than that used on similar commercial equipment.

Base, Bearings, and Frame. Slotted holes in motor bases help to accommodate bolts and mounting. Similarly, mounting plates may have long slots to permit tightening of the bolt. Further, holes in the base may have rubber mounts to cushion the motor against vibration or shock.

The better motors have ball bearings rather than bushings for shaft support. A ball bearing is a set of hardened steel balls in a container or cage that keeps a shaft centered in a frame, figure 34-3. Needle bearings are steel rollers in a cage to keep a shaft centered and useful when extreme pressure is involved. Finally, a bushing is a sleeve in a frame that keeps a shaft centered, figure 34-4.

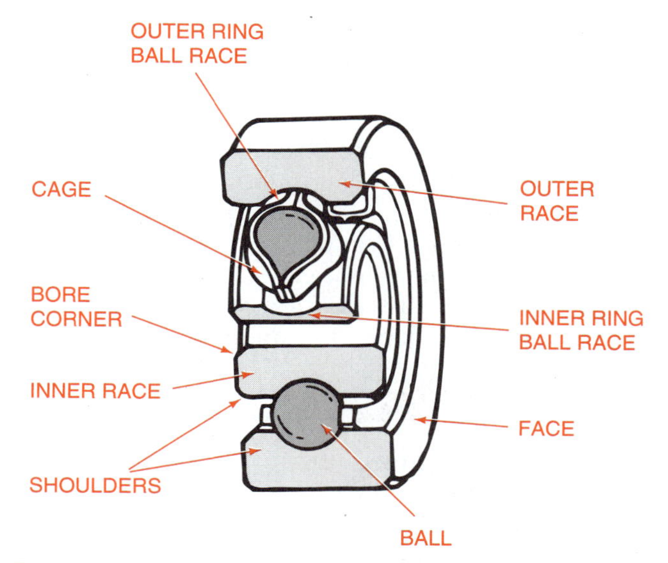

FIGURE 34-3. A ball bearing assembly.

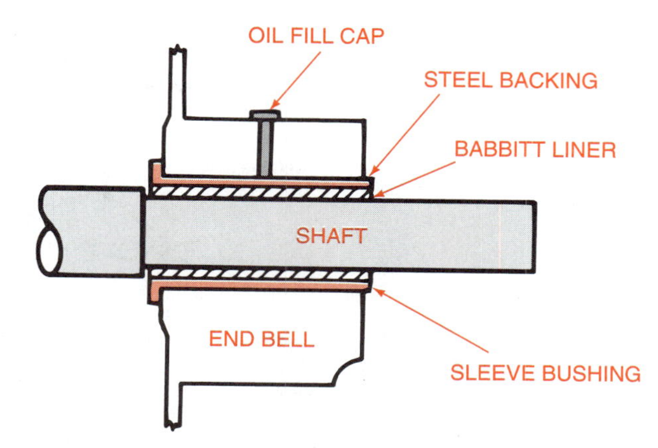

FIGURE 34-4. A bushing-type bearing.

Ball bearings or needle bearings are generally permanently lubricated with grease. However, bushings must be lubricated with small amounts of oil at regular intervals. The oil is generally stored in a felt wick or oil-impregnated bushing material.

Ball bearings add initial cost to a motor but have major advantages over bushings. They can be permanently lubricated and mounted in any position. They also create less friction, withstand more shaft pressure, and rarely need replacing.

The motor frame is the housing or case that contains the other motor components. The frame must be designed to permit the motor to be mounted appropriately; permit adequate ventilation and cooling; and support the bearings, shaft, and other components, figure 34-5.

Starting and Running Characteristics. Electric motors are generally connected directly to the load with no clutch in the drive train. Therefore, they must have sufficient torque to start. A clutch disengages a power source from its load. A drive train is the collective components that transfer power from a power source to a load.

Ease of starting of a load will dictate which types of motors are suitable. Examples of loads that are easy to start are fans, saws, and equipment with free-running lightweight shafts and components. Examples of hard-starting loads are compressors, conveyers, augers, and grinders that must compress air, move weight, or cut material as they start.

Motor type refers to the way current flows in the motor and accessories it has to help it start or run. Type includes a name for starting and a name for running, unless they are the same. For instance, an induction motor that has a capacitor to help in starting, but not when running, is called a "capacitor-start, induction-run" motor. A capacitor manages the

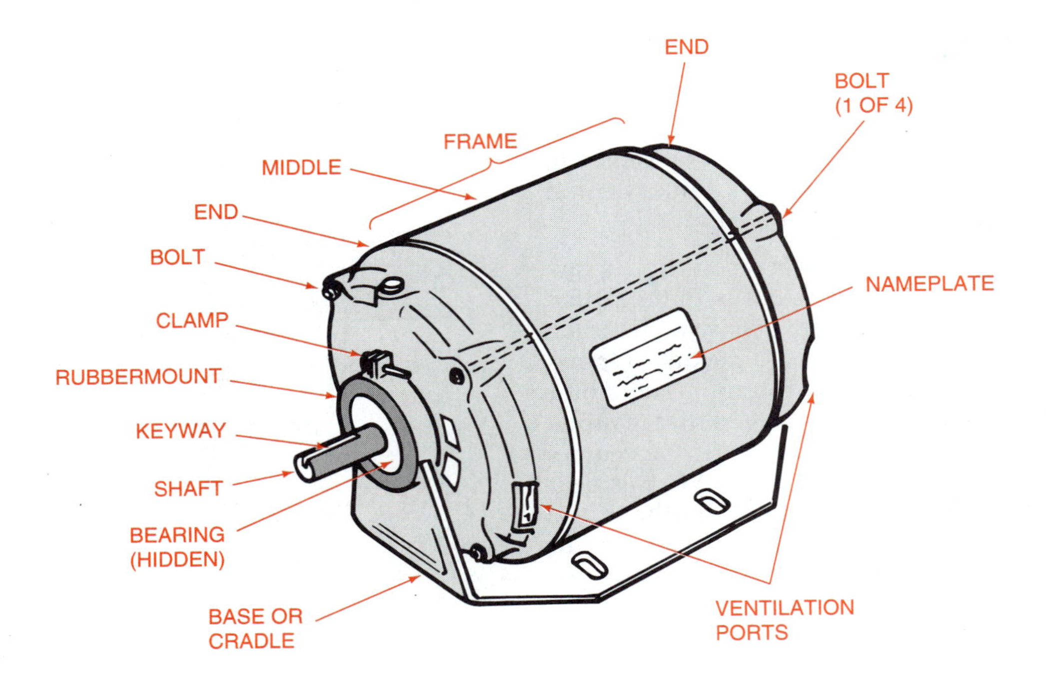

FIGURE 34-5. Exterior parts of a motor.

flow of current to provide increased motor torque. Induction is a type of motor where electricity from the power source does not flow directly to the armature. Motors typically use a centrifugal switch that clicks closed, or turns on, when the motor slows down to stop and clicks open, or turns off, the capacitor or other booster when the motor approaches its running speed on startup. Centrifugal refers to the outward force caused by a spinning object.

Several types of motors have high starting torque. Some of them draw up to six times as much current when starting under a heavy load as they do when running. Such motors cause temporary voltage drop which causes lights in the circuit or building to dim until the motor gets up to speed. Since motor selection is complex, an electrician or engineer should be consulted before purchasing an expensive motor. A selection table may also be used to gain a general understanding of motor types, figure 34-6.

INTERPRETING NAMEPLATE INFORMATION

The nameplate on a motor provides valuable information, figure 34-7. The model and serial number enables the manufacturer to identify the motor and is useful for warranty service or other communications with the supplier. Type, horsepower, rpm, and duty ratios indicate the capabilities of the motor.

One horsepower is the force needed to lift 33,000 pounds one foot in one minute or 550 pounds one foot in one second. The formula for calculating horsepower is:

$$HP = \frac{\text{Pounds lifted x Distance in feet x Time in seconds}}{550}$$

or

$$HP = \frac{W(\text{lbs.}) \times D(\text{ft.}) \times T(\text{sec.})}{550}$$

An efficient motor will generate one horsepower of force while using about 800 watts or 0.8 kilowatt (kw) of electricity. This is about the work done by 8 to 10 men.

Volts, amps, hertz, and phase indicate the type of electricity that must be supplied for the motor to run. Duty rating refers to the percent time a motor may run without overheating. Temperature rise indicates how warm the motor can get without causing internal damage. Other data may be provided on motor nameplates, depending on the manufacturer.

MOTOR DRIVES

Motors may be coupled directly to the equipment they drive, but the speed of the two must be matched. Direct drives include flexible-hose, flange, cushion-flange, and flexible shaft couplings. The first three types are likely to be used to couple motors to water pumps. The flexible shaft is frequently used to

Load Type	Motor Type	Starting Ability (Torque)	Starting Current	Size HP	Electrical Power Requirements		Speed Range	Revers-ible	Relative Cost	Other Characteristics	Typical Uses
					Phase	Voltage					
	(1)	(2)	(3)	(4)	(5)	(6)	(7)	(8)	(9)	(10)	(11)
Easy Starting Loads	(a) Shaded-Pole Induction	Very low. $^1/_2$ to 1 times running torque.	Low.	.035–.2 kW ($^1/_{20}$–1/4 hp)	Single	Usually 120	900 1200 1800 3600	No	Very Low	Light duty, low in efficiency.	Small fans, freezer blowers, arc welder blower, hair dryers.
	(b) Split-Phase	Low. 1 to $1^1/_2$ times running torque.	High. 6 to 8 times running current.	.035–.56 kW ($^1/_{20}$–3/4 hp)	Single	Usually 120	900 1200 1800 3600	Yes	Low	Simple construction.	Fans, furnace blowers, lathes, small shop tools, jet pumps.
	(c) Permanent-Split, Capacitor-Induction	Very low. $^1/_2$ to 1 times running torque.	Low.	.035–.8 kW ($^1/_{20}$–1 hp)	Single	Single voltage 120 or 240	Vari-able 900–1800	Yes	Low	Usually custom-designed for special application.	Air compressors, fans.
	(d) Soft-Start	Very low. $^1/_2$ to 1 times running torque.	Low. 2 to $2^1/_2$ times run-ning current.	5.6–40 kW ($7^1/_2$–50 hp)	Single	240	1800 3600	Yes	High	Used in motor sizes nor-mally served by 3-phase power when 3-phase power not available.	Centrifugal pumps, crop dryer fans. feed grinder.
Difficult Starting Loads	(e) Capacitor-Start, Induction-Run	High. 3 to 4 times running torque.	Medium. 3 to 6 times running current.	.14–8 kW ($^1/_6$–10 hp)	Single	120-240	900 1200 1800 3600	Yes	Moderate	Long service, low maintenance, very popular.	Water systems, air compressors, ventilating fans, grinders, blowers.
	(f) Repulsion-Start, Induction-Run	High. 4 times running torque	Low. $2^1/_2$ to 3 times running current.	.14–16 kW $^1/_6$–20 hp)	Single	120-240	1200 1800 3600	Yes	Moderate to High	Handles large load variations with little variation in current demand.	Grinders, deep-well pumps, silo unloaders, grain conveyors, barn cleaners.
	(g) Capacitor-Start, Capacitor-Run	High. $3^1/_2$ to $4^1/_2$ times running torque.	Medium. 3 to 5 times running current.	.4–20 kW ($^1/_2$–25 hp)	Single	120-240	900 1200 1800 3600	Yes	Moderate	Good starting ability and full-load efficiency.	Pumps, air compressors, dry-ing fans, large conveyors, feed mills.
	(h) Repulsion-Start, Capacitor-Run	High. 4 times running torque.	Low. $2^1/_2$ to 3 times running current.	.8–12 kW (1–15 hp)	Single	Usually 240	1200 1800 3600	Yes	Moderate to High	High efficiency, requires more service than most motors.	Conveyors, deep-well pump, feed mill, silo unloader.
	(i) Three-Phase, General Purpose	Medium. 2 to 3 times running torque.	Low-med. 3 to 4 times running current.	.4–300 kW ($^1/_2$–400 hp)	Three	120-240 240-480 or higher	900 1200 1800 3600	Yes	Very Low	Very simple construction, dependable, service-free.	Conveyors, dryers, elevators, hoists, irrigation pumps.

FIGURE 34-6. Electric motor characteristics. *(Courtesy of American Association for Vocational Instructional Materials—AAVIM—"Electric Motors: Selection, Protection, and Drives")*

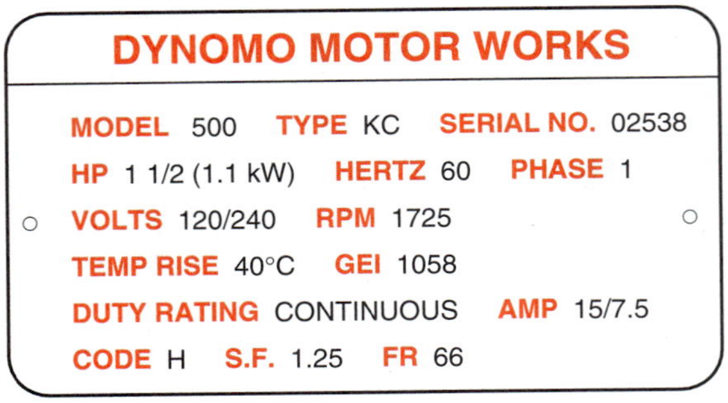

DYNOMO MOTOR WORKS
MODEL 500 TYPE KC SERIAL NO. 02538
HP 1 1/2 (1.1 kW) HERTZ 60 PHASE 1
VOLTS 120/240 RPM 1725
TEMP RISE 40°C GEI 1058
DUTY RATING CONTINUOUS AMP 15/7.5
CODE H S.F. 1.25 FR 66

FIGURE 34-7. The motor nameplate provides a summary of critical information.

couple a motor with a portable tool, such as a grinder.

Belts

The use of belts, gear boxes, and chains permit the motor to run at one speed and the machine at another. Transmissions and variable speed pulley configurations permit the choice of two or more speeds or variable speeds. Since belts are flexible and quiet, they are used in most motor drive systems. Belts are reinforced with nylon or wire cords. They are classified according to shape and size. Popular shapes include the V-belt, multi-V belt, micro-V belt, and flat belt, figure 34-8. Belt length should be determined by measuring around the pulleys, including the space between the motor and the machine.

Pulleys

A pulley is a device attached to a shaft to carry a belt. The pulley is locked to the motor shaft key and/or a set screw. Pulleys may be of the single, step, adjustable, or variable speed type, figure 34-9. Step pulleys are several sizes of pulleys fused together. They permit changing of equipment speeds by moving the belt to a different level on the pulley. Such a move must be accompanied by repositioning the pulley on the mating device unless a second step pulley is part of the system. After moving the belt, the belt must be retightened. An adjustable pulley is one with a movable side locked in place with a nut or set screw. The effective diameter of the pulley is changed by repositioning the movable side. A variable speed pulley is a pulley with a movable side that can be moved while the equipment is running. Such devices permit the changing of speeds of the machine over a certain range frequently and easily with the machine running.

Pulley-Speed Ratio

It is important to understand the relationship between pulley size and belt speed. As the diameter of a motor pulley turning at a given rpm is increased, the resulting belt speed increases proportionally as does the speed of the driven machine. Therefore, the affect of changing the pulley size on the motor or the driven machine can be determined with a mathematical formula.

The formula is: motor rpm (S) × motor pulley diameter (D) = machine rpm (SV) × machine pulley diameter (DV). The formula can be shortened to: $S \times D = S' \times D'$. S = speed of motor, D = diameter of motor pulley, S' = speed of machine, and D' = diameter of machine pulley.

If a motor has a rated speed of 1750 rpm, what size of pulley should be installed on the motor to

*FHP
3/8"
A-SECTION
1/2"
B-SECTION
2 1/32"
MULTI-V BELT
MICRO-V BELT
13/16"
FLAT BELTS

*FRACTIONAL HORSEPOWER

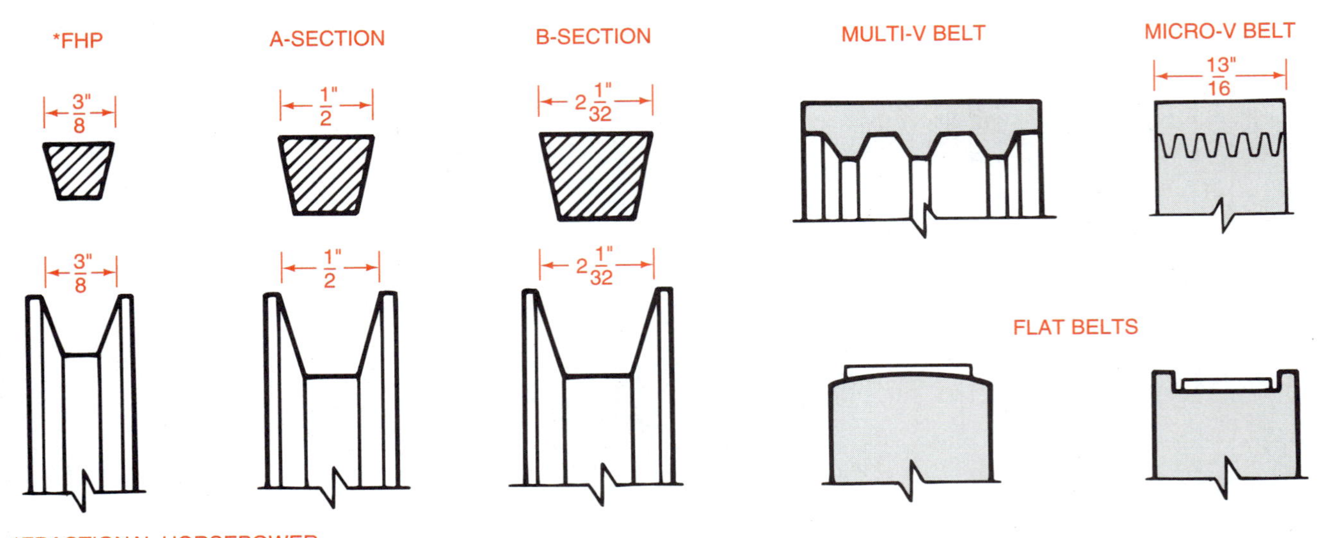

FIGURE 34-8. Some popular types of belts.

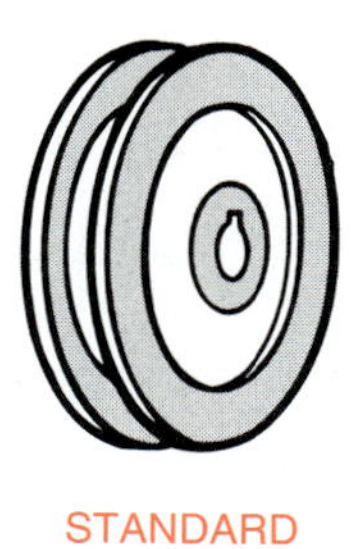

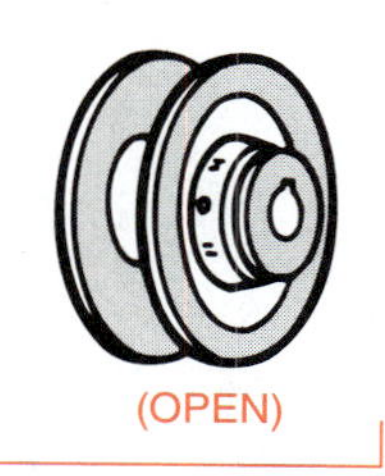

FIGURE 34-9. Common types of V pulleys.

drive a drill press with a 4-inch pulley at 975 rpm? The problem is solved as follows:

$$
\begin{aligned}
S \times D &= S' \times D' \\
1750 \times D &= 875 \times 4 \\
1750 \times D &= 3500 \\
D &= 2
\end{aligned}
$$

Again, suppose the motor was rated at 1150 rpm and came with a 3-inch pulley. What size of pulley should you install on the drill press to run it at 575 rpm? The solution is:

$$
\begin{aligned}
S \times D &= S' \times D' \\
1150 \times 3 &= 575 \times D' \\
3450 \times D &= 575 \times D' \\
6 &= D'
\end{aligned}
$$

The same formula is used to calculate the speed of a machine if the speed of motor and pulley sizes are known. What is the speed of a saw with a 3-inch pulley if it is driven by a motor with a 4-inch pulley turning at 1750 rpm?

$$
\begin{aligned}
S \times D &= S' \times D' \\
1750 \times 4 &= 5 \times 3 \\
7000 &= 5' \times 3 \\
2333 &= S'
\end{aligned}
$$

From the above formula and examples, one can make several observations:

1. To increase the speed of a machine, either increase the size of the motor pulley or decrease the size of the machine pulley.
2. To decrease the speed of a machine, either decrease the size of the motor pulley or increase the size of the machine pulley.
3. The formula applies also to gear sizes and motor and machine speeds when gears are used.

Chains and Sprockets

Drive chains function as belts except they are less flexible and cannot slip. They may be the roller type, like a bicycle chain, or link type. They travel on sprockets, which are wheels with spokelike teeth. Chains and sprocket drives require regular maintenance such as cleaning, oiling, and repair. Their main advantage is strength and positive, nonslip power transfer.

Transmissions

A transmission is a gear box that permits two or more choices of speed, figure 34-10. A gear is a wheel with teeth that mesh with teeth on another wheel to transfer power. Gears are available in many sizes, shapes, and teeth-type, figure 34-11. Some gears are designed for shifting while the shafts are turning when a clutch is used. Others require the machinery to be stopped before the gears can be shifted.

ELECTRIC MOTOR CONTROLS

Electric motors are controlled by monitors, switches, and protective devices. These items enable electric motors to run continuously, on a timed basis, or when conditions call for them to run. They can be started and stopped by humans, or controlled by switches activated by sensors. A sensor is a device that receives and responds to a signal, such as light or pressure.

The Electrical Service

Electricity flows to the user from power plant to local transformer through a system of lines, substations, and switches, figure 34-12. The local transformer converts the current to the type and voltage needed, and the meter measures how many kilowatts of electricity are used, figure 34-13. The service distribution panel contains fuses or circuit breakers to protect circuits going to all electrical appliances, figure 34-14.

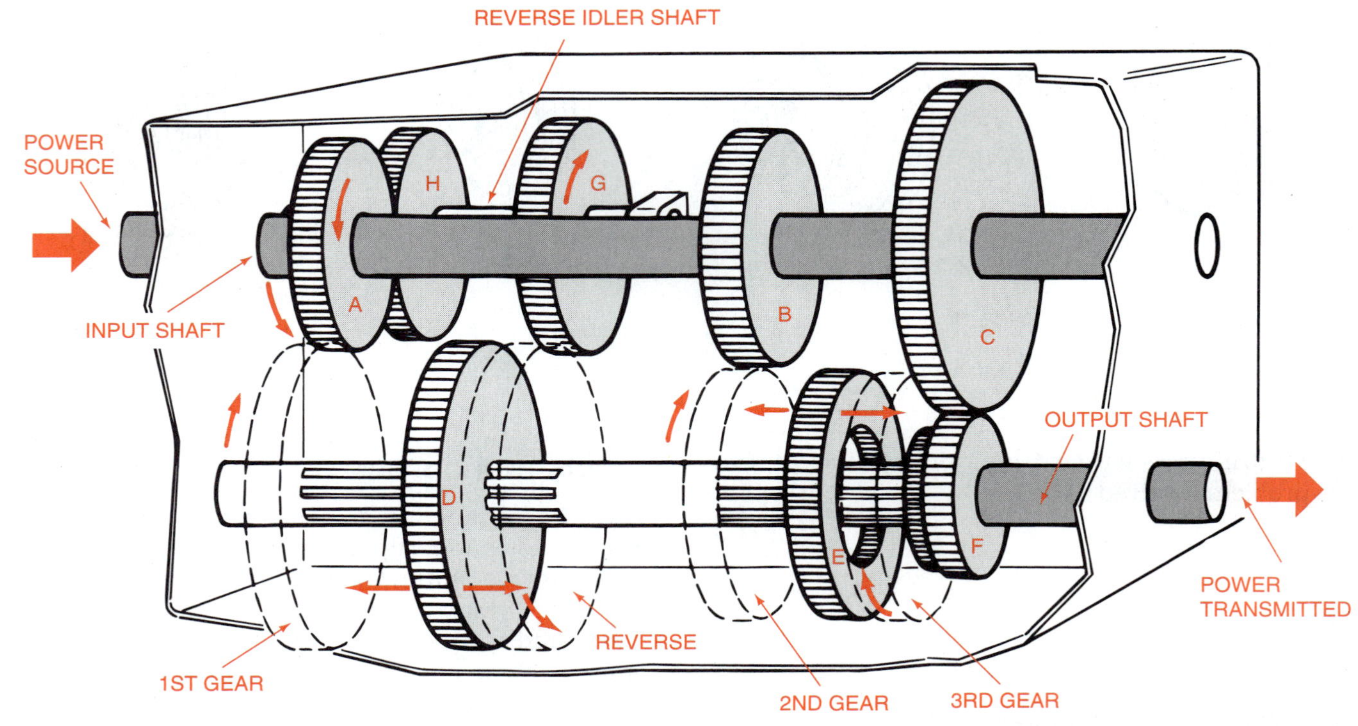

FIGURE 34-10. Example of a three-speed transmission with reverse. (*Courtesy of American Association for Vocational Instructional Materials—AAVIM—"Tractor Transmissions"*)

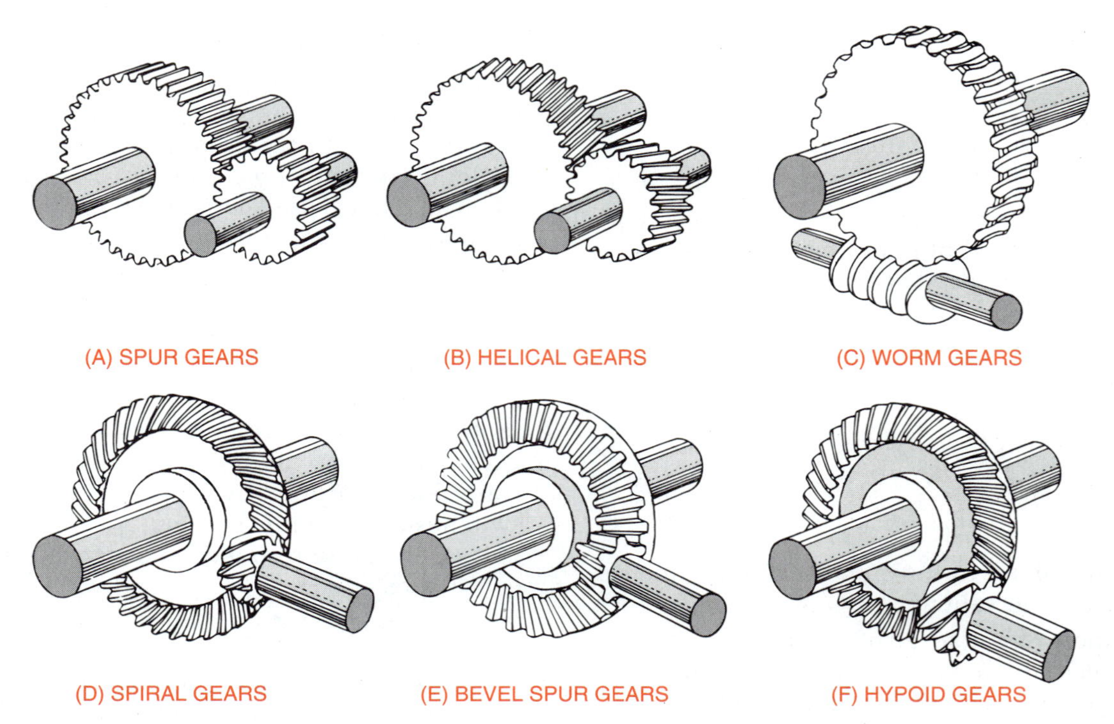

FIGURE 34-11. Common types of gears. (*Courtesy of American Association for Vocational Instructional Materials—AAVIM—"Tractor Transmissions"*)

Wiring Circuits for Motors

Motors may have an independent circuit directly from the distribution panel. Or, they may contain a power cord that plugs into an outlet, figure 34-15. Whichever method is used, the circuit must be properly grounded, figure 34-16, and be protected from overload and overheating by fuses or circuit breakers.

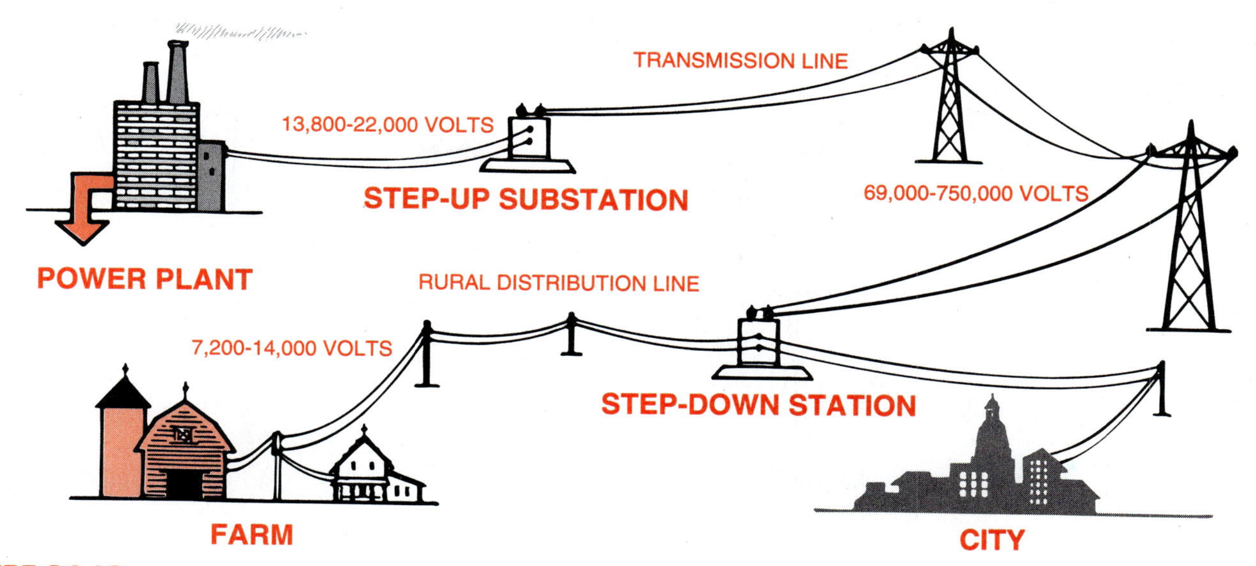

FIGURE 34-12. A power generation and distribution system. (*Courtesy of Instructional Materials Service, Texas A&M University*)

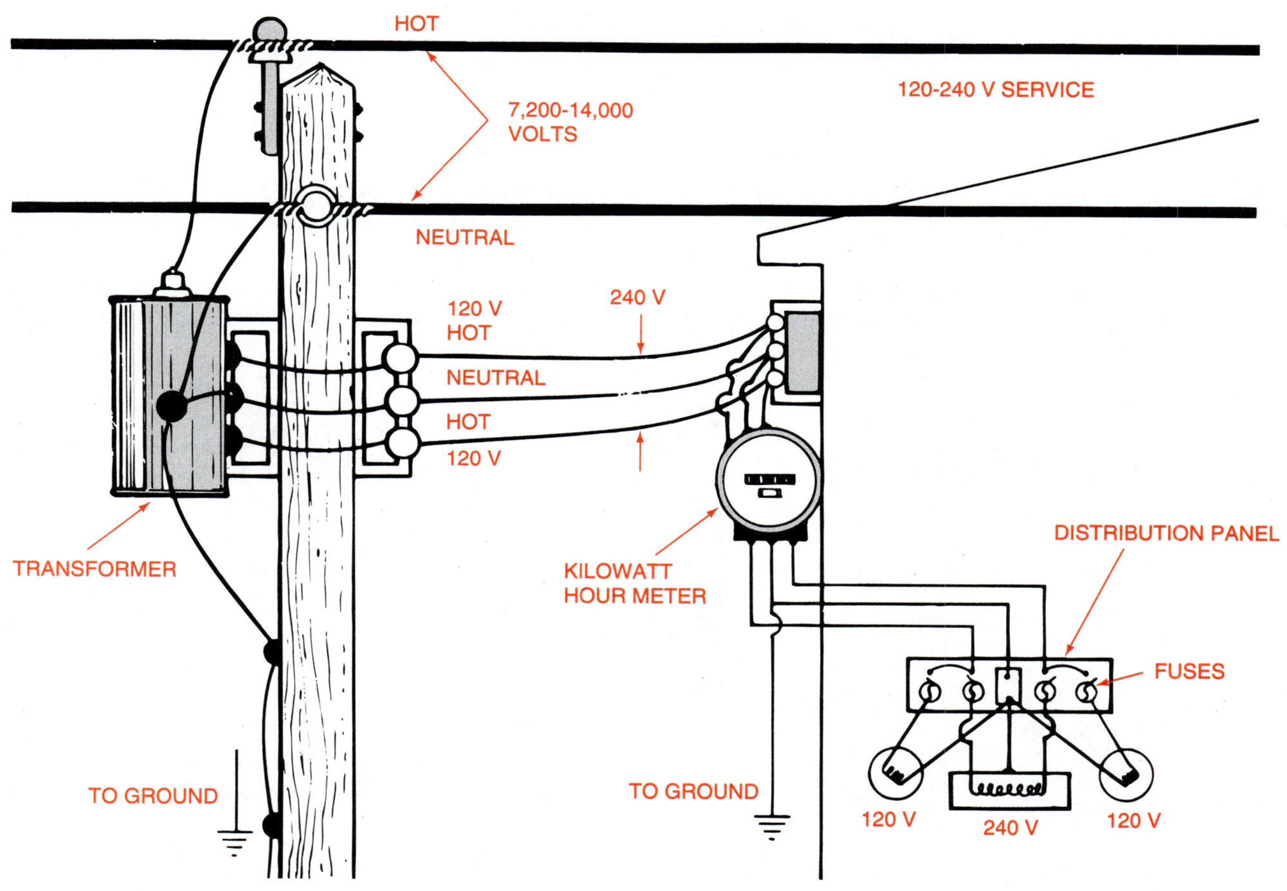

FIGURE 34-13. Movement of current from the local transformer to the distribution panel. (*Courtesy of Instructional Materials Service, Texas A&M University*)

Fuses may be the cartridge type or plug type. They may be the regular type or the slow-blow type that can stand a temporary overload. The fustat is a fuse with threads in its base that permit the fuse to be used only in circuits that match its capacity, figure 34-17.

Circuit breakers protect the circuit, as well as fuses. They are more expensive than fuses, but serve

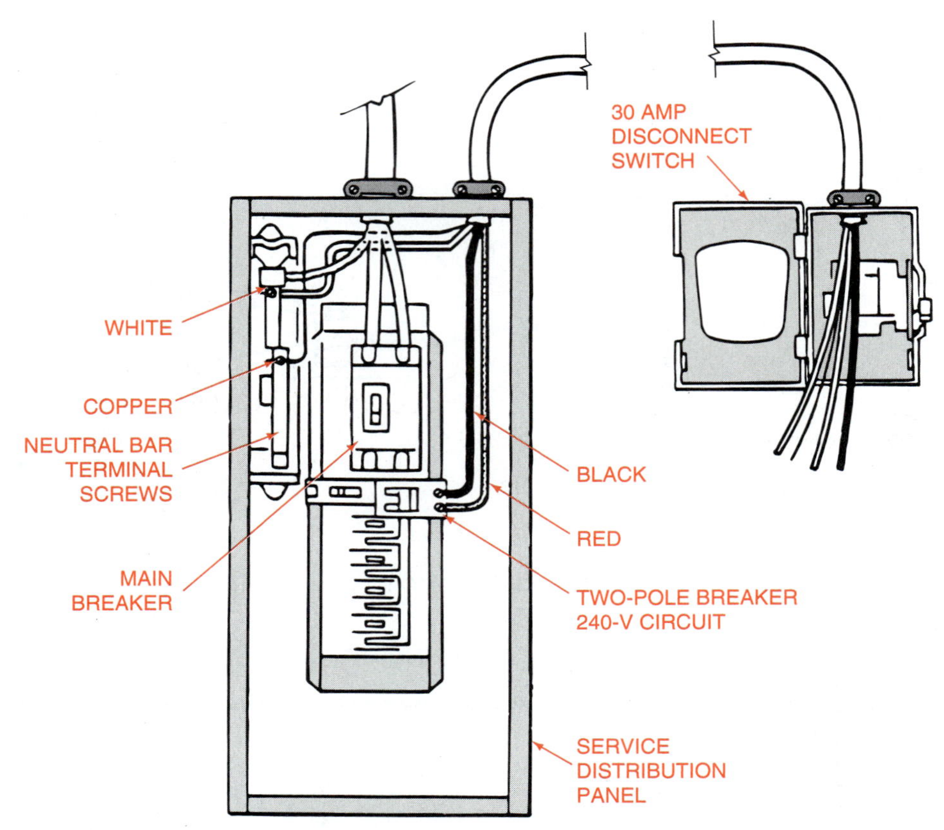

FIGURE 34-14. A service entrance panel and 30 amp disconnect switch located near a motor.

as a switch in the circuit and can be reset if they trip due to circuit overload. Ground-fault interrupters (GFI) add a safety feature for the user. In the event of a faulty circuit that may shock the individual, the GFI

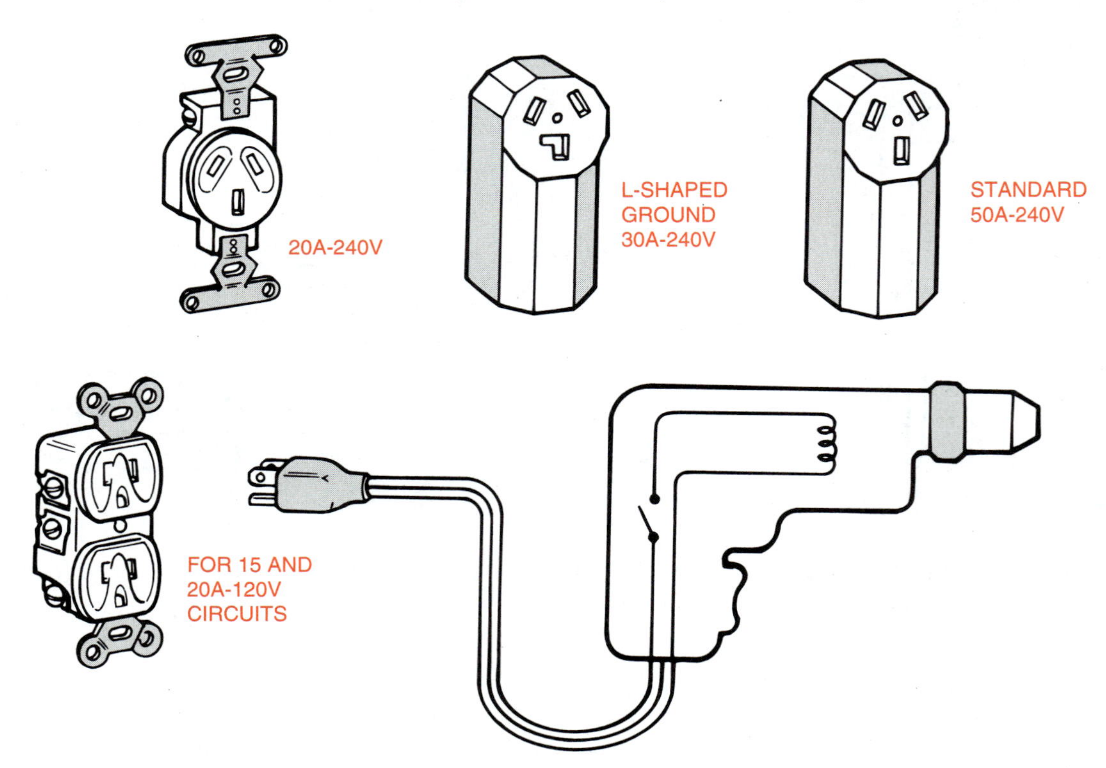

FIGURE 34-15. Motors frequently have power cords that plug into outlets.

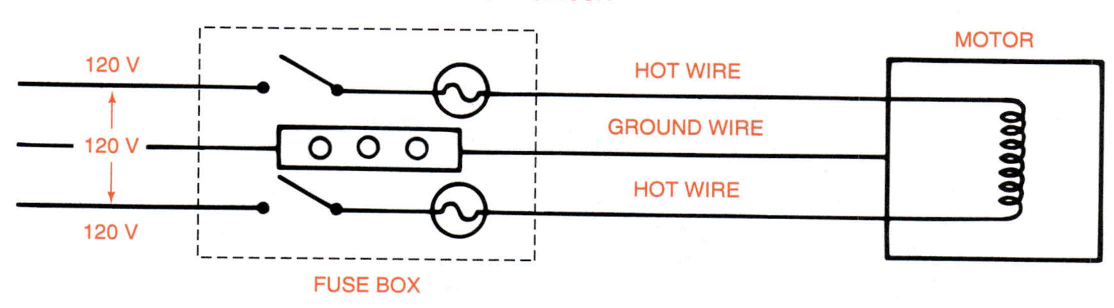

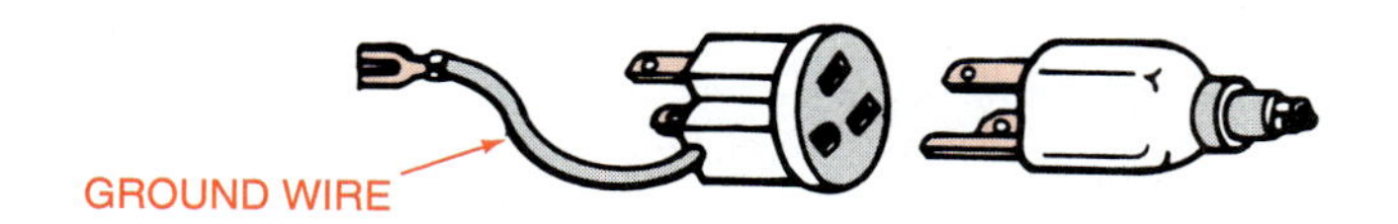

FIGURE 34-16. It is very important to properly ground all motor circuits.

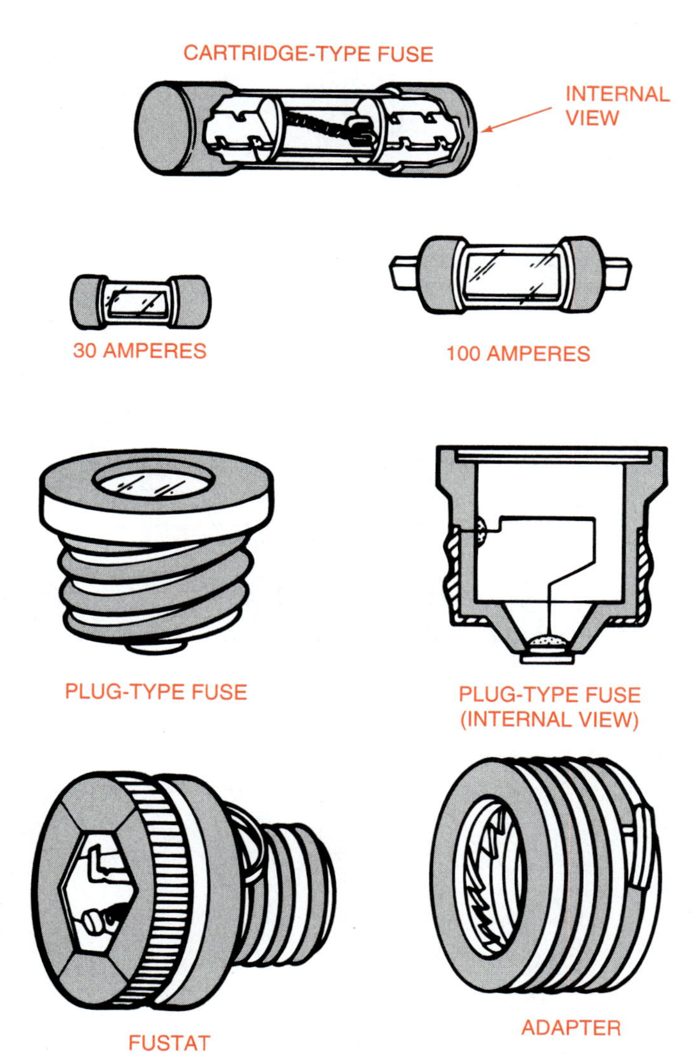

FIGURE 34-17. Fuses are used in many installations to protect the wiring circuit.

will interrupt or stop the flow of current before it causes injury, figure 34-18.

Motor Protection

Fuses and circuit breakers do not provide adequate protection for motors. Motors need overload devices that trip and can be reset if overloaded. These may be mounted on the motor or may be part of the switch assembly, figure 34-19.

Overload devices frequently utilize a bimetallic strip that bends when overheated and opens the circuit, which in turn, trips the motor switch. When the bimetallic element cools, the protector can be reset and the motor will function again, figure 34-20.

CIRCUIT BREAKER
GFI RECEPTACLE FOR CONVENTIONAL OUTLET
ON
R
T
ON-OFF SWITCH

FIGURE 34-18. Circuit breakers and ground fault interrupters provide extra safety and convenience to a circuit.

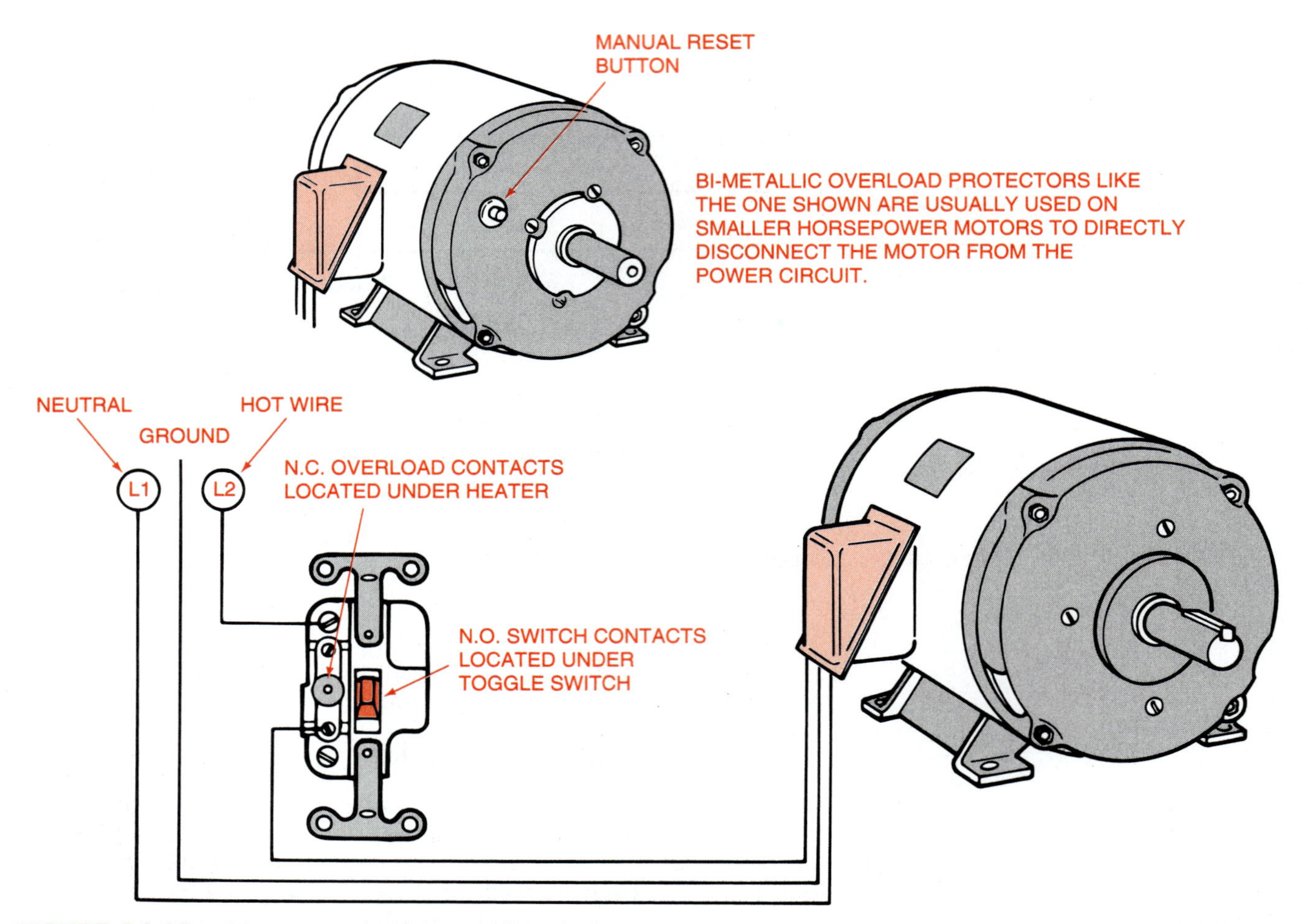

FIGURE 34-19. Motor overload protection devices.

Motor Switches

It should be remembered that switches can only be safely used in the hot wires in a circuit. Never attach neutral or groundwires to switches. A switch generally has one, two, or three pairs of wire terminals.

Types of Switches. Switches may be of the toggle, rotary, push-button, trigger, or magnetic type. All of the switches are useful on motors of about 1 hp or less. However, magnetic switches, which close contact points by electromagnetic solenoids, are recommended for large motors. Such switches close rapidly and can be built to carry more current than the other types, figure 34-21. Such switches are expensive but durable.

Reversing Switches. Some machines must run backwards as well as forwards. A plumber's pipe threading machine is an example. The drum-type switch is useful for this purpose, even when three pairs of wires are needed for three-phase motors, figure 34-22.

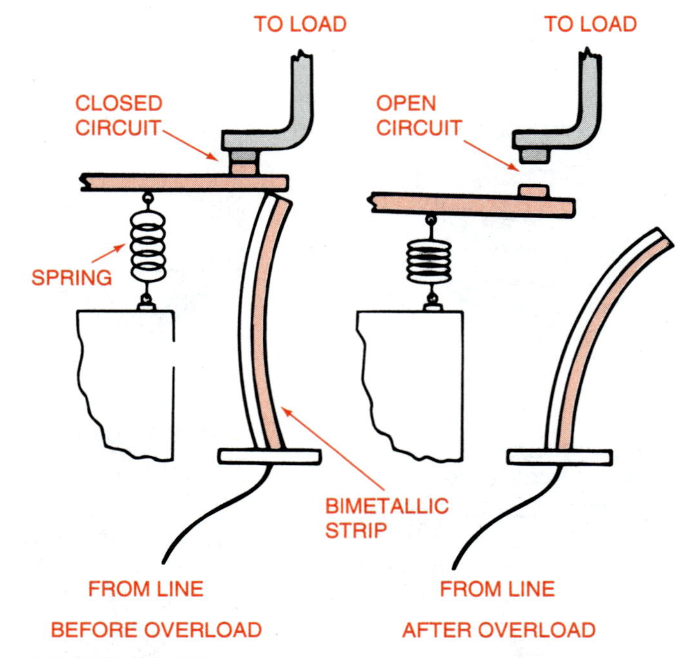

FIGURE 34-20. The bimetallic elements in a motor protector.

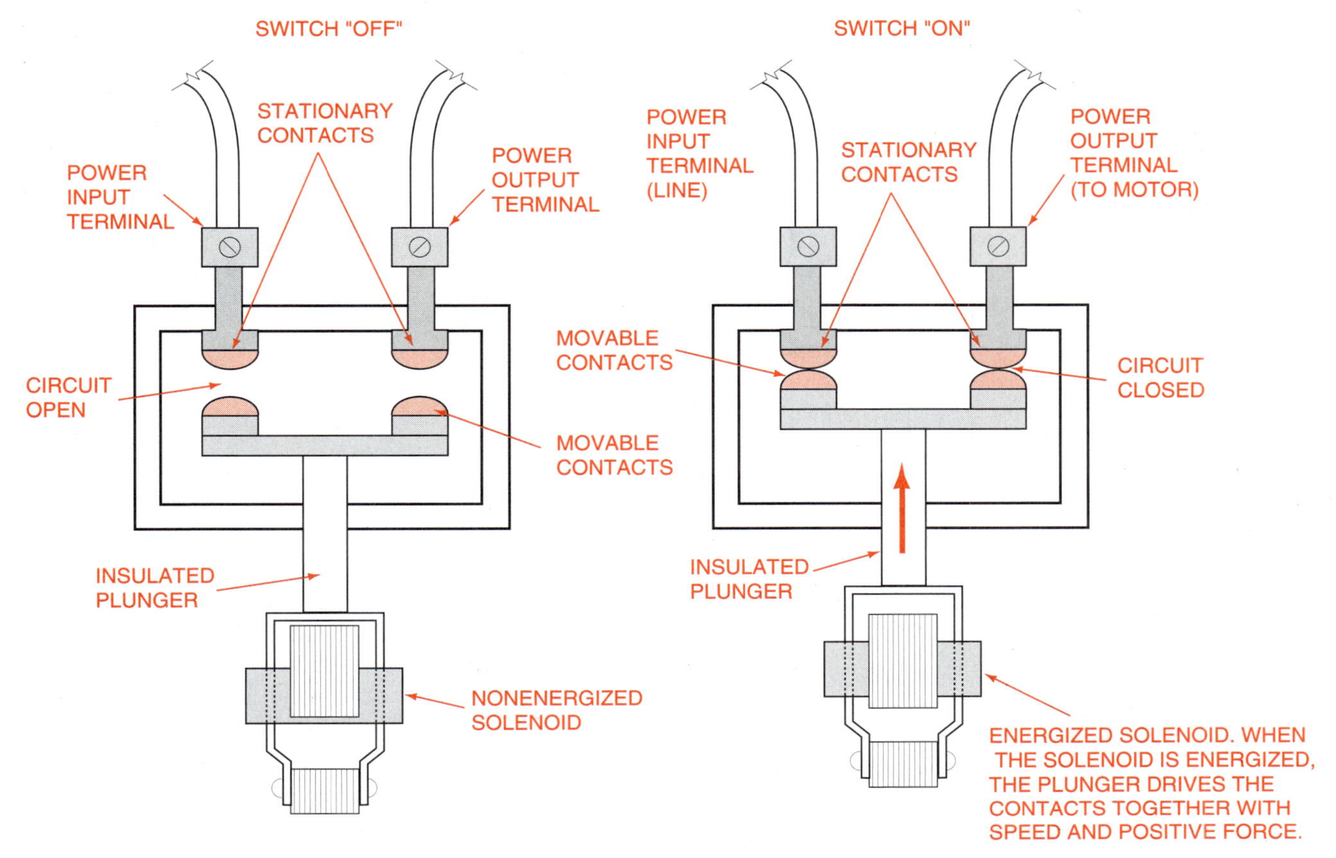

FIGURE 34-21. Magnetic switches are closed by the movement of a solenoid plunger.

Sensors and Other Switch Actuators

There are dozens of types of sensors and switch actuators. An actuator is a device responsible for placing another device in motion. Some of the more common ones will be discussed here.

Float Switch. A float switch is turned on and off by the action of a float. Float switches are used on sump pumps, livestock water tanks, and other devices, where liquid levels must be controlled. A sump is a hole where unwanted water drains and is removed by a pump. As water accumulates and rises to a certain level, a float actuates the switch which turns on the pump. When the pump lowers the water level and the float drops to a certain point, the pump motor is switched off. Circuits with such automatic switching devices also need a manual or override switch for control while repairing equipment, figure 34-23. Manual means "by hand," and override means "to prevail over or take precedence over."

Pilot Light. A pilot light, or indicator light, glows when a circuit is energized. These lights are installed in circuits such as attic fans, sump pumps, and burners on kitchen stoves in order to indicate

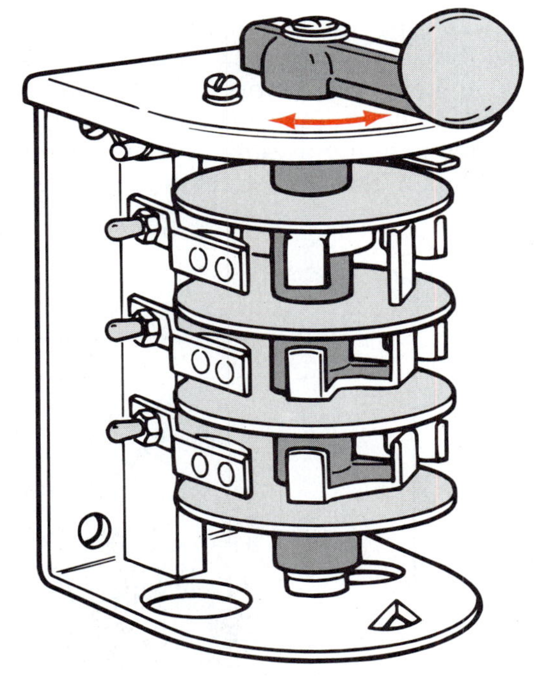

FIGURE 34-22. The drum switch is useful for motors that must run frontwards and backwards.

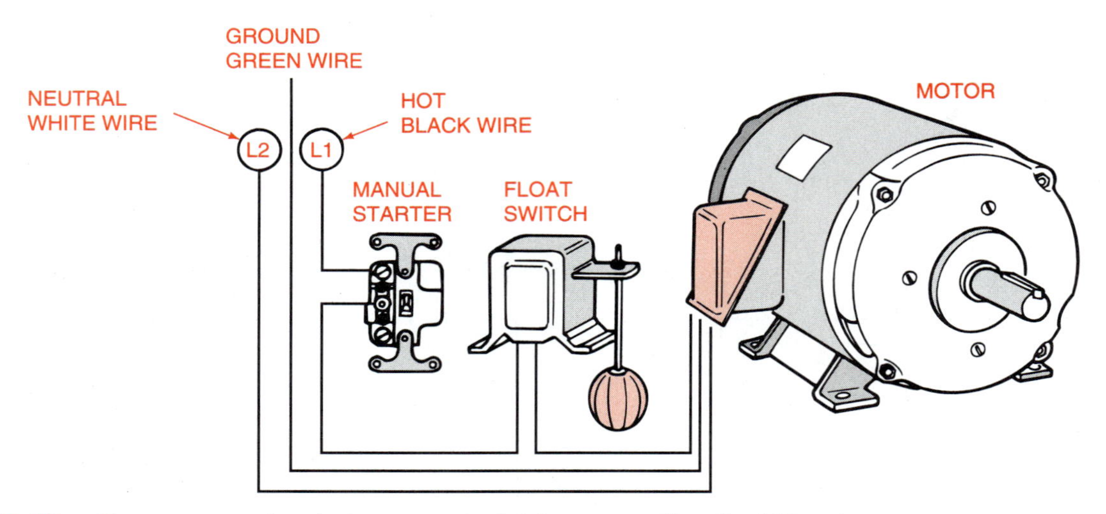

FIGURE 34-23. Float-actuated switches are useful for controlling liquid levels.

when electrical motors and electronic devices are operating, figure 34-24.

Timer. A **timer** is a device that activates a switch after a preset time lapse. Timers are used on stove ovens, clothes dryers, interior and exterior lights, slide projectors, water pumps, mist propagation systems, automatic plant watering systems, and others. Timers may have a manual start and a timed shut-off cycle, and some models can be set for numerous combinations of on-off cycles within a 24-hour period or longer, figure 34-25.

Heat Sensor. **Heat sensors** are switches that respond to temperature changes. They control the electrical flow to oven burners, hot water heaters, boiler equipment, and motors in hot air and liquid systems. They can be designed to turn the current off when the temperature rises to a certain degree and on again when it falls to a certain degree. Other heat sensors may work in reverse. Many heat sensors use a thermocouple to detect heat changes and actuate switches. Thermocouples may use a bimetallic element to sense temperature changes. Others use mercury which expands rapidly in a tube as temperature rises and activates the switch, figure 34-26.

Flow Switch. A **flow switch** reacts to the movement of a gas or liquid. Some are mounted on water lines to indicate when a sprinkler or other liquid system is running. Others are mounted in air ducts to indicate when fans are running, figure 34-27. Flow switches could also be used to start fans if used to sense the absence of natural air movement in a system.

Pressure Switch. A **pressure switch** responds to changes of liquid, gas, or mechanical pressure. For instance, pressure switches on engines activate oil pressure gages or warning lights, figure 34-28.

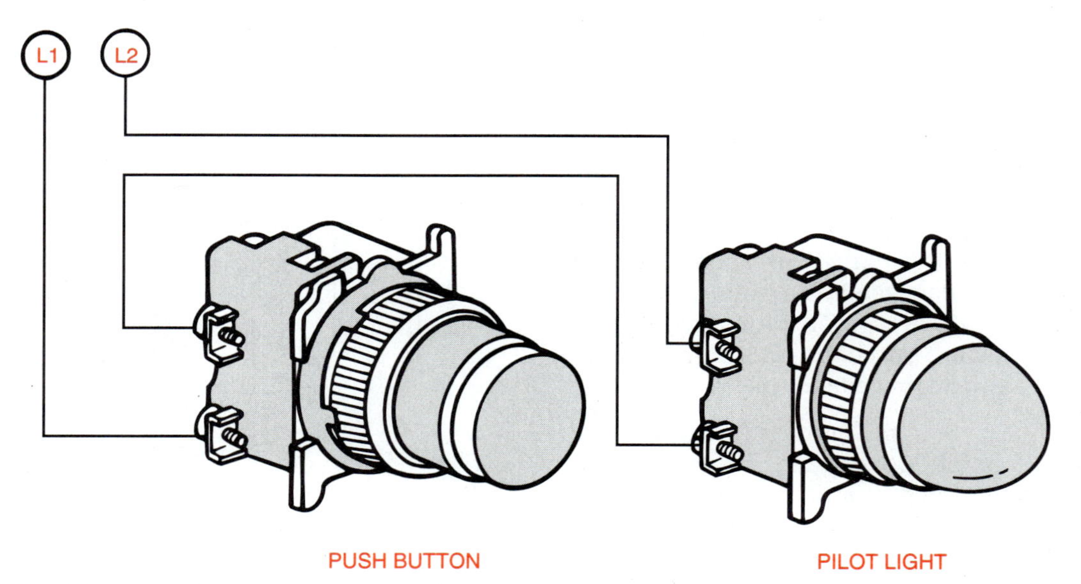

FIGURE 34-24. Pilot or indicator lights are valuable monitoring devices.

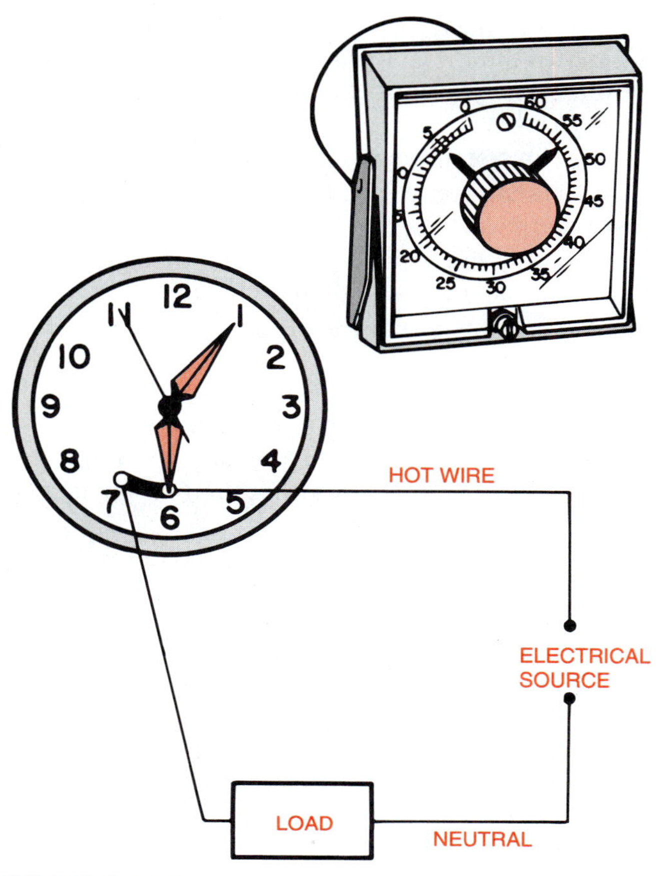

FIGURE 34-25. A timer and how it works.

Pressure switches on water pumps for homes, farms, and ranches start and stop the water pump to maintain pressure between 20 and 50 psi in the water tank. Similarly, air compressors utilize air pressure switches to keep the air in the tank within a narrow pressure range. Pressure switches may also be used to monitor the movement of livestock through gates, chutes, and milking facilities and actuate silo unloaders, feeders, and augers.

Gas Analyzer. A gas analyzer is a device that determines the content of gases such as air and automotive exhaust. An air pollution detector is a type of gas analyzer that senses and presents a warning when air has unacceptable amounts of pollutants, figure 34-29. On the other hand, an automotive gas analyzer has a set of meters that indicate the proportion of various gases in the exhaust and guides the carburetor adjustment process.

Solid-State Devices. Many motor controls are solid-state electronic devices. They are very small and compact. They do not have moving parts to wear out and are quite reliable. For these and other reasons, solid-state sensors and switches are rapidly replacing the traditional ones. Solid-state sensors and switching devices are discussed in Unit 33, Electronics in Agriculture, of this text.

System Controls

Many motors do not operate independently. They are a part of a system and contribute to a larger operation such as heating and moving air or water, operating an assembly line, or controlling traffic lights in a city.

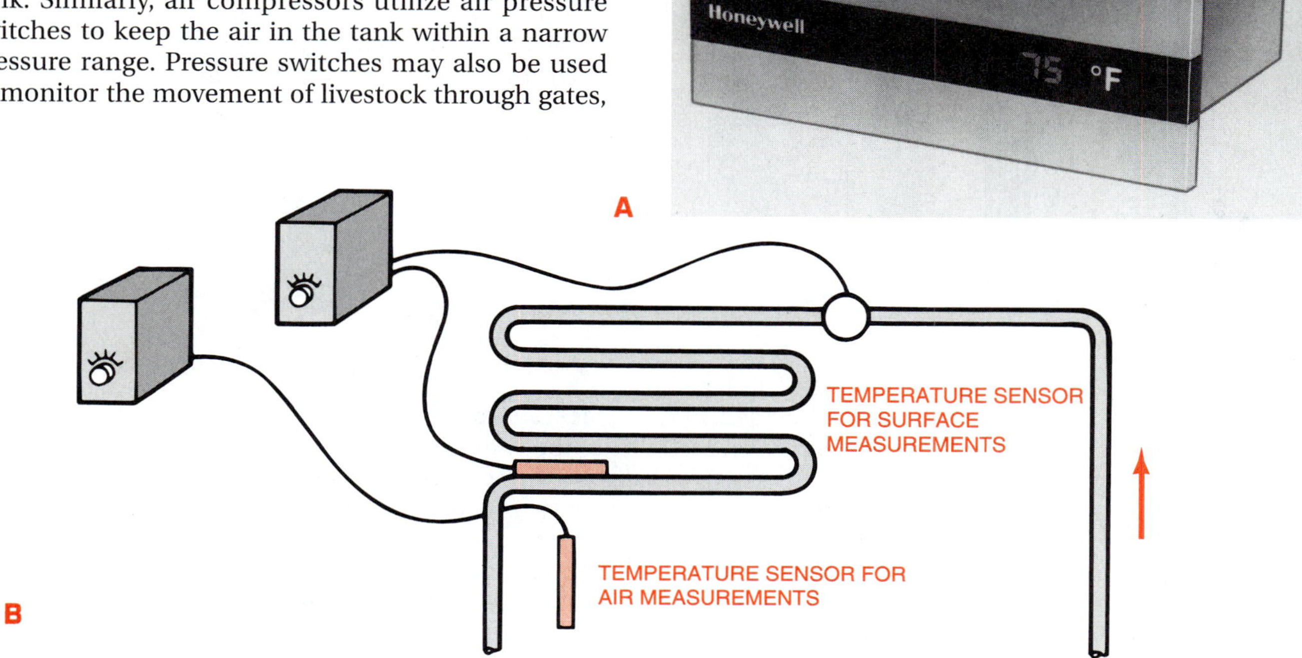

FIGURE 34-26. Examples of heat sensors.

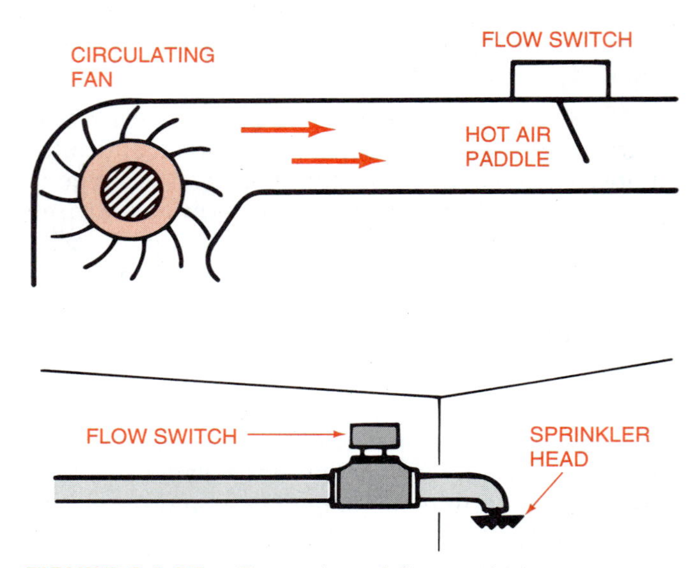

FIGURE 34-27. Examples of flow switches.

FIGURE 34-28. Oil pressure gage.

FIGURE 34-29. An air pollution detector.

The oil-fired, forced-hot-air furnace is a good example of a system with many interrelated sensors and motor controls. It has only two motors, but many controls, figure 34-30.

The oil pump motor in the furnace pumps oil through a nozzle as a highly combustible mist and supplies air to the fire. The fan motor drives a squirrel cage fan which draws cool air from the cool-air return duct and forces the heated air into the living spaces of the building. The following explanation illustrates how many controls direct the action of these two motors to make a furnace work safely and automatically. Briefly, the controls in figure 34-30 work as follows:

1. The system gets its power from a circuit *A*.
2. Disconnect switch *B* is installed on or near the furnace.
3. Room thermostat *C* senses temperature drop and signals the system to start.
4. Oil burner *D* comes on and the fire causes the unit to heat up.
5. When limit control *E* heats up to its "upper-limit" setting, it signals fan motor *G* to start.
6. Fan *G* pulls cold air in from cool-air return duct *I* and pushes warmed air through duct *H* for distribution.
7. When heat warms the space and raises the temperature several degrees, thermostat *C* is "satisfied" and shuts off oil burner *D*.
8. Fan *G* continues to run because there is heat in the furnace.
9. When air being pushed through *H* drops in temperature to the "lower setting," limit control *E* shuts off the current to fan motor *G*. The cycle is complete.

There are several additional safety and protective devices in the system. First, oil burner motor *D* and fan motor *G* have overload protectors that stop them if they get too hot or draw too much current due to overload. Second, if the flue or chimney gases get too hot, stack control *F* will shut down the system. Third, there is a switch in the limit control *E* that will shut down the system if the hot air duct gets too hot. Finally, there is a switch in the system to manually turn on the fan to circulate air continuously if desired, without generating heat.

Items *C, E,* and *F* in the above systems are heat sensors that use bimetallic elements. These sensors measure and respond to temperature changes, figure 34-31.

MAINTAINING MOTORS, DRIVES, AND CONTROLS

Most motors have permanently lubricated bearings. If not, they have oil filler tubes which require only a few drops of electric motor oil periodically. Do not apply oil more frequently or in larger amounts than specified in the owner's manual. Over-oiling can be a problem just as under-oiling. Motors must be kept free of dust, dirt, and moisture.

Shafts need periodic greasing or oiling. When chains are used, they need to be oiled regularly.

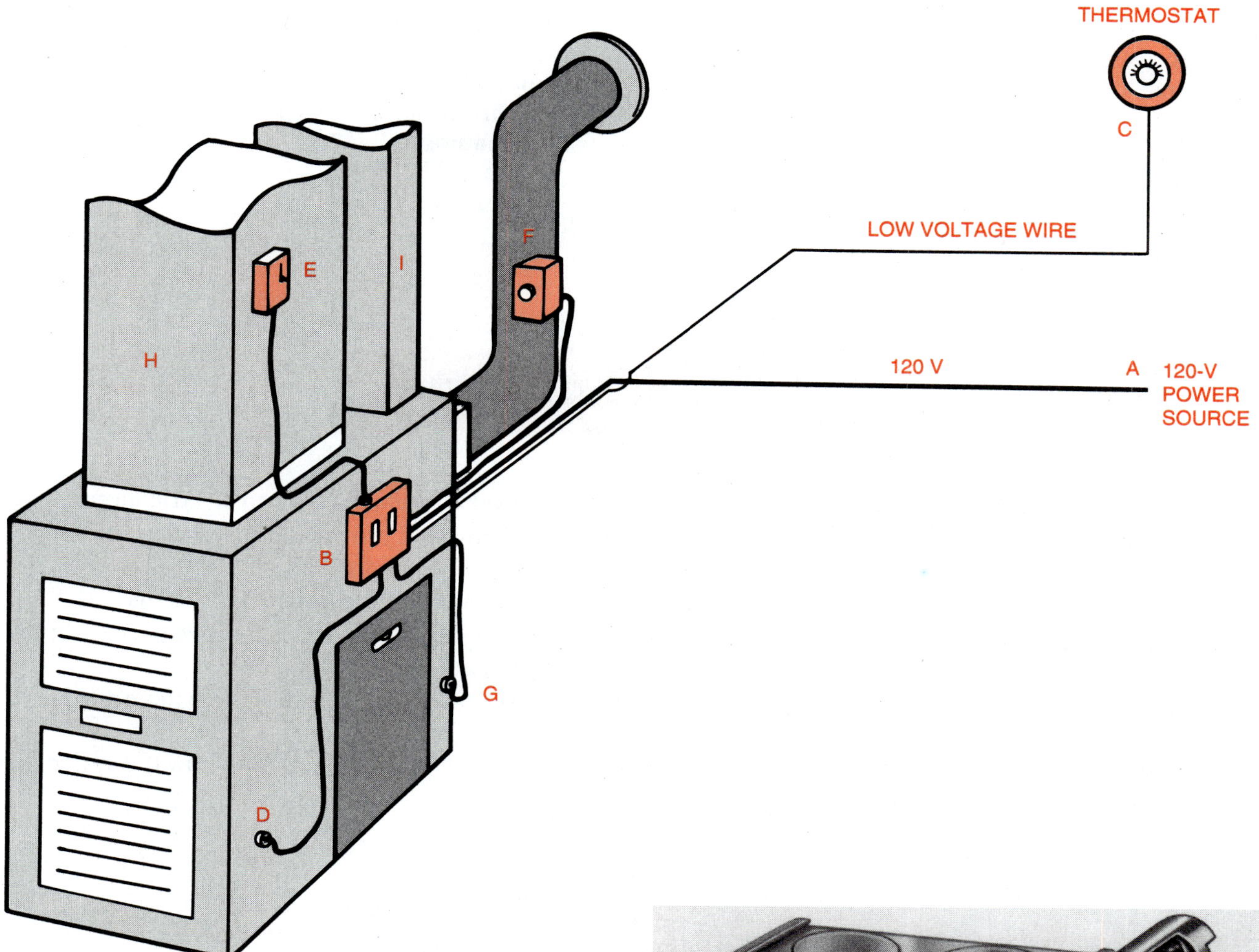

FIGURE 34-30. The oil-fired forced-hot-air furnace utilizes a system of controls.

Additionally, they need to be cleaned in solvent and reoiled periodically to decrease wear caused by dirt and grit.

Belts must be kept clean and dry. Belt tension must be adjusted periodically to prevent slipping and compensate for wear. Whenever the motor is

FIGURE 34-31. Fan/limit control and furnace stack control are examples of heat sensors that are essential motor controls on oil-burning furnaces.

moved, care must be taken to assure proper alignment of belts and pulleys as well as sprockets and chains.

SUMMARY

Electric motors provide low-cost, efficient, and flexible power. Motors are found in all facets of agriculture/agribusiness and renewable natural resources. Sensors and controls are readily used to incorporate motors into mechanical systems. Motors are low in initial cost, economical to operate, and easy to maintain.

STUDENT ACTIVITIES

1. Define the Terms to Know in this unit.
2. Make a survey of your home, and write down the name of every device that has an electric motor.
3. Study the nameplate on a motor, and write down all the information provided. Discuss your findings in class.
4. Collect and make a display of parts and discarded belts, chains, and pulleys. Label the items.
5. Organize a class project in which each student brings in examples of sensors, protective devices, and switches. Organize these and label the items for teaching aids.
6. Choose one type of motor. Do research on how the motor starts, runs, and is designed. Report to the class.
7. Choose a mechanical system, like the example of a furnace in the text, and describe how the system uses sensors, motors, and controls.

SELF-EVALUATION

A. **Multiple Choice**. Select the best answer.

1. The most reliable factor to use when replacing a motor is
   a. time required
   b. specifications of the original motor
   c. personal choice
   d. cost of motors
2. Another name for the case on a motor is
   a. bearing
   b. frame
   c. shield
   d. undercarriage
3. Components that transfer power from motor to machine make up the
   a. clutch assembly
   b. clutch housing
   c. drive train
   d. transmission
4. The device that disconnects a capacitor after a motor is up to speed is a
   a. centrifugal switch
   b. repulsor
   c. sensor
   d. voltage booster
5. A motor running at 1750 rpm with a 3-inch pulley will drive a machine with a 2-inch pulley at
   a. 830 rpm
   b. 1240 rpm
   c. 2630 rpm
   d. 3500 rpm
6. The formula for calculating horsepower is W(Lbs.) × D(ft.) × T(sec.) divided by
   a. 10
   b. 50
   c. 500
   d. 550

7. One type of direct motor drive is
    a. belt and pulleys
    b. differential
    c. flexible shift
    d. transmission
8. The best pulley for changing speeds frequently is
    a. adjustable
    b. standard
    c. step
    d. variable speed
9. The device that permits shifting of gears is
    a. transmission
    b. sensor
    c. differential
    d. clutch
10. Motors are best protected from burnout by
    a. circuit breakers
    b. fuse
    c. overload protectors
    d. all of these

B. Matching. Match the terms in column I with those in column II.

Column I

1. gear box
2. bimetallic
3. magnetic switch
4. reverses motor
5. float switch
6. flow switch
7. pressure switch
8. pollution detector
9. solid state
10. limit control

Column II

a. electromagnetism
b. reacts to liquid levels
c. controls air compressors
d. use in liquid lines
e. controls furnace fans
f. electronic switches
g. gas analyzer
h. drum switch
i. heat sensor
j. transmission

C. Completion. Fill in the blanks with the word or words that make the following statements correct.

1. Calculate the unknown in the following belt systems:
    a. Motor speed is 3450, motor pulley is $3^{1}/_{2}$" diameter, machine pulley is $3^{1}/_{2}$" diameter, machine speed is _____ rpm.
    b. Motor speed is 3450, machine speed is 1725, machine pulley is 3" diameter, motor pulley must be _____.

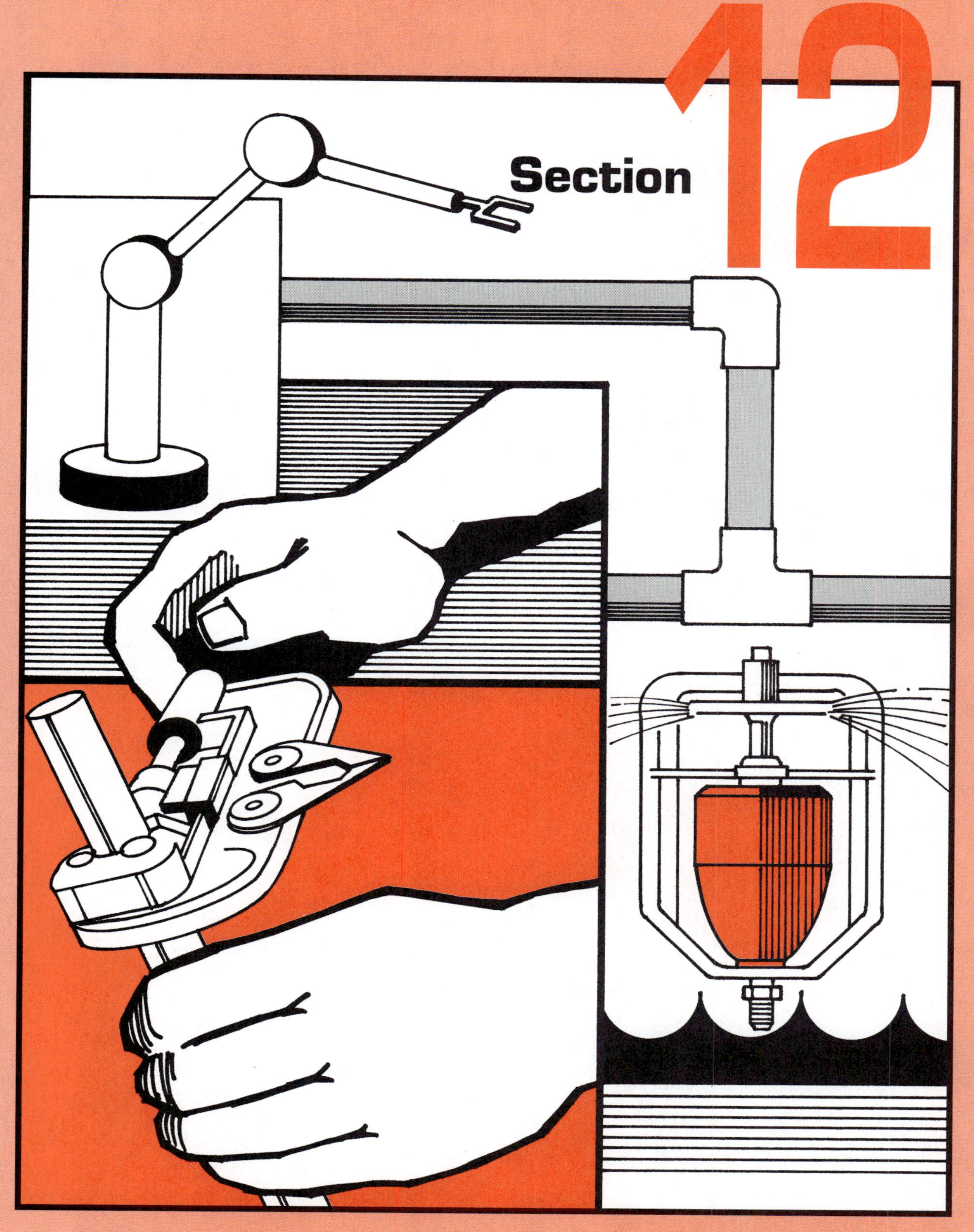

Plumbing, Hydraulic, And Pneumatic Systems

UNIT 35 Plumbing

OBJECTIVE

To identify plumbing materials and perform basic plumbing procedures.

COMPETENCIES TO BE DEVELOPED

After studying this unit, you should be able to:

- Identify plumbing tools.
- Identify and select pipe.
- Identify common pipe fittings.
- Assemble pipe.
- Maintain water systems.

TERMS TO KNOW

plumbing
pipe
fitting
tubing
bench yoke vise
chain vise
chain wrench
tubing cutter
ratchet die stock
flaring tool
flaring block
pipe stand
nominal
ID
OD
black pipe
galvanized pipe
standard pipe
extra heavy
double extra heavy
K
L
M
compression fitting
plastic
polyethlene (PE)
polyvinyl chloride (PVC)
acrylonitrite-butadiene-styrene (ABS)
cast iron soil pipe
oakum
neoprene
adaptor
bushing
reducer
elbow (ell)
tee
Y
coupling
union
plug
cap
nipple
pipe dope
teflon tape
faucet
valve
O ring
packing
faucet washer
seat
diaphragm
aerator
float valve assembly
flush valve

MATERIALS LIST

- ✓ Common pipe fittings:
  - a. bushing
  - b. 90° elbow
  - c. 45° elbow
  - d. coupling
  - e. reducer
  - f. cap
  - g. street elbow
  - h. plug
  - i. tee
  - j. Y
  - k. union
  - l. floor flange
  - m. pipe nipples
- ✓ Examples of steel, copper, and plastic pipe
- ✓ All materials in the bill of materials, figure 35-32

The term plumbing means installing and repairing water pipes and fixtures, including pipes for handling waste water and sewerage. In agricultural mechanics, all work done with pipe and pipe fittings is considered to be a part of plumbing. Pipe refers to rigid tubelike material. A fitting is a part used to connect pieces of pie or to connect other objects to pipe. Tubing is also used in plumbing installations. Tubing generally refers to pipe that is flexible enough to bend.

TOOLS FOR PLUMBING

Many tools used for plumbing have been discussed in previous units. The following additional tools are also commonly required when working with pipe and tubing.

Tools useful for holding pipe and fittings are the bench yoke vise and the chain vise, figure 35-1. The most useful tools for turning pipe and fittings are the chain wrench and pipe wrench, figure 35-2. The hacksaw, file, pipe cutter, reamer, and tubing cutter are useful for cutting and smoothing pipe and tubing, figure 35-3. Taps and dies are used for cutting threads in pipe. A rachet die stock is used to turn the die.

The ends of soft copper tubing may be flared to fit bell-shaped fittings, figure 35-4. Copper tubing is cut to length with a tubing cutter. The pressure and rolling action of the cutter wheel create a burr on the inside of the tubing at the cut. This burr is removed using the burr remover on the tubing cutter. The flaring tool and flaring block are used to hold the tubing and to form a flared end on the tubing, figure 35-5. This end is held against bell-type fittings by a flare nut to create a gas or liquid-tight seal.

A tripod-type pipe stand with a vise makes an excellent work area for cutting, reaming, and threading pipe, figure 35-6.

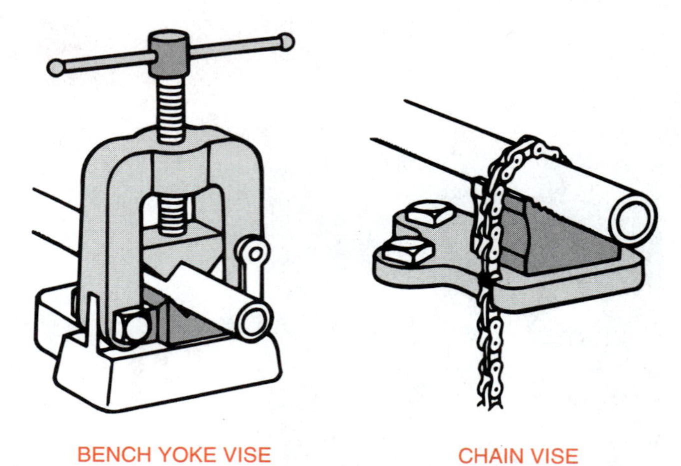

FIGURE 35-1. The bench yoke vise and chain vise are used for holding pipe and pipe fittings.

IDENTIFYING, SELECTING, AND CONNECTING PIPE

The nominal, or identifying, size of pipe is generally based on the inside diameter. Inside diameter is abbreviated ID and outside diameter, OD. Common

FIGURE 35-2. Chain wrenches and pipe wrenches are useful tools for turning pipe and pipe fittings. (*Pipe wrench courtesy of Klein Tools, Inc.; chain wrench courtesy of Mac Tools, Inc.*)

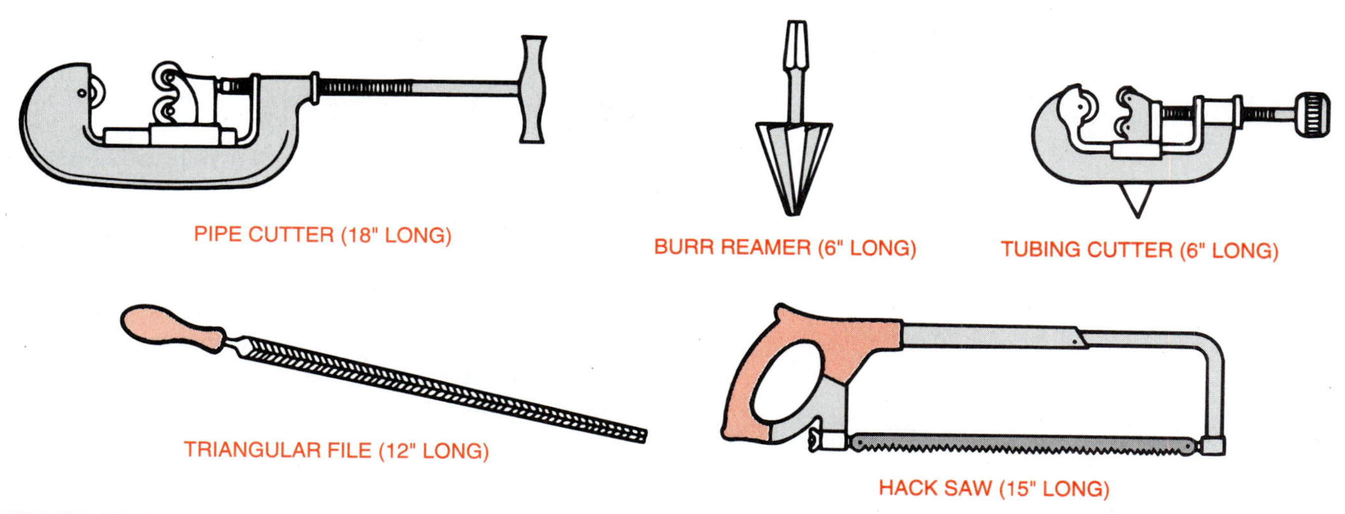

FIGURE 35-3. Tools used for cutting and smoothing pipe and tubing.

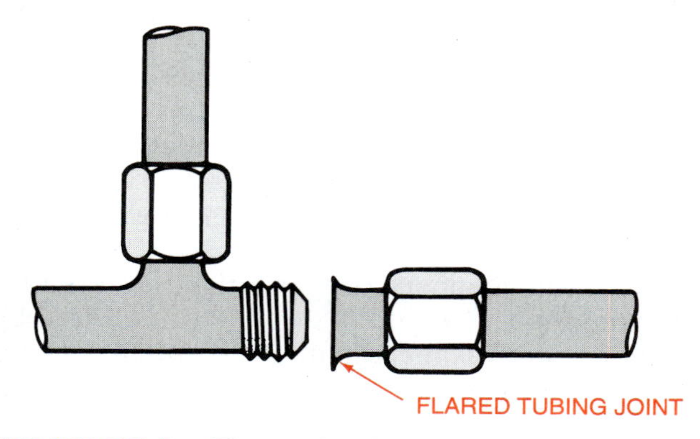

FIGURE 35-4. The ends of soft copper tubing may be flared to fit bell-shaped fittings.

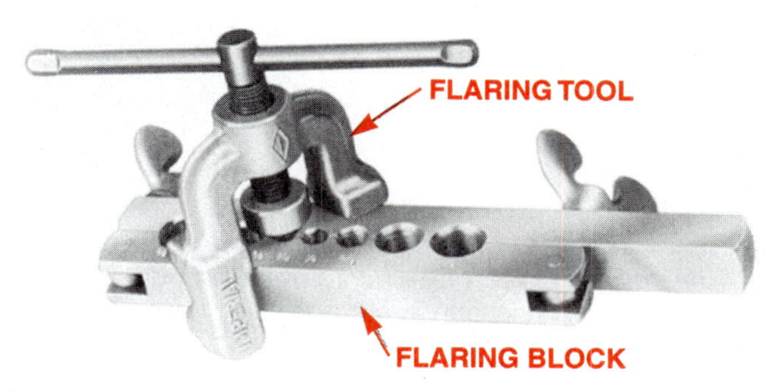

FIGURE 35-5. A flaring tool is used to flare the ends of soft copper tubing. (*Courtesy of Mac Tools, Inc.*)

sizes of pipe for home and farm applications (ID in inches) are $^1/_4$, $^3/_8$, $^1/_2$, $^3/_4$, 1, 1$^1/_4$, 1$^1/_2$, 2, 2$^1/_2$, 3, 4, and 6.

Steel Pipe

Steel pipe may be purchased as black pipe or galvanized pipe. Black pipe is painted black and has little resistance to rusting. Galvanized pipe is coated with zinc inside and outside. It resists rust for many years.

CAUTION:

- **The zinc coating on galvanized pipe gives off toxic fumes when welded or heated with a torch. Special precautions must be taken when using galvanized steel pipe for construction.**

Steel pipe of each nominal size has a specified wall thickness and is classified as standard pipe. Special pipe may be ordered with thicker walls. This pipe is classified as extra heavy or double extra heavy pipe. It is generally used for construction rather than plumbing.

The outside diameter of standard, extra heavy, and double extra heavy pipe is the same. The addi-

FIGURE 35-6. A pipe stand equipped with a chain vise and storage shelf.

tional wall thickness of the extra heavy and double extra heavy pipe results in reduced inside diameters, figure 35-7. Because the outside diameter is consistent, standard pipe fittings can be used with all three types of pipe.

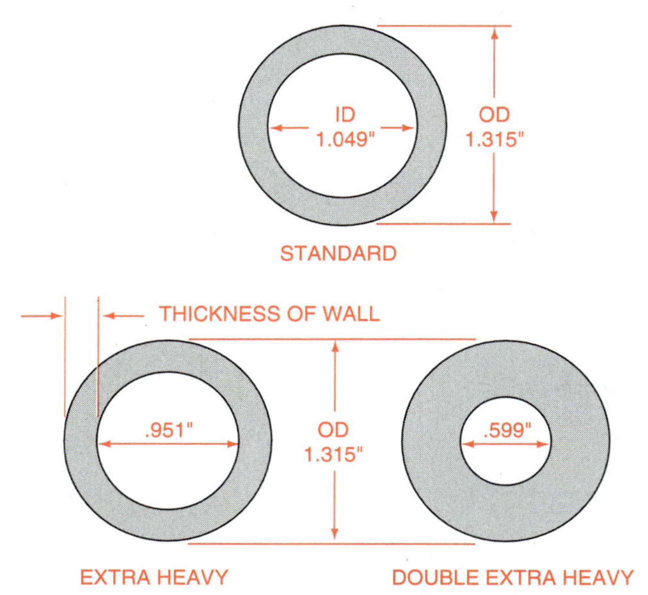

FIGURE 35-7. Standard, extra heavy, and double extra heavy steel pipe all have the same outside diameter for a given size. This permits pipe threading with the same die and the use of the same sized pipe fittings for all three types. (*Courtesy of H. Slater and L. Smith,* Basic Plumbing, 2nd ed., *Delmar Publishers, 1989*)

Copper Tubing and Pipe

Pipe made of copper is frequently referred to as tubing. It may be purchased in the soft, annealed form or the rigid form. For convenience, this unit refers to the soft form as tubing and the rigid form as pipe. The tubing can be bent around irregular parts of buildings. This makes it fast and easy to install. However, tubing does not have the neat appearance of rigid pipe. Therefore, it is used primarily in hidden spaces or where appearance is of little concern.

Copper is frequently preferred to steel for water lines because it resists corrosion, is easy to handle, and has been known to withstand freezing without breaking. Some disadvantages of copper are its high initial cost, high degree of expansion, and the bad taste created if the water is acid. Slightly acid water also reacts with copper to leave green stains in sinks and tubs.

Copper pipe and tubing are available in three types based on wall thickness, figure 35-8. The types are K, L, and M. Type K has the thickest wall, type L has medium wall thickness, and type M has the thinnest wall. Standard copper fittings are used for all three types. Generally, type L is specified by plumbing codes as the minimum acceptable size for use in buildings. Type K is generally required when the pipes are to be buried.

STANDARD WATER TUBE SIZE	ACTUAL OUTSIDE DIAMETER	NOMINAL WALL THICKNESS		
		TYPE K	TYPE L	TYPE M
3/8	.500	.049	.035	.025
1/2	.625	.049	.040	.028
5/8	.750	.049	.042	....
3/4	.875	.065	.045	.032
1	1.125	.065	.050	.035
1 1/4	1.375	.065	.055	.042
1 1/2	1.625	.072	.060	.049
2	2.125	.083	.070	.058
2 1/2	2.625	.095	.080	.065
3	3.125	.109	.090	.072
3 1/2	3.625	.120	.100	.083
4	4.125	.134	.110	.095
5	5.125	.160	.125	.109
6	6.125	.192	.140	.122
8	8.125	.271	.200	.170
10	10.125	.338	.250	.212
12	12.125	.405	.280	.254

Note: All measurements are in inches. Type M is not made in 5/8" size.

FIGURE 35-8. Sizes and weights of copper tubing and pipe. (*Courtesy of H. Slater and L. Smith,* Basic Plumbing, 2nd ed., *Delmar Publishers, 1989*)

Flexible copper tubing may be joined by flaring the ends of the tubing and connecting them with bell-type fittings. Very small tubing can also be joined with compression fittings. A compression fitting grips the pipe by compressing a special collar with a threaded nut. This procedure does not require the use of a flaring tool or soldering torch. A brief review of sweating or soldering copper tubing is shown in figure 35-9.

Plastic Pipe

In addition to steel and copper pipe, plastic pipe is used extensively in water systems. Plastic is the easiest type of pipe to install. The term plastic is used here to include all pipe made from synthetic materials. Such materials have long been used for cold water lines. Some plumbing codes now permit the use of certain plastic pipe materials for hot water

STEP 1

CLEAN THE OUTSIDE OF COPPER TUBE
WITH FINE STEEL WOOL OR EMERY CLOTH
TO A BRIGHT FINISH.

STEP 3

FLUX

APPLY A THIN, EVEN COATING OF FLUX TO
THE OUTSIDE OF THE COPPER TUBE.

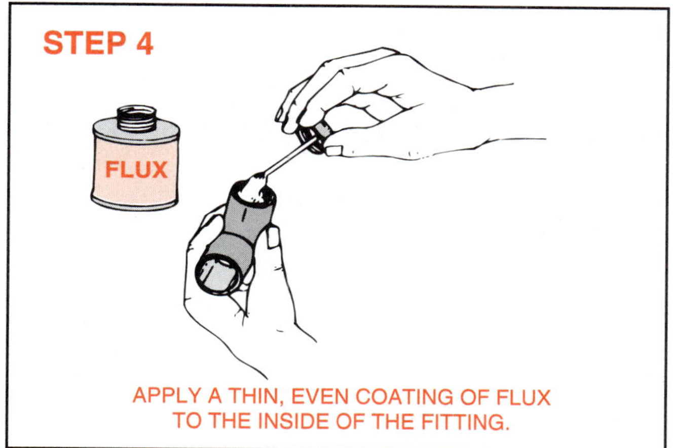

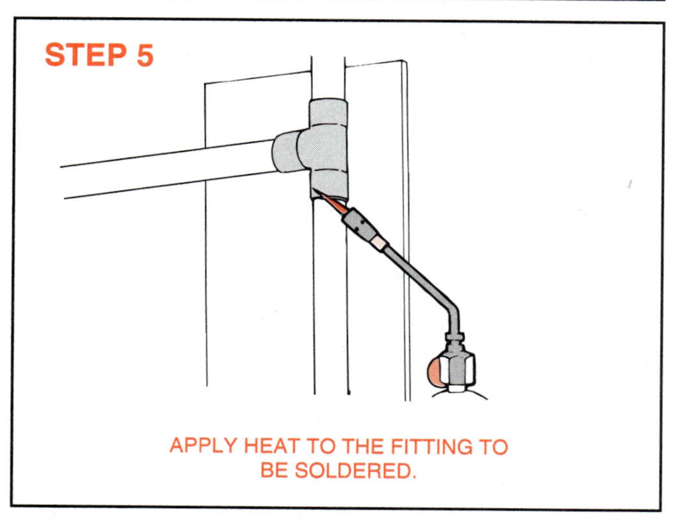

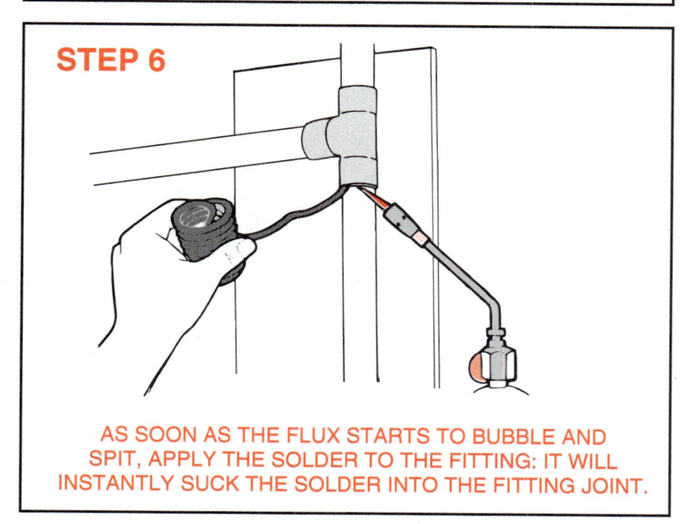

FIGURE 35-9. Six simple steps for soldering copper tubing. (*Courtesy of Step by Step Guide Book Co., 1981*)

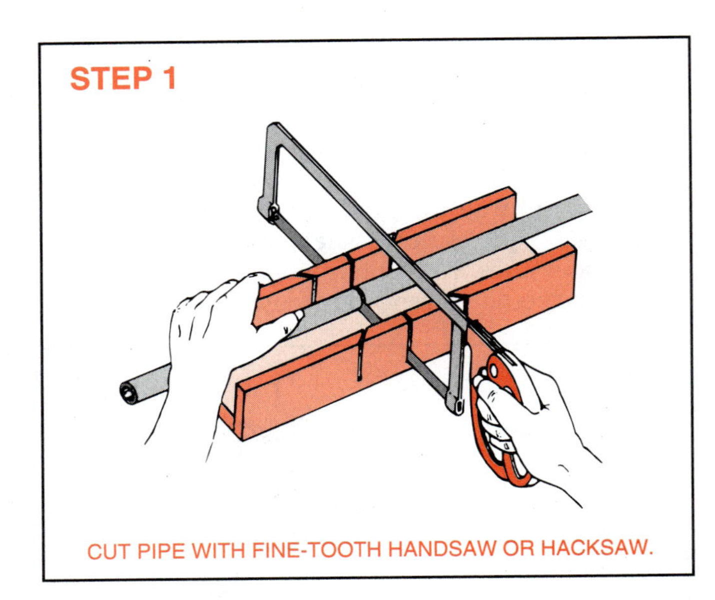

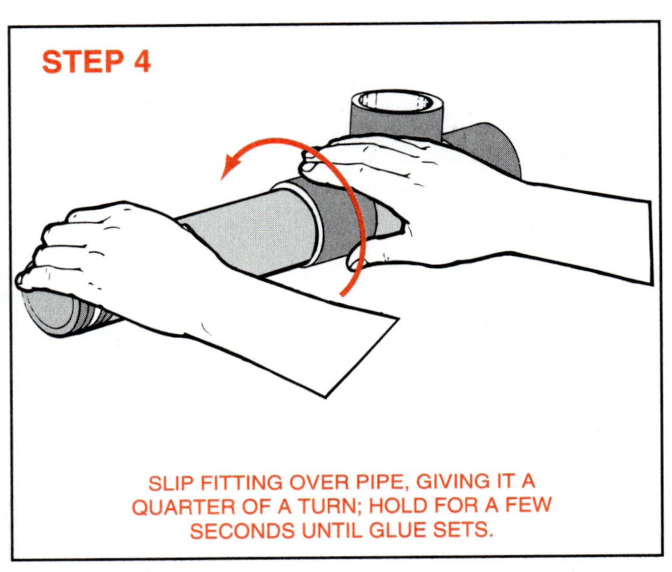

FIGURE 35-10. Steps in cementing plastic pipe. (*Courtesy of Step by Step Guide Book Co., 1981*)

lines. Most plastic pipe is fastened together by the simple procedure shown in figure 35-10. The steps of the procedure follow.

PROCEDURE

1. Use a fine-tooth saw to cut the pipe so the end is square.
2. Remove the rough or ragged edges using a knife or special cutter.
3. Clean the inside of the fitting and the outside of the pipe with a cloth and a special plastic cleaner.
4. Swab plastic cement on the outside of the pipe and in the fitting where the parts will join.
5. Press the fitting onto the pipe and give it a quarter of a turn. The fitting will be difficult to remove after a few seconds. It will set permanently shortly thereafter.

Plastic pipe is made from several different materials. Polyethlene (PE) has been in use for several decades. It is used extensively for cold water lines. PE pipe is usually black, somewhat flexible, and is available in rolls. It is popular for direct burial, pump installations, and surface water lines, figure 35-11. PE pipe is assembled by pushing the pipe over the grooved section of the fittings. The pipe is secured to the fitting using a rustproof stainless steel clamp, figure 35-12. There is no need for cement or pipe sealing compounds. Plastic pipe is easily removed by sawing with a hacksaw.

FIGURE 35-11. Polyethylene pipe is excellent for cold water lines either in underground or surface areas.

FIGURE 35-12. Polyethylene pipe is attached to special grooved pipe fittings with stainless steel bands.

Polyvinyl chloride (PVC) and chlorinated polyvinyl chloride (CPVC) are rigid and generally white or gray in appearance. PVC pipe is used for cold water lines and CPVC is approved by some plumbing codes for hot water lines. Figure 35-13 shows plastic pipe being used for the vent and drain

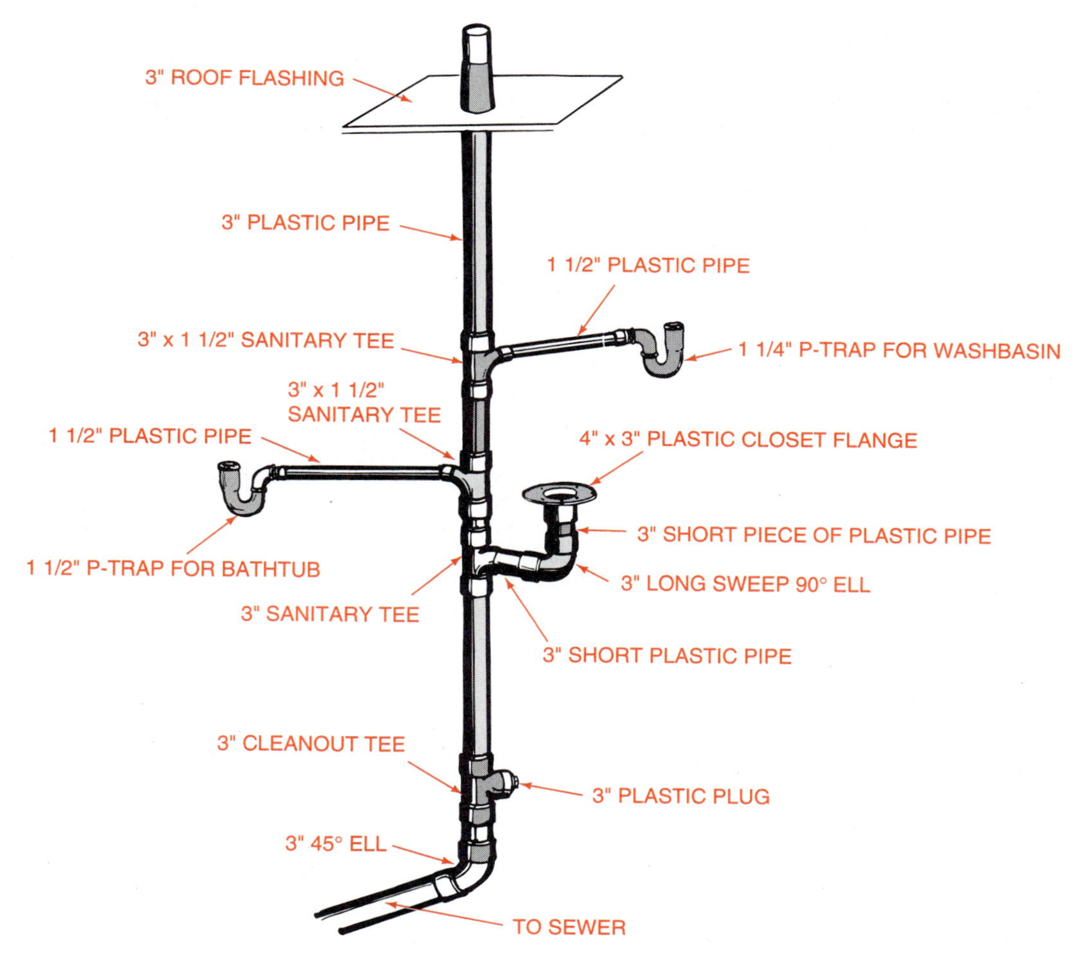

FIGURE 35-13. Plastic vent and drain system for tub, basin, and toilet. (*Courtesy of Step by Step Guide Book Co., 1981*)

system for a tub, basin, and toilet. PVC pipe is especially easy to install. It is light in weight, may be cut with any saw, and is assembled using a liquid cement.

Acrylonitrite-butadiene-styrene (ABS) is used for sewerage and underground applications. Tests seem to indicate that it will last almost indefinitely. Manufacturers reportedly guarantee the material for 50 years. It is assembled by cementing the pipe to the fittings. The light weight of the pipe and the ease of assembly make plastic pipe especially attractive for sewerage installations where large diameters are required.

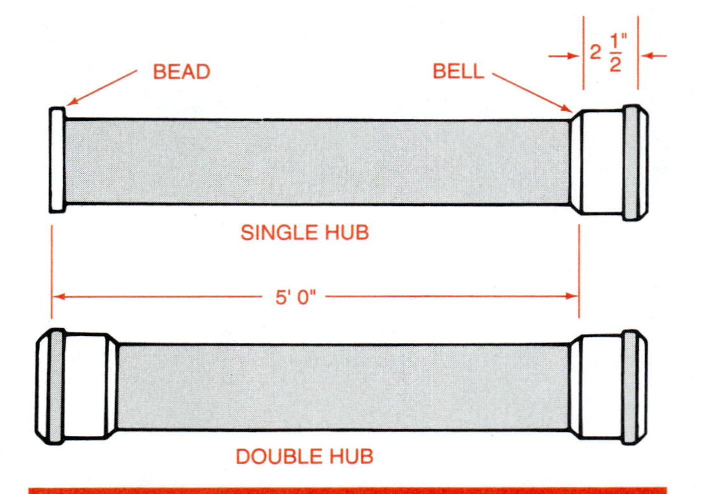

WEIGHT OF SOIL PIPE

Service

Size	Lb Per Ft	Size	Lb Per Ft
2"	4	6"	15
3"	6	7"	20
4"	9	8"	25
5"	12		

Extra Heavy

Size	Lb Per Ft	Size	Lb Per Ft
2"	5	8"	30
3"	9	10"	43
4"	12	12"	54
5"	15	15"	75
6"	19		

FIGURE 35-14. Cast iron soil pipe is available in a variety of diameters and two weights. It may be purchased with straight ends, with a hub on one or both ends, or with a bell-shaped end. (*Courtesy of H. Slater and L. Smith,* Basic Plumbing, 2nd ed., *Delmar Publishers, 1989*)

Cast Iron Soil Pipe

Cast iron soil pipe has thick walls and is used mostly for sewerage systems. Popular sizes range from 2 to 6 inches in diameter. Cast iron pipe is available in 5- and 10-foot lengths and in two weights—service and extra heavy. It can be purchased with straight ends, with an enlarged area called a hub on one or both ends, or with a bell-shaped end, figure 35-14.

Cast iron systems are joined by one of several methods. The oldest method is to place the straight or hub end into the bell end and pack the space with oakum. Oakum is a ropelike material that makes a watertight seal when it is compacted. After the oakum is added, the joint is sealed with molten lead, figure 35-15.

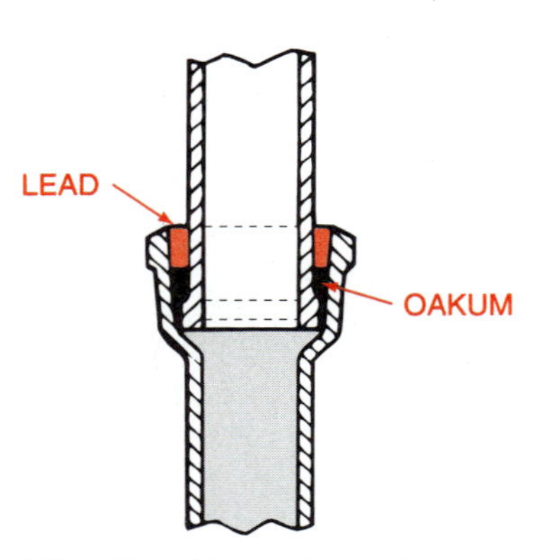

FIGURE 35-15. Cast iron pipe joint made with oakum and lead. (*Courtesy of H. Slater and L. Smith,* Basic Plumbing, 2nd ed., *Delmar Publishers, 1989*)

Two newer methods of connecting cast iron pipe that permit greater speed and ease in assembling joints have become popular. Both methods use a neoprene gasket to make a watertight seal. Neoprene is a tough, rubberlike material that is resistant to moisture and rotting.

One method uses pipe with one bell-shaped end and one straight end. The neoprene gasket is mounted in the bell, and the inside of the gasket is lubricated, figure 35-16. The straight end of the pipe or fitting is then slipped into the gasket. The resulting joint is watertight and flexible.

The other method uses pipe and fittings with straight ends. The straight ends are slipped into a lubricated neoprene gasket. A rustproof stainless steel band is then tightened over the joint, figure 35-17. This method has the advantage of requiring less space than the other methods because there are no bells on the pipe and the bands are thin.

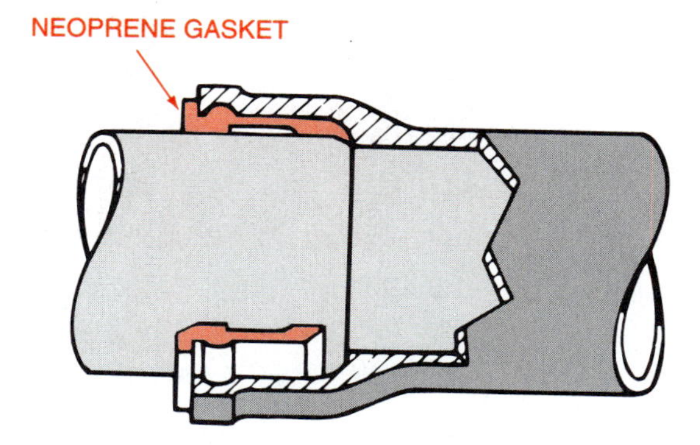

FIGURE 35-16. Cast iron pipe joint made with a neoprene gasket in a bell end. (*Courtesy of H. Slater and L. Smith,* Basic Plumbing, 2nd ed., *Delmar Publishers, 1989*)

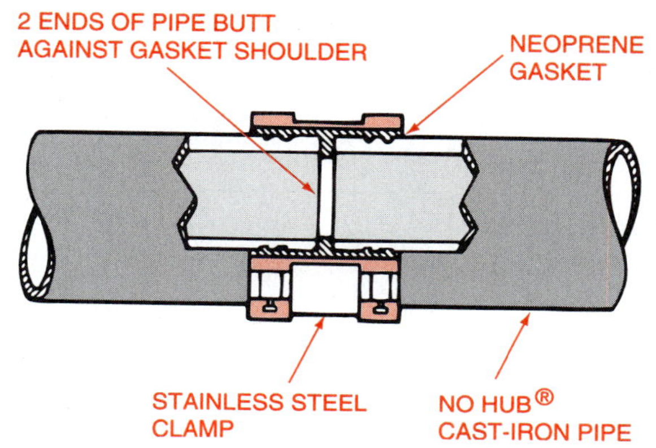

FIGURE 35-17. Cast iron pipe joint made with a neoprene gasket and a stainless steel clamp on straight ends. (*Courtesy of H. Slater and L. Smith,* Basic Plumbing, 2nd ed., *Delmar Publishers, 1989*)

Cast iron is heavy and hard to handle. Since copper and plastic are lightweight, require less space, and are easy to assemble, they are used extensively for indoor sewer lines. However, cast iron is frequently preferred for buried sewer lines. It resists rust well, is durable, and if properly installed is quite resistant to crushing.

Selecting Pipe Material

Each kind of pipe has advantages and disadvantages. When making repairs it is generally easier to repair systems with the kind of pipe used in the original installation. When planning a new system, however, it is important to consider the relative merits of each kind of pipe, figure 35-18.

IDENTIFYING PIPE FITTINGS

Pipe fittings are used to connect pipe. It is important to learn the names of basic pipe fittings in order to plan plumbing systems, order plumbing materials, and communicate with fellow workers about the job. The names of the fittings are the same for all kinds of pipe. Figure 35-19 shows some of these fittings.

Fittings used to connect pipes of different types (such as copper tubing to steel pipe) are called adaptors. Bushings and reducers are used to connect pipes of different sizes (such as $^1/_2$-inch steel pipe to $^3/_4$-inch steel pipe). Pipes coming together from different directions may be connected by an elbow, tee, or Y. The elbow, or ell, is used where a single line changes direction, the tee and Y are used where a line splits in two directions. A coupling is used to connect two pieces of similar pipe. A union also connects two pieces of similar pipe. An advantage of the union is that it permits opening the line without cutting the pipe. A pipe fitting may be closed with a plug; a pipe end may be closed with a cap.

Short pieces of steel pipe are hard to thread on the job. Therefore, they are manufactured and sold in various lengths with both ends threaded. A short piece of pipe is called a nipple. Whenever fittings are threaded onto steel pipe, it is necessary to use pipe dope or teflon tape to make a watertight seal. Teflon tape is also recommended for sealing threads when gas will be used in the lines.

MAINTAINING WATER SYSTEMS

Properly installed water systems are generally quite durable. However, certain types of water react with steel and copper and may shorten the life of pipe, fittings, faucets, valves, and pumps unless the proper water conditioners are used. A faucet is a device that controls the flow of water from a pipe or container. A valve is a device that controls the flow of water in a pipe.

A green stain beneath the faucet in a tub or sink generally indicates a reaction between the water and copper pipes. Reddish water coming from a faucet generally means a rusting condition caused by a reaction between water and steel pipes or tanks.

The green stain or red water problems can generally be corrected by a water treatment system. Such systems are expensive but may be worth the investment. Money spent for water treatment may be offset by savings on repairs to the plumbing system.

Repairing Faucets

Normal maintenance of water systems includes the occasionally tightening of packing nuts and material or changing faucet washers or O rings to stop faucets from dripping. An O ring is a piece of rubber shaped like an O which fits into a groove on a shaft or in a housing. It prevents liquids from passing between a rotating shaft like a faucet stem and the surrounding

FACTORS TO CONSIDER	GALVANIZED STEEL (3 OZ. COATING MIN.)	COPPER		PLASTIC
		TYPE K (HEAVY DUTY)	TYPE L (STANDARD)	
Underground soil corrosion—Probable life expectancy (1)	30 plus yrs. under most soil conditions. (If no-corrosion inside pipe,-life could extend to 100 yrs. or more	40–100 yrs. under-most conditions-	30–80 yrs. under-most conditions-	Experience indicates durability is satisfactory under most soil conditions.
	Waterlogged soils under most conditions —- 12–16 years. May be less than 10 years in very high acid soils	14–20 yrs. in high sulfide conditions. May be less than 10 yrs. in cinders.	12–14 yrs. in high sulfide conditions. May be less than 10 yrs. in cinders.	
Resistance to corrosion inside pipe	Will corrode in acid, alkaline and hard waters or with electrolytic action. (2)	Normally very resistant. May penetrate rapidly-in water containing free carbon dioxide.		Very resistant
Resistance to deposits forming inside pipe	Will accumulate lime deposits from hard water. (2)	Subject to lime scale and encrustation from suspended materials.		Resistant, but occasional deposits will form. (3)
Effect of freezing	Bursts if frozen solidly-	Will stand mild freezes.		PE—will stand some freezing. PVC—will stand mild freezes.
Safe working pressures (lbs. per sq. in.)	Adequate for pressures-developed by small water systems	Adequate for pressures developed by small water systems.		Working pressures at 73°F **PE** 80 to 160 **PVC** 180 to 600
Resistance to puncturing and rodents	Highly resistant to both	Resistant to both		PE—Very limited resistance to puncture and rodents PVC—resistant
Effect of sunlight	No effect	No effect		PE—Weakens with prolonged exposure PVC—High Resistant
Effect on water flavor	Little effect	Very acid water disolves enough copper to cause off flavor.		Little effect (4)
Lengths available	21 ft. lengths	Soft temper: 60-ft.—100-ft. coils up to 1" diameter 60-ft. coils above 1" diameter Hard temper: 12- and 20-ft. lengths		**PE** usually in 100-ft. coils or longer **PVC** usually in 20 ft. lengths
Comparative Weight (Approx. lbs. per foot) Inside diameter				
1/2	.85 lb.	.34 lb.	.29 lb.	.06 lb.
3/4	1.13	.64	.46	.09
1	1.60	.84	.66	.14
1 1/4	2.27	1.04	.88	.24
1 1/2	2.72	1.36	1.14	.30
2	3.65	2.06	1.75	
2 1/2	5.79	2.93	2.48	

FIGURE 35-18. Relative merits of different kinds of piping materials. (*Courtesy of American Association for Vocational Instructional Materials*)

FACTORS TO CONSIDER	GALVANIZED STEEL (3 OZ. COATING MIN.)	COPPER		PLASTIC	
		TYPE K (HEAVY DUTY)	TYPE L (STANDARD)		
Ease of bending	Difficult to bend except for slight bends over long lengths.	Soft temper bends readily, will collapse on short bends. Hard temper difficult to bend except for slight bends over long lengths		**PE** Bends readily, will collapse on short bends	**PVC** Rigid Bends on short bends long radius
Conductor of electricity	Yes	Yes		No	
Comparative cost index				**PE**	**PVC**
Pipe size (in.)					
1/2	10	21	17	3	3
3/4	13	39	28	5	4
1	18	54	42	9	5
1 1/4	24	66	59	14	7
1 1/2	28	87	74	18	10
2	39	130	110	27	15

(1) Derived from studies reported by Dennison, Irving A. and Romanoff, Melvin, "Soil-corrosion Studies, 1946 and 1948; Copper Alloys, Lead and Zinc," Research paper RP2077, Vol. 44, March 1950 and "Corrosion of Galvanized Steel in Soils," Research paper 2366, Vol. 49, No. 5, 1952. National Bureau of Standards, U.S. Dept. of Commerce.

(2) It is possible to greatly reduce corrosion and prevent lime scale in steel pipe by adding a phosphate material. It coats the inside of pipes, as well as the lining of all connected equipment. Prevents further lime scale and greatly reduces corrosion.

(3) Jones, Elmer E., Jr., "New Concepts in Farmstead Water System Design," Am. Society of Agricultural Eng., paper No. 67-216, 1967.

(4) Tiedeman, Walter D., "Studies on Plastic Pipe for Potable Water Supplies," Journal American Water Works Association, Vol. 46, No. 8, Aug. 1954.

FIGURE 35-18. *Continued*

hole. Packing is a soft, slippery, and wear-resistant material. It, too, is used to seal the space between a moving shaft and the surrounding area. A faucet washer is a rubberlike part that creates a seal when pressed against a metal seat. A seat is a nonmoving part that is designed to seal when a moving part is pressed against it.

Toilet tanks and automatic livestock waterers have shut-off valves that need periodic service. Typically the problem is a failure to shut off completely. The replacement of O rings, diaphragms, or rubber seats generally corrects the problem. A diaphragm is a flat piece of rubber that creates a seal between a part moving a short distance, and sealing against a nonmoving part.

The basic parts of valves and single faucets are the handle, stem, packing nut, packing, bonnet, washer, washer screw, seat, and body, figure 35-20. Mixing faucets that control the flow of hot and cold water with one handle have a different design, figure 35-21.

Aerators mix air with the water as it comes from the faucet to create a smooth stream. Aerators have a screen to catch particles of rust or dirt. This screen and other parts of the aerator must be removed and cleaned occasionally to work properly.

Stopping Leaks Around Faucet Stems. To repair a faucet that leaks around the stem the following procedure is suggested.

PROCEDURE

1. Tighten the bonnet or packing nut.
2. If the stem still leaks, turn off the waterline valve, and unscrew the bonnet or packing nut and the stem, figures 35-22 and 35-23.
3. Replace the packing or O ring and reassemble, figure 35-24.
4. Turn on the water to the faucet.
5. Tighten the packing nut or bonnet just enough to prevent leaking.

CAUTION:

- **Overtightening will make the faucet hard to turn on and off.**

Faucet spouts that swing may leak at the base. To correct this problem simply tighten the nut that attaches the spout to the faucet body. If necessary, replace the O ring or packing on the spout.

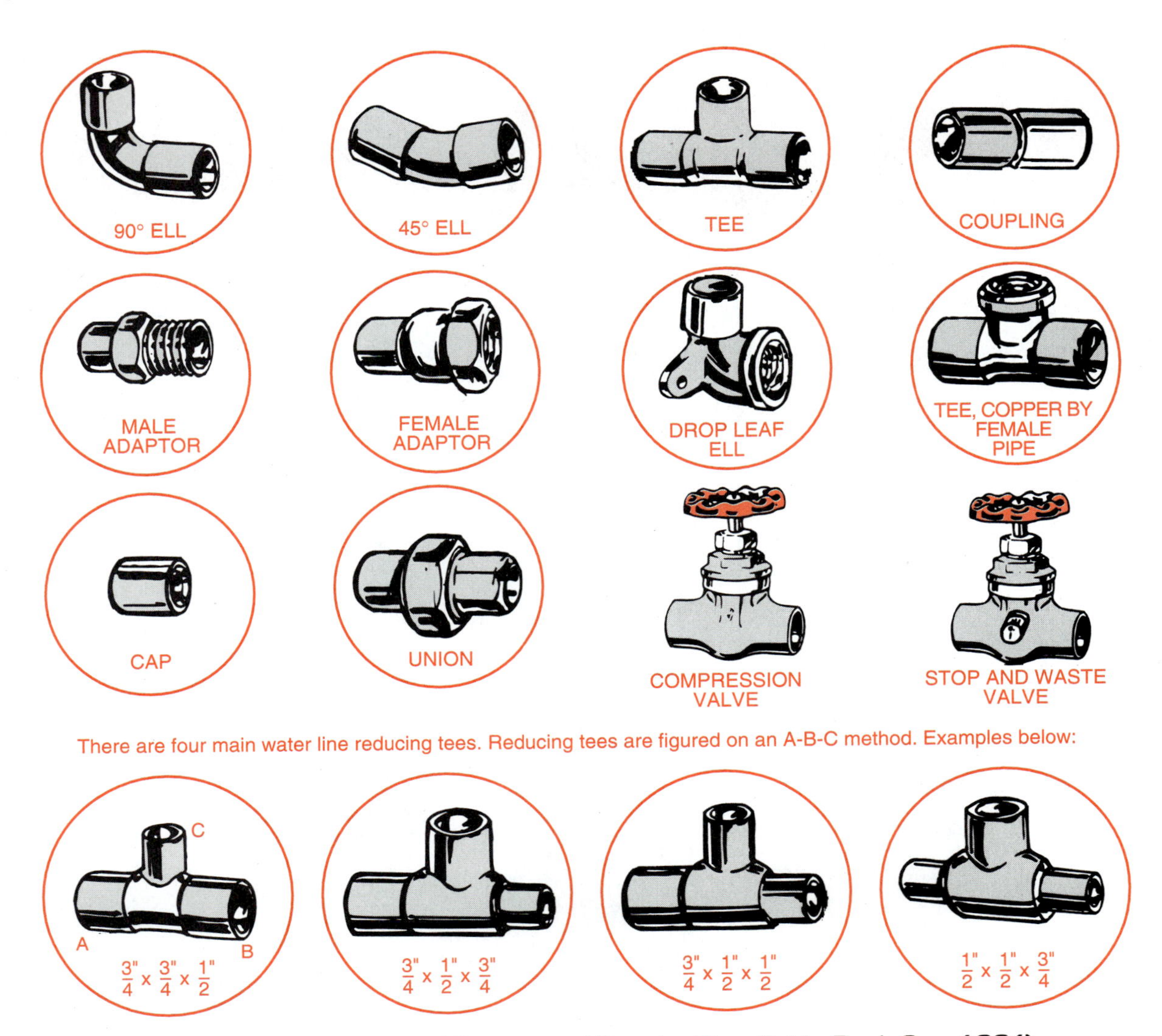

FIGURE 35-19. Names of pipe fittings. (*Courtesy of Step by Step Guide Book Co., 1981*)

Loose handles are tightened by prying off or unscrewing the plug in the handle and tightening the handle screw. If the handle is broken or slips, a replacement handle can generally be purchased. Most faucet parts can be purchased from a hardware store or a plumber.

Stopping Faucets from Hammering or Dripping. A common problem with faucets and valves is a hammering noise when the water is running slowly. This may be caused by worn threads on the stem or a loose washer, figure 35-25. If the problem is worn threads, the stem or entire faucet may need replacing. The problem may be reduced or corrected by tightening the packing nut.

The more likely cause of a hammering faucet is a loose washer. This is corrected by removing the stem and tightening the washer. It is generally advisable to replace the washer while the faucet is open. When replacing washers, be sure to use one of the correct size, shape, and style, figure 35-26. The size is determined by the diameter. When tightening the screw, draw it down snugly but not so tightly that the washer is forced out of shape, figure 35-27.

Replacing Seats. The seats of faucets may become eroded by chemical reaction with the water. This may happen within weeks if the faucet is permitted to drip. The seats in many faucets can be replaced using an allen wrench, figure 35-28.

Repairing Concealed Faucets. Bathtub faucets are generally inside the wall with only the handles in view. To repair the faucet, the handle must be removed first. A large, deep socket is then used to unscrew the bonnet, figure 35-29. The bonnet and stem are removed as an assembly. The washer or seat is then serviced as described previously.

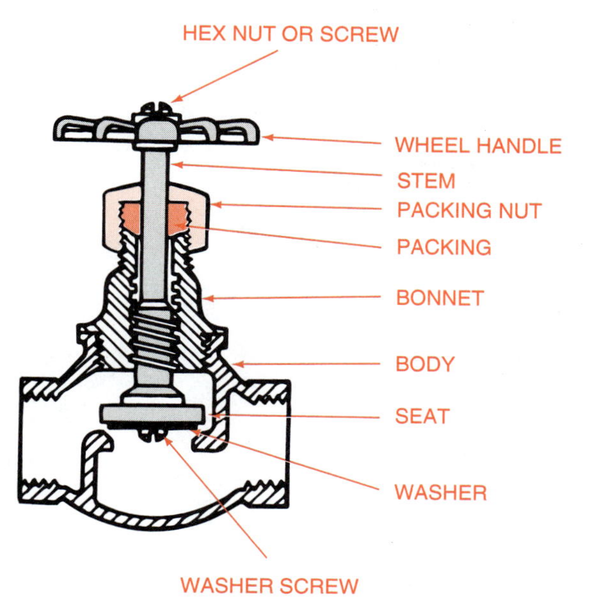

FIGURE 35-20. Parts of typical water valves and single faucets.

FIGURE 35-22. Older type of faucet with the stem removed.

Repairing Float Valve Assemblies

Float valve assemblies are used to control the water flow to flush toilets. Another important use is in livestock watering tanks with automatic filling devices. Float assemblies consist of a body with a diaphragm

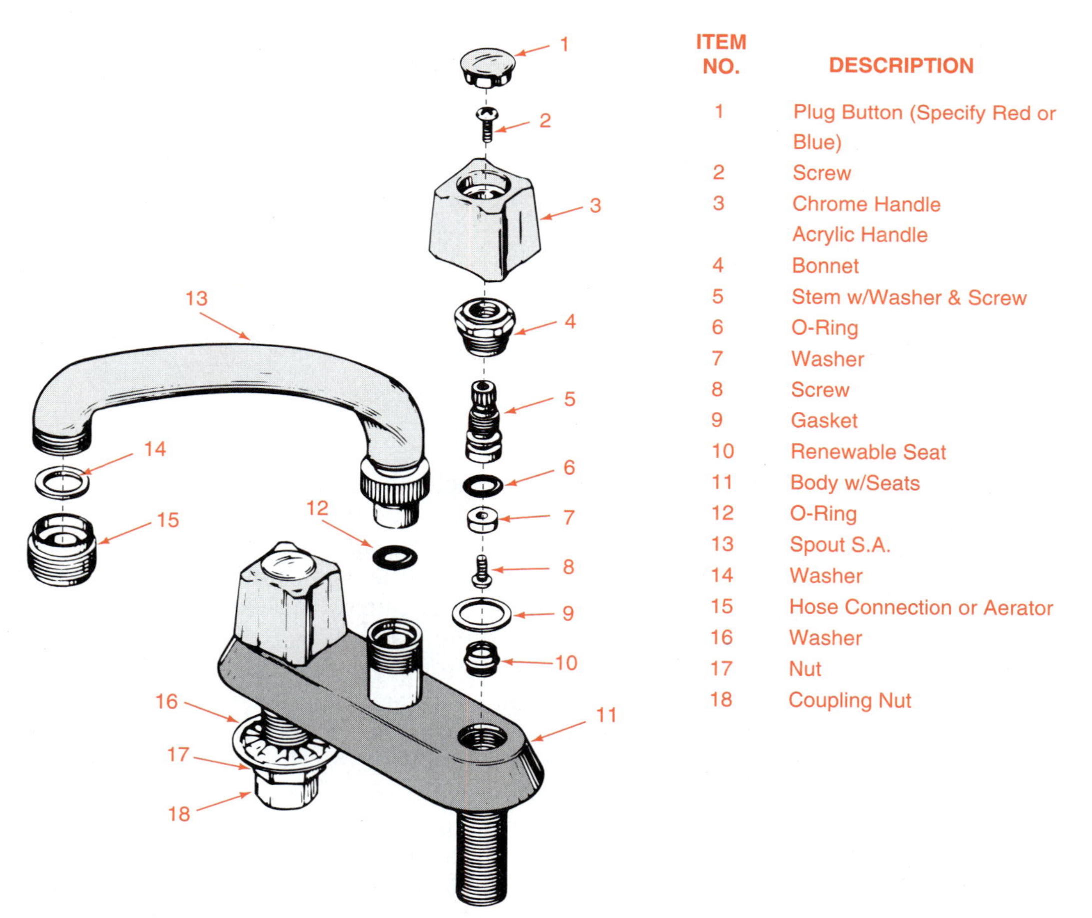

ITEM NO.	DESCRIPTION
1	Plug Button (Specify Red or Blue)
2	Screw
3	Chrome Handle Acrylic Handle
4	Bonnet
5	Stem w/Washer & Screw
6	O-Ring
7	Washer
8	Screw
9	Gasket
10	Renewable Seat
11	Body w/Seats
12	O-Ring
13	Spout S.A.
14	Washer
15	Hose Connection or Aerator
16	Washer
17	Nut
18	Coupling Nut

FIGURE 35-21. Parts of a typical mixing faucet used on kitchen or laundry sinks. (*Courtesy of Kohler Co.*)

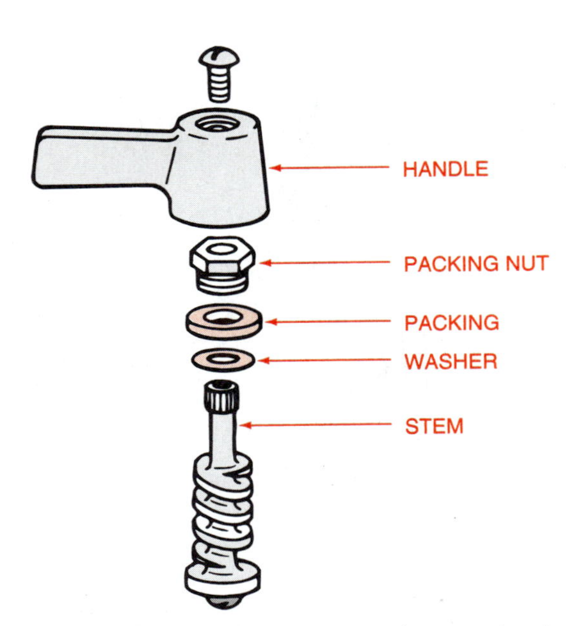

FIGURE 35-23. The handle must be removed to get to the packing nut on this type of faucet. The stem may be removed after the packing nut is unscrewed.

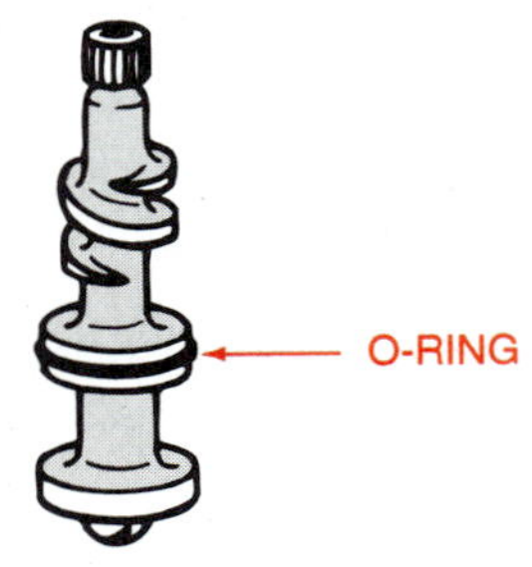

FIGURE 35-24. Some faucet stems have an O ring instead of packing.

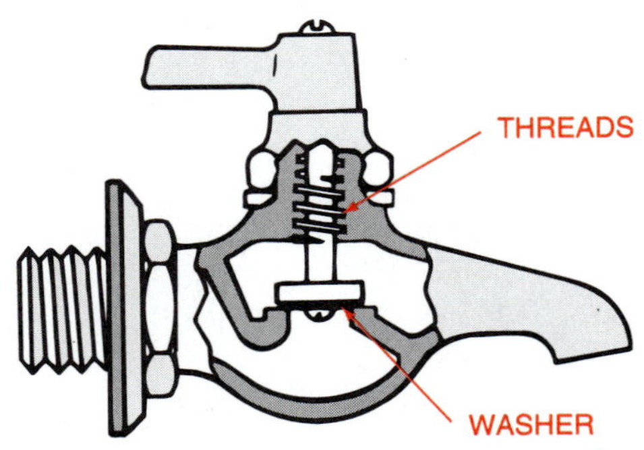

FIGURE 35-25. Worn threads or a loose washer may cause a faucet to make a hammering sound when the water is running slowly.

FIGURE 35-26. It is important to use the correct size and shape of washer.

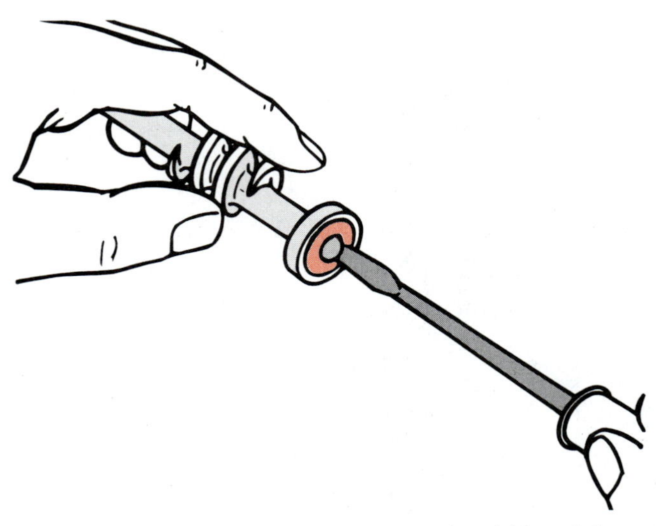

FIGURE 35-27. Faucet washers should be tightened snugly but not so tightly that they are forced out of shape.

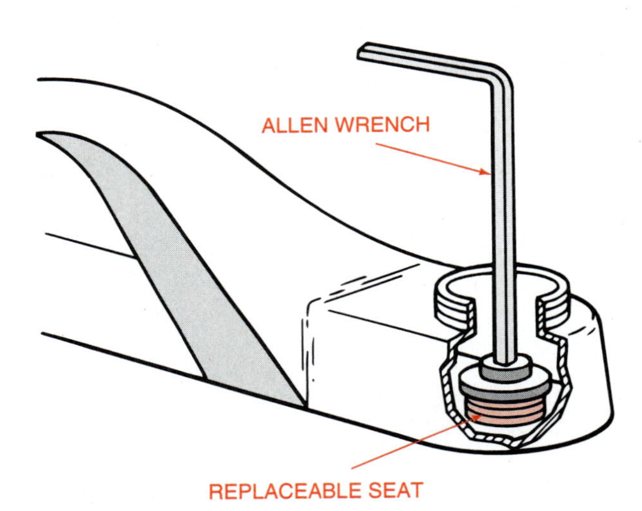

FIGURE 35-28. The seats in many faucets can be replaced.

or washer valve and a float that is lifted by the water as the tank fills, figure 35-30.

The action of the rising float puts pressure on the valve when the water in the tank reaches a specified level. This causes the valve to shut off the flow of incoming water. When the water level drops in the tank, the float drops, opens the valve, and permits water to enter until the tank refills.

Repairs to float valve assemblies may involve tightening the nut or replacing a gasket to stop leaks at the point where the assembly passes through the bottom of the tank; or a washer or diaphragm in the valve may need to be replaced to stop a flooding condition caused by the valve not shutting off completely. Some systems use a hollow ball for a float. If the ball develops a leak, it will fill with water and no

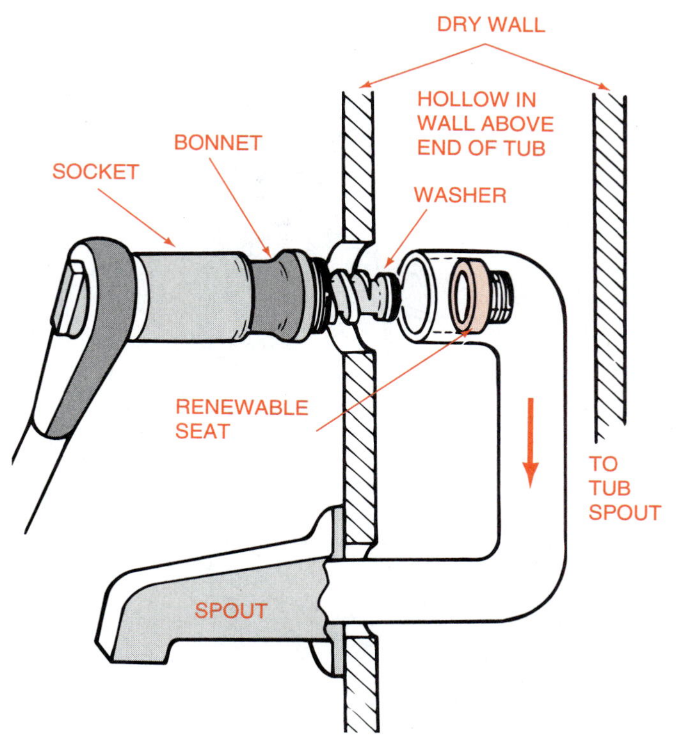

FIGURE 35-29. Bathtub faucets require a large, deep socket wrench to remove the bonnet.

longer float. In this case, the float and its rod are unscrewed from the assembly and replaced.

Repairing Flush Valves

Flush valves on toilets control the water used in the flushing process, figure 35-31. The valves frequently fail to seat properly and permit water to flow continuously from the tank to the toilet bowl, figure 35-32. This causes the filler valve to remain open and water to flow in continuously until an adjustment is made. Such a condition, if not corrected, can cause a well to be pumped dry or a septic system to be flooded, figure 35-33.

In most cases the problem is caused by the chain or linkage getting twisted or caught on another part. This can be corrected by simply flipping the trip level up and down. If this does not permit the valve to reseal, then the chain flapper or ball seal, and other parts should be cleaned of all rust, slime, or other deposits. If this does not correct the problem, replace the flapper and other parts as required.

Maintaining Septic Systems

Some kind of treatment plant is required for all human waste from bathrooms, kitchens, and laundry facilities if diseases are to be controlled. Proper treatment plants will provide conditions for bacteria to decompose solid waste and act upon paper, water,

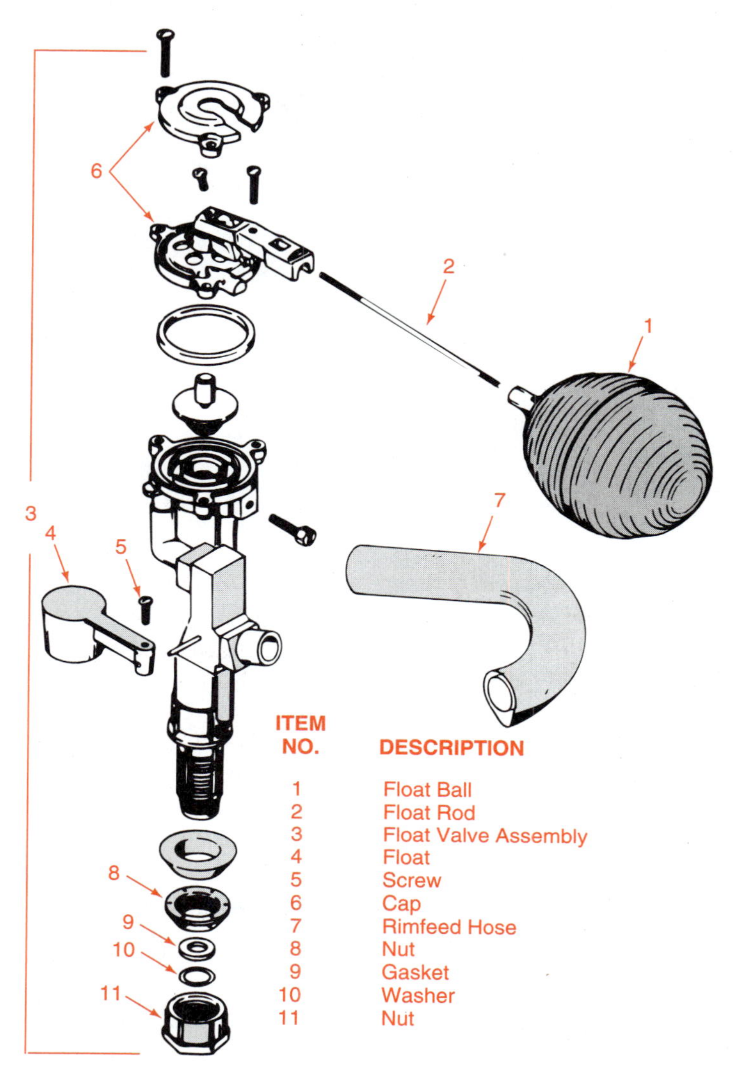

FIGURE 35-30. Parts of a typical float valve assembly used in toilet tanks and livestock watering tanks. (*Courtesy of Kohler Co.*)

and other materials that have been contaminated by harmful micro-organisms. The decomposition of solids in carbon dioxide and water, the filtration of water by soil particles, and oxidation by air will render sewerage relatively free of disease-causing organisms in an effective treatment system. As a further precaution, municipal systems and "package systems" for isolated institutions and housing developments, carefully meter exact amounts of chlorine into the affluent (liquid) from the process. The chlorine will kill any harmful bacteria present.

The septic system of a house or business located in a city or other suburban area will be connected to a street sewer line and feed into a municipal sewerage treatment plant. Such systems are maintained by the municipal government or other public agency. It is the property owner's responsibility to keep the

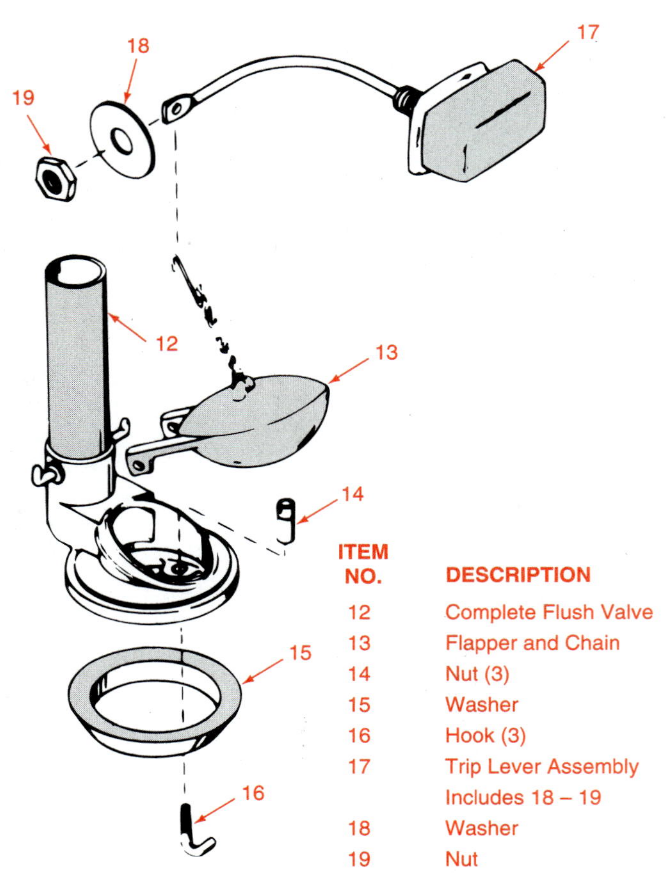

ITEM NO.	DESCRIPTION
12	Complete Flush Valve
13	Flapper and Chain
14	Nut (3)
15	Washer
16	Hook (3)
17	Trip Lever Assembly Includes 18 – 19
18	Washer
19	Nut

FIGURE 35-31. Parts of a flush valve used in toilet tanks. (*Courtesy of Kohler Co.*)

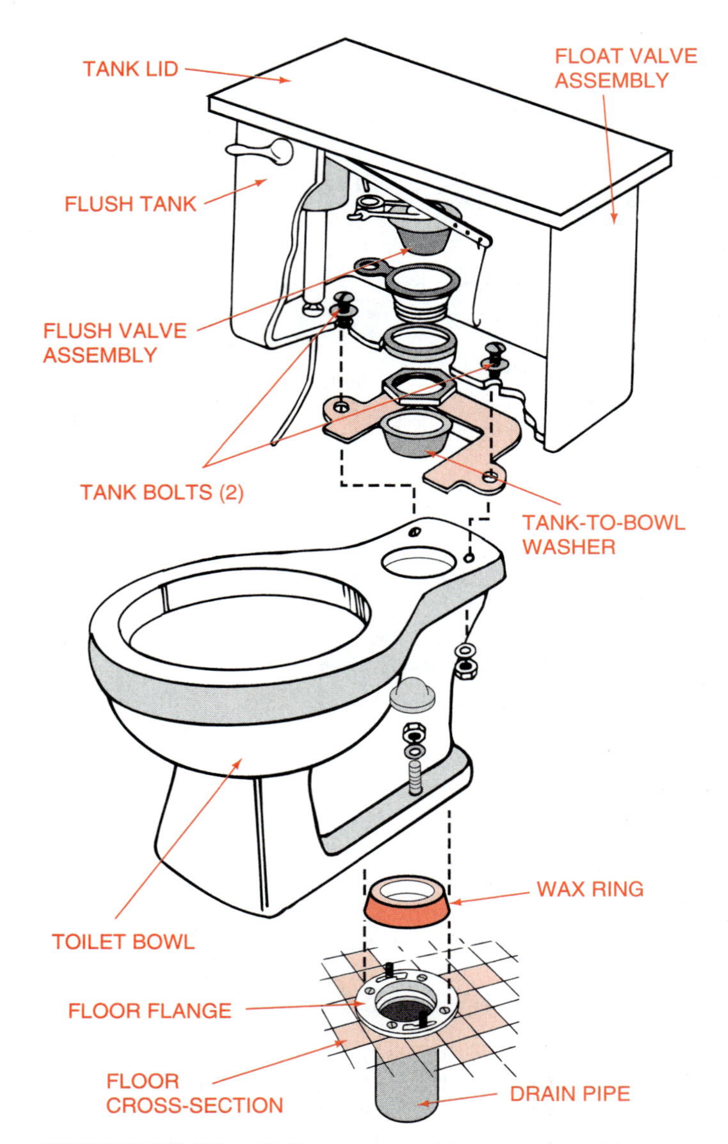

FIGURE 35-32. Toilet tank and bowl assembly.

sewer lines open from the fixtures in the building out to the connection at the street.

The system can accommodate human waste, toilet paper, garbage if shredded by a garbage disposal, and that is about all! Hair, excessive soap, grease, cotton, cloth, rubber, diapers, toys, and most other household objects are likely to plug up sewer lines and require the services of a plumber or specialized equipment to reopen the lines. Sometimes minor stoppages can be opened with chemical drain openers. Exterior lines may also become clogged with tree roots and require the use of a power drain tape or "snake" to cut the roots and restore the drain to good service.

Individual homes outside of cities and towns must have a septic tank to decompose the sewerage and field drains or leach fields to distribute and filter the effluent. Such systems start with a septic tank near the house where solids decompose and settle to the bottom as sludge and the liquid rises to the top. As more sewerage enters the septic tank, the accumulated water flows around a baffle and out to the distribution box. The distribution box has two or more field drains flowing out to the leach fields. Some localities may also permit round dry wells laid up with building blocks without mortar. Such wells have earthen floors and a concrete lid to prevent cave-in. Large gravel or stones about 2 inches in diameter are used to backfill around the structure to further extend the liquid storage area before it drains away through the surrounding soil for filtration and purification.

When field drains are used, a distribution box near the septic tank distributes the liquid to the various field drains, and gravel beds under the drains serve as a reservoir for liquids until they seep away into the soil. Under certain conditions, the drain lines may be placed close to the surface so water can evaporate into the air. When field drains are designed for evaporation, their locations can be observed by the tell-tale plushness of grass over the lines due to the abundance of water in the immediate area.

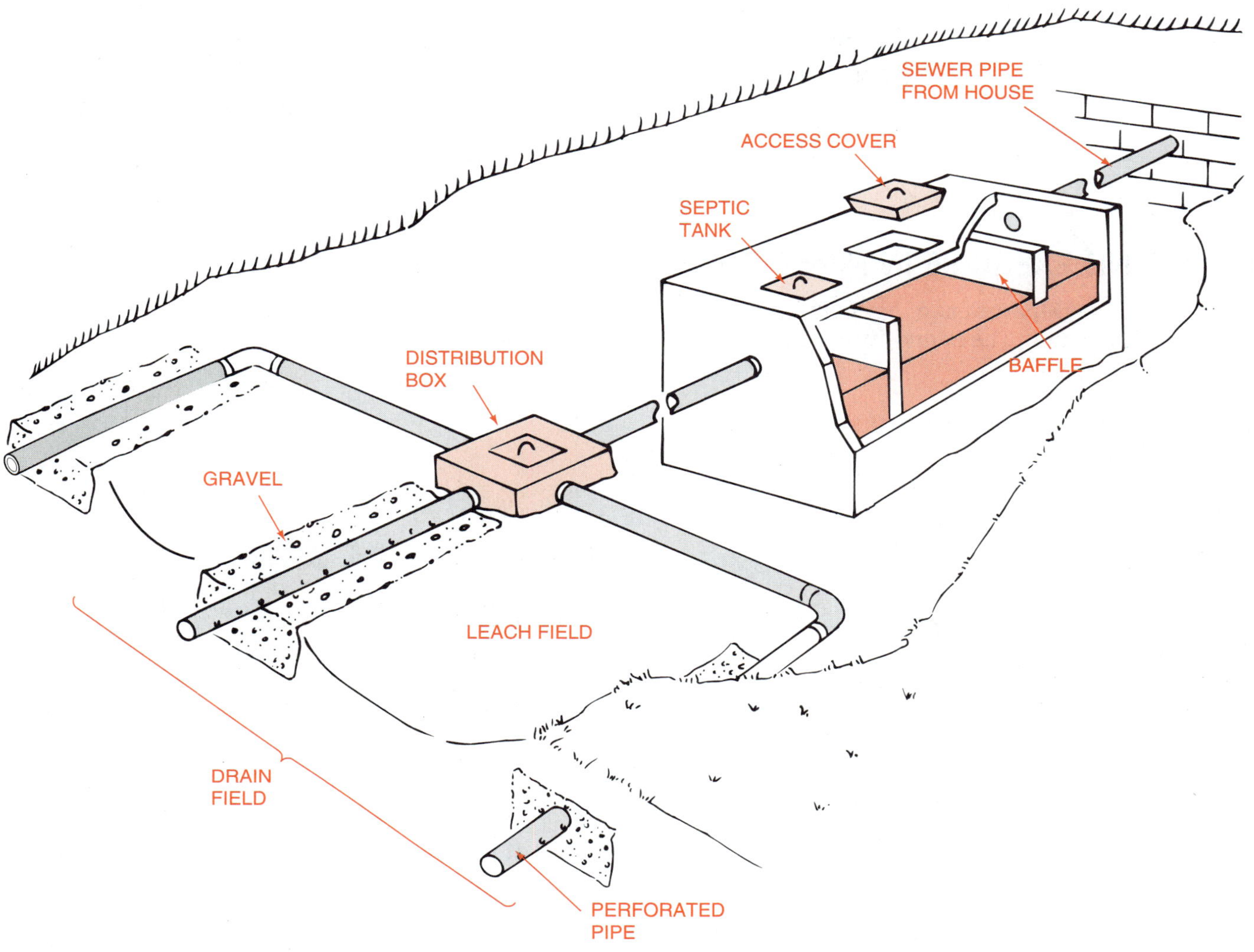

FIGURE 35-33. Typical home sewerage system using field drains.

A properly designed septic system with only appropriate materials being flushed away, will be trouble free for many years. However, depending on the conditions of design, use, and soil conditions, some maintenance may be required. Sludge should be removed from the septic tank periodically and the system inspected by a professional plumber to be sure all components are functioning properly. Once the system is put into service, it is important to keep it in continuous service or the solids will dry on the surfaces of the perforated pipes, building blocks, and stones, and clog the system. Further, any accumulation of surface water with a sewerage odor signals the need for professional attention to the septic system.

Since the water from the septic tank is still contaminated, it is essential to place the septic tank as far as possible from the well. Specifications for such systems are determined by local health authorities and systems must be installed according to state and local health codes.

STUDENT ACTIVITIES

1. Define the Terms to Know in this unit.
2. Sketch and label each pipe fitting illustrated in the unit.
3. Repair a faucet that leaks at the stem.
4. Repair a faucet that makes a hammering noise when the water is running slowly.
5. Repair a dripping faucet.

6. Repair the float mechanism in a water tank.
7. Complete the plumbing exercise in pipe fitting.

PLUMBING EXERCISE IN PIPE FITTING

(Courtesy of the Department of Agriculture and Extension Education, The Pennsylvania State University)

Purpose: The purpose of the pipe fitting exercise is to obtain experience in assembling a watertight unit using different types of plumbing materials. Skills to learn include the following:

- Using various types of adapters and fittings
- Cutting pipe materials to length
- Reaming and cleaning pipe ends

BILL OF MATERIALS

1 1/2-inch hose adapter (F × F)
1 1/2-inch sillcock (F)
1 1/2-inch × 4" nipple (galv.)
1 1/2-inch × 6" nipple (galv.)
1 1/2-inch × 1/2"× 1/2" tee (galv.)
1 1/2-inch coupling (galv.)
1 1/2-inch elbow (galv.)
2 1/2-inch adapters (M × PVC)
1 foot 1/2-inch PVC pipe
2 1/2-inch adapters (M × insert)
2 1/2-inch stainless steel pipe claims
1 foot 1/2-inch polyethylene pipe
1 1/2-inch adapter (F × c)
1 1/2-inch adapting elbow (F × c)
1 1/2-inch tee (c × c × c)
1 1/2-inch union (c × c)
1 1/2-inch adapter (c × M)
15 inches 1/2-inch type L copper pipe
Other materials incude appropriate pipe joint compound, solvent cement, flux and solder, and the necessary tools for performing the job.

Key: F – female threads
M – male threads
c – copper
galv. – galvanized

CAUTION:

- **This exercise involves the use of chemicals and extreme heat. Therefore, follow directions carefully.**

- Threading and making galvanized connections
- Making cemented plastic connections
- Soldering copper connections
- Making polyethylene connections using clamps

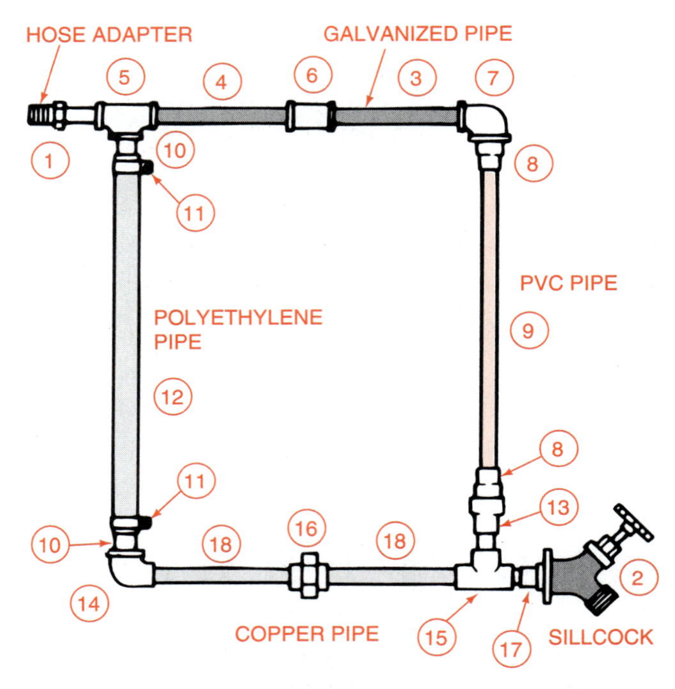

FIGURE 35-34. Plumbing exercise in pipe fitting. (*Courtesy of The Pennsylvania State University, Department of Agricultural and Extension Education*)

PROCEDURE

1. Measure and cut pipe to length as specified in the bill of materials.
2. Ream and clean ends of pipe.
3. Thread galvanized pipe.
4. Apply pipe joint compound to threads.
5. Assemble galvanized pipe and hose adapter as shown, figure 35-34.
6. Working with one connection at a time, apply cement to PVC pipe and its fitting, then assemble quickly using slight turning motion.

7. Assemble polyethylene pipe using insert connectors and clamps.
8. Apply soldering flux to copper pipe and fittings.
9. Assemble and solder copper joints.
10. Cool copper parts before proceeding.
11. Assemble unit as shown, attaching the threaded sillcock. Make use of the union on the copper pipe for assembly purposes.
12. Attach garden hose and test for leaks

SELF-EVALUATION

A. Multiple Choice. Select the best answer.

1. Extra heavy steel pipe has
   a. thicker walls than standard pipe
   b. smaller ID than standard pipe
   c. the same OD as standard pipe
   d. all of these
2. Galvanized pipe is coated with
   a. zinc
   b. paint
   c. oil
   d. galvanoleum
3. Galvanized pipe may be hazardous when
   a. used with acid water
   b. mixed with black pipe in the system
   c. exposed to soil
   d. cut with a torch or welded
4. Pipe that is available in both flexible and rigid form is
   a. steel
   b. PVC
   c. copper
   d. none of these
5. The heaviest copper pipe is type
   a. K
   b. M
   c. L
   d. Z
6. A kind of flexible pipe that is preferred for long runs of outdoor cold water lines is
   a. copper
   b. PE
   c. PVC
   d. steel
7. Green stain in a tub indicates
   a. poor quality finish on the tub
   b. reaction of water and steel
   c. reaction of water and copper
   d. reaction of water and plastic
8. Faucets may drip or leak due to faulty
   a. washers
   b. packing
   c. O rings
   d. any of these
9. A common problem with toilet tanks and automatic livestock waterers is
   a. incomplete shut-off
   b. green stain
   c. dripping
   d. all of these
10. The easiest water or sewerage system to install uses pipe made of
   a. cast iron
   b. copper
   c. plastic
   d. steel

B. Matching. Match the terms in column I with those in column II.

Column I

1. steel pipe
2. copper pipe
3. cast iron pipe
4. plastic pipe
5. ID
6. elbows
7. tee and Y
8. cap and plug
9. coupling and union
10. nipple

Column II

a. nominal pipe size
b. changes direction of single line
c. connect several pipes
d. oakum and lead joints
e. connect similar pipe in line
f. short length of pipe
g. used to close pipe and fittings
h. connected by threading
i. connected by cementing
j. connected by soldering

C. Completion. Fill in the blanks with the word or words that will make the following statements correct.

1. A faucet that leaks around the stem may be corrected by
   a. _____
   b. _____
2. Water reacting with steel pipe may be _____ in color.
3. Two problems that may cause a faucet to make a hammering noise when water runs slowly are:
   a. _____
   b. _____
4. Three parts of a faucet are: a. _____ b. _____ c. _____.
5. The water level in a flush tank or automatic livestock waterer is controlled by a _____.
6. The valve that stops water entering a flush tank or automatic livestock waterer is shut off by the action of a _____________________ .

UNIT 36 Drainage and Irrigation Technology

OBJECTIVE

To select, install, and maintain soil drainage and irrigation systems.

COMPETENCIES TO BE DEVELOPED

After studying this unit, you should be able to:

- State the reasons for draining soils.
- Describe the elements of a basic drainage system.
- State the benefits of irrigation.
- Select an irrigation system.
- Use the soil moisture sensors.
- Relate costs factors involved with irrigation.

TERMS TO KNOW

irrigation
precipitation
drainage
open drainage
reclamation
under drainage
gridiron system
herringbone system
single line system
interceptor
vertical drainage system
hardpan
terra-cotta
tile
grade
land level
transit
laser
causeway
transpiration
pesticide
acre inch
seasonal demand
surface irrigation
subirrigation
trickle irrigation
emitters
wetted zones
humidistat
solenoid valve
microtubing
sprinkler irrigation
sprinkler heads
nozzles
moisture sensors
tensiometer
gypsum blocks with meter
neutron probe

MATERIALS LIST

- ✓ Board $3/4$" x $5\,1/2$" x 36"
- ✓ Discarded one-liter plastic soda bottles (5)
- ✓ Cheesecloth (2 square feet)
- ✓ Samples of different soils (5)
- ✓ Twistees (5)

Drainage and irrigation systems have been used by prominent civilizations since the beginning of recorded history, figure 36-1. The natural flooding of the Nile River Valley and the prosperous agriculture that resulted occurred in Egypt thousands of years B.C. Similarly, the Hanging Gardens of Babylon, one of the seven wonders of the world, and the irrigated rice terraces of ancient China give testimony to the benefits of water management in early times.

DRAINAGE

Drainage and irrigation are discussed in this unit since they are both important soil and water management practices. In some instances both systems may be used on the same land. Irrigation refers to water applied to plants and soil to supplement precipitation. Precipitation is rain and snow. Drainage refers to removal of free water from soil.

Benefits of Drainage

Drainage has been a long standing technique to increase food production in the world. The removal of free-standing water from land has converted hundreds of thousands of acres from marshlands to cropland. This has increased crop and livestock production in a world plagued with starvation.

Some benefits of drainage include:

1. Soft, marshy soil can be made productive.
2. Mosquito breeding areas can be eliminated.
3. Drainage may improve health conditions.
4. Drainage can reduce livestock wallowing and stream pollution.
5. Drained soils absorb more water from rain and snow and thus reduce runoff.
6. Drainage can correct waterlogged conditions in the root zone and improve plant growth.

FIGURE 36-1. Water management was practiced by the early Egyptians, Chinese, Babylonians, Romans, and other civilizations. Here a Roman aqueduct carries water across a deep ravine.

Drainage Systems

Drainage systems may be classified as open and under. Open drainage refers to the use of open ditches or channels to carry excessive water from the soil. This system is used extensively where the land is flat and well drained, except for a high water table. It is also useful to drain water from land depressions where surface water accumulates. Extensive networks of open ditches and canals permit the reclamation of hundreds of thousands of acres of fertile land in Holland. Reclamation means "to reclaim."

Underground systems are called under drainage. There are several types of underdrainage systems as described below.

Gridiron. The gridiron system refers to underground drainage pipes laid parallel and discharging into a main conduit, figure 36-2. The system works well on flatland.

Herringbone. The herringbone system refers to the pipes laid with a center conduit picking up water from laterals coming in from both sides at an angle. It is shaped like the bone structure of a herring fish. The herringbone system works well where water drains from two slopes into a valley, figure 36-3.

Single Line System. The single line system is a single or independent line that discharges directly into a stream, open ditch, or other carrier, figure 36-4. It is suitable for draining narrow areas and carrying water to an outlet. It can also serve as an interceptor, which catches water from spring heads

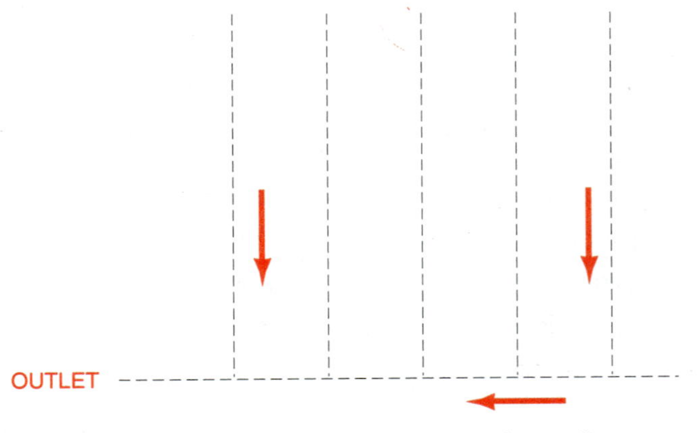

FIGURE 36-2. A gridiron system works well on flat land.

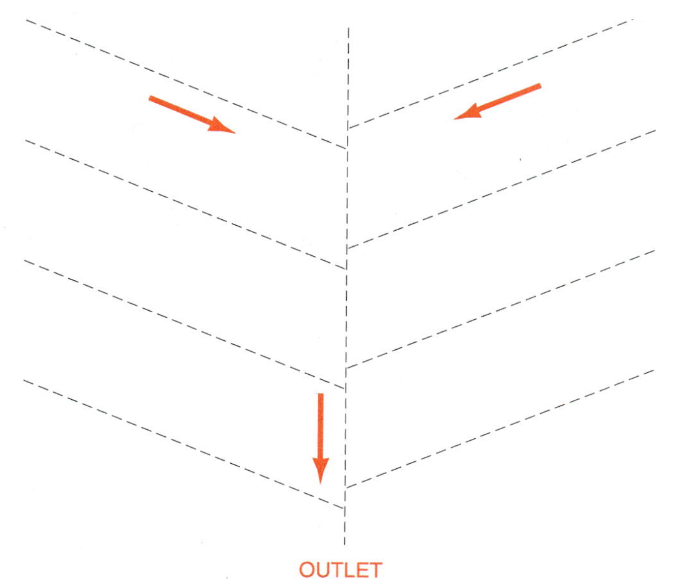

FIGURE 36-3. A herringbone system is useful where two slopes form a valley.

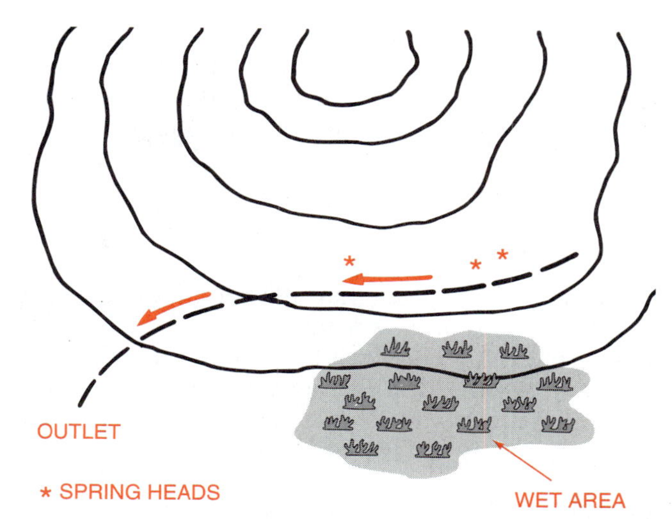

FIGURE 36-5. An interceptor line installed to pick up water on the hillside before it reaches the low area. (*Courtesy of Auburn Consolidated Industries, Inc.*)

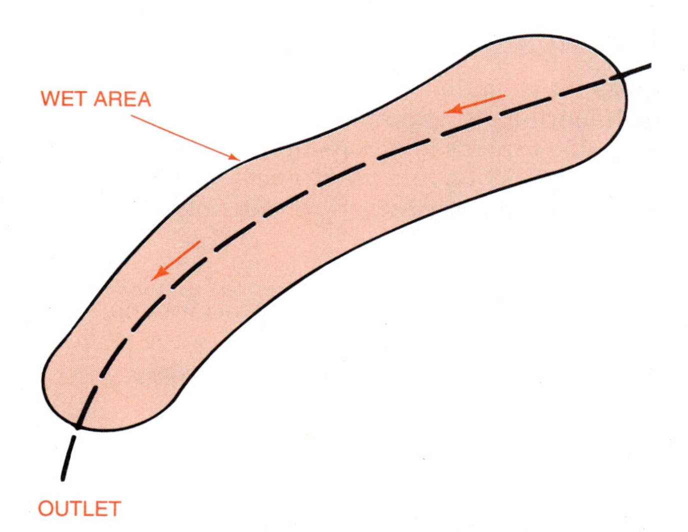

FIGURE 36-4. The single line drainage system.

or underground streams before they create wet places on the surface, figure 36-5.

Vertical Drainage. The vertical drainage system moves water trapped above a hardpan or heavy clay layer to better drainage material below. It consists of a vertical pipe or columns of stone in the soil to permit downward flow. A similar effect may be obtained by using special subsoil chisel plows to break up strips of the clay hardpan.

Installing a Drainage Line

In the past, short lengths of terra-cotta or concrete pipes were laid end-to-end in a trench to form a tile drain line. Terra-cotta is a hard semifired waterproof ceramic clay, also referred to as tile. Most drain lines today are built with special perforated PVC pipe.

To install a drainage line, location of the line is marked with a series of stakes. Then the grade is determined using a land level, transit, or laser equipment. Grade refers to the amount of drop or fall in the line as it progresses downhill. A land level is a hand-held level for sighting level points on the land. A transit is a surveying instrument used to sight horizontal (level) and vertical angles. A laser is an intense beam of light.

Once the grade is calculated on paper, each stake along the line can be marked to show how deep the trench should be at that point. A trenching machine opens the trench according to the depths of cut marked on the stakes or as directed by laser signals, figure 36-6.

FIGURE 36-6. Modern trenching equipment. (*Courtesy of Auburn Consolidated Industries, Inc.*)

Drain pipe is generally laid in a bed of crushed stone. Stone helps collect water and protect the drain pipe. If short lengths of pipe are used instead of perforated PVC pipe, straw or asphalt paper must be used to cover the joints to prevent soil particles from clogging the line.

Drain Outlets

Drain lines are easily damaged when they come onto the surface and discharge into a stream, ditch, canal, or causeway. A causeway is a concrete-lined ditch. To protect the line from damage, a solid steel pipe is recommended with concrete structures to prevent damage by livestock or equipment.

Detrimental Effects of Drainage

While there are many benefits and good reasons for draining land, there are definitely some reasons not to. Scientists are now concluding that draining marshlands has done considerable damage to wildlife habitat and the environment. The benefits of draining land must be weighed against the liabilities.

Marshlands are teeming with living organisms that make up ecosystems of interdependence among microorganisms, plants, and animals. When a wetland is drained, the ecosystem collapses and many of the species are damaged or eliminated.

IRRIGATION PLANNING

Plants take up water through their roots, use some for maintenance and growth, and give some off through transpiration, figure 36-7. Natural precipitation can be supplemented by human-controlled watering called irrigation.

Water may be applied with sprinklers to simulate rain. Water can also be applied by flooding the surface of the land. Water may also be applied directly to the root zone in the soil.

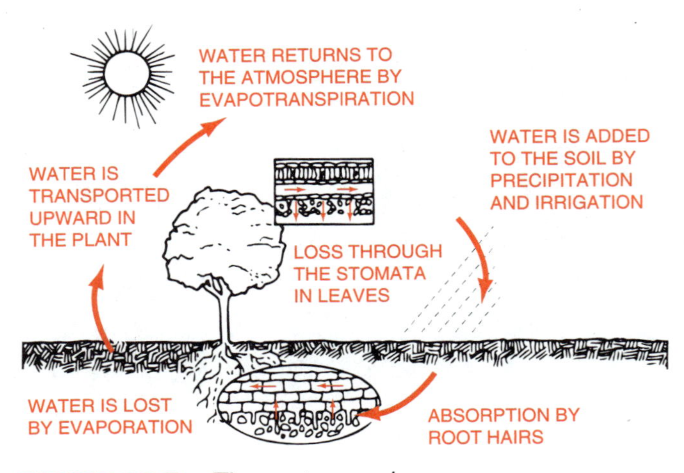

FIGURE 36-7. The water cycle.

Benefits of Irrigation

Irrigation provides many benefits to crops and crop growers. Some of these include:

1. Water may be added to supplement precipitation.
2. Water can be added when precipitation does not match crop needs.
3. Plant nutrients may be applied efficiently in irrigation water.
4. Pesticides may be applied in irrigation water. A pesticide is a substance used to control pests that decrease yields, or damage or destroy desirable plants.
5. Irrigation water can reduce frost damage.
6. Irrigation equipment can be used to distribute liquid manure.
7. Irrigation may be used to cool crops on excessively hot days.

Points to Consider Before Installing An Irrigation System

Costs for irrigation equipment and installation are high. Further, irrigation management requires considerable expertise and operating costs are high for fuel, water, labor, and other items. Some questions that should be asked when considering an irrigation system are:

1. Should I irrigate?
2. Will normal precipitation provide good profits?
3. Will irrigation improve yields? Profits?
4. Can I afford the cost?
5. Can I get the money?
6. What are the energy requirements?
7. Do I have sufficient time, labor, and management expertise?
8. What are the risks involved?
9. Can I get sufficient water?
10. Are my soils suitable for irrigation?
11. What problems might I encounter?
12. Who can help design the system?

WATER REQUIREMENTS

There is no use considering an irrigation system if there is not enough water for efficient crop production. Several steps are necessary to make this determination. A basic measurement of irrigation water use is a unit called an acre inch. An acre inch is the amount of water needed to cover one acre (43,560 square feet) with water 1 inch deep. Assuming no water is running off or seeping into the soil, it takes 27,000 gallons to cover an acre with an inch of water.

Therefore, we say one acre inch equals 27,000 gallons or it takes 27,000 gallons of irrigation water to provide an acre inch of water to the soil.

Seasonal Demand. Scientists have determined the number of acre inches of water taken from the soil in one season by efficiently growing crops in various climates. This is referred to as seasonal demand. The figures for some popular crops are presented in figure 36-8. For example, figure 36-8 indicates that alfalfa growing in southern Arizona requires 44 acre inches per year, but it needs only 26 acre inches per year if growing in New England. Crops in Arizona need more water than in New England because the hot dry climate causes more transpiration and evaporation.

The seasonal demand for one acre of alfalfa in southern Arizona would be 27,000 × 44 or 1,188,000 gallons of water. In New England it would be 27,000 × 26 or 702,000 gallons per acre. In many areas, annual precipitation provides most of the water needed for crops. However, if the precipitation does not occur when the crop needs it, supplemental irrigation is needed. Therefore, irrigation systems may be installed primarily to function during times of maximum or peak demand by the crop. For a system to fulfill its purpose, sufficient water must be available to meet the peak demand or daily peak-use rate.

Peak-use rates needed by certain crops in various areas are presented in figure 36-9. From figure 36-9 we can observe that alfalfa in the California San Joaquin Valley requires 0.25 acre inches, or 0.25 × 27,000 = 6,750 gallons per acre per day, for peak use. In New York State, the same crop requires 0.20 × 27,000 or 5,400 gallons per day. To meet the most demanding needs for irrigation, the system must be capable of supplying peak-use demands every day for extended periods of time.

Sources of Water

Water may be used from streams, rivers, lakes, ponds, reservoirs, irrigation district pipes or canals, or deep wells. Before a decision is made to tap any of these sources, the state, county, and local water laws and ordinances must be checked carefully.

IRRIGATION SYSTEMS

Water may be applied by using one of four systems. These are surface, subirrigation, trickle, and sprinkler irrigation.

	AGRICULTURAL AREAS										
Crop	Lower California	Central Valley California	Washington	Southern Arizona	Western Colorado	Texas High Plains	Western Kansas	Northern Great Plains	Arkansas-Mississippi Bottoms	Piedmont Area	New England Area
	(inches per season)										
Alfalfa	36.0	40.0	34.0	44.0	27.0	45.0	39.0	23.0	39.0	—	26.0
Brome grass	—	—	—	—	—	—	31.0	26.0	—	—	—
Barley	—	—	—	—	—	—	—	16.0	—	—	—
Corn	—	26.0	—	—	20.0	32.0	26.0	17.0	23.0	23.0	17.0
Cotton	25.0	26.0	—	41.0	—	20.0	—	—	—	19.0	—
Grain Sorghum	—	15.0	—	25.0	—	22.0	22.0	—	21.0	20.0	—
Grapefruit	—	—	—	48.0	—	—	—	—	—	—	—
Grapes	—	—	—	20.0	—	—	—	—	—	—	—
Orchard	—	—	—	—	20.0	—	—	—	—	—	—
Oranges	27.0	—	—	39.0	—	—	—	—	—	—	—
Pasture or hay	34.0	36.0	—	—	23.0	—	—	—	—	36.0	—
Potatoes	—	18.0	20.0	24.0	—	—	—	18.0	—	—	12.0
Small grain	—	—	—	23.0	15.0	27.0	27.0	14.0	—	20.0	—
Snap beans	—	—	—	—	—	—	—	—	—	—	9.0
Sugar beets	—	—	28.0	43.0	—	36.0	27.0	23.0	—	—	—
Tomatoes	18.0	20.0	—	—	—	—	—	—	—	—	14.0

FIGURE 36-8. Acre inches of water needed by various crops for efficient production in various regions of the United States.

	AGRICULTURAL AREAS										
Crop	Washington Columbia Crop Basin	Calif. San Joaquin Valley	Texas Southern HighPlains	Arksnsas-Mississippi Bottoms	Nebraska Eastern Part	Colorado Western Part	Wisconsin State	Indiana State	Piedmont Plateau	Virginia Coastal Plains	New York State
	(inches per day)										
Corn	0.27	0.26	0.30	0.24	0.28	0.23	0.30	0.30	0.22	0.18	0.20
Alfalfa	0.25	0.25	0.30	0.24	0.27	0.23	0.30	0.30	0.25	0.22	0.20
Pasture	0.29	0.32	0.30	0.22	0.29	0.23	0.30	0.30	0.25	0.22	—
Grain	0.21	0.17	0.15	0.15	0.26	0.22	0.25	—	0.16	—	—
Sugar Beets	0.26	0.22	—	—	0.26	0.20	0.25	—	—	—	—
Cotton	—	0.22	0.25	0.18	—	—	—	—	0.21	—	—
Potatoes	0.29	0.24	—	—	0.26	0.22	0.20	0.25	0.18	0.18	0.18
Deciduous Orchards	—	0.21	—	—	—	0.18	0.30	0.25	0.25	0.22	0.20
Citrus Orchards	—	0.19	—	—	—	—	—	—	—	—	—
Grapes	—	0.18	—	—	—	—	—	0.25	0.20	—	—
Annual Legumes	—	—	0.18	0.28	—	—	—	—	—	—	—
Soybeans	—	—	—	0.19	0.27	—	0.25	0.30	0.18	—	—
Shallow Truck	—	—	—	—	—	—	0.20	0.20	0.14	—	0.18
Medium Truck	—	—	—	0.12	—	—	0.20	0.20	0.14	0.16	0.18
Deep Truck	—	—	—	—	—	—	0.20	0.20	0.18	—	0.18
Tomatoes	—	—	—	—	—	0.22	0.20	0.20	0.21	0.18	0.18
Tobacco	—	—	—	—	—	—	—	0.25	0.18	0.17	—
Rice	—	—	—	0.17	—	—	—	—	—	—	—

FIGURE 36-9. Inches of water per day needed by various crops for efficient production in various regions of the United States.

Surface Irrigation

The term surface irrigation means adding water to the soil surface by means of gravity. One method consists of flooding an entire area. The area must be level to assure even distribution of water by gravity.

The other method is to plant crops in rows or beds with furrows between them. Then water is released from supply ditches to the furrows. It then flows the length of the field by gravity. Water seeps to the root zone under furrows and rows as it goes, figure 36-10.

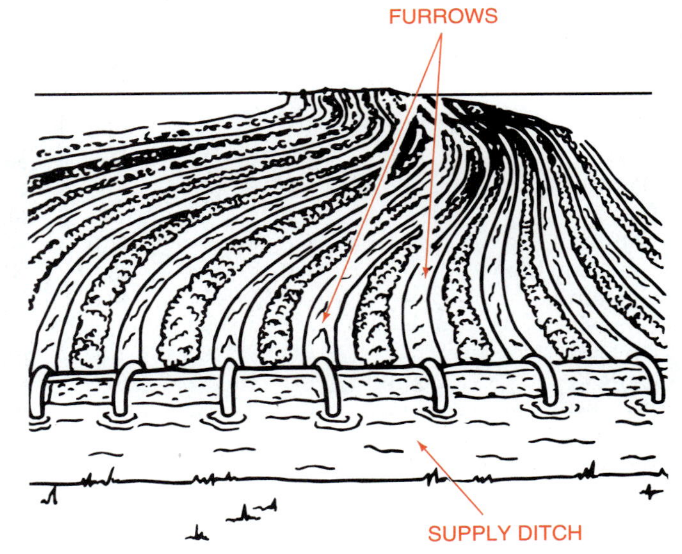

FIGURE 36-10. Surface irrigation by flooding furrows.

Subirrigation

The addition of water to the soil below the surface is called subirrigation. The moisture reaches the root zone area while keeping the surface dry. The open ditch method of subirrigation has water seeping into

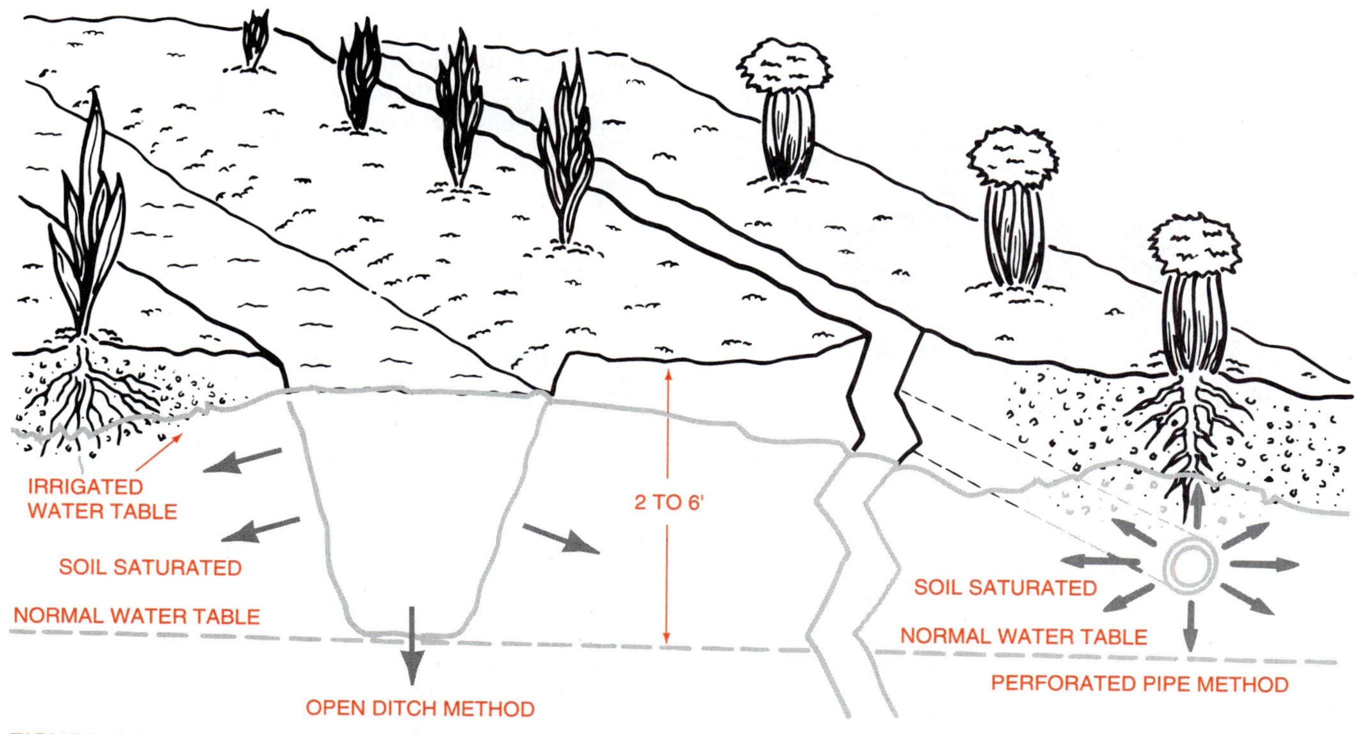

FIGURE 36-11. Subirrigation by the open ditch and perforated pipe method.

the root zone laterally from ditches or canals filled with water, figure 36-11. The other type adds water to the root zone through perforated underground pipes.

Trickle Irrigation

The term trickle irrigation refers to the application of water onto or below the surface of soil through small tubes or pipes. Water is released slowly through special nozzles called emitters. Emitters create small or local watered areas called wetted zones as the water moves through the soil, figure 36-12. Trickle irrigation works especially well for watering individual plants in the field and container-growing plants, figure 36-13.

Trickle irrigation systems require a water source, pump, check valve, gages, filters, control valves, main lines, submain and/or lateral lines, and emitters, figure 36-14. Since the openings in emitters are small, filters should be placed in the system in several locations, figure 36-15.

The use of solenoid valves permit the use of timers, humidistats, and other electronic devices to turn the system on and off. A humidistat measures humidity or moisture in the air. A solenoid valve is a valve actuated by an electromagnet.

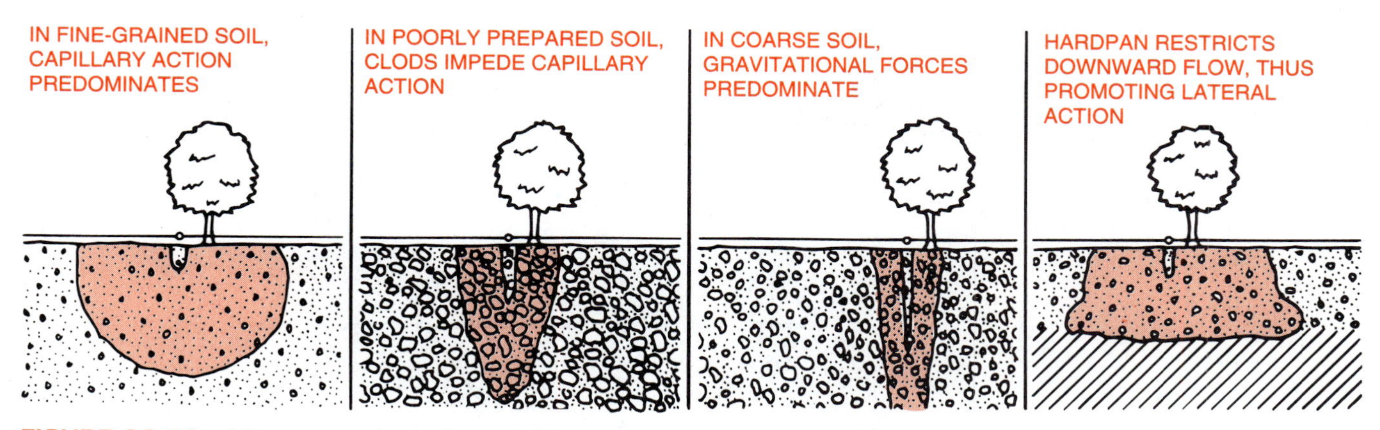

FIGURE 36-12. Wet zones by soil type. (*Courtesy of University of Maryland, Cooperative Extension Service*)

Emitters may be inserted into PVC pipes at any location, figure 36-16. They are quick and economical to install. Emitters are also available in the form of hollow lead devices attached to the end of thin,

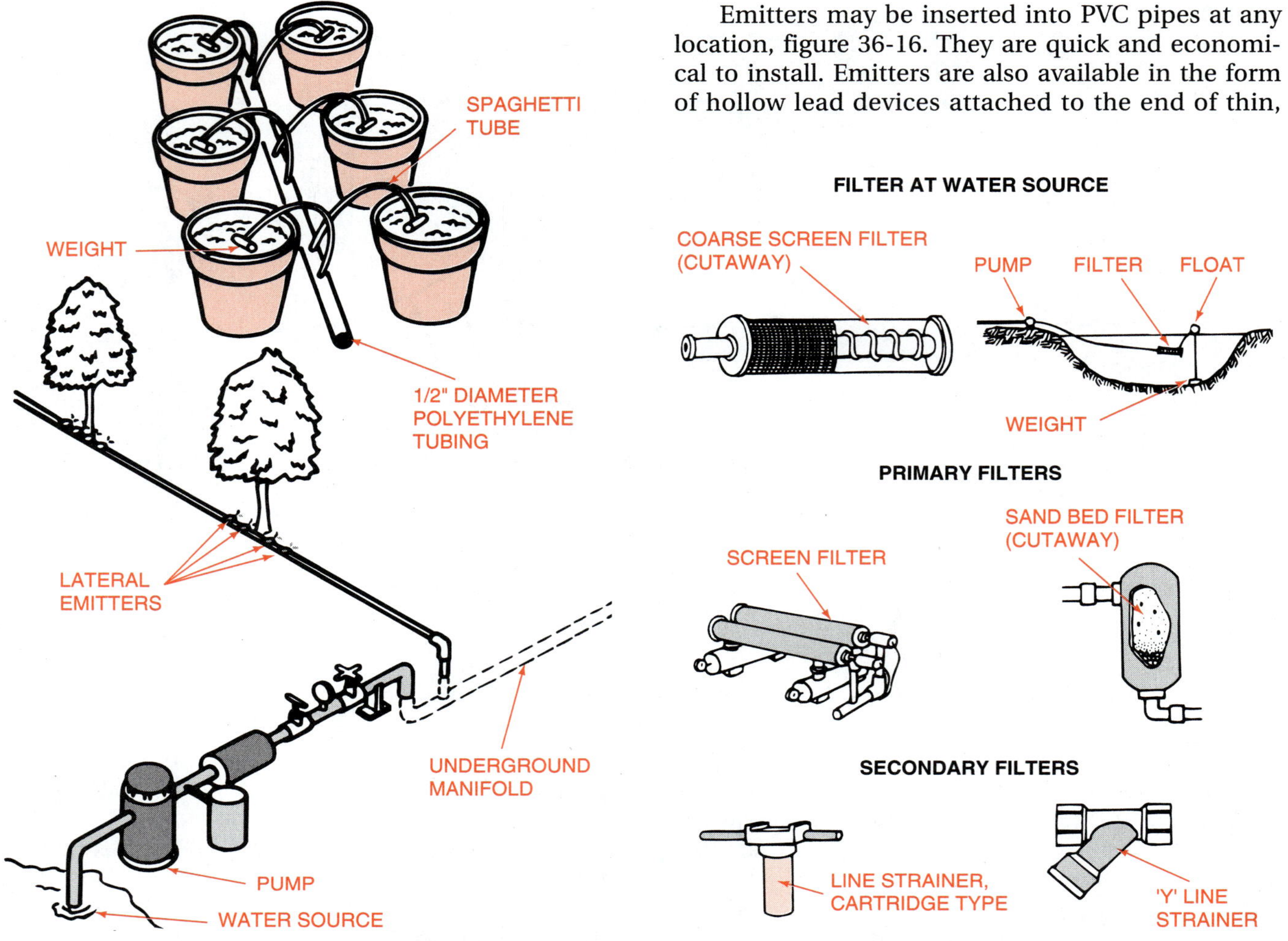

FIGURE 36-13. Trickle irrigation systems are suitable for field, orchard, nursery, greenhouse, shadehouse, and container-growing areas.

FIGURE 36-15. Types of filters. (*Courtesy of University of Maryland, Cooperative Extension Service*)

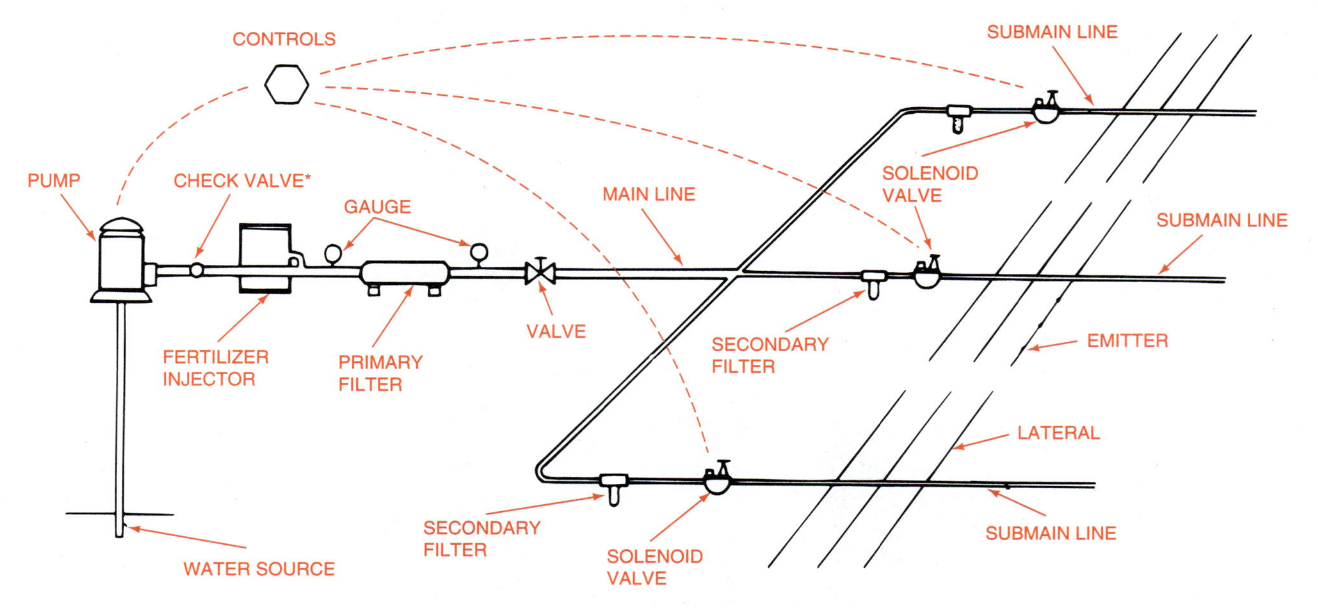

FIGURE 36-14. Trickle irrigation system components. (*Courtesy of University of Maryland, Cooperative Extension Service*)

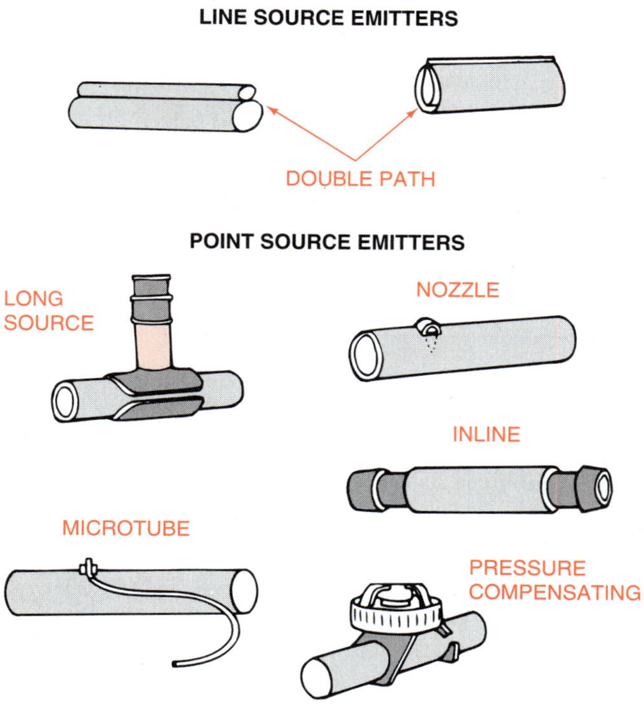

LINE SOURCE EMITTERS

DOUBLE PATH

POINT SOURCE EMITTERS

LONG SOURCE

NOZZLE

INLINE

MICROTUBE

PRESSURE COMPENSATING

FIGURE 36-16. Types of emitters. (*Courtesy of University of Maryland, Cooperative Extension Service*)

flexible, plastic microtubing, sometimes referred to as spaghetti. The spaghetti system is especially handy for container irrigation.

Sprinkler Irrigation

Sprinkler irrigation refers to dispersion of water in droplets to simulate rain. Sprinkler heads or nozzles are devices that break water into droplets and spread them evenly over soil, other media, or crops, figure 36-17. Sprinkler irrigation is popular for lawns, golf courses, nurseries, field crops, vegetables, orchards, citrus groves, and wherever cooling or wetting is desired, figure 36-18.

There are many types of sprinkler irrigation systems. All utilize the same basic equipment as trickle irrigation systems, except sprinkler nozzles can throw water great distances depending on the nozzle type, pipe size, and water pressure. Irrigation may be done by covering large areas with many nozzles or a relatively small area with provisions for moving the nozzle(s) as the job progresses, figures 36-19 and 36-20. Special equipment may be added to the system for adding fertilizers and chemicals, figure 36-21.

Irrigation Wells and Pumps

Since irrigation requires huge volumes of water, special wells are frequently drilled in the center of the area to irrigate. This procedure may save costs of pumping and piping from sources some distance from the crop. Such wells must be designed and constructed with expert assistance, figure 36-22.

FIGURE 36-18. A modern center pivot sprinkler irrigation system. (*Courtesy of E. Plaster,* Soil Science and Management, 2nd ed., *Delmar Publishers, 1992*)

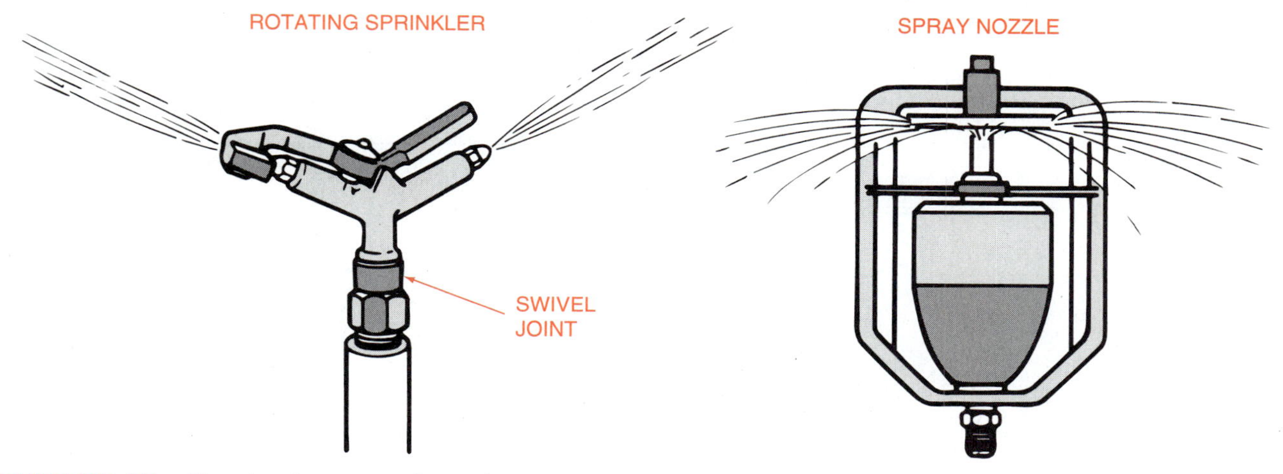

FIGURE 36-17. Two basic types of nozzles.

Type of System	Maximum Slope (Percent)	Water Application Rate Min. (Inches/Hour)	Water Application Rate Max. (Inches/Hour)	Shape of Field	Field Surface Conditions	Max. Height of Crop (Feet)	Labor Required (Hrs./A)	Size of Single System (Acre)	Adaptable To: Cooling and Frost Protect.	Adaptable To: Chemical Application	Adaptable To: Fertilizer Application	Adaptable To: Liquid Animal Waste Distribution
Multi-Sprinkler Permanent	No limit	.05	2	Any shape			.05-.10	1 or more	Yes			Not Recom.
Hand-moved: Portable set	20	.10	2	Rectangular	No Limit	No Limit	.5-1.5	1-40	No			Yes
Solid set	No limit	.05	2	Any shape			.2-.5	1 or more	Yes			Not Recom.
Tractor-moved: Wheel-mounted	10	.10	2		Smooth enough for safe tractor operation		.2-.5	20-50				
Self-moved: Side-wheel-roll	10	.10	2	Rectangular	Reasonably smooth	4	1-3	20-80				
Side-moved	10	.10	2			4-6	1-3					Yes
Self-propelled: Center-pivot	20	.20	1.5	Circular, Square or Rectangular	Clear of obstructions, path for towers	8-10	.05-.15	20-200				
Lateral-move	20	.20	1			8-10	.05-.15	10-100	No	Yes	Yes	
Single-Sprinkler Hand-moved	20	2.5	2				.5-1.5	20-40				Not Recom.
Tractor-moved: Wheel-mounted	5-15	2.5	2	Any shape	Safe operation of tractor	No limit	.2-.4	20-40				
Self-propelled	No limit	2.5	1	Rectangular	Lane for winch and hose		.1-.3	20-40				Yes
Broom-Sprinkler Tractor-moved	5	2.5	1	Any shape	Safe operation of tractor	8-10	2-5	20-40				
Self-propelled	5	2.5	1	Rectangular	Lane for boom and hose	8-10	.1-.5					

*Does not include the cost of water supply, pump, power unit and mainline.

FIGURE 36-19. Factors affecting the selection of sprinkler irrigation systems. (*Courtesy of American Association for Vocational Instructional Materials*)

	GALLONS PER MINUTE FROM EACH SPRINKLER																			
Spacing (feet)	**1**	**2**	**3**	**4**	**5**	**6**	**7**	**8**	**9**	**10**	**11**	**12**	**13**	**14**	**15**	**16**	**17**	**18**	**19**	**20**
20 × 20	0.24	0.48	0.72	0.97	1.21	1.45	1.70	1.94												
20 × 30	.16	.32	.48	.64	.80	.96	1.13	1.29	1.45	1.61	1.77	1.93	2.10							
20 × 40	.12	.24	.36	.48	.60	.72	.85	.97	1.09	1.21	1.33	1.45	1.57	1.70	1.81	1.93	2.06	2.18		
20 × 50	.10	.19	.29	.39	.48	.58	.68	.77	.87	.97	1.06	1.16	1.26	1.36	1.45	1.55	1.64	1.74	1.84	1.94
20 × 60	.08	.16	.24	.32	.40	.48	.56	.64	.72	.81	.88	.97	1.05	1.13	1.21	1.29	1.37	1.45	1.53	1.61
25 × 25	.15	.31	.46	.62	.77	.93	1.09	1.24	1.40	1.55	1.70	1.86	2.02							
30 × 30	.11	.21	.32	.43	.54	.64	.75	.86	.97	1.07	1.18	1.29	1.39	1.50	1.61	1.72	1.83	1.93	2.04	2.15
30 × 40		.16	.24	.32	.40	.48	.56	.64	.72	.81	.89	.97	1.05	1.13	1.21	1.29	1.37	1.45	1.53	1.61
30 × 50		.13	.19	.26	.32	.38	.45	.52	.58	.64	.71	.77	.84	.90	.97	1.03	1.09	1.16	1.22	1.29
30 × 60		.11	.16	.21	.27	.32	.37	.43	.48	.54	.59	.64	.70	.75	.81	.86	.91	.97	1.02	1.07
40 × 40		.12	.18	.24	.30	.36	.42	.48	.54	.60	.66	.72	.78	.84	.90	.96	1.02	1.09	1.14	1.20
40 × 50		.10	.14	.19	.24	.29	.34	.39	.43	.48	.53	.58	.63	.68	.73	.78	.82	.87	.92	.97
40 × 60			.12	.16	.20	.24	.28	.32	.36	.40	.44	.48	.52	.56	.60	.64	.68	.72	.77	.81
40 × 80			.09	.12	.15	.18	.21	.24	.27	.30	.33	.36	.39	.42	.45	.48	.51	.54	.57	.61
50 × 50			.12	.15	.19	.23	.27	.31	.35	.39	.43	.46	.50	.54	.58	.62	.66	.70	.73	.77
50 × 60			.10	.13	.16	.19	.22	.26	.29	.32	.35	.39	.42	.45	.48	.52	.55	.58	.61	.64
50 × 70				.11	.14	.17	.19	.22	.25	.28	.30	.33	.36	.39	.41	.44	.47	.50	.52	.55
50 × 80				.10	.12	.14	.17	.19	.22	.24	.27	.29	.31	.34	.36	.39	.41	.44	.46	.48
60 × 60				.11	.13	.16	.19	.21	.24	.27	.30	.32	.35	.38	.40	.43	.46	.48	.51	.54
60 × 70					.11	.14	.16	.18	.21	.23	.25	.28	.30	.32	.34	.37	.39	.41	.43	.46
60 × 80					.10	.12	.14	.16	.18	.20	.22	.24	.26	.28	.30	.32	.34	.36	.38	.40
70 × 70							.14	.16	.18	.20	.22	.24	.25	.28	.30	.31	.33	.36	.37	.39
70 × 80							.12	.14	.16	.17	.19	.21	.23	.24	.26	.28	.29	.31	.33	.34
70 × 90								.12	.14	.15	.17	.18	.20	.22	.23	.25	.26	.28	.29	.31
80 × 80								.12	.14	.15	.17	.18	.20	.21	.23	.24	.26	.27	.29	.30
80 × 90								.11	.12	.13	.15	.16	.17	.19	.20	.21	.23	.24	.26	.27
80 × 100								.10	.11	.12	.13	.15	.16	.17	.18	.19	.21	.22	.23	.24
100 × 100										.10	.11	.12	.13	.14	.14	.15	.16	.17	.18	.19
120 × 132																	.10	.11	.12	.12

FIGURE 36-20. Average water application rate in inches per hour from sprinklers. (*Courtesy of University of Maryland, Cooperative Extension Service*)

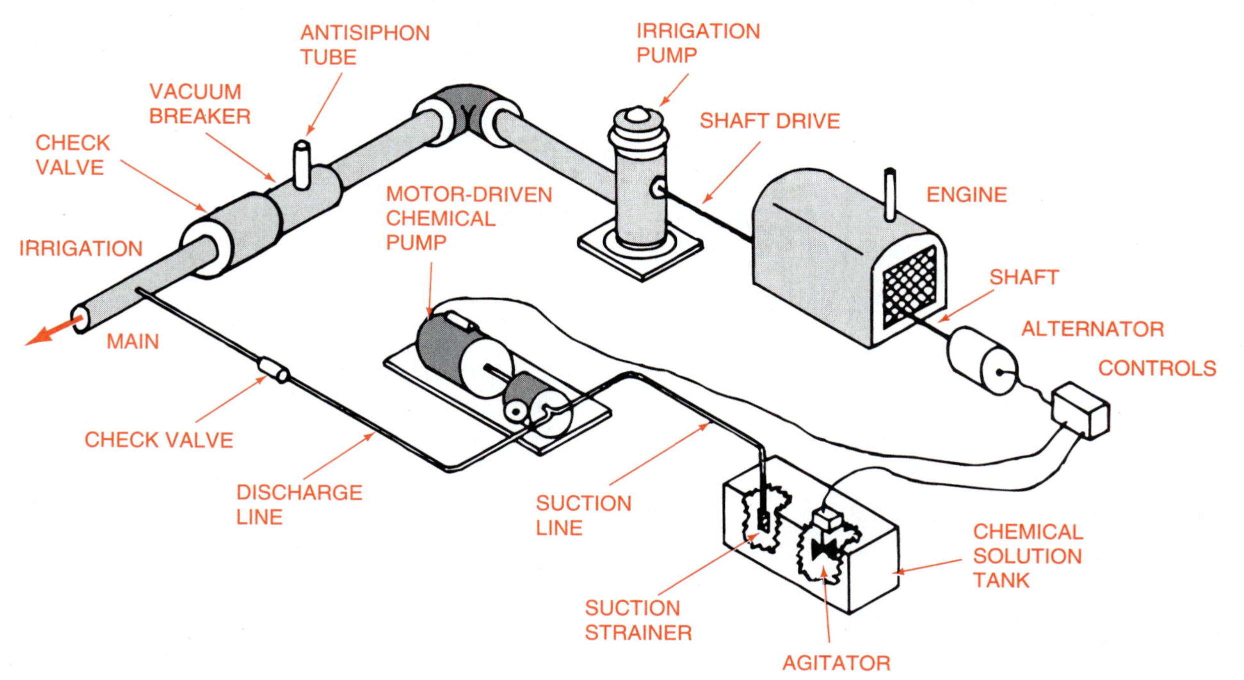

FIGURE 36-21. A sprinkler system with chemical application equipment

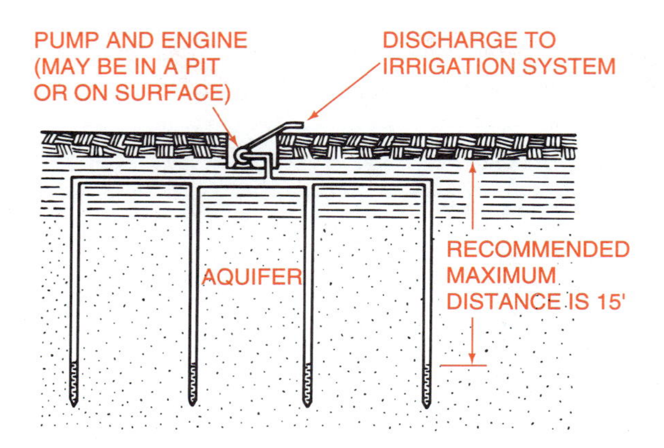

(A) A BATTERY OF DRIVEN WELLS

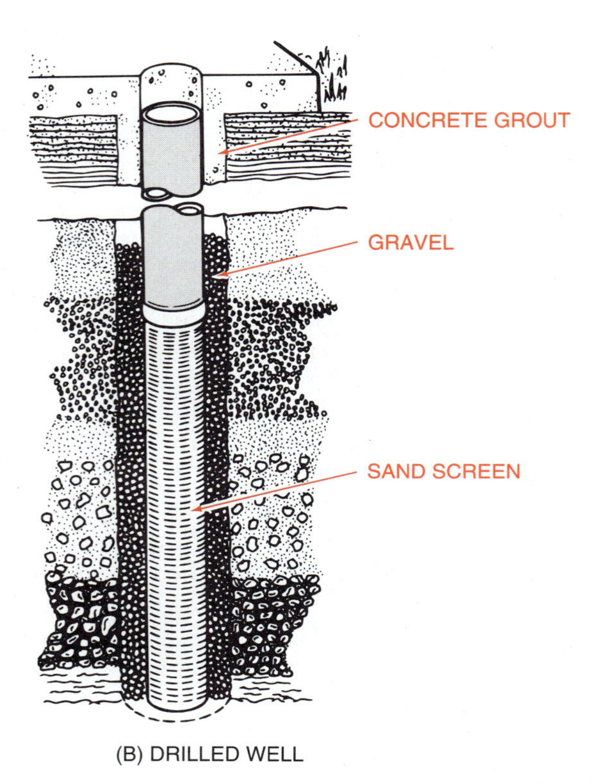

(B) DRILLED WELL

FIGURE 36-22. Cross section of two types of irrigation wells. (*Courtesy of University of Maryland, Cooperative Extension Service*)

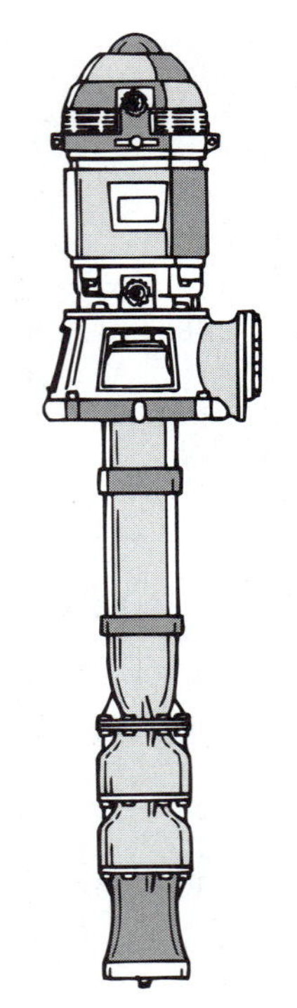

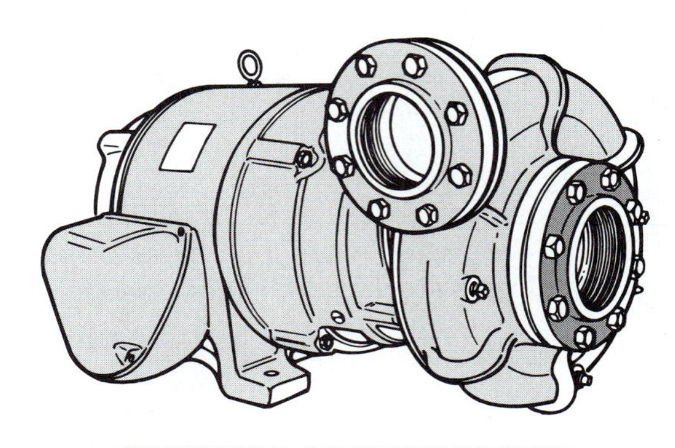

FIGURE 36-23. Two types of electric irrigation pumps.

Similarly, the high cost of pumping, piping, and components dictate the need for a system engineered according to the specific needs, figure 36-23.

MANAGING IRRIGATION

Some areas have such a dry climate that nearly all water requirements must be met through irrigation. Others have sufficient rainfall generally, but need supplemental irrigation occasionally. In all situations, it is important to apply water in accordance with crop needs. Such needs will peak when crops are growing rapidly and producing seed, figure 36-24.

Similarly, it is important to place the water where it does the most good. Some crops, such as azaleas, have shallow root systems and receive most of their water within several inches of the surface. Other crops, like alfalfa, may put roots down 10 or 15 feet, and can survive long dry periods. Most plants fall in between, figure 36-25.

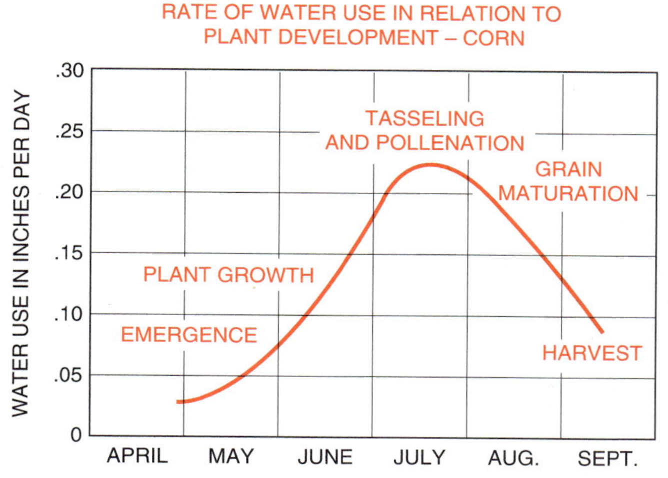

FIGURE 36-24. Seasonal water requirements correlated with growth stage of corn. (*Courtesy of University of Maryland, Cooperative Extension Service*)

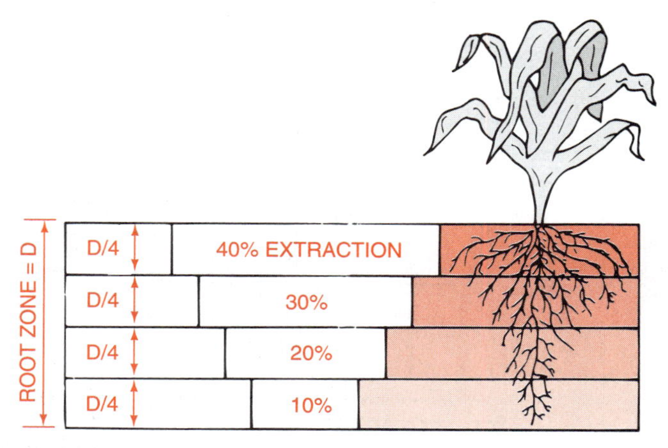

FIGURE 36-25. Moisture use pattern and moisture sensor. (*Courtesy of University of Maryland, Cooperative Extension Service*)

For maximum plant growth, water conservation, and profit, water should be applied only as needed. Moisture sensors are devices placed in the soil to indicate the moisture content. The tensiometer, gypsum blocks with meter, and neutron probe are types of moisture sensors. Their use and placement are illustrated in figure 36-26.

SUMMARY

Crop reduction can be increased and ornamental landscapes can be enhanced with water management practices. Both drainage and irrigation has its place in a productive agriculture. For those needing these practices, the benefits of expert advice should more than offset the cost and should lead to satisfying results.

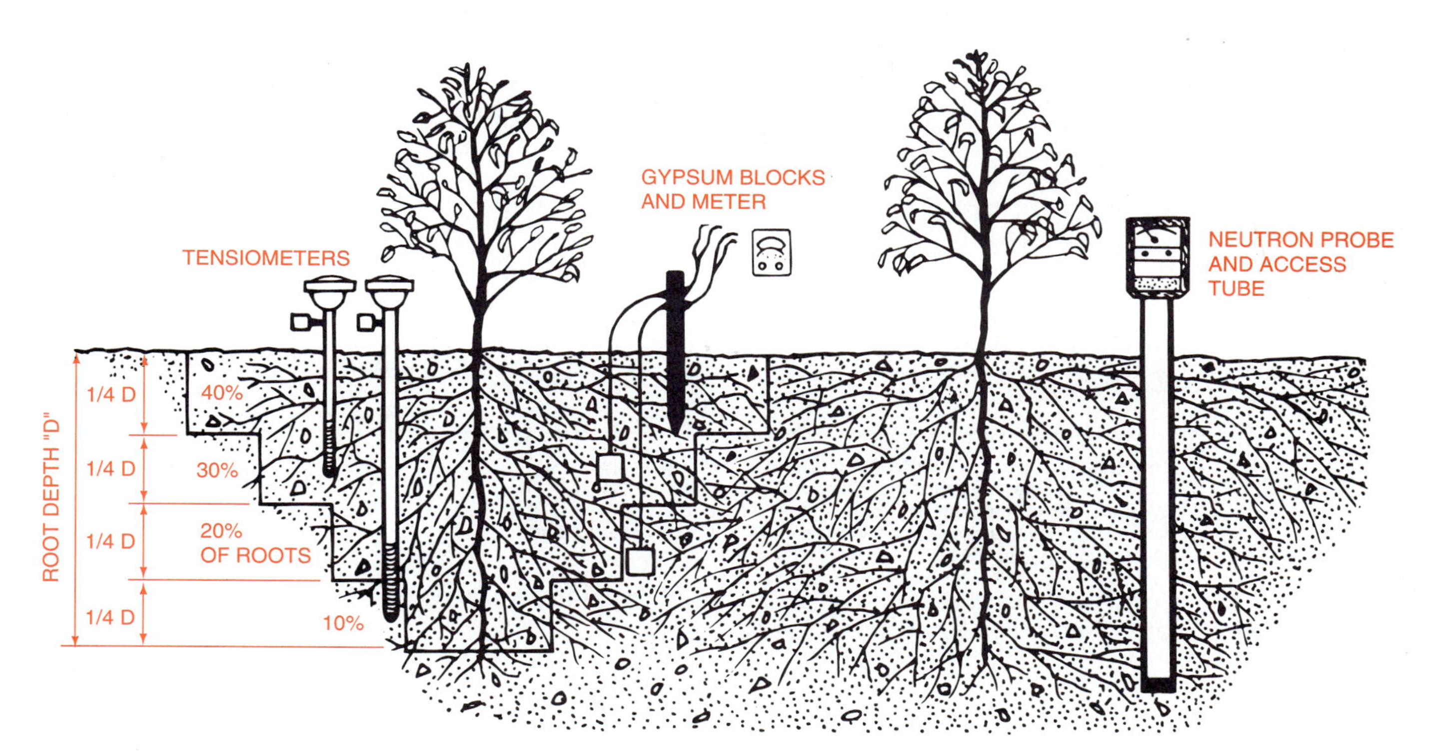

FIGURE 36-26. Plant rooting pattern and moisture sensor placement. (*Courtesy of University of Maryland, Cooperative Extension Service*)

SOIL PERCOLATION EXPERIMENT

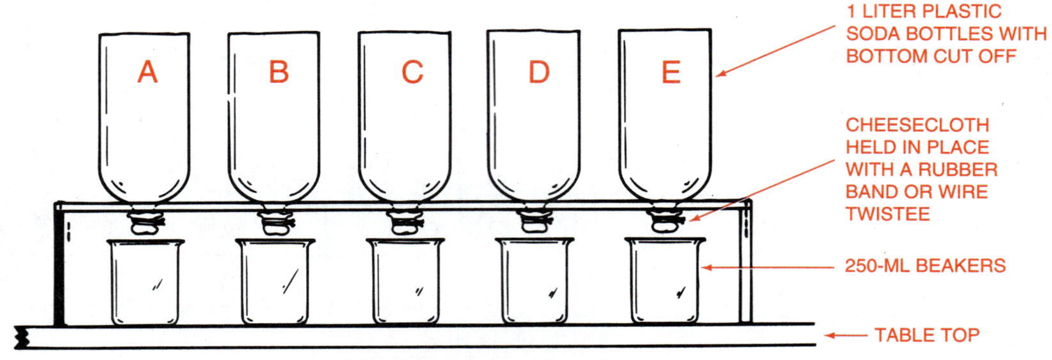

Making the Apparatus

1. Cut two pieces of wood 3/4" x 5 1/2" x 6" long and one piece 3/4" x 5 1/2" x 24" long.
2. Nail or glue the top to the two end pieces.
3. Bore five holes 4" on center (OC) down the centerline of the top board. (Diameter of holes are determined by the bottle sizes.)
4. Cut the bottoms out of five one-liter plastic soda bottles.
5. Invert the bottles into the holes.
6. Place a cheesecloth patch over each opening and secure with a wire twistee or rubber band.
7. Insert 250-ml or larger beakers under each bottle.
8. Mark bottles A through E with a marking pen or label.

Doing the Soil Percolation Experiment

1. Select five samples of soil of 250 milliliters each.
   SAMPLE A – pure sand
   SAMPLE B – a good potting soil
   SAMPLE C – woodland soil
   SAMPLE D – soil from a field where row crops are grown continuously
   SAMPLE E – soil of your choice
2. Break up any particle larger than a pea in the samples, spread the samples out on newspaper, and let them dry overnight.
3. Pour each sample into one of the one-liter bottles and record the bottle number and the source of the soil.
4. Add 250 ml of water to each sample and record the time to the minute and second when you add the water to each bottle.
5. Record the instant the first drop of water flows into each beaker.
6. Compute the time in minutes and seconds for water to start draining from each.
7. Record the amount of water in each beaker after the last one stops dripping.
8. Observe the results and record data in the following chart.

	Source	Time Start	Time of 1st drop	Minutes and seconds for each to start	ml of water released	ml of water retained
SAMPLE A	______	______	______	______	______	______
SAMPLE B	______	______	______	______	______	______
SAMPLE C	______	______	______	______	______	______
SAMPLE D	______	______	______	______	______	______
SAMPLE E	______	______	______	______	______	______

9. Rank your samples from 1 (fastest) to 5 (slowest) in releasing water.
10. Rank your samples from 1 (best) to 5 (worst) in water-holding capacity.
11. What are your observations and conclusions?
12. What are the implications of your findings on selection of drainage systems? Of irrigation systems?

FIGURE 36-27. Soil percolation experiment. (*Courtesy of Steen Westerberg, Agricultural Science and Technology Instructor, Hereford High School, Parkton, MD*)

STUDENT ACTIVITIES

1. Define the Terms To Know in this unit.
2. Set up a soil percolation demonstration as shown in figure 36-22. Present the demonstration to your class.
3. Do research using a variety of soils and add explanatory charts to the soil percolation tests described in figure 36-27. Enter research and findings in your school's science fair.
4. Survey the school grounds and identify areas that are poorly drained. Suggest a suitable drainage system to correct the problem.
5. Read an encyclopedia article on irrigation practices in ancient Egypt, China, Babylonia, or other early civilization. Report your findings to the class.
6. Research the details of one type of irrigation system and report your findings to the class.
7. Design and install an irrigation system for the school greenhouse or grounds or for your lawn.

SELF-EVALUATION

A. **Multiple Choice**. Select the best answer.

1. Which is not a benefit of draining land?
   a. eliminate mosquito breeding areas
   b. improve crop production
   c. reduce livestock wallowing
   d. improve wildlife habitat
2. The drainage system for two slopes with a valley is
   a. gridiron
   b. herringbone
   c. single line
   d. interceptor
3. The best system to drain water from a hillside to prevent marshy flatland below is
   a. gridiron
   b. herringbone
   c. single line
   d. interceptor
4. The gallons of water required to provide one acre inch of water on five acres is
   a. 27,000
   b. 50,000
   c. 100,000
   d. 135,000
5. Peak use is the maximum amount of water needed for
   a. one day
   b. the season
   c. the year
   d. none of these
6. The irrigation system that adds water by flooding furrows is
   a. surface
   b. subirrigation
   c. sprinkler
   d. trickle
7. Valves to start and stop irrigation can be controlled automatically with
   a. solenoids
   b. laddenvalves
   c. filters
   d. emitters
8. Plants require most water
   a. at germination
   b. during early growth
   c. during growth and seed formation
   d. near harvest
9. Most plants get most of their moisture needs from
   a. the top 25 percent of root depth
   b. second 25 percent of root depth
   c. third 25 percent of root depth
   d. lowest 25 percent of root depth
10. Which is not a moisture sensor?
    a. gypsum block and meter
    b. neutron probe
    c. philometer
    d. tensiometer

B. Matching. Match the terms in column I with those in column II.

Column I	Column II
1. terra-cotta	a. surveying instrument
2. drop or fall of land	b. 27,000 gallons
3. transit	c. trickle irrigation
4. land level	d. grade
5. causeway	e. tile
6. one acre inch	f. microtubing
7. emitter	g. creates droplets
8. trickle irrigation	h. water from below
9. irrigation nozzle	i. concrete-lined ditch
10. subirrigation	j. hand-held instrument

UNIT

Hydraulic, Pneumatic, and Robotic Power

OBJECTIVE

To maintain and utilize fluid and robotic power in agricultural applications.

COMPETENCIES TO BE DEVELOPED

After studying this unit, you should be able to:

- Compare hydraulic and pneumatic systems.
- Identify basic theories applying to fluid dynamics.
- Describe fluid power principles.
- Discuss fluid characteristics.
- Identify major components of fluid systems.
- Use and maintain fluid power equipment.
- Discuss some concepts of robotics.

MATERIALS LIST

- ✓ Bulletin board materials
- ✓ Publications with photos of robotics
- ✓ Reference materials on hydraulic and pneumatics
- ✓ Camera and film
- ✓ Materials to demonstrate the principles of fluid dynamics

TERMS TO KNOW

fluid power
hydraulics
pneumatics
force
pressure
Bourdon tube
manometer
viscosity index
additive
viscosity improver
antifoam
corrosion inhibitor
rust inhibitor
antiscuff
extreme-pressure resistor
SAE
API
double action cylinder
positive displacement
eccentric
air compressor
fluid coupling
impeller
pressure regulator
direction control value
check valve
spool valve

tolerance
full-flow system
bypass system
hydraulic cylinder
O ring
hydraulic motor
quick-coupling
robot
robotics
translation
axis
degrees of freedom
cartesian working area
cylindrical working area
hollow
hollow sphere
solid sphere

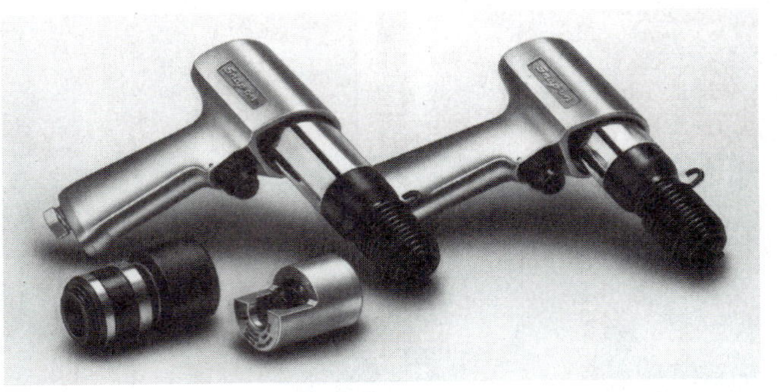

FIGURE 37-1. Pneumatics are used extensively for operating power tools, controlling closures, and moving objects. (*Courtesy of Snap-On Tools*)

Fluid power is an integral part of agricultural and industrial technology. Fluid power is work done utilizing liquids and gases to transfer force. The use of liquids to transfer force is called hydraulics. The use of air or other gases to transfer force is called pneumatics. Together they comprise fluid power.

The use of hydraulic and pneumatic power is extensive in agriculture/agribusiness and the management of renewable natural resources. Pneumatics is used extensively in repair shops to clean parts, drive tools, and operate paint sprayers. It is used in milking operations to open and close gates and direct livestock movements. It is also used extensively in food processing plants to control items on production lines by opening and closing various devices, figure 37-1.

FIGURE 37-2. This log splitter uses a gasoline engine to pump fluid which operates a cylinder to drive logs against the splitting wedge. (*Courtesy of Huss Sales Inc.*)

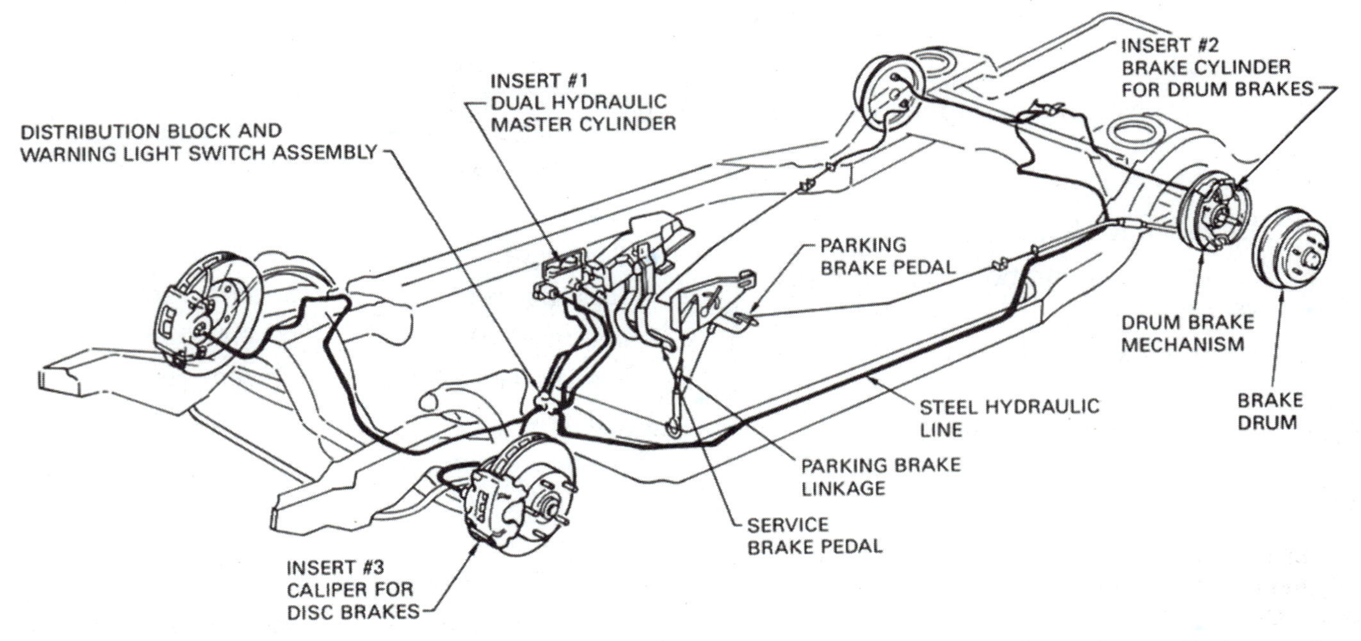

FIGURE 37-3. Automotive braking systems use fluid to transfer foot power to the master cylinder to piston power on the brake shoes and pads. (*Modified from A. Schwaller,* MOTOR Automotive Technology, 2E, *Delmar Publishers, 1993. Courtesy of Delco Moraine Division, General Motors Corp.*)

Hydraulics permit the use of pistons for linear power and hydraulic motors for rotary power on equipment, figure 37-2. Hydraulic power is especially useful in remote areas of automobiles, figure 37-3, and tractors, figure 37-4. Vehicles such as trucks, earth movers, boats, and airplanes rely extensively on hydraulic systems.

PRINCIPLES OF FLUID DYNAMICS

Nature of Hydraulic Fluids and Air

Both liquids and gases are considered fluids. However, they have different characteristics.

1. Hydraulic fluids cannot be compressed—air can be.

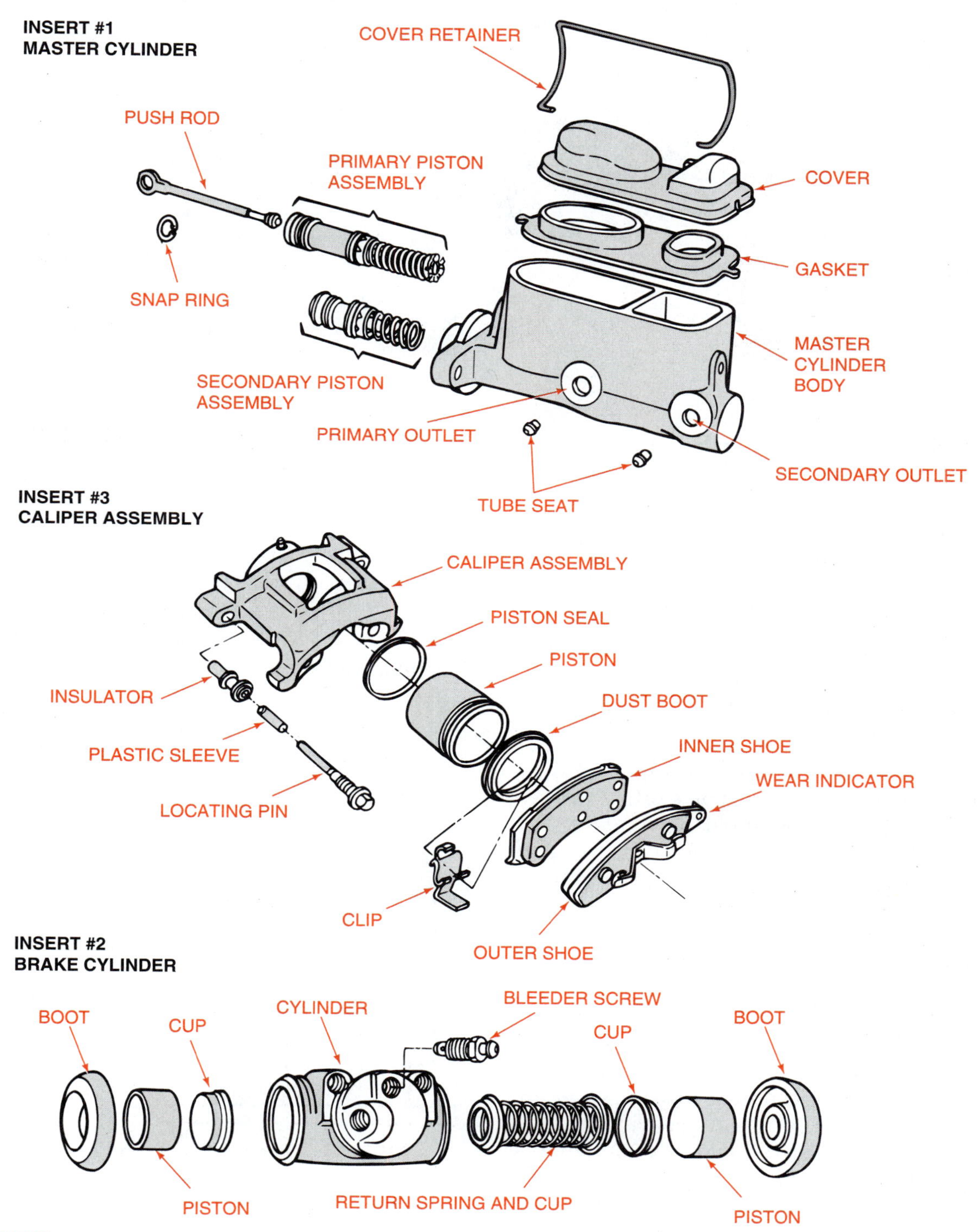

FIGURE 37-3. *Continued*

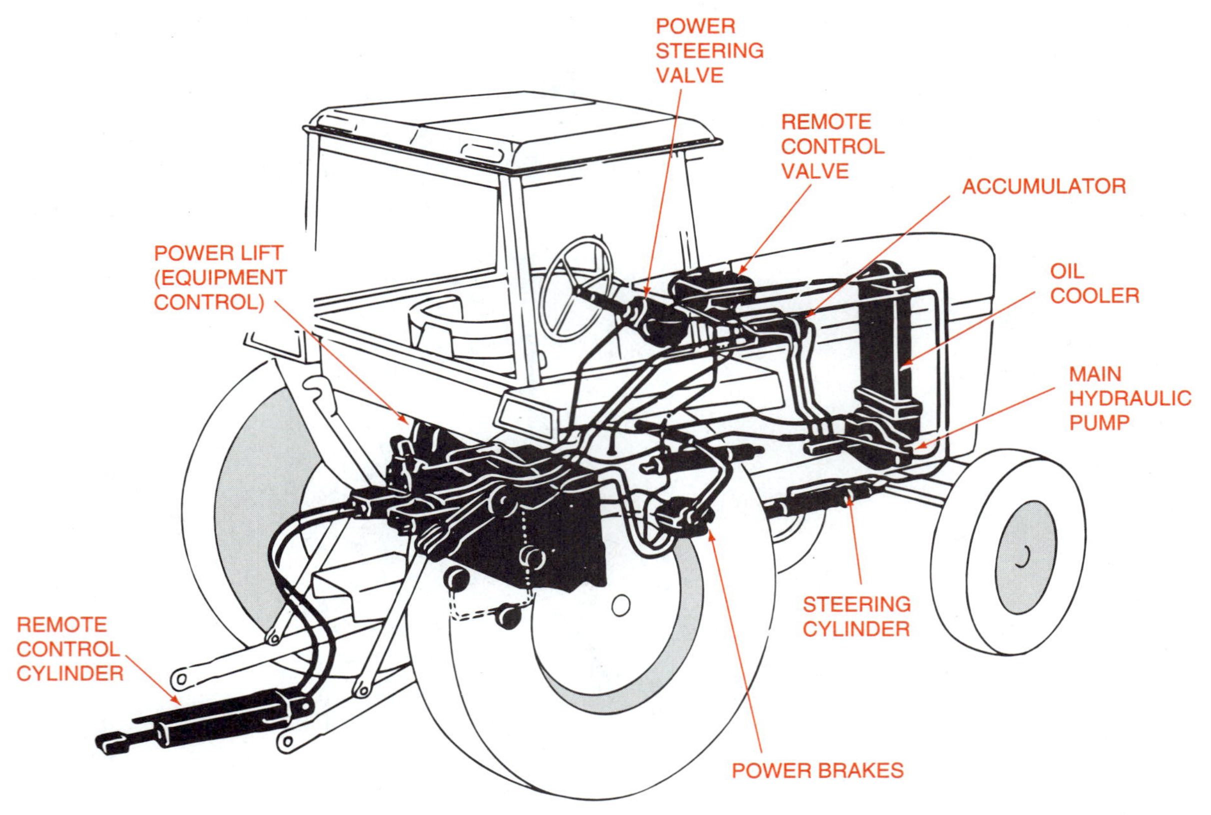

FIGURE 37-4. Hydraulic systems on a modern tractor. (*Courtesy of Instructional Materials Service, Texas A&M University*)

2. Hydraulic fluids must have complete circuits with fluid returning to a system reservoir—air can be pumped in from the atmosphere, used to move components, and be expelled back out into the atmosphere.

Force and Pressure Defined

To understand fluid power, both force and pressure must be defined. Force is the pushing or pulling action of one object on another. Force usually causes objects to move. Force is measured in pounds (lbs.).

On the other hand, pressure is force acting upon an area. For instance, air pressure in a tire may be 30 pounds of pressure per square inch of tire surface. Pressure is measured in pounds per square inch (psi).

Pascal's Law

A pressure applied to a confined fluid is transmitted undiminished to every portion of the surface of the containing vessel. This is Pascal's law and may be shortened to: pressure on a liquid in a container is transferred equally to all surfaces, figure 37-5. Therefore, pressure on a master cylinder piston exerts equal force on all wheel cylinders, figure 37-6.

Also, pressure on a fluid is equal to the force divided by the area or $P = F/A$

where: P = pressure in ps
F = force in pounds
A = area to which force is applied

Therefore, if $F = 100$ and $A = 1$, then $P = 100$.

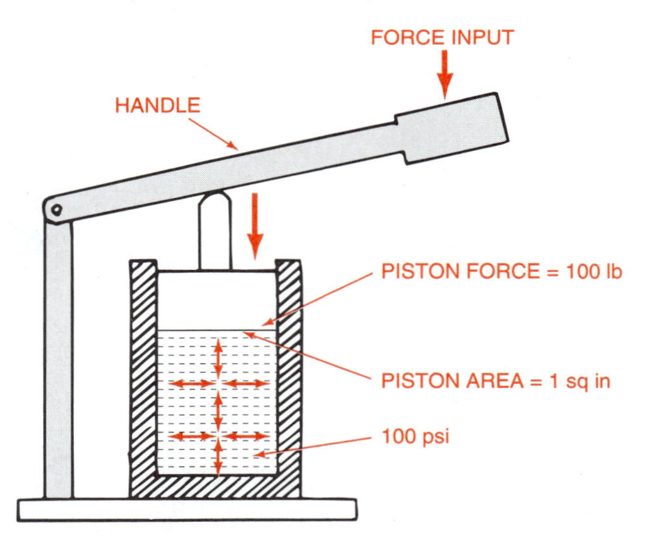

FIGURE 37-5. Pressure exerted on a fluid is transferred equally to all surfaces. (*Courtesy of A. Schwaller,* MOTOR Automotive Technology, 2E, *Delmar Publishers, 1993*)

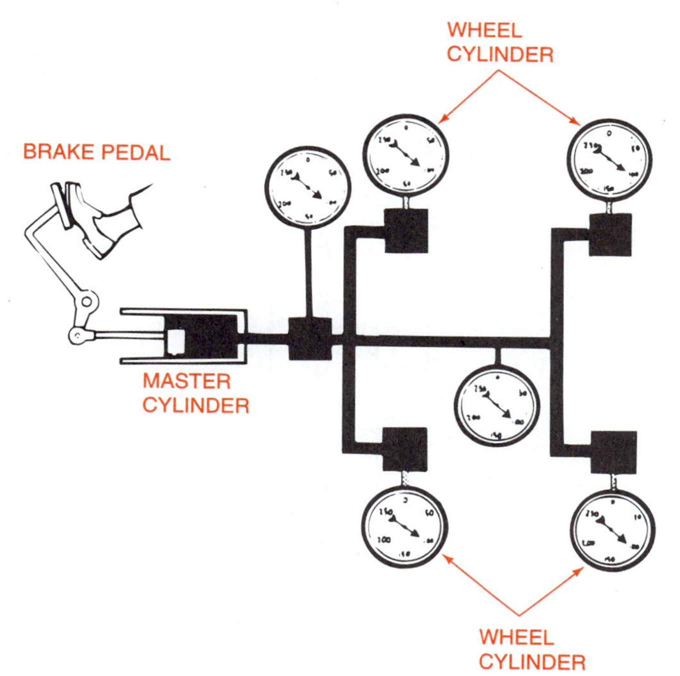

FIGURE 37-6. Pressure transfer in the hydraulic brake system of an automobile. (*From A. Schwaller,* MOTOR Automotive Technology, 2E, *Delmar Publishers, 1993. Courtesy of Delco Marine NDH Division, General Motors Corporation*)

The output or resulting pressure transferred in a system can be changed or manipulated by using different sized pistons. For instance, if 100 pounds of pressure is exerted on a piston with one square inch area, the output would be only 50 psi on a piston with one-half square inch area. But, the pressure exerted on a piston of 2 square inches would be 200 pounds, figure 37-7. Therefore, it can be concluded that P × A (input) = P' ÷ A' (output). From figure 36-6 we observe: 100 × 1 = 50 ÷ $^1/_2$ in cylinder B, 100 × 1 = 100 ÷ 1 in cylinder C, and 100 × 1 = 200 ÷ 2 in cylinder D.

Boyle's Law

A scientist by the name of Robert Boyle made a significant discovery in 1662 about the behavior of gases. Boyle's law states, a volume of gas is inversely proportional to its pressure. That means, as pressure increases volume decreases, and as pressure decreases volume increases. It is like a basketful of dry leaves. Press them with your foot and the leaves decrease in volume. Lift your foot and take the pressure off—the leaves bounce back to their original volume. It is this rebound tendency of gases that makes air a good medium for rubber tires, pressurized water tanks, and aerosol cans.

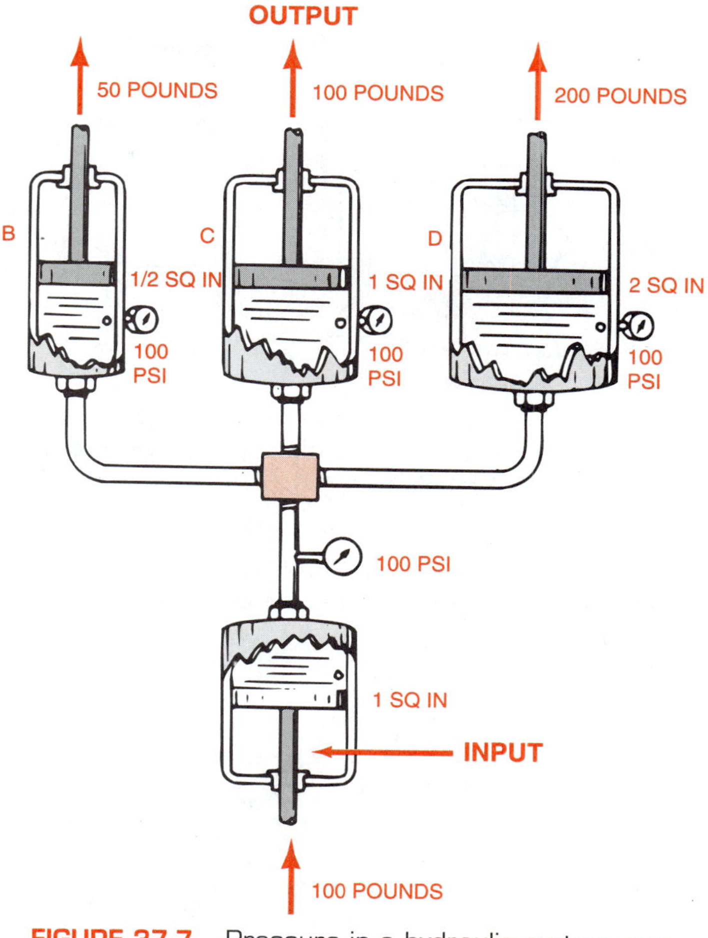

FIGURE 37-7. Pressure in a hydraulic system can be increased or multiplied by changing the size of the piston diameter. (*From A. Schwaller,* MOTOR Automotive Technology, 2E, *Delmar Publishers, 1993. Courtesy of EIS Brake Part Division, Standard Motor Products, Inc.*)

Bernoulli's Theorem

Daniel Bernoulli made an important discovery in 1738 about fluids. Bernoulli's theorem states that, when a fluid flows through a pipe, pressure remains constant unless the diameter of the pipe changes. As the diameter of the pipe decreases, the pressure decreases, figure 37-8.

Measuring Pressure

The **Bourdon tube** is a tube bent in a circular pattern that tends to straighten as internal fluid pressure increases. The outward movement at the end of the loop moves a pointer. The pressure is read directly from a scale on the face of the meter, figure 37-9.

The **manometer** registers pressure as well as vacuum. It uses a glass U-shaped tube filled with water or mercury. Pressure pushes and vacuum

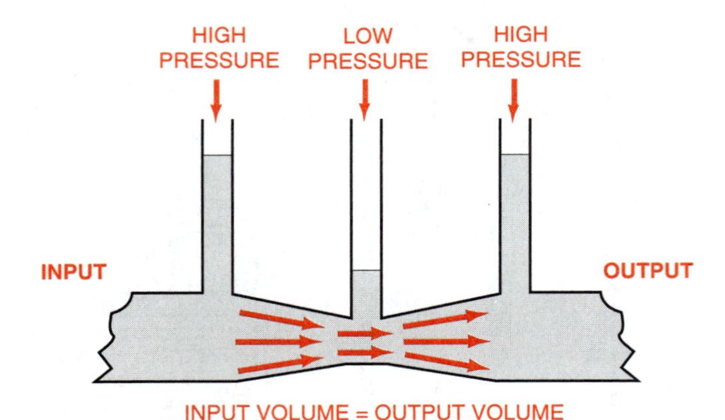

FIGURE 37-8. Bernoulli's theorem indicates that as the velocity of a fluid increases, the pressure at that point decreases.

pulls the liquid column. The pressure or vacuum are read from the scale behind the liquid column, figure 37-10.

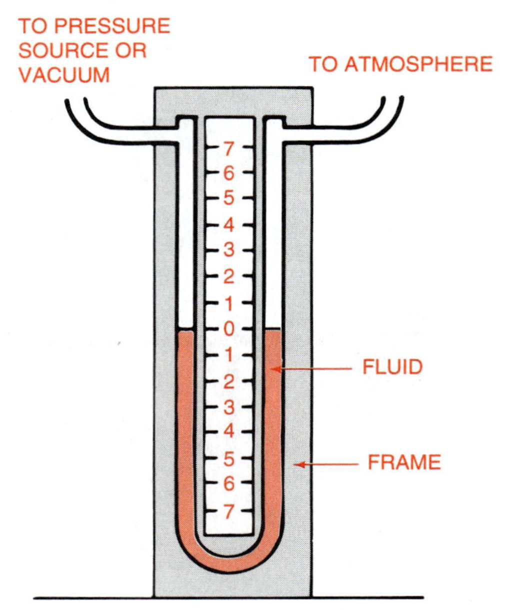

FIGURE 37-10. A manometer measures and reads pressure or vacuum by movement of a liquid column.

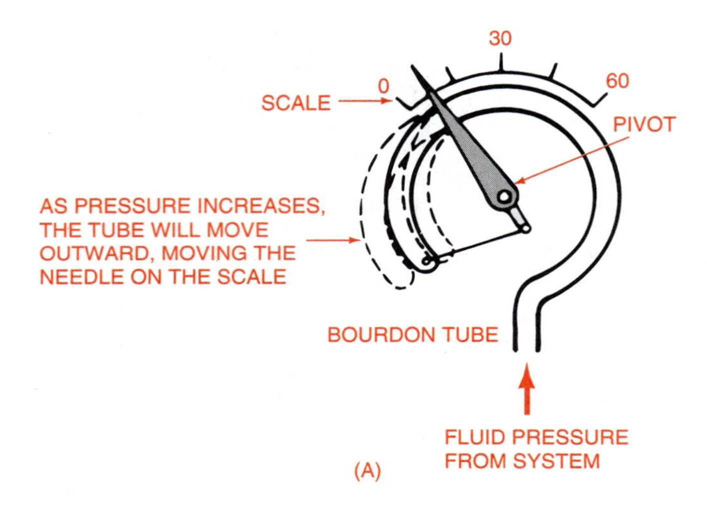

FIGURE 37-9. A Bourdon tube is used to measure and read hydraulic pressure.

Viscosity Index

The viscosity of a liquid refers to its tendency to flow. The viscosity index measures a fluid's tendency to flow at a given temperature. As a fluid becomes colder, it becomes thicker and tends to flow less. As it becomes hotter, it becomes thinner and flows faster.

Fluid Additives

An additive is a material added to a fluid to change its characteristics. Some types of additives used to improve fluids used in hydraulic systems include:

1. Viscosity improver—chemical added to make fluids more stable with temperature changes
2. Antifoam—chemical that reduces foaming in fluids
3. Corrosion inhibitor—chemical that reduces or prevents corrosion
4. Rust inhibitor—chemical that reduces or prevents rusting
5. Antiscuff—chemical that helps fluid polish moving parts
6. Extreme-pressure resistor—additive to prevent splitting of molecules and oil breakdown under high pressure

Fluid Ratings

SAE Rating. Most fluids and lubricating oils are rated by the Society of Automotive Engineers (SAE). The SAE numbers on containers of fluids and oils is a viscosity rating. Fluids and oils must have appropriate viscosity for the climate and the conditions. For instance, fluids used in vehicles in Alaska need different SAE ratings than those used in Florida. SAE ratings of fluids generally range from 5 to 90. SAE 10 flows easily at 0°F, while SAE 40 is too stiff for good lubrication. However, when the engine is operating at 180°F, SAE 10 is too runny, but SAE 40 works fine. The solution is to add viscosity improvers to make the fluid flow like SAE 10 at 0°F and like SAE 40 at 180°F. Such a fluid would be rated SAE 10W-40.

API Rating. The American Petroleum Institute (API) rates oils by the type of service they can tolerate. Oils may be rated SA, SB, SC, SD, or SE for gasoline engines and CA, CB, CC, CD, or CE for the more demanding diesel engine service. In general, the further down the alphabet the better (within the S or C group), the oil. However, it is always very important to use the fluid prescribed by the equipment manufacturer.

HYDRAULIC SYSTEMS

Hydraulic systems must have a reservoir, pump, control valve(s), check valve, filter(s), lines, and cylinder(s). The pump draws the fluid from the reservoir and pushes it through lines and valves to a piston in a cylinder. The pressure on the piston moves the piston and connecting apparatus to move the object. A valve then closes and holds the fluid against the piston. When the object should be lowered, the valve is opened and the piston pushes the fluid back into the reservoir.

Some hydraulic systems are built to apply pressure in either direction. A cylinder that can exert pressure by pushing or pulling is a double action cylinder, figure 37-11.

Hydraulic Pumps

Hydraulic pumps may be of the gear, vane, or piston type. Some gear-type and all piston-type pumps are positive displacement pumps. Positive displacement means that for each revolution an exact volume of fluid will be pumped.

Standard Gear-Type Pumps. The standard gear-type pump consists of two gears in mesh with each other turning in opposite directions. As they turn, fluid is drawn into one side, trapped between the gear teeth and housing, and expelled out the other side, figure 37-12. It is a positive displacement pump.

Eccentric Gear-Type Pump. The eccentric gear-type pump operates on the same principle as the standard gear-type pump. However, the gears move in an eccentric, or off-center, pattern which creates the vacuum and pressure action necessary to move fluids, figure 37-13.

Vane-type Pump. The vane-type pump consists of spring-loaded paddles on an off-center hub. As the hub turns, its body moves close to the housing wall during part of its resolution and away during part of it. The vanes maintain a seal with the wall and move the fluid through the pump, figure 37-14. The vane-type pump is also useful as an air pump.

Piston-Type Pump. Piston-type pumps can be purchased for moving air or fluids. Pumps that increase pressure on air are called air compressors. Some are two-stage air compressors. A two-stage air compressor compresses a large cylinder of air into a small cylinder and then a second piston compresses that air into a tank.

It may be recalled that liquids cannot be compressed. Therefore, piston-type hydraulic pumps are positive displacement pumps. They are relatively expensive, but capable of pumping fluids under great pressure.

Fluid Coupling

Many applications call for a gradual and smooth transfer of power from an engine to the vehicle. Automobile transmissions use fluid couplings for this purpose. A fluid coupling utilizes fluid to transfer power from one turning impeller to drive another as the fluid spins. An impeller is a fan-shaped rotor. The action of a fluid coupling can be demonstrated using a motor-driven fan (air pump) to force air through a tube and drive another fan (turbine), figure 37-15.

Hydraulic Hoses

Hydraulic hoses carry fluid from component to component. They must be strong, flexible, and resistant to pressure and abrasion. Generally, the inner layer or inner tube is made of tough rubber. The intermediate layer(s) is composed of synthetic fibers or wire to withstand high pressure. The outer cover is also rubber for reducing vibration and providing resistance to abrasion, weather, dirt, and moisture, figure 37-16.

Hydraulic Valves

A system of valves is necessary to direct the flow of fluids from pump to cylinders or motors and back to the reservoir. Such valves may determine the routing of fluid or may simply control the direction of flow.

Pressure Regulators. A pressure regulator is a valve that holds pressure in a line at a given value. In hydraulic systems, pressure regulators determine the amount of pressure that can be exerted on the system before fluid is routed back to the reservoir.

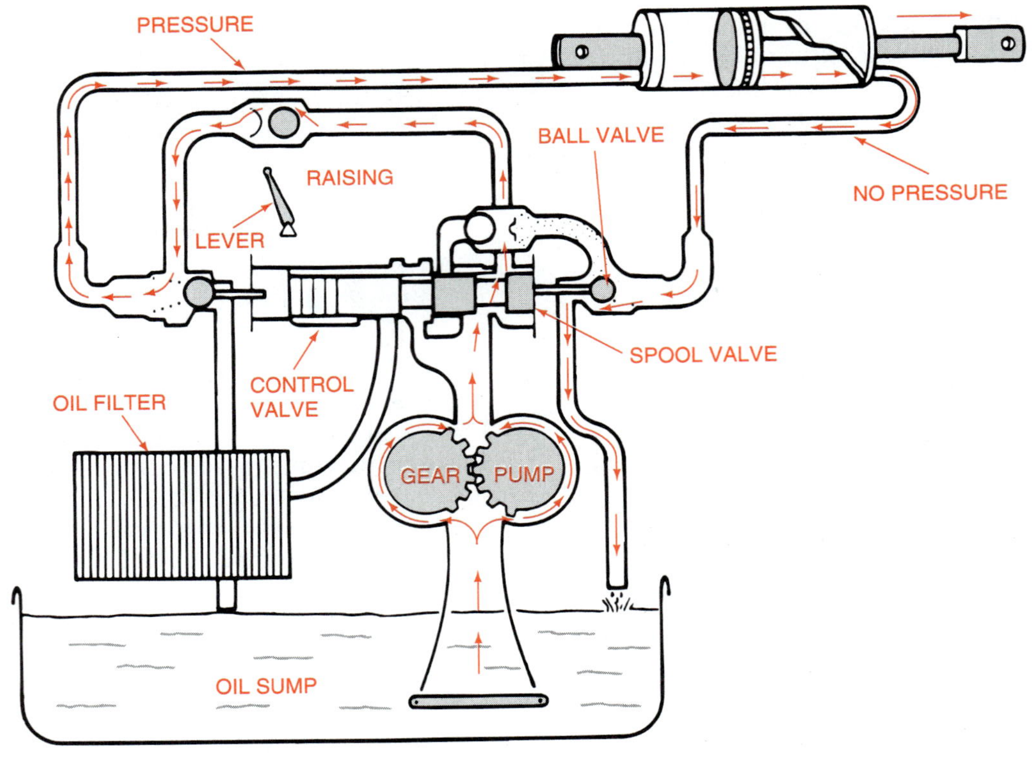

(A) RAISING EQUIPMENT

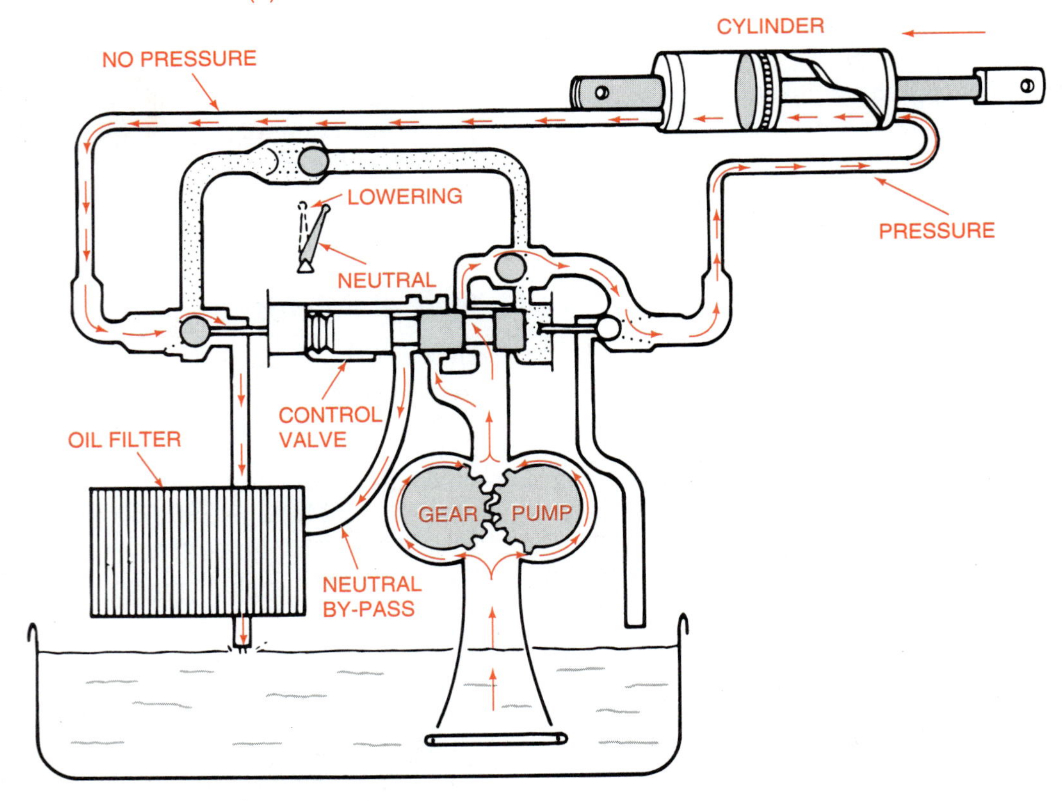

(B) LOWERING EQUIPMENT

FIGURE 37-11. A hydraulic system showing system activity when equipment is being (A) raised and (B) lowered. (*Courtesy of Instructional Materials Services, Texas A&M University*)

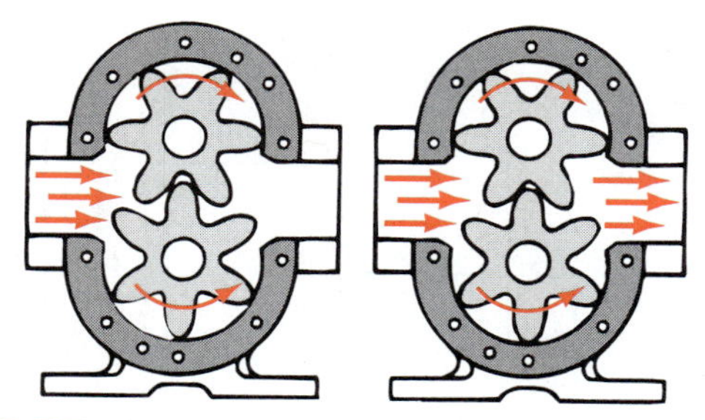

FIGURE 37-12. Standard gear-type pump. (*Courtesy of A. Schwaller,* MOTOR Automotive Technology, 2E, *Delmar Publishers, 1993*)

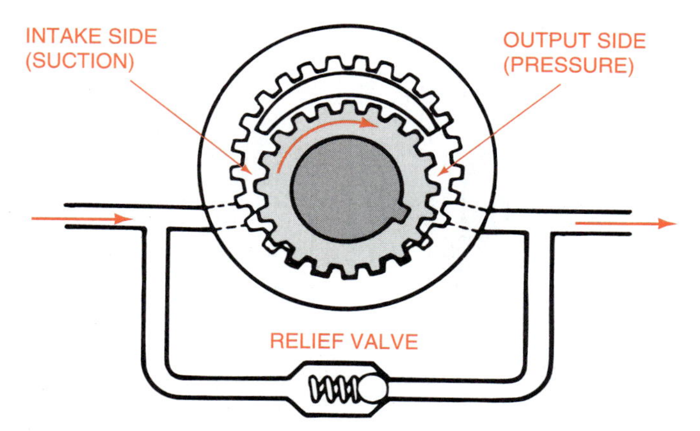

FIGURE 37-13. Eccentric gear-type pump. (*Courtesy of A. Schwaller,* MOTOR Automotive Technology, 2E, *Delmar Publishers, 1993*)

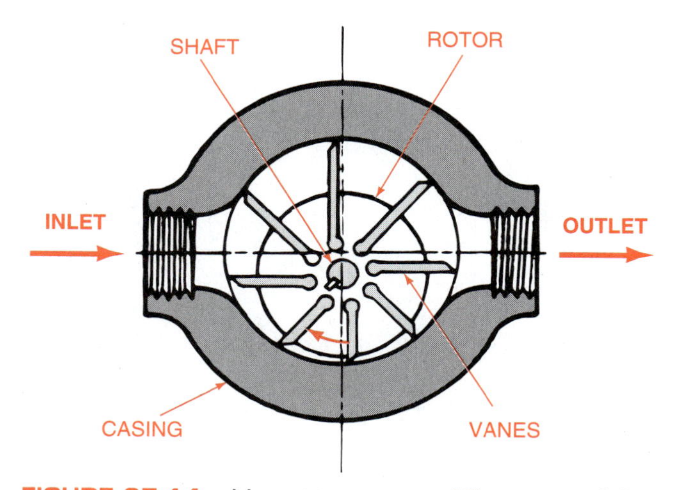

FIGURE 37-14. Vane-type pump. (*Courtesy of A. Schwaller,* MOTOR Automotive Technology, 2E, *Delmar Publishers, 1993*)

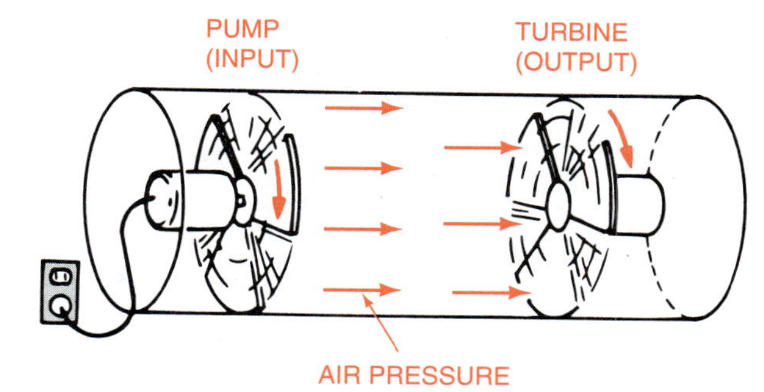

FIGURE 37-15. Fans functioning like a fluid coupling. (*Courtesy of A. Schwaller,* MOTOR Automotive Technology, 2E, *Delmar Publishers, 1993*)

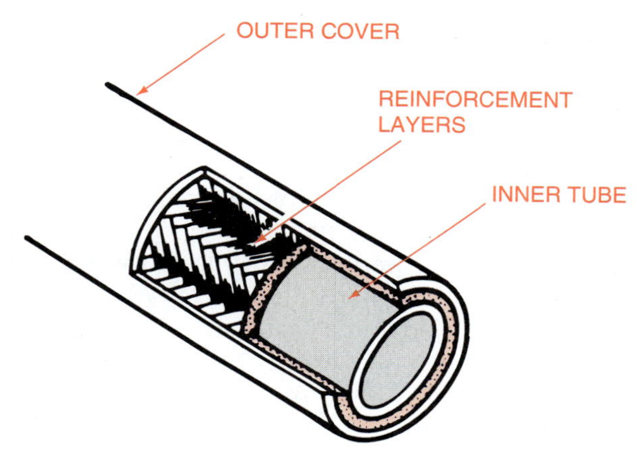

FIGURE 37-16. Construction of hydraulic hoses.

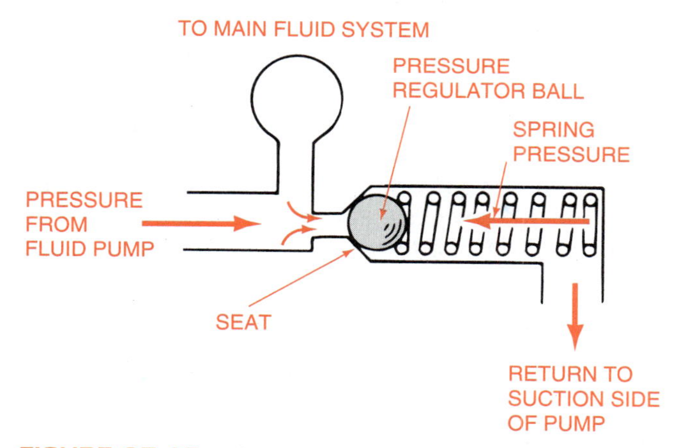

FIGURE 37-17. A pressure regulator protects the system as well as operators. (*Courtesy of A. Schwaller,* MOTOR Automotive Technology, 2E, *Delmar Publishers, 1993*)

This provision protects the system from damage caused by overloading. Therefore, it also protects operators from injury that could result from rupturing hoses or exploding parts, figure 37-17.

Direction Control Valves. A direction control valve permits fluid to flow in one direction only. Such valves are also called check valves. Some check valves consist of a housing with a perfectly round, smooth ball that sits on an opening with a carefully ground seat. Fluid coming in under pressure can lift the ball and flow around it. However, if the pressure drops or is overcome by reverse pressure, the ball drops, seals, and prevents back flow, figure 37-18.

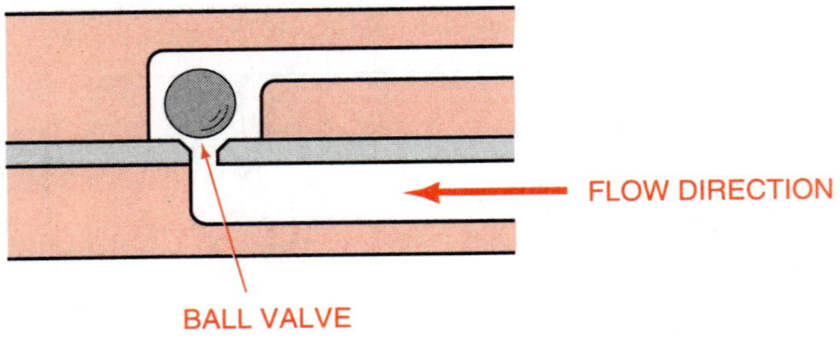

FLOW DIRECTION

BALL VALVE

FIGURE 37-18. The check valve is a direction control valve.

Such valves can operate in any position if the ball is installed under spring tension.

Equipment Control Valves. If you observe an operator of a backhoe, trencher, front-end loader, or aerial bucket perform tree or power line work, you will notice the operator controls the machine by pushing or pulling several levers. These levers operate the equipment control valves.

The most common type of equipment control valve is the spool valve. A spool valve is a valve with one inlet and several outlets controlled by a sliding spool. The spool is constructed so its wheels can block the flow of fluid and its axle can permit fluid to flow through. The position of the sliding spool determines the outlet and equipment that receives the flow and pressure of fluid at any given time, figure 37-19.

FILTERING SYSTEMS

Hydraulic system parts are machined and operated at very close tolerances. Tolerance means leeway or acceptance of variation. Therefore, the smallest speck of dust, dirt, grit, or metal can damage the moving parts of a hydraulic system. A good system of filters must be in place and properly maintained at all times. Most fluid filters are made of special paper folded into pleats that encircle the core. The fluid enters the outside chamber, passes through the paper filter, and leaves the filter through a center tube.

Full-Flow Filter System. A full-flow system directs all fluid leaving the pump through the filter. As long as the filter is changed as needed, the system works well. However, if the filter becomes clogged, the unfiltered fluid will bypass the filter and flow throughout the system. If not corrected promptly, the system will be damaged, figure 37-20.

Bypass Filter System A bypass system is one that filters only a part of the fluid in any given pass through. However, since the fluid circulates frequently in an active system, all fluid is eventually filtered. This system can utilize a filter with less flow capacity than a full-flow system, figure 37-21.

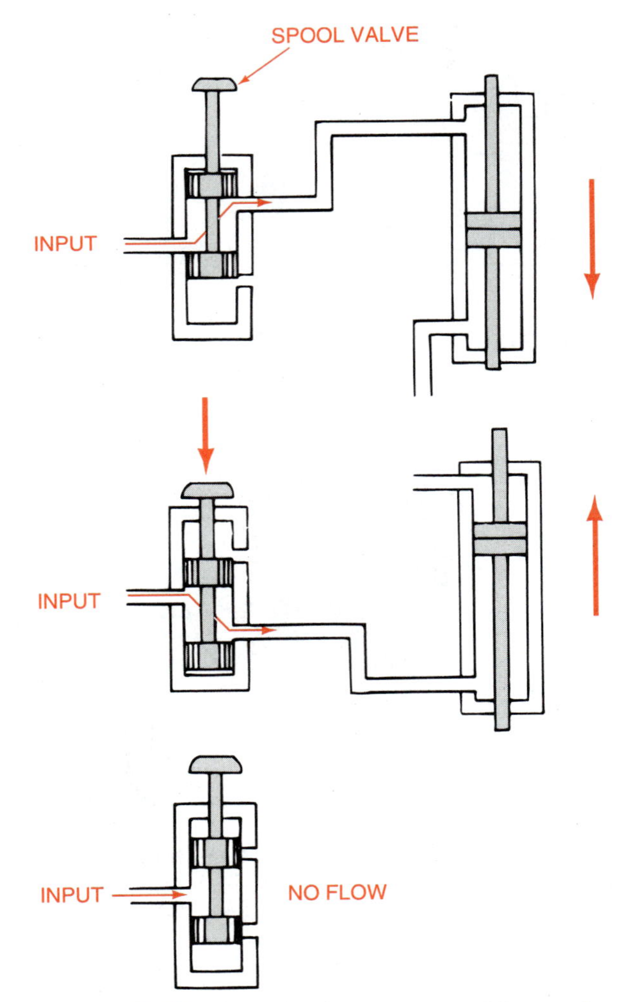

FIGURE 37-19. Spool valves are used by equipment operators to direct the power and movement of hydraulic devices.

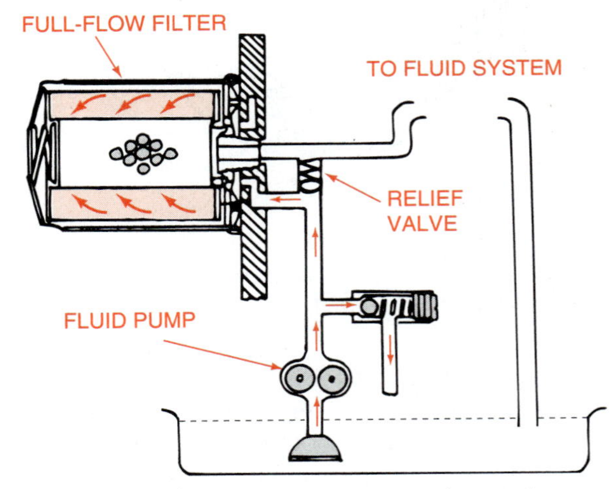

FIGURE 37-20. All fluid flows continuously through the filter of a full flow system. (*Adapted from A. Schwaller,* MOTOR Automotive Technology, 2E, *Delmar Publishers, 1993*)

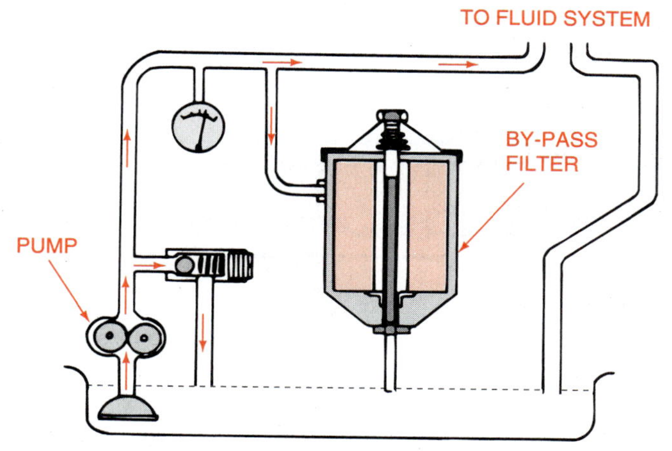

FIGURE 37-21. In a bypass system, only part of the fluid is filtered in any single pass-through. (*Adapted from A. Schwaller,* MOTOR Automotive Technology, 2E, *Delmar Publishers, 1993*)

Hydraulic Cylinder and Motors

Cylinders. A hydraulic cylinder is a tank with a fluid inlet and a close-fitting piston. If it is a double action cylinder, it has a fluid outlet as well. The piston has a connecting rod(s) extending through the end of the cylinder to which movable equipment is attached. The piston has rubber ring or O ring to maintain a fluid seal, figure 37-22.

Hydraulic Motors. A hydraulic motor is a motor that receives its power from moving fluid. Its construction and operation is similar to a gear pump. Fluid, under pressure from the pump, enters the inlet side of the motor, drives the gears as it passes through, and exits under low pressure for return to the reservoir, figure 37-23.

While the hydraulic motor is simple in design, it can convert the tremendous power of fluid under pressure to a rotary power. This quiet and flexible power source is especially useful in remote places on engine-driven machines.

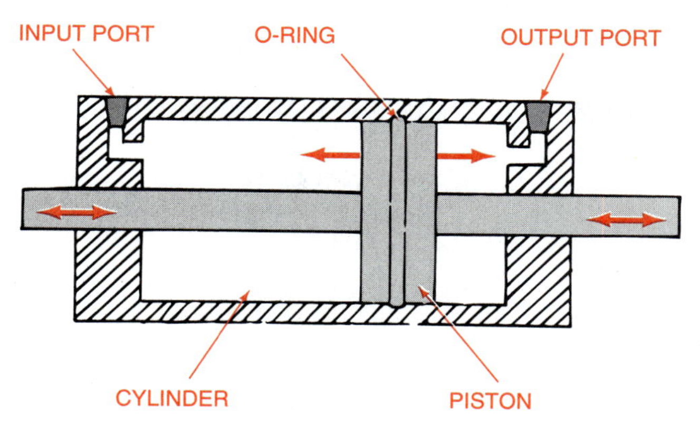

FIGURE 37-22. Parts of a hydraulic cylinder.

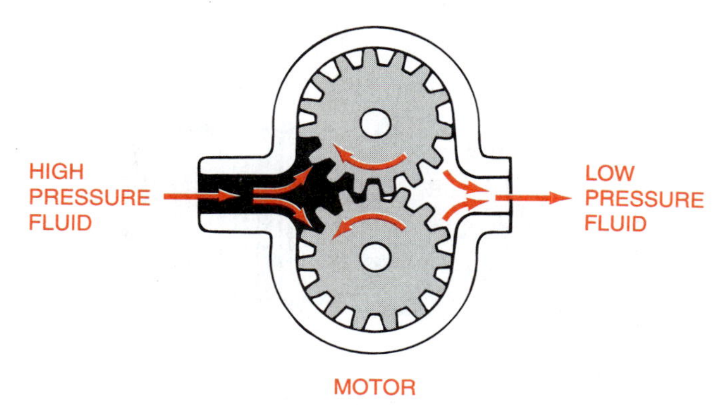

FIGURE 37-23. Hydraulic motor.

Hydraulic System Maintenance

Hydraulic systems must have regular maintenance service to operate trouble-free. Otherwise, unnecessary and expensive repairs will result. Fluid of the proper type and grade must be kept clean and free of dirt, moisture, and contaminants.

Fluid Level. Fluid levels must be checked frequently, and care must be taken not to introduce dirt in the process. Most tractors and other machines with hydraulic systems use a dipstick to monitor fluid levels.

Fluid Filters. Filters must be serviced at regular intervals as prescribed by the manufacturer. This generally involves simply unscrewing a disposable filter canister, wiping the surfaces clean, and screwing on a new filter canister.

Quick-Couplings. Some hydraulic lines go to equipment that can be separated from the machine. In such cases a quick-coupling is generally used in the hoses. A quick-coupling is a device with two connected parts attached to two hose ends. The parts can be connected by retracting a spring-loaded collar on one part, inserting the second part, and releasing the collar to lock the parts together. When quick-couplings are disconnected, the ends should be capped to prevent dirt from entering the hose and when reconnected, the parts must be clean, figure 37-24.

Fluid Change. Fluid systems must be drained and clean fluid and filters installed, as prescribed by the manufacturer. Such service must be done with care to ensure that dirt is not introduced into the system, that proper fluid is put back into the system, and that the old fluid is discarded properly. The latter means it should be recycled and not left to pollute the environment.

If the fluid gets foamy, milky colored, discolored, or dirty, it must be drained and replaced regardless of length of service. Before using the equipment again,

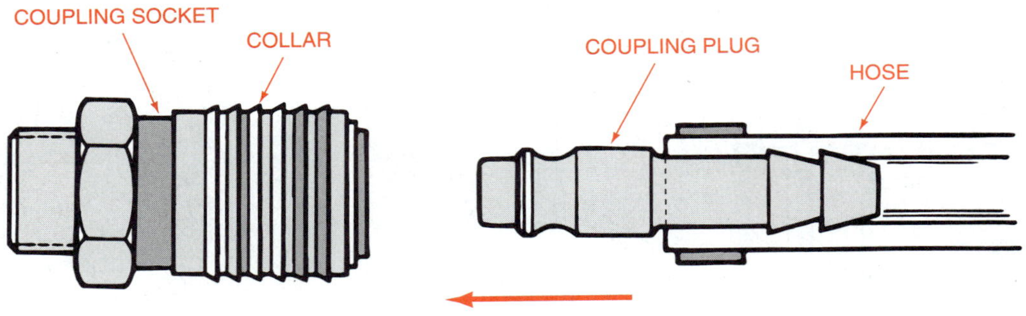

FIGURE 37-24. Quick coupling.

the reasons for fluid contamination or failure must be corrected.

PNEUMATIC SYSTEMS

Pneumatic systems are relatively simple, compact, and have great flexibility. Small electric motor-driven vane pumps and compressors can be very portable. Larger compressors are stationary and driven by large electric motors or gasoline or diesel engines. Service of such units involve frequent draining of condensed water from air tanks and moisture-removing devices in the lines.

Regulators and valves for pneumatic systems are similar in function to those of hydraulic systems. However, air pressure regulators are generally adjustable, figure 37-25. Pneumatic cylinders frequently are single action. These rely on an internal spring to return the piston to its home position when air pressure is released, figure 37-26. Additional information on pneumatic systems is provided in Unit 28, Selecting and Applying Painting Materials.

ROBOTICS

A robot is a mechanical device that is capable of performing human tasks. Robots are used extensively in industry. They are becoming increasingly important in agriculture/agribusiness and the management of renewable natural resources.

The term robotics refers to the study and application of the technology of robots. Robots may be distinguished from other machines in that they:

- are freely computer programmable
- can do a variety of tasks
- have three-dimensional freedom of motion
- are equipped with grippers or tools.

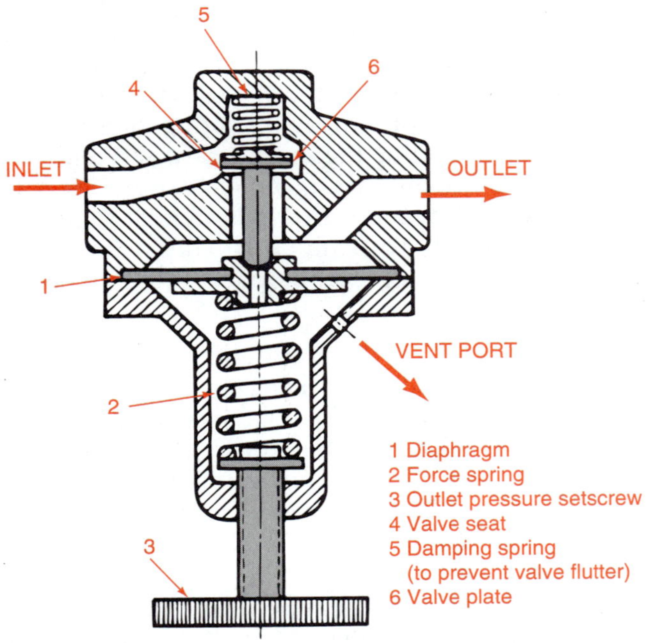

FIGURE 37-25. Air pressure regulator.

COMPONENTS OF A SINGLE-ACTING CYLINDER

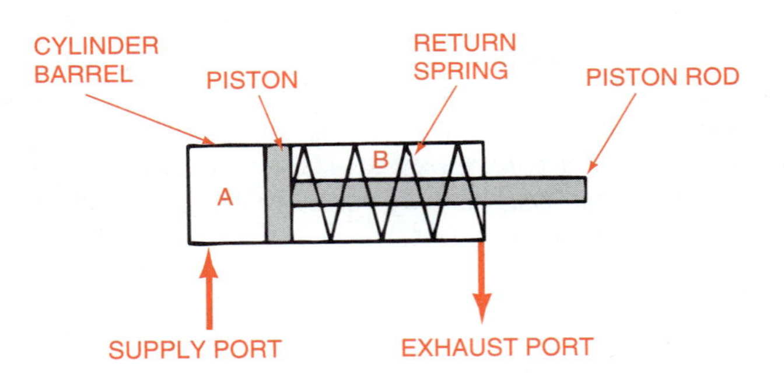

SINGLE-ACTING CYLINDER IN EXTENDED POSITION

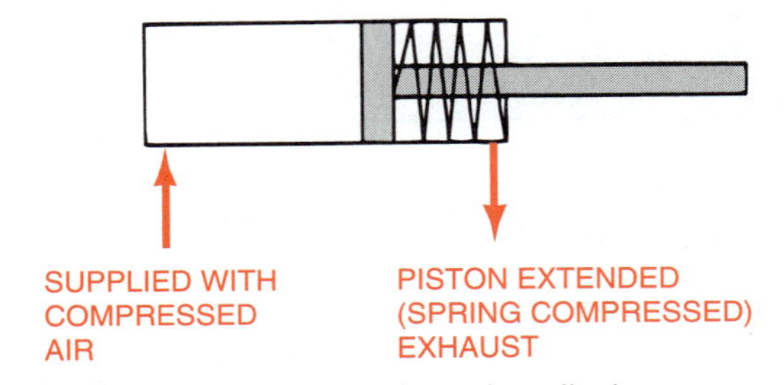

FIGURE 37-26. Single action air cylinder.

Robotic Functions

Robots can be built to perform many tasks faster and more accurately than humans. They are precise in movement and repeat tasks exactly the same for long periods of time. Robots are especially adaptable to:

- arranging parts
- handling parts
- distributing items
- positioning tools and workpieces
- moving tools in predetermined patterns
- gripping, directing, and assembling
- fastening, attaching, and detaching.

Robotic Movements

Robots are built to manage circular and linear motion. Circular motion is referred to as rotation, and linear motion is called translation.

Robots have one or more translational and/or rotational axes. Axes is plural for axis. An axis is a straight line around which a body rotates. The more axes a robot has, the more motions it can perform. Each axis provides the robot with one degree of freedom. A robot has as many degrees of freedom as it has axes.

Cartesian Work Area. A robot with three translational axes can perform translational tasks in a space with box-like characteristics. A box-like work space is known as a cartesian work area, figure 37-27.

Cylindrical Working Area. The work space of the robot in the above example can be changed drastically by changing the axis of the base. If the translational axis is changed to a rotary axis, the working area is cylindrical, or in the shape of a cylinder. However, since the robot now works around itself, there is a hollow (space) in the center where the robot cannot function, figure 37-28.

Hollow Sphere Working Area. A ball-shaped working area is known as a hollow sphere. It can be achieved with two rotary axes and one translational axis. Here the robot is also working around itself and creates a hollow in the center of the work area, figure 37-29.

Solid Sphere Working Area. A solid sphere is like a solid ball. A solid sphere working area can be achieved with three rotational axes on a robot. With this arrangement the robot can work in any part of a round pattern to form a solid sphere work area, figure 37-30.

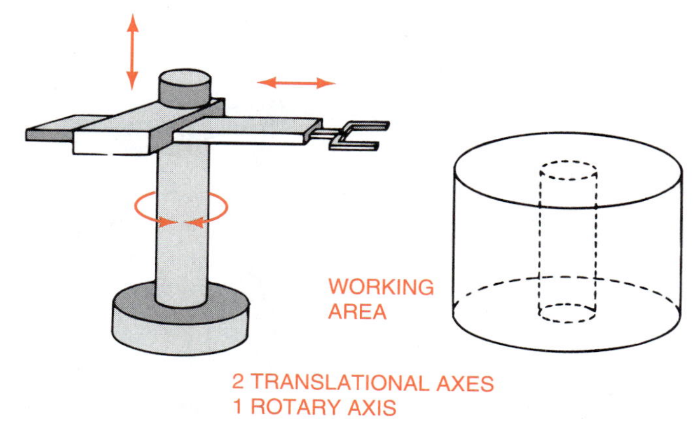

FIGURE 37-28. A cylindrical work area is possible with two translational and one rotational axis.

Robotic Power

Robotic structures are powered by pneumatic, hydraulic, or electrical drives. Both pneumatic and hydraulic power may be utilized with cylinders and motors. However, since air is compressible, the movements of air-driven equipment is not as precise as that of hydraulic equipment. For this reason, pneumatic drives are generally limited to linear point-to-point functions.

Electric power is the choice for many robotic applications. The variety of motors and controls is great, providing an almost infinite number of combinations in equipment design. Electric motors are

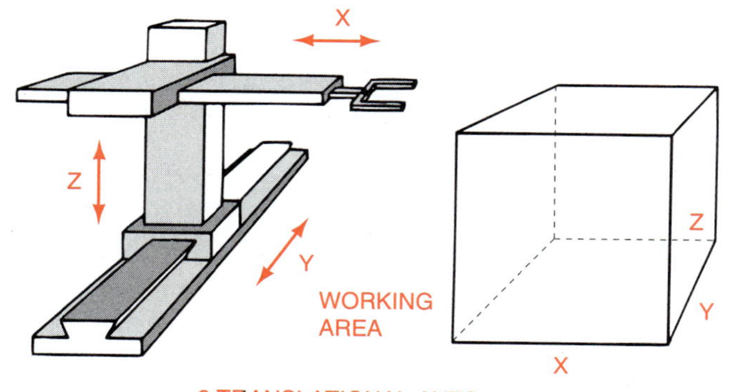

FIGURE 37-27. The cartesian work area of a three-translational robot.

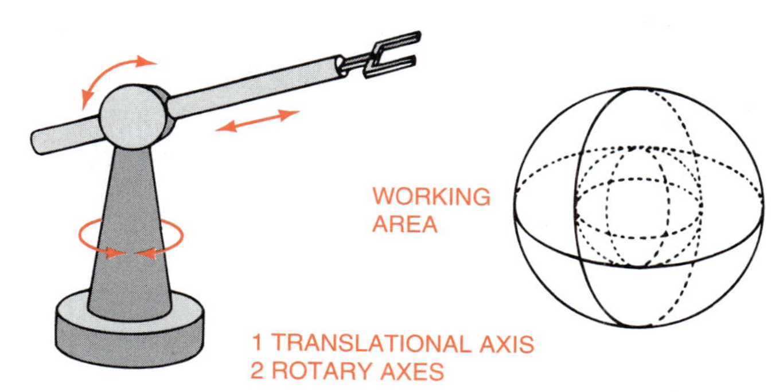

FIGURE 37-29. A hollow sphere work area utilizing two rotary and one translational axis.

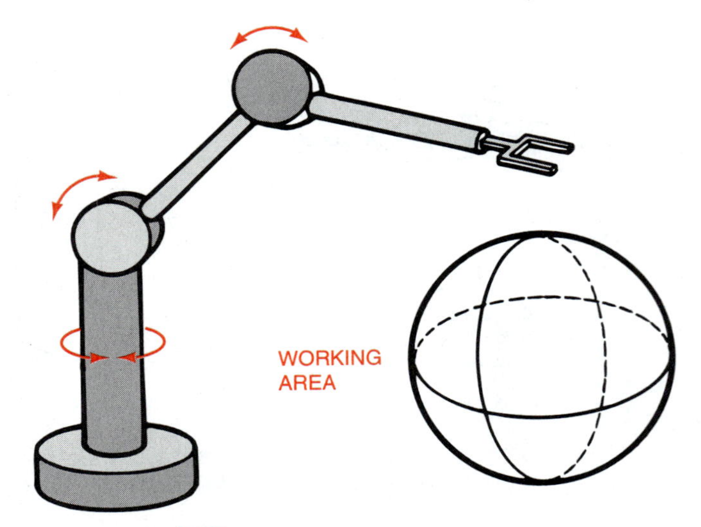

FIGURE 37-30. A solid sphere work area is possible using three rotational axes.

powerful, but compact and electronic controls are tiny. They lend themselves well to computer-controlled operations. The advantages of compactness, ease of getting power to the motor, ease of maintenance, and high reliability all contribute to the advantages of electric power for robotics.

SUMMARY

Hydraulic and pneumatic systems provide flexible power in remote areas. Compressed air is useful as a propellant, can drive motors, and can actuate cylinders. Pressurized liquids provide considerable force when driving hydraulic motors and permit excellent control over hydraulic cylinders.

Hydraulic, pneumatic, and electrical devices are used to operate robots. Robots are performing many tasks faster, cheaper, and better than what can be done by human hands. The use of robots and the expansion of the robotics industry will provide new career opportunities in the future.

STUDENT ACTIVITIES

1. Define the Terms to Know in this unit.
2. Develop a bulletin board illustrating the use of hydraulics and pneumatics in agricultural settings.
3. Make a collage of robots in action.
4. Work with your agricultural mechanics instructor and/or a science teacher, and prepare and present a class demonstration on Boyle's law.
5. Present a class demonstration to illustrate the truth of Bernoulli's theorem.
6. Demonstrate one or more applications of Pascal's law.
7. Prepare and present a report on the use of additives in hydraulic fluids.
8. Do a shop project that requires the servicing of a hydraulic system.
9. Make a shop project that utilizes hydraulics or pneumatics.
10. Visit a firm that uses robots for production. Take pictures of the robots and report to the class how the robots are designed, function, and maintained.

SELF-EVALUATION

A. Multiple Choice. Select the best answer.

1. Fluid power is
   a. hydraulics
   b. pneumatics
   c. pneumatics and hydraulics
   d. pneumatics, hydraulics and electricity
2. Linear power is provided directly by
   a. pistons in cylinders
   b. fluid couplings
   c. motors
   d. rotational robot axes
3. Force acting on an area is
   a. compression
   b. horsepower
   c. measured in pounds
   d. pressure
4. Important laws of force and pressure in hydraulics were formulated by
   a. Bernoulli
   b. Bourdon
   c. Boyle
   d. Pascal
5. Pressure in psi = force in pounds divided by
   a. area
   b. distance
   c. number of pistons
   d. time
6. A volume of gas is inversely proportional to its
   a. weight
   b. temperature
   c. pressure
   d. area
7. When fluid flows through a pipe and encounters a decrease in pipe diameter, pressure in the restriction
   a. cannot be predicted
   b. decreases
   c. increases
   d. remains the same
8. Which is *not* a type of hydraulic pump?
   a. vane
   b. ratchet
   c. piston
   d. gear
9. The full-flow system refers to the use of
   a. valves
   b. spools
   c. lines
   d. filters
10. A hollow sphere working area can be achieved with
    a. three translational axes
    b. two translational and one rotational axis
    c. two translational and one translational axis
    d. none of the above

B. Matching. Match the terms in column I with those in column II.

Column I

1. Bourdon tube
2. manometer
3. viscosity
4. rust inhibitor
5. SAE rating
6. API
7. spool valve
8. quick-coupling
9. cartesian
10. translational

Column II

a. additive
b. service rating
c. measures fluid pressure
d. controls hydraulic equipment
e. measures vacuum
f. tendency to flow
g. uses a sliding collar
h. box like work area
i. linear action
j. viscosity

C. Completion. Fill in the blanks with the word or words that make the following statements correct.

1. In addition to SA and CA, three API oil ratings based on service capability are _____, _____, and _____.
2. Two types of gear pumps are the _____ _____ type and _____ type.
3. The piston in hydraulic cylinders typically contains an _____ to create a moving seal between the piston and cylinder wall.
4. Three important maintenance procedures for hydraulic systems are _____, _____ and _____.
5. A disadvantage of pneumatics in robots is _____.

Section 13

Concrete And Masonry

UNIT

38 Concrete and Masonry

OBJECTIVE

To mix and place concrete and use masonry materials.

COMPETENCIES TO BE DEVELOPED

After studying this unit, you should be able to:

- Identify tools used for concrete work.
- Select ingredients for mixing concrete.
- Make a workable masonry mix.
- Prepare forms for concreting.
- Insulate concrete floors.
- Pour concrete.
- Finish concrete.
- Calculate concrete and block for a job.
- Lay masonry block.

TERMS TO KNOW

masonry
portland cement
concrete
sand
mortar
finishing lime
fine aggregate
gravel
coarse aggregate
silt
clay
washed sand
cement paste
ratio
mix
workable mix
form
footer
footing
insulation
moisture barrier
vapor barrier
construction joint
control joint
reinforced concrete
air-entrained concrete
screeding
bull float
broom finish
curing
masonry units
laying block
mortar bed
core
hollow core block
ears
stretcher block
sash block
jamb block
course
frost line
corner pole

MATERIALS LIST

✓ Portland cement
✓ Sand
✓ Gravel
✓ Mortar mix
✓ Wall reinforcement
✓ Lumber for forms
✓ Stakes
✓ Bull float

- ✓ Hand float
- ✓ Edging tool
- ✓ Grooving tool
- ✓ Push broom
- ✓ Oil for forms
- ✓ Level, 4 foot
- ✓ Mason's trowel
- ✓ Corner poles
- ✓ Wheelbarrow
- ✓ Mortar pan
- ✓ Striking tool
- ✓ Rakes
- ✓ Shovels
- ✓ Quart jar with lid

Concrete and masonry construction is basic to most farm and agribusiness facilities. Masonry refers to anything constructed of brick, stone, tile, or concrete units set or held in place with portland cement. Portland cement is a dry powder made by burning limestone and clay, then grinding and mixing to an even consistency.

Concrete and mortar are two construction materials with portland cement as a component. Concrete is a mixture of stone aggregates, sand, portland cement, and water that hardens as it dries. Sand consists of small particles of stone. Mortar is a mixture of sand, portland cement, water, and finishing lime. Finishing lime is a powder made by grinding and treating limestone. Like concrete, mortar is a mixture of materials mixed with water that hardens as it dries.

Nearly all buildings have concrete and masonry components somewhere in their construction. The ability to use masonry materials is therefore essential to many maintenance and repair operations.

MASONRY CONSTRUCTION

Masonry construction has been used for thousands of years. Its basic materials come from the earth and are plentiful in most areas. The pyramids of Egypt, the colosseum of ancient Rome, the temples of the Mayan Indians, and the skyscrapers of modern America all demonstrate the use of masonry construction. Masonry construction has most of the advantages of concrete. However, the hardness, durability, and moisture-resistant characteristics of the masonry units and mortar will determine the characteristics of the masonry structure.

CONCRETE CONSTRUCTION

Concrete has many advantages for farm use and in other agriculture enterprises. Concrete is:

- fireproof
- insect and rodent proof
- decay resistant
- highly storm resistant
- wear resistant
- waterproof
- strong
- attractive
- easy to make on the job without expensive equipment
- available locally
- low in original and maintenance costs
- sanitary and easy to keep clean
- easily broken up and used as fill material when the structure becomes obsolete.

The strength and durability of concrete depend upon several factors: (1) the strength of the stone particles, (2) the proportion of stone particles by size, (3) the type of portland cement, (4) the purity of the water, (5) the uniformity of the mixture, and (6) the procedures used in placing, finishing, and curing the concrete.

Many types of stone found in fields, streams of water, and solid rock formations are suitable for making concrete. Occasionally the correct type and mixture of stone particles may be found in stream beds or gravel deposits. Usually, however, stone particles must be made by crushing stone and grading it to size. The correct proportion of small particles, called sand or fine aggregate, and large particles, called gravel or coarse aggregate, can then be mixed to meet the requirements of specific jobs.

To make good concrete, aggregates of various sizes should fit together to form a fairly solid mass, figure 38-1. The stone particles must be clean and

FIGURE 38-1. In good concrete, aggregates of various sizes fit together to form a fairly solid mass.

free of clay, silt, chaff, or any other material. Silt is composed of intermediate sized soil particles. Clay consists of the smallest group of soil particles. Both silt and clay particles are too small for use as aggregates in concrete and decrease the quality of concrete if present.

Sand sold commercially for use in concrete generally contains very little clay or silt. If the clay or silt content is above an acceptable level when the material comes from the quarry, the sand is flushed with water to remove the clay or silt. The material is then referred to as washed sand.

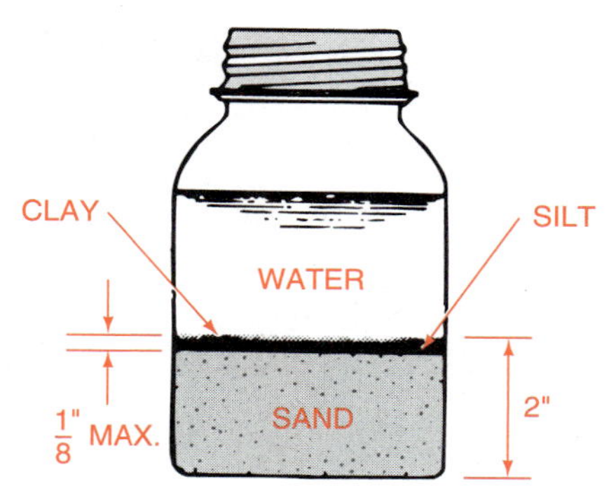

FIGURE 38-2. This simple test will determine the silt content of material that is being considered for use in concrete. (*Courtesy of G. White, Concrete Technology, 4th ed., Delmar Publishers, 1991*)

Testing Sand for Silt or Clay Content

Sand from streambeds or other farm areas may be suitable for making concrete. However, when sand from such places is being considered, a test for clay and silt must be made. The procedure follows.

PROCEDURE

1. Fill a 1-quart glass jar to a depth of 2 inches with the sand to be tested, figure 38-2.
2. Add water until the jar is 3/4 full.
3. Screw on a lid and shake the mixture vigorously for 1 minute to mix all particles with the water.
4. Shake the jar sideways several times to level the sand.
5. Place the jar where it will not be disturbed for 1 hour for a silt test or 12 hours for a clay and silt test.
6. After 1 hour, measure the thickness of the silt layer on top of the sand.
7. If the layer is more than 1/8 inch thick, the sand is not suitable for use in concrete unless the silt is removed by washing.
8. If the layer is not 1/8 inch thick in 1 hour, let the mixture stand for 12 hours. Remeasure the layer(s) that have settled on the sand.
9. If the silt plus clay layer exceeds 1/8 inch, wash the sand before using it in concrete.

CAUTION:

- **A mistake in measuring will ruin the product.**

MIXING CONCRETE

Cement paste is made by mixing portland cement and clean water in precise proportions. The reaction between the cement and the water is a chemical one, and the strength of the concrete is directly affected by the proportions of the materials. As a result, precision in mixing portland cement and water is as important as precision in mixing ingredients when baking a cake.

All sand has some water attached to its particles. This water must be estimated and an allowance made for it when deciding how much water to use in the concrete mix. Sand is described as "damp," "wet," or "very wet." The more moisture there is in the sand, the less water is added from other sources when mixing concrete or mortar. To obtain a more durable concrete to withstand severe weather conditions and/or traffic, add less water, figure 38-3.

The ratio of cement to fine aggregate (sand) and coarse aggregate (gravel) is important. However, the exact proportion will vary with the makeup of each aggregate, figure 38-4. A correct mixture will assure that: (1) each particle of sand and gravel is covered with cement paste, and (2) each particle is bound to others when the cement paste dries and hardens.

The proportion of cement to fine and coarse aggregate is known as the ratio. The ratio is expressed as a three-digit number called a mix. For example, a 1-2-3 mix has one part cement, two parts fine aggregate, and three parts coarse aggregate. Once the mix for a job has been determined, a table can be used to help estimate the amount of material needed for a cubic yard of concrete, figure 38-5. A rule of thumb is that the resulting concrete will be about two-thirds the combined volume of the cement and aggregate used in the mix.

It should be noted that portland cement is sold in bags containing 94 pounds, or exactly 1 cubic foot, of cement. A 1-2-3 mix would consist of 1 cubic foot of cement (one bag), 2 cubic feet of sand, and 3 cubic feet of coarse aggregate. For the purpose of mixing,

INTENDED USE OF CONCRETE	FOR MAXIMUM AGGREGATE SIZE OF	WATER (GALLONS) ADDED TO 1 CU FT OF CEMENT IF SAND IS: Damp	Wet (Average Sand)	Very Wet	SUGGESTED MIXTURE FOR 1 CU FT TRIAL BATCHES* Cement Cu Ft	Aggregates Fine Cu Ft	Aggregates Coarse Cu Ft
Mild Exposure	$1^1/_2$ in.	$6^1/_4$	$5^1/_2$	$4^3/_4$	1	3	4
Normal Exposure	1 in.	$5^1/_2$	5	$4^1/_4$	1	$2^1/_4$	3
Severe Exposure	1 in.	$4^1/_2$	4	$3^1/_2$	1	2	$2^1/_4$

*Mix proportions will vary slightly depending on the gradation of aggregate sizes.
Note: Batches may be made by using 100 pounds (1 cwt) of each ingredient where 1 cubic foot is indicated.

FIGURE 38-3. Ratio of ingredients for mixing concrete. (*Courtesy of G. White,* Concrete Technology, 4th ed., *Delmar Publishers, 1991*)

one bag of cement may be considered as 100 pounds or 1 hundredweight (1 cwt). This is especially useful when the aggregates being used are measured by weight rather than volume. Proportions may be based on either weight or volume.

The term workable mix refers to the consistency of the wet concrete after the various ingredients have been mixed together. A workable mix has the following characteristics:

- All aggregates are clean.
- All portland cement is mixed with water to form a cement paste. No dry powder is present.
- Every particle of aggregate is covered with cement paste.
- Aggregates are distributed evenly throughout the mix.
- No more than the recommended amount of water is added.
- No lumps are present.
- The mixture has uniform color and consistency.
- The mixture can be mixed, moved, and placed with a shovel or spade.

FOR MAXIMUM AGGREGATE SIZE OF:	SUGGESTED MIXTURE FOR 1 CWT TRIAL BATCHES* Cement Cwt** or Cu Ft	Aggregates Fine Cu Ft	Aggregates Coarse Cu Ft
$^3/_4$ in	1	2	$2^1/_4$
1 in	1	$2^1/_4$	3
(preferred mix) $1^1/_2$ in	1	$2^1/_2$	$3^1/_2$
(Optional mix) $1^1/_2$ in	1	3	4

*Mix proportions will vary slightly depending on the gradation of the aggregates. A 10-percent allowance for normal wastage is included in the fine and coarse aggregate values.
**1 cwt of cement equals about 1 cubic foot.

FIGURE 38-4. Materials needed to make trial batches of concrete with separated fine and coarse aggregates. (*Courtesy of G. White,* Concrete Technology, 4th ed., *Delmar Publishers, 1991*)

ESTIMATING MATERIALS FOR A JOB

It is important to know how to estimate the amount of concrete needed for a job. If the estimate is inaccurate, the job will cost more than is necessary and ingredients may be wasted. If insufficient material is available, the job must be stopped and then started again at a considerable increase in cost and time. The following example will show how to determine the amounts of materials needed for a concrete floor 4 inches thick, 14 feet wide, and 24 feet long.

PROCEDURE

1. Determine the cubic yards of concrete needed by multiplying the thickness of the floor by its width and length in feet: Cu Ft = T' × W' × L'. In this case, thickness equals 4 inches which must be converted to $^1/_3$ foot. Therefore, $^1/_3$' × 14' × 24' = 112 cubic foot. There are 27 cubic feet in 1 cubic yard. Thus 112 cubic feet ÷ 27 = 4.15 cubic yards.

MIX RECOMMENDED FOR MAXIMUM AGGREGATE SIZE OF:	MATERIALS NEEDED TO MAKE ONE CU YD OF CONCRETE				
	Cement Cwt or Cu Ft	Aggregates			
		Fine		Coarse	
		Cu Ft	Lb	Cu Ft	Lb
3/4 in 1–2–2 1/4	7	17	1550	19.5	1950
1 in 1–2 1/4–3	5.5	15.5	1400	21	2100
(preferred mix) 1 1/2 in 1–2 1/2–3 1/2	5.25	16.5	1500	23	2300
(optional mix) 1 1/2 in 1–3–4	4.5	16.5	1500	22	2200

Note: 1 cwt of cement or aggregate equals approximately 1 cubic foot.

FIGURE 38-5. Materials needed to make one cubic yard of placed concrete. (*Courtesy of G. White,* Concrete Technology, 4th ed., *Delmar Publishers, 1991*)

2. Select the ratio of the mix needed for the job. In this example, a 1:2 1/4:3 ratio is selected.
3. Determine the amount of each ingredient needed to make 1 cubic yard of concrete. According to figure 38-5, 1 cubic yard of 1:2 1/4:3 mix requires: 5.5 cwt (550 pounds) of cement, 1,400 pounds of fine aggregate, and 2,100 pounds of coarse aggregate.
4. Multiply each of those figures by the number of cubic yards in the project. In this case, multiply each ingredient by 4.15:

4.15 x 550 = 2282.5 lbs of cement
4.15 x 1400 = 5810 lbs of fine aggregate
4.15 x 2100 = 8715 lbs of coarse aggregate

NOTE: If the concrete is to be purchased in ready-mixed form, the estimate is complete once it is determined how many cubic yards are needed. In this case, 4.15 plus a little extra in case the estimate is too low might come to 4.5 cubic yards. If the base is not leveled carefully, 5 cubic yards could be ordered and any excess used in another location.

PREPARING FORMS FOR CONCRETE

A form is a metal or wooden structure that confines the concrete to the desired shape or form until it hardens. Preparing forms for concrete can be as simple as nailing a board in place or as complex as constructing forms for the side of a highway bridge. On farms, around homes, and in landscape projects, form construction is usually required for walks, floors, feedlots, steps, ornamental projects, and building footers. A footer or footing, is the concrete base that provides a solid, level foundation for brick, stone, or block walls. Footers, walks, and concrete slabs are generally constructed with 2" × 4" lumber nailed to 2" × 4" stakes driven into the soil. Forms for walls are generally constructed by nailing 3/4-inch plywood to 2" × 4" materials, figure 38-6. Forms may also be purchased that are made from aluminum plate and struts.

When constructing forms, the following points may be helpful:

- Use soft, clean, straight lumber.
- Sharpen stakes evenly so they can be driven in straight.
- Place stakes about 30 inches apart along the outside of forms for 4-inch-thick concrete. Place the stakes closer when the concrete is more than 4 inches thick.
- Use a transit or level to adjust the height of forms for the desired slope or "fall" of the slab.
- Drive nails through the form and into, but not through, the stakes.
- Be sure the stakes do not stick up above the top of the forms. If they do, saw them off so

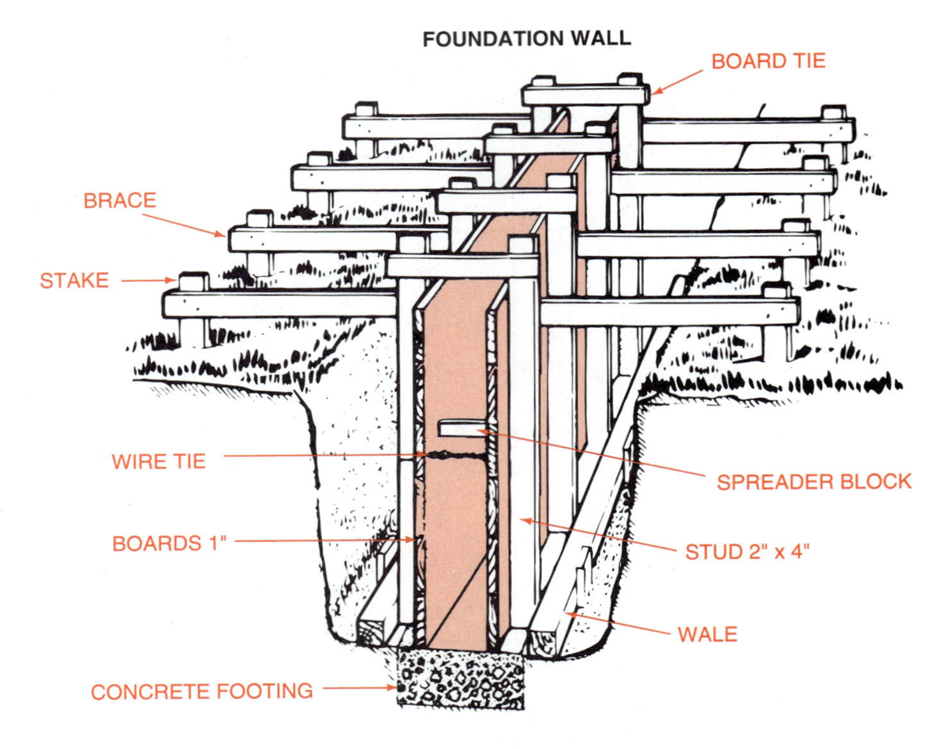

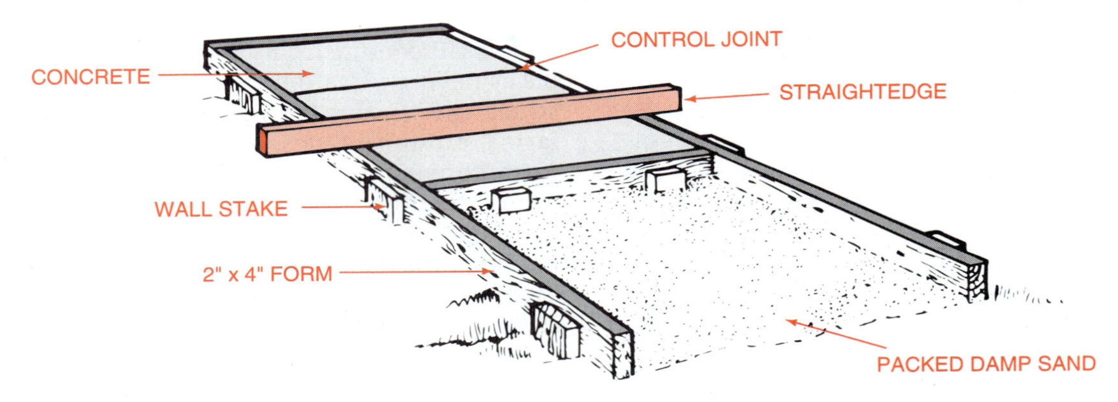

FIGURE 38-6. Proper form construction. (*Courtesy of Instructional Materials, Texas A&M University*)

they are level with or tapered down from the form.

> **CAUTION:**
>
> - Do not saw into the top of the form.

- Construct the inside surfaces of the forms to create the desired shape of the finished concrete. For example, if you want a rounded extension on the front of a concrete step, the shape of the extension is determined by the inside of the form.
- Brush used motor oil on wood surfaces that will be touched by concrete. This is a low-cost product that will prevent the wood from sticking to the concrete and permits easy removal of the forms.

INSULATING CONCRETE FLOORS

In any concrete or masonry structure, consideration should be given to moisture barriers and insulation. Moisture barriers, also known as vapor barriers, prevent moisture from passing through. In climates with either high or low temperatures, energy costs can be reduced with proper insulation. Insulation retards heat movement. Further, moisture barriers and insulation will make floors more comfortable, and waterproofed walls will prevent ground water and precipitation from entering the building, figure 38-7.

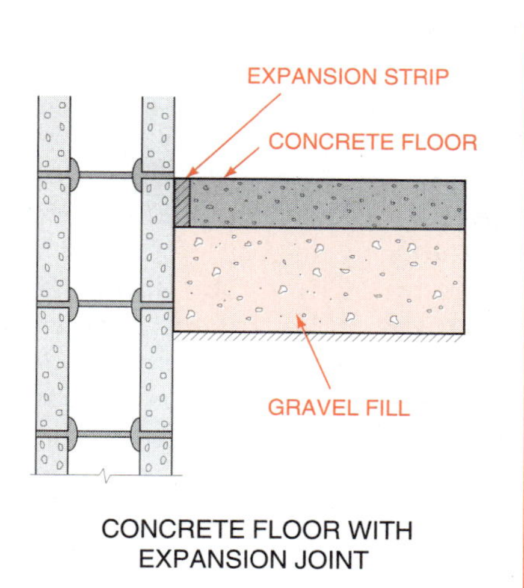

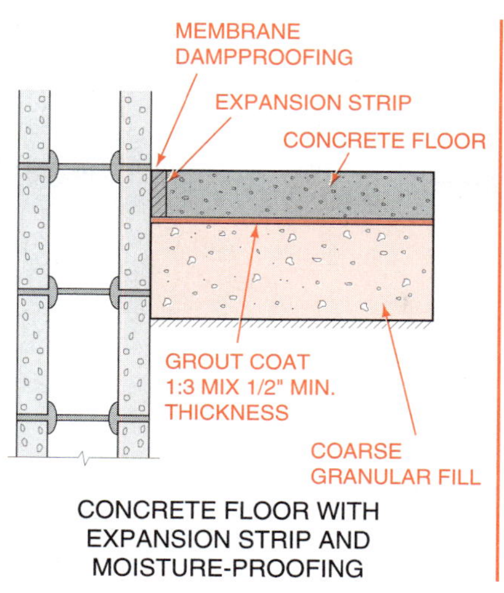

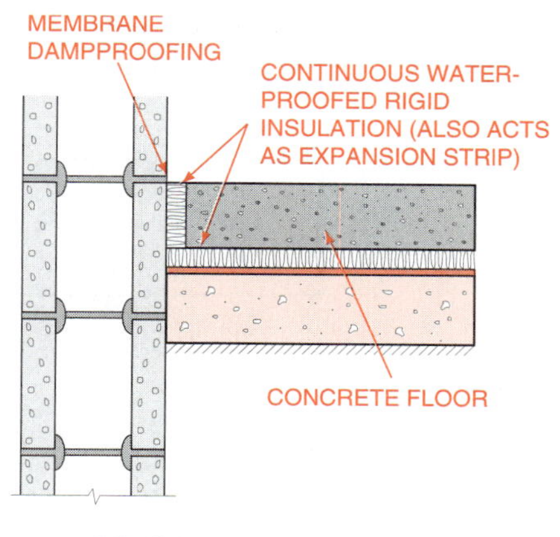

FIGURE 38-7. Moisture-proofing and insulating concrete floors.

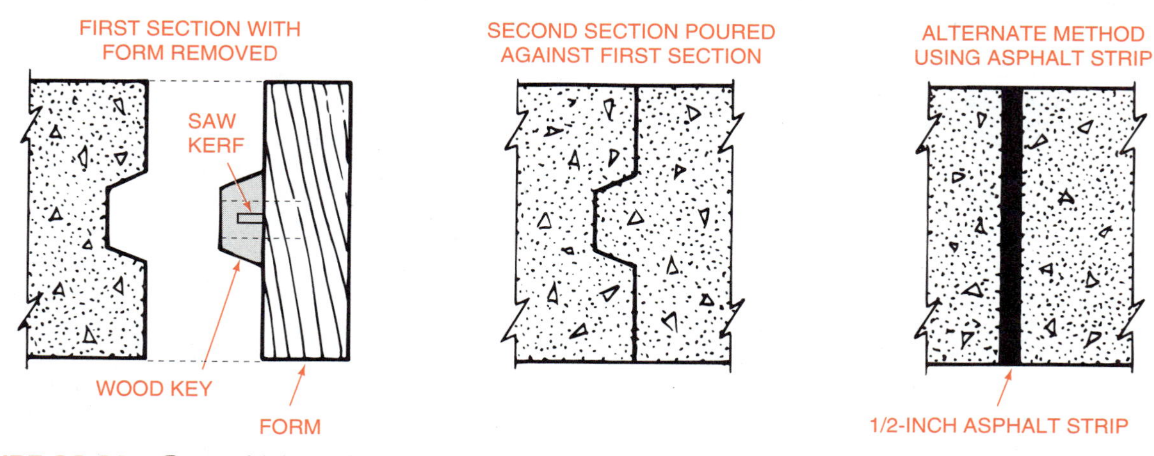

FIGURE 38-8A. Control joints. (*Courtesy of G. White,* Concrete Technology, 4th ed., *Delmar Publishers, 1991*)

Additional information on insulating concrete floors and masonry walls is provided in Unit 39, Planning and Constructing Agriculture Structures. Further, a table showing the insulation values of common building materials is included there.

MAKING JOINTS

Construction joints are needed when slabs of concrete larger than about 10 feet by 10 feet are poured. A construction joint is a place where one pouring of concrete stops and another starts. A control joint is a planned break which permits concrete to expand and contract without cracking. Control joints are placed at equally spaced intervals in the slab.

Control joints are made by placing a wood or galvanized metal "key" on the form, figure 38-8A. A beveled 1" × 2" wood strip is adequate for most slabs. A saw kerf in the key prevents breakage of the concrete when moisture expands the wood strip. The groove left in the concrete after the form is removed provides an indentation into which the next section locks. Before the second section is poured, the first section is coated with a curing compound. This prevents the second section from sticking to the first.

Another popular type of control joint is made by placing asphalt material between sections to absorb the expansion and contraction of the two slabs. This system, however, does not prevent one slab from rising or falling during freezing and thawing, which may leave an uneven surface where the two slabs come together.

Sidewalks need shallow grooves or joints across the surface at 3-foot intervals to control cracking due to expansion. When joints are present the concrete will crack under the joint if stress occurs. A crack under the joint will not create an unsightly appearance.

REINFORCING CONCRETE

Concrete slabs may be greatly strengthened by using steel reinforcing rods or wire mesh, figure 38-8B. Such concrete is referred to as reinforced concrete.

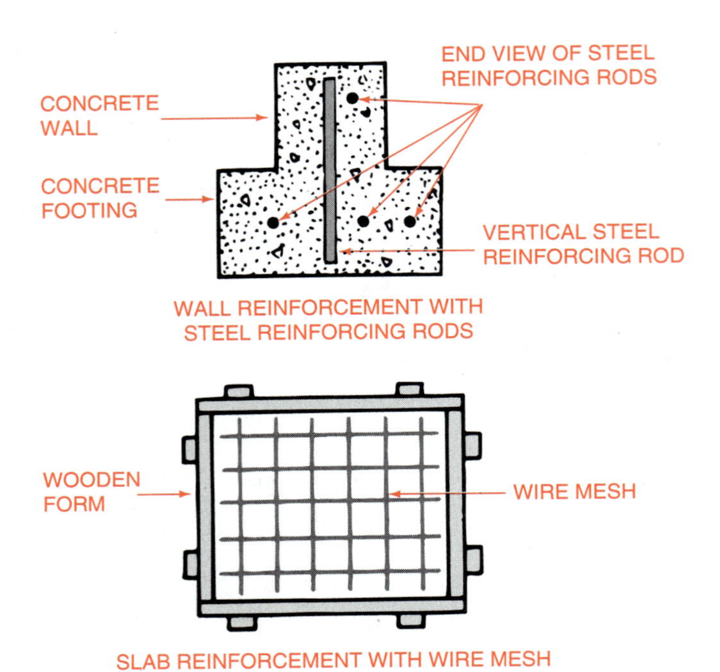

FIGURE 38-8B. Concrete is strengthened greatly by the addition of steel rods or wire mesh.

Any concrete slab that will carry vehicles such as tractors, loaders, or machinery must be reinforced. Otherwise, money is wasted on extra thicknesses of concrete that are still likely to crack under stress. Concrete posts and other thin concrete structures must be reinforced with steel to withstand normal stresses.

Concrete reinforcing rods are made of steel and have a rough, pebbly surface. This surface prevents them from slipping when embedded in concrete. They are classified according to diameter and are available in diameters from 1/4 to 1 inch and over. They may be purchased in 20-, 40-, or 60-foot lengths.

Two or more bars may be joined with wire or by welding to make longer units. These should be wired so that they lap one another by at least a foot. A rule of thumb is to lap bars by 24 times the diameter. Therefore, a 1/2-inch bar should be lapped 12 inches and a 3/4-inch bar lapped 18 inches. Rods should be placed at least 3/4 inch from all surfaces of the concrete. For most applications, 1 1/2 inches or more is preferred.

Rods may be placed in concrete slabs in a cross-sectional pattern. such reinforcement is stronger if the bars are wired together where they cross one another. Reinforcing bars should be free of rust, dirt, oil, or other materials that will reduce adhesion by the concrete.

Concrete slabs are generally reinforced with wire mesh, or fabric. Wire fabric is steel wire welded to form a cross-sectional pattern. It is generally available in a 6-by-6-inch pattern and consists of No. 6, 8, or 10 gauge wire. Both wire mesh and reinforcing rods are more effective if they are not rusted.

Wire fabric is first rolled out on a flat surface to eliminate curves in the material. It is then laid inside the concrete forms. Multiple pieces are wired together to form a continuous piece. Where two or more widths are needed, they should overlap each other by at least one and a half squares. The pieces should be tied securely at regular intervals.

The wire fabric should be supported by stones to keep it off the base material. As the concrete is being

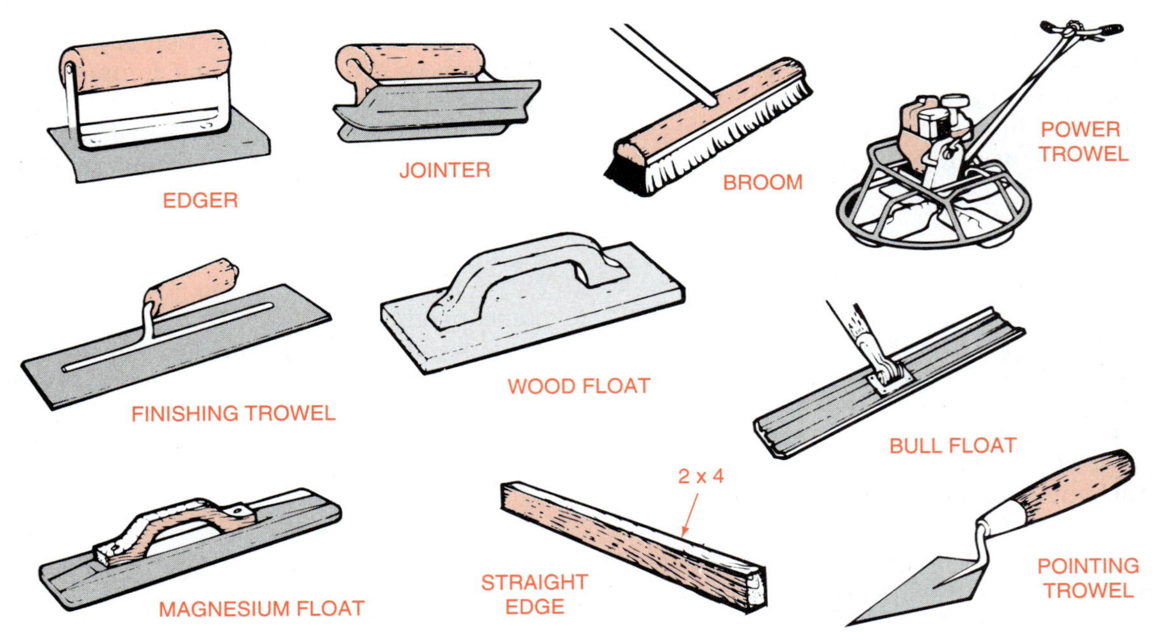

FIGURE 38-9. Tools for placing and finishing concrete. (*Courtesy of Instructional Materials, Texas A&M University*)

poured, it is important to pull the wire up until it is approximately centered in the slab. This ensures maximum reinforcement.

Finally, ready-mixed concrete may be ordered with tiny air bubbles trapped throughout the mixture. Such concrete is known as air-entrained concrete. It is stronger and more resistant to acid, salt, and frost than regular concrete.

POURING, FINISHING, AND CURING CONCRETE

When pouring concrete, common tools such as shovels, spades, and rakes are needed to move and spread the material. Other tools are needed to finish the concrete, figure 38-9. Since concrete is so heavy, it is best to move it downhill with a chute and across level surfaces with a wheelbarrow. This permits the concrete to be pushed, pulled, or lowered rather than lifted. Move a spade up and down in the concrete to help it settle in close to the forms.

Concrete starts to harden about 15 minutes after it is mixed. Thus, it is very important to have all materials on hand before mixing starts.

Before pouring concrete, the stone base should be wetted down with water so that water does not move from the concrete into the dry base. Loss of water from the mixture will weaken the concrete.

Screeding, Floating, and Finishing

After spreading concrete, a straight 2" × 4"or 2" × 6" plank is moved back and forth across the forms to strike off the excess concrete and create a smooth and level surface. This process is called screeding, figure 38-10. When the screeding is finished, a smooth board attached to a handle is pushed and pulled across the surface. This process is called floating, and the tool is called a bull float. The process brings the fine aggregate and cement paste to the surface for a smooth finish. Small areas may be finished with wooden or magnesium floats. Hand floats, edging tools, jointing tools, and trowels are used to finish the edges and/or place grooves or joints in the concrete, figure 38-11. All of these procedures must be done before the concrete begins to set.

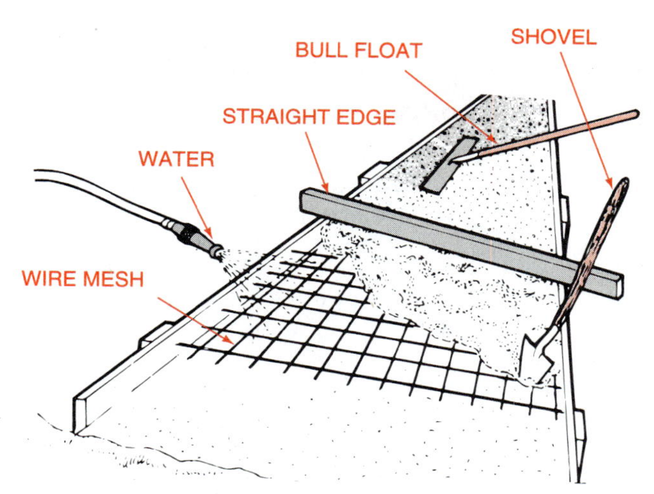

FIGURE 38-10. Pouring, screeding, and floating concrete. (*Courtesy of Instructional Materials, Texas A&M University*)

After the concrete begins to set, a coarse bristled broom may be pulled across it to create a rough surface and improve footing or traction. This is called a broom finish. A very rough, pebbly finish may be obtained by washing away some of the cement paste, leaving the aggregate exposed, figure 38-12. A smooth finish is created by troweling after the concrete starts to harden.

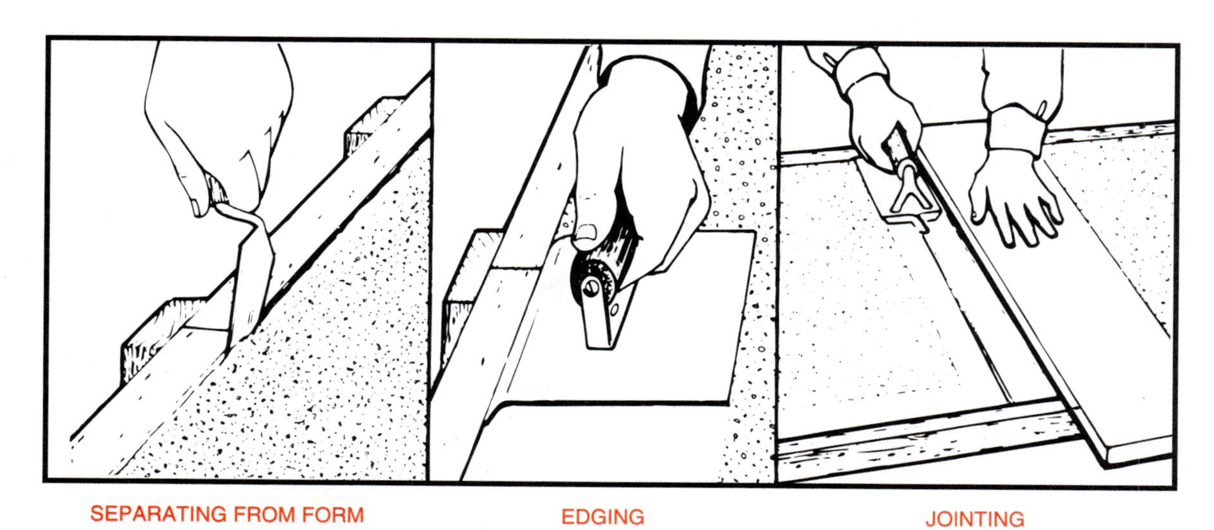

FIGURE 38-11. Separating, edging, and jointing procedures. (*Courtesy of Instructional Materials, Texas A&M University*)

TYPE OF FINISH	USE	OBTAINED BY
Gritty, nonslip surface	Walks, barn floors, driveways, ways, and ramps	Using wood float, stiff broom, or burlap covering
Smooth, troweled finish	Feed troughs, certain type of floors	Using steel trowel
Pea grain or exposed aggregate	Porches and patios	Washing with water hose before curing

FIGURE 38-12. Concrete can be finished in a number of ways. (*Courtesy of Instructional Materials, Texas A&M University*)

When an extremely smooth surface is desirable, the concrete is rubbed with a metal trowel. Generally the trowel is dipped in water and worked over the semihardened surface. This action brings the cement paste to the surface. Excessive troweling will leave the surface weak and easily damaged by frost and chemicals.

Curing

Concrete must dry slowly or it will crack, crumble, and break up long before its intended lifetime has passed. The proper drying of concrete is called curing. To cure concrete, it must be protected from drying air, excessive heat, and freezing temperatures for several days after it is poured.

Concrete is kept moist by covering it with plastic or canvas to prevent the water from evaporating. The surface may also be covered with straw, sawdust, or other insulating material and sprinkled with water occasionally to keep it moist. Insulating materials help protect the concrete from air that is too hot or too cold as well as preventing moisture from escaping. Wooden forms help to protect concrete from drying out. Therefore, forms and other materials used for curing concrete should be left in place for up to a week until curing is complete.

LAYING MASONRY UNITS

Blocks made from concrete, cinders, or other aggregates are used extensively for agricultural buildings. Such blocks are called masonry units. They are referred to as "block" in both singular and plural. Masonry units are held together with mortar. The process of mixing mortar, applying it to block, and placing the block to create walls is called laying block.

Types of Blocks

Concrete and cinder block are available in a variety of sizes and shapes. Standard block are 15 5/8 inches long and 7 5/8 inches high. When laid with 3/8-inch mortar joints they cover an area 16 inches long and 8 inches high. Standard block are available in 4-, 6-, 8-, 10-, and 12-inch widths. The 4-, 8-, and 12-inch sizes are used most frequently.

Block are wider when viewed from the top than when viewed from the bottom. This is because the block is poured into forms that are tapered slightly to permit the block to be removed easily. Block are usually laid with the thicker part up to provide a larger area for the mortar bed. A layer of mortar is called a mortar bed.

Block are generally manufactured with hollow spaces called cores. They are available with two or three cores per block. Such block are referred to as hollow core block. Some block, called stretcher block, have slight extensions on the ends called ears. When laid end to end, the ears create a core. Stretcher block are used in straight wall sections. Corner block have one flat end to create attractive

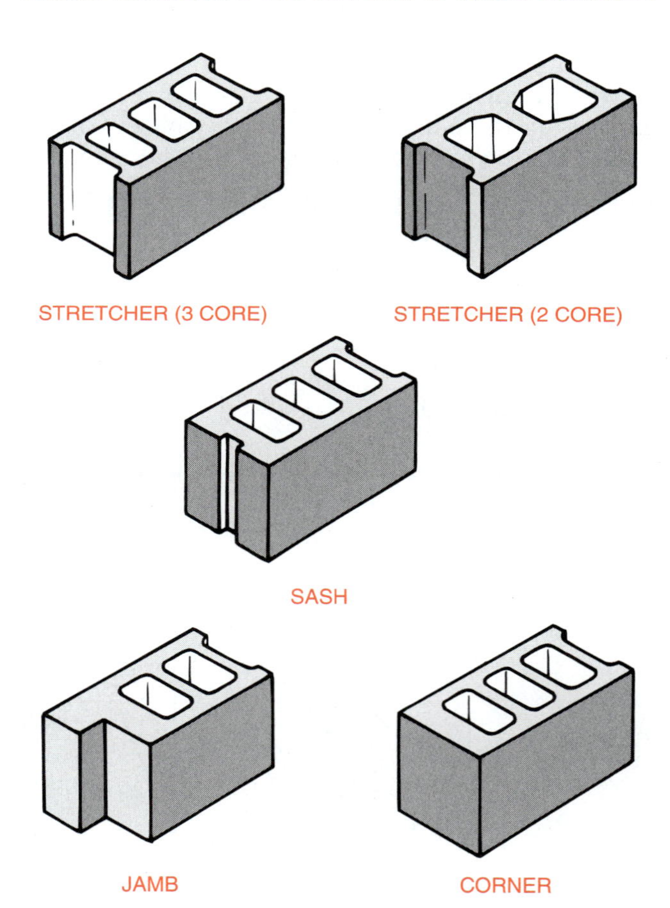

FIGURE 38-13. Stretcher block is used in straight wall sections; sash block hold windows; jamb block hold doors; corner block are used wherever the ends of block are in view.

walls at corners. Block with special grooves can be laid to receive window (sash block) and door (jamb block) parts so the openings are attractive and secure, figure 38-13.

Estimating Block

As stated before, standard block are 8 inches or $2/3$ foot high and 16 inches or $1\,1/3$ foot long when they are laid with a $3/8$-inch mortar joint. One foot, or 12 inches, is $3/4$ the length of one block. Therefore, when estimating the number of block needed for a job, the length of the wall in feet can simply be multiplied by $3/4$. This gives the number of block needed for one row, which is called a course. Similarly, 1 foot is $12/8$ or $3/2$ of the height of a block. Therefore, the height of the wall can be multiplied by $3/2$ to determine the number of courses needed. The number of block per course is then multiplied by the number of courses to obtain the number of block needed for the wall.

For example, to estimate the number of block needed to build a wall 12 feet long and 8 feet high, the procedure is as follows:

12' (length in feet) × 3/4 = 9 block per course
8' (height in feet) × 3/2 = 12 courses
9 (block per course) × 12 (courses) =
108 block needed for the wall

When estimating block for buildings, this procedure is used for each of the four walls. The estimate is then decreased by the area of the doors and windows.

NOTE: When planning a building it is important that the dimensions between corners, windows, and doors utilize full or half length block. Otherwise, large numbers of block must be cut, a procedure that is expensive and time consuming and that results in an unattractive appearance. Figure 38-14 has no cut block.

Constructing Footers

A footer or footing is a continuous slab of concrete that provides a solid, level foundation for block and brick walls. Footers should be at least as deep as the wall is wide, figure 38-15. A wall that is 8 inches wide rests on a footer that is at least 8 inches deep. The width of the footer should be at least twice that of the wall (or more where there is poor soil support). Hence, the 8-inch wall would rest on a footer 16 inches wide. In localities where temperatures drop to 32°F or lower, footers must be placed below the frost line to prevent damage by freezing in cold weather. The frost line is the maximum depth that the soil freezes in cold weather.

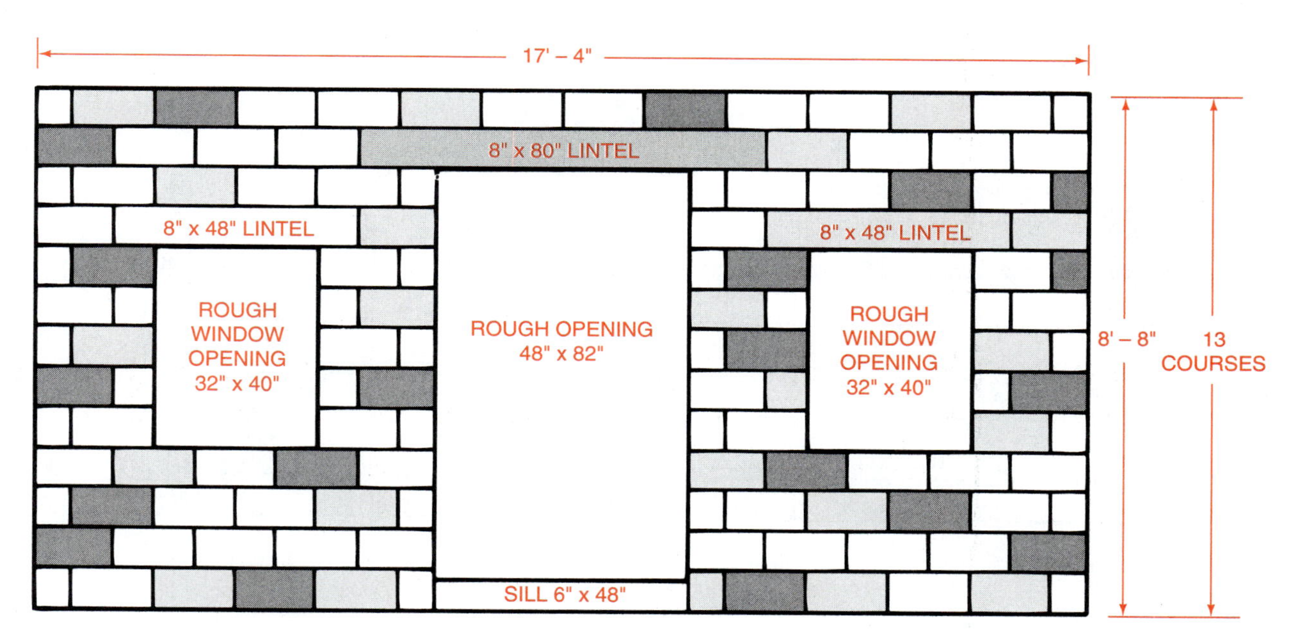

NOTES:
1. DOOR OPENING WILL HAVE 2" JAMBS ON BOTH SIDES AND TOP.
2. DOOR WILL BE A STANDARD HEIGHT (44" x 6'8").
3. DOOR SILL WILL ACCOMMODATE A CONCRETE FLOOR INSIDE.
4. WINDOW OPENINGS WILL ACCOMMODATE STANDARD SIZE WINDOWS.
5. SPACE ABOVE THE TOP OF THE DOOR WILL ACCOMMODATE APPROPRIATE TRIM UNDER THE EAVES OF THE ROOF.
6. REINFORCED CONCRETE LINTELS CARRY THE WEIGHT OVER DOOR AND WINDOW OPENINGS.
7. THERE ARE NO CUT BLOCKS IN THE WALL. THEREFORE, THE WALL CAN BE BUILT WITH A MINIMUM OF COST AND LABOR.

FIGURE 38-14. A building front laid out using only whole and half block.

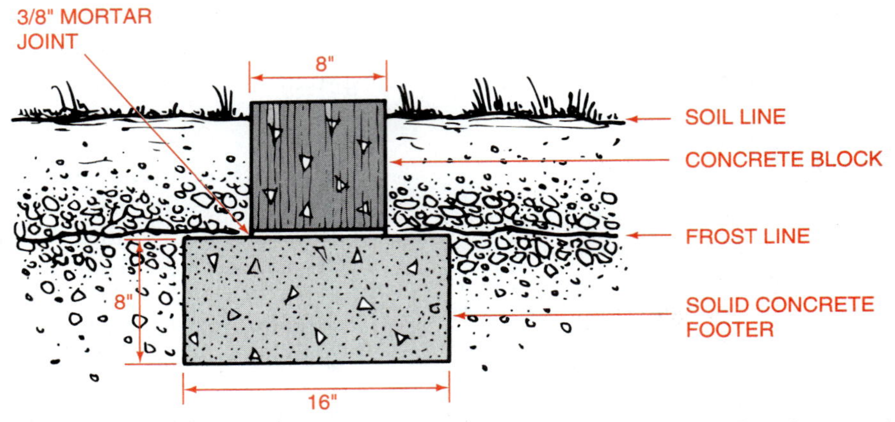

FIGURE 38-15. The footer should be as thick as the wall resting on it is wide. The footer should also be at least twice as wide as the wall. Concrete footers are placed below the frost line to provide a solid base for masonry walls.

When starting a building or wall, a trench is dug by backhoe or shovel. The bottom of the trench should be the width of the footer and as deep as the intended bottom of the footer. Stakes are driven into the bottom of the trench so their tops are level with the top of the intended footer. Concrete is then poured and leveled to the height of the stakes. The stakes are pulled out and the holes filled after the concrete is leveled but is still workable.

Mixing Mortar

Before block can be laid, the mortar must be mixed to a working consistency. Premixed mortar can be purchased that requires only the addition of water. Mortar can also be made at the job site by mixing various combinations of portland cement, finishing lime, sand, and water. The different combinations yield different characteristics, such as greater

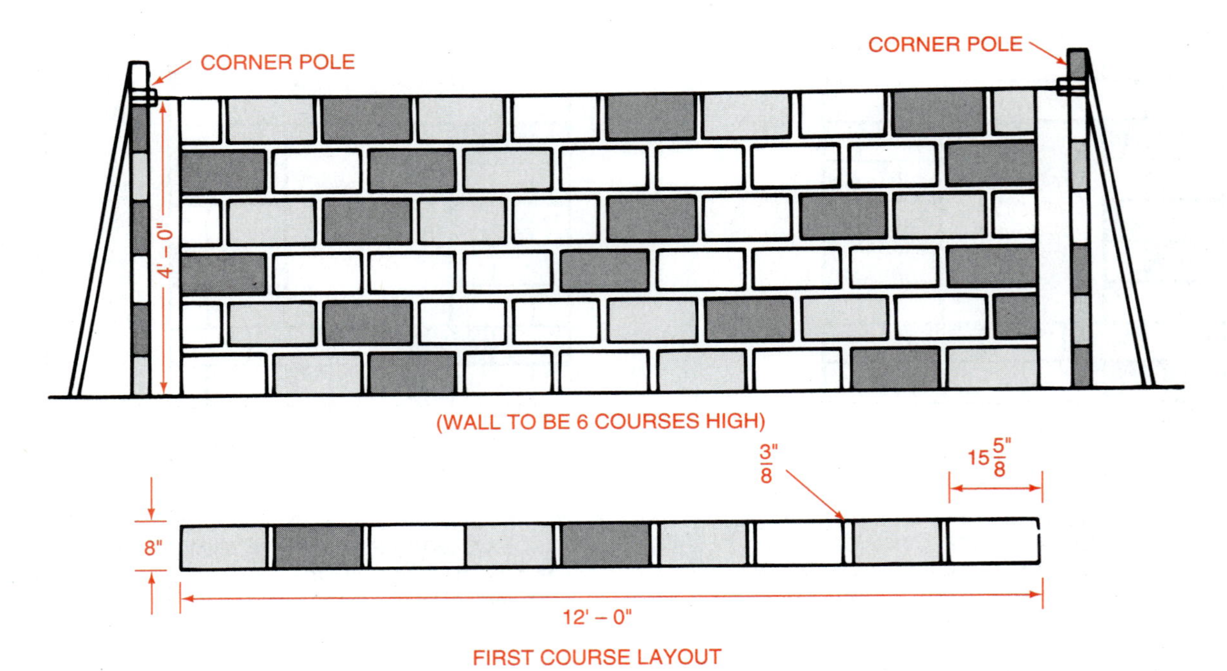

FIGURE 38-16. Placement of corner poles for constructing a block wall. (*Courtesy of R. Kreh,* Masonry Skills, 3rd ed., *Delmar Publishers, 1990*)

strength, higher waterproofness, and greater ease of handling. Unfortunately, no combination has all the ideal qualities.

Small batches of mortar may be mixed with a hoe in a mortar pan or wheelbarrow. Jobs requiring more than a wheelbarrow or two of mortar can be mixed more thoroughly and efficiently in a motor- or engine-driven mixer.

Block must be laid with mortar of the correct wetness. If the mortar is too stiff, it will not bond tightly to the block. If it is too thin, it will be squeezed out of the joint by the weight of the block. In such a case, the joint will be less than 3/8 inch thick and the block must be relaid.

Laying Block

To ensure that block are laid straight and plumb, usually a corner pole is set at each corner of the building, figure 38-16. A corner pole is a straight piece of wood or metal held plumb by diagonal supports. It is used to support the lines to which the wall is built. This process is described further in the procedure that follows.

An alternate method is to lay block corners using a level to guide the construction. Since the level has bubble vials placed in different positions, it can be used to plumb the block as well as to level them. The corners are built, then the areas between them are filled with courses of stretcher block using a line to keep the courses straight, figure 38-17.

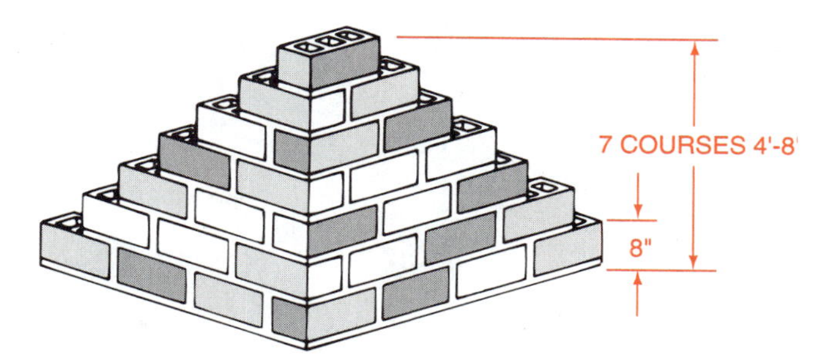

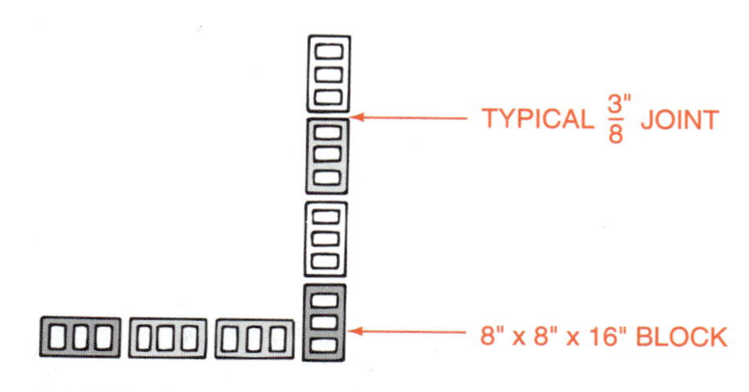

FIGURE 38-17. Construction of corners for block walls. (*Courtesy of R. Kreh,* Masonry Skills, 3rd ed., *Delmar Publishers, 1990*)

The following procedure for laying block is also illustrated in figure 38-18.

SPREAD THE MORTAR BED FOR THE FIRST COURSE.

APPLY MORTAR TO THE EARS OF THE STRETCHER BLOCK. NOTE THE DOWNWARD WIPING OF THE TROWEL.

CHECK THE FIRST COURSE TO BE SURE IT IS PLUMB.

FIGURE 38-18. Procedure for laying block.

LEVEL THE BLOCK.

APPLY THE MORTAR BED FOR THE SECOND COURSE.

CHECK THE FIRST COURSE TO BE SURE IT IS LEVEL.

REMOVE EXCESS MORTAR.

CHECK THE FIRST COURSE TO SEE THAT THE BLOCKS FORM AN EVEN, STRAIGHT LINE.

CUT BLOCK BY MAKING A GROOVE WITH MODERATE BLOWS OF MASON'S HAMMER.

FIGURE 38-18. *Continued*

LAY REINFORCEMENT WIRE IN THE MORTAR BED.

MEASURE A CORNER TO BE SURE THAT EACH COURSE IS 8 INCHES HIGH.

LAY BLOCK TO THE LINE TO ENSURE THE COURSES ARE STRAIGHT. THE LINE IS POSITIONED TO BE LEVEL ALONG THE TOP EDGE OF THE BLOCK.

FINISH A FLUSH OR SMOOTH JOINT WITH A PIECE OF BROKEN BLOCK.

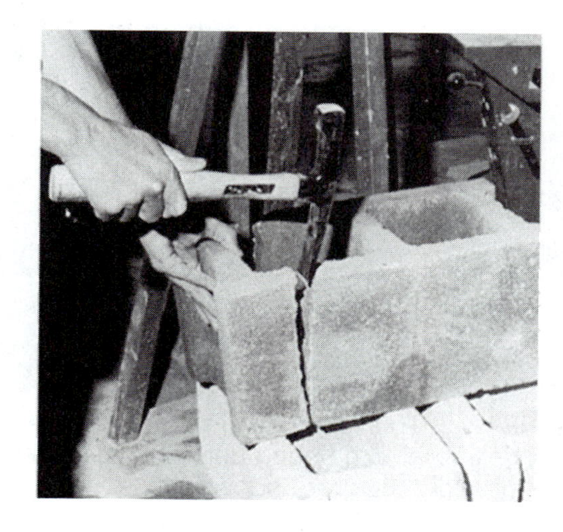

FOLLOW BY A SHARP RAP ON THE EDGE OF THE WEB.

STRIKE A JOINT WITH A SLED RUNNER JOINTER.

FIGURE 38-18. *Continued*

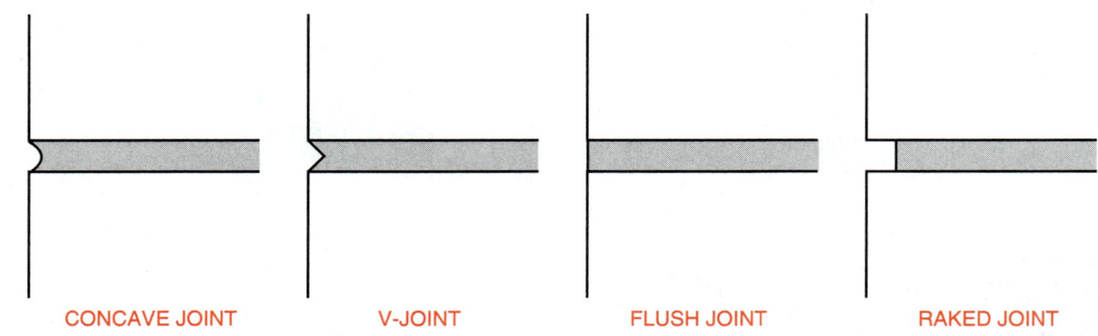

FIGURE 38-19. Types of joint finishes used on block walls. (*Courtesy of R. Kreh,* Masonry Skills, 3rd ed., *Delmar Publishers, 1990*)

PROCEDURE

1. Spread a layer of mortar, called a mortar bed, on the footer.
2. Position the block on the mortar bed so that its outside corner is exactly where the outside corner of the wall should be. Level the block by first placing the level across the block and then lengthwise along the block.
3. Turn several stretcher block on end and apply mortar to the ears with a wiping or swiping stroke of the trowel.
4. Lay several stretcher block in place by working away from the end or corner block.
5. Use the end of the trowel handle to tap the block until each block is plumb, level, and the course is straight.
6. Apply a mortar bed on top of the first course in preparation for the second course.
7. If extra strength is needed in the wall, install reinforcement in the mortar bed.
8. As the block laying progresses, cut off excess mortar with the trowel.
9. Use a line to keep the courses straight. The line is positioned to be level along the top of the block.
10. When a block must be cut, use a mason's hammer and make multiple strikes along the line of cut; then make one sharp strike on the web.
11. Check the height to be sure each new course is an additional 8 inches high.
12. After the mortar dries and hardens slightly, finish the joint by rubbing it with a broken piece of block.
13. If a joint other than a flush joint is desired, use a jointer to compress the mortar and create a watertight joint, figure 38-19. Tools are available to create joints that are concave, V-shaped, flush, or raked.

Walls that are above ground level may be damaged by wind pressure. This is especially true before the mortar cures. Therefore, walls higher than 8 feet should be braced on both sides at 8- to 10-foot intervals, figure 38-20. The braces are removed after the wall is strengthened by curing and by the installation of floor or roof parts.

Mortar joints must be protected from freezing for several days until they are cured. This is generally done by covering the wall with plastic or canvas when temperatures are below 32°F.

FIGURE 38-20. The temporary bracing of this wall will prevent wind damage until additional building parts provide adequate support.

STUDENT ACTIVITIES

1. Define the Terms to Know in this unit.
2. Examine samples of sand and gravel to observe the variety of particle sizes.
3. Conduct a clay and silt test on a sample of sand.
4. Prepare the forms for a small concrete project. Mix and pour the concrete. Finish and cure the concrete.
5. Visit a ready-mix concrete plant to observe the equipment and techniques.
6. Help pour concrete for a project using ready-mixed concrete.
7. Pour a footer.
8. Estimate the block needed for a wall or building.
9. Construct a small block wall.
10. Protect concrete from freezing and drying out until it is cured.
11. Construct a wooden float for use on future concrete projects.

SELF-EVALUATION

A. **Multiple Choice**. Select the best answer.

1. Concrete is a mixture of sand, gravel, water and
   a. clay cement
   b. finishing cement
   c. finishing lime
   d. portland cement
2. Mortar is a mixture of portland cement, sand, water, and
   a. aggregate
   b. clay
   c. finishing cement
   d. finishing lime
3. The strength and durability of concrete are dependent upon the
   a. purity of water
   b. proportion of stone particles by size
   c. type of cement
   d. all of these
4. A concrete slab that is 6 inches deep by 10 feet wide by 20 feet long will contain
   a. 2 cubic yards of concrete
   b. 2.5 cubic yards of concrete
   c. 3.7 cubic yards of concrete
   d. 5.0 cubic yards of concrete
5. Forms for concrete slabs are usually made of
   a. 2" × 4"
   b. 2" × 8"
   c. 2" × 10"
   d. none of these
6. To prevent forms from sticking to the concrete, they are treated with
   a. fat
   b. oil
   c. paint
   d. wax
7. Concrete is reinforced with
   a. air bubbles
   b. aluminum wire
   c. steel bars
   d. wood fibers
8. Concrete is cured by
   a. covering with plastic
   b. protecting from wind
   c. sprinkling with water
   d. all of these
9. A standard sized block when laid will cover an area
   a. 8 inches high by 16 inches long
   b. 4 inches high by 16 inches long
   c. 8 inches high by 12 inches long
   d. none of these
10. Courses of block are laid in a straight line by using a
   a. center point
   b. line
   c. plumb bob
   d. sighting tool

B. **Matching.** Match the terms in column I with those in column II.

Column I

1. sand
2. gravel
3. 1:2:3
4. air
5. drying
6. footing
7. raked joint
8. flat end
9. ears
10. flush joint

Column II

a. course aggregate
b. cement/sand/gravel
c. curing
d. foundation for walls
e. jointing tool
f. fine aggregate
g. entrained
h. corner block
i. stretcher block
j. broken block

C. **Completion.** Fill in the blanks with the word or words that will make the following statements correct.

1. A quart jar with 2 inches of sand for concrete mixtures should not yield more than a _____-inch layer of silt.
2. The amount of water added to a mix is important because _____.
3. A concrete mixture with uniform color and consistency and the correct proportion of ingredients is known as a _____.
4. A block wall is plumbed with a _____.
5. Concrete that is already mixed when it arrives at the job site is called _____.
6. Moisture is prevented from entering or passing through a concrete slab by installing a _____ _____.
7. Heat loss through concrete floors may be reduced by installing _____.

Agricultural Structures

UNIT

Planning and Constructing Agricultural Structures

OBJECTIVE

To plan and construct agricultural structures.

COMPETENCIES TO BE DEVELOPED

After studying this unit, you should be able to:

- Discuss appropriate considerations for site planning.
- Determine storage space requirements for farm machinery.
- Identify major types of buildings used in agricultural settings.
- Name major building parts.
- Describe special features used to make buildings waterproof and wind-resistant.
- Construct agricultural buildings.
- Insulate agriculture structures.
- Tie basic knots and hitches.

MATERIALS LIST

- ✓ Drawing supplies
- ✓ Heavy paper for templates
- ✓ Reference material on agricultural buildings
- ✓ Materials for model agricultural building
- ✓ Rope

TERMS TO KNOW

rafter
girder
truss
pole building
post-and-girder
post-frame
rigid-frame
dressed
plywood
veneer
subfloor
sheathing
interior grade
exterior grade
pressure treated
acid copper chromate (ACC)
ammoniacal copper arsenate (ACA)
chromated copper arsenate (CCA)
ridge
footing pad
girt
purlin
gusset
square
joist
woodframe
foundation
below grade
elevation
batter board
excavate
footer
subgrade
plate
eave
tread
riser
stair stringer
stair horse
square knot
sheet bend
bowline
clove hitch
two half hitches
timber hitch
barrel sling
whip

Buildings of appropriate type, size, and function are an important asset to any business. Agricultural businesses use buildings for protecting field machinery, storing crops, keeping animals and animal products, milking cows, selling commodities, processing crops and animal products, manufacturing commodities, and numerous other activities, figure 39-1.

PLANNING AGRICULTURAL BUILDINGS

Planning saves time and money. Careful consideration is needed when planning buildings for farm, ranch, horticulture, hydroponics, agribusiness sales, aquaculture, or any other agricultural enterprise. Building size, type, design, and placement are all important.

Inadequate buildings can result in crop loss, machinery deterioration, inadequate livestock production, loss of food and related commodities, human discomfort, and wasted energy.

Farmstead Layout and Building Site Selection

Generally, one cannot start from the beginning and build a farmstead. However, as one adds buildings, it is useful to apply some principles of efficient farmstead layout. Some of these are:

1. Place the farmstead in a well-drained area.
2. Since high winds come mostly from the northwest, plant a windbreak on the northwest side of the farmstead.
3. Place the electrical meter pole so it will be as close as possible to all buildings that require a lot of electrical power.
4. Place the livestock facilities downwind from the house to reduce livestock odors around the house.

FIGURE 39-1. Modern farmstead.

5. Face buildings to the south or east for maximum heat and light in the winter and shade in the summer.
6. Position all buildings so they can be enlarged or expanded.
7. Provide a circle to permit traffic to get in and out easily.
8. Hand-surface or gravel the main traffic areas.
9. Provide proper drainage away from each building to avoid polluting streams and wells, figure 39-2.

As many of these principles should be incorporated into building plans as possible. An efficient farmstead is attractive as well as productive, figure 39-3.

Building Use and Size

Most buildings are large, expensive, and specialized. This means professional help is needed for planning the buildings, and probably constructing them. For instance, a milking parlor and free-stall operation

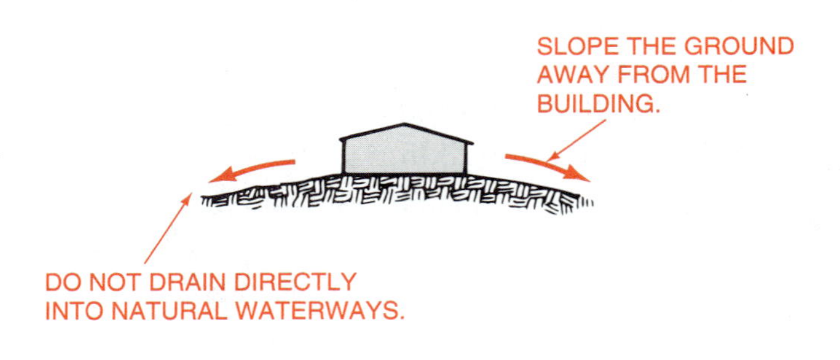

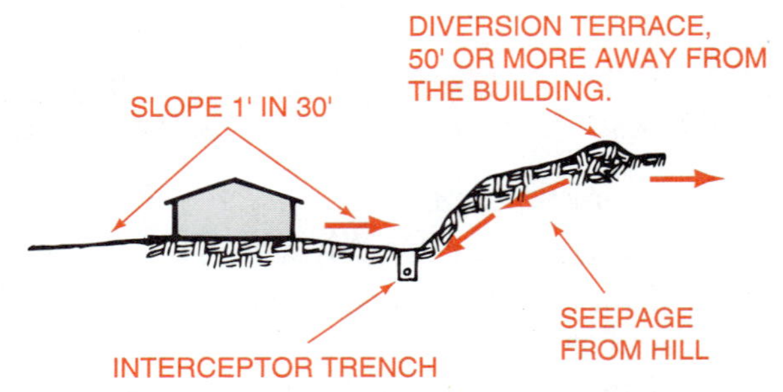

FIGURE 39-2. Building sites should be well drained, but not pollute water sources.

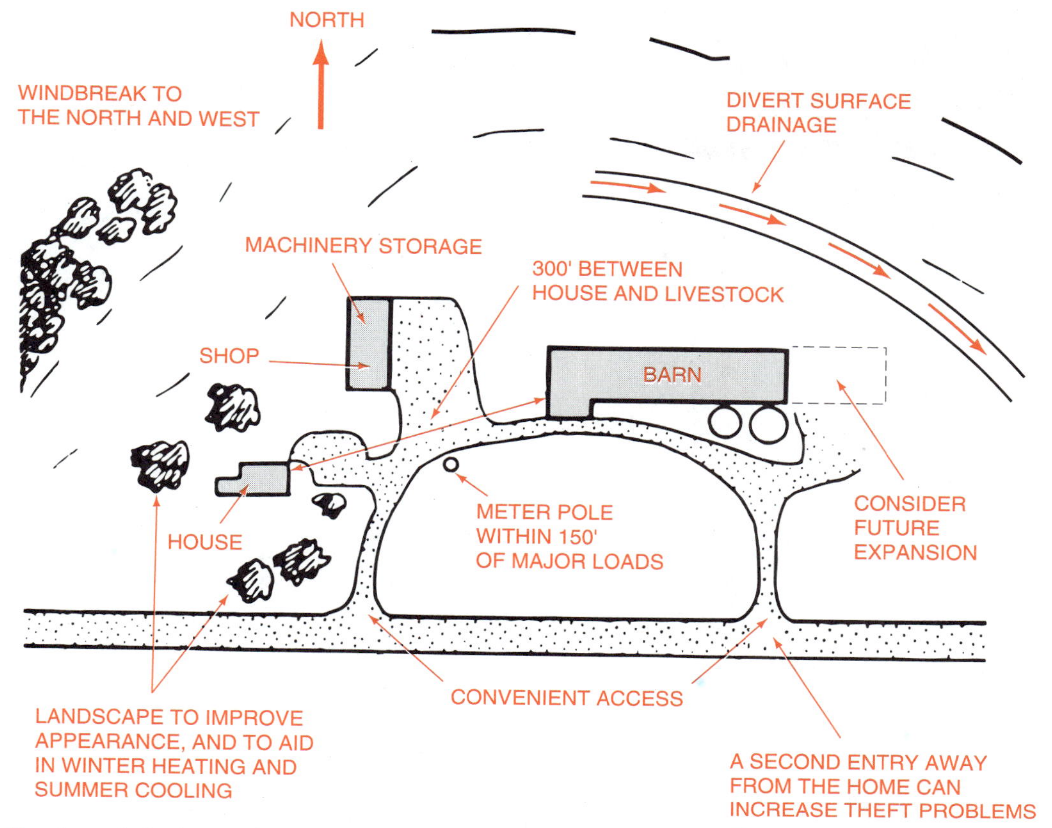

FIGURE 39-3. A well-designed farmstead layout. (*Courtesy of Northeast Regional Agricultural Engineering Service—NRAES*)

must be very carefully planned to be efficient and cost-effective, figure 39-4.

However, some buildings can be planned adequately by the owner or manager and family. Machine storage can be adequately planned with guides on space requirements and building details, figures 39-5 and 39-6. When machinery storage is built, adequate facilities for machinery and farmstead maintenance should be included, figure 39-7.

Some buildings are relatively simple in design and construction. Some buildings such as calf barns, sheep and lambing sheds, horse barns, and roadside stands are constructed by the farm family. In such cases, plans should be obtained from the state agricultural college or USDA, figure 39-8.

FIGURE 39-4. A free-stall dairy operation

Building Types and Shapes

Buildings can be built in almost any shape, but certain shapes have advantages over others. Similarly, some shapes are easier and less expensive to build. For instance, straight gable-type roofs are less expensive to build than hip roofs. For small buildings, shed roofs are less expensive than gable roofs. Still, each roof type has its place for certain uses, figure 39-9.

Types of Trusses. In the past, buildings have been limited in width, without interior posts, due to the limited strength of rafters and girders. A rafter is a single timber that supports a section of the roof. A girder is a timber that carries the weight of floors, and interior walls. However, engineers have designed large, light-weight trusses capable of spanning over 100 feet without posts. A truss is a rigid framework, figure 39-10.

Types of Construction. Buildings may be classified by their construction design. For instance buildings are known as pole buildings if they are supported by poles erected in the soil. If they rest upon posts and rely on heavy girders to carry their sides and roof they are called post-and-girder or post-frame buildings. Buildings that have heavy steel frames are called rigid-frame buildings.

MACHINE	LENGTH (FT.)	WIDTH (FT.)	AREA (SQ. FT.)	HEIGHT (TOTAL IF ABOVE 8 FT.)
Tractors				
Row-crop types,				
1-plow	10 1/2	5*	53	
2 to 3-plow	11 1/2	6*	69	
4 to 5-plow	13	6 1/2*	85	
6-plow	15	8*	112	12
General-purpose types,				
2 to 3-plow	11	6*	66	
4 to 5-plow	12 1/2	6 1/2*	82	
6-plow or larger	14	8*	112	
Tillage Machinery				
Moldboard plow,				
2-bottom	6	3 1/2	21	
3-bottom	9	5	45	
4-bottom	12	6 1/2	78	
5-bottom	15	8	120	
Disk plow,				
2-bottom	11 1/2	6 1/2	75	
3-bottom	15	8	120	
4-bottom	17	10	170	
Disk harrow, single (with end sections folded)				
15-ft.	5 1/2	9	50	
18-ft.	6	10	60	
21-ft.	6	12	72	
Disk harrow, tandem,				
7-ft.	10 1/2	7	74	
8-ft.	11	8	88	
9-ft.	11	9	99	
10-ft.	11 1/2	10	115	
Lister,				
2-row	9 1/2	8 1/2	81	
4-row	11	16	176	
Corn cultivator,				
2-row	8	8	64	
4-row	8 1/2	12	102	
Field cultivator	10 1/2	15	158	
Field cultivator (quack digger), 8-ft.	10	9 1/2	95	
Spring tooth harrow—per section	5	5	25	
Spike tooth or smoothing harrow—lean against wall				
Rotary hoe	7	7 1/2	53	
Corrugated roller—8 ft.	4	9 1/2	38	
Stalk cutter	4	6	24	
Sub-soiler	9	4 1/2	41	
Planting and Seeding Machinery				
Grain drill,				
12 × 6	7	8	56	
20 × 6	7	13	91	
23 × 6	9	16	144	
13 × 7	9	12 1/2	113	
24 × 7 press drill	10	15 1/2	155	
16 × 10 press drill	8 1/2	15	128	
Corn or Cotton planter,				
2-row (with hitch)	10	8	80	
(without hitch)	5	8	40	
4-row (with hitch)	12	14	168	
(without hitch)	6	14	84	
Potato planter,				
1-row	7	4	28	
2-row	11	8	88	
Transplanter	8	5	40	
Harvesting Machinery				
**Grain windrower,				
6-ft.	12	5	60	
8-ft.	14	5	70	
12-ft.	17 1/2	7 1/2	131	
Combine, one-man,				
5-ft. cut	19	10	190	10
6-ft. cut	21	11	231	10
8-ft. cut	24	11	264	11
12-ft. cut	24	12	288	13
Combine, two-man, 14-ft. cut (with header removed)	22 1/2	11	248	13
Combine, self-propelled,				
7-ft. cut	18	9	162	
9-ft. cut	18	10	180	12
12-ft. cut	23 1/2	13 1/2	317	13
14-ft. cut	21	15 1/2	326	13
24-ft. cut	30	24	720	13
Cotton picker,				
1-row	19	10	190	13
2-row	20	11	220	13
Cotton stripper, tractor mounted				
1-row	12	2	24	10
2-row	20	8	160	10
Caneloader	29	9	261	12
Corn harvester	22	7	154	8 1/2
Ensilage harvester, 1-row	13	7	91	
Ensilage cutter & blower	12	5 1/2	66	
Forage crop blower	10-13	6	60	
Corn picker, drawn,				
1-row	15	9 1/2	143	8 1/2
2-row	15	11 1/2	173	8 1/2
Corn picker, mounted,				
2-row	20 1/2	8 1/2	174	8
(with elevator removed)	19	5 1/2	105	
Corn sheller	22	8	176	11
Peanut picker	17	6	102	
Peanut shaker	6	6	36	
Potato digger,				
1-row	17	5	85	
2-row	18	8 1/2	153	
Tobacco harvester	16	9 1/2	152	2-14
Field forage harvester-hay, grass silage and corn silage (with delivery spout removed)	13 1/2	8	108	
Haying Machinery				
Mower,				
6-ft.	7 1/2	6 1/2	49	
7-ft.	7 1/2	7 1/2	56	8 1/2
Side-delivery rake	13	11	143	
Dump rake—10 foot	5	11 1/2	58	
Hay loader	13	8	104	10
Field hay chopper	12	9	108	9
Pickup baler	17	13	221	8
Hauling Machinery				
Manure spreader	15 1/2	6	93	
Tractor trailer	18	7	126	
Truck, 1 1/2-ton	21	8	168	8
Wagons:				
Wagon with box bed	15	6	90	
Wagon with 14-ft. hay rack	16	7-8	112	
Wagon and tractor—tractor backed over tongue	25	7-8	175	
Miscellaneous				
Lime spreader, 8-ft.	4	10 1/2	42	
Manure loader	14	6	84	
Sprayers:				
Orchard	8	6	48	
Weed	5	8	40	
Cut off (buzz) saw	4	4	16	
Rotary cutter (field shredder)	6 1/2	9 1/2	62	

FIGURE 39-5. Approximate space requirements for farm machinery. (*Courtesy of American Association for Vocational Instructional Material—AAVIM*)

EXTRA SPACE
ADD 15% FOR EQUIP. CLEARANCE
9' COTTON & CORN PLANTER 8'
9' STALK CUTTER 4 1/2'
11 1/2' TANDEM DISC HARROW 10'
13' SIDE DELIVERY RAKE 11'
15' CORN PICKER 8 1/2'
19' COTTON PICKER 10'
21' TRUCK 8'
10 1/2' 1-PLOW TRACTOR 7'
16' TRACTOR TRAILER 7'
12' 3-PLOW TRACTOR 7 1/2'
8' COTTON & CORN CULTIVATOR 8'
FIELD FORAGE HARVESTER 7 1/2' 13 1/2'
7 1/2' 7' MOWER
6 1/2' 3-BOTTOM PLOW 12'
13' GRAIN DRILL 7'
18' COMBINE 11'

FIGURE 39-6. Using paper templates to plan machinery storage and determine building size. (*Courtesy of American Association for Vocational Instructional Material—AAVIM*)

FIGURE 39-7. Machine shed with agricultural mechanics shop.

Types of Coverings. Buildings are also classified according to their coverings. They may be known as steel-, aluminum-, or wood-sided buildings.

CONSTRUCTING AGRICULTURAL BUILDINGS

Reference is made to the previous units in this text. Most of the earlier units provide information that is helpful in constructing buildings. Therefore, the reader should be prepared to review previous units if needed.

BUILDING MATERIALS

Lumber. Construction lumber comes in standards sizes from 1" to 6" thick and from 2" to 12" wide. Lengths typically run from 6' to 16' and longer. It is important to remember that thickness and width are stated as nominal dimensions, but when lumber

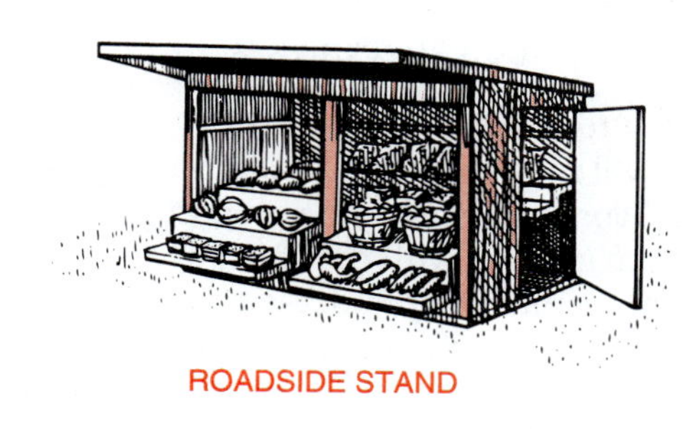

ROADSIDE STAND

SADDLE HORSE BARN

CALF BARN

SHEEP AND LAMBING SHED

FIGURE 39-8. Buildings frequently built by the farm family.

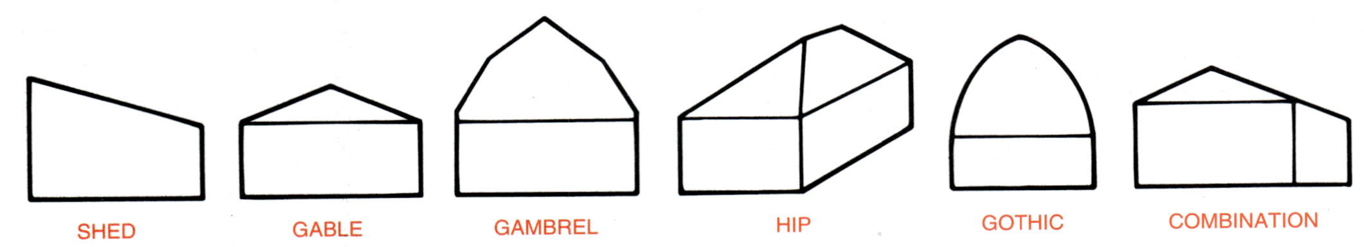

FIGURE 39-9. Common roof designs for agricultural buildings. (*Courtesy of American Association for Vocational Instructional Material—AAVIM*)

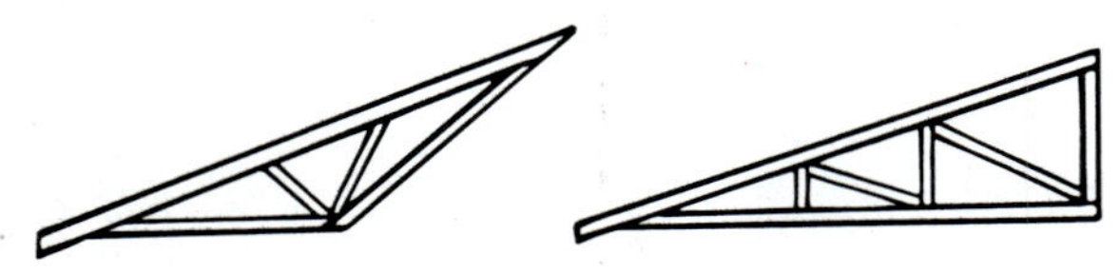

SINGLE SLOPE (MONO PITCH) – SPANS 20' TO 35'.

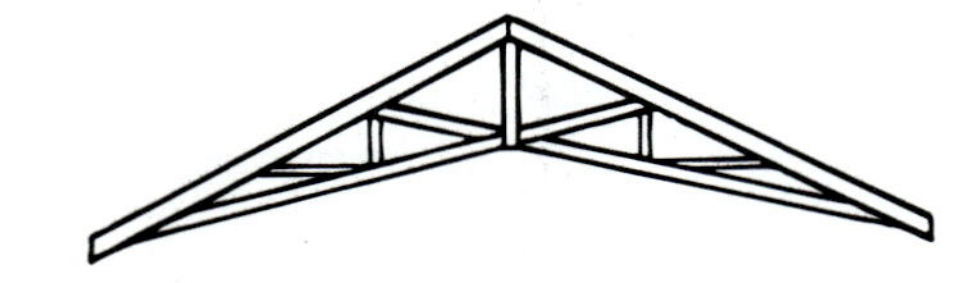

SCISSORS – USED WHERE HIGH CENTER CLEARANCE IS DESIRED. SPANS 20' TO 40'.

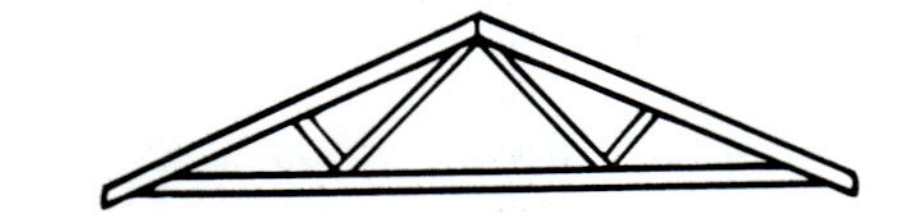

FINK (W) – A POPULAR, EFFICIENT DESIGN FOR SPANS OF 20' TO 50'.

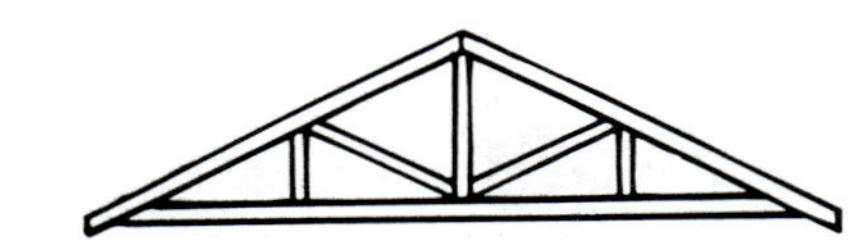

HOWE – SPANS 20' TO 50' FOR HEAVIER CEILING LOADS THAN CARRIED BY THE FINK TRUSS.

PRATT – SPANS 20' TO 60'. FOR USE WITH OR WITHOUT CEILINGS.

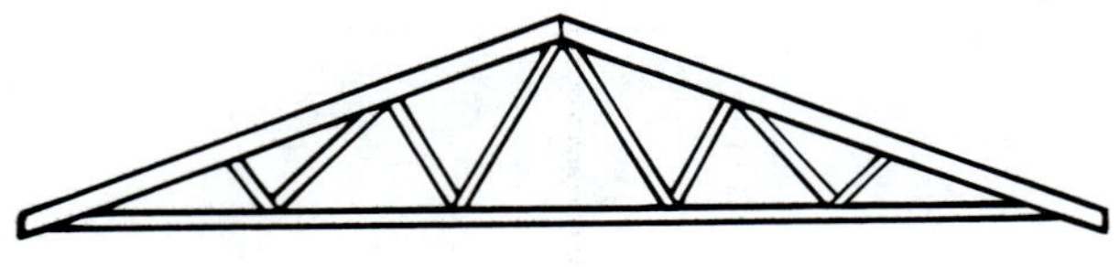

BELGIAN (DOUBLE OR TRIPLE FINK) – SPANS UP TO 80'.

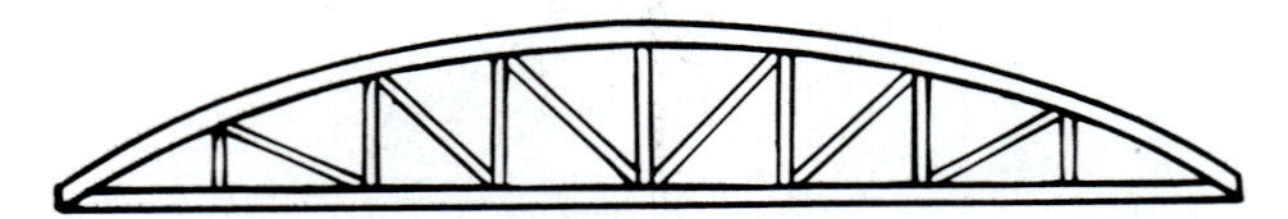

BOWSTRING – GENERALLY USED FOR SPANS FROM 40' TO 120'.

FIGURE 39-10. Common truss designs. (*Courtesy of Northeast Regional Agricultural Engineering Service—NRAES*)

is dressed or planed, the actual thickness is $^{1}/_{4}$" less than nominal thickness and width is $^{1}/_{2}$" less than nominal width. Actual length is the same as nominal length. The dimensions of frequently used lumber is displayed in Table 11 in back of this text. There are also many other tables of information useful for constructing buildings in Appendix B, as well as in this unit.

Plywood. Plywood is made of veneer or thin sheets of wood glued with alternate layers perpendicular to each other. It is used extensively for subfloors, wall sheathing, and roof sheathing. Subfloor refers to the first layer of flooring. Sheathing refers to the first exterior layer of a wall or roof.

Plywood is graded according to the quality of the exterior layers as well as the glue's ability to tolerate moisture. If the glue weakens in dampness, the plywood is classified as interior grade. If the plywood can remain strong and usable when exposed to weather, it is classified as exterior grade.

The top and bottom layers of a sheet of plywood are classified with a letter from A to D. A letter designation is provided for both outside layers. The higher the letter, the better the grade. For instance, a piece of plywood with an AD grade has one side that is smooth and paintable and one side with knots or knotholes up to 2 $^{1}/_{2}$ inches across grain, figure 39-11. Plywood is sold in 4' × 8' sheets of $^{1}/_{4}$", $^{3}/_{8}$", $^{1}/_{2}$", $^{5}/_{8}$", $^{3}/_{4}$, and 1" thicknesses.

Pressure-Treated Lumber. Most species of lumber will rot if exposed to weather for long periods of time. Wood is even more susceptible to rotting if it remains in contact with the earth.

Lumber may be purchased that has been pressure-treated with chemicals to prevent insect damage and decay. Pressure-treated means the chemical is driven into the wood under pressure. The chemicals generally used to treat lumber are creosote, pentachlorophenol, acid copper chromate (ACC), ammoniacal copper arsenate (ACA), and chromated copper arsenate (CCA), figure 39-12. Since treated lumber is toxic to plants and animals, care must be exercised when choosing pressure treated lumber for use in agricultural settings, figure 39-13.

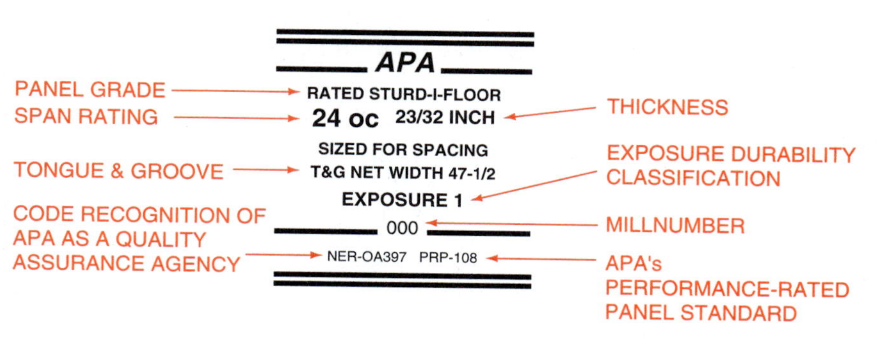

Specially designed as combination subfloor-underlayment. Provides smooth surface for application of carpet and possesess high concentrated and impact load resistance. Can be manufactured as conventional veneerd plywood, as a composite, or as a nonveneer panel. $1^1/_8$" plywood panels marked PS 1 may be used for heavy timber roof construction. Available square edge or tongue-and-groove. EXPOSURE DURABILITY CLASSIFICATION: Exterior, Exposure 1, Exposure 2, COMMON THICKNESS: $^{19}/_{32}$, $^5/_8$, $^{23}/_{32}$, $^3/_4$, $1^1/_8$.

Typical trademarks you may find on plywood and an explanation of what the numbers and terms mean.

American Plywood Association, *APA Design/Construction Guide, Residential & Commerical*, Page 6/7, January 1989

VENEER GRADES

Symbol	Description
A	Smooth, paintable. Not more than 18 neatly made repairs—boat, sled, or router type, and parallel to grain—permitted. May be used for natural finish in less demanding applications. Synthetic repairs permitted.
B	Solid surface. Shims, circular repair plugs, and tight knots to 1 inch across grain permitted. Some minor splits permitted. Synthetic repairs permitted.
C Plugged	Improved C veneer, with splits limited to 1/8 inch width and knotholes and borer holes limited to $^1/_4 \times ^1/_2$ inch. Admits some broken grain. Synthetic repairs permitted.
C	Tight knots to $1^1/_2$ inch. Knotholes to 1 inch across grain and some to $1^1/_2$ inch if total width of knots and knotholes is within specified limits. Synthetic or wood repairs. Discoloration and sanding defects that do not impair strength permitted. Limited splits allowed. Stitching permitted.
D	Knots and knotholes to $2^1/_2$ inch width across grain and $^1/_2$ inch larger within specified limits. Limited splits allowed. Stitching permitted. Limited to Interior, Exposure 1, and Exposure 2 panels.

American Plywood Association, *Grades and Specifications*, Page 5, August 1987.

FIGURE 39-11. Trademark information and grades of plywood. (*Courtesy of the American Plywood Association*)

Roofing and Siding. Steel and aluminum sheets are used extensively to cover agricultural buildings. Most are sold in 24" and wider widths and lengths of 8', 12', or 16'. Steel sheets are strong, but subject to rust. They are galvanized or coated with a layer of zinc equal to 1.25 ounces per square foot. However, better grades of sheet steel are coated with 2 ounces of zinc per square foot, figure 39-14.

Aluminum sheeting will not rust in normal weather, but it is expensive, thin, and easily damaged. Therefore, buildings to be covered with aluminum must have nailing boards with closer spacings than for steel. When using steel or aluminum, the manufacturer's instructions must be followed carefully as to type of nails, nail placement and spacing, overlapping of sheets, use of wicking to improve water lightness, and treatment at the ridge and gables. Ridge refers to the high point of a two-sided roof.

Pole Buildings

Pole buildings are easy to build, fast going up, economical, and flexible. To construct a pole building, a set of plans must be secured. After carefully laying out the building dimensions and digging the holes, pressure-treated poles are set in holes with footing pads. A footing pad is a concrete slab or flat stone in the bottom of a hole to prevent a pole from sinking.

Girders are bolted to the poles and rafters are added. Then girts, or horizontal side nailers, and pressure-treated skirt planking are added. Purlins, or

	Creosote and Creosote Solutions	Pentachlorophenol	Acid Copper Chromate (ACC)	Ammoniacal Copper Arsenate (ACA)	Chromated Copper Arsenate (CCA)
Round Poles as structural members					
Southern Pine, Ponderosa Pine	7.5	0.38	NR	0.6	0.6
Red Pine	10.5	0.53	NR	0.6	0.6
Coastal Douglas-Fir	9.0	0.45	NR	0.6	0.6
Jack Pine, Lodgepole Pine	12.0	0.60	NR	0.6	0.6
Western Red Cedar, Western Larch, Inter Mountain Douglas-Fir	16.0	0.80	NR	0.6	0.6
Posts, Sawn Four Sides as Structural Members					
All softwood species	12.0	0.60	NR	0.60	0.60
Lumber, All Softwood Species					
In contact with soil	10.0	0.5	0.62	0.40	0.40
Not in contact with soil	8.0	0.4	0.25	0.25	0.25
Plywood					
In contact with soil	10.0	0.5	0.62	0.40	0.40
Not in contact with soil	8.0	0.4	0.25	0.25	0.25
Foundation	NR	NR	NR	0.60	0.60
Greenhouses					
Above ground	NR	NR	NR	0.25	0.25
Soil contact	NR	NR	NR	0.40	0.40
Structural Posts	NR	NR	NR	0.60	0.60

[1] As recommended by the American Wood Preservers Association. Standard C 16-82.
NR - Not Recommended

FIGURE 39-12. Recommended minimum preservative retention for wood use on farms in pounds per cubic foot. (*Courtesy of American Wood Preservers Association*)

longitudinal roof nailers, are added, and finally the roofing and siding are installed, figure 39-15.

Care must be taken to brace the building carefully. Wood braces and plywood or metal gussets are used for this purpose. A gusset is a piece of wood or metal used to reinforce a joint, figure 39-16. Windows and skylights may be installed in the structure, figure 39-17.

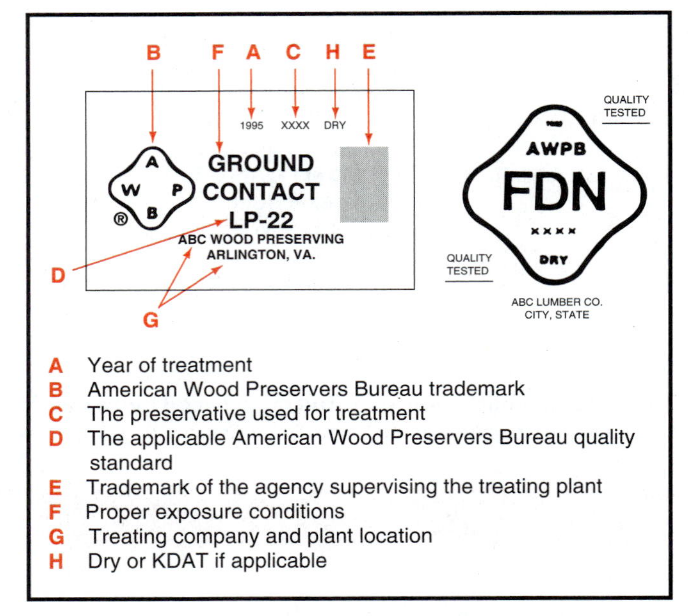

FIGURE 39-13. American Wood Preservers Bureau stamp and its interpretation. (*Courtesy of American Wood Preservers Association*)

Post-Frame Buildings

Post-frame buildings may be built with posts extending to the floor level or to the square. The square is the top of the wall. If posts go to the square, it is easy to add a concrete floor and truss roof. The walls and roof may then be insulated and siding added, figure 39-18.

If a building has a wooden floor, the posts are leveled and cut off at 1' to 3' above the ground. Heavy girders are placed on the posts and floor joists nailed to them. A joist is the timber used to support the flooring. Construction of the walls proceeds after the subfloor is installed. Trusses are generally used to give the building stability and support the roof.

Rigid-Frame Buildings

The use of formed steel for framing parts is called rigid-frame construction. It permits very long roof spans with relatively flat roofs. Rigid-frame buildings must be set on and bolted to extremely strong concrete posts or floors. The downward pressure of roof loads tend to exert an outward pressure on the base of the steel frame.

The frame, sheet coverings, and all accessories are prepunched and ready for assembly upon arrival at the site. Erection is fast and easy for an experienced crew. Steel- and aluminum-covered buildings can be equipped with various styles of doors, windows, ventilators, skylights, gutters, spouting, and

Roofing Materials	Probable life expectancy due to weathering: Years until first rust (approx.)	Probable life expectancy due to weathering: Probable life before roof rusts through *	Fire resistance	Hail resistance	Resistance to heat from sun	Resistance to strong wind	Maintenance	Minimum roof slope with ordinary lap	Approximate wt. per sq. (100 sq. ft.) **	Corrosion resistance to salt and industrial fumes	Relative cost ***
Galvanized Steel (all types)		(years)						(in. per ft.)	(lbs.)		
Gage / Thick-ness (in.)** / Wt. zinc coating (oz. per (sq. ft.)**											
29 / .0172 with coating / 1.25 / 2.00	15 23	40 50+	Highly resistant	Good	Excellent. If painted white is effective for reflecting heat.	Good if well fastened.	If painted with metallic zinc paint when rust starts to appear, life is increased.	Corrugated: 4 V-crimp: 3	80 85	Poor	Low to Medium
28 / .0187 / 1.25 26 / .0217 / 2.00	15 23	45 50+							100		
Aluminum Thickness generally available (inches) .019 .0215 .024 .032	50+ years		Good	Fair with .019 inch unless used with solid decking. Good with heavier sheets.	Excellent. Also excellent for heat reflection.	Fair. Good if well fastened	.019 inch punctures easily. Requires no painting. Sheets must be well nailed.	4	29 36.5 55.2	Poor	Medium
Wood Shingles	8-30 yrs. [26]		Burn readily	Good	Fair	Fair	May damage from strong winds.	6	200	Good	High
Asphalt Shingles	7-25 yrs. Less durable in warmer regions. [26]		Will burn after heat builds up.	Fair	Fair	Fair. Good if well laid.	May damage from strong winds or hail.	5	250 to 300	Good Medium	Low to
Asphalt Roll Roofing: Mineral—surfaced, Selvage edge	About the same as asphalt shingles.		Will burn	Fair	Fair	Fair	May damage from strong winds or hail.	2- to 4-inch lap: 4	100 to 140	Good	Low
Smooth—surfaced	Much less than mineralsurfaced shingles or selvage edge		Will burn	Fair	Fair	Fair	May damage from strong winds or hail.	2- to 4-inch lap: 4	65	Good	Low
Asbestos-Cement Shingles and	50+ yrs.		Good	Good	Excel.	Good	May be damaged from flying objects.	5	260 to 300	Good	Medium to High
Corrugated sheets								3	300		

*Test under rural conditions are not complete. Estimates were derived by projecting rates of corrosion in rural areas and correlating the data with tests already completed under more corrosive conditions. Farms located near industrial areas, or near salt water, will get much less service. Painting with zinc paint when rust first appears will prolong the life of the steel.
**The different types of roofing vary in weight per square according to the thickness and density of the material.
***Cost for roofing material only. Costs will vary with the weights of roofing and also with the location in which it is purchased.
****Standard thicknesses specified by "United States Galvanized Sheet Gage" Sheets imported from other countries may come in under other gage standards that are not of the same thickness as the U.S. gage.

FIGURE 39-14. Characteristics of roofing materials. (*Courtesy of American Association for Vocational Instructional Material—AAVIM*)

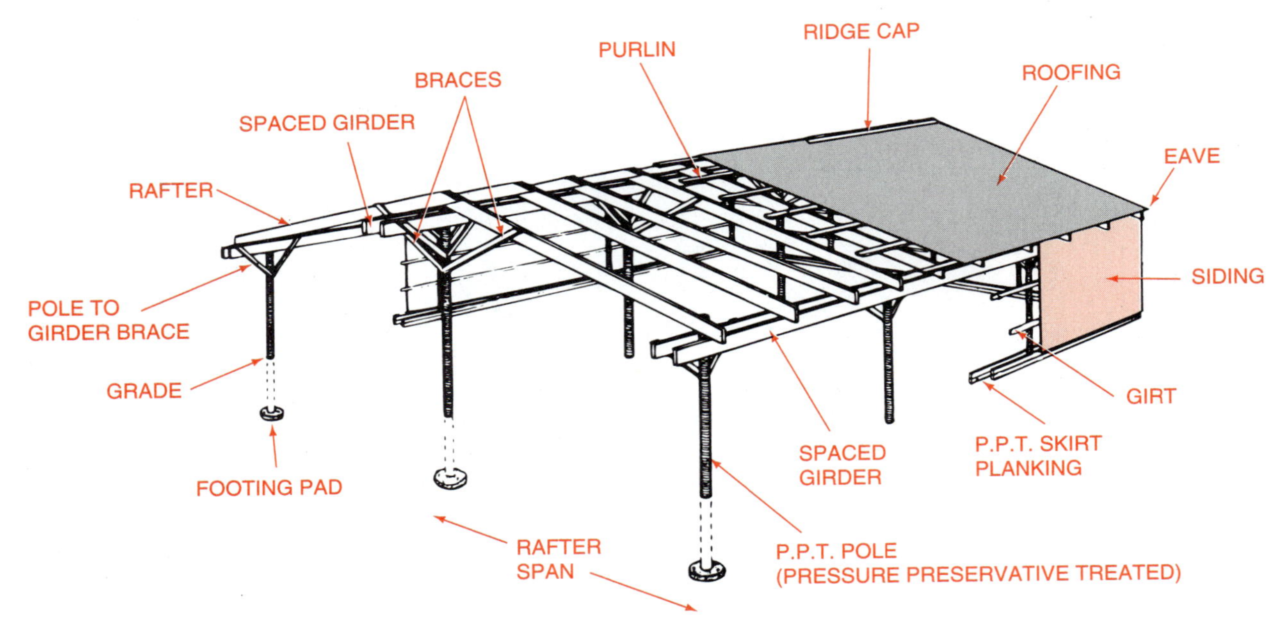

FIGURE 39-15. Pole building.

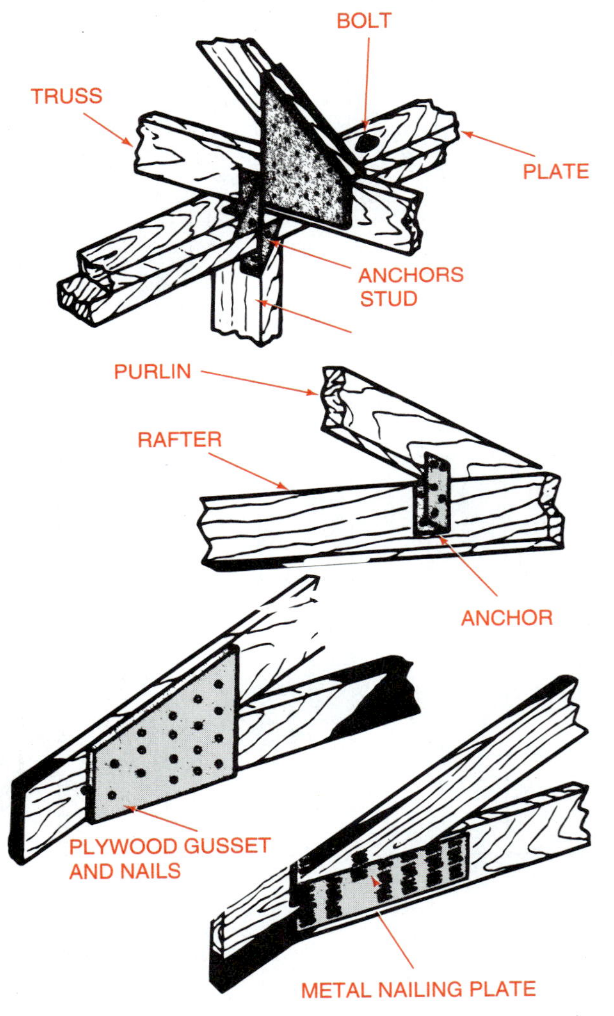

FIGURE 39-16. Securing girders, rafters, and trusses to posts or poles.

other special features. They can be insulated and finished with interior walls for office, manufacturing, warehousing, processing, storage, or other uses, figure 39-19.

Woodframe Buildings

Woodframe buildings have walls constructed of vertical 2" × 4" boards placed 16" on center (OC). Homes, garages, and other one- to three-story buildings with finished interior walls are generally of the woodframe type. These buildings have floors nailed to the floor joists that are carried by the concrete or block walls or foundation and one or more center-

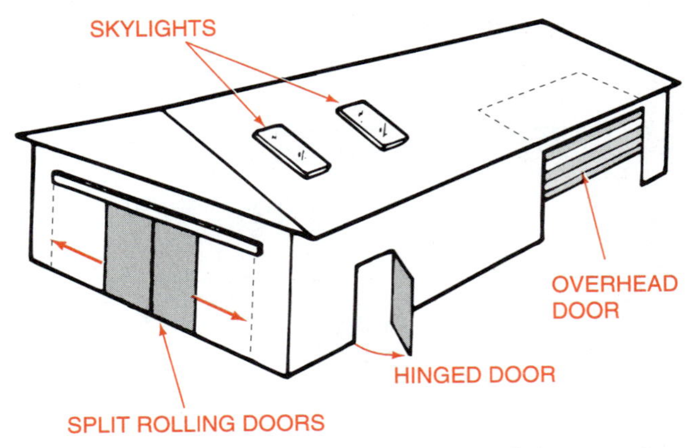

FIGURE 39-17. Doors, windows, skylights, and vents add to the utility of the building.

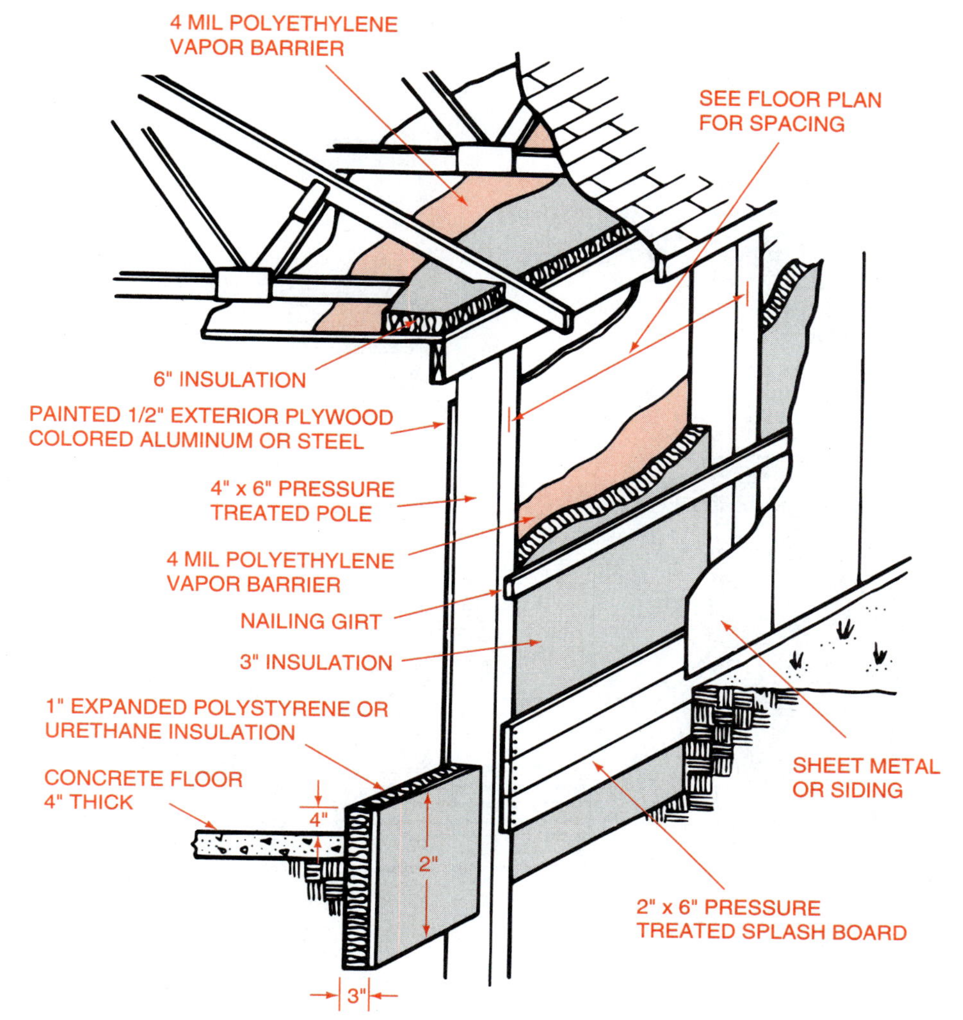

FIGURE 39-18. A post-frame building with truss roof and insulation.

line girders. They rely on solid masonry foundations and one or more posts to bear the weight of the structure, provide anchorage against wind, and prevent moisture from entering below grade. Below grade means below the surface level of the surrounding land. Interior partitions are made of studs and drywall or paneling. They help support upper stories. Woodframe buildings may also be built on concrete slabs in well-drained areas or on posts or masonry pillars in wet areas.

Laying Out a Foundation. The exact locations of corners is necessary for all buildings. In addition, level lines must be stretched along the outer edges of the foundation or posts to establish their placement and elevations. Elevation means relative height and position from sea level. Such lines are attached to batter boards, which are level boards placed 8' to 12' beyond the foundation. The procedure for laying out a building is provided in figure 39-20.

Excavation. To excavate is to dig or remove earth. After a building is laid out, a backhoe and front-end loader is used to dig the trenches for the footer and foundation and remove the soil from the basement area. A footer is a continuous block of concrete under a foundation.

The excavation is made several feet beyond the outside of the layout lines and as deep as the basement subgrade. Subgrade means the level where stone begins for a concrete floor or topsoil begins when grading off land. After the basement excavation is completed, the trenches are dug for the footer. The footer should be at least as deep as the foundation width and $1^1/2$ times as wide. However, engineering specifications and local building codes should be consulted, figure 39-21.

Footers and Foundations. Footers should be poured with concrete made to correct specifications and reinforcement details. The foundation may

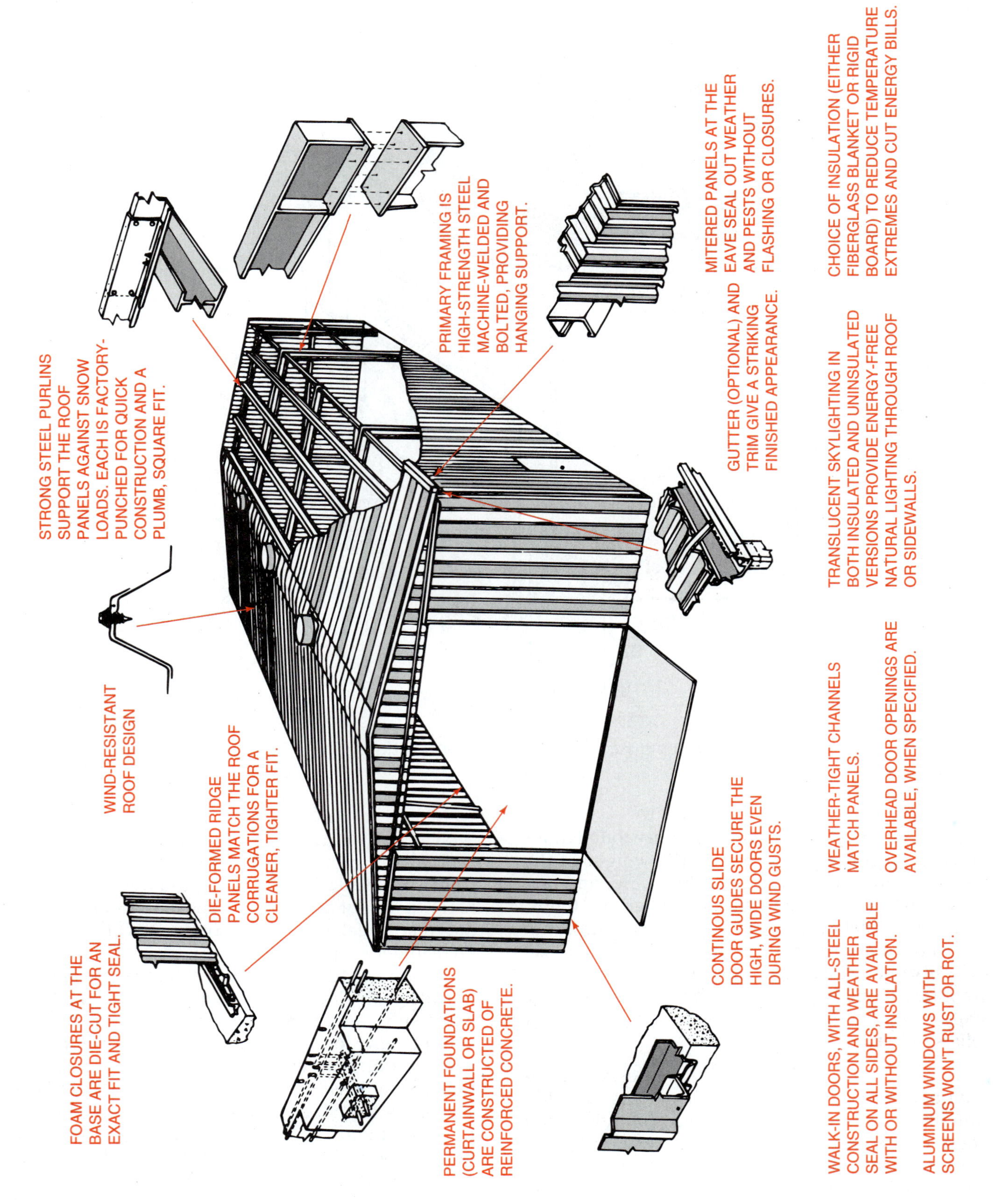

FIGURE 39-19. Rigid-frame steel buildings come in many configurations and are used in many commercial and farm settings. (*Courtesy of Butler Manufacturing Co.*)

Building Layout

The layout of the building establishes exact reference lines and evaluations. Care in layout makes construction easier and helps keep the building square.

Tools and Materials

Two 100' steel tapes, crow bar, hatchet, sledge hammer, level, at least twenty 2 x 4 stakes 35" long, eight 2 x 4 batter boards 8' long, building line or cord (length = building perimeter + 100'), 12d nails, and builder's level and rod.

Procedure

1. Stake out a baseline at the front edge or side of the building.
2. Locate and set front corner stake A on baseline. Drive a nail in the top of the stake as a reference point.
3. Measure the building length along the baseline. Stake A and set corner Stake B. Use a surveyor's level and drive Stake B level with Stake A. Drive a nail in the top of the stake at the exact length of the building.
4. Make the end wall perpendicular to the front wall as follows. Measure 30' along the base line from Stake B and set a temporary stake. The point 50' from this temporary stake and 40' from Stake B is perpendicular to the base line. Set a temporary stake at this point.
5. Measure the width of the building along this line and set the third corner Stake C. Drive Stake C level with Stakes A and B. Drive a nail in top of the stake at the exact width of building.
6. From Stake C measure the building length. From nail in Stake A measure building width. Where these two measurements meet drive fourth corner Stake D. Drive nail at exact corner point.
7. Check tops of all stakes; all should be level. Then measure: (1) baseline length AB, (2) triangulation at second corner (ABC), (3) widths BC & AD, and (4) length CD. Adjust nails or stakes B, C, D as necessary.
8. Check that diagonals AC and BD are equal for a rectangular building. Make diagonals equal by shifting C and D along rear wall line. Keep width BC and CD equal. Check level of shifted stakes.
9. Drive batter board stakes 8' to 12' from all corners. Batter boards provide a level reference plane for the building layout. They should not interfere with excavation or construction and should remain undisturbed until the building frame is complete.
10. Level and fasten batter boards to stakes at same height as tops of corner stakes.
11. Stretch building cord between batter boards to just touch nails on top of stakes. Drive nail or make saw kerf in top of batter boards to line up string. Corner marking stakes can then be removed and corners located where lines cross.

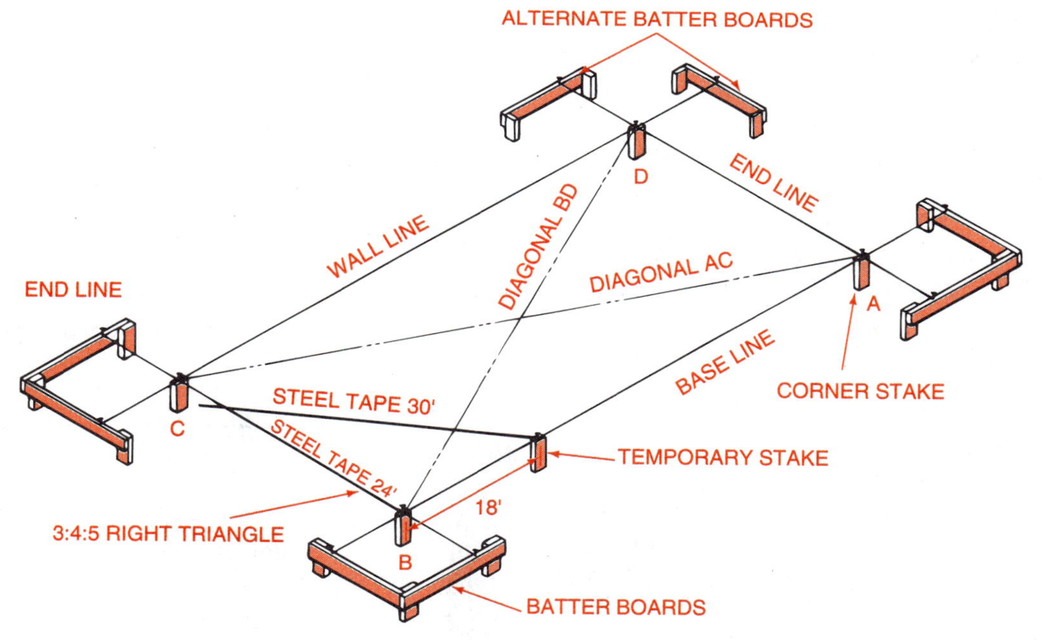

FIGURE 39-20. Building layout procedures. (*Courtesy of Cooperative Extension Northeast Regional Agricultural Engineering Service*)

be poured with solid concrete or laid up with masonry blocks. Block laying is discussed in detail in Unit 38, Concrete and Masonry.

Walls must be specifically treated to be waterproof or water may seep through. The joint where the footer and foundation meet and the foundation wall must be plastered with waterproof cement or special masonry sealer and topped with an asphalt finish. A tile or PVC pipe drainage system should be installed around the total foundation in a stone bed with

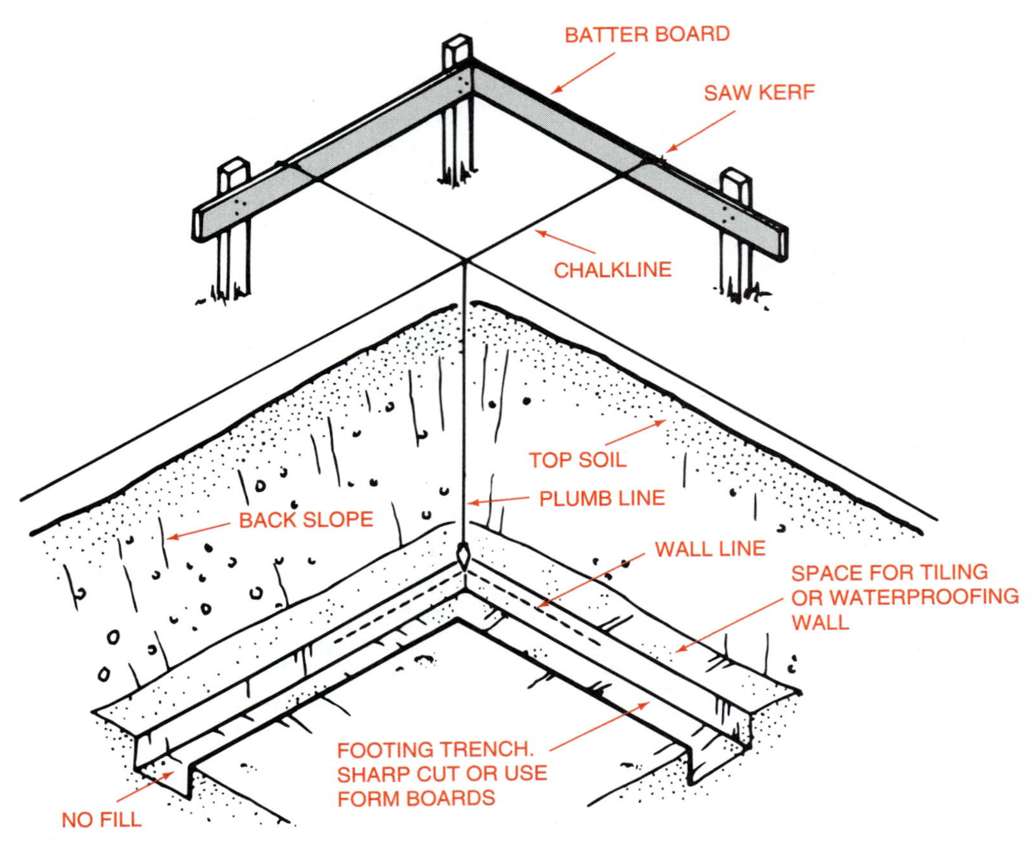

FIGURE 39-21. Excavating for a basement and footing.

stone cover above the footing. The system must have a suitable outlet or pump to carry away water that gets against the foundation, figure 39-22.

Walls and Floors. There are several ways of integrating the floors and walls in a woodframe house. With the modern, platform, and Western methods of building, floor structure and the subflooring are constructed before the wall of the next story is built. This provides a safe working area to construct the walls. Wall and partition framing consisting of 2" × 4" studs with 2" × 4" plates at the top and bottom are laid out and nailed together on the floor. The wall unit is then tilted up and into place, braced, and nailed. After the wall framing is in place, exterior sheathing and siding or brick veneer is installed, figure 39-23.

When a two- or three-story building is desired, one or more floors and wall systems are built on the first. Figure 39-24 shows some additional details of construction. Trusses may be substituted for rafters and ceiling joists Solid plywood sheathing is generally used instead of diagonal sheathing and eliminates the need for diagonal bracing. Also, plywood is frequently used instead of diagonal subflooring. Figure 39-25 shows detailed treatment of eaves of the building. The eave is the overhang.

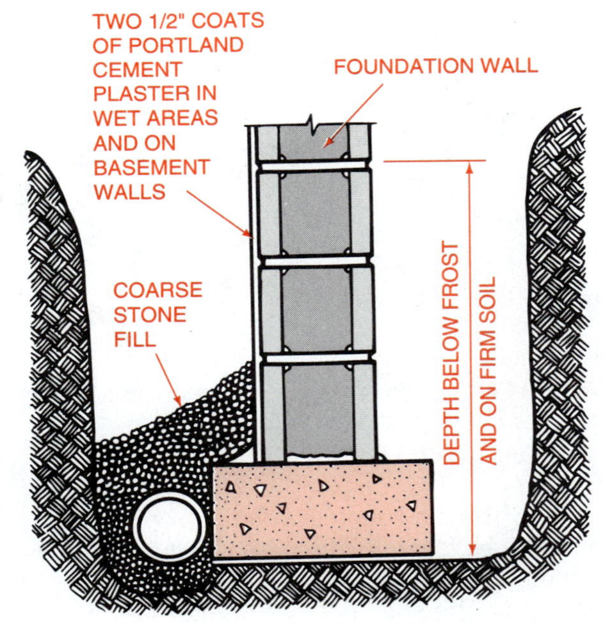

FIGURE 39-22. Footing and foundation with drainage and waterproofing.

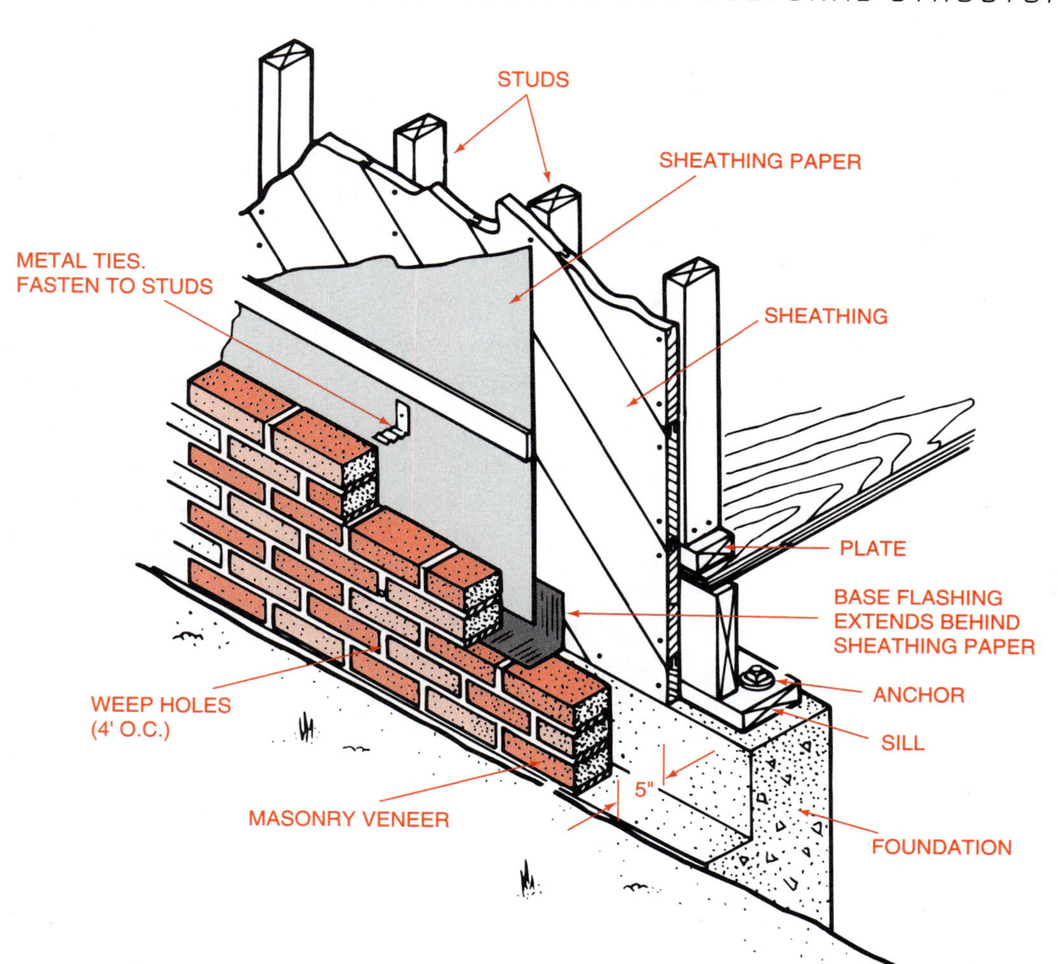

FIGURE 39-23. Wall detail of a modern frame building.

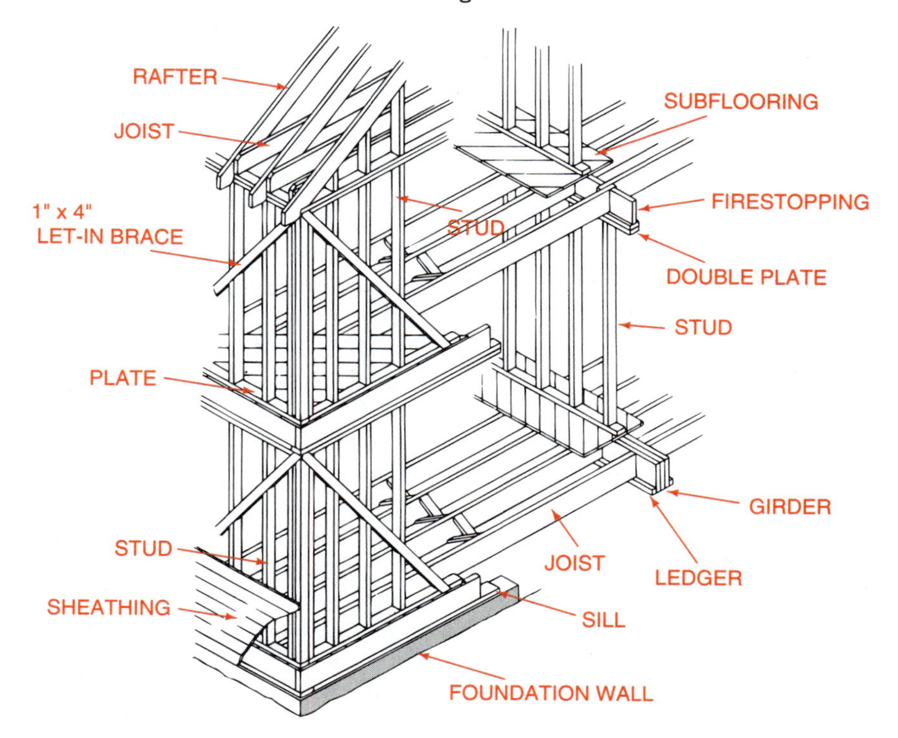

FIGURE 39-24. Framing members of a two-story house.

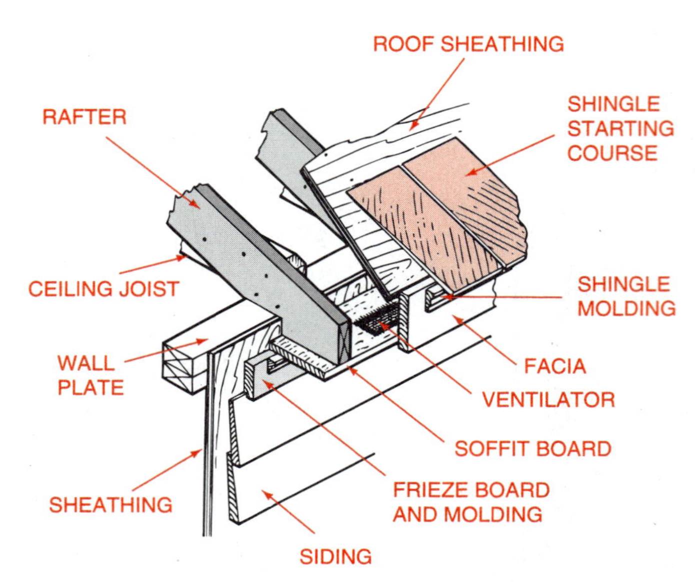

FIGURE 39-25. Finishing of the eaves of a building.

LAYING OUT STAIRS AND RAFTERS

Stairs. Stairs may be needed in original construction or for replacement purposes. When replacing old stairs, remove the trim and treads first. The tread is the part of the stair you step on. Remove the riser or back of the stair step. Finally, remove the stair stringer or stair horse which is the structure that supports the stairs. Be careful not to break the stair horse, and use it as the pattern for cutting replacement parts.

Horses for stairs in new construction must be laid out and sawed out of 2" × 10" or wider material. The framing square is used to lay out and mark the cuts for the bottom and top of the horse and each tread and riser along the way, figure 39-26.

Rafters. There are many kinds of rafters. For straight roofs, trusses are now used in the construction of most new agricultural buildings. Rafter layout and cutting may be needed for small buildings or special roof shapes. The layout of a rafter is discussed in figure 39-27.

INSULATING AGRICULTURE STRUCTURES

Energy conservation and moisture control is of major concern in most buildings. Workers' health, efficiency, and safety are generally improved by comfortable temperatures and humidity. Similarly, plants and animals grow and produce best under optimum conditions of temperature, humidity, and air quality. These are all easier to control when buildings have appropriate insulation and moisture or vapor barriers.

Concrete floors may be protected from moisture penetration by placing a moisture-proof membrane between the fill material and the concrete. The surface of the fill must be relatively smooth to avoid puncturing the film or a thin coat of concrete is recommended over the fill. Four mil polyethylene film is available in large sheets and makes a fast, low-cost, and effective moisture barrier. For small floor areas, 55 lb. roll roofing material over-lapped 6 inches with hot asphalt or bituminous cement in the laps makes a satisfactory vapor barrier. The vapor barrier should be carefully placed across the fill material and up the side walls before the concrete is placed.

Outside surfaces of foundations should be sealed with special masonry products that penetrate the pores and expand as they dry. Further protection against moisture and mechanical damage may be obtained by using a waterproof cement coating followed by a final coat of asphalt. The exteriors of masonry walls can be protected from driving rains by appropriate waterproof paints.

In heated buildings, floors will be warmer and energy conservation better if solid insulation material is placed between the floor and the side walls or foundation. Further, if energy costs are high, consideration should be given to placing solid insulation under the slab.

Masonry and concrete materials have varying insulation values according to the material they contain. For instance, a heavy, high-density concrete block 8" wide may have a low insulation value or resistance (R) of only .98, but a light-aggregate block 8" wide may have an R value as high as 2.3. By filling the cores of such light aggregate block with appropriate insulation material, the R value of the wall may be increased to nearly 7.5. Materials typically used to fill masonry block cores or make concrete with high-insulation value include perlite, vermiculite, and cellulose, figure 39-28.

Exterior masonry or metal walls may also be insulated by building a 4" wood or metal framed wall inside and placing roll or batt insulation in the wall. Woodframe buildings typically have stud walls with 2" or 3" batt or roll insulation between 2" × 4" studs with an R value of 7 or 11 respectively, or 6" insulation between 2" × 6" wall studs with an R value of 19. Ceilings typically have a minimum of 6" insulation or more for an R value of 19 to 30. Popular insulation materials for this purpose include fiber, mineral, fiberglass, and slag.

Metal and pole buildings may be insulated with rigid insulating materials or fiberglass with a poly-

Laying Out Stairs

Although there are many ways to build stairways, only one method is discussed here to illustrate the principles involved. A staircase has a total rise and run, because it travels both horizontally and vertically; but it is also divided into a series of steps, each of which has a rise and run. The width of each tread is its run and the height of each riser makes up the rise of each step.

Determining the width of treads and height of risers

The "magic number" in stair building is 17. This is because the width of the tread and the height of the riser should add together to form 17 inches, or approximately this amount. (Some stair builders try to stay between 16 1/2 and 17 1/2 inches.) Any great departure from 10-inch treads and 7-inch risers is uncomfortable for the average person and may be dangerous.

In laying out a stairway, the total rise is usually a fixed figure. It is important to divide this distance into a number of steps of equal rise. The procedure usually followed is first to determine total rise in inches and divide this distance by 7 in order to determine how many steps there should be. This answer is determined only to the closest whole number. Then divide the total rise by this number and work it out to the closest 1/16 of an inch.

Example: Total rise is 54 inches.

$\frac{54}{7}$ = 7 5/7 or 8 (probable number of steps).

$\frac{54"}{8}$ = 6 6/8, or 6 3/4" (height of each riser)

Width of tread should be 17" - 6 3/4" = 10 1/4"

(Can vary from 9 3/4" to 10 3/4")

Laying out the horses

The side members, which support the treads, are called "horses." There is usually one less tread than the number of risers as figured above. It may be desirable to draw a sketch and number the risers in order to avoid making an error in laying out the stair horses. Such a sketch is shown below.

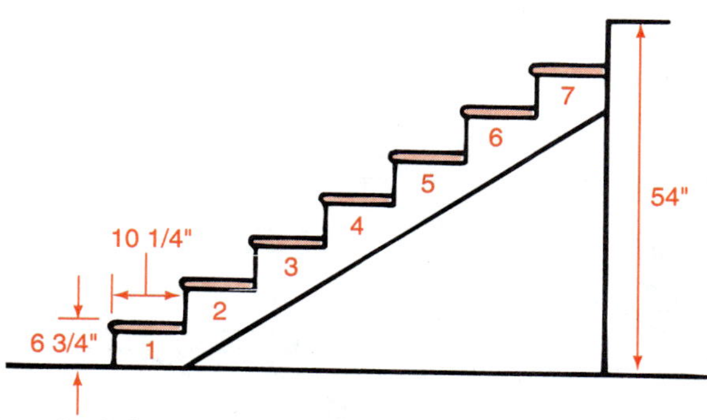

After calculating the number of steps and the height of each riser, draw a sketch and number the risers.

Place the square along the 2" x 10" with the amount of rise on the tongue and the width of tread on the blade of the square. It is more convenient to use clips or wooden fence to avoid the necessity of reading the square each time it is applied.

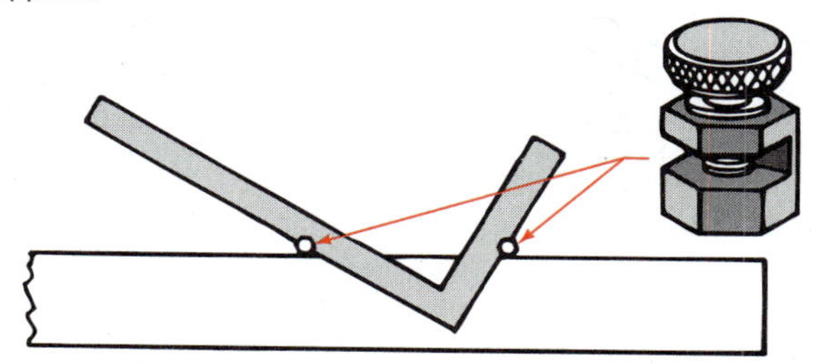

Mark lines as shown in the sketch below where the pieces are to be cut out of the horses. Finally, make a correction at the lower end to shorten the first riser by an amount equal to the thickness of the tread. This amount is usually 1 1/2" or 1 1/16" depending upon whether nominal 2" or 1 1/4" lumber is used. The one correction at the lower end takes care of the top step also as it lowers the entire horse by one tread thickness.

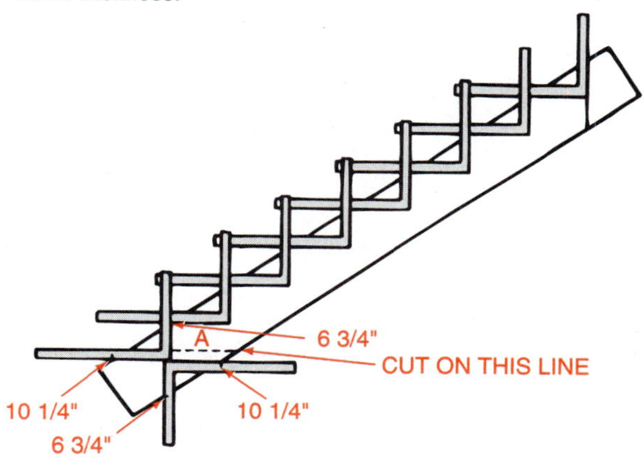

Lay out the stairs horses with the square, making correction at A for the thickness of the stair treads.

FIGURE 39-26. Laying out stairs. (*Courtesy of University of Illinois, Vocational Agricultural Service*)

ethylene vapor barrier. Special attention must be given to the high volume of moisture given off by livestock and appropriate fans must be installed to keep the humidity under control. Further, the ability of birds to pick apart and damage fiberglass and other soft insulating materials must be considered in buildings where birds cannot be excluded.

In all cases, a proper vapor barrier must be installed or the insulation may absorb moisture and lose its insulating ability. Moisture may also lead to

Laying Out a Gable-roof Rafter

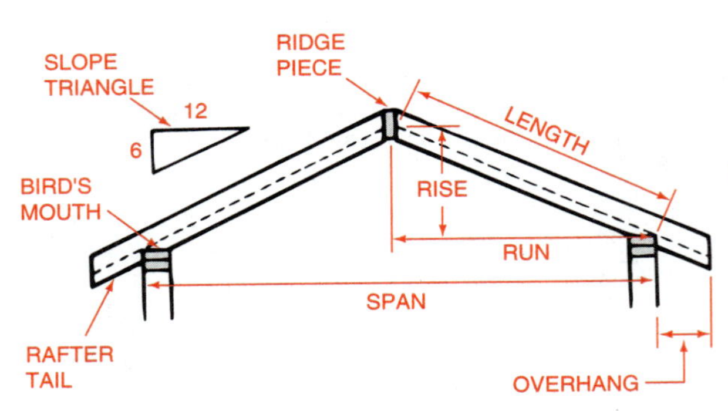

Fig. 1 Standard terms are applied to gable roof buildings and rafters.

Since gable rafters meet in midair, the height of the plate is the only fixed point and it is not necessary to determine the total rise. The only information needed is the total run and the slope, or inches of rise per foot of run. The run is determined by measuring the span from plate to plate (Fig. 1) and dividing by two. Rate of rise is obtained from the slope triangle. In this case the total run is 11 feet and the slope is 6 inches of rise per foot of run.

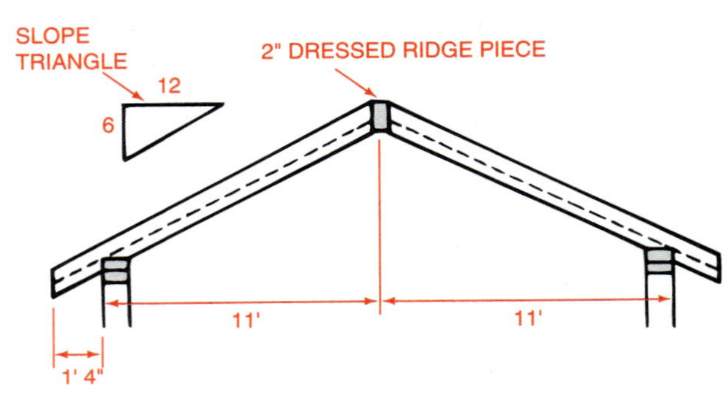

Fig. 2 Total run of a rafter for this building is 22/2" or 11'.

Using a framing square, put a marking line parallel to the edge of the rafter and 2" from the bottom edge. Using 12" on the blade and the number of inches rise per foot of run (6" in this case) on the tongue, lay out the lower end of the rafter first. Mark the lower plumb cut first and then measure with the square in position the desired amount of overhang as a horizontal distance and locate the corner of the bird's mouth on the marking line. Mark out the bird's mouth with a horizontal line where the rafter is to rest on top of the plate and vertical line where it is to contact the edge of the plate.

Calculate the length of the rafter as follows: Find the rafter table on the square and under the 6" mark, since the slope is 6" rise per foot of run, find the figure opposite "Length of Common Rafters per Foot of Run." This figure is 13.42, which means that for each foot of run, the rafter is 13.42" long. Calculate the length for the number of feet of run as follows:

Calculation: 13.42" x 11 = 147.62 inches

$$\frac{147.62"}{12} = 12' \ 3.62" \text{ or}$$

12' 3 5/8" (.625 = 5/8")

Measure out 12' 3 5/8" with a steel tape from the corner of the bird's mouth to a point on the marking line near where the upper plumb cut is to be made. Place the square with the numbers 6 and 12 on the marking line, with the number 12 even with the length mark (Fig. 4). Then measure back along the square a distance equal to half the thickness of the ridge piece. This is 3/4" for a standard-dressed 2" ridge piece, or 3/8"+ for a 1" ridge piece. Draw the line for the plumb cut through this corrected length mark and the rafter is ready to cut. If no ridge piece is used, the upper plumb cut is made to intersect the marking line at the full measured length.

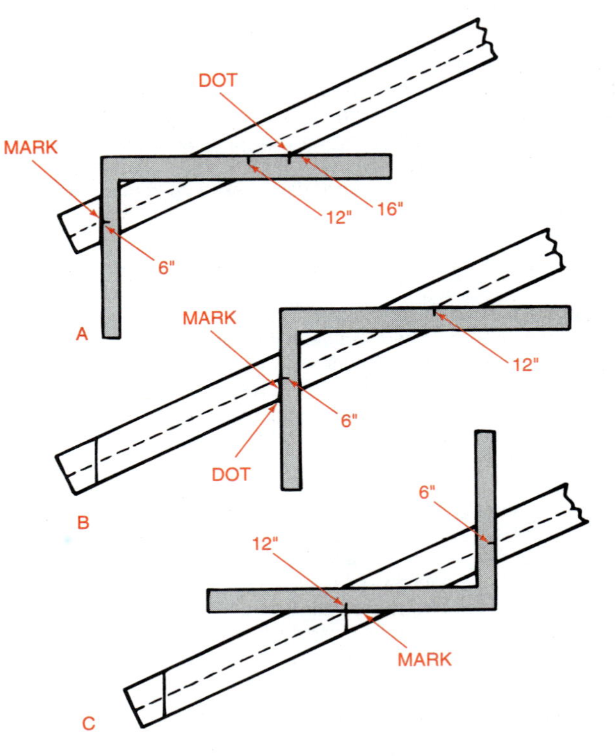

Fig. 3 Lay out lower end of gable rafter as follows:
(A) Place the square so that 6" and 12" are on the marking line. Mark lower plumb cut and place a dot opposite 16" on the blade of the square to indicate length of overhang.
(B) Slide the square up to the rafter and mark the plumb line for the bird's mouth even with the dot.
(C) Rotate the square and mark the level line for the bird's mouth.

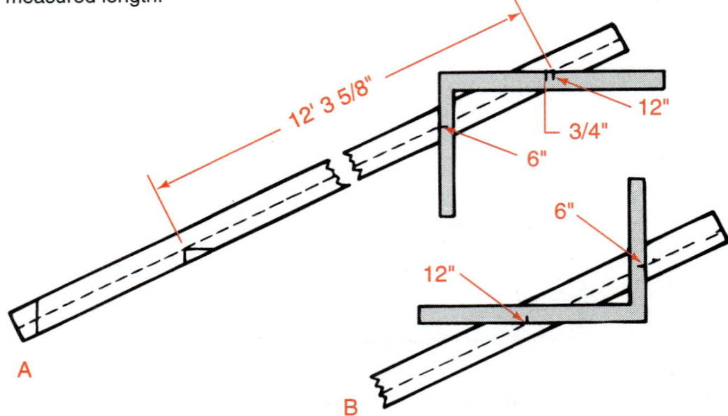

Fig. 4 Lay out upper end of gable rafter as follows:
(A) Measure the calculated length (12' 3 5/8") along the marking line from the corner of the lower bird's mouth and make a dot. Place the square with 6" and 12" on the marking line and with 12" mark even with the dot. Measure back along the blade of the square a distance of 3/4" and make a second dot.
(B) Rotate the square and draw a plumb line through the second dot.

FIGURE 39-27. Laying out a gable-roof rafter. (*Courtesy of University of Illinois, Vocational Agricultural Service*)

MATERIAL	*R VALUE PER INCH OF THICK-NESS	*R VALUE FOR THICKNESS SHOWN
Insulation, rigid		
Fiberglass	4.00	
Mineral fiber, resin binder	3.45	
Polystyrene—cut cell surface	4.00	
Polystyrene—smooth skin surface	5.26	
Polystyrene—molded bead	3.57	
Polyurethane	6.25	
Polyisocyanurate foam, plastic core, and foil each face	7.2	
Insulation—batt fiber type, mineral glass, slag		
2-2.75 in.		7
3-3.50 in.		11
6.5 in.		19
Insulation, fill		
Perlite	2.70	
Vermiculite	2.27	
Cellulose	3.13	
Miscellaneous		
Brick, 4 in.		.44
Gyp board, 1/2 in.		.45
Particle board, 5/8 in.		.82
Plaster, 1/2 in.		.10
Wood, hard	0.9	
Wood, soft	1.2	
Wood, plywood	0.8	

*R = Resistance to heat transfer

FIGURE 39-28. Insulation values of popular building materials.

the decay of wood structures and the rusting of steel components. Vapor barriers generally consist of asphalt impregnated paper, aluminum coated wood or paper products, or polyethylene film. These are generally installed on the warm side of walls, but should be installed according to local climatic conditions, building codes, and manufacturers' recommendations.

ROPE WORK IN AGRICULTURAL SETTINGS

Ropes are used to secure bundles, lift objects, drag materials, secure livestock, rig boats, and other. The practice of splicing is seldom used today, and rope halters for livestock are generally cheaper to buy than make. However, one should know how to tie knots that hold and yet can be untied.

Common Knots and Hitches

Square Knot. The same basic knots are useful in construction as for other agricultural settings. The granny knot is often used, but will pull out and is, therefore, unsafe. The square knot is easy to tie, easy to untie, and will not pull out under tension. It is the preferred knot for attaching ropes of the same size to each other. The sheet bend is good for fastening ropes of different sizes.

Bowline. The bowline is a good knot for creating a loop that will not slip or close up under tension. It is easy to tie and untie.

Clove Hitch and Two Half Hitches. The clove hitch and two half hitches are quick to tie and easy to untie. They are used for fastening ropes to posts.

Timber Hitch. The timber hitch can be used as a single hitch or multiple hitch to lift or drag poles.

Barrel Sling. The barrel sling is useful for rigging barrels and similar objects for lifting. Care must be taken to be certain the rope is across the center of the bottom and cross on the sides at equal distances around the barrel.

The knots and hitches described above are illustrated in figure 39-29.

WHIPPING ROPE ENDS

Most rope will unravel at the end unless treated in some way. Rope ends may be whipped by tightly coiling strong cord around the end for a distance of $^1/_2$" or $^3/_4$", depending on the diameter of the rope. Also friction tape or plastic electrical tape may be tightly wrapped around rope ends for temporary use. Nylon rope ends may be fused into a solid end by burning with a flame.

SUMMARY

Planning, constructing, and maintaining buildings and structures is an important part of agriculture. This unit addressed some principles of planning and

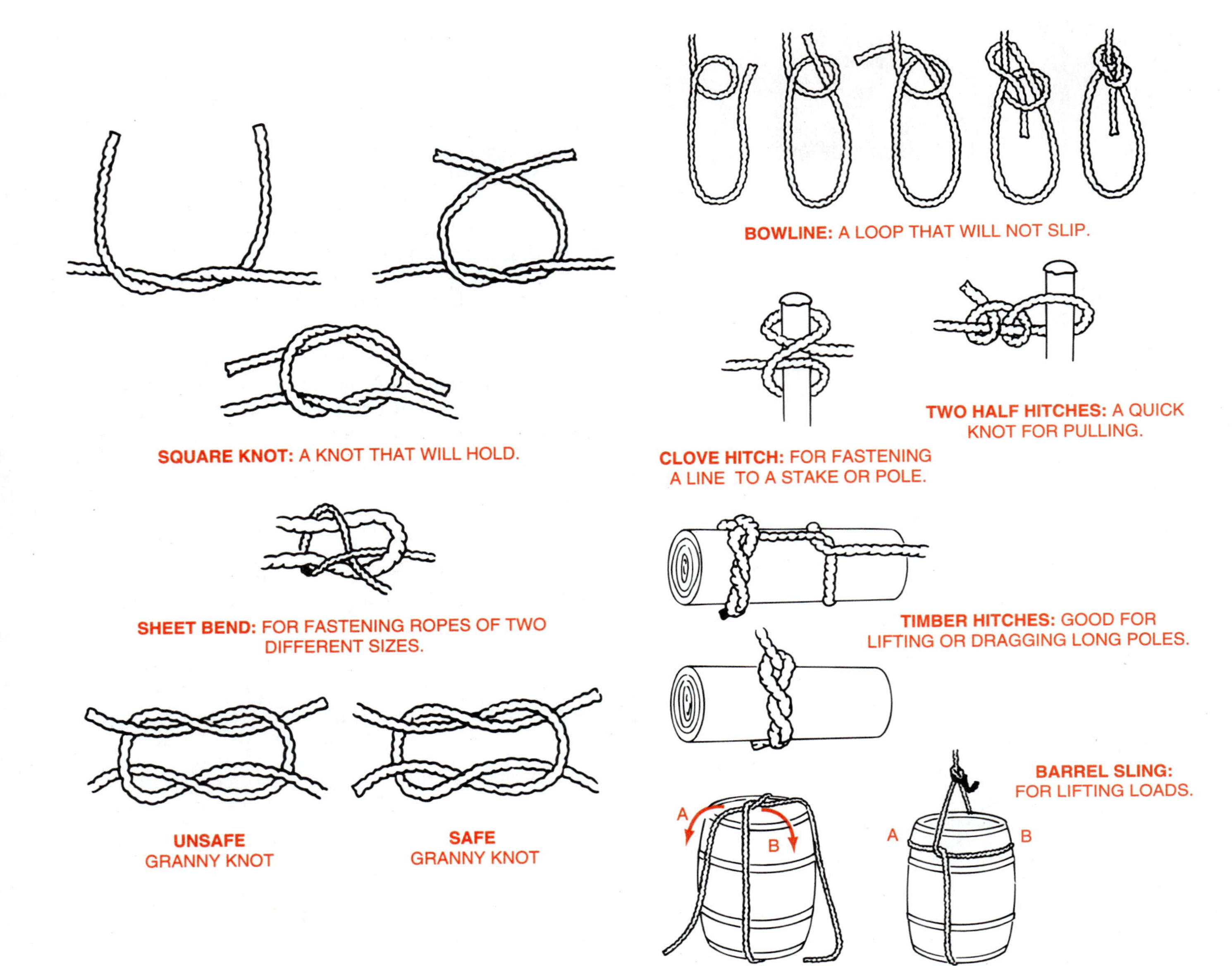

FIGURE 39-29. Useful knots and hitches.

types of structures. While some details of construction are presented, the scope of the book does not permit sufficient detail to proceed without professional advice, plans, and information on local, state, and national building codes.

Related information is presented in previous sections of the text. The unit that follows provides information on some specialized structures.

STUDENT ACTIVITIES

1. Define the Terms to Know in this unit.
2. Visit a farm or other agricultural business. Study the building layout and discuss the advantages and disadvantages of the various building locations.
3. Design and draw a farmstead plan showing electrical meter pole, buildings, windbreak, and roads. Report the advantages of your plan to the class.
4. Study figures 39-5 and 39-6. Draw templates to scale on heavy paper for the major pieces of machinery on a farm with which you are familiar. Cut out the templates and use them to design a proposed machinery storage building to accommodate the machinery.
5. Choose an agricultural enterprise, such as corn, potatoes, poultry, or beef cattle, in which you have special interest. Research the building requirements for the enterprise, and describe in detail the buildings you would recommend for production and/or storage. Report your findings to the class.
6. Build a model of some agricultural structure of your choice.
7. Work with a group of classmates to plan and build a model farmstead to scale.
8. Display the model described in activity #7 in a school showcase, library, and/or local fairs and shows. Consider it for entry in the school science fair.
9. Make a display board showing all the knots described in this unit.

SELF-EVALUATION

A. Multiple Choice. Select the best answer.

1. Prevailing winds generally blow from the
   a. east
   b. north
   c. northwest
   d. west
2. The sun's warmth in winter can be captured best if the building faces
   a. north
   b. northwest
   c. south
   d. west
3. The electrical meter pole for a farmstead should be placed
   a. in a central location
   b. beside the house
   c. close to the highway
   d. at the building using the most electricity
4. Farm building planning should be done by
   a. farm family
   b. farm operator
   c. professional farm or building planner
   d. cooperatively by all the above
5. The item that permits a building to span over 100 feet without posts is a
   a. truss
   b. stud
   c. rafter
   d. girder
6. The actual thickness of 1" × 6" dressed lumber is
   a. 1/8" less than nominal
   b. 1/4" less than nominal
   c. 1/2" less than nominal
   d. Same as nominal
7. An example of a grade of plywood is
   a. AC
   b. S2S
   c. pressure-treated
   d. laminated
8. Which is *not* a material used to pressure treat lumber?
   a. chromated copper arsenate
   b. creosote
   c. pentachlorophenol
   d. sodium biarsenate

9. A building foundation rests upon a
   a. wall
   b. tile drain
   c. girder
   d. footing
10. A woodframe building has walls constructed of
   a. beams
   b. girders
   c. purlins
   d. studs

B. Matching. Match the terms in column I with those in column II.

Column I

1. ACA
2. granny knot
3. galvanized-coated
4. plywood
5. high point of roof
6. joist
7. rigid-frame
8. horse
9. square
10. whipping

Column II

a. made of veneer
b. 1.25 or 2.0 oz. coating
c. ridge
d. supports the floor
e. stair stringer
f. used for pressure treatment
g. safe knot
h. used on rope ends
i. unsafe knot
j. support of steel buildings

C. Completion. Fill in the blanks with the word or words that will make the following statements correct.

1. Outside surfaces of foundations are best waterproofed by sealing with a product that _____ and expands as it dries.
2. Foundation walls may be further protected after being sealed by using _____ and/or _____.
3. The ability of a certain material to restrict the flow of heat and be useful for insulation is referred to as _____ value.
4. Vapor barriers are important in agriculture buildings for
   a. _____
   b. _____
   c. _____
   d. _____
5. Specify the knot or hitch recommended for each of the following:
   a. creating a nonslip loop _____
   b. fastening livestock lead ropes _____
   c. lift poles _____
   d. lift barrels _____
   e. joining ropes of equal size _____

UNIT

Aquaculture, Greenhouse, and Hydroponics Structures

OBJECTIVE

To build and maintain structures for aquaculture, greenhouse, and hydroponics management.

COMPETENCIES TO BE DEVELOPED

After studying this unit, you should be able to:

- Recognize structures used in aquaculture, greenhouse, and hydroponics technology.
- Maintain common structures used in aquaculture, greenhouse, and hydroponics technology.
- Construct simple aquaculture production facilities.
- Erect a small greenhouse.
- Build hydroponic units.

MATERIALS LIST

- ✓ Protective goggles and coveralls
- ✓ General shop tools
- ✓ Specific construction materials are listed for Projects 52 to 54 in Appendix A

TERMS TO KNOW

- aquaculture
- open aquaculture system
- fingerling
- microorganisms
- dissolved oxygen
- rotary air pumps
- diffuser tube
- air stones
- paddle-wheel aerator
- ammonia
- water quality
- submersible pumps
- centrifugal head pumps
- air-lift pump
- net cages
- closed aquaculture systems
- controlled-environment structures
- biological filters
- raceways
- fish feeders
- settling tank
- greenhouse
- translucent
- greenhouse effect
- shadecloth
- freestanding
- evaporative cooling pads
- hydroponics
- nutrient solution
- vermiculite
- bag culture
- trough culture
- tube culture

CAUTION:

- **Read and observe all safety precautions when working with electrical equipment in wet areas.**

ENVIRONMENTAL STRUCTURES USED IN AQUACULTURE

Types of Systems

Aquaculture production systems have been developed to improve the production of aquatic plants and animals used for food, fiber, and recreation. These systems center around bodies of water such as lakes, streams, ponds, or artificial tanks that support the growth and development of economic aquatic species.

Open aquaculture systems are managed with relatively low densities of plants or animals. With low densities, an acceptable balance between the inputs (food, air, or nutrients) and outputs (waste products) may be maintained and pollution of the growing system can be minimized. For example, in the spring, ponds are stocked with small fish called fingerling and managed so that the final number of adult fish in the fall will not pollute the natural ecosystem of bacteria and fungi (microorganisms) that detoxify the fish waste products.

As growers continue to increase the number, or density of fish in the tank, cage, or pond, it is necessary to add inputs, such as feed and oxygen, or remove the accumulated fish wastes by flushing or water treatments. Such systems are called semi-open aquaculture systems and are managed like open aquaculture systems until the fish densities start to limit growth, figure 40-1.

When feed is added to the fish population, an increase in the consumption of oxygen in the water by both fish and microorganisms results. Air must then be mixed with the water to increase the levels of dissolved oxygen so that the fish can continue to grow. Air is pumped into the water using rotary air pumps or air compressors. The air is then spread or diffused by thin-walled diffuser tubes or air stones that break the airstream into fine bubbles, figure 40-2. An alternative method is to use a paddle-wheel aerator, driven by a motor or tractor, that splashes water into the air for additional oxygen to be incorporated, figure 40-3.

FIGURE 40-1. Net pens or cages assist farmers in managing fish growing in natural ponds or lakes.

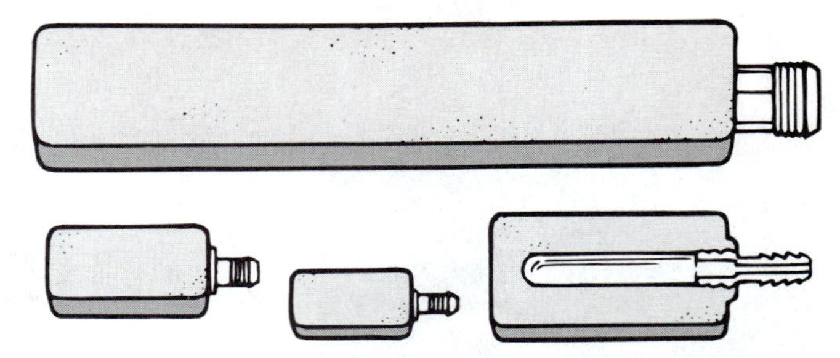

FIGURE 40-2. Ceramic air stones break the airstream into small bubbles.

As additional feed is consumed, fish produce more toxic waste products such as ammonia. Ammonia is a gas that is a by-product of decomposition. Wastes may be removed by microorganisms that live along the bottom and sides of the ponds or by adding new water into the system. Many ponds are flushed with 50 gallons of fresh water/acre/minute to assure good water quality which is free of excessive pollutants. Many growers have designed ponds to use the natural elevation drops to maintain water flows. However, most growers must use pumps to force water flow. Pumps include submersible pumps with submersible motors that can be totally emersed in water, centrifugal head pumps driven by auxiliary motors, and air-lift pumps that use compressed air to lift water.

FIGURE 40-3. Paddle-wheel aerators effectively mix air and water to increase the amount of dissolved oxygen. Portable aerators like the one above can be powered by tractor, PTO, mounted engine, or an electric motor. (*Courtesy of Parker,* Aquaculture Science, *Delmar Publishers, 1995*)

Fish can be grown in net cages or pens to assist in feeding and harvest, figure 40-4. However, because the fish are generally forced into greater density by this method, care must be taken to make sure water quality is maintained within the net cages.

Recirculating systems or closed aquaculture systems are now being developed in many parts of the world where good water quality or land resources are limited. Because of the increased cost of construction, these systems are maintained in controlled-environment structures, figure 40-5. Here temperature, light, and other requirements are managed to maximize fish growth. In this system, fish are grown indoors using tanks. The water flows through production tanks where the fish are held, then through waste treatment facilities or biological filters in which waste products are detoxified and recycled back to the production tanks, figure 40-6. Management includes maintaining the optimum temperature, dissolved oxygen levels, and feed rations for the animals or plants, figure 40-7.

Construction of Indoor Aquaculture Systems

The design of an indoor, recirculating aquaculture production model must include equipment and materials that can operate in constant contact with water and high humidity. Electrical units, switches, and motors must be installed with proper safety

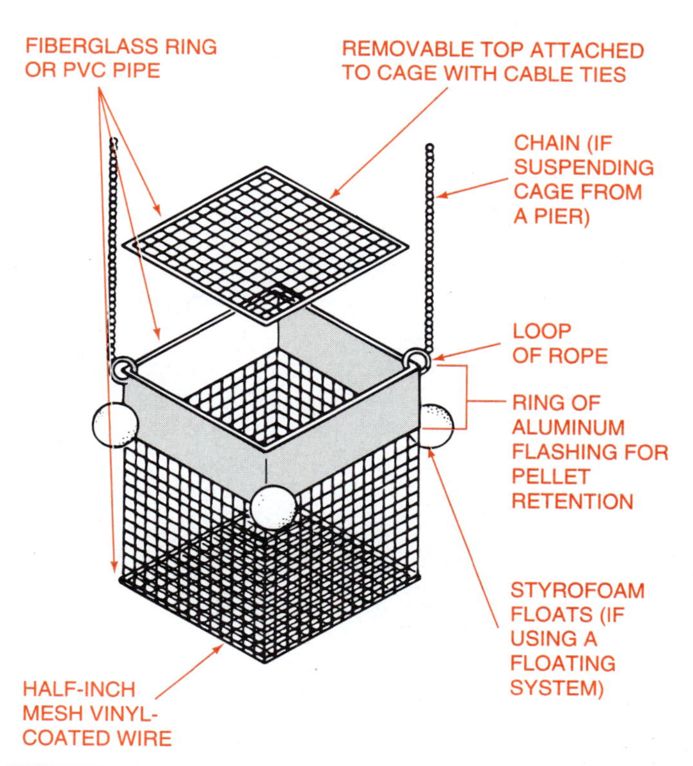

FIGURE 40-4. Net cages or pens are constructed to minimize water movement.

FIGURE 40-5. Greenhouse structures provide year-round production of fish using tank culture. (*Courtesy of Parker,* Aquaculture Science, *Delmar Publishers, 1995;* source: *Douglas Drennan and Dr. Ron Malone, Department of Civil and Environmental Engineering, Louisiana State University*)

FIGURE 40-6. Biological filters provide surface area for microorganisms to grow and detoxify waste products.

FIGURE 40-7. Water quality measurements must be taken every day to maintain optimum growing conditions.

FIGURE 40-8. Rectangular or oval raceways provide economical use of space.

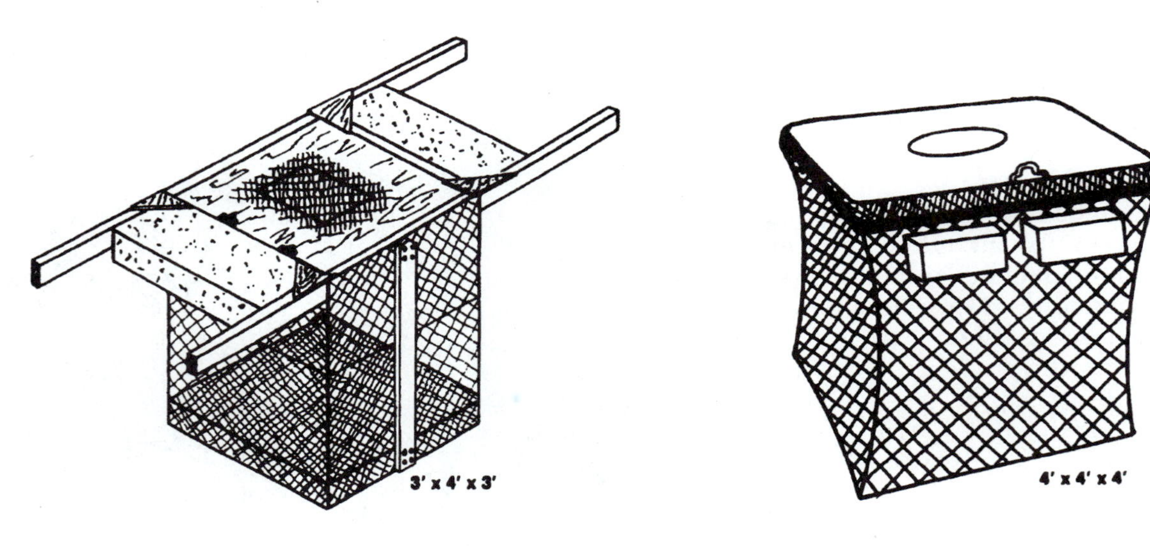

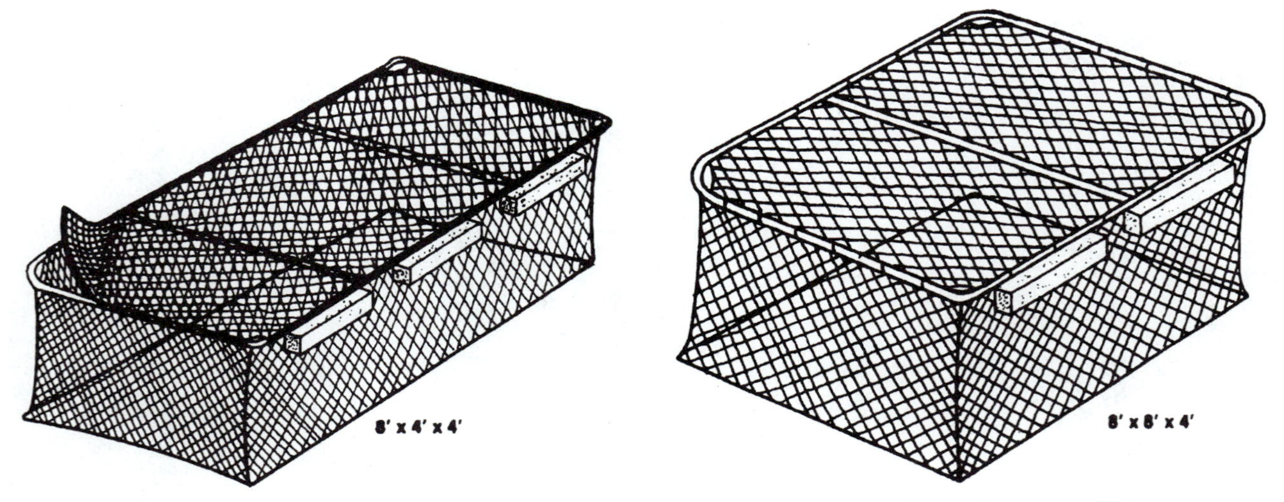

FIGURE 40-9. Some common cage designs. (*Courtesy of Parker,* Aquaculture Science, *Delmar Publishers, 1995; source: Michael P. Masser,* Cage Culture: Cage Construction and Placement, *Southern Regional Aquaculture Center, Publication No. 162*)

precautions. Major design components include the tank and material, air supply, waste treatment, and building.

Tanks may be constructed of fiberglass, polypropylene, epoxy-coated steel, aluminum, or plastic liners supported by wood or steel frames. Most structures average 4' to 6' deep and can be constructed above or below ground level. Round tanks are generally the most economical shape (cost/gallon). Rectangular tanks, figure 40-8, or raceways, figure 40-9, provide economical use of floor space but the corners encourage fighting among fish which stunts overall growth in some species. Use of liners provides the cheapest material cost for initial construction. However, liners must be replaced periodically and create difficulty in connecting equipment to the necessary plumbing fixtures.

The auxiliary equipment necessary to maintain the growing parameters of the fish is based on pounds of fish or density in the system. For example, a 6000 gallon system may support 1200 pounds of fish or more. Density = 0.2 lb of fish/gallon. Fish are typically fed bulk feed with automatic fish feeders that release preset lbs. of fish food during the day, or they may be hand fed, figure 40-10. With a feed conversion rate of 1.5 lbs. of feed per pound of fish, the system should produce 4800 lbs. of fish per year.

Dissolved oxygen may be maintained in water by rotary air pumps which bubble air into the water through a diffuser tube system. Air blowers are typically engineered to provide about 1 pound of oxygen infused into the water per pound of food added, figure 40-11.

A waste treatment component may settle the solid wastes produced by the fish by pumping water sufficient for about seven tank changes per day through a settling tank. The effluent passes through a biological filter which contains the growing surface covered with natural microorganisms that detoxify the wastes and ammonia before the water is recycled to the production tank. The biological filter consists of inert material such as fine gravel, pea gravel, coarse sand, plastic beads, or open-cell foam that provides large amounts of surface area on which the microorganisms grow. The size of the biological filter is determined by:

1. estimating the ammonia produced from the feed added to the system,
2. the surface area per cubic foot of the selected material, and
3. the necessary surface area of microorganisms to remove the ammonia to safe levels.

FIGURE 40-10. Fish feeders deliver feed on a timed schedule to increase the feed conversion ratio. (*Courtesy of Parker,* Aquaculture Science, *Delmar Publishers, 1995*)

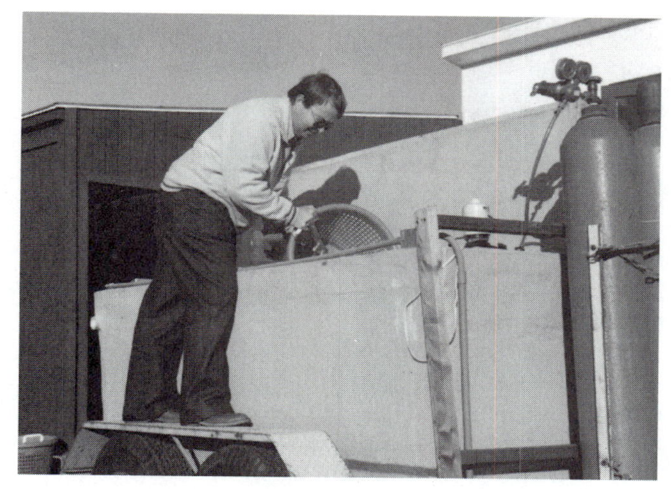

FIGURE 40-11. Cylinder tanks of oxygen are used to maintain dissolved oxygen levels during transport of fingerlings.

Plumbing fixtures and pipes are typically constructed from PVC Schedule 40 grade materials. Copper and iron may be released from other plumbing materials and can become toxic to fish raised in tanks made from these metals.

Climate control in the building is determined by local climate and insulation installed during construction. General building materials must withstand the corrosive humidity and ammonia produced by the system. Recommended materials include fiberglass sheets, lexan sheets, plastic, galvanized tin sheets with vapor barrier, or treated lumber. Care must be taken that no condensation on the building materials comes in contact with the fish water.

General maintenance for such systems requires maintaining the integrity of the water tanks, pump and blower maintenance, and routine flushing of the settling tank. Plans for constructing a closed system aquaculture model are provided in Appendix A, Project 52.

GREENHOUSES

Greenhouse Types and Orientation

Greenhouses are specialized agricultural structures that provide a controlled environment for the production of both plants and animals where either light or heat generated by sunlight create optimal growth, figure 40-12. The greenhouse structure simply supports a transparent or translucent covering that allows between 70 to 90 percent of the sunlight to enter. Such coverings include glass, plastic films, fiberglass sheets, and lexan materials.

Each type of covering has special advantages. Glass is the most durable but often the highest in initial cost. Fiberglass can last up to 15 years but must be cleaned and repaired about every 5 years. Lexan materials have been recently introduced with a 15-year maintenance-free period, but have been reported to be susceptible to hail or other mechanical damage. Plastic films (4 to 6 mm) are economical and are the most commonly used materials on quonset-style greenhouses, figure 40-13. The expected durability of commercial grades of plastic are 2 to 3 years before ultraviolet radiation destroys the integrity of the plastic. Most commercial operations use special ultra-violet-resistant grades of plastics that are more expensive, but last 4 to 5 years before needing replacement.

FIGURE 40-12. Greenhouse production is a major industry in most countries.

FIGURE 40-13. Plastic-covered, quonset-style greenhouses are economical for plant or animal production.

As sunlight enters the greenhouse, the light is absorbed by objects which then re-radiate the energy as heat waves, even on cloudy days. These low-energy waves cannot pass back through the coverings and accumulate inside. This buildup of heat is called the "greenhouse effect." During the summer months, heat must be removed by ventilation or the greenhouse must be shaded to limit the penetration of sunlight. Shading can be applied as a dilute white paint on the outside of the greenhouse that reflects sunlight, or as a woven black shadecloth, which blocks between 30 and 70 percent of the sunlight.

The support work in greenhouses is usually aluminum, galvanized steel pipe, treated wood, or plastic pipe. These members are spaced to provide the mechanical strength to support the weight of the maximum snowfall for the region. Because of the heat, high humidity, and salt from fertilizer solutions, all members must be resistant to a corrosive environment.

Since greenhouses may be low in cost per square foot to build, they are used in a variety of settings from plant production to aquaculture production to recreational activities.

Freestanding greenhouses can be truss-rafter or sawtooth, gothic, or quonset in style, figure 40-14. These shapes support the translucent coverings and provide several alternative methods for ventilation. Vent openings are placed in the top peaks of truss-rafter- or sawtooth-type greenhouses so that as the hot air rises, it is vented to the outside. The gothic- or quonset-style greenhouses are usually covered with single sheets of plastic that stretch from side to side. Vent openings and fans are located on opposite end walls. The fans draw air through the greenhouse and exhaust the heated air outside.

Lean-to or supported greenhouses are connected to existing structures with the translucent covering having a southern exposure. Vent openings and exhaust fans are located on the opposite end wall.

The location of greenhouses for plant growth must have an unrestricted exposure to full sunlight. Long greenhouses are oriented east/west along the long axis so shade from the end walls does not restrict growth. Lean-to greenhouses are located on the southern exposure of support buildings to provide maximum exposure to sunlight.

During the summer months, evaporative cooling pads consisting of porous material saturated with water are installed over the intake vents. As air is pulled through these moist pads, evaporation cools

FIGURE 40-14. Freestanding greenhouses are built following several different construction styles.

the air and helps maintain the optimum growing temperatures inside the greenhouse, figure 40-15.

Heat is provided in the greenhouses using steam, hot water, gas burners, infrared radiation fins, solar collectors, or electrical heaters, figure 40-16. The required output is calculated from:

1. climactic data of the region,
2. growing temperature maintained in the greenhouse, and
3. R factor and square footage of the greenhouse covering.

FIGURE 40-15. Evaporative cooling pads help control the buildup of heat inside the greenhouse during bright days. In this pad and fan cooling system, continuously wet pads of excelsior cool the air drawn through them into the greenhouses. (*Courtesy of Ingels,* Ornamental Horticulture Science, Operations & Management, 2E, *Delmar Publishers, 1995*)

Greenhouse Construction

Once the proper exposure and orientation is established, the site for a proposed greenhouse must be leveled, adequate drainage established, and the necessary water and utilities installed. Following standard construction practices, the building is squared and the foundation lines established.

FIGURE 40-16. Heaters are controlled by thermostats to provide temperature control during cold weather.

Most commercial greenhouses are constructed without a concrete foundation wall unless required by local building codes. Low-cost greenhouses may be made using PVC pipe or aluminum, steel, or wood bows. Plans for building small greenhouses with PVC pipe are shown in Appendix A, Project 54.

HYDROPONICS

Hydroponics is a production technology in plant science that attempts to maximize growth by stimulating nutrient uptake, figure 40-17. It literally means "water culture," a process in which the necessary plant nutrients are dissolved in water. Water with nutrients added, called a nutrient solution, contains all the required elements for plant growth. Therefore, it is not necessary for plants to be planted in soil. Water flow can be maintained so that the dissolved oxygen levels in the nutrient solution are optimal for plant growth.

Hydroponic systems are designed to maintain the nutrient solution in contact with the root systems. This requires a stock tank, pump system, irrigation manifold, plants, and drainage system.

All plumbing should be constructed of PVC plastic pipe and fiberglass because of the corrosive nutrient solution. Copper, steel, or lead pipes affect the nutrient composition of the solution or release toxic compounds. Submersible pumps or centrifugal head pumps used to circulate the nutrient solution must be manufactured to handle saltwater solutions. All electrical equipment and utilities must be installed with safety precautions outlined for wet environments.

FIGURE 40-17. Hydroponic crops are grown using a nutrient solution containing all the essential salts a plant absorbs through its roots.

FIGURE 40-18. Large plants can be grown with all the fertilizer supplied by the nutrient solution. (*Courtesy of Fiesta's Mart, Inc., Dick Scott*)

Hydroponic systems can be installed at ground level or at tabletop level. They operate by either maintaining a constant solution around the roots or by flooding the roots periodically with nutrient solution.

The plants are maintained in various types of bags, troughs, or tubes, figure 40-18. Large plants, like tomatoes and cucumbers, are planted directly into bags filled with vermiculite, which is an artificial soil medium made from mica. In this bag culture system, the bags are laid directly on the ground, punched with holes, and planted with two plants per bag. Trickle irrigation emitters are placed into the bags and the nutrient solution dripped continuously. A drainage system returns the effluent to the stock tank where it is pumped back into the irrigation system.

In a trough culture system, trenches are built and filed with vermiculite or crushed gravel. Plants are

FIGURE 40-19. Plants are grown without soil in tubes or gutters and fed a nutrient solution of dissolved salts.

placed into the media and a pump is used to flood the trough. When the pump turns off, the nutrient solution drains back to the stock tank. The pump is on a timed cycle that prevents the trough media from becoming dry.

Plants can also be held in small diameter gutters or tubes with nutrient solution circulating continuously, figure 40-19. This tube culture maintains a thin stream of nutrient solution in contact with the root system. A plan for constructing this system is provided in Appendix A, Project 53.

Hydroponic systems have been demonstrated to produce the highest production levels of any plant culture system. However, the grower must maintain an intensive inspection schedule that monitors both the irrigation system and the nutrient solution.

The optimum growing conditions of light and temperature are maintained by the environmental controls in the greenhouse structure.

STUDENT ACTIVITIES

1. Define the Terms to Know in this unit.
2. Describe how waste materials are detoxified in a pond or aquarium.
3. Examine a wholesale catalog for aquaculture farmers.
4. Research the permits or requirements for starting an aquaculture industry in your state.
5. Contact the local waste treatment facility and understand how waste products generated in your community are detoxified.
6. Determine what aquatic plants and animals are being grown in your state and locality.
7. Start an aquaculture library for your school.
8. Sketch and label the major structural components of a greenhouse.
9. Outline a detailed procedure for erecting a plastic-covered greenhouse.
10. Determine what vegetables in your local supermarket are being produced in hydroponic systems.
11. Construct models of aquaculture, greenhouse, and hydroponics systems.

SELF-EVALUATION

A. Multiple Choice. Select the best answer.

1. Aquaculture production includes
   a. growing fish for food
   b. growing plants for aquariums
   c. growing fish for recreational fishing
   d. all of the above
2. Farmers can increase the amount of fish they grow by
   a. lowering the water level
   b. decreasing the dissolved oxygen levels
   c. increasing the dissolved oxygen levels
   d. slowing the water flow
3. Fish wastes are detoxified by
   a. birds
   b. small fish
   c. microorganisms
   d. soil
4. Closed aquaculture systems
   a. are expensive to construct
   b. are easy to operate
   c. discharge large amounts of water
   d. are used in ponds
5. Greenhouses built with plastic film coverings
   a. are more expensive to construct than glass greenhouses

b. must be recovered with new plastic every 2 to 3 years
c. cannot handle heavy snow loads
d. cannot be exposed to the sun

6. The "greenhouse effect" occurs
a. on sunny summer days
b. on cloudy summer days
c. on sunny winter days
d. during all of the above days

7. A greenhouse should be built
a. on the southern exposure of a building
b. where the afternoon sun is blocked
c. in a low area where water collects
d. on the northern exposure of a building

8. Heat is generated in the greenhouse from
a. propane heaters
b. electrical heaters
c. sunlight
d. all of the above

9. Dissolved oxygen in a hydroponic nutrient solution is provided by
a. the stock tank
b. the submersible pump
c. the circulating water in contact with air
d. none of the above

10. Greenhouse structures are used for
a. plant production
b. animal production
c. aquaculture production
d. all of the above

B. Matching. Match the terms in column I with those in column II.

Column I
1. air-lift pump
2. hydroponics
3. biological filter
4. raceway
5. gothic style
6. shadecloth
7. wind brace
8. aquaculture
9. nutrient solution
10. ammonia

Column II
a. oval fish tank
b. waste product
c. freestanding greenhouse
d. triangle brace or greenhouse
e. production of fish
f. mixture of plant nutrients
g. "water culture"
h. water pump using bubbles
i. light reflecting cover
j. mixture of microorganisms

C. Completion. Fill in the blanks with the word or words that will make the following statements correct.

1. Closed aquaculture systems require farmers to add extra feed, dissolved oxygen, and remove _____ products in order to grow fish.
2. Copper plumbing is not used in aquaculture systems because high levels of copper are toxic to _____.
3. Dissolved oxygen is increased in water by mixing water with _____.
4. Diffuser tubes and _____ _____ are used to generate small bubbles that help mix the air in water.
5. The microorganisms in the _____ _____ or growing on the bottom of a pond detoxify waste products.
6. A controlled-environment structure, like a _____, is used to grow plants.
7. Glass, plastic films, and fiberglass sheets are _____ coverings that let sunlight enter the greenhouse.
8. The "greenhouse effect" causes the temperature in the greenhouse to _____.
9. A supported greenhouse is best built on the _____ side of the house.
10. A hydroponic system recirculates the _____ _____ that provides the root system with water, oxygen, and nutrients.

Appendices

A. PROJECTS

B. DATA TABLES

APPENDIX

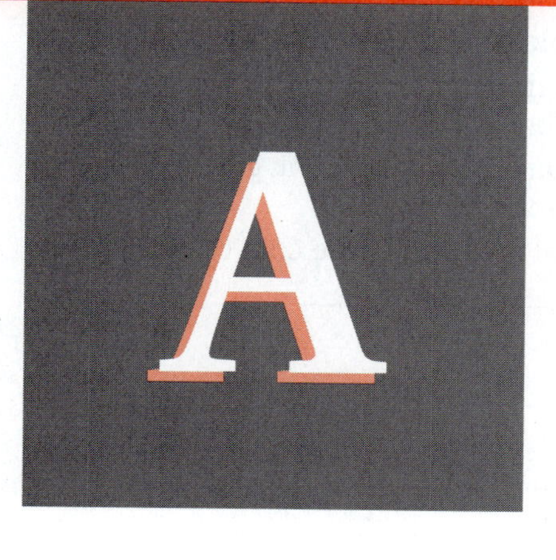

A Projects

PROJECT 1: CONCRETE FLOAT

PLAN FOR CONCRETE FLOAT

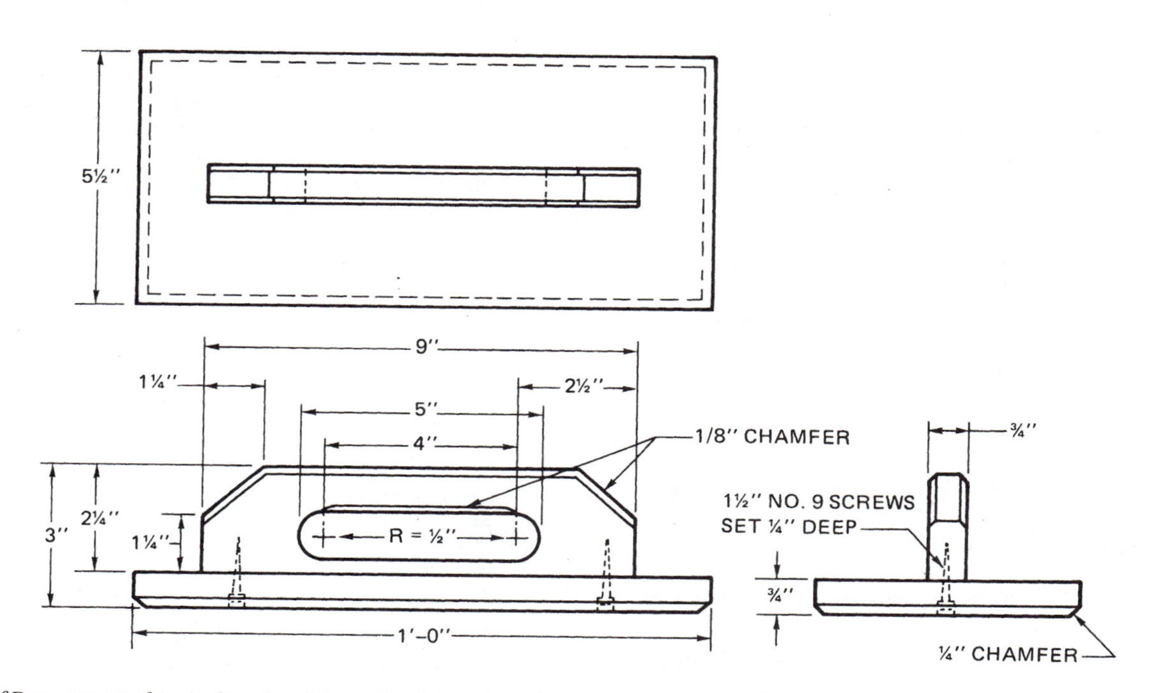

Courtesy of Department of Agricultural and Extension Education, The Pennsylvania State University, University Park, PA

BILL OF MATERIALS			
Materials Needed	**Quantity**	**Dimensions**	**Description**
Lumber (hard)	1 piece	$^3/_4$" × 5$^1/_2$" × 1'0"	Base
Lumber (hard)	1 piece	$^3/_4$" × 2$^1/_4$" × 9"	Handle

CONSTRUCTION PROCEDURE FOR CONCRETE FLOAT

1. Lay out and cut base to the dimensions given in the plan drawing.
2. Chamfer the base edges as shown in the plan.
3. Lay out and cut the outside dimensions of the handle.
4. Locate position of 2 one-inch holes in the handle.
5. Bore the 2 one-inch holes in the handle and saw out the piece between the holes.
6. Chamfer the edges of the handle as noted in the plan.
7. Drill pilot holes for the screws.
8. Center the handle on the base; locate and drill all screw holes.
9. Fasten handle to the base using glue and two $2^1/_2$" No. 9 wood screws.
10. Sand the float and apply one coat of linseed oil.

PROJECT 2: PUSH STICK

PLAN FOR PUSH STICK

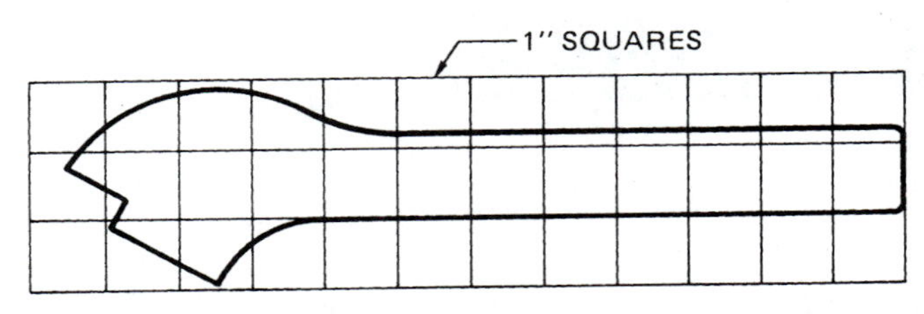

Courtesy of T. J. Wakeman and V. L. McCoy, The Farm Shop, *The Macmillan Company, 1960*

BILL OF MATERIALS

Materials Needed	Quantity	Dimensions	Description
Lumber			
Unsurfaced	1 piece	1" × 3" × 12"'	
or			
Surfaced	1 piece	$^3/_4$" × 3" × 12"	
or			
Plywood	1 piece	$^1/_2$" × 3" × 12"	

CONSTRUCTION PROCEDURE FOR PUSH STICK

1. Surface the 1" board to $^3/_4$" thickness with a hand plane or sander.
2. Lay off 1" squares on a piece of heavy paper or cardboard, and sketch an outline of the push stick, as shown in the drawing.
3. Cut out the pattern to use a template.
4. Lay out the push stick by marking around the template.
5. Cut out the push stick by sawing along the outside edges of the mark with a band saw, jig saw, coping saw, or compass saw.
6. Sand and finish.

PROJECT 3: MITER BOX

PLAN FOR MITER BOX

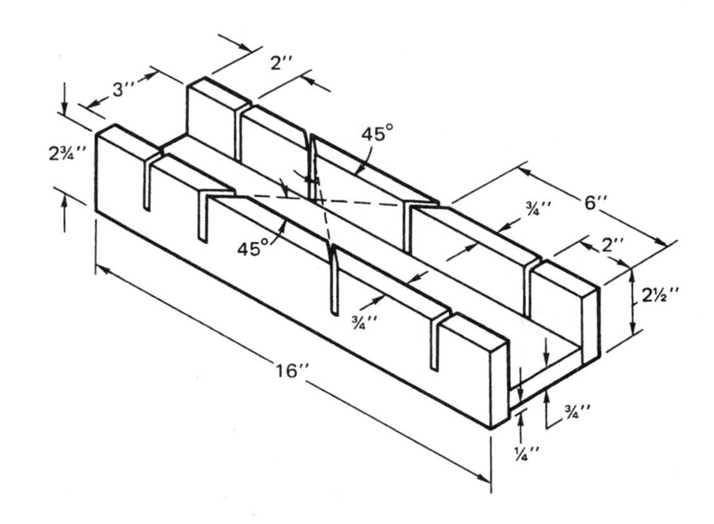

BILL OF MATERIALS

Materials Needed	Quantity	Dimensions	Description or Use
Lumber (hard)	1 piece	$^3/_4$" × 3" × 16"	Bottom
Lumber (hard)	1 piece	$^3/_4$" × 2$^3/_4$" × 16"	Front
Lumber (hard)	1 piece	$^3/_4$" × 2$^1/_2$" × 16"	Back
Wood Screws	6	No. 8 × 2" Flat Head	Attach

CONSTRUCTION PROCEDURE FOR MITER BOX

1. Cut a kiln-dried piece of $^3/_4$" hard wood to 49" long.
2. Joint one edge of the board.
3. Rip the board to 3 $^1/_{16}$" wide.
4. Joint the board to 3" wide.
5. Saw board into three pieces 16" long.
6. Rip pieces to $^1/_{16}$" wider than is shown in the materials list. Then set the jointer to remove $^1/_{16}$" and joint to exact width.
7. Lay off, bore, and countersink holes for wood screws. Space a hole 1" from each end on each side. Space the other hole in the middle on each side. Fasten side pieces to bottom piece with three wood screws on each side.

NOTE: The front side projects $^1/_4$" below the bottom. This forms a ledge to hold the box against the edge of a table when it is being used.

8. Lay off and saw a square cut in the side pieces 2" from the ends, with a miter box saw. Saw both side pieces at the same time.
9. Lay off and saw 45° cuts in the side pieces 6" from the right end, as shown in the drawing. These cuts serve as a guide for the saw when square or miter cuts are desired. In some instances, it may be desirable to cut other angles in the miter box to fit a particular situation. If so, the desired angle cut may be substituted for one of the 45° cuts.

PROJECT 4: GUN RACK

PLAN FOR GUN RACK

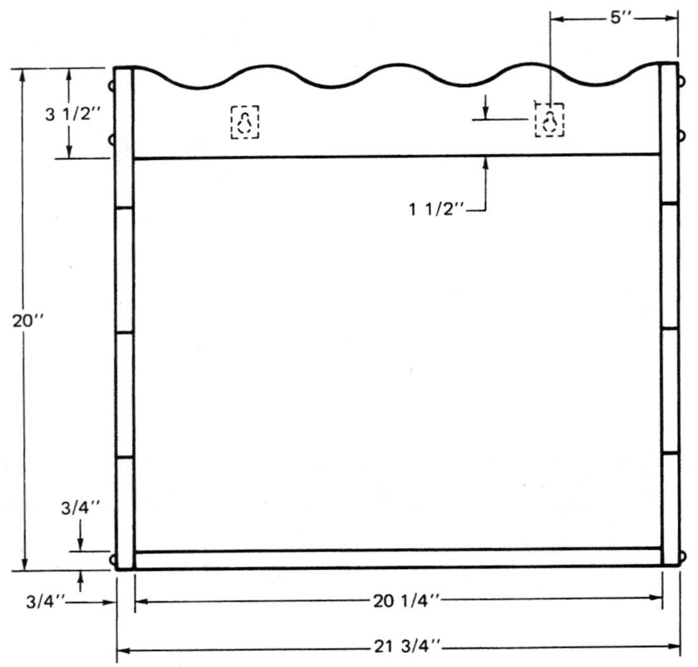

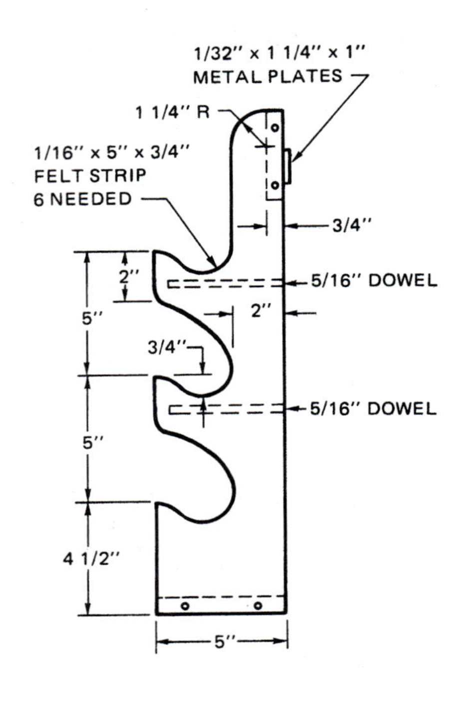

Courtesy of Agricultural Education and Agricultural Engineering, Virginia Polytechnic Institute and State University, Blacksburg, VA

BILL OF MATERIALS

Materials Needed	Quantity	Dimensions	Description
Lumber, surfaced	1 piece	$^3/_4$" × $3^1/_2$" × $20^1/_4$"	Back
Lumber, surfaced	2 pieces	$^3/_4$"' × 5" × 20"	Ends
Lumber, surfaced	1 piece	$^3/_4$" × 5" × $20^1/_4$"	Bottom
Lumber, surfaced	2 pieces	$^5/_{16}$" × $4^1/_4$"	Dowels
Screws	6 each	No. 8 × $1^1/_2$"	Round or flat head
Metal Plates	2 each	$^1/_{32}$" × $1^1/_4$" × 1"	Hangers

CONSTRUCTION PROCEDURE FOR GUN RACK

1. Cut a piece of heavy paper $3^1/_2$" × $20^1/_4$". Fold the paper in half end-to-end and cut a pattern for the back.
2. Draw the pattern topline onto a piece of stock $^3/_4$" × $3^1/_2$" × $20^1/_4$".
3. Use a piece of heavy paper 5" × 20" to cut a pattern for the sides.
4. Cut the topline of the back using a coping, jig, band, or bayonet saw.
5. Cut two 5" × 20" end pieces from $^3/_4$" stock.
6. Use the pattern to draw the gun supports on the end pieces.
7. Lay out the $^3/_4$" × $3^1/_2$" cut on each piece to receive the back.
8. Saw out the ends.
9. Cut a $^3/_4$" × 5" × $20^1/_4$" piece for the bottom.
10. Carefully drill holes and install dowel pins in the ends.
11. Carefully and thoroughly sand all surfaces, being careful not to round edges where the joints are made.
12. Glue and screw all parts together.
13. Apply a suitable finish.
14. Use stick-on felt strips in the curves where the guns are cradled.
15. Install hangers or plan to hang the rack by two nails or screws.

PROJECT 5: FLICKER/ WOODPECKER HOUSES

PLAN FOR FLICKER/WOODPECKER HOUSES

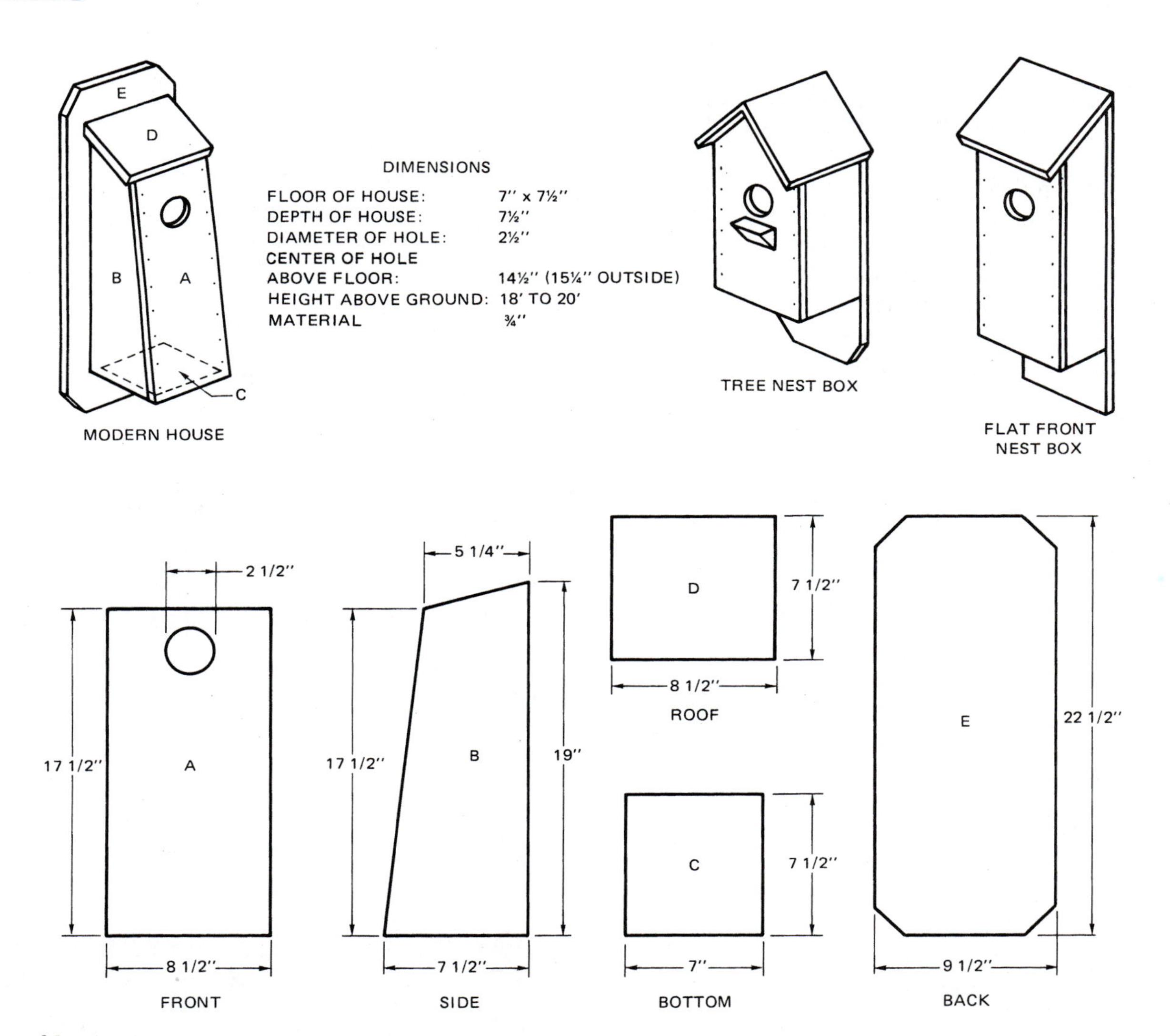

Courtesy of Cooperative Extension Service, Virginia Polytechnic Institute and State University, Blacksburg, VA

BILL OF MATERIALS (for modern house)

Materials Needed	Quantity	Dimensions	Description
Lumber (rough and rustic)	1	3/4" × 8 1/2" × 17 1/2"	Front (A)
"	2	3/4" × 7 1/2" × 19"	Sides (B)
"	1	3/4" × 7" × 7 1/2"	Bottom (C)
"	1	3/4" × 7 1/2" × 8 1/2"	Roof (D)
"	1	3/4" × 9 1/2" × 22 1/2"	Back (E)
Box nails or finishing nails		4d or 6d	

The flicker will nest readily in a well-placed birdhouse of the proper dimensions. Roughen the interior of the nest box to assist the young woodpeckers to reach the entrance hole. Cover the bottom with sawdust so that the mother bird can shape the nest for eggs and the young birds. Material may be $^{1}/_{2}$ inch instead of $^{3}/_{4}$ inch thick, in which case the width of the front and roof will be 8 inches.

CONSTRUCTION PROCEDURE FOR FLICKER/WOODPECKER HOUSES

1. Select pine, redwood, or other lumber that will withstand weather without paint. Rough lumber is generally preferred and will provide a rustic appearance. Planed lumber should be used if the house is to be painted for decorative purposes, but this is not recommended for most species except Martin houses.
2. Cut all parts from a $^{3}/_{4}$" × 9 $^{1}/_{2}$" × 8' board.
3. Cut the sides to the proper angle.
4. Bevel the top edge of the front to conform to the angle of the sides.
5. Bevel the upper edge of the roof to conform to the back and sides.
6. Nail the front, sides, and bottom together.
7. Nail the back to the sides and bottom. (Center the assembly on the back for a good appearance. Leave 1"at bottom and 2 $^{1}/_{2}$"at the top.)
8. Attach the roof with two hinges or nail it lightly so the roof is easily removed for periodic cleaning of the nest area.

PROJECT 6: NESTING AND DEN BOXES

HOUSES FOR SONGBIRDS

SHEET METAL RIDGE TACKED TO ONE SIDE OF ROOF

WREN BOXES

ONLY BOXES INTENDED FOR WRENS SHOULD BE HUNG FROM A LIMB; OTHERS SHOULD BE SECURELY FASTENED TO A POST OR TREE TRUNK

BLUEBIRD BOX WITH HINGED ROOF FOR EASY CLEANING

AIR DUCT

SCREENED VENTILATOR HOLE

GUIDE BLOCKS

VENTILATION HOLE 5/16"

"ADD-A-STORY" MARTIN HOUSE (ADDITIONAL SECTIONS ADDED AS COLONY GROWS)

Courtesy of Maryland Department of Natural Resources, Public Information Services

PLAN FOR NESTING AND DEN BOXES

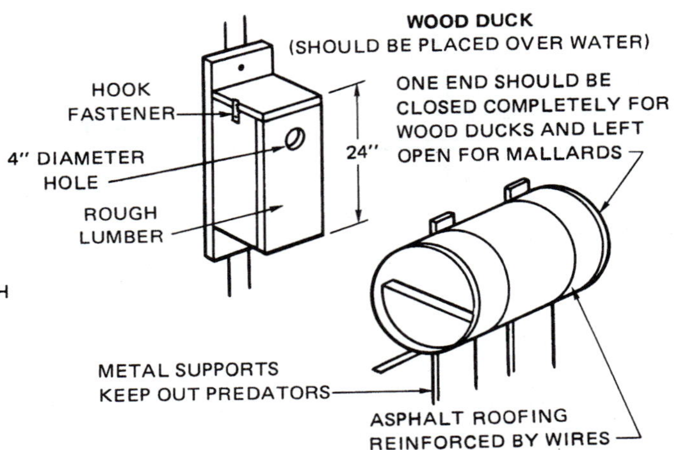

BILL OF MATERIALS

Bills of materials are not provided with these nesting and den boxes due to the variety of types pictured. Nesting and den boxes should be planned from the information provided. An appropriate bill of materials may then be developed for the plan.

CONSTRUCTION PROCEDURE FOR NESTING AND DEN BOXES

Construction procedures will vary with the nesting or den box being constructed. However, the procedure outlined for Project 5, Flicker/Woodpecker Houses, may be helpful.

DIMENSIONS FOR EACH SPECIES

Species	Diameter of Entrance	Floor of Cavity	Depth of Cavity	Entrance Above Floor
Bluebird	$1^1/_2$"	5" × 5"	8"	6"
Chickadee	$1^1/_8$"	4" × 4"	8"-10"	6"-8"
Titmouse	$1^1/_4$"	4" × 4"	8"-10"	6"-8"
Nuthatches	$1^1/_4$"	4" × 4"	8"-10"	6"-8"
House Wren	$^7/_8$"	4" × 4"	6"-8"	1"-6"
Carolina Wren	$1^1/_8$"	4" × 4"	6"-8"	1"-6"
Crested Flycatcher	2"	6" × 6"	8"-10"	6"-8"
Flicker	$2^1/_2$"	7" × $7^1/_2$"	16"-18"	14"-16"
Purple Martin	$2^1/_2$"	6" × 6"	6"	1"
Tree Swallow	$1^1/_2$"	5" × 5"	6"	1"-5"
Sparrow Hawk	3"	8" × 8"	12"-15"	9"-12"
Barn Owl	3"	8" × 8"	12"-15"	9"-12"

PROJECT 7: SMALL ENGINE STAND—WOOD

PLAN FOR SMALL ENGINE STAND—WOOD

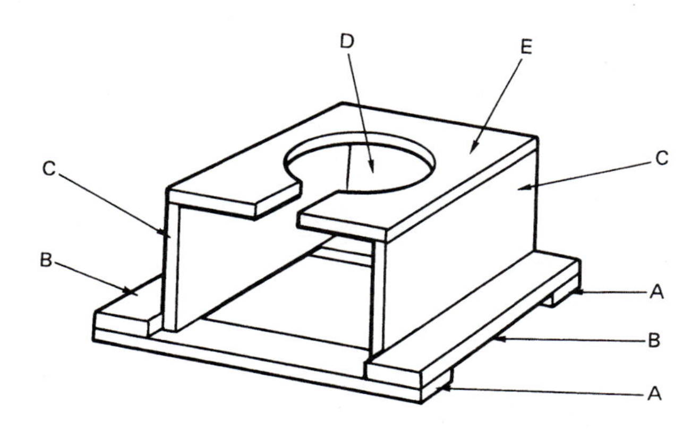

Courtesy of Department of Agricultural and Extension Education, The Pennsylvania State University, University Park, PA

CONSTRUCTION PROCEDURE FOR SMALL ENGINE STAND—WOOD

1. All wooden joints should be made with glue and nails, or glue and screws.
2. Use pieces "A" and "B" to construct the frame base as shown in the drawing, making sure the base is square.
3. Fasten the two "C" pieces into position. One end of each "C" piece should be flush to the rear edge of the base frame; the front ends will fall short of the front edge of the base frame by $^1/_2$ inch.
4. Place the "D" piece or back of the stand between the sides of the stand and flush with the back ends.
5. Determine the center of the top plywood piece "E" and draw a 7-inch diameter circle around it. Cut a 7-inch hole and a 2-inch front slot (centered on the front) before fastening the top to the stand.

BILL OF MATERIALS

Materials Needed	Quantity	Dimensions	Description
Lumber, No. 2 White Pine, surfaced	2 pieces	$^3/_4$" × $2^1/_4$" × 16"	—Front and back of base frame (A)
Lumber, No. 2 White Pine, surfaced	2 pieces	$^3/_4$" × $2^1/_4$" × 12"	—Sides of base frame (B)
Lumber, No. 2 White Pine, surfaced	2 pieces	$^3/_4$" × $5^1/_2$" × $11^1/_2$"	—Sides of stand (C)
Lumber, No. 2 White Pine, surfaced	1 piece	$^3/_4$" × $5^1/_2$" × 10"	—Back of stand (D)
Plywood, Good one side	1 piece	$^3/_4$" × $11^1/_2$" × $11^1/_2$"	—Top of stand (E)

6. For the front of the stand a *safety door,* not shown, should be constructed from 1/2-inch material. It may be fitted to the front and secured in place by small hooks and eyes or hinges.

7. All wooden parts should be painted with an enamel paint. Recommended colors are grey, green, or blue.

PROJECT 8: NAIL BOX

PLAN FOR NAIL BOX

CONSTRUCTION PROCEDURE FOR NAIL BOX

1. Surface the 1" × 6" × 10'-0" board to 3/4" thickness with a thickness planer or a hand plane.

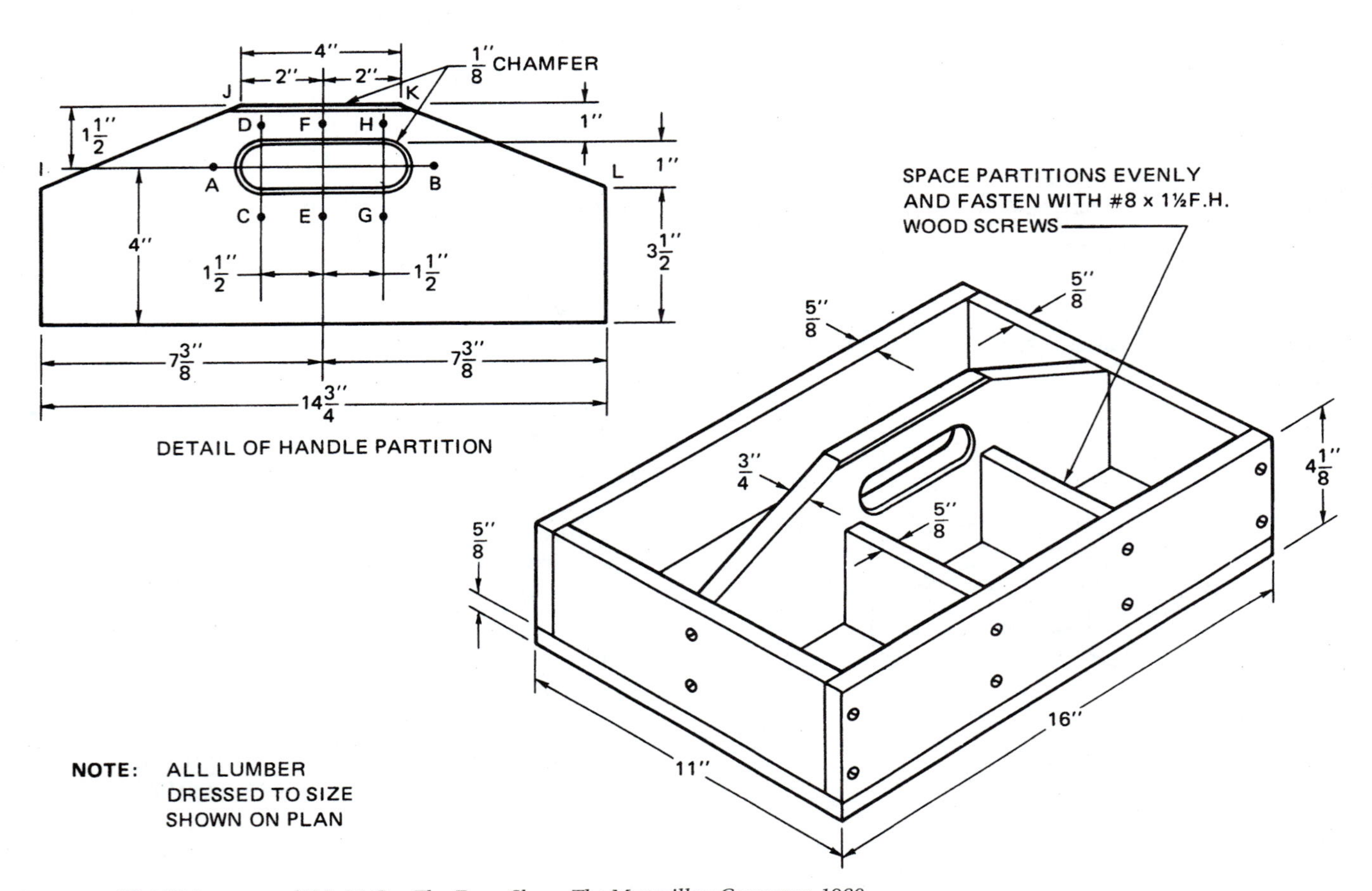

Courtesy of T. J. Wakeman and V. L. McCoy, The Farm Shop, *The Macmillan Company, 1960*

BILL OF MATERIALS

Materials Needed	Quantity	Dimensions	Description or Use
Lumber, unsurfaced*	1 piece	1" × 6" × 10'	
Lumber, surfaced	1 piece	3/4" × 5 1/2" × 14 3/4"	Handle partition
Lumber, surfaced	2 pieces	5/8" × 5 1/2" × 16"	Bottom
Lumber, surfaced	2 pieces	5/8" × 3 1/2" × 16"	Sides
Lumber, surfaced	2 pieces	5/8" × 3 1/2" × 9 3/4"	Ends
Lumber, surfaced	2 pieces	5/8" × 3 1/2" × 4 1/2"	Partitions
Wood screws	34	1 1/2"	No. 8 flat head

*All surfaced pieces are to be cut from this board.

2. Cut one piece $14^{3}/_{4}$" long for the handle partition.
3. Surface the remainder of the board to $^{5}/_{8}$" thickness, and cut bottom, side, end, and partition pieces to size, as shown in the bill of materials.
4. Lay off, bore, and countersink holes for screws in the side pieces. Fasten the sides to the ends with the two screws at each joint starting $^{3}/_{4}$" from each edge.
5. Fasten the bottom to the side and end pieces. Use three screws in the side pieces, two spaced 1" from each end and one in the center. Use four screws on each end; place one screw 2" from each corner and one $4^{1}/_{2}$" from each corner.
6. Cut the handle partition to size as shown in the bill of materials. (Check distance between end pieces before cutting handle partition to exact length.)
7. Draw lines AB, EF, CD, GH, as shown in the detail of the handle partition in the drawing.
8. Using a 1" auger, bore holes where lines CD and GH cross line AB. Bore until the feed screw starts through the opposite side of the board; then complete the hole from the opposite side.
9. Locate points I, J, K, and L.
10. Mark and cut lines IJ and KL.
11. Slightly round off the corners of the slot and top of handle with a wood rasp, and sand smooth.
12. Space the partitions evenly on one side of the handle, fastening each with two screws.

NOTE: Partitions are left out of one side so that it may be used for small tools. If the box is to be used for nails or bolts only, partitions should be inserted in the other side.

13. Place handle and partitions in box. Square, drill holes for screws, and fasten partitions to side with two screws at each joint. Fasten handle to ends with two screws at each end.
14. Sand and finish.

PROJECT 9: TOOL BOX

PLAN FOR TOOL BOX

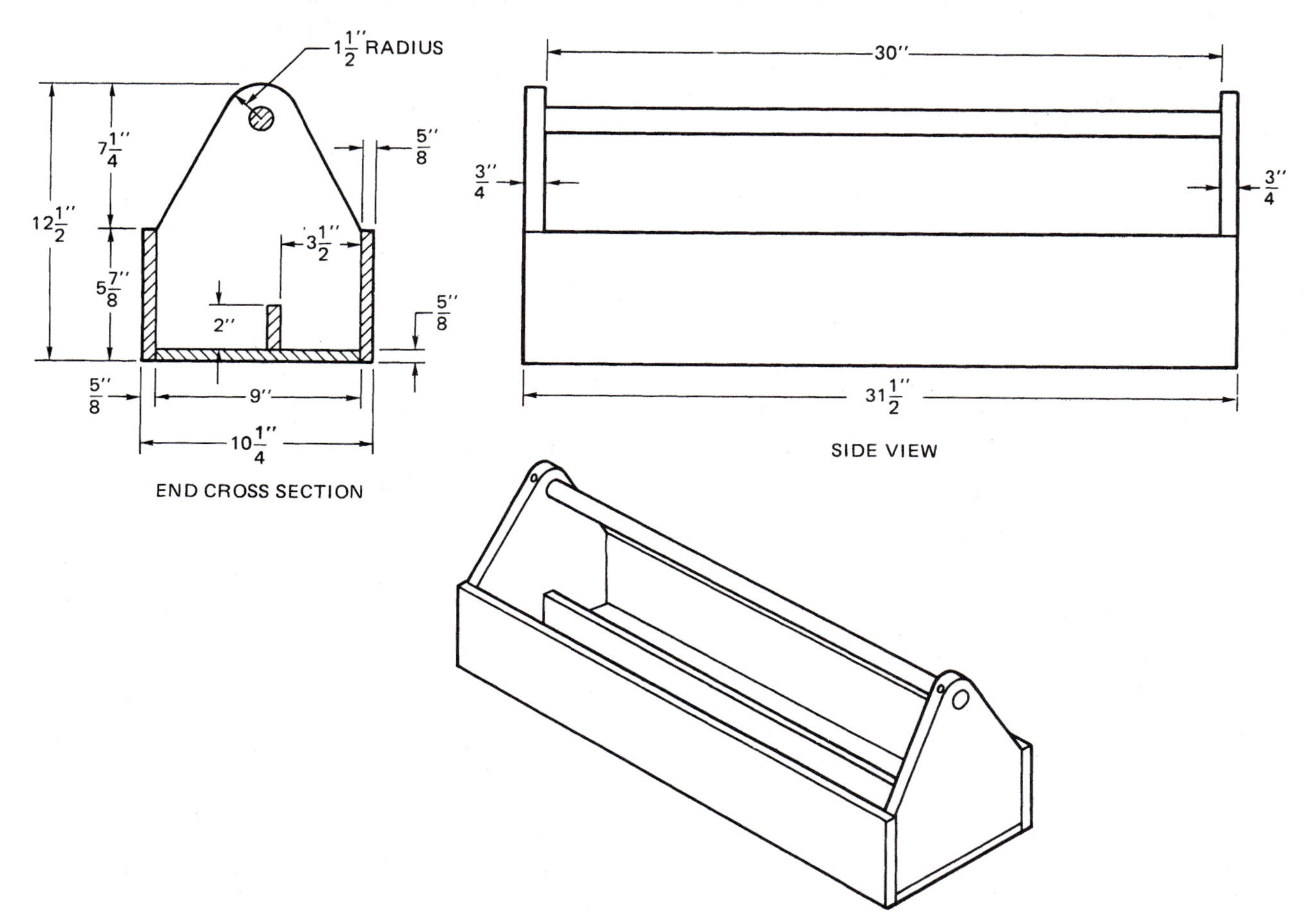

Courtesy of T. J. Wakeman and V. L. McCoy, The Farm Shop, *The Macmillan Company, 1960*

BILL OF MATERIALS

Materials Needed	Quantity	Dimensions	Description or Use
Lumber, unsurfaced*	1 piece	1" × 12" × 8'	
Lumber, surfaced	2 pieces	$^3/_4$" × 9" × 12$^1/_2$"	Ends
Lumber, surfaced	1 piece	$^5/_8$" × 9" × 31$^1/_2$"	Bottom
Lumber, surfaced	1 piece	$^5/_8$" × 2" × 30"	Partition
Lumber, surfaced	2 pieces	$^5/_8$" × 5$^7/_8$" × 31$^1/_2$"	Side boards
Round wood stock	1 piece	1" diameter × 31$^1/_2$"	Handle (cut from broom handle or dowel)
Box of finishing nails	$^1/_4$ pound	6d	
Wood screws	2	2"	No. 6 flat head

*All surfaced pieces are to be cut from this board.

CONSTRUCTION PROCEDURE FOR TOOL BOX

1. Surface the 1" board to $^3/_4$" thickness with a thickness planer.
2. Cut ends to size; shape and bore holes for handle, as shown in end cross section. Use an auger bit the same size as the ends of the handle.
3. Surface remainder of board to $^5/_8$" thickness with a thickness planer.
4. Cut partition, bottom, and side boards to size, as shown in the drawing.
5. Cut handle to length, and place in holes in end pieces. Glue and nail bottom, side boards, and partition in place.
6. Bore holes in end pieces and handle for the wood screws. Countersink and place screws to hold the handle in place.
7. Sand and finish.

NOTE: $^3/_4$" lumber may be substituted where $^5/_8$" material is specified. Flat head wood screws may be used in place of nails.

PROJECT 10: SADDLE RACK

PLAN FOR SADDLE RACK

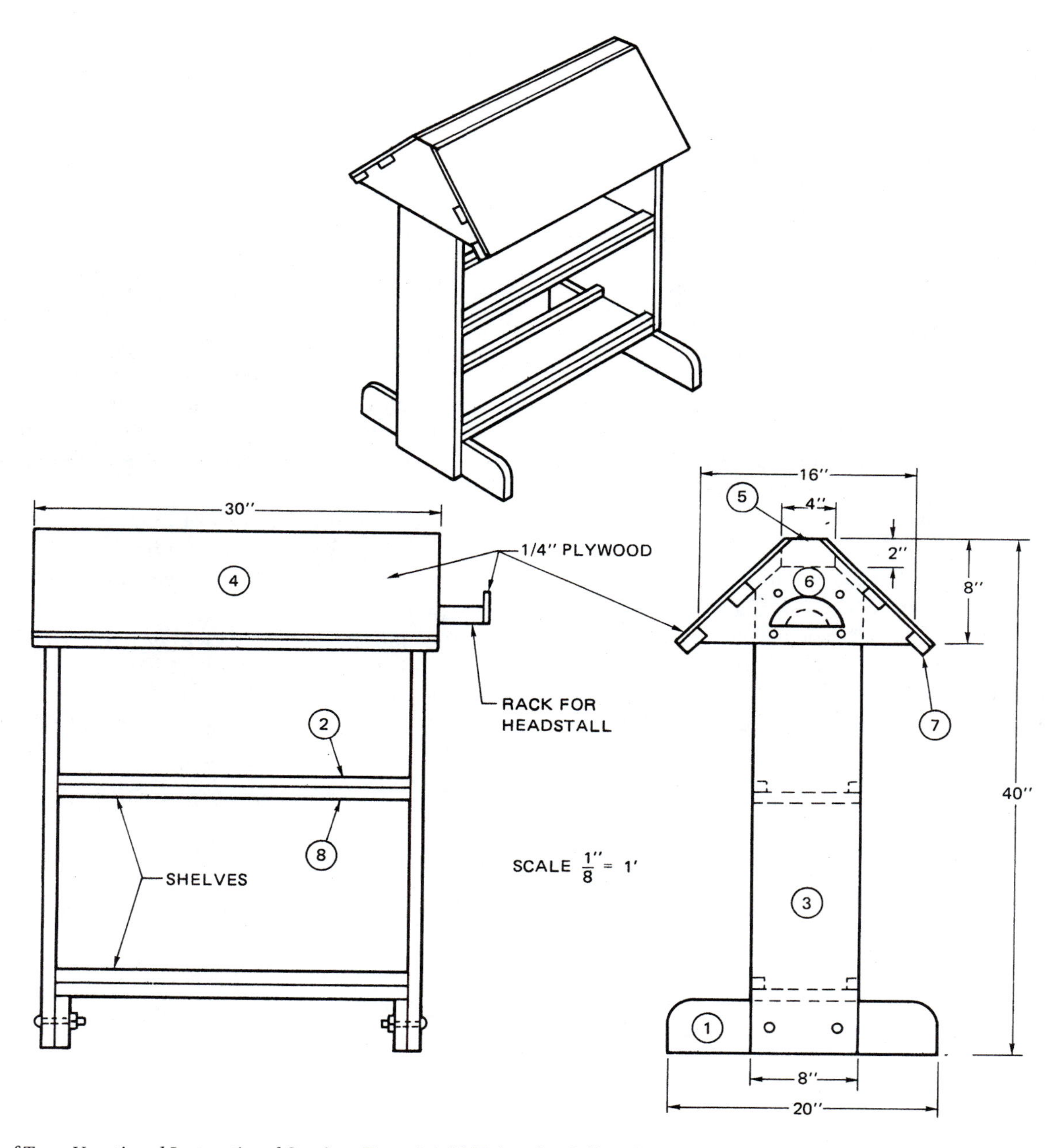

Courtesy of Texas Vocational Instructional Services, Texas A & M University, College Station, TX

BILL OF MATERIALS

Materials Needed	Quantity	Dimensions	Description
Lumber	2 pieces	1" × 4" × 20"	Base (1)
Lumber	4 pieces	1" × 1" × 26"	Cleats (2)
Lumber	2 pieces	1" × 8" × 40"	End Supports (3)
Lumber	2 pieces	1/4" × 12" × 30"	Top (4)
Lumber	1 piece	2" × 4" × 30"	Center (Ridge) (5)
Lumber	2 pieces	1" × 8" × 16"	Ends (6)
Lumber	4 pieces	1" × 2" × 30"	Ribs (7)
Lumber	2 pieces	1" × 8" × 26"	Shelves (8)

Note: 3/4" stock may be used in place of 1" material.

CONSTRUCTION PROCEDURE FOR SADDLE RACK

1. Cut two pieces 1" × 8" × 16" for the top ends. Set the boards on edge and mark the center on the top edge. Mark locations 1" out in both directions from the centers. Draw a line from each of these points to the lower corners on each end of the boards. Saw the ends to these lines.
2. Cut the ridge to length and joint two corners so its shape matches the top of the end pieces.
3. Cut all other pieces to size as listed in the bill of materials, except the shelves and cleats. These should be cut about 1" longer than indicated.
4. Use glue and nails or screws to assemble the top.
5. Use glue and 1/4" or 5/16" carriage bolts with washers to attach the sides to the base pieces.
6. Use glue and eight 1/4" carriage bolts with washers to attach the sides to the top. Be very careful to attach the ends so the unit is straight and sits squarely on its feet on a flat surface.
7. Attach the cleats to the shelf pieces with glue and nails.
8. Measure the distance from inside to inside of the two ends at the top of the rack. Cut the length of the shelf and cleat assemblies to this dimension.
9. Install the shelf and cleat assemblies with glue and 10d finishing nails or No. 8 × 2 1/2" wood screws.
10. Smooth all surfaces and apply a suitable finish.

PROJECT 11: SAW HORSE—LIGHTWEIGHT

PLAN FOR SAW HORSE—LIGHTWEIGHT

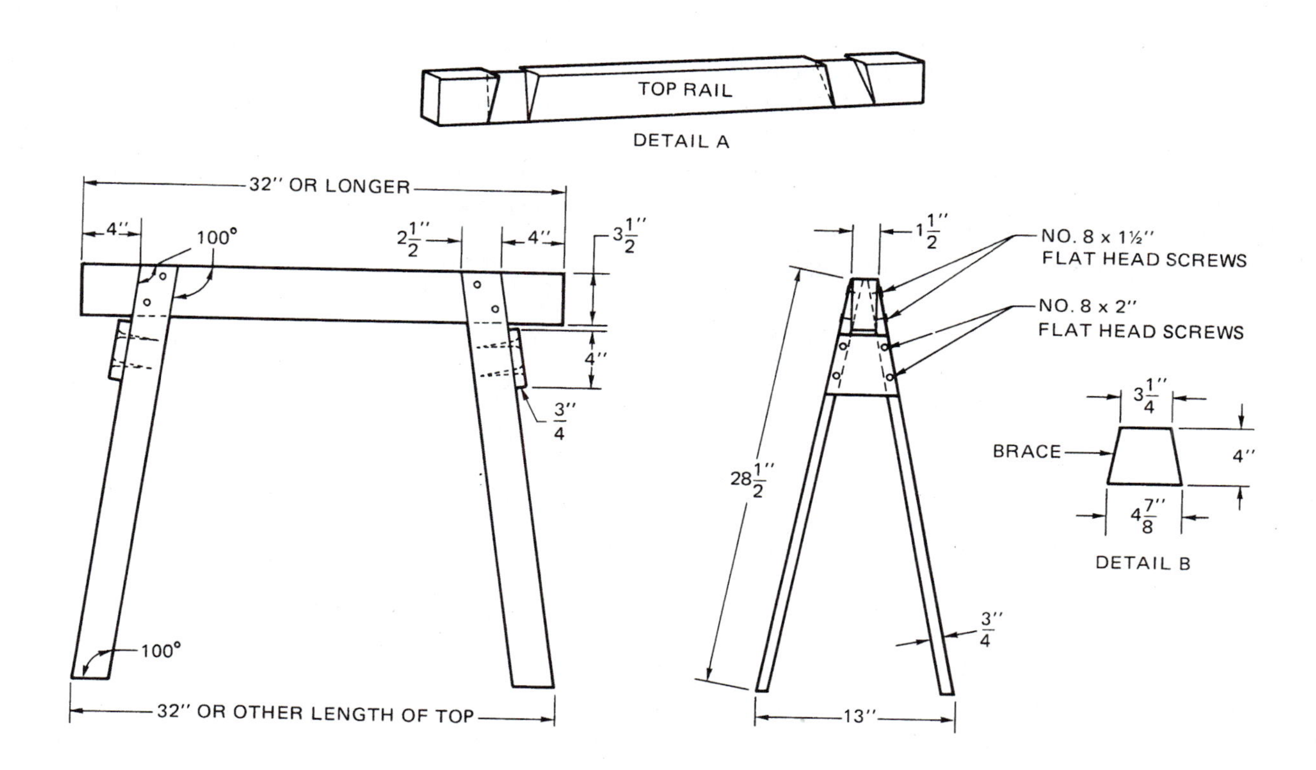

BILL OF MATERIALS

Materials Needed	Quantity	Dimensions	Description or Use
Lumber, surfaced	1 piece	$1^1/_2$" × $3^1/_2$" × 32"	Top rail
*Lumber, surfaced	4 pieces	$^3/_4$" × $2^1/_2$" × 29"	Legs
Lumber, surfaced	2 pieces	$^3/_4$" × 4" × 5"	Braces
Wood screws	8	No. 8 × $1^1/_2$"	Flat head, legs
Wood screws	8	No. 8 × 2"	Flat head, braces

*Use clear lumber, preferably hardwood.

CONSTRUCTION PROCEDURE FOR SAW HORSE—LIGHTWEIGHT

1. Cut the top rail from a surfaced 2 × 4.
2. Cut one leg from clear lumber so the finished size is $^3/_4$" × $2^1/_2$" × $28^1/_2$" with ends cut at 100° angles as shown in the drawing. Tough wood such as oak or yellow pine is recommended.
3. Use the first leg as a pattern to cut the other three legs.
4. Mark "top" on one of the $1^1/_2$" edges of the top rail.
5. Mark the top of the rail 4" in from each end and square a line across the $1^1/_2$" surface at each mark. (See Detail A for illustration of steps 5–8.)
6. Use a T-bevel to draw a line at 100° from each end of each line down the sides of the rail.

7. Lay the leg pattern against each line and draw a second line parallel to each. The lines for all four legs are now drawn along the sides of the top rail. These should be such that when legs are attached, the outside toe of each leg will be in line with the end of the top rail.
8. Draw a line down the center of the top of the rail between each pair of lines.
9. Use a hand saw to make multiple cuts inside the sets of parallel lines as the first step in cutting a dado for each leg. The depth of the cuts is 3/4" on the top of the rail and 0" on the bottom of the rail.
10. Use a wood chisel to cut out the dados. Smooth the bottom of each dado with a coarse file.
11. Drill and countersink holes in the legs for wood screws.
12. Cut out two leg braces 3/4" × 4" wide by 3 1/4" and 4 7/8". See Detail B.
13. Drill and countersink for the wood screws.
14. Assemble the saw horse with glue and screws.
15. If the saw horse rocks when placed on a flat surface, place a thin block of wood on the floor against the leg. Draw a line at the top of the block and one on each side and on each edge of the leg. Repeat the process on the other three legs. Be careful that the saw horse does not rock during this procedure. Resaw the legs on the lines so established.
16. Smooth the ends of the legs on top of the rail.
17. Apply a suitable finish.

PROJECT 12: STANDARD SAW HORSE

PLAN FOR STANDARD SAW HORSE

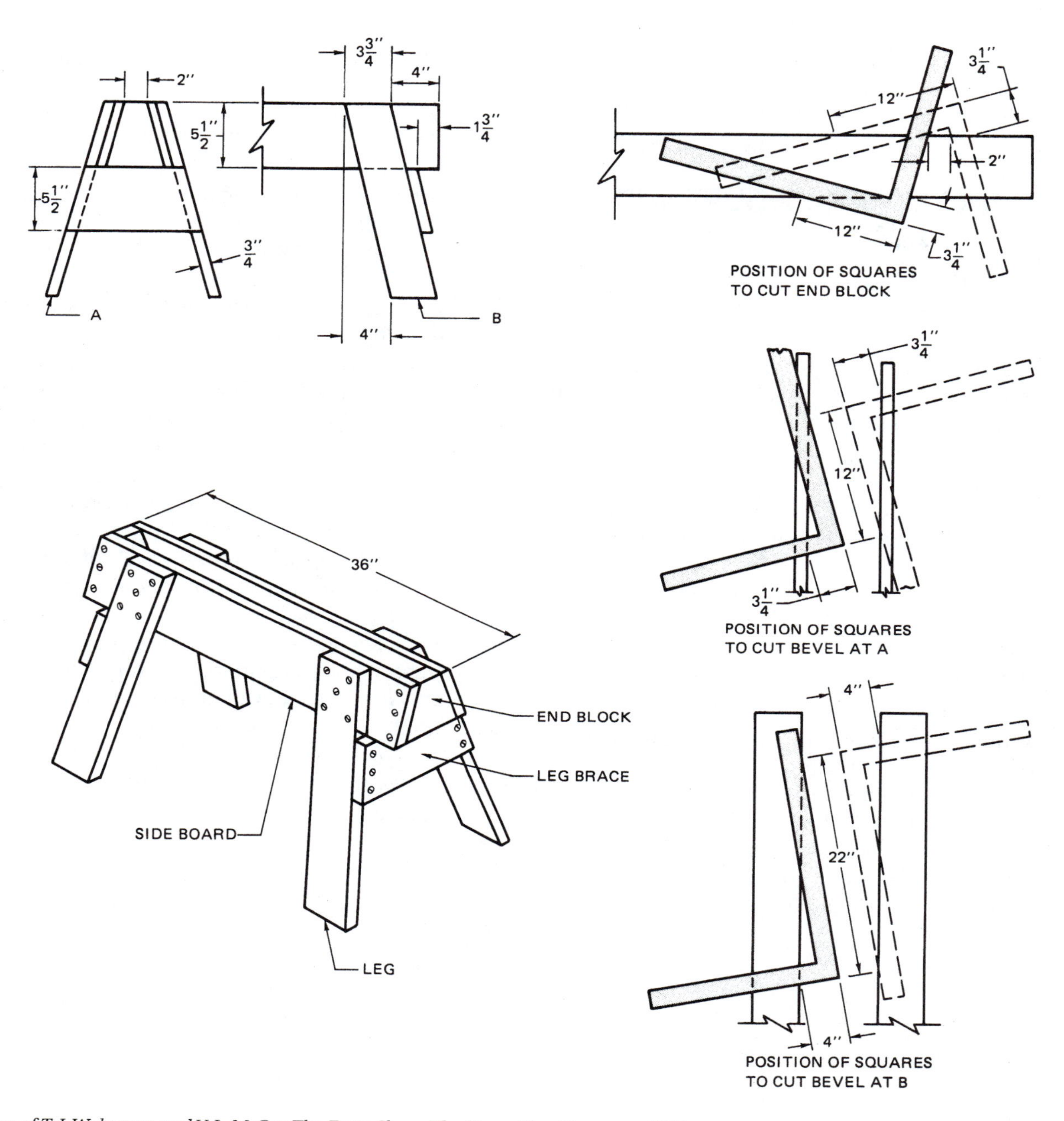

Courtesy of T. J. Wakeman and V. L. McCoy, The Farm Shop, *The Macmillan Company, 1960*

BILL OF MATERIALS

Materials Needed	Quantity	Dimensions	Description or Use
Lumber, surfaced	2 pieces	$^3/_4$" × $5^1/_2$" × 36"	Side boards
Lumber, surfaced	2 pieces	$^3/_4$" × $5^1/_2$" × 14"	Leg braces
Lumber, surfaced	2 pieces	$1^3/_4$" × $5^1/_2$" × 6"	End blocks
Lumber, surfaced	4 pieces	$^3/_4$" × $3^3/_4$" × 24"	Legs
Wood screws	44	$1^1/_2$"	No. 12 flat head

Note: 1 inch lumber may be used when $^3/_4$ inch material is specified.

CONSTRUCTION PROCEDURE FOR STANDARD SAW HORSE

1. Surface 1" material for side boards, leg braces, and legs to $^3/_4$" thickness using a thickness planer.
2. Cut side boards to length, as shown in the drawing.
3. Rip side boards to $5^1/_2$" width.
4. Set T-bevel at the same degree as the end blocks (see drawing), and use a template for setting the jointer fence at the proper angle for beveling the top and bottom edges of the side boards.
5. Make the top and bottom bevels of the side boards parallel by jointing each bevel from the same end.
6. Surface the end blocks to $1^3/_4$" thickness.
7. Joint one edge of each end block.
8. Lay out and mark the end blocks, as shown in the drawing. Mark the length of slope on each end block the same as the width of the side boards; then joint the bottom edge of each block to this width.
9. Fasten side boards to the end blocks with glue and No. 12 × $1^1/_2$" flat head wood screws, as shown in the drawing. Bore holes for the wood screws in the side boards and end blocks. Countersink the side boards for the screw heads.
10. Lay out and cut the legs to length, as shown in detailed drawing.
11. Bore holes, and fasten legs to side with glue and No. 12 × $1^1/_2$" flat head wood screws, as shown.
12. Set the T-bevel to the same degree as the side cuts of the legs (4" and 22"on the framing square), and use a template for setting the jointer fence at the proper angle for beveling the top and bottom of the leg braces.
13. Make the top and bottom bevels of the leg braces parallel by jointing each bevel from the same end.
14. Mark leg braces to length by holding the braces tight in place and marking along the outside edges of the legs. Cut at these marks.
15. Bore holes, and fasten the leg braces to legs with glue and No. 12 × $1^1/_2$" flat head wood screws as shown in the drawing.
16. Sand and finish.

PROJECT 13: WORKBENCH

PLAN FOR WORKBENCH

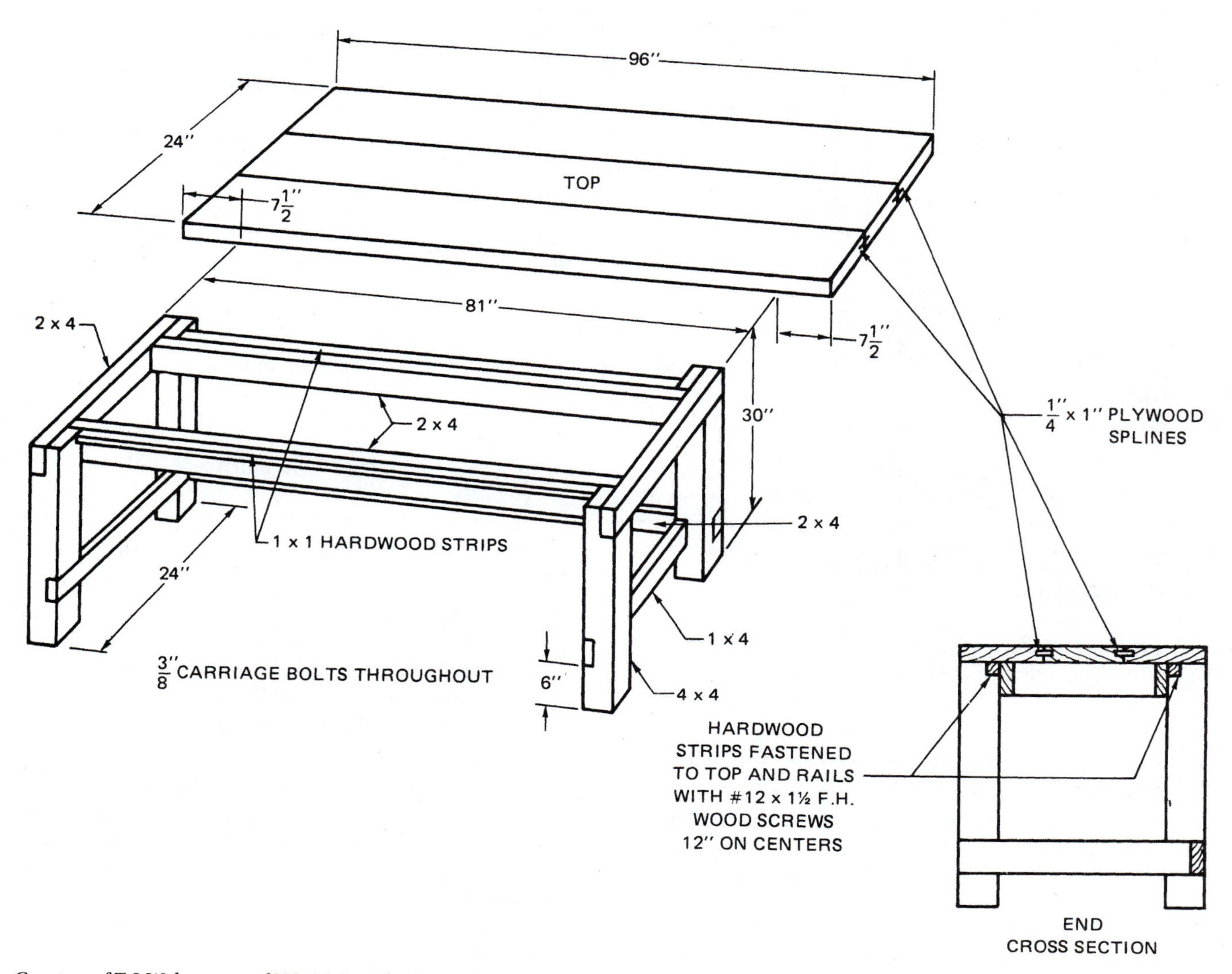

Courtesy of T. J. Wakeman and V. L. McCoy, The Farm Shop, *The Macmillan Company, 1960*

BILL OF MATERIALS

Materials Needed	Quantity	Dimensions	Description or Use
Lumber, surfaced	4 pieces	$3^3/_4" \times 3^3/_4" \times 30"$	Legs
Lumber, surfaced	2 pieces	$1^3/_4" \times 3^3/_4" \times 6'5^1/_2"$	Side rails
Lumber, surfaced	2 pieces	$1^3/_4" \times 3^3/_4" \times 24"$	End rails
Lumber, surfaced	1 piece	$1^3/_4" \times 3^3/_4" \times 6'9"$	Side braces
Lumber, surfaced	2 pieces	$^3/_4" \times 3^3/_4" \times 22^1/_4"$	End braces
Hardwood, surfaced	2 pieces	$^3/_4" \times ^3/_4" \times 6'1^1/_2"$	Strips to fasten frame to top
Hardwood, surfaced	2 pieces	$1^3/_4" \times 7^1/_2" \times 8'0"$	Top, outside pieces
Hardwood, surfaced	1 piece	$1^3/_4" \times 9^1/_2" \times 8'0"$	Top, middle piece
Plywood, 1/4" thick	2 pieces	$^1/_4" \times 1" \times 8'0"$	Top, splines
Carriage bolts	10	$^3/_8" \times 4^1/_2"$	
Carriage bolts	4	$^3/_8"' \times 6"$	
Washers	14	$^3/_8"$	Flat
Wood screws	28	$1^1/_2"$	No. 12 flat head
Wood glue	3 ounces		
Paint	1 quart		For frame
Thinner	$^1/_2$ pint		For frame
Floor or gym seal	1 pint		For top

Note: Commercially surfaced lumber will be $1^1/_2" \times 3^1/_2"$ for 2 × 4s; $1^1/_2" \times 5^1/_2"$ for 2 × 6s; and $3^1/_2"' \times 3^1/_2"$ for 4 × 4s. Use shorter bolts if commercial lumber is used.

CONSTRUCTION PROCEDURE FOR WORKBENCH

1. With a thickness planer, surface the lumber for the legs, side and end rails, and side and end braces to the thickness shown in bill of materials.
2. Cut legs, side and end rails, and side and end braces to length, as shown in bill of materials.
3. Lay off and cut notches in legs for rails and leg braces. (This can be done by hand, or with a dado head on a power saw.) Remember that each leg must be made to fit a specific corner. See the drawing for the location of the notches. Be sure the notches are cut so that the end and side rails and the leg braces will fit tightly.
4. Tack the frame together. (Be sure that the nails are not placed at points where bolt holes are to be made, as described in step 5.)
5. Bore a $^3/_8"$ hole at each joint in the frame and fasten the frame together with $^3/_8"$ carriage bolts. Insert bolts in holes from outside, and tighten nuts.
6. Surface lumber for top pieces to $1^3/_4"$ thickness using a thickness planer. Then joint both edges of the middle piece and one edge of the outside pieces. The jointed edge must be straight and square. Clamp the top pieces together with the best sides up. If the edges of the boards do not match perfectly, repeat the process until they do.
7. Mark boards on the top side with chalk or pencil so that they may be placed in the same position for gluing.
8. Rip splines 1" wide and 96" long from a piece of $^1/_4"$ plywood. (The splines can be ripped from short pieces of plywood if a 96" piece is not available. The short pieces can be placed end to end in the groove.)
9. Adjust the dado head to cut the proper size grooves for the $^1/_4" \times 1"$ splines. Establish a distance between the fence and the center of the dado that is equal to $^1/_2$ the thickness of the lumber to be grooved. To be sure of the proper setting, cut a groove in two pieces of scrap wood that are the same thickness as the table top. Place the spline in the grooves of the scrap pieces to see if they are grooved correctly.
10. Cut a groove in the jointed edge of the two outside pieces and in each edge of the middle piece. The marked side of each piece must be held tightly (use a feather board) against the fence while the groove is being cut. This will make all the grooves the same distance from the top surface. If a power saw is not available, the top may be held together with glue and with $^1/_2" \times 2"$ dowel pins placed about 8" apart.
11. Glue the top together.
12. After the glue has set properly, joint one edge of the table top straight; then cut the top to 24" wide by ripping and jointing the other edge.
13. Cut the top to 96" in length by sawing both ends square.

14. Lay off, bore, and countersink holes in the strips to fasten the hardwood strips to the top rail. Bore one hole about 2" from each end of each strip. Bore the other holes approximately 12" apart.
15. Bore and countersink holes in the hardwood strips for fastening the top to the frame (base). Space vertical holes to avoid hitting the horizontal ones.
16. Bore anchor holes in rails. Fasten strips to rails with No. 12 × $1^1/_2$" flat head wood screws.
17. Place the table top on a smooth surface with the finished side down. Place the frame over the table top, and adjust to the proper distance from both ends and sides. Fasten the strips to the table top with wood screws.
18. Finish the frame with a clear finish or with paint.
19. Sand the table top and edges. Apply two coats of floor seal or gym seal.

PROJECT 14: PICNIC TABLE

PLAN FOR PICNIC TABLE

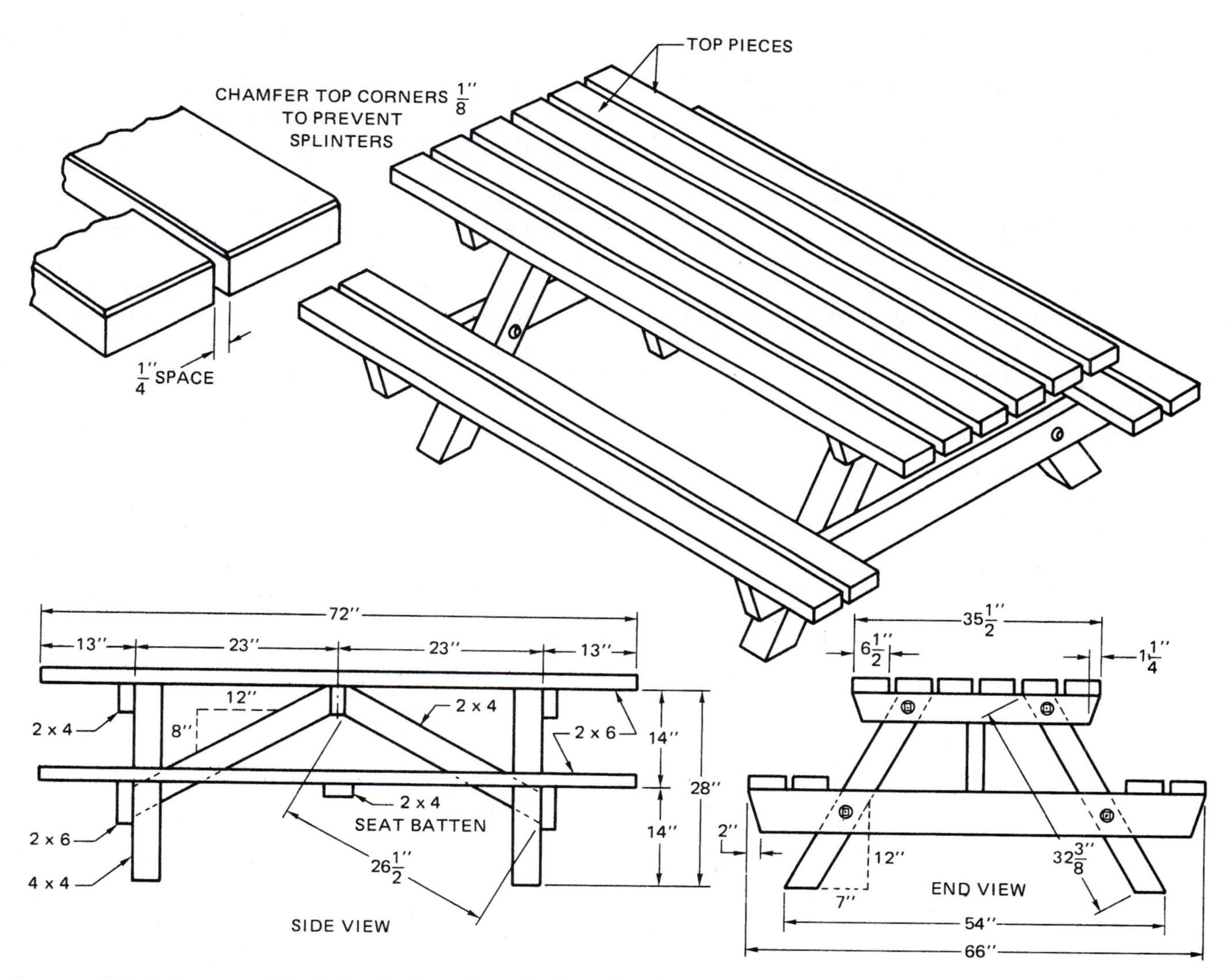

Courtesy of T. J. Wakeman and V. L. McCoy, The Farm Shop, *The Macmillan Company, 1960*

BILL OF MATERIALS

Materials Needed	Quantity	Dimensions	Description or Use
Lumber, surfaced	10 pieces	$1^3/_4$" × $5^3/_4$" × 72"	Top and seats
Lumber, surfaced	2 pieces	$1^3/_4$" × $5^3/_4$"' × 66"	Seat supports
Lumber, surfaced	3 pieces	$1^3/_4$" × $3^3/_4$" × $35^1/_2$"	Top supports
Lumber, surfaced	4 pieces	$3^3/_4$" × $3^3/_4$" × 34"	Legs
Lumber, surfaced	2 pieces	$1^3/_4$"' × $3^3/_4$" × 28"	Braces
Lumber, surfaced	2 pieces	$1^3/_4$" × $3^3/_4$" × 12"	Seat battens
Carriage bolts	8	$^1/_2$" × $6^1/_2$"	Ends
Washers	16	$^1/_2$"	Flat
Common nails	2 pounds	16d	Galvanized finish

Note: Commercially surfaced lumber will be $1^1/_2$" × $3^1/_2$" for 2 × 4s; $1^1/_2$" × $5^1/_2$" for 2 × 6s; and $3^1/_2$" × 3 $^1/_2$" for 4 × 4s. Use shorter bolts if commercially dressed lumber is used.

CONSTRUCTION PROCEDURE FOR PICNIC TABLE

1. Surface the lumber to size, as listed in the bill of materials using a thickness planer. Joint the edges smooth.
2. Cut the top and seat pieces to length, and chamfer the top corners with a hand plane (see detail drawing).
3. Cut the three top supports to length and shape. Place them on a level surface with the long edge up, and space as shown in the side view.
4. Nail the two outside top pieces to each top support with one 16d nail.
5. Square the top, and finish nailing the two outside pieces with 16d nails.
6. Space the remaining top pieces approximately $^1/_4$" apart, and nail them in place.
7. Cut the legs and seat supports to length and angle, as shown in the end view.
8. Turn the top over, and tack the legs and seat supports in place.
9. Bore holes with a $^1/_2$" auger bit, and bolt the legs and seat supports in place with $^1/_2$" carriage bolts. Place a $^1/_2$" flat washer under the nut of each bolt.
10. Square legs with the top of the table. Cut the braces to length and shape; nail in place, as shown in the side view.
11. Cut 2" × 4" seat battens to length, as listed in bill of materials. Turn the table upright, and nail on seats and seat battens.
12. Apply a suitable clear, stain, or paint finish for continuous exterior exposure.

PROJECT 15: HOG-SHIPPING CRATE

PLAN FOR HOG-SHIPPING CRATE

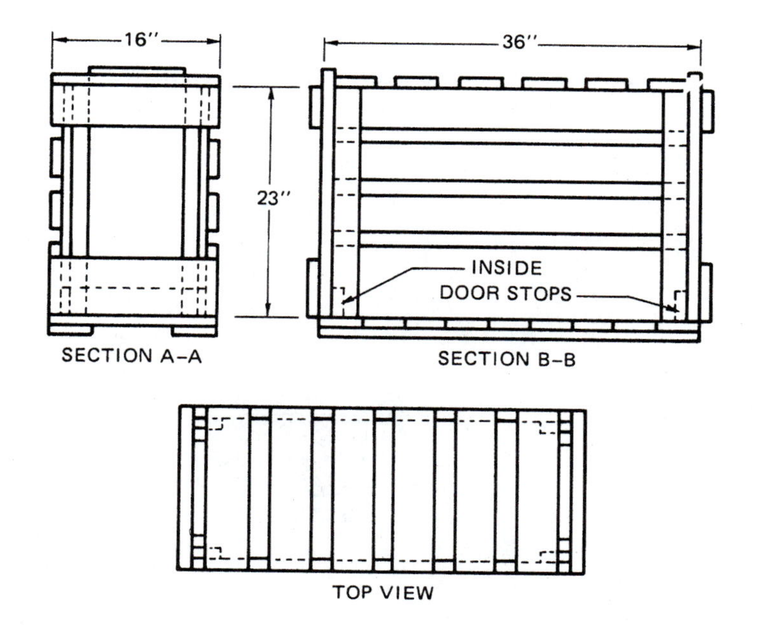

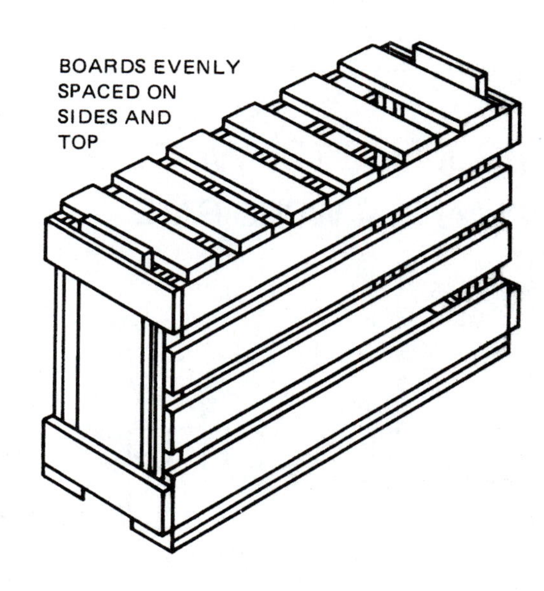

Courtesy of T. J. Wakeman and V. L. McCoy, The Farm Shop, *The Macmillan Company, 1960*

BILL OF MATERIALS

Materials Needed	Quantity	Dimensions	Description or Use
Lumber, unsurfaced	2 pieces	1" × 6" × 36"	Bottom side pieces
	2 pieces	1" × 5" × 16"	Bottom end pieces
	2 pieces	1" × 4" × 36"	Runners
	9 pieces	1" × 4" × 16"	Flooring
	4 pieces	1" × 4" × 23"	Corner uprights
	6 pieces	1" × 4" × 36"	Top and middle side pieces
	6 pieces	1" × 4" × 16"	Top crosspieces
	2 pieces	1" × 4" × 12"	Inside doorstop
	2 pieces	1" × 4" × 16"	Top end pieces
	2 pieces	1" × 10"× 24"	Doors
Machine bolts	2	$^{1}/_{4}$" × 2$^{1}/_{2}$"	
Wing nuts	2	$^{1}/_{4}$"	
Box nails	3 pounds	8d	

CONSTRUCTION PROCEDURE FOR HOG-SHIPPING CRATE

1. Cut all the lumber pieces to length as shown in the bill of materials. If the crate is to be painted, surface the 1" boards to $^{7}/_{8}$" thick using a thickness planer before cutting to length. Joint the edges smooth after the pieces are cut to length.
2. Nail the flooring to the two bottom side pieces (see drawing), and nail on the two runners, using 8d box or common nails.
3. Nail the two bottom end pieces to the ends of the bottom side pieces and floor. Nail the corner uprights to the bottom side and end pieces, as shown.

4. Nail the two top side pieces and the two end pieces to the top of the corner upright pieces. Nail on the middle side pieces.
5. Fit the doors in place, and nail the bottom inside door stops to the floor. Leave 1/8"clearance between the door and the inside door stop.
6. Nail on the top crosspieces. Leave 1/8" clearance between the door and the first top crosspiece.
7. Bore a 5/16" hole in the top end pieces and through the door on each end so the door can be bolted with a 1/4" machine bolt.

NOTE: The size of the crate may be increased for large hogs. Carriage bolts may be used in high-stress joints for added strength.

PROJECT 16: WOOD GATE

PLAN FOR WOOD GATE

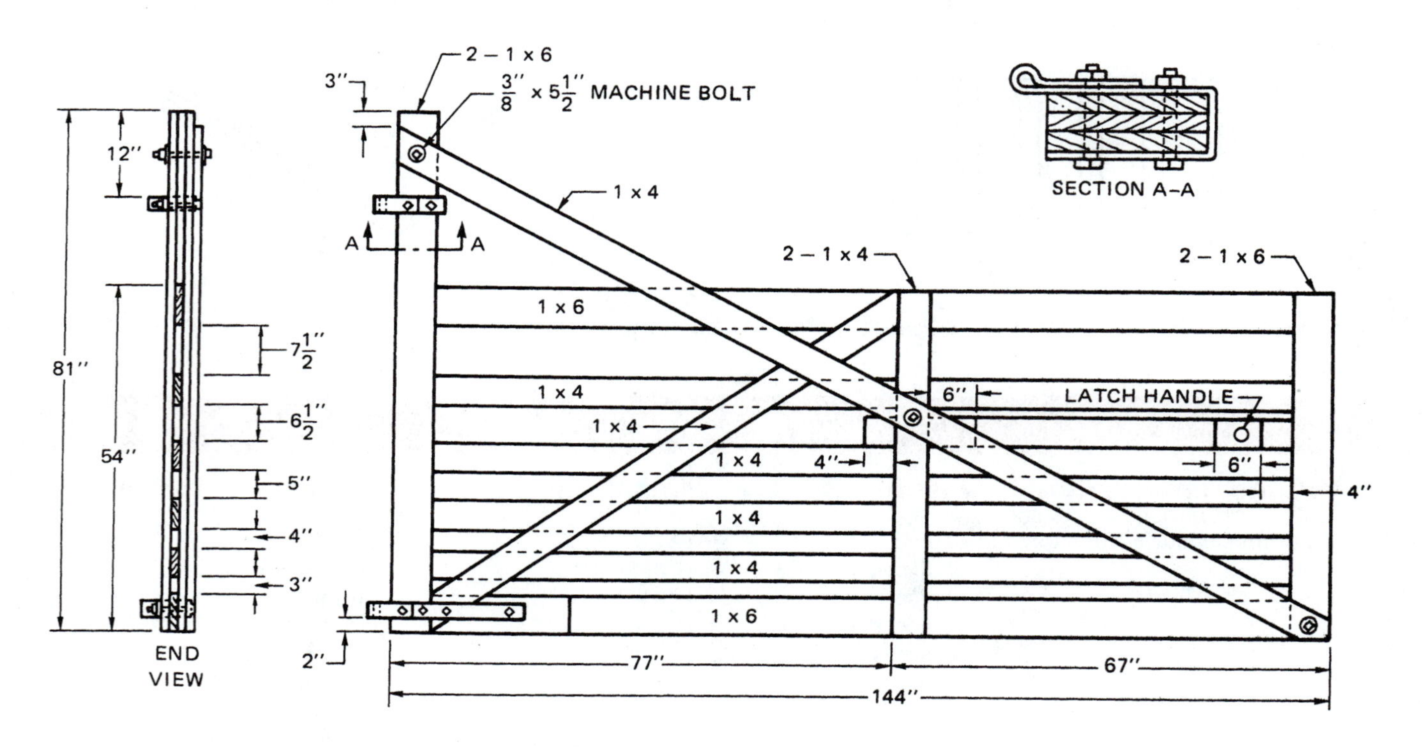

Courtesy of T. J. Wakeman and V. L. McCoy, The Farm Shop, *The Macmillan Company, 1960*

BILL OF MATERIALS

Materials Needed	Quantity	Dimensions	Description or Use
Lumber, unsurfaced*	2 pieces	1" × 6" × 81"	Hinge battens
	2 pieces	1" × 6" × 54"	Front battens
	1 piece	1" × 6" × 36"	Hinge filler blocks
	2 pieces	1" × 4" × 54"	Center battens
	1 piece	1" × 4" × 36"	Latch
	2 pieces	1" × 4" × 8'0"	Short braces
	2 pieces	1" × 6" × 12'0"	Top and bottom slats
	4 pieces	1" × 4" × 12'0"	Inside slats
	2 pieces	1" × 4" × 14'0"	Long braces
Carriage bolts	3	3/8" × 5 1/2"	
Carriage bolts	6	3/8" × 3 1/2"	
Washers	8	3/8"	Flat
Common nails	2 pounds	8d	
Common nails	1/2 pound	6d	
Hinges	2		Make or purchase

*Use surfaced lumber if gate is to be painted.

CONSTRUCTION PROCEDURE FOR WOOD GATE

1. Cut all battens and slats to length, as shown in the drawing. (If the gate is to be painted, surface all boards to 7/8" thickness.)
2. Place on front batten and one hinge batten on a flat surface. Tack the top and bottom slats in place with a 6d common nail at each joint.
3. Square gate with a framing square. Tack the inside slats at the proper spacing. The spacing may vary slightly, depending on the width of the slats.
4. Place the two remaining end battens directly over the other two battens. Fasten permanently with 8d common nails.
5. Place the two center battens in position. Fasten permanently with 8d common nails.
6. Place the short brace for this side in position. Mark ends in line with the inside edge of the battens. Cut brace and tack it in position with 6d common nails.
7. Mark and cut the long brace and tack it in position on the brace and battens.
8. Turn the gate over. Cut braces for this side in the same manner, and fasten all joints on this side permanently with 8d common nails, except where long brace touches battens.
9. Turn the gate over. Fasten braces from this side in like manner.
10. Use 3/8" carriage bolts in joints where long braces cross the battens.
11. Place the latch in position and nail on stop blocks. Drill hole in latch and insert a short piece of dowel or broom handle for the latch handle.
12. Nail the bottom and top hinge filler blocks on the same side of the gate that the hinges are to be placed. Fasten the hinges to the gate with carriage bolts.
13. Prime the gate and paint it with two coats of exterior paint, or finish with a wood preservative.

NOTE: If additional rigidity and permanence are desired, add a 5/16" carriage bolt at every joint in the gate.

PROJECT 17: SALT AND MINERAL BOX—PORTABLE

PLAN FOR SALT AND MINERAL BOX—PORTABLE

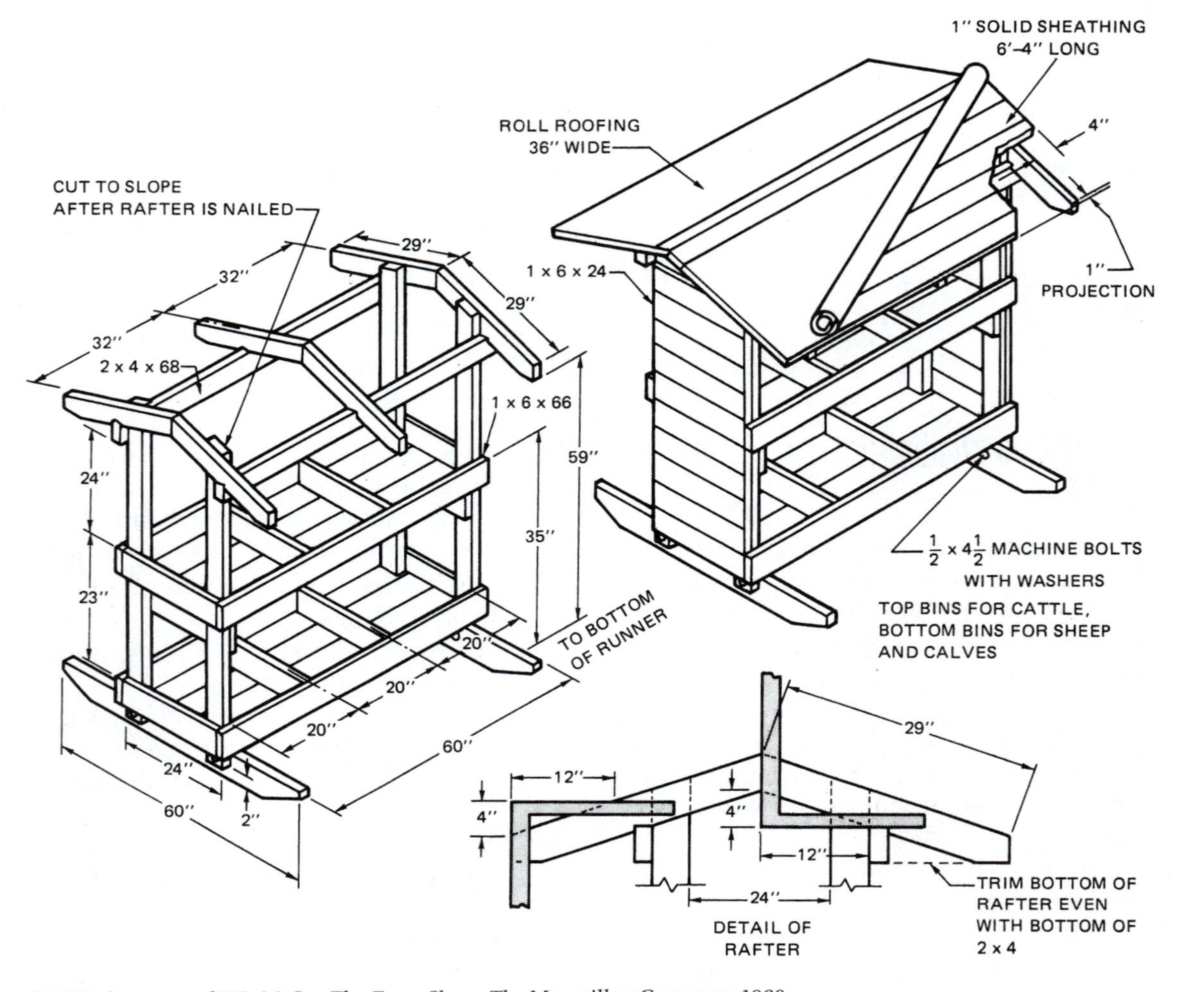

Courtesy of T. J. Wakeman and V. L. McCoy, The Farm Shop, *The Macmillan Company, 1960*

BILL OF MATERIALS

Materials Needed	Quantity	Dimensions	Description or Use
Lumber, unsurfaced*	2 pieces	2" × 6" × 60"	Runners**
	4 pieces	2" × 4" × 72"	Uprights
	6 pieces	2" × 4" × 29"	Rafters
	2 pieces	2" × 4" × 58"	Rafter supports
	16 pieces	1" × 6" × 12'0" stock	Siding, roof sheathing, trough supports, floors, sides, pieces, partitions
Roofing nails	1/2 pound	3/4"	
Common nails	1/4 pound	16d	
Box nails	3 pounds	8d	
Machine bolts	4	1/2" × 4 1/2"	
Washers	8	1/2"	Flat
Mineral roofing	1 piece	13'0" × 36"	Roll roofing

*Surface pieces if box is to be painted.
**Pressure treated material to prevent rotting is recommended.

CONSTRUCTION PROCEDURE FOR SALT AND MINERAL BOX—PORTABLE

1. Cut the two runners to length and shape, as shown in the drawing.
2. Cut the four uprights to length as listed in the bill of materials.
3. Lay the runners down flat, and tack uprights to the sides of the runners. See the drawing for the correct position of the uprights.
4. Square the uprights with the runners, bore holes, and bolt them to the runners with 1/2" × 4 1/2" machine bolts. Use 1/2" washers under the bolt heads and nuts. Cut the end pieces (1" × 6" × 24") to length, and nail to the outside of the uprights with 8d box nails.
5. Cut the rafter supports (2" × 4" × 68") to length and nail to the uprights with 16d common nails.
6. Cut the two trough supports (1" × 6" × 24") and nail to the inside of the uprights, as shown in the drawing. Cut flooring for the two troughs to length. Fit and nail in place with 8d box nails.
7. Square the uprights with the floor. Cut the 1" × 6" side pieces to length, and nail to the uprights and floor. Cut the trough partitions to length and nail to the trough sides, as shown in the drawing. Turn the salt box on its side and nail flooring to trough partitions, using 8d box nails.
8. Lay off, mark, and cut the rafters. (See the detail drawing for the position of the square in laying off the rafter.) After the first rafter is cut, it can be used as a template in marking the remaining rafters.
9. Nail the rafters in place, and saw off the ends of the uprights even with the top of the rafters.
10. Cut the 1" × 6" sheathing to length and nail to the rafters with 8d box nails. Cut the mineral roll roofing and nail to the sheathing with 3/4" roofing nails.
11. Apply safe paint or preservative to appropriate wood surfaces. Creosote the runners if pressure treated material is not used.

PROJECT 18: CATTLE FEED TROUGH

PLAN FOR CATTLE FEED TROUGH

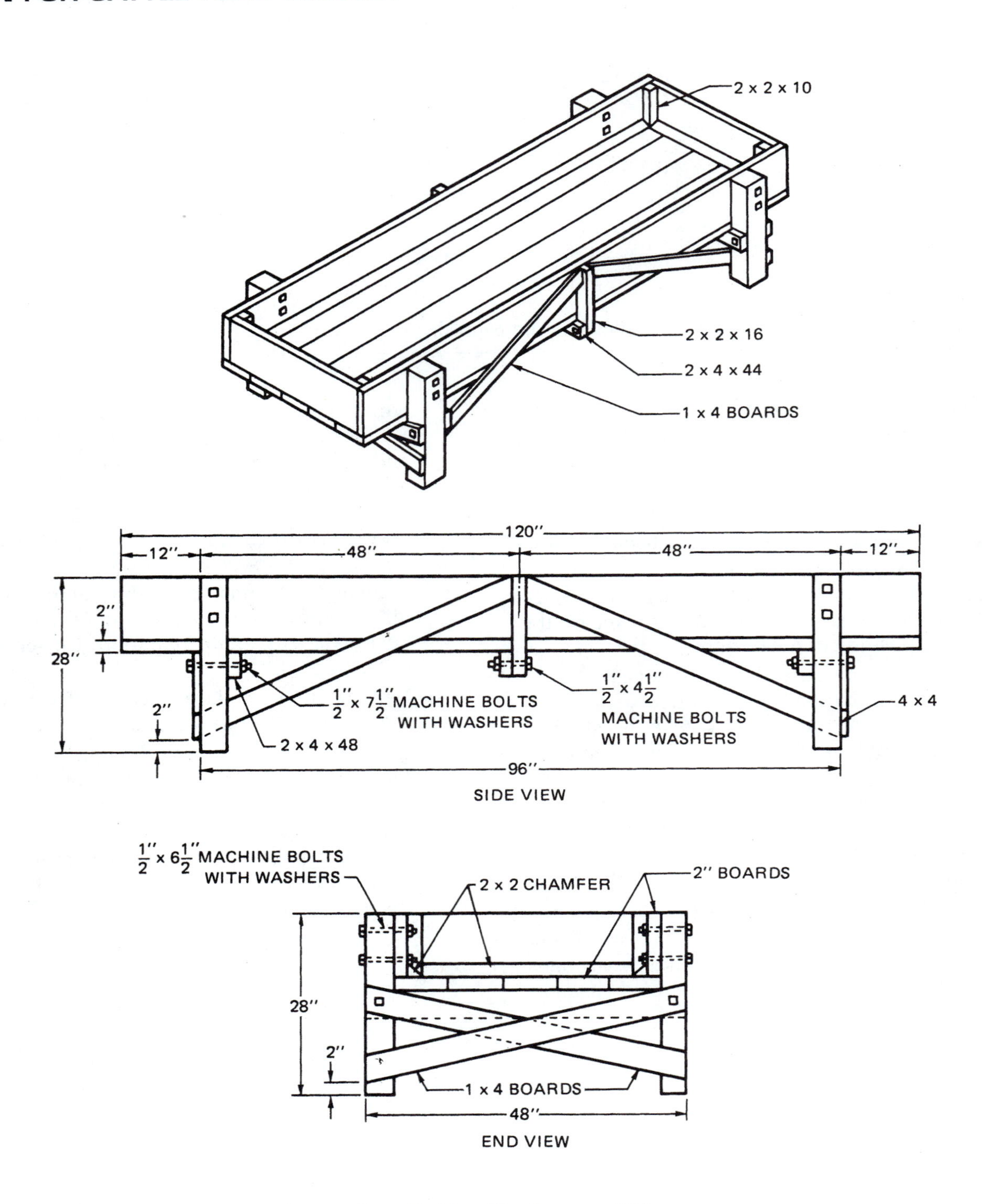

Courtesy of T. J. Wakeman and V. L. McCoy, The Farm Shop, *The Macmillan Company, 1960*

BILL OF MATERIALS

Materials Needed	Quantity	Dimensions	Description or Use
Lumber, unsurfaced	4 pieces	4" × 4" × 28"	Legs
	2 pieces	2" × 4" × 48"	End trough supports
	1 piece	2" × 4" × 44"	Center trough supports
	4 pieces	1" × 4" × 51"	End braces
	4 pieces	2" × 8" × 10'0"	Trough bottom*
	3 pieces	2" × 10" × 10'0"	Trough bottom and sides
	2 pieces	2" × 10" × 36"	Trough ends
	4 pieces	2" × 2" × 10"	Corner blocks
	2 pieces	2" × 2" × 16"	Center side supports
	2 pieces	2" × 2" × 9'5"	Chamfer strips
	2 pieces	2" × 2" × 33"	Chamfer strips
	4 pieces	1" × 4" × 54"	Side braces
Common nails	4 pounds	20d	
Common nails	1 pound	16d	
Common nails	1 pound	8d	
Machine bolts	8	1/2" × 6 1/2"	
Machine bolts	4	1/2" × 7 1/2"	
Machine bolts	6	1/2" × 4 1/2"	
Washers	36	1/2"	Flat
Creosote	1 gallon		

*Use surfaced or tongue and groove lumber for the bottom, if ground feed is to be fed in the trough.

CONSTRUCTION PROCEDURE FOR CATTLE FEED TROUGH

1. Cut the 4" × 4" legs to length as shown in the bill of materials.
2. Cut the end trough supports and end braces to length; square and tack them to legs with 20d common nails. See drawings. Place nails where they will not interfere with the bolt holes.
3. Bore holes, and bolt end trough supports and top ends of braces to legs with 1/2" × 7 1/2" machine bolts.
4. Cut the pieces for the trough bottom to length; joint and nail to the trough supports with 20d common nails.
5. Cut the center trough support to length. Nail the trough bottom to the support with 20d common nails.
6. Cut the trough ends and sides to length; bore holes, and bolt sides to legs with 1/2" × 6 1/2" machine bolts. Nail ends in place as shown in the drawings.
7. Turn trough over and nail bottom to side and end pieces with 20d common nails. Turn trough back over.
8. Cut the 2" × 2" center side supports to length. Bolt the side supports to the side pieces and to the center trough support with 1/2" × 4 1/2" machine bolts.
9. Cut the four side braces to length. Square the legs with the bottom of the trough, and nail all braces to sides and legs with 8d common nails.
10. Cut the 2" × 2" corner blocks to length. Drill nail holes in the corner blocks and nail to end and side pieces with 16d common nails, or use 5/16" carriage bolts.
11. Cut the chamfer strips to shape and length, and nail in place with 8d common nails.
12. Apply a safe wood preservative.

PROJECT 19: PORTABLE RANGE FEEDER

PLAN FOR PORTABLE RANGE FEEDER

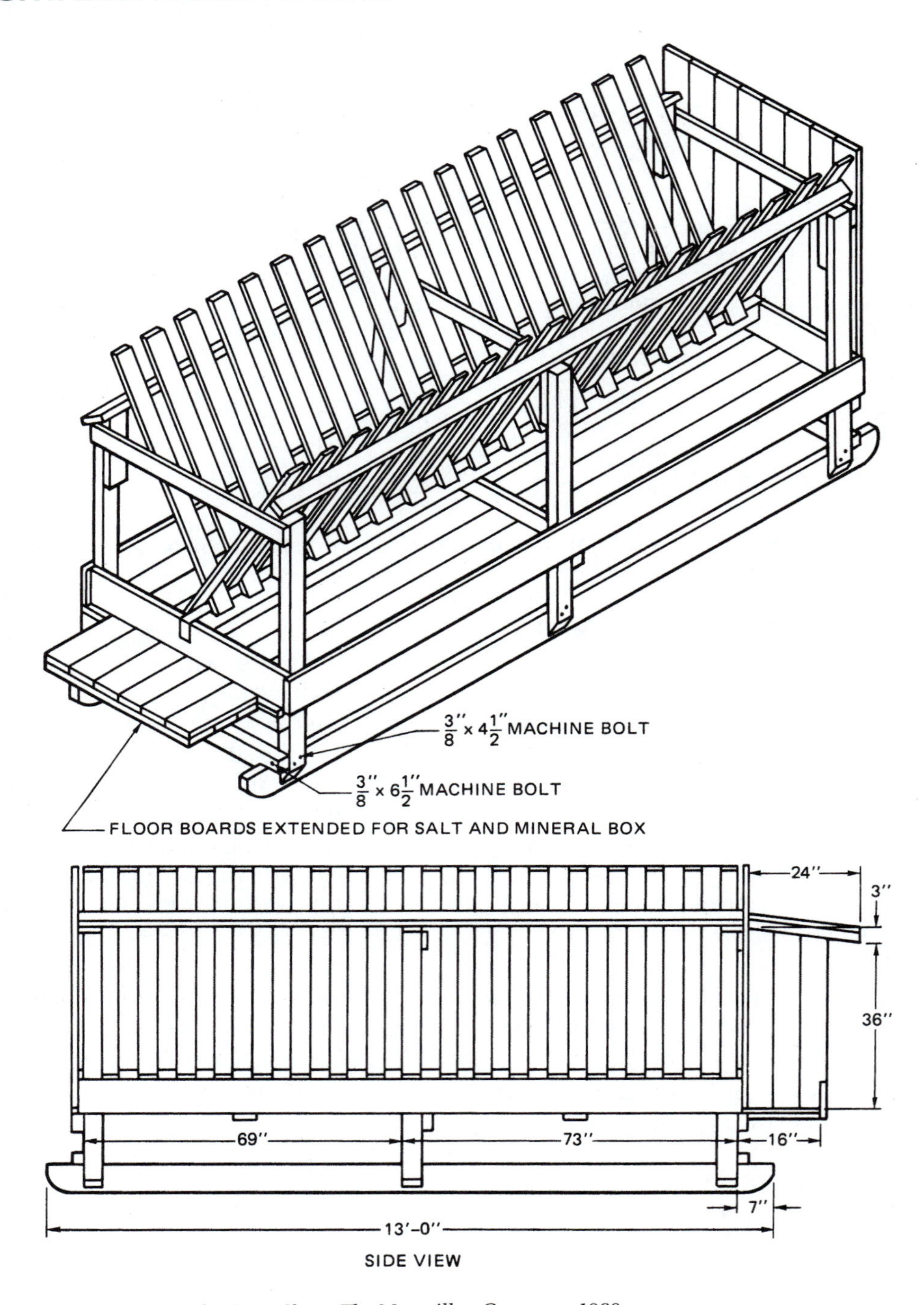

Courtesy of T. J. Wakeman and V. L. McCoy, The Farm Shop, *The Macmillan Company, 1960*

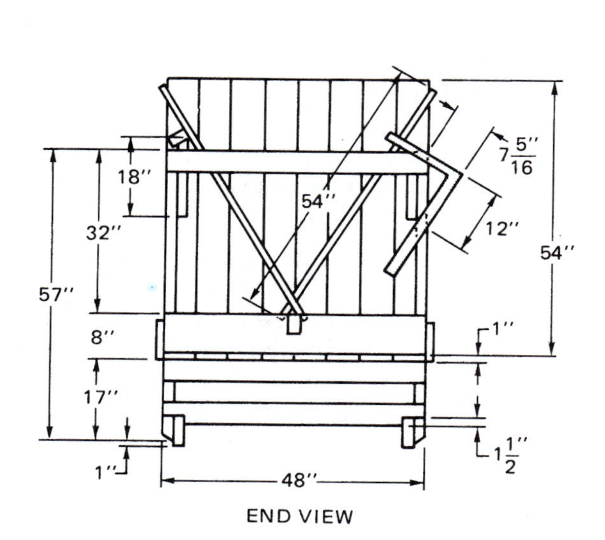

END VIEW

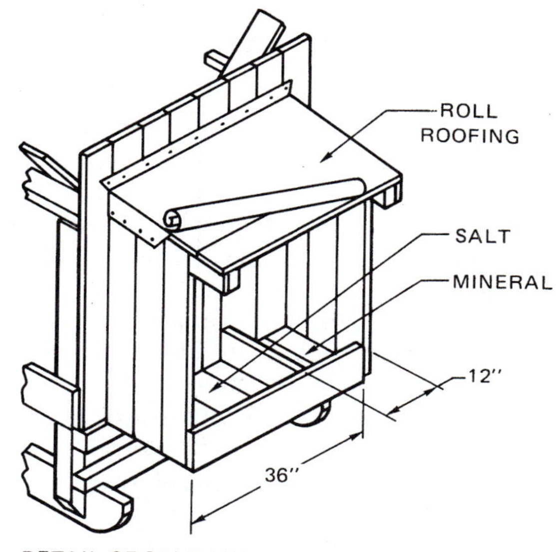

DETAIL OF SALT AND MINERAL BOX

BILL OF MATERIALS

Materials Needed	Quantity	Dimensions	Description or Use
Lumber, unsurfaced	2 pieces	2" × 6" × 13'0"	Runners
	6 pieces	2" × 4" × 60"	Uprights
	6 pieces	2" × 4" × 18"	Upright scabs
	3 pieces	2" × 4" × 48"	Trough supports
	2 pieces	2" × 4" × 48"	Struts between runners
	2 pieces	2" × 4" × 24"	Rafters
	2 pieces	2" × 4" × 12'0"	Top slat supports
	1 piece	2" × 4" × 12'0"	Bottom slat support
	3 pieces	1" × 4" × 48"	Top crosspieces
	1 piece	1" × 4" × 48"	Center crosspiece
	2 pieces	1" × 4" × 6"	Rafter filler block
	2 pieces	1" × 6" × 12'2"	Outside floor pieces
	6 pieces	1" × 6" × 13'6"	Inside floor pieces
	16 pieces	1" × 6" × 54"	End boards, vertical
	4 pieces	1" × 6" × 38"	Sheathing
	2 pieces	1" × 8" × 12'0"	Trough sides
	2 pieces	1" × 8" × 48"	Trough ends
	35 pieces	1" × 4" × 54"	Slats
	2 pieces	1" × 8" × 36"	Salt box subfloor
	1 piece	1" × 8" × 36"	Salt box end
	6 pieces	1" × 6" × 42"	Siding; salt box partitions
Machine bolts	12	3/8" × 4 1/2"	
Machine bolts	4	3/8" × 6 1/2"	
Washers	32	3/8"	Flat
Common nails	4 pounds	8d	
Common nails	2 pounds	20d	
Roofing nails	1 pound	3/4"	
Mineral roofing	1 piece	48" × 36"	Roll roofing
Preservative	2 gallons		

CONSTRUCTION PROCEDURE FOR PORTABLE RANGE FEEDER

1. Cut the runners to length and shape, as shown in the drawings.
2. Cut the uprights to length, using 7 5/16" on the tongue of a framing square and 12" on the body for marking cuts, as shown in the end view.
3. Space one upright 7" from each end and one in the center of the runners, and tack in position. Be sure uprights are at right angles to runners. Bore holes, and bolt each upright to runners with two 3/8" × 4 1/2" machine bolts.
4. Cut upright scabs, and nail to the inside top of the uprights.
5. Cut, bore holes, and bolt struts to the corner uprights with 3/8" × 6 1/2" machine bolts. Be sure the notches fit right between the runners.
6. Cut trough supports, and nail to corner uprights with 20d common nails, as shown in the drawing.
7. Cut the three top crosspieces, and nail to uprights as shown.
8. Cut top slat supports 12'0" long.
9. Place top slat supports on top of the uprights, and nail in place, as shown in the drawing. The support should project 1" beyond the uprights at each end.
10. Cut the two outside floor pieces long enough (approximately 12'2" long) to fit even with the outside edge of the trough supports. Notch one edge of each piece to fit around the 2" × 4" upright pieces, and nail to trough supports.
11. Cut the inside floor pieces long enough to extend 16" past one end of the trough support to form the bottom for the salt box. Fit the boards tightly together, and nail in place.
12. Cut the trough ends and center crosspiece, and nail in place.
13. Cut the trough side pieces the same length as the outside bottom pieces, and nail in place.
14. Cut a notch in the center of the trough ends to fit the 2" × 4" bottom slat support.
15. Cut the bottom slat support, and nail in position, as shown in the drawing.
16. Cut 35 slats 54" long.
17. Place the first slat tight against the trough end and top crosspiece, and fasten with 8d common nails. The bottom end of the slat should project 2" below the top corner of the bottom slat support. Fasten the remaining slats to alternate sides with 8d common nails as shown in the drawing. Be sure the slats are spaced the same distance (4") apart at the top and bottom.
18. Cut end boards 54" long, and nail in place.
19. Cut subflooring for the salt box, and nail in place.
20. Mark the 2" × 4" rafters, using 3" and 24" on a framing square. Cut rafters, and nail in place, as shown in the detail of the salt and mineral box.
21. Cut and fasten the siding to rafters and floor with 8d common nails.
22. Cut and fasten the sheathing to rafters with 8d common nails.
23. Cut and nail the rafter filler blocks in position, as shown.
24. Nail roofing to sheathing.
25. Cut the salt box end, and nail to the floor and bottom edge of the side pieces.
26. Cut the salt box partition and fasten in place, as shown in the drawing.
27. Apply a suitable preservative to all surfaces.

PROJECT 20: TOOL SHARPENING GAUGE

PLAN FOR TOOL SHARPENING GAUGE

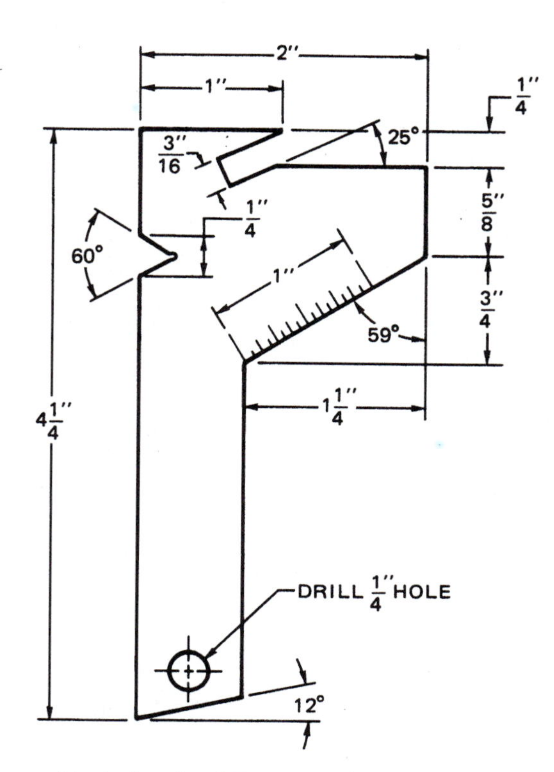

GAUGES
- PLANE IRON OR WOOD CHISEL
- COLD CHISEL OR CENTER PUNCH
- TWIST DRILL, CUTTING EDGE ANGLE AND LENGTH
- TWIST DRILL, LIP CLEARANCE

Courtesy of Department of Agricultural and Extension Education, The Pennsylvania State University, University Park, PA

BILL OF MATERIALS

Materials Needed	Quantity	Dimensions	Description
Sheet metal, galvanized, 24 gauge	1	2" × 4 1/4"	

CONSTRUCTION PROCEDURE FOR TOOL SHARPENING GAUGE

1. Measure and scribe outline on the stock using a scratch awl.
2. Cut out tool gauge with snips.
3. Use snips and a flat file to cut the 25° wood chisel slot.
4. Use a taper file to cut the 60° cold chisel vee.
5. Position and drill a 1/4" hole at the narrow end of the gauge.
6. Measure and scribe a 1" rule by 1/16" graduations.
7. Cut rule indicator marks with a cold chisel or awl.
8. Smooth all edges with a flat file.

PROJECT 21: FEED SCOOP

PLAN FOR FEED SCOOP

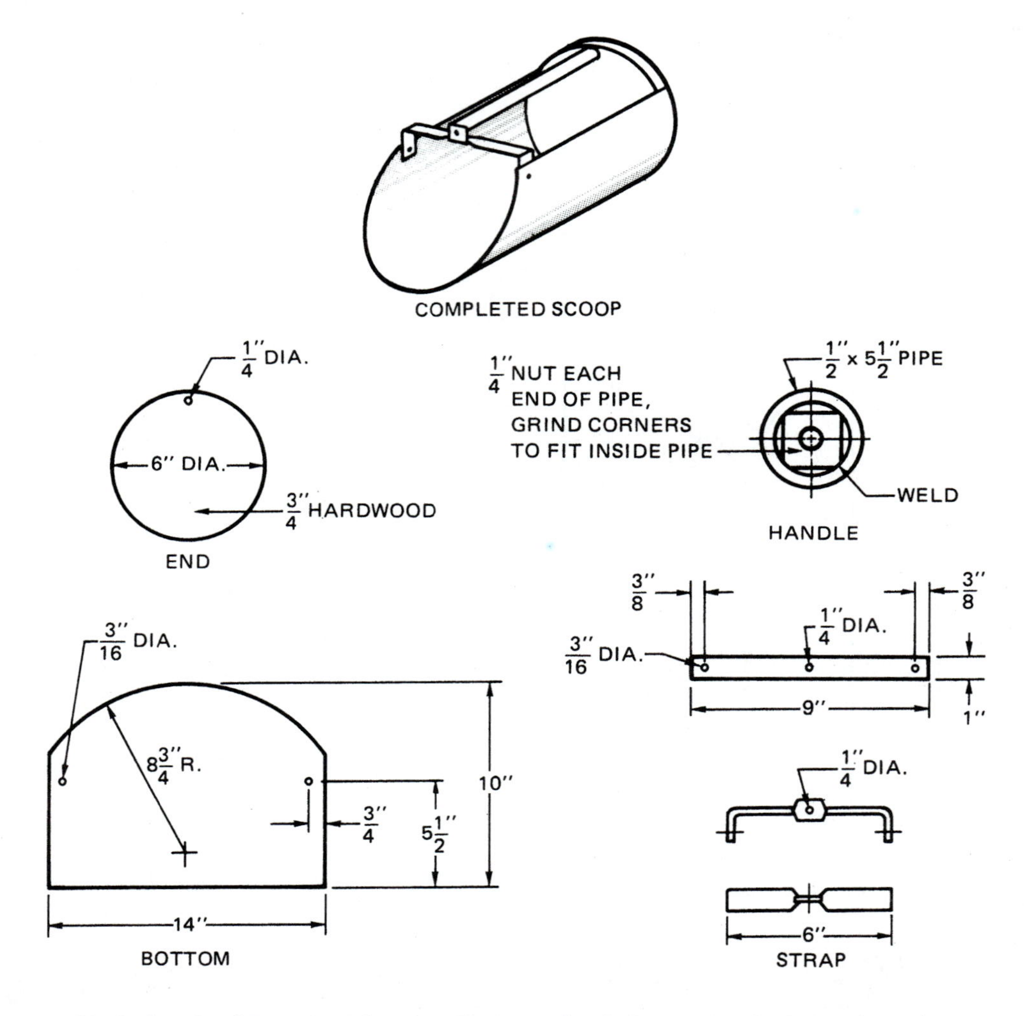

Courtesy of Department of Agricultural and Extension Education, The Pennsylvania State University, University Park, PA

BILL OF MATERIALS

Materials Needed	Quantity	Dimensions	Description or Use
Sheet metal, galvanized, 24 gauge	1	10" × 14"	Bottom
Lumber, hardwood	1	3/4" × 6" × 6"	End
Mild steel, hot rolled	1	1/8" × 1" × 9"	Strap
Steel tubing	1	1/2" I.D. × 5 1/2"	Handle
Hardware			
Stove bolt	1	1/4" × 6 1/2"	
Nut	2	1/4"	
Rivet	2	3/16"	
Screw	6	#6 × 3/4"	Pan head
Paint (nonlead)	1/2 pint		For end, handle, and hand bracket

CONSTRUCTION PROCEDURE FOR FEED SCOOP

1. Lay out and cut the sheet metal and wood end.
2. Cut and bore holes in the strap iron, 1/8" × 1" × 9".
3. Cut the light steel tubing, 1/2" × 5 1/2" long for the handle.

   NOTE: A 3/4" dowel may be used instead of steel tubing.

4. Bend the bottom sheet metal to fit the end.
5. Drill holes for screws and handle bolt.
6. Screw the metal to the wood end.

   NOTE: Screw-shank or other improved nails may be used in place of screws.

7. Heat and bend the iron strap.
8. Attach the iron strap to the sheet metal using rivets.
9. Weld nuts into ends of the tubing.

   NOTE: Before welding, run the bolt through both nuts to check alignment.

10. Attach the handle with 1/4" × 6 1/2" bolt.
11. Finish and paint the handle, bracket, and wood end.

PROJECT 22: STAPLE

PLAN FOR STAPLE

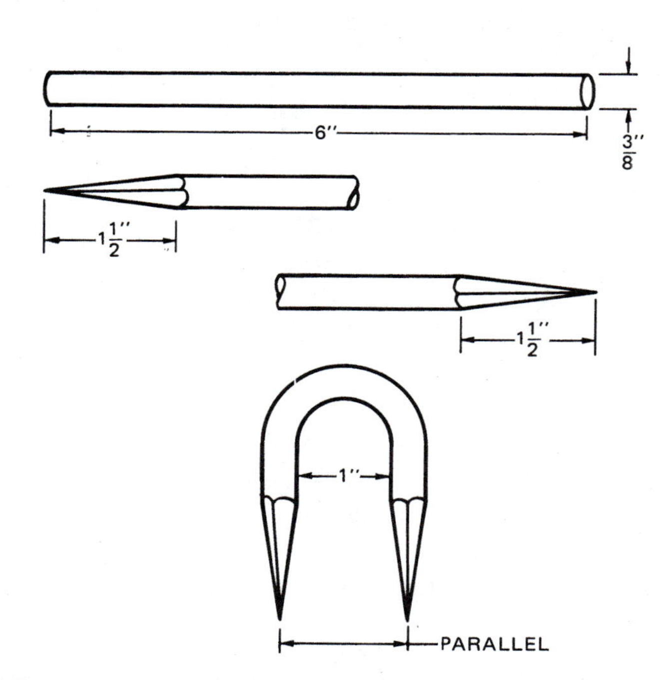

BILL OF MATERIALS			
Materials Needed	**Quantity**	**Dimensions**	**Description**
Steel, round hot or cold rolled	1	3/8" × 6"	

CONSTRUCTION PROCEDURE FOR STAPLE

1. Cut a piece of 3/8" round rod 6" long.
2. Heat and put a long, round point on one end.
3. Heat and put a long, square point on the other end.
4. Bend to form a staple:
   a. One inch width between staple legs
   b. Legs of equal length

PROJECT 23: COLD CHISEL

PLAN FOR COLD CHISEL

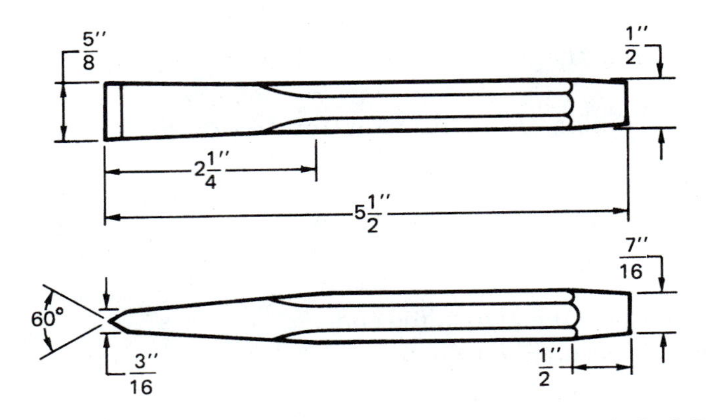

Courtesy of Department of Agricultural and Extension Education, The Pennsylvania State University, University Park, PA

BILL OF MATERIALS

Materials Needed	Quantity	Dimensions	Description
Tool steel	1	1/2" × 5 1/2"	Octagonal

CONSTRUCTION PROCEDURE FOR COLD CHISEL

1. Heat 2 1/4" of one end of the stock to a uniform cherry red color.
2. Place one side against the anvil face. Using drawing blows, work to shape rapidly starting at the end and working back to 2 1/4" taper.
3. Finish to 3/16" thickness and 5/8" width at the tip. Keep area hot (from a dull red to a cherry red color) when working it.
4. Anneal by heating to cherry red and cooling slowly (12 to 24 hours) in lime or sand.
5. File and polish the forged faces. Do not grind.
6. Temper with water (practice on an old cold chisel)
   a. Heat 2" to 3" of the tip to a uniform cherry red color.
   b. Cool 3/4" to 1" in water until drops cling to the tip when it is removed from the water.
   c. Move the tip to avoid cracks at the water line.
   d. Quickly remove scale using a steel brush or file.
   e. Observe the color changes and quench lower 1/4" on purple color. Color order is light straw, dark straw, brown, purple, dark blue, and light blue.
7. Grind the cutting edge to a 60° angle. Use a tool gauge to check the angle.
8. Chamfer the opposite end to approximately 1/2" by 7/16" to prevent mushrooming.

PROJECT 24: DRAWBAR HITCH PIN

PLAN FOR DRAWBAR HITCH PIN

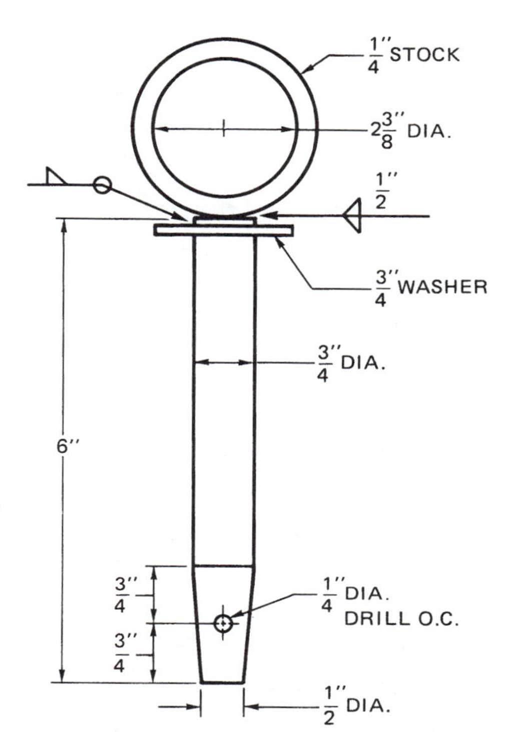

Courtesy of Department of Agricultural and Extension Education, The Pennsylvania State University, University Park, PA

BILL OF MATERIALS			
Materials Needed	**Quantity**	**Dimensions**	**Description or Use**
Hot rolled steel, M1020	1	1/4" × 8"	Handle
C1042 cold rolled or C1045 hot rolled steel, round	1	3/4" × 6"	Pin
Washer, flat	1	3/4"	
Hair pin	1	1/4" × 2"	Safety pin

CONSTRUCTION PROCEDURE FOR DRAWBAR HITCH PIN

1. Bend 1/4" × 8" round stock around 2" pipe, using a vise as a bending aid.
2. Cut the pin to length.
3. Drill a 1/4" hole 3/4" from one end of the pin.
4. Shape the end as indicated in the plan.
5. Weld the washer to the other end of the pin.
6. Place handle on pin and weld.
7. Remove slag and clean with a steel brush.

PROJECT 25: HAY HOOK

PLAN FOR HAY HOOK

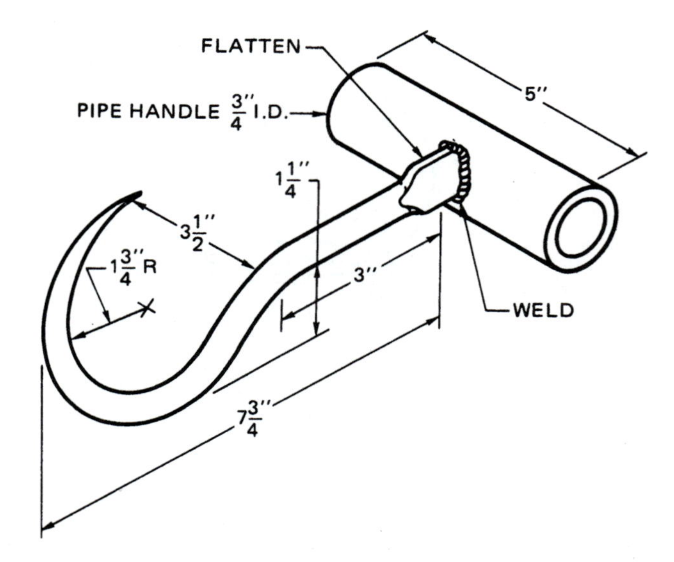

Courtesy of Texas Vocational Instructional Services, Texas A & M University, College Station, TX

BILL OF MATERIALS

Materials Needed	Quantity	Dimensions	Description or Use
Pipe, steel	1	$^3/_4$" ID × 5"	Handle
Rod	1	$^7/_{16}$" dia × 12"	Hook

CONSTRUCTION PROCEDURE FOR HAY HOOK

1. Heat one end of the rod and forge it into a long, slender point.
2. Finish shaping the point with a grinder.
3. Reheat the pointed end and shape it to the dimensions given in the plan drawing.
4. Heat the other end and forge a flat area about $^1/_4$" thick and in line with the curved section.
5. Cut a piece of $^3/_4$" ID steel pipe 5" long.
6. Weld the hook to the center of the handle, being careful to keep the hook square to the handle.
7. Grind and sand all surfaces smooth and round for a comfortable feel.

PROJECT 26: CANT HOOK

PLAN FOR CANT HOOK

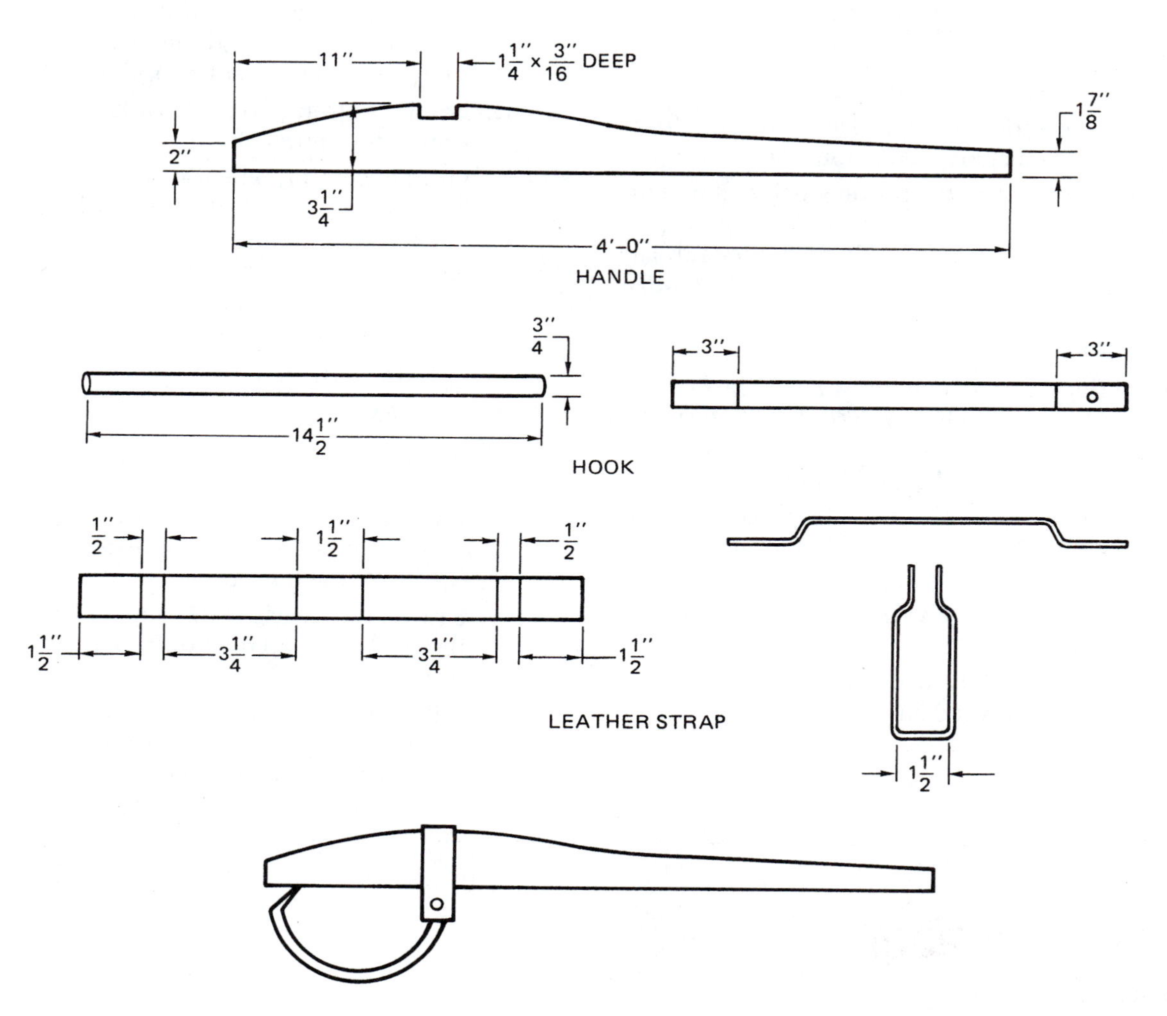

Courtesy of Douglas Hering

BILL OF MATERIALS			
Materials Needed	**Quantity**	**Dimensions**	**Description or Use**
Lumber, oak	1 piece	2" × 4" × 4'	Handle
Round stock	1	$^3/_4$" dia × 14$^1/_2$"	Hook
Metal	1 piece	$^3/_{16}$" × 1$^1/_4$" × 12"	Strap
Bolt, hex head	1	$^3/_8$" × 1$^1/_4$"	
Nut, hex	2	$^3/_8$"	

CONSTRUCTION PROCEDURE FOR CANT HOOK

A. THE HANDLE

1. Plane the 48" oak 2 × 4 to a thickness of 1$^1/_2$".
2. Joint the 2 × 4 to a width of 3$^1/_2$".
3. Cut one side of the 2 × 4 on a band saw to make the handle. Be sure to leave 2" of width at the bottom, and at least 1$^7/_8$" at the top. It is also important to leave 3$^1/_4$" of width 12 inches from the bottom to support the strap.

4. Notch the curved side of the handle to receive the strap.
5. Chamfer all edges to the desired roundness. The straight side can be jointed 1/4" at 45 degrees.

B. THE HOOK

1. Cut a piece of 3/4" round stock 14 1/2" long.
2. Place one end of the round stock in a gas forge and heat to a cherry red color.
3. Flatten the end of the round stock as if making a cold chisel. Stop when the end is tapered back about 3" and is 3/8" thick on the end. This will be the hinged end of the cant hook. Drill a 3/8" hole after rounding the end on the grinder.
4. Heat the opposite end in the gas forge and shape it in a similar way *perpendicular* to the other end.
5. Flatten this end to 1/4", cool and sharpen like a cold chisel.
6. Reheat the sharp end until cherry red.
7. Place 1 1/2" of the point into a vice and bend the remainder over at an 80 degree angle forming the hook.
8. Reheat the round stock and shape the entire piece to the desired curvature.

C. THE STRAP

1. Obtain a piece of 3/16" × 1 1/4" strap 12" long. Mark off the dimensions with soap stone as shown.
2. Using an oxyacetylene unit, a vise and a blacksmith's hammer, bend the two ends into the shape shown (procedure optional).
3. Bend the strap into a "U" with the 1 1/2" dimension in the center.
4. From this point, reheating and reshaping may be necessary to fit different handles. Custom fit each strap before drilling the 3/8" holes to receive the bolt that hinges the hook.
5. The strap should fit neatly into the 3/16" × 1 1/4" notch in the handle.
6. Put the hook on the straight side of the handle.

D. THE ASSEMBLY

1. Assemble by putting the strap and hook in place.
2. Using a 3/8" bolt, install a hex nut and lock washer and tighten until the desired control of the hook is achieved.
3. Saw off the excess end of the bolt and hammer it slightly to keep the nut from coming off during use.
4. If the strap is not tight on the handle, drill a 1/4" hole through the strap and handle and install a flat-headed rivet or bolt to hold the strap securely to the handle.

PROJECT 27: ROUND METAL JIG

PLAN FOR ROUND METAL JIG

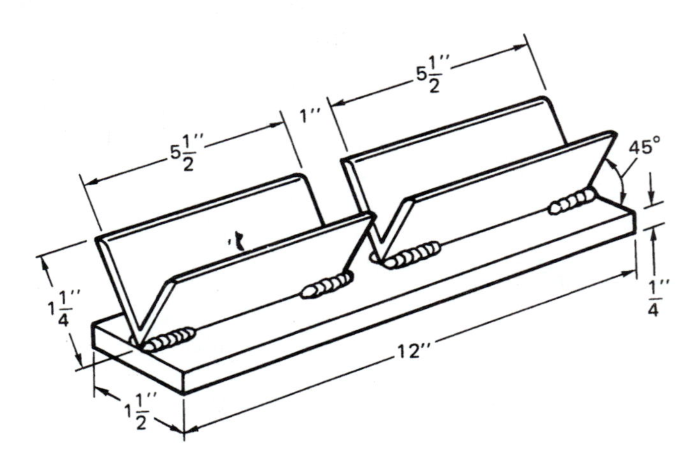

Courtesy of T. J. Wakeman and V. L. McCoy, The Farm Shop, *The Macmillan Company, 1960*

BILL OF MATERIALS			
Materials Needed	**Quantity**	**Dimensions**	**Description or Use**
Angle iron	2 pieces	$^3/_{16}$" × $1^1/_4$" × $1^1/_4$" × $5^1/_2$"	Trough
Flat iron	1 piece	$^1/_4$" × $1^1/_2$" × 12"	Base

CONSTRUCTION PROCEDURE FOR ROUND METAL JIG

1. Cut the materials to length, as listed in the bill of materials.
2. Place both pieces of angle iron on top of the flat iron, as shown. Place a piece of round iron or angle iron in the V, and clamp it to the flat iron to hold the two pieces of angle iron in line while welding.
3. Electric weld the angle iron to the base by welding on both sides at each end of both pieces.

PROJECT 28: FOOT SCRAPER

PLAN FOR FOOT SCRAPER

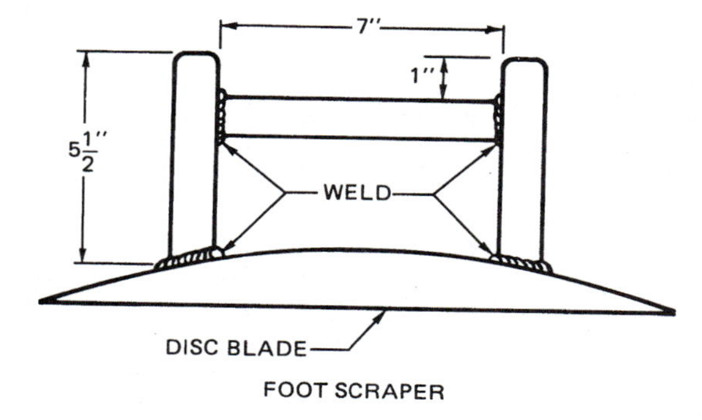

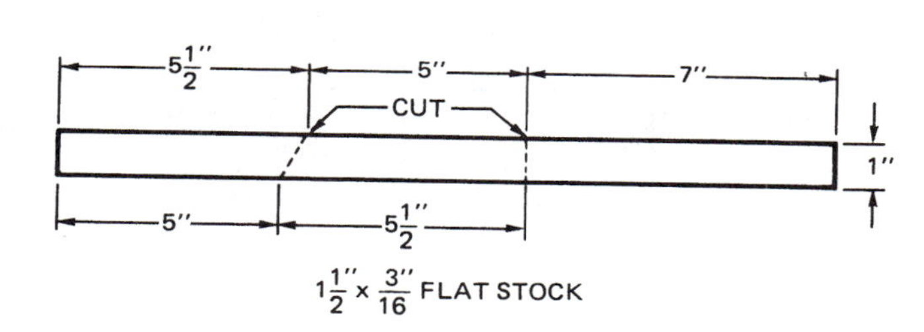

Courtesy of Vocational Agriculture Service, University of Illinois, Urbana, IL

BILL OF MATERIALS			
Materials Needed	**Quantity**	**Dimensions**	**Description**
Disc blade	1	15" diameter	Base
Band, steel	2	$^3/_{16}$" × $1^1/_2$" × $5^1/_2$"	Legs
	1	$^3/_{16}$" × $1^1/_2$" × 7"	Scraper

CONSTRUCTION PROCEDURE FOR FOOT SCRAPER

1. Mark and cut the three pieces of flat stock with a hacksaw or torch.
2. Square the ends of the 7" piece and round the upper ends of the upright pieces on grinder.
3. Place the scraper piece on the edge of the welding table with the angled ends extending over the edge.
4. Clamp firmly and weld the crosspiece to the uprights.
5. Fit the angled ends of the uprights to the contour of the disc blade by grinding if necessary.
6. Clean the disc blade with a wire brush or portable grinder.
7. Weld the scraper to the disc blade carefully. Use low heat. Since the disc blade is high-carbon steel, either preheat or use a low-hydrogen electrode.

PROJECT 29: SMALL ENGINE STAND—METAL

PLAN FOR SMALL ENGINE STAND—METAL

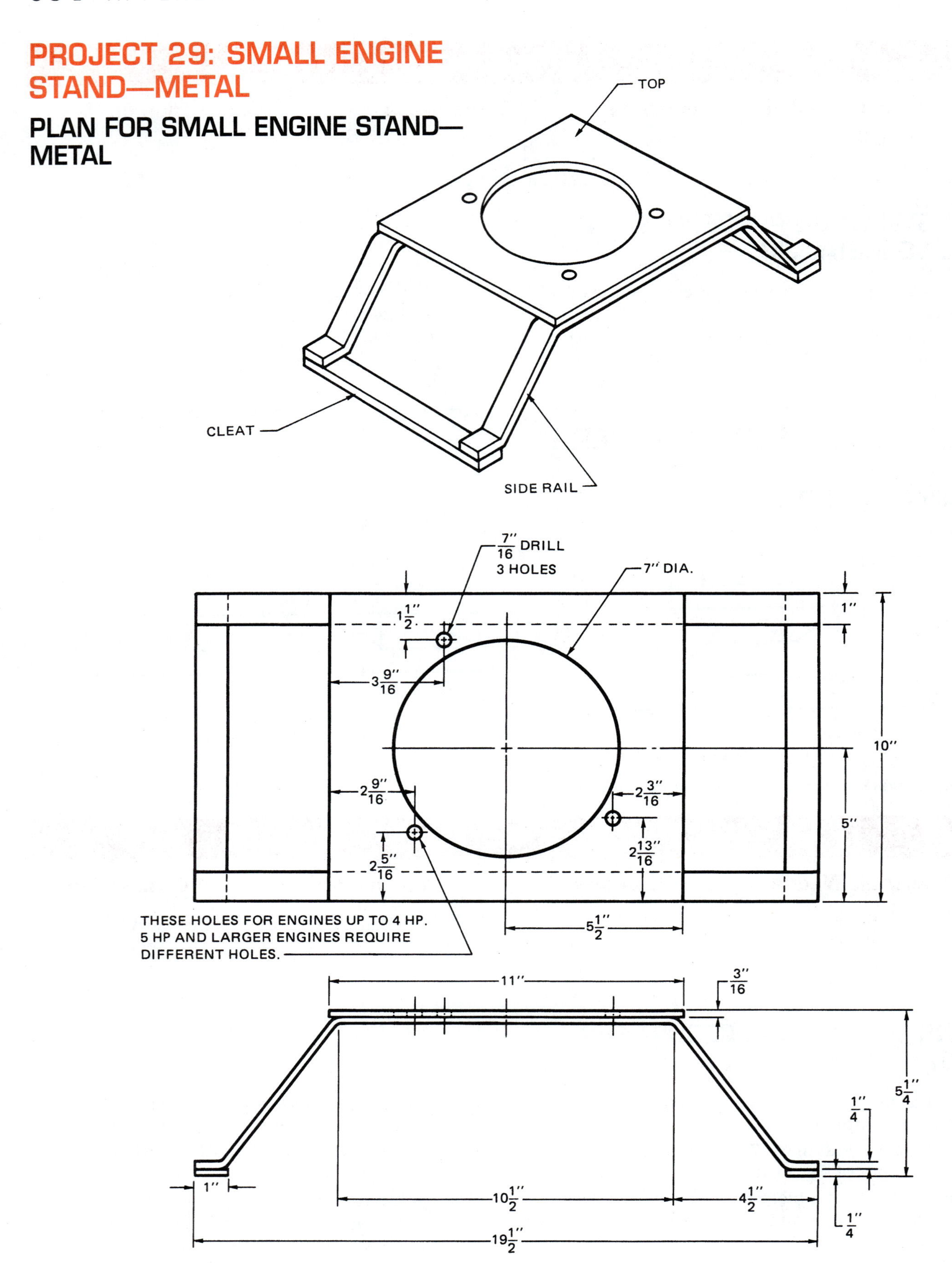

Courtesy of Agricultural Education and Agricultural Engineering, Virginia Polytechnic Institute and State University, Blacksburg, VA

BILL OF MATERIALS

Materials Needed	Quantity	Dimensions	Description
Band, steel	2	$^1/_4$" × 1" × 24"	Side rails
Band, steel	2	$^1/_4$" × 1" × 10"	Cleats
Plate, steel	1	$^3/_{16}$" × 10" × 11"	Top

CONSTRUCTION PROCEDURE FOR SMALL ENGINE STAND—METAL

1. Locate and mark the midpoint of the two side pieces.
2. Measure and mark points 5$^1/_2$" in both directions from the midpoint.
3. Measure and mark points 1" from both ends of the side pieces.
4. Bend the side pieces so they are 19$^1/_2$" from end to end and 4$^{13}/_{16}$" from the bottom to the upper surface. Place them side by side and be sure their shapes are identical.
5. Weld the cleats to the sides at 90° angles.
6. Lay out and center punch for all holes in the top.
7. Drill the $^7/_{16}$" holes.
8. Use a divider to lay out the 7" diameter hole.
9. Drill a hole inside the 7" circle to insert a metal cutting bayonet saw blade and cut out the circle; or, use a cutting torch to make the cut.
10. Weld the top to the sides.
11. Apply a suitable finish.

PROJECT 30: SHOP STAND—ADJUSTABLE

PLAN FOR SHOP STAND—ADJUSTABLE

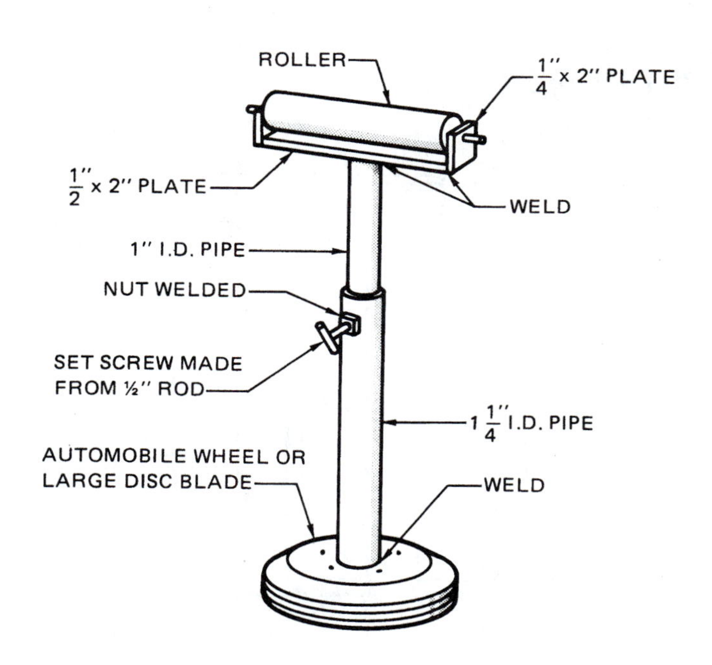

BILL OF MATERIALS

Materials Needed	Quantity	Dimensions	Description
Steel automobile or truck rim		According to need	Base
Pipe, black steel	1	$1^1/_4$" ID × *A	External post
Pipe, black steel	1	1" ID × *B	Internal post
Band, steel	1	$^1/_2$" × 2" × *C	Base plate
Band, steel	2	$^1/_4$" × 2" × 2"	Bearing ends
Wood or metal or rubber	1	Approx. 2" dia × 12"	Roller

*A—Determined by the desired minimum height of the stand minus the height of the base and roller.
*B—Same length as A.
*C—Same length as the roller plus $^1/_4$".
Note: Standard 1" ID pipe will fit well in standard $1^1/_4$" ID pipe. 2" ID pipe will fit well in $2^1/_2$" ID pipe.

CONSTRUCTION PROCEDURE FOR SHOP STAND—ADJUSTABLE

1. Obtain a rim of suitable size and weight to support the height of the stand needed. A large steel disc or other heavy metal object will also work well.
2. Determine the length of pipes needed to provide the desired height of the stand.
3. Drill a $^5/_8$" hole 1" from one end of the external post. Weld a $^1/_2$" nut over the hole. Make a setscrew from $^1/_2$" cold rolled round stock or use a standard bolt with a cross handle welded to it.
4. Weld the external post to the rim.
5. Obtain a standard roller or make one from 2" wood stock or $1^1/_2$" ID pipe 12" long.
6. Cut the base plate $^1/_4$" longer than the roller.
7. Lay out and drill bearing holes in the bearing ends to hold the top of the roller slightly above the tops of the ends.
8. Weld the bearing ends with the roller in place.
9. Apply a suitable finish.

PROJECT 31: STEEL POST DRIVER

PLAN FOR STEEL POST DRIVER

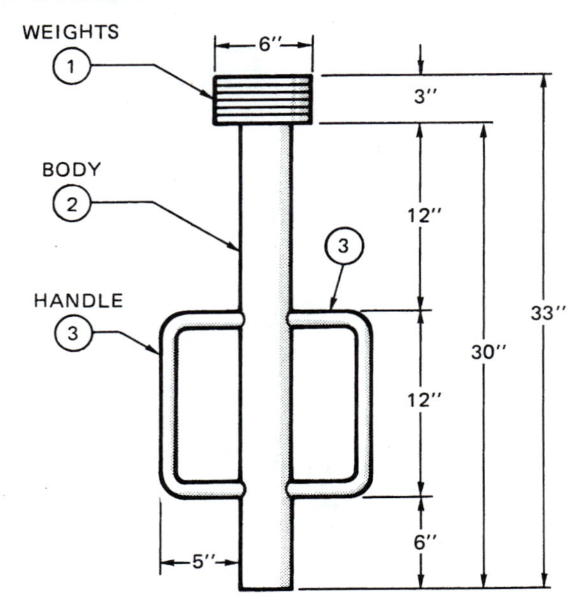

Courtesy of Texas Vocational Instructional Services, Texas A & M University, College Station, TX

BILL OF MATERIALS

Materials Needed	Quantity	Dimensions	Description
Flat or plate steel	6	1/2" × 6" × 6" or 6" dia	Weights
Pipe	1	3" ID × 30"	Body
Round stock	2	1/2" × 20"	Handles

CONSTRUCTION PROCEDURE FOR STEEL POST DRIVER

1. Cut six pieces for weights and weld them to form a mass 3" thick × 6" × 6", or 6" in diameter.
2. Weld the weight mass securely to the body.
3. Bend the two handles so they are identical.
4. Weld the handles to the body.
5. Apply a suitable finish.

PROJECT 32: EXTENSION CORD STAND

PLAN FOR EXTENSION CORD STAND

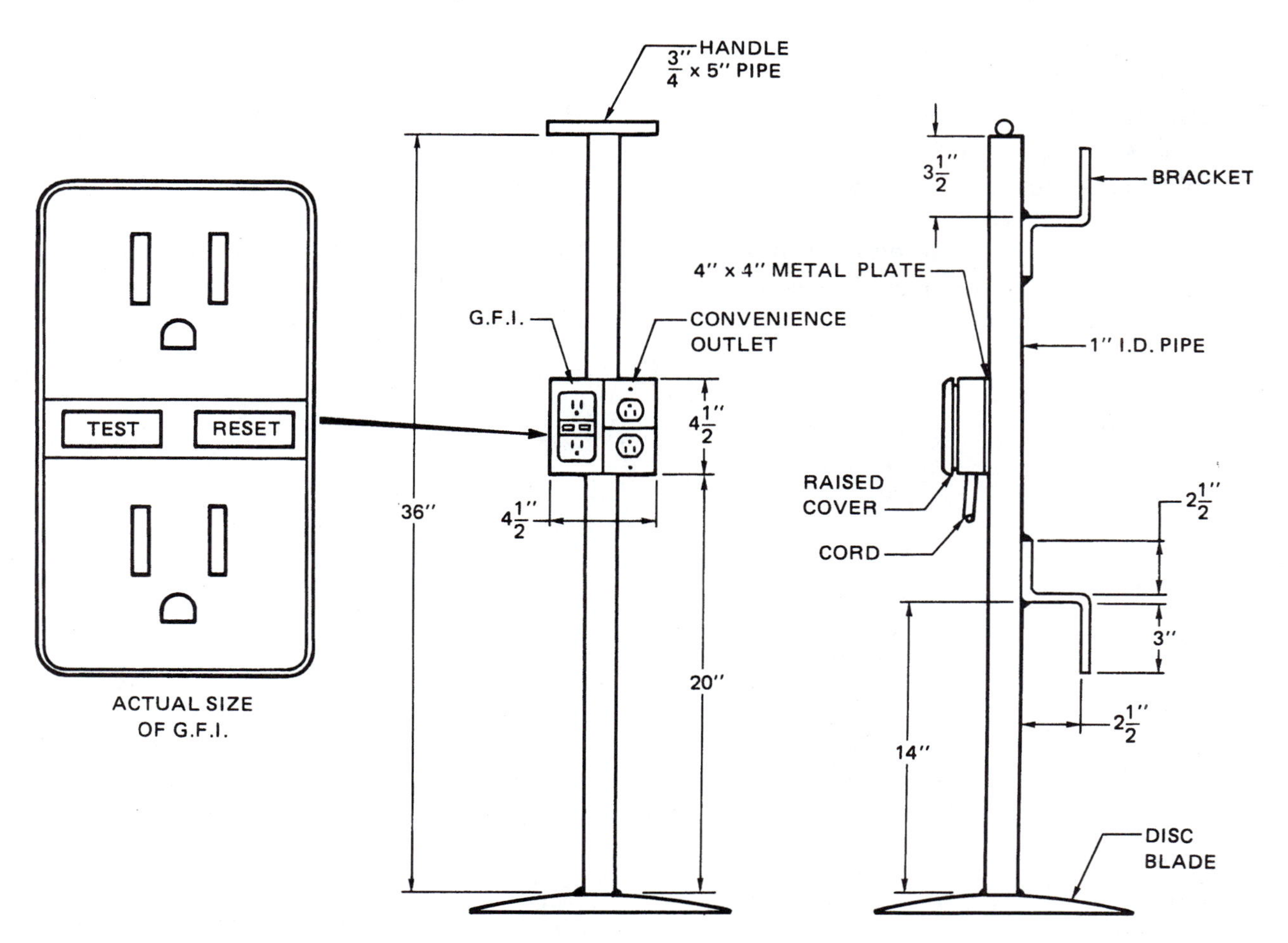

Courtesy of Department of Agricultural and Extension Education, The Pennsylvania State University, University Park, PA

BILL OF MATERIALS

Materials Needed	Quantity	Dimensions	Description or Use
Pipe, black iron	1	1" ID × 36"	Upright
Pipe, black iron	1	3/4" × 5"	Handle
Band, iron	2	1/8" × 3/4" × 8"	Brackets
Disc blade	1	12" dia	Base
Plate, steel	1	1/8" × 4" × 4"	Outlet box base
Outlet box, square	1	4" square, 1 1/2" deep	
*Ground fault interrupter	1		
Duplex receptacle	1		
Grounded cap, male plug	1		
Raised cover	1		
Box connector	1	3/4"	
Solderless connectors	3		
Cable	100 feet		Neoprene or plastic covered portable cable, AWG 12-2 stranded conductor with ground

*Substitute a duplex receptable if the circuit is already protected by a ground fault interrupter.

CONSTRUCTION PROCEDURE FOR EXTENSION CORD STAND

1. Cut the handle 5" long.
2. Cut the bracket pieces 8" long.
3. Shape the brackets to hold the electrical cable.
4. Cut 1" pipe to 36" length.
5. Weld the pipe to the disc blade, using a low hydrogen electrode.
6. Weld the handle and brackets to the pipe.
7. Drill 1/8" × 4" × 4" steel plate for electrical boxes; weld the plate to the pipe.
8. Bolt the electrical box to the plate.
9. Install one end of conductor into the box.
10. Install the ground fault interrupter (GFI).
11. Install the duplex receptacle so it is protected by the GFI.
12. Add the cover plate.
13. Connect the male plug to the electrical conductor.
14. Apply a suitable finish.

PROJECT 33: JACK STAND—AUTOMOBILE

PLAN FOR JACK STAND—AUTOMOBILE

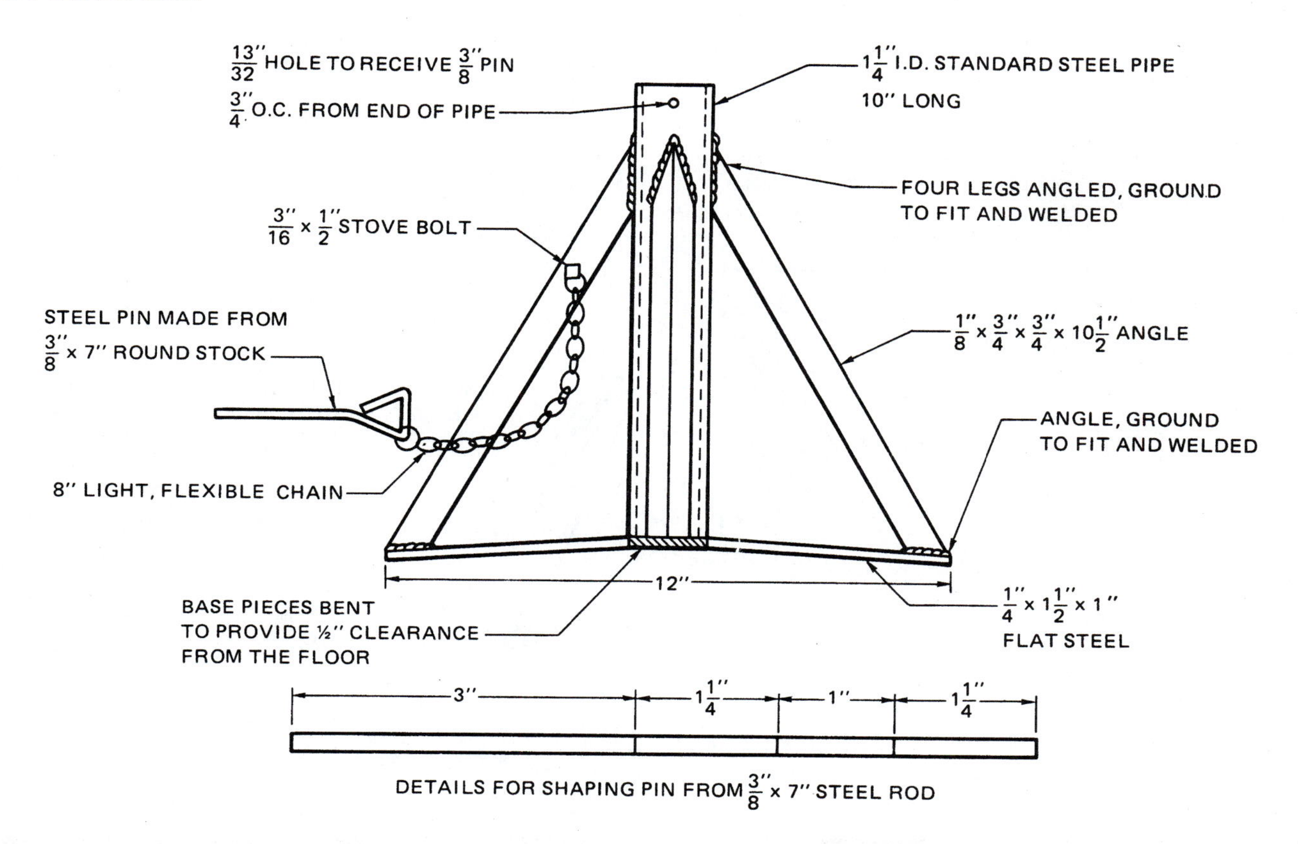

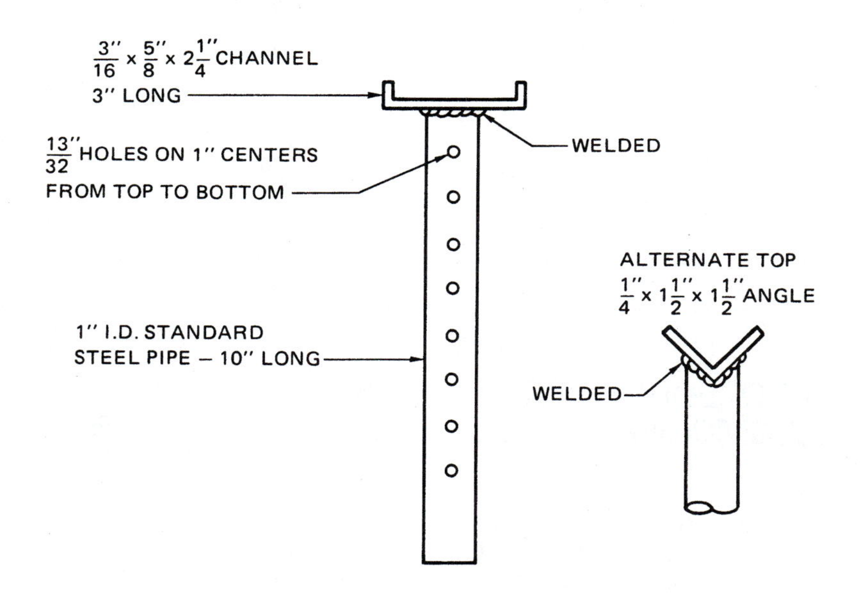

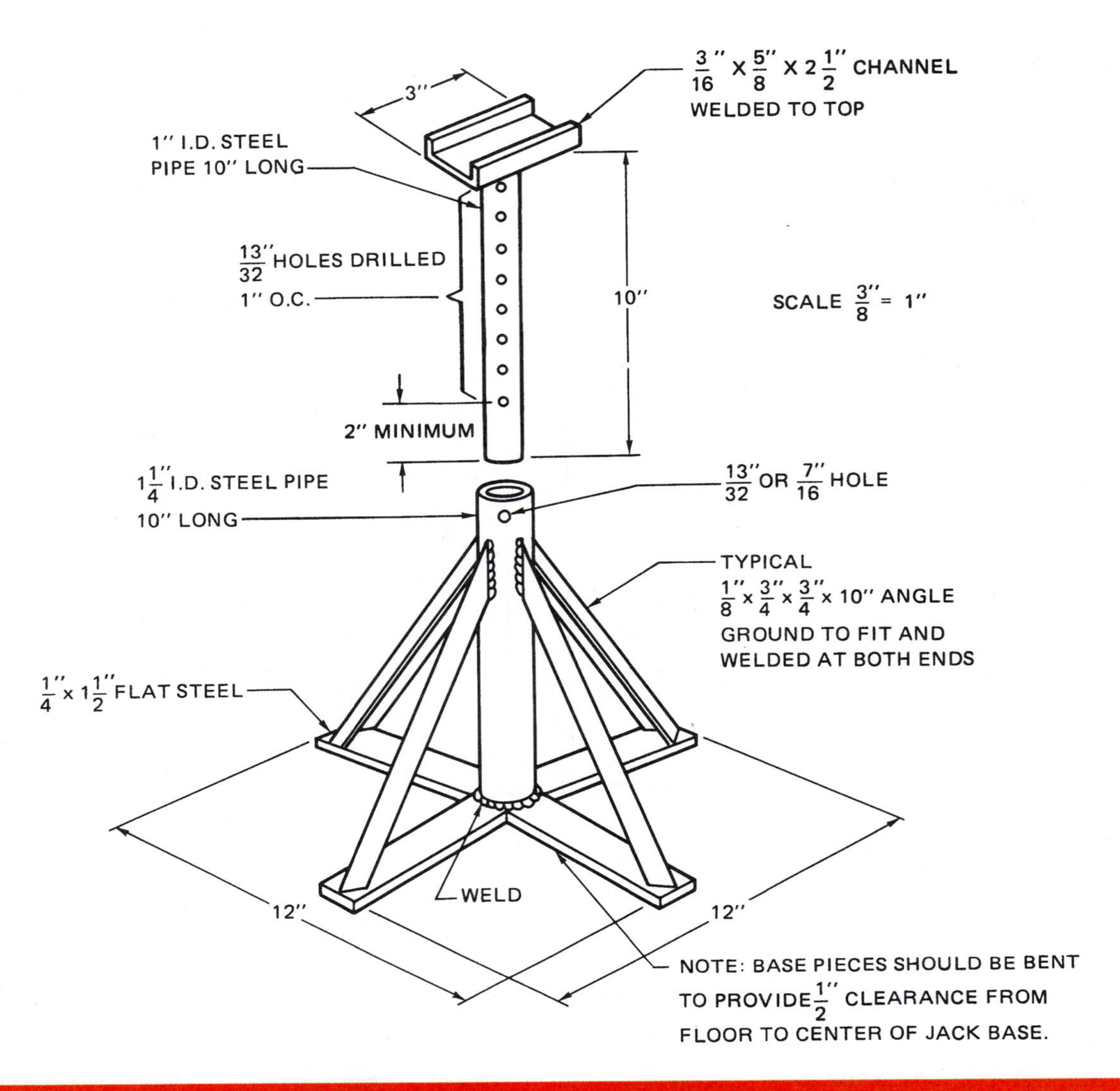

BILL OF MATERIALS

Materials Needed	Quantity	Dimensions	Description
Band, steel	1	1/4" × 1 1/2" × 12"	Base
Band, steel	2	1/4" × 1 1/2" × 5 1/4"	Base
Angle, steel	4	1/8" × 3/4" × 3/4" × 10"	Legs
Pipe, steel	1	1 1/4" ID × 10"	External post
Pipe, steel	1	1" ID × 10"	Internal post
Channel, steel	1	3/16" × 5/8" × 2 1/2" × 3"	Top
Round stock	1	3/8" dia × 7"	Pin, hardened
Chain	1	Light weight, 8"	Pin chain
Bolt, stove	1	3/16" × 1/2"	Chain fastener

CONSTRUCTION PROCEDURE FOR JACK STAND—AUTOMOBILE

1. Cut all pieces to length.
2. Weld two short base pieces to the longer piece to form a 90 degree cross. Grind welds smooth and slightly round off the exposed ends.
3. Bend the pieces of the base slightly so there will be 1/2" clearance between the floor and the center of the cross.
4. Ream the inside of the ends of the external post.
5. Center punch 3/4" from one end of the external post.

6. Scribe a mark from end to end down the center axis of the interior post.
7. Center punch at 1" intervals on the line from end to end of the interior post.
8. Place the exterior post in a drill press vise or clamp it in a vee block and drill a $^{13}/_{32}$" hole through both walls of the pipe.
9. Without unclamping the exterior post, slide the interior post into the exterior post and run the drill through the exterior post and on through both walls of the interior post. Drill all holes in the interior post in like fashion.
10. Weld the exterior post to the center of the base so the assembly sits with the post perpendicular to the floor.
11. Grind the legs so they fit the post and base. Tack weld the legs to the base and to the post, being careful to keep the parts and spacings equal and the post perpendicular. Weld all joints completely.
12. Weld the top to the interior post.
13. Bend a loop in the pin and attach the chain to the pin.
14. Drill a $^{3}/_{16}$" hole $1^{1}/_{2}$" from the top of one leg and anchor the chain with a $^{3}/_{16}$" × $^{1}/_{2}$" stove bolt.
15. Chip all welds, and remove all burrs.
16. Apply a suitable finish.

NOTE: All welds in this project must be high quality and a full-size pin must pass through all four pipe walls when stand is in use. For added safety, insert a hardened bolt in the first hole above the external post.

CAUTION:

- Maximum capacity is one-half ton.

PROJECT 34: JACK STAND—TRACTOR

PLAN FOR JACK STAND—TRACTOR

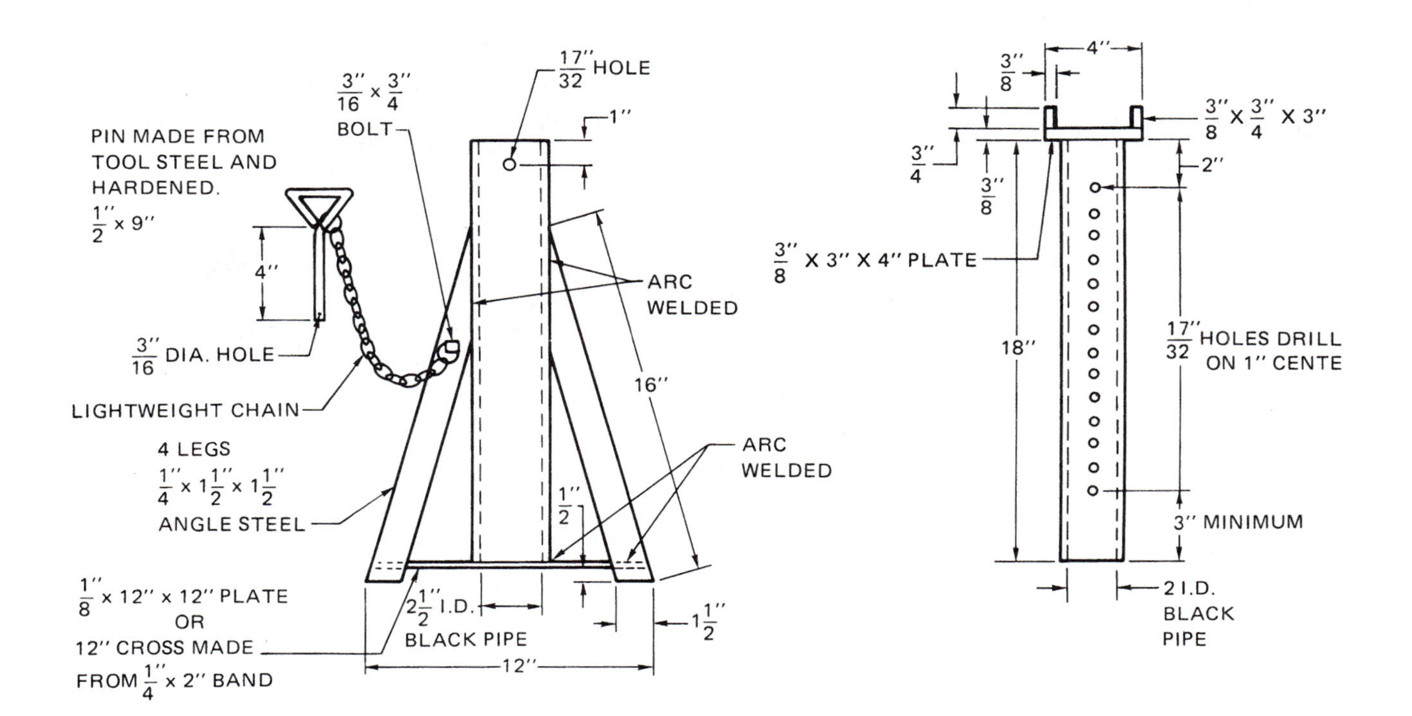

BILL OF MATERIALS

Materials Needed	Quantity	Dimensions	Description
Band, steel	1	$1/4$" × 2"× 15"	Base
Band, steel	2	$1/4$" × 2"× $6^1/2$"	Base
Angle, steel	4	$1/4$" × $1^1/2$" × $1^1/2$" × 16"	Legs
Pipe, steel	1	$2^1/2$" ID × 18"	External post
Pipe, steel	1	2" ID × 18"	Internal post
Plate	1	$3/8$" × 3" × 4"	Top base
Plate	2	$3/8$" × $3/4$" × 3"	Top rails
Round stock	1	$1/2$" dia × 9"	Pin, hardened
Chain	1	Light weight, 12"	Pin chain
Bolt, stove	1	$3/16$" × $3/4$"	Chain fastener

CONSTRUCTION PROCEDURE FOR JACK STAND—TRACTOR

1. Cut all pieces to length.
2. Make a cross by welding two pieces of $1/4$" × $6\,1/2$" band to a piece of $1/4$" × 15" band to form a 90° cross. Grind the welds smooth.
3. Cut all four ends with two 45° angles so the ends will fit inside the legs.
4. Ream the inside of the ends of the external post.
5. Center punch 1" from one end of the external post.
6. Scribe a mark from end to end down the center axis of the interior post.
7. Center punch 2" from one end and at 1" intervals on the line from that end to a point 3" from the other end of the interior post.
8. Place the exterior post in a drill press vise or clamp it in a vee block and drill the $17/32$" hole through both walls of the pipe.
9. Without unclamping the exterior post, slide the interior post into the exterior post and run the drill through the exterior post and on through both walls of the interior post. Drill all holes in the interior post in like fashion.
10. Weld the exterior post to the center of the base so the assembly sits with the post perpendicular to the floor.
11. Grind the legs so they fit the post. Tack weld the legs to the base and to the post, being careful to keep the parts and spacing equal and the post perpendicular. Weld all joints completely.
12. Weld the top on the interior post.
13. Bend a loop in the pin, harden the pin, and attach the chain to the pin.
14. Drill a $3/16$" hole $1^1/2$" from the top of one leg and anchor the chain with a $3/16$" × $3/4$" stove bolt.
15. Chip all welds and remove all burrs.
16. Apply a suitable finish.

NOTE: All welds in this project must be high quality and a full-size pin must pass through all four pipe walls when stand is in use. For added safety, insert a second $1/2$" hardened bolt in the first hole above the external post.

CAUTION:

- Maximum capacity is one ton.

PROJECT 35: CAR RAMP

PLAN FOR CAR RAMP

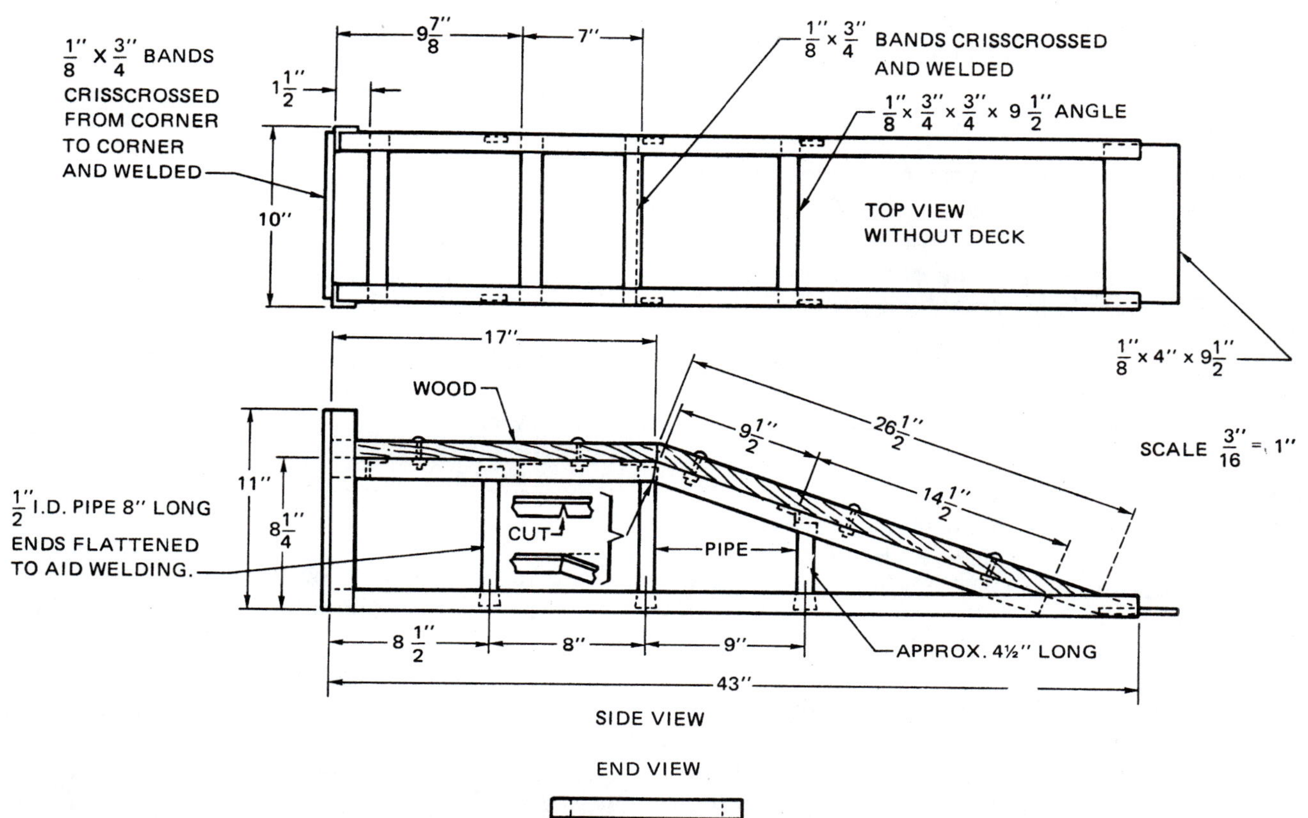

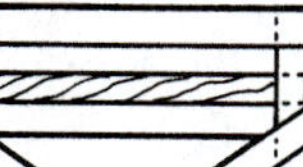

BILL OF MATERIALS

Materials Needed	Quantity	Dimensions	Description
Angle, steel	2	1/8" × 3/4" × 3/4" × 43"	Bottom runners
Angle, steel	2	1/8" × 3/4" × 3/4" × 41"	Top rails
Angle, steel	4	1/8" × 3/4" × 3/4" × 9 1/2"	Cross supports
Flat (band), steel	2	1/8" × 3/4" × 3/4" × 11"	Corner standards
Flat (band), steel	1	1/8" × 4"× 9 1/2"	Toe plate
Flat (band), steel	4	1/8" × 3/4" × 12" (approx)	Diagonal braces
Flat (band), steel	1	1/8" × 3/4" × 10"	Stop bar
*Pipe, standard, black steel	4	1/4" ID × 8"	Posts
Pipe, standard, black steel	2	1/4" ID × 4 1/2" (approx)	Ramp posts
**Plywood, exterior	1	3/4" × 9 3/4" × 17" (approx)	Deck
Plywood, exterior	1	3/4" × 9 3/4" × 26 1/2" (approx)	Ramp deck
Bolts, carriage or stove	10	1/4" × 1 1/2"	Ramp fasteners

*Flatten ends to improve welds. Angles (1/8" ×3/4" × 3/4") may be used in place of pipe.
**Sawed lumber may be used if it is placed with the grain running across the ramp, or sheet steel may be welded on for decking.

CONSTRUCTION PROCEDURE FOR CAR RAMP

1. Cut to length, exactly as specified, the bottom runners, top rails, cross supports, corner standards, toe plate, stop bar, and posts.
2. Assemble the parts and tack weld one side frame.

> **CAUTION:**
> - **Sides must be welded so the corner standard will be on the outside of the runner and rail. When fully assembled, runners will have one flat side down and one flat side out. Rails have one flat side up and one flat side out.**

The following procedure is suggested:

a) Lay a runner and rail on a flat surface parallel to each other. One leg of each points towards the opposite piece and the other leg of each points upwards.
b) Slip one leg of a standard under the ends of the runner and rail.
c) Lay two 8" posts in place between the runner and rail.
d) Mark the standard where the rail intersects. The top of the rail should be 8 1/4" from the bottom of the runner.
e) Starting at the standard, measure out the rail 17" and lay out and cut a notch in the rail to permit a 25° bend.
f) Place the rail in a vise and bend it to close the notch.
g) Clamp the runner, standard, and rail assembly to a flat metal surface. Square the joints carefully and tack weld.
h) Place the 8" posts in position and tack weld.
i) Tack weld the rail to the runner at the bottom of the ramp.
j) Position the 4 1/2" post as shown in the plan. If it's too long, shorten it as needed. Tack weld.

3. Assemble and tack weld the opposite side frame.

> **CAUTION:**
> - **Make the unit match the first one except be careful to set it up so it faces the first one.**

4. Install and tack weld cross members.

> **CAUTION:**
> - **Be sure the two sides are square to one another across the front and parallel from end to end.**

5. Recheck the unit to be sure it is square and parallel. Install and tack weld the toe plate, stop bar, and cross braces.
6. Make permanent welds at all points.

> **CAUTION:**
> - **Select a welding sequence that minimizes distortion.**

7. Chip all welds. Prime and paint the unit.
8. Measure the unit for the exact cutting of the deck pieces.
9. Cut, drill holes, and install the deck pieces.

> **CAUTION:**
> - **Cut angles on the ends of the ramp pieces so they fit well for maximum strength at the floor and where they meet at the top of the incline. Drill bolt holes in the frame and deck so there is room to apply nuts when bolts are installed.**

10. Prime and paint the deck parts.

    NOTE: All welds in this project must be high quality and fuse all parts of the joint.

> **CAUTION:**
> - **Maximum capacity is one-half ton.**

PROJECT 36: CREEPER

PLAN FOR CREEPER

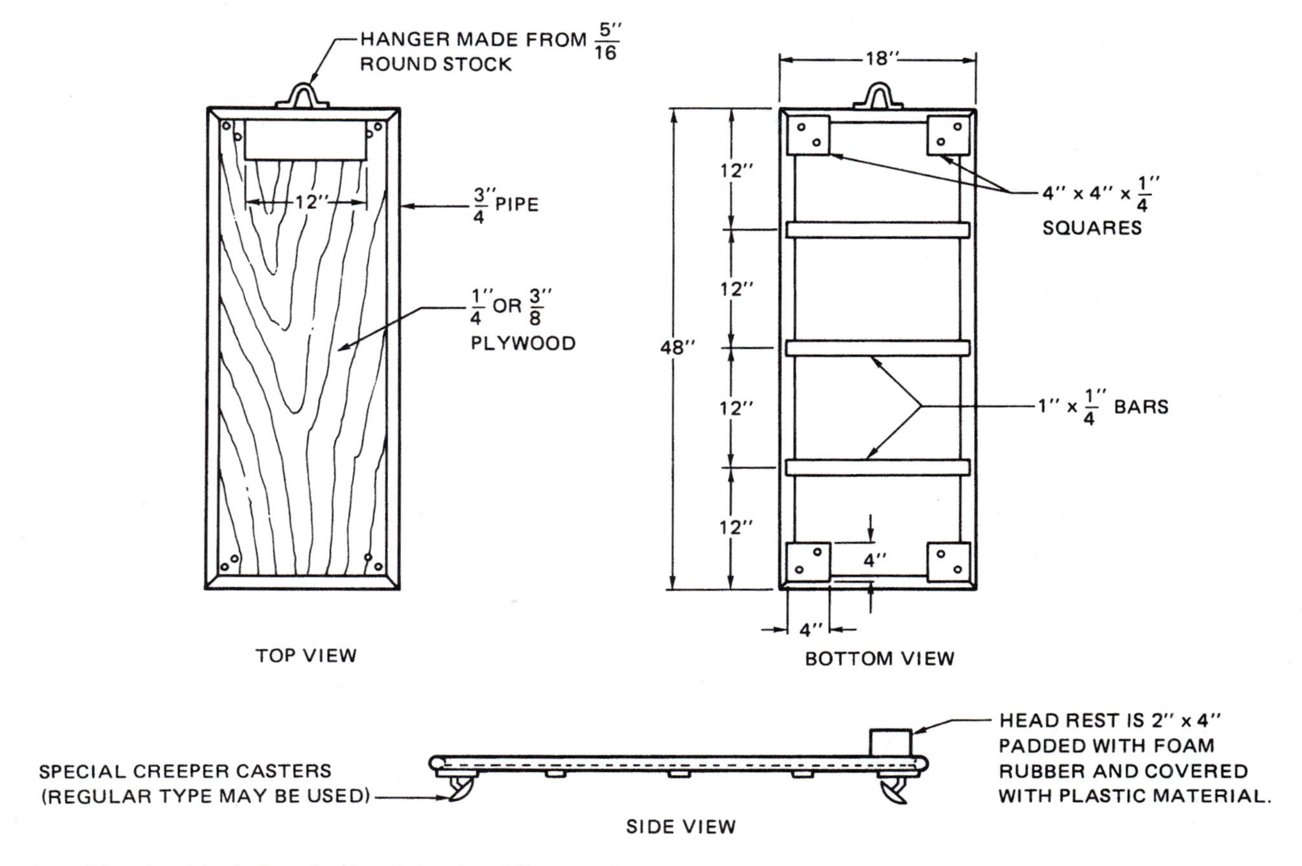

Courtesy of Vocational Agriculture Service, University of Illinois, Urbana, IL

BILL OF MATERIALS

Materials Needed	Quantity	Dimensions	Description
*Pipe, black steel	2	3/4" ID × 48"	Side rails
*Pipe, black steel	2	3/4" ID × 18"	End rails
Band, steel	3	1/4" × 1" × 17"	Bars
Band or plate, steel	4	1/8" or 1/4" × 4" × 4"	Square plates
Round stock, steel	1	5/16" × 4"	Hangers
Plywood	1	1/4" or 3/8" × 15" × 46"	Deck
Lumber	1	2" × 4" × 12"	Head rest
Foam	1	As needed	Head rest
Plastic cover	1	As needed	Head rest
Bolt, carriage	8	1/4" × 1"	Deck and caster fasteners
Creeper casters	4	As needed	Casters

*Frame may also be constructed of hard wood or angle steel.

CONSTRUCTION PROCEDURE FOR CREEPER

1. Cut all metal pieces to size.
2. Assemble and weld the rails, bars, plates, and hanger.
3. Cut the deck to size.
4. Install the deck and casters using bolts with lock washers. Saw off excess bolt material.
5. Apply a suitable finish.
6. Construct and install the head rest using wood screws.

PROJECT 37: SAW HORSE—PIPE

PLAN FOR SAW HORSE—PIPE

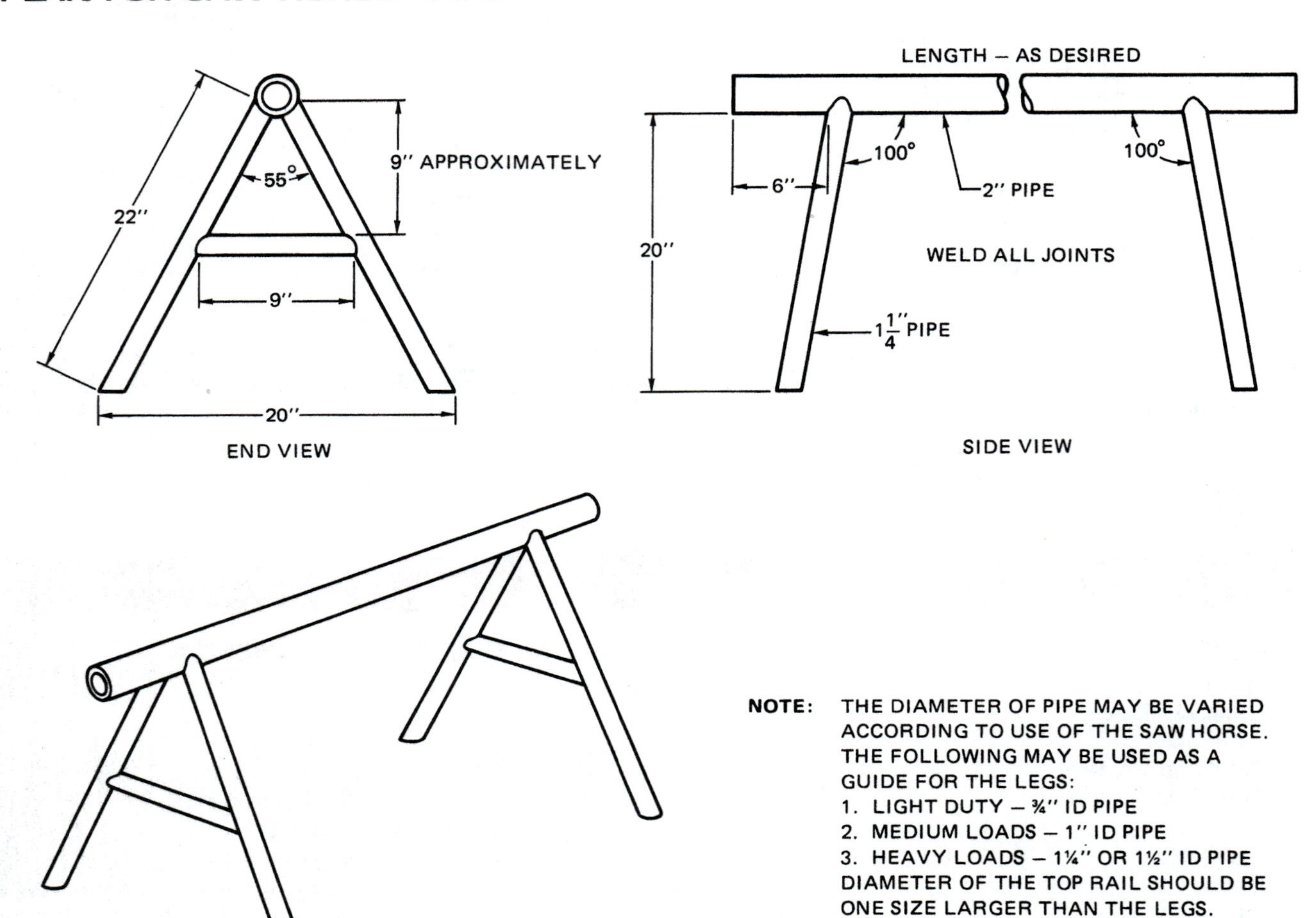

NOTE: THE DIAMETER OF PIPE MAY BE VARIED ACCORDING TO USE OF THE SAW HORSE. THE FOLLOWING MAY BE USED AS A GUIDE FOR THE LEGS:
1. LIGHT DUTY – ¾" ID PIPE
2. MEDIUM LOADS – 1" ID PIPE
3. HEAVY LOADS – 1¼" OR 1½" ID PIPE
DIAMETER OF THE TOP RAIL SHOULD BE ONE SIZE LARGER THAN THE LEGS.

BILL OF MATERIALS

Materials Needed	Quantity	Dimensions	Description
Pipe, black steel	1	$1^1/_4$" ID × 36"	Top rail
Pipe, black steel	4	1" ID × 20"	Legs
Pipe, black steel	2	1" ID × 9"	Braces

CONSTRUCTION PROCEDURE FOR SAW HORSE—PIPE

1. Cut all pipe to length.
2. Grind the ends of the legs to fit the top rail.

   NOTE: Welding will be easier if the end of each leg is flattened slightly where it will be fitted to the top rail.

3. Tack weld the legs to the top rail.
4. Adjust the leg positions so the saw horse stands squarely on a flat surface and all spacings are correct.
5. Weld the legs permanently.
6. Slightly flatten the ends of the braces and grind them to fit the legs.
7. Weld the braces in place.
8. Grind the bottoms of the legs so they are flat on the floor.

PROJECT 38: WORKBENCH BRACKETS

PLAN FOR WORKBENCH BRACKETS

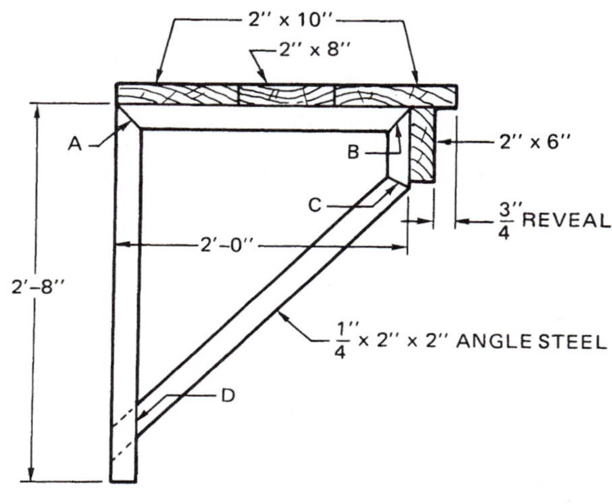

ANGLE BRACKET

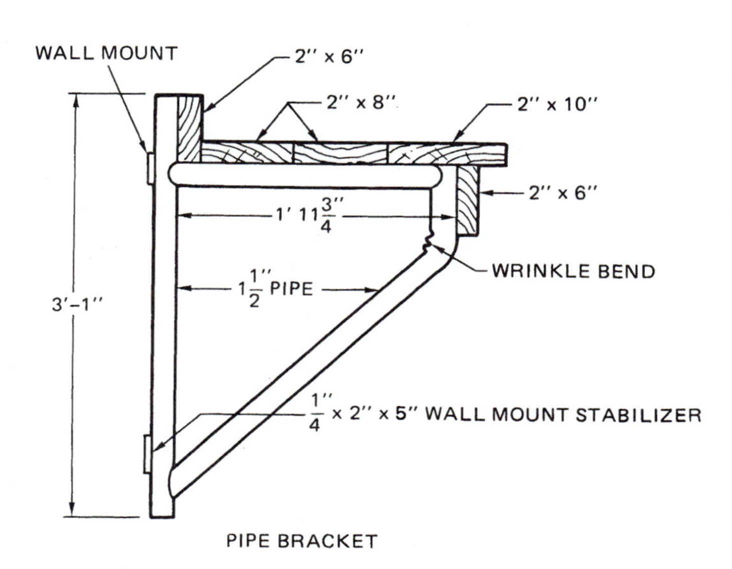

PIPE BRACKET

GENERAL

THREE BRACKETS ARE USUALLY REQUIRED FOR A 10- OR 12- FOOT BENCH. TWO MAY BE SUFFICIENT FOR AN 8- FOOT BENCH.

BOLT BRACKETS TO OR THROUGH THE WALL.

BOLT ALL WOOD PIECES TO BRACKETS, USING 3/8" CARRIAGE BOLTS. PLANKS SHOULD BE JOINTED FOR A COMBINED TOTAL BENCH WIDTH TO PROVIDE A 3/4" REVEAL ON THE FRONT. EDGES OF PLANKS MAY BE GLUED AND CLAMPED TO PROVIDE A CONTINUOUS BENCH TOP.
HOLES ON TOP ARE COUNTERBORED 1 INCH AND FILLED WITH DOWEL PLUGS.

Courtesy of Vocational Agriculture Service, University of Illinois, Urbana, IL

BILL OF MATERIALS			
Materials Needed	**Quantity**	**Dimensions**	**Description**
ANGLE BRACKET (EACH):			
Angle, steel	2	$^1/_4$" × 2" × 2" × 2'8"	Legs
	1	$^1/_4$" × 2" × 2" × 2'0"	Horizontal support
or			
PIPE BRACKET (EACH):			
Pipe, black	2	$1^1/_2$" ID × 3'1"	Standards
Pipe, black	1	$1^1/_2$" ID × 1'10"	Horizontal support
Band, steel	2	$^1/_4$" × 2" × 5"	Wall mounts

CONSTRUCTION PROCEDURE FOR WORKBENCH BRACKETS

ANGLE IRON BRACKET

1. Cut two standards each with a 45 degree angle for the upper joint. Cut the horizontal support with a 45 degree angle on each end.
2. Cut a 30 degree wedge out at point C.
3. Square and tack weld A.
4. Heat and bend at C to create correct fits at B and D.
5. Tack weld B and D, being careful to keep A and B square.
6. Weld all joints.

PIPE BRACKET

1. Cut pipes to length, make wrinkle bend in one standard, and weld together.
2. Cut, drill, and weld on wall brackets.

PROJECT 39: WORKBENCH BRACKET—ALTERNATE PLAN

PLAN FOR WORKBENCH BRACKET—ALTERNATE PLAN

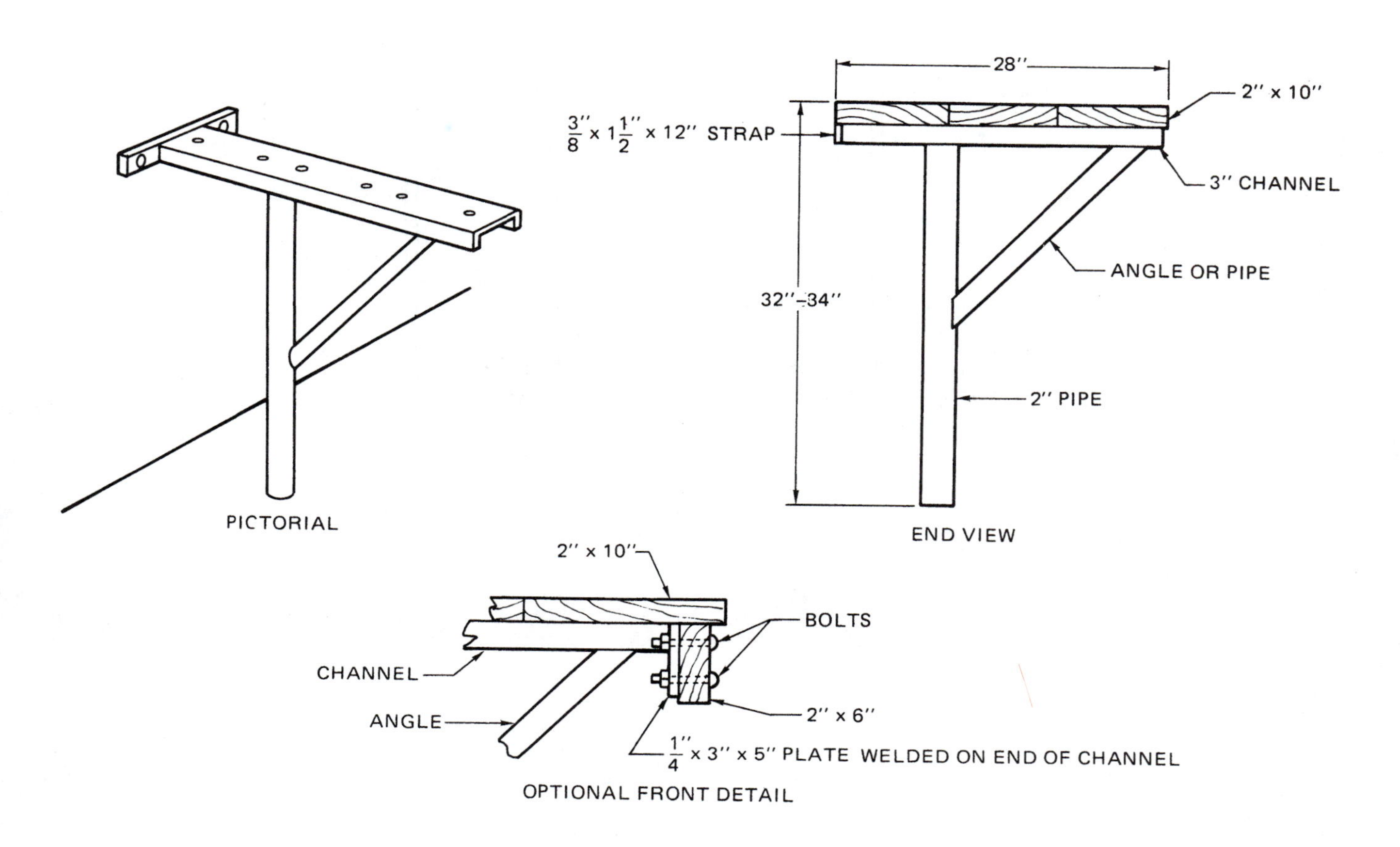

Courtesy of Vocational Agriculture Service, University of Illinois, Urbana, IL

BILL OF MATERIALS

Materials Needed	Quantity	Dimensions	Description
Channel, steel	1	3" × 27"	Top
Pipe, black steel	1	2" ID × 32"	Leg
Pipe, black steel	1	1" ID × 15"	Brace
Band, steel	1	3/8" × 1 1/2" × 12"	Bracket

CONSTRUCTION PROCEDURE FOR WORKBENCH BRACKET—ALTERNATE PLAN

1. Cut all pieces to length.
2. Drill 7/16" holes in the bracket.
3. Drill holes in the top as needed.
4. Round the corners of the channel that will protrude forward.
5. Weld the bracket to the top.
6. Tack weld the top to the leg and square the assembly.
7. Grind the ends of the brace to fit. Tack weld the brace.
8. Weld all joints permanently.
9. Apply a suitable finish.

PROJECT 40: HAND CART

PLAN FOR HAND CART

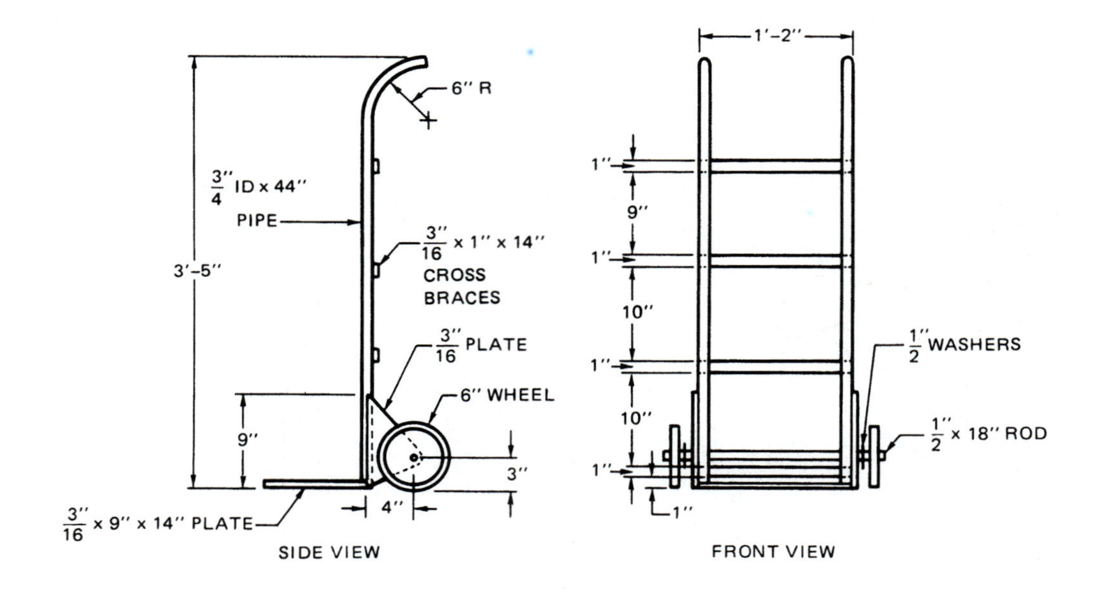

Courtesy of Department of Agricultural and Extension Education, The Pennsylvania State University, University Park, PA

BILL OF MATERIALS

Materials Needed	Quantity	Dimensions	Description
Wheels, heavy duty	2	6" × 1.50	For 1/2" axle
Round, cold rolled	1	1/2" × 18" long	Axle*
Plate, steel	1	3/16" × 9" × 14"	Base
Pipe, black	2	3/4" × 44" long	Handles
Plate, steel	2	3/16" × 4" × 9"	For axle supports
Band, steel	4	3/16" × 1" × 14"	Cross braces
Washers	4	1/2"	
Cotter pins	2	1/8" × 1"	
Metal primer	1 pint		
Enamel	1 pint		

*Axle length may vary with different wheels.

CONSTRUCTION PROCEDURE FOR HAND CART

1. Measure and cut the pipe to the dimension given in the bill of materials.
2. Bend the handles to a 6" radius.
3. Cut the axle supports using an oxyacetylene torch. Grind to identical shape.
4. Drill 1/2" holes with centers 1/2" from the point of the axle supports. Insert the axle.
5. Cut the four cross braces and weld in place.
6. Tack weld the base in position, handles in air.
7. Position the axle supports with the wheels mounted so that the base and wheels are level with the floor. Tack weld in place.

8. Drill the axle for the $^1/_8$" cotter pins to keep wheels on the cart and weld the axle in place.
9. Weld the $^1/_2$" washers to the axle on the inside of each wheel to serve as stops.
10. Complete welding all weld joints.
11. Paint the finished product with metal primer and a finish coat of enamel.

NOTE: If cart is to be used as a bag cart, add fenders by welding $^1/_8$" × 2" × 6" steel plate to the pipe over each wheel.

PROJECT 41: WELDING STAND—ACETYLENE

PLAN FOR WELDING STAND—ACETYLENE

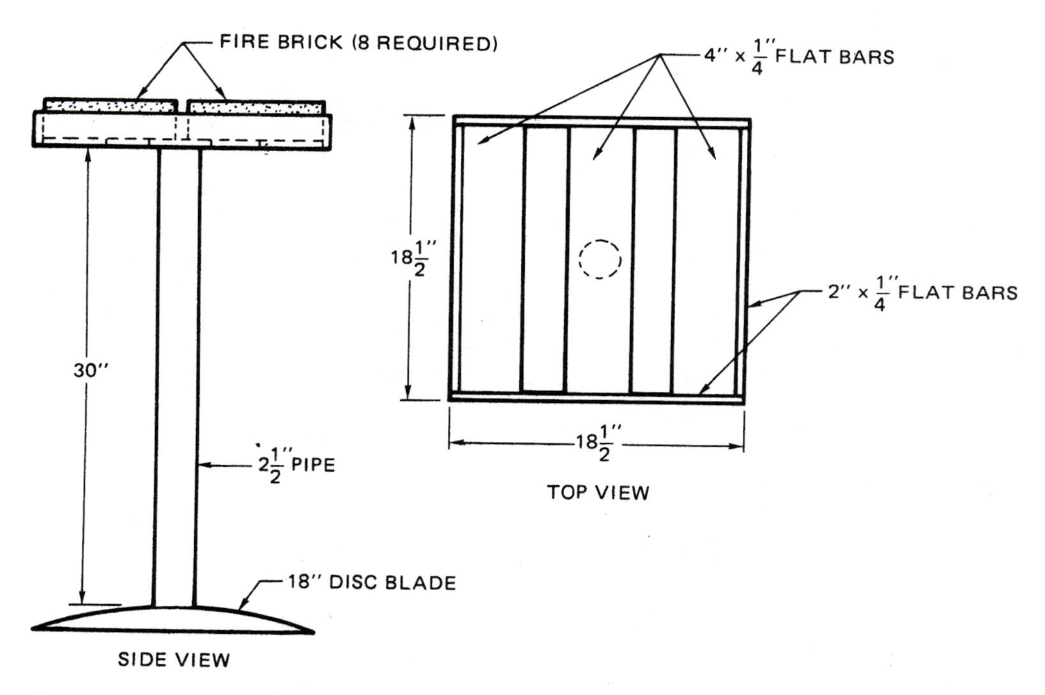

Courtesy of Vocational Agriculture Service, University of Illinois, Urbana, IL

BILL OF MATERIALS

Materials Needed	Quantity	Dimensions	Description
Disc blade	1	18" diameter	Base
Pipe, steel	1	22$^1/_2$" ID × 30"	Post
Flat (band) steel	3	$^1/_4$" × 4" × 18"	Bottom
	2	$^1/_4$" × 2" × 18$^1/_2$"	Sides
	1	$^1/_4$" × 2" × 18"	Ends

CONSTRUCTION PROCEDURE FOR WELDING STAND—ACETYLENE

1. Set two rows of four fire bricks each on a flat surface. Check to see if they form a bed 18" × 18". If they do not, determine the changes needed in the plan to accommodate your bricks. Modify the bill of materials accordingly.
2. Cut all pieces to length.
3. Tack weld the sides on the outside corners.
4. Tack weld the bottom pieces to the sides.
5. Invert the table on a flat surface and weld all joints from the underside.
6. Weld the post to the center bottom piece.
7. Weld the post to the disc base using a low-hydrogen electrode.
8. Apply a suitable finish.
9. Add the fire brick.

PROJECT 42: WELDING STAND—ARC

PLAN FOR WELDING STAND—ARC

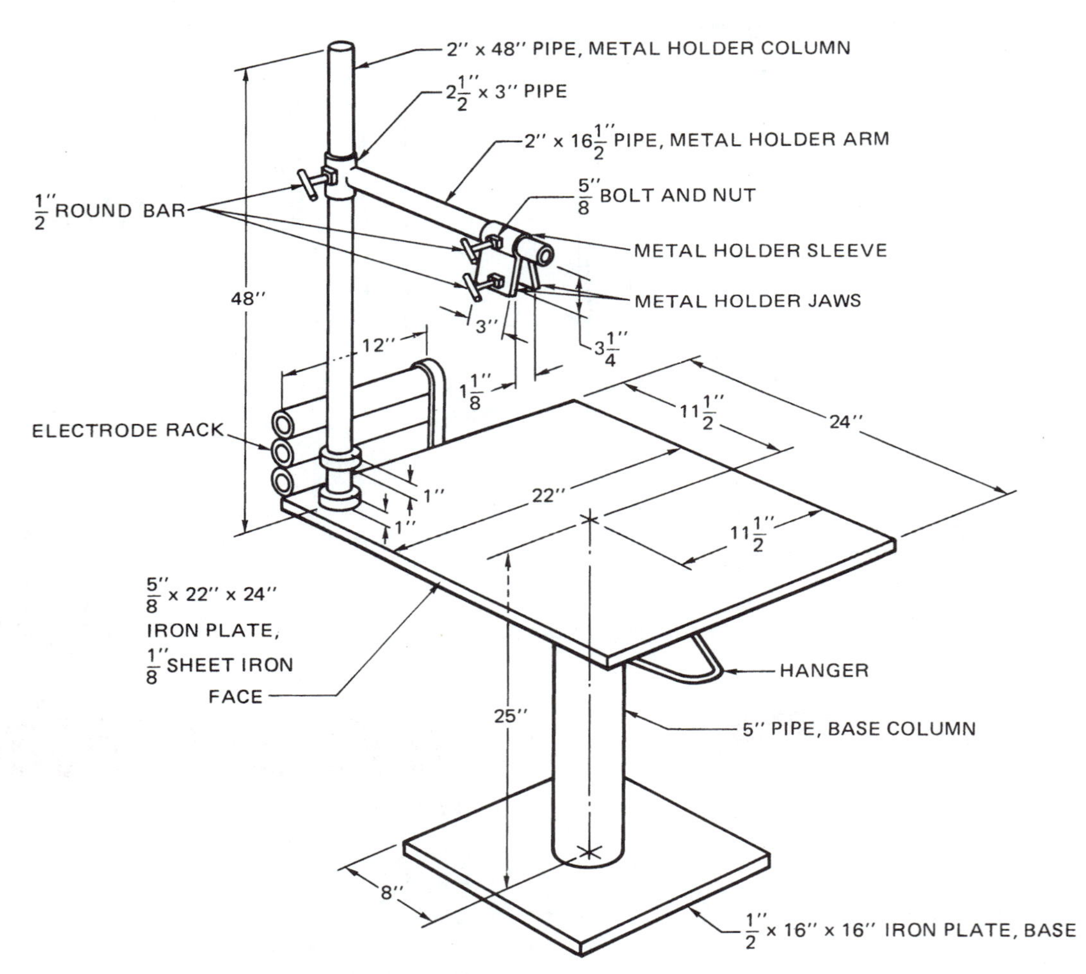

Courtesy of T. J. Wakeman and V. L. McCoy, The Farm Shop, *The Macmillan Company, 1960*

BILL OF MATERIALS

Materials Needed	Quantity	Dimensions	Description or Use
Plate iron	1 piece	$1/2" \times 16" \times 16"$	Base
	1 piece	$5/8" \times 2" \times 24"$	Top
	1 piece	$1/8" \times 22" \times 24"$	Top cover
	1 piece	$1/4" \times 2^1/2" \times 7^1/2"$	Electrode rack end
	2 pieces	$1/4" \times 3" \times 3^1/4"$	Clamp jaws
Black iron pipe	1 piece	$5" \times 26"$	Base column
	3 pieces	$2" \times 11^3/4"$	Electrode rack
	1 piece	$2^1/2" \times 2"$	Rack supports
	1 piece	$2" \times 48"$	Metal holder column
	1 piece	$2" \times 16^1/2"$	Metal holder arm
	2 pieces	$2^1/2" \times 3"$	Sleeves for arm
Round iron	1 piece	$1/2" \times 10"$	Hanger
	3 pieces	$1/2" \times 4"$	Adjustment screw bars
Coupling	1	$2^1/2" \times 2"$	Arm anchor
Machine bolts, nuts	3 each	$5/8" \times 2^1/2"$ bolts	NC threads

CONSTRUCTION PROCEDURE FOR WELDING STAND—ARC

1. Center and square the 5" pipe on the base plate, and weld them together.

CAUTION:

- **Tack weld and check for squareness before welding permanently.**

2. Place the top on a flat surface. Weld the other end of the 5" pipe to the top, as shown. Set the table upright.

CAUTION:

- **Tack weld and check for squareness before welding permanently.**

3. Cut the 2" coupling into two short pieces and weld the cut end of one to the table top, being careful to keep it level.
4. Thread one end of the 2" × 48" pipe. Screw it into the coupling welded to the table top.
5. Clamp the 2" × 11" pipes, ends even, on a flat surface. Weld 1" at each end between the pipes. Turn the pipes over and repeat the 1" welds.
6. Place both rack supports over the 48" column; weld the rack to them.
7. Drill a $3/4"$ hole in the center of one side of both $2^1/2"$ sleeves. Weld a $5/8"$ nut directly over the hole.
8. Shape one end of the 2" × 16 $1/2"$ pipe to fit the sleeve. Weld as shown.
9. In one metal-holder jaw, drill a $3/4"$ hole and weld on a $5/8"$ nut as in step 7.
10. Weld both jaws to the sleeve.
11. Weld a $1/2" \times 3^1/4"$ round rod to the head of each bolt, as shown. Round the ends of all handles with a grinder or file.
12. Bend a $1/2"$ rod for the electrode-holder hanger into a 2" wide U. Cover all but 1" of each end with rubber hose. Weld the ends to the underside of the table.

NOTE: The weldor sits at or works from the side of the table between the metal-holder column and the hanger.

PROJECT 43: WELDING TABLE—COMBINATION

PLAN FOR WELDING TABLE—COMBINATION

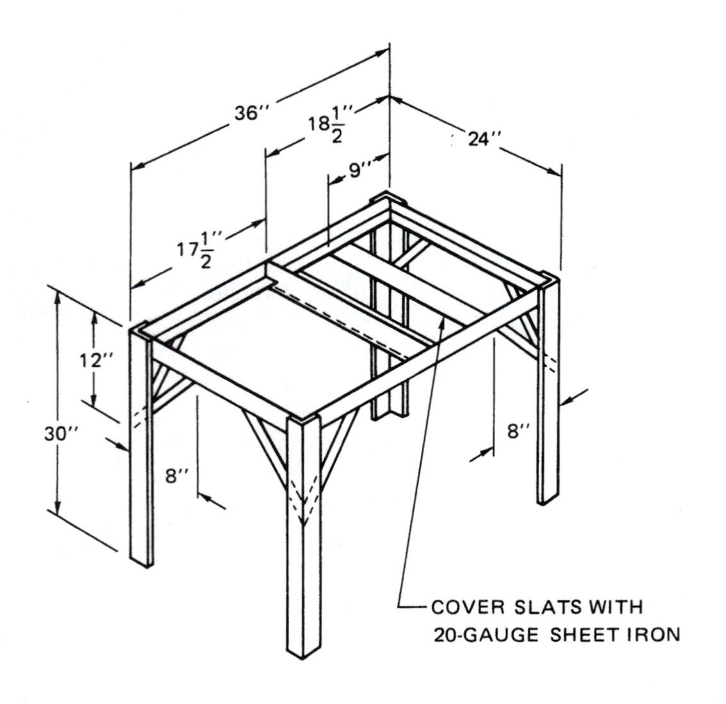

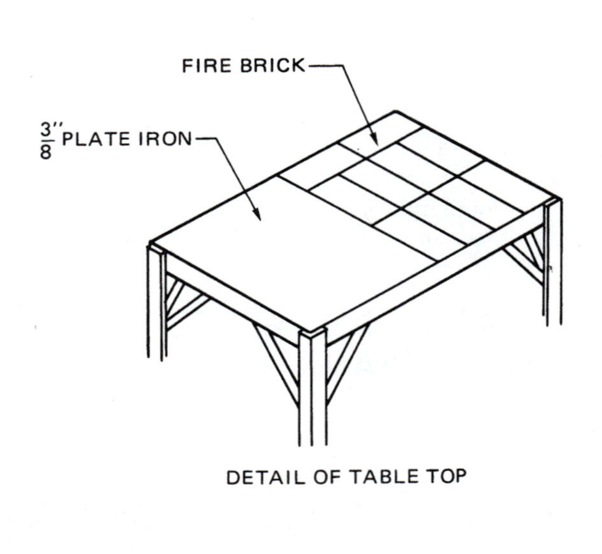

Courtesy of T. J. Wakeman and V. L. McCoy, The Farm Shop, *The Macmillan Company, 1960*

BILL OF MATERIALS

Materials Needed	Quantity	Dimensions	Description or Use
Angle steel	4 pieces	1/8" × 1 1/2" × 1 1/2" × 30"	Legs
	2 pieces	1/8" × 1 1/2" × 2" × 36"	Frame
	2 pieces	1/8" × 1 1/2" × 2" × 24"	Frame
Flat steel	8 pieces	3/16" × 1" × 14 1/2"	Braces
	2 pieces	3/16" × 2" × 21"	Bottom slats
	1 piece	3/16" × 2" × 23 3/4"	Partition
Plate steel	1 piece	3/8" × 17 1/2" × 24"	Top for arc welding
Sheet steel	1 piece	18" × 23 3/4"	20 gauge
Firebrick	10	2 3/8" × 4 1/2" × 9"	

CONSTRUCTION PROCEDURE FOR WELDING TABLE—COMBINATION

1. Cut all pieces to length as listed in the bill of materials. Cut the pieces for the frame at 45 degree angles with the 2" outsides of the angle steel pointing upward, as shown in the drawing.

2. Place the pieces for the frame on a smooth surface. Square and tack weld the frame on the outside and underside. Check for squareness, and then weld the outside and bottom side of corners.

3. Place the partition in place, with the top edge even with the top edge of the frame, as shown in the drawing. Weld the partition to the frame, avoiding any excess weld buildup.
4. Weld one slat on the under side against the bottom edge of the partition, with the ends welded to the inside leg of the angle-iron frame. Place the other slat in the center of the space as shown in the drawing and weld on the under side to the frame.
5. Square and tack the legs to the frame.
6. Clamp the leg braces in place and weld them to the frame and legs. Finish welding the legs to the frame.
7. Place 3/8" plate iron in position, and weld it to the top of the frame.
8. Place the sheet iron on top of the slats, as indicated in the drawing.
9. Place firebrick on top of the sheet iron.

PROJECT 44: GATE—METAL, SMALL

PLAN FOR GATE—METAL, SMALL

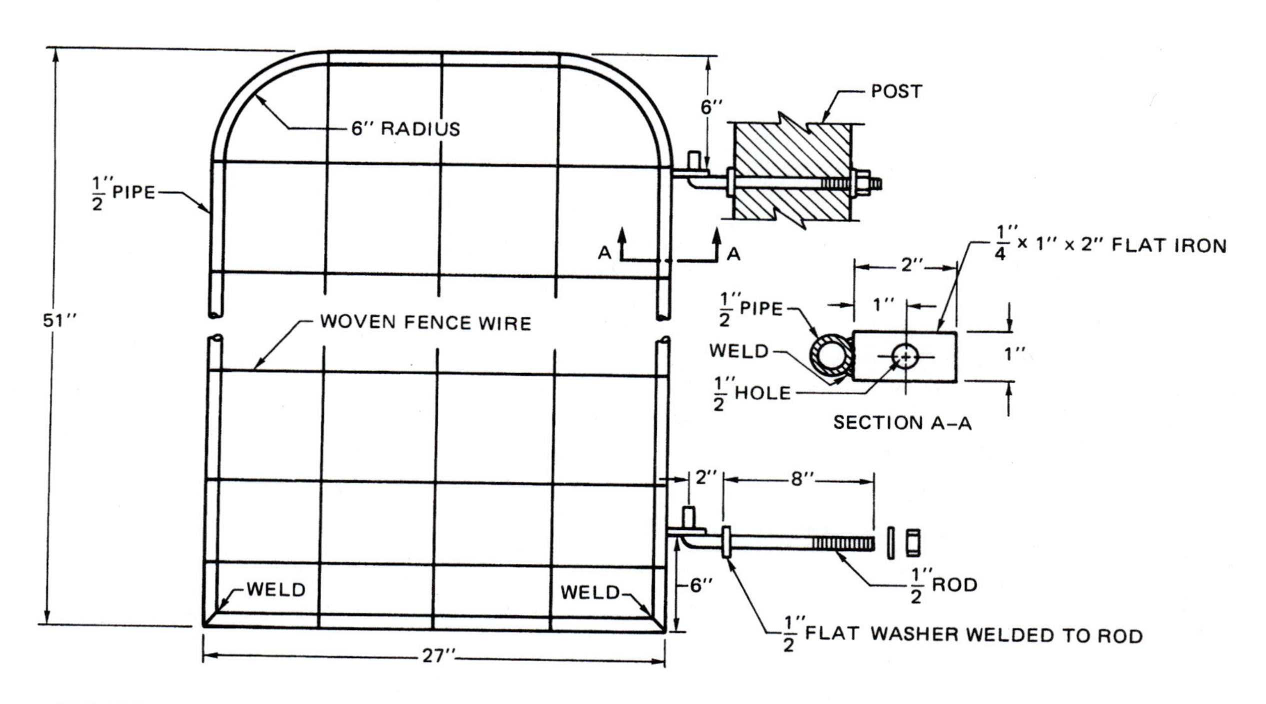

Courtesy of T. J. Wakeman and V. L. McCoy, The Farm Shop, *The Macmillan Company, 1960*

BILL OF MATERIALS

Materials Needed	Quantity	Dimensions	Description or Use
Pipe, steel, black	1 piece	1/2" × 13'0"	Frame
Flat iron	2 pieces	1/4" × 1" × 2"	Hinge
Round iron	2 pieces	1/2" × 11"	Hinge hooks
Washers	2	1/2"	Flat
	2	7/16"	Flat
Woven wire	1 piece	36" × 48"	

CONSTRUCTION PROCEDURE FOR GATE—METAL, SMALL

1. Bend pipe for the frame as shown in the drawing, starting the first bend 45" from one end.
2. Cut off the long end 51" from the top of the gate at a 45 degree angle. Make a cut of the same angle on the opposite corner.
3. Cut the bottom piece 27" long (or the width of the gate near the top), and weld in place.
4. Cut the flat iron for hinges to the length listed in the bill of materials. Lay off, center punch, and drill 1/2" holes. Weld to the gate frame, as shown in detail section A-A.
5. Cut the hinge hooks to the length listed in the bill of materials. Make a 90 degree bend 1" from one end. (The length will vary with the size of the gate post.)
6. Cut 2" of NC threads on the other end of the hooks.
7. Drive a 7/16" washer 2" from the bend in each hook. Weld in place, as shown in the drawing.
8. Fasten woven wire to the gate by passing the wire ends around the frame and wrapping around themselves. Use short pieces of wire to fasten the woven wire to the top and bottom of the gate.

NOTE: The height and width of the gate and the type of wire may be varied to suit a particular situation.

PROJECT 45: GATE—METAL, 12-FOOT OR AS NEEDED

PLAN FOR GATE—METAL, 12-FOOT OR AS NEEDED

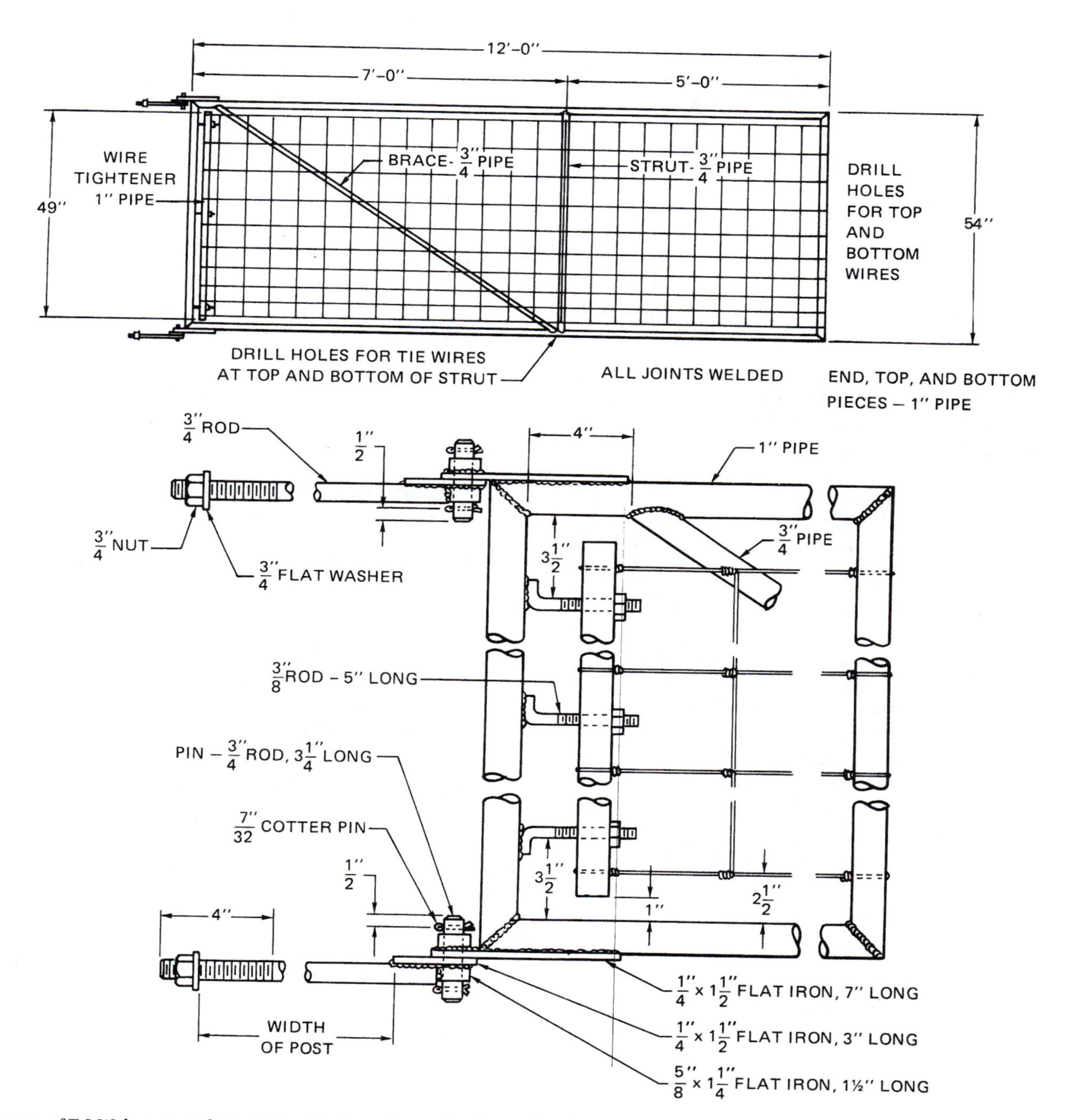

Courtesy of T. J. Wakeman and V. L. McCoy, The Farm Shop, *The Macmillan Company, 1960*

BILL OF MATERIALS

Materials Needed	Quantity	Dimensions	Description or Use
Pipe, steel, black	2 pieces	1" × 12'0"	Top and bottom
	2 pieces	1" × 54"	Ends
	1 piece	1" × 49"	Wire tightener
	1 piece	$^{3}/_{4}$" × 51$^{1}/_{2}$"'	Strut
	1 piece	$^{3}/_{4}$" × 7'9$^{1}/_{4}$"	Brace
Flat iron	2 pieces	$^{1}/_{4}$" × 1$^{1}/_{2}$" × 7"	Hinges
	2 pieces	$^{1}/_{4}$" × 1$^{1}/_{2}$" × 3"	Hinges
	4 pieces	$^{5}/_{8}$" × 1$^{1}/_{4}$" × 1$^{1}/_{2}$"	Hinges
Round iron	3 pieces	$^{3}/_{8}$" × 5"	Tightener rods
	2 pieces	$^{3}/_{4}$" × width of post + 2$^{1}/_{2}$"	Hinge bolts
	2 pieces	$^{3}/_{4}$" × 3$^{1}/_{4}$"	Hinge pins
Nuts	3	$^{3}/_{8}$"	NC
	2	$^{3}/_{4}$"	NC
Washers	2	$^{3}/_{4}$"	Flat
Cotter pins	4	$^{7}/_{32}$" × 1$^{1}/_{2}$"	
Woven wire	1 piece	47" × 13'6"	10 strands
Wire, galvanized	2 pieces	6" long	10 gauge, tie wires

CONSTRUCTION PROCEDURE FOR GATE—METAL, 12-FOOT OR AS NEEDED

1. Cut the pipe for the outside frame to the length as listed in the bill of materials, with corners mitered at 45 degree. Cut the wire tightener and strut to the length listed in the bill of materials. A hand or power hacksaw may be used to cut the pipe.
2. Center punch and drill the $^{1}/_{4}$" holes in the wire tightener, strut, and latch end of the gate for the top and bottom wires, as shown in the detail. Locate the bottom hole 2$^{1}/_{2}$" from the bottom piece of the gate. Measure the height of the wire to determine the location of the top holes.
3. Place the ends, top, and bottom pieces together, as shown in the drawing. Place on a level surface; square and tack weld each corner on the top side.
4. Place the strut in position, and tack weld.
5. Cut the brace to length and proper angle and tack weld.
6. Turn the gate over, and weld all joints. Place the gate in the appropriate positions and complete all welds.
7. Center punch and drill three $^{7}/_{16}$" holes through the wire tightener for $^{3}/_{8}$" rods, as shown in the drawing. After the first hole is drilled, place a 12" or longer $^{3}/_{8}$" rod in the hole, and hold it in line with the drill press spindle while drilling the other two holes. This will help to drill the holes in line with each other.
8. Bend a 1" leg on one end of the three $^{3}/_{8}$" × 5" tightener rods. Cut 3"of NC threads on the other end.
9. Place nuts on the three tightener rods, then insert the rods through the $^{7}/_{16}$" holes in the tightener. Place nuts on the other side and tighten to hold the tightener rods in line while welding to the frame.
10. Lay the gate and tightener on a flat surface. Weld the legs of the tightener rods to the hinge end of the gate, as shown in the detail.
11. Remove the nuts from the tightener rods, then replace the wire tightener and the three outside nuts. Leave the nuts as near the ends of the rods as possible to allow maximum adjustment for tightening the wire.
12. Cut flat iron for hinges to the sizes listed in the bill of materials. Locate the center of the four pieces of $^{5}/_{8}$" flat iron, and mark with a center punch for drilling.
13. Place one piece of $^{5}/_{8}$" flat iron lengthwise $^{1}/_{4}$" from the end of each piece of 1$^{1}/_{4}$" flat iron. Be sure that the marked side is up. Clamp the two pieces together in this position, and weld all four sides.

14. Cut two pieces of $3/4$" round iron to the length shown in the drawing. Weld one $3/4$" rod to each of the $1/4" \times 1\,1/2" \times 3"$ pieces of flat iron. Hold the end of the rod against the $5/8$" piece welded in step 13. Cut 4" of NC threads on the other end of each rod.
15. Grind the corners of the $1/4$" pieces round on the reinforced ends of the hinges. Drill a $25/32$" hole through the four hinge parts at the center-punched mark.
16. Cut two pieces of $3/4$" round iron $3\,1/4$" long for the hinge pins. Center punch and drill a $1/4$" hole $1/2$" from each end.
17. Weld the 7" pieces of the hinges to the top and bottom of the gate with the reinforced side up in both cases, as shown in the drawing. A piece of $3/4$" round iron about 60" long should be inserted in the holes of the hinges to assist in keeping the holes in line and at the correct distance from the end of the gate while welding.
18. Attach woven wire to the latch end of the gate with the vertical stay wire parallel to the end of the gate. Place the top and bottom wires through holes in the end piece and wrap each wire around the pipe and back around itself.
19. Pull the wire tight by hand. Attach the wire to the tightener in the same manner as in step 18. Tighten nuts against the tightening bar until about one-half of the tension curve remains in the wires.
20. Tie the top and bottom wires to the strut with tie wires.
21. Apply a suitable finish.

NOTE: Commercial gate hinges may be used to simplify the project.

PROJECT 46: FARROWING CRATE

PLAN FOR FARROWING CRATE

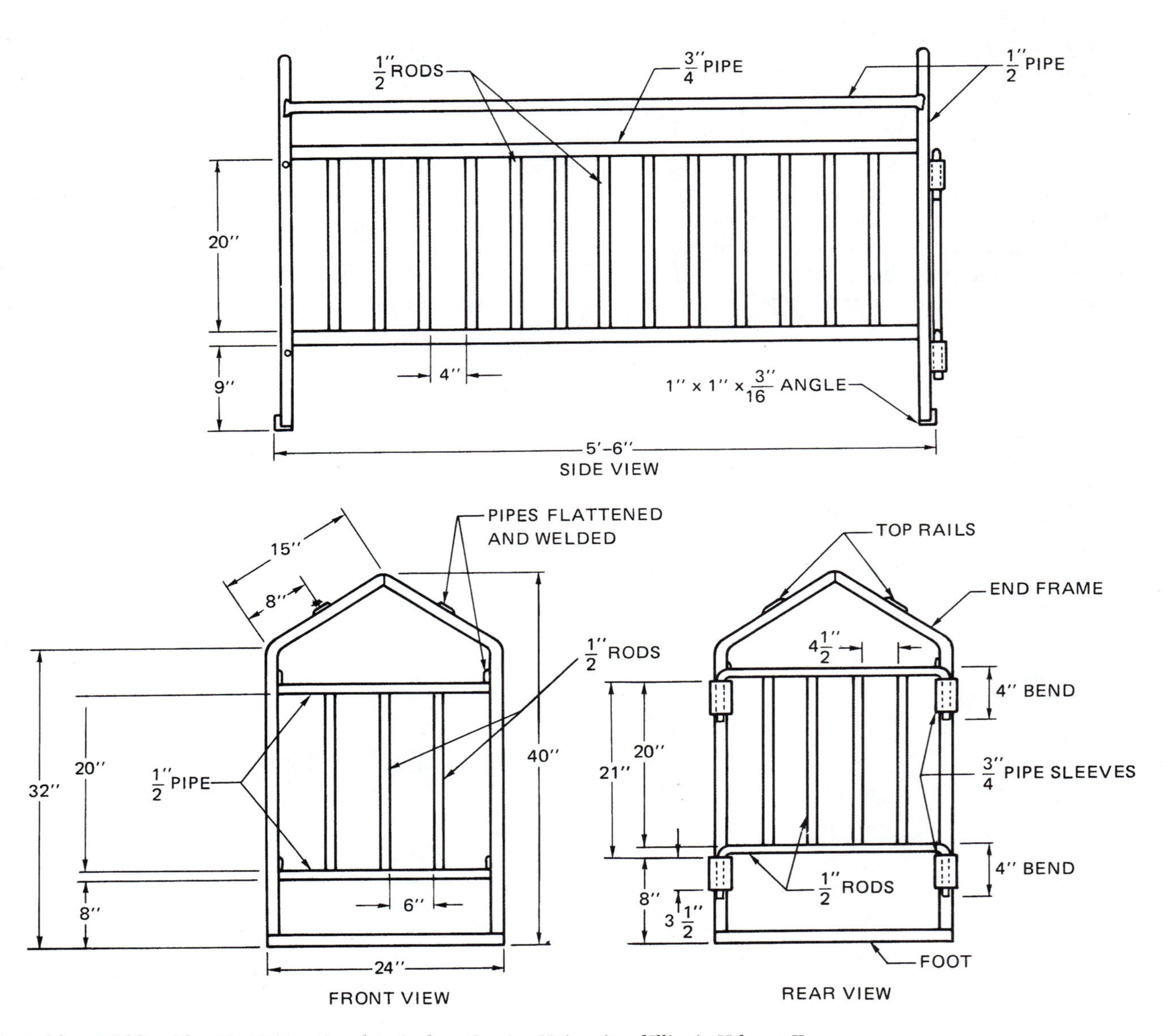

Adapted from Welding Plan No. 20, Vocational Agriculture Service, University of Illinois, Urbana, IL

BILL OF MATERIALS

Materials Needed	Quantity	Dimensions	Description
Pipe, steel	2	1/2" ID × 5'6"	Top side rails
	4	3/4" ID × 5'4"	Upper and lower side rails
Rod, steel	26	1/2" × 20"	Vertical slats for sides
Pipe, steel	4	1/2" ID × 4'	Half of end frame
	4	3/4" ID × 3 1/2"	Sleeves
	2	1/2" ID × 22"	Top and bottom rails of front
Rod, steel	7	1/2" × 20"	Vertical slats for ends
	2	1/2" × 32"	Top and bottom rails of rear gate
Angle, steel	2	3/16" × 1" × 1" × 24"	Feet

CONSTRUCTION PROCEDURE FOR FARROWING CRATE

1. Cut the pieces to be used in the end frame halves and feet.
2. Measure in 15" from one end and bend the four half frames to match each other.
3. Saw or grind the ridge angle on the half frames.
4. Weld the half frames at the top and weld them to the feet to form the ends.
5. Cut and weld the front rails and slats in place.
6. Cut and weld the side rails and slats in place.
7. Set up the ends and weld the side rails to the ends.
8. Make up the rear gate.
9. Cut four 3 1/2" lengths of 3/4" ID pipe for the sleeves and weld them into place.
10. Apply a suitable finish.

PROJECT 47: FEED/SILAGE CART

PLAN FOR FEED/SILAGE CART

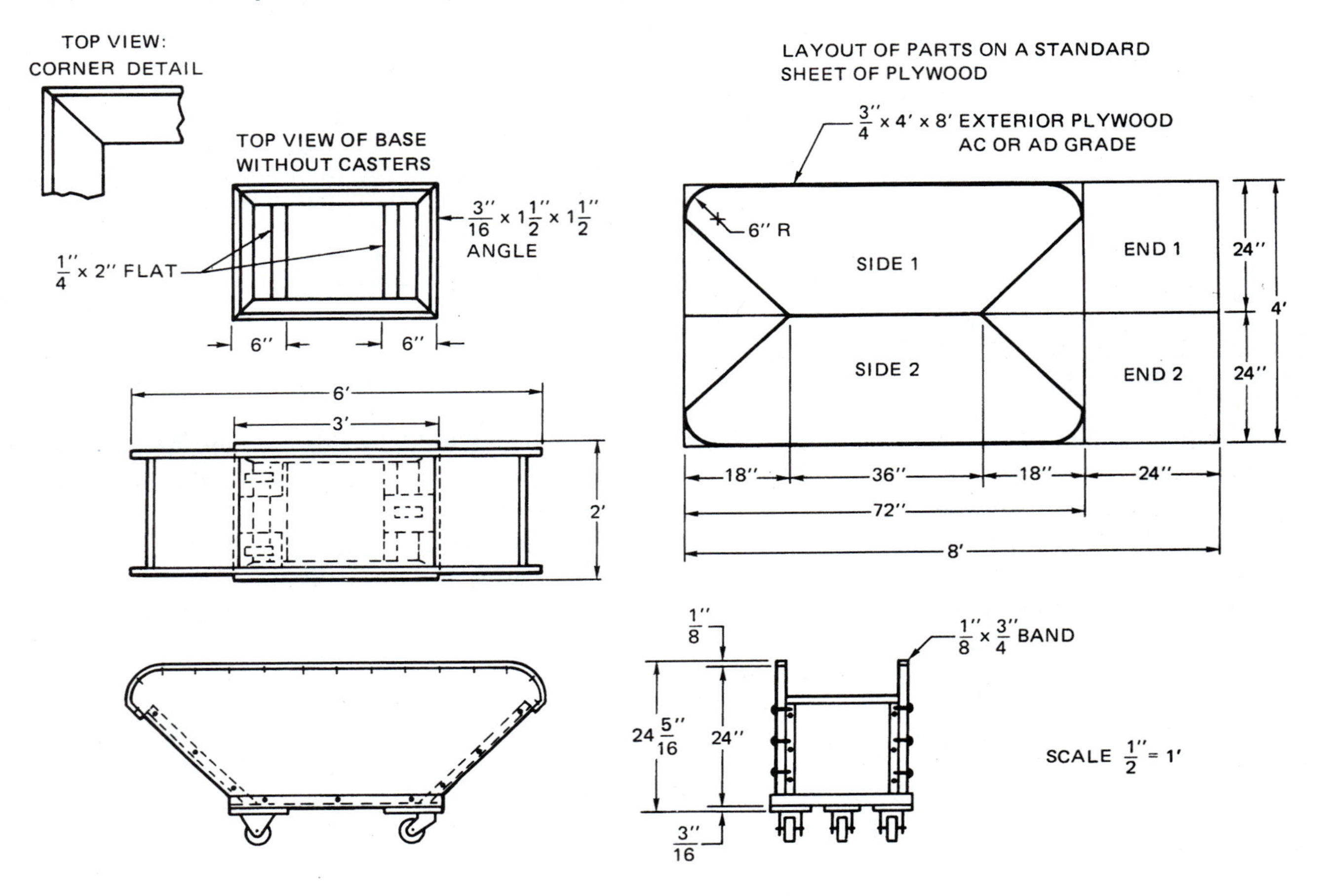

BILL OF MATERIALS

Materials Needed	Quantity	Dimensions	Description
Plywood, exterior AC or better recommended	2	3/4" × 2' × 6'	Sides
	2	3/4" × 2' × 2'	Ends
	1	3/4" × 2' × 3'	Bottom
Steel, angle	2	3/16" × 1 1/2" × 1 1/2" × 36"	Sides of base
	2	3/16" × 1 1/2" × 1 1/2" × 24"	Ends of base
	4	1/8" × 1" × 1" × 22"	Corner angles
Steel, band	2	1/8" × 1" × 90"	Edge bands
	2	3/16" × 2" × 21"	Caster supports
Bolts, carriage, with nuts and lock washers	24	1/4" × 1 1/4"	Bolt plywood to angles
Bolts, stove, flat head	6	1/4" × 1 1/4"	Bolt sides to base
Bolts, stove, flat head	12	3/8" × 1"	Attach casters
*Caster, straight	2	6" or 8" dia	Rigid wheels on one end
Caster, swivel	1	6" or 8"	Swivel wheel in center on other end
Screws, wood, flat	30	No. 6 × 1"	Fasten edge bands

*Two 6", 8", or 10" wheels with supports and axle may be used in place of the straight casters. A metal support is then fabricated on the other end to level the cart on the swivel caster.

CONSTRUCTION PROCEDURE FOR FEED/SILAGE CART

1. Using a $^3/_4$" × 4' × 8' sheet of plywood, cut the sides by cutting two pieces 2' × 6'. Then lay out the other lines and finish cutting out the sides. See plan for most efficient layout.
2. Cut the ends by sawing the remaining piece in half.
3. Cut two pieces of $^3/_{16}$" × 1$^1/_2$" × 1$^1/_2$" angle 36" long and two pieces of $^3/_{16}$" × 1$^1/_2$" × 1 $^1/_2$" angle 24" long with 45 degree angles on all ends. Weld them into a rectangular frame to form the base.
4. Cut two $^1/_4$" × 2" bands 21" long.
5. Place the base on a flat surface. Weld the caster supports into place according to the size of the base of the casters.
6. Bolt the casters to the base using flat head stove bolts passing downward through countersunk holes and through the base of each caster.
7. Cut four pieces of $^1/_8$" × 1" × 1" angle 22" long. Drill three $^9/_{32}$" holes in each leg of each piece. Holes should be spaced so they are 1" to 1$^1/_2$" from each end with one in the middle. Holes must be placed so the bolts do not collide.
8. Notch the ends of the sides so they fit down into the base. Bolt the sides to the base using $^1/_4$" × 1$^1/_4$" flat head stove bolts passing through countersunk holes in the frame and then through the wood sides. Install $^1/_4$" flat washers and draw them slightly into the wood. Saw off any excess bolt length.
9. Using the four pieces of angle, bolt the ends to the sides at each corner. The angle steel and bolt nuts should be on the exterior of the ends and interior of the sides.
10. Cut a piece of $^3/_4$" plywood to drop in to form the bottom. Small notches may be needed to clear the nuts at the bottom of the sides. The bottom must fit tight to prevent feed from running through. Caulk any cracks that can permit feed to pass through.
11. Cut two pieces of $^1/_8$" × $^3/_4$" band 90" long. Drill and countersink $^1/_2$" from each end and at 6" intervals for No. 6 screws.
12. Shape the bands to fit the top of the sides.
13. Apply suitable finishes to all wood and metal parts.
14. Install the bands on the top of each side of the cart.

PROJECT 48: HOIST FRAME—PORTABLE

PLAN FOR HOIST FRAME—PORTABLE

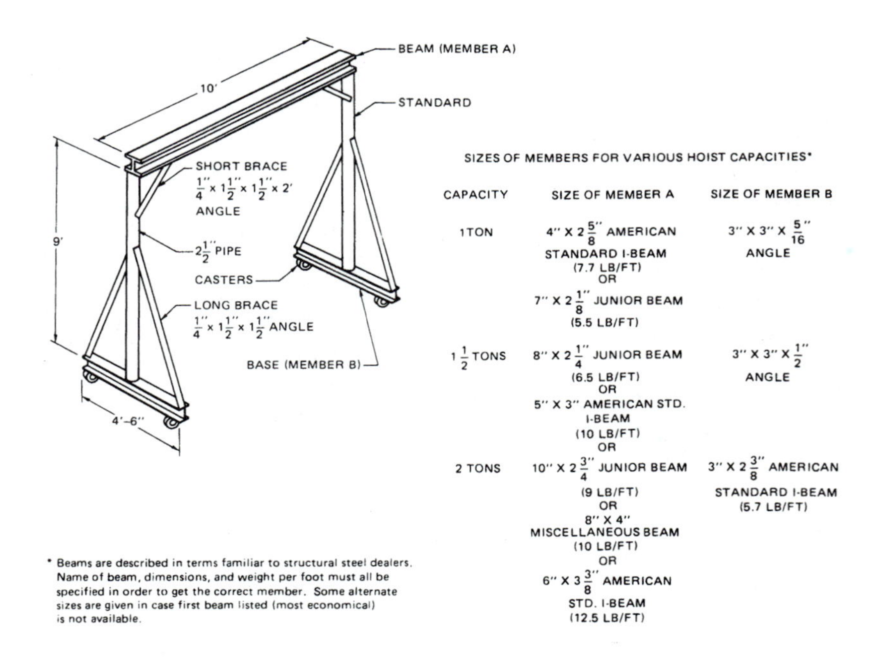

SIZES OF MEMBERS FOR VARIOUS HOIST CAPACITIES*

CAPACITY	SIZE OF MEMBER A	SIZE OF MEMBER B
1 TON	4" X $2\frac{5}{8}$" AMERICAN STANDARD I-BEAM (7.7 LB/FT) OR 7" X $2\frac{1}{8}$" JUNIOR BEAM (5.5 LB/FT)	3" X 3" X $\frac{5}{16}$" ANGLE
$1\frac{1}{2}$ TONS	8" X $2\frac{1}{4}$" JUNIOR BEAM (6.5 LB/FT) OR 5" X 3" AMERICAN STD. I-BEAM (10 LB/FT) OR	3" X 3" X $\frac{1}{2}$" ANGLE
2 TONS	10" X $2\frac{3}{4}$" JUNIOR BEAM (9 LB/FT) OR 8" X 4" MISCELLANEOUS BEAM (10 LB/FT) OR 6" X $3\frac{3}{8}$" AMERICAN STD. I-BEAM (12.5 LB/FT)	3" X $2\frac{3}{8}$" AMERICAN STANDARD I-BEAM (5.7 LB/FT)

* Beams are described in terms familiar to structural steel dealers. Name of beam, dimensions, and weight per foot must all be specified in order to get the correct member. Some alternate sizes are given in case first beam listed (most economical) is not available.

Courtesy of Vocational Agriculture Service, University of Illinois, Urbana, IL

BILL OF MATERIALS

Materials Needed	Quantity	Dimensions	Description
1 beam, standard, steel	1	4" × $2^5/_8$" × 10' (7.7 lb/ft)	Beam (member A)
Pipe, steel, standard	2	$2^1/_2$" × 9'	Standards
Angle, steel	2	$^5/_{16}$" × 3" × 3" × 4'6"	Base (member B)
Angle, steel	4	$^1/_4$" × $1^1/_2$" × $1^1/_2$" × 6'6"	Long braces
Angle, steel	2	$^1/_4$" × $1^1/_2$" × $1^1/_2$" × 2'	Short braces

*A 7" × $2^1/_8$" × 10' (5.5 lb/ft) junior beam should be cheaper, if available, and may be substituted. See plan for specifications of beams for capacity greater than one ton.

CONSTRUCTION PROCEDURE FOR HOIST FRAME—PORTABLE

NOTE: Be sure the casters used are designed to carry the combined weight of the frame, hoist, and load for which the hoist is designed.

1. Cut all pieces to length.
2. Lay the standards and member A on a level floor. Square and tack weld them.
3. Check each joint between the standards and member A, and tack weld the braces into place. Weld all joints permanently.
4. Set the assembly up on the B members.

5. Check for squareness and parallelism between the B members, then tack weld the standards to the B members.
6. Check for squareness and weld the braces between the standards and B members.
7. Weld all joints permanently.
8. Attach casters to the B members.

CAUTION:

- **Bill of materials is for a hoist with a maximum capacity of 1 ton.**

PROJECT 49: UTILITY TRAILER

PLAN FOR UTILITY TRAILER

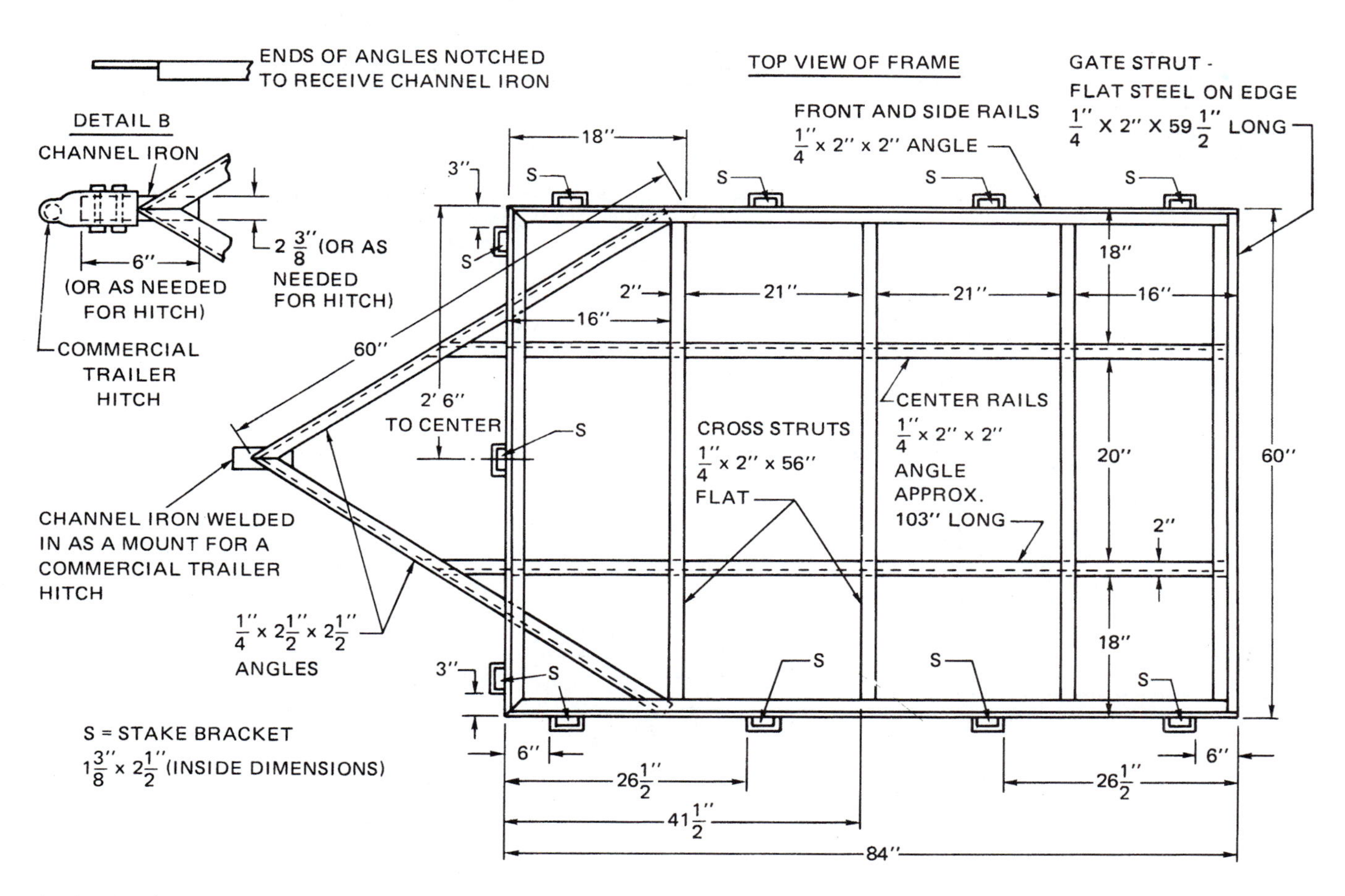

Plan by Elmer L. and Timothy B. Cooper

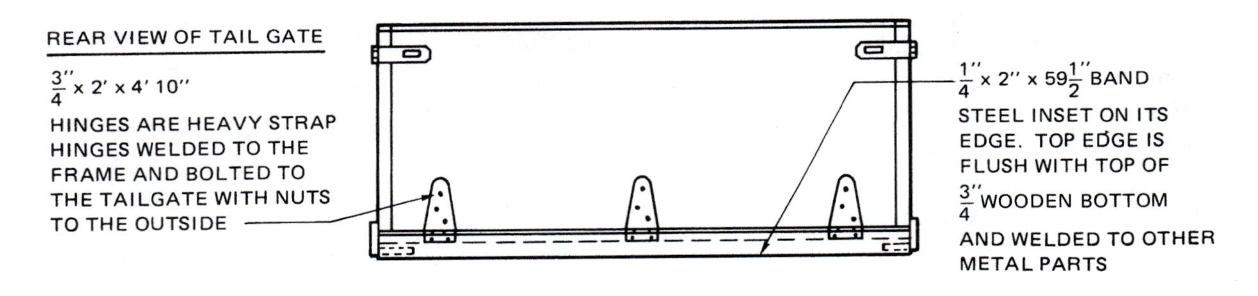
REAR VIEW OF TAIL GATE
$\frac{3}{4}$'' x 2' x 4' 10''
HINGES ARE HEAVY STRAP HINGES WELDED TO THE FRAME AND BOLTED TO THE TAILGATE WITH NUTS TO THE OUTSIDE
$\frac{1}{4}$'' x 2'' x 59$\frac{1}{2}$'' BAND STEEL INSET ON ITS EDGE. TOP EDGE IS FLUSH WITH TOP OF $\frac{3}{4}$'' WOODEN BOTTOM AND WELDED TO OTHER METAL PARTS

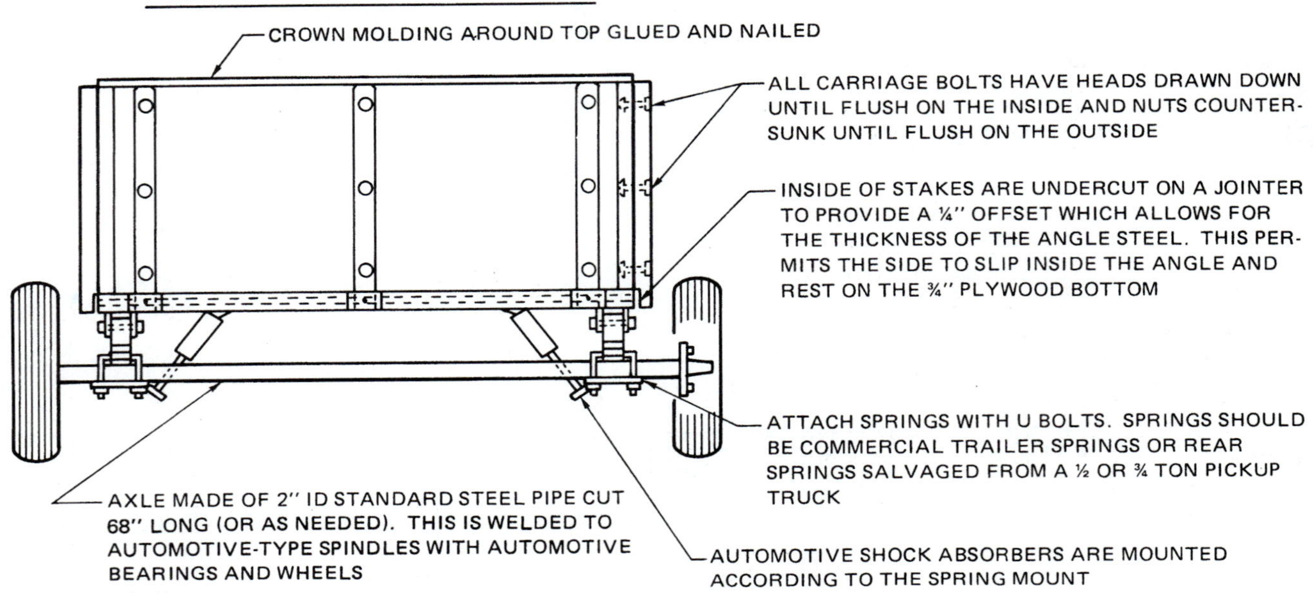
FRONT VIEW OF FRONT AND SIDE BOARDS
CROWN MOLDING AROUND TOP GLUED AND NAILED
ALL CARRIAGE BOLTS HAVE HEADS DRAWN DOWN UNTIL FLUSH ON THE INSIDE AND NUTS COUNTER-SUNK UNTIL FLUSH ON THE OUTSIDE
INSIDE OF STAKES ARE UNDERCUT ON A JOINTER TO PROVIDE A ¼'' OFFSET WHICH ALLOWS FOR THE THICKNESS OF THE ANGLE STEEL. THIS PERMITS THE SIDE TO SLIP INSIDE THE ANGLE AND REST ON THE ¾'' PLYWOOD BOTTOM
ATTACH SPRINGS WITH U BOLTS. SPRINGS SHOULD BE COMMERCIAL TRAILER SPRINGS OR REAR SPRINGS SALVAGED FROM A ½ OR ¾ TON PICKUP TRUCK
AXLE MADE OF 2'' ID STANDARD STEEL PIPE CUT 68'' LONG (OR AS NEEDED). THIS IS WELDED TO AUTOMOTIVE-TYPE SPINDLES WITH AUTOMOTIVE BEARINGS AND WHEELS
AUTOMOTIVE SHOCK ABSORBERS ARE MOUNTED ACCORDING TO THE SPRING MOUNT

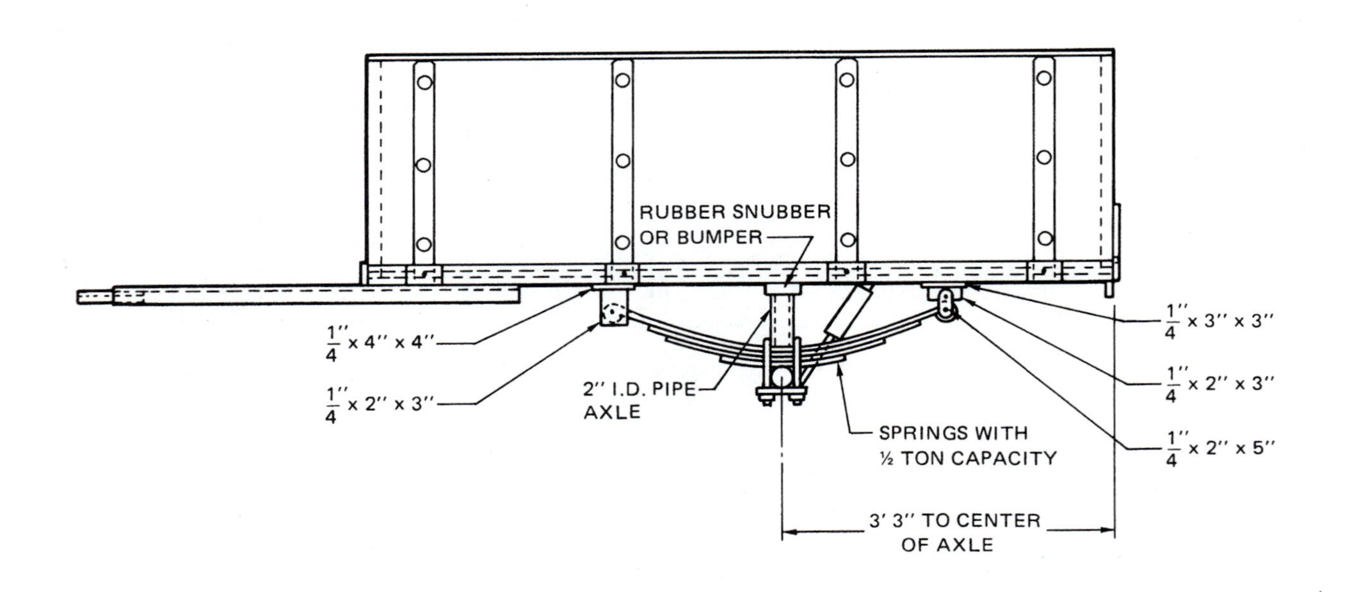
RUBBER SNUBBER OR BUMPER
$\frac{1}{4}$'' x 4'' x 4''
$\frac{1}{4}$'' x 2'' x 3''
2'' I.D. PIPE AXLE
SPRINGS WITH ½ TON CAPACITY
$\frac{1}{4}$'' x 3'' x 3''
$\frac{1}{4}$'' x 2'' x 3''
$\frac{1}{4}$'' x 2'' x 5''
3' 3'' TO CENTER OF AXLE

BILL OF MATERIALS

Materials Needed	Quantity	Dimensions	Description
Angle, steel	2	1/4" × 2" × 2" × 7'	Side rails
	1	1/4" × 2" × 2" × 5'	Front rail
	2	1/4" × 2" × 2" × 8'7" (approx)	Center rails
	2	1/4" × 2 1/2" × 2 1/2" × 5'	Tongue rails
Band (flat), steel	4	1/4" × 2" × 4'8"	Cross struts
	1	1/4" × 2" × 4'11 1/2"	Gate strut
	1	1/8" or 1/4" × 2" × 4'	Materials for pockets (or use 11 commercial pockets) 1 3/8" × 2 1/2" inside dimensions
	2	1/4" × 4" × 4"	Front spring mount base
	4	1/4" × 2" × 3"	Front spring mount
	2	1/4" × 3" × 3"	Rear spring mount base
	4	1/4" × 2" × 3"	Rear spring mount
	4	1/4" × 2" × 5"	Rear spring shackle
Channel, steel	1	2 3/8" wide × 6" long (or as needed) for the trailer hitch	Hitch mount
Trailer hitch	1	Commercial	Ball type, heavy duty
Pipe, steel, standard	1	2" ID × 5'8" (approx)	Axle
Springs, leaf type	2	Rated to carry 1/2 ton (rear truck or trailer type)	Springs with matching base mounts and U bolts
Spindles, automotive type	2	To carry 1/2 ton	With matching wheels and tires
Shock absorbers	2	As needed	Automotive type
Fenders	2	As needed	Commercial boat or utility trailer fenders
*Plywood, exterior type AC or better	1	3/4" × 5' × 7'	Bottom
	2	3/4" × 2' × 7'	Side boards
	2	3/4" × 2' × 4'10"	Front and back boards
Lumber	11	1 1/2" × 2 1/2" × 2'1"	Stakes
Crown molding	2	1/2" × 3/4" × 7'	Top of side boards
	2	1/2" × 3/4" × 5'	Top of front and gate boards
T hinge, steel	3	8" heavy duty	Back gate
Hasps, steel	4	4" or as needed	Corner fasteners
Bolts, carriage	33	1/4" × 2 1/2" with hex nuts and small washers	Stake bolts
	12	1/4" × 1 1/4"	Hinge bolts

*Plywood may be special ordered in five foot widths, and lengths other than eight feet. Local prices should be explored to determine the most economical sizes to purchase. The trailer is designed with a five-foot width to accommodate a large lawn tractor with a four-foot mower.
All dimensions for plywood pieces must be adjusted according to the dimensions of the actual frame.

CONSTRUCTION PROCEDURE FOR UTILITY TRAILER

1. Cut the two side rails to length with a 45 degree cut in the front end. Cut 1/4" off the bottom leg in the rear.
2. Cut the front rail with 45 degree cuts on both ends.

CAUTION:

- **Be careful to lay out the cuts with the parts as they will be positioned on the completed trailer.**

3. Weld the front and side rails together.

4. Cut the cross struts and weld them in place.
5. Cut the gate strut and weld it in place. Be careful to position the gate strut so it extends 3/4" above the top edge of the back cross strut to protect the edge of the 3/4" plywood bottom.
6. Make up and install the stake brackets. Drill a 3/16" hole in the face of each bracket that is located adjacent to each corner for insertion of wood screws into the stakes.
7. Install the tongue pieces and hitch.
8. Install the center rails and weld them at all contact points.
9. Make up the spring and shock absorber mounts and shackles according to the springs being used. Place the spring mounts so the axle will be 3 to 4 inches slightly to the rear of center to provide appropriate balance in the completed trailer.
10. Make up the axle and spindle assembly to allow approximately 2" clearance between the body and tires.
11. Decide if the axle should be mounted above or below the springs to obtain the desired clearance to permit loading of the springs and to provide the desired height of the trailer to the towing vehicle.
12. Mount the axle assembly and shock absorbers.
13. Cut the plywood bottom to fit the frame and install it temporarily. Carefully lay out and drill six 1/4" holes through the plywood floor and into the frame for carriage bolts to anchor the floor.
14. Cut 11 stakes for the front and sides. Cut a 1/4" × 2" rabbet on the flat side of one end of each stake. The resulting end should fit into the brackets and the inside of the stake should be in line with the inside of the metal side rail. Chamfer (1/4") the appropriate three edges on the opposite end of each stake to provide a finished appearance.
15. Cut and install the side boards with the stakes in their brackets so the side boards fit inside the side rails.
16. Cut and install the front board between the side boards.
17. Tack weld the T hinges to the gate strut so the hinge pins are in an absolutely straight line and parallel to the gate strut edge.
18. Cut and install the back gate. Install one bolt in each of the outside hinges. Move the gate up and down to be sure it does not bind. Install a bolt in the center hinge. Check again for binding. Weld all hinges permanently. Install all other hinge bolts.
19. Install corner hasps. The front corners may be secured with rigid corner brackets rather than hasps if desired.
20. Drill a 1/2" hole through the floor near each front corner for drainage.
21. Remove all wooden parts for painting on all sides and edges.
22. Chip all welds thoroughly. Smooth all rough areas and prepare all parts for painting.
23. Install commercial fenders if desired.
24. Prime all wood and metal parts with suitable primers.
25. Apply two coats of machinery enamel to all parts.
26. Re-install all wooden parts. Install No. 10 screws in the stakes to secure the sides.

CAUTION:

- **Trailer is designed for maximum load of 1/2 ton.**

PROJECT 50: PICKUP TRUCK RACKS

PLAN FOR PICKUP TRUCK RACKS

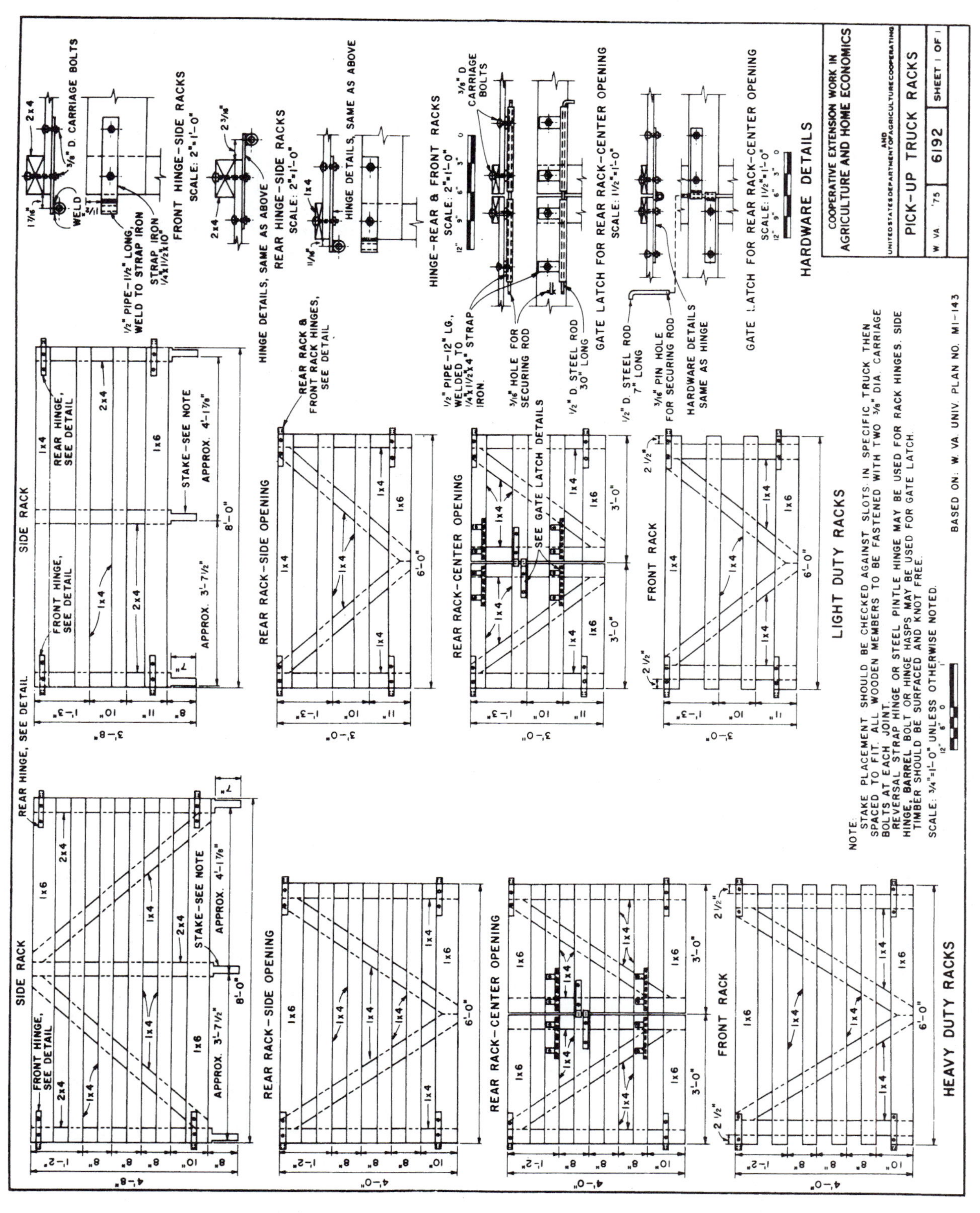

Courtesy of Department of Agricultural Engineering, University of Maryland, College Park, MD

BILL OF MATERIALS

This plan provides construction details for several types of racks. These include plans for heavy-duty and light-duty racks with center-opening or side-opening rear racks. The user must determine which weight and style of racks are preferred and then make up a bill of mterials accordingly.

CONSTRUCTION PROCEDURE FOR PICKUP TRUCK RACKS

1. Examine the truck to determine if the truck body and pockets are such that the plans can be used as provided.
2. Select the racks from the plans that are to be used and modify them as needed to provide a custom fit.
3. Make the wooden racks.
4. Make or purchase the appropriate hardware for the corners and latches.
5. Install the hardware.
6. Apply a primer and two coats of enamel, or apply a polyurethane finish.

PROJECT 51: WAGON RACK FOR TOSSED BALES

PLAN FOR WAGON RACK FOR TOSSED BALES

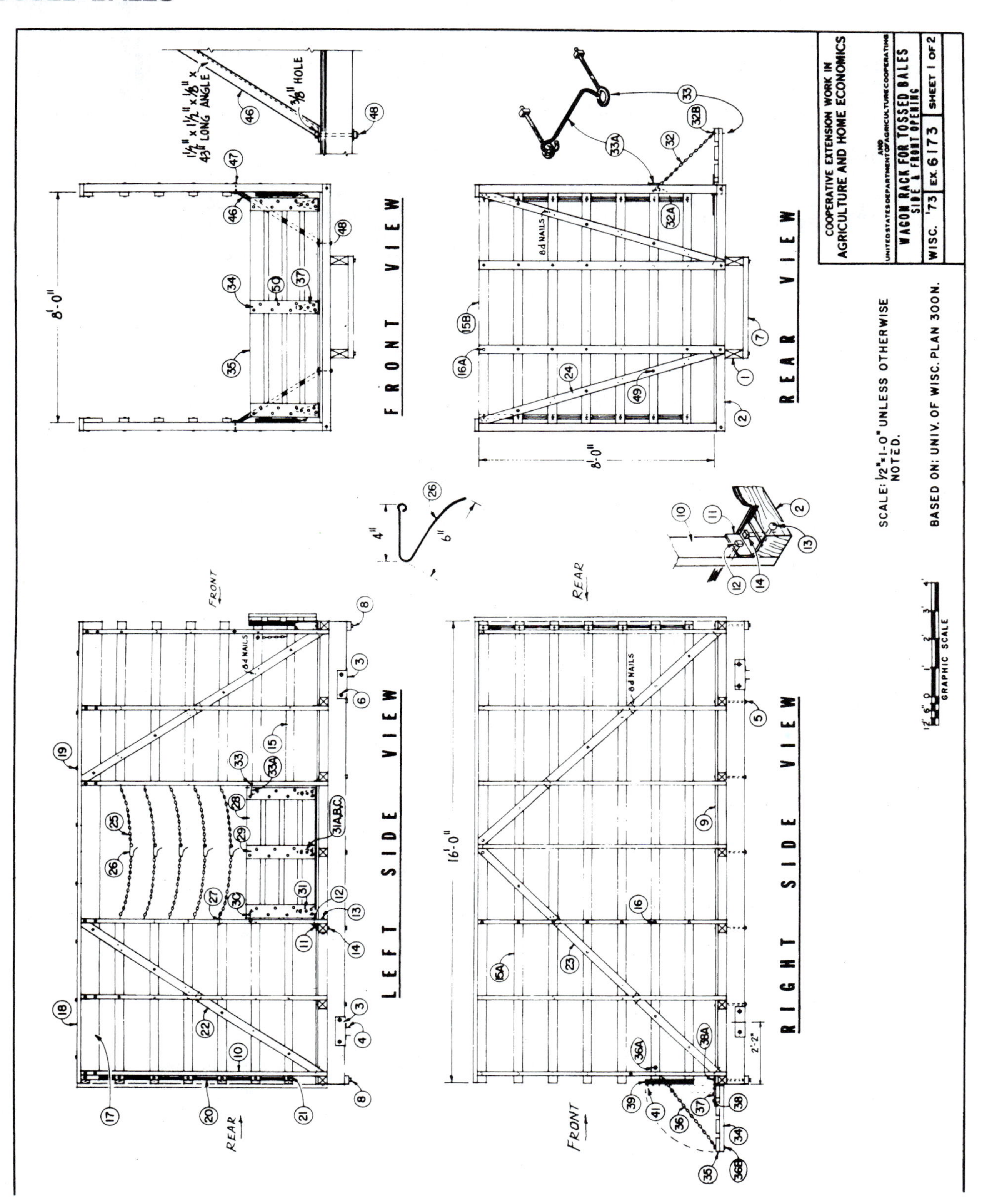

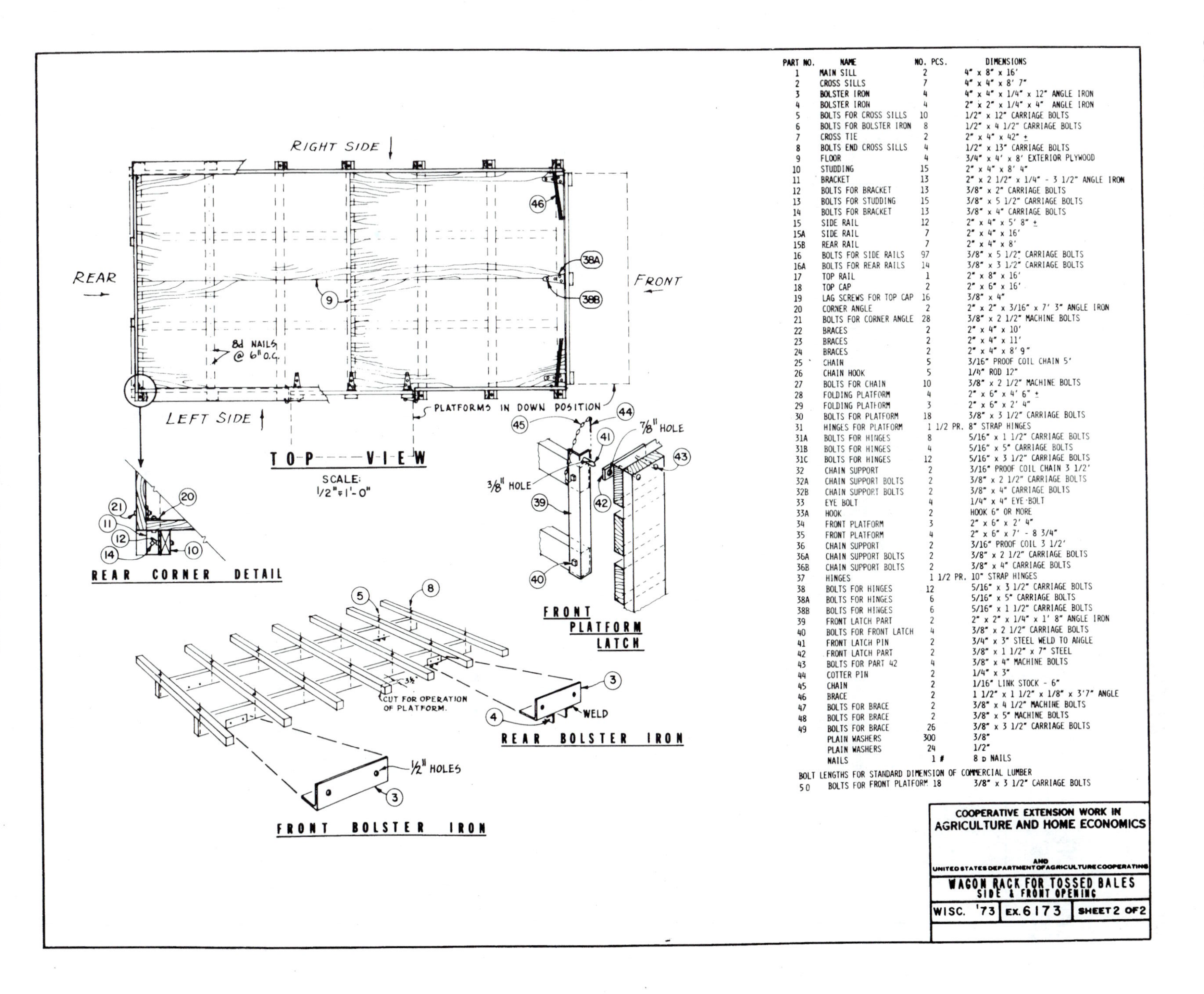

PART NO.	NAME	NO. PCS.	DIMENSIONS
1	MAIN SILL	2	4" x 8" x 16'
2	CROSS SILLS	7	4" x 4" x 8' 7"
3	BOLSTER IRON	4	4" x 4" x 1/4" x 12" ANGLE IRON
4	BOLSTER IRON	4	2" x 2" x 1/4" x 4" ANGLE IRON
5	BOLTS FOR CROSS SILLS	10	1/2" x 12" CARRIAGE BOLTS
6	BOLTS FOR BOLSTER IRON	8	1/2" x 4 1/2" CARRIAGE BOLTS
7	CROSS TIE	2	2" x 4" x 42" ±
8	BOLTS END CROSS SILLS	4	1/2" x 13" CARRIAGE BOLTS
9	FLOOR	4	3/4" x 4' x 8' EXTERIOR PLYWOOD
10	STUDDING	15	2" x 4" x 8' 4"
11	BRACKET	13	2" x 2 1/2" x 1/4" - 3 1/2" ANGLE IRON
12	BOLTS FOR BRACKET	13	3/8" x 2" CARRIAGE BOLTS
13	BOLTS FOR STUDDING	15	3/8" x 5 1/2" CARRIAGE BOLTS
14	BOLTS FOR BRACKET	13	3/8" x 4" CARRIAGE BOLTS
15	SIDE RAIL	12	2" x 4" x 5' 8" ±
15A	SIDE RAIL	7	2" x 4" x 16'
15B	REAR RAIL	7	2" x 4" x 8'
16	BOLTS FOR SIDE RAILS	97	3/8" x 5 1/2" CARRIAGE BOLTS
16A	BOLTS FOR REAR RAILS	14	3/8" x 3 1/2" CARRIAGE BOLTS
17	TOP RAIL	1	2" x 8" x 16'
18	TOP CAP	2	2" x 6" x 16'
19	LAG SCREWS FOR TOP CAP	16	3/8" x 4"
20	CORNER ANGLE	2	2" x 2" x 3/16" x 7' 3" ANGLE IRON
21	BOLTS FOR CORNER ANGLE	28	3/8" x 2 1/2" MACHINE BOLTS
22	BRACES	2	2" x 4" x 10'
23	BRACES	2	2" x 4" x 11'
24	BRACES	2	2" x 4" x 8' 9"
25	CHAIN	5	3/16" PROOF COIL CHAIN 5'
26	CHAIN HOOK	5	1/4" ROD 12"
27	BOLTS FOR CHAIN	10	3/8" x 2 1/2" MACHINE BOLTS
28	FOLDING PLATFORM	4	2" x 6" x 4' 6" ±
29	FOLDING PLATFORM	3	2" x 6" x 2' 4"
30	BOLTS FOR PLATFORM	18	3/8" x 3 1/2" CARRIAGE BOLTS
31	HINGES FOR PLATFORM	1 1/2 PR.	8" STRAP HINGES
31A	BOLTS FOR HINGES	8	5/16" x 1 1/2" CARRIAGE BOLTS
31B	BOLTS FOR HINGES	4	5/16" x 5" CARRIAGE BOLTS
31C	BOLTS FOR HINGES	12	5/16" x 3 1/2" CARRIAGE BOLTS
32	CHAIN SUPPORT	2	3/16" PROOF COIL CHAIN 3 1/2'
32A	CHAIN SUPPORT BOLTS	2	3/8" x 2 1/2" CARRIAGE BOLTS
32B	CHAIN SUPPORT BOLTS	2	3/8" x 4" CARRIAGE BOLTS
33	EYE BOLT	4	1/4" x 4" EYE-BOLT
33A	HOOK	2	HOOK 6" OR MORE
34	FRONT PLATFORM	3	2" x 6" x 2' 4"
35	FRONT PLATFORM	4	2" x 6" x 7' - 8 3/4"
36	CHAIN SUPPORT	2	3/16" PROOF COIL 3 1/2'
36A	CHAIN SUPPORT BOLTS	2	3/8" x 2 1/2" CARRIAGE BOLTS
36B	CHAIN SUPPORT BOLTS	2	3/8" x 4" CARRIAGE BOLTS
37	HINGES	1 1/2 PR.	10" STRAP HINGES
38	BOLTS FOR HINGES	12	5/16" x 3 1/2" CARRIAGE BOLTS
38A	BOLTS FOR HINGES	6	5/16" x 5" CARRIAGE BOLTS
38B	BOLTS FOR HINGES	6	5/16" x 1 1/2" CARRIAGE BOLTS
39	FRONT LATCH PART	2	2" x 2" x 1/4" x 1' 8" ANGLE IRON
40	BOLTS FOR FRONT LATCH	4	3/8" x 2 1/2" CARRIAGE BOLTS
41	FRONT LATCH PIN	2	3/4" x 3" STEEL WELD TO ANGLE
42	FRONT LATCH PART	2	3/8" x 1 1/2" x 7" STEEL
43	BOLTS FOR PART 42	4	3/8" x 4" MACHINE BOLTS
44	COTTER PIN	2	1/4" x 3"
45	CHAIN	2	1/16" LINK STOCK - 6"
46	BRACE	2	1 1/2" x 1 1/2" x 1/8" x 3'7" ANGLE
47	BOLTS FOR BRACE	2	3/8" x 4 1/2" MACHINE BOLTS
48	BOLTS FOR BRACE	2	3/8" x 5" MACHINE BOLTS
49	BOLTS FOR BRACE	26	3/8" x 3 1/2" CARRIAGE BOLTS
	PLAIN WASHERS	300	3/8"
	PLAIN WASHERS	24	1/2"
	NAILS	1 #	8 D NAILS

BOLT LENGTHS FOR STANDARD DIMENSION OF COMMERCIAL LUMBER

50	BOLTS FOR FRONT PLATFORM	18	3/8" x 3 1/2" CARRIAGE BOLTS

Courtesy of Department of Agricultural Engineering, University of Maryland, College Park, MD

BILL OF MATERIALS

Part No.	Materials Needed	Quantity	Dimensions	Description
1	Main sill	2	4" × 8" × 16'	
2	Cross sills	7	4" × 4" × 8'7"	
3	Bolster iron	4	4" × 4" × $^{1}/_{4}$" × 12" angle iron	
4	Bolster iron	4	2" × 2" × $^{1}/_{4}$" × 4" angle iron	
5	Bolts for cross sills	10	$^{1}/_{2}$" × 12" carriage bolts	
6	Bolts for bolster iron	8	$^{1}/_{2}$" × 4$^{1}/_{2}$" carriage bolts	
7	Cross tie	2	2" × 4" × 42" ±	
8	Bolts end cross sills	4	$^{1}/_{2}$" × 13" carriage bolts	
9	Floor	4	$^{3}/_{4}$" × 4' × 8' exterior plywood	
10	Studding	15	2" × 4" × 8'4"	
11	Bracket	13	2" × 2$^{1}/_{2}$" × $^{1}/_{4}$" — 3$^{1}/_{2}$" angle iron	
12	Bolts for bracket	13	$^{3}/_{8}$" × 2" carriage bolts	
13	Bolts for studding	15	$^{3}/_{8}$" × 5$^{1}/_{2}$" carriage bolts	
14	Bolts for brackets	13	$^{3}/_{8}$" × 4" carriage bolts	
15	Side rail	12	2" × 4" × 5'8" ±	
15A	Side rail	7	2" × 4" × 16'	
15B	Rear rail	7	2" × 4" × 8'	
16	Bolts for side rails	97	$^{3}/_{8}$" × 5$^{1}/_{2}$" carriage bolts	
16A	Bolts for rear rails	14	$^{3}/_{8}$" × 3$^{1}/_{2}$" carriage bolts	
17	Top rail	1	2" × 8" × 16'	
18	Top cap	2	2" × 6" × 16'	
19	Lag screws for top cap	16	$^{3}/_{8}$" × 4"	
20	Corner angle	2	2" × 2" × $^{3}/_{16}$" × 7'3" angle iron	
21	Bolts for corner angle	28	$^{3}/_{8}$" × 2 $^{1}/_{2}$" machine bolts	
22	Braces	2	2" × 4" × 10'	
23	Braces	2	2" × 4" × 11'	
24	Braces	2	2" × 4" × 8'9"	
25	Chain	5	$^{3}/_{16}$" proof coil chain 5'	
26	Chain hook	5	$^{1}/_{4}$" rod 12"	
27	Bolts for chain	10	$^{3}/_{8}$" × 2$^{1}/_{2}$" machine bolts	
28	Folding platform	4	2" × 6" × 4'6" ±	
29	Folding platform	3	2" × 6" × 2'4"	
30	Bolts for platform	18	$^{3}/_{8}$" × 3 $^{1}/_{2}$" carriage bolts	
31	Hinges for platform	3	8" strap hinges	
31A	Bolts for hinges	8	$^{5}/_{16}$" × 1$^{1}/_{2}$" carriage bolts	
31B	Bolts for hinges	4	$^{5}/_{16}$" × 5" carriage bolts	
31C	Bolts for hinges	12	$^{5}/_{16}$" × 3$^{1}/_{2}$" carriage bolts	
32	Chain support	2	$^{3}/_{16}$" proof coil chain 3$^{1}/_{2}$'	
32A	Chain support bolts	2	$^{3}/_{8}$" × 2 $^{1}/_{2}$" carriage bolts	
32B	Chain support bolts	2	$^{3}/_{8}$" × 4" carriage bolts	
33	Eye bolt	4	$^{1}/_{4}$" × 4" eye bolt	
33A	Hook	2	Hook 6" or more	
34	Front platform	3	2" × 6" × 2'4"	
35	Front platform	4	2" × 6" × 7'8$^{3}/_{4}$"	
36	Chain support	2	$^{3}/_{16}$" proof coil 3$^{1}/_{2}$'	
36A	Chain support bolts	2	$^{3}/_{8}$" × 2$^{1}/_{2}$" carriage bolts	
36B	Chain support bolts	2	$^{3}/_{8}$" × 4" carriage bolts	
37	Hinges	3	10" strap hinges	
38	Bolts for hinges	12	$^{5}/_{16}$" × 3$^{1}/_{2}$" carriage bolts	
38A	Bolts for hinges	6	$^{5}/_{16}$" × 5" carriage bolts	
38B	Bolts for hinges	6	$^{5}/_{16}$" × 1$^{1}/_{2}$" carriage bolts	
39	Front latch part	2	2" × 2" × $^{1}/_{4}$" × 1'8" angle iron	
40	Bolts for front latch	4	$^{3}/_{8}$" × 2$^{1}/_{2}$" carriage bolts	
41	Front latch pin	2	$^{3}/_{4}$" × 3" steel weld to angle	

BILL OF MATERIALS (Cont.)

Part No.	Materials Needed	Quantity	Dimensions	Description
42	Front latch part	2	3/8" × 1 1/2" × 7" steel	
43	Bolts for part 42	4	3/4" × 4" machine bolts	
44	Cotter pin	2	1/4" × 3"	
45	Chain	2	1/16" link stock—6"	
46	Brace	2	1 1/2" × 1 1/2" × 1/8" × 3'7" angle	
47	Bolts for brace	2	3/8" × 4 1/2" machine bolts	
48	Bolts for brace	2	3/8" × 5" machine bolts	
49	Bolts for brace	26	3/8" × 3 1/2" carriage bolts	
	Plain washers	300	3/8"	
	Plain washers	24	1/2"	
	Nails	1 #	8d nails	
50	Bolts for front platform	18	3/8" × 3 1/2" carriage bolts	

Note: Bolt lengths are specified for standard dimensions of commercial lumber.

CONSTRUCTION PROCEDURE FOR WAGON RACK FOR TOSSED BALES

1. Cut the main sills to length, install the bolster irons, and mount the sills on a suitable running gear.
2. Cut the cross sills to length.
3. Paint or use a wood preservative on the main sills and cross sills. (This procedure will delay decay of the wood due to moisture accumulation in the joints.)
4. Attach the cross sills to the main sills.
5. Saw 3 1/2" off one end of the center cross sill as shown if a center platform is to be installed.
6. Cut the side studding (uprights) to length and bolt them to the cross sills.
7. Nail the floor boards in place and saw them off flush.
8. Cut the side rails to length and bolt them to the studding.
9. Install the top caps.
10. Install the diagonal bracing.
11. Cut the two rear studs to length and bolt them in place.
12. Bolt the rear rails in place.
13. Install the rear diagonal braces.
14. Install the corner angles between the rear and side rails.
15. Make and install the front platform.
16. Install the front angle braces.
17. Make and install the side platform.
18. Install chains and all remaining hardware.
19. Apply paint or other suitable finish.

PROJECT 52: AQUACULTURE MODEL—CLOSED SYSTEM

PLAN FOR AQUACULTURE MODEL—CLOSED SYSTEM (BIOLOGICAL FILTER)

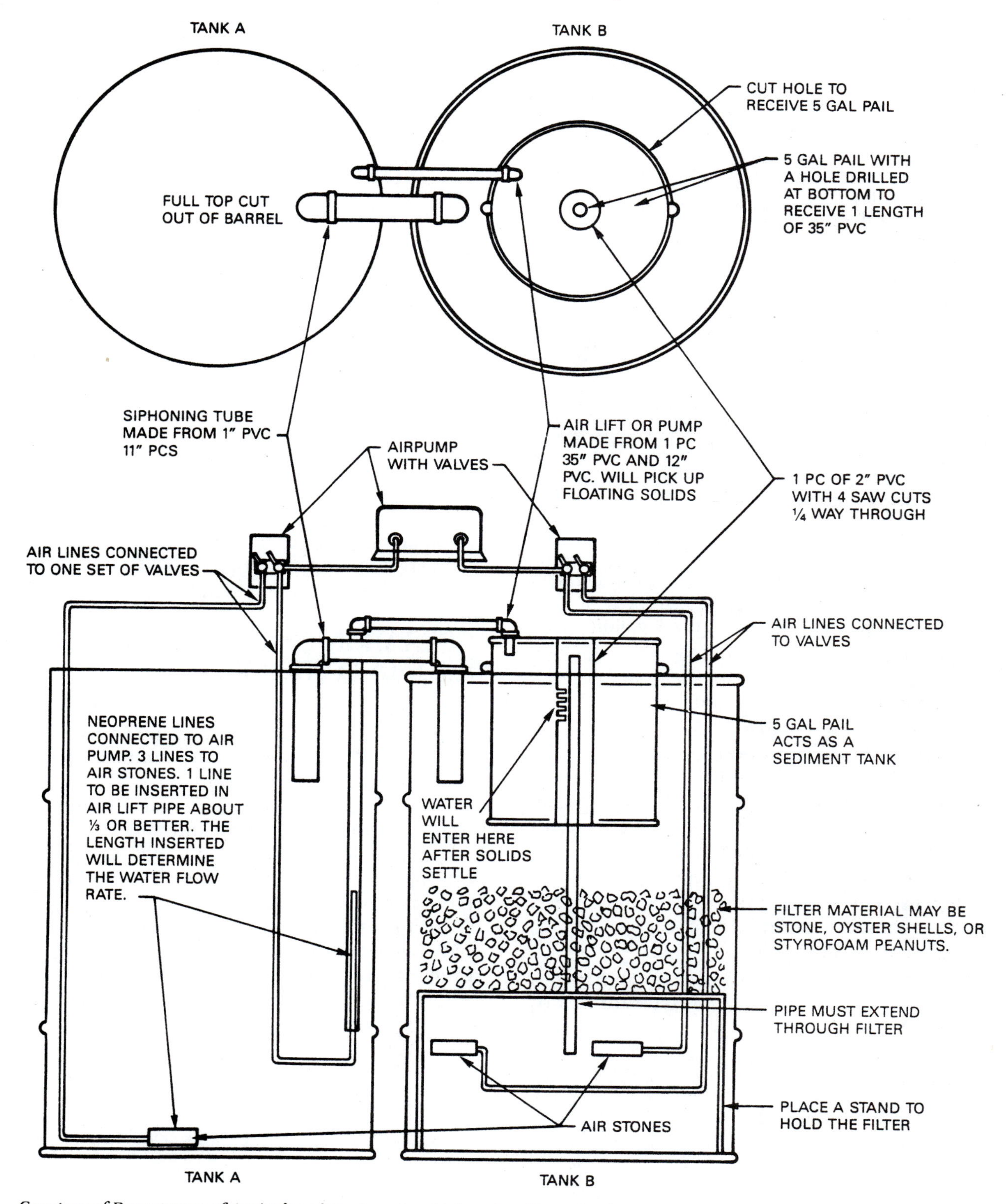

Courtesy of Department of Agricultural Engineering, University of Maryland, Eastern Shore

Courtesy of Department of Agricultural Engineering, University of Maryland, Eastern Shore

BILL OF MATERIALS

Materials Needed	Quantity	Dimensions	Description
55 gal. drums	2		
5 gal. pail	1		
1/2" PVC 35"	2 pieces		
1/2" PVC 12"	1 piece		
90° PVC 1/2" elbows	2		
1" PVC 11"	3 pieces		
90° PCV 1" elbows	2		
2" PVC 12"	1 piece		
neoprene tubing 1/4"	15 ft		
air pump with 2 outlets	1		
air stones	3		
Two-way valves	2		
Filter material	1 bag		
stand	1		(to hold filter material off bottom of barrel)

CONSTRUCTION PROCEDURE FOR AQUACULTURE MODEL—CLOSED SYSTEM

1. Cut the top out of tank A to hold fish.
2. Cut section out of tank B to support the 5-gallon sedimentation tank by the top rim.
3. Drill a hole in center of sediment tank and insert 35" of 1" PVC pipe. The top of this pipe will determine the water height in the 5-gallon pail.
4. Prepare a bag of filter material (crushed small gravel, oyster shell, or glass beads) so that the water that runs down the sediment pipe will flow out below the filter matrix.
5. Position both barrels together and construct an air-lift pump using the 1/2" PVC pipe. An air line will be inserted into the bottom of the lift in tank A so that water flows up and over into the outer ring of the 5-gallon sediment tank.
6. Cut a 12" piece of 2" PVC pipe and cut several saw slits about 4" from the end. This slips over the 1" center PVC pipe and acts as a sleeve to calm the water flow.
7. Construct a siphon tube using 1" PVC pipe as illustrated to return the clean water from tank B to tank A.
8. Connect the air pump to control valves so that two lines with 2" air stones can be placed in tank B under the biological filter. Two other lines are used in tank A, one for the air-lift and one with a 2" air stone suspended in the tank.

OPERATION OF THE CLOSED AQUACULTURE MODEL

It will take about three to four weeks for the microorganisms to become established in the biological filter. This can be aided by adding 1 ounce of feed every two days until fish are added.

Fish are kept in tank A. The air-lift pump moves water into the sediment tank in tank B. It flows into the inner chamber, down the 1" PVC stand pipe and up through the biological filter. The siphon returns the water back to tank A.

If fish are not eating, the water quality might have deteriorated: stop feeding, clean sediment tank, exchange 10 percent of the water, and observe the fish behavior.

PROJECT 53: HYDROPONICS MODEL—TUBE CULTURE

PLAN FOR HYDROPONICS MODEL—TUBE CULTURE

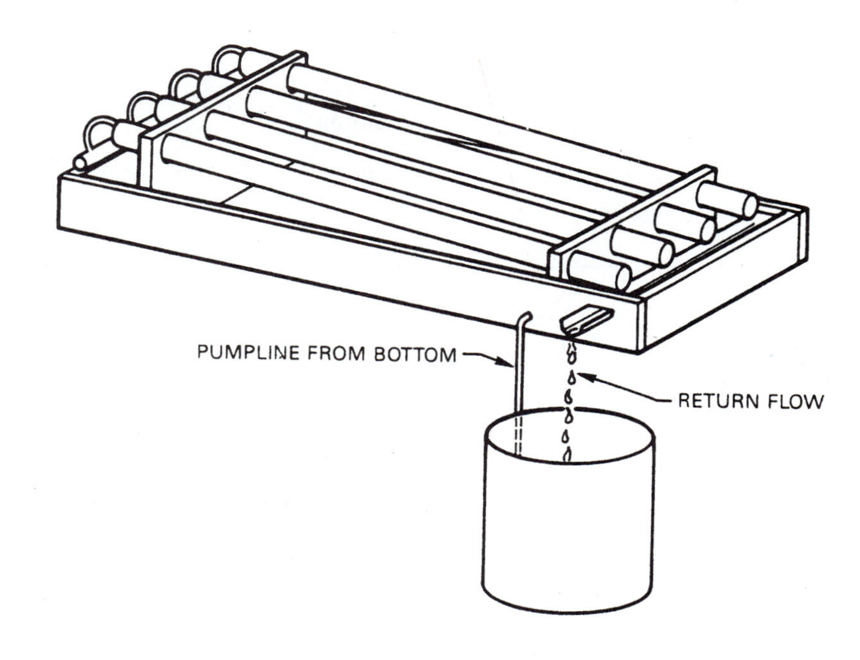

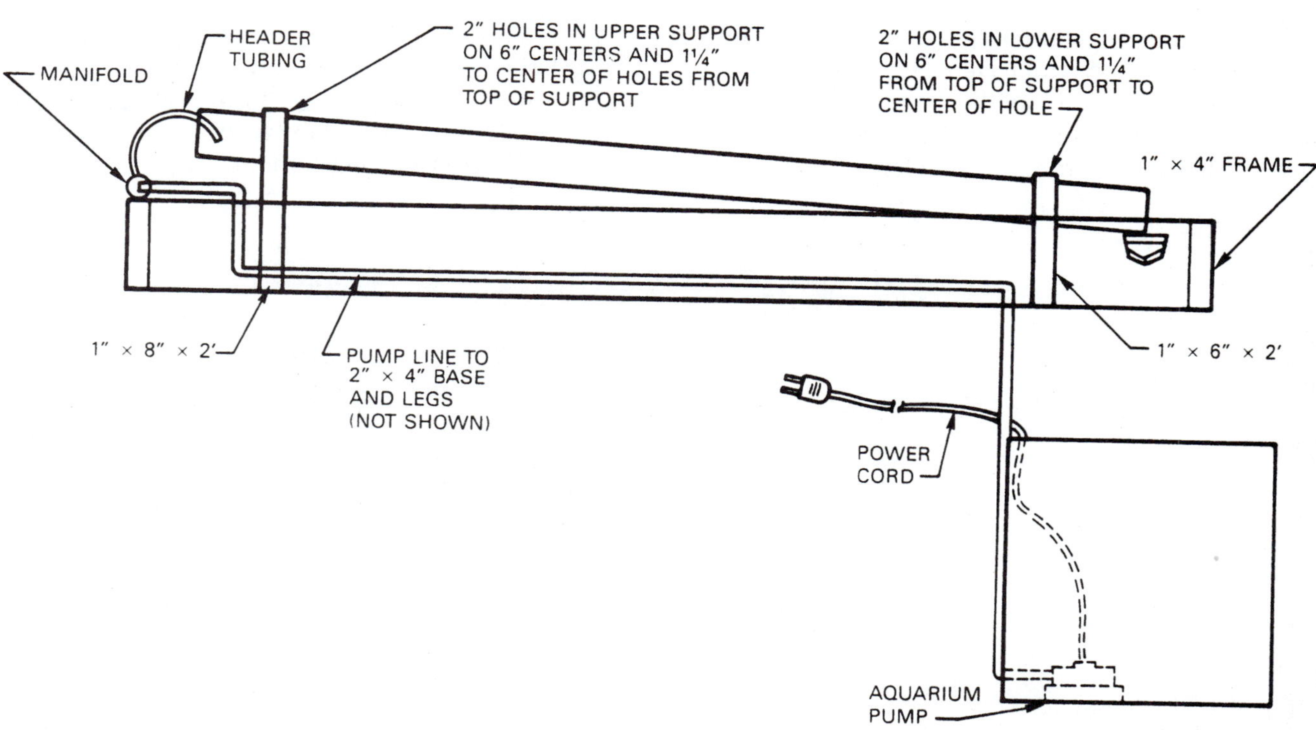

Courtesy University of Maryland at Princess Anne, Dr. Thomas Handwerker

BILL OF MATERIALS

Materials Needed	Quantity	Dimensions	Description
Wood	12'	1" × 4"	Frame
Wood	2'	1" × 8"	Upper support
Wood	2'	1" × 6"	Lower support
PVC pipe Sch. 40	20'	2"	Growing tube
PVC cap	5, 1	2", 1"	Caps
Plastic pipe	2'	1"	Manifold
Plastic tubing and fittings	10'	$^{5}/_{16}$"	Supply line
Header tubing	4'	$^{1}/_{4}$"	Header tubing
Submersible pump	1	Aquarium size	Pump
Reducing coupling	1	1" <> $^{5}/_{16}$"	Connection
Plastic pail	1	5 gallon	Stock tank
Miscellaneous: wood screws, paint and nails			

CONSTRUCTION PROCEDURE FOR HYDROPONICS MODEL—TUBE CULTURE

1. Construct a 2' × 4' base frame (1" × 4" material).
2. Drill 2$^{1}/_{4}$" holes, 1$^{1}/_{2}$" to center from the top edge of the 1" × 8" and 1" × 6" supports. Space four holes across each member on 6" centers.
3. Attach supports within the framework 6" from the upper end and 10" from the lower end.
4. Cut 2" PVC pipe into four 3'6" lengths. Insert into framework and fasten with small wood screws through top of the wooden supports.
5. Drill 2" holes along the top axis of each pipe, spaced 6" on center to hold plants.
6. Saw a 2 $^{1}/_{2}$" length of PVC pipe in half (end to end). Glue a 2" cap on one end and install as drain trough at the lower end of the tubes.
7. Place submersible pump in pail, attach tubing and adjust length to reach manifold connection.
8. Attach 1" manifold along the upper framework. Cap one end and connect to other tubing.
9. Drill manifold and install irrigation header tubes.
10. Friction fit 2" PVC caps to upper ends of tubes. Drill $^{1}/_{4}$" hole in each and insert header tubes.
11. Apply a suitable finish. Observe all safety precautions when working with electrical equipment in a wet environment.

PROJECT 54: GREENHOUSE—PIPE FRAME

PLAN FOR GREENHOUSE—PIPE FRAME

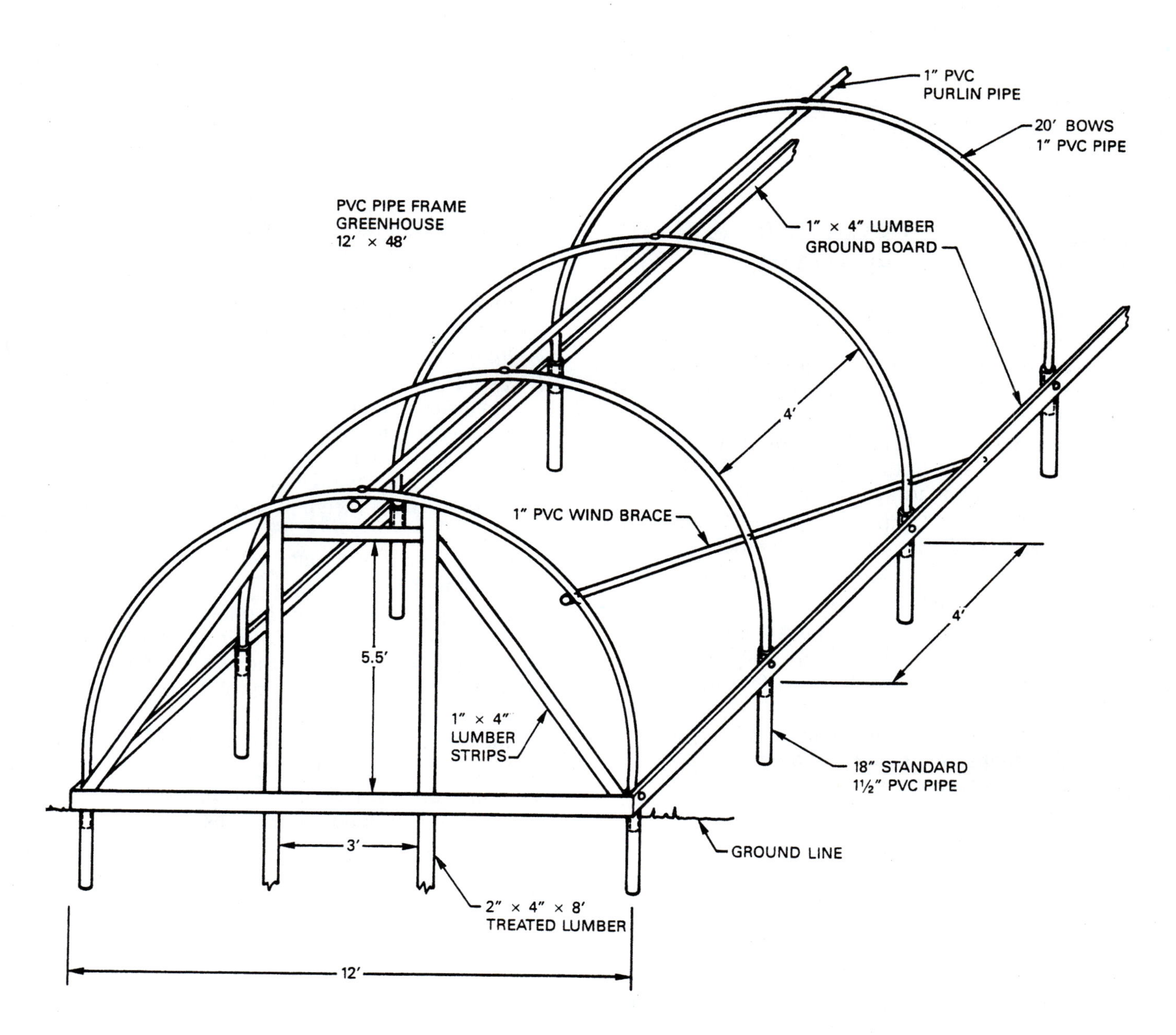

Courtesy University of Maryland at Princess Anne, Dr. Thomas Handwerker

BILL OF MATERIALS

Materials Needed	Quantity	Dimensions	Description
Wood	4	1" × 4" × 12'	Ground board
Wood	4	2" × 4" × 8'	Door frame
Plywood, exterior grade	2	4' × 8' A/C	Doors
PVC pipe Sch. 40	7	1" × 20'	Bows and braces
PVC pipe Sch. 40	8	$1^1/_2$" × 18"	Standards
#9 Wire	10'	#9 × 10'	Straps
Hex bolts and nuts	12	$^1/_4$" × 2"	Attach bows
Plastic film	1	20' × 30'	6 mil thick
Miscellaneous: Door hinges, staples, nails, and paint			

CONSTRUCTION PROCEDURE FOR GREENHOUSE—PIPE FRAME

1. Prepare a level site and ensure adequate drainage.
2. Square the structure. Drive or set in concrete the 18" × $1^1/_2$" PVC standards so only 4" are visible.
3. Drive or set remaining standards on 4' centers down each side until only 4"are visible. The tops of all standards should be level.
4. Mark four 1" PVC pipes with a line 6" from each end and at the midpoint of each pipe.
5. Place the 1" × 4" ground boards around the outside of all the standards.
6. Insert one end of a 1" PVC pipe into the first corner standard to the 6" mark and drill through the ground board, bow, and standard. Insert a $^1/_4$" bolt and tighten.
7. Insert the opposite end of the 1" PVC pipe into the corner standard across from the first one and secure as described in #6.
8. Complete installation of all bows.
9. Cut 1" PVC pipe 12' long and mark every 4'. Hold this purlin up to the inside of the arched bows. Drill and insert $^1/_4$" bolts with purlin marks aligned with the center of all bows. Saw off any protruding bolts to prevent head injuries.
10. Take 1" PVC pipe and bolt end about 4' high on the inside of end bow. Allow the pipe to triangulate down to the ground board. Bolt the pipe to all bows that it passes and to the ground board. This is a wind brace and is an important member of the greenhouse. Be sure bows are plumb.
11. Construct the end walls and door frames as illustrated.
12. Drape plastic across the framework and staple the leading edge to the door frame. Proceed down each side, pulling the plastic tight to the ground board. Roll the edge of the plastic several times and staple the roll to the wooden members. Finish with the trailing end attached to the opposite door frame.
13. Doors, hinges, and utilities are constructed according to standard practices.

APPENDIX

Data Tables

TABLE 1 WEIGHTS AND MEASURES

Liquid/Dry Measure

16 fluid ounces or 1 pound (water)	1 pint
2 pints	1 quart
4 quarts	1 gallon
8 quarts	1 peck
4 pecks	1 bushel
1 bushel	$1\frac{1}{4}$ cu ft
2 barrels	1 hogshead
$31\frac{1}{2}$ gallons (US)	1 barrel
7.48 gallons (US)	1 cu ft
1 gallon (US)	231 cu in
1 inch of water per acre	27,154 gallons
1 inch of water per hectare	67,885 gallons

(1 hectare = 2.5 acres)

Weight Measure

Gram	15.432 Grains
Gram	0353 Ounce
Kilogram	2.2046 Lb
Kilogram	0011 Ton (Short)
Met. Ton	1.1025 Ton (Short)
Grain	.064 Gram
Ounce	28.35 Grams
16 ounces	1 pound
Lb	453.5 Grams
100 pounds	1 hundredweight
20 hundredweight	1 ton
Ton (Short)	907.18 Kilgms.
Ton (Short)	.907 Met. Ton
Ton (Short)	2,000 Lb

1 gallon of water weighs 8.34 pounds
1 gallon of milk weighs 8.6 pounds

Temperature

° C. $= (°F. - 32) \times 5/9$
° F. $= (9/5\ °C) + 32$

Linear Measure

12 inches	1 foot
3 feet	1 yard
$5\frac{1}{2}$ yards ($16\frac{1}{2}$ feet)	1 rod
320 rods	1 mile
1760 yards (5,280 feet)	1 mile

Square Measure

144 square inches	1 square foot
9 square feet	1 square yard
$30\frac{1}{4}$ square yards	1 square rod
160 square rods	1 acre (43,560 sq ft)
640 acres	1 square mile
1 square mile	1 section

Cubic Measure

1728 cubic inches	1 cu ft
27 cubic feet	1 cu yd
128 cu ft (8' × 4' × 4')	1 cord
1' × 1' × 1"	1 bd ft

Land Measure

To find number of acres:
Divide no. of sq ft by 43,560
Divide no. of sq yds by 4,840
Divide no. of sq rods by 160
Divide no. of sq chains by 10

Height of Tree or Building

To estimate the height of a tree or building:

1. Measure the height (H_1) of a nearby object which is vertical.
2. Measure the length of the shadow (S_1) cast by that object.
3. Measure the length of the shadow (S_2) cast by the tree or building.
4. Length of shadow (S_2) of tree or building times the height (H_1) of the object divided by length of shadow (S_1) of the object equals the height (H_2) of the tree or building ($H_2 = S_2 \times H_1 \div S_1$)

TABLE 2 METRIC CONVERSIONS AND MEASUREMENTS

Comparison of the International Metric System and the English System of Measurement

1 Centimeter	= .3937 inches	1 Kilometer	= 1000 meters	1 Gallon	= 3.785 liters
1 Inch	= 2.54 centimeter	1 Kilometer	= .62137 miles	1 Gram	= 15.43 grains
1 Foot	= 30.48 centimeter	1 Sq. Centimeter	= .155 Sq. inches	1 Ounce	= 28.35 grams
1 Meter	= 39.37 inches	1 Sq. Decimeter	= 100 cu. centimeters	1 Kilogram	= 1000 grams
1 Meter	= 100 centimeters	1 Cu. Centimeter	= .061 cu. inches	1 Kilogram	= 2.205 lbs.
1 Meter	= 1.094 yards	1 Cu. Decimeter	= 1000 Cu. centimeter	1 Pound	= 7000 grains
1 Meter	= 1000 millimeters	1 Cu. Meter	= 100 liters	1 Pound	= .4536 kiolgrams
1 Millimeter	= .001 meter	1 Fluid Ounce	= 29.54 milliliters	1 Kilogram	= 1000 milliliters
1 Yard	= .9144 meter	1 Liter	= 1000 cu. centimeters	1 Kilogram	= 1 liter
1 Mile	= 1609.344 meters	1 Liter	= 1.057 quarts		

Approximate Conversion of Common Units

U.S. to Metric

LENGTH

1 inch	= 25.0 millimeters (mm)
1 foot	= 0.3 meter (m)
1 yard	= 0.9 meter
1 mile	= 1.6 kilometers (km)

AREA

1 sq. inch	= 6.5 sq. centimeters (cm^2)
1 sq. foot	= 0.09 sq. meter (m^2)
1 sq. yard	= 0.8 sq. meter
1 acre	= 0.4 hectare*
1 sq. mile	= 2.6 sq. kilometers

*1 hectare equals 10,000 sq. meters

MASS

1 grain	= 64.8 milligrams (mg)
1 ounce (dry)	= 28.0 grams (g)
1 pound	= 0.45 kilogram (kg)
1 short ton	= 9.071 kilograms

VOLUME

1 cubic inch	= 16.0 cubic centimeters (cm^3)
1 cubic foot	= 0.03 cubic meter (m^3)
1 cubic yard	= 0.76 cubic meter
1 teaspoon	= 5.0 milliliters (ml)
1 tablespoon	= 15.0 milliliters
1 fl. ounce	= 30.0 milliliters
1 cup	= 0.24 liter (l)**
1 pint	= 0.47 liter
1 quart (liq.)	= 0.95 liter
1 gallon (liq.)	= 0.004 cubic meter
1 peck	= 0.009 cubic meter
1 bushel	= 0.04 cubic meter

**1 liter equals 1 cubic decimeter (dm^3)

Metric to U.S.

LENGTH

1 millimeter (mm)	= 0.04 inch
1 meter (m)	= 3.3 feet
1 meter	= 1.1 yards
1 kilometer (km)	= 0.6 mile

AREA

1 sq. centimeter (cm^2)	= 0.16 sq. inch
1 sq. meter (m^2)	= 11.0 sq. feet
1 sq. meter	= 1.2 sq. yards
1 hectare	= 2.5 acres
1 sq. kilometer	= 0.39 sq. mile

MASS

1 milligram (mg)	= 0.015 grain
1 gram (g)	= 0.035 ounce
1 kilogram (kg)	= 2.2 pounds
1 metric ton	= 1.102 tons (short)

VOLUME

1 cubic centimeter (cm^3)	= 0.06 cubic inch
1 cubic meter (m^3)	= 35.0 cubic feet
1 cubic meter	= 1.3 cubic yards
1 milliliter (ml)	= 0.2 teaspoon
1 milliliter	= 0.07 tablespoon
1 milliliter	= 0.03 ounce
1 liter (l)	= 4.2 cups
1 liter	= 2.1 pints
1 liter	= 1.1 quarts
1 cubic meter	= 264.0 gallons
1 cubic meter	= 113.0 pecks
1 cubic meter	= 28.0 bushels

Courtesy of The Maryland State Department of Education, Division of Vocational Technical Education, Baltimore, MD

TABLE 3 APPROXIMATE DILUTION RATIOS AND PROPORTIONS

Dilution of Liquid Pesticides at Various Concentrations

Dilution	1 Gal.	3 Gal.	5 Gal.	15 Gal.
1-100	2 tbs. + 2 tsp.	1/2 cup	3/4 cup + 2 tsp.	2 cups + 6 1/2 tbs.
1-200	4 tsp.	1/4 cup	6 1/2 tbs.	1 cup + 3 1/3 tbs.
1-400	2 tsp.	2 tbs.	3 tbs.	1/2 cup + 1 tbs.
1-800	1 tsp.	1 tbs.	1 tbs. + 2 tsp.	4 tbs. + 2 tsp.
1-1000	3/4 tsp.	2 1/4 tsp.	1 tbs. + 2 tsp.	4 tbs.

NOTE: 1 gal. = 16 cups or 256 tbs. or 768 tsp.
1 cup = 16 tbs.; 1 tbs. = 3 tsp.

Water	Quantity of Material					
100 gals.	1 lb.	2 lb.	3 lb.	4 lb.	5 lb.	6 lb.
25 gals.	4 oz.	8 oz.	12 oz.	1 lb.	1 1/4 lb.	1 1/2 lb.
5 gals.	3 tbs.	1 1/2 oz.	2 1/2 oz.	3 1/4 oz.	4 oz.	5 oz.
1 gal.	1 tsp.	2 tsp.	1 tbs.	4 tsp.	5 tsp.	2 tbs.

Equivalent quantities of liquid materials (emulsion concentrates, etc.) for various quantities of water based on pints per 100 gallons.

Water	Quantity of Material					
100 gals.	1/2 pint	1 pint	2 pints	3 pints	4 pints	5 pints
25 gals.	2 fl. oz.	4 fl. oz.	8 fl. oz.	12 fl. oz.	1 pint	1 1/4 pints
5 gals.	1 tbs.	1 fl. oz.	2 fl. oz.	2 1/2 fl. oz.	3 fl. oz.	4 fl. oz.
1 gal.	1/2 tsp.	1 tsp.	2 tsp.	3 tsp.	4 tsp.	5 tsp.

TABLE 4 CUBIC AIR CONTENT OF A GREENHOUSE

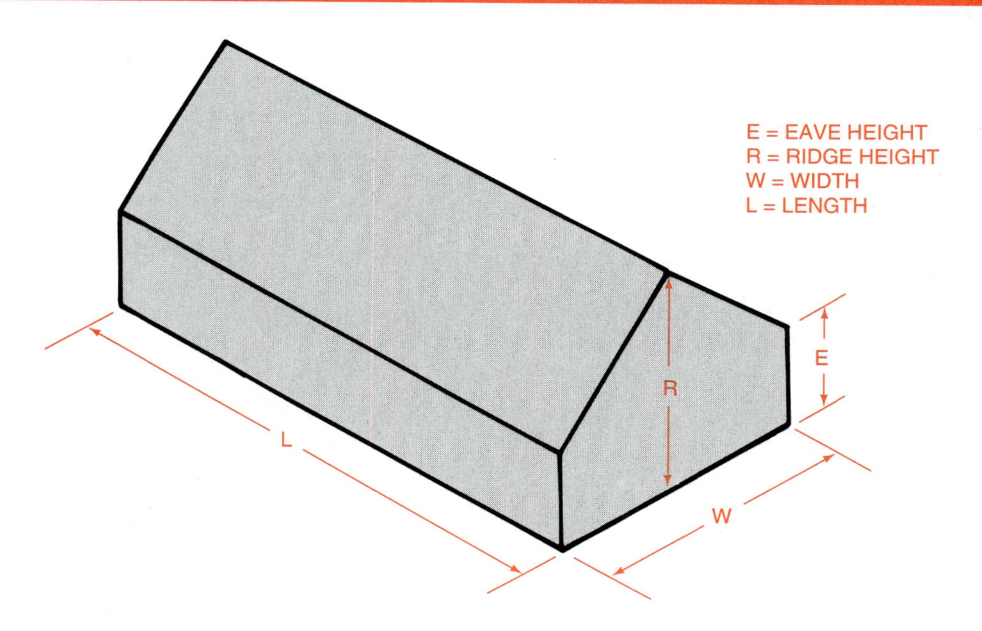

FORMULA: $\frac{E + R}{2} \times W \times L$ = CUBIC CONTENT

Courtesy of The Maryland State Department of Education, Division of Vocational Technical Education, Baltimore, MD

TABLE 5 WEIGHTS PER BUSHEL OF COMMODITIES

Commodity	Weight Per Bushel (Pounds)
FIELD CROPS	
Barley	48
Cottonseed	30
Corn, ear	70
Corn, shelled	56
Oats	32
Sorghum seed	56
Soybeans	60
Wheat	60
LEGUMES	
Lespedeza, common	25
Lespedeza, Korean	40
Lespedeza, Kobe	25
Clover, crimson	60
Clover, white	60
Alfalfa	60
Vetch	60
GRASSES	
Bermuda	40
Dallis	15
Bahia	15–17
Fescue	10–30
Orchard	14
Rye	53
FRUITS	
Apples	48
Peaches	48
Pears	50
VEGETABLES	
Potatoes, Irish	56
Potatoes, Sweet	50–55
Cabbage	50
Cucumbers	48
Okra	32–36
Green beans	30
English peas	30
Dried peas	60
Turnips	54

TABLE 6 MOISTURE CONTENT OF GRAINS FOR LONG-TERM STORAGE

Crop	Maximum Moisture Content
Wheat	12%
Oats	13%
Barley	13%
Grain sorghums	12%
Shelled corn	13%
Soybeans	11%
Rice	12%

TABLE 7 QUICK REFERENCE FRACTION TO DECIMAL EQUIVALENTS

Fraction	Decimal
1/64	.015625
1/32	.03125
3/64	.046875
1/16	.0625
5/64	.078125
3/32	.09375
7/64	.109375
1/8	.125
9/64	.140625
5/32	.15625
11/64	.171875
3/16	.1875
13/64	.203125
7/32	.21875
15/64	.234375
1/4	.250
17/64	.265625
9/32	.28125
19/64	.296875
5/16	.3125
21/64	.328125
11/32	.34375
23/64	.359375
3/8	.375
25/64	.390625
13/32	.40625
27/64	.421875
7/16	.4375
29/64	.453125
15/32	.46875
31/64	.484375
1/2	.500

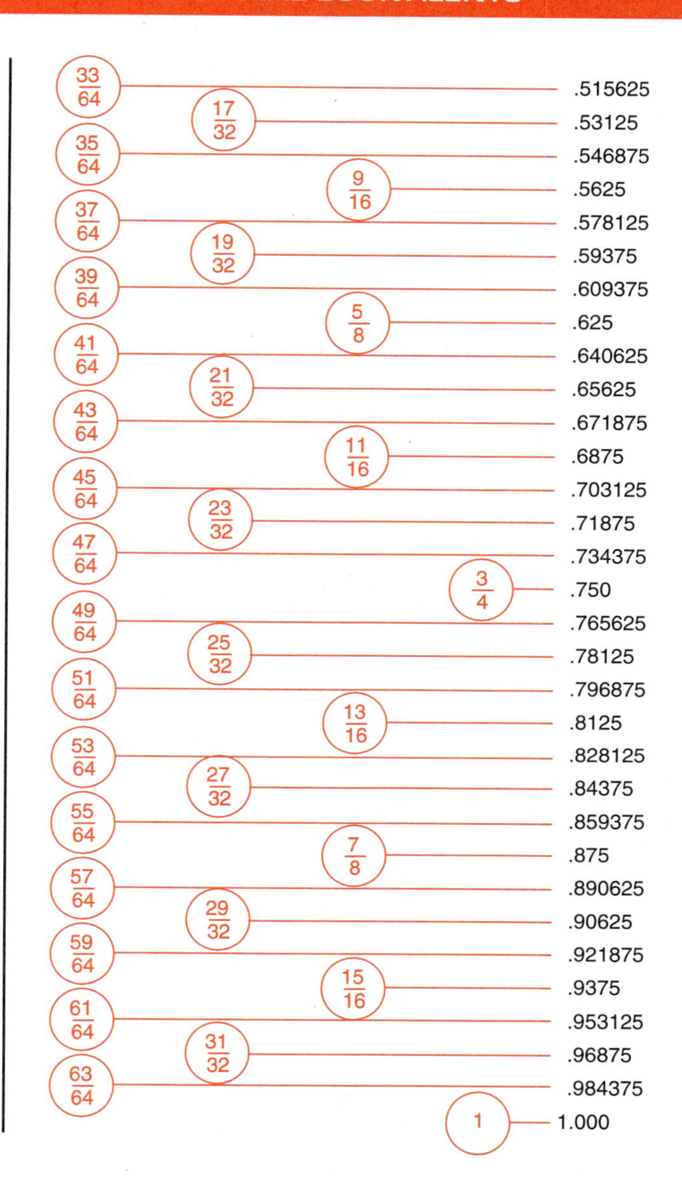

Fraction	Decimal
33/64	.515625
17/32	.53125
35/64	.546875
9/16	.5625
37/64	.578125
19/32	.59375
39/64	.609375
5/8	.625
41/64	.640625
21/32	.65625
43/64	.671875
11/16	.6875
45/64	.703125
23/32	.71875
47/64	.734375
3/4	.750
49/64	.765625
25/32	.78125
51/64	.796875
13/16	.8125
53/64	.828125
27/32	.84375
55/64	.859375
7/8	.875
57/64	.890625
29/32	.90625
59/64	.921875
15/16	.9375
61/64	.953125
31/32	.96875
63/64	.984375
1	1.000

TABLE 8 NAIL DATA

Penny	Length (in.)	Gauge No. of Shank			Approx. No. Per Lb.		
		Common	Box	Finish	Common	Box	Finish
2d	1	15	15 1/2	16 1/2	830	1010	1351
3d	1 1/4	14	14 1/2	15 1/2	528	635	807
4d	1 1/2	12 1/2	14	15	316	473	584
6d	2	11 1/2	12 1/2	13	168	236	309
8d	2 1/2	10 1/4	11 1/2	12 1/2	106	145	189
10d	3	9	10 1/2	11 1/2	69	94	121
16d	3 1/2	8	10	11	49	71	90
20d	4	6	9	10	31	52	62
30d	4 1/2	5			24		
40d	5	4			18		
60d	6	2			11		

TABLE 9 WOOD SCREW INFORMATION

Wood Screw Size	Shank Hole Size	Pilot Hole Size		Auger Bit Sizes by 16th for Counterbore Hole
	All Woods	Hardwood	Softwood	
	Nearest Fractional Size Drill	Nearest Fractional Size Drill	Nearest Fractional Size Drill	
0	1/16	1/32	—	—
1	5/64	1/32	1/32	—
2	3/32	3/64	1/32	3
3	7/64	1/16	3/64	4
4	7/64	1/16	3/64	4
5	1/8	5/64	1/16	4
6	9/64	5/64	1/16	5
7	5/32	3/32	1/16	5
8	11/64	3/32	5/64	6
9	3/16	7/64	5/64	6
10	3/16	7/64	3/32	6
11	13/64	1/8	3/32	7
12	7/32	1/8	7/64	7

Standard screw lengths in inches are 3/16, 1/4, 3/8, 1/2, 5/8, 3/4, 7/8, 1, 1 1/4, 1 1/2, 1 3/4, 2, 2 1/4, 2 1/2, 2 3/4, 3, 3 1/2, and up.

TABLE 10 GRADES OF HARDWOOD LUMBER*

Grades	Clear cut ** (%)	Minimum Dimensions (W × L)	Principal Farm Uses
Firsts	91.7	6" × 8'	Flooring, trim, furniture, implement & tool handles
Seconds	83.3	6" × 8'	
Selects	91.7	4" × 6'	
No. 1 Common	66.7	3" × 4'	Permanent structures (barns, pens, feeders, troughs, fences)
No. 2 Common	50.0	3" × 4'	
Sound & Wormy	33.3	3" × 4'	Temporary structures (sheds, fences, feeders, crates)
No. 3A Common	33.3	3" × 4'	
No. 3B Common	25.0	3" × 4'	

*Adopted by National Hardwood Lumber Association.
**Percentage of face surface that is free of all defects.

TABLE 11 LUMBER DIMENSIONS

Type of Lumber	Rough Size (in.) (Sawmill Cut)	Actual Size (in.) (S 4 S)*	Bd. Ft. Per Ft. of Length	Amount to Add to Cover**
Square Edge Boards	1 × 4	$3/4 \times 3\,1/2$	$1/3$	$1/10$
	1 × 6	$3/4 \times 5\,1/2$	$1/2$	$1/16$
	1 × 8	$3/4 \times 7\,1/4$	$2/3$	$1/16$
	1 × 10	$3/4 \times 9\,1/4$	$5/6$	$1/20$
	1 × 12	$3/4 \times 11\,1/4$	1	$1/24$
Framing	2 × 4	$1\,1/2 \times 3\,1/2$	$2/3$	—
	2 × 6	$1\,1/2 \times 5\,1/2$	1	—
	2 × 8	$1\,1/2 \times 7\,1/4$	$1\,1/3$	—
	2 × 10	$1\,1/2 \times 9\,1/4$	$1\,2/3$	—
	2 × 12	$1\,1/2 \times 11\,1/4$	2	—
	4 × 4	$3\,1/2 \times 3\,1/2$	$1\,1/3$	—
	6 × 6	$5\,1/2 \times 5\,1/2$	3	—

*Surfaced on all four sides.
**Amount to add to total surface to allow for surfacing and matching of lumber.
(Does not include carpenter's waste.)

TABLE 12 WOOD MEASURE

A board foot of lumber = A Unit 1' long × 1' wide × 1" thick or its equivalent

1 U.S. cord of wood = A Pile 4' wide × 4' high × 8' long or its equivalent

1 cord foot of wood = 4' wide × 4' high × 1' long or its equivalent

Board feet = Number of pieces of wood × thickness in inches × width in inches × length in feet ÷ 12

TABLE 13 POSITIONS OF FRAMING SQUARE FOR SOME COMMON ANGLES

Position of Body	Position of Tongue	For Angle at Body	For Angle at Tongue
12"	1 1/16"	5°	85°
12"	2 1/8"	10°	80°
12"	3 3/16"	15°	75°
12"	4 3/8"	20°	70°
12"	5 9/16"	25°	65°
12"	6 15/16"	30°	60°
12"	8 3/8"	35°	55°
12"	10 1/16"	40°	50°
12"	12"	45°	45°

TABLE 14 RECOMMENDED RPM FOR WOOD TURNING

Stock Diam.	Rounding Stock	General Cutting	Finishing
1"	900–1400	2400–3000	3000–4400
2"	800–1200	2000–2700	2700–4000
3"	700–1050	1800–2400	2400–3500
4"	600– 900	1600–2100	2100–3000
5"	550– 750	1400–1800	1850–2600
6"	500– 650	1200–1550	1600–2200
7"	450– 550	1000–1300	1350–1650
8"	400– 500	800–1050	1100–1400
9"	350– 450	600– 700	750–1150
10"	300– 400	400– 500	600– 950

TABLE 15 ABRASIVE PAPER GRADES

Grit	Equiv. "O" Series	General Description	Remarks
600	—	Super Fine	Range of abrasive papers used for wet sanding
500	—		
400	10/0		
360	—		
320	9/0		
280	8/0	Very Fine	Used for dry sanding all finishing undercoats
240	7/0		
220	6/0		
180	5/0	Fine	For final sanding of bare wood. Good for smoothing old paint
150	4/0		
120	3/0	Medium	Use for general wood sanding. Good for first smoothing of old paint, plaster patches
100	2/0		
80	1/0		
60	1/2	Coarse	For rough-wood sanding
50	1		
40	1 1/2		
36	2	Very Coarse	Too coarse for pad sanders. Heavy machines and high speed recommended
30	2 1/2		
24	3		

TABLE 16 RECOMMENDED RPM OF HIGH-SPEED DRILLS IN VARIOUS METALS*

Drill Size**	Wrought Iron, Low-Carbon Steel	Medium-Carbon Steel	High-Carbon Tool Steel	Cast Iron	Aluminum & Brass
1/16	5000–6700	4000–4900	3000–3600	4200–6100	12,000–18,000
1/8	2500–3350	2100–2450	1500–1800	2100–3000	6000–9000
3/16	1600–2200	1400–1600	1000–1200	1400–2000	4000–6000
1/4	1200–1700	1000–1200	750–900	1100–1500	3000–4500
5/16	1000–1300	850–950	600–725	850–1200	2400–3600
3/8	800–1100	700–800	500–600	700–1000	2000–3000
7/16	700–950	600–700	425–525	600–875	1700–2500
1/2	600–850	450–600	375–450	525–750	1500–2250
9/16	550–750	425–550	340–400	475–675	1325–2000
5/8	500–650	400–500	300–350	425–600	1200–1800
11/16	450–600	375–450	275–325	375–550	1100–1650
3/4	400–550	350–400	250–300	350–500	1000–1500
13/16	375–500	325–375	235–275	325–460	900–1375
7/8	350–450	300–350	225–350	300–425	850–1275
15/16	325–425	275–325	210–235	275–400	800–1200
1	300–400	250–300	200–225	250–375	750–1125

*Reduce rpm one-half for carbon drills.
**For number and letter drills, use speed of nearest fractional-sized drill.

TABLE 17 DATA FOR SCREW EXTRACTORS

Screw or Bolt (In.)	Twist Drill (In.)	Extractor (No.)	Pipe (In.)
3/16–1/4	5/64	1	
1/4–5/16	7/64	2	
5/16–7/16	5/32	3	
7/16–9/16	1/4	4	1/8
9/16–3/4	17/64	5	1/4
3/4–1	13/32	6	3/8
1–1 3/8	17/32	7	1/2
1 3/8–1 3/4	13/16	8	3/4
1 3/4–2 1/8	1 1/16	9	1
2 1/8–2 1/2	1 5/16	10	1 1/4
2 1/2–3	1 9/16	11	1 1/2
3–3 1/2	1 15/16	12	2

TABLE 18 SHEET METAL RIVETING DATA

Metal Gauge	Rivets			Drill, Punch	Rivet Set
	Weight	Diam.	Length		
28	14 oz.	.109"	3/16"	7/64"	#8
26	1 lb.	.112"	13/64"	1/8"	#7
24	2 lb.	.144"	17/64"	9/64"	#5
22	2 1/2 lb.	.148"	9/32"	5/32"	#4
20	3 lb.	.160"	5/16"	11/64"	#4

TABLE 19 SELF-TAPPING SHEET METAL SCREW DATA (POINTED TYPE)

Screw No.	Screw Lengths*	Metal Gauge	Drill Size	
			No.	Inches
4	1/4–3/4	29–22	42	3/32
6	3/8–1	29–20	38	7/64
8	3/8–1 1/2	26–18	32	1/8
10	3/8–2	24–18	29	9/64
12	1/2–2	24–18	25	5/32
14	1/2–2	24–18	12	3/16

*Standard screw lengths in inches are: 1/4, 3/8, 1/2, 5/8, 3/4, 7/8, 1, 1 1/4, 1 1/2, and 2. This column shows range of screw lengths available—example, for No. 4, lengths are 1/4, 3/8, 1/2, 5/8, and 3/4.

TABLE 20 FLUXES FOR SOFT SOLDERING

Metal	Flux	Chemical Name
Aluminum	Special compounds	
Brass, copper	Cut acid, rosin, sal ammoniac	Zinc chloride, colophony, ammonium chloride
Galvanized iron	Muriatic acid	Hydrochloric acid
Iron, steel	Cut acid, sal ammoniac	Zinc chloride, ammonium chloride
Tin	Rosin, cut acid	Colophony, zinc chloride

TABLE 21 MELTING POINTS OF SOFT SOLDERS

Tin	Lead	Liquifies at *
63%	37%	360°F
50	50	415
45	55	437
40	60	459
33	67	621

*First melting point for each type is 360°F.

TABLE 22 RECOMMENDED COLORS FOR TEMPERING VARIOUS TOOLS*

Degrees Fahrenheit	Color	Kind of Tool
430	Yellow	Scrapers, lathe cutting tools, hammers
470	Straw	Punches, dies, hack-saw blades, drills, taps, knives, reamers
500	Brown	Axes, wood chisels, drifts, shears
530	Purple	Cold chisels, center punches, rivet sets
560	Blue	Screw drivers, springs, gears, picks, saws
700	Grey	Soft (annealed), must harden again

*Because of the difference in the quality of steel used in tool manufacture, the color recommendations will not always apply. *E.g.*, it may be necessary to cool a cold chisel when at a straw or blue color instead of purple to get the desired temper.

TABLE 23 HACK SAW BLADES

Teeth per In.	Stock	Material
14	1" & over	cast iron, machine steel, brass, copper, aluminum, bronze, slate
18	1/4" to 1"	annealed toolsteel, high-speed steel, rails, bronze, copper, aluminum
24	1/8" to 1/4"	iron, steel, drill rod, brass & copper tubing, wrought-iron pipe, conduit, trim
32	1/8" & less	same as for 24 teeth per in.

TABLE 24 FUEL-OIL MIXTURES

	U.S. Gallons		Imperial Gallons		Metric	
Mix.	**Fuel**	**Oil**	**Fuel**	**Oil**	**Petrol**	**Oil**
16:1	1 Gallon 3 Gallons 5 Gallons 6 Gallons	8 oz. 24 oz. 40 oz. 48 oz.	1 Gallon 3 Gallons 5 Gallons 6 Gallons	10 oz. 30 oz. 50 oz. 60 oz.	4 Liters 12 Liters 20 Liters 24 Liters	.250 Liter .750 Liter 1.250 Liter 1.500 Liter
24:1	1 Gallon 3 Gallons 5 Gallons 6 Gallons	5.33 oz. 16 oz. 26.66 oz. 32 oz.	1 Gallon 3 Gallons 5 Gallons 6 Gallons	6.4 oz. 19.2 oz. 32 oz. 38.4 oz.	4 Liters 12 Liters 20 Liters 24 Liters	.160 Liter .470 Liter .790 Liter .940 Liter
32:1	1 Gallon 3 Gallons 5 Gallons 6 Gallons	4 oz. 12 oz. 20 oz. 24 oz.	1 Gallon 3 Gallons 5 Gallons 6 Gallons	5 oz. 15 oz. 25 oz. 30 oz.	4 Liters 12 Liters 20 Liters 24 Liters	.125 Liter .375 Liter .625 Liter .750 Liter
50:1	1 Gallon 3 Gallons 5 Gallons 6 Gallons	2.5 oz. 8 oz. 13 oz. 15.5 oz.	1 Gallon 3 Gallons 5 Gallons 6 Gallons	3 oz. 9 oz. 15 oz. 18.5 oz.	4 Liters 12 Liters 20 Liters 24 Liters	.080 Liter .240 Liter .400 Liter .480 Liter

TABLE 25 ELECTRODES FOR FARM WELDING

Type of Work	Electrode		Current & Polarity (S-straight) (R-reversed)	Welding Position	Pene-tration	Characteristics	Uses
	AWS No.	NEMA Color					
Mild Steel	E6010	None	DC-R	All	Deep	Digging affects it; leaves rough, rippled surface; slow burn off; a strong weld with much spatter.	All mild-steel welding; $^{1}/_{8}$" very good for holes and cutting; best farm electrode.
	E6011	Blue spot	All	All	Deep		
	E6012	White spot	All	All	Medium	Melted metal is gummy.	Fill in gaps and poor fit ups; medium arc keeps sagging metal from closing the circuit.
	E6013	Brown spot	All	All	Shallow		Very good for down welds on thin sheet metal.
	E6013	Brown spot	All	All	Shallow	General-purpose, easy to use; less burn through owing to shallower penetration.	All mild steel; for extra strength, lay a heavy first bead or two light beads.
	E6014	Brown spot	All	All	Shallow	Cross between E6013 and E6024; easy to use; fast burn off.	Second best for farm; same uses as above.
	E6024	Yellow spot	All	Flat, Horizontal	Medium	Fast deposit; slag removes itself if amperage is right and edges not pinned by poor electrode motion.	All downhand welds, in fillets and horizontal; $^{3}/_{32}$" good as spray rod on vertical down welds.
	E6027	Silver spot	All	Flat, Horizontal	Medium	Less undercut than E6024; better weld; slag not as easily removed; less penetration on rusty metal.	Same as E6024, but see Character-istics (to the left).
Low Alloy Metal	E7016	(See Char-acter.)	AC, DC-R	All	Varies	Requires clean surface*; NEMA code; green group, orange spot, blue end.	All difficult welds; rather hard for beginners to use.
Non-Machin-able Cast Iron	No AWS; non-machin-able cast iron	Orange end	AC, DC-S	All	Medium	Requires clean surface*; low heat needed reduces cracking of weld or work; *do not permit work to become dark red.*	For shielded arc welds if not to be machined; hold close arc, but coating must not touch molten metal; intermittent beads not over 3" long; peen lightly after each bead; cool and clean before next bead.
Cast Iron	No AWS; cast i.	Orange spot	AC, DC-R	All	Medium	Same as above, except machinable.	All cast iron, machinery gears, housings, parts, mold boards, etc.
Stain-less Steel	No AWS; s. steel 308-16	Yellow group & end	All	All	Medium	Requires short arc, electrode held 15° in direction of travel.	High-to-medium carbon-alloy steel and most nonferrous metals; auto bumpers, mold boards, etc.
Hard Sur-facing	No AWS numbers or NEMA color. Consult manufac-urers' catalogs according to uses.		AC, DC-R	All	Light	For all hardsurfacing; requires clean surface*; not for joining parts; build up worn parts with a strength rod, then hardsurface; for wear resistance, lay straight or weavy bead not over 3/4" wide; remove slag before new bead.	Metal-to-metal wear: building-up gives moderate hardness to resist shock and abrasion.
			AC, DC-S	All	Light		Metal in rocky soil: build-up for resistance to impact and severe abrasion; for mild or carbon steel, low-alloy or high-manganese steel.
			AC, DC-S	All	Light		Metal in sandy soil: to resist any kind of abrasion, mild impact; for carbon, alloy, or manganese steel.
			AC, Arc torch	All	Light		Knife edges: fine-grain alloy pow-der applied with carbon arc gives smooth, abrasion-resistance surface.

*This means that surfaces must be cleaned of rust, grease, oil, moisture, and other foreign matter by wire brushing or grinding

TABLE 26 ELECTRICAL WIRE SIZES FOR COPPER AT 115–120 VOLTS

Minimum Allowable Size of Conductor

COPPER up to 200 Amperes, *115=120 Volts*, Single Phrase, Based on 2% Voltage Drop

Length of Run in Feet

Compare size shown below with size shown to left of double line. Use the larger size.

In Cable, Conduit, Earth: Load in Amps	Types R, T, TW	Types RH, RHW, THW	Overhead In Air*: Bare & Covered Conductors	30	40	50	60	75	100	125	150	175	200	225	250	275	300	350	400	450	500	550	600	650	700
5	12	12	10	12	12	12	12	12	12	12	10	10	10	10	8	8	8	8	6	6	6	6	4	4	4
7	12	12	10	12	12	12	12	12	12	10	10	8	8	8	8	6	6	6	6	4	4	4	4	4	3
10	12	12	10	12	12	12	12	10	10	8	8	8	6	6	6	6	4	4	4	4	3	3	2	2	2
15	12	12	10	12	12	10	10	10	8	6	6	6	4	4	4	4	4	3	2	2	1	1	1	0	0
20	12	12	10	12	10	10	8	8	6	6	4	4	4	4	3	3	2	2	1	1	0	0	00	00	00
25	10	10	10	10	10	8	8	6	6	4	4	4	3	3	2	2	1	1	0	0	00	00	000	000	000
30	10	10	10	10	8	8	8	6	4	4	4	3	2	2	1	1	1	0	00	00	000	000	000	4/0	4/0
35	8	8	10	10	8	8	6	6	4	4	3	2	2	1	1	0	0	00	00	000	000	4/0	4/0	4/0	250
40	8	8	10	8	8	6	6	4	4	3	2	2	1	1	0	0	00	00	000	000	4/0	4/0	250	250	300
45	6	8	10	8	8	6	6	4	4	3	2	1	1	0	0	00	00	000	000	4/0	4/0	250	250	300	300
50	6	6	10	8	6	6	4	4	3	2	1	1	0	0	00	00	000	000	4/0	4/0	250	250	300	300	350
60	4	6	8	8	6	4	4	4	2	1	1	0	00	00	000	000	000	4/0	250	250	300	300	350	400	400
70	4	4	8	6	6	4	4	3	2	1	0	00	00	000	000	4/0	4/0	250	300	300	350	400	400	500	500
80	2	4	6	6	4	4	3	2	1	0	00	00	000	000	4/0	4/0	250	300	300	350	400	400	500	500	600
90	2	3	6	6	4	4	3	2	1	0	00	000	000	4/0	4/0	250	250	300	350	400	500	500	500	600	600
100	1	3	6	4	4	3	2	1	0	00	000	000	4/0	4/0	250	250	300	350	400	500	500	500	600	600	700
115	0	2	4	4	4	3	2	1	0	00	000	4/0	4/0	250	300	300	350	400	500	500	600	600	700	700	750
130	00	1	4	4	3	2	1	0	00	000	4/0	4/0	250	300	300	350	400	500	500	600	600	700	750	800	900
150	000	0	2	4	2	1	1	0	000	4/0	4/0	250	300	350	350	400	500	500	600	700	700	800	900	900	1M
175	4/0	00	2	3	2	1	0	00	000	4/0	250	300	350	400	400	500	500	600	700	750	800	900	1M		
200	250	000	1	2	1	0	00	000	4/0	250	300	350	400	500	500	500	600	700	750	900	1M				

TABLE 27 ELECTRICAL WIRE SIZES FOR ALUMINUM AT 115–120 VOLTS

Minimum Allowable Size of Conductor

ALUMINUM up to 200 Amperes, *115=120 Volts*, Single Phrase, Based on 2% Voltage Drop

Length of Run in Feet

Compare size shown below with size shown to left of double line. Use the larger size

In Cable, Conduit, Earth: Load in Amps	Types R, T, TW	Types RH, RHW, THW	Overhead In Air*: Bare & Covered Conductors Single Triplex	30	40	50	60	75	100	125	150	175	200	225	250	275	300	350	400	450	500	550	600	650	700
5	12	12	10	12	12	12	12	12	10	10	8	8	8	8	6	6	6	6	4	4	4	4	3	3	3
7	12	12	10	12	12	12	12	10	10	8	8	6	6	6	6	4	4	4	4	3	3	2	2	2	1
10	12	12	10	12	12	10	10	8	8	6	6	6	4	4	4	4	3	3	2	2	1	1	0	0	0
15	12	12	10	12	10	8	8	8	6	4	4	4	3	3	2	2	2	1	0	0	00	00	00	000	000
20	10	10	10	10	8	8	6	6	4	4	3	3	2	2	1	1	0	0	00	00	000	000	4/0	4/0	4/0
25	10	10	10	8	8	6	6	4	4	3	2	2	1	1	0	0	00	00	000	000	4/0	4/0	250	250	300
30	8	8	10	8	6	6	6	4	3	2	2	1	0	0	00	00	00	000	4/0	4/0	250	250	300	300	350
35	6	8	10	8	6	6	4	4	3	2	1	0	0	00	00	000	000	4/0	4/0	250	300	300	350	350	400
40	6	8	10	6	6	4	4	3	2	1	0	0	00	00	000	000	4/0	4/0	250	300	300	350	350	400	500
45	4	6	10	6	6	4	4	3	2	1	0	00	00	000	000	4/0	4/0	250	300	300	350	400	400	500	500
50	4	6	8	6	4	4	3	2	1	0	00	00	000	000	4/0	4/0	250	300	300	350	400	400	500	500	600
60	2	4	6	6	4	3	3	2	0	00	00	000	4/0	4/0	250	250	300	350	350	400	500	500	600	600	700
70	2	2	6	4	4	3	2	1	0	00	000	4/0	4/0	250	300	300	350	400	500	500	600	600	700	700	750
80	1	2	6	4	3	2	1	0	00	000	4/0	4/0	250	300	300	350	350	500	500	600	600	700	750	800	900
90	0	2	4	4	3	2	1	0	00	000	4/0	250	300	300	350	400	400	500	600	600	700	750	800	900	1M
100	0	1	4	3	2	1	0	00	000	4/0	250	300	300	350	400	400	500	600	600	700	750	900	900	1M	
115	00	0	2	3	1	1	0	00	000	4/0	300	300	350	400	500	500	600	600	700	800	900	1M			
130	000	00	2	2	1	0	00	000	4/0	250	300	350	400	500	500	600	600	700	800	900	1M				
150	4/0	000	1	2	0	00	00	000	250	300	350	400	500	500	600	600	700	800	900	1M					
175	300	4/0	0	1	0	00	000	4/0	300	350	400	500	600	600	700	750	800	900							
200	350	250	00	0	00	000	4/0	250	300	400	500	600	600	700	750	900	900								

TABLE 28 ELECTRICAL WIRE SIZES FOR COPPER AT 230-240 VOLTS

Minimum Allowable Size of Conductor — **COPPER up to 400 Amperes, *230=240 Volts*, Single Phrase, Based on 2% Voltage Drop**

In Cable, Conduit, Earth			Overhead In Air*	Length of Run in Feet																					
				Compare size shown below with size shown to left of double line. Use the larger size.																					
Load in Amps	**Types R, T, TW**	**Types RH, RHW, THW**	**Bare & Covered Conductors**	**50**	**60**	**75**	**100**	**125**	**150**	**175**	**200**	**225**	**250**	**275**	**300**	**350**	**400**	**450**	**500**	**550**	**600**	**650**	**700**	**750**	**800**
5	12	12	10	12	12	12	12	12	12	12	12	12	12	12	10	10	10	10	8	8	8	8	8	6	6
7	12	12	10	12	12	12	12	12	12	12	12	10	10	10	10	8	8	8	8	6	6	6	6	6	6
10	12	12	10	12	12	12	12	12	10	10	10	10	8	8	8	8	6	6	6	6	4	4	4	4	4
15	12	12	10	12	12	12	10	10	10	8	8	8	6	6	6	6	4	4	4	4	4	3	3	3	2
20	12	12	10	12	12	10	10	8	8	8	6	6	6	6	4	4	4	4	3	3	2	2	2	1	1
25	10	10	10	12	10	10	8	8	6	6	6	6	4	4	4	4	3	3	2	2	1	1	1	0	0
30	10	10	10	10	10	10	8	6	6	6	4	4	4	4	4	3	2	2	1	1	1	0	0	0	00
35	8	8	10	10	10	8	8	6	6	4	4	4	4	3	3	2	2	1	1	0	0	0	00	00	00
40	8	8	10	10	8	8	6	6	4	4	4	4	3	3	2	2	1	1	0	0	00	00	00	000	000
45	6	8	10	10	8	8	6	6	4	4	4	3	3	2	2	1	1	0	0	00	00	00	000	000	000
50	6	6	10	8	8	6	6	4	4	4	3	3	2	2	1	1	0	0	00	00	000	000	000	4/0	4/0
60	4	6	8	8	8	6	4	4	4	3	2	2	1	1	1	0	00	00	000	000	000	4/0	4/0	4/0	250
70	4	4	8	8	6	6	4	4	3	2	2	1	1	0	0	00	00	000	000	4/0	4/0	4/0	250	250	300
80	2	4	6	6	6	4	4	3	2	2	1	1	0	0	00	00	000	000	4/0	4/0	250	250	300	300	300
90	2	3	6	6	6	4	4	3	2	1	1	0	0	00	00	000	000	4/0	4/0	250	250	300	300	350	350
100	1	3	6	6	4	4	3	2	1	1	0	0	00	00	000	000	4/0	4/0	250	250	300	300	350	350	400
115	0	2	4	6	4	4	3	2	1	0	0	00	00	000	000	4/0	4/0	250	300	300	350	350	400	400	500
130	00	1	4	4	4	3	2	1	0	0	00	00	000	000	4/0	4/0	250	300	300	350	400	400	500	500	500
150	000	0	2	4	4	3	1	0	0	00	000	000	4/0	4/0	4/0	250	300	350	350	400	500	500	500	600	600
175	4/0	00	2	4	3	2	1	0	00	000	000	4/0	4/0	250	250	300	350	400	400	500	500	600	600	600	700
200	250	000	1	3	2	1	0	00	000	000	4/0	4/0	250	250	300	350	400	500	500	500	600	600	700	700	750
225	300	4/0	0	3	2	1	0	00	000	4/0	4/0	250	300	300	350	400	500	500	600	600	700	700	750	800	900
250	350	250	00	2	1	0	00	000	4/0	4/0	250	300	300	350	350	400	500	600	600	700	700	750	800	900	1M
275	400	300	00	2	1	0	00	000	4/0	250	250	300	350	350	400	500	500	600	700	700	800	900	900	1M	
300	500	350	000	1	1	0	000	4/0	4/0	250	300	350	350	400	500	500	600	700	700	800	900	900	1M		
325	600	400	4/0	1	0	00	000	4/0	250	300	300	350	400	500	500	600	600	700	750	900	900	1M			
350	600	500	4/0	1	0	00	000	4/0	250	300	350	400	400	500	500	600	700	750	800	900	1M				
375	700	500	250	0	0	00	4/0	250	300	300	350	400	500	500	600	600	700	800	900	1M					
400	750	600	250	0	00	000	4/0	250	300	350	400	500	500	500	600	700	750	900	1M						

Conductors in overhead spans must be at least No. 10 for spans up to 50 feet and No. 8 for longer spans. See NEC, Sec. 225-6(a).
See Note 3 of NEC "Notes to Tables 310-16 through 310-19".

TABLE 29 ELECTRICAL WIRE SIZES FOR ALUMINUM AT 230-240 VOLTS

Minimum Allowable Size of Conductor (In Cable, Conduit, Earth; Overhead In Air*)

ALUMINUM up to 400 Amperes, *230=240 Volts*, Single Phrase, Based on 2% Voltage Drop

Length of Run in Feet

Compare size shown below with size shown to left of double line. Use the larger size.

Load in Amps	Types R, T, TW	Types RH, RHW, THW	Bare & Covered Conductors Single	Bare & Covered Conductors Triplex	50	60	75	100	125	150	175	200	225	250	275	300	350	400	450	500	550	600	650	700	750	800
5	12	12	10		12	12	12	12	12	12	12	10	10	10	10	8	8	8	8	6	6	6	6	6	4	4
7	12	12	10		12	12	12	12	12	10	10	10	8	8	8	8	6	6	6	6	4	4	4	4	4	4
10	12	12	10		12	12	12	10	10	8	8	8	8	6	6	6	6	4	4	4	4	3	3	3	2	2
15	12	12	10		12	12	10	8	8	8	6	6	6	4	4	4	4	3	3	2	2	2	1	1	1	0
20	10	10	10		10	10	8	8	6	6	6	4	4	4	4	3	3	2	2	1	1	0	0	0	00	00
25	10	10	10		10	8	8	6	6	4	4	4	4	3	3	2	2	1	1	0	0	00	00	00	000	000
30	8	8	10		8	8	8	6	4	4	4	3	3	2	2	2	1	0	0	00	00	00	000	000	000	4/0
35	6	8	10		8	8	6	6	4	4	3	3	2	2	1	1	0	0	00	00	000	000	000	4/0	4/0	4/0
40	6	8	10		8	6	6	4	4	3	3	2	2	1	1	0	0	00	00	000	000	4/0	4/0	4/0	250	250
45	4	6	10		8	6	6	4	4	3	2	2	1	1	0	0	00	00	000	000	4/0	4/0	250	250	250	300
50	4	6	8		6	6	4	4	3	2	2	1	1	0	0	00	00	000	000	4/0	4/0	250	250	300	300	300
60	2	4	6	6	6	6	4	3	2	2	1	0	0	00	00	00	000	4/0	4/0	250	250	300	300	350	350	350
70	2	2(a)	6	4	6	4	4	3	2	1	0	0	00	00	000	000	4/0	4/0	250	300	300	350	350	400	400	500
80	1	2(a)	6	4	4	4	3	2	1	0	0	00	00	000	000	4/0	4/0	250	300	300	350	350	400	500	500	500
90	0	2(a)	4	2	4	4	3	2	1	0	00	00	000	000	4/0	4/0	250	300	300	350	400	400	500	500	500	600
100	0	1(a)	4	2	4	3	2	1	0	00	00	000	000	4/0	4/0	250	300	300	350	400	400	500	500	600	600	600
115	00	0(a)	2	1	4	3	2	1	0	00	000	000	4/0	4/0	250	300	350	400	500	500	600	600	600	700	700	
130	000	00(a)	2	0	3	2	1	0	00	000	000	4/0	250	250	300	300	400	500	500	600	600	700	700	750	800	
150	4/0	000(a)	1	00	2	2	1	00	000	000	4/0	250	250	300	300	350	500	500	600	600	700	750	800	900	900	
175	300	4/0(a)	0	000	2	1	0	00	000	4/0	250	300	300	350	400	400	600	600	700	750	800	900	900	1M		
200	350	250	00	4/0	1	0	00	000	4/0	250	300	300	350	400	400	500	600	700	750	900	900	1M				
225	400	300	000		1	0	00	000	4/0	250	300	350	400	500	500	500	600	700	750	900	1M	1M				
250	500	350	000		0	00	000	4/0	250	300	350	400	500	500	500	600	700	750	900	1M						
275	600	500	4/0		0	00	000	4/0	250	300	400	400	500	500	600	600	750	900	1M							
300	700	500	250		00	00	000	250	300	350	400	500	500	600	600	700	800	900	1M							
325	800	600	300		00	000	4/0	250	300	400	500	500	600	600	700	750	900	1M								
350	900	700	300		00	000	4/0	300	350	400	500	600	600	700	750	800	900									
375	1M	700	350		000	000	4/0	300	350	500	500	600	700	700	800	900	1M									
400		900	350		000	4/0	250	300	400	500	600	600	700	750	900	900										

Conductors in overhead spans must be at least No. 10 for spans up to 50 feet and No. 8 for longer spans. See NEC, Sec. 225-6(a).
See Note 3 of NEC "Notes to Tables 310-16 through 310-19".

TABLE 30 AVERAGE DAILY WATER REQUIREMENTS OF FARM ANIMALS

Horse	10-12 gallons
Beef Cattle	8-12 gallons
Dairy Cows (dry)	8-12 gallons
Dairy Cows	12-20 gallons
Hogs	4-11 gallons
Sheep	1-2 gallons
Layers (100)	5 gallons
Broilers (100)	1-10 gallons

TABLE 31 STEEL AND WROUGHT-IRON PIPE DATA*

Nom. Pipe Size	Outside Diam.	Standard		Extra Strong		Threads Per Inch	Tap Drill Size	Hole Size **	Length of Thread	Dist. Pipe Goes into Fitting
		Inside Diam.	Weight Lb. Per Ft.	Inside Diam.	Weight Lb. Per Ft.					
1/8	0.405	0.269	0.24	0.215	0.31	27	5/16	7/16	13/32	1/4
1/4	0.540	0.364	0.42	0.302	0.53	18	7/32	9/16	5/8	3/8
3/8	0.675	0.493	0.57	0.423	0.74	18	19/32	11/16	5/8	3/8
1/2	0.840	0.622	0.85	0.546	1.09	14	23/32	7/8	13/16	1/2
3/4	1.050	0.824	1.13	0.742	1.47	14	15/16	1 1/16	7/8	1/2
1	1.315	1.049	1.68	0.957	2.17	11 1/2	1 3/16	1 11/32	1 1/32	9/16
1 1/4	1.660	1.380	2.27	1.278	3.00	11 1/2	1 15/16	1 11/16	1 1/16	5/8
1 1/2	1.900	1.610	2.72	1.500	3.63	11 1/2	1 25/32	1 15/16	1 1/16	11/16
2	2.375	2.067	3.65	1.937	5.04	11 1/2	2 3/16	2 7/16	1 1/8	3/4
2 1/2	2.875	2.469	5.79	2.323	7.66	8	2 5/8	2 15/16	1 5/8	7/8
3	3.500	3.068	7.58	2.900	10.25	8	3 1/4	3 9/16	1 11/16	15/16

*All measurements are stated in fractional or decimal inches.
**Size of hole to drill to insert pipe through walls, floors, and other materials.

TABLE 32 STANDARD MALLEABLE FITTINGS

Nominal Size (in.)	General Dimensions (inches)					
	A	B	C	D	E	F
1/8	11/16	11/16	15/16	—	1	13/16
1/4	13/16	3/4	1 1/16	1	1 3/16	15/16
3/8	15/16	13/16	1 3/16	1 1/8	1 7/16	1
1/2	1 1/8	7/8	1 5/16	1 1/4	1 5/8	1 1/8
3/4	1 5/16	1	1 1/2	1 1/2	1 7/8	1 5/16
1	1 1/2	1 1/8	1 11/16	1 11/16	2 1/8	1 7/16
1 1/4	1 3/4	1 5/16	1 15/16	2 1/16	2 7/16	1 11/16
1 1/2	1 15/16	1 7/16	2 1/8	2 5/16	2 11/16	1 7/8
2	2 1/4	1 11/16	2 1/2	2 13/16	3 1/4	2 1/4
2 1/2	2 11/16	1 15/16	2 7/8	3 1/4	3 7/8	—
3	3 1/16	2 3/16	3 3/16	3 11/16	4 1/2	—

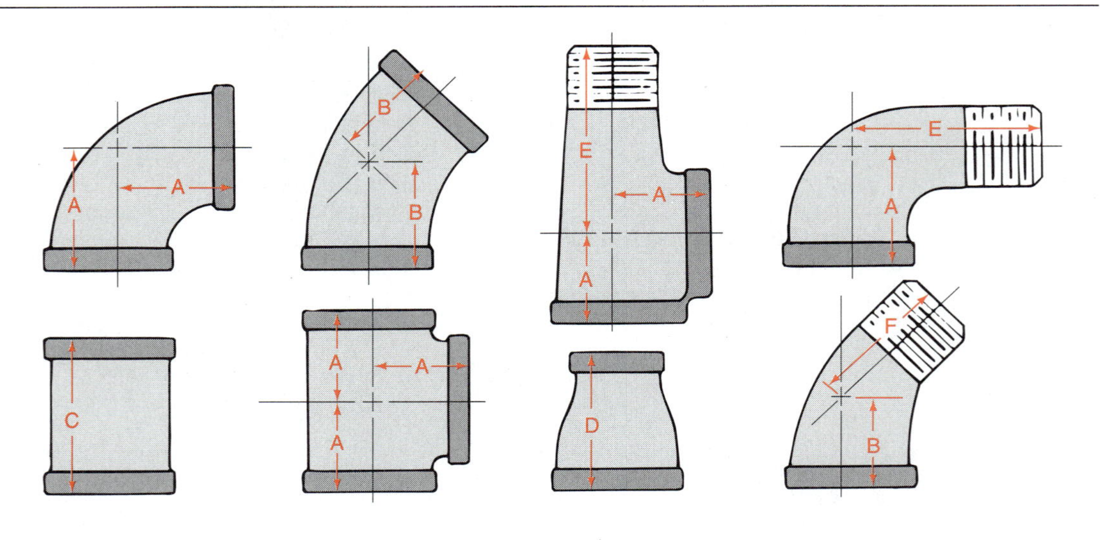

TABLE 33 RECOMMENDED MIXTURES FOR CONCRETE

Uses of Concrete	Water (gal.) for 1 Sack Cement—Sand				Trial Batch 1 Sack Cement plus		Space Trial Batch Fills (ft³)	Materials for 1 yd³**		
	Dry	Damp	Wet	Very Wet	Sand (ft³)	Gravel (ft³)		Cement (sack)	Sand (ft³)	Gravel (ft³)
Acid-, alkali-resistant; dairy, creamery floors	5*	4 3/4	4 1/2	4 1/4	1 3/4	2	3.38	8	14	16
Medium wear; reinforced; water-tight; floors, tanks, etc.	6*	5 1/2	5	4 1/4	2 1/4	3	4.32	6 1/4	14	19
Medium wear, indoor, underground, no water pressure	7*	6 1/4	5 1/2	4 3/4	2 3/4	4	5.4	5	14	20

*Called 5-, 6-, or 7-gallon pastes, indicating that there are 5, 6, or 7 gallons of water to each sack of cement, including the amount of water in the sand.

**To calculate materials for 1 cu. ft., divide these figures by 27.

TABLE 34 APPROXIMATE MATERIALS REQUIRED FOR ONE CUBIC YARD OF CONCRETE

Proportions of Concrete or Mortar			Quantities of Materials		
Cement	Sand	Gravel	Sacks Cement	Cu. Yds. Sand	Cu. Yds. Gravel
1	1.5	—	15.5	0.86	—
1	2.0	—	12.8	.95	—
1	2.5	—	11.0	1.02	—
1	3.0	—	9.6	1.07	—
1	1.5	3	7.6	.42	0.85
1	2.0	2	8.2	.60	.60
1	2.0	3	7.0	.52	.78
1	2.0	4	6.0	.44	.89
1	2.5	3.5	5.9	.55	.77
1	2.5	4	5.6	.52	.83
1	2.5	5	5.0	.46	.92
1	3.0	5	4.6	.51	.85
1	3.0	6	4.2	.47	.94

TABLE 35
AREAS COVERED BY ONE CUBIC YARD OF MIXED CONCRETE

Depth, in.	Sq. ft.	Depth, in.	Sq. ft.	Depth, in.	Sq. ft.
1	324	$4^3/_4$	68	$8^1/_2$	38
$1^1/_4$	259	5	65	$8^3/_4$	37
$1^1/_2$	216	$5^1/_4$	62	9	36
$1^3/_4$	185	$5^1/_2$	59	$9^1/_4$	35
2	162	$5^3/_4$	56	$9^1/_2$	34
$2^1/_4$	144	6	54	$9^3/_4$	33
$2^1/_2$	130	$6^1/_4$	52	10	32.5
$2^3/_4$	118	$6^1/_2$	50	$10^1/_4$	31.5
3	108	$6^3/_4$	48	$10^1/_2$	31
$3^1/_4$	100	7	46	$10^3/_4$	30
$3^1/_2$	93	$7^1/_4$	45	11	29.5
$3^3/_4$	86	$7^1/_2$	43	$11^1/_4$	29
4	81	$7^3/_4$	42	$11^1/_2$	28
$4^1/_4$	76	8	40	$11^3/_4$	27.5
$4^1/_2$	72	$8^1/_4$	39	12	27

TABLE 36
PIGMENTS FOR COLORING CONCRETE

Color Desired	Pigment to Use
Blues	Cobalt oxide
Browns	Brown oxide of iron
Buffs	Synthetic yellow oxide of iron
Greens	Chromium oxide
Reds	Red oxide of iron
Grays & slate effects	Black iron oxide or carbon black (never use common lamp black)

GLOSSARY

A

abuse—to use wrongly, make bad use of, or misuse.

abusar—utilizar injustamente, usa mal de una cosa, o maltratar.

ACA—ammoniacal copper arsenate.

ACA—arsenato de cobre amoniacal.

ACC—acid copper chromate.

CCA–cromato de cobre ácido.

acid copper chromate—wood preservative.

cromato de cobre ácido—preservativo de la madera.

acre inch—amount needed to cover one acre of land one inch deep in water.

pulgada acre—cantidad de agua que se necesita para que sea sumergido un acre de tierra con agua de una pulgada de profundidad.

acrylonitrite-butadiene-styrene (ABS)—a type of plastic pipe used for sewerage and underground applications.

acrilonitrito-butadiene-estireno—un tipo de tubo plástico que se usa para el alcantarillado y utilizaciones subterráneas.

actuator—device that places another device in motion.

impulsor—dispositivo que da impulso a otro dispositivo, poniéndolo en movimiento.

adaptor—a fitting used to connect pipes of different types.

adaptor—un ajuste que se usa para conectar tubos de tipos diferentes.

additive—changes characteristics of a fluid.

aditivo–cambia las características de un fluido.

adhesive—sticky substance used to bind two materials.

adhesivo—sustancia pegajosa que se usa para ligar dos materiales.

adjust—to set a part or parts to function as designed.

ajustar—encajar una pieza o unas piezas para que funcionen así como deben según su diseño.

adjustable pulley—pulley with a movable side.

polea ajustable–polea cuyo canto es movible.

aerator—device that mixes air with water as it comes from a faucet.

aireador—aparato que mezcla el aire con el agua mientras sale ésta de un grifo.

aerosol—high-pressure container with a valve and spray nozzle.

aerosol—recipiente de alta presión con una válvula y un pulverizador.

agribusiness—broad range of activities associated with agriculture.

agroindustria—una gran variedad de actividades relativas a la agricultura.

agricultural mechanics—selection, operation, maintenance, service, sale, and use of power units, machinery, equipment, structures, and utilities in agriculture.

mecánica agrícola—elección, operación, mantenimiento, servicio, venta, y uso de motores, maquinaria, equipo, estructuras y herramientas de la agricultura.

agriculture—enterprises involving the production of plants and animals, along with supplies, services, mechanics, products, processing, and marketing related to those enterprises.

agricultura—empresas que comprenden la producción de plantas y animales, juntos con los artículos, servicios, mecanismos, productos, el elaborar, y la venta relativos a esas empresas.

air atomization—paint split into tiny droplets using compressed air.

atomización de aire—pintura pulverizada en gotitas por usar aire comprimido.

air compressor—pump that increases pressure on air.

compresor de aire—bomba que aumenta la presión sobre el aire.

air-dried lumber—sawed lumber separated with wooden strips and protected from rain and snow for six months or more.

madera de construcción secado al aire libre—madera aserrada separada con listones y protegida de la lluvia y la nieve por seis meses or más.

air-entrained concrete—ready-mixed concrete with tiny bubbles of air trapped throughout the mixture to strengthen it.

hormigón de oclusión de aire—hormigón hecho con pequeñas burbujas de aire retenidas por la mixtura para fortalecerlo.

air-lift pump—pumps or lifts water using rising air bubbles.

bomba elevadora de líquidos por aire comprimido—saca o levanta el agua usando burbujas de aire que se levantan.

air stones—ceramic tips that produce small air bubbles.

puntas de soplado—puntas arcillosas que producen pequeñas burbujas de aire.

air tool—a tool powered by compressed air.

herramienta aérea—herramienta accionada por el aire comprimido.

air vane—type of governor used on small engines.

aspa—un tipo de manostato regulador usado en motores pequeños.

alkyd—oil-base paint.

alquídica—pintura derivada de petroleo.

Allen screw—a screw with a six-sided hole in the head.

tornillo de cabeza allen—un tornillo cuyo cabeza consiste en agujero hexagonal.

alloy—mixture of two or more metals.

aleación—mezcla de dos metales o más.

alternating current (AC)—current that reverses its direction 60 times per second.

corriente alterna (CA)—corriente que recorre ya en un sentido ya en otro, sesenta veces por segundo.

alternator—device that produces alternating electrical current but usually equipped with diodes to change current to direct current.

alternador—dispositivo que produce la corriente eléctrica alterna pero usualmente equipado con diodos para convertir la corriente en corriente continua.

aluminum—tough, light, and durable metal; an element associated with high-quality paint.

aluminio—metal duro, ligero y resistente; elemento asociado con pintura de buena calidad.

American Welding Society (AWS)—an organization that supports education in welding processes and that developed a system of numerical classification of electrodes.

Sociedad Americana de Soldadura (SAS)—una organización que apoya la instrucción en los procesos de soldadura y que ha realizado un sistema de clasificación numérica de electrodos.

ammeter—measures current in amperes.

amperímetro—mide la corriente en amperios.

ammonia—nitrogen waste product by animals.

amoníaco—desecho de nitrógeno producido por animales.

ammoniacal copper arsenate—wood preservative.

arsenato de cobre amoniaco—preservativo de la madera.

ampere (A) (amp)—a measure of the rate of flow of current in a conductor.

amperio—una medida del régimen de la corriente que atraviesa un conductor.

amplifier—circuit used to increase the amplitude of a signal.

amplificador—circuito que se usa para aumentar la amplitud de una señal.

analog meter—instrument with a graduated scale and a pointer.

medidor análogo—instrumento que tiene una regla graduada y un índice.

anchor hole—pilot hole.

agujero de áncora—guía.

anneal—to cool steel slowly so as to make it soft and malleable.

recocer—dejar enfriar lentamente el acero para que se haga maleable y dulce.

annual rings—pattern in a tree caused by the hardening and disuse of tubes in the woody parts.

capa cortical—diseño en un árbol resultando del endurecimiento y desuso de las partes leñosas.

antifoam—reduces foaming tendency.

agente antiespumante—reduce la tendencia de hacer espuma.

antiscuff—helps polish moving parts.

antipatinador—ayuda en bruñir las partes móviles.

anvil—heavy steel object used to help bend, cut, and shape metal.

yunke—objeto pesado de acero empleado para ayudar en remachar, embutir, y cortar el metal.

API—American Petroleum Institute.

IAP—Instituto Americano de Petróleo.

apparatus—equipment necessary to carry out a function.

aparato—equipo necesario para llevar a cabo una función.

aquaculture—production of aquatic plants and animals.

piscicultura—producción de plantas y animales acuáticos.

arc—the discharge of electricity through an air space.

arco—la descarga de electricidad por el espacio entre polos y armadura.

arc welder—a machine that produces current for welding.

transformador para soldadura por arco—una máquina que produce la corriente eléctrica para soldadura.

arc welding—*See* shielded metal arc welding.

soldadura por arco—*V.* soldadura por arco protegido.

armature—the rotating part of a motor. Also the iron core portion of a magneto.

armadura—la parte de rotación del motor. También la parte del núcleo de hierro de un magneto.

armored cable—flexible metal sheath with individual wires inside.

cable armado—forro flexible de metal que tiene hilos individuos adentro.

asphalt—black paint pigment with excellent hiding power.

asfalto–pigmento de pintura negra con gran poder cubriente.

B

backfire—a loud snap or popping noise in a gas torch that generally blows out the flame.

retorno de llama—un ruido de explosión o un chasquido fuerte del soplete oxídrico, que generalmente sopla la llama.

back saw—a saw that has very fine teeth and a stiff metal back.

cola de zorro—una sierra de dientes muy finos y una banda dura de metal.

back-stepping—series of short welds each made in the backward direction while the bead progresses forward.

soldadura de retroceso—una serie de soldaduras cortas, cada una hecha en la dirección retroceso mientras el cordón sigue adelante.

bag culture—plant culture with artificial media contained in bags fed with a nutrient solution.

cultivo sin suelo, bolso para plantas—cultivo vegetal con medios artificiales contenidos en bolsas y alimentado con una solución nutritiva.

balance—weight equally distributed on all sides of center.

equilibrio—el peso distribuido igualmente por todos lados del centro.

ball bearing—set of hardened steel balls in a container.

cojinete de bolas—juego de bolas duras de acero en un contenedor (que se apoya un eje).

band saw—power tool with saw teeth on a continuous blade or band.

sierra de cinta tronzadora de trozos—máquina herramienta cuyos dientes son partes de una hoja continua o una cinta.

barrel sling—procedure for attaching rope to barrels.

nudo de barril—procedimiento de atar cuerda a los barriles.

base metal—main piece of metal or main component.

metal prima—pieza de metal principal o componente mayor.

batter board—board placed to carry level guide lines.

tabla marcadora para establecer la línea de excavación—tabla colocada para llevar líneas de guía a nivel.

battery—produces electricity by chemical action.

acumulador—produce la electricidad por acción química.

bead—continuous and uniform line of filler metal.

cordón—raya continuosa y uniforme de masilla de metal de aporte.

below-grade—below surface level of land.

por debajo de la superficie—por debajo de la superficie de la tierra.

belt sander—power tool with a moving sanding belt.

lijadora de cinta—máquina–herramienta con una cinta móvil para lijar.

bench stone—sharpening stone designed to rest on a bench.

piedra de banco—agudización hecha para descansar sobre un banco.

bench yoke vise—a tool useful for holding pipe.

tornillo de banco—una herramienta útil para sujetar tubería.

bevel—a sloping edge.

bisel—un borde pendiente.

bill of materials—a listing of materials with specifications that are needed in a project.

lista de materiales—una lista de los materiales y sus especificaciones que se necesitan para un proyecto.

binary—having two parts or components.

binario—que tiene dos partes o componentes.

binary system—contains two digits, 0 and 1.

sistema de numeración binaria—tiene solamente dos dígitos, cero y úno.

biological filters—colony of microorganisms that detoxify waste products.

filtros biológicos—colonia de microorganismos que desintoxican los desechos.

biosensor—device that measures processes of the body.

biosensor—dispositivo que mide los procesos del cuerpo.

bit—one digit, one number, or a yes/no answer to a question.

bitio—un dígito, un número, o una respuesta sí/no a una cuestión.

bit brace—a device with a large crank-type handle used to turn wood cutting bits.

berbiquí—dispositivo que tiene un gran manubrio empleado para tornear los taladros de cortamaderos.

black pipe—steel pipe painted black that has little resistance to rusting.

tubo de hierro negro—tubo de acero, pintado negro, que tiene poca resistencia al oxidarse.

blade guide—metal blocks or wheels that support the blade of a band saw.

guías para la sierra—bloques o ruedas de metal que sostienen la cinta de una cierra de cinta tronzadora de trozos.

bleeding the lines—removing gas pressure from all lines and equipment.

purgar las líneas—quitar la presión de gas de todos líneas y equipo.

blind cut—cut made by piercing a hole with a sabre saw blade or by slowly lowering a portable hand saw blade into the material.

picadura inferior—corte hecho por perforar con hoja de sierra de sable o por lentamente bajar la hoja de un serrucho en el material.

blow-by—compression leakage past pistons.

oclusión de aire—pérdida o escape de compresión más allá de los émbolos.

blue—the safety color used for signs of warning or caution.

azul—el color de seguridad que se usa para señales de peligro o aviso.

board foot—amount of wood equal to a board 1 inch thick, 1 foot wide, and 1 foot long, or 144 cubic inches in volume.

«board foot» (b.ft.)—cantidad de madera igual a un madero de una pulgada de espresor, un pie de ancho, y un pie de largo, o sea de 144 pulgadas cúbicas de volumen.

border—heavy, solid, black line drawn close to the outer edges of paper used for drawing plans.

orladura margenal—línea gruesa negra y continua, dibujada cerca de los bordes de papel usado para los planos.

border line—*See* border.

filete de encuadramiento—*V.* orladura margenal.

bore—to make or drill a hole.

taladrar—hacer un agujero por medio de penetrar o atravesar.

bottom dead center (BDC)—piston at its lowest point (point nearest the crankshaft).

punto muerto inferior (PMI)—cuando el émbolo está en su punto más bajo (punto más cercana al cigüeñal).

Bourdon tube—measures fluid pressure.

manómetro de Bourdon—mide la presión de los fluidos.

bowline—useful as a nonslip loop in a rope.

nudo marinero—útil como lazo de no desatarse en una cuerda.

bows—a framework that supports the greenhouse covering.

arcos—armazón que sostiene el cobertizo de un invernadero.

branch circuit—electrical wiring, switches, and outlets that extend out from a service entrance panel.

circuito derivado—cableado, interruptores y toma de corriente que extienden desde un panel de acometida de energía.

brass—mixture of copper and zinc.

latón—una mixtura de cobre y cinc.

braze welding—bonding with metals and alloys that melt at or above 840°F when capillary action does not occur.

soldeo oxigás con metal de aportación—unir piezas con metales y aleaciones que se funden a los 840°F o más, cuando no ocurre la acción capilar.

brazing—bonding with metals and alloys that melt at or above 840° when capillary action occurs.

soldadura fuerte—unir piezas con metales y aleaciones que se funden a los 840°F o más, cuando sí ocurre la acción capilar.

break line—solid, zigzag line used to show that the illustration stops but the object does not.

filete tremente con zigzag—una línea continua en zigzag usada para mostrar el límite de la ilustración aunque todavía continuye el objeto.

brick set—wide chisel used for breaking masonry units.

cincel de ladrillería—buril ancho empleado para quebrar los elementos de albañilería.

broom finish—rough surface formed by pushing a broom over concrete before it hardens to improve footing or traction.

acabado de escoba—superficie aspera formada por pasar una escoba sobre el hormigón antes de endurecerse para mejorar el pie o la fricción.

bull float—smooth board attached to a long handle used to smooth out newly poured concrete.

llana para hormigón—madero liso ligado a un mango largo, que se usa para extender el hormigón que se acaba de derramar.

bushing—a fitting used to connect pipe to pipe fittings or larger sizes. Also a device used to fill the space between a shaft and a larger hole in a blade or wheel. Also a sleeve in a frame.

aislador de entrada—ajuste empleado para conectar tubos a los accesorios para tubos o los de tamaños más grandes. También un dispositivo empleado para llenar el vacío entre el eje y el agujero en una rueda. También un manguito de un bastidor.

business—work done for profit.

negocio—trabajo hecho por ganancia.

butt joint—joint formed by placing two pieces end to end in line or at a 90 degree angle.

empalme plano—ensambladura formada por unir dos piezas por sus extremos o a un ángulo de 90 grados.

butt weld—bead laid between two pieces of metal set edge to edge or end to end.

costadura transversal soldada a tope—cordón soldado entre dos piezas de metal puestas por sus extremos o por sus bordes.

bypass system—the part of a fluid system that bypasses the filter.

válvula de desvío—parte del sistema de fluidos que desvía el filtro.

C

cadmium—metal used for plating of fasteners and other steel products for rust-resistance.

cadmio—metal empleado en el revestimiento de cierres y otros productos de acero para que sean resistentes al orín.

calcium—a low-quality paint pigment.

calcio—pigmento de pintura de mala calidad.

caliper—an instrument used to measure the diameter or thickness of an object.

calibre—un instrumento que se usa para medir el diámetro o espresor de un objeto.

cam—a raised area on a shaft.

leva—un área elvada en un eje.

cap—a fitting used to close a pipe end.

tapa de recubrimiento—un encastre usado para tapar la boca de un tubo.

capacitance—ability of a device to store electrical energy in an electrostatic field.

capacitancia—la capacidad de un dispositivo de almacenar la energía eléctrica en un campo electrostático.

capacitor—two conductors separated by an insulator.

capacitor—dos conductores separados por un aislador.

capillary action—rising of the surface of a liquid at the point of contact with a solid.

acción capilar—el subir a la superficie de un líquido a la punta de contacto con un sólido.

cap screw—a hex-head screw that is threaded over its entire length and is generally 2 inches or shorter.

perno de máquina—un tornillo de cabeza hexagonal que es roscado a lo largo de su cuerpo y generalmente mide 2 pulgadas o menos.

carbon arc torch—a device that holds two carbon sticks and produces a flame from the energy of an electric welder.

linterna de arco de carbón—un dispositivo que tiene dos barras de carbono y produce una llama de la energía de una soldadora eléctrica.

carbonizing flame—a flame with an excess of acetylene.

llama carbonizada—una llama con demasiado de acetileno.

carburetor—provides fuel and air to the engine in appropriate portions and volume.

carburado—proporciona el combustible y el aire al motor en proporciones y volumen adecuados.

cardiopulmonary resuscitation (CPR)—a first-aid technique to provide oxygen to the body and circulate blood when breathing and heartbeat stop.

resucitación cadiopulminar—una técnica de primeros auxilios que suministra el oxígeno al cuerpo y circula la sangre cuando arrestan la respiración y el latido de corazón.

carriage bolt—a threaded fastener with a round head over square shoulders.

perno de carrocería—cierre roscado con una cabeza redonda encima de hombros cuadrados.

cartesian working area—boxlike shape.

área de trabajo cartesiano—en forma de caja.

casting—a mold that holds molten metal.

colada—matriz que guarda el metal fundido.

cast iron soil pipe—iron pipe with thick walls used mostly for sewerage systems.

caño de desagüe de hierro colado—caño de hierro colado que tiene paredes gruesas, empleado principalmente en los alcantrillados.

caulk—material that stretches, compresses, and rebounds to maintain a tight seal between materials as they expand and contract.

calafateado—materia que se estira, se comprime, y regota para mantener una junta estanca entre materiales mientras se expanden y se contraen.

causeway—concrete-lined ditch.

cuneta de hormigón—zanja pavimentada con hormigón.

CCA—chromated copper arsenate.

ACC—arsenato de cobre cromado.

cement paste—a paste made by mixing portland cement and clean water in exact proportions.

pasta de cemento—una masa hecha por mezclar el cemento pórtland y agua limpia en proporciones exactas.

center line—a long-short-long line used to indicate the center of a found object on a diagram or plan.

línea central—un punto de referencia dibujado en forma de raya-punto-raya, usado para indicar el centro de un objeto hallado en un diagrama o plano.

center punch—steel punch with a sharp point.

sacabocados—perforadora de acero con punzón agudo.

centrifugal—outward force cause by a spinning object.

centrífugo—dícese de la fuerza que se aleja del centro, causado por un objeto dando vueltas.

centrifugal head pumps—self-contained unit powered by an auxiliary motor.

bombas centrífugas—grupo autosuficiente impulsado por un motor auxiliar.

CFM—cubic feet per minute.

pcm—pie cúbico por minuto.

chain vise—a tool useful for holding pipe.

prensa de cadena—herramienta útil para fijar tubería.

chain wrench—wrench that utilizes a chain to grip and turn pipe.

llave de cadena—tornillo portátil para tubería con una cadena que permite fijar el tubo y tornearlo.

chalking—a process whereby paint gradually washes off in the rain to stay bright.

ayesamiento—un proceso por el cual la pintura gradualmente se quita con la lluvia para que permanezca viva.

chalk line—a cotton cord with chalk applied used to create long, straight lines.

cordel entizado—una cuerda de algodón entizado, que se usa para hacer líneas largas y derechas.

chamfer—cutting down of a corner between the edge and face of a board.

chaflanar—planar una esquina entre el borde y la superficie de un madero.

check valve—direction control valve.

válvula de retención—válvula de control de dirección.

chip—tiny electronic processing circuit in a computer.

chip—circuito pequeñito del procesamiento electrónico empleado en un computador.

chipping hammer—a hammer with a sharp edge and/or point used to remove slag from a welding bead.

martillo ancelador—martillo con un corte o una punta afilado que se usa para extirpar escoria de un cordón de soldadura.

chlorinated polyvinyl chloride (CPVC)—plastic pipe suitable for hot water lines.

cloruro de polivilino clorado (CPVC)—tubo plástico adecuado para líneas de agua caliente.

chromated copper arsenate—wood preservative.

arsenato de cobre cromado—preservativo de la madera.

chuck—a devise with jaws that open and close to receive and hold bits or other objects.

mandril—dispositivo con quijadas que abren y cierran para asegurar brocas y otros objetos.

circuit—an electrical source and wires connected to a light, heater, or motor.

circuito—conjunto de conductores y cables eléctricos conectados a una luz, un filamento, o un motor.

circuit breaker—a switch that trips and breaks the circuit when more than a specified amount of current passes through it.

ruptor de circuito—un interruptor que interrumpe y corta el circuito cuando más que la cantidad especificada de corriente pasa por él.

claw—the cup of a milking machine.

distribuidor filtrante—el gancho, o garfio de la ordeñadora.

clay—the smallest group of soil particles.

arcilla—el grupo más pequeño de partículas de tierra.

clearance hole—shank hole.

agujero de holgura—agujero del mango.

clinch—bending a nail over and driving the flattened end down into the material.

remachar—curvar un clavo y machacar la cabeza o punta allanada en el material.

clockwise—the direction that hands turn on a clock.

en el sentido de las agujas del reloj—de derecha a izquierda.

closed aquaculture systems—intensive production systems using oxygen, feed, and waste treatment to maximize economic return.

sistemas cerrados de piscicultura—sistemas de producción intensivos que usan el oxígeno, pienso y tratamiento de residuos para llevar al máximo el rendimiento económico.

cloth—fabric material used in some measuring tapes.

tela—un tejido usado en algunas cintas métricas.

clove hitch—knot useful for tying livestock.

ballestrinque—nudo útil para atar ganado.

coarse aggregate—gravel; larger particles of stone used in concrete.

árido grueso—grava; partículas más grandes de piedra que se usan en el hormigón.

cold chisel—a piece of tool steel shaped, tempered, and sharpened to cut mild steel when driven with a hammer.

cortafrío—una herramienta de acero labrada, templada, y afilada para cortar el acero suave cuando se clava con un martillo.

combination square—type of square that combines many tools in one.

escuadra de combinación—tipo de escuadra que combina muchos instrumentos en uno.

combustion—burning.

combustión—acción de quemar.

commutator—the part of a motor that changes the flow of current at appropriate times.

colector de delgas—la parte de un motor que cambia la corriente en el momento apropiado.

compatible—going together with no undesirable reactions.

compatible—que se llevan bien sin reacciones indeseables.

compress—to reduce in volume by pressure.

comprimir—reducir el volumen de algo por hacer presión.

compressed air—air pumped under high pressure that may be carried by pipe, tubing, or reinforced hose.

aire comprimido—aire inyectado bajo alta presión que puede ser llevado por medio de tubo o manguera reforzada.

compression fitting—a fitting that grips copper or steel tubing by compressing a special collar with a tapered, threaded nut.

ajuste de compresión—un ajuste que agarra entubado de cobre o acero por comprimir una abrazadera especial con una tuerca afilada y roscada.

compression ratio—the relationship between the volume of an engine cylinder at the beginning and end of the compression stroke.

relación de compresión—la relación entre el volumen de un cilindro de motor al comienzo y a término de la carrera de émbolo.

compression stroke—movement of an engine piston to squeeze or compress the air-fuel mixture.

carrera de émbolo—movimiento del émbolo de motor para comprimir la mezcla carburante.

concave—hollow or curved in.

cóncavo—hueco o curvado hacia el interior.

concrete—a mixture of stone aggregates, sand, portland cement, and water that hardens as it dries.

hormigón—una mixtura de áridos de piedra, arena, pórtland, y agua, que se fragua mientras seca.

condensor—stores and releases current to boost current in the primary circuit.

condensador—almacena y suelta la corriente para aumentar la corriente del circuito primario.

conductance (G)—ability of a material to carry electrons.

conductancia (G)—capacidad de un material para llevar electrones.

conductor—any material that will permit electrons (electricity) to move through it.

conductor—cualquier material que permite que los electrones (electricidad) lo atraviesen.

conduit—metal tube with individual insulated wires inside.

tubo aislante—tubo de metal que contiene los cables individuales aislados adentro.

construction joint—place where one pouring of concrete stops and another starts.

junta de construcción—lugar en donde ún bloque de hormigón termina y otro empieza.

contaminant—any material that does not belong in a substance.

contaminador—cualquier material que no debe ser parte de la sustancia.

continuity—connectedness.

continuidad—ser conectado.

continuity tester—device used to determine if electricity can flow between two points.

óhmetro—dispositivo empleado para comprobar si se puede recorrer la corriente eléctrica entre dos polos.

continuous duty—tool that can be used all the time for a 6- or 8-hour day.

servicio continuo—dícese de una herramienta que se puede usar todo el tiempo por un día de trabajo de 6 o 8 horas.

control joint—planned break that permits concrete to expand and contract without cracking.

junta de dilatación, junta de retracción—una abertura o rotura intencional que permite la expansión y la contracción del hormigón sin que se rompa.

controlled environment structures—production facilities where temperature, light, and humidity are maintained for agricultural production.

crianza bajo ambiente controlado—instalaciones de producción en donde se mantienen la temperatura, la luz y la humedad para la producción agrícola.

convex—curved out.

convexo—curvado hacia el exterior.

coolant—liquid used to cool parts or assemblies.

refrigerante—líquido usado para congelar piezas o montajes.

coping saw—a saw that has a very thin and narrow blade supported by a spring steel frame.

sierra bastidor—una sierra que tiene una hoja cortante muy delgada y estrecha, sostenida por un bastido de muelle aceroso.

cordless—tool containing a rechargeable battery pack to drive the unit when not plugged into an electrical outlet.

sin cordones—herramienta que incluye una batería recargable para impulsar el grupo cuando no está enchufado a un enchufe eléctrico.

core—the hollow space in a masonry block.

núcleo hueco—el hueco de un bloque de albañilería.

corner pole—a straight piece of wood or metal held plumb by diagonal supports and used to support a line when laying block or brick.

poste de ángulo—un poste recto de madera o metal aplomado por tirantes de extensión y empleado para sostener una línea en poniendo ladrillos o bloque.

corrosion—reaction of metal to liquids and gases that causes them to deteriorate or break down.

corrosión—una reacción de metal a líquidos y gases que los causa deterior o corroer.

corrosion inhibitor—reduces or prevents corrosion.

sustancia anticorrosiva—reduce o evita la corrosión.

counterclockwise—opposite of the way hands move around a clock.

en sentido opuesto a las agujas del reloj—de derecha a izquierda.

countersink—tapered hole for a screw head.

contrafresadura—bisel practicado en el borde de un agujero para recibir tornillos.

coupling—a fitting used to connect two pieces of similar pipe.

acoplamiento—un ajuste que se usa para unir entre sí dos piezas semejantes de entubado.

course—a row of masonry units.

hilada de ladrillos—una fila de elementos de albañilería.

crack a cylinder—turn gas on and off quickly to blow dust from the opening.

destaponar un cilindro—apenas abrir el cilindro de gas y en seguida muy rapidamente cerrarlo para soplar el polvo de la abertura.

craftsman—skilled worker.

artesano—obrero cualifcado.

crankshaft—a shaft with an offset projection that converts circular motion to reciprocal motion, or vise versa.

cigüeñal—una leva con una proyección desviada que convierte el movimento circular en el movimiento rectilíneo o vice versa.

crater—a low spot in metal where the force of a flame has pushed out molten metal.

cráter—depresión en el metal donde la fuerza de la llama ha abollado el metal fundido.

creosote—popular wood preservative that is black in color.

creosota—preservativo de madera popular que es de color negro.

crosscut—to cut across the grain of wood.

picadura en cruz—cortar a través del hilo de la madera.

crosscut saw—saw with teeth cut and filed to a point, used to cut across the grain of wood.

sierra de picadura en cruz—sierra de dientes cortados y afilados a puntas, usado para cortar a través del hilo de madera.

crown—the part of a tool that receives the blow of a hammer.

cabeza—parte de la herramienta que recibe el golpe del martillo.

cuprinol—popular wood preservative used when plants will be near the wood.

cuprinol—preservativo de madera popular que se usa cuando hay plantas cerca de la madera.

curing—proper drying of concrete.

fraguarse—método correcto de secar y endurecer la masa de hormigón.

cutoff saw—motor-driven circular blade for angle cuts that is fed down into the material.

sierra de cortes angulares—hoja circular de motor que se emplea para cortes angulares y que se baja en el material al cortar.

cutting tool—tool used to cut, chop, saw, or otherwise remove material.

herramienta cortadura—herramienta usada para cortar, tronchar, serrar, o de otro modo extirpar material.

cycle—all the events that take place as an engine takes in air and fuel, compresses the air-fuel mixture, burns the fuel, and expels the burned gasses. Also the current produced by one turn of a generator armature.

ciclo—todos los sucesos que toman lugar mientras el motor toma aire y combustible, comprime la mezcla carburante, quema el combustible, y expela los gases quemados. También la corriente producido por una vuelta de la armadura del generador.

cylinder—long round tank with extremely thick walls built to hold gases under great pressure. Also, engine cavity containing a piston.

cilindro—cámara larga y redonda con paredes extremamente fruesas, construida para contener los gases bajo gran presión. También, la cavidad de motor conteniendo el émbolo.

cylinder hone—tool for restoring cylinders.

rectificador de cilindros—herramienta usada para restaurar los cilindros.

cylindrical working area—working area in shape of a cylinder.

área de trabajar cilíndrica—area de trabajar en forma de cilindro.

dado—square or rectangular groove in a board.

dado—ranura cuadrada o rectangular en un madero.

dado head—special blade that is adjustable to cut kerfs from 1/8 to 3/4 inch in one pass.

fresa rotativa de ranurar—hoja especial que se ajuste para cortar cortes de 1/8 a 3/4 de pulgadas en una pasada.

dado joint—rectangular groove cut in a board with the end or edge of another board inserted.

mortaja filete—ranura rectangular cortada de un madero en cuyo borde otro listón o madero está encajado.

decibel (dB)—standard unit of sound.

decibelio (dB)—unidad estándar de sonido.

degrees of freedom—number of axes.

grados de libertad—número de cortes de hacha.

diameter—distance across the center of a circle or round object.

diámetro—la distancia a través del centro de un círculo u objeto redondo.

diaphragm—a flat piece of rubber that creates a seal between moving and nonmoving parts.

diafragma—una pieza llana de hule que forma una obturación entre piezas móviles y inmóviles.

die—an instrument used to cut threads onto a rod or bolt.

troquel—instrumento usado para cortar roscas en una biela o un perno.

die stock—handle used to turn a die.

terraja de anillo—mango empleada para tornear un troquel.

dielectric—insulated part of a capacitor.

dieléctrico—pieza aislante de un capacitor.

diffuser tube—porous underwater tube for producing small bubbles.

difusor—tubo poroso submarino que produce pequeñas burbujas.

digging tool—device used to move soil or rock.

herramienta de cava—dispositivo empleado para remover la tierra o piedra.

digital computer—electronic processor using digital techniques.

calculadora digital—máquina electrónica de proceso de datos usando técnicas digitales.

digital meter—device providing a direct numerical readout.

medidador digital—dispositivo que da presentación numérica directa.

digits—the numbers 0 through 9.

dígitos—los números 0 a 9.

dimension—measurement of length, width, or thickness.

dimensión–medida de lo largo, lo ancho, y lo alto.

dimension line—solid line with arrowheads at the ends to indicate the length, width, or height of an object or part.

línea de dimensión—línea continua con puntas de flecha en los extremos para indicar lo largo, lo ancho, y lo alto de un objeto o una pieza.

diode—simplest type of semiconductor.

diodo—tipo más sencillo del semiconductor.

direct current (DC)—current that flows in one direction continuously.

corriente continua (CC)—corriente cuya dirección es más o menos constante.

direction control valve—restricts fluid movement to one direction.

válvula de regulación—restringe el movimiento de fluidos a una sóla dirección.

disc sander—tool with sanding grit on a revolving plate.

esmeriladora de disco—herramienta que tiene arenisca de grano en una placa giratoria.

discharge—loss of power from a battery.

descarga—pérdida de potencia del acumulador.

dissolved oxygen—level of oxygen dissolved in water used by aquatic plants, animals, and microorganisms.

disociación de gases—índice de oxígeno disuelto en agua que se usan plantas, animales, y microorganismos acuáticos.

distilled—free from ions or other impurities.

destilado—libre de iones u otros impurezas.

dividers—two sharp steel legs connected on one end that are used to make arcs and circles.

compás de división—dos piernas agudas de acero articuladas en un extremo que se emplean para dibujar arcos y círculos.

double action cylinder—works two ways.

cilindro de doble-acción—*ésta* acciona en dos sentidos.

double cut file—file that has teeth laid out in two directions.

lima de segundo corte—lima cuyas aristas cortantes son talladas en dos direcciones.

double extra heavy—steel pipe with walls thicker than standard and extra heavy pipe.

extragrueso doble—tubo de acero cuyas paredes son más gruesas que las de los tubos estándares y extragruesos.

double insulated—motor with electrical parts insulated or separated from the user by special insulation inside the motor and by its insulating plastic motor housing.

doble-aislado—motor con piezas eléctricas que son aisladas o separadas del usuario por un aislamiento especial dentro del motor y por su cárter plástico de motor que también aisla.

dovetail joint—a joint formed by interlocking parts of two pieces.

ranura en cola de milano—ensambladura formado por las partes entrelazadas de dos piezas.

dowel—a round piece of wood.

pasador—pieza cilíndrica de madera.

drainage—removal of free water from soil.

avenamiento—eliminación del agua libre del suelo.

draw filing—a filing procedure done by placing the file at a 90 degree angle to the metal.

limar a rectángulo—una acción de limar realizada por poner la lima a un ángulo de 90 grados al metal.

draw the temper—modify the temper of steel to render it softer.

revenir—modifica el templado de acero para renderlo más blando.

drawing—a picture or likeness made with a pencil, pen, chalk, crayon, or other instrument.

dibujo—figura o retrato hecho con lápiz, pluma, tiza, crayola, u otro instrumento.

dress—to remove material and leave a smooth surface.

raboter—quitar el material y dejar una superficie lisa.

drill press—stationary tool used for making holes in metal and other materials.

taladradora radial—herramienta fijada empleada para hacer agujas en el metal u otros materiales.

drive train—collective parts that transfer power from power source to load.

tren de propulsión—conjunto de piezas que transfieren la potencia desde la punta de potencia hasta la carga.

driving tool—tool used to move another tool or object.

herramienta de impulso—herramienta que se emplea para poner en moción otra herramienta u objeto.

drop cloth—material used to protect floors, furniture, shrubbery, and the like from falling paint.

cuentagotas—tela que se usa para proteger los pisos, las muebles, los arbustos, y otras cosas semejantes de las gotitas de pintura que se caen.

drywall screw—hardened steel self-threading screw.

tornillo de muro de piedra—tornillo autoroscante de acero colado.

dual—two.

dual—de dos.

ductile—able to be bent slightly without breaking.

dúctil—que puede estirarse sin romperse.

duplex receptacle—double receptacle wired so that both outlets are on the same circuit.

receptáculo dúplex—receptáculo doble contectado para que las dos tomas corran en el mismo circuito.

duty cycle—the proportion of time a motor or welder can run without overheating.

ciclo de servicio—la proporción de tiempo en que se puede marchar un motor o soldador sin calentar demasiado.

duty rating—percent of time a motor may run without overheating.

relación de trabajo—el porcentaje de tiempo en que se puede marchar un motor sin calentar demasiado.

dysfunctional—does not work the way it should.

disfuncional—que no funciona en la manera a que debe.

E

ear—short extensions on the ends of masonry block.

asa—salientes cortos que sobresalen de los extremos de ladrillos.

eccentric—off-center.

excéntrico—muy alejado del centro.

efficient—able to produce with a minimum of time, energy, and expense.

de buen rendimiento—que puede producir dentro del mínimo de tiempo, energía, y gasto.

elbow (ell)—a fitting used where a single pipe line changes direction.

codo—un ajuste que es empleado donde la tubería cambia de sentido.

electrical meter—a device that measures the activities of electrons.

electrómetro—un dispositivo que mide las actividades de electrones.

electricity—form of energy that can produce light, heat, magnetism, and chemical changes.

electricidad—forma de energía que puede producir luz, calentura, magnetismo, y reacciones químicas.

electrode—a metal welding rod coated with flux and used with an electric welder.

electrodo—una varilla de soldadura de metal, cubierta con pasta para soldar y usada con una soldadora eléctrica.

electrode holder—a spring-loaded device with insulated handles used to grip welding electrodes.

portaelectrodo—un dispositivo de válvula de descarga de muelle que tiene mangos aislantes y que se usa para fijar electrodos de soldadura.

electrolyte—acid solution in a battery.

electrólito—solución acídico en un acumulador.

electromagnet—a core of magnetic material surrounded by a coil of wire through which an electric current is passed to magnetize the core.

electroimán—un núcleo de materia magnética rodeado por una bobina de alambre a través de que pasa una corriente eléctrica para imantar el núcleo.

electromagnetic induction—conversion of low-voltage current to high-voltage current with a coil.

inducción electromagnético—la conversión de corriente de baja tensión a la de alta tensión usando una bobina.

elevation—relative height from sea level.

elevación—altura relativa sobre el nivel del mar.

EMF—electromotive force.

FEM—fuerza electromotriz.

emitters—nozzles that release water slowly.

emisores—boquillas que emiten el aqua lentamente.

employment—work done for which one is paid by the hour, day, week, month, or year.

empleo—trabajo hecho, por lo cual se paga a uno por hora, día, semana, mes o año.

enamel—paint with a gloss or semigloss finish.

pintura esmaltada—pintura con acabado brillante o semi-brillante.

end marking—color on the end of an electrode.

marca del terminal—color que es pintado en la punta de un electrodo.

end nailing—nailing through the thickness of one piece and into the end of another piece.

clavar contrahilo—clavar por el espesor de una pieza y por el borde de otra pieza.

endwalls—wall construction at the end of greenhouses supporting doors, fans, and vents.

pared del extremo—pared en el extremo del invernadero que sostiene las puertas, albanicas y rejillas de ventilación.

epoxy—a synthetic paint material with exceptional adhesion and wear-resistant qualities.

epoxi—un material sintético de pintura que tiene adhesión excepcional y cualidades resistentes al desgaste.

evaporative cooling pads—porous excelsior or material kept moist to help cool incoming air into the greenhouse.

almohadillas evaporativas de enfriamiento —excelsior poroso u otro material que se mantiene húmedo para enfriar el aire entrante al invernadero.

excavate—to dig.

excavar—cavar.

exhaust—burned gases removed by the motion of a piston.

escape—gases quemados traslados por la moción de un émbolo.

exhaust stroke—movement of a piston that expels burned gases from a cylinder.

tiempo de escape—movimiento del émbolo que expulsa los gases quemados del cilindro.

extension line—solid line showing the exact area specified by a dimension.

línea de extensión—línea continua que representa el área exacta que se especifica una dimensión.

exterior—paint able to withstand moisture and outside weather conditions.

pintura exterior—pintura que puede aguantar la precipitación y el clima.

exterior grade—plywood suitable for outdoor use.

cualidad para exterior—madera contra-chapada apropiada para el uso al aire libre.

extinguish—to put out a fire by cooling, smothering, or removing fuel.

extinguir—hacer que cese un fuego por congelar, apagar, o quitar el combustible.

extractor—device that removes moisture and oil from compressed air.

extractor—dispositivo que extrae el agua y aceite del aire comprimido.

extra heavy—pipe with thicker than standard walls.

extragrueso—tubería cuyas paredes son más gruesas que las estándares.

extreme-pressure resistor—reduces oil breakdown.

lubricante para presión extrema—reduce la descomposición del aceite.

eye—a piece of metal bent into a small circle.

ojo—trozo de metal remachado, formando un círculo pequeño.

F

face—surface that is intended for use. Also tapered section of a valve head.

cara—superficie que se intenta usar.

cara de válvula—también la parte ahusada de la cabeza de válvula.

farmstead layout—arrangement of buildings on a farm.

disposición de las instalaciones de la finca—arreglo de los edificios en una finca.

fastener—device used to hold two or more pieces of material together.

sujetador, broche—dispositivo que se usa para juntar dos tejidos.

faucet—device that controls the flow of water from a pipe or container.

grifa—dispositivo que controla el flujo del agua de un tubo o contenedor.

faucet washer—rubberlike part that creates a seal when pressed against a metal seat.

anillo obturador—pieza de tipo de hule que forma una obturación cuando aprieta fuerte a una hoja de metal.

feather—to sand so that a chipped edge is tapered.

recubrimiento solapado—lijar hasta que se haga ahusado un filo desportillado.

ferrous—metal that comes from iron ore.

ferroso—dícese del metal que es de mineral de hierro.

ferrule—a metal collar fitted on a handle to prevent splitting of the wood.

virola—una abrazadera que encaja sobre un mango para evitar el hendimiento de la madera.

field—magnetic force around an electric wire or iron core.

campo—fuerza magnética que rodea un hilo eléctrico o un núcleo de hierro.

field use—location where repairs and other operations are done.

estación de reparación y servicio técnico—sitio en donde se realizan las reparaciones y otras operaciones.

50-50 solder—solder composed of 50 percent tin and 50 percent lead.

soldadura mitad y mitad—soldadura compuesta de 50 por ciento de estaño y 50 por ciento de plomo.

filament—special metal element in a vacuum that produces light when electricity flows through it.

filamento—elemento especial de metal en un vacío que produce luz al pasar la corriente.

file—flat, round, half-round, square, or three-sided piece of metal with fine teeth.

lima—pieza de metal plana, redonda, de media caña, de cuatro cuartes, o triangular, tallada de dientes muy finos.

filler rod—metal in the form of a long, thin rod used to add to or fill joints when brazing and welding.

varilla de metal de aportación—metal en la forma de una válvula larga y estrecha empleada para llenar junturas en soldando.

fillet weld—a weld placed in a joint created by a 90 degree angle.

soldadura de filete—soldadura hecha en una unión formada por un ángulo recto.

filter—converts pulsating DC voltage to smooth voltage.

filtro—convierte corriente continua modulada en voltaje uniforme.

fine aggregate—sand and other small particles of stone.

árido fino—arena y otras partículas pequeñas de piedra.

fingering—dividing and clustering of the bristles of a paintbrush.

«fingering»—dividir y poner en grupitos las cerditas de una brocha.

finish—chemical layer that protects the surface of a material.

acabado—capa química que protege la superficie de un material.

finishing lime—powder made by grinding and treating limestone for use in masonry materials.

cal apagada en polvo—polvillo hecho por pulverizar y tratar el cal para uso en materiales de albañilería.

finishing sander—tool with a small sanding pad driven in a forward-backward or circular pattern.

enarenador de acabado—herramienta con placa pequeña de lijar, operada en movimientos circulares o hacia adelante hacia atrás.

fire—flame; to make a spark jump across an air gap.

fuego—llama; hacer que brinque una chispa a través de un espacio de aire.

fire triangle—the three conditions–fuel, oxygen, and heat–that must be present to produce a fire.

triángulo de fuego—las tres condiciones—combustible, oxígeno, y calentura—que tienen que ser presente para producir un fuego.

fish feeders—built feed holders calibrated to disperse feed at timed intervals.

canales de alimentación para los peces—construidos y calibrados para disponer alimentación a intervalos determinados.

fitting—a part used to connect pieces of pipe or to connect other objects to pipe.

ajuste—una pieza empleada para conectar pedazos de tubería o para conectar otros objetos a la tubería.

fixture—a base or housing for a light bulb, fan motor, or other electrical device.

instalación—un fundamento o cárter para una bombilla, un ventilador, u otro dispositivo eléctrico.

flammable—capable of burning easily.

inflamable—que puede encenderse facil-mente.

flaring block—tool used to hold tubing when flaring.

bloque abocinadora—herramienta emple-ada para expandir la boca de la tubería en ensanchando.

flaring tool—tool used to expand the opening in tubing to a bell shape.

herramienta abocinadora—herramienta empleada para expandir la boca de la tubería para que tome la forma de una campana.

flashback—burning inside an oxyfuel torch that causes a squealing or hissing noise.

retroceso de la llama—lo ardiendo dentro de la llama de soplete que causa un ruido chirriante o silbante.

flat finish—dull or without shine.

acabado mate—sin relieve o sin lustre.

flat head screw—screw that has a flat head with a tapered underside designed to fit down into the material being secured.

tornillo de cabeza plana—tornillo que tiene cabeza llana con cara inferior ahusada, hecha para encajar el material que se asegura.

flat nailing—two flat pieces nailed to each other.

clavar sobre costero—dos piezas planas clavadas una a la otra.

float switch—a switch turned on and off by the action of a float.

interruptor de flotador—un interruptor encendido y apagado por la acción de un flotador.

float valve assembly—a device used to control liquid flow into a reservoir, such as a flush tank in a toilet.

montaje de válvula de flotador—un dispositivo empleado para controlar el flujo de líquido a un embalse, como el tanque de un lavabo.

floppy disk (diskette)—flexible disk for computer data storage.

disco flexible—disco blando donde se almacena la información.

flow switch—reacts to the movement of a gas or liquid.

controlador de circulación de fluídos—reacciona al movimiento de un gas o un líquido.

fluid coupling—uses fluid to transfer motion.

acoplamiento flúido—usa fluidos para transferir movimiento.

fluid power—transfer of force with liquids to gases.

potencia flúida—transmisión de la fuerza de líquidos a los gases.

fluorescent light—light tube that glows as a result of electricity flowing through gases.

lámpara fluorescente—tubo que se enciende como resulta de la corriente eléctrica que corre por los gases.

flush plate—thin device fastened across a joint to provide support.

placa para embutir a línea recta—dispositivo delgado asegurado a través de una ensambladura para sostenerla.

flush valve—used to control water used in the flushing process.

válvua de chorreo—aparato empleado para regular el agua durante el proceso de chorrear.

flux—material that removes tarnish or corrosion, prevents corrosion from developing, and acts as an agent to help solder spread over metal.

fundente—material que quita la empañadura o corrosión, evita que ocurra la corrosión, y actua como agente para que la soldadura pueda esparcirse sobre el metal.

folding rule—rigid rule of 2 to 8 feet in length that folds into a compact unit.

metro plegable—una regla rígida de 2 a 8 de pies de largo que se dobla para ponerse compacta.

footer (footing)—a continuous slab of concrete that provides a solid, level foundation for block and other masonry.

zócalo—un bloque continuo de hormigón que sirve como cimientos sólidos y planos para el bloque y otra albañilería.

footing—concrete base under a wall.

zócalo—fundamento de hormigón debajo de una pared.

footing pad—concrete or stone under a vertical pole.

zapata del soporte—hormigón o piedra bajo un poste vertical.

force—pushing or pulling action.

fuerza—acción de empujar o jalear.

form—a metal or wooden structure that contains and shapes concrete until it hardens.

molde—una estructura de metal o de madera que contiene y moldea el hormigón hasta que se fragüe.

formulate—to put together according to a formula.

formular—recetar conforme a una fórmula.

foundation—block or concrete wall supporting a building.

cimientos—bloque o pared de hormigón que sostiene un edificio.

four-cycle engine—*See* four-stroke-cycle engine.

motor de ciclo de cuatro tiempos—un motor que tiene cuatro tiempos por ciclo.

four-stroke-cycle engine—an engine with four strokes per cycle.

motor de ciclo de cuatro tiempos—un motor que tiene cuatro tiempos por ciclo.

frame—the housing or case for motor components.

bastidor—el cárter o armazón para componentes del motor.

framing square (carpenter's square)—a flat square with a body and tongue.

escuadra—un instrumento plano con cuerpo y lengüeta, usado para segurar ensambladuras.

freestanding—construction-style greenhouse requiring no support structures.

autoportante—invernadero de tipo de construcción que no requiere armazones para sostenerlo.

frost line—the maximum depth that the soil freezes in a give locality.

depósito de escarcha—la parte más profundo del suelo que se hiela en un sitio determinado.

fuel—any material that will burn.

combustible—cualquier material que se quema.

full-flow system—all fluid flows through filter.

sistema de flujo completo—en que todo fluido recorre por un filtro.

full scale—a drawing the same size as the object it represents.

tamaño natural—una representación del mismo tamaño que el objeto a que representa.

fuse—plug or cartridge containing a strip of metal that melts when more than a specified amount of current passes through it.

fusible—encufe o cartucho que contiene una chapa metálica que se funde al pasar una cantidad excesiva de corriente eléctrica.

fusion—joining by melting.

fusión—unir por fundición.

fusion welding—joining parts by melting them together.

soldadura por fusión—unir piezas por fundirlas.

fustat—a fuse with threads that permit it to be used only in circuits that match its capacity.

fustato—un fusible con filamentos que le permiten emplearse solamente en circuitos de la misma capacidad.

G

galvanized—coated with zinc for rust resistance.

galvanizado—cubierto con una capa de cinc para protección contra la corrosión.

gas—any fluid substance that can expand without limit.

gas—cualquiera sustancia fluida que puede expandirse sin límite.

gas analyzer—device that determines the content of gases.

analizador de gas—aparato que determina el contenido de gases.

gas forge—gas-burning unit used to heat pieces of metal and temper tools.

forjador de gas—quemador de gas empleado para fundir piezas de metal y templar las herramientas.

gas furnace—gas-burning unit used to heat objects.

horno de gas—quemador de gas empleado para calentar objetos.

gauge—a device used to determine the dimension of materials or space; a device used to measure and indicate pressure in a hose, pipe, or tank.

medida—un dispositivo empleado para medir la dimensión de materiales o del vacío; un dispositivo empleado para medir e indicar la presión en una manga, un tubo, o un tanque.

gear—wheel with teeth that mesh to another wheel.

engranaje—rueda dentada que penetra entre los dientes de otra rueda.

generator—a device that produces direct current.

generador—un dispositivo que produce corriente continua.

girder—carrie floors and interior walls.

viga—sostiene pisos y paredes interiores.

girt—horizontal side nailers in a building.

riostra—vigas de rigidez horizontales en un edificio.

glazing compound—special puttylike material used to install window glass.

masilla—materia especial de tipo de yeso usada para instalar los cristales en los bastidores de las ventanas.

gloss—paint with a shiny finish.

pintura brillante—pintura con un acabado brillante.

glue—sticky liquid used to hold things together.

pegamento—líquido pegajoso que se usa para unir, o pegar cosas.

governor—speed control device.

regulador—dispositivo que controla la velocidad de algo.

gradations—numbers and lines stamped or painted on measuring devices.

gradaciones—números y líneas embutidos o pintados sobre los dispositivos de medida.

grade—amount or drop or fall in the land.

grado—medida del declive del terreno.

grain—lines on lumber caused by the annual rings in a tree.

grano—rayas o estrías en los maderos causadas por la capa cortical del árbol.

graph paper—paper laid out in squares of equal size.

papel cuadriculado—papel en el cual son imprimidas cuadrículas.

gravel—particle of stone larger than sand, also called coarse aggregate.

gravilla—partículas de piedra más grande que arena, también se llama árido grueso.

green—the safety color used to indicate the presence of safety equipment, safety areas, first-aid, and medical practice.

verde—el color de seguridad usado para indicar la presencia de equipo de seguridad, áreas de seguridad, primeros auxilios, y tratamiento médico.

greenhouse—structure that provides light and heat from solar energy.

invernadero—una estructura que proporciona luz y calor derividos de energía solar.

greenhouse effect—the buildup of heat waves accumulated inside a translucent structure or greenhouse.

efecto invernadero—el aumento de ondas de calor acumuladas dentro de una estructura translúcida o un invernadero.

gridiron—underground drainage pipes carrying water to a conduit.

red—tubos subterráneos de desagüe llevando el agua a un conducto.

grinding wheel—abrasive cutting particles formed into a wheel by a bonding agent.

rueda de muela abrasiva—partículas abrasivas para cortar que son formadas para hacerse una rueda, por un agente de encolamiento.

ground board—wooden or aluminum construction member for attaching plastic or fiberglass coverings on greenhouses.

«ground board»—pieza de construcción de madera o aluminio que sirve para atar cubiertas plásticas o de fibra de vidrio a los invernaderos.

ground clamp—connector used to attach a cable, wire, or object to a ground source.

abrazadera de toma de masa—conectador usado para conectar un cable, alambre, u objeto a una fuente a tierra.

ground (gee) clip—a device by which groundwires are attached to electrical boxes.

abrazadera de toma de masa—dispositivo por lo cual los cables de toma de tierra son conectados a cajas eléctricas.

ground-fault interrupter (GFI)—a device that cuts off the electricity if even very tiny amounts of current leave the normal circuit.

interruptor de la falta de tierra—un dispositivo que interrumpe la corriente eléctrica incluso si cantidades muy pequeñas de corriente eléctrica salgan del circuito normal.

grounding—making an electrical connection between a piece of equipment and the earth.

conectar con tierra—establecer una conexión eléctrica entre un aparato y la tierra.

gusset—wood or metal used to reinforce a joint.

esquinero—chapa de madera o metal empleada para reforzar una juntura.

H

hacksaw—a device that holds a blade designed for cutting metal.

sierra de arco—un dispositivo que tiene una hoja creada para cortar metal.

hammer—striking action.

martillazo—acción de golpear.

hand drill—a manually operated device with gears that drive a bit faster than the handle turns.

taladradora manual—un dispositivo operado manualmente con engranajes que impulsan el mandril más rápido que tornea la manivela.

hand saw—a saw used to cut across boards or to rip boards and panels.

sierra de mano—una sierra usada para cortar a través de maderos o hender maderos y tableros.

hand stone—a sharpening stone designed to be held in the hand.

piedra de aguzamiento de mano—piedra de aguzamiento hecho para sujetarse en la mano.

hand tool—a tool operated by hand to do work.

herramienta manual—una herramienta manejada a mano para hacer una obra.

hands-on—actually doing something rather than reading or hearing others tell about it.

comando—verdaderamente hacer algo en vez de leer o escuchar a otros enseñarlo.

hard disk—rigid disk for computer data storage.

disco duro—disco rígido empleado para almacenaje.

hardpan—impervious clay.

arcilla dura—arcilla impermeable.

hardware—fasteners; objects made from metal.

ferretería—sujetadores; objetos hechos de metal.

head—flat part of a valve; cylinder cover containing the spark plug and combustion chamber.

cabeza—la parte llana de la válvula.

culata—parte superior del cilindro, conteniendo la bujía y la cámara de combustión.

head gasket—seal between the head and the cylinder block.

junta de culata—juntura entre la culata y el bloque del cilindro.

heat—type of energy that causes the temperature to rise.

calor—forma de energía que resulta en el aumento de la temperatura.

heat sensor—switch that responds to temperature change.

ruptor térmico—interruptor que responde al cambio de la temperatura.

henry (H)—unit of inductance.

henrio—unidad de inductancia.

herringbone—pipes laid with a center conduit.

doble helicoidal—tubos tendidos con un conducto central.

hertz (Hz)—one cycle per second.

hertz (Hz)—un ciclo por segundo.

hidden line—a series of dashes that indicates the presence of unseen edges.

línea oculta—una serie de rayas que indican la presencia de bordes ocultos.

hiding power—the ability of a material to create color and mask out the presence of colors over which it is spread.

poder cubriente—la capacidad que tiene una materia para crear color y ocultar la vista de los colores sobre los cuales cubre.

high-speed drill—a twist drill made and tempered specially to drill metal.

barrena de alta velocidad—una broca helicoidal hecha y templada especificamente para taladrar metal.

high-tension wire–high-voltage wire in a secondary circuit.

alambre de alta tensión—alambre de alto voltaje en un circuito secundario.

hinge—an object that pivots and permits a door or other object to swing back and forth or up and down.

bisagra—un objeto que da vueltas sobre su eje y permite que una puerta u otro objeto gire de acá para allá o de arriba abajo.

holding tool—a tool used to grip wood, metal, plastic, and other materials.

herramienta fijadora—herramienta empleada para agarrar madera, metal, plástico, y otros materiales.

hollow core block—masonry block with two or three holes per block.

núcleo de aire—bloque de albañilería con dos o tres cavidades por bloque.

hollow core solder—solder with flux inside.

soldadura con núcleo de fundente—soldadura envolviendo el fundente que está en el centro.

hollow ground—a saw blade with teeth wider at the points than at the base.

afilado con cara cóncava—una hoja de sierra con dientes más anchos de los puntos que de la base.

hollow sphere—work area in the shape of a hollow ball.

esfera hueca—área de trabajo en la forma de una bola hueca.

horizontal—flat or level.

horizontal—llano o a nivel.

horizontal shaft engine—an engine with a crankshaft that lies crossways for normal operation.

motor de eje horizontal—un motor que tiene un cigüeñal que yace transversalmente para funcionamiento normal.

horsepower—force needed to lift 550 pounds one foot in one second.

caballo de vapor—la potencia necesario para levantar 550 libras un pie por segundo.

hose—flexible line that carries gases or liquids.

manga—tubo largo flexible que lleva gases o líquidos.

hydraulic cylinder—tank with inlet and piston.

cilindro hidráulico—tanque con válvula de admisión y émbolo.

hydraulic motor—receives power from moving fluid.

motor hidráulico—recibe potencia de fluido en movimiento.

hydraulics—use of liquids to transfer force.

hidráulica—uso de líquidos para transferir potencia.

hydroponics—production of plants using dissolved nutrients in a recirculating system.

hidropónica—cultivación de plantas usando alimentos nutritivos disueltos en un sistema de recirculación.

I

ID—inside diameter.

DI—diámetro interior.

ignition—a spark igniting an air-fuel mixture.

encendido—una chispa encendiendo una mezcla carburante.

ignition spark—hot electrical arc across an air gap.

chispa de encendido—arco eléctrico caliente a través de un espacio de aire.

impeller—fan-shaped rotor.

propulsor—rueda de aletas.

impurity—any material other than the base metal.

impureza—cualquiera materia extraña al metal principal.

inclined plane—a surface at an angle to another surface; one of the six simple machines.

plano inclinado—una superficie de un ángulo a otra superficie; una de las seis máquinas sencillas.

inductance (L)—ability to draw energy and store it in a magnetic field.

inductancia (L)—capacidad de sacar energía y almacenarla en un campo magnético.

induction—motor where power does not flow directly to the armature.

inducción—motor a que la potencia no corre directamente a la armadura.

inductor—a device that provides inductance.

inductor—un dispositivo que sirve como fuente de inductancia.

infrared light sensors—used to detect heat or motion.

detector de rayos infrarrojos—usado para detectar calor o movimiento.

inside micrometer—telescoping gauge used to measure inside surfaces of hollow objects.

micrómetro interior—medida telescópica empleada para medir las caras interiores de objetos huecos.

insulator—material that provides great resistance to the flow of electricity.

aislador—material que da gran resistencia al paso de la electricidad.

intake stroke—engine process of taking fuel and air into the combustion chamber.

carrera de admisión—el proceso motriz de tomar combustible y aire a la cámara de combustión.

integrated circuit—a miniature electronic circuit.

circuito integrado—un circuito electrónico pequeño.

interceptor—line that catches water from a spring head.

línea de interceptación—línea que coge el agua de la cabecera del manantial.

interior—paint that will not hold up if exposed to weather.

pintura interior—pintura que desgasta cuando está expuesta a clima.

interior grade—plywood suitable for indoor use only.

cualidad para interior—madera contrachapada apropiada solamente para uso dentro de la casa.

internal combustion engine—device that burns fuel inside a cylinder to create a force that drives a piston.

motor de combustión interna—dispositivo que quema el combustible dentro de un cilindro para crear una fuerza que impulsa el émbolo.

iron—ferrous; an element associated with high-quality paint.

hierro—ferroso; un elemento asociado con pintura de buena calidad.

irrigation—human-controlled water system applied to plants and soils.

irrigación—abastecimiento de agua controlado por los seres humanos que se aplica a los cultivos y los suelos.

item—separate thing.

artículo—cosa distinta.

J

jet or seat—a hole shaped to receive the needle and control the flow of fuel.

chicler—un agujero formado para recibir el agujo y controlar el paso de combustible.

jigsaw—a saw that cuts very short curves by reciprocal action.

sierra de vaivén—una sierra que corta en curvas muy cortas por movimiento alternativo.

jobbers-length drill—a drill of longer than standard length.

taladradora para agujos profundos—taladradora cuyo mandril es más largo que lo estándar.

joint—the place where two pieces come together.

juntura—sitio en donde se juntan dos piezas.

ensambladura—sitio en donde dos piezas de madera se juntan.

Note: In Spanish, "ensambladura" is specifically used when referring to joints of wood.

jointer—a machine with rotating knives used to straighten and smooth edges of boards; a masonry tool used to shape joints.

garlopa—para igualar y lisar las superficies de madera.

palustre—una herramienta de albañilería empleado para formar junturas.

K

K—classification of thick wall copper pipe or tubing.

K–clasificación de tubos o cañerías de pared gruesa de cobre gruesa.

kerf—opening in board made by a saw; opening in steel made by oxyfuel cutting.

aserradura—corte hecho en un madero por una sierra.

corte—abertura hecha en el acero por el soldeo con soplete oxiacetilénico.

keyhole saw—a saw designed for starting in holes.

serrucho de punta—una sierra creada para empezar en agujeros.

kick starter—foot-operated lever and cog mechanism to crank an engine.

pedal de arranque—mecanismo de palanca y rueda dentada en que pisa para arrancar el motor.

kiln-dried—lumber that has been dried by heat in a special oven called a kiln.

secado al horno—dícese de los maderos que han sido secados por la calentura en un gran horno especial que se llama «tostadero».

kilo—one thousand.

kilo—mil.

kilowatthour—the use of 1000 watts for one hour.

kilovatio-hora—el uso de 1000 vatios por una hora.

knockout—partially punched impression in electrical boxes.

cortacircuitos—un molde en las cajas eléctricas parcialmente perforado.

L

L—classification of medium wall copper pipe or tubing.

L—clasificación de tubos o cañerías de pared mediana de cobre.

labor—to struggle or work hard to keep running.

funcionar con dificultad—luchar o avanzar penosamente para que siga en marcha.

lag screws (lag bolts)—screws with coarse threads designed for use in structural timber or lead anchors.

tirafondo—tornillos con roscas gruesas creados para usarse en maderos de construcción o áncoras de plomo.

laminate—to fasten two or more flat pieces together with an adhesive.

contrachapear—unir dos chapas planas o más con un adhesivo.

land level—hand-held level used for sighting level points.

escuadra de agrimensor—nivel usado para averigüar si la tierra es plana u horizontal a ciertos puntos.

lap joint—joint formed by fastening one member face-to-face on another member of an assembly.

juntura imbricante—juntura formada por sujetar una pieza cara-a-cara a otra pieza, de una ensambladura.

lapping compound—gritty material used for lapping in valves.

compuesto para pulir—materia arenosa usada para lapear las válvulas.

lapping in—grinding valves to fit the seat for a perfect seal.

lapear las válvulas—esmerilar las válvulas para encajar el asiento para hacer una obturación perfecta.

laser—intense beam of light.

láser—rayo de luz intensa.

latex—water-base paint.

látex—pintura de base de agua.

laying block—the process of mixing mortar, applying it to masonry block, and placing the block to create walls.

tender bloque—el proceso de mezclar mortero, aplicándolo a bloques de albañilería y colocando los ladrillos o piedras para crear paredes.

layout—a plan, map, or pattern for future operations.

trazado—un plano, mapa, o dibujo para operaciones futuras.

layout tool—tool used to measure or mark wood, metal, and other materials.

instrumento para trazar—instrumento usado para medir o marcar madera, metal, y otras materiales.

lead—heavy metal element that stays in the body once it is ingested; one component of solder.

plomo—elemento metal pesado que permanece en el cuerpo una vez que es ingerido; un componente de soldadura.

leader line—solid line with an arrow used with an explanatory note to point to a specific feature.

línea de guía—línea continua con una flecha usada con comentario explanatorio para señalar una característica específica.

leaner—a greater proportion of air and a lesser proportion of fuel in an air-fuel mixture.

mezcla pobre—una proporción más grande de aire y una proporción menor de combustible en la mezcla carburante.

level—a device used to determine if an object has the same height at two or more points. *See* spirit level.

nivel—un instrumento empleado para determinar si un objeto tiene la misma altura entre dos puntos o más. *V.* nivel de alcohol, nivel de burbuja.

light-emitting diode (LED)—produces light when subjected to a current flow.

diodo emiso de luz (DEL)—produce luz al sujetarse a la corriente eléctrica.

line—small diameter material stretched tightly between two or more points.

cordón flexible—material de diámetro que se puede estirar tensamente entre dos puntos o más.

linear—in a straight line.

lineal—de línea recta.

loading the brush—the process of dipping bristles into paint.

preparar la brocha—el proceso de zambullir las cerdas en la pintura.

lodestone—magnetized5black iron oxide.

magnetita—piedra de imán de óxido de hierro negro.

M

M—classification of thin wall copper pipe or tubing.

M—clasificación de tubos o cañerías de pared delgada de cobre.

machine bolt—fastener with a square or hex head on one end and threads on the last inch or so on the other end.

perno hecho a torno—cierre de cabeza cuadrada o hexagonal de un extremo y roscado de la última pulgada del otro extremo.

magnesium—a low-quality paint pigment.

magnesio—pigmento de pintura de mala calidad.

magnetic field—area around an iron core or electric wire influenced by the presence of magnetism.

campo magnético—región alrededor del núcleo de hierro o alambre eléctrico sometida a la influencia de magnetismo.

magnetic flux—lines of magnetic force that occur in patterns between poles of a magnet.

flujo magnético—rayos de fuerza magnética que ocurren en patrones de campo entre los polos de un imán.

magnetic tape—vinyl material used for storage of electronic impulses.

cinta magnética—material de vinilo empleado en el almacenamiento de impulsos electrónicos.

magnetism—a force that attracts or repels iron or steel.

magnetismo—una fuerza que atrae o repele el hierro o acero.

magneto—produces electricity by magnetism.

magneto—produce electricidad por el magnetismo.

mainframe—a very large capacity computer.

computadora principal—una computadora muy grande de mucha capacidad.

maintenance—doing the tasks that keep a machine in good condition.

mantenimiento—hacer las tareas para el cuidado de una máquina, preservándola en buen estado.

malleable—workable.

maleable—que puede labrarse.

malleable cast iron—combination of cast iron core and ductible metal.

fundición maleable—mezcla de núcleo de arrabio y metal dúctil.

manifold—device with openings to serve two or more units.

tubería multiple—dispositivo con aberturas que puede acometer a dos unidades o más.

manometer—measures fluid pressure and vacuum.

manómetro—mide las presiones de los líquidos y de lo neumático.

manual—by hand.

manual—a mano.

margin—outer edge of a valve head.

margen—borde externo de la cabeza de válvula.

mask—to cover so paint will not touch.

ocultar—cubrir para que no se manche con pintura.

masonry—anything constructed of brick, stone, tile, or concrete units held in place with portland cement.

albañilería—cualquiera cosa construida de unidades de ladrillo, piedra, azulego, u hormigón, unidas por pórtland.

masonry units—block made from concrete or cinders.

unidades de albañilería—bloques hechos de hormigón o escoria.

mechanic—a person specifically trained to perform mechanical tasks.

mecánico—persona especializada en las obras o tareas mecánicas.

mechanical—of or having to do with a machine, mechanism, or machinery.

mecánico—de o que tiene que ver con una máquina, un mecanismo o la mecánica.

mechanics—the branch of physics dealing with motion and the action of forces on bodies or fluids.

mecánica—el ramo de la física que trata del movimiento y la acción de las fuerzas sobre cuerpos o líquidos.

meter—a device that measures electricity, gas, or liquid that passes through a system.

metro—un dispositivo que mide la corriente eléctrica, el gas, o el líquido que pasa por un sistema.

meter (m)—a metric unit of linear measure equalling 39.37 inches.

metro (m)—unidad métrica de medida longitudinal que es igual a 39,37 de pulgadas.

metric system—a decimal system of measures and weights.

sistema métrico—sistema decimal de medidas y pesas.

microcomputer—self-contained small computer.

microcomputadora—computadora pequeña y autosuficiente.

micrometer—used to measure outside surfaces of round objects.

micrómetro—usado para medir las superficies de objetos redondos.

microorganisms—colony of microscopic bacteria and fungi.

microorganismos—colonia de bacterias y mohos microscópicos.

microprocessor—central processing unit (CPU) of small computer.

microprocesador—unidad central de proceso (CPU) de una computadora pequeña.

milliampere—thousandth of an ampere.

miliamperio—milésima parte de un amperio.

millimeter (mm)—thousandth of a meter.

milímetro (mm)—milésima parte de un metro.

minicomputer—intermediate capacity computer.

minicomputadora—computadora de capacidad intermedia.

miter—an angle.

inglete—un ángulo.

miter box—a device used to cut molding and other narrow boards at angles.

caja de ingletes—dispositivo empleado para cortar molduras y otras tablas estrechas a ángulos.

miter gauge—adjustable, sliding device to guide stock into a saw at the desired angle.

medida de inglete—dispositivo ajustible, móvil, para guiar materia prima a una sierra al ángulo determinado.

miter joint—joint formed by cutting the ends of two pieces at a 45 degree angle.

inglete—ensambladura formada por cortar los bordes de dos piezas a un ángulo de 45 grados.

mix—the ratio of materials in concrete or mortar.

mezcla—la ración de materias en el hormigón o el mortero.

moisture barrier—prevents movement of water, steam, or vapor.

barrera de humedad—impide el movimiento de agua o vapor.

molding head—a device that holds knives to shape wood into moldings of various types.

cabeza de limadora—un dispositivo que sujeta cuchillas que moldean la madera en molduras de varios tipos.

momentum—turning force of the flywheel and other moving parts that carries an engine through non-power strokes.

velocidad—fuerza giratoria del volante y otras partes móviles que lleva un motor por carreras no motrices.

mortar—a mixture of portland cement, finishing lime, water, and sand.

mortero—una argamasa de pórtland, cal apagada en polvo, agua y arena.

mortar bed—a layer of mortar.

capa de mortero—capa de mortero.

Note: In Spanish, "mortar bed" is referred to as a "layer of mortar."

multimeter—combination volt-ohm-milliammeter.

aparato medidor multiple—una combinación voltiohm-miliamperímetro.

mushroom—to spread out over an edge.

tomar forma de una seta—desbordar un borde.

mushroomed—a spread or pushed over condition caused by being struck repeatedly.

tomar forma de seta—una condición de rebosar o hacer caer causada por ser golpeada repetidamente.

N

nail—fastener that is driven into the material it holds.

clavo—sujetador que se hinca en el material a que sujeta.

nail-set—punchlike tool with a cupped end.

embutidera—herramienta de tipo perforador con un extremo en forma de bocina.

nap—soft, woolly, threadlike surface.

lanilla—superficie de lana finos, suave.

National Electrical Manufacturers Association (NEMA)—the group that developed the system of color coding electrodes.

Asociación Nacional de Fabricantes Electrotécnicos (ANFE)—el grupo que desarrolló el sistema de codificar electrodos por asignarlos colores específicos.

neat's-foot oil—light yellow oil obtained by boiling the feet and shinbones of cattle and used to soften and preserve leather.

aceite de pie de buey—aceite amarillo blanco obtenido por hervir las patas y corvejones del ganado y usado para ablandar y proteger el cuero.

needle—a long, tapered shaft.

aguja—eje largo y afilado.

needle bearing—steel rollers in a cage to keep a shaft centered.

cojinete de agujas—rodillos de acero en un cojinete en el cual se apoya y gira un eje.

neoprene—tough rubberlike material that is resistant to moisture and rotting.

neopreno—material duro de tipo de hule que es resistente de precipitación y pudrimiento.

net cages—pens or confinement structures for herding fish.

trasmallos—redes o estructuras de restricción para cazar peces.

neutral flame—flame with a balance of acetylene and oxygen.

llama neutra—llama con un equilibrio de acetileno y oxígeno.

neutral wires—conductors with white insulation that carry current from an appliance back to the source.

hilos neutros—conductores con aislamiento blanco que llevan la corriente desde un aparato eléctrica hasta el origen.

new wood—wood that has never been painted or sealed.

madera virgen—madera que nunca ha sido pintada ni chapada.

new work—original construction, including new wood.

obra original—construcción original, inclusive la madera virgen.

nipple—a short piece of steel pipe threaded at both ends.

boquilla de unión—un trozo de tubo de acero roscado en ambos extremos.

noise intensity—energy in the sound waves.

intensidad de ruido—energía de las ondas acústicas.

nominal—the size of material as used in the name for trade and building purposes.

nominal—el tamaño del material como lo usado en la nomenclatura para fines de comercio y construcción.

nonmetallic sheathed cable—cable composed of copper or aluminum wires covered with paper, rubber, or vinyl for insulation and protection.

cable con cubierta de material no metálico—cable compuesto de alambres de cobre o aluminio cubiertas con papel, hule, o vinilo para aislamiento y protección.

north pole—end of a magnet opposite the south pole.

polo positivo—terminal del imán opuesto del polo negativo.

nut—threaded device with shoulders.

tuerca—dispositivo roscado que tiene hombros.

nutrient solution—dissolved nutrients and dissolved oxygen needed for plant growth.

solución nutritiva—alimentos nutritivos disueltos y oxígeno disuelto necesarios para el crecimiento de plantas.

O

oakum—ropelike material that makes a watertight seal when compacted.

estopa—material de tipo de soga que forma una junta estanca cuando comprimido.

object line—solid line in a drawing that shows visible edges of an object.

(no translation)—línea continua en un dibujo que representa los bordes visibles de un objeto.

occupation—business, employment, or trade.

ocupación—negocio, empleo, o empresa.

occupational cluster—group of related jobs.

grupo profesional—grupo de empleos relacionados.

occupational division—group of occupations or jobs within a cluster that require similar skills.

división profesional–grupo de ocupaciones o trabajos dentro de un grupo que requieren oficios semejantes.

OD—outside diameter.

DO—diámetro exterior.

off-the-farm agricultural jobs—those jobs requiring agricultural skills, but not regarded as farming or ranching.

trabajos agrícolos fuera de la finca—aquellos trabajos que exigen técnicas agrícolas, pero que no se consideran cultivar la tierra o llevar una hacienda.

ohm—a measure of the resistance of a material to the flow of electrical current.

ohmio—medida de la resistencia de un material a la corriente eléctrica.

Ohm's law—the relationship between electric current (I), electromotive force (E), and resistance (R): E=IR.

la ley de Ohm—la relación entre la corriente eléctrica (I), la fuerza electromotriz (E), y la resistencia (R): E=IR.

oil-base—paint containing some type of oil as the vehicle.

colores al óleo—pintura que contiene algún tipo de óleo como el vehículo.

old work—wood that was previous painted or finished.

obra vieja—maderos que fueron pintados o acabados anteriormente.

open aquaculture system—natural water environment for growing plants and animals.

sistema abierto de piscicultura—ambiente acuático natural para cultivar plantas y animales.

open circuit—an electrical circuit in which there is at least one place where current cannot flow.

circuito abierto—un circuito eléctrico en que hay por lo menos un sitio donde no puede pasar la corriente.

open drainage—use of open ditches to carry excessive water.

avenamiento destapado—utilización de canales descubiertos para llevar el agua excesiva.

optoelectric device—designed to interact with light energy in the visible, infrared, and ultraviolet ranges.

dispositivo optoelectrónico—dispositivo diseñado para actuar recíprocamente con energía de luz en las escalas visibles infrarrojas, y ultravioletas.

orange—safety color used to designate machine hazards, such as edges and openings.

anaranjado—color de seguridad usado para señalar peligros de máquina, como bordes y bocas.

orbital—circular or egg-shaped pattern.

orbital—diseño circular o de curva elíptica.

orbital sander—a finishing sander that travels in a circular pattern.

enarenador orbital—un enarenador de acabado que se mueve en una manera circular.

O ring—piece of rubber shaped like an O that fits in a groove on a shaft and prevents liquids from passing.

junta tórica—pieza de hule en forma del «O» que cabe en una ranura sobre un eje y previene que pasen líquidos.

oscillator—nonrotating device for producing alternating current.

oscilador—aparato que no gira, empleado para producir corriente alterna.

oscilloscope—provides visual display of frequency, duration, phase, wave form, and amplitude of signals.

osciloscopio—da representación visual de la frecuencia, duración, fase, ondulación y amplitud de sintonías.

oval head screw—screw with a head that extends both above and below the surface of the material being held.

tornillo de cabeza ovalada—tornillo con cabeza que asoma encima de y está hincada debajo de la superficie del material a que sujete.

overhaul—complete disassembly with cleaning and reconditioning or replacement of most moving parts.

reparar totalmente; poner en buen estado—desmontaje completo con limpieza y arreglo o reemplazo de la mayoría de las piezas móviles.

override—take precedence over.

dominar—tener prioridad sobre algo.

over-voltage protection circuit—circuit protection device.

circuito de protección de máxima tensión—protección contra sobretensiones.

oxidation—combining with oxygen.

oxidación—el combinar del oxígeno con algo.

oxide—the product resulting from oxidation of metal.

óxido—el producto que resulta de la oxidación de metal.

oxidizing flame—flame with an excess of oxygen; hot-test type of oxyfuel flame.

llama oxidante—llama con demasiado oxígeno; tipo de llama oxifuel.

oxyacetylene—oxygen and acetylene combined.

oxiacetilénico—la combinación de oxígeno con acetileno.

oxyfuel—the combination of nearly pure oxygen and a combustible gas to produce a flame.

oxifuel—la combinación de oxígeno casi puro y un gas combustible para producir una llama.

oxyfuel cutting—process in which steel is heated to the point when it burns and is removed in a thin line called a kerf.

soldeo con soplete oxiacetilénico—proceso en que se caldea el acero hasta que se funda y se corte en un corte delgado.

oxygen—gas in the atmosphere that is necessary to support combustion.

oxígeno—el gas en la atmosfera que es necesario para sostener la combustión.

P

packing—soft, slippery, wear-resistant material used to seal space between a moving and nonmoving part.

embalaje—material suave, escurridizo, resistente al desgaste, que sirve para llenar el vacío entre las partes móviles e inmóviles.

pad—metal for practicing welding.

placa de soldadura—chapa de metal para entrenamiento en soldadura.

paddle-wheel aerator—splashing paddle wheel used to mix air and water.

aireador de rueda de paletas—álabe salpicante que se usa para mezclar aire y agua.

paint—substance consisting of pigments and a vehicle.

pintura—sustancia que consiste en pigmentos y un vehículo.

paint film—material left after paint has dried.

película de pintura—material residuo al secarse la pintura.

paint roller—a cylinder that turns on a handle and is used to apply paint.

rodillo—un cilindro que gira por medio de un mango y que se usa para pintar.

pallet—portable platform for storing or moving cargo.

paleta—plataforma portátil para almacenar o trasladar la carga.

pan head screw—screw that has a head that looks like a frying pan turned upside down.

tornillo de cabeza troncocónica—tornillo cuya cabeza es parecida a una sartén puesta boca abajo.

parallel—two edges or lines the same distance apart at all points.

paralelo—dos bordes o líneas equidistantes entre sí a todos puntos.

parallel circuit—provides two or more paths for current flow.

circuito en paralelo—proporciona dos senderos o más para el paso de corriente.

pass—one bead of metal or other material.

pasada—un cordón de metal u otra materia.

pattern—a model or guide for something to be made.

muestra—un modelo o guía para algo que se ha de hacer.

penny—a unit of measure used to designate the length of most nails.

penique—unidad de medida usada para determinar el longitud de la mayoría de los clavos.

pentachlorophenol—a popular wood preserva-tive sold in a nearly clear vehicle.

pentachlorofenol—preservativo popular de la madera que se vende en un vehículo casi claro.

permanent magnet—a piece of iron or steel that holds its magnetism.

imán permanente—un trozo de hierro o acero imantado.

perpendicular—at a 90 degree angle to some object.

perpendicular—algo que forma ángulo recto con otro objeto.

pesticide—substance used to control pests.

pesticida—sustancia empleada para controlar los animales e insectos nocivos.

phillips head screw—a screw with screwdriver slots in the shape of a plus sign.

tornillo de cabeza phillips—un tornillo con muescas en forma del signo más.

phillips screwdriver—a screwdriver with the tip shaped like a plus sign (+).

destornillador phillips—un destornillador con punta en forma del signo más (+).

photoconductive cell—light-sensitive device.

célula fotoconductor—dispositivo sensible a la luz.

photodiode—light-sensitive device that controls current flow.

fotodiodo—dispositivo sensible a la luz que controla el paso de la corriente.

phototransistor—produces higher output than a photodiode.

fototransistor—produce más energía que un fotodiodo.

photovoltaic cell (solar cell)—converts light energy into electrical energy.

célula fotovoltaica (célula solar)—transforma energía solar en energía eléctrica.

pictorial drawing—a drawing that shows three views in one.

representación esquemática—un dibujo que presenta tres vistas.

pierce—to make a hole by pushing through.

agujerear—hacer agujeros por perforar.

pigment—a solid coloring substance suspended in a liquid or vehicle.

pigmento—una sustancia colorante sólida que es suspendida en un líquido o vehículo.

pilot hole—a small hole drilled in material to guide the center point of larger drills; hole drilled to receive the threaded part of a screw.

guía—un agujero pequeño taladrado en el material par guiar el punto central de taladradoras más grandes; agujero taladrado para recibir la parte roscada de un tornillo.

pilot light—glows when a circuit is energized.

lámpara indicadora—está encendida cuando se activa un circuito.

pipe—rigid tubelike material.

tubo—material cilíndrico rígido.

pipe dope—substance applied to threaded fittings to prevent leakage at pipe connections.

composición impermeable para tubos—sustancia aplicada a los ajustes roscados para evitar la salida en las uniones.

pipe fitting—*See* fitting.

ajuste de tubería—*V.* ajuste.

pipe stand—work area for cutting and threading pipe.

soporte de tubería—area de trabajo usado para cortar y roscar tubería.

piston—a sliding cylinder fitting within a cylindrical vessel that receives the force of combusting fuel.

émbolo—un ajuste cilíndrico móvil dentro de un recipiente cilíndrico que recibe la fuerza de carburante ardiendo.

piston ring—completes the seal between a piston and cylinder wall.

aro del émbolo—termina la junta entre el émbolo y la superficie del cilindro.

piston ring compressor—used to force piston rings into their grooves.

compresor de aro de émbolo—empleado para meter a la fuerza los aros del émbolo en las gargantas.

pivot—to turn or swing on.

hacer giras—dar vueltas o girar sobre algo.

plan reading—reading or interpreting scale drawings.

leer los planes—leer o interpretar los dibujos a escala.

plane—a tool that shaves off small amounts of wood and leaves the surface smooth.

cepillo—una herramienta que cepilla los maderos y deja lisa la superficie.

plane iron—the cutting part of a plane.

pletina de cepillo—la pieza cortante del cepillo.

planer—a machine with turning knives that dress the sides of boards to a uniform thickness.

cepillo—una máquina con cuchillas de tornear que cepillan los lados de los tableros a un espesor uniforme.

plastic—term used for a group of synthetic materials made from chemicals and molded into objects.

plástica—término aplicado al grupo de materias sintéticas hechas de sustancias químicas y moldeado en objetos.

plastic wood—a soft wood filler that dries and hardens very quickly.

madera plástica—un relleno blando para maderos que se seca y se endurezca muy rapidamente.

Plastigage—carefully designed material that flattens out uniformly when pressed.

«Plastigage»—material diseñado cuidados amente que se aplasta uniformemente cuando prensado.

plate—2" × 4" on the top or bottom of a stud wall.

viga horiozontal—2" × 4" encima de o debajo del montante.

plates—the conducting parts of a capacitor.

placas—las piezas conductivas del capacitor.

play the flame—to alternately move a flame into and out of an area to carefully control temperatures.

jugar la llama—trasladar una llama alternativamente aquí dentro allá fuera del área para controlar las temperaturas cuidadosamente.

pliable—able to move without separating if pushed or pulled.

flexible—capaz de moverse sin partirse al ser empujado o arrastrado.

plow bolt—a bolt with a square tapered head that is flush with the surface when installed.

perno de arado—un perno con cabeza cuadrada ahusada que se pone embutido cuando instalado.

plug tap—a device used to cut pipe threads in a hole.

macho de aterrajar—un dispositivo que se emplea para aterrajar las roscas de tubo en un agujero o una boca.

plumb—vertical to the axis of the earth or in line with the pull of gravity.

aplomo—vertical al eje de la tierra o alineado con la fuerza de gravedad.

plumb bob—a round, tapered piece of metal attached to a plumb line.

pesa de plomo—un trozo de metal redondo, ahusado, atado a una plomada.

plumb line—a string with a plumb bob attached.

plomada—una pesa de plomo colgada de un hilo.

plumbing—installing and repairing water pipes and fixtures.

fontanería—la instalación y reparación de cañerías y conductos domésticos de agua.

plywood—sheets of lumber made from veneer.

madera contrachapada—tablas de maderos hechas de capas de madera terciada.

pneumatics—use of air to transfer force.

neumática—uso de aire para transmitir potencia.

polarity—direction of electrical flow in a circuit.

polaridad—dirección o sentido del flujo eléctrico de un circuito.

pole building—building supported by erect poles in the soil.

«pole building»—edificio apoyado de postes derechos en el suelo.

poles—opposite ends of a magnet.

polos—terminales opuestos del imán.

policy—firm rules that govern management.

poliza—reglas estrictas que regulan administraciones.

pollution detector—gas analyzer that senses and warns when air is unacceptable.

detector de contaminación del aire—analizador de gas que detecta y avisa cuando está inaceptable la contaminación del aire.

polyethylene (PE)—plastic pipe used extensively for cold water lines.

polietileno—tubo plástico que se usa extensivamente para cañería de agua fría.

polyurethane—a clear, durable, water-resistant finish.

poliuretano—un acabado claro, duradero, e impermeable al agua.

polyvinyl chloride (PVC)—a relatively new type of plastic pipe suitable for interior plumbing.

cloruro de polivinilo—un tipo relativamente recién de tubo plástico que es adecuado para instalación sanitaria dentro de la casa.

pop rivet—a tubular rivet that is enlarged by pulling a ball-ended stem in it until the stem breaks, leaving the rivet secured.

roblón de salto—un roblón tubular que se amplia por tirar del vástago con bola en su extremo, hasta romperse, dejándo seguro al roblón.

poppet valve—a valve that controls the flow of air and gases by moving up and down.

válvula de movimiento vertical—una válvula que controla el movimiento de aire y gases por moverse verticalmente en sentido alternativo.

port—a special hole in the cylinder wall of a two-cycle engine to permit gases to flow in or out of the cylinder.

portillo—una abertura especial en la pared de cilindro de un motor de ciclo de dos tiempos que permite que los gases entren a y salgan del cilindro.

portable circular saw (power hand saw)—a lightweight, motor-driven, round-bladed saw.

sierra circular portátil—una sierra ligera, propulsada por motor, y de hoja circular.

portland cement—dry powder made by burning limestone and clay followed by grinding and mixing.

pórtland—polvo seco obtenido por calcinación de caliza y arcilla, y en seguida pulverizándolo y mezclándolo.

positive displacement—consistent volume pumped.

desplazamiento positivo—cuando un volumen consistente es bombeado.

positive (hot) wires—conductors with black, red, or blue insulation that carry current to an appliance.

alambres de alta tensión—conductores con aislamiento negro, rojo, o azul, que elevan la corriente a un aparato.

post-and-girder building—building resting on posts with a heavy timber frame.

edificio de poste y viga—edificio que descansa sobre postes con una armazón pesada de maderos.

post-frame—another name for post-and-girder building.

«post-frame»—otro nombre por edificio de poste y viga.

potentiometer (pot)—variable resistor used to control voltage.

potenciómetro—resistencia variable empleada para controlar la tensión.

power hand saw—*See* portable circular saw.

sierra mecánica portátil—*V.* sierra circular portátil.

power machine—a tool driven by electric motor, hydraulics, air, gas engine, or some force other than or in addition to human power.

motor mecánico—una herramienta accionada por un motor eléctrico, por hidráulica, por motor de gas o por alguna fuerza distinta o en adición a la fuerza humana.

power stroke—the engine process in which burning fuel expands rapidly but evenly to drive the piston down.

carrera motriz—el proceso motriz en que el combustible ardiente se expansiona rápidamente pero uniformemente para apretar el émbolo.

power supply—provides electricity to circuits.

suministro de electricidad—suministra la corriente eléctrica a los circuitos.

power tool—a tool operated by some source other than human power.

herramienta mecánica—una herramienta accionada por alguna cosa diferente de la fuerza humana.

precipitation—rain or snow.

precipitación—lluvia o nieve.

precleaner—a device that removes large particles from air entering an air cleaner.

filtro—un dispositivo que, al entrar una limpiadora de aire, quita las partículas grandes al aire.

pressure—force acting upon an area.

presión—la fuerza ejerciéndose sobre un área.

pressure feed system—a spray painting system that delivers paint under pressure to the mixing nozzle of the spray gun.

sistema de alimentación bajo presión—un sistema de pintar con esprai que entrega la pintura bajo presión a la tobera mezclante de la pistola.

pressure regulator—maintains constant line pressure.

regulador de presión—mantiene constante la presión de línea.

pressure switch—responds to changes of liquid, gas, or mechanical pressure.

interruptor automático—responde a cambios de presión líquida, gaseosa, o mecánica.

pressure-treated—chemicals injected under pressure.

manipulado bajo presión—sustancias químicas inyectadas bajo presión.

primary circuit—low-voltage circuit of an ignition system.

circuito primario—circuito de baja tensión de un sistema de encendido.

primer—a special paint used to seal bare wood and prepare metal surfaces for high-quality top coats.

primer mano—una pintura especial usada para chapar madera descubierta y preparar las superficies de metal para mano de buena calidad.

procedure—a sequence of steps for doing something; a particular course of action.

procedimiento—una sucesión de medidas para hacer algo; una línea particular de acción.

profit—income made from the sale of goods or services.

beneficios—ingresos ganados de la venta de productos o servicios.

project—special activity planned and conducted with the purpose of learning.

proyecto—actividad especial planeado y conducido con el objetivo de aprender.

propellant—the gas that propels paint toward the object being painted.

propulsor—el gas que propela la pintura hacia el objeto que se ha de pintar.

protractor—an instrument for drawing or measuring angles.

transportador—un instrumento empleado para trazar o medir triángulos.

PTO—power take-off.

TF—toma de fuerza.

puddle—a small pool of liquid metal.

mezcla—goteo de metal líquido.

pulley—device attached to a shaft to carry a belt.

polea—dispositivo atado a un eje para llevar una correa.

purge the lines—remove undesirable gases.

purgar las líneas—quitar los gases no deseosos.

purlin—longitudinal roof nailers.

correa—claves longitudinales del techo.

purloins—greenhouse construction member maintains distance between bows.

correa superior—pieza de construcción del invernadero que mantiene la distancia entre los arcos.

purple—the safety color used to designate radiation hazards.

morado—el color de seguridad que se usa para señalar peligros de radiación.

push drill (automatic drill)—a drill with a spiral shaft that turns a chuck when pushed against an object.

taladradora automática—una taladradora con eje espiral que tornea un mandril cuando se empuja contra el objeto.

push stick—a wooden device with a notch in the end to push or guide stock on the table of a power tool.

varilla de guiar—un instrumento de madera con una muesca en el extremo para empujar o guiar la materia prima que está en la mesa de una herramienta-máquina.

putty—soft material containing oils that keep it pliable over a long period of time, used to seal joints or fill holes.

masilla—materia consistente en aceites que la mantienen flexible durante mucho tiempo; se usa para tapar junturas o tapar agujeros.

Q

quick-coupling—hose ends that snap together.

empalme automático—bocas de manga que cierran de golpe.

rabbet—a cut or groove at the end of a board made to receive another board and form a joint.

ranura—un corte o hendidura en el borde del tablero hecho para que se encaje con otro tablero, formando una ensambladura.

rabbet joint—a joint formed by setting the end or edge of one board into a groove in the end or edge of another board.

ensambladura de ranura—una ensambladura formada por encajar el borde de un tablero en la ranura o borde de otro tablero.

raceways—rectangular or oval fish tanks.

«raceways»—peceras elípticas o rectangulares.

radial arm saw—a power circular saw that rolls along a horizontal arm.

sierra de brazo de movimiento radial—una sierra circular mecánica que se roda a lo largo de un brazo horizontal.

rafter—single timber supporting a roof section.

par—una viga única que apoya una sección del techo.

rasp—a file with very coarse teeth.

escofina—lima con dientes muy gruesos.

ratio—proportion of one component to another by weight or volume.

proporción—razón del peso o volumen de ún componente a otro.

receptacle—a device for receiving an electrical plug.

enchufe hembra—un dispositivo que recibe un enchufe (macho).

reciprocal—back and forth or opposite.

recíprico—de acá para allá o al contrario.

reciprocate—return; move back and forth.

reciprocar—regresar; tener movimiento alterno.

reciprocating saw—a saw with a stiff blade that moves back and forth.

sierra de movimiento alternativo—sierra con una banda rígida que tiene movimiento alterno.

reclamation—to reclaim or make usable.

regeneración—regenerar o hacer útil.

recondition—to do what is needed to put back into good condition.

arreglar—hacer lo que sea necesario para que se ponga de nuevo.

rectifier—converts AC to DC current.

rectificador—convierte la corriente CA en la corriente CC.

red—the safety color used to designate areas or items of danger or emergency.

rojo—el color de seguridad usado para señalar áreas o artículos de peligro o de emergencia.

reducer—a fitting used to connect pipes of different sizes.

reductor—un ajuste empleado para conectar tubos de tamaños distintos.

reed valve—a flat, flexible plate that permits air or liquid to pass in one direction but seals when the flow reverses.

válvula de lengüeta—una placa llana flexible que permite que el aire o líquido pase en un sentido pero que cierre cuando pase en otro sentido.

regulator—a device that keeps pressure at a set level or controls the rate of flow of a gas or liquid.

regulador—un dispositivo que mantiene la presión a un nivel determinado o controla la velocidad del flujo de un gas o líquido.

reinforced concrete—concrete slabs or structures that are strengthened with embedded steel rods or wire mesh.

hormigón armado—bloques o armazones de hormigón que son armados entre su masa con alambres y barras de acero.

renewable natural resources—resources provided by nature that can replace or renew themselves.

recursos naturales renovables—recursos proporcionados por la naturaleza que pueden volver a llenarse o renoverse sí mismos.

repair—to replace a faulty part or make it work correctly.

reparar—reemplazar una pieza defectuosa o hacer que ésta funcione correctamente.

repel—to push away.

repeler—rechazar.

represent—to stand for or to be a sign or symbol of.

representar—hacer presente algo o servir como señal de algo.

resistance—any tendency of a material to prevent electrical flow.

resistencia—cualquiera tendencia de un material de impedir el flujo eléctrico.

resistivity—the ability to prevent flow of electricity.

resistividad—la capacidad de impedir el flujo de electricidad.

resistor—device that offers specific resistance to a current.

reóstato—dispositivo que ofrece resistencia específica a una corriente eléctrica.

reverse (positive) polarity (RP)—DC current flowing in the opposite direction from straight polarity; to reverse the direction of current.

polaridad inversa (positiva)—corriente continua que fluye en sentido contrario a la polaridad negativa.

invertir la polaridad—invertir el sentido de la corriente.

reversible—capable of running backward as well as forward.

reversible—dícese de un aparato que es capaz de funcionar en los dos sentidos contrarios.

revolution—one complete turn of 360º.

revolución—una rotación, o un giro completo de 360º.

rheostat—variable resistor used to control current.

reóstato—resistencia variable empleado para controlar la corriente.

richer—a mixture with an increased proportion of fuel to air.

mezcla rica—una mezcla que tiene más combustible en proporción con aire.

ridge—highest point of a roof.

caballete—el lomo de un techo.

rig—a piece of apparatus assembled to conduct an operation.

aparejo—conjunto de cosas necesarias para conducir una operación.

right triangle—a three-sided figure with one 90 degree angle.

rectángulo—figura delimitada por tres líneas rectas que tiene un ángulo recto.

rigid-frame—building with a steel frame.

de armazón rígida—edificio con armazón de acero.

ring expander—tool used to remove and install piston rings.

llave de aro—herramienta empleada para instalar y quitar los aros de émbolo.

rip—to cut the long way on a board or with the grain.

serrar al hilo—cortar según la dirección de los hilos, o del grano.

rip fence—a guide that helps keep work in a straight line with a saw blade.

guía de aserrar al hilo—una guía que ayuda mantener los materiales alineados con la hoja de sierra.

rip saw—a saw with teeth filled to a knifelike edge and used to cut with the grain.

sierra de hender—una sierra cuyos dientes son afilados a corte como cuchillo y que se emplea para cortar según la dirección de los hilos.

riser—vertical part of a step.

contrahuella—parte vertical del peldaño.

rivet—to spread or shape by hammering; a fastening device held in place by spreading one or both ends.

remachar—triturar o cambiar la forma por martillazos.

roblón—un sujetador que se sujete por machacar o partir la punta.

robot—mechanical device capable of humanlike movements.

robot—aparato mecánico capaz de movimientos movimientos así como los de los seres humanos.

robotics—study and application of technology to robots.

robótica—estudio y aplicación de tecnología a los robotes.

roller cover—a hollow, fabric-covered cylinder that slides onto the handle of a paint roller.

cilindro de rodillo—un cilindro hueco y cubierto con tela que se introduce en el mango del rodillo para pintar.

root—the deepest point in a weld.

raíz—el punto más hondo de una soldadura.

root pass—the first welding pass that is made in a joint.

pasada de raíz—la primera pasada que se hace en una soldadura.

rope starter—rope wrapped around a pulley for turning power to start an engine.

arranque de soga—soga que corre por una polea para potencia de torno para poner en marcha un motor.

rot—to decay or break down into other substances.

pudrir—descomponer.

rotary air pumps—high-volume, low-pressure air pumps.

bombas aspirantes giratorias—bombas de aire de gran cantidad y presión baja.

rough lumber—lumber as it comes from the sawmill.

madera en bruto—maderos así como llegan del aserradero.

round head screw—screw that extends in an even curve above the surface of the material being held.

tornillo de cabeza redonda—tornillo que asoma en una curva uniforme encima del material sujetado.

rounded up or down—going to the next higher or lower whole number.

redondear—convertir la cantidad en el número completo que la precede o sigue.

RPM—revolutions per minute.

RPM—revoluciones por minuto.

rust—reddish brown or orange coating that results when iron reacts with air and moisture.

orín—capa castaña y anaranjada que resulta cuando el hierro reacciona con el aire y la humedad.

rust inhibitor—reduces or prevents rusting.

inhibidor de oxidación—reduce o impide la oxidación.

S

sabre saw—a reciprocal saw used primarily for cutting curves or holes in wood, metal, cardboard, and similar materials.

sierra de sable—una sierra de movimiento alternativo que se emplea mayormente para cortar curvas o agujeros en madera, metal, cartón, y materiales semejantes.

saddle soap—a product used to clean, soften, and preserve leather.

(no translation other than definition)—producto que se usa para limpiar, ablandar, y curar el cuero.

SAE—Society of Automotive Engineers.

SIA—Sociedad de Ingenieros Automotrices.

safe—free from harm or danger.

seguro—libre de daño o peligro.

safety—freedom from accidents.

seguridad—libre de los accidentes.

sag—shifting of a large area of paint downward.

caída—movedizo de una gran área de pintura hacia abajo.

sal ammoniac—a cube or block of special flux for cleaning soldering coppers.

sal amoníaco—un cubo o bloque de fundente especial para limpiar los soldadores.

sand—small particles of stone.

arena—partículas pequeñas de piedra.

sandblasting—cleaning by sand particles thrown by compressed air.

limpiar con chorro de arena—limpiar con partículas por el aire comprimido.

satin—a wood or metal finish with low sheen or shine.

raso—un acabado para madera o metal con poco lustre.

saturate—to add a substance until the excess starts to run out.

empapar—añadir una sustancia hasta que lo exceso empiece a rebosar.

sawhorse—trestle; a wood or metal bar with legs, used for temporary support of materials.

burro—caballete; madero o metal horizontal con patas empleado para el soporte temporario de materiales.

scale—an instrument with all increments shortened according to proportion; numbers and gradations on measuring tools; a rigid steel or wooden measuring device; the size of a plan compared with that of the object it represents.

escala—instrumento cuyos incrementos son graduados según de la proporción; números y gradaciones en las herramientas de medida; un dispositivo rígido de acero o madera; la representación de un plan relacionado a la dimensión del objeto que representa.

scale drawing — a drawing that represents an object in exact proportion although the object is larger or smaller than the drawing itself.

dibujo a escala—un trazado que representa un objeto en proporción exacta, aunque sea el objeto más grande o más pequeño que el trazado si mismo.

scope—size and complexity.

amplitud—tamaño y complejidad.

score—to scratch.

rayar—raspar (a metal, madera, etc.)

scratch awl—a sharply pointed tool with a wooden or plastic handle used to mark metal.

punzón marcador—un instrumento con una punta muy aguda y un mango de madera o plástica, que se emplea para raspar metal.

screeding—striking off excess concrete to create a smooth and level surface.

descantillar—usar un escantillón para enrasar el hormigón exceso para crear una superficie lisa y plana.

screw—a fastener with threads that bite into material as it is turned.

tornillo—un sujetador con filetes que se introducen en el material cuando da vueltas.

screw plate—a set of taps, dies, and handles used for making threads.

dispositivo de roscado por fresa—un juego de machos, troqueles y mangos usado para hacer filetes, o roscados.

screwdriver—a turning tool with a straight tip, phillips tip, or special tip.

destornillador—una herramienta que da vueltas y que tiene punta llana, de phillips, o especial.

scriber—a very small, metal, sharp-tipped marker.

punta de trazar—un marcador muy pequeña, de metal, y puntiaguda.

seal—to apply a coating that fills or blocks the pores so that no material can pass through the surface.

impermeablizar—aplicar una capa que tapa los agrietos o poros para que un material no pueda penetrar la superficie.

sealer—a coating used to fill pores and prevent material from passing through a surface.

producto de selladura—una capa usada para tapar los poros y impedir que penetren materiales por la superficie.

seasonal demand—amount of acre inches of water for growing a crop.

demanda de temporada—cantidad de pulgadas acre de agua que se exige la cultivación de un cultivo determinado.

seat—a stationary or nonmoving part that is designed to seal when a moving part is pressed against it.

asiento—una pieza fija que es hecha a obturar cuando se la aprieta una pieza en movimiento.

secondary circuit—high-voltage circuit of an ignition system.

circuito secundario—circuito de alta tensión del sistema de encendido.

semiconductor—a material with characteristics that fall between those of insulators and conductors.

semiconductor—un material cuyas características caen entre los de los aisladores y de los conductores.

semigloss—paint with a slight shine or gloss.

pintura semi-brillante—pintura con un acabado de poco lustre o brilla.

sensor—device that receives and responds to a signal.

sensor—dispositivo que recibe y responde a una señal.

series circuit—provides single path for current flow.

circuito en serie—da un sendero sólo para el flujo de corriente eléctrica.

series-parallel circuit—combines properties of series and parallel circuits.

circuito serie-paralelo—combina propiedades de los circuitos en serie y los circuitos paralelos.

serrated—notched.

dentado—entallado.

service drop—an assembly of electrical wires, connectors, and fasteners used to transmit electricity from a transformer to a service entrance panel.

conductor de bajada del poste—un conjunto de alambres, conectadores y abrazaderas eléctricos usados para transmitir la electricidad de un transformador a una caja de línea de acometida de energía.

service entrance panel—a box with fuses or circuit breakers where electricity enters a building.

caja de línea de acometida de energía—una caja con fusibles o cortacircuitos donde la electricidad entra el edificio.

set—to drive the head of a nail below the surface.

embutir—hincar la cabeza del clavo bajo la superficie.

settling tank—low-pressure tank with slow water velocity that allows solid waste particles to settle to the bottom.

depósito de sedimentación—depósito con velocidad lenta de agua que deja asentarse al fondo a las partículas sólidas residuales.

shadecloth—woven plastic fabric that reflects 30 to 70 percent of the sunlight hitting it.

pantalla para sombra—tejido plástico que refleja 30 a 70 por ciento de la luz del sol que lo golpea.

shank—the nonthreaded part of a nail or screw.

pie—la parte no roscada del tornillo o clave.

shank hole—the hole provided for the shank of a screw.

agujero del mango—el agujero a que se introduce el pie del tornillo.

shear—to cut by action of opposed cutting edges.

cizallar—cortar por medio de acción contraria de dos filos cortantes.

shears—large scissorlike tools for cutting sheet metal and fabrics.

cizallas—instrumentos grandes parecidos a tijeras, que se emplean para cortar chapas de metal y tejidos.

sheathing—first exterior layer of wall or roof.

revestimiento—primera capa exterior de pared o techo.

sheet bend—knot for connecting ropes of different sizes.

nudo de escota—nudo para conectar sogas de tamaños diferentes.

sheet metal screw—a screw that has threads wide enough to permit thin metal to fit between the ridges of the threads.

tornillo de chapa—un tornillo cuyos filetes son bastante anchos para que chapa delgada de metal pueda encajar entre las partes salientes de los filetes.

shellac—a natural material that comes from an insect and is used to seal wood.

goma laca—un material natural que origina de un insecto y que se usa para chapar madera.

shielded metal arc welding (stick welding)—welding with electrical power as a source of heat and rods covered with flux which form a gaseous shield around the molten metal until it solidifies.

soldadura por arco protegido—soldar con energía eléctrica como la fuente de caldeo y con electrodos cubiertos con fundente que forman campo gaseoso alrededor del metal fundido hasta que se fragüe.

shock—the body's reaction to an electric current.

sacudida—la reacción del cuerpo a una corriente eléctrica.

short circuit—a condition that occurs when electricity flows back to its source too rapidly and blows fuses, burns wires, and drains batteries.

cortocircuito—fenómeno que ocurre cuando la corriente eléctrica pasa al retroceso a su fuente tan rapidamente y funde los fusibles y los cables, y drena las baterías o pilas.

shroud—to cover; a cover.

amortajar—envolver, cubrir.

mortaja—lienzo para envolver.

Note: In Spanish, the noun "shroud" is only used when referring to the covering of the dead.

siemen (S)—unit for measuring conductance.

siemen (S)—unidad para medir la conductancia.

silicon—a low-quality pigment. Also the material most commonly used for semiconductors.

silicio—un pigmento de mala calidad. También el material lo más usado en los semiconductores.

silt—a substance composed of intermediate size soil particles.

légamo—una sustancia compuesta de partículas de tamaño intermedio de tierra.

sine wave—created by the flow of current through one cycle.

onda sinusoidal—formada por el paso de corriente por un ciclo.

single line system—single or independent line.

vía sencilla—línea individua o independiente.

single-pole switch—the one switch that controls one or more lights and/or outlets.

interruptor unipolar—el único interruptor que controla úno o más de luces y/o tomas.

siphon system—a spray painting system that creates a vacuum at the nozzle of the gun and utilizes atmospheric pressure to push the paint to the nozzle.

aparato de sifón—un sistema de pintar con esprai que crea un vacío en la boquilla de la pistola y utiliza la presión atmosférica para avanzar la pintura a la boquilla.

60-cycle current—electricity that alternates direction of flow 60 times per second.

corriente de ciclo 60—corriente eléctrica que cambia el sentido 60 veces por segundo.

sketch—a rough drawing of an idea, object, or procedure.

croquis—un trazado de una idea, objeto, o procedimiento.

slag—the product formed when burning steel combines with oxygen.

escoria—el producto que se forma cuando acero ardiente combina con oxígeno.

slag box—a container of water or sand placed to catch hot slag and metal from the cutting process.

depósito de escorias—un recipiente de agua o arena colocado para recibir escoria y metal calientes del proceso de soldar.

sliding T bevel—a device used to draw angles on boards or metal.

saltarregla móvil—un instrumento usado para trazar ángulos sobre maderos o metal.

slotted head screw—a screw with one straight slot across the head.

tornillo de cabeza ranurada—un tornillo con una sóla ranura derecha a través de la cabeza.

slow moving vehicle (SMV) emblem—reflective emblem consisting of an orange triangle with a red strip on each of the three sides.

señal de vehículo en marcha lenta—señal reflectiva que consiste en un triángulo anaranjado con tres ribetes en cada de los tres lados.

snips—a large scissorlike tool for cutting sheet metal and fabrics.

cizallas—gran instrumento parecido a tijeras, que se emplea para cortar chapa de metal y tejidos.

soapstone—a soft, gray rock that shows up well when marked on most metals.

esteatita—una piedra blanda, gris que se muestra bien al marcarse sobre la mayoría de los metales.

solar cell (photovoltaic cell)—converts light energy into electrical energy.

célula solar (célula fotovoltáica)—transforma energía solar en energía eléctrica.

solder—mixture of tin and lead.

soldadura—mixtura de estaño y plomo.

soldering—bonding with metals and alloys that melt at temperatures below 840°F.

soldadura—unir con metales y aleaciones que se funden a temperaturas menos de 840°F.

soldering copper—a tool consisting of a handle, steel shank, and copper tip used to heat metal for soldering.

soldador—un instrumento consistente en mango, pierna de acero, y punta de cobre usado para caldear el metal a que se ha de soldar.

solderless connector—*See* wire nut.

conectador sin soldadura—*V.* conectador de alambres.

solenoid valve—valve actuated by an electromagnet.

válvula de soleinoide—válvula accionada por un electro-imán.

solid sphere—work area in the shape of a solid ball.

sfera sólida—área de trabajo en forma de una bola sólida.

solidify—to harden or change from a liquid to a solid.

solidificarse—endurecerse o pasar de líquido a sólido.

south pole—end of magnet opposite the north pole.

polo negativo—terminal opuesto del polo positivo del imán.

spar varnish—special finish for wood exposed to high moisture conditions.

barniz de espato—acabado especial para madera expuesta a condiciones de mucha precipitación.

species—plants or animals with the same permanent characteristics.

especies—conjunto de plantas o animales que tienen las mismas características permanentes.

speed indicator—a device used to measure revolutions per minute (RPM) of a turning shaft or part.

contador de velocidad—un dispositivo usado para medir las revoluciones por minutos (rpm) de un eje o pieza girándose.

spirit level—a tool containing alcohol in a sealed, curved tube with a small air space or bubble, used to determine whether an object has the same height at two or more points.

nivel de alcohol, nivel de burbuja—instrumento que contiene alcohol en un tubo tapado y con una burbuja de aire, empleado para determinar si un objeto tenga la misma altura a dos o más puntos.

spool valve—one inlet and multiple outlets.

válvula de carrete—una admisión y salidas multiples.

spot marking—color on the surface of the wire of an electrode.

«spot marking»—color en la superficie del alambre de un electrodo.

spot welder, spot welding—a device and process used when sheet metal must be fastened in many places near the edges.

soldadura eléctrica por puntos—un dispositivo y proceso empleado cuando chapa de metal tiene que ser juntado en muchos puntos cerca de los bordes.

sprinkler irrigation—dispersion of water in droplets.

riego por aspersión—dispersión del agua en gotitas.

sprocket—a wheel with spiked teeth.

catalina—una rueda con dientes estacas.

square—a device used to draw angles for cutting and to check the cuts for accuracy. Also the top of a wall.

escuadra—instrumento que se emplea en trazar ángulos para cortar y para revisar la precisión de los cortes.

square head—a head with four equal sides.

cabeza cuadrada—cabeza cuadrilátera de lados.

square knot—safe knot for connecting two ropes of equal size.

nudo de envergue—nudo seguro para conectar dos sogas de tamaños iguales.

stair horse—stair stringer.

zanca—estructura que apoya una escalera.

stair stringer—structure that supports stairs.

zanca—estructura que apoya una escalera.

standard pipe—steel pipe sold or used unless extra heavy or double extra heavy is specified.

tubo estándar—tubo de acero que se vende o se usa a menos que se especifique extragrueso o doble extragrueso.

staple—a piece of wire with both ends sharpened and bent to form two legs of equal length.

grapa—una pieza de alambre con ambos extremos afilados y combados para formar dos piernas del mismo longitud.

stationary—having a fixed position.

estacionario—teniendo una posición fija.

steam cleaner—a portable machine that uses water, a pump, and a burner to produce steam.

limpiadora de vapor—una máquina portátil que utiliza el agua, una bomba, y un quemador para producir el vapor.

steel screws—screws made of steel with a blued, galvanized, cadmium, nickel, chromium, or brass finish.

tornillos de acero—tornillos hechos de acero con un acabado pavonado, galvanizado, cadmio, de níquel, chromado, o de latón.

stem—long, round section of a valve.

vástago—parte larga y cilíndrica de la válvula.

step pulley—a pulley with several sizes.

motón—una polea de varios tamaños.

stick welding—*See* shielded metal arc welding.

(no translation)—*V.* soldadura por arco protegido.

stock—a piece of material such as wood or metal.

materia prima—un trozo de material como madera o metal.

stove bolt—a round head bolt with a straight screwdriver slot, threaded its entire length.

tornillo de estufa—un perno de cabeza redonda ranurada, y roscado por su longitud entero.

straight (negative) polarity (SP)—DC current flowing in one direction, the opposite of reverse polarity.

polaridad negativa—corriente continua pasando en una sóla dirección, al contrario a la polaridad inversa.

strands—small wires placed with others to form bundles to improve flexibility and conductivity.

filamentos—cables pequeños colocados con otros para mejorar la flexibilidad y la conductividad.

stringer bead—weld bead produced without weaving.

cordón longitudinal—cordón producido sin pasada pendular.

stroke—the movement of a piston from top to bottom or from bottom to top.

carrera—el movimiento de un émbolo de arriba abajo o de abajo arriba.

subfloor—first layer of flooring.

subpiso—primera capa de suelo.

subirrigation—water supplied below the ground surface.

riego subterráneo—agua suministrado subterráneamente.

submersible pumps—low-pressure, high-volume immersible pumps.

bombas sumergibles—bombas de baja presión, y gran cantidad que se pueden sumergir.

sump—hole where unwanted water drains and is removed by a pump.

pozo negro—hoyo en que el agua drena y de que se recoge por una bomba aspirante impelete.

supercomputer—enormous capacity computer.

supercomputadora—computadora de capacidad enorme.

supervised occupational experience program (SOEP)—activities of the student outside the agricultural class or laboratory done to develop agricultural skills.

programa de experiencia ocupacional supervisado (PEOS)–actividades del estudiante fuera del programa o laboratorio de agricultura llevadas a cabo para desarrollar técnicas agrícolas.

surface irrigation—adding water to the soil by gravity.

riego por aspersión—suministrar agua al suelo por la gravedad.

sweating—process of soldering a piece of copper pipe into a fitting.

suelda por fusión—el proceso de soldar un trozo de tubo de cobre a un ajuste.

switch—a device used to stop the flow of electricity.

interruptor—un dispositivo usado para interrumpir el flujo de una corriente eléctrica.

switching wires (traveler wires)—pair of wires attached to the light-colored screws on three-way or four-way switches.

circuitos selectores—par de cables atado a los tornillos de colores blancos en los interruptores de tres o cuatro direcciones.

T

table saw (bench saw)—a stationary circular saw with either a tilting arbor or a tilting table.

sierra de mesa—una sierra circular fija con mesa o árbol inclinado.

tack rag—a rag dampened with solvent.

(no translation)—trapo humedecido con disolvente.

tacking—making a small weld to hold metal parts temporarily.

soldadura discontinua—el hacer de una soldadura pequeña para unir piezas metales temporalmente.

tacky—sticky.

(no translation)—pegajoso.

tang—tapered end on a file.

cola—parte extremo ahusado de una lima.

tap—a hardened, brittle, fluted tool used to cut threads into holes in metal.

macho de aterrajar—instrumento endurecido, quebradizo y estriado que se usa para labrar la rosca en los agujeros de metal.

tap wrench—a device used to turn a tap.

llave—un instrumento empleado para abrir o cerrar un grifo.

tape—flexible measuring device that rolls onto a spool.

cinta métrica—un instrumento flexible de medida que bobina a una bobina.

tee—a fitting used where a pipe line splits into two directions.

unión en T—un ajuste empleado donde la tubería se bifurca.

teflon tape—tape used on threaded fittings to prevent water or gas leakage at pipe connections.

cinta de teflón—cinta aplicado a los ajustes roscados para evitar escape de agua o gas de los conexiones de tubos.

temper—to heat a piece of tool steel followed by controlled cooling so as to control the degree of hardness.

templar—caldear una pieza de acero de herramientas y seguir con enfriamento controlado para controlar el grado de dureza.

temperature rise—how warm the motor can get without causing internal damage.

elevación de temperatura—la temperatura a que puede alcanzar un motor sin causar daño interno.

tensile strength—the amount of tension or pull a weld can withstand.

resistencia a la tracción—la tensión o fuerza máxima que una soldadura puede resistir.

terra cotta—hard, semifired, weatherproof ceramic clay.

terracota—barro cocido, duro, y que resiste a la intemperie.

threads—grooves of even shape and taper that wrap continuously around a shank or hole.

filetes—ranuras de forma y estriamiento uniforme que se enrollan continuamente en un mango o agujero.

three-view drawing—a drawing that shows three views or sides of an object in three separate representations.

representación esquemática—un dibujo que muestra tres vistas o caras de un objeto en tres representaciones distintas.

three-way switch—a switch that permits a light or receptacle to be controlled from two different locations by a pair of switches.

conmutador de tres direcciones—un interruptor que permite que una luz o receptáculo sea controlado desde dos sitios distintos por un par de interruptores.

throughput—rate of flow of grain and straw through a combine.

salida—el índice del paso de granos y hechaduras o granzos por una segadora trilladora.

thyristor—a semiconductor used to electronically control switching.

tiristor—un semiconductor que se emplea para controlar electronicamente el interrumpir de corriente.

tile—terra cotta.

teja—terracota.

tilting arbor—a motor, belt, pulley, shaft, and blade assembly that tilts as a unit in a power saw.

árbol inclinable—un conjunto de motor, cinta, polea, eje y banda que se inclina como grupo en una sierra de máquina.

tilting table—a table that can be set at various angles.

mesa incinable—una mesa que se puede fijar a varias inclinaciones.

timber hitch—knot for attaching rope to poles.

vuelta de braza—nudo para atar la soga a los postes.

timer—device that activates a switch for a preset time.

temporizador electrónico—dispositivo que activa un interruptor programado a una hora específica.

tin—one component of solder.

estaño—un componente de soldadura.

tinning—bonding filler material to a base metal.

estañar—uniendo materia de revestimiento a un metal de base.

tip cleaners—rods with rough edges designed to remove soot, dirt, or metal residue from the hole in the tip of a torch.

limpiadoras de la boquilla del soplete—varillas con puntos ásperos hechos para quitar residuos de metal, mugre, u hollín de la cavidad de la boquilla del soplete.

titanium dioxide—a high-quality pigment used in many paints.

dióxido de titanio—un pigmento de buena calidad que se usa en muchas pinturas.

title block—the section of a drawing reserved for information about the drawing in general.

viñeta de título—la parte de un dibjuo reservada para información sobre el dibujo en general.

toe nail—to drive a nail at an angle near the end of one piece and into the face of another piece.

clavar en cruz—clavar un clavo a un ángulo cerca del borde de úna pieza y por la cara de otra pieza.

tolerance—acceptable variation.

tolerancia—variación aceptable.

tongue and groove (T&G)—board with a tonguelike edge on one side and a groove on the other which permits boards to be locked together.

ranura y lengüeta—tablero con un borde de tipo lengüeta de un extremo y una ranura en el otro que permite que se encajen los tableros.

tool—any instrument used in doing work.

herramienta—cualquier instrumento empleado para trabajar.

tool fitting—to clean, reshape, repair, or resharpen a tool.

ajustar una herramienta—limpiar, reformar, reparar, o reafilar una herramienta.

tool steel—steel with a specific carbon content that allows the tool to be annealed and tempered.

acero de herramientas—acero con un índice específico de carbono que permite que se rocozca y se temple el acero.

top dead center (TDC)—position of a piston when at its highest point (furthest from the crankshaft).

punto muerto superior—posición del émbolo cuando está a su punto más alto (más lejano del cigüeñal).

torch—an assembly that mixes gases and discharges them to support a controllable flame.

soplete—un montaje que mezcla y descarga gases para sostener una llama controlable.

torque—a twisting force; to twist.

momento de torsión—una fuerza de torsión; torcer.

trade—specific kinds of work or businesses, especially those that require skilled mechanical work.

industria—empresas o negocios de tipo específico, especialmente aquellos que requieren una fuerza laboral experta.

transformer—a device that converts high voltage from high-power lines to 230 volts for home and farm installations; device used to step current up or down.

transformador—un aparato que transforma la tensión alta de las líneas de alta tensión en 230 voltios para instalaciones de casa y de finca; un aparato empleado para aumentar o reducir la corriente eléctrica.

transistor—a three-element, two-junction device used to control electron flow.

transistor—un dispositivo de tres elementos y dos empalmes que sirve para controlar el flujo de electrones.

transit—surveying instrument.

escala de agrimensura—instrumento para medir tierras.

translation—linear movement.

traslado—movimiento linear.

translucent—allows sunlight to pass through.

translúcido—que deja pasar la luz del sol.

transmission—great box that permits two or more choices of speed.

transmisión—gran caja que permite dos opciones de velocidad.

traveler wires—*See* switching wires.

cable flexible de ascensor—*V.* circuitos selectores.

tread—stair step.

peldaño—escalón de un escalera.

trestle—*See* sawhorse.

caballete—caballete.

Note: In Spanish, the two words are the same.

trickle irrigation—tubes putting water onto or below the soil.

irrigación por infiltración—tubos suministrando el agua sobre o por debajo de la tierra.

trim and shutter paint—paint formulated to stay bright without chalking.

pintura para marcos y contravientos—pintura formulada para permanecer brillante sin ponerse gredoso.

troubleshooting—determining what causes a malfunction in a machine or process.

localizador de averías—determina lo que causa un funcionamiento defectuoso de una máquina o proceso.

trough culture—plant production where the root zone is periodically flooded with nutrient solution.

«trough culture»—producción de cultivos en donde el rizoma es inundado periodicamente con solución nutritiva.

truss—rigid framework.

cercha—armazón rígida.

try square—a tool used to try or test the accuracy of cuts that have been made.

escuadra de comprobación—un instrumento utilizado para probar la precisión de los cortes de se han hecho.

tube culture—plant production by which the root zone is immersed in a constant flow of nutrient solution.

«tube culture»—producción de cultivos por la cual el rizoma es sumergido en un flujo constante de solución nutritiva.

tubing—pipe that is flexible enough to bend.

tubos flexibles—que son bastante flexibles para curvar.

tubing cutter—a tool used for cutting and smoothing pipe or tubing.

trefiladora de tubos—una herramienta que se emplea para cortar y lisar tubos y cañería.

turbine—a high-speed rotary engine driven by water, steam, or other gases.

turbina—un motor giratorio de alta velocidad, impulsado por agua, vapor, u otros gases.

turning tool—a tool used to turn nuts, bolts, or screws.

herramienta de tornear—una herramienta empleada para tornear tuercas, pernos, o tornillos.

two half hitches—knot useful for tying livestock.

dos vueltas de cabo—nudo útil para atar ganado.

two-cycle engine (two-stroke-cycle engine)—an engine with two strokes per cycle.

motor de dos tiempos—un motor que tiene dos carreras por ciclo.

type—the way current flows in a motor.

tipo—la manera en que pasa la corriente en un motor.

U

undercoater—special paint used to prepare surfaces for high-quality top coats.

pintura de primera mano—pintura especial usado para preparar las superficies para capas de buena calidad.

underdrainage—underground drainage system.

drenaje subterráneo—sistema de drenaje subterráneo.

union—a fitting that can be easily opened between two pieces of similar threaded pipe.

unión de tuerca—un accesorio que puede abrirse facilmente entre dos trozos de tubo roscado de la misma manera.

unique—having no like or equal.

único—no teniendo ni parecido ni igual.

V

valve—a device that controls the flow of water or gas.

válvula—un dispositivo que regula el flujo de agua o gas.

valve grinding tool—rotates valve back and forth when lapping.

rectificadora para machos de válvula—gira de movimiento alternativo mientras esmerila.

valve guide—holds valve stems in alignment.

soporte de válvula—sujeta los vástagos de las válvulas en alineación.

valve keepers or valve pin—transfers spring force to valve stem.

retenes de las válvulas—transmite el empuje de muelle al vástago de la válvula.

valve spring compressor—compresses valve spring to remove valve keepers.

compresor de muelle de válvula—comprime el muelle de válvula para quitar los retenes.

valve stem clearance—air gap between valve and push rod.

aire libre del vástago-guía de la válvula—espacio de aire entre válvula e impulsador.

vapor barrier—moisture barrier.

barrera de vapor—barrera de humedad.

variable speed motor—a motor that has speed that can be controlled by the operator.

motor de velocidad regulable—un motor que tiene velocidad que puede ser controlada por el operador.

variable speed pulley—pulley with a movable side that can be moved while the equipment is running.

polea de velocidad regulable—polea con canto móvil que puede moverse mientras está puesto en marcha el equipo.

varnish—clear, tough, water-resistant finish.

barniz—acabado claro, duro, e impermeable al agua.

vehicle—a device for carrying something.

vehículo—un instrumento para cargar algo.

veneer—thin sheets of wood.

chapa de madera—maderos muy delgados.

vermiculite—artificial planting media manufactured from heated mica.

vermiculita—medios artificiales de planta fabricados de mica calentada.

vertical down—a weld made by moving downward across the metal.

vertical y hacia abajo—una soldadura realizada por mover hacia abajo a través del metal.

vertical drainage system—moves water through impervious clay.

drenaje longitudinal—transporta el agua por la arcilla impermeable.

vertical shaft engine—an engine with a vertical crankshaft for normal operation.

motor de eje vertical—un motor con un cigüeñal rectolineo para funcionamiento normal.

vertical up—a weld made by moving upward on the metal.

vertical y hacia arriba—una soldadura realizada por mover hacia arriba a través del metal.

video amplifier—used to amplify pictorial signals.

amplificador de imagen—se utiliza para amplificar señales pictoriales.

viscosimeter—an instrument used to measure the rate of flow of a liquid.

viscosímetro—un instrumento usado para medir el índice del flujo de un líquido.

viscosity—tendency to flow.

viscosidad—tendencia para fluir.

vista green—special shade of green used as a focal color.

verde vista—tono especial de verde que se usa como color de foco.

volt (V), voltage—a measure of electrical pressure.

voltio (V)—una medida de la tensión eléctrica.

voltage drop—loss of voltage as electricity travels through a wire.

caída de tensión—pérdida de voltaje mientras corre la corriente eléctrica por un alambre.

voltage multiplier—used to step up DC voltage.

multiplicador de tensión—utilizado para aumentar la tensión CC.

voltage regulator—controls voltage output.

regulador de tensión—controla la salida de tensión.

voltmeter—measures voltage.

voltímetro—mide la tensión.

volt-ohm-milliampere meter (VOM)—a popular combination meter used in electronics.

voltiohmímetro—metro combinado popular empleado en la electrónica.

warp—to bend or twist out of shape.

alabear—combar o torcear hasta que se pierda la forma o el pérfil original.

washed sand—sand flushed with water to remove clay or silt.

arena lavada—arena limpiada con agua para quitar tierra arcillosa y légamo.

washer—a device with a hole used to increase the holding power or prevent loosening of a bolt, nut, or screw.

arandela—un aparato con agujera que se usa para aumentar la retención o evitar el aflojamiento de perno, tuerca, o tornillo.

water quality—levels of dissolved oxygen, pH, salts, and waste products that affect aquatic plants and animals.

calidad de agua—niveles de oxígeno disuelto, el pH, y desechos que afectan plantas y animales acuáticos.

water-base—paint containing water as the vehicle.

colores al agua—pintura que contiene agua como su vehículo.

waterproof—material that is sealed against the entrance of water.

impenetrable al agua—materia hermética a que no puede penetrar el agua.

watt (W)—a measure of energy available or work that can be done using one ampere at one volt.

vatio—una medida de la potencia disponible o del trabajo producido que se puede hacer usando un amperio a un voltio.

watthour—the use of one watt of electricity for one hour.

vatio-hora—el uso de un vatio de corriente eléctrica durante una hora.

weaving—moving an electrode sideways to create a wider bead.

pasada pendular—pasando el electrodo a un lado para crear un cordón más ancho.

wedge—a piece of wood or metal that is thick on one end and tapers down to a thin edge on the other; to pack tightly.

calzo—una pieza de madera o metal que es gruesa de un lado y termina en ángulo agudo del otro.

weld—to join by fusion; the seam created by fusion.

soldar—unir por fusión.

soldadura—la juntura formada por fusión.

welder—a machine that welds.

soldadora—una máquina que solda.

weldor—a person who welds.

soldador— una persona que solda.

wetted zones—watered areas.

zonas mojadas—regiones mojadas.

whet—to sharpen by rubbing on a stone.

afilar con «piedra de agua»—afilar por frotar sobre una piedra.

whip—end wrap on a rope.

látigo—extremo cónico de una cuerda.

white—the color used to mark traffic areas; color of insulation on neutral wires.

blanco—el color que se usa para marcar áreas de tráfico; color de aislante de cables eléctricos neutros.

white and black stripes—stripes used as traffic markings.

rayas blancas y negras—rayas utilizadas como señalización de tráfico.

wind brace—construction member forming a triangle brace that strengthens the greenhouse structure.

contraviento—pieza de construcción que forma una riostra triangular que fortalece el armazón del invernadero.

wind-up starter—uses a lever to coil a spring for cranking an engine.

desvanado de puesta en marcha—usa una palanca para devanar una muella para arrancar un motor con la manivela.

wire nut—an insulated solderless connector used to connect the ends of electrical wires.

conectador de alambre—un conducto aislado sin soldadura usado para conectar los terminales de hilos eléctricos.

wood—the hard, compact, fibrous material that comes from the stems and branches of trees.

madera—el material fibroso, duro y compacto que viene de los tallos y ramos de los árboles.

wood fillers—a thick material used to fill the pores of open-grained woods.

auxiliar—materia gruesa usada para llenar los poros de maderas porosas.

wood preservatives—liquids used on wood to prevent rotting and insect damage.

preservativo de la madera—sustancias líquidas usadas sobre la madera para evitar pudrimiento y daño de los insectos.

wood screws—screws with threads designed to bite into wood fibers and draw down when turned.

tornillo de rosca de madera—tornillos con filetes que penetran las fibras de madera y que se bajan al tornearse.

wood-frame—walls constructed of vertical 2" × 4"s.

entramado—paredes construidas de los 2" × 4".

workable mix—the consistency of wet concrete when the various ingredients are mixed together correctly.

mezcla buena—la consistencia de la masa de hormigón cuando los ingredientes son mezclados correctamente.

wrist pin—pin between a connecting rod and piston.

pasador de articulación—pasador entre la biela y el émbolo.

Y—a fitting used where a pipe line splits in two directions at an angle of less than 90 degrees.

unión en Y—un montaje empleado donde tubería se bifurca a un ángulo de menos de 90 grados.

yellow—the safety color meaning caution, used to identify parts of machines such as wheels, levers, and knobs that control or adjust the machine.

amarillo—el color de seguridad, que significa cuidado, usado para identificar piezas de máquinas como ruedas, volantes, palancas, y bultos que controlan o ajustan la máquina.

Z

zinc—element associated with high-quality paint; material used for galvanizing steel.

cinc—elemento asociado con pintura de buena calidad; materia que se usa para galvanizar el acero.

INDEX

A

B

C

F

G

H

I

J

K

V

W

Y

Z